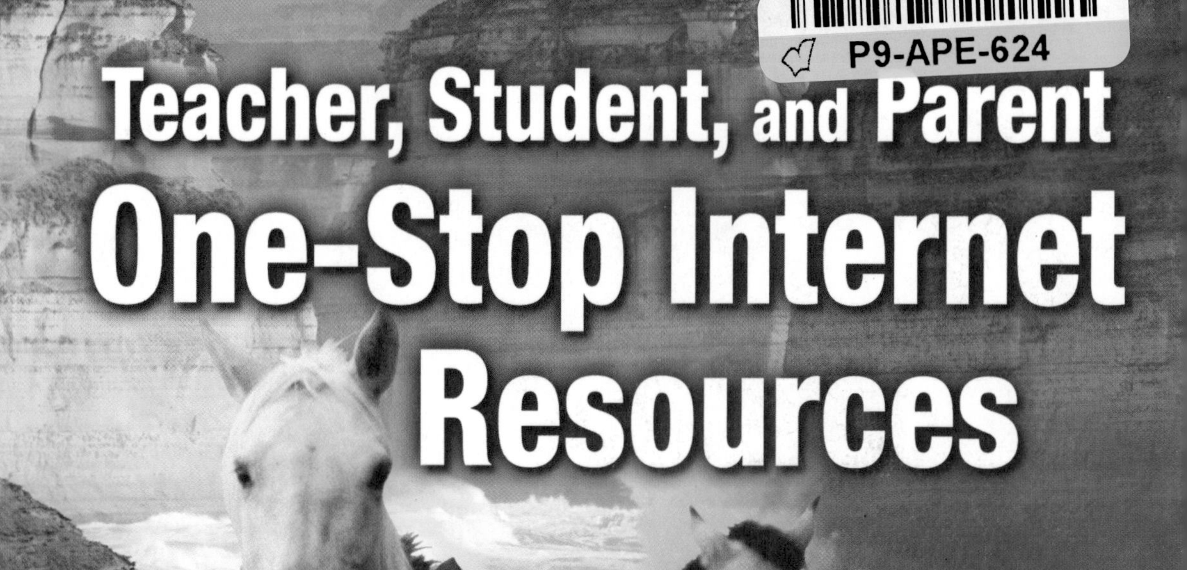

Teacher, Student, and Parent One-Stop Internet Resources

Log on to green.msscience.com

ONLINE STUDY TOOLS

- Section Self-Check Quizzes
- Interactive Tutor
- Chapter Review Tests
- Standardized Test Practice
- Vocabulary PuzzleMaker

ONLINE RESEARCH

- WebQuest Projects
- Prescreened Web Links
- Career Links
- Internet Labs

INTERACTIVE ONLINE STUDENT EDITION

- Complete Interactive Student Edition available at mhln.com

FOR TEACHERS

- Teacher Bulletin Board
- Teaching Today—Professional Development

SAFETY SYMBOLS	HAZARD	EXAMPLES	PRECAUTION	REMEDY
DISPOSAL	Special disposal procedures need to be followed.	certain chemicals, living organisms	Do not dispose of these materials in the sink or trash can.	Dispose of wastes as directed by your teacher.
BIOLOGICAL	Organisms or other biological materials that might be harmful to humans	bacteria, fungi, blood, unpreserved tissues, plant materials	Avoid skin contact with these materials. Wear mask or gloves.	Notify your teacher if you suspect contact with material. Wash hands thoroughly.
EXTREME TEMPERATURE	Objects that can burn skin by being too cold or too hot	boiling liquids, hot plates, dry ice, liquid nitrogen	Use proper protection when handling.	Go to your teacher for first aid.
SHARP OBJECT	Use of tools or glassware that can easily puncture or slice skin	razor blades, pins, scalpels, pointed tools, dissecting probes, broken glass	Practice common-sense behavior and follow guidelines for use of the tool.	Go to your teacher for first aid.
FUME	Possible danger to respiratory tract from fumes	ammonia, acetone, nail polish remover, heated sulfur, moth balls	Make sure there is good ventilation. Never smell fumes directly. Wear a mask.	Leave foul area and notify your teacher immediately.
ELECTRICAL	Possible danger from electrical shock or burn	improper grounding, liquid spills, short circuits, exposed wires	Double-check setup with teacher. Check condition of wires and apparatus.	Do not attempt to fix electrical problems. Notify your teacher immediately.
IRRITANT	Substances that can irritate the skin or mucous membranes of the respiratory tract	pollen, moth balls, steel wool, fiberglass, potassium permanganate	Wear dust mask and gloves. Practice extra care when handling these materials.	Go to your teacher for first aid.
CHEMICAL	Chemicals can react with and destroy tissue and other materials	bleaches such as hydrogen peroxide; acids such as sulfuric acid, hydrochloric acid; bases such as ammonia, sodium hydroxide	Wear goggles, gloves, and an apron.	Immediately flush the affected area with water and notify your teacher.
TOXIC	Substance may be poisonous if touched, inhaled, or swallowed.	mercury, many metal compounds, iodine, poinsettia plant parts	Follow your teacher's instructions.	Always wash hands thoroughly after use. Go to your teacher for first aid.
FLAMMABLE	Flammable chemicals may be ignited by open flame, spark, or exposed heat.	alcohol, kerosene, potassium permanganate	Avoid open flames and heat when using flammable chemicals.	Notify your teacher immediately. Use fire safety equipment if applicable.
OPEN FLAME	Open flame in use, may cause fire.	hair, clothing, paper, synthetic materials	Tie back hair and loose clothing. Follow teacher's instruction on lighting and extinguishing flames.	Notify your teacher immediately. Use fire safety equipment if applicable.

Eye Safety
Proper eye protection should be worn at all times by anyone performing or observing science activities.

Clothing Protection
This symbol appears when substances could stain or burn clothing.

Animal Safety
This symbol appears when safety of animals and students must be ensured.

Handwashing
After the lab, wash hands with soap and water before removing goggles.

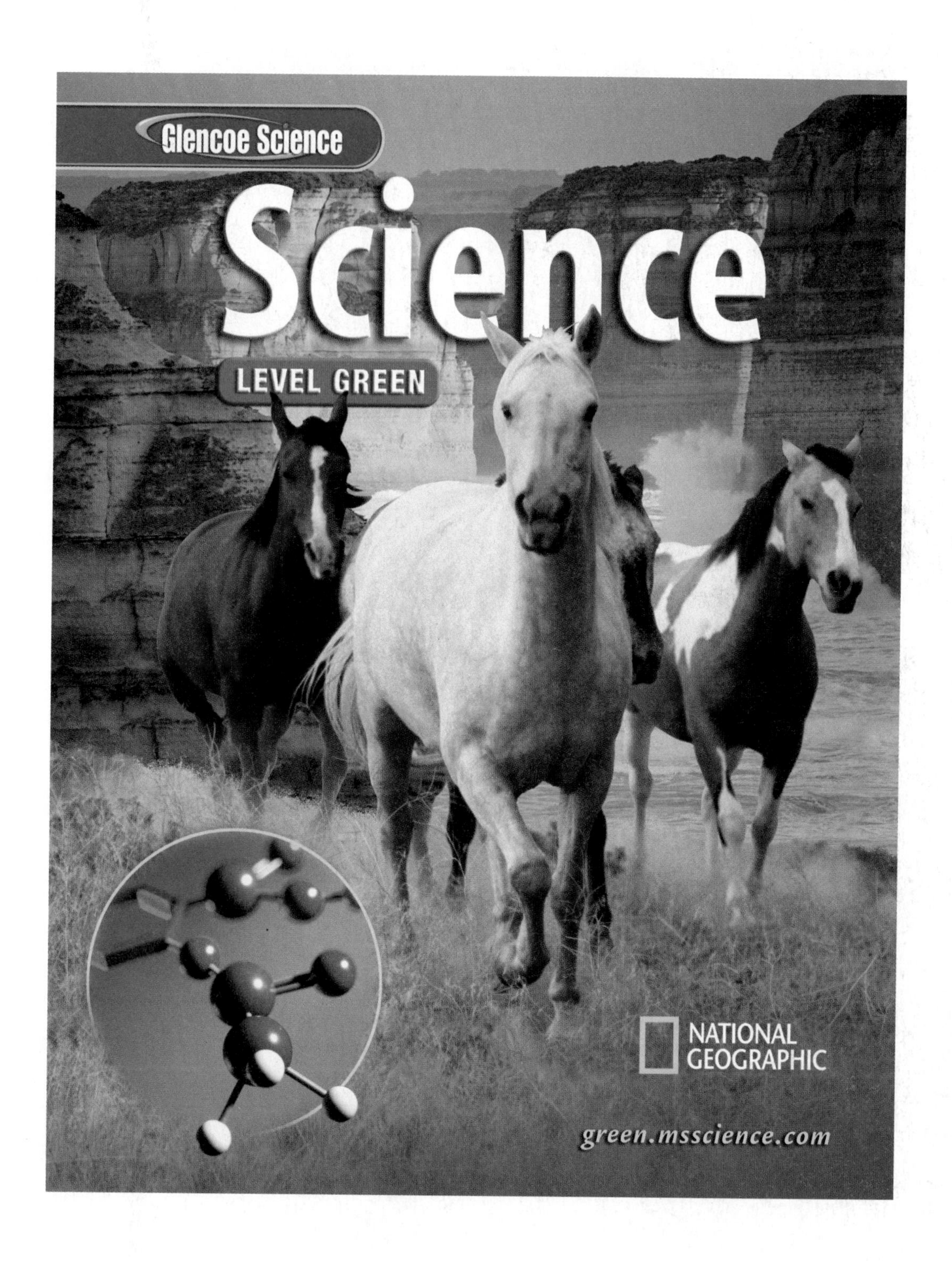

New York, New York Columbus, Ohio Chicago, Illinois Peoria, Illinois Woodland Hills, California

Glencoe Science

Level Green

These ocean cliffs on the Victoria coast in Australia are called the Twelve Apostles. They form a background for wild horses running through a field. The inset photo is a model of an aspirin molecule. These elements represent the Earth, life, and physical sciences, respectively.

Glencoe

The McGraw-Hill Companies

Send all inquiries to:
Glencoe/McGraw-Hill
8787 Orion Place
Columbus, OH 43240-4027

ISBN: 0-07-860047-2

Printed in the United States of America.

5 6 7 8 9 10 071/055 09 08 07 06

Contents In Brief

Authors

Education Division
Washington, D.C.

Alton Biggs
Biology Teacher
Allen High School
Allen, TX

Lucy Daniel, PhD
Teacher/Consultant
Rutherford County Schools
Rutherfordton, NC

Ralph M. Feather Jr., PhD
Assistant Professor
Geoscience Department
Indiana University of Pennsylvania
Indiana, PA

Edward Ortleb
Science Consultant
St. Louis, MO

Peter Rillero, PhD
Associate Professor of Science Education
Arizona State University West
Phoenix, AZ

Susan Leach Snyder
Retired Teacher/Consultant
Jones Middle School
Upper Arlington, OH

Dinah Zike
Educational Consultant
Dinah-Might Activities, Inc.
San Antonio, TX

Series Consultants

CONTENT

Alton J. Banks, PhD
Director of the Faculty Center
for Teaching and Learning
North Carolina State University
Raleigh, NC

Jack Cooper
Ennis High School
Ennis, TX

Sandra K. Enger, PhD
Associate Director,
Associate Professor
UAH Institute for Science Education
Huntsville, AL

David G. Haase, PhD
North Carolina State University
Raleigh, NC

Michael A. Hoggarth, PhD
Department of Life and
Earth Sciences
Otterbein College
Westerville, OH

Jerome A. Jackson, PhD
Whitaker Eminent Scholar in Science
Program Director
Center for Science, Mathematics,
and Technology Education
Florida Gulf Coast University
Fort Meyers, FL

William C. Keel, PhD
Department of Physics
and Astronomy
University of Alabama
Tuscaloosa, AL

Linda McGaw
Science Program Coordinator
Advanced Placement Strategies, Inc.
Dallas, TX

Madelaine Meek
Physics Consultant Editor
Lebanon, OH

Robert Nierste
Science Department Head
Hendrick Middle School, Plano ISD
Plano, TX

Connie Rizzo, MD, PhD
Department of Science/Math
Marymount Manhattan College
New York, NY

Dominic Salinas, PhD
Middle School Science Supervisor
Caddo Parish Schools
Shreveport, LA

Cheryl Wistrom
St. Joseph's College
Rensselaer, IN

Carl Zorn, PhD
Staff Scientist
Jefferson Laboratory
Newport News, VA

MATH

Michael Hopper, DEng.
Manager of Aircraft Certification
L-3 Communications
Greenville, TX

Teri Willard, EdD
Mathematics Curriculum Writer
Belgrade, MT

READING

Elizabeth Babich
Special Education Teacher
Mashpee Public Schools
Mashpee, MA

Barry Barto
Special Education Teacher
John F. Kennedy Elementary
Manistee, MI

Carol A. Senf, PhD
School of Literature, Communication, and Culture
Georgia Institute of Technology
Atlanta, GA

Rachel Swaters-Kissinger
Science Teacher
John Boise Middle School
Warsaw, MO

SAFETY

Aileen Duc, PhD
Science 8 Teacher
Hendrick Middle School, Plano ISD
Plano, TX

Sandra West, PhD
Department of Biology
Texas State University-San Marcos
San Marcos, TX

ACTIVITY TESTERS

Nerma Coats Henderson
Pickerington Lakeview Jr. High School
Pickerington, OH

Mary Helen Mariscal-Cholka
William D. Slider Middle School
El Paso, TX

Science Kit and Boreal Laboratories
Tonawanda, NY

Reviewers

Deidre Adams
West Vigo Middle School
West Terre Haute, IN

Sharla Adams
IPC Teacher
Allen High School
Allen, TX

Maureen Barrett
Thomas E. Harrington Middle School
Mt. Laurel, NJ

John Barry
Seeger Jr.-Sr. High School
West Lebanon, IN

Desiree Bishop
Environmental Studies Center
Mobile County Public Schools
Mobile, AL

William Blair
Retired Teacher
J. Marshall Middle School
Billerica, MA

Tom Bright
Concord High School
Charlotte, NC

Lois Burdette
Green Bank Elementary-Middle School
Green Bank, WV

Marcia Chackan
Pine Crest School
Boca Raton, FL

Obioma Chukwu
J.H. Rose High School
Greenville, NC

Karen Curry
East Wake Middle School
Raleigh, NC

Joanne Davis
Murphy High School
Murphy, NC

Robin Dillon
Hanover Central High School
Cedar Lake, IN

Anthony J. DiSipio, Jr.
8th Grade Science
Octorana Middle School
Atglen, PA

Sandra Everhart
Dauphin/Enterprise Jr. High Schools
Enterprise, AL

Mary Ferneau
Westview Middle School
Goose Creek, SC

Cory Fish
Burkholder Middle School
Henderson, NV

Linda V. Forsyth
Retired Teacher
Merrill Middle School
Denver, CO

George Gabb
Great Bridge Middle School
Chesapeake Public Schools
Chesapeake, VA

Annette D'Urso Garcia
Kearney Middle School
Commerce City, CO

Nerma Coats Henderson
Pickerington Lakeview Jr.
High School
Pickerington, OH

Lynne Huskey
Chase Middle School
Forest City, NC

Maria E. Kelly
Principal
Nativity School
Catholic Diocese of Arlington
Burke, VA

Michael Mansour
Board Member
National Middle Level Science
Teacher's Association
John Page Middle School
Madison Heights, MI

Mary Helen Mariscal-Cholka
William D. Slider Middle School
El Paso, TX

Michelle Mazeika-Simmons
Whiting Middle School
Whiting, IN

Joe McConnell
Speedway Jr. High School
Indianapolis, IN

Sharon Mitchell
William D. Slider Middle School
El Paso, TX

Amy Morgan
Berry Middle School
Hoover, AL

Norma Neely, EdD
Associate Director for Regional
Projects
Texas Rural Systemic Initiative
Austin, TX

Annette Parrott
Lakeside High School
Atlanta, GA

Nora M. Prestinari Burchett
Saint Luke School
McLean, VA

Mark Sailer
Pioneer Jr.-Sr. High School
Royal Center, IN

Joanne Stickney
Monticello Middle School
Monticello, NY

Dee Stout
Penn State University
University Park, PA

Darcy Vetro-Ravndal
Hillsborough High School
Tampa, FL

Karen Watkins
Perry Meridian Middle School
Indianapolis, IN

Clabe Webb
Permian High School
Ector County ISD
Odessa, TX

Alison Welch
William D. Slider Middle School
El Paso, TX

Kim Wimpey
North Gwinnett High School
Suwanee, GA

Kate Ziegler
Durant Road Middle School
Raleigh, NC

Teacher Advisory Board

The Teacher Advisory Board gave the editorial staff and design team feedback on the content and design of the Student Edition. They provided valuable input in the development of the 2005 edition of ***Glencoe Science Level Green.***

John Gonzales
Challenger Middle School
Tucson, AZ

Rachel Shively
Aptakisic Jr. High School
Buffalo Grove, IL

Roger Pratt
Manistique High School
Manistique, MI

Kirtina Hile
Northmor Jr. High/High School
Galion, OH

Marie Renner
Diley Middle School
Pickerington, OH

Nelson Farrier
Hamlin Middle School
Springfield, OR

Jeff Remington
Palmyra Middle School
Palmyra, PA

Erin Peters
Williamsburg Middle School
Arlington, VA

Rubidel Peoples
Meacham Middle School
Fort Worth, TX

Kristi Ramsey
Navasota Jr. High School
Navasota, TX

Student Advisory Board

The Student Advisory Board gave the editorial staff and design team feedback on the design of the Student Edition. We thank these students for their hard work and creative suggestions in making the 2005 edition of ***Glencoe Science Level Green*** student friendly.

Jack Andrews
Reynoldsburg Jr. High School
Reynoldsburg, OH

Peter Arnold
Hastings Middle School
Upper Arlington, OH

Emily Barbe
Perry Middle School
Worthington, OH

Kirsty Bateman
Hilliard Heritage Middle School
Hilliard, OH

Andre Brown
Spanish Emersion Academy
Columbus, OH

Chris Dundon
Heritage Middle School
Westerville, OH

Ryan Manafee
Monroe Middle School
Columbus, OH

Addison Owen
Davis Middle School
Dublin, OH

Teriana Patrick
Eastmoor Middle School
Columbus, OH

Ashley Ruz
Karrar Middle School
Dublin, OH

The Glencoe middle school science Student Advisory Board taking a timeout at COSI, a science museum in Columbus, Ohio.

HOW TO...

Use Your Science Book

Why do I need my science book?

Have you ever been in class and not understood all of what was presented? Or, you understood everything in class, but at home, got stuck on how to answer a question? Maybe you just wondered when you were ever going to use this stuff?

These next few pages are designed to help you understand everything your science book can be used for . . . besides a paperweight!

Before You Read

- **Chapter Opener** Science is occurring all around you, and the opening photo of each chapter will preview the science you will be learning about. The **Chapter Preview** will give you an idea of what you will be learning about, and you can try the **Launch Lab** to help get your brain headed in the right direction. The **Foldables** exercise is a fun way to keep you organized.
- **Section Opener** Chapters are divided into two to four sections. The **As You Read** in the margin of the first page of each section will let you know what is most important in the section. It is divided into four parts. **What You'll Learn** will tell you the major topics you will be covering. **Why It's Important** will remind you why you are studying this in the first place! The **Review Vocabulary** word is a word you already know, either from your science studies or your prior knowledge. The **New Vocabulary** words are words that you need to learn to understand this section. These words will be in **boldfaced** print and highlighted in the section. Make a note to yourself to recognize these words as you are reading the section.

As You Read

- **Headings** Each section has a title in large red letters, and is further divided into blue titles and small red titles at the beginnings of some paragraphs. To help you study, make an outline of the headings and subheadings.
- **Margins** In the margins of your text, you will find many helpful resources. The **Science Online** exercises and **Integrate** activities help you explore the topics you are studying. **MiniLabs** reinforce the science concepts you have learned.
- **Building Skills** You also will find an **Applying Math** or **Applying Science** activity in each chapter. This gives you extra practice using your new knowledge, and helps prepare you for standardized tests.
- **Student Resources** At the end of the book you will find **Student Resources** to help you throughout your studies. These include **Science, Technology,** and **Math Skill Handbooks,** an **English/Spanish Glossary,** and an **Index.** Also, use your **Foldables** as a resource. It will help you organize information, and review before a test.
- **In Class** Remember, you can always ask your teacher to explain anything you don't understand.

Science Vocabulary Make the following Foldable to help you understand the vocabulary terms in this chapter.

STEP 1 **Fold** a vertical sheet of notebook paper from side to side.

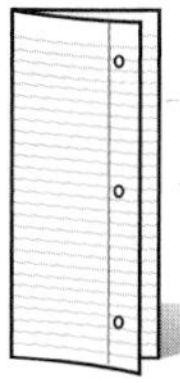

STEP 2 **Cut** along every third line of only the top layer to form tabs.

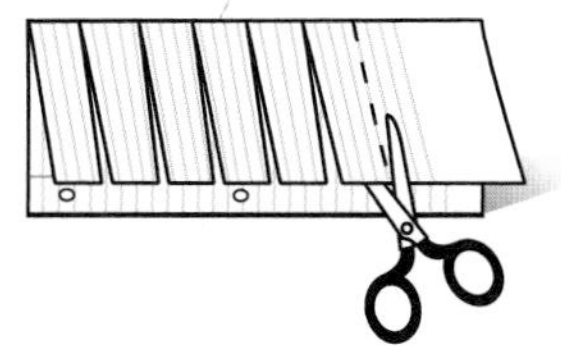

STEP 3 **Label** each tab with a vocabulary word from the chapter.

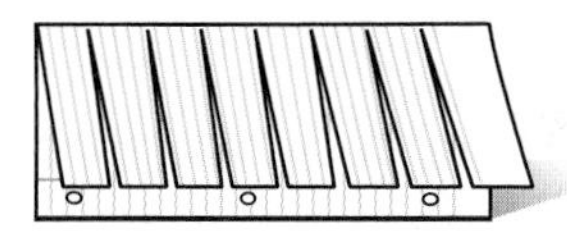

Build Vocabulary As you read the chapter, list the vocabulary words on the tabs. As you learn the definitions, write them under the tab for each vocabulary word.

Look For...

FOLDABLES™

At the beginning of every section.

In Lab

Working in the laboratory is one of the best ways to understand the concepts you are studying. Your book will be your guide through your laboratory experiences, and help you begin to think like a scientist. In it, you not only will find the steps necessary to follow the investigations, but you also will find helpful tips to make the most of your time.

- Each lab provides you with a **Real-World Question** to remind you that science is something you use every day, not just in class. This may lead to many more questions about how things happen in your world.
- Remember, experiments do not always produce the result you expect. Scientists have made many discoveries based on investigations with unexpected results. You can try the experiment again to make sure your results were accurate, or perhaps form a new hypothesis to test.
- Keeping a **Science Journal** is how scientists keep accurate records of observations and data. In your journal, you also can write any questions that may arise during your investigation. This is a great method of reminding yourself to find the answers later.

Before a Test

Admit it! You don't like to take tests! However, there *are* ways to review that make them less painful. Your book will help you be more successful taking tests if you use the resources provided to you.

- Review all of the **New Vocabulary** words and be sure you understand their definitions.
- Review the notes you've taken on your **Foldables,** in class, and in lab. Write down any question that you still need answered.
- Review the **Summaries** and **Self Check questions** at the end of each section.
- Study the concepts presented in the chapter by reading the **Study Guide** and answering the questions in the **Chapter Review.**

a or b?

?

T or F?

Look For...

- **Reading Checks** and **caption questions** throughout the text.
- the **Summaries** and **Self Check questions** at the end of each section.
- the **Study Guide** and **Review** at the end of each chapter.
- the **Standardized Test Practice** after each chapter.

Let's Get Started

To help you find the information you need quickly, use the Scavenger Hunt below to learn where things are located in Chapter 1.

1. What is the title of this chapter?
2. What will you learn in Section 1?
3. Sometimes you may ask, "Why am I learning this?" State a reason why the concepts from Section 2 are important.
4. What is the main topic presented in Section 2?
5. How many reading checks are in Section 1?
6. What is the Web address where you can find extra information?
7. What is the main heading above the sixth paragraph in Section 2?
8. There is an integration with another subject mentioned in one of the margins of the chapter. What subject is it?
9. List the new vocabulary words presented in Section 2.
10. List the safety symbols presented in the first Lab.
11. Where would you find a Self Check to be sure you understand the section?
12. Suppose you're doing the Self Check and you have a question about concept mapping. Where could you find help?
13. On what pages are the Chapter Study Guide and Chapter Review?
14. Look in the Table of Contents to find out on which page Section 2 of the chapter begins.
15. You complete the Chapter Review to study for your chapter test. Where could you find another quiz for more practice?

Earth's Materials—2

In each chapter, look for these opportunities for review and assessment:
- Reading Checks
- Caption Questions
- Section Review
- Chapter Study Guide
- Chapter Review
- Standardized Test Practice
- Online practice at green.msscience.com

Contents

unit 2 Earth's Atmosphere and Beyond—86

chapter 4 Atmosphere—88

chapter 5 Weather—116

In each chapter, look for these opportunities for review and assessment:

- Reading Checks
- Caption Questions
- Section Review
- Chapter Study Guide
- Chapter Review
- Standardized Test Practice
- Online practice at green.msscience.com

chapter 6 Climate—146

chapter 7 Earth in Space—176

unit 3 The Basis of Life—210

chapter 8 Life's Structure and Classification—212

Contents

Contents

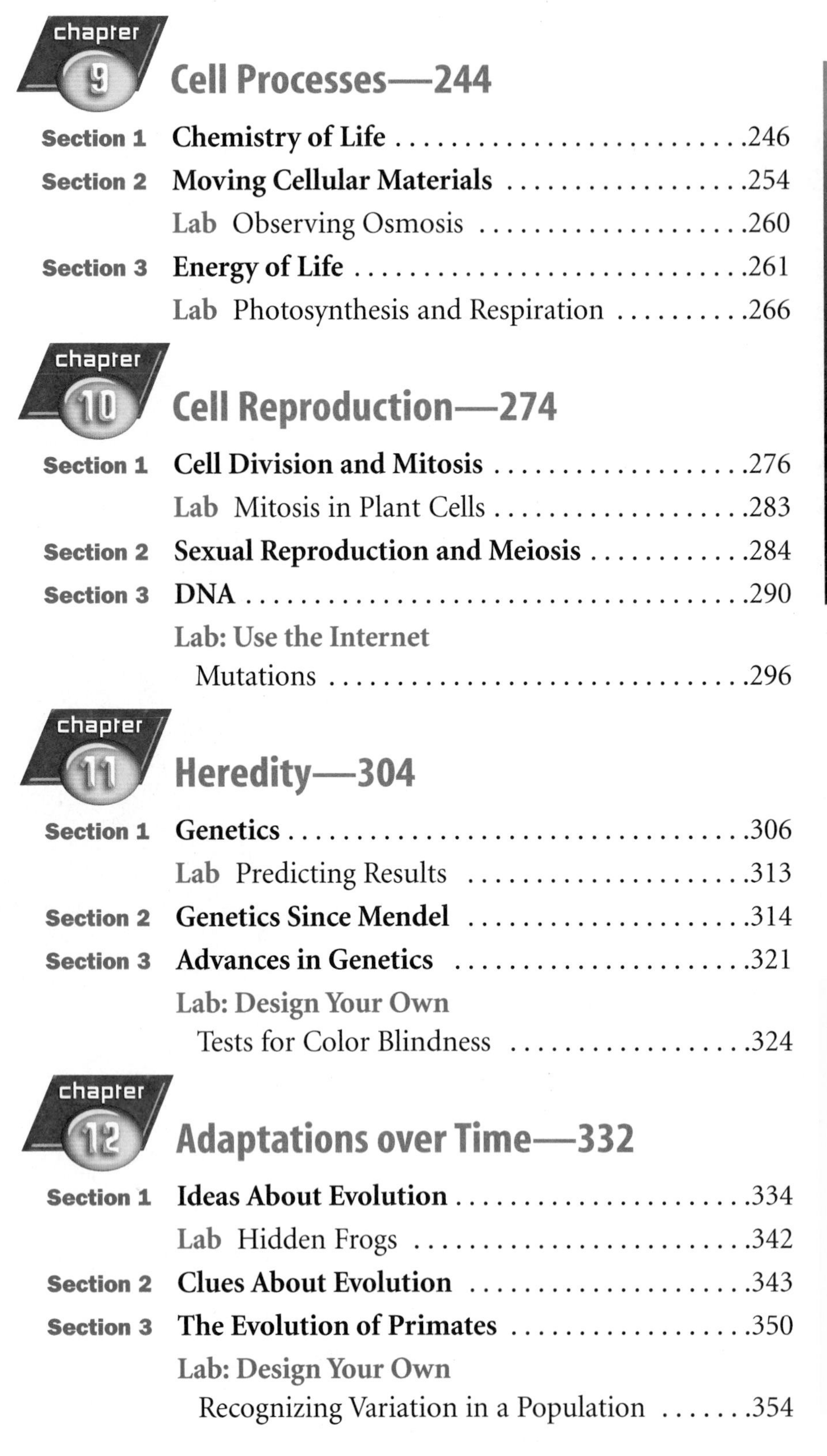

In each chapter, look for these opportunities for review and assessment:
- Reading Checks
- Caption Questions
- Section Review
- Chapter Study Guide
- Chapter Review
- Standardized Test Practice
- Online practice at **green.msscience.com**

Human Body Systems—362

Circulation and Immunity—364

Digestion, Respiration, and Excretion—398

Support, Movement, and Responses—432

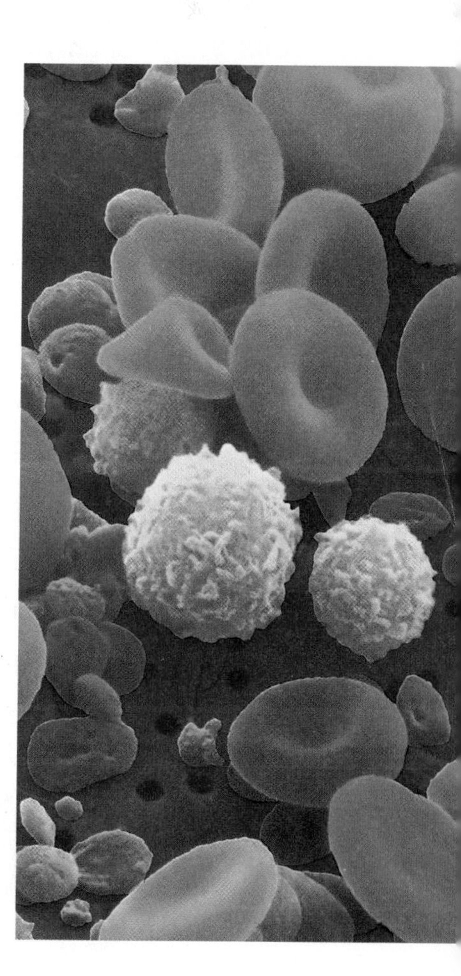

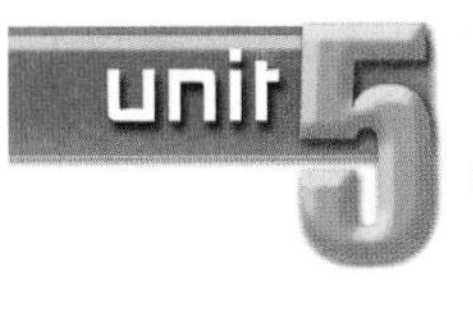

unit 5 The Interdependence of Life—496

Conserving Resources—558

In each chapter, look for these opportunities for review and assessment:
- Reading Checks
- Caption Questions
- Section Review
- Chapter Study Guide
- Chapter Review
- Standardized Test Practice
- Online practice at green.msscience.com

unit 6

Matter and Energy—590

Properties and Changes of Matter—592

Substances, Mixtures, and Solubility—618

Contents

Contents

In each chapter, look for these opportunities for review and assessment:
- Reading Checks
- Caption Questions
- Section Review
- Chapter Study Guide
- Chapter Review
- Standardized Test Practice
- Online practice at green.msscience.com

Student Resources—746

Science Skill Handbook—748

Extra Try at Home Labs—760

Technology Skill Handbook—772

Math Skill Handbook—776

Reference Handbooks—791

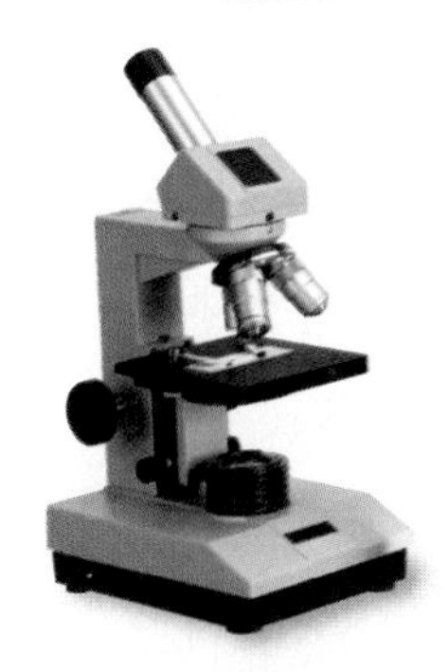

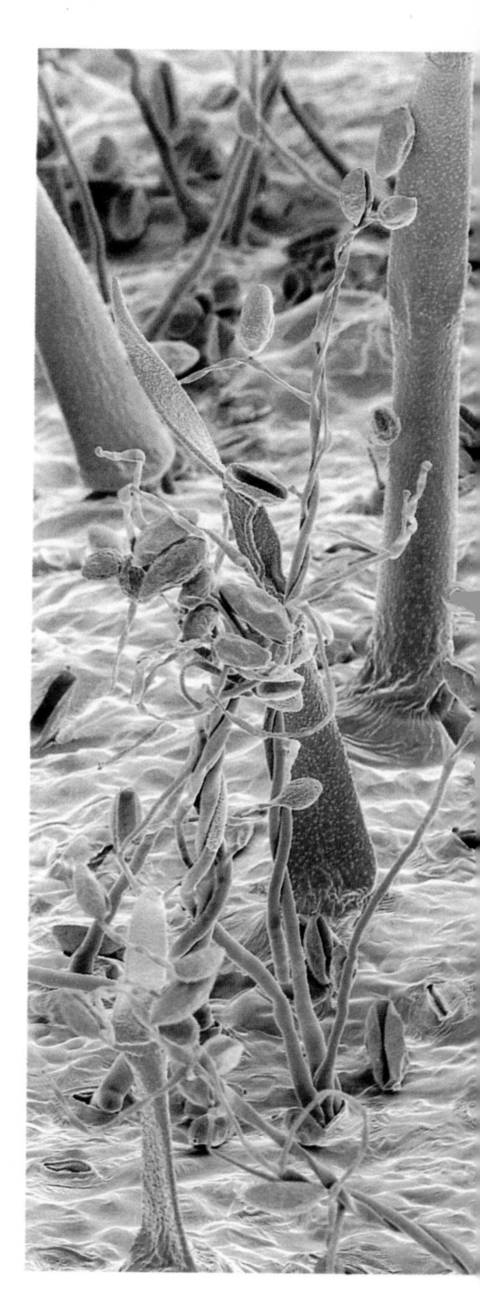

English/Spanish Glossary—802

Index—825

Credits—845

Cross-Curricular Readings

NATIONAL GEOGRAPHIC Unit Openers

NATIONAL GEOGRAPHIC VISUALIZING

TIME Science and Society

TIME Science and History

Oops! Accidents in Science

Science and Language Arts

Science Stats

available as a video lab

Mini LAB

Mini LAB Try at Home

available as a video lab

One-Page Labs

Two-Page Labs

Design Your Own Labs

Model and Invent Labs

Use the Internet Labs

Activities

Content Details

Applying Math

Applying Science

INTEGRATE

Standardized Test Practice

How Are Canals & the Paleozoic Era Connected?

Before the invention of the locomotive, canals, such as the one at upper right, were an important means of transportation. In the 1790s, an engineer traveled around England to study new canals. The engineer noticed something odd: All across the country, certain types of rocks seemed to lie in predictable layers, or strata. And the same strata always had the same kinds of fossils in them. Since each layer of sedimentary rock typically forms on top of the previous one, scientists realized that the strata recorded the history of life on Earth. By the mid-1800s, the known rock strata had been organized into a system that we now know as the geologic time scale. In this system, Earth's history is divided into units called eras, which in turn are divided into periods. Many of the rock layers in the Grand Canyon (background) date from the Paleozoic, or "ancient life," Era.

unit projects

Visit green.msscience.com/unit_project to find project ideas and resources. Projects include:

- **History** Discover some of Earth's inhabitants of different time periods using the fossil record. Create a drawing of a scene in Earth's history.
- **Technology** Choose an extinct animal to investigate. How has technology allowed paleontologists to learn about how it lived?
- **Model** As a group, design a wall mural or diorama depicting the layers of the geologic time scale, or a particular scene of interest from an era.

WebQuest Use online resources to form your own opinion concerning plate tectonics. Investigate the *Fossils of Antarctica* and what they could tell us about its ancient climate and location.

chapter 1

The Nature of Science

chapter preview

sections

How did they live?

Digging in a cave in Southern France, these researchers are unearthing ancient human relics—some dating back to about 200,000 years before present. Notice the string grid layout on the cave floor.

Science Journal Write about any human artifact you know of that was discovered in an area near your home, or that was unearthed in another region.

Start-Up Activities

Model an Excavation

Excavations to unearth tools or other clues of past human life often are slow processes. Care must be taken so the remains are not broken or destroyed as they are removed from the soil. Try to excavate a cookie without destroying the treasures within.

WARNING: *Never eat or drink in the science lab, and never use lab glassware as a food or drink container.*

1. Obtain an oatmeal cookie with raisins and walnuts from your teacher.
2. Place the cookie on a large paper towel.
3. Use a biology probe to remove the raisins and walnuts from the cookie without damaging either one.
4. Give all pieces of the excavated cookie to your teacher for disposal.
5. Wash your hands with soap and water when you have finished.
6. **Think Critically** In your Science Journal, write a paragraph that explains how probing the cookie might be similar to removing bones, tools, or other evidence of ancient life from Earth's crust.

FOLDABLES™ Study Organizer

Science and Technology Make the following Foldable to compare and contrast science and technology.

STEP 1 **Fold** one sheet of paper lengthwise.

STEP 2 **Fold** into thirds.

STEP 3 **Unfold** and **draw** overlapping ovals. **Cut** the top sheet along the folds.

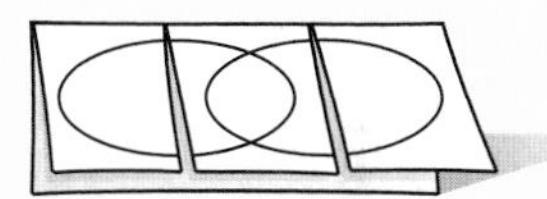

STEP 4 **Label** the ovals as shown.

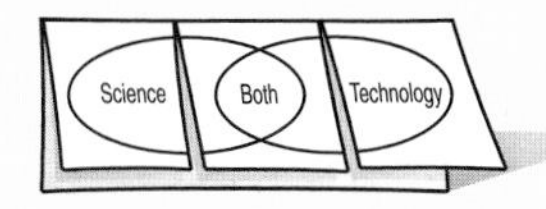

Construct a Venn Diagram As you read the chapter, list the aspects unique to science under the left tab, those unique to technology under the right tab, and those characteristics common to both under the middle tab.

Preview this chapter's content and activities at green.msscience.com

section 1

How Science Works

as you read

What You'll Learn

- **Explain** the science of archaeology.
- **Compare and contrast** science and technology.

Why It's Important

Science and technology are important parts of your everyday world.

Review Vocabulary
artifact: object of historical interest produced by humans, such as a tool or weapon

New Vocabulary
- science
- technology

Groundbreaking News

It was Friday morning, and the students in Ms. Garcia's science lab were waiting eagerly for class to start. Unlike most days in science class at York Middle School, this meeting would be a field trip to the north end of the school. Students were eager to observe work that would result in the long-awaited, new gymnasium. The students in group 4—Ben, Emily, Maria, and Juan—peered out the windows. They saw construction equipment, including bulldozers and trucks much like the ones shown in **Figure 1,** pull up to the school. With pencils and notebooks in hand, the interested students hiked out to the site. They watched as massive shovels moved hundreds of kilograms of dirt from one spot to another.

Buried treasure? All of a sudden, the power-shovel operator stopped the giant scoop in midair. He looked curiously into the hole he was making, and then he climbed from his seat high above the ground. He called some of the other workers over. They all stared into the pit. One of the workers motioned for Ms. Garcia and her students to come a little closer. Everyone was surprised at what they saw. A piece of broken pottery was sticking out from the loosened soil.

Figure 1 Construction efforts sometimes unearth prehistoric sites.

Figure 2 Much can be learned about ancient cultures from materials they left behind.

Archaeologists study pottery and other items to learn more about ancient humans.

Archaeologists work in the field to gather data.

Science in Action One worker suggested that the pottery might be only one of thousands of pieces of trash that were buried long before the school was built. Another worker, however, wasn't so sure. He thought that the pottery could perhaps be an ancient piece of art, such as the one shown in **Figure 2.** Nonetheless, a decision was made to stop the excavation, at least for the moment.

Back in the classroom, the students talked excitedly about the find. This, they all agreed, was real science. **Science,** they knew, is the process of trying to understand the world.

Calling in the Experts Although the discovery was exciting, Ms. Garcia reminded the students that the piece of pottery might be something that was thrown out only decades ago. To be sure, however, the school's principal called an archaeologist at the local college. Archaeologists, such as the two shown in **Figure 2,** are scientists who study the cultural remains of ancient humans. Cultural remains, known as artifacts, might be tools, weapons, rock drawings, buildings, or pottery, such as that found at the school. Dr. Lum, the students were told, would be at the school on Monday to examine the pottery.

Ms. Garcia suggested that her students research more about the history of their area. This would help the students evaluate how this pottery might have originated from ancient cultures that once lived in the area. Ben and the others in his group quickly began their research. Maria thought that it would be a good idea to take notes on their findings. That way, they could compare what they found with what Dr. Lum told them on Monday. The others in the group agreed and put their science notebooks into their backpacks before heading to the library.

Topic: Artifacts and Human History

Visit green.msscience.com for Web links to information about how artifacts provide clues about human behavior and history.

Activity Select a human artifact, such as a tool, art piece, or waste material. List three specific examples of this type of artifact, and include the location where each was found. Summarize the knowledge gained about the humans who produced the artifacts.

Figure 3 Archaeologists study artifacts of ancient people like those from ancient Egypt or like those who came to North America about 12,000 years ago.

One branch of archaeology studies the cultural remains of people who lived before history was written.

Researching the Past

At the library, Juan used an encyclopedia to begin his research. He found out that archaeology is a branch of science that studies the tools and other cultural remains of humans. There are two major branches of archaeology, as shown in **Figure 3.** One focuses on groups of people who lived before history was written. The other studies civilizations that developed since people began writing things down. To his surprise, Juan also discovered that archaeology covers a time span of more than 3 million years. About 3.5 million years ago, he read, the first ancestors of humans are thought to have appeared on Earth.

Reading Check ***What are the two major branches of archaeology?***

The other students took turns finding out about the history of their area. Ben found that many scientists hypothesize that the first people came to North America from Asia about 12,000 years ago. Over thousands of years, these people migrated to different parts of the continent. Emily and Maria discovered that the area around their city was settled about 2,000 years ago. After locating a few more sources of information, the students took notes on all the information they had gathered. Emily suggested that they also write any questions they had about the pottery or the science of archaeology. Juan, Ben, and Maria agreed, and each wrote a few questions. The group left the library eager to hear how its findings would compare with what Dr. Lum would tell them on Monday.

Another branch of archaeology studies civilizations that have developed since written history began.

Think Critically *How would you define written history when distinguishing rock drawings from hieroglyphics?*

Figure 4 Computers and radar are two examples of technology used in archaeological research.
List *at least three other forms of technology.*

Dr. Lum's Visit Dr. Lum arrived before nine o'clock. When the bell rang, Emily's hand shot up. She was hoping to be the first to ask about the pottery. However, before calling on her, Dr. Lum said she wanted to give the students some background information and then she would answer their questions.

Dr. Lum explained how important it is to preserve prehistoric sites for present and future generations. She also said that many archaeological sites, like the possible one on the school grounds, are found by accident. More scientific work would have to be done before construction on the site could continue.

Technology Several kinds of technology would be used to study the area, such as computers and cameras. **Technology** is the use of knowledge gained through science to make new products or tools people can use. **Figure 4** shows some common types of technology. Dr. Lum told the students that a radar survey would be conducted to help study the find at the school. This type of technology, Dr. Lum explained, helps scientists "see" what's beneath the ground without disturbing the site. Experts from other fields of science probably would be called upon to help evaluate the site. For instance, geologists, scientists who study Earth processes, might be contacted to help with soil studies.

Relative Ages Archaeologists and geologists determine the relative ages of layers of sediment and artifacts by examining where they lie in comparison to other layers. For example, in undisturbed sediment, the bottom layer is the oldest and the layer on top is the youngest. Model this concept with a stack of books.

Working Together Dr. Lum ended her talk by suggesting that the students go back to the site with her. There, she would examine what had been found. She also would try to answer any questions the students might have about the find.

Maria and Emily led the group of curious students toward the north end of the school yard. Dr. Lum used her hand lens to observe the piece of pottery carefully. After examining the piece for awhile, she announced that she thought the pottery was old and that an archaeological dig, or excavation of the site, was in order. The students asked if they could participate in the dig. Dr. Lum said she would welcome all the help they could give.

Digging In

Weeks passed before the radar surveys were complete. The students in Ms. Garcia's class spent most of their time learning about how an archaeological excavation is done. Maria reported to the class that the holes and ditches that were being dug around the site would help determine its size. She also added that it was important that the site be disturbed as little as possible. By keeping the site intact, much of its history could be retold.

Finally, the day came when the students could participate in the dig. Each was given a small hand shovel, a soft paintbrush, and a pair of gardening gloves. Each student was paired with an amateur archaeologist. All of those involved were instructed to work slowly and carefully to excavate this important piece of their city's past. **Figure 5** shows a piece of pottery recovered from a similar archaeological dig site. *(4)

Artifacts are carefully mapped before they are excavated.

Figure 5 This paint brush, along with other tools, such as dental probes and toothbrushes, are commonly used to remove artifacts.
Explain *why ancient sites must be excavated carefully.*

Clues to the Past Many pieces of pottery, along with some tools, were found at the school site. Before the artifacts were removed from the soil, college students working with Dr. Lum took pictures or made drawings of each piece. These were used to make maps showing the exact location of each artifact before it was removed. The maps also would be used to show vertical and horizontal differences in the site.

Figure 6 After artifacts are mapped and excavated, they're taken to a laboratory where they are cleaned and tagged for further study.

Lab Work Each artifact was given a number and its location and orientation in the soil was recorded. After the artifacts were cataloged, they were removed from the site. Dr. Lum told the students that she would take the finds back to her lab. There, they would be cleaned, studied, and stored for future analysis, as shown in **Figure 6.**

Chemical analyses of certain artifacts would be used to determine their approximate age. Based on her knowledge of the area, Dr. Lum thought that the site was at least several thousand years old.

section 1 review

Summary

Groundbreaking News

- Science is the process of trying to understand the world.
- Important discoveries in science sometimes happen by accident.
- Discoveries must be subject to scientific testing in order to be validated.

Researching the Past

- Background research is an important part of any scientific study.
- Technology applies knowledge that is gained from doing science.

Digging In

- Both field and lab methods are used during scientific studies.

Self Check

1. **Explain** what archaeology is.
2. **Describe** several common forms of technology used in science.
3. **Explain** why scientists conduct radar surveys of archaeological sites.
4. **List** some examples of cultural remains studied by archaeologists.
5. **Think Critically** Why are maps of prehistoric sites often made before removing the artifacts?

Applying Skills

6. **Compare and contrast** science and technology. Include a discussion of how progress in science can lead to progress in technology, and vice versa.

section 2

Scientific Problem Solving

as you read

What You'll Learn

- **Explain** the steps taken in scientific methods.
- **Compare and contrast** scientific variables and constants.
- **Explain** how a control is used during an experiment.

Why It's Important

Scientific methods can help you solve many types of problems.

Review Vocabulary
analyze: to separate and study something as parts or basic principles in order to understand it as a whole

New Vocabulary
- scientific methods
- observation
- inference
- hypothesis
- independent variable
- dependent variable
- constant
- control

Scientific Methods

Several steps were taken to learn about the pottery found at York Middle School. When the pottery was found, a decision was made to stop construction at the site. One adult guessed that the pottery was old. An expert was called to verify the guess made about the pottery. Based on prior knowledge and further testing, it was concluded that the pottery was from a prehistoric culture.

Think about the last time you had a problem that took several steps or actions to solve. Step-by-step procedures of scientific problem solving are called **scientific methods.** Solving any problem scientifically involves several steps. The basic steps in a commonly used scientific approach are shown in **Figure 7.** The steps used can vary from situation to situation and aren't always done in the same order. But for now, take a look at each step in turn.

Reading Check *What are scientific methods used for?*

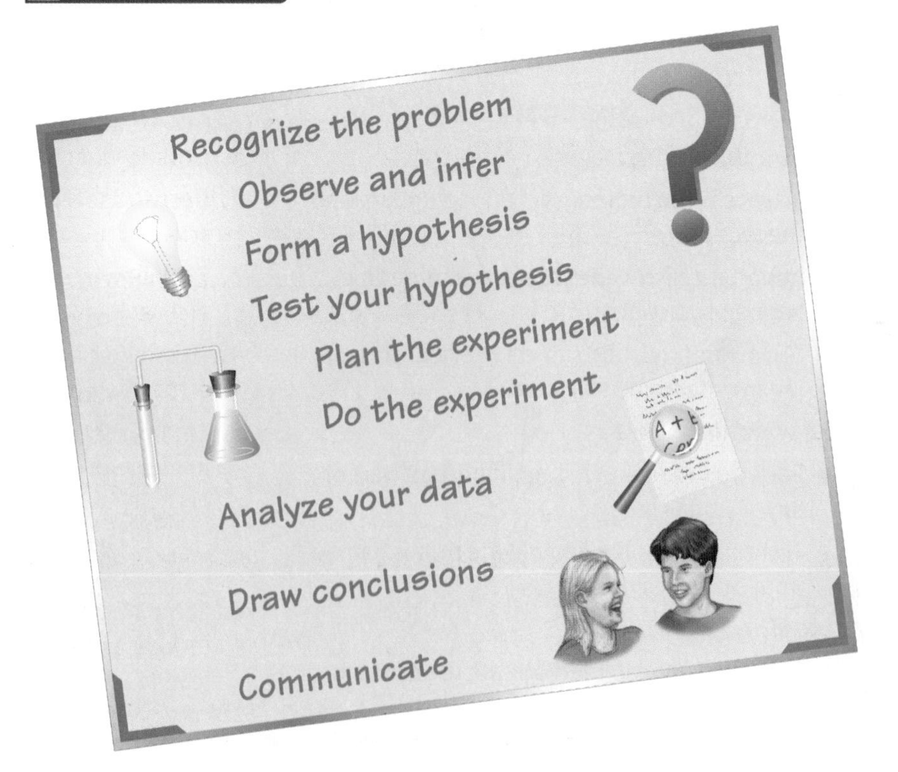

Figure 7 This illustration shows one way to solve a problem or find an answer to a question.

Recognize the Problem

Ben thought about all the science he had learned over the past few months. He was eager to find out more about the world around him. As he looked around his bedroom, he wondered what he could explore. It was then that Ben noticed that the plant on his windowsill was droopy. He quickly watered the wilting plant. Later in the day, Ben observed that the droopy plant had perked up. He concluded that he should water the plant on a regular basis. Every day after school, he watered the plant in his room.

After a few weeks, Ben noticed that the leaves on his plant had turned yellow and brown. He knew from science class that plants need water, so why was this plant not doing well? He talked to his teacher about the plant. She suggested that Ben use what he learned in science class to solve his problem. She pointed out that this problem might make a good project for the upcoming science fair.

Ben already has completed the first step in using a scientific approach to solving a problem—he recognizes a problem. A scientific problem is a question that can be answered using scientific methods. To solve his problem, Ben must do research about his plant. Using sources of information such as those shown in **Figure 8,** Ben identified his plant as a fig. In his Science Journal, he drew a picture of the plant and listed some facts about it.

Reading Check ***What is the first step in any scientific approach to solving a problem?***

Figure 8 Gathering information in the library or on the Internet can make your problem-solving tasks easier.
List *Besides books and computers, what other resources can you use to gather information?*

Observing and Inferring

Procedure

1. Look at the **illustration** above. It is a part of a larger illustration.
2. Record in your Science Journal everything you observe about the illustration.

Analysis

1. Infer what might be happening in the illustration.
2. Compare your inferences with the illustration on the left-hand page of the Chapter Review. How close were your inferences to the actual illustration?

Try at Home

Observe In order for Ben to be able to answer the question about why his plant was not thriving, he needed to plan and carry out an experiment. First, he made and recorded careful observations about his plant. **Observations** can be bits of information you gather with your senses. Most scientific observations are made with your eyes and ears. You also can observe with your senses of touch, taste, and smell. Ben observed that many of the leaves had fallen off his plant. The stem, in places, was peeling. Ben also noticed that some white, powdery, smelly stuff was covering the soil in the pot. He stuck his finger into the soil. It was wet.

Infer Observations like Ben's often lead to inferences. An **inference** is a conclusion about an observation. Ben inferred that perhaps he was watering his plant too often. Can you make any other inferences about why Ben's plant wasn't thriving? To learn more about observing and inferring, do the Try At Home MiniLab on this page.

Form a Hypothesis

After a problem is identified, a scientist might make a hypothesis. A **hypothesis** (hi PAH thuh sus) is a statement that can be tested. **Figure 9** illustrates how hypotheses are based on observations, research, and prior knowledge of a problem. Sometimes more than one hypothesis can be developed.

Table 1 compares and contrasts hypotheses with two other scientific statements—scientific theories and scientific laws. Ben decided to use his inference about watering too often as his hypothesis. His hypothesis was the following: Fig plants grow best when they are watered only once a week.

Table 1 Scientific Statements

Hypothesis	A hypothesis is a statement that is tested with experiments. Hypotheses supported by repeated tests are used to form theories.
Theory	A theory is an explanation supported by results of many experiments. Theories attempt to explain why something happens.
Law	A scientific law describes the behavior of something in nature. Generally, laws predict or describe what will happen in a given situation, but they don't always explain why it happens.

NATIONAL GEOGRAPHIC **VISUALIZING A HYPOTHESIS**

Figure 9

Many hypotheses begin when people notice something interesting. A chance observation led Dr. Katharine Payne to form and test a hypothesis after watching elephant behavior in a zoo. While observing the animals, Payne felt a throbbing sensation in the air. She hypothesized that elephants use sounds that are below the range of human hearing to communicate over long distances.

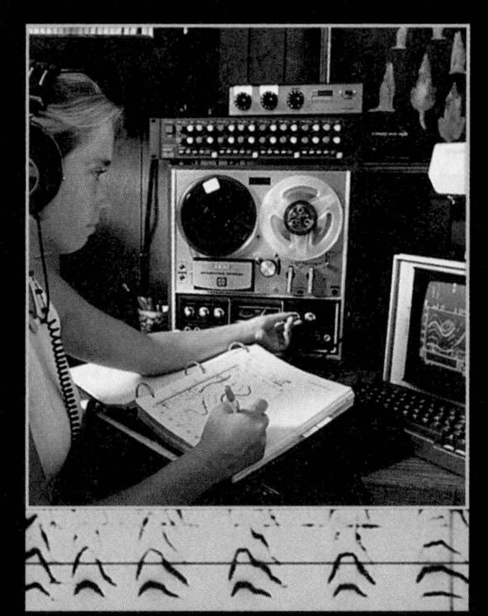

Ⓐ To test her hypothesis, Payne recorded the zoo elephants with special audio equipment, like that used by the researcher at right. Electronic printouts of the recordings revealed that the elephants were indeed making very low frequency sounds.

Ⓑ Dr. Payne and her research team traveled to Africa to further test the hypothesis. They made audio recordings of infrasonic vibrations emitted by elephants in the wild.

Ⓒ From a distance of two miles, Payne played the recordings to a herd of elephants. With raised ears, they stood motionless to locate the source of the sound. Payne used a second group of elephants as a control group. This group was not exposed to the recorded sounds and displayed no unusual behavior. The results supported Payne's hypothesis that elephants use infrasound to communicate.

Figure 10 The number of times each plant is watered in one month is the independent variable in this experiment.

Throughout the experiment, similar-sized fig plants received the same amount of sunlight. They were planted in similar containers with the same type and amount of soil.

Test Your Hypothesis

To test his hypothesis, Ben will carry out an experiment using three plants. An experimental investigation is a series of carefully planned steps used to test a hypothesis. In any experiment, it's important to carefully consider what resources you will use and how to conserve them. It's also important to keep everything the same except for the item or variable you are testing so that you'll know which variable caused the results. The one factor that you change in an experiment is called the **independent variable.**

In Ben's proposed experiment, the independent variable is the number of times he waters each plant in a week. He then will observe how well each plant grows based on how frequently it receives water. The growth of the plants is the dependent variable in Ben's experiment. A **dependent variable** is the factor, or outcome, that will be measured in an experiment. **Figure 10** shows an experiment that tests the effects of water on plants.

What is a dependent variable in an experiment?

Topic: Scientific Methods
Visit green.msscience.com for Web links to information about how scientific research projects are carried out.

Activity Select three science projects that interest you. Make an outline of the steps used in each of the three scientific methods carried out in the projects. Compare and contrast the methods used in the three studies.

Plan the Experiment In order to test only one variable at a time, scientists often use constants. **Constants** are factors in an experiment that stay the same. In his experiment, Ben will use the same species and size of plants, which will be potted with the same kinds and amounts of soil in identical containers. Another constant will be the amount of light each plant will get.

Some experimental investigations also have a control. A **control** is a standard used for comparison. For example, suppose a scientist wished to study the chemical makeup of a soil sample. A control soil—one of known chemistry—could be analyzed first. That way, data from the sample of interest could be directly compared to data from the control soil.

One month later, by keeping other factors constant and changing only one variable—the results of the experiment show the effect of watering frequency on the growth of fig plants.

Do the Experiment Ben gathered all the materials he would need to test his hypothesis. Before he started, Ben knew from Ms. Garcia's labs that he must write down a plan to follow. In his Science Journal, he wrote that he would use three fig plants. Plant A would only be watered once, at the beginning of the experiment. A second fig plant, Plant B, would get watered every day. The third fig plant, Plant C, would get watered once each week. His experiment would last one month.

Ben then made a table for recording his observations. He listed each plant and the number of times it was to be watered. Ben made room in the table for his measurements. He also made a plan to record his observations, which would include the height of each plant, the color of its leaves, and the number of leaves it dropped, if any.

Analyze Your Data

Data are collected during any scientific study. Some data are numeric values such as the length of an object or the temperature of a liquid. Other data you collect may include observations that use adjectives and phrases such as *faster, smaller, not as well as,* and *greener.* An experimenter must record and study the data collected, as shown in **Figure 11,** before he or she can draw conclusions about an experiment.

By the end of the month, Ben observed that the leaves still left on the plant that was watered only once were brown and shriveled. It had lost most of its leaves. The plant that was watered every day had a few leaves left on its branches, but these leaves didn't look too healthy. A white, smelly substance covered the soil. Ben noticed that the plant that was watered once each week had grown the tallest. Many healthy green-and-white leaves extended from its branches.

Figure 11 Observations and other data taken during an experiment must be logged carefully so that they are analyzed and properly interpreted later.

Draw Conclusions and Communicate

After studying his data, Ben was ready to draw some conclusions. A conclusion is a statement based on what is observed. Ben concluded that not watering a plant enough causes the leaves to dry out and die. Watering a plant too much also causes the leaves to die. Watering the plant once a week seems to be the best schedule, of those tested, for a fig plant.

Ben told his teacher about his results. She reminded him that in order to make sure his conclusions were valid, he should repeat his experiment. Ben agreed and did the same experiment again. Based on the results of his second experiment, Ben was able to conclude confidently that watering a fig plant once a week made it grow well in the conditions he used. His hypothesis was supported. An important step in the scientific process is to communicate the results of an investigation. Ben entered his project in his school's science fair, much like the students shown in **Figure 12.**

Figure 12 When results of experiments are communicated, it helps other researchers decide what to do next to help solve a problem.

These students are preparing for their school's science fair.

Plant Heights (cm)			
Week	Plant A	Plant B	Plant C
1	10.5	10.3	10.8
2	10.7	11.2	12.6
3	9.2	12.0	14.6
4	5.1	12.4	15.5

This table shows the results of an experiment similar to Ben's.

section 2 review

Summary

Scientific Methods

- Step-by-step procedures used during scientific investigations are called scientific methods.

Recognize the Problem

- Defining a problem usually is the first step.

Form and Test a Hypothesis

- A hypothesis is a statement that can be tested.
- Hypotheses are tested by controlling, manipulating, and measuring variables.

Analyze Your Data and Draw Conclusions

- Data analysis includes calculations and graphs.
- After data analysis, an investigator draws conclusions and communicates results.

Self Check

1. **List** the steps followed in a typical scientific method.
2. **Explain** how observations are different from inferences.
3. **Evaluate** why experiments should be repeated.
4. **Compare and contrast** a control and a constant.
5. **Think Critically** Isaac Newton once said, "If I have seen further, it is by standing on the shoulders of giants." What do you think Newton meant by this statement?

Applying Skills

6. **Use Variables and Controls** Describe another independent variable that might influence the growth of Ben's figs. How could Ben set up an experiment to test the effect of this different variable?

green.msscience.com/self_check_quiz

Advertising Inferences

Imagine you're reading a magazine and you see an ad for a pest control service. The ad states that 8 out of 10 homes have a problem with carpenter ants. Would you infer that your home might have ants? In this lab, you'll use advertisements to practice the science skills of observing and inferring. How do service providers get their data? Are the data correct?

Real-World Question

What observations and inferences can you make from advertisements?

Goals

- **Make** inferences based on observations.
- **Recognize** the limits of observations.

Materials

magazine advertisements
paper (1 sheet)
colored pencils or markers

Procedure

1. Select three magazine advertisements from those supplied by your teacher. In your Science Journal, make a table like the one shown.

Ad Data

Type of Ad	Observation	Inference
Ad 1	Do not write in this book.	
Ad 2		
Ad 3		

2. For each magazine advertisement, list your observations. For example, you might observe that large, ferocious looking insects are pictured in a pest control ad.
3. What inferences does the magazine advertiser want you to make? Make inferences that relate your observations to the service or product being provided. The pest control advertisement, for example, might lead you to infer that if you don't want to be invaded by insects, you should hire their service.
4. Share your magazine advertisements and inferences with others in your class.

Conclude and Apply

1. **Compare and contrast** your classmates' inferences about the advertisements with your own. Are there other explanations for the things you observed in the advertisements?
2. **Create** your own magazine advertisement to sell a product. Think about what people will observe in the ad and what you want them to infer from it.
3. **Infer** Have a classmate make inferences about your magazine advertisement. What did your classmate infer about the magazine advertisement you created? Is this what you wanted the classmate to infer? Explain.

Communicating Your Data

Describe a new product or service to your class. As a group, brainstorm ideas for an ad to sell your product. **For more help, refer to the** Science Skill Handbook.

Model an Archaeological Dig

Goals

- In this lab you will use the skills, patience, and tools of a scientist while modeling and uncovering your example of a prehistoric site.

Possible Materials

craft sticks
toothpicks
plastic shovels
small paintbrushes
small stones
bits of black tissue paper
sand
interlocking building blocks
clear-plastic storage box
ruler, pencil, and paper

Safety Precautions

Real-World Question

Scientists often use models to study objects that are too large or too small to observe directly. In this lab, your group will construct a model of a prehistoric site. After you cover the site with sand, you'll exchange it with another group for them to unearth. Using materials provided by your teacher, you will create a miniature archaeological dig. What can be learned from an archaeological excavation? How do models help us learn about science?

Procedure

1. Obtain a storage box in which to build your site.
2. Using the materials provided by your teacher, begin by planning what remains your site will contain and where they will be placed.

3. While designing your site, keep in mind that this is an area where people once lived. Artifacts you might want to include are a hearth used for cooking, a trash pit, some sort of shelter, a protective wall, a burial site, a water source, and tools.
4. Now that you've placed your artifacts, create a map of your site. Draw your map to scale.
5. Cover your site with sand so that another group from your class can excavate the artifacts.
6. Exchange your model with one made by another group. Keep the map of your site for now.
7. Using the paintbrushes and shovels given to you, begin by slowly excavating the site your group received.
8. As you excavate, be sure to accurately record the locations of all of your discoveries. Draw a map as you excavate. Be sure to use the measurements you made as you unearthed the site.

Conclude and Apply

1. **Compare and Contrast** How was this experience similar to a real archaeological dig? Did any of the excavating tools damage or disturb the site? How do archaeologists avoid damaging the site?
2. **Infer** How do you think archaeologists recognize findings that aren't familiar to them? What clues do they use?
3. **Explain** Why did you make maps of your site and the site you excavated? How would maps help scientists after they have excavated a site?
4. **Compare and contrast** your map of the site you excavated to the map of the student who made it. How are your maps similar? How are they different? Do the same with the map you made for your own model of your archaeological dig.
5. **Identify** what other things do scientists study using models? Think of a scientific concept you've learned already this year, either small or large, that scientists study by using models.

Communicating Your Data

Make an enlarged version of the map you created while uncovering your findings. Display your maps on a poster like a scientist would. **For more help, refer to the** **Science Skill Handbook.**

Science and Language Arts

Mama Solves a Murder

a mystery novel

by Nora DeLoach

My mama's name is Grace. Everybody calls her "Candi," like candied sweet potatoes, because of her skin, a golden brown color with yellow undertones that looks as smooth as silk.

. . . Mama can be shrewd and cunning. She has uncanny perception and self control. Her mind is formidable[1], her beauty is enticing, but I've seen her use either to get in and out of places.

Mama is fifty-three and she works as a social worker in the small town of Hampton, South Carolina. While most of the things that arouse the mind don't excite Mama, I've seen her absolutely euphoric when her mind is deducing. She's a self-styled private investigator who sees herself as the romantic loner. Mama enjoys the tedious jobs of digging up bits and pieces until she's solved a mystery. Long ago, I don't remember when, Mama decided that if she could get at the truth of a problem, she would have made a contribution to humankind.

Who am I? My name is Simone, I'm Mama's one and only daughter....

...To be perfectly honest, Mama is my Sherlock Holmes and I'm her Dr. Watson.

1 admirable or awe inspiring

Understanding Literature

Characterization An author's method of developing the personality of a character is called characterization. Simone, who is the narrator, talks directly to the reader about her mother. Her own observations help the reader understand the character of Mama. The author also uses an allusion—a reference to a well-known character, place, or situation—to describe her relationship to her mother. Can you find it?

Respond to the Reading

1. Based on this passage, how does Simone feel about her mother?
2. How is Mama like a scientist?
3. **Linking Science and Writing** Create an idea for a protagonist who is a scientist who solves a mystery involving a clue of soil particles on the victim's clothing.

In a mystery novel, readers are entertained by a protagonist, or main character, who employs a scientific method to solve a mystery. In *Mama Solves a Murder,* Mama and Simone create a scientific method as they gather information, make inferences, propose a hypothesis, and then test their hypothesis. Mystery novels usually include descriptions of technology and other resources used to solve the mystery.

Reviewing Main Ideas

Section 1 How Science Works

1. Science is a process of understanding the world around you. Technology is the use of knowledge gained through scientific thinking and problem solving. Archaeologists use science and technology to study the artifacts of ancient people.
2. Many archaeological sites and the artifacts they contain are found by accident. The excavation of an archaeological site is done slowly and carefully so the artifacts and the site itself are not damaged or destroyed.
3. Some artifacts, such as bones and charcoal, can be dated using chemical analyses.

Section 2 Scientific Problem Solving

1. Many scientific experiments involve variables, or factors that can change. An independent variable is a factor that the experimenter changes. The dependent variable is the factor that changes as a result of the independent variable. Independent variables should be changed one at a time, so the experimenter can determine what influenced the dependent variable's change.
2. Constants are factors in an experiment that don't change. A control, when one is included, is a standard used for comparison.
3. Scientific methods are step-by-step approaches to solving problems. These can include identifying the problem, forming and testing a hypothesis, analyzing the results of the test, and drawing conclusions.

Visualizing Main Ideas

Copy and complete the following concept map on methods used in an archaeological study. Use the following words and phrases: library, field studies, cleaning and storing, research, Internet, radar, *and* chemical analysis.

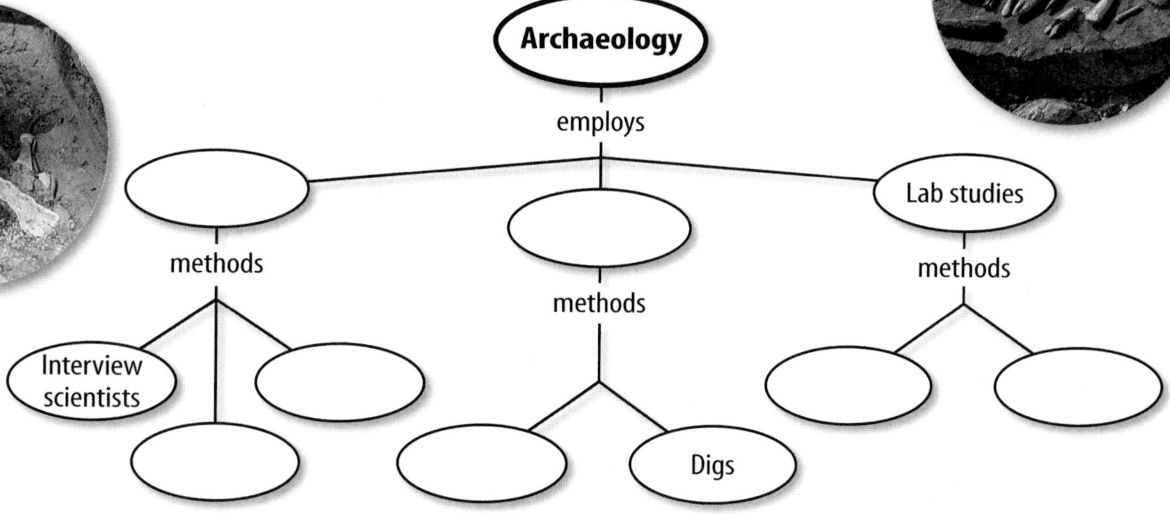

Archaeology
employs
Lab studies
methods
methods
methods
Interview scientists
Digs

chapter 1 Review

Using Vocabulary

constant p. 16
control p. 16
dependent variable p. 16
hypothesis p. 14
independent variable p. 16
inference p. 14
observation p. 14
science p. 7
scientific methods p. 12
technology p. 9

Each phrase below describes a science term from the list. Write the term that matches the phrase describing it.

1. variable changed by the experimenter
2. a statement that can be tested
3. step-by-step approach to solving problems
4. the process of understanding the world
5. a factor that remains the same phases of an experiment
6. new products or tools made because of knowledge gained through science
7. variable measured during an experiment
8. standard used for comparison

Checking Concepts

Choose the word or phrase that best answers the question.

9. What should an experimenter do after analyzing data?
   A) carry out the experiment
   B) draw conclusions
   C) observe and infer
   D) identify the problem

10. Why do scientists make maps of archaeological sites?
   A) to photograph artifacts
   B) to calculate the exact age of artifacts
   C) to record where the artifacts were found
   D) to discover artifacts

Use the illustration below to answer question 11.

11. What is a conclusion that is based on an observation?
   A) a control **C)** an inference
   B) a hypothesis **D)** a variable

12. A scientist publishes the results of an experiment. Which science skill is this?
   A) observing **C)** communicating
   B) inferring **D)** hypothesizing

13. What is a series of carefully planned steps used to test a hypothesis?
   A) a constant **C)** an experiment
   B) an observation **D)** a conclusion

14. Why should an experiment be repeated?
   A) to form a hypothesis
   B) to reduce the chance of error
   C) to change controls
   D) to identify the problem

15. What technology can help an archaelologist "see" a buried site before he or she begins to excavate it?
   A) computer **C)** radar
   B) mapmaker **D)** camera

16. What is the first step in a commonly used scientific method?
   A) digging for artifacts
   B) drawing conclusions
   C) controlling variables
   D) recognizing a problem

Science Online green.msscience.com/vocabulary_puzzlemaker

Thinking Critically

17. **Draw Conclusions** An archaeologist finds a site that contains many different layers of artifacts. What might he or she conclude about the people who lived at the site?

18. **Explain** why the following statement is incorrect: Scientists do all of their work in laboratories.

19. **Identify and Manipulate Variables and Controls** Identify all variables in Ben's fig experiment. Give an example of how Ben kept some variables constant.

20. **Explain** whether every scientific problem is solved using the same steps.

21. **Evaluate** Why is it important to keep good records during a scientific investigation?

22. **Concept Map** Copy the concept map shown on this page in your Science Journal. Then, use the following terms to complete the concept map of a commonly used scientific method: *perform the experiment, analyze data, form a hypothesis,* and *observe and infer.*

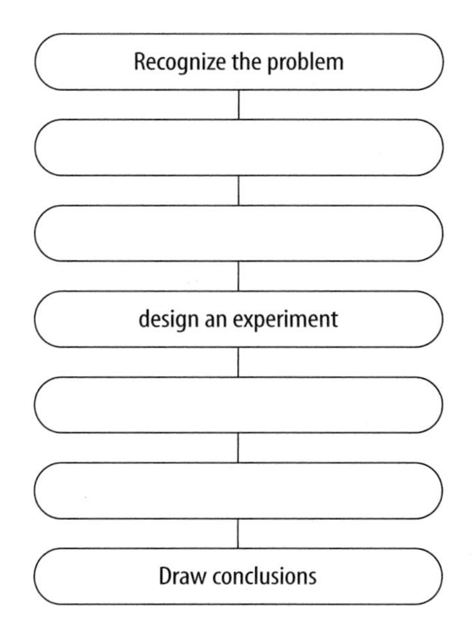

Performance Activities

23. **Design an Experiment** Describe how you might test which laundry soap cleans the best. Be sure to include variables, constants, and a control.

24. **Oral Presentation** Research how technology has been used to study ancient human artifacts in your area. Present your findings in a speech to your class.

Applying Math

25. **Soil Sample** A geologist collected a 2.5-kg soil sample for analysis. If she only needs 20 grams of sample to perform the analysis, what percentage of the soil sample will be tested?

Use the table below to answer questions 26–28.

Roman Units of Measure (Volume)		
Unit	Conversion	Volume (L)
1 congius	= 6 sextarii	3.281
1 modius	= 16 sextarii	8.751
1 amphora	= 3 modii	26.261

26. **Classical Units** In the Classical Period, an amphora was a container used to transport liquids such as wine and olive oil. The amphora also was the name of a unit of volume used during this period. According to the table, about how many sextarii are in one amphora?

27. **Soda Serving** Suppose you were serving your friends a soft drink using the units in the table. Which of the units represents the likeliest single serving of soda? One sextarius is equivalent to how many liters?

28. **Modius Conversion** According to the table, one modius is equivalent to how many liters? How many congii make one modius?

chapter 1 Standardized Test Practice

Part 1 Multiple Choice

Record your answers on the answer sheet provided by your teacher or on a sheet of paper.

1. What predicts or describes the behavior of something in nature?
   A. hypothesis **C.** law
   B. inference **D.** theory

Use the photos below to answer question 2.

2. What step of a scientific method is illustrated in the photos?
   A. observation
   B. hypothesis
   C. conclusion
   D. data collection and analysis

3. Hearing, sight, taste, and smell all are used for which aspects of science?
   A. hypotheses **C.** observations
   B. laws **D.** theories

4. Which of the following is a process of trying to understand things about the world?
   A. science **C.** bias
   B. technology **D.** conclusion

5. Which of the following would not be considered an artifact?
   A. tool **C.** drawing
   B. weapon **D.** soil

6. Which is used as a background research tool?
   A. Reciting a law
   B. Inventing a microchip
   C. Forming a hypothesis
   D. Studying a periodical

Use the figure below to answer question 7.

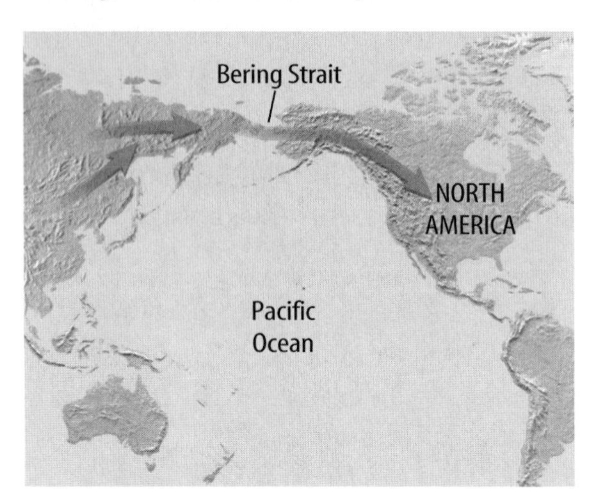

7. According to this map, what best describes a path of ancient people entering North America?
   A. from Asia to North America
   B. from Australia to North America
   C. from the Bering Strait to the Atlantic
   D. from South America to North America

8. What are computers and microscopes examples of?
   A. hypotheses
   B. variables
   C. technology
   D. constants

9. Select a tool that could be used to collect data during an archaeological dig.
   A. law
   B. radar
   C. periodical
   D. artifact

10. Which of the following can be used to test a hypothesis?
   A. experiment
   B. theory
   C. law
   D. variable

Part 2 Short Response/Grid In

Record your answers on the answer sheet provided by your teacher or on a sheet of paper.

Use the photo below to answer question 11.

11. Complete the table by determining whether each statement about the plant is an observation, an inference or a hypothesis.

Statement	Observation, Inference or Hypothesis
The plant needs more water.	
The plant has big leaves.	
The plant does not have flowers. Something is eating the plant.	
If the plant is moved to a different location, then it will be healthier.	
The plant may need more sunlight.	
If an insecticide is used on the plant, then it will become healthier.	

12. How does a hypothesis become a theory?

13. What happens if data are not recorded properly?

14. What is the difference between an inference and a hypothesis?

15. Explain how independent and dependent variables are used in an experiment.

16. Why is it important to have constants in an experiment?

Part 3 Open Ended

Record your answers on a sheet of paper.

17. List several reasons why experiments with plants could be less problematic than those involving humans.

18. Suppose a scientist is studying a disease, such as cancer. Why is it important to communicate his or her findings? List several ways that the scientist could communicate his or her data.

Use the photo below to answer questions 19 and 20.

19. List several observations and inferences and formulate a hypothesis about changing the condition of this plant.

20. Design an experiment to test your hypothesis about the plant. What are your independent and dependent variables? What would you include as constants?

21. Explain how an archaeologist should approach a new site.

chapter 2

Minerals

chapter preview

sections

1 Minerals
Lab *Crystal Formation*

2 Mineral Identification

3 Uses of Minerals
Lab *Mineral Identification*

Virtual Lab *How can minerals be defined by their properties?*

Nature's Beautiful Creation

Although cut by gemologists to enhance their beauty, these gorgeous diamonds formed naturally—deep within Earth. One requirement for a substance to be a mineral is that it must occur in nature. Human-made diamonds serve their purpose in industry but are not considered minerals.

Science Journal Write two questions you would ask a gemologist about the minerals that he or she works with.

Start-Up Activities

Distinguish Rocks from Minerals

When examining rocks, you'll notice that many of them are made of more than one material. Some rocks are made of many different crystals of mostly the same mineral. A mineral, however, will appear more like a pure substance and will tend to look the same throughout. Can you tell a rock from a mineral?

1. Use a magnifying lens to observe a quartz crystal, salt grains, and samples of sandstone, granite, calcite, mica, and schist (SHIHST).
2. Draw a sketch of each sample.
3. Infer which samples are made of one type of material and should be classified as minerals.
4. Infer which samples should be classified as rocks.
5. **Think Critically** In your Science Journal, compile a list of descriptions for the minerals you examined and a second list of descriptions for the rocks. Compare and contrast your observations of minerals and rocks.

Minerals Make the following Foldable to help you better understand minerals.

STEP 1 **Fold** a vertical sheet of notebook paper from side to side.

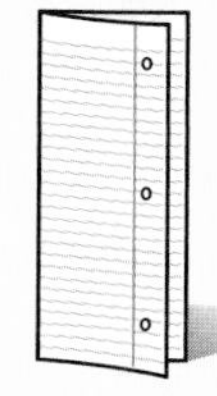

STEP 2 **Cut** along every third line of only the top layer to form tabs.

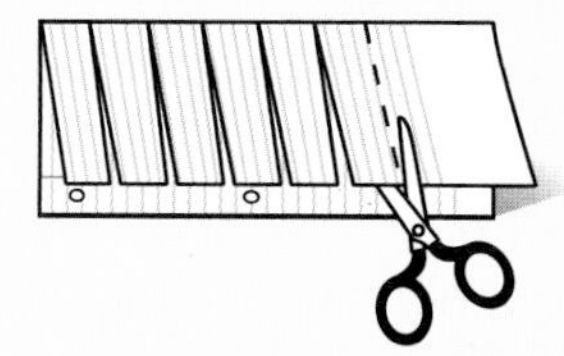

STEP 3 **Label** each tab with a question.

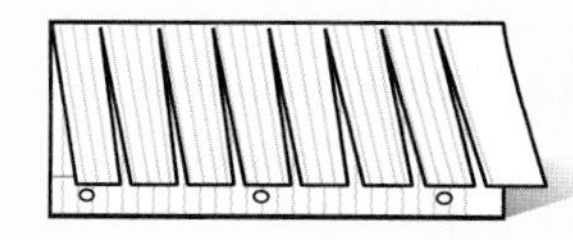

Ask Questions Before you read the chapter, write questions you have about minerals on the front of the tabs. As you read the chapter, add more questions and write answers under the appropriate tabs.

Preview this chapter's content and activities at green.msscience.com

section 1

Minerals

as you read

What You'll Learn

- **Describe** characteristics that all minerals share.
- **Explain** how minerals form.

Why It's Important

You use minerals and products made from them every day.

Review Vocabulary

atoms: tiny particles that make up matter; composed of protons, electrons, and neutrons

New Vocabulary

- mineral
- crystal
- magma
- silicate

What is a mineral?

How important are minerals to you? Very important? You actually own or encounter many things made from minerals every day. Ceramic, metallic, and even some paper items are examples of products that are derived from or include minerals. **Figure 1** shows just a few of these things. Metal bicycle racks, bricks, and the glass in windows would not exist if it weren't for minerals. A **mineral** is a naturally occurring, inorganic solid with a definite chemical composition and an orderly arrangement of atoms. About 4,000 different minerals are found on Earth, but they all share these four characteristics.

Mineral Characteristics First, all minerals are formed by natural processes. These are processes that occur on or inside Earth with no input from humans. For example, salt formed by the natural evaporation of seawater is the mineral halite, but salt formed by evaporation of saltwater solutions in laboratories is not a mineral. Second, minerals are inorganic. This means that they aren't made by life processes. Third, every mineral is an element or compound with a definite chemical composition. For example, halite's composition, NaCl, gives it a distinctive taste that adds flavor to many foods. Fourth, minerals are crystalline solids. All solids have a definite volume and shape. Gases and liquids like air and water have no definite shape, and they aren't crystalline. Only a solid can be a mineral, but not all solids are minerals.

Figure 1 You probably use minerals or materials made from minerals every day without thinking about it.
Infer *How many objects in this picture might be made from minerals?*

Atom Patterns The word *crystalline* means that atoms are arranged in a pattern that is repeated over and over again. For example, graphite's atoms are arranged in layers. Opal, on the other hand, is not a mineral in the strictest sense because its atoms are not all arranged in a definite, repeating pattern, even though it is a naturally occurring, inorganic solid.

Figure 2 More than 200 years ago, the smooth, flat surfaces on crystals led scientists to infer that minerals had an orderly structure inside.

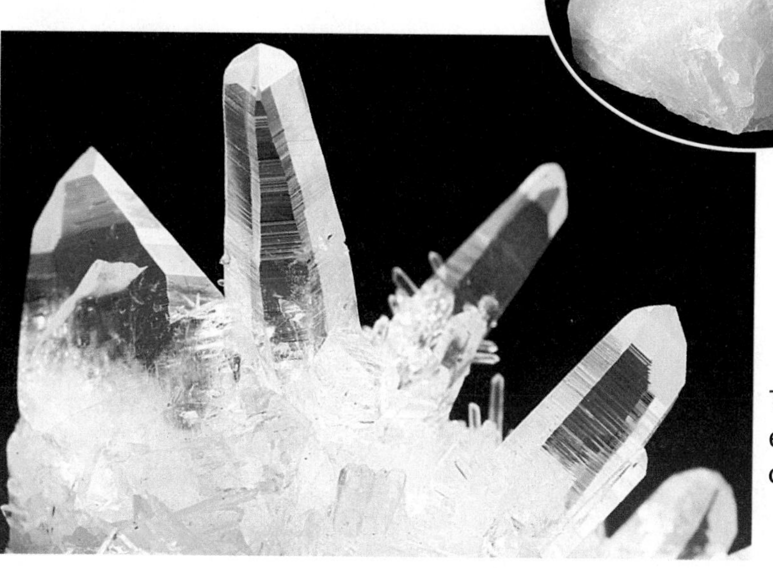

Even though this rose quartz looks uneven on the outside, its atoms have an orderly arrangement on the inside.

The well-formed crystal shapes exhibited by these clear quartz crystals suggest an orderly structure.

The Structure of Minerals

Do you have a favorite mineral sample or gemstone? If so, perhaps it contains well-formed crystals. A **crystal** is a solid in which the atoms are arranged in orderly, repeating patterns. You can see evidence for this orderly arrangement of atoms when you observe the smooth, flat outside surfaces of crystals. A crystal system is a group of crystals that have similar atomic arrangements and therefore similar external crystal shapes.

Reading Check *What is a crystal?*

Crystals Not all mineral crystals have smooth surfaces and regular shapes like the clear quartz crystals in **Figure 2.** The rose quartz in the smaller photo of **Figure 2** has atoms arranged in repeating patterns, but you can't see the crystal shape on the outside of the mineral. This is because the rose quartz crystals developed in a tight space, while the clear quartz crystals developed freely in an open space. The six-sided, or hexagonal crystal shape of the clear quartz crystals in **Figure 2,** and other forms of quartz can be seen in some samples of the mineral. **Figure 3** illustrates the six major crystal systems, which classify minerals according to their crystal structures. The hexagonal system to which quartz belongs is one example of a crystal system.

Crystals form by many processes. Next, you'll learn about two of these processes—crystals that form from magma and crystals that form from solutions of salts.

Mini LAB

Inferring Salt's Crystal System

Procedure

1. Use a **magnifying lens** to observe grains of common **table salt** on a dark sheet of **construction paper.** Sketch the shape of a salt grain. **WARNING:** *Do not taste or eat mineral samples. Keep hands away from your face.*
2. Compare the shapes of the salt crystals with the shapes of crystals shown in **Figure 3.**

Analysis

1. Which characteristics do all the grains have in common?
2. Research another mineral with the same crystal system as salt. What is this crystal system called?

Try at Home

NATIONAL GEOGRAPHIC VISUALIZING CRYSTAL SYSTEMS

Figure 3

A crystal's shape depends on how its atoms are arranged. Crystal shapes can be organized into groups known as crystal systems—shown here in 3-D with geometric models (in blue). Knowing a mineral's crystal system helps researchers understand its atomic structure and physical properties.

▲ CUBIC **Fluorite is an example of a mineral that forms cubic crystals. Minerals in the cubic crystal system are equal in size along all three principal dimensions.**

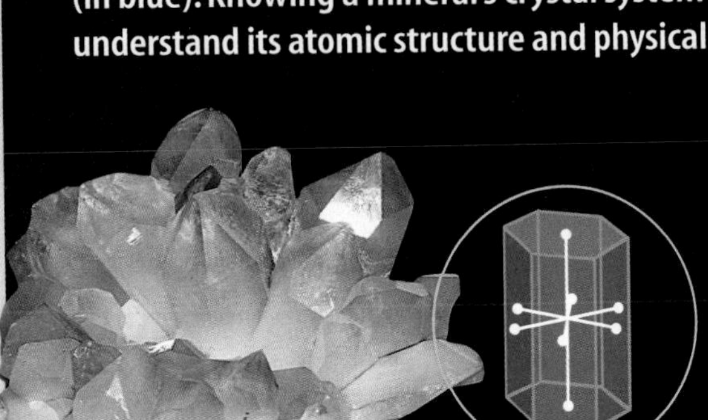

▲ HEXAGONAL **(hek SA guh nul) In hexagonal crystals, horizontal distances between opposite crystal surfaces are equal. These crystal surfaces intersect to form 60° or 120° angles. The vertical length is longer or shorter than the horizontal lengths.**

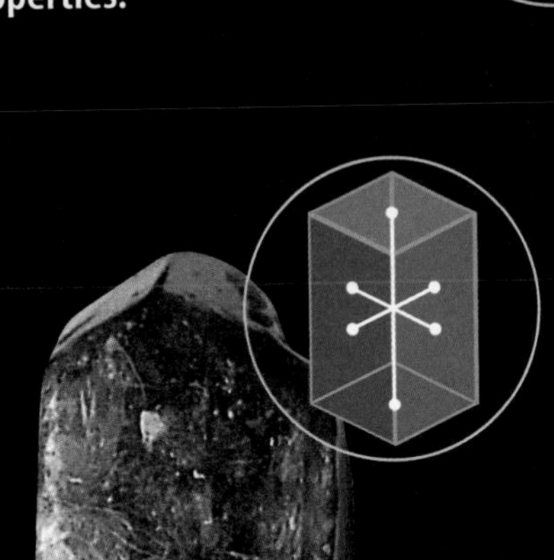

◀ TETRAGONAL **(te TRA guh nul) Zircon crystals are tetragonal. Tetragonal crystals are much like cubic crystals, except that one of the principal dimensions is longer or shorter than the other two dimensions.**

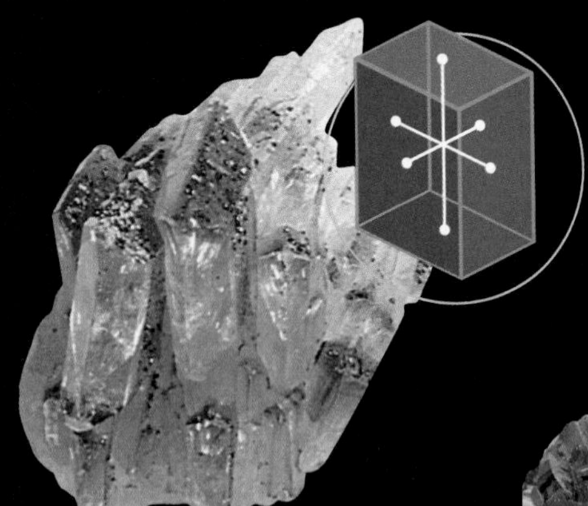

▲ ORTHORHOMBIC **(awr thuh RAHM bihk) Minerals with orthorhombic structure, such as barite, have dimensions that are unequal in length, resulting in crystals with a brick-like shape.**

▲ MONOCLINIC **(mah nuh KLIH nihk) Minerals in the monoclinic system, such as orthoclase, also exhibit unequal dimensions in their crystal structure. Only one right angle forms where crystal surfaces meet. The other angles are oblique, which means they don't form 90° angles where they intersect.**

▲ TRICLINIC **(tri KLIH nihk) The triclinic crystal system includes minerals exhibiting the least symmetry. Triclinic crystals, such as rhodonite (ROH dun ite), are unequal in all dimensions, and all angles where crystal surfaces meet are oblique.**

Figure 4 Minerals form by many natural processes.

A This rock formed as magma cooled slowly, allowing large mineral grains to form.

Labradorite

B Some minerals form when salt water evaporates, such as these white crystals of halite in Death Valley, California.

Crystals from Magma Natural processes form minerals in many ways. For example, hot melted rock material, called **magma,** cools when it reaches Earth's surface, or even if it's trapped below the surface. As magma cools, its atoms lose heat energy, move closer together, and begin to combine into compounds. During this process, atoms of the different compounds arrange themselves into orderly, repeating patterns. The type and amount of elements present in a magma partly determine which minerals will form. Also, the size of the crystals that form depends partly on how rapidly the magma cools.

When magma cools slowly, the crystals that form are generally large enough to see with the unaided eye, as shown in **Figure 4A.** This is because the atoms have enough time to move together and form into larger crystals. When magma cools rapidly, the crystals that form will be small. In such cases, you can't easily see individual mineral crystals.

Crystals from Solution Crystals also can form from minerals dissolved in water. When water evaporates, as in a dry climate, ions that are left behind can come together to form crystals like the halite crystals in **Figure 4B.** Or, if too much of a substance is dissolved in water, ions can come together and crystals of that substance can begin to form in the solution. Minerals can form from a solution in this way without the need for evaporation.

Crystal Formation Evaporites commonly form in dry climates. Research the changes that take place when a saline lake or shallow sea evaporates and halite or gypsum forms.

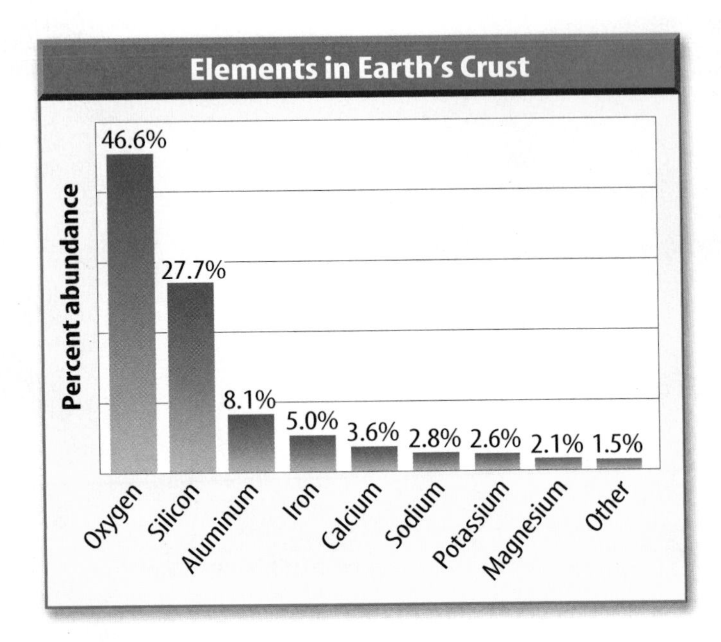

Figure 5 Most of Earth's crust is composed of eight elements.

Mineral Compositions and Groups

Ninety elements occur naturally in Earth's crust. Approximately 98 percent (by weight) of the crust is made of only eight of these elements, as shown in **Figure 5.** Of the thousands of known minerals, only a few dozen are common, and these are mostly composed of the eight most common elements in Earth's crust.

Most of the common rock-forming minerals belong to a group called the silicates. **Silicates** (SIH luh kayts) are minerals that contain silicon (Si) and oxygen (O) and usually one or more other elements. Silicon and oxygen are the two most abundant elements in Earth's crust. These two elements alone combine to form the basic building blocks of most of the minerals in Earth's crust and mantle. Feldspar and quartz, which are silicates, and calcite, which is a carbonate, are examples of common, rock-forming minerals. Other mineral groups also are defined according to their compositions.

section 1 review

Summary

What is a mineral?

- Many products used by humans are made from minerals.
- Minerals are defined by four main characteristics.

The Structure of Minerals

- The crystal shape of a mineral reflects the way in which its atoms are arranged.
- Minerals are classified according to the types of atoms in their structures and the way that the atoms are arranged.

Mineral Compositions and Groups

- Only eight elements form approximately 98 percent (by weight) of Earth's crust.
- The majority of Earth's crust is composed of silicate minerals.

Self Check

1. **List** four characteristics that all minerals share.
2. **Describe** two ways that minerals can form from solution.
3. **Explain** whether diamonds made in the laboratory are considered to be minerals.
4. **Describe** how crystals of minerals are classified.
5. **Think Critically** The mineral dolomite, a rock-forming mineral, contains oxygen, carbon, magnesium, and calcium. Is dolomite a silicate? Explain.

Applying Skills

6. **Graph** Make a graph of your own design that shows the relative percentages of the eight most common elements in Earth's crust. Then determine the approximate percentage of the crust that is made up of iron and aluminum. If one is available, you may use an electronic spreadsheet program to make your graph and perform the calculation.

Science Online green.msscience.com/self_check_quiz

Crystal Formation

In this lab, you'll have a chance to learn how crystals form from solutions.

Real-World Question

How do crystals form from solution?

Goals

- **Compare and contrast** the crystals that form from salt and sugar solutions.
- **Observe** crystals and infer how they formed.

Materials

250-mL beakers (2)
cardboard
large paper clip
table salt
flat wooden stick
granulated sugar
cotton string
hot plate
magnifying lens
thermal mitt
shallow pan
spoon

Safety Precautions

WARNING: *Never taste or eat any lab materials.*

Procedure

1. Gently mix separate solutions of salt in water and sugar in water in the two beakers. Keep stirring the solutions as you add salt or sugar to the water. Stop mixing when no more salt or sugar will dissolve in the solutions. Label each beaker.
2. Place the sugar solution beaker on a hot plate. Use the hot plate to heat the sugar solution gently. **WARNING:** *Do not touch the hot beaker without protecting your hands.*
3. Tie one end of the thread to the middle of the wooden stick. Tie a large paper clip to the free end of the string for weight. Place the stick across the opening of the sugar beaker so the thread dangles in the sugar solution.
4. Remove the beaker from the hot plate and cover it with cardboard. Place it in a location where it won't be disturbed.
5. Pour a thin layer of the salt solution into the shallow pan.
6. Leave the beaker and the shallow pan undisturbed for at least one week.
7. After one week, examine each solution with a magnifying lens to see whether crystals have formed.

Conclude and Apply

1. **Compare and contrast** the crystals that formed from the salt and the sugar solutions. How do they compare with samples of table salt and sugar?
2. **Describe** what happened to the saltwater solution in the shallow pan.
3. Did this same process occur in the sugar solution? Explain.

Communicating Your Data

Make a poster that describes your methods of growing salt and sugar crystals. Present your results to your class.

section 2

Mineral Identification

as you read

What You'll Learn

- **Describe** physical properties used to identify minerals.
- **Identify** minerals using physical properties such as hardness and streak.

Why It's Important

Identifying minerals helps you recognize valuable mineral resources.

Review Vocabulary
physical property: any characteristic of a material that you can observe without changing the identity of the material

New Vocabulary
- hardness
- luster
- specific gravity
- streak
- cleavage
- fracture

Physical Properties

Why can you recognize a classmate when you see him or her in a crowd away from school? A person's height or the shape of his or her face helps you tell that person from the rest of your class. Height and facial shape are two properties unique to individuals. Individual minerals also have unique properties that distinguish them.

Mineral Appearance Just like height and facial characteristics help you recognize someone, mineral properties can help you recognize and distinguish minerals. Color and appearance are two obvious clues that can be used to identify minerals.

However, these clues alone aren't enough to recognize most minerals. The minerals pyrite and gold are gold in color and can appear similar, as shown in **Figure 6.** As a matter of fact, pyrite often is called fool's gold. Gold is worth a lot of money, whereas pyrite has little value. You need to look at other properties of minerals to tell them apart. Some other properties to study include how hard a mineral is, how it breaks, and its color when crushed into a powder. Every property you observe in a mineral is a clue to its identity.

Figure 6 The general appearance of a mineral often is not enough to identify it.

Using only color, observers can be fooled when trying to distinguish between pyrite and gold.

The mineral azurite is identified readily by its striking blue color.

Hardness A measure of how easily a mineral can be scratched is its **hardness.** The mineral talc is so soft you can scratch it loose with your fingernail. Talcum powder is made from this soft mineral. Diamonds, on the other hand, are the hardest mineral. Some diamonds are used as cutting tools, as shown in **Figure 7.** A diamond can be scratched only by another diamond. Diamonds can be broken, however.

Why is hardness sometimes referred to as scratchability?

Figure 7 Some saw blades have diamonds embedded in them to help slice through materials, such as this limestone. Blades are kept cool by running water over them.

Sometimes the concept of hardness is confused with whether or not a mineral will break. It is important to understand that even though a diamond is extremely hard, it can shatter if given a hard enough blow in the right direction along the crystal.

Mohs Scale In 1824, the Austrian scientist Friedrich Mohs developed a list of common minerals to compare their hardnesses. This list is called Mohs scale of hardness, as seen in **Table 1.** The scale lists the hardness of ten minerals. Talc, the softest mineral, has a hardness value of one, and diamond, the hardest mineral, has a value of ten.

Here's how the scale works. Imagine that you have a clear or whitish-colored mineral that you know is either fluorite or quartz. You try to scratch it with your fingernail and then with an iron nail. You can't scratch it with your fingernail but you can scratch it with the iron nail. Because the hardness of your fingernail is 2.5 and that of the iron nail is 4.5, you can determine the unknown mineral's hardness to be somewhere around 3 or 4. Because it is known that quartz has a hardness of 7 and fluorite has a hardness of 4, the mystery mineral must be fluorite.

Some minerals have a hardness range rather than a single hardness value. This is because atoms are arranged differently in different directions in their crystal structures.

Table 1 Mineral Hardness

Mohs Scale	Hardness	Hardness of Common Objects	
Talc (softest)	1		
Gypsum	2	fingernail	(2.5)
Calcite	3	piece of copper	(2.5 to 3.0)
Fluorite	4	iron nail	(4.5)
Apatite	5	glass	(5.5)
Feldspar	6	steel file	(6.5)
Quartz	7	streak plate	(7.0)
Topaz	8		
Corundum	9		
Diamond (hardest)	10		

Figure 8 Luster is an important physical property that is used to distinguish minerals. Graphite has a metallic luster. Fluorite has a nonmetallic, glassy luster.

Luster The way a mineral reflects light is known as **luster.** Luster can be metallic or nonmetallic. Minerals with a metallic luster, like the graphite shown in **Figure 8,** shine like metal. Metallic luster can be compared to the shine of a metal belt buckle, the shiny chrome trim on some cars, or the shine of metallic cooking utensils. When a mineral does not shine like metal, its luster is nonmetallic. Examples of terms for nonmetallic luster include dull, pearly, silky, and glassy. Common examples of minerals with glassy luster are quartz, calcite, halite, and fluorite.

Specific Gravity Minerals also can be distinguished by comparing the weights of equal-sized samples. The **specific gravity** of a mineral is the ratio of its weight compared with the weight of an equal volume of water. Like hardness, specific gravity is expressed as a number. If you were to research the specific gravities of gold and pyrite, you'd find that gold's specific gravity is about 19, and pyrite's is 5. This means that gold is about 19 times heavier than water and pyrite is 5 times heavier than water. You could experience this by comparing equal-sized samples of gold and pyrite in your hands—the pyrite would feel much lighter. The term *heft* is sometimes used to describe how heavy a mineral sample feels.

Applying Science

How can you identify minerals?

Properties of Minerals

Mineral	Hardness	Streak
Copper	2.5–3	copper-red
Galena	2.5	dark gray
Gold	2.5–3	yellow
Hematite	5.5–6.5	red to brown
Magnetite	6–6.5	black
Silver	2.5–3	silver-white

You have learned that minerals are identified by their physical properties, such as streak, hardness, cleavage, and color. Use your knowledge of mineral properties and your ability to read a table to solve the following problems.

Identifying the Problem

The table includes hardnesses and streak colors for several minerals. How can you use these data to distinguish minerals?

Solving the Problem

1. What test would you perform to distinguish hematite from copper? How would you carry out this test?
2. How could you distinguish copper from galena? What tool would you use?
3. What would you do if two minerals had the same hardness and the same streak color?

Streak When a mineral is rubbed across a piece of unglazed porcelain tile, as in **Figure 9,** a streak of powdered mineral is left behind. **Streak** is the color of a mineral when it is in a powdered form. The streak test works only for minerals that are softer than the streak plate. Gold and pyrite can be distinguished by a streak test. Gold has a yellow streak and pyrite has a greenish-black or brownish-black streak.

Some soft minerals will leave a streak even on paper. The last time you used a pencil to write on paper, you left a streak of the mineral graphite. One reason that graphite is used in pencil lead is because it is soft enough to leave a streak on paper.

Figure 9 Streak is more useful for mineral identification than is mineral color. Hematite, for example, can be dark red, gray, or silver in color. However, its streak is always dark reddish-brown.

Reading Check ***Why do gold and pyrite leave a streak, but quartz does not?***

Cleavage and Fracture The way a mineral breaks is another clue to its identity. Minerals that break along smooth, flat surfaces have **cleavage** (KLEE vihj). Cleavage, like hardness, is determined partly by the arrangement of the mineral's atoms. Mica is a mineral that has one perfect cleavage. **Figure 10** shows how mica breaks along smooth, flat planes. If you were to take a layer cake and separate its layers, you would show that the cake has cleavage. Not all minerals have cleavage. Minerals that break with uneven, rough, or jagged surfaces have **fracture.** Quartz is a mineral with fracture. If you were to grab a chunk out of the side of that cake, it would be like breaking a mineral that has fracture.

Figure 10 Weak or fewer bonds within the structures of mica and halite allow them to be broken along smooth, flat cleavage planes.
Infer *If you broke quartz, would it look the same?*

Observing Mineral Properties

Procedure

1. Obtain samples of some of the following clear minerals: **gypsum, muscovite mica, halite,** and **calcite.**
2. Place each sample over the print on this page and observe the letters.

Analysis

1. Which mineral can be identified by observing the print's double image?
2. What other special property is used to identify this mineral?

Figure 11 Some minerals are natural magnets, such as this lodestone, which is a variety of magnetite.

Other Properties Some minerals have unique properties. Magnetite, as you can guess by its name, is attracted to magnets. Lodestone, a form of magnetite, will pick up iron filings like a magnet, as shown in **Figure 11.** Light forms two separate rays when it passes through calcite, causing you to see a double image when viewed through transparent specimens. Calcite also can be identified because it fizzes when hydrochloric acid is put on it.

Now you know that you sometimes need more information than color and appearance to identify a mineral. You also might need to test its streak, hardness, luster, and cleavage or fracture. Although the overall appearance of a mineral can be different from sample to sample, its physical properties remain the same.

section 2 review

Summary

Physical Properties

- Minerals are identified by observing their physical properties.
- Hardness is a measure of how easily a mineral can be scratched.
- Luster describes how a mineral reflects light.
- Specific gravity is the ratio of the weight of a mineral sample compared to the weight of an equal volume of water.
- Streak is the color of a powdered mineral.
- Minerals with cleavage break along smooth, flat surfaces in one or more directions.
- Fracture describes any uneven manner in which a mineral breaks.
- Some minerals react readily with acid, form a double image, or are magnetic.

Self Check

1. **Compare and contrast** a mineral fragment that has one cleavage direction with one that has only fracture.
2. **Explain** how an unglazed porcelain tile can be used to identify a mineral.
3. **Explain** why streak often is more useful for mineral identification than color.
4. **Determine** What hardness does a mineral have if it does not scratch glass but it scratches an iron nail?
5. **Think Critically** What does the presence of cleavage planes within a mineral tell you about the chemical bonds that hold the mineral together?

Applying Skills

6. **Draw Conclusions** A large piece of the mineral halite is broken repeatedly into several perfect cubes. How can this be explained?

Science Online green.msscience.com/self_check_quiz

section 3

Uses of Minerals

Gems

Walking past the window of a jewelry store, you notice a large selection of beautiful jewelry—a watch sparkling with diamonds, a necklace holding a brilliant red ruby, and a gold ring. For thousands of years, people have worn and prized minerals in their jewelry. What makes some minerals special? What unusual properties do they have that make them so valuable?

Properties of Gems As you can see in **Figure 12, gems** or gemstones are highly prized minerals because they are rare and beautiful. Most gems are special varieties of a particular mineral. They are clearer, brighter, or more colorful than common samples of that mineral. The difference between a gem and the common form of the same mineral can be slight. Amethyst is a gem form of quartz that contains just traces of iron in its structure. This small amount of iron gives amethyst a desirable purple color. Sometimes a gem has a crystal structure that allows it to be cut and polished to a higher quality than that of a nongem mineral. **Table 2** lists popular gems and some locations where they have been collected.

as you read

What You'll Learn

- **Describe** characteristics of gems that make them more valuable than other minerals.
- **Identify** useful elements that are contained in minerals.

Why It's Important

Minerals are necessary materials for decorative items and many manufactured products.

Review Vocabulary
metal: element that typically is a shiny, malleable solid that conducts heat and electricity well

New Vocabulary
- gem
- ore

Figure 12 It is easy to see why gems are prized for their beauty and rarity. Shown here is The Imperial State Crown, made for Queen Victoria of England in 1838. It contains thousands of jewels, including diamonds, rubies, sapphires, and emeralds.

Table 2 Minerals and Their Gems

Fun Facts	Mineral	Gem Example	Some Important Locations
Beryl is named for the element beryllium, which it contains. Some crystals reach several meters in length.	Beryl		Colombia, Brazil, South Africa, North Carolina
A red spinel in the British crown jewels has a mass of 352 carats. A carat is 0.2 g.	Spinel		Sri Lanka, Thailand, Myanmar (Burma)
Purplish-blue examples of zoisite were discovered in 1967 near Arusha, Tanzania.	Zoisite		Tanzania
The most valuable examples are yellow, pink, and blue varieties.	Topaz (uncut)		Siberia, Germany, Japan, Mexico, Brazil, Colorado, Utah, Texas, California, Maine, Virginia, South Carolina

Fun Facts	Mineral	Gem Example	Some Important Locations
Olivine composes a large part of Earth's upper mantle. It is also present in moon rocks.		Peridot	Myanmar (Burma), Zebirget (Saint John's Island, located in the Red Sea), Arizona, New Mexico
Garnet is a common mineral found in a wide variety of rock types. The red color of the variety almandine is caused by iron in its crystal structure.		Almandine	Ural Mountains, Italy, Madagascar, Czech Republic, India, Sri Lanka, Brazil, North Carolina, Arizona, New Mexico
Quartz makes up about 30 percent of Earth's continental crust.		Amethyst	Colorless varieties in Hot Springs, Arkansas; Amethyst in Brazil, Uruguay, Madagascar, Montana, North Carolina, California, Maine
The blue color of sapphire is caused by iron or titanium in corundum. Chromium in corundum produces the red color of ruby.		Blue sapphire	Thailand, Cambodia, Sri Lanka, Kashmir

Important Gems All gems are prized, but some are truly spectacular and have played an important role in history. For example, the Cullinan diamond, found in South Africa in 1905, was the largest uncut diamond ever discovered. Its mass was 3,106.75 carats (about 621 g). The Cullinan diamond was cut into 9 main stones and 96 smaller ones. The largest of these is called the Cullinan 1 or Great Star of Africa. Its mass is 530.20 carats (about 106 g), and it is now part of the British monarchy's crown jewels, shown in **Figure 13A.**

Another well-known diamond is the blue Hope diamond, shown in **Figure 13B.** This is perhaps the most notorious of all diamonds. It was purchased by Henry Philip Hope around 1830, after whom it is named. Because his entire family as well as a later owner suffered misfortune, the Hope diamond has gained a reputation for bringing its owner bad luck. The Hope diamond's mass is 45.52 carats (about 9 g). Currently it is displayed in the Smithsonian Institution in Washington, D.C.

Topic: Gemstone Data
Visit green.msscience.com for Web links to information about gems at the Smithsonian Museum of Natural History.

Activity List three important examples of gems other than those described on this page. Prepare a data table with the heads *Gem Name/Type, Weight (carats/grams), Mineral,* and *Location.* Fill in the table entries for the gemstones you selected.

Useful Gems In addition to their beauty, some gems serve useful purposes. You learned earlier that diamonds have a hardness of 10 on Mohs scale. They can scratch almost any material—a property that makes them useful as industrial abrasives and cutting tools. Other useful gems include rubies, which are used to produce specific types of laser light. Quartz crystals are used in electronics and as timepieces. When subjected to an electric field, quartz vibrates steadily, which helps control frequencies in electronic devices and allows for accurate timekeeping.

Most industrial diamonds and other gems are synthetic, which means that humans make them. However, the study of natural gems led to their synthesis, allowing the synthetic varieties to be used by humans readily.

Figure 13 These gems are among the most famous examples of precious stones.

A The Great Star of Africa is part of a sceptre in the collection of British crown jewels.

B Beginning in 1668, the Hope diamond was part of the French crown jewels. Then known as the French Blue, it was stolen in 1792 and later surfaced in London, England in 1812.

Figure 14 Bauxite, an ore of aluminum, is processed to make pure aluminum metal for useful products.

Useful Elements in Minerals

Gemstones are perhaps the best-known use of minerals, but they are not the most important. Look around your home. How many things made from minerals can you name? Can you find anything made from iron?

Ores Iron, used in everything from frying pans to ships, is obtained from its ore, hematite. A mineral or rock is an **ore** if it contains a useful substance that can be mined at a profit. Magnetite is another mineral that contains iron.

When is a mineral also an ore?

INTEGRATE Chemistry

Aluminum sometimes is refined, or purified, from the ore bauxite, shown in **Figure 14.** In the process of refining aluminum, aluminum oxide powder is separated from unwanted materials that are present in the original bauxite. After this, the aluminum oxide powder is converted to molten aluminum by a process called smelting.

During smelting, a substance is melted to separate it from any unwanted materials that may remain. Aluminum can be made into useful products like bicycles, soft-drink cans, foil, and lightweight parts for airplanes and cars. The plane flown by the Wright brothers during the first flight at Kitty Hawk had an engine made partly of aluminum.

Historical Mineralogy An early scientific description of minerals was published by Georgius Agricola in 1556. Use print and online resources to research the mining techniques discussed by Agricola in his work *De Re Metallica.*

Figure 15 The mineral sphalerite (greenish when nearly pure) is an important source of zinc. Iron often is coated with zinc to prevent rust in a process called galvanization.

Vein Minerals Under certain conditions, metallic elements can dissolve in fluids. These fluids then travel through weaknesses in rocks and form mineral deposits. Weaknesses in rocks include natural fractures or cracks, faults, and surfaces between layered rock formations. Mineral deposits left behind that fill in the open spaces created by the weaknesses are called vein mineral deposits.

Reading Check *How do fluids move through rocks?*

Sometimes vein mineral deposits fill in the empty spaces after rocks collapse. An example of a mineral that can form in this way is shown in **Figure 15.** This is the shiny mineral sphalerite, a source of the element zinc, which is used in batteries. Sphalerite sometimes fills spaces in collapsed limestone.

Minerals Containing Titanium You might own golf clubs with titanium shafts or a racing bicycle containing titanium. Perhaps you know someone who has a titanium hip or knee replacement. Titanium is a durable, lightweight, metallic element derived from minerals that contain this metal in their crystal structures. Two minerals that are sources of the element titanium are ilmenite (IHL muh nite) and rutile (rew TEEL), shown in **Figure 16.** Ilmenite and rutile are common in rocks that form when magma cools and solidifies. They also occur as vein mineral deposits and in beach sands.

Figure 16 Rutile and ilmenite are common ore minerals of the element titanium.

Figure 17 Wheelchairs used for racing and playing basketball often have parts made from titanium.

Uses for Titanium Titanium is used in automobile body parts, such as connecting rods, valves, and suspension springs. Low density and durability make it useful in the manufacture of aircraft, eyeglass frames, and sports equipment such as tennis rackets and bicycles. Wheelchairs used by people who want to race or play basketball often are made from titanium, as shown in **Figure 17.** Titanium is one of many examples of useful materials that come from minerals and that enrich humans' lives.

section 3 review

Summary

Gems

- Gems are highly prized mineral specimens often used as decorative pieces in jewelry or other items.
- Some gems, especially synthetic ones, have industrial uses.

Useful Elements in Minerals

- Economically important quantities of useful elements or compounds are present in ores.
- Ores generally must be processed to extract the desired material.
- Iron, aluminum, zinc, and titanium are common metals that are extracted from minerals.

Self Check

1. **Explain** why the Cullinan diamond is an important gem.
2. **Identify** Examine **Table 2.** What do rubies and sapphires have in common?
3. **Describe** how vein minerals form.
4. **Explain** why bauxite is considered to be a useful rock.
5. **Think Critically** Titanium is nontoxic. Why is this important in the manufacture of artificial body parts?

Applying Skills

6. **Use Percentages** Earth's average continental crust contains 5 percent iron and 0.007 percent zinc. How many times more iron than zinc is present in average continental crust?

Science Online green.msscience.com/self_check_quiz

Mineral Identification

Goals

- **Hypothesize** which properties of each mineral are most useful for identification purposes.
- **Test** your hypothesis as you attempt to identify unknown mineral samples.

Materials

mineral samples
magnifying lens
pan balance
graduated cylinder
water
piece of copper
**copper penny*
glass plate
small iron nail
steel file
streak plate
5% HCl with dropper
Mohs scale of hardness
Minerals Appendix
**minerals field guide*
safety goggles
**Alternate materials*

Safety Precautions

WARNING: *If an HCl spill occurs, notify your teacher and rinse with cool water until you are told to stop. Do not taste, eat, or drink any lab materials.*

Real-World Question

Although certain minerals can be identified by observing only one property, others require testing several properties to identify them. How can you identify unknown minerals?

Procedure

1. Copy the data table into your Science Journal. Obtain a set of unknown minerals.
2. Observe a numbered mineral specimen carefully. Write a star in the table entry that represents what you hypothesize is an important physical property. Choose one or two properties that you think will help most in identifying the sample.
3. Perform tests to observe your chosen properties first.
   a. To estimate hardness:
      - Rub the sample firmly against objects of known hardness and observe whether it leaves a scratch on the objects.
      - Estimate a hardness range based on which items the mineral scratches.

   b. To estimate specific gravity: Perform a density measurement.
      - Use the pan balance to determine the sample's mass, in grams.

- Measure its volume using a graduated cylinder partially filled with water. The amount of water displaced by the immersed sample, in mL, is an estimate of its volume in cm^3.
- Divide mass by volume to determine density. This number, without units, is comparable to specific gravity.

4. With the help of the Mineral Appendix or a field guide, attempt to identify the sample using the properties from step 2. Perform more physical property observations until you can identify the sample. Repeat steps 2 through 4 for each unknown.

Physical Properties of Minerals

Sample Number	Hardness	Cleavage or Fracture	Color	Specific Gravity	Luster and Streak	Crystal Shape	Other Properties	Mineral Name
1								
2			Do not write in this book.					
etc.								

Analyze Your Data

1. Which properties were most useful in identifying your samples? Which properties were least useful?
2. **Compare** the properties that worked best for you with those that worked best for other students.

Conclude and Apply

1. **Determine** two properties that distinguish clear, transparent quartz from clear, transparent calcite. Explain your choice of properties.
2. Which physical properties would be easiest to determine if you found a mineral specimen in the field?

Communicating Your Data

For three minerals, list physical properties that were important for their identification. **For more help, refer to the** Science Skill Handbook.

TIME **SCIENCE AND HISTORY** SCIENCE CAN CHANGE THE COURSE OF HISTORY!

Dr. Dorothy Crowfoot Hodgkin

Like X rays, electrons are diffracted by crystalline substances, revealing information about their internal structures and symmetry. This electron diffraction pattern of titanium was obtained with an electron beam focused along a specific direction in the crystal.

Trailblazing scientist and humanitarian

What contributions did Dorothy Crowfoot Hodgkin make to science?

Dr. Hodgkin used a method called X-ray crystallography (kris tuh LAH gruh fee) to figure out the structures of crystalline substances, including vitamin B^{12}, vitamin D, penicillin, and insulin.

What's X-ray crystallography?

Scientists expose a crystalline sample to X rays. As X rays travel through a crystal, the crystal diffracts, or scatters, the X rays into a regular pattern. Like an individual's fingerprints, each crystalline substance has a unique diffraction pattern. Crystallography has applications in the life, Earth, and physical sciences. For example, geologists use X-ray crystallography to identify and study minerals found in rocks.

1910–1994

What were some obstacles Hodgkin overcame?

During the 1930s, there were few women scientists. Hodgkin was not even allowed to attend meetings of the chemistry faculty where she taught because she was a woman. Eventually, she won over her colleagues with her intelligence and tenacity.

How does Hodgkin's research help people today?

Dr. Hodgkin's discovery of the structure of insulin helped scientists learn how to control diabetes, a disease that affects more than 15 million Americans. Diabetics' bodies are unable to process sugar efficiently. Diabetes can be fatal. Fortunately, Dr. Hodgkin's research with insulin has saved many lives.

Research Look in reference books or go to the Glencoe Science Web site for information on how X-ray crystallography is used to study minerals. Write your findings and share them with your class.

For more information, visit green.msscience.com/time

Reviewing Main Ideas

Section 1 Minerals

1. Much of what you use each day is made at least in some part from minerals.
2. All minerals are formed by natural processes and are inorganic solids with definite chemical compositions and orderly arrangements of atoms.
3. Minerals have crystal structures in one of six major crystal systems.

Section 2 Mineral Identification

1. Hardness is a measure of how easily a mineral can be scratched.
2. Luster describes how light reflects from a mineral's surface.
3. Streak is the color of the powder left by a mineral on an unglazed porcelain tile.
4. Minerals that break along smooth, flat surfaces have cleavage. When minerals break with rough or jagged surfaces, they are displaying fracture.
5. Some minerals have special properties that aid in identifying them. For example, magnetite is identified by its attraction to a magnet.

Section 3 Uses of Minerals

1. Gems are minerals that are more rare and beautiful than common minerals.
2. Minerals are useful for their physical properties and for the elements they contain.

Visualizing Main Ideas

Copy and complete the following concept map about minerals. Use the following words and phrases: the way a mineral breaks, the way a mineral reflects light, ore, a rare and beautiful mineral, how easily a mineral is scratched, streak, *and* a useful substance mined for profit.

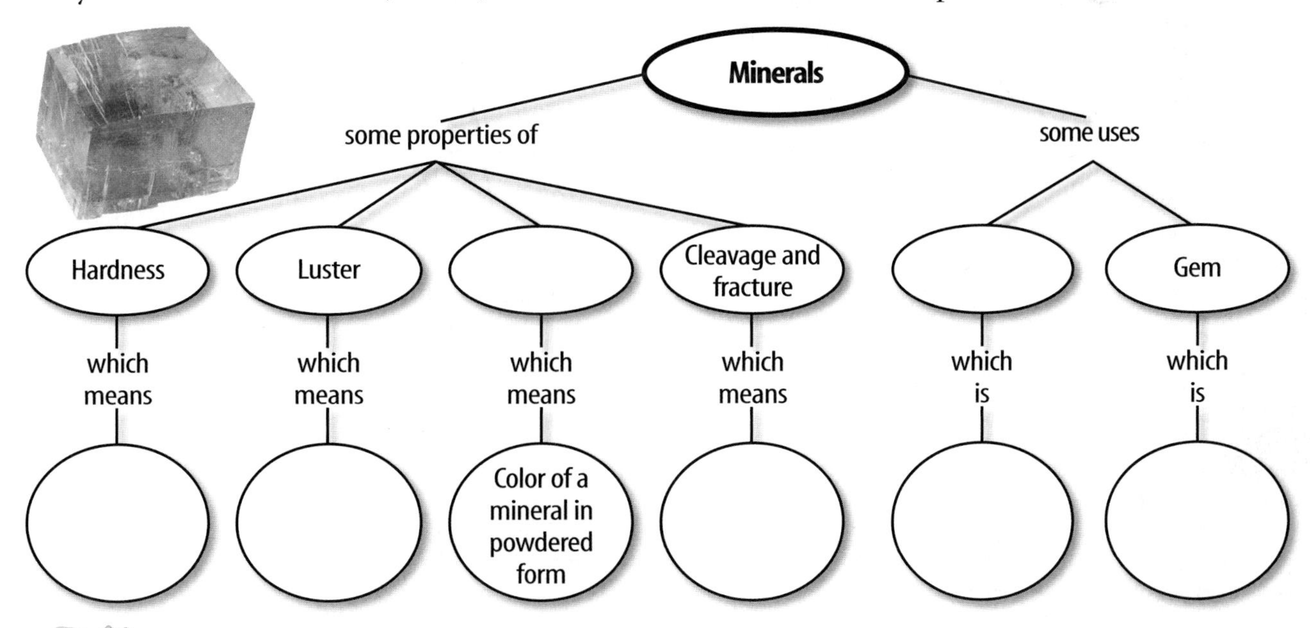

Using Vocabulary

cleavage p. 39
crystal p. 31
fracture p. 39
gem p. 41
hardness p. 37
luster p. 38
magma p. 33
mineral p. 30
ore p. 45
silicate p. 34
specific gravity p. 38
streak p. 39

Explain the difference between the vocabulary words in each of the following sets.

1. cleavage—fracture
2. crystal—mineral
3. luster—streak
4. magma—crystal
5. hardness—specific gravity
6. ore—mineral
7. crystal—luster
8. mineral—silicate
9. gem—crystal
10. streak—specific gravity

Checking Concepts

Choose the word or phrase that best answers the question.

11. Which is a characteristic of a mineral?
    A) It can be a liquid.
    B) It is organic.
    C) It has no crystal structure.
    D) It is inorganic.

12. What must all silicates contain?
    A) magnesium
    B) silicon and oxygen
    C) silicon and aluminum
    D) oxygen and carbon

13. What is the measure of how easily a mineral can be scratched?
    A) luster
    B) hardness
    C) cleavage
    D) fracture

Use the photo below to answer question 14.

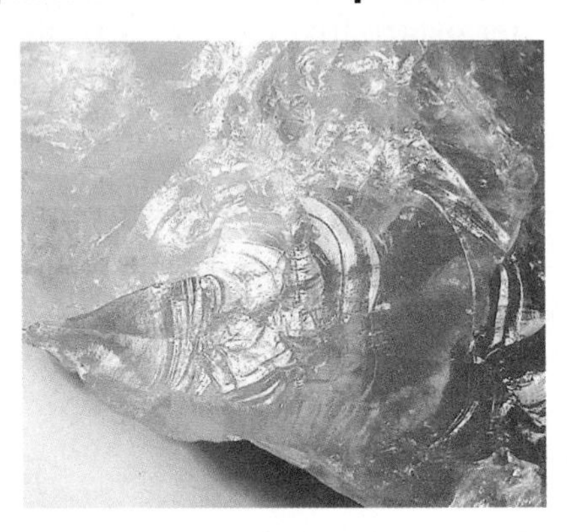

14. Examine the photo of quartz above. In what way does quartz break?
    A) cleavage
    B) fracture
    C) luster
    D) flat planes

15. Which of the following must crystalline solids have?
    A) carbonates
    B) cubic structures
    C) orderly arrangement of atoms
    D) cleavage

16. What is the color of a powdered mineral formed when rubbing it against an unglazed porcelain tile?
    A) luster
    B) density
    C) hardness
    D) streak

17. Which is hardest on Mohs scale?
    A) talc
    B) quartz
    C) diamond
    D) feldspar

Science Online green.msscience.com/vocabulary_puzzlemaker

Thinking Critically

18. Classify Water is an inorganic substance that is formed by natural processes on Earth. It has a unique composition. Sometimes water is a mineral and other times it is not. Explain.

19. Determine how many sides a perfect salt crystal has.

20. Apply Suppose you let a sugar solution evaporate, leaving sugar crystals behind. Are these crystals minerals? Explain.

21. Predict Will a diamond leave a streak on a streak plate? Explain.

22. Collect Data Make an outline of how at least seven physical properties can be used to identify unknown minerals.

23. Explain how you would use **Table 1** to determine the hardness of any mineral.

24. Concept Map Copy and complete the concept map below, which includes two crystal systems and two examples from each system. Use the following words and phrases: *hexagonal, corundum, halite, fluorite,* and *quartz.*

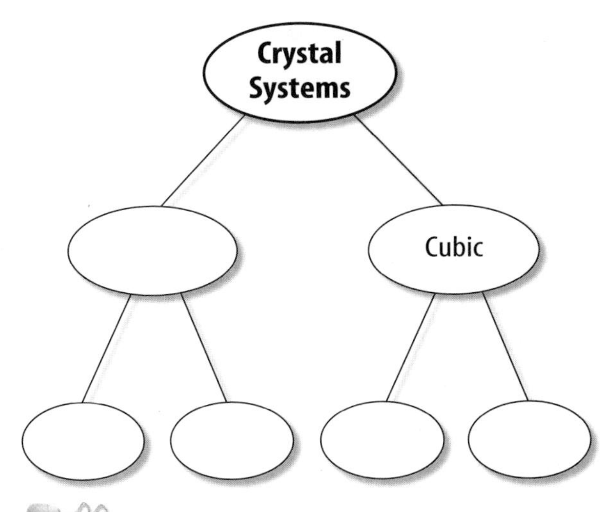

Performance Activities

25. Display Make a display that shows the six crystal systems of minerals. Research the crystal systems of minerals and give three examples for each crystal system. Indicate whether any of the minerals are found in your state. Describe any important uses of these minerals. Present your display to the class.

Applying Math

26. Mineral Volume Recall that 1 mL = 1 cm^3. Suppose that the volume of water in a graduated cylinder is 107.5 mL. A specimen of quartz, tied to a piece of string, is immersed in the water. The new water level reads 186 mL. What is the volume, in cm^3, of the piece of quartz?

Use the graph below to answer questions 27 and 28.

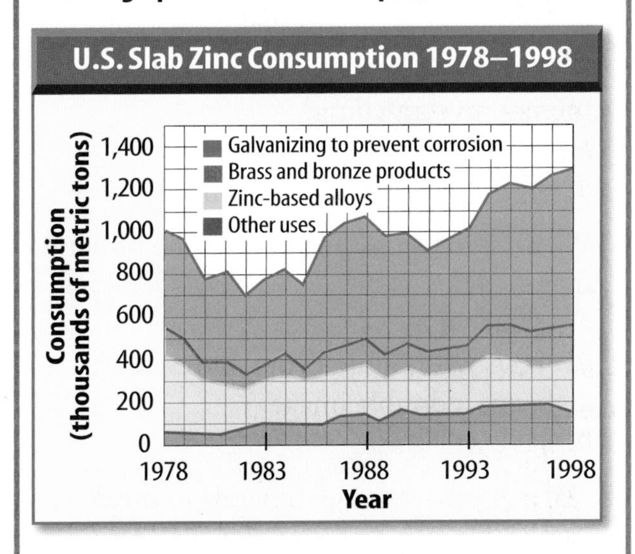

27. Zinc Use According to the graph above, what was the main use of zinc consumed in the United States between 1978 and 1998?

28. Metal Products According to the graph, approximately how many thousand metric tons of zinc were used to make brass and bronze products in 1998?

Standardized Test Practice

Part 1 Multiple Choice

Record your answers on the answer sheet provided by your teacher or on a sheet of paper.

Use the photo below to answer question 1.

1. To which crystal system does the crystal shown above belong?
   A. hexagonal
   B. cubic
   C. triclinic
   D. monoclinic

2. Which of the following is a common rock-forming mineral?
   A. azurite
   B. gold
   C. quartz
   D. diamond

3. Which term refers to the resistance of a mineral to scratching?
   A. hardness
   B. specific gravity
   C. luster
   D. fracture

4. Which is a special property of the mineral magnetite?
   A. attracted by a magnet
   B. fizzes with dilute hydrochloric acid
   C. forms a double image
   D. has a salty taste

5. Which causes some minerals to break along smooth, flat surfaces?
   A. streak
   B. cleavage
   C. luster
   D. fracture

Test-Taking Tip

If you are taking a timed test, keep track of time during the test. If you find that you're spending too much time on a multiple-choice question, mark your best guess and move on.

6. Which of these forms in cracks or along faults?
   A. bauxite
   B. silicates
   C. vein minerals
   D. rock-forming minerals

7. Which is the most abundant element in Earth's crust?
   A. silicon
   B. manganese
   C. iron
   D. oxygen

Use the table below to answer questions 8–10.

Mineral	Hardness
Talc	1
Gypsum	2
Calcite	3
Fluorite	4
Apatite	5
Feldspar	6
Quartz	7
Topaz	8
Corundum	9
Diamond	10

8. Which mineral in the table is softest?
   A. diamond
   B. feldspar
   C. talc
   D. gypsum

9. Which mineral will scratch feldspar but not topaz?
   A. quartz
   B. calcite
   C. apatite
   D. diamond

10. After whom is the scale shown above named?
    A. Neil Armstrong
    B. Friedrich Mohs
    C. Alfred Wegener
    D. Isaac Newton

Part 2 Short Response/Grid In

Record your answers on the answer sheet provided by your teacher or on a sheet of paper.

11. What is the definition of a mineral?

12. Why are gems valuable?

13. Explain the difference between fracture and cleavage.

14. Why is mineral color sometimes not helpful for identifying minerals?

Use the conversion factor and table below to answer questions 15–17.

1.0 carat = 0.2 grams

Diamond	Carats	Grams
Uncle Sam: largest diamond found in United States	40.4	?
Punch Jones: second largest U.S. diamond; named after boy who discovered it	?	6.89
Theresa: discovered in Wisconsin in 1888	21.5	4.3
2001 diamond production from western Australia	21,679,930	?

15. How many grams is the *Uncle Sam* diamond?

16. How many carats is the *Punch Jones* diamond?

17. How many grams of diamond were produced in western Australia in 2001?

18. What is the source of most of the diamonds that are used for industrial purposes?

19. Explain how minerals are useful to society. Describe some of their uses.

Part 3 Open Ended

Record your answers on a sheet of paper.

Use the photo below to answer question 20.

20. The mineral crystals in the rock above formed when magma cooled and are visible with the unaided eye. Hypothesize about how fast the magma cooled.

21. What is a crystal system? Why is it useful to classify mineral crystals this way?

22. How can a mineral be identified using its physical properties?

23. What is a crystal? Do all crystals have smooth crystal faces? Explain.

24. Are gases that are given off by volcanoes minerals? Why or why not?

25. What is the most abundant mineral group in Earth's crust? What elements always are found in the minerals included in this group?

26. Several layers are peeled from a piece of muscovite mica? What property of minerals does this illustrate? Describe this property in mica.

chapter 3

Rocks

chapter preview

sections

1 The Rock Cycle
2 Igneous Rocks
Lab *Igneous Rock Clues*
3 Metamorphic Rocks
4 Sedimentary Rocks
Lab *Sedimentary Rocks*
Virtual Lab *How are rocks classified?*

How did it get there?

The giant rocky peak of El Capitan towers majestically in Yosemite National Park. Surrounded by flat landscape, it seems out of place. How did this expanse of granite rock come to be?

Science Journal Are you a rock collector? If so, write two sentences about your favorite rock. If not, describe the rocks you see in the photo in enough detail that a nonsighted person could visualize them.

Start-Up Activities

Observe and Describe Rocks

Some rocks are made of small mineral grains that lock together, like pieces of a puzzle. Others are grains of sand tightly held together or solidified lava that once flowed from a volcano. If you examine rocks closely, you sometimes can tell what they are made of.

1. Collect three different rock samples near your home or school.
2. Draw a picture of the details you see in each rock.
3. Use a magnifying lens to look for different types of materials within the same rock.
4. Describe the characteristics of each rock. Compare your drawings and descriptions with photos, drawings, and descriptions in a rocks and minerals field guide.
5. Use the field guide to try to identify each rock.
6. **Think Critically** Decide whether you think your rocks are mixtures. If so, infer or suggest what these mixtures might contain. Write your explanations in your Science Journal.

Major Rock Types Make the following Foldable to help you organize facts about types of rocks.

STEP 1 **Fold** a sheet of paper in half lengthwise. Make the back edge about 5 cm longer than the front edge.

STEP 2 **Turn** the paper so the fold is on the bottom. Then fold it into thirds.

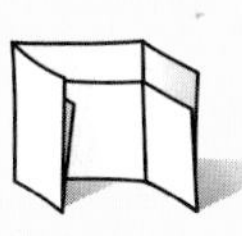

STEP 3 **Unfold and cut** only the top layer along both folds to make three tabs.

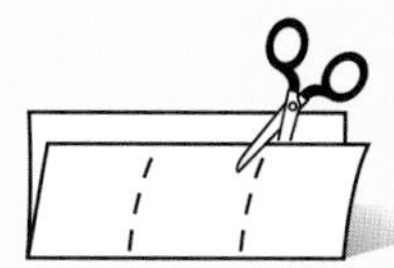

STEP 4 **Label** the Foldable as shown.

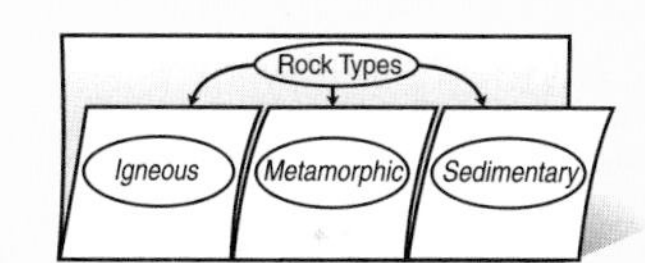

Make an Organizational Study Fold As you read the chapter, write and illustrate what you learn about the three main types of rocks in your study fold.

Preview this chapter's content and activities at green.msscience.com

section 1

The Rock Cycle

as you read

What You'll Learn

- **Distinguish** between a rock and a mineral.
- **Describe** the rock cycle and some changes that a rock could undergo.

Why It's Important

Rocks exist everywhere, from under deep oceans and in high mountain ranges, to the landscape beneath your feet.

Review Vocabulary
mineral: a naturally occurring, inorganic solid with a definite chemical composition and an orderly arrangement of atoms

New Vocabulary
- rock
- rock cycle

What is a rock?

Imagine you and some friends are exploring a creek. Your eye catches a glint from a piece of rock at the edge of the water. As you wander over to pick up the rock, you notice that it is made of different-colored materials. Some of the colors reflect light, while others are dull. You put the rock in your pocket for closer inspection in science lab.

Common Rocks The next time you walk past a large building or monument, stop and take a close look at it. Chances are that it is made out of common rock. In fact, most rock used for building stone contains one or more common minerals, called rock-forming minerals, such as quartz, feldspar, mica, or calcite. When you look closely, the sparkles you see are individual crystals of minerals. A **rock** is a mixture of such minerals, rock fragments, volcanic glass, organic matter, or other natural materials. **Figure 1** shows minerals mixed together to form the rock granite. You might even find granite near your home.

Figure 1 Mount Rushmore, in South Dakota, is made of granite. Granite is a mixture of feldspar, quartz, mica, hornblende, and other minerals.

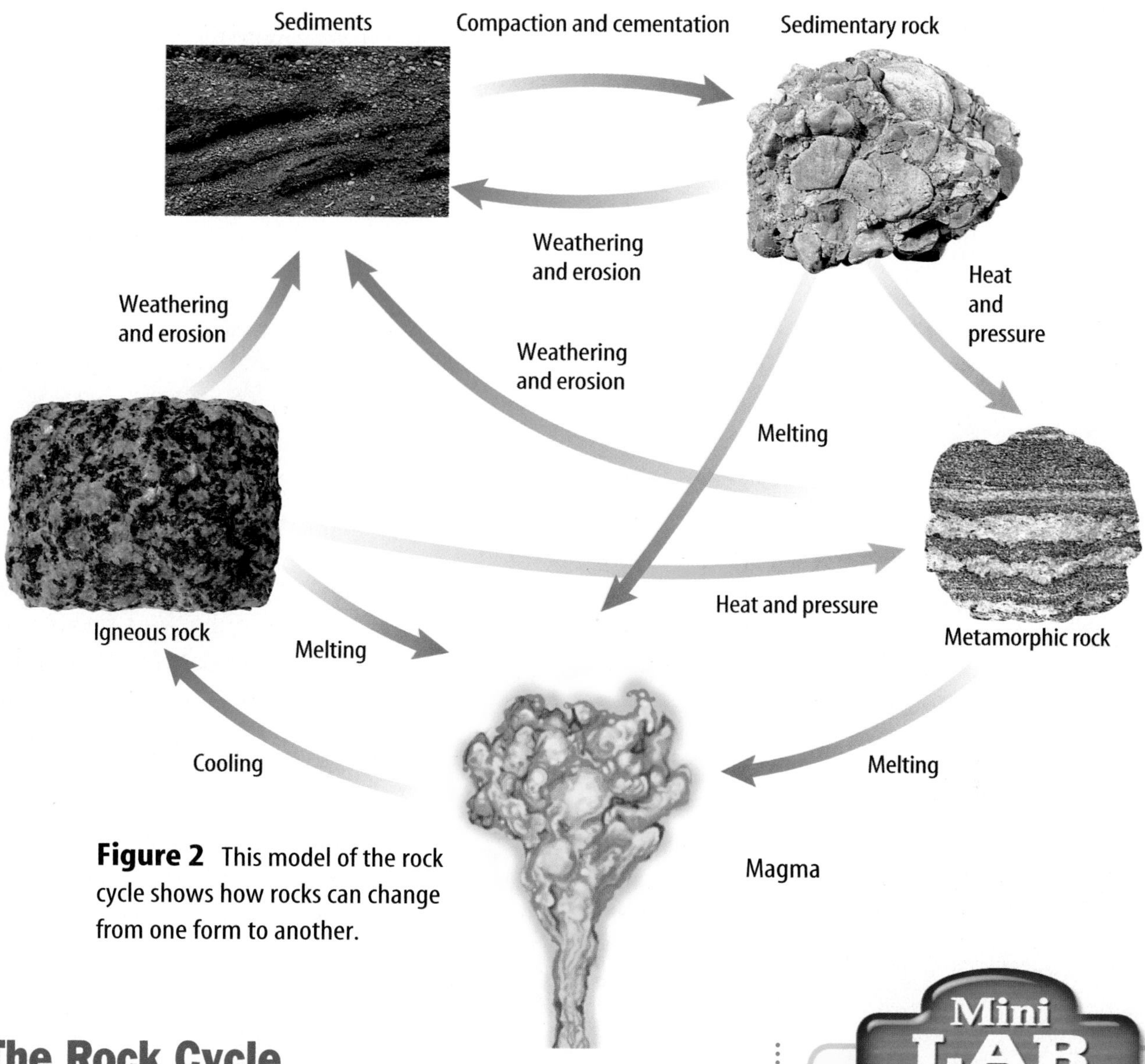

Figure 2 This model of the rock cycle shows how rocks can change from one form to another.

The Rock Cycle

To show how rocks slowly change through time, scientists have created a model called the **rock cycle,** shown in **Figure 2.** It illustrates the processes that create and change rocks. The rock cycle shows the three types of rock—igneous, metamorphic, and sedimentary—and the processes that form them.

Look at the rock cycle and notice that rocks change by many processes. For example, a sedimentary rock can change by heat and pressure to form a metamorphic rock. The metamorphic rock then can melt and later cool to form an igneous rock. The igneous rock then could be broken into fragments by weathering and erode away. The fragments might later compact and cement together to form another sedimentary rock. Any given rock can change into any of the three major rock types. A rock even can transform into another rock of the same type.

Reading Check ***What is illustrated by the rock cycle?***

Mini LAB

Modeling Rock

Procedure

1. Mix about 10 mL of **white glue** with about 7 g of **dirt** or **sand** in a **small paper cup.**
2. Stir the mixture and then allow it to harden overnight.
3. Tear away the paper cup carefully from your mixture.

Analysis

1. Which rock type is similar to your hardened mixture?
2. Which part of the rock cycle did you model?

Try at Home

VISUALIZING THE ROCK CYCLE

Figure 3

Rocks continuously form and transform in a process that geologists call the rock cycle. For example, molten rock—from volcanoes such as Washington's Mount Rainier, background—cools and solidifies to form igneous rock. It slowly breaks down when exposed to air and water to form sediments. These sediments are compacted or cemented into sedimentary rock. Heat and pressure might transform sedimentary rock into metamorphic rock. When metamorphic rock melts and hardens, igneous rock forms again. There is no distinct beginning, nor is there an end, to the rock cycle.

▲ The black sand beach of this Polynesian island is sediment weathered and eroded from the igneous rock of a volcano nearby.

▲ This alluvial fan on the edge of Death Valley, California, was formed when gravel, sand, and finer sediments were deposited by a stream emerging from a mountain canyon.

▲ Layers of shale and chalk form Kansas's Monument Rocks. They are remnants of sediments deposited on the floor of the ancient sea that once covered much of this region.

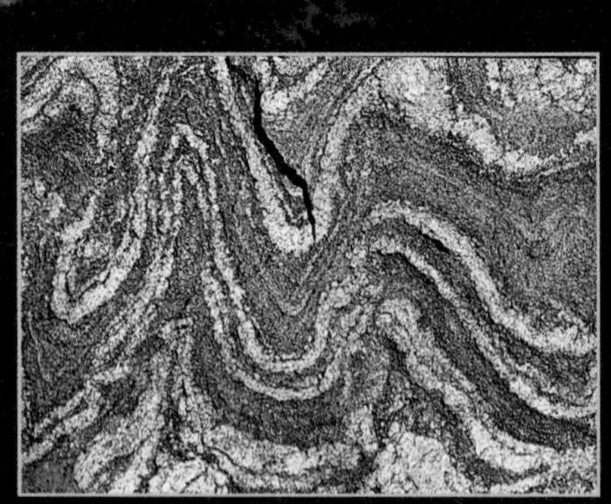

▲ Heat and pressure deep below Earth's surface can change rock into metamorphic rock, like this banded gneiss.

Matter and the Rock Cycle

The rock cycle, illustrated in **Figure 3,** shows how rock can be weathered to small rock and mineral grains. This material then can be eroded and carried away by wind, water, or ice. When you think of erosion, it might seem that the material is somehow destroyed and lost from the cycle. This is not the case. The chemical elements that make up minerals and rocks are not destroyed. This fact illustrates the principle of conservation of matter. The changes that take place in the rock cycle never destroy or create matter. The elements are just redistributed in other forms.

Figure 4 The rock formations at Siccar Point, Scotland, show that rocks undergo constant change.

What is the principle of conservation of matter?

Discovering the Rock Cycle James Hutton, a Scottish physician and naturalist, first recognized in 1788 that rocks undergo profound changes. Hutton noticed, among other things, that some layers of solid rock in Siccar Point, shown in **Figure 4,** had been altered since they formed. Instead of showing a continuous pattern of horizontal layering, some of the rock layers at Siccar Point are tilted and partly eroded. However, the younger rocks above them are nearly horizontal.

Hutton published these and other observations, which proved that rocks are subject to constant change. Hutton's early recognition of the rock cycle continues to influence geologists.

section 1 review

Summary

What is a rock?

- Rocks are mixtures of minerals, rock fragments, organic matter, volcanic glass, and other materials found in nature.

The Rock Cycle

- The three major types of rock are igneous, metamorphic, and sedimentary.
- Rock cycle processes do not create or destroy matter.
- Processes that are part of the rock cycle change rocks slowly over time.
- In the late eighteenth century, James Hutton recognized some rock cycle processes by observing rocks in the field.
- Some of Hutton's ideas continue to influence geologic thinking today.

Self Check

1. **Explain** how rocks differ from minerals.
2. **Compare and contrast** igneous and metamorphic rock formation.
3. **Describe** the major processes of the rock cycle.
4. **Explain** one way that the rock cycle can illustrate the principle of conservation of matter.
5. **Think Critically** How would you define magma based on the illustration in **Figure 2?** How would you define sediment and sedimentary rock?

Applying Skills

6. **Communicate** Review the model of the rock cycle in **Figure 2.** In your Science Journal, write a story or poem that explains what can happen to a sedimentary rock as it changes throughout the rock cycle.

green.msscience.com/self_check_quiz

section 2

Igneous Rocks

as you read

What You'll Learn

- **Recognize** magma and lava as the materials that cool to form igneous rocks.
- **Contrast** the formation of intrusive and extrusive igneous rocks.
- **Contrast** granitic and basaltic igneous rocks.

Why It's Important

Igneous rocks are the most abundant kind of rock in Earth's crust. They contain many valuable resources.

Review Vocabulary
element: substance made of one type of atom that cannot be broken down by ordinary chemical or physical means

New Vocabulary
- igneous rock
- lava
- intrusive
- extrusive
- basaltic
- granitic

Formation of Igneous Rocks

Perhaps you've heard of recent volcanic eruptions in the news. When some volcanoes erupt, they eject a flow of molten rock material, as shown in **Figure 5.** Molten rock material, called magma, flows when it is hot and becomes solid when it cools. When hot magma cools and hardens, it forms **igneous** (IHG nee us) **rock.** Why do volcanoes erupt, and where does the molten material come from?

Magma In certain places within Earth, the temperature and pressure are just right for rocks to melt and form magma. Most magmas come from deep below Earth's surface. Magma is located at depths ranging from near the surface to about 150 km below the surface. Temperatures of magmas range from about 650°C to 1,200°C, depending on their chemical compositions and pressures exerted on them.

The heat that melts rocks comes from sources within Earth's interior. One source is the decay of radioactive elements within Earth. Some heat is left over from the formation of the planet, which originally was molten. Radioactive decay of elements contained in rocks balances some heat loss as Earth continues to cool.

Because magma is less dense than surrounding solid rock, it is forced upward toward the surface, as shown in **Figure 6.** When magma reaches Earth's surface and flows from volcanoes, it is called **lava.**

Figure 5 Some lava is highly fluid and free-flowing, as shown by this spectacular lava fall in Volcano National Park, East Rift, Kilauea, Hawaii.

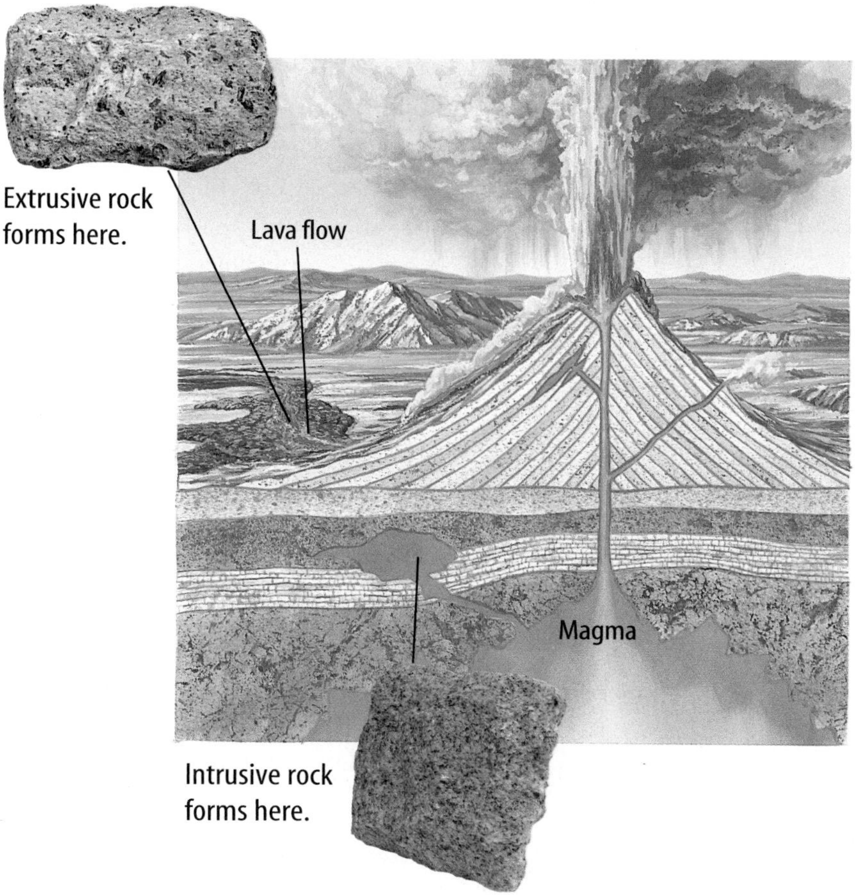

Extrusive rock forms here.

Lava flow

Magma

Intrusive rock forms here.

Figure 6 Intrusive rocks form from magma trapped below Earth's surface. Extrusive rocks form from lava flowing at the surface.

Intrusive Rocks Magma is melted rock material composed of common elements and fluids. As magma cools, atoms and compounds in the liquid rearrange themselves into new crystals called mineral grains. Rocks form as these mineral grains grow together. Rocks that form from magma below the surface, as illustrated in **Figure 6,** are called **intrusive** igneous rocks. Intrusive rocks are found at the surface only after the layers of rock and soil that once covered them have been removed by erosion. Erosion occurs when the rocks are pushed up by forces within Earth. Because intrusive rocks form at depth and they are surrounded by other rocks, it takes a long time for them to cool. Slowly cooled magma produces individual mineral grains that are large enough to be observed with the unaided eye.

Extrusive Rocks **Extrusive** igneous rocks are formed as lava cools on the surface of Earth. When lava flows on the surface, as illustrated in **Figure 6,** it is exposed to air and water. Lava, such as the basaltic lava shown in **Figure 5,** cools quickly under these conditions. The quick cooling rate keeps mineral grains from growing large, because the atoms in the liquid don't have the time to arrange into large crystals. Therefore, extrusive igneous rocks are fine grained.

What controls the grain size of an igneous rock?

Table 1 Common Igneous Rocks

Magma Type	Basaltic	Andesitic	Granitic
Intrusive	Gabbro	Diorite	Granite
Extrusive	Basalt Scoria	Andesite	Rhyolite Obsidian Pumice

Topic: Rock Formation
Visit green.msscience.com for Web links to information about intrusive and extrusive rocks.

Activity List several geographic settings where intrusive or extrusive rocks are found. Select one setting for intrusive rocks, and one for extrusive rocks. Describe how igneous rocks form in the two settings, and locate an example of each on a map.

Volcanic Glass Pumice, obsidian, and scoria are examples of volcanic glass. These rocks cooled so quickly that few or no mineral grains formed. Most of the atoms in these rocks are not arranged in orderly patterns, and few crystals are present.

In the case of pumice and scoria, gases become trapped in the gooey molten material as it cools. Some of these gases eventually escape, but holes are left behind where the rock formed around the pockets of gas.

Classifying Igneous Rocks

Igneous rocks are intrusive or extrusive depending on how they are formed. A way to further classify these rocks is by the magma from which they form. As shown in **Table 1,** an igneous rock can form from basaltic, andesitic, or granitic magma. The type of magma that cools to form an igneous rock determines important chemical and physical properties of that rock. These include mineral composition, density, color, and melting temperature.

Reading Check *Name two ways igneous rocks are classified.*

Basaltic Rocks **Basaltic** (buh SAWL tihk) igneous rocks are dense, dark-colored rocks. They form from magma that is rich in iron and magnesium and poor in silica, which is the compound SiO_2. The presence of iron and magnesium in minerals in basalt gives basalt its dark color. Basaltic lava is fluid and flows freely from volcanoes in Hawaii, such as Kilauea. How does this explain the black beach sand common in Hawaii?

Granitic Rocks **Granitic** igneous rocks are light-colored rocks of a lower density than basaltic rocks. Granitic magma is thick and stiff and contains lots of silica but lesser amounts of iron and magnesium. Because granitic magma is stiff, it can build up a great deal of gas pressure, which is released explosively during violent volcanic eruptions.

Melting Rock Inside Earth, materials contained in rocks can melt. In your Science Journal, describe what is happening to the atoms and molecules to cause this change of state.

Andesitic Rocks Andesitic igneous rocks have mineral compositions between those of basaltic and granitic rocks. Many volcanoes around the rim of the Pacific Ocean formed from andesitic magmas. Like volcanoes that erupt granitic magma, these volcanoes also can erupt violently.

Take another look at **Table 1.** Basalt forms at the surface of Earth because it is an extrusive rock. Granite forms below Earth's surface from magma with a high concentration of silica. When you identify an igneous rock, you can infer how it formed and the type of magma that it formed from.

section 2 review

Summary

Formation of Igneous Rocks

- When molten rock material, called magma, cools and hardens, igneous rock forms.
- Intrusive igneous rocks form as magma cools and hardens slowly, beneath Earth's surface.
- Extrusive igneous rocks form as lava cools and hardens rapidly, at or above Earth's surface.

Classifying Igneous Rocks

- Igneous rocks are further classified according to their mineral compositions.
- The violent nature of some volcanic eruptions is partly explained by the composition of the magma that feeds them.

Self Check

1. **Explain** why some types of magma form igneous rocks that are dark colored and dense.
2. **Identify** the property of magma that causes it to be forced upward toward Earth's surface.
3. **Explain** The texture of obsidian is best described as glassy. Why does obsidian contain few or no mineral grains?
4. **Think Critically** Study the photos in **Table 1.** How are granite and rhyolite similar? How are they different?

Applying Skills

5. **Make and Use Graphs** Four elements make up most of the rocks in Earth's crust. They are: *oxygen—46.6 percent, aluminum—8.1 percent, silicon—27.7 percent,* and *iron—5.0 percent.* Make a bar graph of these data. What might you infer from the low amount of iron?

Igneous Rock Clues

You've learned how color often is used to estimate the composition of an igneous rock. The texture of an igneous rock describes its overall appearance, including mineral grain sizes and the presence or absence of bubble holes, for example. In most cases, grain size relates to how quickly the magma or lava cooled. Crystals you can see without a magnifying lens indicate slower cooling. Smaller, fine-grained crystals indicate quicker cooling, possibly due to volcanic activity. Rocks with glassy textures cooled so quickly that there was no time to form mineral grains.

Real-World Question

What does an igneous rock's texture and color indicate about its formation history?

Goals

- **Classify** different samples of igneous rocks by color and infer their composition.
- **Observe** the textures of igneous rocks and infer how they formed.

Materials

rhyolite
basalt
vesicular basalt
pumice
granite
obsidian
gabbro
magnifying lens

Safety Precautions

WARNING: *Some rock samples might have sharp edges. Always use caution while handling samples.*

Procedure

1. **Arrange** rocks according to color (light or dark). Record your observations in your Science Journal.
2. **Arrange** rocks according to similar texture. Consider grain sizes and shapes, presence of holes, etc. Use your magnifying lens to see small features more clearly. Record your observations.

Conclude and Apply

1. **Infer** which rocks are granitic based on color.
2. **Infer** which rocks cooled quickly. What observations led you to this inference?
3. **Identify** any samples that suggest gases were escaping from them as they cooled.
4. **Describe** Which samples have a glassy appearance? How did these rocks form?
5. **Infer** which samples are not volcanic. Explain.

Communicating Your Data

Research the compositions of each of your samples. Did the colors of any samples lead you to infer the wrong compositions? Communicate to your class what you learned.

section 3

Metamorphic Rocks

Formation of Metamorphic Rocks

Have you ever packed your lunch in the morning and not been able to recognize it at lunchtime? You might have packed a sandwich, banana, and a large bottle of water. You know you didn't smash your lunch on the way to school. However, you didn't think about how the heavy water bottle would damage your food if the bottle was allowed to rest on the food all day. The heat in your locker and the pressure from the heavy water bottle changed your sandwich. Like your lunch, rocks can be affected by changes in temperature and pressure.

Metamorphic Rocks Rocks that have changed because of changes in temperature and pressure or the presence of hot, watery fluids are called **metamorphic rocks.** Changes that occur can be in the form of the rock, shown in **Figure 7,** the composition of the rock, or both. Metamorphic rocks can form from igneous, sedimentary, or other metamorphic rocks. What Earth processes can change these rocks?

as you read

What You'll Learn

- **Describe** the conditions in Earth that cause metamorphic rocks to form.
- **Classify** metamorphic rocks as foliated or nonfoliated.

Why It's Important

Metamorphic rocks are useful because of their unique properties.

Review Vocabulary
pressure: the amount of force exerted per unit of area

New Vocabulary
- metamorphic rock
- foliated
- nonfoliated

+ pressure

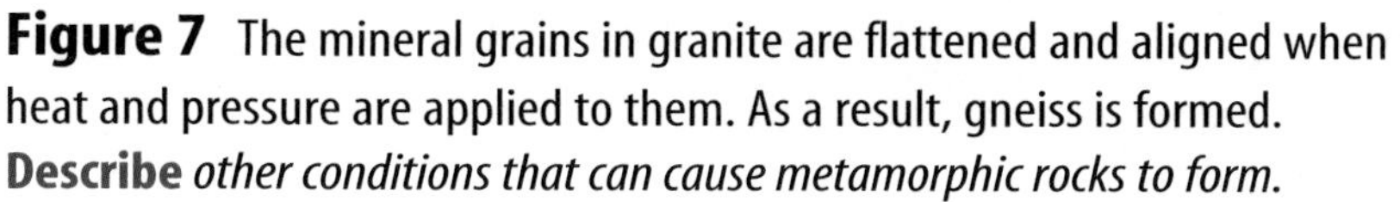

Figure 7 The mineral grains in granite are flattened and aligned when heat and pressure are applied to them. As a result, gneiss is formed. **Describe** *other conditions that can cause metamorphic rocks to form.*

Topic: Shale Metamorphism

Visit green.msscience.com for Web links to information about the metamorphism of shale. Communicate to your class what you learn.

Activity Make a table with headings that are major rock types that form from shale metamorphism. Under each rock heading, make a list of minerals that can occur in the rock.

Heat and Pressure Rocks beneath Earth's surface are under great pressure from rock layers above them. Temperature also increases with depth in Earth. In some places, the heat and pressure are just right to cause rocks to melt and magma to form. In other areas where melting doesn't occur, some mineral grains can change by dissolving and recrystallizing—especially in the presence of fluids. Sometimes, under these conditions, minerals exchange atoms with surrounding minerals and new, bigger minerals form.

Depending upon the amount of pressure and temperature applied, one type of rock can change into several different metamorphic rocks, and each type of metamorphic rock can come from several kinds of parent rocks. For example, the sedimentary rock shale will change into slate. As increasing pressure and temperature are applied, the slate can change into phyllite, then schist, and eventually gneiss. Schist also can form when basalt is metamorphosed, or changed, and gneiss can come from granite.

How can one type of rock change into several different metamorphic rocks?

Hot Fluids Did you know that fluids can move through rock? These fluids, which are mostly water with dissolved elements and compounds, can react chemically with a rock and change its composition, especially when the fluids are hot. That's what happens when rock surrounding a hot magma body reacts with hot fluids from the magma, as shown in **Figure 8.** Most fluids that transform rocks during metamorphic processes are hot and mainly are comprised of water and carbon dioxide.

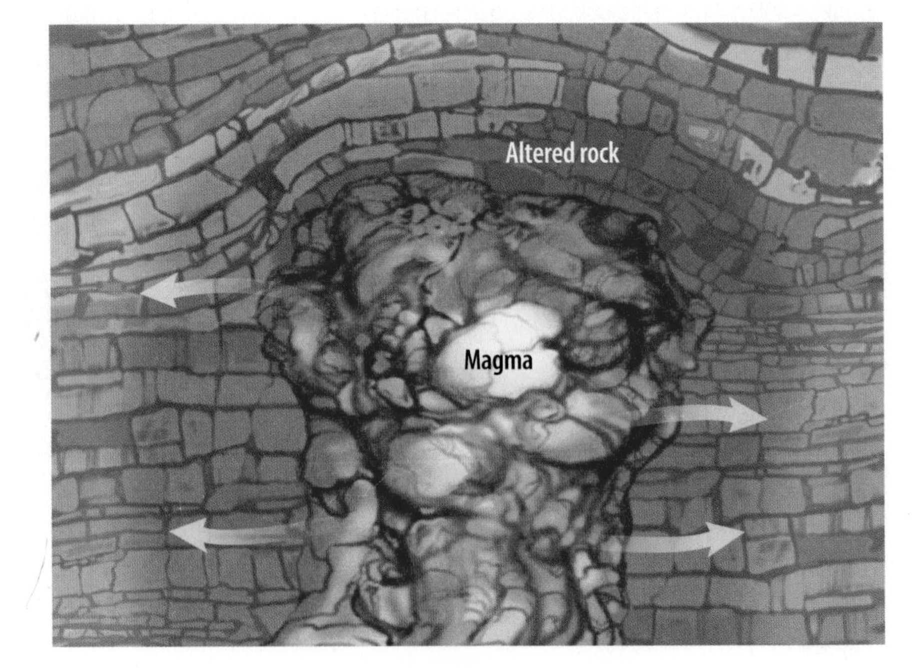

Figure 8 In the presence of hot, water-rich fluids, solid rock can change in mineral composition without having to melt.

Classifying Metamorphic Rocks

Metamorphic rocks form from igneous, sedimentary, or other metamorphic rocks. Heat, pressure, and hot fluids trigger the changes. Each resulting rock can be classified according to its composition and texture.

Foliated Rocks When mineral grains line up in parallel layers, the metamorphic rock is said to have a **foliated** texture. Two examples of foliated rocks are slate and gneiss. Slate forms from the sedimentary rock shale. The minerals in shale arrange into layers when they are exposed to heat and pressure. As **Figure 9** shows, slate separates easily along these foliation layers.

The minerals in slate are pressed together so tightly that water can't pass between them easily. Because it's watertight, slate is ideal for paving around pools and patios. The naturally flat nature of slate and the fact that it splits easily make it useful for roofing and tiling many surfaces.

Gneiss (NISE), another foliated rock, forms when granite and other rocks are changed. Foliation in gneiss shows up as alternating light and dark bands. Movement of atoms has separated the dark minerals, such as biotite mica, from the light minerals, which are mainly quartz and feldspar.

Reading Check *What type of metamorphic rock is composed of mineral grains arranged in parallel layers?*

Figure 9 Slate often is used as a building or landscaping material. **Identify** *the properties that make slate so useful for these purposes.*

Figure 10 This exhibit in Vermont shows the beauty of carved marble.

Nonfoliated Rocks In some metamorphic rocks, layering does not occur. The mineral grains grow and rearrange, but they don't form layers. This process produces a **nonfoliated** texture.

Sandstone is a sedimentary rock that's often composed mostly of quartz grains. When sandstone is heated under a lot of pressure, the grains of quartz grow in size and become interlocking, like the pieces of a jigsaw puzzle. The resulting rock is called quartzite.

Marble is another nonfoliated metamorphic rock. Marble forms from the sedimentary rock limestone, which is composed of the mineral calcite. Usually, marble contains several other minerals besides calcite. For example, hornblende and serpentine give marble a black or greenish tone, whereas hematite makes it red. As **Figure 10** shows, marble is a popular material for artists to sculpt because it is not as hard as other rocks.

So far, you've investigated only a portion of the rock cycle. You still haven't observed how sedimentary rocks are formed and how igneous and metamorphic rocks evolve from them. The next section will complete your investigation of the rock cycle.

section 3 review

Summary

Formation of Metamorphic Rocks

- Changes in pressure, temperature, or the presence of fluids can cause metamorphic rocks to form.
- Rock, altered by metamorphic processes at high temperatures and pressures, changes in the solid state without melting.
- Hot fluids that move through and react with preexisting rock are composed mainly of water and carbon dioxide.
- One source of hot, watery fluids is magma bodies close to the changing rock.
- Any parent rock type—igneous, metamorphic, or sedimentary—can become a metamorphic rock.

Classifying Metamorphic Rocks

- Texture and mineral composition determine how a metamorphic rock is classified.
- Physical properties of metamorphic rocks, such as the watertight nature of slate, make them useful for many purposes.

Self Check

1. **Explain** what role fluids play in rock metamorphism.
2. **Describe** how metamorphic rocks are classified. What are the characteristics of rocks in each of these classifications?
3. **Identify** Give an example of a foliated and a nonfoliated metamorphic rock. Name one of their possible parent rocks.
4. **Think Critically** Marble is a common material used to make sculptures, but not just because it's a beautiful stone. What properties of marble make it useful for this purpose?

Applying Skills

5. **Concept Map** Put the following events in an events-chain concept map that explains how a metamorphic rock might form from an igneous rock. *Hint: Start with "Igneous Rock Forms."* Use each event just once.

   Events: *sedimentary rock forms, weathering occurs, heat and pressure are applied, igneous rock forms, metamorphic rock forms, erosion occurs, sediments are formed, deposition occurs*

Science Online green.msscience.com/self_check_quiz

section 4

Sedimentary Rocks

Formation of Sedimentary Rocks

Igneous rocks are the most common rocks on Earth, but because most of them exist below the surface, you might not have seen too many of them. That's because 75 percent of the rocks exposed at the surface are sedimentary rocks.

Sediments are loose materials such as rock fragments, mineral grains, and bits of shell that have been moved by wind, water, ice, or gravity. If you look at the model of the rock cycle, you will see that sediments come from already-existing rocks that are weathered and eroded. **Sedimentary rock** forms when sediments are pressed and cemented together, or when minerals form from solutions.

Stacked Rocks Sedimentary rocks often form as layers. The older layers are on the bottom because they were deposited first. Sedimentary rock layers are a lot like the books and papers in your locker. Last week's homework is on the bottom, and today's notes will be deposited on top of the stack. However, if you disturb the stack, the order in which the books and papers are stacked will change, as shown in **Figure 11.** Sometimes, forces within Earth overturn layers of rock, and the oldest are no longer on the bottom.

as you read

What You'll Learn

- **Explain** how sedimentary rocks form from sediments.
- **Classify** sedimentary rocks as detrital, chemical, or organic in origin.
- **Summarize** the rock cycle.

Why It's Important

Some sedimentary rocks, like coal, are important sources of energy.

Review Vocabulary
weathering: surface processes that work to break down rock mechanically or chemically

New Vocabulary
- sediment
- sedimentary rock
- compaction
- cementation

Figure 11 Like sedimentary rock layers, the oldest paper is at the bottom of the stack. If the stack is disturbed, then it is no longer in order.

Classifying Sediments

Procedure

WARNING: *Use care when handling sharp objects.*

1. Collect different samples of **sediment.**
2. Spread them on a sheet of **paper.**
3. Use **Table 2** to determine the size range of gravel-sized sediment.
4. Use **tweezers or a dissecting probe** and a **magnifying lens** to separate the gravel-sized sediments.
5. Separate the gravel into piles—rounded or angular.

Analysis

1. Describe the grains in both piles.
2. Determine what rock could form from each type of sediment you have.

Classifying Sedimentary Rocks

Sedimentary rocks can be made of just about any material found in nature. Sediments come from weathered and eroded igneous, metamorphic, and sedimentary rocks. Sediments also come from the remains of some organisms. The composition of a sedimentary rock depends upon the composition of the sediments from which it formed.

Like igneous and metamorphic rocks, sedimentary rocks are classified by their composition and by the manner in which they formed. Sedimentary rocks usually are classified as detrital, chemical, or organic.

Detrital Sedimentary Rocks

The word *detrital* (dih TRI tul) comes from the Latin word *detritus,* which means "to wear away." Detrital sedimentary rocks, such as those shown in **Table 2,** are made from the broken fragments of other rocks. These loose sediments are compacted and cemented together to form solid rock.

Weathering and Erosion When rock is exposed to air, water, or ice, it is unstable and breaks down chemically and mechanically. This process, which breaks rocks into smaller pieces, is called weathering. **Table 2** shows how these pieces are classified by size. The movement of weathered material is called erosion.

Compaction Erosion moves sediments to a new location, where they then are deposited. Here, layer upon layer of sediment builds up. Pressure from the upper layers pushes down on the lower layers. If the sediments are small, they can stick together and form solid rock. This process, shown in **Figure 12,** is called **compaction.**

Reading Check *How do rocks form through compaction?*

Figure 12 During compaction, pore space between sediments decreases, causing them to become packed together more tightly.

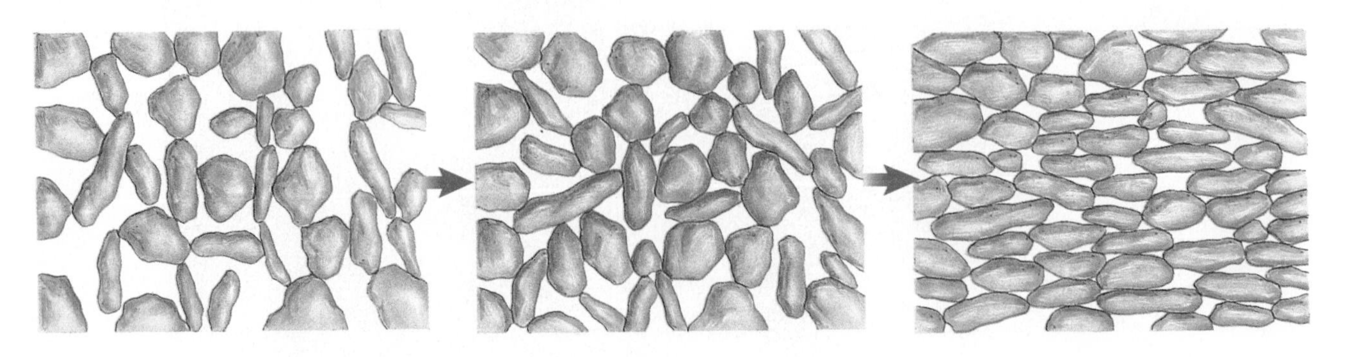

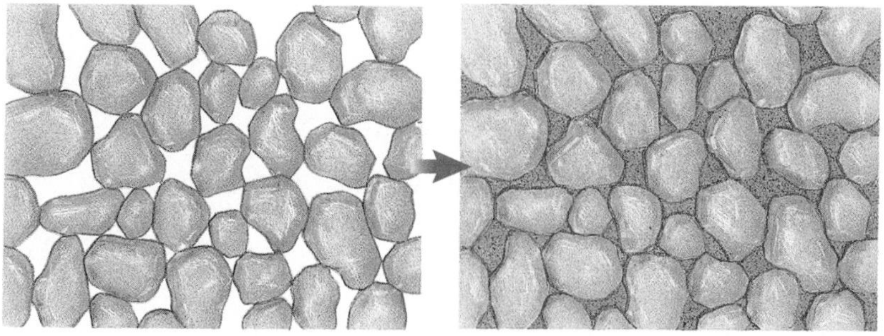

Figure 13 Sediments are cemented together as minerals crystallize between grains.

Cementation If sediments are large, like sand and pebbles, pressure alone can't make them stick together. Large sediments have to be cemented together. As water moves through soil and rock, it picks up materials released from minerals during weathering. The resulting solution of water and dissolved materials moves through open spaces between sediments. **Cementation,** which is shown in **Figure 13,** occurs when minerals such as quartz, calcite, and hematite are deposited between the pieces of sediment. These minerals, acting as natural cements, hold the sediment together like glue, making a detrital sedimentary rock.

Shape and Size of Sediments Detrital rocks have granular textures, much like granulated sugar. They are named according to the shapes and sizes of the sediments that form them. For example, conglomerate and breccia both form from large sediments, as shown in **Table 2.** If the sediments are rounded, the rock is called conglomerate. If the sediments have sharp angles, the rock is called breccia. The roundness of sediment particles depends on how far they have been moved by wind or water.

Table 2 Sediment Sizes and Detrital Rocks

Sediment	Clay	Silt	Sand	Gravel
Size Range	<0.004 mm	0.004–0.063 mm	0.063–2 mm	>2 mm
Example	Shale 	Siltstone 	Sandstone 	Conglomerate (shown) or Breccia 

Figure 14 Although concrete strongly resembles conglomerate, concrete is not a rock because it does not occur in nature.

Materials Found in Sedimentary Rocks The gravel-sized sediments in conglomerate and breccia can consist of any type of rock or mineral. Often, they are composed of chunks of the minerals quartz and feldspar. They also can be pieces of rocks such as gneiss, granite, or limestone. The cement that holds the sediments together usually is made of quartz or calcite.

Have you ever looked at the concrete in sidewalks, driveways, and stepping stones? The concrete in **Figure 14** is made of gravel and sand grains that have been cemented together. Although the structure is similar to that of naturally occurring conglomerate, it cannot be considered a rock.

Sandstone is formed from smaller particles than conglomerates and breccias. Its sand-sized sediments can be just about any mineral, but they are usually grains of minerals such as quartz and feldspar that are resistant to weathering. Siltstone is similar to sandstone except it is made of smaller, silt-sized particles. Shale is a detrital sedimentary rock that is made mainly of clay-sized particles. Clay-sized sediments are compacted together by pressure from overlying layers.

Sedimentary Petrology Research the work done by sedimentary petrologists. Include examples of careers in academia and in industry.

Chemical Sedimentary Rocks

Chemical sedimentary rocks form when dissolved minerals come out of solution. You can show that salt is deposited in the bottom of a glass or pan when saltwater solution evaporates. In a similar way, minerals collect when seas or lakes evaporate. The deposits of minerals that come out of solution form sediments and rocks. For example, the sediment making up New Mexico's White Sands desert consists of pieces of a chemical sedimentary rock called rock gypsum. Chemical sedimentary rocks are different. They are not made from pieces of preexisting rocks.

Reading Check *How do chemical sedimentary rocks form?*

Limestone Calcium carbonate is carried in solution in ocean water. When calcium carbonate ($CaCO_3$) comes out of solution as calcite and its many crystals grow together, limestone forms. Limestone also can contain other minerals and sediments, but it must be at least 50 percent calcite. Limestone usually is deposited on the bottom of lakes or shallow seas. Large areas of the central United States have limestone bedrock because seas covered much of the country for millions of years. It is hard to imagine Kansas being covered by ocean water, but it has happened several times throughout geological history.

Rock Salt When water that is rich in dissolved salt evaporates, it often deposits the mineral halite. Halite forms rock salt, shown in **Figure 15.** Rock salt deposits can range in thickness from a few meters to more than 400 m. Companies mine these deposits because rock salt is an important resource. It's used in the manufacturing of glass, paper, soap, and dairy products. The halite in rock salt is processed and used as table salt.

Figure 15 Rock salt is extracted from this mine in Germany. The same salt can be processed and used to season your favorite foods.

Organic Sedimentary Rocks

Rocks made of the remains of once-living things are called organic sedimentary rocks. One of the most common organic sedimentary rocks is fossil-rich limestone. Like chemical limestone, fossil-rich limestone is made of the mineral calcite. However, fossil-rich limestone mostly contains remains of once-living ocean organisms instead of only calcite that formed directly from ocean water.

Animals such as mussels, clams, corals, and snails make their shells from $CaCO_3$ that eventually becomes calcite. When they die, their shells accumulate on the ocean floor. When these shells are cemented together, fossil-rich limestone forms. If a rock is made completely of shell fragments that you can see, the rock is called coquina (koh KEE nuh).

Chalk Chalk is another organic sedimentary rock that is made of microscopic shells. When you write with naturally occurring chalk, you're crushing and smearing the calcite-shell remains of once-living ocean organisms.

Coal Another useful organic sedimentary rock is coal, shown in **Figure 16.** Coal forms when pieces of dead plants are buried under other sediments in swamps. These plant materials are chemically changed by microorganisms. The resulting sediments are compacted over millions of years to form coal, an important source of energy. Much of the coal in North America and Europe formed during a period of geologic time that is so named because of this important reason. The Carboniferous Period, which spans from approximately 360 to 286 million years ago, was named in Europe. So much coal formed during this interval of time that coal's composition—primarily carbon—was the basis for naming a geologic period.

Applying Math — Calculate Thickness

COAL FORMATION It took 300 million years for a layer of plant matter about 0.9 m thick to produce a bed of bituminous coal 0.3 m thick. Estimate the thickness of plant matter that produced a bed of coal 0.15 m thick.

Solution

1. *This is what you know:*
   - original thickness of plant matter = 0.9 m
   - original coal thickness = 0.3 m
   - new coal thickness = 0.15 m
2. *This is what you need to know:* thickness of plant matter needed to form 0.15 m of coal
3. *This is the equation you need to use:* (thickness of plant matter)/(new coal thickness) = (original thickness of plant matter)/(original coal thickness)
4. *Substitute the known values:* (? m plant matter)/(0.15 m coal) = (0.9 m plant matter)/(0.3 m coal)
5. *Solve the equation:* (? m plant matter) = (0.9 m plant matter)(0.15 m coal)/(0.3 m coal) = 0.45 m plant matter
6. *Check your answer:* Multiply your answer by the original coal thickness. Divide by the original plant matter thickness to get the new coal thickness.

Practice Problems

1. Estimate the thickness of plant matter that produced a bed of coal 0.6 m thick.
2. About how much coal would have been produced from a layer of plant matter 0.50 m thick?

For more practice, visit green.msscience.com/math_practice

Figure 16 This coal layer in Alaska is easily identified by its jet-black color, as compared with other sedimentary layers.

Another Look at the Rock Cycle

You have seen that the rock cycle has no beginning and no end. Rocks change continually from one form to another. Sediments can become so deeply buried that they eventually become metamorphic or igneous rocks. These reformed rocks later can be uplifted and exposed to the surface—possibly as mountains to be worn away again by erosion.

All of the rocks that you've learned about in this chapter formed through some process within the rock cycle. All of the rocks around you, including those used to build houses and monuments, are part of the rock cycle. Slowly, they are all changing, because the rock cycle is a continuous, dynamic process.

section 4 review

Summary

Formation of Sedimentary Rocks

- Sedimentary rocks form as layers, with older layers near the bottom of an undisturbed stack.

Classifying Sedimentary Rocks

- To classify a sedimentary rock, determine its composition and texture.

Detrital Sedimentary Rocks

- Rock and mineral fragments make up detrital rocks.

Chemical Sedimentary Rocks

- Chemical sedimentary rocks form from solutions of dissolved minerals.

Organic Sedimentary Rocks

- The remains of once-living organisms make up organic sedimentary rocks.

Self Check

1. **Identify** where sediments come from.
2. **Explain** how compaction is important in the formation of coal.
3. **Compare and contrast** detrital and chemical sedimentary rock.
4. **List** chemical sedimentary rocks that are essential to your health or that are used to make life more convenient. How is each used?
5. **Think Critically** Explain how pieces of granite and slate could both be found in the same conglomerate. How would the granite and slate pieces be held together?

Applying Math

6. **Calculate Ratios** Use information in **Table 2** to estimate how many times larger the largest grains of silt and sand are compared to the largest clay grains.

Sedimentary Rocks

Goals
- **Observe** sedimentary rock characteristics.
- **Compare and contrast** sedimentary rock textures.
- **Classify** sedimentary rocks as detrital, chemical, or organic.

Materials
unknown sedimentary rock samples
marking pen
5% hydrochloric acid (HCl) solution
dropper
paper towels
water
magnifying lens
metric ruler

Safety Precautions

WARNING: *HCl is an acid and can cause burns. Wear goggles and a lab apron. Rinse spills with water and wash hands afterward.*

Sedimentary rocks are formed by compaction and cementation of sediment. Because sediment is found in all shapes and sizes, do you think these characteristics could be used to classify detrital sedimentary rocks? Sedimentary rocks also can be classified as chemical or organic.

Real-World Question

How are rock characteristics used to classify sedimentary rocks as detrital, chemical, or organic?

Procedure

1. Make a Sedimentary Rock Samples chart in your Science Journal similar to the one shown on the next page.
2. **Determine** the sizes of sediments in each sample, using a magnifying lens and a metric ruler. Using **Table 2,** classify any grains of sediment in the rocks as gravel, sand, silt, or clay. In general, the sediment is silt if it is gritty and just barely visible, and clay if it is smooth and if individual grains are not visible.
3. Place a few drops of 5% HCl solution on each rock sample. Bubbling on a rock indicates the presence of calcite.
4. **Examine** each sample for fossils and describe any that are present.
5. **Determine** whether each sample has a granular or nongranular texture.

Sedimentary Rock Samples

Sample	Observations	Minerals or Fossils Present	Sediment Size	Detrital, Chemical, or Organic	Rock Name
A					
B		Do not write in this book.			
C					
D					
E					

Analyze Your Data

1. **Classify** your samples as detrital, chemical, or organic.
2. **Identify** each rock sample.

Conclude and Apply

1. **Explain** why you tested the rocks with acid. What minerals react with acid?
2. **Compare and contrast** sedimentary rocks that have a granular texture with sedimentary rocks that have a nongranular texture.

Communicating Your Data

Compare your conclusions with those of other students in your class. **For more help, refer to the** Science Skill Handbook.

Australia's controversial rock star

One of the most famous rocks in the world is causing serious problems for Australians

Uluru (yew LEW rew), also known as Ayers Rock, is one of the most popular tourist destinations in Australia. This sandstone skyscraper is more than 8 km around, over 300 m high, and extends as much as 4.8 km below the surface. One writer describes it as an iceberg in the desert. Geologists hypothesize that the mighty Uluru rock began forming 550 million years ago during Precambrian time. That's when large mountain ranges started to form in Central Australia.

For more than 25,000 years, this geological wonder has played an important role in the lives of the Aboriginal peoples, the Anangu (a NA noo). These native Australians are the original owners of the rock and have spiritual explanations for its many caves, holes, and scars.

Athlete Nova Benis-Kneebone had the honor of receiving the Olympic torch near the sacred Uluru and carried it partway to the Olympic stadium.

Tourists Take Over

In the 1980s, some 100,000 tourists visited—and many climbed—Uluru. In 2000, the rock attracted about 400,000 tourists. The Anangu take offense at anyone climbing their sacred rock. However, if climbing the rock were outlawed, tourism would be seriously hurt. That would mean less income for Australians.

To respect the Anangu's wishes, the Australian government returned Ayers Rock to the Anangu in 1985 and agreed to call it by its traditional name. The Anangu leased back the rock to the Australian government until the year 2084, when its management will return to the Anangu. Until then, the Anangu will collect 25 percent of the money people pay to visit the rock.

The Aboriginal people encourage tourists to respect their beliefs. They offer a walking tour around the rock, and they show videos about Aboriginal traditions. The Anangu sell T-shirts that say "I *didn't* climb Uluru." They hope visitors to Uluru will wear the T-shirt with pride and respect.

Write **Research a natural landmark or large natural land or water formation in your area. What is the geology behind it? When was it formed? How was it formed? Write a folktale that explains its formation. Share your folktale with the class.**

For more information, visit green.msscience.com/time

Reviewing Main Ideas

Section 1 The Rock Cycle

1. A rock is a mixture of one or more minerals, rock fragments, organic matter, or volcanic glass.
2. The rock cycle includes all processes by which rocks form.

Section 2 Igneous Rocks

1. Magma and lava are molten materials that harden to form igneous rocks.
2. Intrusive igneous rocks form when magma cools slowly below Earth's surface. Extrusive igneous rocks form when lava cools rapidly at the surface.
3. The compositions of most igneous rocks range from granitic to andesitic to basaltic.

Section 3 Metamorphic Rocks

1. Heat, pressure, and fluids can cause metamorphic rocks to form.
2. Slate and gneiss are examples of foliated metamorphic rocks. Quartzite and marble are examples of nonfoliated metamorphic rocks.

Section 4 Sedimentary Rocks

1. Detrital sedimentary rocks form when fragments of rocks and minerals are compacted and cemented together.
2. Chemical sedimentary rocks come out of solution or are left behind by evaporation.
3. Organic sedimentary rocks contain the remains of once-living organisms.

Visualizing Main Ideas

Copy and complete the following concept map on rocks. Use the following terms: organic, metamorphic, foliated, extrusive, igneous, *and* chemical.

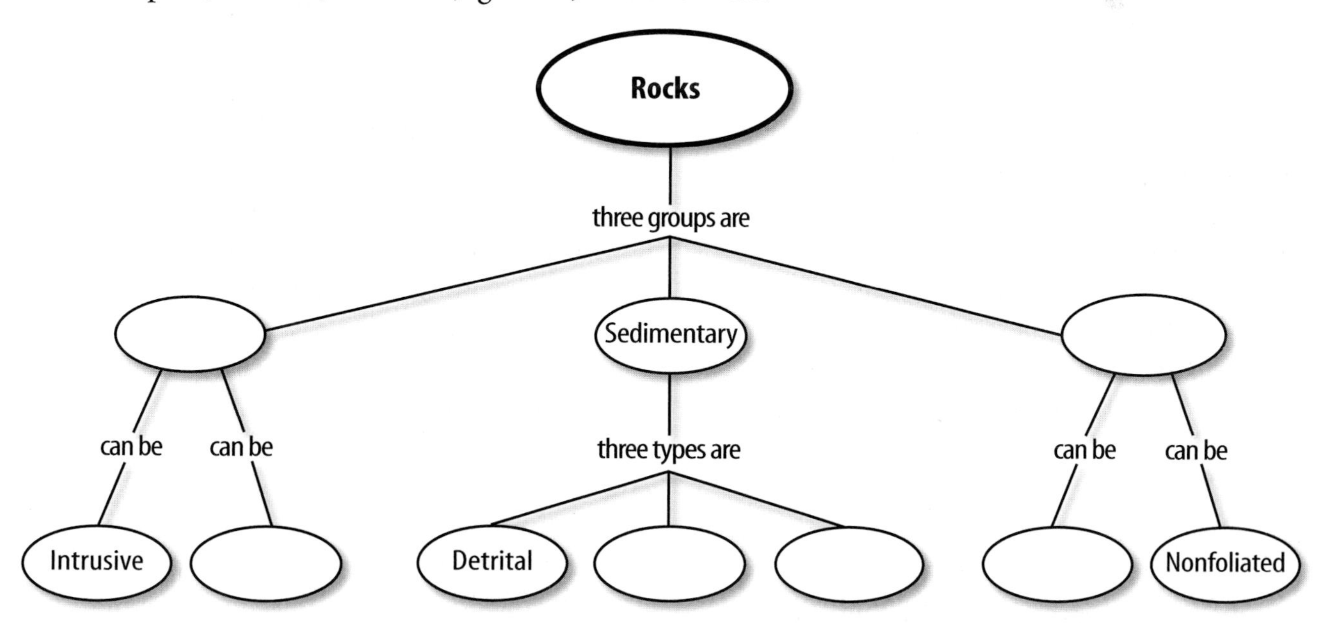

chapter 3 Review

Using Vocabulary

basaltic p. 65
cementation p. 73
compaction p. 72
extrusive p. 63
foliated p. 69
granitic p. 65
igneous rock p. 62
intrusive p. 63
lava p. 62
metamorphic rock p. 67
nonfoliated p. 70
rock p. 58
rock cycle p. 59
sediment p. 71
sedimentary rock p. 71

Explain the difference between the vocabulary words in each of the following sets.

1. foliated—nonfoliated
2. cementation—compaction
3. sediment—lava
4. extrusive—intrusive
5. rock—rock cycle
6. metamorphic rock—igneous rock—sedimentary rock
7. sediment—sedimentary rock
8. lava—igneous rock
9. rock—sediment
10. basaltic—granitic

Checking Concepts

Choose the word or phrase that best answers the question.

11. Why does magma tend to rise toward Earth's surface?
    A) It is more dense than surrounding rocks.
    B) It is more massive than surrounding rocks.
    C) It is cooler than surrounding rocks.
    D) It is less dense than surrounding rocks.

12. During metamorphism of granite into gneiss, what happens to minerals?
    A) They partly melt.
    B) They become new sediments.
    C) They grow smaller.
    D) They align into layers.

13. Which rock has large mineral grains?
    A) granite **C)** obsidian
    B) basalt **D)** pumice

14. Which type of rock is shown in this photo?
    A) foliated
    B) nonfoliated
    C) intrusive
    D) extrusive

15. What do igneous rocks form from?
    A) sediments **C)** gravel
    B) mud **D)** magma

16. What sedimentary rock is made of large, angular pieces of sediments?
    A) conglomerate **C)** limestone
    B) breccia **D)** chalk

17. Which of the following is an example of a detrital sedimentary rock?
    A) limestone **C)** breccia
    B) evaporite **D)** chalk

18. What is molten material at Earth's surface called?
    A) limestone **C)** breccia
    B) lava **D)** granite

19. Which of these is an organic sedimentary rock?
    A) coquina **C)** rock salt
    B) sandstone **D)** conglomerate

Science Online green.msscience.com/vocabulary_puzzlemaker

Thinking Critically

20. Infer Granite, pumice, and scoria are igneous rocks. Why doesn't granite have airholes like the other two?

21. Infer why marble rarely contains fossils.

22. Predict Would you expect quartzite or sandstone to break more easily? Explain your answer.

23. Compare and contrast basaltic and granitic magmas.

24. Form Hypotheses A geologist was studying rocks in a mountain range. She found a layer of sedimentary rock that had formed in the ocean. Hypothesize how this could happen.

25. Concept Map Copy and complete the concept map shown below. Use the following terms and phrases: *magma, sediments, igneous rock, sedimentary rock, metamorphic rock.* Add and label any missing arrows.

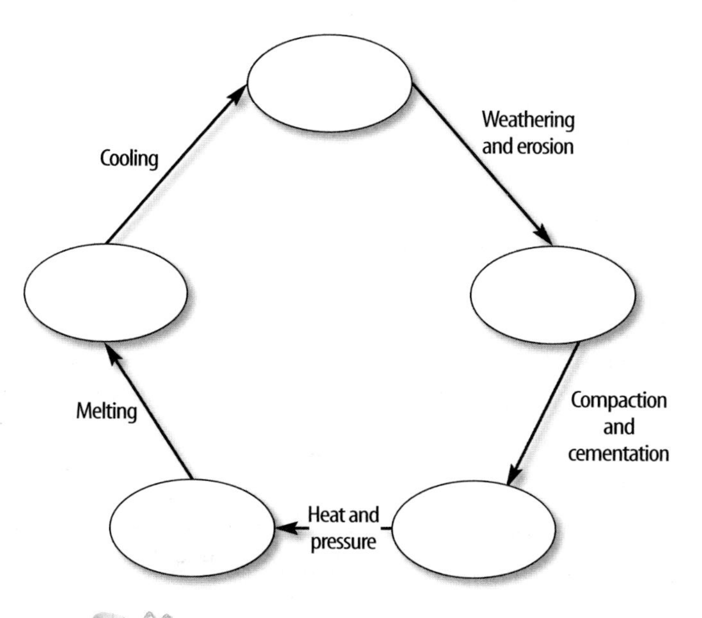

Performance Activities

26. Poster Collect a group of rocks. Make a poster that shows the classifications of rocks, and glue your rocks to the poster under the proper headings. Describe your rocks and explain where you found them.

Applying Math

27. Grain Size Assume that the conglomerate shown on the second page of the "Sedimentary Rocks" lab is one-half of its actual size. Determine the average length of the gravel in the rock.

28. Plant Matter Suppose that a 4-m layer of plant matter was compacted to form a coal layer 1 m thick. By what percent has the thickness of organic material been reduced?

Use the graph below to answer questions 29 and 30.

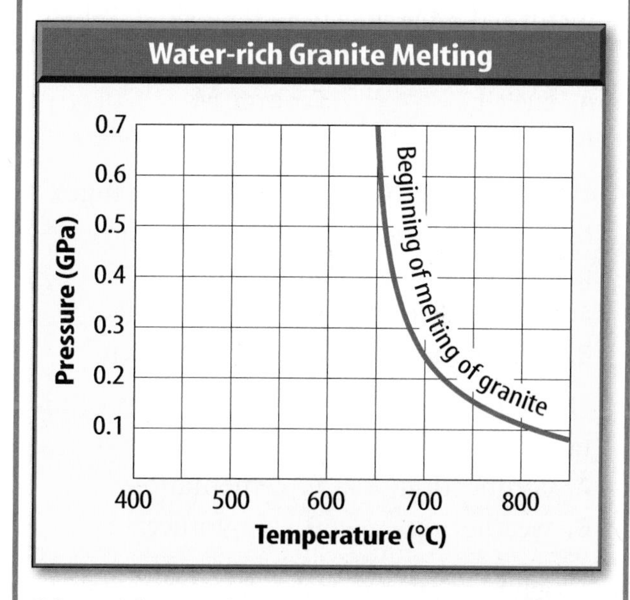

29. Melting Granite Determine the melting temperature of a water-rich granite at a pressure of 0.2 GPa.

Pressure conversions:
1 GPa, or gigapascal, = 10,000 bars
1 bar = 0.9869 atmospheres

30. Melting Pressure At about what pressure will a water-rich granite melt at 680°C?

Part 1 Multiple Choice

Record your answers on the answer sheet provided by your teacher or on a sheet of paper.

Use the illustration below to answer question 1.

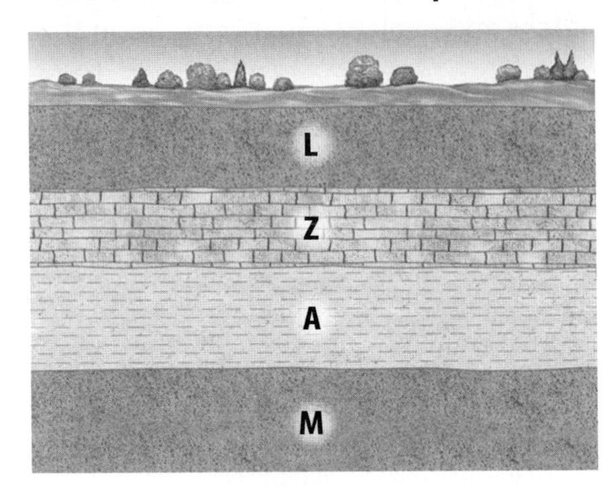

1. These layers of sedimentary rock were not disturbed after they were deposited. Which layer was deposited first?
   A. layer L **C.** layer M
   B. layer Z **D.** layer A

2. Who realized that rocks undergo changes through long periods of time after observing rocks at Siccar Point, Scotland?
   A. James Hutton **C.** Galileo Galilei
   B. Neil Armstrong **D.** Albert Einstein

3. During which process do minerals precipitate in the spaces between sediment grains?
   A. compaction **C.** cementation
   B. weathering **D.** conglomerate

4. Which rock often is sculpted to create statues?
   A. shale **C.** coquina
   B. marble **D.** conglomerate

Test-Taking Tip

Careful Reading Read each question carefully for full understanding.

5. Which of the following rocks is a metamorphic rock?
   A. shale **C.** slate
   B. granite **D.** pumice

6. Which rock consists mostly of pieces of seashell?
   A. sandstone **C.** pumice
   B. coquina **D.** granite

Use the diagram below to answer questions 7–9.

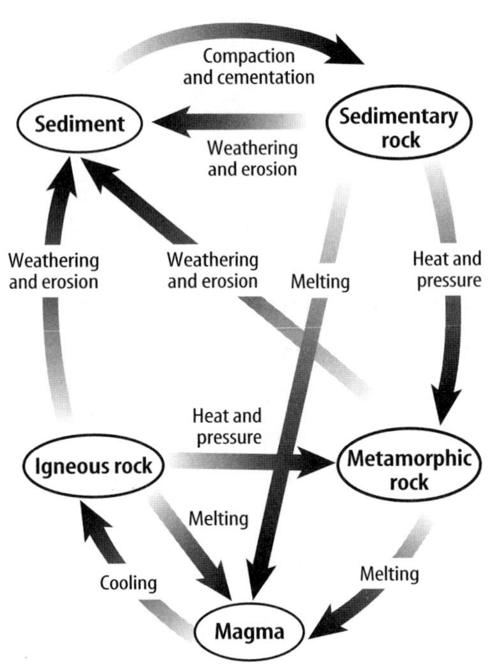

7. Which process in the rock cycle causes magma to form?
   A. melting **C.** weathering
   B. erosion **D.** cooling

8. What forms when rocks are weathered and eroded?
   A. igneous rock **C.** sedimentary rock
   B. sediment **D.** metamorphic rock

9. Which type of rock forms because of high heat and pressure without melting?
   A. igneous rock **C.** sedimentary rock
   B. intrusive rock **D.** metamorphic rock

Part 2 | Short Response/Grid In

Record your answers on the answer sheet provided by your teacher or on a sheet of paper.

10. What is a rock? How is a rock different from a mineral?

11. Explain why some igneous rocks are coarse and others are fine.

12. What is foliation? How does it form?

13. How do chemical sedimentary rocks, such as rock salt, form?

14. Why do some rocks contain fossils?

15. How is the formation of chemical sedimentary rocks similar to the formation of cement in detrital sedimentary rocks?

Use the graph below to answer questions 16–17.

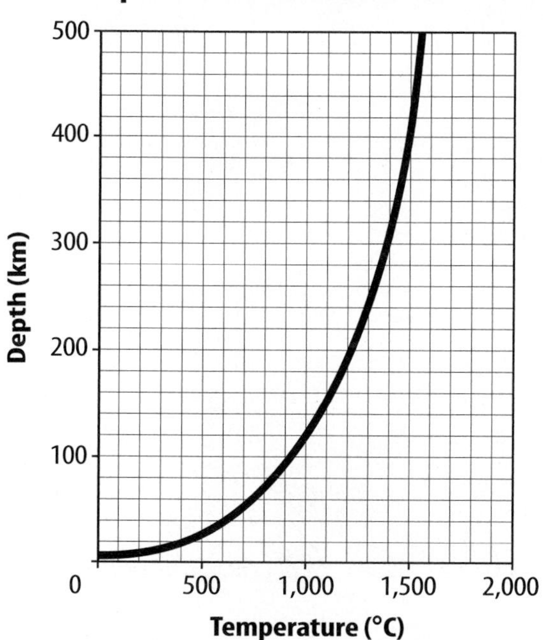

16. According to the graph, about how deep below a continent does the temperature reach 1,000°C?

17. In general, what happens to temperature as depth below Earth's surface increases?

Part 3 | Open Ended

Record your answers on a sheet of paper.

Use the table below to answer questions 18 and 19.

Magma Type	Basaltic	Andesitic	Granitic
Intrusive		Do not write in this book.	
Extrusive			

18. Copy the table on your paper. Then, fill in the empty squares with a correct rock name.

19. Explain how igneous rocks are classified.

20. Explain how loose sediment can become sedimentary rock.

21. Why does pressure increase with depth in Earth? How does higher pressure affect rocks?

22. Why is slate sometimes used as shingles for roofs? What other rocks are used for important purposes in society?

23. How are organic sedimentary rocks different from other rocks? List an example of an organic sedimentary rock.

24. Why is the rock cycle called a cycle?

25. A geologist found a sequence of rocks in which 200-million-year-old shales were on top of 100-million-year-old sandstones. Hypothesize how this could happen.

26. Explain why coquina could be classified in more than one way.

unit 2 Earth's Atmosphere and Beyond

How Are Bats & Tornadoes Connected?

Bats are able to find food and avoid obstacles without using their vision. They do this by producing high-frequency sound waves which bounce off objects and return to the bat. From these echoes, the bat is able to locate obstacles and prey. This process is called echolocation. If the reflected waves have a higher frequency than the emitted waves, the bat senses the object is getting closer. If the reflected waves have a lower frequency, the object is moving away. This change in frequency is called the Doppler effect. Like echolocation, sonar technology uses sound waves and the Doppler effect to determine the position and motion of objects. Doppler radar also uses the Doppler effect, but with radar waves instead of sound waves. Higher frequency waves indicate if an object, such as a storm, is coming closer, while lower frequencies indicate if it is moving away. Meteorologists use frequency shifts indicated by Doppler radar to detect the formation of tornadoes and to predict where they will strike.

unit projects

Visit green.msscience.com/unit_project to find project ideas and resources. Projects include:

- **Technology** Predict and track the weather of a city in a different part of the world, and compare it to your local weather pattern.
- **Career** Explore weather-related careers while investigating different types of storms. Compare and contrast career characteristics and history.
- **Model** Research animal behavior to discover if animals are able to predict the weather. Present your samples of weather-predicting proverbs as a collection, or use them in a folklore tale.

Hurricanes! investigates a variety of tropical storms, their source of energy, classifications, and destructive forces.

chapter 4

Atmosphere

chapter preview

sections

Fresh mountain air?

On top of Mt. Everest the air is a bit thin. Without breathing equipment, an average person quickly would become dizzy, then unconscious, and eventually would die. In this chapter you'll learn what makes the atmosphere at high altitudes different from the atmosphere we are used to.

Science Journal Write a short article describing how you might prepare to climb Mt. Everest.

Start-Up Activities

Observe Air Pressure

The air around you is made of billions of molecules. These molecules are constantly moving in all directions and bouncing into every object in the room, including you. Air pressure is the result of the billions of collisions of molecules into these objects. Because you usually do not feel molecules in air hitting you, do the lab below to see the effect of air pressure.

1. Cut out a square of cardboard about 10 cm from the side of a cereal box.
2. Fill a glass to the brim with water.
3. Hold the cardboard firmly over the top of the glass, covering the water, and invert the glass.
4. Slowly remove your hand holding the cardboard in place and observe.
5. **Think Critically** Write a paragraph in your Science Journal describing what happened to the cardboard when you inverted the glass and removed your hand. How does air pressure explain what happened?

Earth's Atmospheric Layers Make the following Foldable to help you visualize the five layers of Earth's atmosphere.

STEP 1 **Collect** 3 sheets of paper and layer them about 1.25 cm apart vertically. Keep the edges level.

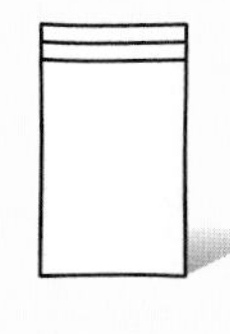

STEP 2 **Fold** up the bottom edges of the paper to form 6 equal tabs.

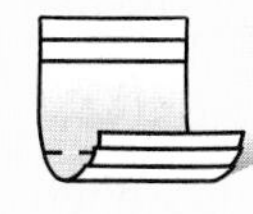

STEP 3 **Fold** the paper and crease well to hold the tabs in place. Staple along the fold. **Label** each tab.

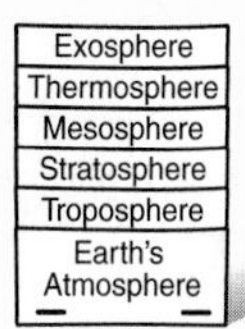

Find Main Ideas Label the tabs *Earth's Atmosphere, Troposphere, Stratosphere, Mesosphere, Thermosphere,* and *Exosphere* from bottom to top as shown. As you read the chapter, write information about each layer of Earth's atmosphere under the appropriate tab.

Preview this chapter's content and activities at green.msscience.com

section 1

Earth's Atmosphere

as you read

What You'll Learn

- **Identify** the gases in Earth's atmosphere.
- **Describe** the structure of Earth's atmosphere.
- **Explain** what causes air pressure.

Why It's Important

The atmosphere makes life on Earth possible.

Review Vocabulary
pressure: force exerted on an area

New Vocabulary
- atmosphere
- troposphere
- ionosphere
- ozone layer
- ultraviolet radiation
- chlorofluorocarbon

Importance of the Atmosphere

Earth's **atmosphere,** shown in **Figure 1,** is a thin layer of air that forms a protective covering around the planet. If Earth had no atmosphere, days would be extremely hot and nights would be extremely cold. Earth's atmosphere maintains a balance between the amount of heat absorbed from the Sun and the amount of heat that escapes back into space. It also protects life-forms from some of the Sun's harmful rays.

Makeup of the Atmosphere

Earth's atmosphere is a mixture of gases, solids, and liquids that surrounds the planet. It extends from Earth's surface to outer space. The atmosphere is much different today from what it was when Earth was young.

Earth's early atmosphere, produced by erupting volcanoes, contained nitrogen and carbon dioxide, but little oxygen. Then, more than 2 billon years ago, Earth's early organisms released oxygen into the atmosphere as they made food with the aid of sunlight. These early organisms, however, were limited to layers of ocean water deep enough to be shielded from the Sun's harmful rays, yet close enough to the surface to receive sunlight. Eventually, a layer rich in ozone (O_3) that protects Earth from the Sun's harmful rays formed in the upper atmosphere. This protective layer eventually allowed green plants to flourish all over Earth, releasing even more oxygen. Today, a variety of life forms, including you, depends on a certain amount of oxygen in Earth's atmosphere.

Figure 1 Earth's atmosphere, as viewed from space, is a thin layer of gases. The atmosphere keeps Earth's temperature in a range that can support life.

Gases in the Atmosphere Today's atmosphere is a mixture of the gases shown in **Figure 2.** Nitrogen is the most abundant gas, making up 78 percent of the atmosphere. Oxygen actually makes up only 21 percent of Earth's atmosphere. As much as four percent of the atmosphere is water vapor. Other gases that make up Earth's atmosphere include argon and carbon dioxide.

The composition of the atmosphere is changing in small but important ways. For example, car exhaust emits gases into the air. These pollutants mix with oxygen and other chemicals in the presence of sunlight and form a brown haze called smog. Humans burn fuel for energy. As fuel is burned, carbon dioxide is released as a by-product into Earth's atmosphere. Increasing energy use may increase the amount of carbon dioxide in the atmosphere.

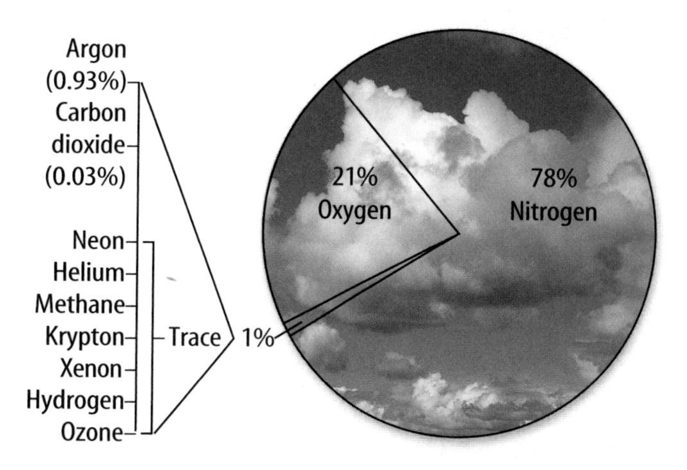

Figure 2 This circle graph shows the percentages of the gases, excluding water vapor, that make up Earth's atmosphere.
Determine *Approximately what fraction of Earth's atmosphere is oxygen?*

Solids and Liquids in Earth's Atmosphere In addition to gases, Earth's atmosphere contains small, solid particles such as dust, salt, and pollen. Dust particles get into the atmosphere when wind picks them up off the ground and carries them along. Salt is picked up from ocean spray. Plants give off pollen that becomes mixed throughout part of the atmosphere.

The atmosphere also contains small liquid droplets other than water droplets in clouds. The atmosphere constantly moves these liquid droplets and solids from one region to another. For example, the atmosphere above you may contain liquid droplets and solids from an erupting volcano thousands of kilometers from your home, as illustrated in **Figure 3.**

Figure 3 Solids and liquids can travel large distances in Earth's atmosphere, affecting regions far from their source.

On June 12, 1991, Mount Pinatubo in the Philippines erupted, causing liquid droplets to form in Earth's atmosphere.

Droplets of sulfuric acid from volcanoes can produce spectacular sunrises.

Layers of the Atmosphere

What would happen if you left a glass of chocolate milk on the kitchen counter for a while? Eventually, you would see a lower layer with more chocolate separating from upper layers with less chocolate. Like a glass of chocolate milk, Earth's atmosphere has layers. There are five layers in Earth's atmosphere, each with its own properties, as shown in **Figure 4.** The lower layers include the troposphere and stratosphere. The upper atmospheric layers are the mesosphere, thermosphere, and exosphere. The troposphere and stratosphere contain most of the air.

Topic: Earth's Atmospheric Layers

Visit green.msscience.com for Web links to information about layers of Earth's atmosphere.

Activity Locate data on recent ozone layer depletion. Graph your data.

Lower Layers of the Atmosphere You study, eat, sleep, and play in the **troposphere** which is the lowest of Earth's atmospheric layers. It contains 99 percent of the water vapor and 75 percent of the atmospheric gases. Rain, snow, and clouds occur in the troposphere, which extends up to about 10 km.

The stratosphere, the layer directly above the troposphere, extends from 10 km above Earth's surface to about 50 km. As **Figure 4** shows, a portion of the stratosphere contains higher levels of a gas called ozone. Each molecule of ozone is made up of three oxygen atoms bonded together. Later in this section you will learn how ozone protects Earth from the Sun's harmful rays.

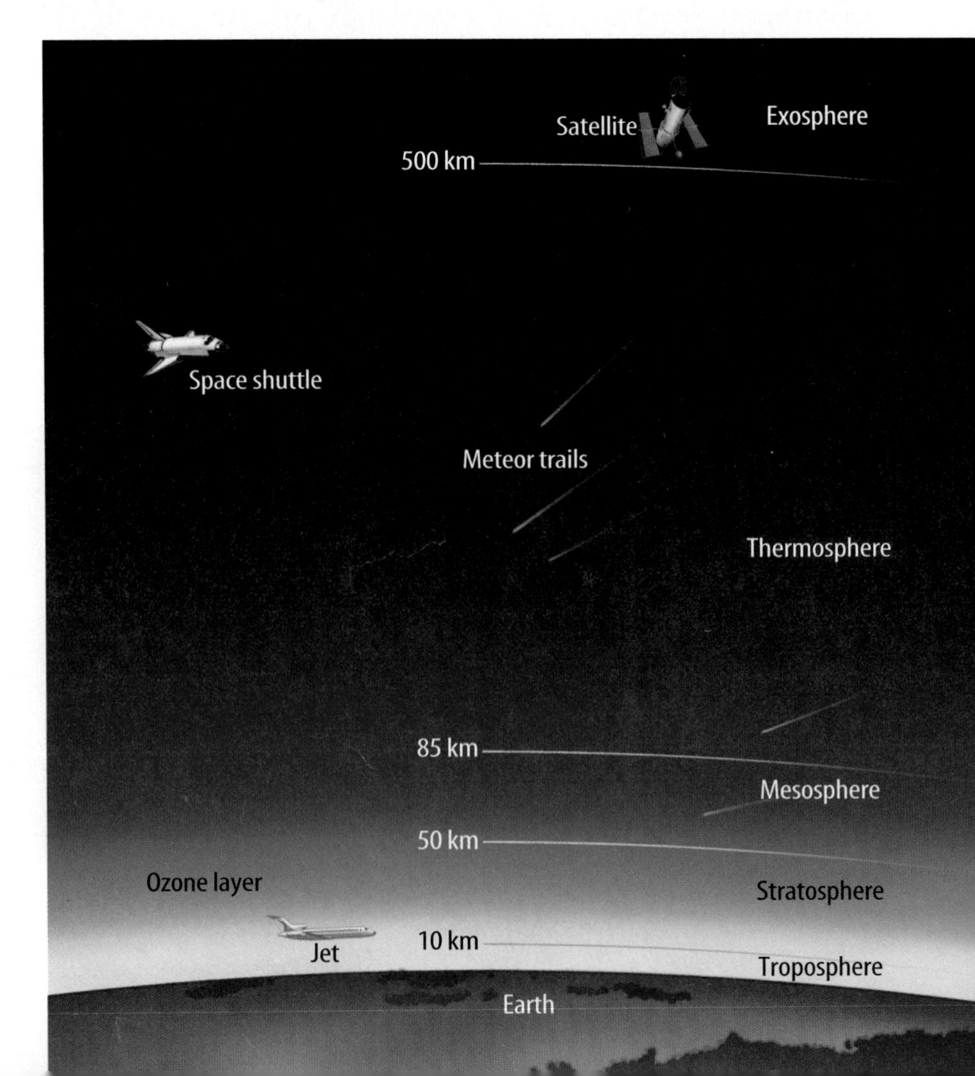

Figure 4 Earth's atmosphere is divided into five layers.
Describe *the layer of the atmosphere in which you live.*

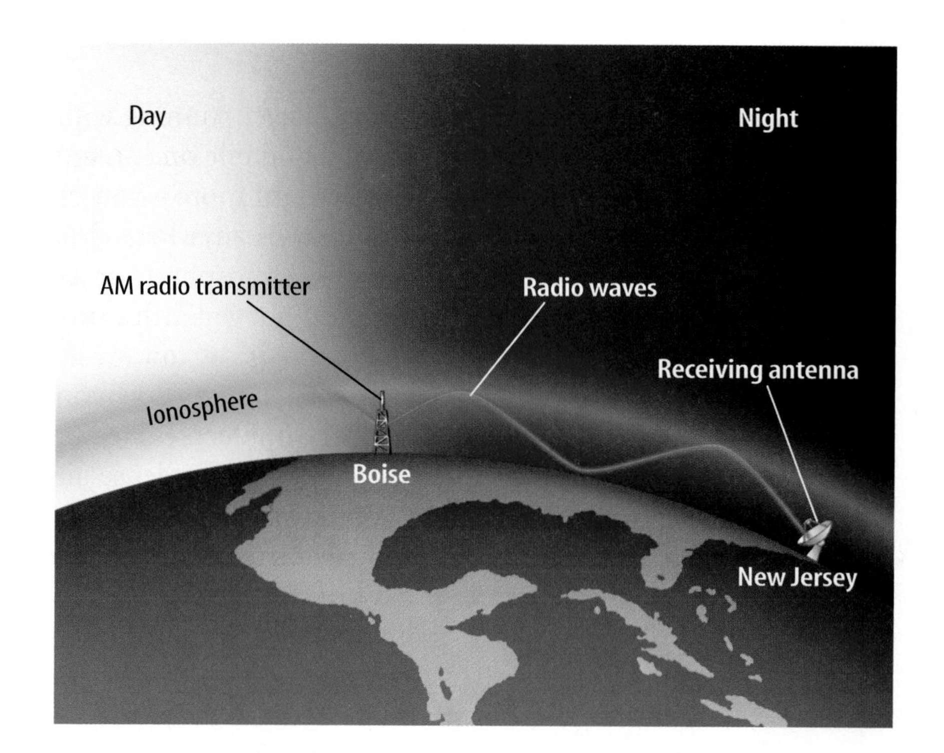

Figure 5 During the day, the ionosphere absorbs radio transmissions. This prevents you from hearing distant radio stations. At night, the ionosphere reflects radio waves. The reflected waves can travel to distant cities.
Describe *what causes the ionosphere to change between day and night.*

Upper Layers of the Atmosphere Beyond the stratosphere are the mesosphere, thermosphere, and exosphere. The mesosphere extends from the top of the stratosphere to about 85 km above Earth. If you've ever seen a shooting star, you might have witnessed a meteor in the mesosphere.

The thermosphere is named for its high temperatures. This is the thickest atmospheric layer and is found between 85 km and 500 km above Earth's surface.

Within the mesosphere and thermosphere is a layer of electrically charged particles called the **ionosphere** (i AH nuh sfihr). If you live in New Jersey and listen to the radio at night, you might pick up a station from Boise, Idaho. The ionosphere allows radio waves to travel across the country to another city, as shown in **Figure 5.** During the day, energy from the Sun interacts with the particles in the ionosphere, causing them to absorb AM radio frequencies. At night, without solar energy, AM radio transmissions reflect off the ionosphere, allowing radio transmissions to be received at greater distances.

The space shuttle in **Figure 6** orbits Earth in the exosphere. In contrast to the troposphere, the layer you live in, the exosphere has so few molecules that the wings of the shuttle are useless. In the exosphere, the spacecraft relies on bursts from small rocket thrusters to move around. Beyond the exosphere is outer space.

Reading Check *How does the space shuttle maneuver in the exosphere?*

Figure 6 Wings help move aircraft in lower layers of the atmosphere. The space shuttle can't use its wings to maneuver in the exosphere because so few molecules are present.

Figure 7 Air pressure decreases as you go higher in Earth's atmosphere.

Atmospheric Pressure

Imagine you're a football player running with the ball. Six players tackle you and pile one on top of the other. Who feels the weight more—you or the player on top? Like molecules anywhere else, atmospheric gases have mass. Atmospheric gases extend hundreds of kilometers above Earth's surface. As Earth's gravity pulls the gases toward its surface, the weight of these gases presses down on the air below. As a result, the molecules nearer Earth's surface are closer together. This dense air exerts more force than the less dense air near the top of the atmosphere. Force exerted on an area is known as pressure.

Like the pile of football players, air pressure is greater near Earth's surface and decreases higher in the atmosphere, as shown in **Figure 7.** People find it difficult to breathe in high mountains because fewer molecules of air exist there. Jets that fly in the stratosphere must maintain pressurized cabins so that people can breathe.

Where is air pressure greater—in the exosphere or in the troposphere?

Applying Science

How does altitude affect air pressure?

Atmospheric gases extend hundreds of kilometers above Earth's surface, but the molecules that make up these gases are fewer and fewer in number as you go higher. This means that air pressure decreases with altitude.

Identifying the Problem

The graph on the right shows these changes in air pressure. Note that altitude on the graph goes up only to 50 km. The troposphere and the stratosphere are represented on the graph, but other layers of the atmosphere are not. By examining the graph, can you understand the relationship between altitude and pressure?

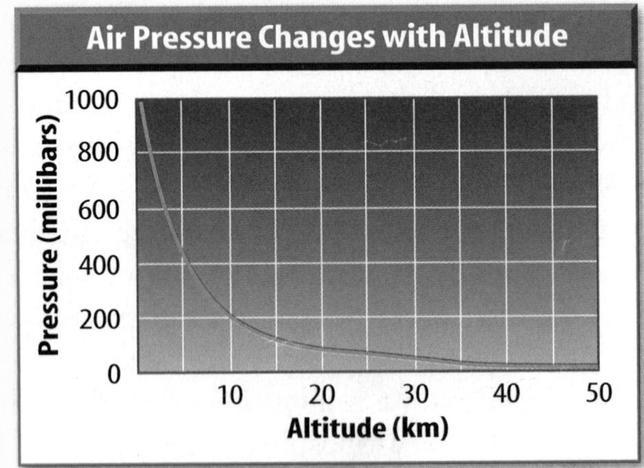

Solving the Problem

1. Estimate the air pressure at an altitude of 5 km.
2. Does air pressure change more quickly at higher altitudes or at lower altitudes?

Temperature in Atmospheric Layers

The Sun is the source of most of the energy on Earth. Before it reaches Earth's surface, energy from the Sun must pass through the atmosphere. Because some layers contain gases that easily absorb the Sun's energy while other layers do not, the various layers have different temperatures, illustrated by the red line in **Figure 8.**

Molecules that make up air in the troposphere are warmed mostly by heat from Earth's surface. The Sun warms Earth's surface, which then warms the air above it. When you climb a mountain, the air at the top is usually cooler than the air at the bottom. Every kilometer you climb, the air temperature decreases about 6.5°C.

Molecules of ozone in the stratosphere absorb some of the Sun's energy. Energy absorbed by ozone molecules raises the temperature. Because more ozone molecules are in the upper portion of the stratosphere, the temperature in this layer rises with increasing altitude.

Like the troposphere, the temperature in the mesosphere decreases with altitude. The thermosphere and exosphere are the first layers to receive the Sun's rays. Few molecules are in these layers, but each molecule has a great deal of energy. Temperatures here are high.

Determining if Air Has Mass

Procedure

1. On a **pan balance,** find the mass of an **inflatable ball** that is completely deflated.
2. Hypothesize about the change in the mass of the ball when it is inflated.
3. Inflate the ball to its maximum recommended inflation pressure.
4. Determine the mass of the fully inflated ball.

Analysis

1. What change occurs in the mass of the ball when it is inflated?
2. Infer from your data whether air has mass.

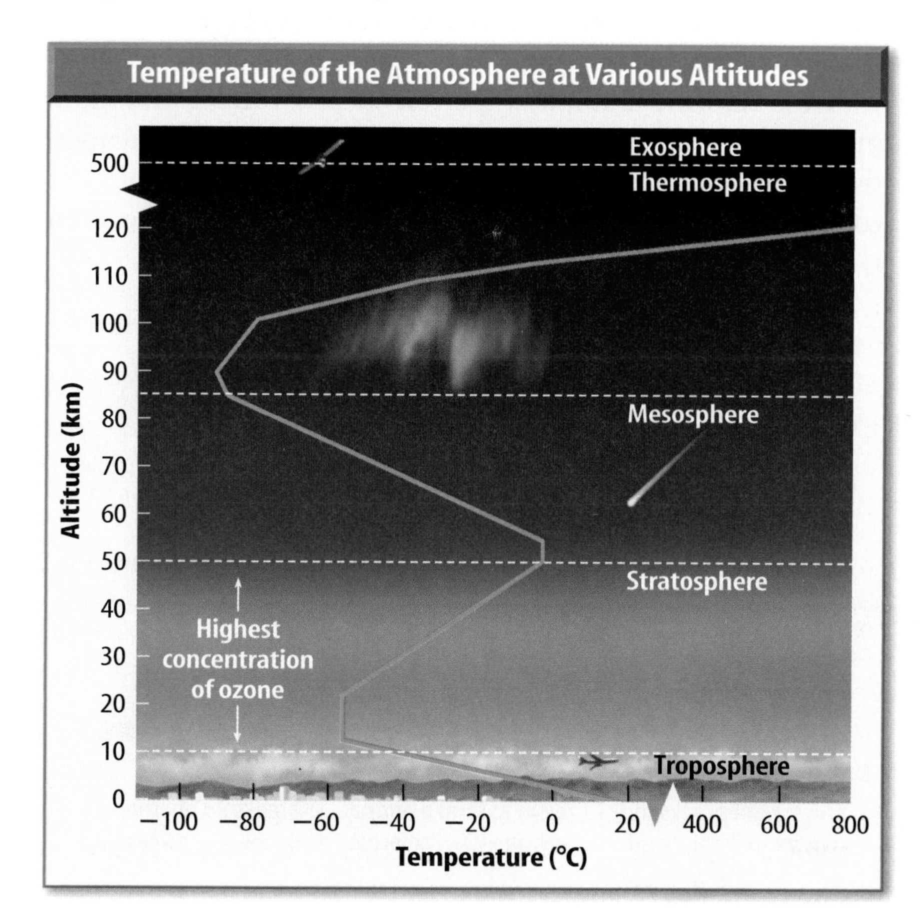

Figure 8 The division of the atmosphere into layers is based mainly on differences in temperature.
Determine *Does the temperature increase or decrease with altitude in the mesosphere?*

Effects of UV Light on Algae Algae are organisms that use sunlight to make their own food. This process releases oxygen to Earth's atmosphere. Some scientists suggest that growth is reduced when algae are exposed to ultraviolet radiation. Infer what might happen to the oxygen level of the atmosphere if increased ultraviolet radiation damages some algae.

The Ozone Layer

Within the stratosphere, about 19 km to 48 km above your head, lies an atmospheric layer called the **ozone layer.** Ozone is made of oxygen. Although you cannot see the ozone layer, your life depends on it.

The oxygen you breathe has two atoms per molecule, but an ozone molecule is made up of three oxygen atoms bound together. The ozone layer contains a high concentration of ozone and shields you from the Sun's harmful energy. Ozone absorbs most of the ultraviolet radiation that enters the atmosphere. **Ultraviolet radiation** is one of the many types of energy that come to Earth from the Sun. Too much exposure to ultraviolet radiation can damage your skin and cause cancer.

CFCs Evidence exists that some air pollutants are destroying the ozone layer. Blame has fallen on **chlorofluorocarbons** (CFCs), chemical compounds used in some refrigerators, air conditioners, and aerosol sprays, and in the production of some foam packaging. CFCs can enter the atmosphere if these appliances leak or if they and other products containing CFCs are improperly discarded.

Recall that an ozone molecule is made of three oxygen atoms bonded together. Chlorofluorocarbon molecules, shown in **Figure 9,** destroy ozone. When a chlorine atom from a chlorofluorocarbon molecule comes near a molecule of ozone, the ozone molecule breaks apart. One of the oxygen atoms combines with the chlorine atom, and the rest form a regular, two-atom molecule. These compounds don't absorb ultraviolet radiation the way ozone can. In addition, the original chlorine atom can continue to break apart thousands of ozone molecules. The result is that more ultraviolet radiation reaches Earth's surface.

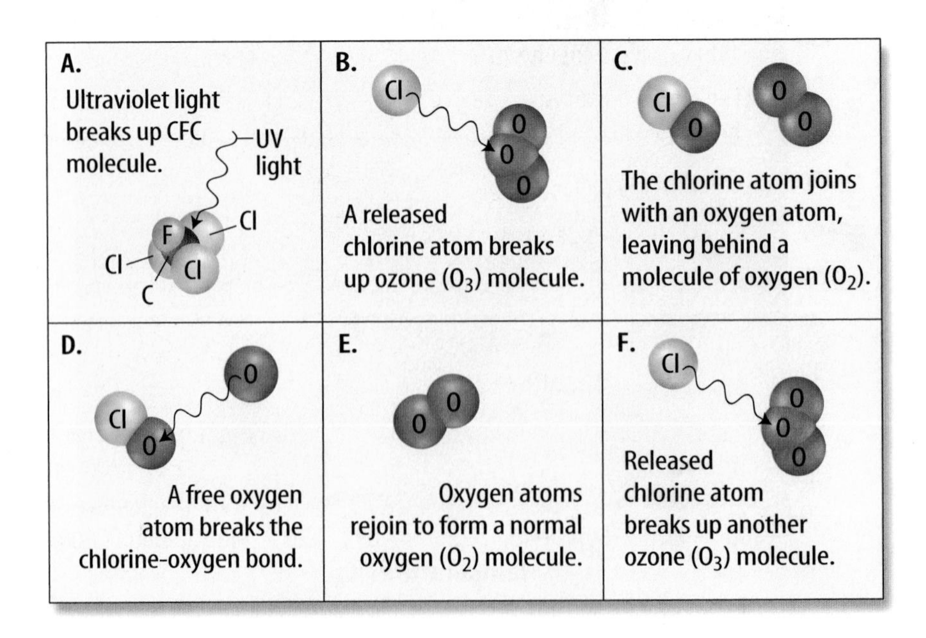

Figure 9 Chlorofluorocarbon (CFC) molecules once were used in refrigerators and air conditioners. Each CFC molecule has three chlorine atoms. One atom of chlorine can destroy approximately 100,000 ozone molecules.

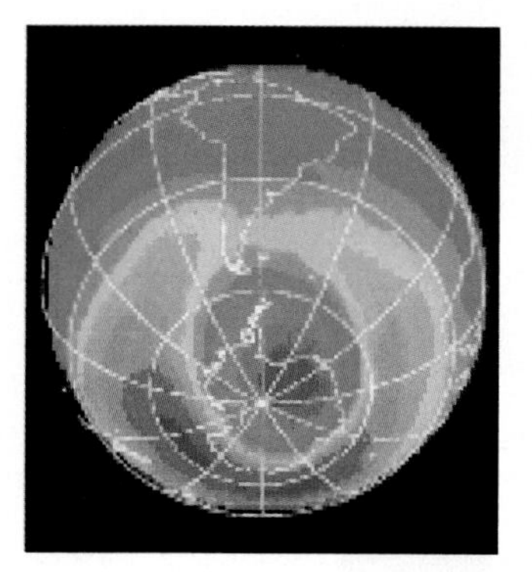

October 1980

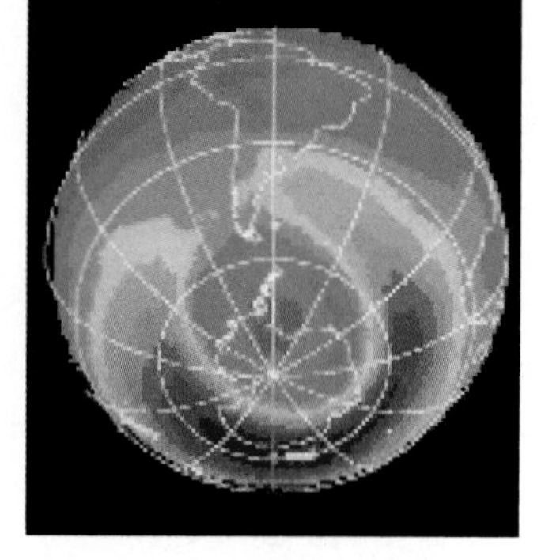

October 1988

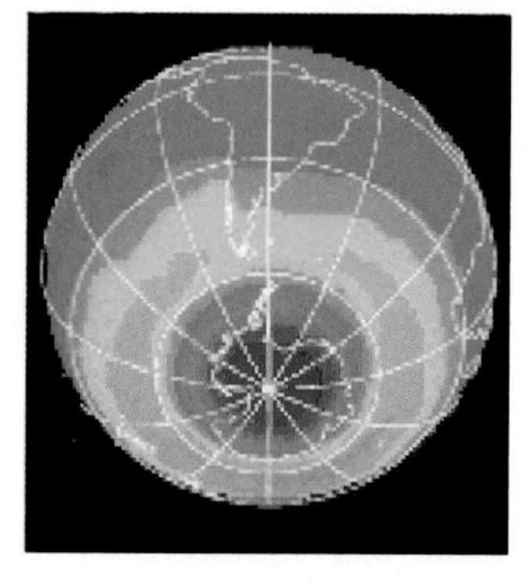

October 1990

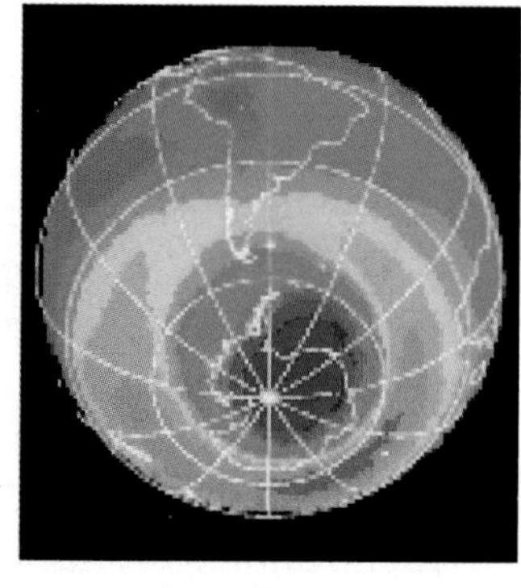

September 1999

The Ozone Hole The destruction of ozone molecules by CFCs seems to cause a seasonal reduction in ozone over Antarctica called the ozone hole. Every year beginning in late August or early September the amount of ozone in the atmosphere over Antarctica begins to decrease. By October, the ozone concentration reaches its lowest values and then begins to increase again. By December, the ozone hole disappears. **Figure 10** shows how the ozone hole over Antarctica has changed. In the mid-1990s, many governments banned the production and use of CFCs. Since then, the concentration of CFCs in the atmosphere has started to decrease.

Figure 10 These images of Antarctica were produced using data from a NASA satellite. The lowest values of ozone concentration are shown in dark blue and purple. These data show that the size of the seasonal ozone hole over Antarctica has grown larger over time.

section 1 review

Summary

Layers of the Atmosphere

- The atmosphere is a mixture of gases, solids, and liquids.
- The atmosphere has five layers—troposphere, stratosphere, mesosphere, thermosphere, and exosphere.
- The ionosphere is made up of electrically charged particles.

Atmospheric Pressure and Temperature

- Atmospheric pressure decreases with distance from Earth.
- Because some layers absorb the Sun's energy more easily than others, the various layers have different temperatures.

Ozone Layer

- The ozone layer absorbs most UV light.
- Chlorofluorocarbons (CFCs) break down the ozone layer.

Self Check

1. **Describe** How did oxygen come to make up 21 percent of Earth's present atmosphere?
2. **Infer** While hiking in the mountains, you notice that it is harder to breathe as you climb higher. Explain.
3. **State** some effects of a thinning ozone layer.
4. **Think Critically** Explain why, during the day, the radio only receives AM stations from a nearby city, while at night, you're able to hear a distant city's stations.

Applying Skills

5. **Interpret Scientific Illustrations** Using **Figure 2,** determine the total percentage of nitrogen and oxygen in the atmosphere. What is the total percentage of argon and carbon dioxide?
6. **Communicate** The names of the atmospheric layers end with the suffix *-sphere,* a word that means "ball." Find out what *tropo-, meso-, thermo-,* and *exo-* mean. Write their meanings in your Science Journal and explain if the layers are appropriately named.

Evaluating Sunscreens

Without protection, sun exposure can damage your health. Sunscreens protect your skin from UV radiation. In this lab, you will draw inferences using different sunscreen labels.

Real-World Question

How effective are various brands of sunscreens?

Goals

- **Draw inferences** based on labels on sunscreen brands.
- **Compare** the effectiveness of different sunscreen brands for protection against the Sun.
- **Compare** the cost of several sunscreen brands.

Materials

variety of sunscreens of different brand names

Safety Precautions

Procedure

1. Make a data table in your Science Journal using the following headings: *Brand Name, SPF, Cost per Milliliter,* and *Misleading Terms.*
2. The Sun Protection Factor (SPF) tells you how long the sunscreen will protect you. For example, an SPF of 4 allows you to stay in the Sun four times longer than if you did not use sunscreen. Record the SPF of each sunscreen on your data table.
3. **Calculate** the cost per milliliter of each sunscreen brand.
4. Government guidelines say that terms like *sunblock* and *waterproof* are misleading because sunscreens can't block the Sun's rays, and they do wash off in water. List misleading terms in your data table for each brand.

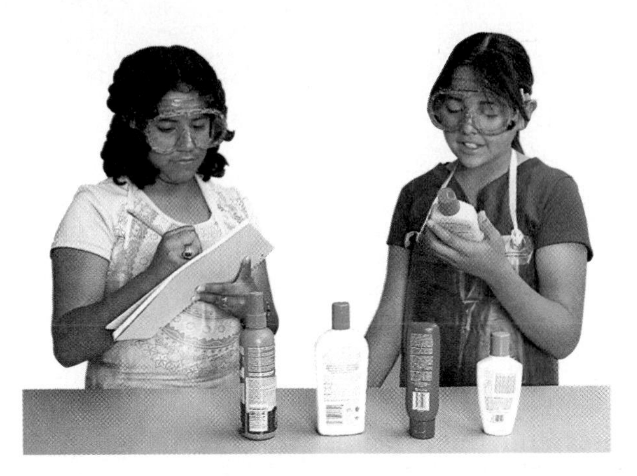

Sunscreen Assessment			
Brand Name			
SPF			
Cost per Milliliter	Do not write in this book.		
Misleading Terms			

Conclude and Apply

1. **Explain** why you need to use sunscreen.
2. **Evaluate** A minimum of SPF 15 is considered adequate protection for a sunscreen. An SPF greater than 30 is considered by government guidelines to be misleading because sunscreens wash or wear off. Evaluate the SPF of each sunscreen brand.
3. **Discuss** Considering the cost and effectiveness of all the sunscreen brands, discuss which you consider to be the best buy.

Create a poster on the proper use of sunscreens, and provide guidelines for selecting the safest product.

section 2

Energy Transfer in the Atmosphere

Energy from the Sun

The Sun provides most of Earth's energy. This energy drives winds and ocean currents and allows plants to grow and produce food, providing nutrition for many animals. When Earth receives energy from the Sun, three different things can happen to that energy, as shown in **Figure 11.** Some energy is reflected back into space by clouds, particles, and Earth's surface. Some is absorbed by the atmosphere or by land and water on Earth's surface.

Heat

Heat is energy that flows from an object with a higher temperature to an object with a lower temperature. Energy from the Sun reaches Earth's surface and heats it. Heat then is transferred through the atmosphere in three ways—radiation, conduction, and convection, as shown in **Figure 12.**

as you read

***What* You'll Learn**

- **Describe** what happens to the energy Earth receives from the Sun.
- **Compare and contrast** radiation, conduction, and convection.
- **Explain** the water cycle and its effect on weather patterns and climate.

***Why* It's Important**

The Sun provides energy to Earth's atmosphere, allowing life to exist.

Review Vocabulary

evaporation: when a liquid changes to a gas at a temperature below the liquid's boiling point

New Vocabulary

- radiation
- conduction
- convection
- hydrosphere
- condensation

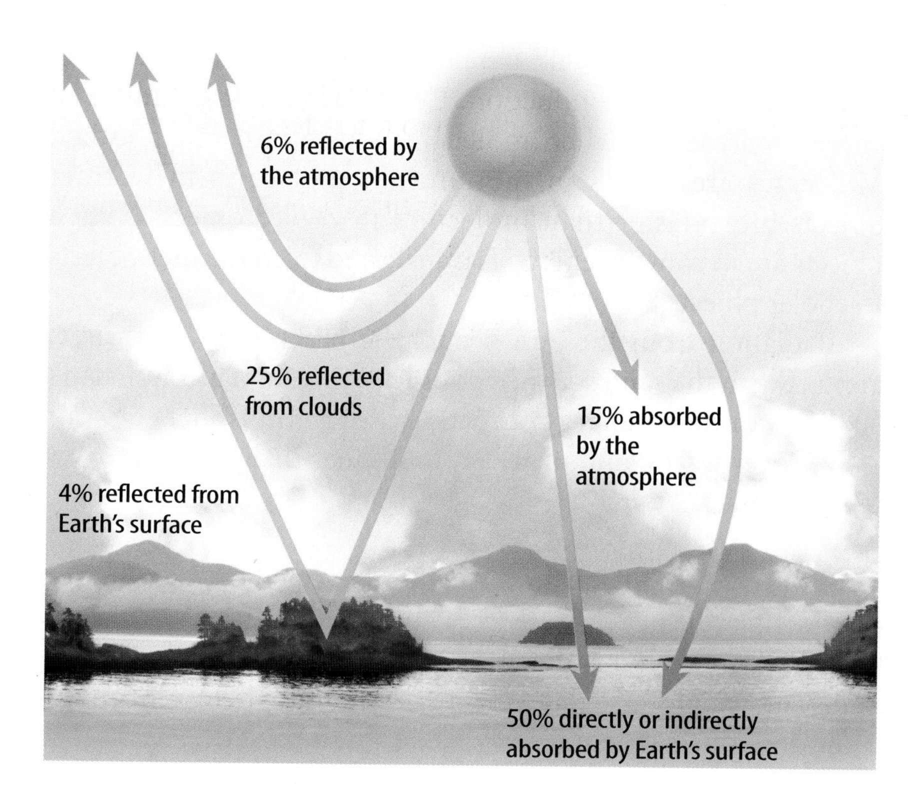

Figure 11 The Sun is the source of energy for Earth's atmosphere. Thirty-five percent of incoming solar radiation is reflected back into space.
Infer *how much is absorbed by Earth's surface and atmosphere.*

Figure 12 Heat is transferred within Earth's atmosphere by radiation, conduction, and convection.

Radiation Sitting on the beach, you feel the Sun's warmth on your face. How can you feel the Sun's heat even though you aren't in direct contact with it? Energy from the Sun reaches Earth in the form of radiant energy, or radiation. **Radiation** is energy that is transferred in the form of rays or waves. Earth radiates some of the energy it absorbs from the Sun back toward space. Radiant energy from the Sun warms your face.

Reading Check *How does the Sun warm your skin?*

Conduction If you walk barefoot on a hot beach, your feet heat up because of conduction. **Conduction** is the transfer of energy that occurs when molecules bump into one another. Molecules are always in motion, but molecules in warmer objects move faster than molecules in cooler objects. When objects are in contact, energy is transferred from warmer objects to cooler objects.

Radiation from the Sun heated the beach sand, but direct contact with the sand warmed your feet. In a similar way, Earth's surface conducts energy directly to the atmosphere. As air moves over warm land or water, molecules in air are heated by direct contact.

Convection After the atmosphere is warmed by radiation or conduction, the heat is transferred by a third process called convection. **Convection** is the transfer of heat by the flow of material. Convection circulates heat throughout the atmosphere. How does this happen?

Specific Heat Specific heat is the amount of heat required to raise the temperature of one kilogram of a substance one degree Celsius. Substances with high specific heat absorb a lot of heat for a small increase in temperature. Land warms faster than water does. Infer whether soil or water has a higher specific heat value.

When air is warmed, the molecules in it move apart and the air becomes less dense. Air pressure decreases because fewer molecules are in the same space. In cold air, molecules move closer together. The air becomes more dense and air pressure increases. Cooler, denser air sinks while warmer, less dense air rises, forming a convection current. As **Figure 12** shows, radiation, conduction, and convection together distribute the Sun's heat throughout Earth's atmosphere.

The Water Cycle

Hydrosphere is a term that describes all the waters of Earth. The constant cycling of water within the atmosphere and the hydrosphere, as shown in **Figure 13,** plays an important role in determining weather patterns and climate types.

Energy from the Sun causes water to change from a liquid to a gas by a process called evaporation. Water that evaporates from lakes, streams, and oceans enters Earth's atmosphere. If water vapor in the atmosphere cools enough, it changes back into a liquid. This process of water vapor changing to a liquid is called **condensation.**

Clouds form when condensation occurs high in the atmosphere. Clouds are made up of tiny water droplets that can collide to form larger drops. As the drops grow, they fall to Earth as precipitation. This completes the water cycle within the hydrosphere. Classification of world climates is commonly based on annual and monthly averages of temperature and precipitation that are strongly affected by the water cycle.

Modeling Heat Transfer

Procedure

1. Cover the outside of an empty **soup can,** with **black construction paper.**
2. Fill the can with **cold water** and feel it with your fingers.
3. Place the can in sunlight for 1 h, then pour the water over your fingers.

Analysis

1. Does the water in the can feel warmer or cooler after placing the can in sunlight?
2. What types of heat transfer did you model?

Try at Home

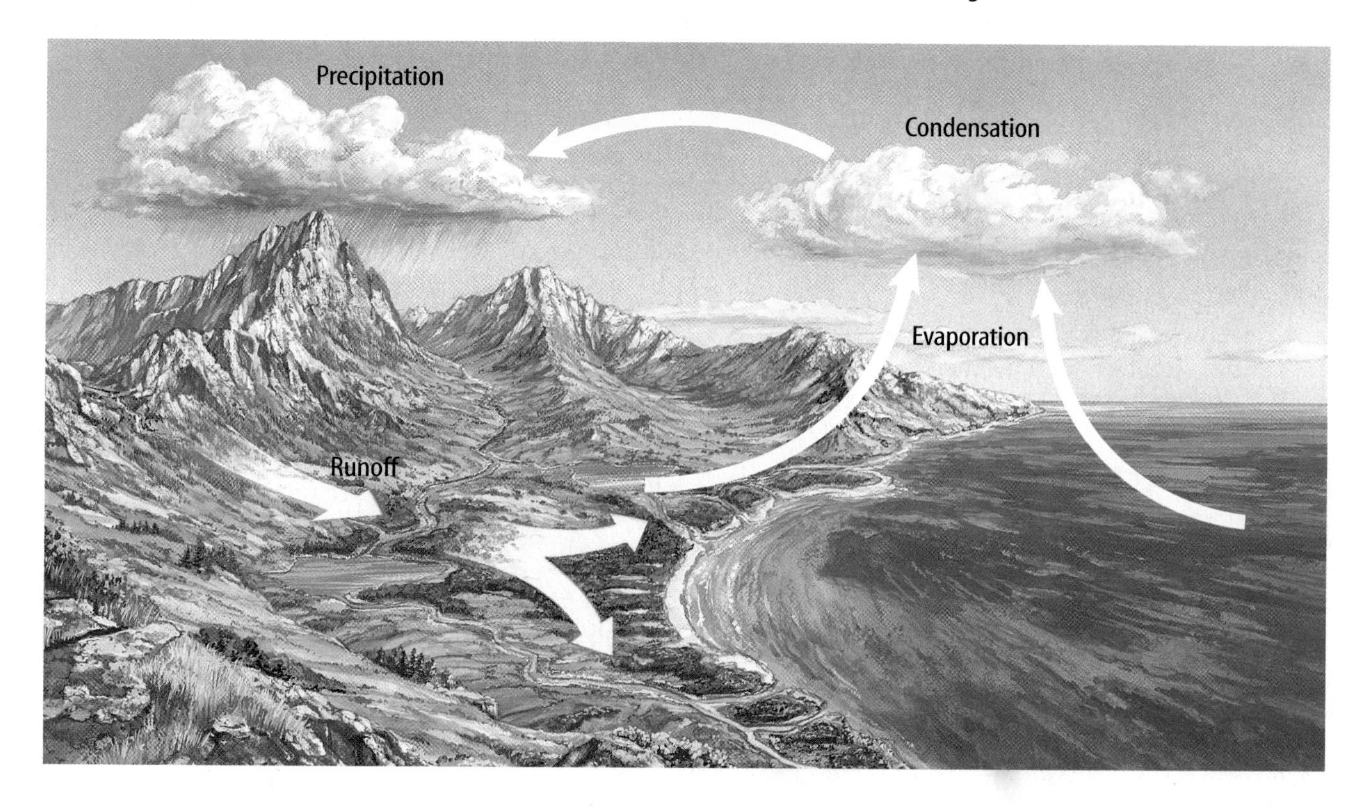

Figure 13 In the water cycle, water moves from Earth to the atmosphere and back to Earth again.

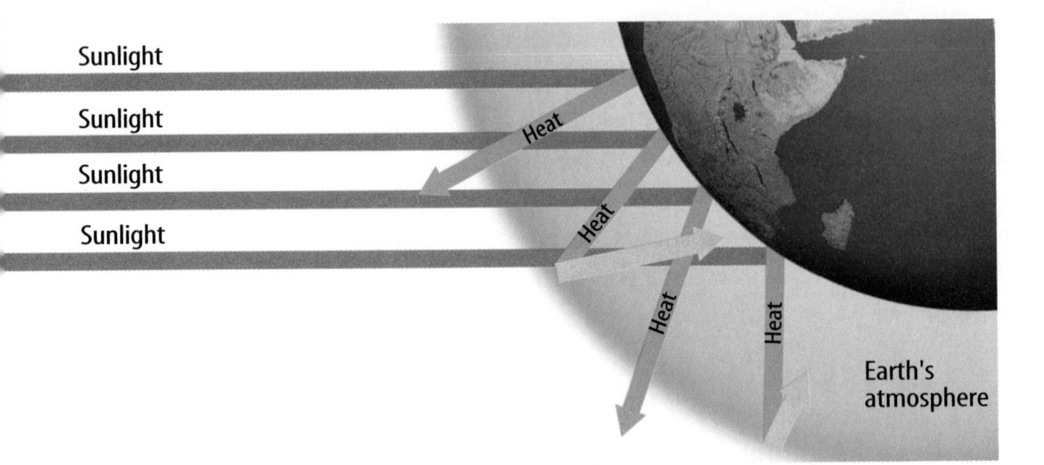

Figure 14 Earth's atmosphere creates a delicate balance between energy received and energy lost. **Infer** *What could happen if the balance is tipped toward receiving more energy than it does now?*

Earth's Atmosphere is Unique

On Earth, radiation from the Sun can be reflected into space, absorbed by the atmosphere, or absorbed by land and water. Once it is absorbed, heat can be transferred by radiation, conduction, or convection. Earth's atmosphere, shown in **Figure 14,** helps control how much of the Sun's radiation is absorbed or lost.

Reading Check *What helps control how much of the Sun's radiation is absorbed on Earth?*

Why doesn't life exist on Mars or Venus? Mars is a cold, lifeless world because its atmosphere is too thin to support life or to hold much of the Sun's heat. Temperatures on the surface of Mars range from 35°C to −170°C. On the other hand, gases in Venus's dense atmosphere trap heat coming from the Sun. The temperature on the surface of Venus is 470°C. Living things would burn instantly if they were placed on Venus's surface. Life on Earth exists because the atmosphere holds just the right amount of the Sun's energy.

section 2 review

Summary

Energy From the Sun

- The Sun's radiation is either absorbed or reflected by Earth.
- Heat is transferred by radiation (waves), conduction (contact), or convection (flow).

The Water Cycle

- The water cycle affects climate.
- Water moves between the hydrosphere and the atmosphere through a continual process of evaporation and condensation.

Earth's Atmosphere is Unique

- Earth's atmosphere controls the amount of solar radiation that reaches Earth's surface.

Self Check

1. **State** how the Sun transfers energy to Earth.
2. **Contrast** the atmospheres of Earth and Mars.
3. **Describe** briefly the steps included in the water cycle.
4. **Explain** how the water cycle is related to weather patterns and climate.
5. **Think Critically** What would happen to temperatures on Earth if the Sun's heat were not distributed throughout the atmosphere?

Applying Math

6. **Solve One-Step Equations** Earth is about 150 million km from the Sun. The radiation coming from the Sun travels at 300,000 km/s. How long does it take for radiation from the Sun to reach Earth?

Science Online green.msscience.com/self_check_quiz

section 3

Air Movement

Forming Wind

Earth is mostly rock or land, with three-fourths of its surface covered by a relatively thin layer of water, the oceans. These two areas strongly influence global wind systems. Uneven heating of Earth's surface by the Sun causes some areas to be warmer than others. Recall that warmer air expands, becoming lower in density than the colder air. This causes air pressure to be generally lower where air is heated. Wind is the movement of air from an area of higher pressure to an area of lower pressure.

Heated Air Areas of Earth receive different amounts of radiation from the Sun because Earth is curved. **Figure 15** illustrates why the equator receives more radiation than areas to the north or south. The heated air at the equator is less dense, so it is displaced by denser, colder air, creating convection currents.

This cold, denser air comes from the poles, which receive less radiation from the Sun, making air at the poles much cooler. The resulting dense, high-pressure air sinks and moves along Earth's surface. However, dense air sinking as less-dense air rises does not explain everything about wind.

as you read

***What* You'll Learn**

- **Explain** why different latitudes on Earth receive different amounts of solar energy.
- **Describe** the Coriolis effect.
- **Explain** how land and water surfaces affect the overlying air.

***Why* It's Important**

Wind systems determine major weather patterns on Earth.

Review Vocabulary

density: mass per unit volume

New Vocabulary

- Coriolis effect
- jet stream
- sea breeze
- land breeze

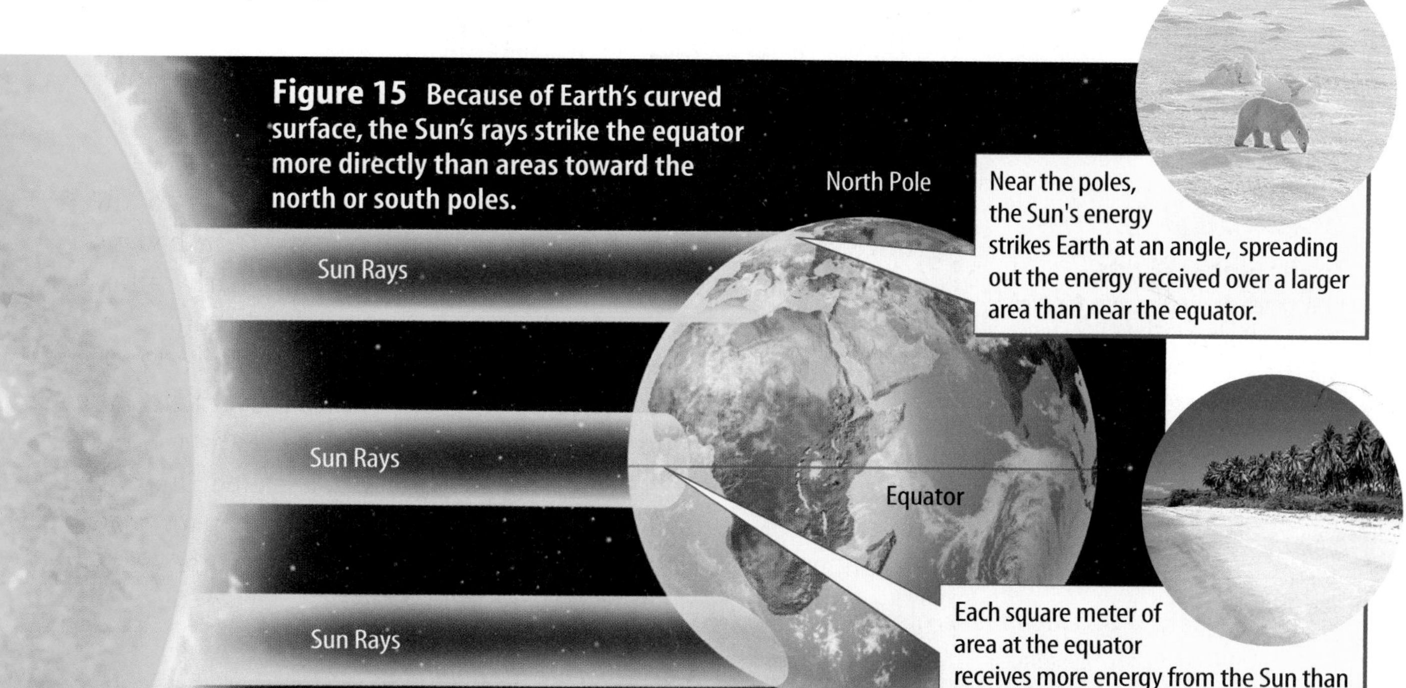

Figure 15 Because of Earth's curved surface, the Sun's rays strike the equator more directly than areas toward the north or south poles.

Figure 16 The Coriolis effect causes moving air to turn to the right in the northern hemisphere and to the left in the southern hemisphere.
Explain *What causes this to happen?*

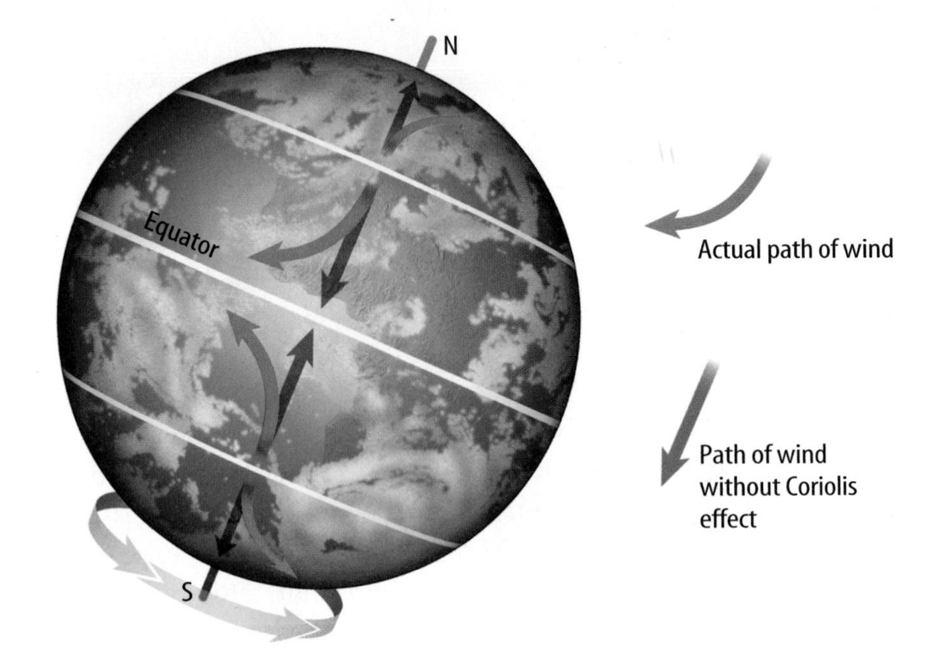

The Coriolis Effect What would happen if you threw a ball to someone sitting directly across from you on a moving merry-go-round? Would the ball go to your friend? By the time the ball got to the opposite side, your friend would have moved and the ball would appear to have curved.

Like the merry-go-round, the rotation of Earth causes moving air and water to appear to turn to the right north of the equator and to the left south of the equator. This is called the **Coriolis** (kohr ee OH lus) **effect.** It is illustrated in **Figure 16.** The flow of air caused by differences in the amount of solar radiation received on Earth's surface and by the Coriolis effect creates distinct wind patterns on Earth's surface. These wind systems not only influence the weather, they also determine when and where ships and planes travel most efficiently.

Topic: Global Winds
Visit green.msscience.com for Web links to information about global winds.

Activity Make a model of Earth showing the locations of global wind patterns.

Global Winds

How did Christopher Columbus get from Spain to the Americas? The *Nina,* the *Pinta,* and the *Santa Maria* had no source of power other than the wind in their sails. Early sailors discovered that the wind patterns on Earth helped them navigate the oceans. These wind systems are shown in **Figure 17.**

Sometimes sailors found little or no wind to move their sailing ships near the equator. It also rained nearly every afternoon. This windless, rainy zone near the equator is called the doldrums. Look again at **Figure 17.** Near the equator, the Sun heats the air and causes it to rise, creating low pressure and little wind. The rising air then cools, causing rain.

What are the doldrums?

NATIONAL GEOGRAPHIC

VISUALIZING GLOBAL WINDS

Figure 17

The Sun's uneven heating of Earth's surface forms giant loops, or cells, of moving air. The Coriolis effect deflects the surface winds to the west or east, setting up belts of prevailing winds that distribute heat and moisture around the globe.

A WESTERLIES Near 30° north and south latitude, Earth's rotation deflects air from west to east as air moves toward the polar regions. In the United States, the westerlies move weather systems, such as this one along the Oklahoma-Texas border, from west to east.

B DOLDRUMS Along the equator, heating causes air to expand, creating a zone of low pressure. Cloudy, rainy weather, as shown here, develops almost every afternoon.

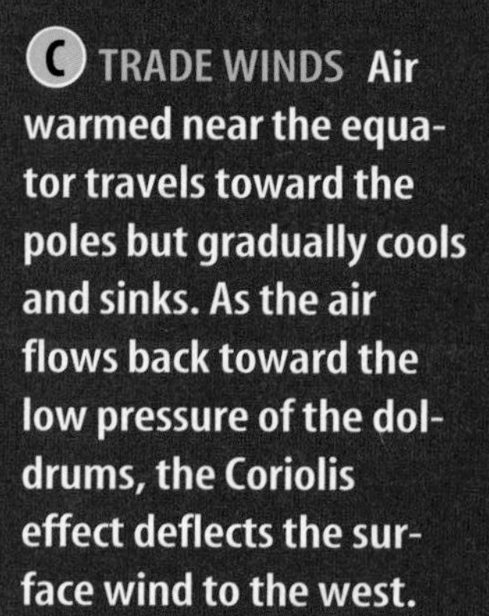

C TRADE WINDS Air warmed near the equator travels toward the poles but gradually cools and sinks. As the air flows back toward the low pressure of the doldrums, the Coriolis effect deflects the surface wind to the west. Early sailors, in ships like the one above, relied on these winds to navigate global trade routes.

D POLAR EASTERLIES In the polar regions, cold, dense air sinks and moves away from the poles. Earth's rotation deflects this wind from east to west.

Surface Winds Air descending to Earth's surface near 30° north and south latitude creates steady winds that blow in tropical regions. These are called trade winds because early sailors used their dependability to establish trade routes.

Between 30° and 60° latitude, winds called the prevailing westerlies blow in the opposite direction from the trade winds. Prevailing westerlies are responsible for much of the movement of weather across North America.

Polar easterlies are found near the poles. Near the north pole, easterlies blow from northeast to southwest. Near the south pole, polar easterlies blow from the southeast to the northwest.

Winds in the Upper Troposphere Narrow belts of strong winds, called **jet streams,** blow near the top of the troposphere. The polar jet stream forms at the boundary of cold, dry polar air to the north and warmer, more moist air to the south, as shown in **Figure 18.** The jet stream moves faster in the winter because the difference between cold air and warm air is greater. The jet stream helps move storms across the country.

Jet pilots take advantage of the jet streams. When flying eastward, planes save time and fuel. Going west, planes fly at different altitudes to avoid the jet streams.

Local Wind Systems

Global wind systems determine the major weather patterns for the entire planet. Smaller wind systems affect local weather. If you live near a large body of water, you're familiar with two such wind systems—sea breezes and land breezes.

Figure 18 The polar jet stream affecting North America forms along a boundary where colder air lies to the north and warmer air lies to the south. It is a swiftly flowing current of air that moves in a wavy west-to-east direction and is usually found between 10 km and 15 km above Earth's surface.

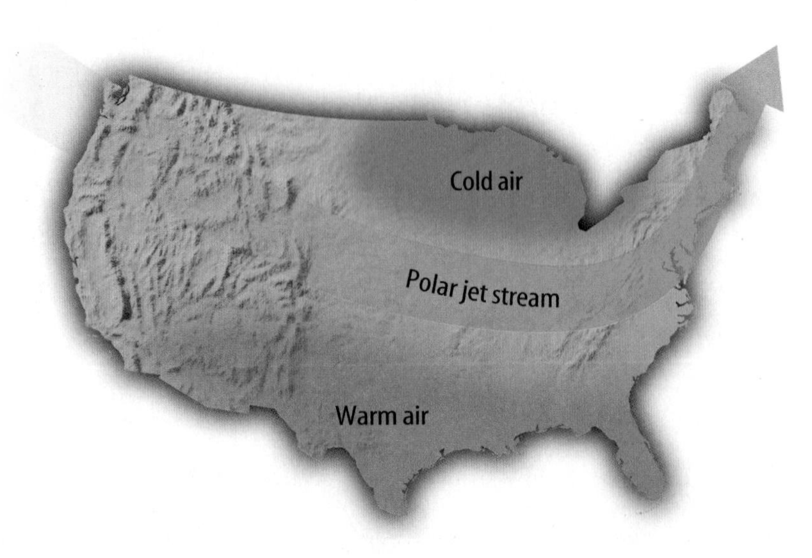

Flying from Boston to Seattle may take 30 min longer than flying from Seattle to Boston.
Think Critically *Why would it take longer to fly from east to west than it would from west to east?*

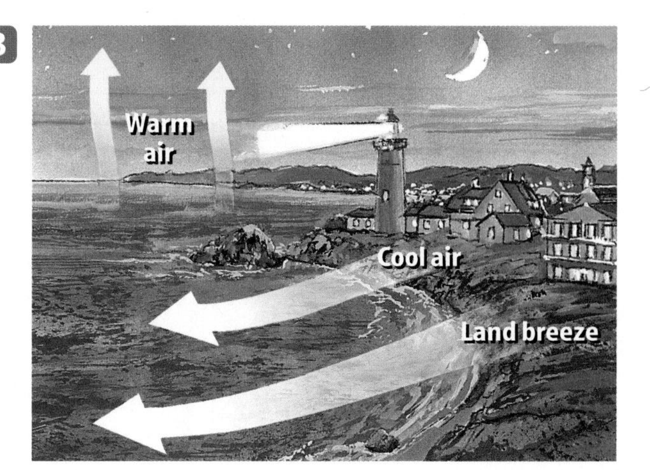

Figure 19 These daily winds occur because land heats up and cools off faster than water does. **A** During the day, cool air from the water moves over the land, creating a sea breeze. **B** At night, cool air over the land moves toward the warmer air over the water, creating a land breeze.

Sea and Land Breezes Convection currents over areas where the land meets the sea can cause wind. A **sea breeze,** shown in **Figure 19,** is created during the day because solar radiation warms the land more than the water. Air over the land is heated by conduction. This heated air is less dense and has lower pressure. Cooler, denser air over the water has higher pressure and flows toward the warmer, less dense air. A convection current results, and wind blows from the sea toward the land. The reverse occurs at night, when land cools much more rapidly than ocean water. Air over the land becomes cooler than air over the ocean. Cooler, denser air above the land moves over the water, as the warm air over the water rises. Movement of air toward the water from the land is called a **land breeze.**

Reading Check *How does a sea breeze form?*

section 3 review

Summary

Forming Wind

- Warm air is less dense than cool air.
- Differences in density and pressure cause air movement and wind.
- The Coriolis effect causes moving air to appear to turn right north of the equator and left south of the equator.

Wind Systems

- Wind patterns are affected by latitude.
- High-altitude belts of wind, called jet streams, can be found near the top of the troposphere.
- Sea breezes blow from large bodies of water toward land, while land breezes blow from land toward water.

Self Check

1. **Conclude** why some parts of Earth's surface, such as the equator, receive more of the Sun's heat than other regions.
2. **Explain** how the Coriolis effect influences winds.
3. **Analyze** why little wind and much afternoon rain occur in the doldrums.
4. **Infer** which wind system helped early sailors navigate Earth's oceans.
5. **Think Critically** How does the jet stream help move storms across North America?

Applying Skills

6. **Compare and contrast** sea breezes and land breezes.

Design Your Own

The Heat Is On

Goals

- **Design** an experiment to compare heat absorption and release for soil and water.
- **Observe** how heat release affects the air above soil and above water.

Possible Materials

ring stand
soil
metric ruler
water
masking tape
clear-plastic boxes (2)
overhead light
 with reflector
thermometers (4)
colored pencils (4)

Safety Precautions

WARNING: *Be careful when handling the hot overhead light. Do not let the light or its cord make contact with water.*

Real-World Question

Sometimes, a plunge in a pool or lake on a hot summer day feels cool and refreshing. Why does the beach sand get so hot when the water remains cool? A few hours later, the water feels warmer than the land does. How do soil and water compare in their abilities to absorb and emit heat?

Form a Hypothesis

Form a hypothesis about how soil and water compare in their abilities to absorb and release heat. Write another hypothesis about how air temperatures above soil and above water differ during the day and night.

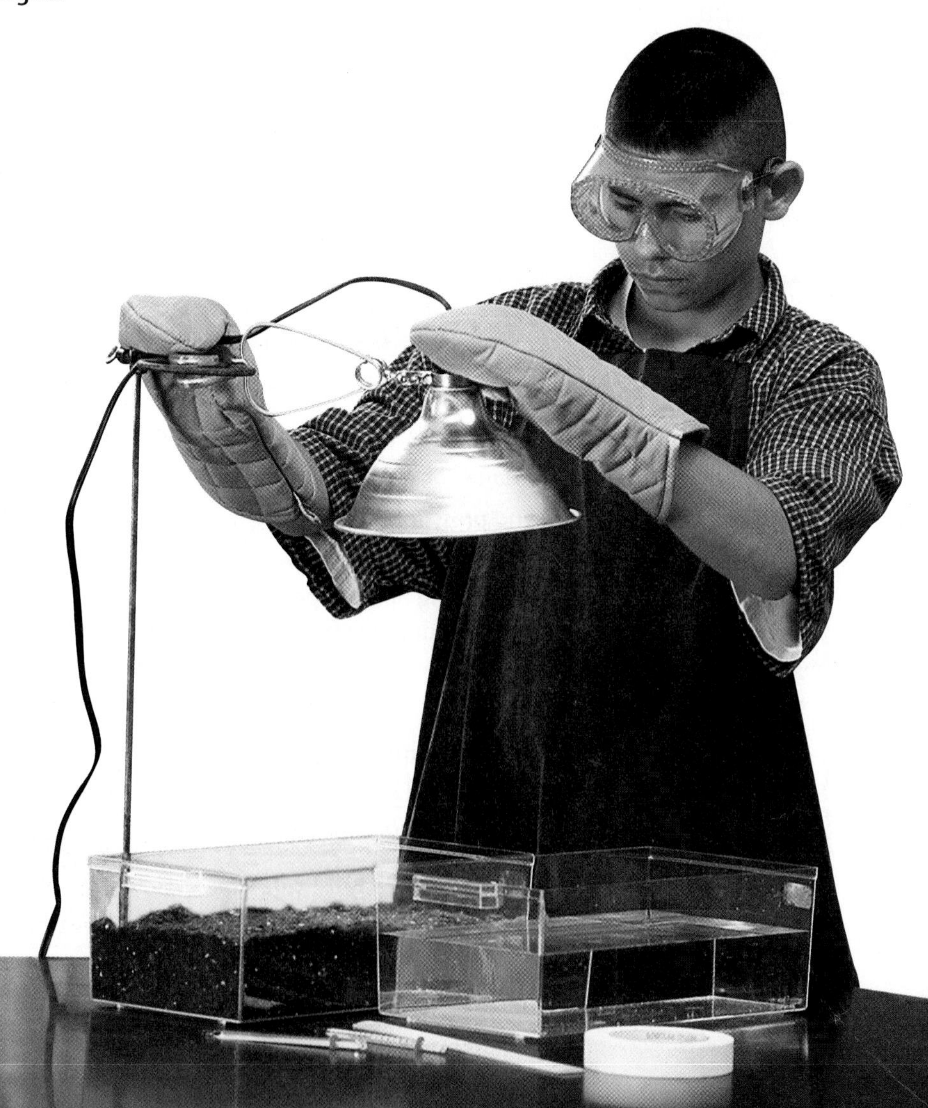

Test Your Hypothesis

Make a Plan

1. As a group, agree upon and write your hypothesis.
2. **List** the steps that you need to take to test your hypothesis. Include in your plan a description of how you will use your equipment to compare heat absorption and release for water and soil.
3. **Design** a data table in your Science Journal for both parts of your experiment—when the light is on and energy can be absorbed and when the light is off and energy is released to the environment.

Follow Your Plan

1. Make sure your teacher approves your plan and your data table before you start.
2. Carry out the experiment as planned.
3. During the experiment, record your observations and complete the data table in your Science Journal.
4. Include the temperatures of the soil and the water in your measurements. Also compare heat release for water and soil. Include the temperatures of the air immediately above both of the substances. Allow 15 min for each test.

Analyze Your Data

1. Use your colored pencils and the information in your data tables to make line graphs. Show the rate of temperature increase for soil and water. Graph the rate of temperature decrease for soil and water after you turn the light off.
2. **Analyze** your graphs. When the light was on, which heated up faster—the soil or the water?
3. **Compare** how fast the air temperature over the water changed with how fast the temperature over the land changed after the light was turned off.

Conclude and Apply

1. Were your hypotheses supported or not? Explain.
2. **Infer** from your graphs which cooled faster—the water or the soil.
3. **Compare** the temperatures of the air above the water and above the soil 15 minutes after the light was turned off. How do water and soil compare in their abilities to absorb and release heat?

Make a poster showing the steps you followed for your experiment. Include graphs of your data. Display your poster in the classroom.

Science and Language Arts

Song of the Sky Loom[1]

Brian Swann, ed.

This Native American prayer probably comes from the Tewa-speaking Pueblo village of San Juan, New Mexico. The poem is actually a chanted prayer used in ceremonial rituals.

Mother Earth Father Sky

we are your children
With tired backs we bring you gifts you love
Then weave for us a garment of brightness
its warp[2] the white light of morning,
weft[3] the red light of evening,
fringes the falling rain,
its border the standing rainbow.
Thus weave for us a garment of brightness
So we may walk fittingly where birds sing,
So we may walk fittingly where grass is green.

Mother Earth Father Sky

1 a machine or device from which cloth is produced

2 threads that run lengthwise in a piece of cloth

3 horizontal threads interlaced through the warp in a piece of cloth

Understanding Literature

Metaphor A metaphor is a figure of speech that compares seemingly unlike things. Unlike a simile, a metaphor does not use the connecting words *like* or *as*. Why does the song use the image of a garment to describe Earth's atmosphere?

Respond to Reading

1. What metaphor does the song use to describe Earth's atmosphere?
2. Why do the words *Mother Earth* and *Father Sky* appear on either side and above and below the rest of the words?
3. **Linking Science and Writing** Write a four-line poem that uses a metaphor to describe rain.

In this chapter, you learned about the composition of Earth's atmosphere. The atmosphere maintains the proper balance between the amount of heat absorbed from the Sun and the amount of heat that escapes back into space. The water cycle explains how water evaporates from Earth's surface back into the atmosphere. Using metaphor instead of scientific facts, the Tewa song conveys to the reader how the relationship between Earth and its atmosphere is important to all living things.

Reviewing Main Ideas

Section 1 Earth's Atmosphere

1. Earth's atmosphere is made up mostly of gases, with some suspended solids and liquids. The unique atmosphere allows life on Earth to exist.
2. The atmosphere is divided into five layers with different characteristics.
3. The ozone layer protects Earth from too much ultraviolet radiation, which can be harmful.

Section 2 Energy Transfer in the Atmosphere

1. Earth receives its energy from the Sun. Some of this energy is reflected back into space, and some is absorbed.
2. Heat is distributed in Earth's atmosphere by radiation, conduction, and convection.
3. Energy from the Sun powers the water cycle between the atmosphere and Earth's surface.
4. Unlike the atmosphere on Mars or Venus, Earth's unique atmosphere maintains a balance between energy received and energy lost that keeps temperatures mild. This delicate balance allows life on Earth to exist.

Section 3 Air Movement

1. Because Earth's surface is curved, not all areas receive the same amount of solar radiation. This uneven heating causes temperature differences at Earth's surface.
2. Convection currents modified by the Coriolis effect produce Earth's global winds.
3. The polar jet stream is a strong current of wind found in the upper troposphere. It forms at the boundary between cold, polar air and warm, tropical air.
4. Land breezes and sea breezes occur near the ocean.

Visualizing Main Ideas

Copy and complete the following cycle map on the water cycle.

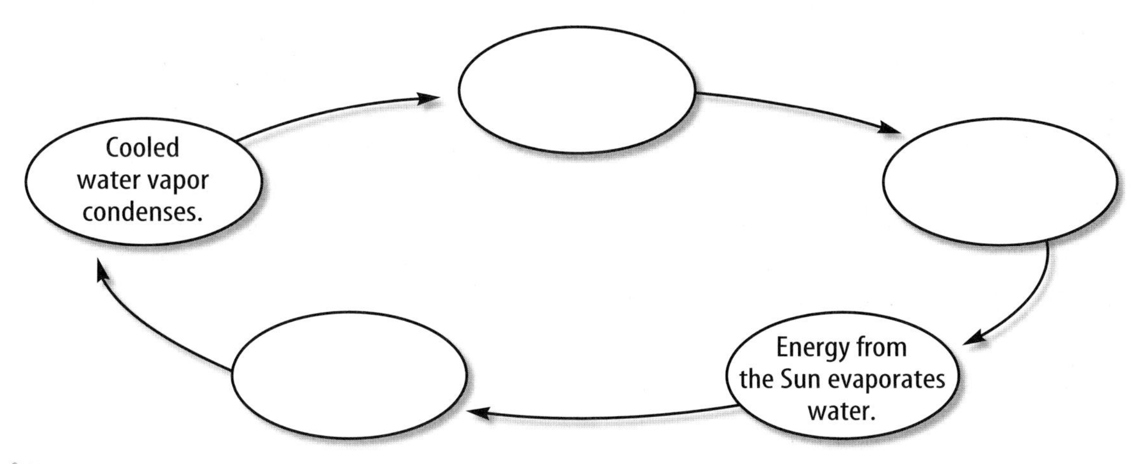

Using Vocabulary

atmosphere p. 90
chlorofluorocarbon p. 96
condensation p. 101
conduction p. 100
convection p. 100
Coriolis effect p. 104
hydrosphere p. 101
ionosphere p. 93
jet stream p. 106
land breeze p. 107
ozone layer p. 96
radiation p. 100
sea breeze p. 107
troposphere p. 92
ultraviolet radiation p. 96

Fill in the blanks below with the correct vocabulary word or words.

1. Chlorofluorocarbons are dangerous because they destroy the ________.
2. Narrow belts of strong winds called ________ blow near the top of the troposphere.
3. The thin layer of air that surrounds Earth is called the ________.
4. Heat energy transferred in the form of waves is called ________.
5. The ozone layer helps protect us from ________.

Checking Concepts

Choose the word or phrase that best answers the question.

6. Nitrogen makes up what percentage of the atmosphere?
   A) 21%
   B) 1%
   C) 78%
   D) 90%
7. What causes a brown haze near cities?
   A) conduction
   B) mud
   C) car exhaust
   D) wind
8. Which is the uppermost layer of the atmosphere?
   A) troposphere
   B) stratosphere
   C) exosphere
   D) thermosphere
9. What layer of the atmosphere has the most water?
   A) troposphere
   B) stratosphere
   C) mesosphere
   D) exosphere
10. What protects living things from too much ultraviolet radiation?
    A) the ozone layer
    B) oxygen
    C) nitrogen
    D) argon
11. Where is air pressure least?
    A) troposphere
    B) stratosphere
    C) exosphere
    D) thermosphere
12. How is energy transferred when objects are in contact?
    A) trade winds
    B) convection
    C) radiation
    D) conduction
13. Which surface winds are responsible for most of the weather movement across the United States?
    A) polar easterlies
    B) sea breeze
    C) prevailing westerlies
    D) trade winds
14. What type of wind is a movement of air toward water?
    A) sea breeze
    B) polar easterlies
    C) land breeze
    D) trade winds
15. What are narrow belts of strong winds near the top of the troposphere called?
    A) doldrums
    B) jet streams
    C) polar easterlies
    D) trade winds

Science Online green.msscience.com/vocabulary_puzzlemaker

Thinking Critically

16. **Explain** why there are few or no clouds in the stratosphere.
17. **Describe** It is thought that life could not have existed on land until the ozone layer formed about 2 billion years ago. Why does life on land require an ozone layer?
18. **Diagram** Why do sea breezes occur during the day but not at night?
19. **Describe** what happens when water vapor rises and cools.
20. **Explain** why air pressure decreases with an increase in altitude.
21. **Concept Map** Copy and complete the cycle concept map below using the following phrases to explain how air moves to form a convection current: *Cool air moves toward warm air, warm air is lifted and cools, and cool air sinks.*

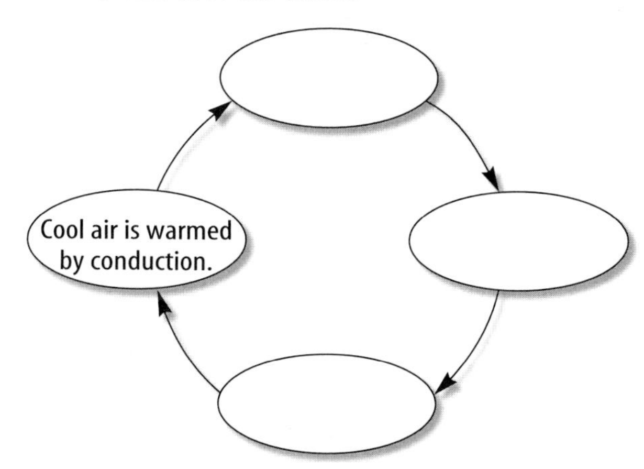

22. **Form Hypotheses** Carbon dioxide in the atmosphere prevents some radiation from Earth's surface from escaping to space. Hypothesize how the temperature on Earth might change if more carbon dioxide were released from burning fossil fuels.
23. **Identify and Manipulate Variables and Controls** Design an experiment to find out how plants are affected by differing amounts of ultraviolet radiation. In the design, use filtering film made for car windows. What is the variable you are testing? What are your constants? Your controls?
24. **Recognize Cause and Effect** Why is the inside of a car hotter than the outdoor temperature on a sunny summer day?

Performance Activities

25. **Make a Poster** Find newspaper and magazine photos that illustrate how the water cycle affects weather patterns and climate around the world.
26. **Experiment** Design and conduct an experiment to find out how different surfaces such as asphalt, soil, sand, and grass absorb and reflect solar energy. Share the results with your class.

Applying Math

Use the graph below to answer questions 27–28.

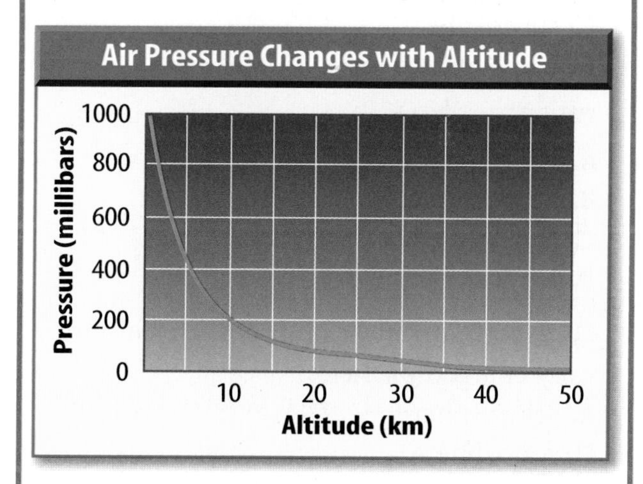

27. **Altitude and Air Pressure** What is the altitude at which air pressure is about 1,000 millibars? What is it at 200 millibars?
28. **Mt. Everest** Assume the altitude on Mt. Everest is about 10 km high. How many times greater is air pressure at sea level than on top of Mt. Everest?

Standardized Test Practice

Part 1 Multiple Choice

Record your answers on the answer sheet provided by your teacher or on a sheet of paper.

Use the illustration below to answer questions 1–3.

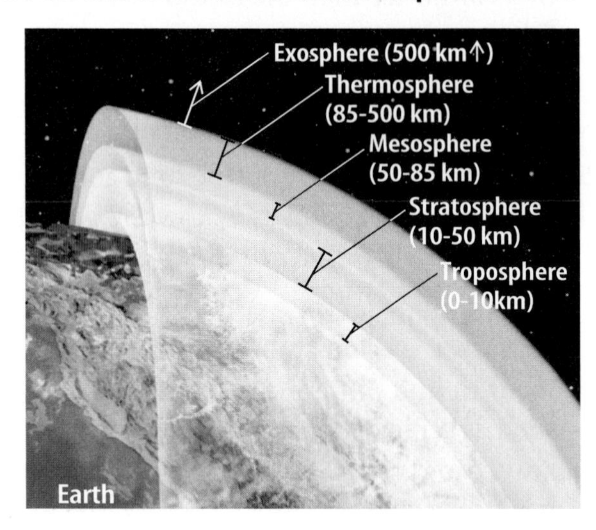

1. Which layer of the atmosphere contains the ozone layer?
A. exosphere
B. mesosphere
C. stratosphere
D. troposphere

2. Which atmospheric layer contains weather?
A. mesosphere
B. stratosphere
C. thermosphere
D. troposphere

3. Which atmospheric layer contains electrically charged particles?
A. stratosphere
B. ionosphere
C. exosphere
D. troposphere

4. What process changes water vapor to a liquid?
A. condensation
B. evaporation
C. infiltration
D. precipitation

5. Which process transfers heat by contact?
A. conduction
B. convection
C. evaporation
D. radiation

6. Which global wind affects weather in the U.S.?
A. doldrums
B. easterlies
C. trade winds
D. westerlies

Use the illustration below to answer question 7.

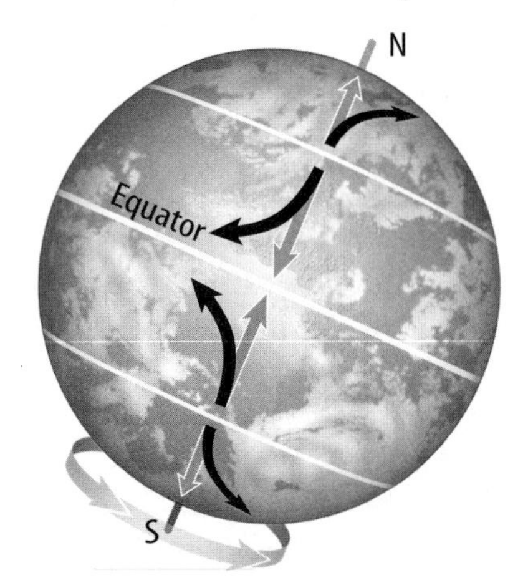

7. Which deflects winds to the west or east?
A. convection
B. Coriolis effect
C. jet stream
D. radiation

8. Which forms during the day because water heats slower than land?
A. easterlies
B. westerlies
C. land breeze
D. sea breeze

9. Which is the most abundant gas in Earth's atmosphere?
A. carbon dioxide
B. nitrogen
C. oxygen
D. water vapor

Part 2 Short Response/Grid In

Record your answers on the answer sheet provided by your teacher or on a sheet of paper.

10. Why does pressure drop as you travel upward from Earth's surface?

11. Why does the equator receive more radiation than areas to the north or south?

12. Why does a land breeze form at night?

13. Why does the jet stream move faster in the winter?

14. Why is one global wind pattern known as the trade winds?

Use the illustration below to answer questions 15–17.

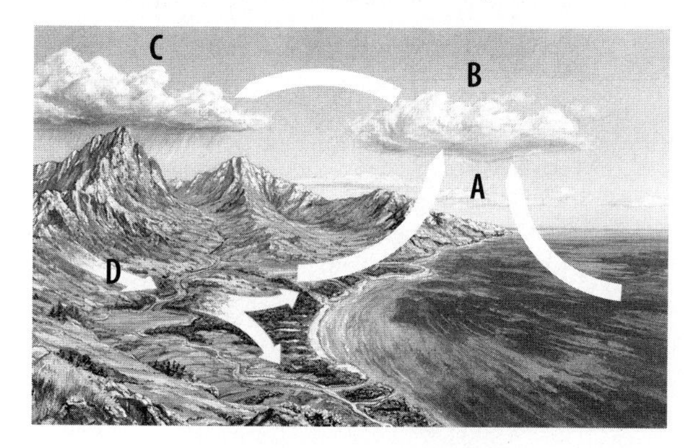

15. What process is illustrated?

16. Explain how this cycle affects weather patterns and climate.

17. What happens to water that falls as precipitation and does not runoff and flow into streams?

18. How do solid particles become part of Earth's atmosphere?

19. Why can flying from Seattle to Boston take less time than flying from Boston to Seattle in the same aircraft?

Part 3 Open Ended

Record your answers on a sheet of paper.

20. Explain how ozone is destroyed by chlorofluorocarbons.

21. Explain how Earth can heat the air by conduction.

22. Explain how humans influence the composition of Earth's atmosphere.

23. Draw three diagrams to demonstrate radiation, convection, and conduction.

24. Explain why the doldrums form over the equator.

Use the graph below to answer question 25.

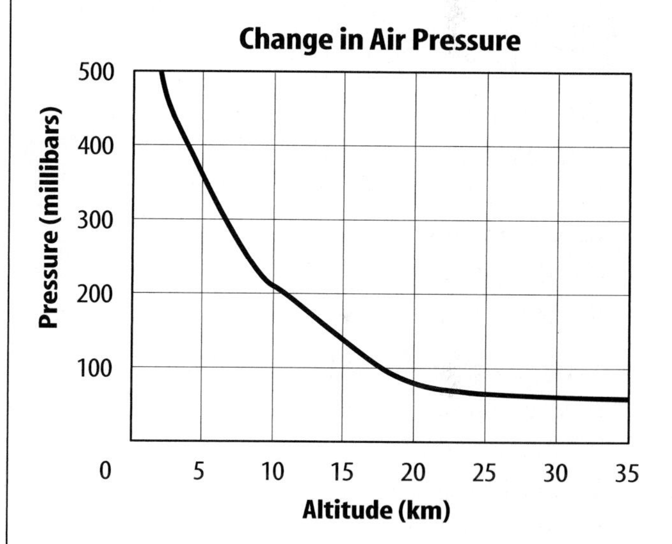

25. As you increase in altitude what happens to the air pressure? How might this affect people who move to the mountains?

Test-Taking Tip

Trends in Graphs When analyzing data in a table or graph, look for a trend. Questions about the pattern may use words like *increase, decrease, hypothesis,* or *summary.*

Question 25 The word "increase" indicates that you should look for the trend in air pressure as altitude increases.

chapter 5

Weather

chapter preview

sections

To play or not to play?

Will this approaching storm be over before the game begins? New weather technology can provide information that allows us to make plans based on predicted weather conditions, such as whether or not to delay the start of a baseball game.

Science Journal Write three questions you would ask a meteorologist about weather.

Start-Up Activities

What causes rain?

How can it rain one day and be sunny the next? Powered by heat from the Sun, the air that surrounds you stirs and swirls. This constant mixing produces storms, calm weather, and everything in between. What causes rain and where does the water come from? Do the lab below to find out. **WARNING:** *Boiling water and steam can cause burns.*

1. Bring a pan of water to a boil on a hot plate.
2. Carefully hold another pan containing ice cubes about 20 cm above the boiling water. Be sure to keep your hands and face away from the steam.
3. Keep the pan with the ice cubes in place until you see drops of water dripping from the bottom.
4. **Think Critically** In your Science Journal, describe how the droplets formed. Infer where the water on the bottom of the pan came from.

Weather When information is grouped into clear categories, it is easier to make sense of what you are learning. Make the following Foldable to help you organize your thoughts about weather.

STEP 1 **Collect** 2 sheets of paper and layer them about 1.25 cm apart vertically. Keep the edges level.

STEP 2 **Fold** up the bottom edges of the paper to form 4 equal tabs.

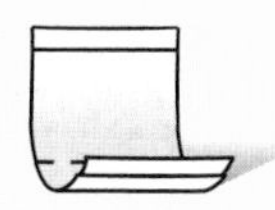

STEP 3 **Fold** the papers and crease well to hold the tabs in place. Staple along the fold.

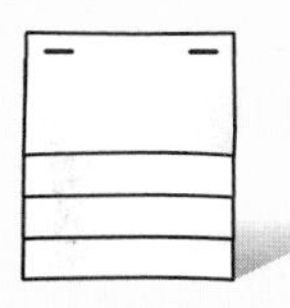

STEP 4 Label the tabs *Weather, What is weather?, Weather Patterns,* and *Forecasting Weather* as shown.

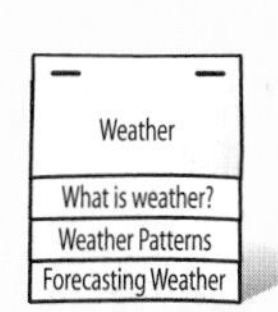

Summarize As you read the chapter, summarize what you learn under the appropriate tabs.

Preview this chapter's content and activities at green.msscience.com

section 1

What is weather?

as you read

What You'll Learn

- **Explain** how solar heating and water vapor in the atmosphere affect weather.
- **Discuss** how clouds form and how they are classified.
- **Describe** how rain, hail, sleet, and snow develop.

Why It's Important

Weather changes affect your daily activities.

Review Vocabulary
factor: something that influences a result

New Vocabulary
- weather
- humidity
- relative humidity
- dew point
- fog
- precipitation

Weather Factors

It might seem like small talk to you, but for farmers, truck drivers, pilots, and construction workers, the weather can have a huge impact on their livelihoods. Even professional athletes, especially golfers, follow weather patterns closely. You can describe what happens in different kinds of weather, but can you explain how it happens?

Weather refers to the state of the atmosphere at a specific time and place. Weather describes conditions such as air pressure, wind, temperature, and the amount of moisture in the air.

The Sun provides almost all of Earth's energy. Energy from the Sun evaporates water into the atmosphere where it forms clouds. Eventually, the water falls back to Earth as rain or snow. However, the Sun does more than evaporate water. It is also a source of heat energy. Heat from the Sun is absorbed by Earth's surface, which then heats the air above it. Differences in Earth's surface lead to uneven heating of Earth's atmosphere. Heat is eventually redistributed by air and water currents. Weather, as shown in **Figure 1,** is the result of heat and Earth's air and water.

Figure 1 The Sun provides the energy that drives Earth's weather.
Identify *storms in this image.*

When air is heated, it expands and becomes less dense. This creates lower pressure.

Molecules making up air are closer together in cooler temperatures, creating high pressure. Wind blows from higher pressure toward lower pressure.

Figure 2 The temperature of air can affect air pressure. Wind is air moving from high pressure to low pressure.

Infer *In the above picture, which way would the wind move at night if the land cooled?*

Air Temperature During the summer when the Sun is hot and the air is still, a swim can be refreshing. But would a swim seem refreshing on a cold, winter day? The temperature of air influences your daily activities.

Air is made up of molecules that are always moving randomly, even when there's no wind. Temperature is a measure of the average amount of motion of molecules. When the temperature is high, molecules in air move rapidly and it feels warm. When the temperature is low, molecules in air move less rapidly, and it feels cold.

Wind Why can you fly a kite on some days but not others? Kites fly because air is moving. Air moving in a specific direction is called wind. As the Sun warms the air, the air expands and becomes less dense. Warm, expanding air has low atmospheric pressure. Cooler air is denser and tends to sink, bringing about high atmospheric pressure. Wind results because air moves from regions of high pressure to regions of low pressure. You may have experienced this on a small scale if you've ever spent time along a beach, as in **Figure 2.**

Many instruments are used to measure wind direction and speed. Wind direction can be measured using a wind vane. A wind vane has an arrow that points in the direction from which the wind is blowing. A wind sock has one open end that catches the wind, causing the sock to point in the direction toward which the wind is blowing. Wind speed can be measured using an anemometer (a nuh MAH muh tur). Anemometers have rotating cups that spin faster when the wind is strong.

Body Temperature Birds and mammals maintain a fairly constant internal temperature, even when the temperature outside their bodies changes. On the other hand, the internal temperature of fish and reptiles changes when the temperature around them changes. Infer from this which group is more likely to survive a quick change in the weather.

Figure 3 Warmer air can have more water vapor than cooler air can because water vapor doesn't easily condense in warm air.

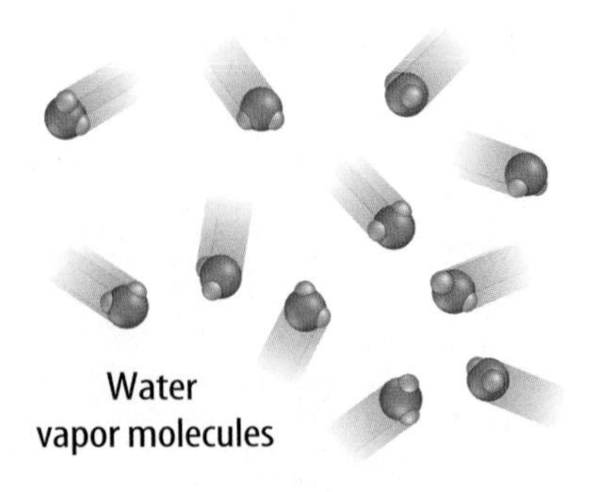

Water vapor molecules in warm air move rapidly. The molecules can't easily come together and condense.

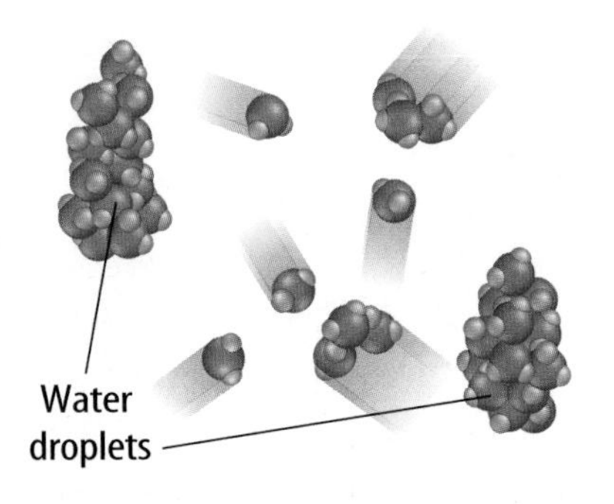

As air cools, water molecules in air move closer together. Some of them collide, allowing condensation to take place.

Determining Dew Point

Procedure

1. Partially fill a **metal can** with **room-temperature water**. Dry the outer surface of the can.
2. Place a **stirring rod** in the water.
3. Slowly stir the water and add small amounts of **ice.**
4. Make a data table in your **Science Journal.** With a **thermometer,** note the exact water temperature at which a thin film of moisture first begins to form on the outside of the metal can.
5. Repeat steps 1 through 4 two more times.
6. The average of the three temperatures at which the moisture begins to appear is the dew point temperature of the air surrounding the metal container.

Analysis

1. What determines the dew point temperature?
2. Will the dew point change with increasing temperature if the amount of moisture in the air doesn't change? Explain.

Humidity Heat evaporates water into the atmosphere. Where does the water go? Water vapor molecules fit into spaces among the molecules that make up air. The amount of water vapor present in the air is called **humidity.**

Air doesn't always contain the same amount of water vapor. As you can see in **Figure 3,** more water vapor can be present when the air is warm than when it is cool. At warmer temperatures, the molecules of water vapor in air move quickly and don't easily come together. At cooler temperatures, molecules in air move more slowly. The slower movement allows water vapor molecules to stick together and form droplets of liquid water. The formation of liquid water from water vapor is called condensation. When enough water vapor is present in air for condensation to take place, the air is saturated.

Why can more water vapor be present in warm air than in cold air?

Relative Humidity On a hot, sticky afternoon, the weather forecaster reports that the humidity is 50 percent. How can the humidity be low when it feels so humid? Weather forecasters report the amount of moisture in the air as relative humidity. **Relative humidity** is a measure of the amount of water vapor present in the air compared to the amount needed for saturation at a specific temperature.

If you hear a weather forecaster say that the relative humidity is 50 percent, it means that the air contains 50 percent of the water needed for the air to be saturated.

As shown in **Figure 4,** air at 25°C is saturated when it contains 22 g of water vapor per cubic meter of air. The relative humidity is 100 percent. If air at 25°C contains 11 g of water vapor per cubic meter, the relative humidity is 50 percent.

Dew Point

When the temperature drops, less water vapor can be present in air. The water vapor in air will condense to a liquid or form ice crystals. The temperature at which air is saturated and condensation forms is the dew point. The **dew point** changes with the amount of water vapor in the air.

You've probably seen water droplets form on the outside of a glass of cold milk. The cold glass cooled the air next to it to its dew point. The water vapor in the surrounding air condensed and formed water droplets on the glass. In a similar way, when air near the ground cools to its dew point, water vapor condenses and forms dew. Frost may form when temperatures are near 0°C.

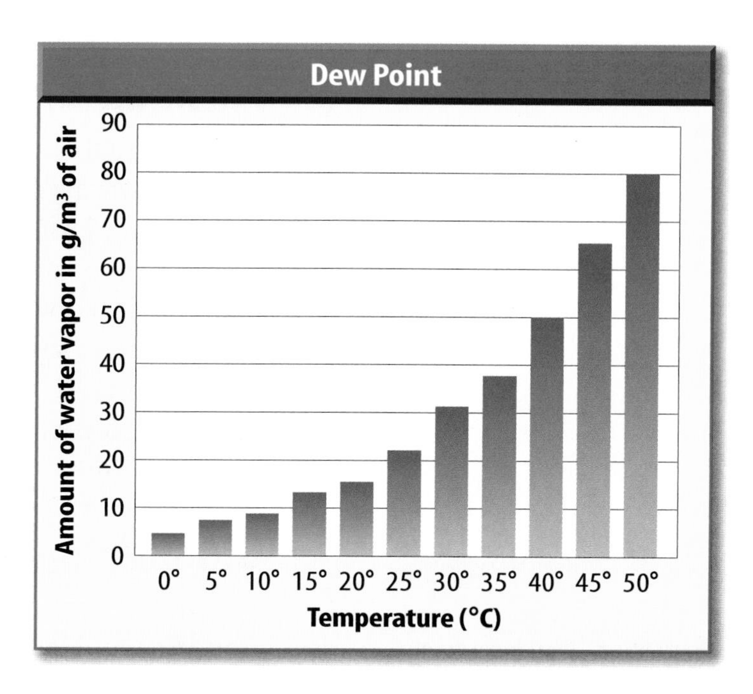

Figure 4 This graph shows that as the temperature of air increases, more water vapor can be present in the air.

Applying Math — Calculate Percent

DEW POINT One summer day, the relative humidity is 80 percent and the temperature is 35°C. Use **Figure 4** to find the dew point reached if the temperature falls to 25°C?

Solution

1 *This is what you know:*

Air Temperature (°C)	Amount of Water Vapor Needed for Saturation (g/m^3)
35	37
25	22
15	14

2 *This is what you need to find out:*

x = amount of water vapor in 35°C air at 80 percent relative humidity. Is $x > 22$ g/m^3 or is $x < 22$ g/m^3?

3 *This is how you solve the problem:*

$x = .80\ (37\ g/m^3)$

$x = 29.6\ g/m^3$ of water vapor

$29.6\ g/m^3 > 22\ g/m^3$, so the dew point is reached and dew will form.

Practice Problems

1. If the relative humidity is 50 percent and the air temperature is 35°C, will the dew point be reached if the temperature falls to 20°C?
2. If the air temperature is 25°C and the relative humidity is 30 percent, will the dew point be reached if the temperature drops to 15°C?

For more practice, visit green.msscience.com/math_practice

Forming Clouds

Why are there clouds in the sky? Clouds form as warm air is forced upward, expands, and cools. **Figure 5** shows several ways that warm, moist air forms clouds. As the air cools, the amount of water vapor needed for saturation decreases and the relative humidity increases. When the relative humidity reaches 100 percent, the air is saturated. Water vapor soon begins to condense in tiny droplets around small particles such as dust and salt. These droplets of water are so small that they remain suspended in the air. Billions of these droplets form a cloud.

Classifying Clouds

Clouds are classified mainly by shape and height. Some clouds extend high into the sky, and others are low and flat. Some dense clouds bring rain or snow, while thin, wispy clouds appear on mostly sunny days. The shape and height of clouds vary with temperature, pressure, and the amount of water vapor in the atmosphere.

Figure 5 Clouds form when moist air is lifted and cools. This occurs where air is heated, at mountain ranges, and where cold air meets warm air.

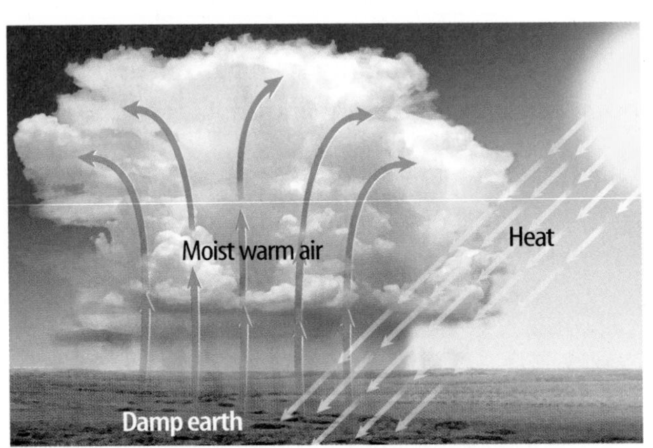

Rays from the Sun heat the ground and the air next to it. The warm air rises and cools. If the air is moist, some water vapor condenses and forms clouds.

As moist air moves over mountains, it is lifted and cools. Clouds formed in this way can cover mountains for long periods of time.

When cool air meets warm, moist air, the warm air is lifted and cools.

Explain *what happens to the water vapor when the dew point is reached.*

Shape The three main cloud types are stratus, cumulus, and cirrus. Stratus clouds form layers, or smooth, even sheets in the sky. Stratus clouds usually form at low altitudes and may be associated with fair weather or rain or snow. When air is cooled to its dew point near the ground, it forms a stratus cloud called **fog,** as shown in **Figure 6.**

Figure 6 Fog surrounds the Golden Gate Bridge, San Francisco. Fog is a stratus cloud near the ground.
Think Critically *Why do you think fog is found in San Francisco Bay?*

Cumulus (KYEW myuh lus) clouds are masses of puffy, white clouds, often with flat bases. They sometimes tower to great heights and can be associated with fair weather or thunderstorms.

Cirrus (SIHR us) clouds appear fibrous or curly. They are high, thin, white, feathery clouds made of ice crystals. Cirrus clouds are associated with fair weather, but they can indicate approaching storms.

Height Some prefixes of cloud names describe the height of the cloud base. The prefix *cirro-* describes high clouds, *alto-* describes middle-elevation clouds, and *strato-* refers to clouds at low elevations. Some clouds' names combine the altitude prefix with the term *stratus* or *cumulus.*

Cirrostratus clouds are high clouds, like those in **Figure 7.** Usually, cirrostratus clouds indicate fair weather, but they also can signal an approaching storm. Altostratus clouds form at middle levels. If the clouds are not too thick, sunlight can filter through them.

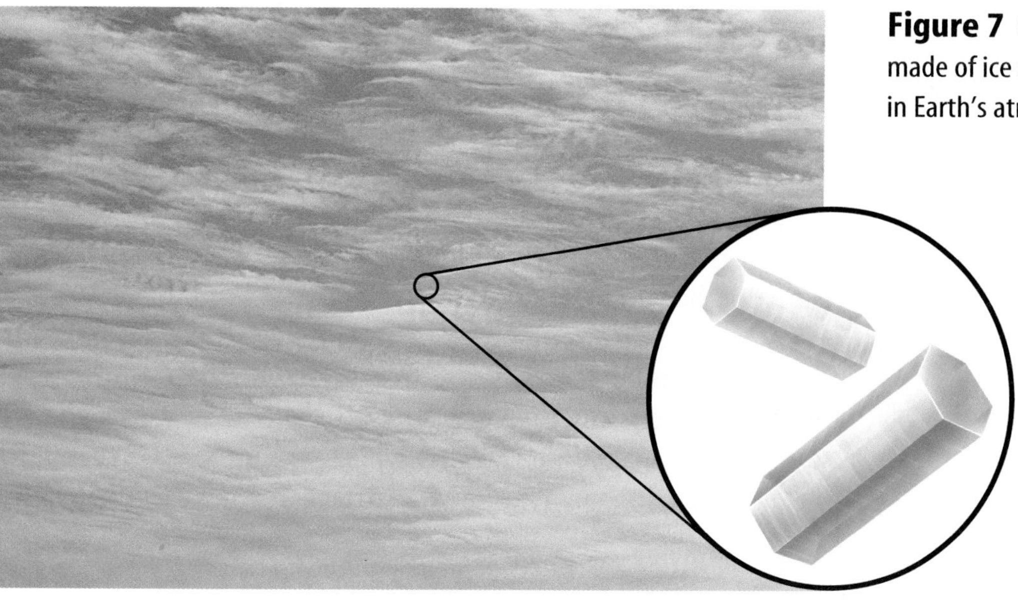

Figure 7 Cirrostratus clouds are made of ice crystals and form high in Earth's atmosphere.

Figure 8 Water vapor in air collects on particles to form water droplets or ice crystals. The type of precipitation that is received on the ground depends on the temperature of the air.

When the air is warm, water vapor forms raindrops that fall as rain.

When the air is cold, water vapor forms snowflakes.

Rain- or Snow-Producing Clouds Clouds associated with rain or snow often have the word nimbus attached to them. The term *nimbus* is Latin for "dark rain cloud" and this is a good description, because the water content of these clouds is so high that little sunlight can pass through them. When a cumulus cloud grows into a thunderstorm, it is called a cumulonimbus (kyew myuh loh NIHM bus) cloud. These clouds can tower to nearly 18 km. Nimbostratus clouds are layered clouds that can bring long, steady rain or snowfall.

Precipitation

Water falling from clouds is called **precipitation.** Precipitation occurs when cloud droplets combine and grow large enough to fall to Earth. The cloud droplets form around small particles, such as salt and dust. These particles are so small that a puff of smoke can contain millions of them.

You might have noticed that raindrops are not all the same size. The size of raindrops depends on several factors. One factor is the strength of updrafts in a cloud. Strong updrafts can keep drops suspended in the air where they can combine with other drops and grow larger. The rate of evaporation as a drop falls to Earth also can affect its size. If the air is dry, the size of raindrops can be reduced or they can completely evaporate before reaching the ground.

Air temperature determines whether water forms rain, snow, sleet, or hail—the four main types of precipitation. **Figure 8** shows these different types of precipitation. Drops of water falling in temperatures above freezing fall as rain. Snow forms when the air temperature is so cold that water vapor changes directly to a solid. Sleet forms when raindrops pass through a layer of freezing air near Earth's surface, forming ice pellets.

Reading Check ***What are the four main types of precipitation?***

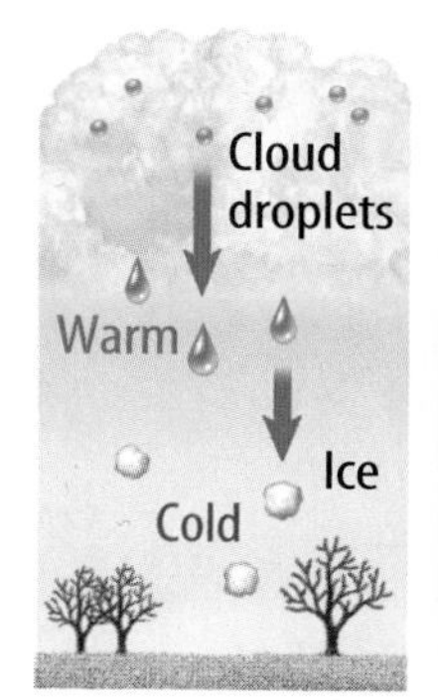

When the air near the ground is cold, sleet, which is made up of many small ice pellets, falls.

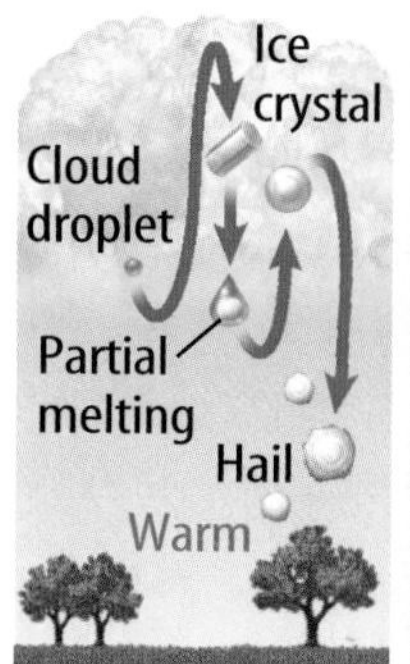

Hailstones are pellets of ice that form inside a cloud.

Hail Hail is precipitation in the form of lumps of ice. Hail forms in cumulonimbus clouds of a thunderstorm when water freezes in layers around a small nucleus of ice. Hailstones grow larger as they're tossed up and down by rising and falling air. Most hailstones are smaller than 2.5 cm but can grow larger than a softball. Of all forms of precipitation, hail produces the most damage immediately, especially if winds blow during a hailstorm. Falling hailstones can break windows and destroy crops.

If you understand the role of water vapor in the atmosphere, you can begin to understand weather. The relative humidity of the air helps determine whether a location will have a dry day or experience some form of precipitation. The temperature of the atmosphere determines the form of precipitation. Studying clouds can add to your ability to forecast weather.

section 1 review

Summary

Weather Factors

- Weather is the state of the atmosphere at a specific time and place.
- Temperature, wind, air pressure, dew point, and humidity describe weather.

Clouds

- Warm, moist air rises, forming clouds.
- The main types of clouds are stratus, cumulus, and cirrus.

Precipitation

- Water falling from clouds is called precipitation.
- Air temperature determines whether water forms rain, snow, sleet, or hail.

Self Check

1. **Explain** When does water vapor in air condense?
2. **Compare and contrast** humidity and relative humidity.
3. **Summarize** how clouds form.
4. **Describe** How does precipitation occur and what determines the type of precipitation that falls to Earth?
5. **Think Critically** Cumulonimbus clouds form when warm, moist air is suddenly lifted. How can the same cumulonimbus cloud produce rain and hail?

Applying Math

6. **Use Graphs** If the air temperature is 30°C and the relative humidity is 60 percent, will the dew point be reached if the temperature drops to 25°C? Use the graph in **Figure 4** to explain your answer.

section 2

Weather Patterns

as you read

What You'll Learn

- **Describe** how weather is associated with fronts and high- and low-pressure areas.
- **Explain** how tornadoes develop from thunderstorms.
- **Discuss** the dangers of severe weather.

Why It's Important

Air masses, pressure systems, and fronts cause weather to change.

Review Vocabulary
barometer: instrument used to measure atmospheric pressure

New Vocabulary
- air mass
- front
- tornado
- hurricane
- blizzard

Weather Changes

When you leave for school in the morning, the weather might be different from what it is when you head home in the afternoon. Because of the movement of air and moisture in the atmosphere, weather constantly changes.

Air Masses An **air mass** is a large body of air that has properties similar to the part of Earth's surface over which it develops. For example, an air mass that develops over land is dry compared with one that develops over water. An air mass that develops in the tropics is warmer than one that develops over northern regions. An air mass can cover thousands of square kilometers. When you observe a change in the weather from one day to the next, it is due to the movement of air masses. **Figure 9** shows air masses that affect the United States.

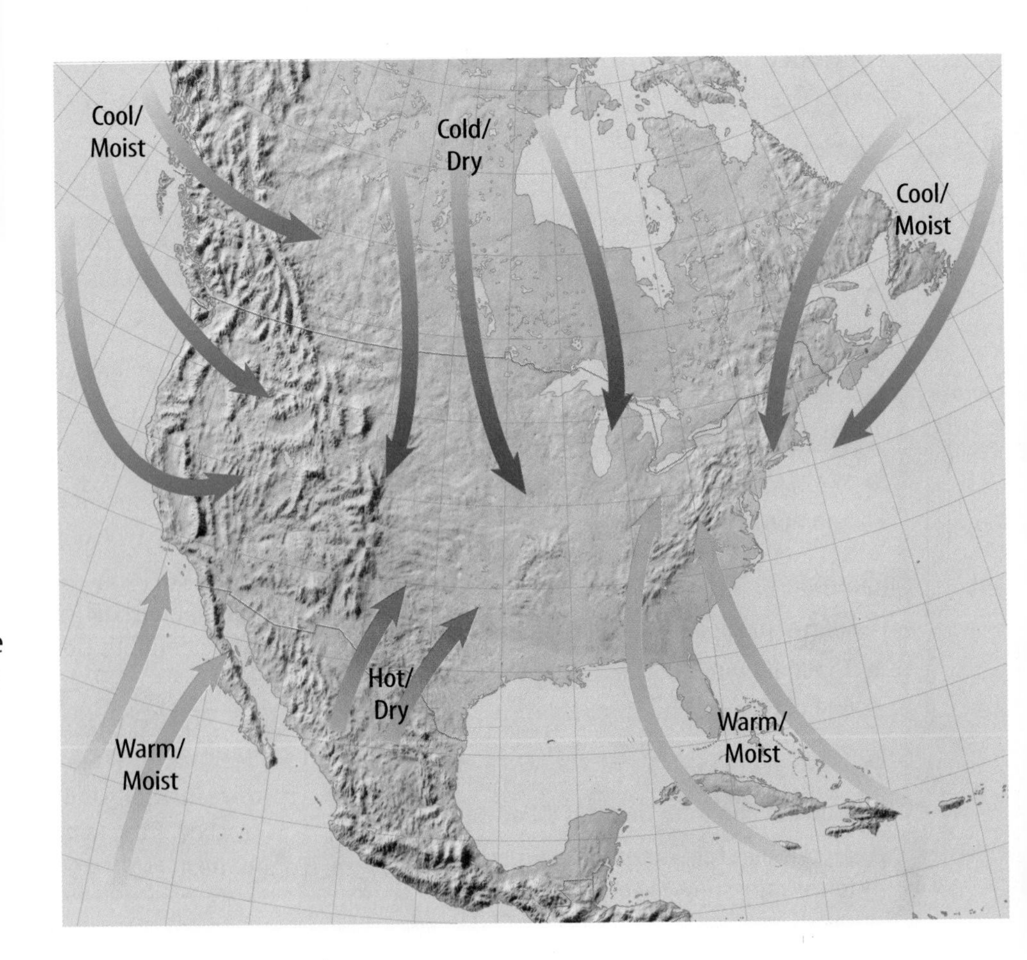

Figure 9 Six major air masses affect weather in the United States. Each air mass has the same characteristics of temperature and moisture content as the area over which it formed.

Highs and Lows Atmospheric pressure varies over Earth's surface. Anyone who has watched a weather report on television has heard about high- and low-pressure systems. Recall that winds blow from areas of high pressure to areas of low pressure. As winds blow into a low-pressure area in the northern hemisphere, Earth's rotation causes these winds to swirl in a counterclockwise direction. Large, swirling areas of low pressure are called cyclones and are associated with stormy weather.

Reading Check *How do winds move in a cyclone?*

Winds blow away from a center of high pressure. Earth's rotation causes these winds to spiral clockwise in the northern hemisphere. High-pressure areas are associated with fair weather and are called anticyclones. Air pressure is measured using a barometer, like the one shown in **Figure 10.**

Figure 10 A barometer measures atmospheric pressure. The red pointer points to the current pressure. Watch how atmospheric pressure changes over time when you line up the white pointer to the one indicating the current pressure each day.

Variation in atmospheric pressure affects the weather. Low pressure systems at Earth's surface are regions of rising air. Clouds form when air is lifted and cools. Areas of low pressure usually have cloudy weather. Sinking motion in high-pressure air masses makes it difficult for air to rise and clouds to form. That's why high pressure usually means good weather.

Fronts

A boundary between two air masses of different density, moisture, or temperature is called a **front.** If you've seen a weather map in the newspaper or on the evening news, you've seen fronts represented by various types of curving lines.

Cloudiness, precipitation, and storms sometimes occur at frontal boundaries. Four types of fronts include cold, warm, occluded, and stationary.

Cold and Warm Fronts A cold front, shown on a map as a blue line with triangles, occurs when colder air advances toward warm air. The cold air wedges under the warm air like a plow. As the warm air is lifted, it cools and water vapor condenses, forming clouds. When the temperature difference between the cold and warm air is large, thunderstorms and even tornadoes may form.

Warm fronts form when lighter, warmer air advances over heavier, colder air. A warm front is drawn on weather maps as a red line with red semicircles.

Topic: Atmospheric Pressure

Visit green.msscience.com for Web links to information about the current atmospheric pressure of your town or nearest city.

Activity Look up the pressure of a city west of your town and the pressure of a city to the east. Compare the pressures to local weather conditions. Share your information with the class.

Occluded and Stationary Fronts An occluded front involves three air masses of different temperatures—colder air, cool air, and warm air. An occluded front may form when a cold air mass moves toward cool air with warm air between the two. The colder air forces the warm air upward, closing off the warm air from the surface. Occluded fronts are shown on maps as purple lines with triangles and semicircles.

A stationary front occurs when a boundary between air masses stops advancing. Stationary fronts may remain in the same place for several days, producing light wind and precipitation. A stationary front is drawn on a weather map as an alternating red and blue line. Red semicircles point toward the cold air and blue triangles point toward the warm air. **Figure 11** summarizes the four types of fronts.

Figure 11 Cold, warm, occluded, and stationary fronts occur at the boundaries of air masses.
Describe *what type of weather occurs at front boundaries.*

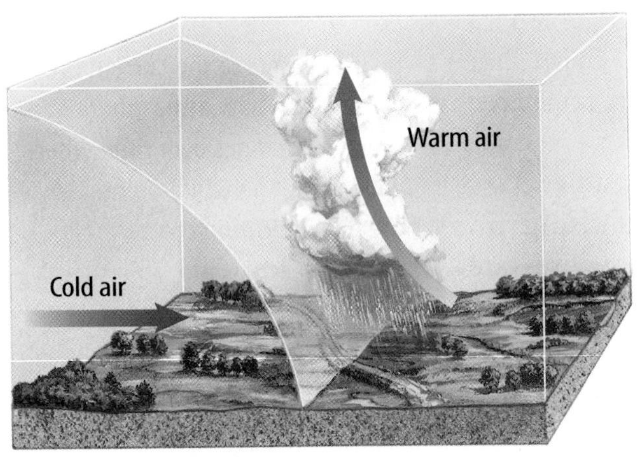

A cold front can advance rapidly. Thunderstorms often form as warm air is suddenly lifted up over the cold air.

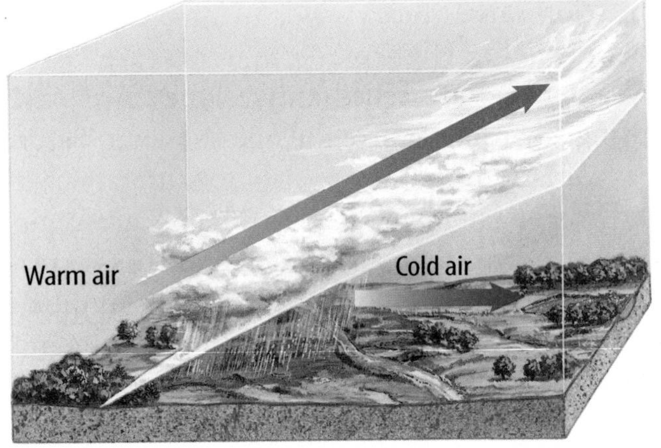

Warm air slides over colder air along a warm front, forming a boundary with a gentle slope. This can lead to hours, if not days, of wet weather.

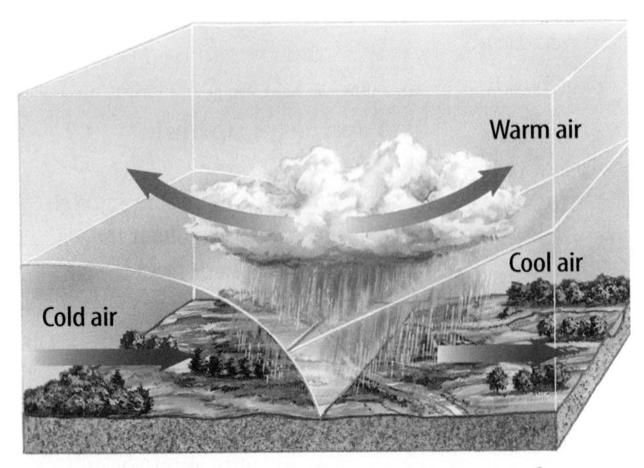

The term *occlusion* means "closure." Colder air forces warm air upward, forming an occluded front that closes off the warm air from the surface.

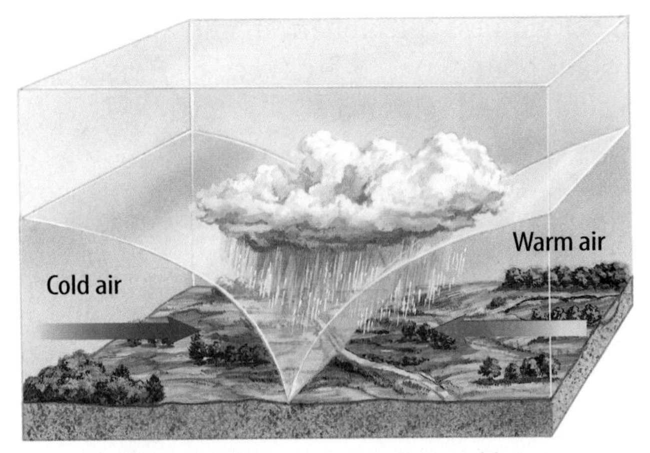

A stationary front results when neither cold air nor warm air advances.

Severe Weather

Despite the weather, you usually can do your daily activities. If it's raining, you still go to school. You can still get there even if it snows a little. However, some weather conditions, such as those caused by thunderstorms, tornadoes, and blizzards, prevent you from going about your normal routine. Severe weather poses danger to people, structures, and animals.

Thunderstorms In a thunderstorm, heavy rain falls, lightning flashes, thunder roars, and hail might fall. What forces cause such extreme weather conditions?

Thunderstorms occur in warm, moist air masses and along fronts. Warm, moist air can be forced upward where it cools and condensation occurs, forming cumulonimbus clouds that can reach heights of 18 km, like the one in **Figure 12.** When rising air cools, water vapor condenses into water droplets or ice crystals. Smaller droplets collide to form larger ones, and the droplets fall through the cloud toward Earth's surface. The falling droplets collide with still more droplets and grow larger. Raindrops cool the air around them. This cool, dense air then sinks and spreads over Earth's surface. Sinking, rain-cooled air and strong updrafts of warmer air cause the strong winds associated with thunderstorms. Hail also may form as ice crystals alternately fall to warmer layers and are lifted into colder layers by the strong updrafts inside cumulonimbus clouds.

Figure 12 Tall cumulonimbus clouds may form quickly as warm, moist air rapidly rises.
Identify *some things these clouds are known to produce.*

Thunderstorm Damage Sometimes thunderstorms can stall over a region, causing rain to fall heavily for a period of time. When streams cannot contain all the water running into them, flash flooding can occur. Flash floods can be dangerous because they occur with little warning.

Strong winds generated by thunderstorms also can cause damage. If a thunderstorm is accompanied by winds traveling faster than 89 km/h, it is classified as a severe thunderstorm. Hail from a thunderstorm can dent cars and the aluminum siding on houses. Although rain from thunderstorms helps crops grow, hail has been known to flatten and destroy entire crops in a matter of minutes.

Figure 13 This time-elapsed photo shows a thunderstorm over Arizona.

Lightning and Thunder

What are lightning and thunder? Inside a storm cloud, warm air is lifted rapidly as cooler air sinks. This movement of air can cause different parts of a cloud to become oppositely charged. When current flows between regions of opposite electrical charge, lightning flashes. Lightning, as shown in **Figure 13,** can occur within a cloud, between clouds, or between a cloud and the ground.

Thunder results from the rapid heating of air around a bolt of lightning. Lightning can reach temperatures of about 30,000°C, which is more than five times the temperature of the surface of the Sun. This extreme heat causes air around the lightning to expand rapidly. Then it cools quickly and contracts. The rapid movement of the molecules forms sound waves heard as thunder.

Tornadoes Some of the most severe thunderstorms produce tornadoes. A **tornado** is a violently rotating column of air in contact with the ground. In severe thunderstorms, wind at different heights blows in different directions and at different speeds. This difference in wind speed and direction, called wind shear, creates a rotating column parallel to the ground. A thunderstorm's updraft can tilt the rotating column upward into the thunderstorm creating a funnel cloud. If the funnel comes into contact with Earth's surface, it is called a tornado.

Reading Check *What causes a tornado to form?*

A tornado's destructive winds can rip apart buildings and uproot trees. High winds can blow through broken windows. When winds blow inside a house, they can lift off the roof and blow out the walls, making it look as though the building exploded. The updraft in the center of a powerful tornado can lift animals, cars, and even houses into the air. Although tornadoes rarely exceed 200 m in diameter and usually last only a few minutes, they often are extremely destructive. In May 1999, multiple thunderstorms produced more than 70 tornadoes in Kansas, Oklahoma, and Texas. This severe tornado outbreak caused 40 deaths, 100 injuries, and more than $1.2 billion in property damage.

Science Online

Topic: Lightning
Visit green.msscience.com for Web links to research the number of lightning strikes in your state during the last year.

Activity Compare your findings with data from previous years. Communicate to your class what you learn.

VISUALIZING TORNADOES

NATIONAL GEOGRAPHIC

Figure 14

Tornadoes are extremely rapid, rotating winds that form at the base of cumulonimbus clouds. Smaller tornadoes may even form inside larger ones. Luckily, most tornadoes remain on the ground for just a few minutes. During that time, however, they can cause considerable—and sometimes strange—damage, such as driving a fork into a tree.

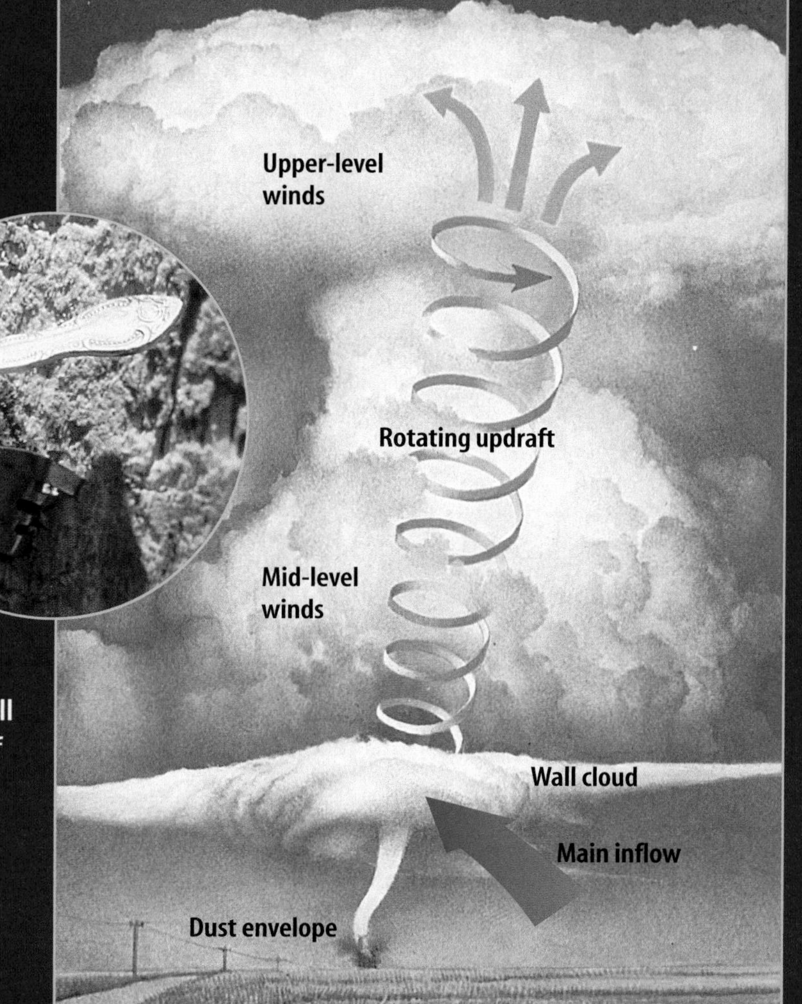

Tornadoes often form from a type of cumulonimbus cloud called a wall cloud. Strong, spiraling updrafts of warm, moist air may form in these clouds. As air spins upward, a low-pressure area forms, and the cloud descends to the ground in a funnel. The tornado sucks up debris as it moves along the ground, forming a dust envelope.

The Fujita Scale

	Wind speed (km/h)	Damage
F0	<116	Light: broken branches and chimneys
F1	116–180	Moderate: roofs damaged, mobile homes upturned
F2	181–253	Considerable: roofs torn off homes, large trees uprooted
F3	254–332	Severe: trains overturned, roofs and walls torn off
F4	333–419	Devastating: houses completely destroyed, cars picked up and carried elsewhere
F5	420–512	Incredible: total demolition

The Fujita scale, named after tornado expert Theodore Fujita, ranks tornadoes according to how much damage they cause. Fortunately, only one percent of tornadoes are classified as violent (F4 and F5).

Global Warming Some scientists hypothesize that Earth's ocean temperatures are increasing due to global warming. In your Science Journal, predict what might happen to the strength of hurricanes if Earth's oceans become warmer.

Hurricanes The most powerful storm is the hurricane. A **hurricane,** illustrated in **Figure 15,** is a large, swirling, low-pressure system that forms over the warm Atlantic Ocean. It is like a machine that turns heat energy from the ocean into wind. A storm must have winds of at least 119 km/h to be called a hurricane. Similar storms are called typhoons in the Pacific Ocean and cyclones in the Indian Ocean.

Hurricanes are similar to low-pressure systems on land, but they are much stronger. In the Atlantic and Pacific Oceans, low pressure sometimes develops near the equator. In the northern hemisphere, winds around this low pressure begin rotating counterclockwise. The strongest hurricanes affecting North America usually begin as a low-pressure system west of Africa. Steered by surface winds, these storms can travel west, gaining strength from the heat and moisture of warm ocean water.

When a hurricane strikes land, high winds, tornadoes, heavy rains, and high waves can cause a lot of damage. Floods from the heavy rains can cause additional damage. Hurricane weather can destroy crops, demolish buildings, and kill people and other animals. As long as a hurricane is over water, the warm, moist air rises and provides energy for the storm. When a hurricane reaches land, however, its supply of energy disappears and the storm loses power.

Figure 15 In this hurricane cross section, the small, red arrows indicate rising, warm, moist air. This air forms cumulus and cumulonimbus clouds in bands around the eye. The green arrows indicate cool, dry air sinking in the eye and between the cloud bands.

Blizzards Severe storms also can occur in winter. If you live in the northern United States, you may have awakened from a winter night's sleep to a cold, howling wind and blowing snow, like the storm in **Figure 16.** The National Weather Service classifies a winter storm as a **blizzard** if the winds are 56 km/h, the temperature is low, the visibility is less than 400 m in falling or blowing snow, and if these conditions persist for three hours or more.

Figure 16 Blizzards can be extremely dangerous because of their high winds, low temperatures, and poor visibility.

Severe Weather Safety When severe weather threatens, the National Weather Service issues a watch or warning. Watches are issued when conditions are favorable for severe thunderstorms, tornadoes, floods, blizzards, and hurricanes. During a watch, stay tuned to a radio or television station reporting the weather. When a warning is issued, severe weather conditions already exist. You should take immediate action. During a severe thunderstorm or tornado warning, take shelter in the basement or a room in the middle of the house away from windows. When a hurricane or flood watch is issued, be prepared to leave your home and move farther inland.

Blizzards can be blinding and have dangerously low temperatures with high winds. During a blizzard, stay indoors. Spending too much time outside can result in severe frostbite.

section 2 review

Summary

Weather Changes

- Air masses tend to have temperature and moisture properties similar to Earth's surface.
- Winds blow from areas of high pressure to areas of lower pressure.

Fronts

- A boundary between different air masses is called a front.

Severe Weather

- The National Weather Service issues watches or warnings, depending on the severity of the storm, for people's safety.

Self Check

1. **Draw Conclusions** Why is fair weather common during periods of high pressure?
2. **Describe** how a cold front affects weather.
3. **Explain** what causes lightning and thunder.
4. **Compare and contrast** a watch and a warning. How can you keep safe during a tornado warning?
5. **Think Critically** Explain why some fronts produce stronger storms than others.

Applying Skills

6. **Recognize Cause and Effect** Describe how an occluded front may form over your city and what effects it can have on the weather.

section 3

Weather Forecasts

as you read

What You'll Learn

- **Explain** how data are collected for weather maps and forecasts.
- **Identify** the symbols used in a weather station model.

Why It's Important

Weather observations help you predict future weather events.

Review Vocabulary
forecast: to predict a condition or event on the basis of observations

New Vocabulary
- meteorologist
- station model
- isotherm
- isobar

Weather Observations

You can determine current weather conditions by checking the thermometer and looking to see whether clouds are in the sky. You know when it's raining. You have a general idea of the weather because you are familiar with the typical weather where you live. If you live in Florida, you don't expect snow in the forecast. If you live in Maine, you assume it will snow every winter. What weather concerns do you have in your region?

A **meteorologist** (mee tee uh RAH luh jist) is a person who studies the weather. Meteorologists take measurements of temperature, air pressure, winds, humidity, and precipitation. Computers, weather satellites, Doppler radar shown in **Figure 17,** and instruments attached to balloons are used to gather data. Such instruments improve meteorologists' ability to predict the weather. Meteorologists use the information provided by weather instruments to make weather maps. These maps are used to make weather forecasts.

Forecasting Weather

Meteorologists gather information about current weather and use computers to make predictions about future weather patterns. Because storms can be dangerous, you do not want to be unprepared for threatening weather. However, meteorologists cannot always predict the weather exactly because conditions can change rapidly.

The National Weather Service depends on two sources for its information—data collected from the upper atmosphere and data collected on Earth's surface. Meteorologists of the National Weather Service collect information recorded by satellites, instruments attached to weather balloons, and from radar. This information is used to describe weather conditions in the atmosphere above Earth's surface.

Figure 17 A meteorologist uses Doppler radar to track a tornado. Since the nineteenth century, technology has greatly improved weather forecasting.

Station Models When meteorologists gather data from Earth's surface, it is recorded on a map using a combination of symbols, forming a **station model.** A station model, like the one in **Figure 18,** shows the weather conditions at a specific location on Earth's surface. Information provided by station models and instruments in the upper atmosphere is entered into computers and used to forecast weather.

Temperature and Pressure In addition to station models, weather maps have lines that connect locations of equal temperature or pressure. A line that connects points of equal temperature is called an **isotherm** (I suh thurm). *Iso* means "same" and *therm* means "temperature." You probably have seen isotherms on weather maps on TV or in the newspaper.

An **isobar** is a line drawn to connect points of equal atmospheric pressure. You can tell how fast wind is blowing in an area by noting how closely isobars are spaced. Isobars that are close together indicate a large pressure difference over a small area. A large pressure difference causes strong winds. Isobars that are spread apart indicate a smaller difference in pressure. Winds in this area are gentler. Isobars also indicate the locations of high- and low-pressure areas.

Reading Check *How do isobars indicate wind speed?*

Measuring Rain

Procedure

1. You will need a **straight-sided container,** such as a soup or coffee can, **duct tape,** and a **ruler.**
2. Tape the ruler to the inner wall of your container.
3. Place the container on a level surface outdoors, away from buildings or plants.
4. Measure the amount of water in your container after it rains. Continue to take measurements for a week.

Analysis

1. What was the average daily rainfall?
2. Why is it necessary to use containers with straight sides?

Try at Home

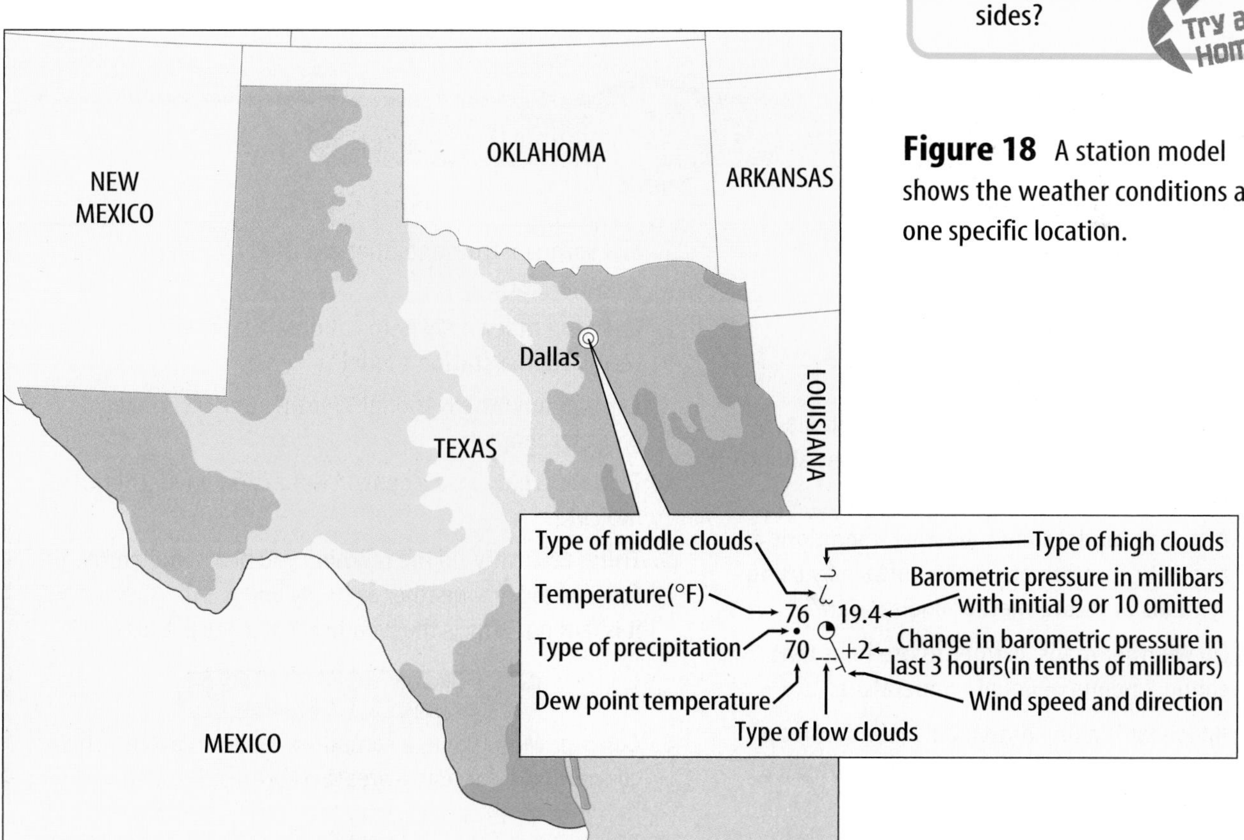

Figure 18 A station model shows the weather conditions at one specific location.

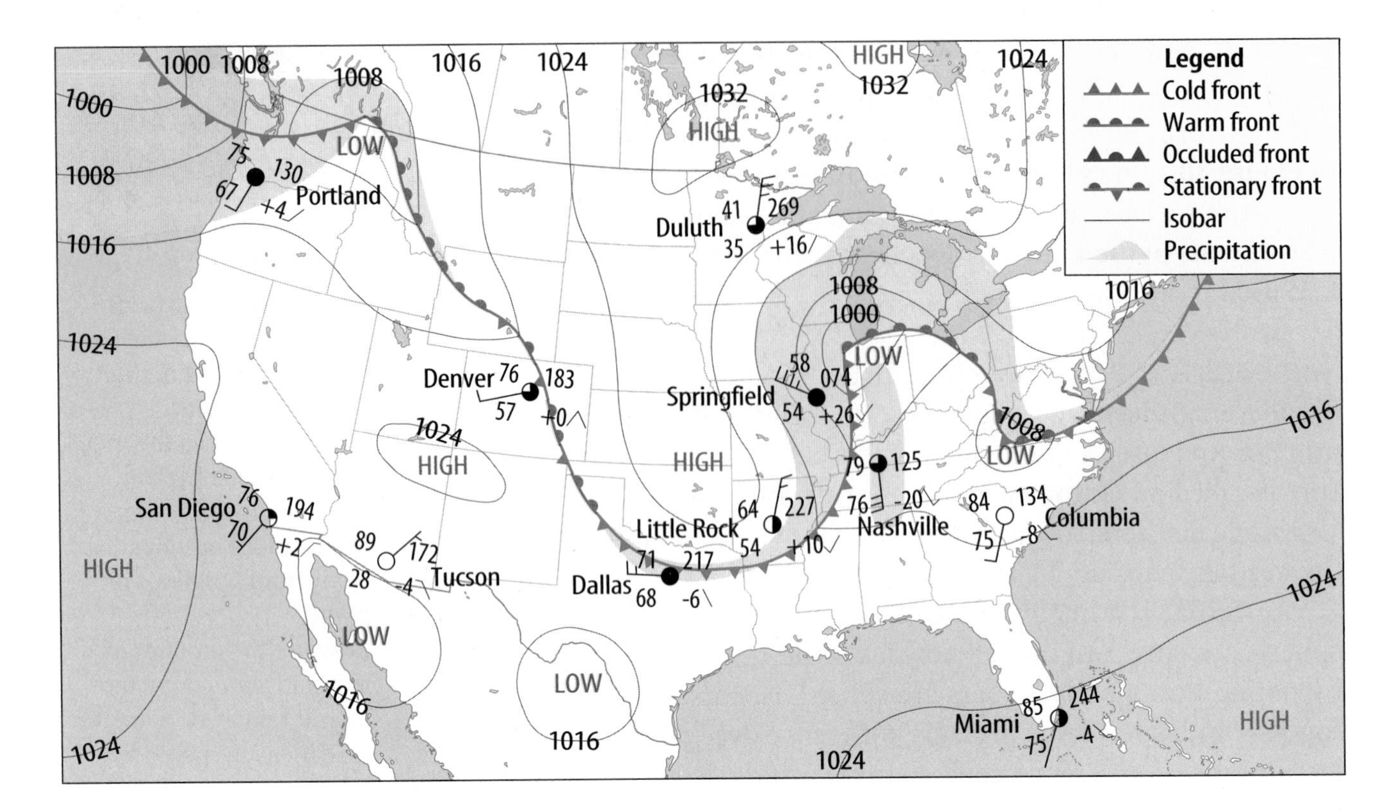

Figure 19 Highs, lows, isobars, and fronts on this weather map help meteorologists forecast the weather.

Weather Maps On a weather map like the one in **Figure 19,** pressure areas are drawn as circles with the word High or Low in the middle of the circle. Fronts are drawn as lines and symbols. When you watch weather forecasts on television, notice how weather fronts move from west to east. This is a pattern that meteorologists depend on to forecast weather.

section 3 review

Summary

Weather Observations

- Meteorologists are people who study the weather and make weather maps.

Forecasting Weather

- Meteorologists gather information about current weather and make predictions about future weather patterns.
- A station model shows weather conditions at a specific location on Earth's surface by using symbols to record meteorological data.
- On weather maps, isotherms are lines that connect points of equal temperature.
- An isobar is a line drawn on a weather map that connects points of equal atmospheric pressure.

Self Check

1. **List** some instruments that are used to collect weather data.
2. **Describe** at least six items of data that might be recorded in a station model.
3. **Explain** how the National Weather Service makes weather maps.
4. **Explain** what closely spaced isobars on a weather map indicate.
5. **Think Critically** In the morning you hear a meteorologist forecast today's weather as sunny and warm. After school, it is raining. Why is the weather so hard to predict?

Applying Skills

6. **Concept Map** Using a computer, make an events-chain concept map for how a weather forecast is made.

Science Online green.msscience.com/self_check_quiz

Reading a Weather Map

Meteorologists use a series of symbols to provide a picture of local and national weather conditions. With what you know, can you interpret weather information from weather map symbols?

Real-World Question

How do you read a weather map?

Materials

magnifying lens
Weather Map Symbols Appendix
Figure 19 (Weather Map)

Goals

- **Learn** how to read a weather map.
- **Use** information from a station model and a weather map to forecast weather.

Procedure

Use the information provided in the questions below and the Weather Map Symbols Appendix to learn how to read a weather map.

1. Find the station models on the map for Portland, Oregon, and Miami, Florida. Find the dew point, wind direction, barometric pressure, and temperature at each location.
2. Looking at the placement of the isobars, determine whether the wind would be stronger at Springfield, Illinois, or at San Diego, California. Record your answer. What is another way to determine the wind speed at these locations?
3. **Determine** the type of front near Dallas, Texas. Record your answer.
4. The triangles or half-circles are on the side of the line toward the direction the front is moving. In which direction is the cold front located over Washington state moving?

Conclude and Apply

1. Locate the pressure system over southeast Kansas. Predict what will happen to the weather of Nashville, Tennessee, if this pressure system moves there.
2. Prevailing westerlies are winds responsible for the movement of much of the weather across the United States. Based on this, would you expect Columbia, South Carolina, to continue to have clear skies? Explain.
3. The direction line on the station model indicates the direction from which the wind blows. The wind is named for that direction. Infer from this the name of the wind blowing at Little Rock, Arkansas.

Communicating Your Data

Pretend you are a meteorologist for a local TV news station. Make a poster of your weather data and present a weather forecast to your class.

Model and Invent

Measuring Wind Speed

Goals
- **Invent** an instrument or devise a system for measuring wind speeds using common materials.
- **Devise** a method for using your invention or system to compare different wind speeds.

Possible Materials
paper
scissors
confetti
grass clippings
meterstick
**measuring tape*
**Alternate materials*

Safety Precautions

Data Source
Refer to Section 1 for more information about anemometers and other wind speed instruments. Consult the data table for information about Beaufort's wind speed scale.

Real-World Question

When you watch a gust of wind blow leaves down the street, do you wonder how fast the wind is moving? For centuries, people could only guess at wind speeds, but in 1805, Admiral Beaufort of the British navy invented a method for estimating wind speeds based on their effect on sails. Later, Beaufort's system was modified for use on land. Meteorologists use a simple instrument called an anemometer to measure wind speeds, and they still use Beaufort's system to estimate the speed of the wind. What type of instrument or system can you invent to measure wind speed? How could you use simple materials to invent an instrument or system for measuring wind speeds? What observations do you use to estimate the speed of the wind?

Make a Model

1. Scan the list of possible materials and choose the materials you will need to devise your system.
2. **Devise** a system to measure different wind speeds. Be certain the materials you use are light enough to be moved by slight breezes.

Check the Model Plans

1. **Describe** your plan to your teacher. Provide a sketch of your instrument or system and ask your teacher how you might improve its design.
2. Present your idea for measuring wind speed to the class in the form of a diagram or poster. Ask your classmates to suggest improvements in your design that will make your system more accurate or easier to use.

Test Your Model

1. Confetti or grass clippings that are all the same size can be used to measure wind speed by dropping them from a specific height. Measuring the distances they travel in different strength winds will provide data for devising a wind speed scale.
2. Different sizes and shapes of paper also could be dropped into the wind, and the strength of the wind would be determined by measuring the distances traveled by these different types of paper.

Analyze Your Data

1. **Develop** a scale for your method.
2. **Compare** your results with Beaufort's wind speed scale.
3. **Analyze** what problems may exist in the design of your system and suggest steps you could take to improve your design.

Conclude and Apply

1. **Explain** why it is important for meteorologists to measure wind speeds.
2. **Evaluate** how well your system worked in gentle breezes and strong winds.

Beaufort's Wind Speed Scale

Description	Wind Speed (km/h)
calm—smoke drifts up	less than 1
light air—smoke drifts with wind	1–5
light breeze—leaves rustle	6–11
gentle breeze—leaves move constantly	12–19
moderate breeze—branches move	20–29
fresh breeze—small trees sway	30–39
strong breeze—large branches move	40–50
moderate gale—whole trees move	51–61
fresh gale—twigs break	62–74
strong gale—slight damage to houses	75–87
whole gale—much damage to houses	88–101
storm—extensive damage	102–120
hurricane—extreme damage	more than 120

Communicating Your Data

Demonstrate your system for the class. Compare your results and measurements with the results of other classmates.

TIME SCIENCE AND Society

SCIENCE ISSUES THAT AFFECT YOU!

Rainmakers

Cloud seeding is an inexact science

You listen to a meteorologist give the long-term weather forecast. Another week with no rain in sight. As a farmer, you are concerned that your crops are withering in the fields. Home owners' lawns are turning brown. Wildfires are possible. Cattle are starving. And, if farmers' crops die, there could be a shortage of food and prices will go up for consumers.

Meanwhile, several states away, another farmer is listening to the weather report calling for another week of rain. Her crops are getting so water soaked that they are beginning to rot.

Weather. Can't scientists find a way to better control it? The answer is...not exactly. Scientists have been experimenting with methods to control our weather since the 1940s. And nothing really works.

Cloud seeding is one such attempt. It uses technology to enhance the natural rainfall process. The idea has been used to create rain where it is needed or to reduce hail damage. Government officials also use cloud seeding or weather modification to try to reduce the force of a severe storm.

Flares are lodged under a plane. The pilot will drop them into potential rain clouds.

Some people seed a cloud by flying a plane above it and releasing highway-type flares with chemicals, such as silver iodide. Another method is to fly beneath the cloud and spray a chemical that can be carried into the cloud by air currents.

Cloud seeding doesn't work with clouds that have little water vapor or are not near the dew point. Seeding chemicals must be released into potential rain clouds. The chemicals provide nuclei for water molecules to cluster around. Water then falls to Earth as precipitation.

Cloud seeding does have its critics. If you seed clouds and cause rain for your area, aren't you preventing rain from falling in another area? Would that be considered "rain theft" by people who live in places where the cloudburst would naturally occur? What about those cloud-seeding agents? Could the cloud-seeding chemicals, such as silver iodide and acetone, affect the environment in a harmful way? Are humans meddling with nature and creating problems in ways that haven't been determined?

Debate **Learn more about cloud seeding and other methods of changing weather. Then debate whether or not cloud seeding can be considered "rain theft."**

For more information, visit green.msscience.com/time

Reviewing Main Ideas

Section 1 What is weather?

1. Factors that determine weather include air pressure, wind, temperature, and the amount of moisture in the air.
2. More water vapor can be present in warm air than in cold air. Water vapor condenses when the dew point is reached. Clouds are formed when warm, moist air rises and cools to its dew point.
3. Rain, hail, sleet, and snow are types of precipitation.

Section 2 Weather Patterns

1. Fronts form when air masses with different characteristics meet. Types of fronts include cold, warm, occluded, and stationary fronts.
2. High atmospheric pressure at Earth's surface usually means good weather. Cloudy and stormy weather occurs under low pressure.
3. Tornadoes, thunderstorms, hurricanes, and blizzards are examples of severe weather.

Section 3 Weather Forecasts

1. Meteorologists use information from radar, satellites, computers, and other weather instruments to forecast the weather.
2. Weather maps include information about temperature and air pressure. Station models indicate weather at a particular location.

Visualizing Main Ideas

Copy and complete the following concept map about air temperature, water vapor, and pressure.

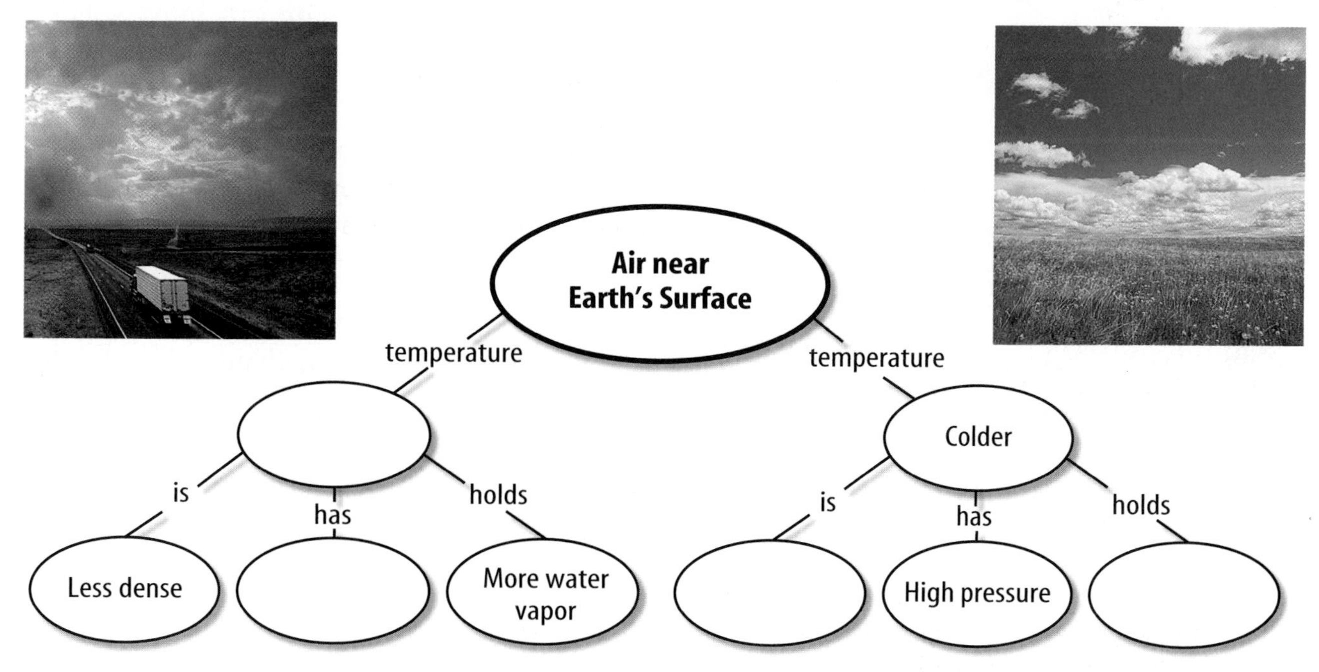

Using Vocabulary

air mass p. 126	isotherm p. 135
blizzard p. 133	meteorologist p. 134
dew point p. 121	precipitation p. 124
fog p. 123	relative humidity p. 120
front p. 127	station model p. 135
humidity p. 120	tornado p. 130
hurricane p. 132	weather p. 118
isobar p. 135	

Explain the differences between the vocabulary words in each of the following sets.

1. air mass—front
2. humidity—relative humidity
3. relative humidity—dew point
4. dew point—precipitation
5. hurricane—tornado
6. blizzard—fog
7. meteorologist—station model
8. precipitation—fog
9. isobar—isotherm
10. isobar—front

Checking Concepts

Choose the word or phrase that best answers the question.

11. Which term refers to the amount of water vapor in the air?
    A) dew point **C)** humidity
    B) precipitation **D)** relative humidity
12. What does an anemometer measure?
    A) wind speed **C)** air pressure
    B) precipitation **D)** relative humidity
13. Which type of air has a relative humidity of 100 percent?
    A) humid **C)** dry
    B) temperate **D)** saturated

Use the photo below to answer question 14.

14. Which type of the following clouds are high feathery clouds made of ice crystals?
    A) cirrus **C)** cumulus
    B) nimbus **D)** stratus
15. What is a large body of air that has the same properties as the area over which it formed called?
    A) air mass **C)** front
    B) station model **D)** isotherm
16. At what temperature does water vapor in air condense?
    A) dew point **C)** front
    B) station model **D)** isobar
17. Which type of precipitation forms when water vapor changes directly into a solid?
    A) rain **C)** sleet
    B) hail **D)** snow
18. Which type of front may form when cool air, cold air, and warm air meet?
    A) warm **C)** stationary
    B) cold **D)** occluded
19. Which is issued when severe weather conditions exist and immediate action should be taken?
    A) front **C)** station model
    B) watch **D)** warning
20. What is a large, swirling storm that forms over warm, tropical water called?
    A) hurricane **C)** blizzard
    B) tornado **D)** hailstorm

Science Online green.msscience.com/vocabulary_puzzlemaker

Thinking Critically

21. **Explain** the relationship between temperature and relative humidity.
22. **Describe** how air, water, and the Sun interact to cause weather.
23. **Explain** why northwest Washington often has rainy weather and southwest Texas is dry.
24. **Determine** What does it mean if the relative humidity is 79 percent?
25. **Infer** Why don't hurricanes form in Earth's polar regions?
26. **Compare and contrast** the weather at a cold front and the weather at a warm front.
27. **Interpret Scientific Illustrations** Use the cloud descriptions in this chapter to describe the weather at your location today. Then try to predict tomorrow's weather.
28. **Compare and contrast** tornadoes and thunderstorms. Include information about wind location and direction.
29. **Concept Map** Copy and complete the sequence map below showing how precipitation forms.

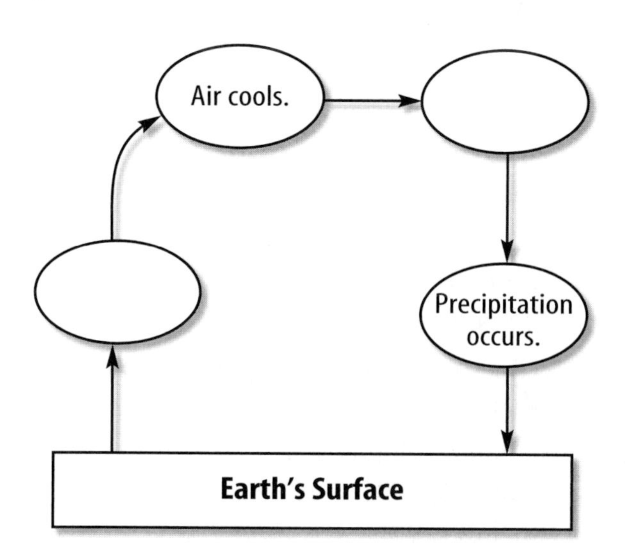

30. **Observe and Infer** You take a hot shower. The mirror in the bathroom fogs up. Infer from this information what has happened.

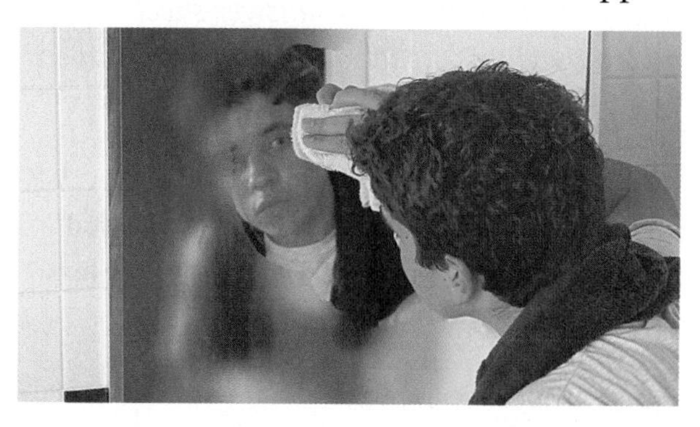

Performance Activities

31. **Board Game** Make a board game using weather terms. You could make cards to advance or retreat a token.
32. **Design** your own weather station. Record temperature, precipitation, and wind speed for one week.

Applying Math

Use the table below to answer question 33.

Air Temperature (°C)	Amount of Water Vapor Needed for Saturation (g/m^3)
25	22
20	15

33. **Dew Point** If the air temperature is 25°C and the relative humidity is 55 percent, will the dew point be reached if the temperature drops to 20°C?
34. **Rising Temperature** If the air temperature is 30°C and the relative humidity is 60 percent, will the dew point be reached if the temperature rises to 35°C? Use the graph in **Figure 4** to explain your answer.

Part 1 Multiple Choice

Record your answers on the answer sheet provided by your teacher or on a sheet of paper.

Use the table and paragraph below to answer questions 1 and 2.

Hurricanes are rated on a scale based on their wind speed and barometric pressure. The table below lists hurricane categories.

Hurricane Rating Scale		
Category	Wind Speed (km/h)	Barometric Pressure (millibars)
1	119–154	>980
2	155–178	965–980
3	179–210	945–964
4	211–250	920–944
5	>250	<920

1. Hurricane Mitch, with winds of 313 km/h and a pressure of 907 mb, struck the east coast of Central America in 1998. What category was Hurricane Mitch?
   A. 2 **C.** 4
   B. 3 **D.** 5

2. Which of the following is true when categorizing a hurricane?
   A. Storm category increases as wind increases and pressure decreases.
   B. Storm category increases as wind decreases and pressure increases.
   C. Storm category increases as wind and pressure increase.
   D. Storm category decreases as wind and pressure decrease.

Test-Taking Tip

Fill In All Blanks Never leave any answer blank.

3. Which of the following instruments is used to measure air pressure?
   A. anemometer **C.** barometer
   B. thermometer **D.** rain gauge

4. Which of the following is a description of a tornado?
   A. a large, swirling, low-pressure system that forms over the warm Atlantic Ocean
   B. a winter storm with winds at least 56 km/h and low visibility
   C. a violently rotating column of air in contact with the ground
   D. a boundary between two air masses of different density, moisture, or temperature

Use the figure below to answer questions 5 and 6.

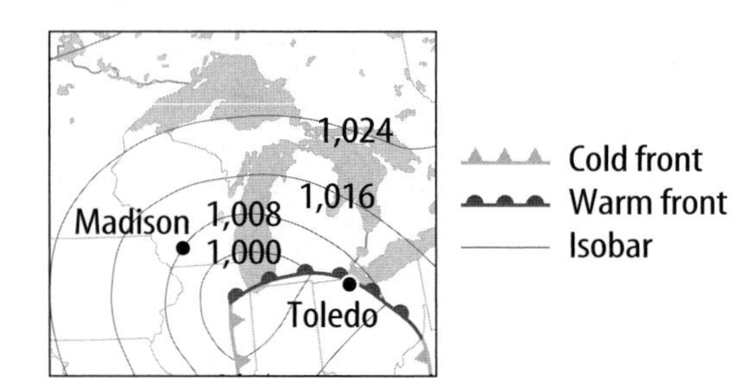

5. What is the atmospheric pressure in the city of Madison, Wisconsin?
   A. 1024 mb **C.** 1008 mb
   B. 1016 mb **D.** 1000 mb

6. What type of front is near Toledo, Ohio?
   A. cold front
   B. warm front
   C. stationary front
   D. occluded front

7. Which of the following terms is used to describe a person studies the weather?
   A. meteorologist
   B. geologist
   C. biologist
   D. paleontologist

Part 2 Short Response/Grid In

Record your answer on the answer sheet provided by your teacher or on a sheet of paper.

8. Compare and contrast the formation of stratus clouds, cumulus clouds and cirrus clouds.

Use the graph below to answer questions 9 and 10.

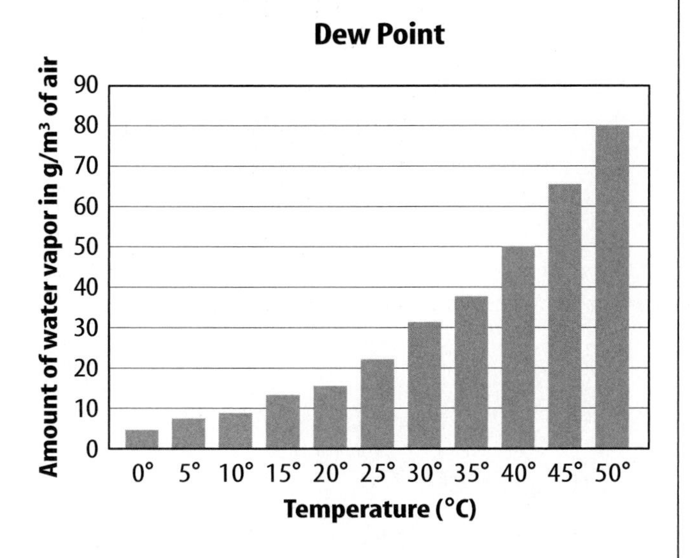

9. What amount of water vapor is in air at a temperature of 45°C?

10. On a fall day, the relative humidity is 72 percent and the temperature is 30°C. Will the dew point be reached if the temperature drops to 20°C? Why or why not?

11. Describe weather conditions during which hailstones form and the process by which they form.

12. What effects do high-pressure systems have on air circulation and weather? What effects do low-pressure systems have on weather?

13. Explain the relationship between differences in atmospheric pressure and wind speed.

Part 3 Open Ended

Record your answers on a sheet of paper.

14. Explain the relationship between lightning and thunder.

15. Describe how a hurricane in the Northern Hemisphere forms.

16. Explain why hurricanes lose power once they reach land.

Use the figure below to answer question 17.

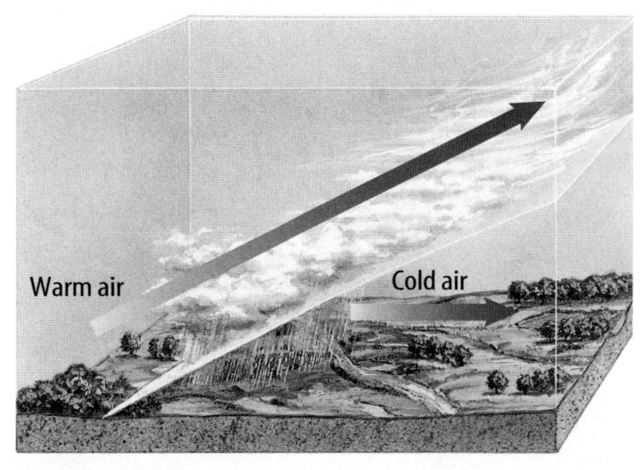

17. What type of front is shown? How does this type of front form?

18. Explain what type of weather occurs at front boundaries.

19. List the safety precautions you should take during a severe weather alert, including tornado warnings, flood watches, and blizzards, respectively.

20. Explain how the Sun's heat energy creates Earth's weather.

21. What are the four main types of precipitation? Describe the differences between each type.

chapter 6

Climate

chapter preview

sections

Why do seasons change?

Why do some places have four distinct seasons, while others have only a wet and dry season? In this chapter, you will learn what climate is and how climates are classified. You will also learn what causes climate changes and how humans and animals adapt to different climates.

Science Journal Write a paragraph explaining what you already know about the causes of seasons.

Start-Up Activities

Tracking World Climates

You wouldn't go to Alaska to swim or to Jamaica to snow ski. You know the climates in these places aren't suited for these sports. In this lab, you'll explore the climates in different parts of the world.

1. Obtain a world atlas, globe, or large classroom map. Select several cities from different parts of the world.
2. Record the longitude and latitude of your cities. Note if they are near mountains or an ocean.
3. Research the average temperature of your cities. In what months are they hottest? Coldest? What is the average yearly rainfall? What kinds of plants and animals live in the region? Record your findings.
4. Compare your findings with those of the rest of your class. Can you see any relationship between latitude and climate? Do cities near an ocean or a mountain range have different climatic characteristics?
5. **Think Critically** Keep track of the daily weather conditions in your cities. Are these representative of the kind of climates your cities are supposed to have? Suggest reasons why day-to-day weather conditions may vary.

FOLDABLES™ Study Organizer

Classifying Climates Make the following Foldable to help you compare climatic types.

STEP 1 **Fold** two pieces of paper lengthwise into thirds.

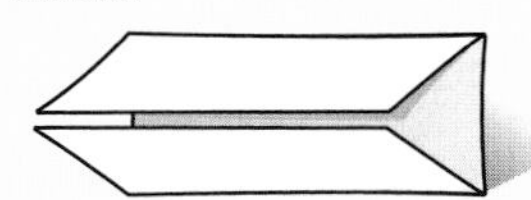

STEP 2 **Fold** the papers widthwise into fourths.

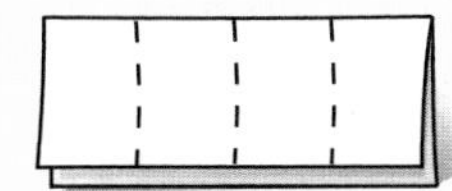

STEP 3 **Unfold,** lay the papers lengthwise, and draw lines along the folds as shown.

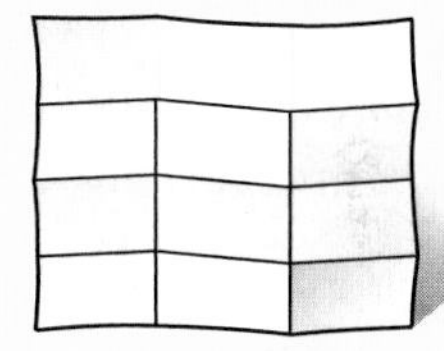

STEP 4 **Label** your tables as shown.

Climate Classification		
Tropical		
Mild		
Dry		

Climate Classification		
Continental		
Polar		
High elevation		

Make a Table As you read the chapter, define each type of climate and write notes on its weather characteristics.

Preview this chapter's content and activities at green.msscience.com

What is climate?

as you read

What You'll Learn

- **Describe** what determines climate.
- **Explain** how latitude, oceans, and other factors affect the climate of a region.

Why It's Important

Climate affects the way you live.

Review Vocabulary
latitudes: distance in degrees north or south of the equator

New Vocabulary
- climate
- tropics
- polar zone
- temperate zone

Climate

If you wandered through a tropical rain forest, you would see beautiful plants flowering in shades of pink and purple beneath a canopy of towering trees. A variety of exotic birds and other animals would dart among the tree branches and across the forest floor. The sounds of singing birds and croaking frogs would surround you. All of these organisms thrive in hot temperatures and abundant rainfall. Rain forests have a hot, wet climate. **Climate** is the pattern of weather that occurs in an area over many years. It determines the types of plants or animals that can survive, and it influences how people live.

Climate is determined by averaging the weather of a region over a long period of time, such as 30 years. Scientists average temperature, precipitation, air pressure, humidity, and number of days of sunshine to determine an area's climate. Some factors that affect the climate of a region include latitude, landforms, location of lakes and oceans, and ocean currents.

Latitude and Climate

As you can see in **Figure 1,** regions close to the equator receive the most solar radiation. Latitude, a measure of distance north or south of the equator, affects climate. **Figure 2** compares cities at different latitudes. The **tropics**—the region between latitudes 23.5°N and 23.5°S—receive the most solar radiation because the Sun shines almost directly over these areas. The tropics have temperatures that are always hot, except at high elevations. The **polar zones** extend from 66.5°N and 66.5°S latitude to the poles. Solar radiation hits these zones at a low angle, spreading energy over a large area. During winter, polar regions receive little or no solar radiation. Polar regions are never warm.

Reading Check *How does latitude affect climate?*

Between the tropics and the polar zones are the **temperate zones.** Temperatures here are moderate. Most of the United States is in a temperate zone.

Figure 1 The tropics are warmer than the temperate zones and the polar zones because the tropics receive the most direct solar energy.

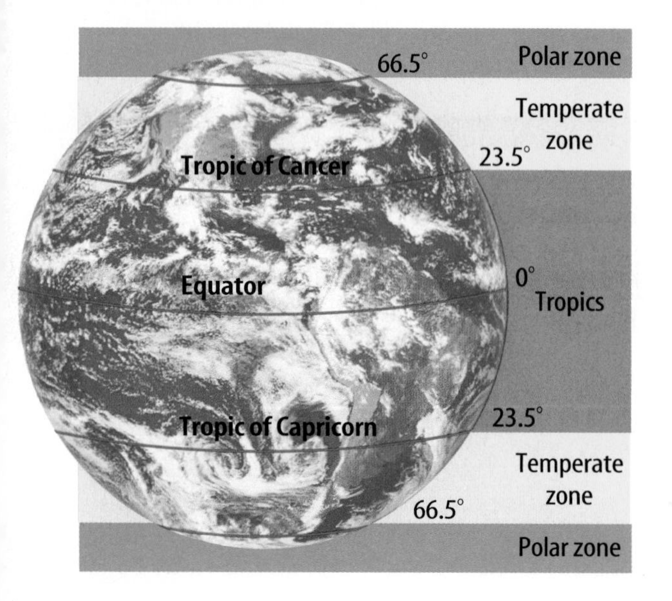

Other Factors

In addition to the general climate divisions of polar, temperate, and tropical, natural features such as large bodies of water, ocean currents, and mountains affect climate within each zone. Large cities also change weather patterns and influence the local climate.

Large Bodies of Water If you live or have vacationed near an ocean, you may have noticed that water heats up and cools down more slowly than land does. This is because it takes a lot more heat to increase the temperature of water than it takes to increase the temperature of land. In addition, water must give up more heat than land does for it to cool. Large bodies of water can affect the climate of coastal areas by absorbing or giving off heat. This causes many coastal regions to be warmer in the winter and cooler in the summer than inland areas at similar latitude. Look at **Figure 2** again. You can see the effect of an ocean on climate by comparing the average temperatures in a coastal city and an inland city, both located at 37°N latitude.

Observing Solar Radiation

Procedure

1. Darken the room.
2. Hold a **flashlight** about 30 cm from a **globe.** Shine the light directly on the equator. With your finger, trace around the light.
3. Now, tilt the flashlight to shine on 30°N latitude. The size of the lighted area should increase. Repeat at 60°N latitude.

Analysis

1. How did the size and shape of the light beam change as you directed the light toward higher latitudes?
2. How does Earth's tilt affect the solar radiation received by different latitudes?

Figure 2 This map shows average daily low temperatures in four cities during January and July. It also shows average yearly precipitation.

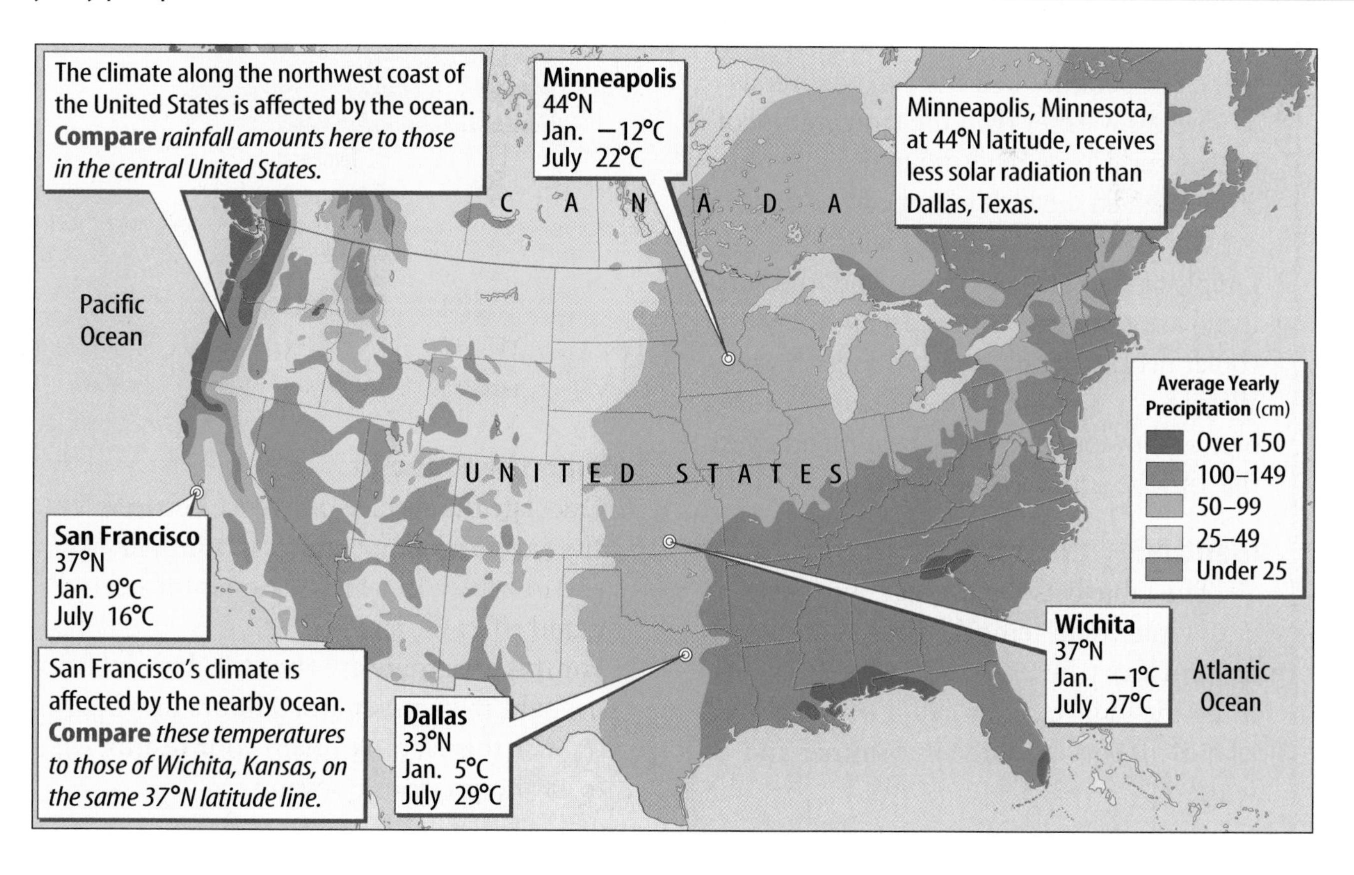

Mountain Air When air rises over a mountain, the air expands and its temperature decreases, causing water vapor to condense and form rain. Temperature changes caused by air expanding or contracting also occur in some machines. Why does the air coming out of a bicycle pump feel cold?

Ocean Currents Ocean currents affect coastal climates. Warm currents begin near the equator and flow toward higher latitudes, warming the land regions they pass. When the currents cool off and flow back toward the equator, they cool the air and climates of nearby land.

Reading Check *How do ocean currents affect climate?*

Winds blowing from the sea are often moister than those blowing from land. Therefore, some coastal areas have wetter climates than places farther inland. Look at the northwest coast of the United States shown in **Figure 2.** The large amounts of precipitation in Washington, Oregon, and northern California can be explained by this moist ocean air.

Mountains At the same latitude, the climate is colder in the mountains than at sea level. When radiation from the Sun is absorbed by Earth's surface, it heats the land. Heat from Earth then warms the atmosphere. Because Earth's atmosphere gets thinner at higher altitudes, the air in the mountains has fewer molecules to absorb heat.

Applying Science

How do cities influence temperature?

The temperature in a city can be several degrees warmer than the temperature of nearby rural areas. This difference in temperature is called the heat island effect. Cities contain asphalt and concrete which heat up rapidly as they absorb energy from the Sun. Rural areas covered with vegetation stay cooler because plants and soil contain water. Water heats up more slowly and carries away heat as it evaporates. Is the heat island effect the same in summer and winter?

Identifying the Problem

The table lists the average summer and winter high temperatures in and around a city in 1996 and 1997. By examining the data, can you tell if the heat island effect is the same in summer and winter?

Average Seasonal Temperatures

Season	Temperature (°C)	
	City	Rural
Winter 1996	2.0	0.6
Summer 1996	28.5	25.9
Winter 1997	4.9	3.2
Summer 1997	28.6	26.2

Solving the Problem

1. Calculate the average difference between city and rural temperatures in summer and in winter. In which season is the heat island effect the largest?
2. For this area there are about 15 hours of daylight in summer and 9 hours in winter. Use this fact to explain your results from the previous question.

Rain Shadows Mountains also affect regional climates, as shown in **Figure 3.** On the windward side of a mountain range, air rises, cools, and drops its moisture. On the leeward side of a mountain range air descends, heats up, and dries the land. Deserts are common on the leeward sides of mountains.

Figure 3 Large mountain ranges can affect climate by forcing air to rise over the windward side, cooling and bringing precipitation. The air descends with little or no moisture, creating desertlike conditions on the leeward side.

Cities Large cities affect local climates. Streets, parking lots, and buildings heat up, in turn heating the air. Air pollution traps this heat, creating what is known as the heat-island effect. Temperatures in a city can be 5°C higher than in surrounding rural areas.

section 1 review

Summary

Latitude and Climate

- Climate is the pattern of weather that occurs in an area over many years.
- The tropics receive the most solar radiation because the Sun shines most directly there.
- The polar zones receive the least solar energy due to the low-angled rays.
- Temperate zones, located between the tropics and the polar zones, have moderate temperatures.

Other Factors

- Natural features such as large bodies of water, ocean currents, and mountains can affect local and regional climates.
- Large cities can change weather patterns and influence local climates.

Self Check

1. **Explain** how two cities located at the same latitude can have different climates.
2. **Describe** how mountains affect climate.
3. **Define** the heat island effect.
4. **Compare and contrast** tropical and polar climates.
5. **Think Critically** Explain why plants found at different elevations on a mountain might differ. How can latitude affect the elevation at which some plants are found?

Applying Math

6. **Solve One-Step Equations** The coolest average summer temperature in the United States is 2°C at Barrow, Alaska, and the warmest is 37°C at Death Valley, California. Calculate the range of average summer temperatures in the United States.

section 2

Climate Types

as you read

What You'll Learn

- **Describe** a climate classification system.
- **Explain** how organisms adapt to particular climates.

Why It's Important

Many organisms can survive only in climates to which they are adapted.

Review Vocabulary
regions: places united by specific characteristics

New Vocabulary
- adaptation
- hibernation

Classifying Climates

What is the climate like where you live? Would you call it generally warm? Usually wet and cold? Or different depending on the time of year? How would you classify the climate in your region? Life is full of familiar classification systems—from musical categories to food groups. Classifications help to organize your thoughts and to make your life more efficient. That's why Earth's climates also are classified and are organized into the various types that exist. Climatologists—people who study climates—usually use a system developed in 1918 by Wladimir Köppen to classify climates. Köppen observed that the types of plants found in a region depended on the climate of the area. **Figure 4** shows one type of region Köppen might have observed. He classified world climates by using the annual and monthly averages of temperature and precipitation of different regions. He then related the types and distribution of native vegetation to the various climates.

The climate classification system shown in **Figure 5** separates climates into six groups—tropical, mild, dry, continental, polar, and high elevation. These groups are further separated into types. For example, the dry climate classification is separated into semiarid and arid.

Figure 4 The type of vegetation in a region depends on the climate. **Describe** *what these plants tell you about the climate shown here.*

Adaptations

Climates vary around the world, and as Köppen observed, the type of climate that exists in an area determines the vegetation found there. Fir trees aren't found in deserts, nor are cacti found in rain forests. In fact, all organisms are best suited for certain climates. Organisms are adapted to their environment. An **adaptation** is any structure or behavior that helps an organism survive in its environment. Structural adaptations are inherited. They develop in a population over a long period of time. Once adapted to a particular climate, organisms may not be able to survive in other climates.

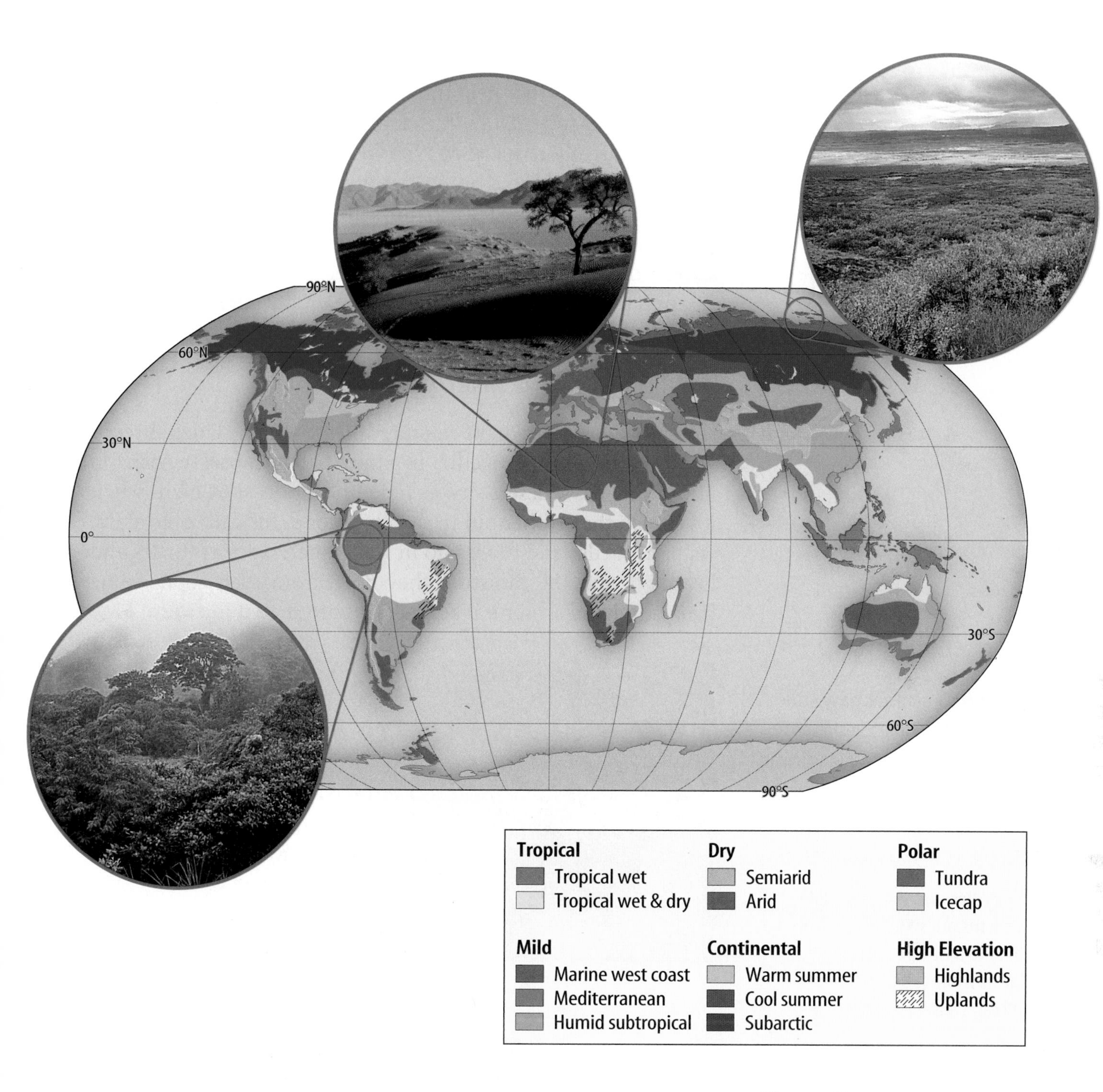

Figure 5 This map shows a climate classification system similar to the one developed by Köppen. **Describe** *the patterns you can see in the locations of certain climate types.*

INTEGRATE Life Science

Structural Adaptations Some organisms have body structures that help them survive in certain climates. The fur of mammals is really hair that insulates them from cold temperatures. A cactus has a thick, fleshy stem. This structural adaptation helps a cactus hold water. The waxy stem covering prevents water inside the cactus from evaporating. Instead of broad leaves, these plants have spiny leaves, called needles, that further reduce water loss.

Reading Check *How do cacti conserve water?*

Behavioral Adaptations Some organisms display behavioral adaptations that help them survive in a particular climate. For example, rodents and certain other mammals undergo a period of greatly reduced activity in winter called **hibernation.** During hibernation, body temperature drops and body processes are reduced to a minimum. Some of the factors thought to trigger hibernation include cooler temperatures, shorter days, and lack of adequate food. The length of time that an animal hibernates varies depending on the particular species of animal and the environmental conditions.

Reading Check *What is hibernation?*

Other animals have adapted differently. During cold weather, bees cluster together in a tight ball to conserve heat. On hot, sunny days, desert snakes hide under rocks. At night when it's cooler, they slither out in search of food. Instead of drinking water as turtles and lizards do in wet climates, desert turtles and lizards obtain the moisture they need from their food. Some behavioral and structural adaptations are shown in **Figure 6.**

Figure 6 Organisms have structural and behavioral adaptations that help them survive in particular climates.

These hibernating bats have adapted their behavior to survive winter.

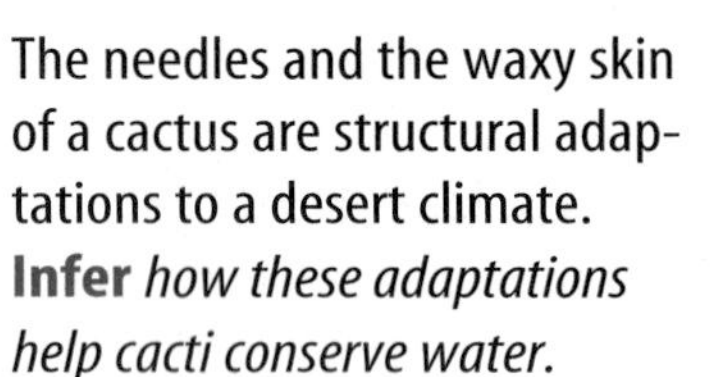

The needles and the waxy skin of a cactus are structural adaptations to a desert climate. **Infer** *how these adaptations help cacti conserve water.*

Polar bears have structural adaptations to keep them warm. The hairs of their fur trap air and heat.

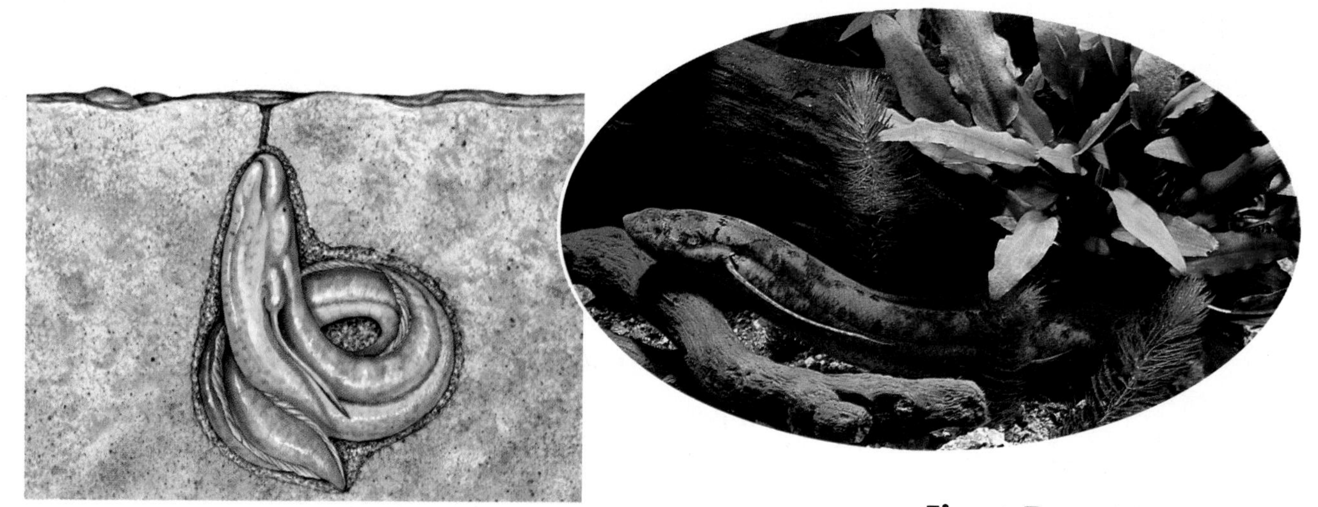

Figure 7 Lungfish survive periods of intense heat and drought by going into an inactive state called estivation. During the dry season when water evaporates, lungfish dig into the mud and curl up in a small chamber they make at the lake's bottom. During the wet season, lungfish reemerge to live in small lakes and pools.

Estivation Lungfish, shown in **Figure 7,** survive periods of intense heat by entering an inactive state called estivation (es tuh VAY shun). As the weather gets hot and water evaporates, the fish burrows into mud and covers itself in a leathery mixture of mud and mucus. It lives this way until the warm, dry months pass.

Like other organisms, you have adaptations that help you adjust to climate. In hot weather, your sweat glands release water onto your skin. The water evaporates, taking some heat with it. As a result, you become cooler. In cold weather, you may shiver to help your body stay warm. When you shiver, the rapid muscle movements produce some heat. What other adaptations to climate do people have?

section 2 review

Summary

Classifying Climates

- Climatologists classify climates into six main groups: tropical, mild, dry, continental, polar, and high elevation.

Adaptations

- Adaptations are any structures or behaviors that help an organism to survive.
- Structural adaptations such as fur, hair, and spiny needles help an organism to survive in certain climates.
- Behavioral adaptations include hibernation, a period of greatly reduced activity in winter; estivation, an inactive state during intense heat; clustering together in the cold; and obtaining water from food when water is not found elsewhere.

Self Check

1. **List** Use **Figure 5** and a world map to identify the climate type for each of the following locations: Cuba, North Korea, Egypt, and Uruguay.
2. **Compare and contrast** hibernation and estivation.
3. **Think Critically** What adaptations help dogs keep cool during hot weather?

Applying Skills

4. **Form Hypotheses** Some scientists have suggested that Earth's climate is getting warmer. What effects might this have on vegetation and animal life in various parts of the United States?
5. **Communicate** Research the types of vegetation found in the six climate regions shown in **Figure 5.** Write a paragraph in your Science Journal describing why vegetation can be used to help define climate boundaries.

section 3

Climatic Changes

as you read

What You'll Learn

- **Explain** what causes seasons.
- **Describe** how El Niño affects climate.
- **Explore** possible causes of climatic change.

Why It's Important

Changing climates could affect sea level and life on Earth.

Review Vocabulary
solar radiation: energy from the Sun transferred by waves or rays

New Vocabulary
- season
- El Niño
- greenhouse effect
- global warming
- deforestation

Earth's Seasons

In temperate zones, you can play softball under the summer Sun and in the winter go sledding with friends. Weather changes with the season. **Seasons** are short periods of climatic change caused by changes in the amount of solar radiation an area receives. **Figure 8** shows Earth revolving around the Sun. Because Earth is tilted, different areas of Earth receive changing amounts of solar radiation throughout the year.

Seasonal Changes Because of fairly constant solar radiation near the equator, the tropics do not have much seasonal temperature change. However, they do experience dry and rainy seasons. The middle latitudes, or temperate zones, have warm summers and cool winters. Spring and fall are usually mild.

Reading Check *What are seasons like in the tropics?*

Figure 8 As Earth revolves around the Sun, different areas of Earth are tilted toward the Sun, which causes different seasons.
Identify *During which northern hemisphere season is Earth closer to the Sun?*

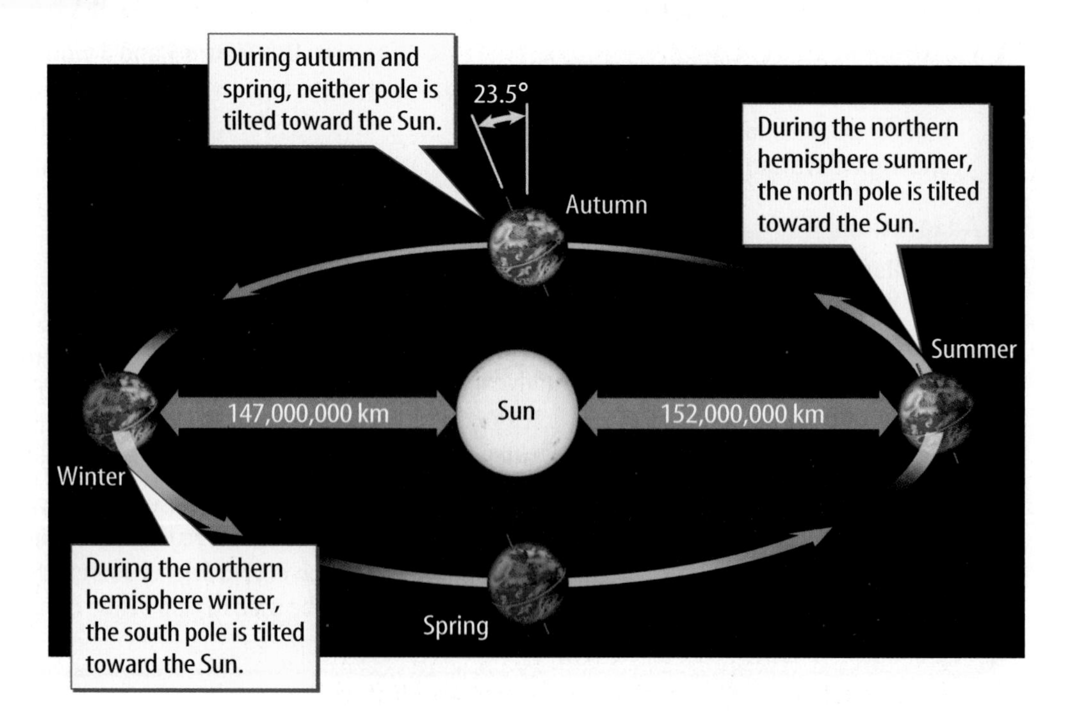

Figure 9 A strong El Niño, like the one that occurred in 1998, can affect weather patterns around the world.

A severe drought struck Indonesia, contributing to forest fires.

California was plagued by large storms that produced pounding surf and shoreline erosion.

High Latitudes During the year, the high latitudes near the poles have great differences in temperature and number of daylight hours. As shown in **Figure 8,** during summer in the northern hemisphere, the north pole is tilted toward the Sun. During summer at the north pole, the Sun doesn't set for nearly six months. During that same time, the Sun never rises at the south pole. At the equator days are about the same length all year long.

El Niño and La Niña

El Niño (el NEEN yoh) is a climatic event that involves the tropical Pacific Ocean and the atmosphere. During normal years, strong trade winds that blow east to west along the equator push warm surface water toward the western Pacific Ocean. Cold, deep water then is forced up from below along the coast of South America. During El Niño years, these winds weaken and sometimes reverse. The change in the winds allows warm, tropical water in the upper layers of the Pacific to flow back eastward to South America. Cold, deep water is no longer forced up from below. Ocean temperatures increase by 1°C to 7°C off the coast of Peru.

El Niño can affect weather patterns. It can alter the position and strength of one of the jet streams. This changes the atmospheric pressure off California and wind and precipitation patterns around the world. This can cause drought in Australia and Africa. This also affects monsoon rains in Indonesia and causes storms in California, as shown in **Figure 9.**

The opposite of El Niño is La Niña, shown in **Figure 10.** During La Niña, the winds blowing across the Pacific are stronger than normal, causing warm water to accumulate in the western Pacific. The water in the eastern Pacific near Peru is cooler than normal. La Niña may cause droughts in the southern United States and excess rainfall in the northwestern United States.

Mini LAB

Modeling El Niño

Procedure

1. During El Niño, trade winds blowing across the Pacific Ocean from east to west slacken or even reverse. Surface waters move back toward the coast of Peru.
2. Add **warm water** to a **9-in × 13-in baking pan** until it is two-thirds full. Place the pan on a smooth countertop.
3. Blow as hard as you can across the surface of the water along the length of the pan. Next, blow with less force. Then, blow in the opposite direction.

Analysis

1. What happened to the water as you blew across its surface? What was different when you blew with less force and when you blew from the opposite direction?
2. Explain how this is similar to what happens during an El Niño event.

Try at Home

NATIONAL GEOGRAPHIC **VISUALIZING EL NIÑO AND LA NIÑA**

Figure 10

Weather in the United States can be affected by changes that occur thousands of kilometers away. Out in the middle of the Pacific Ocean, periodic warming and cooling of a huge mass of seawater—phenomena known as El Niño and La Niña, respectively—can impact weather across North America. During normal years (right), when neither El Niño nor La Niña is in effect, strong winds tend to keep warm surface waters contained in the western Pacific while cooler water wells up to the surface in the eastern Pacific.

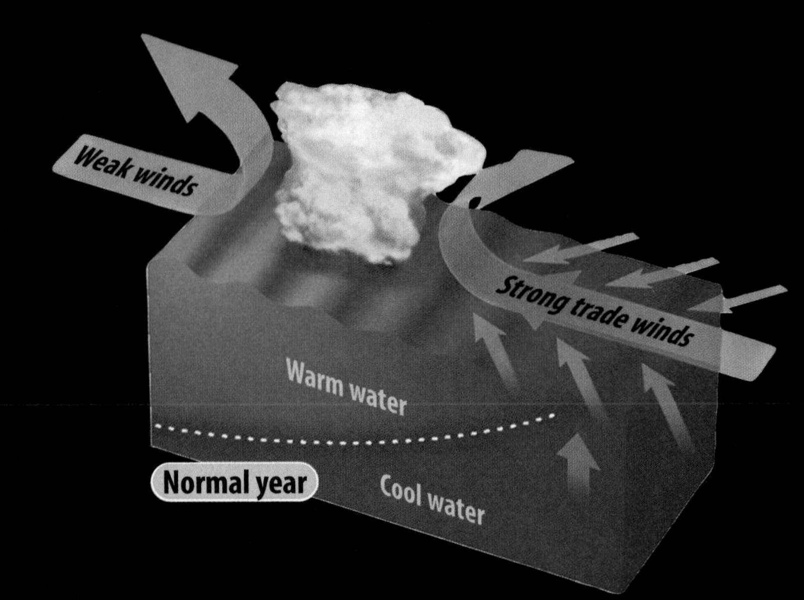

EL NIÑO During El Niño years, winds blowing west weaken and may even reverse. When this happens, warm waters in the western Pacific move eastward, preventing cold water from upwelling. These changes can alter global weather patterns and trigger heavier-than-normal precipitation across much of the United States.

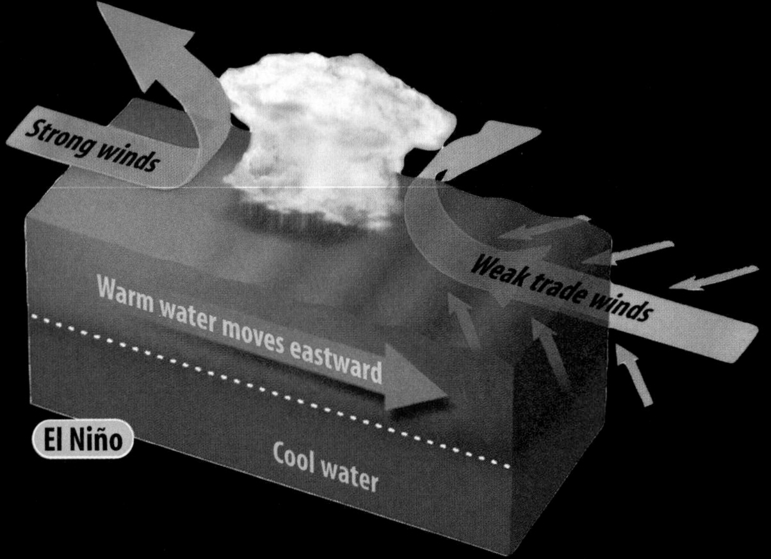

LA NIÑA During La Niña years, stronger-than-normal winds push warm Pacific waters farther west, toward Asia. Cold, deep-sea waters then well up strongly in the eastern Pacific, bringing cooler and often drier weather to many parts of the United States.

El Niño

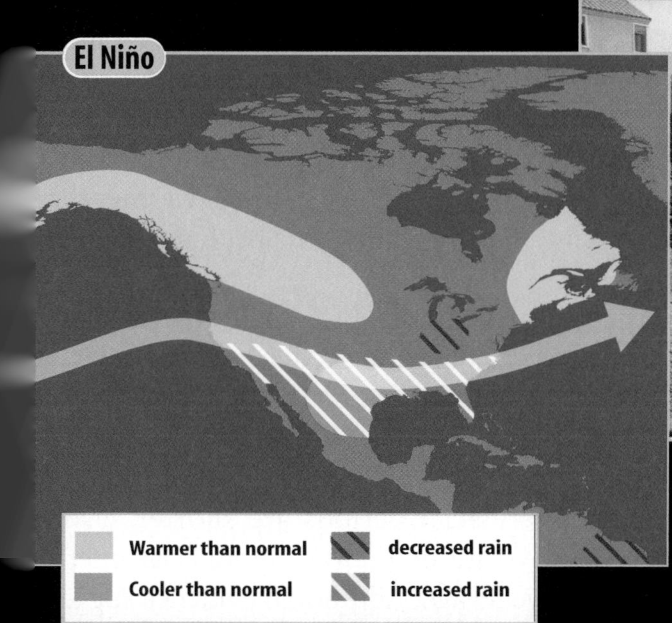

Sun-warmed surface water spans the Pacific Ocean during El Niño years. Clouds form above the warm ocean, carrying moisture aloft. The jet stream, shown by the white arrow above, helps bring some of this warm, moist air to the United States.

▲ LANDSLIDE Heavy rains in California resulting from El Niño can lead to landslides. This upended house in Laguna Niguel, California, took a ride downhill during the El Niño storms of 1998.

La Niña

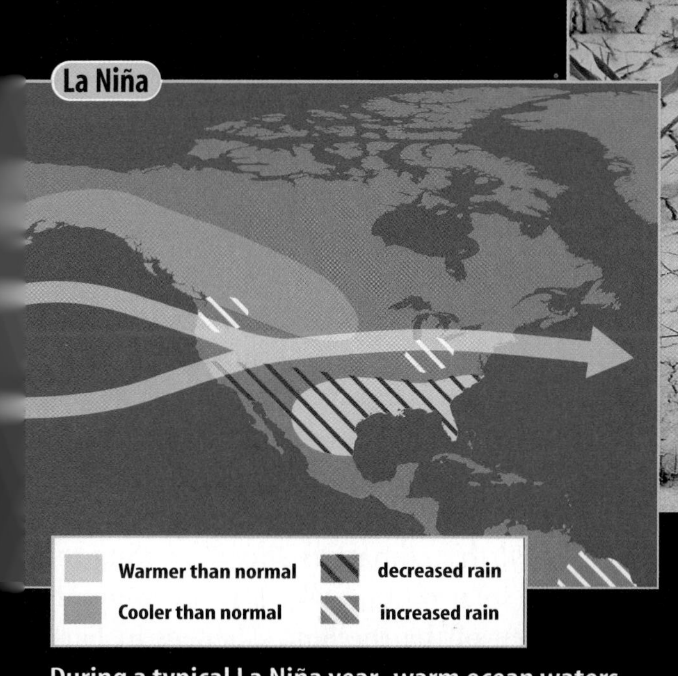

During a typical La Niña year, warm ocean waters, clouds, and moisture are pushed away from North America. A weaker jet stream often brings cooler weather to the northern parts of the continent and hot, dry weather to southern areas.

▲ PARCHED LAND The Southeast may experience drought conditions, like those that struck the cornfields of Montgomery County, Maryland, during the La Niña summer of 1988.

Climatic Change

If you were exploring in Antarctica near Earth's south pole and found a 3-million-year-old fossil of a warm-weather plant or animal, what would it tell you? You might conclude that the climate of that region changed because Antarctica is much too cold for similar plants and animals to survive today. Some warm-weather fossils found in polar regions indicate that at times in Earth's past, worldwide climate was much warmer than at present. At other times Earth's climate has been much colder than it is today.

Sediments in many parts of the world show that at several different times in the past 2 million years, glaciers covered large parts of Earth's surface. These times are called ice ages. During the past 2 million years, ice ages have alternated with warm periods called interglacial intervals. Ice ages seem to last 60,000 to 100,000 years. Most interglacial periods are shorter, lasting 10,000 to 15,000 years. We are now in an interglacial interval that began about 11,500 years ago. Additional evidence suggests that climate can change even more quickly. Ice cores record climate in a way similar to tree rings. Cores drilled in Greenland show that during the last ice age, colder times lasting 1,000 to 2,000 years changed quickly to warmer spells that lasted about as long. **Figure 11** shows a scientist working with ice cores.

Figure 11 Some ice cores consist of layers of ice that record detailed climate information for individual years. These ice cores can cover more than 300,000 years. **Describe** *how this is helpful.*

What causes climatic change?

Climatic change has many varied causes. These causes of climatic change can operate over short periods of time or very long periods of time. Catastrophic events, including meteorite collisions and large volcanic eruptions, can affect climate over short periods of time, such as a year or several years. These events add solid particles and liquid droplets to the upper atmosphere, which can change climate. Another factor that can alter Earth's climate is short- or long-term changes in solar output, which is the amount of energy given off by the Sun. Changes in Earth's movements in space affect climate over many thousands of years, and movement of Earth's crustal plates can change climate over millions of years. All of these things can work separately or together to alter Earth's climate.

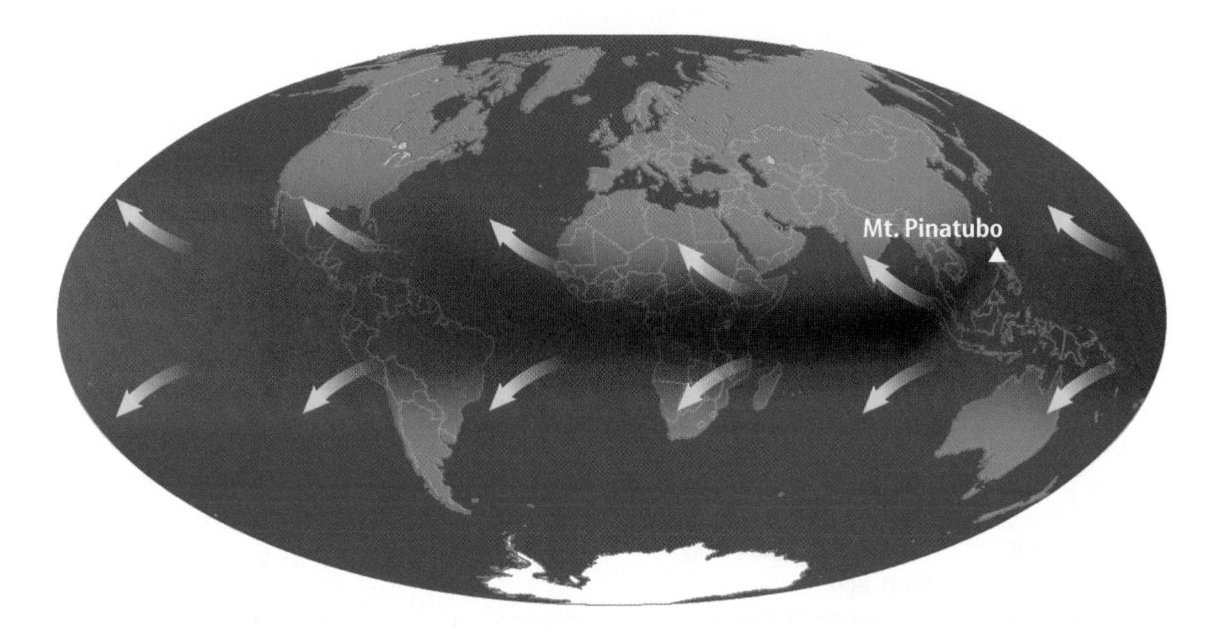

Figure 12 Mount Pinatubo in the Philippines erupted in 1991. During the eruption, particles were spread high into the atmosphere and circled the globe. Over time, particles spread around the world, blocking some of the Sun's energy from reaching Earth. The gray areas show how particles from the eruption moved around the world.

Atmospheric Solids and Liquids Small solid and liquid particles always are present in Earth's atmosphere. These particles can enter the atmosphere naturally or be added to the atmosphere by humans as pollution. Some ways that particles enter the atmosphere naturally include volcanic eruptions, soot from fires, and wind erosion of soil particles. Humans add particles to the atmosphere through automobile exhaust and smokestack emissions. These small particles can affect climate.

Catastrophic events such as meteorite collisions and volcanic eruptions put enormous volumes of dust, ash, and other particles into the atmosphere. These particles block so much solar radiation that they can cool the planet. **Figure 12** shows how a major volcanic eruption affected Earth's atmosphere.

In cities, particles put into the atmosphere as pollution can change the local climate. These particles can increase the amount of cloud cover downwind from the city. Some studies have even suggested that rainfall amounts can be reduced in these areas. This may happen because many small cloud droplets form rather than larger droplets that could produce rain.

Energy from the Sun Solar radiation provides Earth's energy. If the output of radiation from the Sun varies, Earth's climate could change. Some changes in the amount of energy given off by the Sun seem to be related to the presence of sunspots. Sunspots are dark spots on the surface of the Sun. **WARNING:** *Never look directly at the Sun.* Evidence supporting the link between sunspots and climate includes an extremely cold period in Europe between 1645 and 1715. During this time, very few sunspots appeared on the Sun.

Air Quality Control/Monitor
Atmospheric particles from pollution can affect human health as well as climate. These small particles, often called particulates, can enter the lungs and cause tissue damage. The Department of Environmental Protection employs people to monitor air pollution and its causes. Research what types of laws air quality control monitors must enforce.

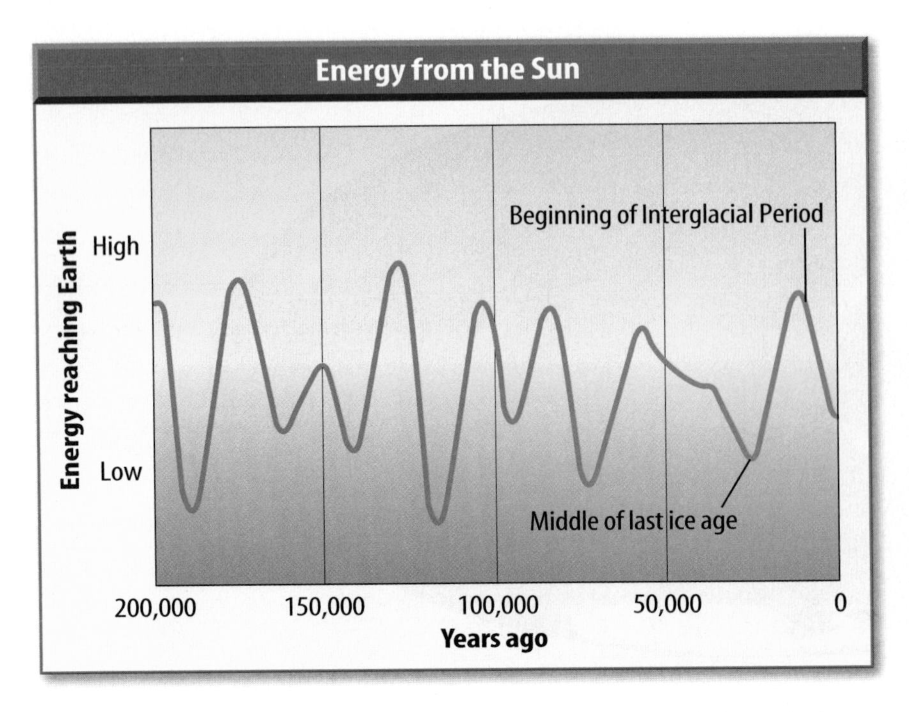

Figure 13 The curving line shows how the amount of the Sun's energy that strikes the northern hemisphere changed over the last 200,000 years.
Describe *the amount of energy that reached the northern hemisphere during the last ice age.*

Earth Movements Another explanation for some climatic changes involves Earth's movements in space. Earth's axis currently is tilted 23.5° from perpendicular to the plane of its orbit around the Sun. In the past, this tilt has increased to 24.5° and has decreased to 21.5°. When this tilt is at its maximum, the change between summer and winter is probably greater. Earth's tilt changes about every 41,000 years. Some scientists hypothesize that the change in tilt affects climate.

Two additional Earth movements also cause climatic change. Earth's axis wobbles in space just like the axis of a top wobbles when it begins to spin more slowly. This can affect the amount of solar energy received by different parts of Earth. Also, the shape of Earth's orbit changes. Sometimes it is more circular than at present and sometimes it is more flattened. The shape of Earth's orbit changes over a 100,000-year cycle.

Amount of Solar Energy These movements of Earth cause the amount of solar energy reaching different parts of Earth to vary over time, as shown in **Figure 13.** These changes might have caused glaciers to grow and shrink over the last few million years. However, they do not explain why glaciers have occurred so rarely over long spans of geologic time.

Crustal Plate Movement Another explanation for major climatic change over tens or hundreds of millions of years concerns the movement of Earth's crustal plates. The movement of continents and oceans affects the transfer of heat on Earth, which in turn affects wind and precipitation patterns. Through time, these altered patterns can change climate. One example of this is when movement of Earth's plates created the Himalaya about 40 million years ago. The growth of these mountains changed climate over much of Earth.

As you've learned, many theories attempt to answer questions about why Earth's climate has changed through the ages. Probably all of these things play some role in changing climates. More study needs to be done before all the factors that affect climate will be understood.

Climatic Changes Today

Beginning in 1992, representatives from many countries have met to discuss the greenhouse effect and global climate change. These subjects also have appeared frequently in the headlines of newspapers and magazines. Some people are concerned that the greenhouse effect could be responsible for some present-day warming of Earth's atmosphere and oceans.

The **greenhouse effect** is a natural heating process that occurs when certain gases in Earth's atmosphere trap heat. Radiation from the Sun strikes Earth's surface and causes it to warm. Some of this heat then is radiated back toward space. Some gases in the atmosphere, known as greenhouse gases, absorb a portion of this heat and then radiate heat back toward Earth, as shown in **Figure 14.** This keeps Earth warmer than it would be otherwise.

There are many natural greenhouse gases in Earth's atmosphere. Water vapor, carbon dioxide, and methane are some of the most important ones. Without these greenhouse gases, life would not be possible on Earth. Like Mars, Earth would be too cold. However, if the greenhouse effect is too strong, Earth could get too warm. High levels of carbon dioxide in its atmosphere indicate that this has happened on the planet Venus.

Topic: Greenhouse Effect
Visit green.msscience.com for Web links to information about the greenhouse effect.

Activity Research changes in the greenhouse effect over the last 200 years. Infer what might be causing the changes.

Figure 14 The Sun's radiation travels through Earth's atmosphere and heats the surface. Gases in our atmosphere trap the heat. **Compare and contrast** *this to the way a greenhouse works.*

Global Warming

Over the past 100 years, the average global surface temperature on Earth has increased by about 0.6°C. This increase in temperature is known as **global warming.** Over the same time period, atmospheric carbon dioxide has increased by about 20 percent. As a result, researchers hypothesize that the increase in global temperatures may be related to the increase in atmospheric carbon dioxide. Other hypotheses include the possibility that global warming might be caused by changes in the energy emitted by the Sun.

If Earth's average temperature continues to rise, many glaciers could melt. When glaciers melt, the extra water causes sea levels to rise. Low-lying coastal areas could experience increased flooding. Already some ice caps and small glaciers are beginning to melt and recede, as shown in **Figure 15.** Sea level is rising in some places. Some scientific studies show that these events are related to Earth's increased temperature.

You learned in the previous section that organisms are adapted to their environments. When environments change, can organisms cope? In some tropical waters around the world, corals are dying. Many people think these deaths are caused by warmer water to which the corals are not adapted.

Some climate models show that in the future, Earth's temperatures will increase faster than they have in the last 100 years. However, these predictions might change because of uncertainties in the climate models and in estimating future increases in atmospheric carbon dioxide.

Figure 15 This glacier in Greenland might have receded from its previous position because of global warming. The pile of sediment in front shows how far the glacier once reached.

Figure 16 When forests are cleared or burned, carbon dioxide levels increase in the atmosphere.

Human Activities

Human activities affect the air in Earth's atmosphere. Burning fossil fuels and removing vegetation increase the amount of carbon dioxide in the atmosphere. Because carbon dioxide is a greenhouse gas, it might contribute to global warming. Each year, the amount of carbon dioxide in the atmosphere continues to increase.

Burning Fossil Fuels When natural gas, oil, and coal are burned for energy, the carbon in these fossil fuels combines with atmospheric oxygen to form carbon dioxide. This increases the amount of carbon dioxide in Earth's atmosphere. Studies indicate that humans have increased carbon dioxide levels in the atmosphere by about 25 percent over the last 150 years.

Deforestation Destroying and cutting down trees, called **deforestation**, also affects the amount of carbon dioxide in the atmosphere. Forests, such as the one shown in **Figure 16,** are cleared for mining, roads, buildings, and grazing cattle. Large tracts of forest have been cleared in every country on Earth. Tropical forests have been decreasing at a rate of about one percent each year for the past two decades.

As trees grow, they take in carbon dioxide from the atmosphere. Trees use this carbon dioxide to produce wood and leaves. When trees are cut down, the carbon dioxide they could have removed from the atmosphere remains in the atmosphere. Cut-down trees often are burned for fuel or to clear the land. Burning trees produces even more carbon dioxide.

What can humans do to slow carbon dioxide increases in the atmosphere?

Topic: Deforestation
Visit green.msscience.com for Web links to information about deforestation.

Activity Collect data on the world's decline in forests. Infer what the world's forests will be like in 100 years.

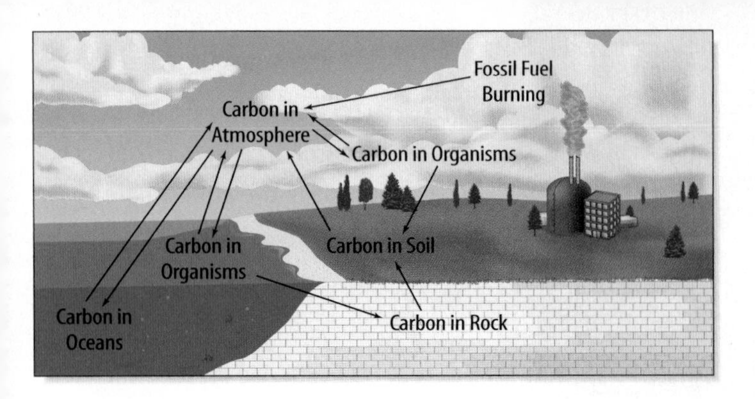

Figure 17 Carbon constantly is cycled among the atmosphere, oceans, solid earth, and biosphere.

The Carbon Cycle

Carbon, primarily as carbon dioxide, is constantly recycled in nature among the atmosphere, Earth's oceans, and organisms that inhabit the land. Organisms that undergo photosynthesis on land and in the water take in carbon dioxide and produce and store carbon-based food. This food is consumed by non-photosynthetic organisms. Carbon dioxide is released as food is broken down to release energy. When organisms die and decay, some carbon is stored as humus in soil and some carbon is released as carbon dioxide. This carbon cycle is illustrated in **Figure 17.**

Some carbon dioxide in the atmosphere dissolves in the oceans, and is used by algae and other photosynthetic, aquatic organisms. Just as on land, aquatic organisms give off carbon dioxide. However, Earth's oceans currently absorb more carbon dioxide from the atmosphere than they give off.

When Earth's climate changes, the amount of carbon dioxide that cycles among atmosphere, ocean, and land also can change. Some people hypothesize that if Earth's climate continues to warm, more carbon dioxide may be absorbed by oceans and land. Scientists continue to collect data to study any changes in the global carbon cycle.

section 3 review

Summary

Earth's Seasons

- Seasons are short periods of climatic changes due to Earth's tilt on its axis while revolving around the Sun, causing differing amounts of solar energy to reach areas of Earth.

El Niño and La Niña

- El Niño begins in the tropical Pacific Ocean when trade winds weaken or reverse directions, disrupting the normal temperature and precipitation patterns around the globe.

Climatic Changes Today

- The greenhouse effect is a natural heating process that occurs when certain gases in Earth's atmosphere trap heat.
- Burning fossil fuels increases the amount of carbon dioxide in the air.
- Deforestation increases the amount of carbon dioxide in the atmosphere.

Self Check

1. **Explain** how Earth's tilted axis is responsible for seasons.
2. **Compare and contrast** El Niño and La Niña. What climate changes do they demonstrate?
3. **List** factors that can cause Earth's climate to change.
4. **Explain** how people are adding carbon dioxide to the atmosphere.
5. **Think Critically** If Earth's climate continues to warm, how might your community be affected?

Applying Skills

6. **Use Models** Using a globe, model the three movements of Earth in space that can cause climatic change.
7. **Use a word processor** to make a table that lists the different processes that might cause Earth's climate to change. Include in your table a description of the process and how it causes climate to change.

Science Online green.msscience.com/self_check_quiz

The Greenhouse Effect

Do you remember climbing into the car on a warm, sunny day? Why was it so hot inside the car when it wasn't that hot outside? It was hotter in the car because the car functioned like a greenhouse. You experienced the greenhouse effect.

Real-World Question

How can you demonstrate the greenhouse effect?

Goals

- **Model** the greenhouse effect.
- **Measure and graph** temperature changes.

Materials

identical large, empty glass jars (2)
lid for one jar
nonmercury thermometers (3)

Safety Precautions

WARNING: *Be careful when you handle glass thermometers. If a thermometer breaks, do not touch it. Have your teacher dispose of the glass safely.*

Procedure

1. Lay a thermometer inside each jar.
2. Place the jars next to each other by a sunny window. Lay the third thermometer between the jars.
3. **Record** the temperatures of the three thermometers. They should be the same.
4. Place the lid on one jar.
5. **Record** the temperatures of all three thermometers at the end of 5, 10, and 15 min.
6. Make a line graph that shows the temperatures of the three thermometers for the 15 min of the experiment.

Conclude and Apply

1. **Explain** why you placed a thermometer between the two jars.
2. **List** the constants in this experiment. What was the variable?
3. **Identify** which thermometer showed the greatest temperature change during your experiment.
4. **Analyze** what occurred in this experiment. How was the lid in this experiment like the greenhouse gases in the atmosphere?
5. **Infer** from this experiment why you should never leave a pet inside a closed car in warm weather.

Communicating Your Data

Give a brief speech describing your conclusions to your class.

MICROCLIMATES

Goals

- **Observe** temperature, wind speed, relative humidity, and precipitation in areas outside your school.
- **Identify** local microclimates.

Materials

thermometers
psychrometer
paper strip or wind sock
large cans (4 or 5)
* *beakers or rain gauges (4 or 5)*
unlined paper
**Alternate materials*

Safety Precautions

WARNING: *If a thermometer breaks, do not touch it. Have your teacher dispose of the glass safely.*

Real-World Question

A microclimate is a localized climate that differs from the main climate of a region. Buildings in a city, for instance, can affect the climate of the surrounding area. Large buildings, such as the Bank of America Plaza in Dallas, Texas, can create microclimates by blocking the Sun or changing wind patterns. Does your school create microclimates?

Procedure

1. Select four or five sites around your school building. Also, select a control site well away from the school.
2. Attach a thermometer to an object near each of the locations you selected. Set up a rain gauge, beaker, or can to collect precipitation.
3. Visit each site at two predetermined times, one in the morning and one in the afternoon, each day for a week. Record the temperature and measure any precipitation that might have fallen. Use a wind sock or paper strip to determine wind direction.

Relative Humidity

Dry Bulb Temperature (°C)	Dry Bulb Temperature Minus Wet Bulb Temperature (°C)									
	1	2	3	4	5	6	7	8	9	10
14	90	79	70	60	51	42	34	26	18	10
15	90	80	71	61	53	44	36	27	20	13
16	90	81	71	63	54	46	38	30	23	15
17	90	81	72	64	55	47	40	32	25	18
18	91	82	73	65	57	49	41	34	27	20
19	91	82	74	65	58	50	43	36	29	22
20	91	83	74	66	59	51	44	37	31	24
21	91	83	75	67	60	53	46	39	32	26
22	92	83	76	68	61	54	47	40	34	28
23	92	84	76	69	62	55	48	42	36	30
24	92	84	77	69	62	56	49	43	37	31
25	92	84	77	70	63	57	50	44	39	33

4. To find relative humidity, you'll need to use a psychrometer. A psychrometer is an instrument with two thermometers—one wet and one dry. As moisture from the wet thermometer evaporates, it takes heat energy from its environment, and the environment immediately around the wet thermometer cools. The thermometer records a lower temperature. Relative humidity can be found by finding the difference between the wet thermometer and the dry thermometer and by using the chart on the previous page. Record all of your weather data.

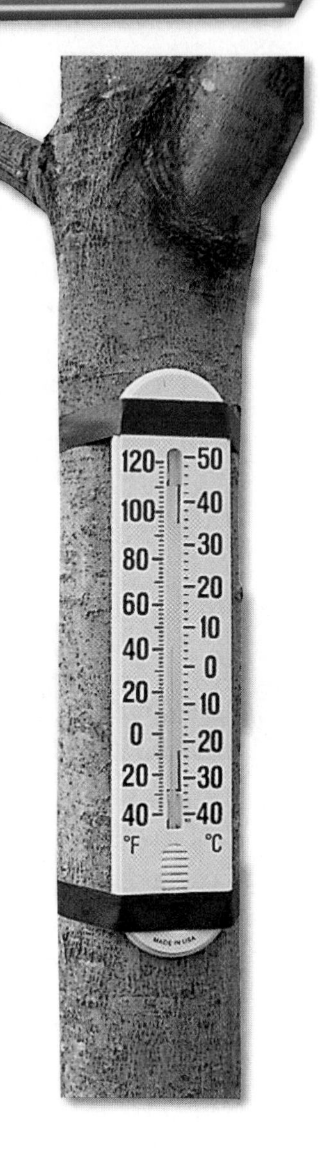

Analyze Your Data

1. Make separate line graphs for temperature, relative humidity, and precipitation for your morning and afternoon data. Make a table showing wind direction data.
2. **Compare and contrast** weather data for each of your sites. What microclimates did you identify around your school building? How did these climates differ from the control site? How did they differ from each other?

Conclude and Apply

1. **Explain** Why did you take weather data at a control site away from the school building? How did the control help you analyze and interpret your data?
2. **Infer** what conditions could have caused the microclimates that you identified. Are your microclimates similar to those that might exist in a large city? Explain.

Use your graphs to make a large poster explaining your conclusions. Display your poster in the school building. **For more help, refer to the** Science Skill Handbook.

TIME SCIENCE AND HISTORY

SCIENCE CAN CHANGE THE COURSE OF HISTORY!

The Year there was No Summer

You've seen pictures of erupting volcanoes. One kind of volcano sends smoke, rock, and ash high into the air above the crater. Another kind of volcano erupts with fiery, red-hot rivers of lava snaking down its sides. Erupting volcanoes are nature's forces at their mightiest, causing destruction and death. But not everyone realizes how far-reaching the destruction can be. Large volcanic eruptions can affect people thousands of kilometers away. In fact, major volcanic eruptions can have effects that reach around the globe.

An erupting volcano can temporarily change Earth's climate. The ash a volcano ejects into the atmosphere can create day after day without sunshine. Other particles move high into the atmosphere and are carried all the way around Earth, sometimes causing global temperatures to drop for several months.

The Summer That Never Came

An example of a volcanic eruption with wide-ranging effects occurred in 1783 in Iceland, an island nation in the North Atlantic Ocean. Winds carried a black cloud of ash from an erupting volcano in Iceland westward across northern Canada, Alaska, and across the Pacific Ocean to Japan. The summer turned bitterly cold in these places. Water froze, and heavy snowstorms pelted the land. Sulfurous gases from the erupting volcano combined with water to form particles of acid that reflected solar energy back into space. This "blanket" in the atmosphere kept the Sun's rays from heating up part of Earth.

The most tragic result of this eruption was the death of many Kauwerak people, who lived in western Alaska. Only a handful of Kauwerak survived the summer that never came. They had no opportunity to catch needed foods to keep them alive through the following winter.

Locate Using an atlas, locate Indonesia and Iceland. Using reference materials, find five facts about each place. Make a map of each nation and illustrate the map with your five facts.

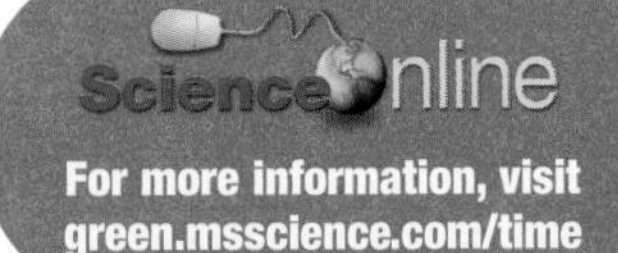

For more information, visit green.msscience.com/time

Reviewing Main Ideas

Section 1 What is climate?

1. An area's climate is the average weather over a long period of time, such as 30 years.
2. The three main climate zones are tropical, polar, and temperate.
3. Features such as oceans, mountains, and even large cities affect climate.

Section 2 Climate Types

1. Climates can be classified by various characteristics, such as temperature, precipitation, and vegetation. World climates commonly are separated into six major groups.
2. Organisms have structural and behavioral adaptations that help them survive in particular climates. Many organisms can survive only in the climate they are adapted to.
3. Adaptations develop in a population over a long period of time.

Section 3 Climatic Changes

1. Seasons are caused by the tilt of Earth's axis as Earth revolves around the Sun.
2. El Niño disrupts normal temperature and precipitation patterns around the world.
3. Geological records show that over the past few million years, Earth's climate has alternated between ice ages and warmer periods.
4. The greenhouse effect occurs when certain gases trap heat in Earth's atmosphere.
5. Carbon dioxide enters the atmosphere when fossil fuels such as oil and coal are burned.

Visualizing Main Ideas

Copy and complete the following concept map on climate.

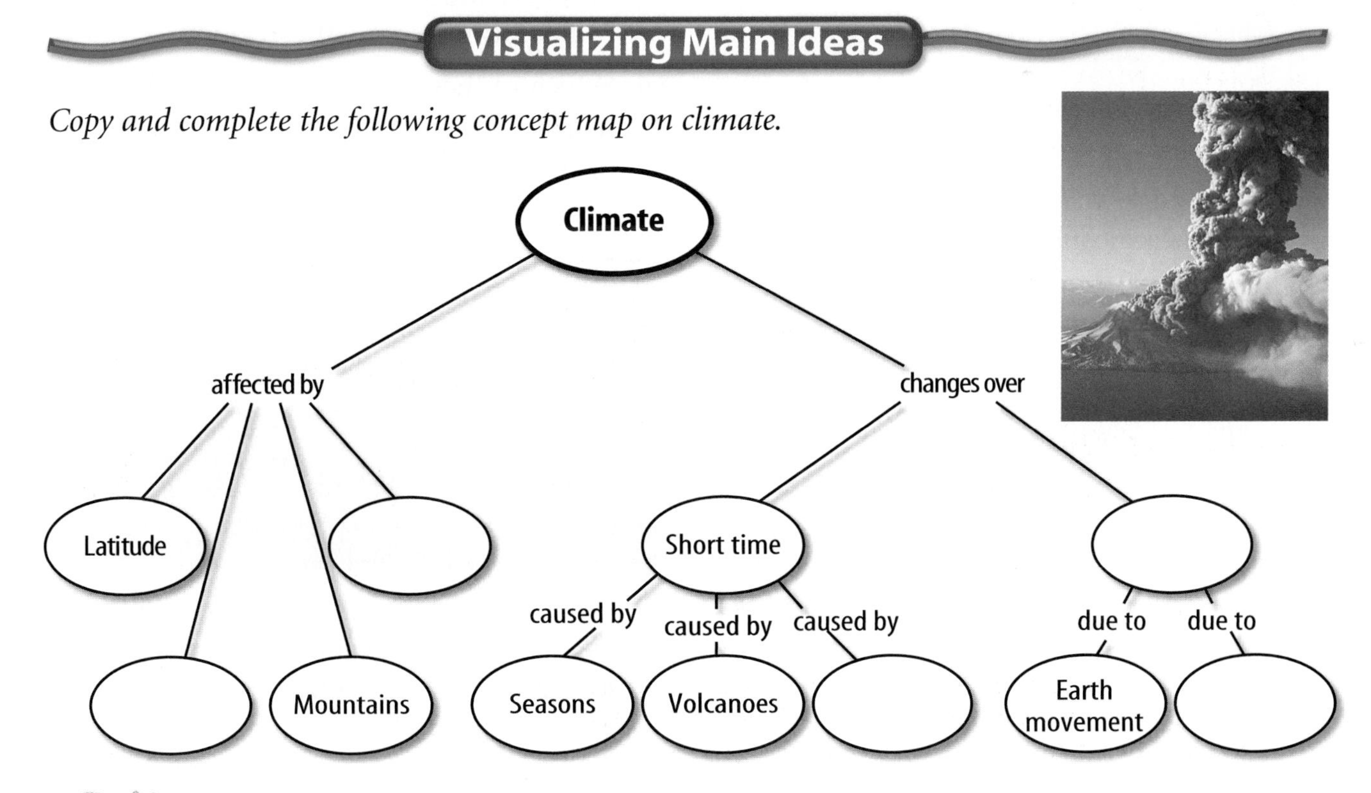

chapter 6 Review

Using Vocabulary

adaptation p. 152	hibernation p. 154
climate p. 148	polar zone p. 148
deforestation p. 165	season p. 156
El Niño p. 157	temperate zone p. 148
global warming p. 164	tropics p. 148
greenhouse effect p. 163	

Fill in the blanks with the correct vocabulary word or words.

1. Earth's north pole is in the ____________.
2. ____________ causes the Pacific Ocean to become warmer off the coast of Peru.
3. During ____________, an animal's body temperature drops.
4. ____________ is the pattern of weather that occurs over many years.
5. ____________ means global temperatures are rising.

Checking Concepts

Choose the word or phrase that best answers the question.

6. Which of the following is a greenhouse gas in Earth's atmosphere?
   A) helium **C)** hydrogen
   B) carbon dioxide **D)** oxygen

7. During which of the following is the eastern Pacific warmer than normal?
   A) El Niño **C)** summer
   B) La Niña **D)** spring

8. Which latitude receives the most direct rays of the Sun year-round?
   A) 60°N **C)** 30°S
   B) 90°N **D)** 0°

9. What happens as you climb a mountain?
   A) temperature decreases
   B) temperature increases
   C) air pressure increases
   D) air pressure remains constant

10. Which of the following is true of El Niño?
    A) It cools the Pacific Ocean near Peru.
    B) It causes flooding in Australia.
    C) It cools the waters off Alaska.
    D) It may occur when the trade winds slacken or reverse.

11. What do changes in Earth's orbit affect?
    A) Earth's shape **C)** Earth's rotation
    B) Earth's climate **D)** Earth's tilt

12. The Köppen climate classification system includes categories based on precipitation and what other factor?
    A) temperature **C)** winds
    B) air pressure **D)** latitude

13. Which of the following is an example of structural adaptation?
    A) hibernation **C)** fur
    B) migration **D)** estivation

14. Which of these can people do in order to help reduce global warming?
    A) burn coal **C)** conserve energy
    B) remove trees **D)** produce methane

Use the illustration below to answer question 15.

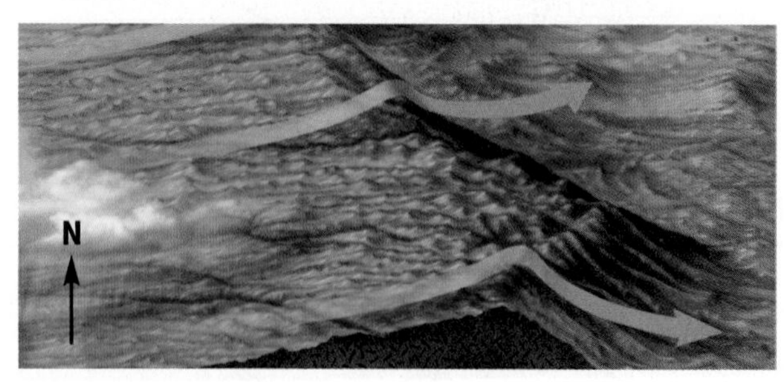

15. What would you most likely find on the leeward side of this mountain range?
    A) lakes **C)** deserts
    B) rain forests **D)** glaciers

Science Online green.msscience.com/vocabulary_puzzlemaker

Thinking Critically

16. **Draw a Conclusion** How could climate change cause the types of organisms in an area to change?
17. **Infer** What might you infer if you find fossils of tropical plants in a desert?
18. **Describe** On a summer day, why would a Florida beach be cooler than an orange grove that is 2 km inland?
19. **Infer** what would happen to global climates if the Sun emitted more energy.
20. **Explain** why it will be cooler if you climb to a higher elevation in a desert.
21. **Communicate** Explain how atmospheric pressure over the Pacific Ocean might affect how the trade winds blow.
22. **Predict** Make a chain-of-events chart to explain the effect of a major volcanic eruption on climate.
23. **Form Hypotheses** A mountain glacier in South America has been getting smaller over several decades. What hypotheses should a scientist consider to explain why this is occurring?
24. **Concept Map** Copy and complete the concept map using the following: *tropics, 0°–23.5° latitude, polar, temperate, 23.5°–66.5° latitude,* and *66.5° latitude to poles.*

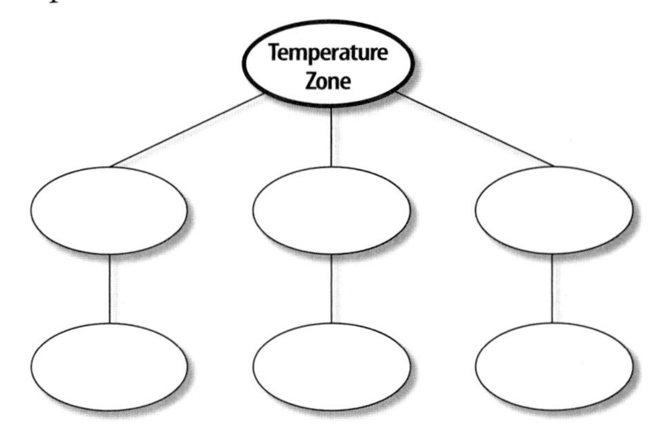

25. **Explain** how global warming might lead to the extinction of some organisms.
26. **Describe** how dust and ash from large volcanoes can change the atmosphere.
27. **Explain** how heat energy carried by ocean currents influences climate.
28. **Describe** how sediments, fossils, and ice cores record Earth's geologic history.
29. **Describe** how volcanic eruptions or meteorite collisions have changed past climates.

Performance Activities

30. **Science Display** Make a display illustrating different factors that can affect climate. Be sure to include detailed diagrams and descriptions for each factor in your display. Present your display to the class.

Applying Math

Use the table below to answer questions 31 and 32.

Precipitation in Phoenix, Arizona

Season	Precipitation (cm)
Winter	5.7
Spring	1.2
Summer	6.7
Autumn	5.9
Total	19.5

31. **Precipitation Amounts** The following table gives average precipitation amounts for Phoenix, Arizona. Make a bar graph of these data. Which climate type do you think Phoenix represents?
32. **Local Precipitation** Use the table above to help estimate seasonal precipitation for your city or one that you choose. Create a bar graph for that data.

Part 1 Multiple Choice

Record your answers on the answer sheet provided by your teacher or on a sheet of paper.

Use the graph below to answer questions 1 and 2.

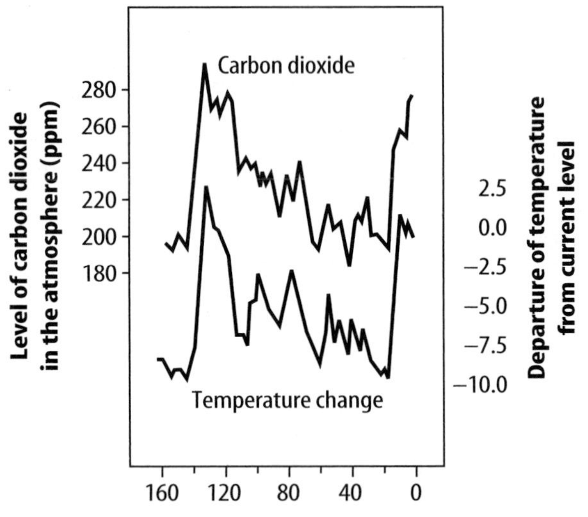

1. Which of these statements is true according to the graph?

A. Earth's mean temperature has never been hotter than it is today.
B. The level of CO_2 has never been higher than today.
C. The mean global temperature 60,000 years ago was less than today.
D. The level of CO_2 in the atmosphere 80,000 years ago was 280 parts per million.

2. Which of the following statements best describes this graph?

A. As CO_2 levels have increased, so has global temperature.
B. As CO_2 levels have increased, global temperature has decreased.
C. As global temperature has increased, CO_2 levels have decreased.
D. No relationship exists between CO_2 and global temperatures.

Use the table below to answer question 3.

Apparent Temperature Index				
	Relative Humidity (%)			
Air Temperature (°F)	80	85	90	95
85	97	99	102	105
80	86	87	88	89
75	78	78	79	79
70	71	71	71	71

3. The National Weather Service created the Apparent Temperature Index to show the temperature the human body feels when heat and humidity are combined. If the relative humidity is 85% and the temperature is 75°F, what is the apparent temperature?

A. 78°F
B. 79°F
C. 88°F
D. 89°F

4. What is the most likely reason that the air temperature is warmest at the tropical latitudes?

A. These latitudes receive the most solar radiation because there are no clouds.
B. These latitudes receive the most solar radiation because the sun's angle is high.
C. These latitudes receive the least solar radiation because the sun's angle is low.
D. These latitudes receive the least solar radiation because of heavy cloud cover.

Test-Taking Tip

Qualifiers Look for quantifiers in a question. Such questions are not looking for absolute answers. Qualifiers would be words such as *most likely, most common,* or *least common.*

Question 4 Look for the most likely scientific explanation.

Part 2 | Short Response/Grid In

Record your answers on the answer sheet provided by your teacher or on a sheet of paper.

5. Explain how a large body of water can affect the climate of a nearby area.
6. Describe the relationship between ocean currents and precipitation in a coastal region.
7. The city of Redmond, Oregon is near the Cascade Mountain Range. The average annual rainfall for the Redmond, OR area is about 8 inches. Infer whether Redmond, OR is located on the windward side or leeward side of the mountain range. Explain your answer.

Use the figure below to answer question 8.

8. What is the greenhouse effect?
9. List three greenhouse gases.
10. How does the greenhouse effect positively affect life on Earth? How could it negatively affect life on Earth?
11. Explain why the temperature of a city can be up to 5°C warmer than the surrounding rural areas.
12. What are the different ways in which solid and liquid particles enter the Earth's atmosphere?

Part 3 | Open Ended

Record your answers on a sheet of paper.

Use the figure below to answer questions 13–15.

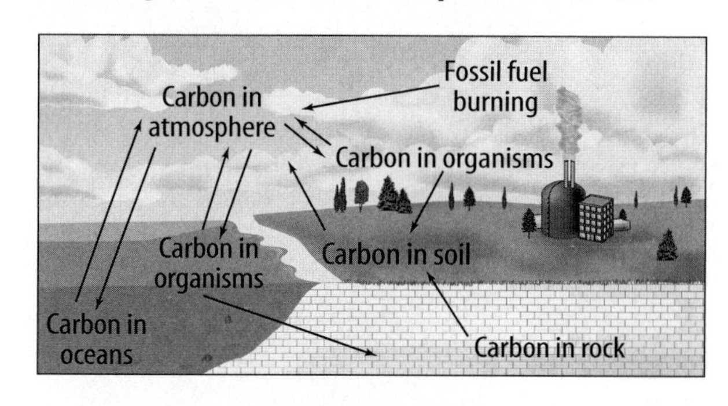

13. Describe the carbon cycle. Explain how carbon is transferred from organisms to soil.
14. How does the burning of fossil fuels affect the amount of carbon dioxide entering the carbon cycle?
15. How does deforestation affect the amount of carbon dioxide entering the carbon cycle?
16. In 1991 Mt. Pinatubo erupted, releasing volcanic particulates into the atmosphere. Temperatures around the world fell by as much as 0.7°C below average during 1992. How was this global temperature change related to the volcanic eruption?
17. What is global warming? What hypotheses help explain global warming? Explain the relationship between global warming and the level of seawater.
18. A scientist analyzes the pollen of ancient plants found preserved in lake sediments. The pollen is determined to be from a plant that needs moisture and year-round warm temperatures to grow. Make an inference about the type of climate that area experienced during the time the plant lived.

chapter 7

Earth in Space

chapter preview

sections

Light Up the Night

For thousands of years humans have looked at the night sky, but only during the past 50 years have we been able to view Earth from space. This photo combines views of Earth taken from the *International Space Station.*

Science Journal Describe how our view of Earth has changed in the past 50 years.

Start-Up Activities

Model Earth's Shape

Could you prove that Earth is round? What can you see from Earth's surface that proves Earth is round?

1. Cut a strip of cardboard about 8 cm long and 8 cm tall into a triangle. Fold up about 2 cm on one side and tape it to a basketball so that the peak of the triangle sticks straight up resembling a sailboat.
2. Place the basketball on a table at eye level so that the sailboat sticks out horizontally, parallel to the table, pointing away from you.
3. Roll the ball toward you slowly so that the sail comes into view over the top of the ball.
4. **Think Critically** Write a paragraph that explains how the shape of the basketball affected your view of the sailboat. How does this model show how Earth's shape affects your view over long distances.

Earth and the Moon Make the following Foldable to help you see how Earth and the Moon are similar and different.

STEP 1 **Fold** a vertical sheet of paper in half from top to bottom.

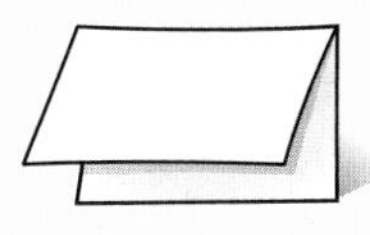

STEP 2 **Fold** in half from side to side with the fold at the top.

STEP 3 **Unfold** the paper once. **Cut** only the fold of the top flap to make two tabs.

STEP 4 **Turn** the paper vertically and **draw** on the from tabs as shown.

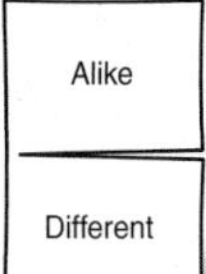

Compare and Contrast Before you read the chapter, list the ways you think Earth and the Moon are alike and different under the appropriate tabs. As you read, correct or add to this information.

Preview this chapter's content and activities at green.msscience.com

section 1

Earth's Motion and Seasons

as you read

What You'll Learn

- **Identify** Earth's shape and other physical properties.
- **Compare and contrast** Earth's rotation and revolution.
- **Explain** the causes of Earth's seasons.

Why It's Important

Movements of Earth regulate life patterns.

Review Vocabulary
satellite: a natural or artificial object that revolves around another object in space

New Vocabulary
- axis
- rotation
- revolution
- orbit
- solstice
- equinox

Earth's Physical Data

Think about the last time you saw a beautiful sunset. Late in the day, you may have noticed the Sun sinking lower and lower in the western sky. Eventually, as the Sun went below the horizon, the sky became darker. Was the Sun actually traveling out of view, or were you?

In the past, some people thought that the Sun, the Moon, and other objects in space moved around Earth each day. Now it is known that some of the motions of these objects, as observed from Earth, are really caused by Earth's movements.

Also, many people used to think that Earth was flat. They thought that if you sailed far enough out to sea, you eventually would fall off. It is now known that this is not true. What general shape does Earth have?

Spherical Earth As shown in **Figure 1,** pictures from space show that Earth is shaped like a ball, or a sphere. A sphere (SFIHR) is a three-dimensional object whose surface at all points is the same distance from its center. What other evidence can you think of that reveals Earth's shape?

Figure 1 Earth's nearly spherical shape was first observed directly by images taken from a spacecraft. **Describe** *observations from Earth's surface that also suggest that it is spherical.*

Evidence for Earth's Shape Have you ever stood on a dock and watched a sailboat come in? If so, you may have noticed that the first thing you see is the top of the boat's sail. This occurs because Earth's curved shape hides the rest of the boat from view until it is closer to you. As the boat slowly comes closer to you, more and more of its sail is visible. Finally, the entire boat is in view.

More proof of Earth's shape is that Earth casts a curved shadow on the Moon during a lunar eclipse, like the one shown in **Figure 2.** Something flat, like a book, casts a straight shadow, whereas objects with curved surfaces cast curved shadows.

Reading Check ***What object casts a shadow on the Moon during a lunar eclipse?***

Influence of Gravity The spherical shape of Earth and other planets is because of gravity. Gravity is a force that attracts all objects toward each other. The farther away the objects are, the weaker the pull of gravity is. Also, the more massive an object is, the stronger its gravitational pull is. Large objects in space, such as planets and moons often are spherical because of how they formed. At first, particles collide and stick together randomly. However, as the mass increases, gravity plays a role. Particles are attracted to the center of the growing mass, making it spherical.

Even though Earth is round, it may seem flat to you. This is because Earth's surface is so large compared to your size.

Figure 2 Earth's spherical shape also is indicated by the curved shadow it casts on the Moon during a partial lunar eclipse.

Table 1 Physical Properties of Earth	
Diameter (pole to pole)	12,714 km
Diameter (equator)	12,756 km
Circumference (poles) *(distance around Earth through N and S poles)*	40,008 km
Circumference (equator) *(distance around Earth at the equator)*	40,075 km
Mass	5.98×10^{24} kg
Average density *(average mass per unit volume)*	5.52 g/cm^3
Average distance from the Sun	149,600,000 km
Period of rotation relative to stars (1 day)	23h, 56 min
Solar day	24 h
Period of revolution (1 year) *(path around the Sun)*	365 days, 6 h, 9 min

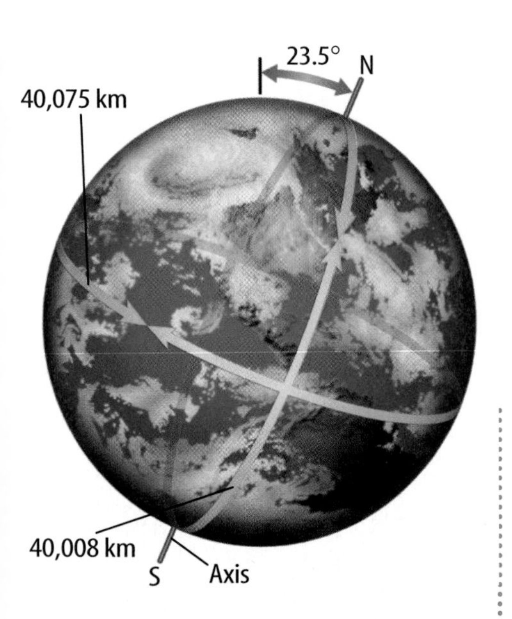

Figure 3 Earth is almost a sphere, but its circumference measurements vary slightly. The north-south circumference of Earth is smaller than the east-west circumference.

Almost a Sphere Earth's shape is not a perfect sphere. It bulges slightly at the equator and is somewhat flattened around the poles. As shown in **Figure 3,** this causes Earth's circumference at the equator to be a bit larger than Earth's circumference as measured through the north and south poles. The circumference of Earth and some other physical properties are listed in **Table 1.**

Motions of Earth

Why the Sun appears to rise and set each day and why the Moon and other objects in the sky appear to move from east to west is illustrated in **Figure 4.** Earth's geographic poles are located at the north and south ends of Earth's axis. Earth's **axis** is the imaginary line drawn from the north geographic pole through Earth to the south geographic pole. Earth spins around this imaginary line. The spinning of Earth on its axis, called **rotation,** causes you to experience day and night.

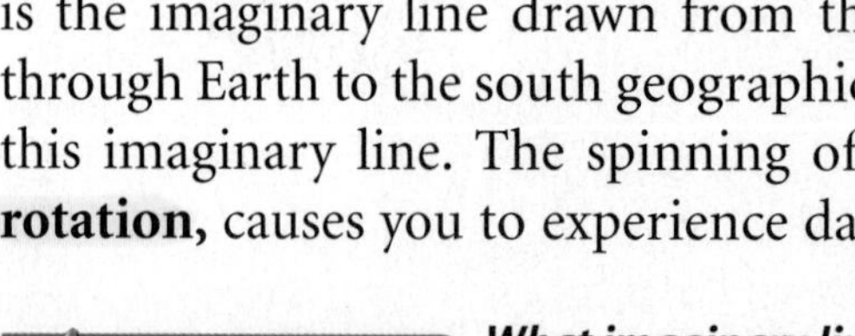

What imaginary line runs through Earth's north and south geographic poles?

Earth's Orbit Earth has another type of motion. As it rotates on its axis each day, Earth also moves along a path around the Sun. This motion of Earth around the Sun, shown in **Figure 4,** is called **revolution.** How many times does Earth rotate on its axis during one complete revolution around the Sun? Just as day and night are caused by rotation, what happens on Earth that is caused by its revolution?

Seasons A new year has begun. As days and weeks pass, you notice that the Sun remains in the sky later and later each day. You look forward to spring when you will be able to stay outside longer in the evening because the number of daylight hours gradually increases. What is causing this change?

You learned earlier that Earth's rotation causes day and night. Earth also moves around the Sun, completing one revolution each year. Earth is really a satellite of the Sun, moving around it along a curved path called an **orbit.** The shape of Earth's orbit is an ellipse, which is rounded like a circle but somewhat flattened. As Earth moves along in its orbit, the way in which the Sun's light strikes Earth's surface changes.

Earth's elliptical orbit causes it to be closer to the Sun in January and farther from the Sun in July. But, the total amount of energy Earth receives from the Sun changes little during a year. However, the amount of energy that specific places on Earth receive varies quite a lot.

Topic: Earth's Rotation and Revolution

Visit green.msscience.com for Web links to information about Earth's rotation and revolution and the seasons.

Activity Build a model, perhaps using a plastic foam sphere and toothpicks, showing how the portion of the surface lit by a small flashlight varies at different angles of tilt.

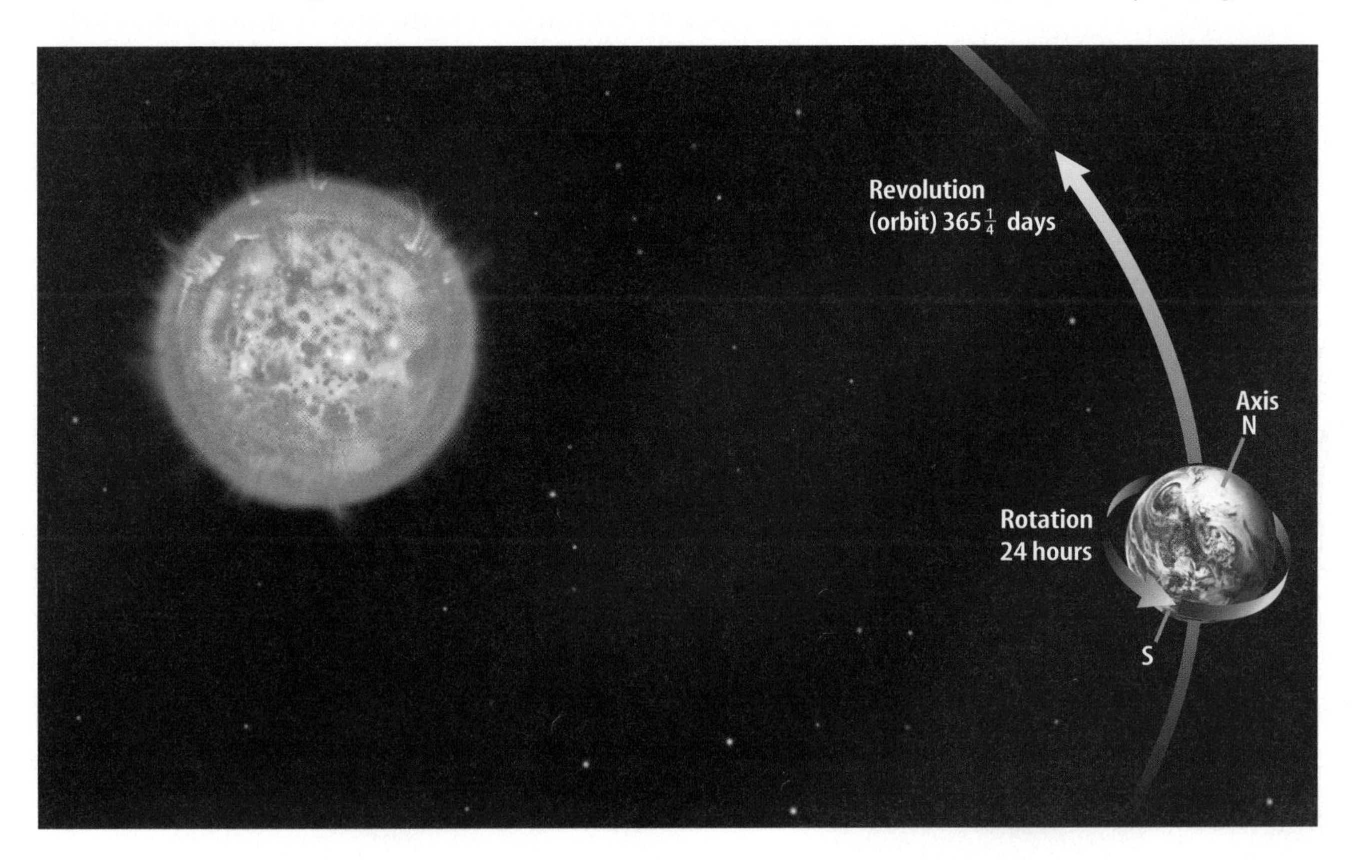

Figure 4 Earth's west-to-east rotation causes day and night.

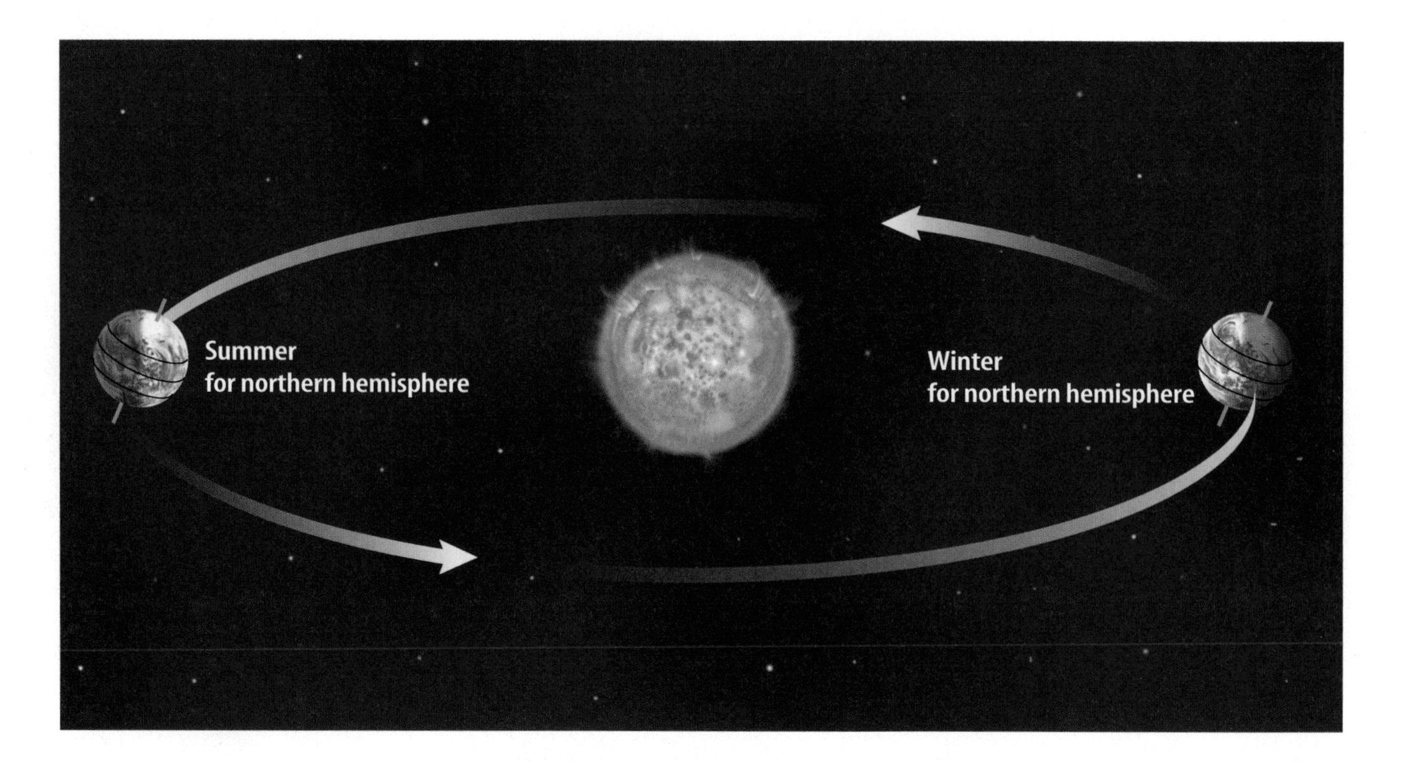

Figure 5 When the northern hemisphere is tilted toward the Sun, it experiences summer. **Explain** *why days are longer during the summer.*

Earth's Tilt You can observe one reason why the amount of energy from the Sun varies by moving a globe slowly around a light source. If you keep the globe tilted in one direction, as shown in **Figure 5,** you will see that the top half of the globe is tilted toward the light during part of its orbit and tilted away from the light during another.

Earth's axis forms a 23.5-degree angle with a line perpendicular to the plane of its orbit around the Sun. It always points to the North Star. Because of this, there are more daylight hours for the hemisphere tilted toward the Sun. Think about when it gets dark outside at different times of the year. More hours of sunlight in summer is one reason why summer is warmer than winter. Another reason is that because of Earth's tilt, sunlight strikes the hemisphere tilted toward the Sun at a higher angle, that is, closer to 90 degrees. Sunlight strikes the hemisphere tilted away from the Sun at a lower angle. This lessens solar radiation and brings winter.

Solstices Because of the tilt of Earth's axis, the Sun's position relative to Earth's equator changes. Twice during the year, the Sun reaches its greatest distance north or south of the equator and is directly over either the Tropic of Cancer or the Tropic of Capricorn, as shown in **Figure 6.** These times are known as the summer and winter **solstices.** Summer solstice, which is when the Sun is highest in the sky at noon, happens on June 21 or 22 for the northern hemisphere and on December 21 or 22 for the southern hemisphere. The opposite of this for each hemisphere is winter solstice, which is when the Sun is lowest at noon.

Equinoxes At an **equinox,** (EE kwuh nahks) when the Sun is directly above Earth's equator, the lengths of day and night are nearly equal all over the world. During equinox, Earth's tilt is neither toward nor away from the Sun. In the northern hemisphere, spring equinox is March 21 or 22 and fall equinox is September 21 or 22. As you saw in **Table 1,** the time it takes for Earth to revolve around the Sun is not a whole number of days. Because of this, the dates for solstices and equinoxes change slightly over time.

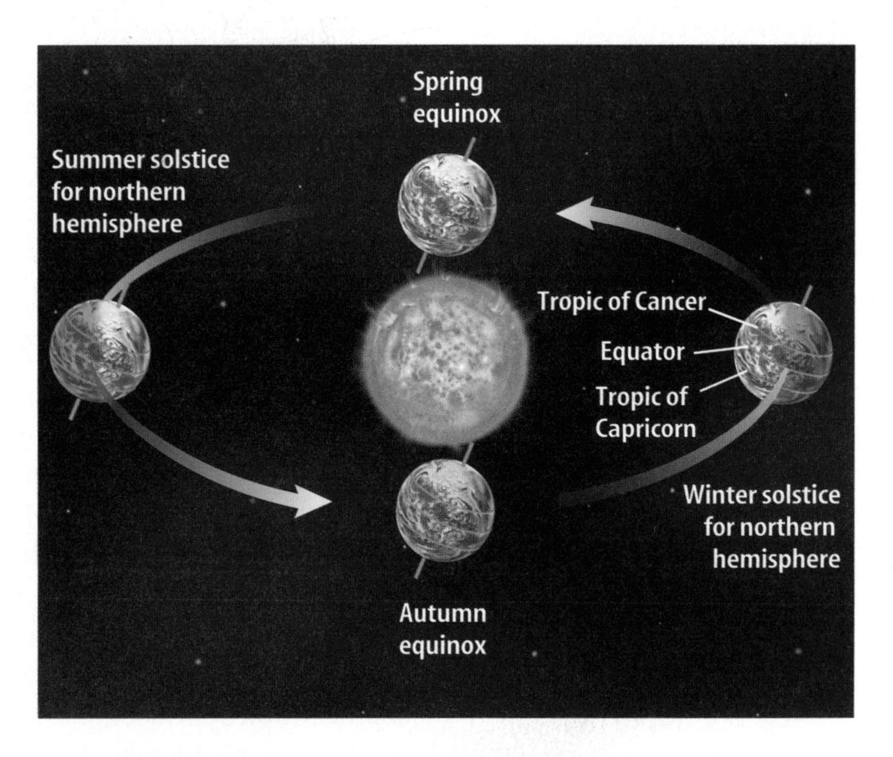

Figure 6 When the Sun is directly above the equator, day and night have nearly equal lengths.

Earth's Place in Space Earth is shaped much like a sphere. As Earth rotates on its axis, the Sun appears to rise and set in the sky. Earth's tilt and revolution around the Sun cause seasons to occur. In the next section, you will learn about Earth's nearest neighbor in space, the Moon. Later, you will learn about other planets in our solar system and how they compare with Earth.

section 1 review

Summary

Earth's Physical Data

- Earth has a spherical shape.
- The evidence for Earth's shape is the way objects appear on the horizon and its curved shadow during a lunar eclipse.

Motions of Earth

- Earth rotates on its axis, causing day and night.
- Earth revolves around the Sun in an elliptical orbit.
- Earth's rotational axis is tilted at an angle of 23.5°.
- The tilt of Earth's axis as it moves around the Sun causes the seasons.
- At a solstice, the Sun is directly over the Tropic of Capricorn or the Tropic of Cancer.
- At an equinox the Sun is directly over the equator.

Self Check

1. **Explain** how planets develop their spherical shapes.
2. **Explain** Although Earth receives the same total amount of energy from the Sun throughout the year, why is it so much warmer in summer than it is in winter?
3. **Describe** whether the shape of Earth's orbit affects or does not affect the seasons.
4. **Explain** why the dates of the solstices and equinoxes vary.
5. **Think Critically** In **Table 1,** why is Earth's distance from the Sun reported as an average distance?

Applying Skills

6. **Use Models** Use a globe and an unshaded light source to illustrate how the tilt of the Earth on its axis, as it rotates and revolves around the Sun, causes changes in the number of daylight hours.

section 2

Earth's Moon

as you read

What You'll Learn

- **Identify** the Moon's surface features and interior.
- **Explain** the Moon's phases.
- **Explain** the causes of solar and lunar eclipses.
- **Identify** the origin of the Moon.

Why It's Important

Learning about the Moon can help you understand Earth.

Review Vocabulary
density: the amount of matter in a given volume of a substance

New Vocabulary
- crater
- moon phase
- solar eclipse
- lunar eclipse

The Moon's Surface and Interior

Take a good look at the surface of the Moon during the next full moon. You can see some of its large surface features, especially if you use binoculars or a small telescope. You will see dark-colored maria (MAR ee uh) and lighter-colored highland areas, as illustrated in **Figure 7.** Galileo first named the dark-colored regions *maria,* the Latin word for seas. They reminded Galileo of the oceans. Maria probably formed when lava flows from the Moon's interior flooded into large, bowl-like regions on the Moon's surface. These depressions may have formed early in the Moon's history. Collected during *Apollo* missions and then analyzed in laboratories on Earth, rocks from the maria are about 3.2 billion to 3.7 billion years old. They are the youngest rocks found on the Moon thus far.

The oldest moon rocks analyzed so far—dating to about 4.6 billion years old—were found in the lunar highlands. The lunar highlands are areas of the lunar surface with an elevation that is several kilometers higher than the maria. Some lunar highlands are located in the south-central region of the Moon.

Figure 7 On a clear night and especially during a full moon, you can observe some of the Moon's surface features.
Describe *how you can recognize the maria, lunar highlands, and craters.*

Craters As you look at the Moon's surface features, you will see craters. **Craters,** also shown in **Figure 7,** are depressions formed by large meteorites—space objects that strike the surface. As meteorites struck the Moon, cracks could have formed in the Moon's crust, allowing lava flows to fill in the large depressions. Craters are useful for determining how old parts of a moon's or a planet's surface are compared to other parts. The more abundant the craters are in a region, the older the surface is.

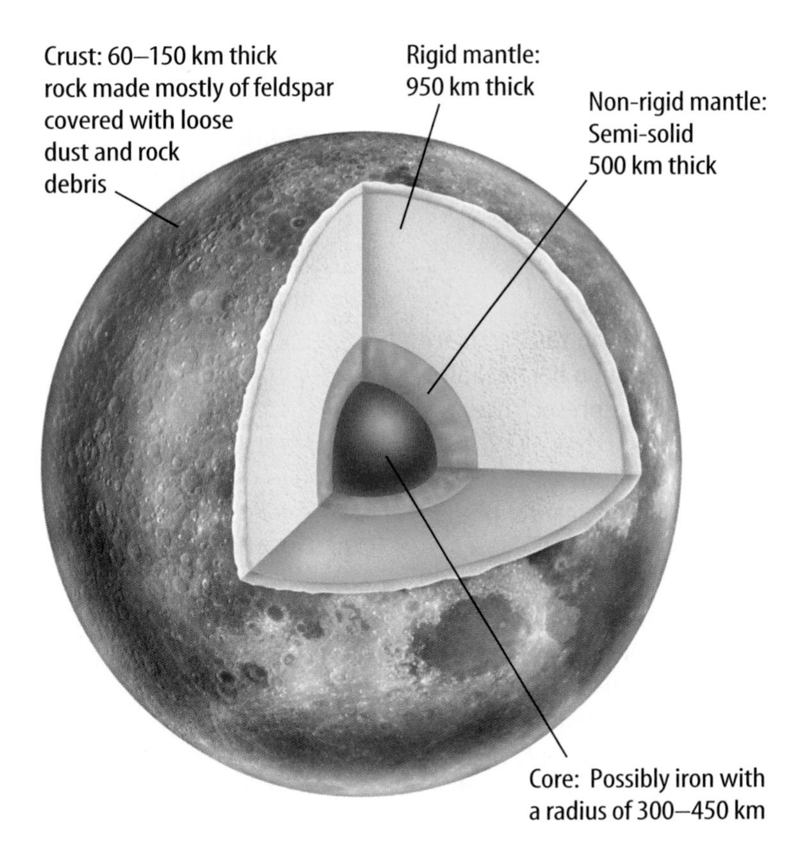

Figure 8 The small size of the Moon's core suggests the Moon formed from a part of the crust and mantle of Earth.

The Moon's Interior During the *Apollo* space program, astronauts left several seismographs (size muh grafs) on the Moon. A seismograph is an instrument that detects tremors, or seismic vibrations. On Earth, seismographs are used to measure earthquake activity. On the Moon, they are used to study moonquakes. Based on the study of moonquakes, a model of the Moon's interior has been proposed, as illustrated in **Figure 8.** The Moon's crust is about 60 km thick on the side facing Earth and about 150 km thick on the far side. The difference in thickness is probably the reason fewer lava flows occurred on the far side of the Moon. Below the crust, a solid layer called the mantle may extend 900 km to 950 km farther down. A soft layer of mantle may continue another 500 km deeper still. Below this may be an iron-rich, solid core with a radius of about 300–450 km.

Like the Moon, Earth also has a dense, iron core. However, the Moon's core is small compared to its total volume. Compared with Earth, the Moon is most like Earth's outer two layers—the mantle and the crust—in density. This supports a hypothesis that the Moon may have formed primarily from material ejected from Earth's mantle and crust.

Moon Colonies Research from space probes indicates that conditions on some areas of the Moon might make moon colonies possible someday. Why might a moon colony prove useful?

Motions of the Moon

The same side of the Moon is always facing Earth. You can verify this by examining the Moon in the sky night after night. You'll see that bright and dark surface features remain in the same positions. Does this mean that the Moon doesn't turn on an axis as it moves around Earth? Next, explore why the same side of the Moon always faces Earth.

Modeling the Moon's Rotation

Procedure

1. Use **masking tape** to place a large X on a **basketball** that will represent the Moon.
2. Ask two students to sit in **chairs** in the center of the room.
3. Place other students around the outer edge of the room.
4. Slowly walk completely around the two students in the center while holding the basketball so that the side with the X always faces the two students.

Analysis

1. Ask the two students in the center whether they think the basketball turned around as you circled them. Then ask several students along the outer edge of the room whether they think the basketball turned around.
2. Based on these observations, infer whether or not the Moon rotates as it moves around Earth. Explain your answer.

Revolution and Rotation of the Moon The Moon revolves around Earth at an average distance of about 384,000 km. It takes 27.3 days for the Moon to complete one orbit around Earth. The Moon also takes 27.3 days to rotate once on its axis. Because these two motions of the Moon take the same amount of time, the same side of the Moon is always facing Earth. Examine **Figure 9** to see how this works.

Reading Check *Why does the same side of the Moon always face Earth?*

However, these two lunar motions aren't exactly the same during the Moon's 27.3-day rotation-and-revolution period. Because the Moon's orbit is an ellipse, it moves faster when it's closer to Earth and slower when it's farther away. During one orbit, observers are able to see a little more of the eastern side of the Moon and then a little more of the western side.

Moon Phases If you ever watched the Moon for several days in a row, you probably noticed how its shape and position in the sky change. You learned that the Moon rotates on its axis and revolves around Earth. Motions of the Moon cause the regular cycle of change in the way the Moon looks to an observer on Earth.

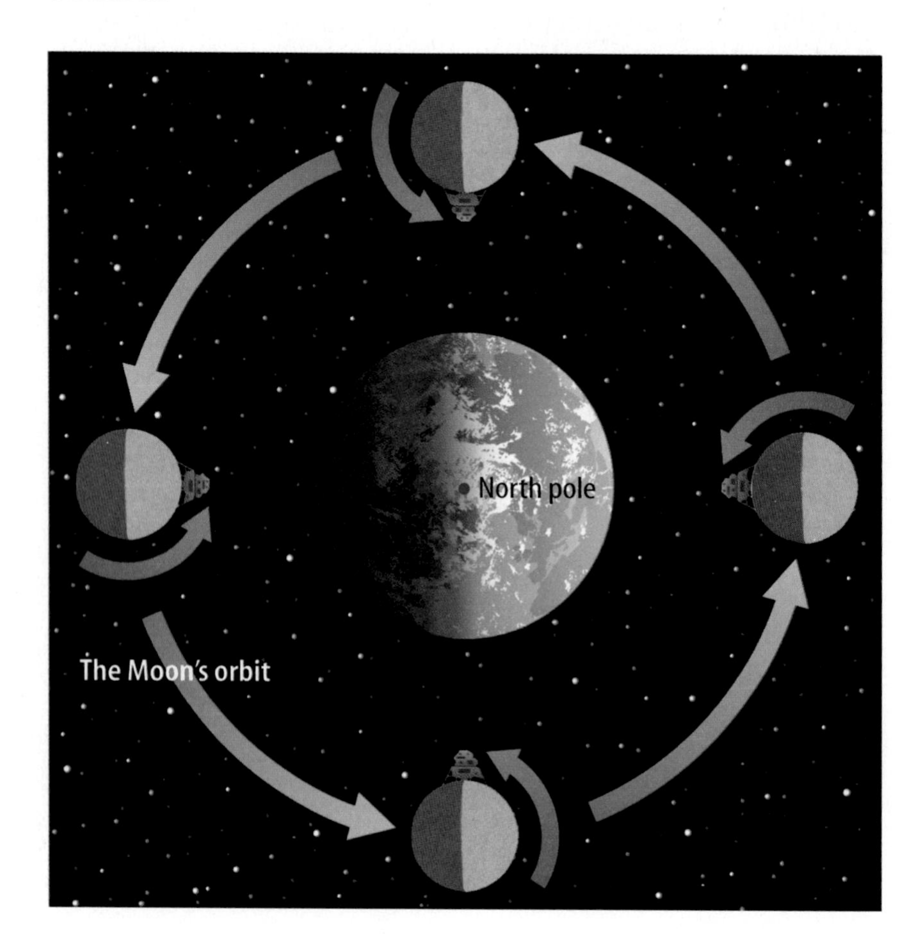

Figure 9 Observers viewing the Moon from Earth always see the same side of the Moon. This is caused by two separate motions of the Moon that take the same amount of time.

The Sun Lights the Moon You see the Moon because it reflects sunlight. As the Moon revolves around Earth, the Sun always lights one half of it. However, you don't always see the entire lighted part of the Moon. What you do see are phases, or different portions of the lighted part. **Moon phases,** illustrated in **Figure 10,** are the changing views of the Moon as seen from Earth.

New Moon and Waxing Phases New moon occurs when the Moon is positioned between Earth and the Sun. You can't see any of a new moon, because the lighted half of the Moon is facing the Sun. The new moon rises and sets with the Sun and never appears in the night sky.

Figure 10 The amount of the Moon's surface that looks bright to observers on Earth changes during a complete cycle of the Moon's phases.
Explain *what makes the Moon's surface appear so bright.*

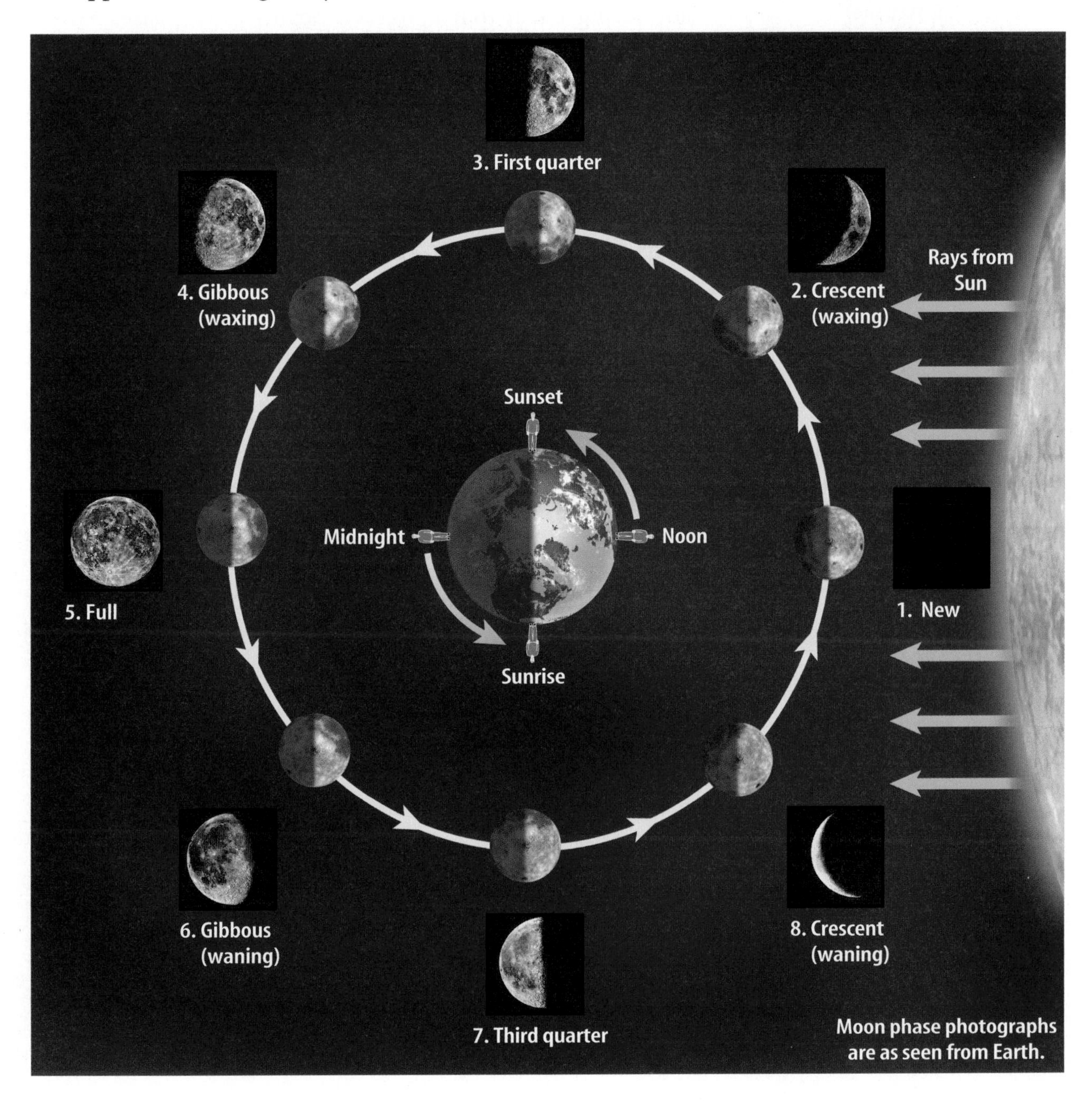

Topic: Moon Phases

Visit green.msscience.com for Web links to information about moon phases.

Activity When weather permits, take photos or make drawings of the Moon every three or four days. Record the date of each photo and make a poster showing the Moon's phases. How do your moon pictures correspond to what you viewed on the Web site?

Waxing Moon Shortly after new moon, more and more of the side facing Earth is lighted. The phases are said to be waxing, or growing in size. About 24 hours after new moon, a thin sliver on the side facing Earth is lighted. This phase is called waxing crescent. As the Moon continues its trip around Earth, half of that side is lighted. This phase is first quarter and occurs about a week after new moon.

The Moon's phases continue to wax through waxing gibbous (GIH bus) and then full moon—the phase when you can see all of the side facing Earth lighted. At full moon, Earth is between the Sun and the Moon.

Waning Moon After passing full moon, the amount of the side facing Earth that is lighted begins to decrease. Now the phases are said to be waning. Waning gibbous occurs just after full moon. Next comes third quarter when you can see only half of the side facing Earth lighted, followed by waning crescent, the final phase before the next new moon.

Reading Check *What are the waning phases of the Moon?*

The complete cycle of the Moon's phases takes about 29.5 days. However, you will recall that the Moon takes only 27.3 days to revolve once around Earth. **Figure 11** explains the time difference between these two lunar cycles. Earth's revolution around the Sun causes the time lag. It takes the Moon about two days longer to align itself again between Earth and the Sun at new moon.

Figure 11 It takes about two days longer for a complete Moon phase cycle than for the Moon to orbit Earth.
Explain *how the revolution of the Earth-Moon system around the Sun causes this difference in time.*

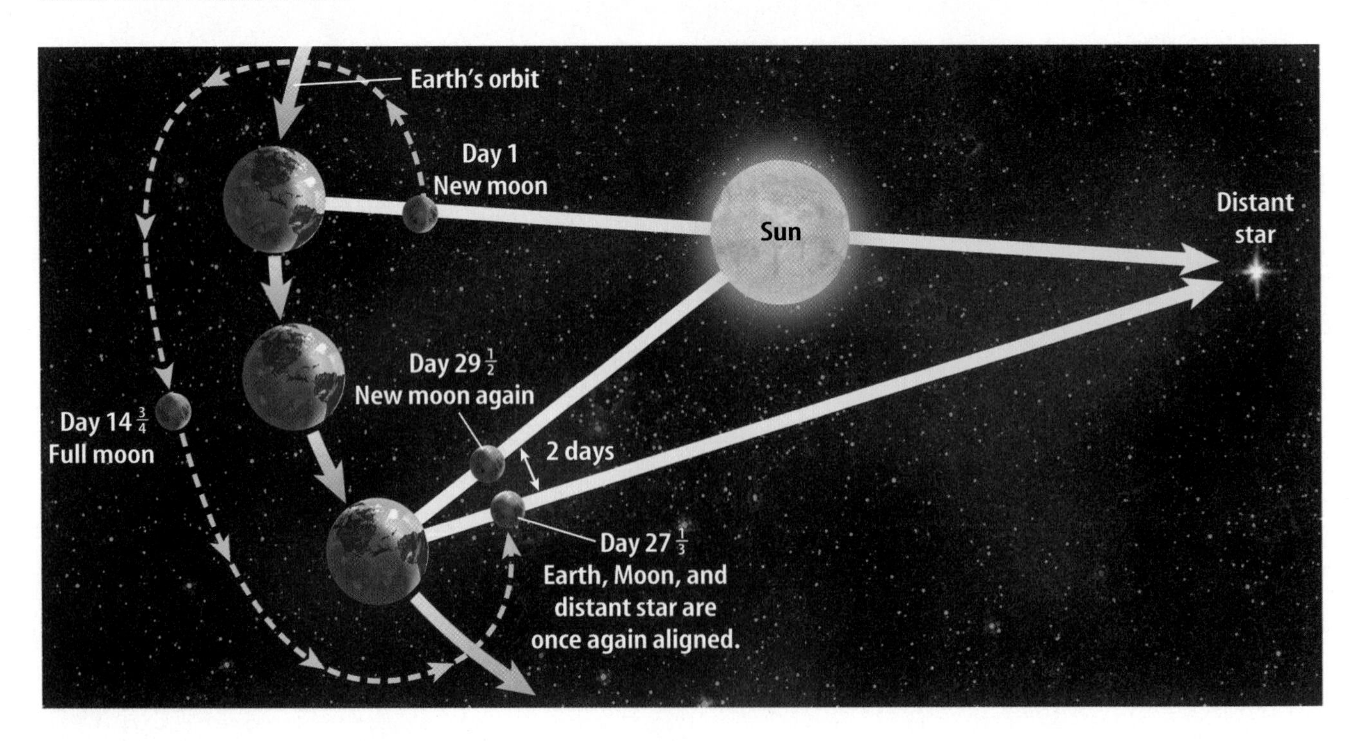

Eclipses

You can see other effects of the Moon's revolution than just the changes in its phases. Sometimes during new and full moon, shadows cast by one object will fall on another. While walking along on a sunny day, have you ever noticed how a passing airplane can cast a shadow on you? On a much larger scale, the Moon can do this too, when it lines up directly with the Sun. When this happens, the Moon can cast its shadow all the way to Earth. Earth also can cast its shadow onto the Moon during a full moon. When shadows are cast in these ways, eclipses occur.

Eclipses occur only when the Sun, the Moon, and Earth are lined up perfectly. Because the Moon's orbit is tilted at an angle from Earth's orbit, the Moon's shadow most often misses Earth, and eclipses happen only a few times each year.

Solar Eclipses During new moon, if Earth moves into the Moon's shadow, a **solar eclipse** occurs. As shown in **Figure 12,** the Moon blocks sunlight from reaching a portion of Earth's surface. Only areas on Earth in the Moon's umbra, or the darkest part of its shadow, experience a total solar eclipse. Those areas in the penumbra, or lighter part of the shadow, experience a partial solar eclipse. During a total solar eclipse, the sky becomes dark and stars can be seen easily. Because Earth rotates and the Moon is moving in its orbit, a solar eclipse lasts only a few minutes in any one location.

Figure 12 During a total solar eclipse, viewers on Earth within the Moon's umbra will see the Moon cover the Sun completely. Only the Sun's outer atmosphere is visible as a halo.

Viewers on Earth within the Moon's penumbra will see only a portion of the Sun covered. **WARNING:** *Never look directly at a solar eclipse. Only observe solar eclipses indirectly.*

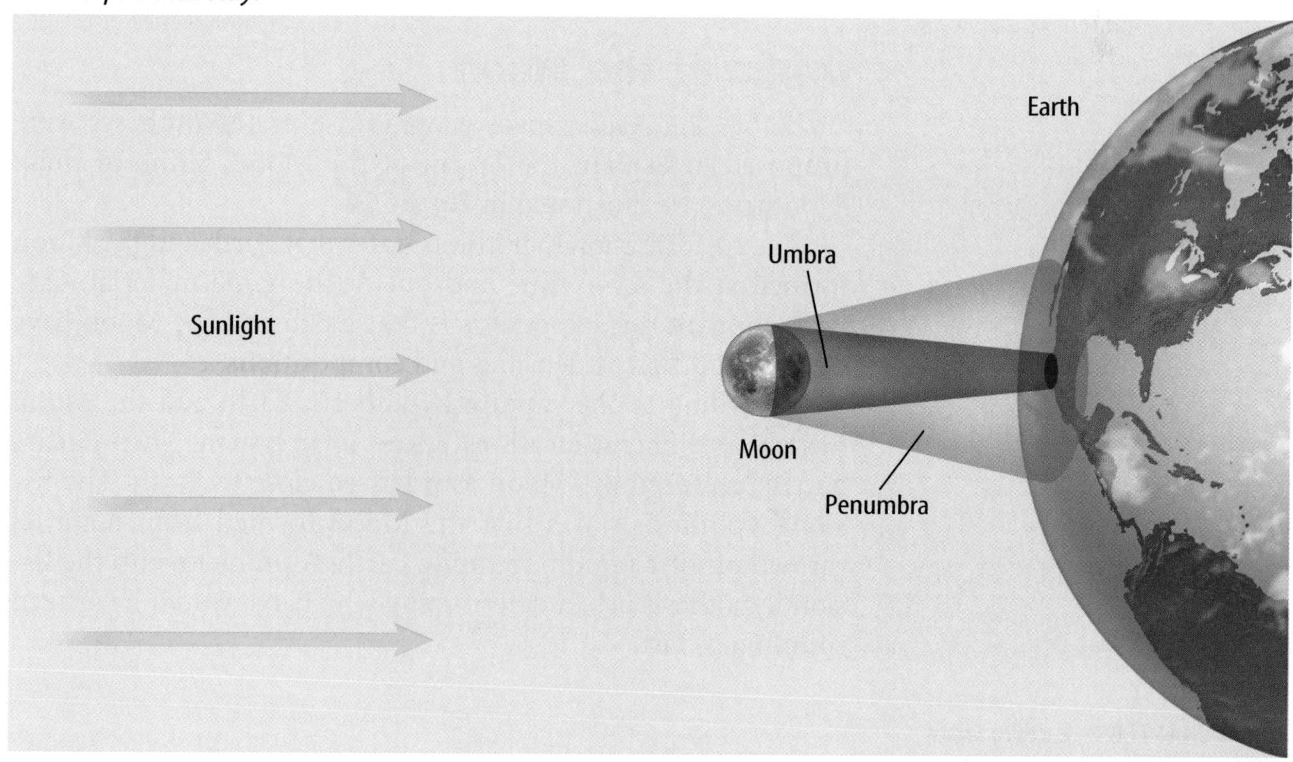

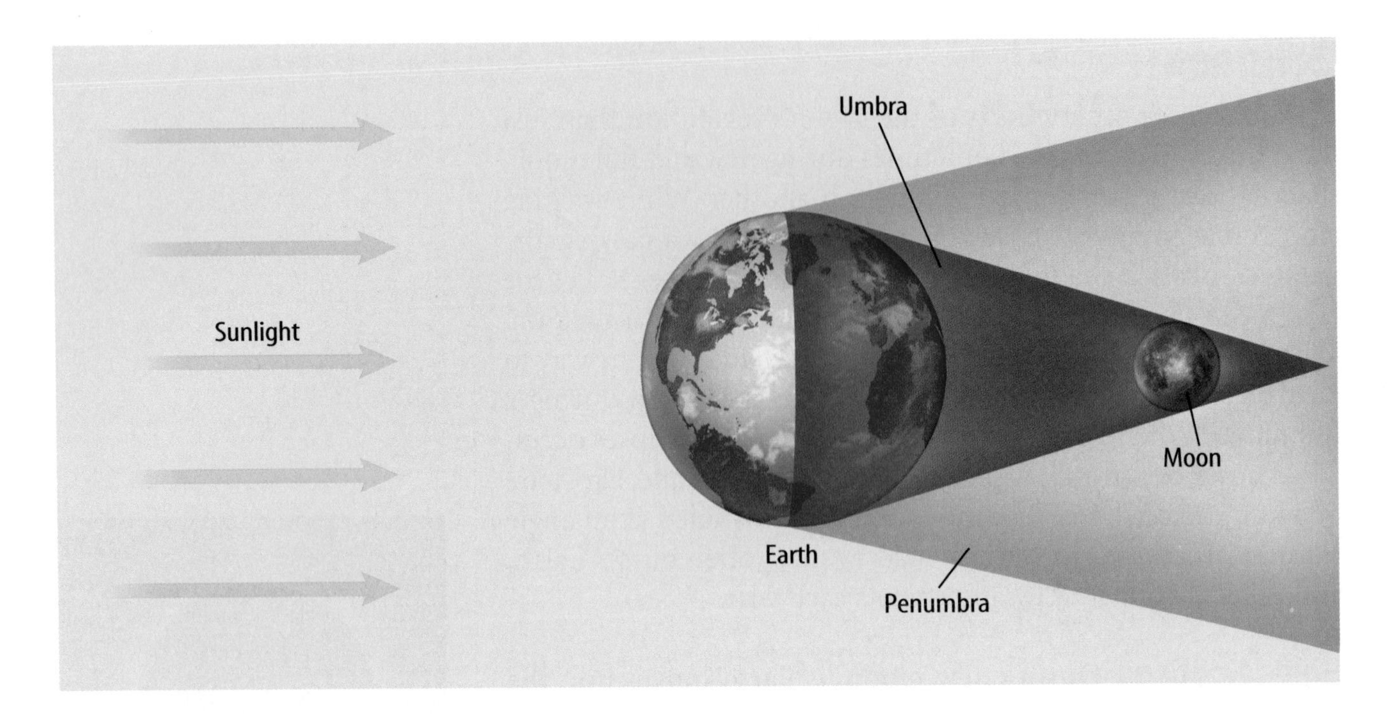

Figure 13 Lunar eclipses occur when the Sun, the Moon, and Earth line up so that Earth's shadow is cast upon a full moon. When the entire Moon is eclipsed, anyone on Earth who can see a full moon can see the lunar eclipse.

Lunar Eclipses A **lunar eclipse,** illustrated in **Figure 13,** occurs when the Sun, Earth, and the Moon are lined up so that the full moon moves into Earth's shadow. Direct sunlight is blocked from reaching the Moon. When the Moon is in the darkest part of Earth's shadow, a total lunar eclipse occurs.

During a total lunar eclipse, the full moon darkens. Because some sunlight refracts through Earth's atmosphere, the Moon appears to be deep red. As the Moon moves out of the umbra and into the penumbra, or lighter shadow, you can see the curved shadow of Earth move across the Moon's surface. When the Moon passes partly through Earth's umbra, a partial lunar eclipse occurs.

Origin of the Moon

Before the *Apollo* space program, several hypotheses were proposed to explain the origin of the Moon. Some of these hypotheses are illustrated in **Figure 14.**

The co-formation hypothesis states that Earth and the Moon formed at the same time and out of the same material. One problem with this hypothesis is that Earth and the Moon have somewhat different densities and compositions.

According to the capture hypothesis, Earth and the Moon formed at different locations in the solar system. Then Earth's gravity captured the Moon as it passed close to Earth. The fission hypothesis states that the Moon formed from material thrown off of a rapidly spinning Earth. A problem with the fission hypothesis lies in determining why Earth would have been spinning so fast.

NATIONAL GEOGRAPHIC VISUALIZING HOW THE MOON FORMED

Figure 14

Scientists have proposed several possible explanations, or hypotheses, to account for the formation of Earth's Moon. As shown below, these include the co-formation, fission, capture, and collision hypotheses. The latter—sometimes known as the giant impact hypothesis—is the most widely accepted today.

▲ CO-FORMATION Earth and the Moon form at the same time from a vast cloud of cosmic matter that condenses into the bodies of the solar system.

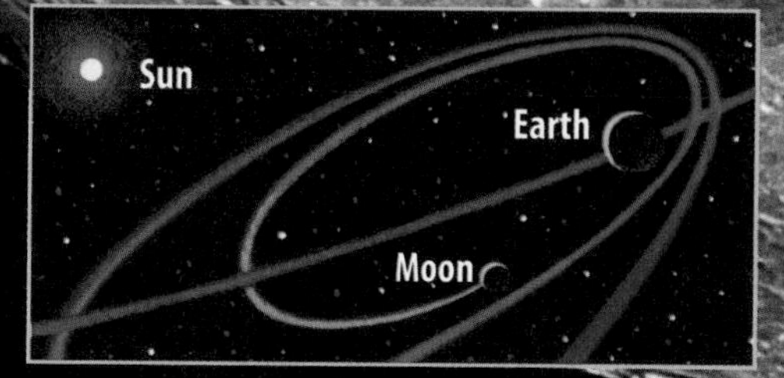

▲ CAPTURE Earth's gravity captures the Moon into Earth orbit as the Moon passes close to Earth.

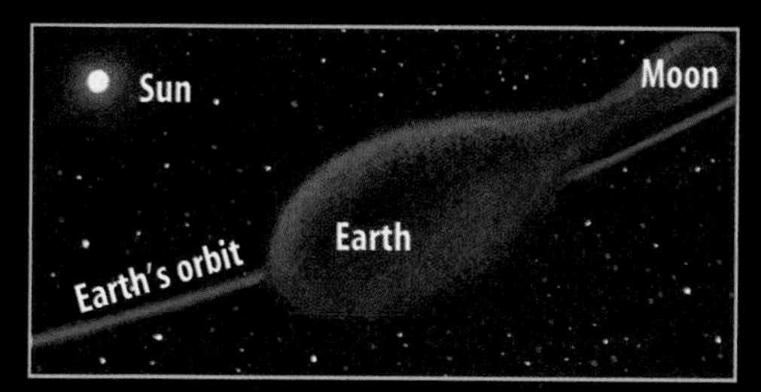

▲ FISSION A rapidly spinning molten Earth tears in two. The smaller blob of matter enters into orbit as the Moon.

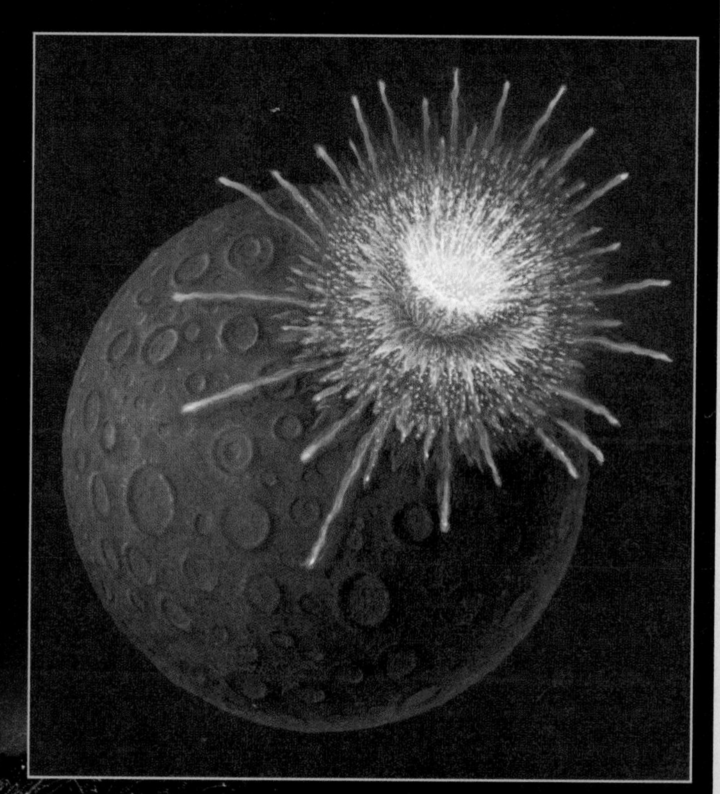

▲ COLLISION A Mars-sized body collides with the primordial Earth. The colossal impact smashes off sufficient debris from Earth to form the Moon.

Figure 15 Moon rocks collected during the *Apollo* space program provide clues about how the Moon formed.

Collision Hypothesis A lot of uncertainty still exists about the origin of the Moon. However, the collection and study of moon rocks, shown in **Figure 15,** brought evidence to support one recent hypothesis. This hypothesis, summarized in **Figure 14,** involves a great collision. When Earth was about 100 million years old, a Mars-sized space object may have collided with Earth. Such an object would have broken through Earth's crust and plunged toward the core. This collision would have thrown large amounts of gas and debris into orbit around Earth. Within about 1,000 years the gas and debris then could have condensed to form the Moon. The collision hypothesis is strengthened by the fact that Earth and the Moon have different densities. The Moon's density is similar to material that would have been thrown off Earth's mantle and crust when the object collided with Earth.

Earth is the third planet from the Sun. Along with the Moon, Earth could be considered a double planet. In the next section you will learn about other planets in the solar system. Some have properties similar to Earth's—others are different from Earth.

section 2 review

Summary

The Moon's Surface and Interior

- Surface features include highlands, maria, and craters.
- The Moon has an outer crust, a rigid mantle, a non-rigid mantle, and possibly a small iron core.

Motions of the Moon

- The Moon revolves around Earth and rotates on its axis once every 27.3 days.
- The varying portions of the Moon's near side that are lit by the Sun are called phases.

Eclipses

- A solar eclipse takes place when the Moon passes between Earth and the Sun.
- A lunar eclipse takes place when Earth passes between the Sun and the Moon.

Origin of the Moon

- Evidence obtained from moon rocks supports the collision hypothesis of lunar formation.

Self Check

1. **Name** the phase of the Moon that occurs when Earth is located between the Moon and the Sun.
2. **Explain** how maria formed on the Moon and how this name for them originated.
3. **Infer** Based on what you know about the Moon's crust, explain why there are fewer maria on the far side of the Moon.
4. **Explain** why the fact that Earth and the Moon have different densities favors the collision theory of lunar formation.
5. **Think Critically** Explain why more people observe a total lunar eclipse than a total solar eclipse.

Applying Skills

6. **Recognize Cause and Effect** How does the Moon's orbit around Earth cause the observed cyclical phases of the Moon? What role does the Sun play?
7. **Use a Spreadsheet** to compare the four hypotheses of lunar formation in terms of modern factual evidence of the Moon's density.

Science Online green.msscience.com/self_check_quiz

Viewing the Moon

The position of the Moon in the sky varies as the phases of the Moon change. Do you know when you might be able to see the Moon during daylight hours? How will viewing the Moon through a telescope be different from viewing it with the unaided eye?

Real-World Question

What features of the Moon are visible when viewed through a telescope?

Goals

- **Determine** when you may be able to observe the Moon during the day.
- Use a telescope to observe the Moon.
- **Draw** a picture of the Moon's features as seen through the telescope.

Materials

telescope
drawing pencils
drawing paper

Safety Precautions

WARNING: *Never look directly at the Sun. It can damage your eyes.*

Procedure

1. Using your own observations, books about astronomy, or other resource materials, determine when the Moon may be visible to you during the day. You will need to find out during which phases the Moon is up during daylight hours, and where in the sky you likely will be able to view it. You will also need to find out when the Moon will be in those phases in the near future.
2. **Observe** the Moon with your unaided eye. Draw the features that you are able to see.
3. Using a telescope, observe the Moon again. Adjust the focus of the telescope so that you can see as many features as possible.
4. **Draw** a new picture of the Moon's features.

Conclude and Apply

1. **Describe** what you learned about when the Moon is visible in the sky. If a friend wanted to know when to try to see the Moon during the day next month, what would you say?
2. **Describe** the differences between how the Moon looked with the naked eye and through the telescope. Did the Moon appear to be the same size when you looked at it both ways?
3. **Determine** what features you were able to see through the telescope that were not visible with the unaided eye.
4. **Observe** Was there anything else different about the way the Moon looked through the telescope? Explain your answer.
5. **Identify** some of the types of features that you included in your drawings.

Communicating Your Data

The next time you notice the Moon when you are with your family or friends, talk about when the Moon is visible in the sky and the different features that are visible.

section 3

Our Solar System

as you read

What You'll Learn

- **List** the important characteristics of inner planets.
- **Identify** how other inner planets compare and contrast with Earth.
- **List** the important characteristics of outer planets.

Why It's Important

Learning about other planets helps you understand Earth and the formation of our solar system.

Review Vocabulary
atmosphere: the gaseous layers surrounding a planet or moon

New Vocabulary
- solar system
- astronomical unit
- asteroid
- comet
- nebula

Size of the Solar System

Measurements in space are difficult to make because space is so vast. Even our own solar system is extremely large. Our **solar system,** illustrated in **Figure 16,** is composed of the Sun, planets, asteroids, comets, and other objects in orbit around the Sun. How would you begin to measure something this large? If you are measuring distances on Earth, kilometers work fine, but not for measuring huge distances in space. Earth, for example, is about 150,000,000 km from the Sun. This distance is referred to as 1 **astronomical unit,** or 1 AU. Jupiter, the largest planet in the solar system, is more than 5 AU from the Sun. Astronomical units can be used to measure distances between objects within the solar system. Even larger units are used to measure distances between stars.

Located at the center of the solar system is a star you know as the Sun. The Sun is an enormous ball of gas that produces energy by fusing hydrogen into helium in its core. More than 99 percent of all matter in the solar system is contained in the Sun.

Figure 16 Our solar system is composed of the Sun, planets and their moons, and smaller bodies that revolve around the Sun, such as asteroids and comets.

The asteroid belt is composed of rocky bodies that are smaller than the planets.
Evaluate *About how many AU are between the Sun and the middle of the asteroid belt?*

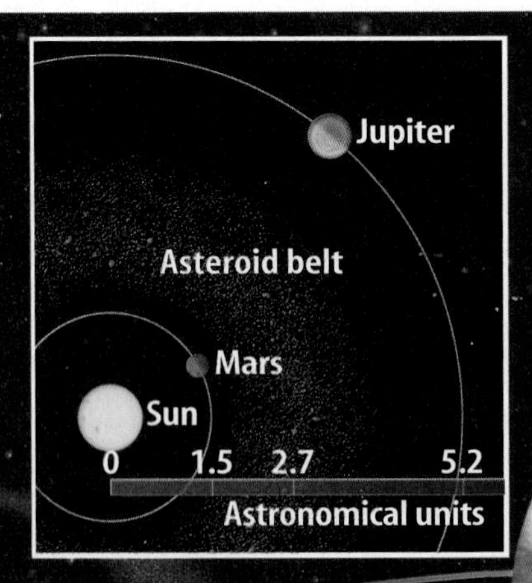

An Average Star Although the Sun is important to life on Earth, it is much like many other stars in the universe. The Sun is middle-aged and about average in the amount of light it gives off.

The Planets

The planets in our solar system can be classified as inner or outer planets. Inner planets have orbits that lie inside the orbit of the asteroid belt. The inner planets are mostly solid, rocky bodies with thin atmospheres compared with the atmospheres of outer planets. Outer planets have orbits that lie outside the orbit of the asteroid belt. Four of these are known as gas giants, and one is a small ice/rock planet that seems to be out of place.

Inner Planets

The inner planets are Mercury, Venus, Earth, and Mars. Known as the terrestrial planets, after the Latin word *terra* (earth), they are similar in size to Earth and are made up mainly of rock.

Mercury Mercury is the closest planet to the Sun. It is covered by craters formed when meteorites crashed into its surface. The surface of Mercury also has cliffs, as shown in **Figure 17**, some of which are 3 km high. These cliffs may have formed when Mercury's molten, iron-rich core cooled and contracted, causing the outer solid crust to shrink. The planet seems to have shrunk about 2 km in diameter. It has no atmosphere.

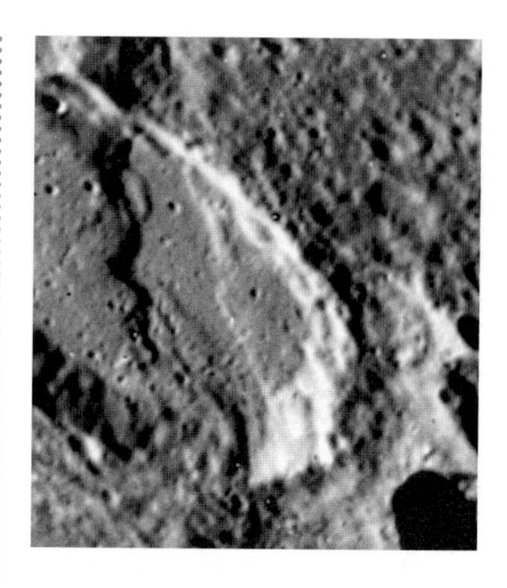

Figure 17 The Discovery Rupes Scarp is a huge cliff that may have formed as Mercury cooled and contracted.

Figure 18 Clouds in Venus's atmosphere are composed partly of sulfuric acid droplets.
Describe *What are clouds on Earth composed of?*

Venus Venus, the second inner planet from the Sun, shown in **Figure 18,** often has been referred to as Earth's twin, but only because of their similar sizes and masses. Otherwise, the surface conditions and atmospheres of Earth and Venus are extremely different. Thick clouds surround Venus and trap energy from the Sun, causing Venus's surface temperature to reach about 472°C. The process is similar to what occurs in a greenhouse.

Earth Earth is the third inner planet from the Sun. It is unique because surface temperatures enable water to exist in three states—solid, liquid, and gas. Ozone, a molecule of three oxygen atoms bound together, exists in the layer of Earth's atmosphere known as the stratosphere. This ozone protects life from the Sun's harmful ultraviolet radiation.

Mars Mars is the fourth inner planet from the Sun. It often is called the red planet. Iron oxide, the same material found in rust, exists in Mars's weathered surface rocks, giving the planet a reddish color. The Martian surface is shown in **Figure 19.** The rocks shown here are similar in composition to some volcanic rocks on Earth. The largest volcano in the solar system, Olympus Mons, is found on Mars.

Mars has two polar ice caps that change in size between Martian winter and summer. Recent data from *Mars Odyssey* indicate that both ice caps are made of frozen water covered by a layer of frozen carbon dioxide. Mars has two Moons, Phobos (FOH buhs) and Deimos (DI mos). Also, long channels exist on Mars. The channels on Mars are hypothesized to have been carved by flowing water sometime in the past. Mars's atmosphere, made up mostly of carbon dioxide with some nitrogen and argon, is much thinner than Earth's.

Planetology The space age has opened many new careers. One of these is planetology, the study of other planets. NASA advises those interested in this field to start with basic chemistry, physics, and math. Most planetary scientists first earn advanced degrees in areas such as geology, hydrology, meteorology, and biology and then apply their knowledge to the study of other planets.

Figure 19 Just as a metal toy left outside on Earth rusts, red rocks on Mars's surface show that iron in the rocks has combined with oxygen to form iron oxide.

Outer Planets

The outer planets are Jupiter, Saturn, Uranus, Neptune, and Pluto. Except for Pluto, they all are gaseous giant planets with dense atmospheres. The outer planets are mainly made up of light elements such as hydrogen and helium.

Jupiter The giant planet Jupiter, shown in **Figure 20,** is the largest planet in the solar system. It is the fifth planet from the Sun. Jupiter's atmosphere is made mostly of hydrogen and helium and contains many huge storms. The largest and most prominent of these storms is the Great Red Spot, which has raged for more than 300 years. It is about twice the width of Earth and rotates once every six days. With its 61 moons, Jupiter is like a miniature solar system.

Applying Science

What influences a planet's atmosphere?

The inner planets are small and dense and have thin atmospheres. The outer planets are large and gaseous. Do a planet's gravity and distance from the Sun affect what kinds of gases its atmosphere contains? Use your ability to interpret a data table to find out.

Identifying the Problem

The table below lists the main gases in the atmospheres of two inner and two outer planets. Each planet's atmosphere also contains many other gases, but these are only present in small amounts. Looking at the table, what conclusions can you draw? How do you think a planet's distance from the Sun and the size of the planet contribute to the kind of atmosphere it has?

Solving the Problem

1. What gases do the atmospheres of the inner and outer planets contain? What is special about Earth's atmosphere that makes it able to support modern life?
2. Can you think of any reasons why the outer planets have the gaseous atmospheres they do? *Hint: Hydrogen and Helium are the two lightest elements.*

Atmospheric Composition of the Planets	
Earth	78.1% nitrogen; 20.9% oxygen; traces of argon and carbon dioxide
Mars	95.3% carbon dioxide; 2.7% nitrogen; 1.6% argon; 0.13% oxygen; 0.08% carbon monoxide
Jupiter	86.1% molecular hydrogen; 13.8% helium
Uranus	82.5% molecular hydrogen; 15.2% helium; about 2.3% methane

Jupiter's atmosphere is dense because of its gravity and great distance from the Sun.

Saturn's rings include seven main divisions—each of which is composed of particles of ice and rock.

Figure 20 The first four outer planets—Jupiter, Saturn, Uranus, and Neptune are known also as the gas giants.

Jupiter's Moons The four largest moons of Jupiter are Io, Europa, Ganymede, and Callisto. They are called the Galilean satellites after Galileo Galilei, who discovered them in 1610. Volcanoes continually erupt on Io, the most volcanically active body in the solar system. An ocean of liquid water is hypothesized to exist beneath the cracked, frozen ice crust of Europa. Does this mean that life could exist on Europa? The National Aeronautics and Space Administration (NASA) is studying a mission to launch an orbiting spacecraft in 2008 to study this moon.

Topic: The Planets
Visit green.msscience.com for Web links to information about the planets.

Activity Decide which planet interests you most and make a list of three questions you would like to ask an astronomer about it.

Saturn The next outer planet is Saturn, shown in **Figure 20.** The gases in Saturn's atmosphere are made up in large part of hydrogen and helium. Saturn is the sixth planet from the Sun. It often is called the ringed planet because of its striking ring system. Although all of the gaseous giant planets have ring systems, Saturn's rings are by far the most spectacular. Saturn is known to have seven major ring divisions made up of hundreds of smaller rings. Each ring is made up of pieces of ice and rock.

Saturn has at least 31 moons, the largest of which is Titan. The atmosphere surrounding Titan is denser than the atmospheres of either Earth or Mars. The environment on Titan might be similar to the environment on Earth before oxygen became a major atmospheric gas.

The bluish-green color of Uranus is thought to be caused by methane in its atmosphere.

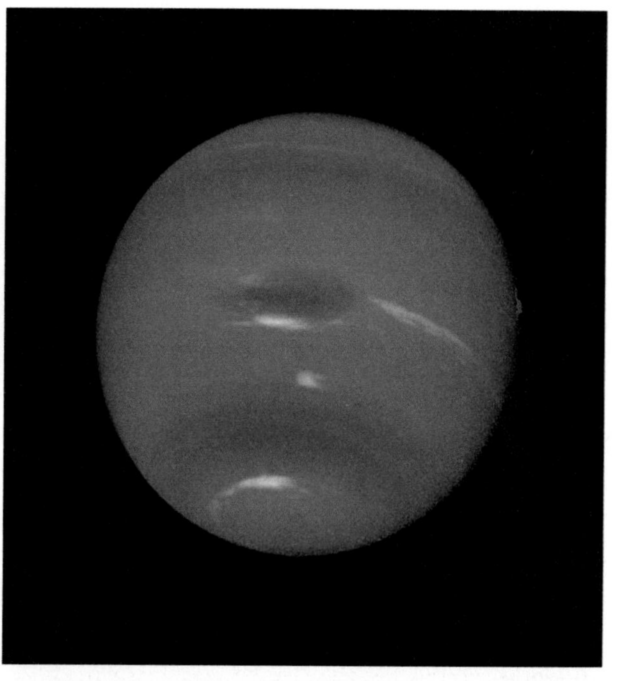

Like Uranus, the blue color of Neptune is also thought to be caused by methane. The Great Dark Spot was a storm system similar in size to the diameter of Earth.

Uranus Shown in **Figure 20,** the seventh planet from the Sun is Uranus. The atmosphere of Uranus, made up mostly of hydrogen, also contains helium and methane. The methane gives the planet a distinctive, bluish-green color. This is because methane gas reflects blue light and absorbs red light. Uranus is thought to have at least 27 moons. However, additional satellites might exist.

Neptune The eighth planet from the Sun is Neptune. It is thought that Neptune's atmosphere of hydrogen, helium, and methane gradually changes into a slushlike layer, comprised partially of water and other melted ices. Toward the interior, this slushy material is thought to change into an icy solid. In turn, this icy layer may surround a central, rocky core that is about the size of Earth.

As with Uranus, the methane in Neptune's atmosphere gives the planet its bluish color, as shown in **Figure 20.** Winds in the gaseous portion of Neptune exceed speeds of 2,400 km per hour—faster than winds on any other planet.

At least eleven natural satellites of Neptune have been discovered so far and many more probably exist. The largest of these, Triton, has great geysers that shoot gaseous nitrogen into space. A lack of craters on Triton's surface suggests that the surface of Triton is fairly young.

Mini LAB

Interpreting Your Creature Feature

Procedure

1. Select any one of the planets or moons in our solar system except Earth.
2. Research its surface conditions.
3. Imagine a life-form that might have developed on your chosen planet or moon. Be sure to indicate how your creature would eat, breathe, and reproduce.

Analysis

1. How is this creature protected from its environment?
2. How does your creature obtain nourishment for survival?

Try at Home

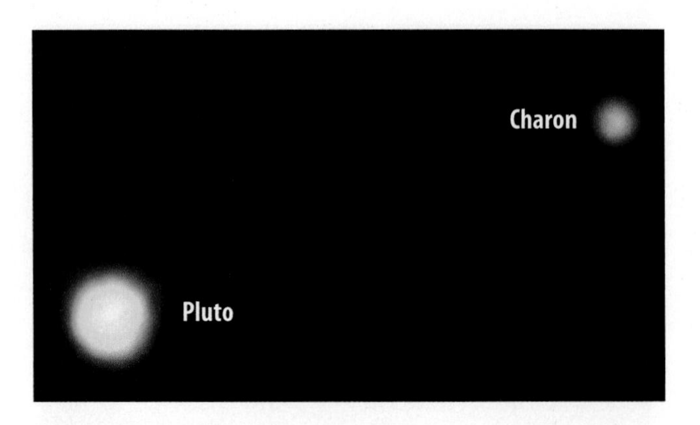

Figure 21 Pluto and its moon, Charon, are so close together that they usually can't be detected separately using ground-based telescopes. Because of this, Charon was discovered nearly 50 years after Pluto.

Pluto Pluto, shown in **Figure 21,** is so far from the Sun that it has completed less than 20 percent of one revolution around the Sun since its discovery in 1930. Pluto is totally different from the other outer planets. It is a planet that is thought to be made partly of ice and partly of rock. Apparently, a frozen layer of methane, nitrogen, and carbon monoxide sometimes covers Pluto's surface. At times, however, when Pluto is at its closest point to the Sun, these materials thaw into their gaseous states and rise, forming a temporary atmosphere. The surface of Charon, Pluto's moon, appears to be covered by water ice.

Other Objects in the Solar System

Other objects that exist in the solar system include asteroids, comets, and meteoroids. **Asteroids** are small, rocky objects that mostly lie in a belt located between the orbits of Mars and Jupiter. The asteroid belt is used by astronomers as a dividing line that separates the inner and outer planets. Jupiter's tremendous gravity probably kept a planet from forming from the matter contained in the asteroid belt.

Comets are made mainly of rocky particles and water ice. As their orbits approach the Sun, parts of comets vaporize and form tails. Comet tails, shown in **Figure 22,** always point away from the Sun. Almost all of the solar system's comets are located in the Kuiper Belt and the Oort Cloud. The Kuiper Belt is located beyond Neptune's orbit, and the Oort Cloud is located far beyond Pluto's orbit.

When comets break up, some of the resulting particles remain in orbit. When asteroids collide, small pieces break off. Both of these processes produce small objects in the solar system known as meteoroids. If meteoroids enter Earth's atmosphere, they are called meteors, and when they fall to Earth, they are called meteorites.

How are meteoroids related to meteorites?

Figure 22 The brilliant head of a comet, called a coma, glows as its ices vaporize upon approaching the Sun.

Origin of the Solar System

How did the solar system begin? One hypothesis is that the Sun and all the planets and other objects condensed from a large cloud of gas, ice, and dust about 5 billion years ago, as illustrated in **Figure 23.** This large **nebula** (NEB yuh luh), or cloud of material, was rotating slowly in space. Shock waves, perhaps from a nearby exploding star, might have caused the cloud to start condensing. As it condensed, it started rotating faster and flattened into a disk. Most of the condensing material was pulled by gravity toward the center to form an early Sun. The remaining gas, ice, and dust in the outer areas of the nebula condensed, collided, and stuck together forming planets, moons, and other components of the solar system. Conditions in the inner part of the cloud caused small, solid planets to form, whereas conditions in the outer part were better for the formation of gaseous giant planets. Comets are thought to be made up of material left over from the original condensation of the cloud.

Figure 23 The solar system is thought to have formed from a cloud of rotating gases and dust particles.

section 3 review

Summary

Solar System

- An astronomical unit or AU is used to measure distance in the solar system.
- The solar system includes nine planets and other objects, such as asteroids and comets.

Planets

- Mercury, Venus, Earth, and Mars are the inner planets.
- They are similar in size to Earth.
- Jupiter, Saturn, Uranus, Neptune, and Pluto are the outer planets.

Origin of the Solar System

- One hypothesis is that the Sun, planets, and other objects formed from a large cloud of gas, ice, and dust about 5 billion years ago.

Self Check

1. **Explain** why astronomers do not use kilometers to measure distances in the solar system.
2. **Explain** why Venus is called Earth's twin planet.
3. **Explain** how 3-km-high cliffs could have formed on the surface of Mercury.
4. **Think Critically** How can the composition of the planets in the solar system be explained by the hypothesis described on this page?

Applying Math

5. **Calculate Ratios** Research the equatorial diameters of Earth's Moon and the four Galilean moons of Jupiter. Calculate the ratio of each moon's diameter to that of Earth's Moon.

Model and Invent

The Slant of the Sun's Rays

Goals

- **Design** a model for simulating the effect of changing angles of the Sun's rays on Earth's surface temperatures.

Materials

shallow baking pans lined with cardboard
**paper, boxes, or box lids*
thermometers
wood blocks
**bricks or textbooks*
protractor
clock
**stopwatch*
**Alternate materials*

Data Source

Copy the data table into your Science Journal and fill it in, providing angles of the Sun's rays for your area. Go to green.msscience.com to collect this data.

Safety Precautions

Use thermometers as directed by teacher. Do not use "shake down" lab thermometers.

WARNING: *Never look directly at the Sun at any time during your experiment.*

During winter in the northern hemisphere, the north pole is positioned away from the Sun. This causes the angle of the Sun's rays striking Earth to be smaller in winter than in summer, and there are fewer hours of sunlight. The reverse is true during the summer months. The Sun's rays strike Earth at higher angles that are closer to 90°.

Real-World Question

How does the angle of the Sun's rays affect Earth's surface temperature?

Angle of the Sun's Rays at Noon at Your Latitude

Date	Angle
December 22 (winter solstice)	
January 22	
February 22	
March 21 (vernal equinox)	
April 21	
May 21	
June 21 (summer solstice)	

Equinoxes
Mar. 21
Sept. 22
Summer
solstice
June 21
Winter
solstice
Dec. 22
Vertical
pole
Dallas, Texas 32°45' N 12:00 P.M.
79°
57°
34°

Using Scientific Methods

Plan the Model

1. **Design** a model that will duplicate the angle of the Sun's rays during different seasons of the year.
2. Choose the materials you will need to construct your model. Be certain to provide identical conditions for each angle of the Sun's rays that you seek to duplicate.

Check the Model Plans

1. **Present** your model design to the class in the form of diagrams, poster, slide show, or video. Ask your classmates how your group's model design could be adjusted to make it more accurate.
2. Decide on a location that will provide direct sunlight and will allow your classmates to easily observe your model.

Make a Model

1. Create a model that demonstrates the effects different angles of the Sun's rays have on the temperature of Earth's surface.
2. **Demonstrate** your model during the morning, when the Sun's rays will hit the flat tray at an angle similar to the Sun's rays during winter solstice. Measure the angle of the Sun's rays by laying the protractor flat on the tray. Then sight the angle of the Sun's rays with respect to the tray.
3. Tilt other trays forward to simulate the Sun's rays striking Earth at higher angles during different times of the year.

Conclude and Apply

1. **Determine** Which angle had the greatest effect on the surface temperature of your trays? Which angle had the least effect?
2. **Predict** how each of the seasons in your area would change if the tilt of Earth's axis changed suddenly from 23.5 degrees to 40 degrees.

Demonstrate your model for your class. Explain how your model replicated the angle of the Sun's rays for each of the four seasons in your area.

TIME SCIENCE AND Society

SCIENCE ISSUES THAT AFFECT YOU!

Collision Course

Will an asteroid collide with Earth?

Asteroids have been the basis for several disaster movies. Can asteroids really threaten Earth? "Absolutely!" say many scientists who study space. In fact, asteroids have hit our planet many times.

Earth is scarred with about 120 recognizable craters and others may be covered by erosion, plant growth, and other processes. Visitors to Meteor Crater, Arizona, can see a 1.2-km-wide depression caused by an asteroid that impacted Earth about 49,000 years ago. A much older crater lies in Mexico's Yucatan Peninsula. This depression is about 195 km wide and was created about 65 million years ago.

Some scientists believe the Yucatan asteroid created a giant dust cloud that blocked the Sun's rays from reaching Earth for about six months and led to freezing temperatures. This would have put an end to much of Earth's early plant life, and may have led to the extinction of the dinosaurs and many other species.

Meteor Crater, near Winslow, Arizona, was formed about 49,000 years ago. It is about 200 m deep.

Rocks in Space

Can we protect ourselves from such an impact? Astronomer/geologist Eugene Shoemaker thinks so, and is responsible for alerting the world to the dangers of asteroid impact. In 1973, he and geologist Eleanor Helin began the first Near Earth Objects (NEO) watch at the Mount Palomar Observatory in California. But few others were concerned.

Then, in 1996, all that changed. An asteroid, about 0.5 km wide, came within 450,800 km of Earth. Scientists said this was a close call! Today, groups of scientists are working on creating systems to track NEOs. As of 2003, they recorded 2,565 NEOs. Of those, nearly 700 were at least 1.0 km in diameter.

Don't worry, though: there is little chance of an asteroid hitting Earth anytime soon. When an asteroid hits Earth's atmosphere, it is usually vaporized completely. Just in case, some physicists and astronomers are thinking about ways to defend our planet from NEOs.

Brainstorm **Working in small groups, come up with as many ways as you can to blast an asteroid to pieces or make it change course before hitting Earth. Present your reports to the class.**

For more information, visit green.msscience.com/time

Reviewing Main Ideas

Section 1 Earth's Motion

1. Earth's shape is nearly spherical.
2. Earth's motions include rotation around its axis and revolution around the Sun.
3. Earth's rotation causes day and night. Its tilt and revolution cause the seasons.

Section 2 Earth's Moon

1. Surface features on the Moon include maria, craters, and lunar highlands.
2. The Moon rotates once and revolves around Earth once in 27.3 days.
3. Phases of Earth's Moon, solar eclipses, and lunar eclipses are caused by the Moon's revolution around Earth.
4. One hypothesis concerning the origin of Earth's Moon is that a Mars-sized body collided with Earth, throwing off material that later condensed to form the Moon.

Section 3 Our Solar System

1. The solar system includes the Sun, planets, moons, asteroids, comets, and meteoroids.
2. Planets can be classified as inner or outer.
3. Inner planets are small and rocky with thin atmospheres (or none). Outer planets are generally large and gaseous with thick atmospheres.
4. One hypothesis on the origin of the solar system states that it condensed from a large cloud of gas, ice, and dust.

Visualizing Main Ideas

Copy and complete the following concept map to complete the moon phase cycle.

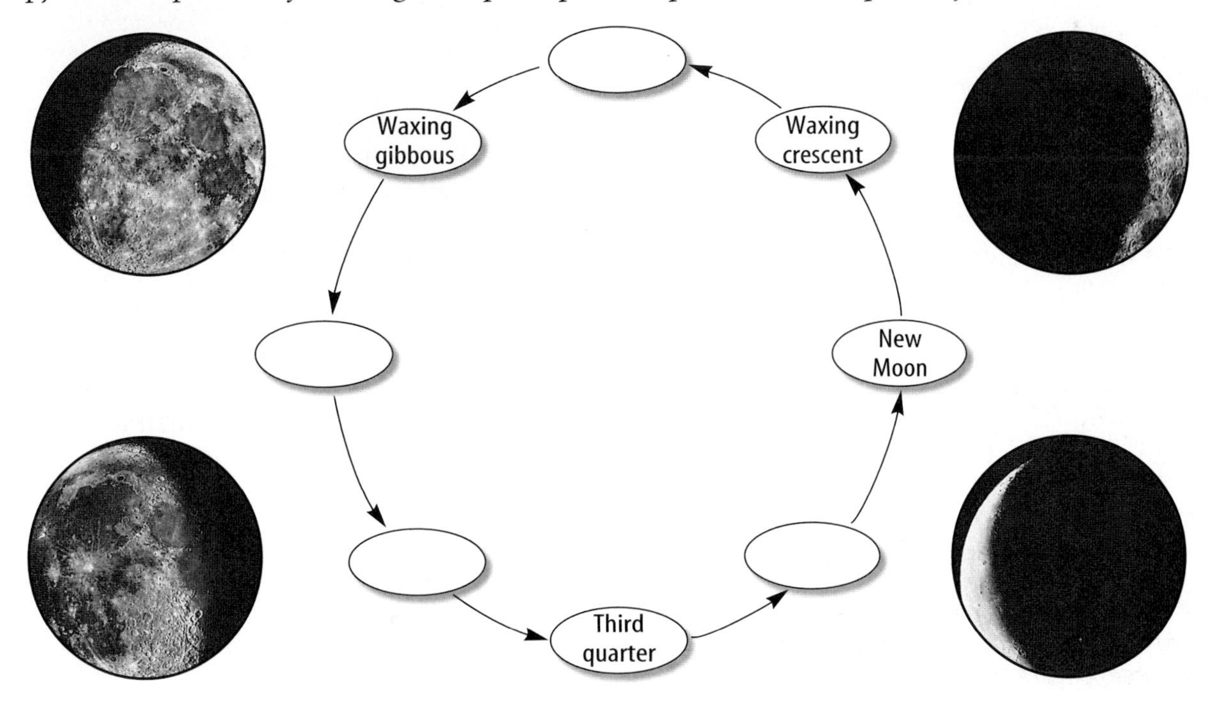

Using Vocabulary

asteroid p. 200
astronomical unit p. 194
axis p. 180
comet p. 200
crater p. 185
equinox p. 183
lunar eclipse p. 190
moon phase p. 187
nebula p. 201
orbit p. 181
revolution p. 181
rotation p. 180
solar eclipse p. 189
solar system p. 194
solstice p. 182

Fill in the blanks with the correct word or words.

1. A(n) ________ is a chunk of rock that circles the sun between the orbits of Mars and Jupiter.
2. The motion that describes Earth's orbit around the Sun is called ________.
3. The times when the Sun reaches its greatest distance north or south of the equator is called a(n) ________.
4. When Earth passes between the Sun and the Moon, a(n) ________ can result.
5. The changing views of the Moon seen from Earth are called ________.

Checking Concepts

Choose the word or phrase that best answers the question.

6. Which motion refers to Earth's spinning on its axis?
   A) rotation
   B) waxing
   C) revolution
   D) waning
7. What occurs twice each year when Earth's tilt is neither toward nor away from the Sun?
   A) orbit **C)** solstice
   B) equinox **D)** axis
8. What is the imaginary line around which Earth spins?
   A) orbit **C)** solstice
   B) equinox **D)** axis
9. Which moon surface features probably formed when lava flows filled large basins?
   A) maria **C)** highlands
   B) craters **D)** volcanoes
10. Meteorites that strike the Moon's surface cause which surface feature?
    A) maria **C)** highlands
    B) craters **D)** volcanoes
11. How long is the Moon's period of revolution?
    A) 27.3 hours **C)** 27.3 days
    B) 29.5 hours **D)** 29.5 days
12. How long does it take for the Moon to rotate once on its axis?
    A) 27.3 hours **C)** 27.3 days
    B) 29.5 hours **D)** 29.5 days

Use the illustration below to answer question 13.

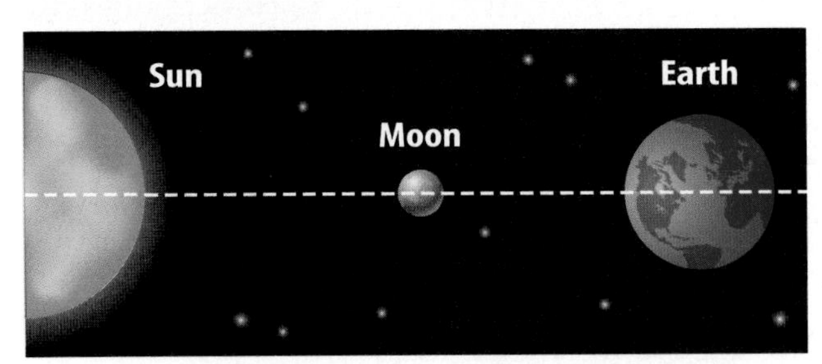

13. What could result from the alignment shown above?
    A) lunar eclipse **C)** full moon
    B) solar eclipse **D)** waxing crescent
14. Which planet is most like Earth in size and mass?
    A) Mercury **C)** Saturn
    B) Mars **D)** Venus
15. Europa is a satellite of which planet?
    A) Uranus **C)** Jupiter
    B) Saturn **D)** Mars

Science Online green.msscience.com/vocabulary_puzzlemaker

Thinking Critically

16. **Compare and contrast** the inner and outer planets.
17. **Explain** why more maria are found on the near side of the Moon than on the far side.
18. **Explain** why scientists hypothesize that life might exist on Europa.
19. **Describe** the Sun's positions relative to the equator during winter and summer solstices and equinox.
20. **Classify** A new planet is found circling the Sun, and you are given the job of classifying it. The new planet has a thick, dense atmosphere and no apparent solid surface. It lies beyond the orbit of Pluto. How would you classify this newly discovered planet?
21. **Make and Use Tables** Copy and complete the table of outer planets. Show how many satellites each planet has and what gases are found in each planet's atmosphere.

Planetary Facts

Planet	Number of Satellites	Major Atmospheric Gases
Jupiter		
Saturn		
Uranus	Do not write in this book.	
Neptune		
Pluto		

22. **Sequence** the phases of the moon starting and ending with the new moon phase, and explain why we can see only these lighted portions of the Moon. Consider the fact that the Sun lights one half of the Moon at all times.

Performance Activities

23. **Observe** the moons of Jupiter. You can see the four Galilean moons using binoculars of at least 7× power. Check the newspaper, almanac, or other reference to see whether Jupiter is visible in your area and at what times. Depending on where they are in their orbits and how clear a night it is, you may be able to see all of them.
24. **Display** Design a chart showing the nine planets, using the appropriate colors for each. Include rings and moons and any other characteristics mentioned in this chapter.

Applying Math

Use the table below to answer question 25.

Some Planet Diameters

Planet	Kilometers (km)
Earth	12,700
Jupiter	143,000
Saturn	120,000
Uranus	50,800
Neptune	50,450

25. **How Giant Are They?** Using the data in the table above calculate the ratios between the diameter of Earth and each of the gas giants.
26. **Jupiter's Orbit** The orbit of Jupiter lies about 4.2 AU from that of Earth. Express this distance in kilometers.
27. **Martian Mass** The mass of Mars is about 0.11 times that of Earth. Assuming that the mass of Earth is 5.98×10^{24} kg, calculate the mass of Mars.

chapter 7 Standardized Test Practice

Part 1 Multiple Choice

Record your answers on the answer sheet provided by your teacher or on a sheet of paper.

Use the illustration below to answer questions 1 and 2.

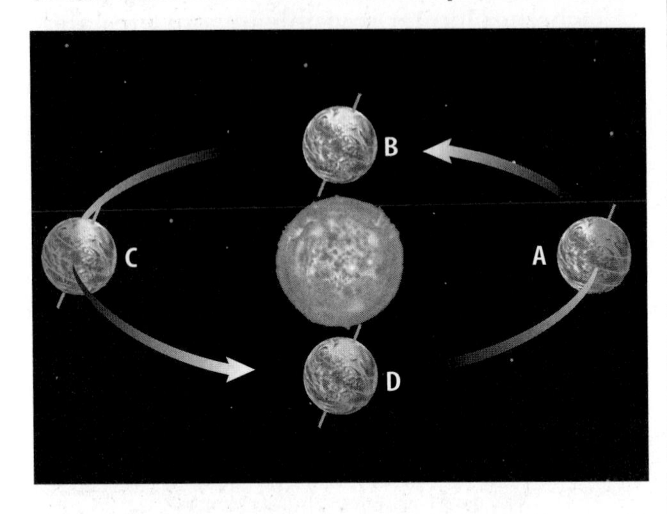

1. At which point in Earth's orbit does summer begin in the southern hemisphere?
   A. A **C.** C
   B. B **D.** D

2. Which two positions indicate equinoxes, when day and night lengths are nearly equal all over the world?
   A. A and C **C.** B and D
   B. B and C **D.** A and D

3. Which of these instruments monitors moonquakes to yield information about the Moon's interior?
   A. telescope **C.** seismograph
   B. spectroscope **D.** centrifuge

Test-Taking Tip

Consider All Choices Be sure you understand the question before you read the answer choices. Make special note of words like NOT or EXCEPT. Read and consider all the answer choices before you mark your answer sheet.

Question 2 Be sure that both letters represent equinoxes.

4. Which of these is referred to as a terrestrial planet?
   A. Saturn **C.** Neptune
   B. Uranus **D.** Venus

Use the illustration below to answer questions 5 and 6.

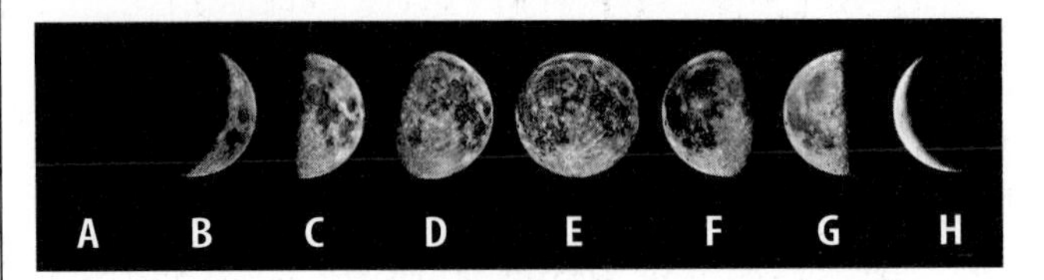

5. Which letter corresponds to the moon phase waning crescent?
   A. D **C.** B
   B. H **D.** F

6. During which moon phase could a solar eclipse occur?
   A. A **C.** C
   B. E **D.** G

7. Which of these planets has a large, permanent storm in its atmosphere known as the Great Red Spot?
   A. Saturn **C.** Jupiter
   B. Venus **D.** Neptune

8. Between the orbits of which two planets is the asteroid belt found?
   A. Earth and Mars
   B. Mars and Jupiter
   C. Jupiter and Saturn
   D. Mercury and Venus

9. Which of the following terms is used to describe the motion of Earth as it orbits the Sun?
   A. revolution **C.** rotation
   B. ellipse **D.** solstice

10. Which is a satellite of Saturn?
    A. Charon **C.** Phobos
    B. Europa **D.** Titan

Part 2 | Short Response/Grid In

Record your answers on the answer sheet provided by your teacher or on a sheet of paper.

11. Explain how Earth's shape differs somewhat from a perfect sphere.

12. Explain how mass and the distance of an object from other objects affect its gravitational force.

13. Why does the same side of the Moon always face Earth?

Use the illustration below to answer questions 14 and 15.

14. During what type of eclipse might you observe something like the image in this photograph? What causes the dark circle in the center of the image?

15. Identify the three solar system members which line up to cause this type of eclipse. Describe the order in which they align.

16. Define the unit used to describe distances in the solar system. Why is this unit used rather than kilometers?

17. Compare the atmospheres of Venus, Earth, and Mars.

18. Why do Neptune and Uranus appear blue or bluish-green?

19. How can lunar craters help to determine the relative age of a particular area on the Moon's surface?

Part 3 | Open Ended

Record your answers on a sheet of paper.

20. Explain how the tilt of Earth's axis causes warmer summer than winter temperatures in the northern hemisphere.

21. Scientists have proposed several hypotheses to explain the Moon's origin. Choose two hypotheses and describe how they differ.

22. Describe two observations which can be made with the unaided eye from Earth that show Earth is spherical.

Use the illustration below to answer question 23.

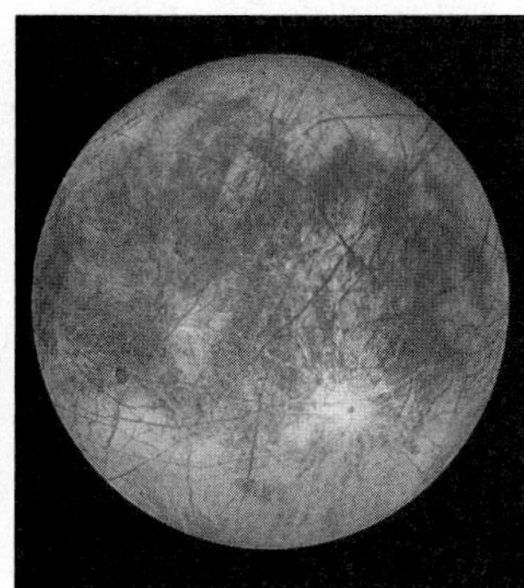

23. The photos show Jupiter's moons Io and Europa, respectively. Could life exist on these moons? Why or why not?

24. Compare and contrast the motions of Earth and the Moon in space.

25. Imagine your spacecraft has landed somewhere on the planet Mars. Describe your observations of the surface.

26. Explain the differences among meteoroids, meteors, and meteorites.

27. Describe the composition of comets and explain what happens as they come closer to the Sun.

28. Describe the asteroid belt and explain what may have caused it to form.

How Are Cargo Ships & Cancer Cells Connected?

Below, a present-day cargo ship glides through a harbor in New Jersey. In 1943, during World War II, another cargo ship floated in an Italian harbor. The ship carried a certain type of chemical. When a bomb struck the ship, the chemical accidentally was released. Later, when doctors examined the sailors who were exposed to the chemical, they noticed that the sailors had low numbers of white blood cells. The chemical had interfered with the genetic material in certain cells, preventing the cells from reproducing. Since cancer cells (like the ones at lower left) are cells that reproduce without control, scientists wondered whether this chemical could be used to fight cancer. A compound related to the chemical became the first drug developed to fight cancer. Since then, many other cancer-fighting drugs have been developed.

unit projects

Visit green.msscience.com/unit_project to find project ideas and resources.
Projects include:

- **History** Create a Moment in History to present the lives of the two famous scientists credited with discovering the structure of DNA.
- **Technology** Using online resources, research the adaptations of a species and design a flow chart showing four changes through time.
- **Model** Using a penny and a three-generation family tree, determine the genotype and phenotype of each generation.

WebQuest *New Research on Cells* provides an opportunity to explore current research on cells, and how different cells work in the human body.

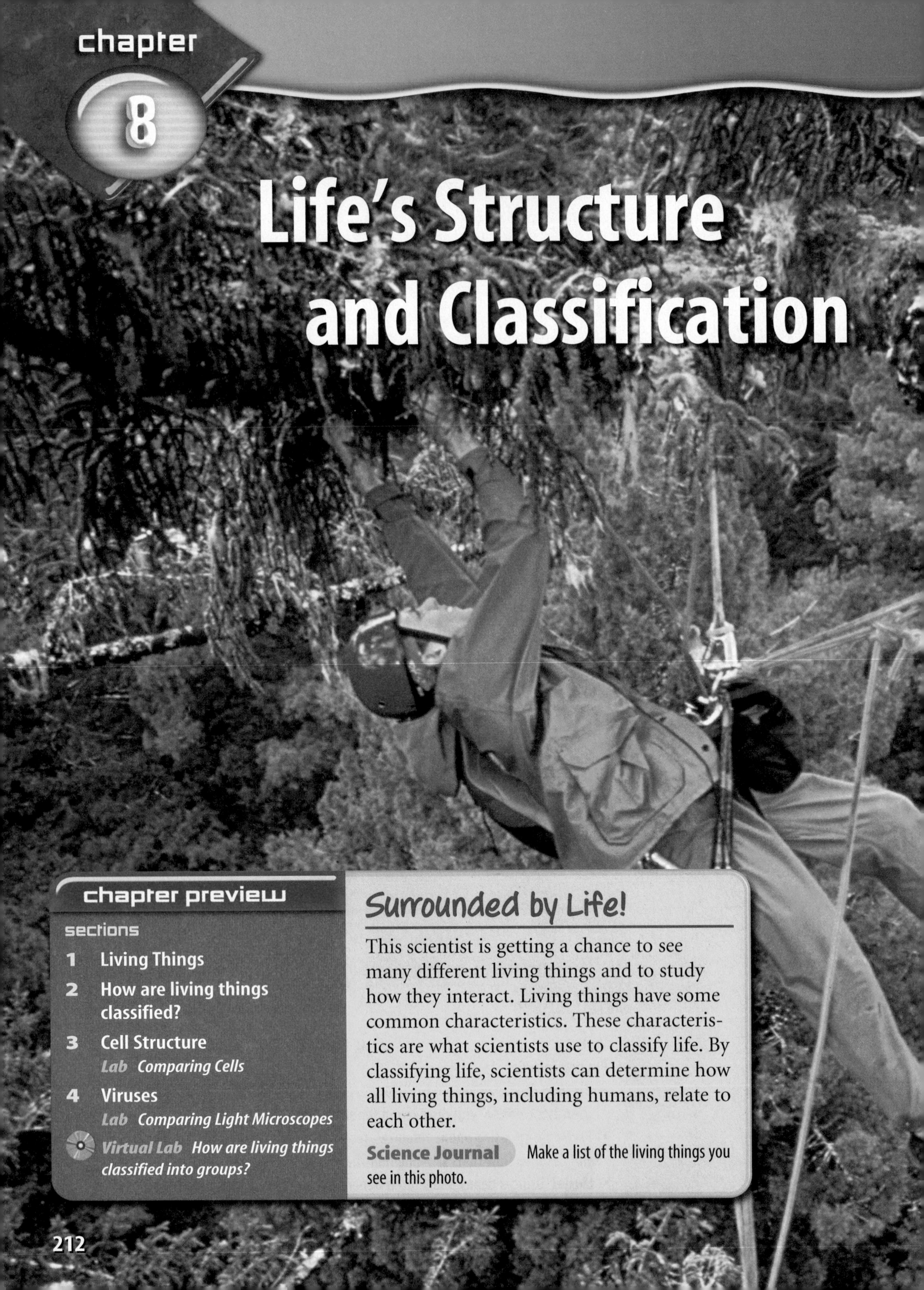

chapter

8

Life's Structure and Classification

chapter preview

sections

1 Living Things
2 How are living things classified?
3 Cell Structure
   Lab *Comparing Cells*
4 Viruses
   Lab *Comparing Light Microscopes*
   Virtual Lab *How are living things classified into groups?*

Surrounded by Life!

This scientist is getting a chance to see many different living things and to study how they interact. Living things have some common characteristics. These characteristics are what scientists use to classify life. By classifying life, scientists can determine how all living things, including humans, relate to each other.

Science Journal Make a list of the living things you see in this photo.

Start-Up Activities

Classifying Life

Life scientists discover, describe, and name hundreds of organisms every year. What methods do they use to decide if an insect is more like a grasshopper or a beetle?

1. Observe an insect collection or a collection of other organisms.
2. Decide which feature could be used to separate the organisms into two groups, then sort the organisms into the two groups.
3. Continue to make new groups using different features until each organism is in a category by itself.
4. **Think Critically** List the features you would use to classify the living thing in the photo above. How do you think scientists classify living things?

Preview this chapter's content and activities at green.msscience.com

Life's Structure and Classification Make the following Foldable to help you understand the vocabulary terms in this chapter. Use this Foldable to review for the chapter test.

STEP 1 **Fold** a sheet of notebook paper in half lengthwise.

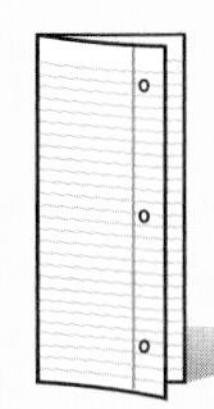

STEP 2 **Cut** along every third line of only the top layer to form tabs.

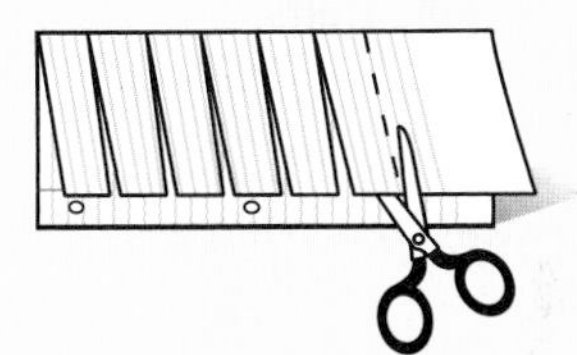

STEP 3 **Turn** vertically and label each tab as described below.

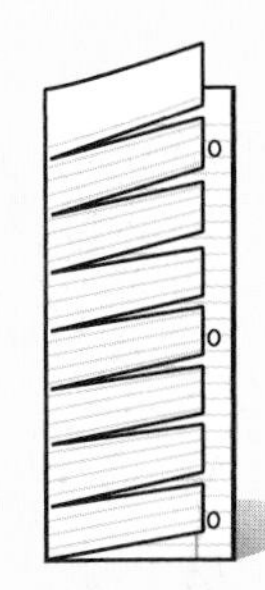

Build Vocabulary As you read the chapter, list the vocabulary words about the classification and structure of life on the tabs. As you learn the definitions, write them under the tab for each vocabulary word.

section 1

Living Things

as you read

What You'll Learn

- **Distinguish** between living and nonliving things.
- **Identify** what living things need to survive.

Why It's Important

All living things, including you, have many of the same traits.

Review Vocabulary
trait: a feature of a living thing

New Vocabulary
- organism
- cell
- homeostasis

What are living things like?

What does it mean to be alive? If you walked down your street after a thunderstorm, you'd probably see birds flying and clouds moving across the sky. You'd see living and nonliving things that are alike in some ways. For example, birds and clouds move. Yet clouds are nonliving things, and birds are living things. Any living thing is called an **organism.**

Organisms vary in size—from the microscopic bacteria in mud puddles to gigantic oak trees—and are found just about everywhere. They have different behaviors and food requirements. In spite of these differences, all organisms have similar traits. These traits determine what it means to be alive.

Living Things Are Organized Imagine looking at almost any part of an organism, such as a plant leaf or your skin, under a microscope. You would see that it is made up of small units called cells, such as the ones pictured in **Figure 1.** A **cell** is the smallest unit of an organism that carries on the functions of life. Cells take in materials from their surroundings and use them in complex ways. Some organisms are composed of just one cell while others are composed of many cells. Each cell has an orderly structure and contains the instructions for cellular organization and function in its hereditary material. All the things an organism can do are possible because of what their cells can do.

Reading Check *How are living things organized?*

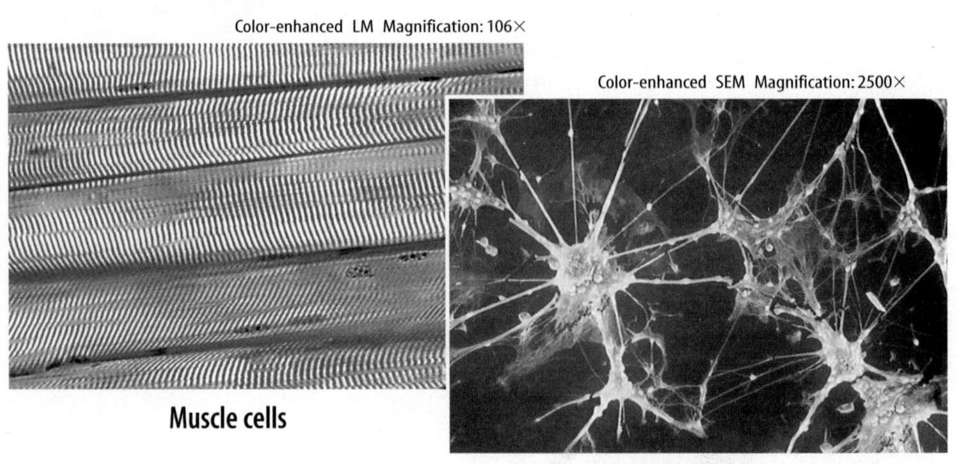

Figure 1 The human body is organized into many different types of cells. Two types are shown to the right.

A dog

A human

A pea plant

A butterfly

Figure 2 Complete development of an organism can take a few days or several years.

Living Things Grow and Develop When a puppy is born, it might be small enough to hold in one hand. After the same dog is fully grown, you might not be able to hold it at all. How does this happen? Growth of a many-celled organism, such as a puppy, is mostly due to an increase in the number of cells. In one-celled organisms, growth is due to an increase in the size of the cell.

Organisms change as they grow. Puppies can't see or walk when they are born. In eight or nine days, their eyes open, and their legs become strong enough to hold them up. All of the changes that take place during the life of an organism are called development. **Figure 2** shows how four different organisms changed as they grew.

The length of time an organism is expected to live is its life span. Some dogs can live for 20 years. Other organisms have a short life span. For example, mayflies live only one day. Yet, others have a much longer life span. A land tortoise can live for more than 180 years, and some bristlecone pine trees have been alive for more than 4,600 years! A human's life span is about 80 years.

Human Stages of Development Human infants can't take care of themselves at birth. Research to find out what human infants can do at different stages of development. Make a chart that shows changes from birth to one year of age.

Topic: Homeostasis

Visit green.msscience.com for Web links to information about homeostasis.

Activity Choose two living things and describe a method each uses to maintain homeostasis.

Living Things Respond Living things must interact with their surroundings. Anything that causes some change in an organism is a stimulus (plural, *stimuli*). The reaction to a stimulus is a response. Often, that response results in movement. An organism must respond to stimuli to carry on its daily activity and to survive.

Living Things Maintain Homeostasis Living things also must respond to stimuli that occur inside them. For example, water or food levels in an organism's cells can increase or decrease. The organism then makes internal changes to maintain the right amounts of water and food in its cells. The regulation of an organism's internal, life-maintaining condition despite changes in its environment is called **homeostasis.** Homeostasis is a trait of all living things.

Reading Check *What is homeostasis?*

Living Things Use Energy Staying organized and carrying on activities like finding food requires energy. The energy used by most organisms comes either directly or indirectly from the Sun. Plants and some other organisms can use the energy in sunlight, carbon dioxide, and water to make food. You and many other organisms can't use the energy in sunlight directly. Instead, you take in and use food as a source of energy. In order to release the energy in food, you and many other organisms must take in oxygen.

Some bacteria live at the bottom of the oceans and in other areas where sunlight cannot reach. They can't use the Sun's energy to produce food; instead, the bacteria use energy stored in some chemical compounds and carbon dioxide to make food. Unlike many other organisms, most of these bacteria do not need oxygen to release the energy that is found in their food.

Figure 3 Living things reproduce themselves in many different ways.

Beetles, like most insects, reproduce by laying eggs.

Color-enhanced LM Magnification: 400×

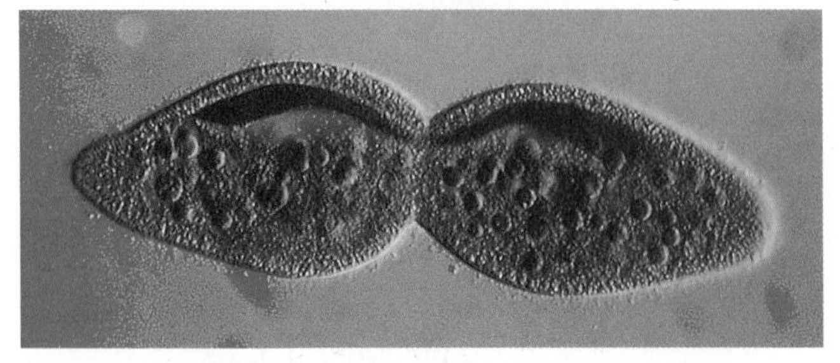

Paramecium species reproduce by dividing into two.

Living Things Reproduce All living things eventually reproduce, to make more of their own kind. Some bacteria reproduce every 20 minutes, while it might take a pine tree two years to produce seeds. **Figure 3** shows some ways organisms reproduce.

Without reproduction, living things would not exist to replace those individuals that die. An individual cat can live its entire life without reproducing. However, if cats never reproduced, all cats soon would disappear.

What do living things need?

Do you have any needs that are different from those of other living things? All living things need a place to live, water, and food source to survive.

A Place to Live All organisms need a place to live that is suited to their unique needs. Could a cactus survive in Antarctica, or a penguin in the Sahara? A place to live also provides enough space for the organism. When weeds grow in a flower bed, they crowd out the flowers, taking up the space that the flowers need to grow. The flower bed is no longer a suitable place for the flower to live.

Water Water is important for all living things. All organisms take in water from their surroundings, as shown in **Figure 4.** Most organisms are composed of more that 50 percent water. You are made up of 60 to 70 percent water. Water performs many functions, such as transporting materials within a cell and between cells. Organisms also give off large amounts of water each day. Homeostasis balances the amount of water exchanged.

Food Sources Living things are made up of substances such as proteins, fats, and sugars. Animals take in these substances as part of the foods that they eat. Plants and some bacteria make their own food. When organisms die, substances in their bodies are broken down and released into the environment. The substances can then be used again by other living organisms. Some of the substances in your body might once have been part of a butterfly or an apple tree!

Figure 4 Most animals, including humans, drink water. They also take in water through the foods they eat, such as corn.
Infer *how a corn plant takes in water.*

section 1 review

Summary

What are living things like?

- An organism is any living thing.
- All living things are organized into cells, which are the smallest units that carry on the functions of life.
- Living things also grow and develop, respond to stimuli, maintain homeostasis, reproduce, and use energy.

What do living things need?

- All living things need a place to live, a food source, and water to survive.

Self Check

1. **Identify** the main source of energy used by most organisms.
2. **Infer** why you would expect to see cells if you looked at a section of a mushroom cap under a microscope.
3. **Think Critically** Why is homeostasis important to organisms?

Applying Skills

4. **Compare and contrast** the similarities and differences between a goldfish and the flame of a burning candle.

section 2

How are living things classified?

as you read

What You'll Learn

- **Describe** how early scientists classified living things.
- **Explain** the system of binomial nomenclature.
- **Demonstrate** how to use a dichotomous key.

Why It's Important

Knowing how living things are classified will help you understand the relationships that exist between all living things.

Review Vocabulary
hereditary: relating to the passing of traits from parent to offspring

New Vocabulary
- binomial nomenclature
- genus
- phylogeny
- kingdom

Classification

People have grouped together similar organisms, or classified them, for thousands of years. Many different systems were used until the late eighteenth century. Carolus Linnaeus, a Swedish naturalist, developed a new system of grouping organisms that was accepted and used by most scientists. His classification system was based on looking for organisms with similar structures. For example, plants that had a similar flower structure were grouped together. Linnaeus also developed a scientific naming system that still is used today.

Binomial Nomenclature The two-word naming system that Linnaeus used to name various organisms is called **binomial nomenclature** (bi NOH mee ul • NOH mun klay chur). This two-word name is an organism's species. Organisms that can mate and produce fertile offspring belong to the same species.

The first word of the two-word name identifies the genus of the organism. A **genus** is a group of similar species. For example, a salamander's genus is *Ambystoma.* The second word of the name usually describes a feature. The Eastern tiger salamander is *Ambystoma tigrinum,* shown in **Figure 5.** Latin is the language used for scientific names.

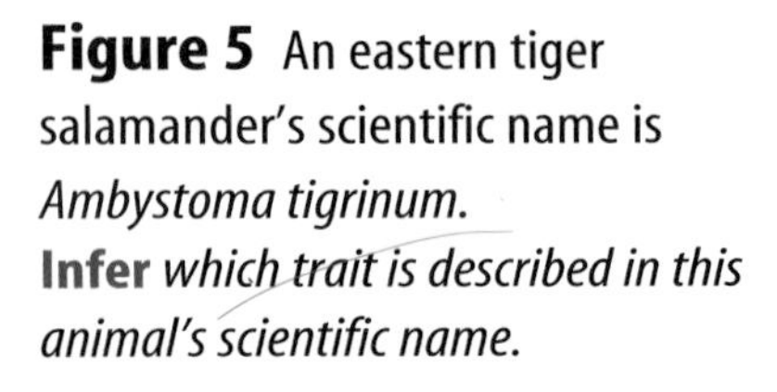

Figure 5 An eastern tiger salamander's scientific name is *Ambystoma tigrinum.*
Infer *which trait is described in this animal's scientific name.*

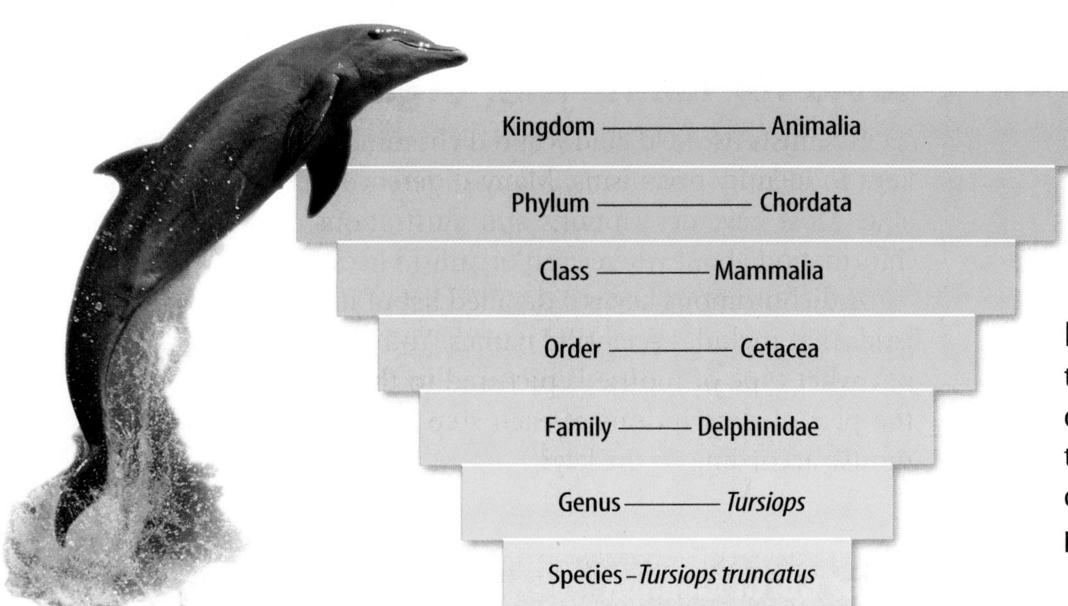

Figure 6 The classification of the bottle-nosed dolphin shows that it is in the order Cetacea. This order includes whales and porpoises.

Uses of Scientific Names Why use scientific names at all? Scientific names are used for four reasons. First, they help avoid mistakes. Often, common names for two different organisms are the same, or one organism has many different common names. Scientific names help distinguish between those organisms. Second, organisms with similar evolutionary histories are classified together. Because of this, you know that organisms in the same genus are related. Third, scientific names give descriptive information about the species, like the salamander mentioned earlier. Fourth, scientific names allow information about organisms to be organized easily and efficiently. Such information may be found in a book or a pamphlet that lists related organisms and gives their scientific names.

Reading Check *What are the functions of scientific names?*

Modern Classification Like Linnaeus, modern scientists use similarities in structure to classify organisms. They also study fossils, hereditary information, and early stages of development. Scientists use all of this information to determine an organism's phylogeny. **Phylogeny** (fi LAH juh nee) is the evolutionary history of an organism, that is, how the organism has changed over time. Today, it is the basis for the classification of many organisms.

In the classification system used today, the smallest group is a species. There are broader groups preceding species, the largest of which is a **kingdom.** To understand how an organism is classified, look at the classification of the bottle-nosed dolphin in **Figure 6.** Some scientists have proposed that before organisms are grouped into kingdoms, they should be placed in larger groups called domains. One proposed system groups all organisms into three domains.

Communicating Ideas

Procedure

1. Find a **magazine picture of a piece of furniture** that can be used as a place to sit and to lie down.
2. Show the picture to ten people and ask them to tell you what word they use for this piece of furniture.
3. Keep a record of the answers in your **Science Journal.**

Analysis

1. In your Science Journal, infer how using common names can be confusing.
2. How do scientific names make communication among scientists easier?

Try at Home

Tools for Identifying Organisms

Scientists use field guides and dichotomous (di KAH tuh mus) keys to identify organisms. Many different field guides are available. Most have descriptions and illustrations of organisms and information about where each organism lives.

A dichotomous key is a detailed list of identifying characteristics that includes scientific names. You can use **Table 1** to find out what type of mouse is pictured to the left. Choose between the pair of descriptions at each step. What is the name of this mouse according to the key?

Table 1 Mice of North America

1. Tail hair	**a.** no tail hair; scales show plainly; house mouse, *Mus musculus* **b.** hair on tail, go to 2
2. Ear size	**a.** ears small and nearly hidden in fur, go to 3 **b.** ears large and not hidden in fur, go to 4
3. Tail length	**a.** less than 25 mm; woodland vole, *Microtus pinetorum* **b.** more than 25 mm; prairie vole, *Microtus ochrogaster*
4. Tail coloration	**a.** sharply dark above; deer mouse, *Peromyscus maniculatus* **b.** slightly dark above; white-footed mouse, *Peromyscus leucopus*

section 2 review

Summary

Classification

- Linnaeus developed the first widely accepted method of classification based on similar structures.
- Binomial nomenclature is the two-name system that scientists use today. It consists of the genus name and another identifying name.
- Scientists study fossils, hereditary information, and early stages of development to determine classification.
- Scientists use field guides and dichotomous keys to identify organisms.

Self Check

1. **Explain** the purpose of classification.
2. **Identify** what information a scientist would use to determine an organism's phylogeny.
3. **Think Critically** Why would a field guide have common names as well as scientific names?

Applying Skills

4. **Classify** Create a dichotomous key that identifies types of cars.
5. **Communicate** Select a field guide for trees, insects, or mammals. Select two organisms in the field guide that closely resemble each other. Use labeled diagrams to show how they are different.

green.msscience.com/self_check_quiz

section 3

Cell Structure

Viewing Cells

Four hundred years ago, scientists did not know what cells looked like, or that they even existed. That changed in the late 1500s, when the first microscope was made by a Dutch optometrist. He put two magnifying lenses together in a tube and got an image that was larger than the image that was made by either lens alone. In the mid 1600s, Antonie van Leeuwenhoek, a Dutch fabric merchant, made a simple microscope with a tiny glass bead for a lens, as shown in **Figure 7** on the next page. These microscopes eventually led to the types of microscopes that scientists use today.

Development of the Cell Theory

In the seventeenth century the microscope was improved, which led to the discovery of cells. In 1665, Robert Hooke cut a thin slice of cork and looked at it under his microscope. To Hooke, the cork seemed to be made up of empty little boxes, which he named cells.

In the 1830s, Matthias Schleiden used a microscope to study plant parts. He concluded that all plants are made of cells. Theodor Schwann, after observing many different animal cells, concluded that all animals also are made up of cells. Eventually, they combined their ideas and became convinced that all living things are made of cells.

Several years later, Rudolf Virchow hypothesized that cells divide to form new cells. Virchow proposed that every cell came from a cell that already existed. His observations and conclusions and those of others are summarized in the **cell theory,** as described in **Table 2.**

as you read

What You'll Learn

- **Describe** the development of the cell theory.
- **Identify** names and functions of each part of a cell.
- **Explain** how important a nucleus is in a cell.
- **Compare** tissues, organs, and organ systems.

Why It's Important

Humans are like other living things because we are made of cells.

Review Vocabulary
theory: an explanation of events or circumstances based on scientific knowledge resulting from repeated observations and tests

New Vocabulary
- cell theory
- cell wall
- cell membrane
- cytoplasm
- ribosome
- organelle
- nucleus
- chloroplast
- mitochondrion
- endoplasmic reticulum
- Golgi body
- tissue
- organ
- organ system

Table 2 The Cell Theory

All organisms are made up of one or more cells.	An organism can be one cell or many cells like most plants and animals.
The cell is the basic unit of organization in organisms.	Even in complex organisms, the cell is the basic unit of structure and function.
All cells come from cells.	Most cells can divide to form two new, identical cells.

VISUALIZING MICROSCOPES

Figure 7

Microscopes give us a glimpse into a previously invisible world. Improvements have vastly increased their range of visibility, allowing researchers to study life at the molecular level. A selection of these powerful tools—and their magnification power—is shown here.

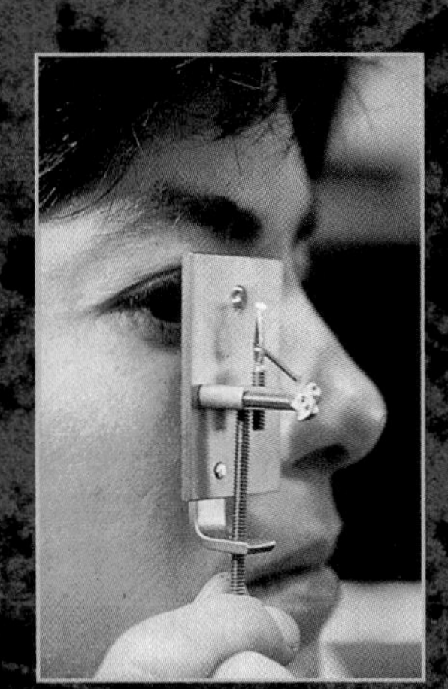

▶ Up to 250× LEEUWENHOEK MICROSCOPE Held by a modern researcher, this historic microscope allowed Leeuwenhoek to see clear images of tiny freshwater organisms that he called "beasties."

▼ Up to 2,000× BRIGHTFIELD / DARKFIELD MICROSCOPE The light microscope is often called the brightfield microscope because the image is viewed against a bright background. A brightfield microscope is the tool most often used in laboratories to study cells. Placing a thin metal disc beneath the stage, between the light source and the objective lenses, converts a brightfield microscope to a darkfield microscope. The image seen using a darkfield microscope is bright against a dark background. This makes details more visible than with a brightfield microscope. Below are images of a *Paramecium* as seen using both processes.

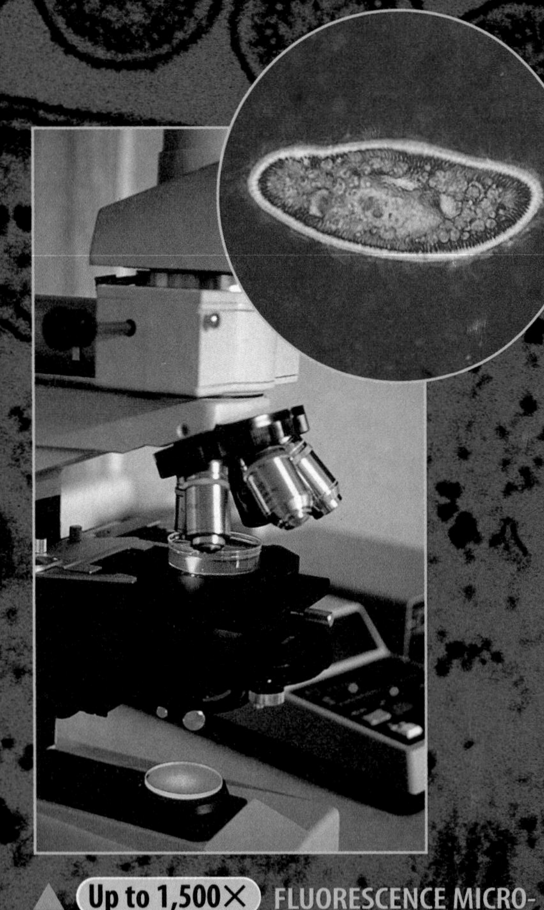

▲ Up to 1,500× FLUORESCENCE MICROSCOPE This type of microscope requires that the specimen be treated with special fluorescent stains. When viewed through this microscope, certain cell structures or types of substances glow, as seen in the image of a *Paramecium* above.

Up to 1,000,000× TRANSMISSION ELECTRON MICROSCOPE A TEM aims a beam of electrons through a specimen. Denser portions of the specimen allow fewer electrons to pass through and appear darker in the image. Organisms, such as the *Paramecium* at right, can only be seen when the image is photographed or shown on a monitor. A TEM can magnify hundreds of thousands of times.

Up to 1,500× PHASE-CONTRAST MICROSCOPE A phase-contrast microscope emphasizes slight differences in a specimen's capacity to bend light waves, thereby enhancing light and dark regions without the use of stains. This type of microscope is especially good for viewing living cells, like the *Paramecium* above left. The images from a phase-contrast microscope can only be seen when the specimen is photographed or shown on a monitor.

Up to 200,000× SCANNING ELECTRON MICROSCOPE An SEM sweeps a beam of electrons over a specimen's surface, causing other electrons to be emitted from the specimen. SEMs produce realistic, three-dimensional images, which can only be viewed as photographs or on a monitor, as in the image of the *Paramecium* at right. Here a researcher compares an SEM picture to a computer monitor showing an enhanced image.

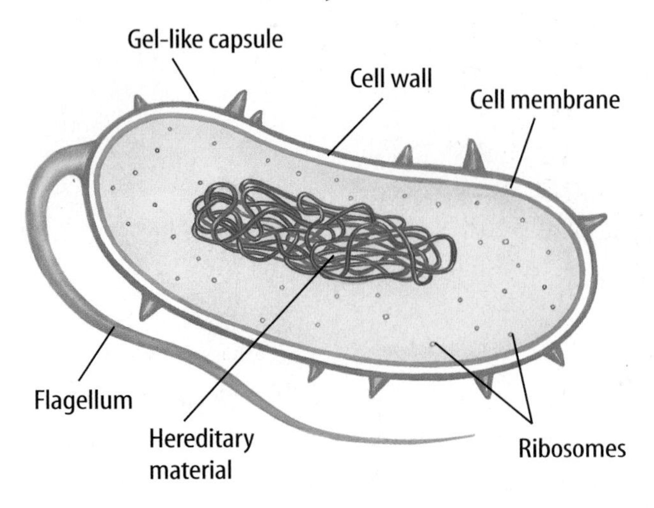

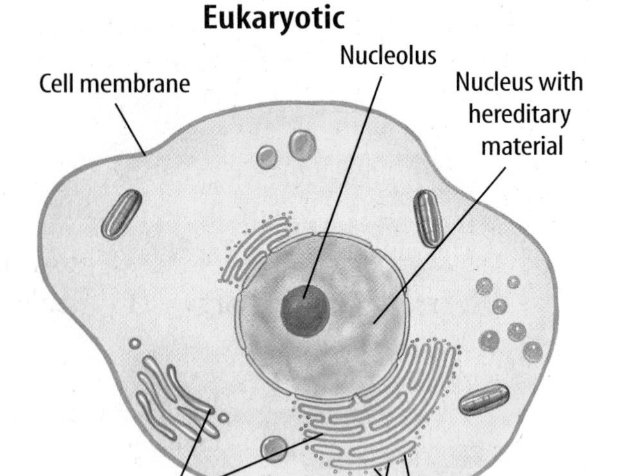

Figure 8 Examine these drawings of cells. Prokaryotic cells only are found in one-celled organisms, such as bacteria. Protists, fungi, plants and animals are made of eukaryotic cells.
Describe *the differences between these cells.*

Cellular Organization

Scientists have found that cells can be separated into two groups. One group has membrane-bound structures inside the cell and the other group does not, as shown in **Figure 8.** Cells without membrane-bound structures are called prokaryotic (proh kayr ee AH tihk) cells. Cells with membrane-bound structures are called eukaryotic (yew kayr ee AH tihk) cells. Each cell performs specific functions; however, all cells must constantly take in nutrients, store, produce, and breakdown substances, and take in and use energy.

Cell Wall The cells of plants, algae, fungi, and most bacteria are enclosed in a cell wall. **Cell walls** are tough, rigid outer coverings that protect cells and give them shape.

A plant cell wall mostly is made up of a carbohydrate called cellulose. The long, threadlike fibers of cellulose form a thick mesh that allows water and dissolved materials to pass through. Plant cell walls also may contain pectin and lignin. Pectin aids in cell growth, development, defense, and strength, among other functions. It is also what makes jams and jellies have a thick texture. Lignin is a compound that makes cell walls rigid. Plant cells responsible for support have more lignin in their walls than other plant cells.

Cell Membrane The protective layer surrounding every cell is the **cell membrane,** as shown in **Figure 9.** The cell membrane is the outermost covering of a cell unless a cell wall is present. The cell membrane regulates interactions between the cell and its environment. The cell membrane allows nutrients to move into the cell, while waste products leave.

Figure 9 The cell membrane is made up of a double layer of fat-like molecules.

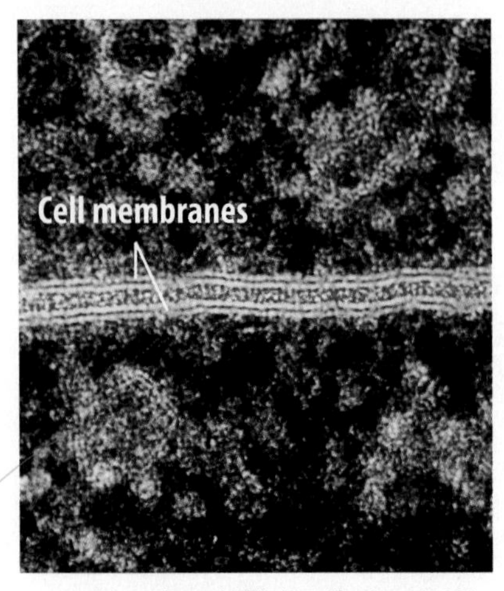

Color-enhanced TEM Magnification: 125,000×

Cytoplasm Cells are filled with a gelatinlike substance called **cytoplasm** (SI toh pla zuhm) that constantly flows inside the cell membrane. Most of a cell's life processes occur in the cytoplasm. In prokaryotic cells, the hereditary material is found here. Throughout the cytoplasm is a framework called the cytoskeleton, shown in **Figure 10,** which helps the cell maintain or change its shape and enables some cells to move. The cytoskeleton is made up of thin, hollow tubes of protein and thin, solid protein fibers. Proteins are organic molecules made up of amino acids.

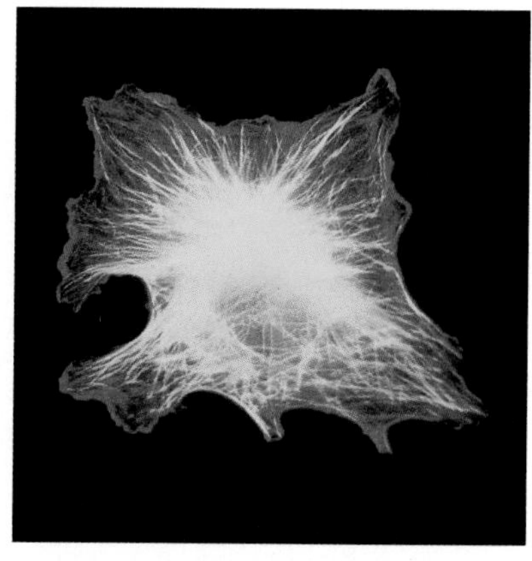

Figure 10 Cytoskeleton, a network of fibers in the cytoplasm, gives cells structure and helps them maintain shape.
Describe *what the cytoskeleton is made of.*

Reading Check *How might an amoeba move using its cytoskeleton?*

Manufacturing Proteins One substance that takes part in nearly every cell activity is protein. Proteins are part of cell membranes and are needed for chemical reactions that take place in the cytoplasm. Cells make their own proteins on small structures called **ribosomes.** All ribosomes in prokaryotic cells, and some in eukaryotic cells, are made and float freely in the cytoplasm, like the ribosomes in **Figure 11.** Ribosomes receive directions from the hereditary material on how, when, and in what order to make specific proteins.

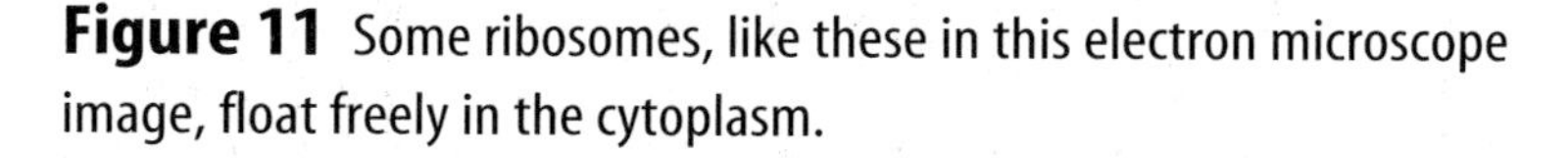

Figure 11 Some ribosomes, like these in this electron microscope image, float freely in the cytoplasm.

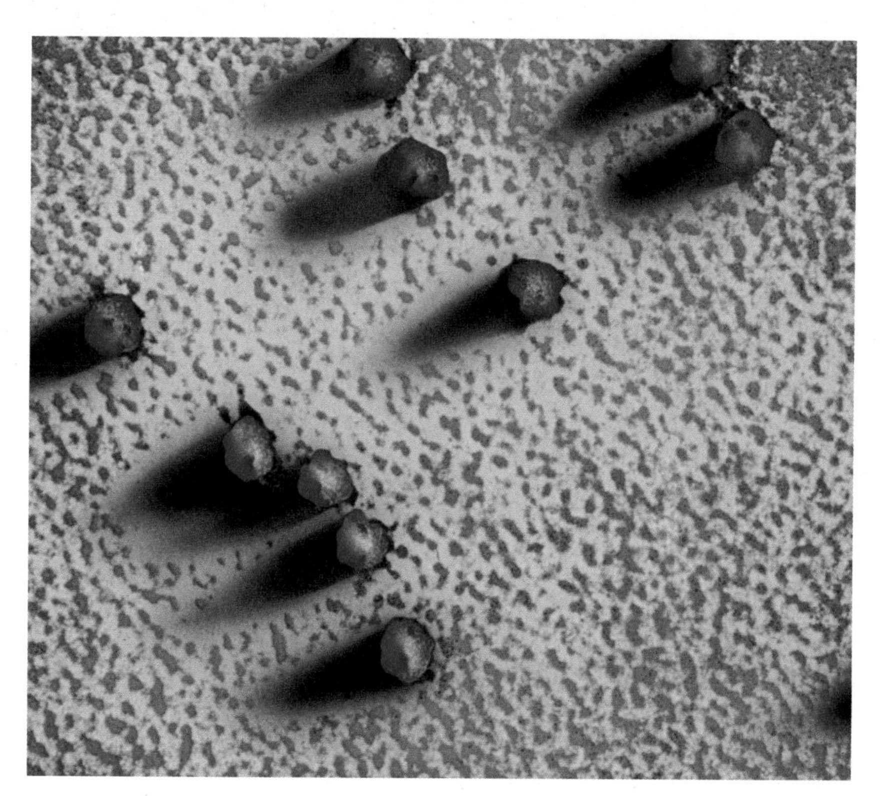

Modeling Cytoplasm

Procedure

1. Add 100 mL **water** to a **clear container.**
2. Add **unflavored gelatin** and stir.
3. Shine a **flashlight** through the solution.

Analysis

1. Describe what you see.
2. How does a model help you understand what cytoplasm might be like?

Membrane-Bound Organelles Within the cytoplasm of eukaryotic cells are structures called **organelles,** the largest of which is usually the nucleus. Some organelles process energy and others manufacture substances needed by the cell or other cells. Certain organelles move materials, while others act as storage sites. Most organelles are surrounded by a membrane. Ribosomes are considered organelles, but are not membrane-bound.

Nucleus All cellular activities are directed by the **nucleus.** Materials enter and leave the nucleus through openings in its membrane. The nucleus contains long, threadlike, hereditary material made of DNA. DNA is the chemical that contains the code for the cell's structure and activities. A structure called a nucleolus also is found in the nucleus, and is where most ribosomes are made in a eukaryotic cell.

INTEGRATE Physics

Organelles That Process Energy Cells require a continuous supply of energy to process food, make new substances, eliminate wastes, and communicate with each other. In plant cells, food is made in green organelles in the cytoplasm called **chloroplasts** (KLOR uh plasts), shown in **Figure 12.** Chloroplasts contain the green pigment chlorophyll, which gives many leaves and stems their color. Chlorophyll captures light energy that is used to make a sugar called glucose. This light energy is changed and stored in glucose as chemical energy. Many cells do not have chloroplasts for making food and must get food from their environment.

The energy in food usually is released by mitochondria. **Mitochondria** (mi tuh KAHN dree uh) (singular, *mitochondrion*), also shown in **Figure 12,** are organelles where energy is released when food is broken down into carbon dioxide and water. Some types of cells, such as muscle cells, are more active than other cells and have larger numbers of mitochondria. Both chloroplasts and mitochondria contain ribosomes and hereditary material. Both plant and animal cell structures can be found in **Figure 13.**

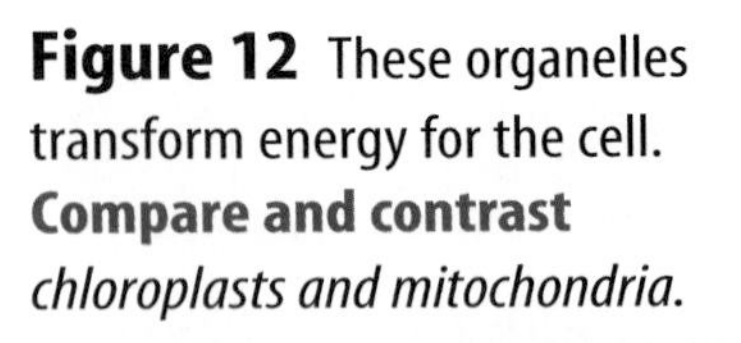

Figure 12 These organelles transform energy for the cell. **Compare and contrast** *chloroplasts and mitochondria.*

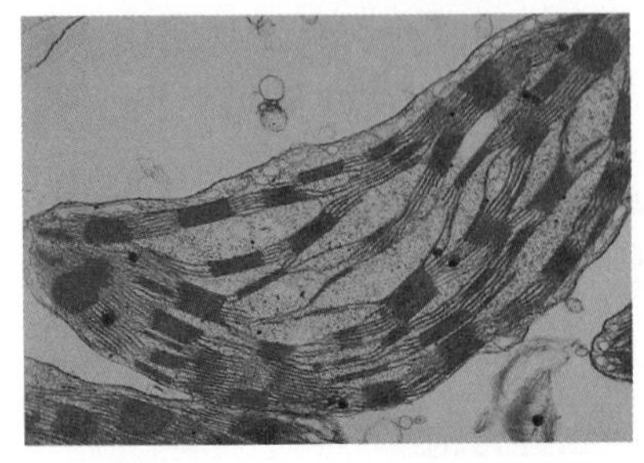

Color-enhanced TEM Magnification: 37,000×

Chloroplasts are organelles in organisms, such as plants, that use light to make sugar from carbon dioxide and water.

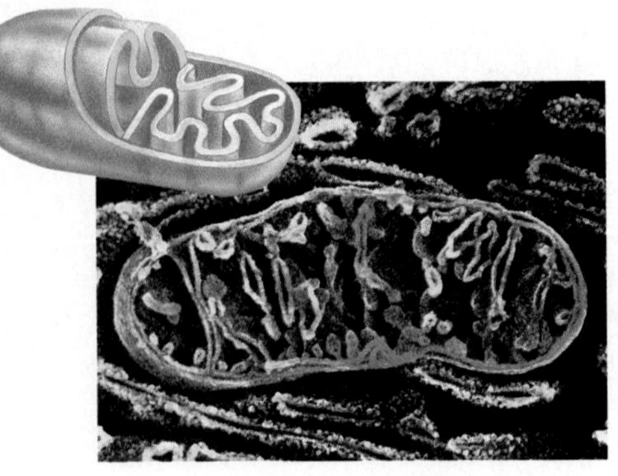

Color-enhanced SEM Magnification: 48,000×

Mitochondria are known as the powerhouses of the cell because they release energy that is needed by the cell.

Figure 13 Refer to these diagrams of a typical animal cell (top) and plant cell (bottom) as you read about cell structures and their functions.

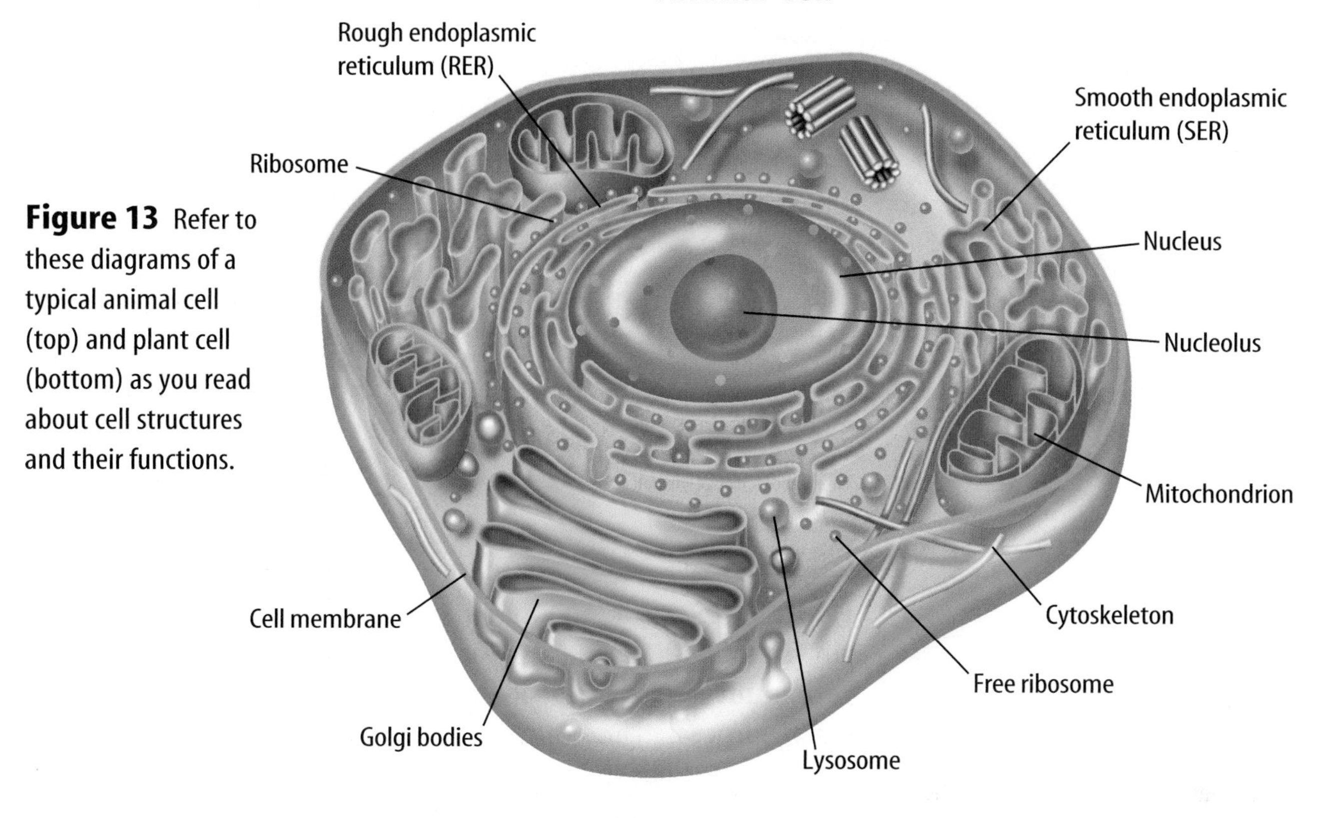

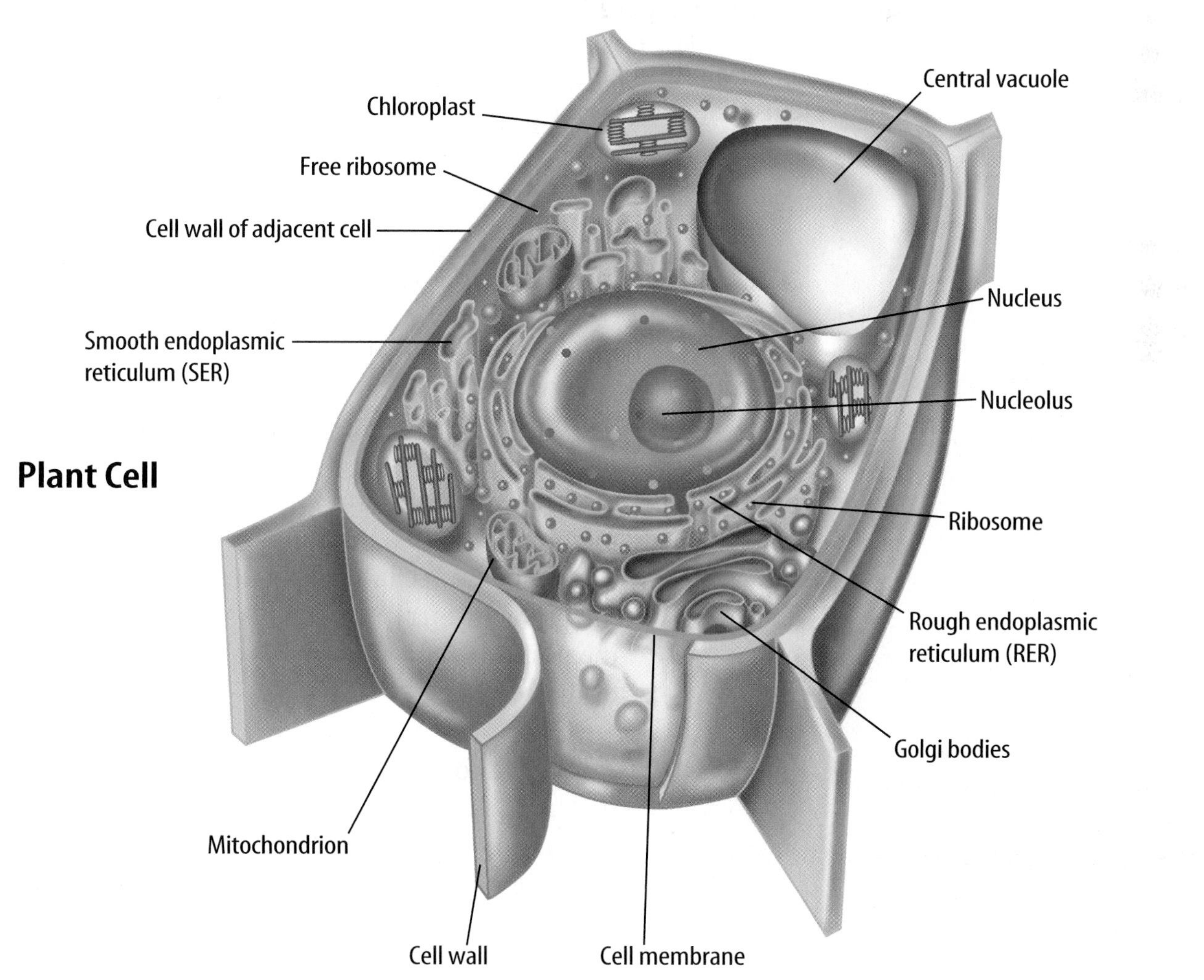

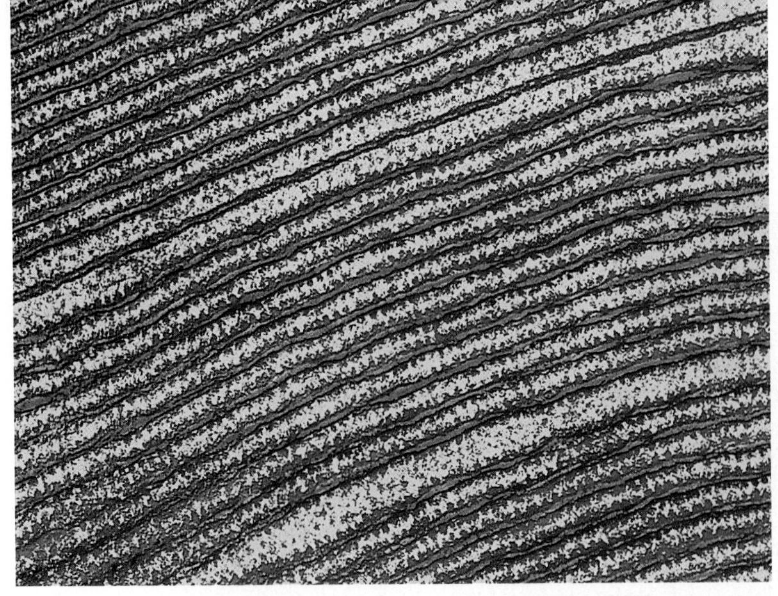

Color-enhanced TEM Magnification: 65,000×

Figure 14 Endoplasmic reticulum (ER) is a complex series of membranes in the cytoplasm of the cell.
Describe *smooth ER.*

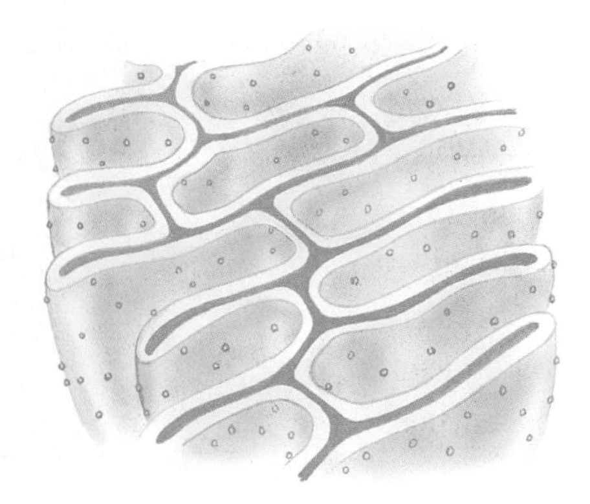

Organelles That Process, Transport, and Store

Figure 14 shows the endoplasmic reticulum (en duh PLAZ mihk • rih TIHK yuh lum), also called the ER. The **endoplasmic reticulum** is a series of folded membranes in which materials can be processed and moved around inside of the cell. It extends from the nucleus to the cell membrane and takes up a considerable amount of space in some cells.

The ER may be "rough" or "smooth." Ribosomes are attached to areas on the rough ER. There they carry out their job of making proteins that are moved out of the cell or used within the cell. ER that does not have attached ribosomes is called smooth ER. This type of ER processes cellular substances such as lipids that store energy.

Reading Check *What is the difference between rough ER and smooth ER?*

After proteins are made in a cell, they are transferred to another type of cell organelle called the Golgi (GAWL jee) bodies. The **Golgi bodies,** shown in **Figure 15,** are stacked, flattened membranes. The Golgi bodies sort proteins and other cellular substances and package them into membrane-bound structures called vesicles. The vesicles deliver cellular substances to areas inside the cell, and carry cellular substances to the cell membrane where they are released to the outside of the cell.

Cells also have membrane-bound spaces called vacuoles for the temporary storage of materials. A vacuole can store water, waste products, food, and other cellular materials. In plant cells, the vacuole may make up most of the cell's volume.

Figure 15 Golgi bodies package materials.
Infer *why some materials are removed from the cell.*

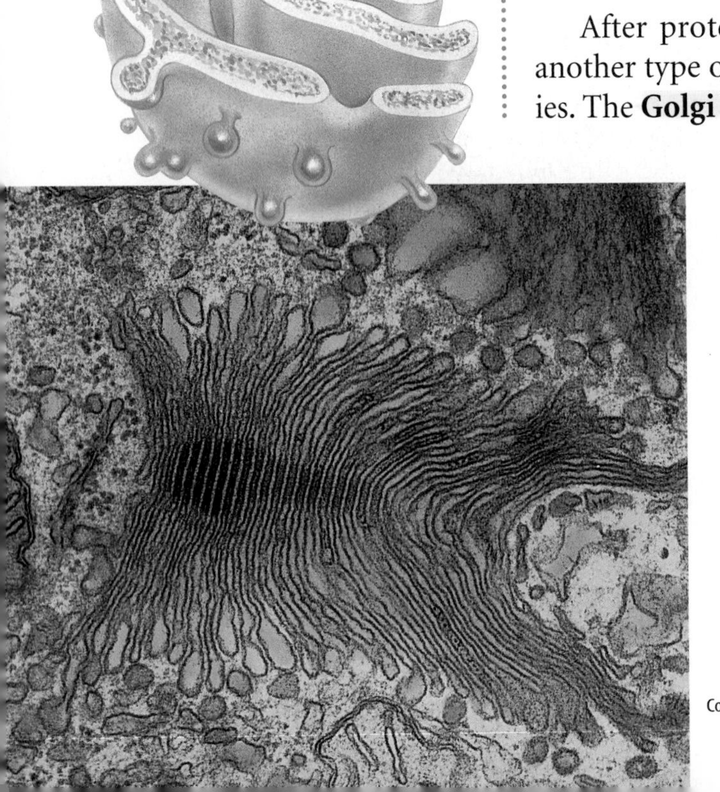

Color-enhanced TEM Magnification: 28,000×

Organelles That Recycle Active cells break down and recycle substances. Organelles called lysosomes (LI suh sohmz) contain digestive chemicals that help break down food molecules, cell wastes, worn-out cell parts, and viruses and bacteria that enter a cell. Chemicals can be released into vacuoles when needed to break down its contents. The lysosome's membrane prevents the digestive chemicals inside from leaking into the cytoplasm and destroying the cell.

When a cell dies, the lysosome's membrane disintegrates, releasing digestive chemicals that allow the quick breakdown of the cell's contents. This is what happens when a frog tadpole becomes an adult. The lysosomes digest the cells in the tadpole's tail. These cellular molecules then are used to make other cells, such as those in the legs of the frog.

Recycle Just like a cell, you can recycle materials. Paper, plastic, aluminum, and glass can be recycled into usable items. Many other countries require recycling because of a lack of space for landfills. Research recycling in other countries and make a bar graph that compares recycling rates in other countries to the rate in the United States.

Applying Math Find a Ratio

CELL SURFACE AREA AND VOLUME You can find the surface area and volume of a cell in the same way you find the surface area and volume of a cube, if you assume that a cell is like a cube with six equal sides. Find the ratio of surface area to volume for a cell that is 4 cm high.

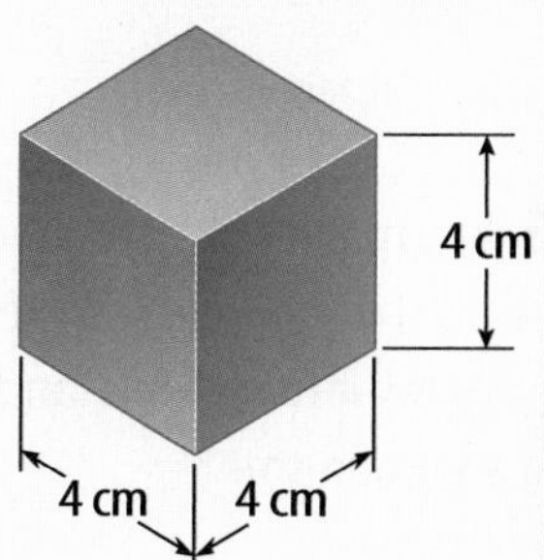

Solution

1 *This is what you know:* A cell has six equal sides of 4 cm × 4 cm

2 *This is what you want to find out:* What is the ratio (R) of surface area to volume for each cell?

3 *This is the procedure you need to use:*

- Use the following equations:
  surface area (A) = width × length × 6
  volume (V) = length × width × height
  $R = A/V$
- Substitute in known values and solve.
  $A = 4\text{ cm} \times 4\text{ cm} \times 6 = 96\text{ cm}^2$
  $V = 4\text{ cm} \times 4\text{ cm} \times 4\text{ cm} = 64\text{ cm}^3$
  $R = 96\text{ cm}^3/64\text{ cm}^2 = 1.5\text{ cm}^2/\text{cm}^3$

4 *Check your answer:* Multiply $1.5\text{ cm}^2/\text{cm}^3 \times 64\text{ cm}^3$. You should get 96 cm^2.

Practice Problems

1. Calculate the ratio of surface area to volume for a cell that is 2 cm high. What happens to this ratio as the size of the cell decreases?
2. If a 4-cm cell doubled just one of its dimensions—length, width, or height—what would happen to the ratio of surface area to volume?

For more practice, visit green.msscience.com/math_practice

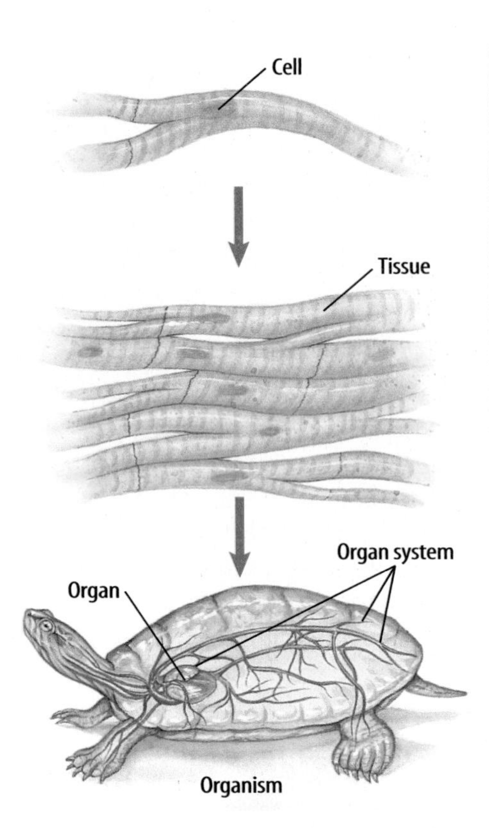

Figure 16 In a many-celled organism, cells are organized into tissues, tissues into organs, organs into systems, and systems into an organism.

Many-Celled Organisms

Cells in a many-celled organism do not work alone. Each cell carries on its own life functions while depending in some way on other cells in the organism. Many one-celled organisms, however, perform all life functions on their own.

In **Figure 16,** you can see cardiac muscle cells grouped together to form a tissue. A **tissue** is a group of similar cells that work together to do one job. Each cell in a tissue does its part to keep the tissue alive.

Tissues are organized into organs. An **organ** is a structure made up of two or more different types of tissues that work together. Your heart is an organ made up of nerve, blood, and cardiac muscle tissues. The cardiac muscle tissue contracts, making the heart pump. The nerve tissue brings messages that tell the heart how fast to beat. The blood tissue is carried from the heart to other organs of the body.

A group of organs working together to perform a certain function is an **organ system.** Your heart, arteries, veins, and capillaries make up your cardiovascular system. Organ systems work together to make up a many-celled organism.

section 3 review

Summary

Viewing Cells and the Cell Theory

- Cells could be seen after the invention of the microscope by van Leeuwenhoek.
- The cell theory states that all organisms are made up of cells, the cell is the basic unit of organization, and all cells come from other cells.

Cellular Organization

- Cells are prokaryotic or eukaryotic.
- All cells have a cell membrane, cytoplasm, and ribosomes. Some have a cell wall.
- All eukaryotic cells have membrane-bound organelles to carry out life processes.

Many-Celled Organisms

- Cells in many-celled organisms cannot function alone. Cells group together to form tissues, tissues form organs, organs form organ systems, and organ systems form an organism.

Self Check

1. **Describe** the events leading to the discovery of the cell theory.
2. **Identify** the role of the nucleus in the life of a cell.
3. **Compare and contrast** the differences between plant and animal cells.
4. **Explain** how a cell is like your school or town. What is the nucleus of your town? The mitochondria?
5. **Think Critically** How is the cell of a one-celled organism different from the cells in many-celled organisms?

Applying Math

6. **Solve One-Step Equations** The magnification of a light microscope can be found by multiplying the power of the eyepiece by the power of the objective lens. Calculate the magnifications of a microscope that has an 8× eyepiece, and 10× and 40× objectives.

Comparing Cells

Real-World Question

If you compared a goldfish to a rose, you would find them unlike each other. Are their individual cells also different? Try this lab to compare plant and animal cells. How do human cheek cells and plant cells compare?

Goals

- **Compare and contrast** an animal and a plant cell.

Materials

microscope
microscope slide
coverslip
forceps
tap water
dropper
Elodea plant
prepared slide of human cheek cells

Safety Precautions

Procedure

1. Copy the data table in your Science Journal. Check off the cell parts as you observe them.

Cell Observations		
Cell Part	Cheek	*Elodea*
Cytoplasm		
Nucleus	Do not write in this book.	
Chloroplasts		
Cell wall		
Cell membrane		

2. Using forceps, make a wet-mount slide of a young leaf from the tip of an *Elodea* plant.
3. Observe the leaf on low power. Focus on the top layer of cells.
4. Switch to high power and focus on one cell. In the center of the cell is a membrane-bound organelle called the central vacuole. Observe the chloroplasts—the green, disk-shaped objects moving around the central vacuole. Try to find the cell nucleus. It looks like a clear ball.
5. Draw the *Elodea* cell. Label the cell wall, cytoplasm, chloroplasts, central vacuole, and nucleus. Return to low power and remove the slide. Properly dispose of the slide.
6. Observe the prepared slide of cheek cells under low power.
7. Switch to high power and observe the cell nucleus. Draw and label the cell membrane, cytoplasm, and nucleus. Return to low power and remove the slide. Properly dispose of the slide.

Conclude and Apply

1. **Compare and contrast** the shapes of the cheek cell and the *Elodea* cell.
2. What can you conclude about the differences between plant and animal cells?

Communicating Your Data

Draw the two kinds of cells on one sheet of paper. Use a green pencil to label the organelles found only in plants, a red pencil to label the organelles found only in animals, and a blue pencil to label the organelles found in both.

section 4

Viruses

as you read

What You'll Learn

- **Explain** how a virus makes copies of itself.
- **Identify** the benefits of vaccines.
- **Investigate** some uses of viruses.

Why It's Important

Viruses can infect nearly all organisms, including humans.

Review Vocabulary
bacteria: a prokaryotic, one-celled organism

New Vocabulary
- virus
- host cell

What are viruses?

Cold sores, measles, chicken pox, colds, the flu, and AIDS are diseases caused by nonliving particles called viruses. A **virus** is a strand of hereditary material surrounded by a protein coating. A virus multiplies by making copies of itself with the help of a living cell called a **host cell.** Viruses don't have a nucleus, other organelles, or a cell membrane. They have a variety of shapes and are too small to be seen with a light microscope. They were discovered only after the electron microscope was invented. Before that time, scientists only hypothesized about viruses.

Active Viruses When a virus enters a cell and is active, it causes the host cell to make new viruses. This process destroys the host cell. Follow the steps in **Figure 17** to see one way that an active virus functions inside a cell.

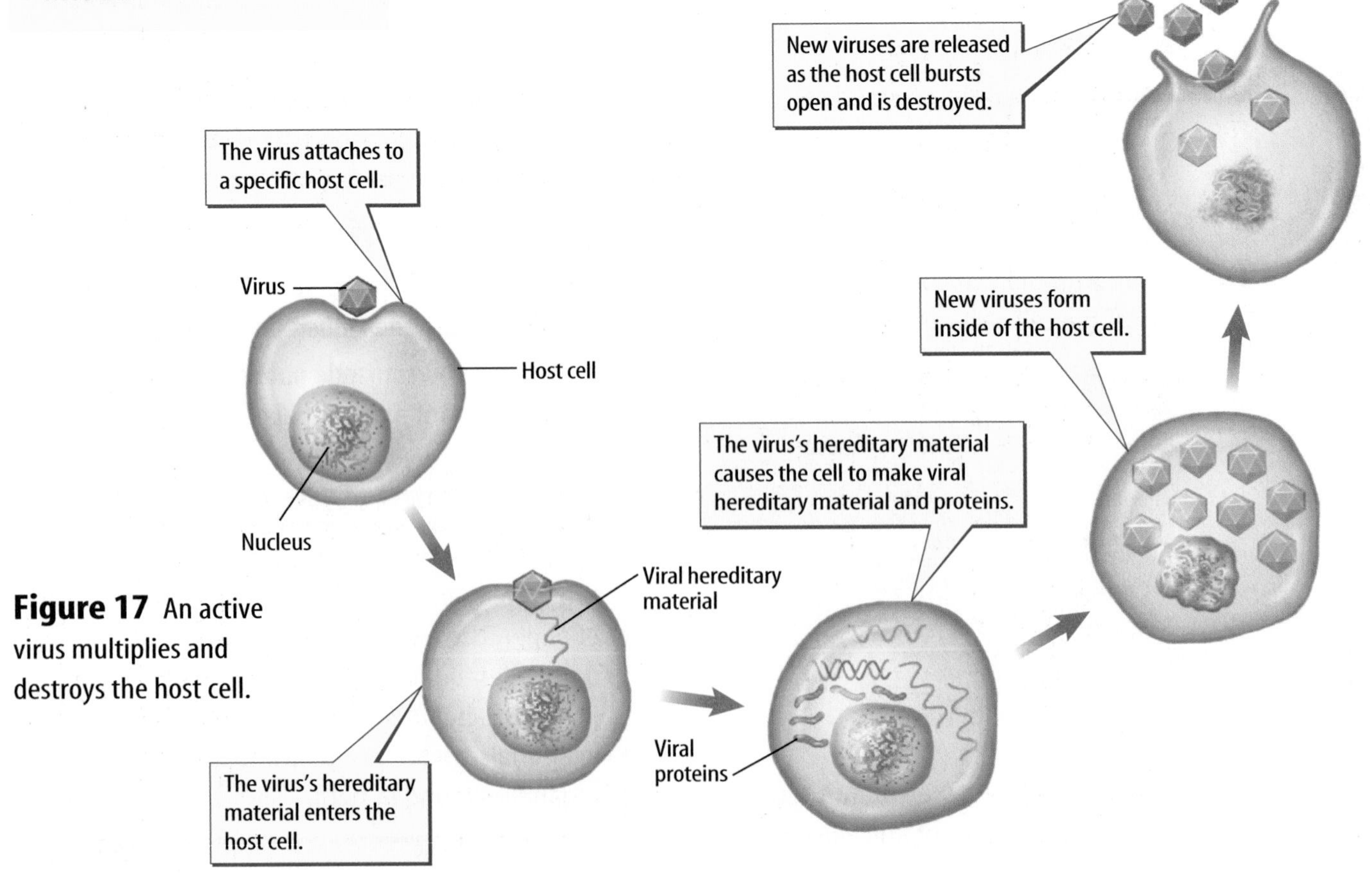

Figure 17 An active virus multiplies and destroys the host cell.

Latent Viruses Some viruses can be inactive, and are called latent. This means that after the virus enters a cell, its hereditary material becomes part of the cell's hereditary material. It does not immediately make new viruses or destroy the cell. As the host cell reproduces, the hereditary material of the virus is copied. A virus can be latent for many years. Then, at any time, certain conditions, either inside or outside the body, can activate the virus.

If you have had a cold sore on your lip, a latent virus in your body has become active. The cold sore is a sign that the virus is active and destroying cells in your lip. When the cold sore disappears, the virus has become latent again. The virus, however, is still in your body's cells.

Topic: Latent Viruses
Visit green.msscience.com for Web links to information about viruses.

Activity Name three environmental stimuli that may activate a latent virus.

How do viruses affect organisms?

Viruses can infect animals, plants, fungi, protists, and all bacteria. Most viruses can infect only specific kinds of cells. For instance, many viruses, such as the potato leafroll virus in **Figure 18,** are limited to one host species or to one type of tissue within a species. A few viruses affect a broad range of hosts. An example of this is the rabies virus. Rabies can infect humans and many other animal hosts.

A virus cannot move by itself, but it can reach a host's body in several ways. For example, it can be carried onto a plant's surface by the wind or it can be inhaled by an animal. In a viral infection, the virus first attaches to the surface of the host cell. The virus and the place where it attaches must fit together exactly, also shown in **Figure 18,** which is why most viruses attack only one kind of host cell. Viruses that infect bacteria are called bacteriophages (bak TIHR ee uh fay juhz).

Figure 18 Viruses and the attachment sites of the host cell must match exactly, like a puzzle. That's why most viruses infect only one kind of host cell.

The potato leafroll virus, PLRV, damages potato crops worldwide, and is specific to only potatoes.

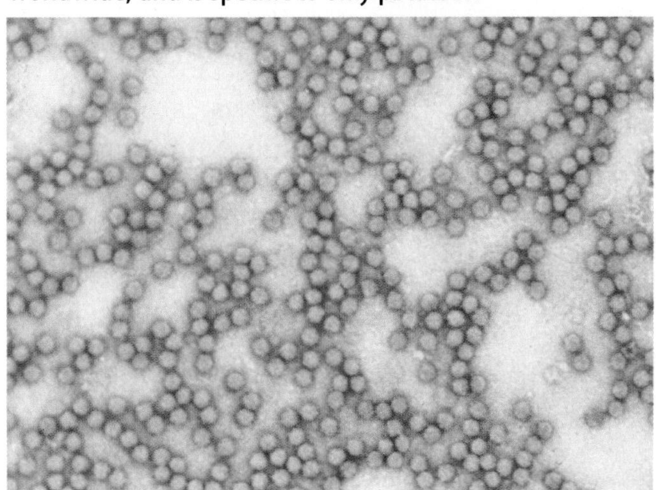

Magnification: unknown

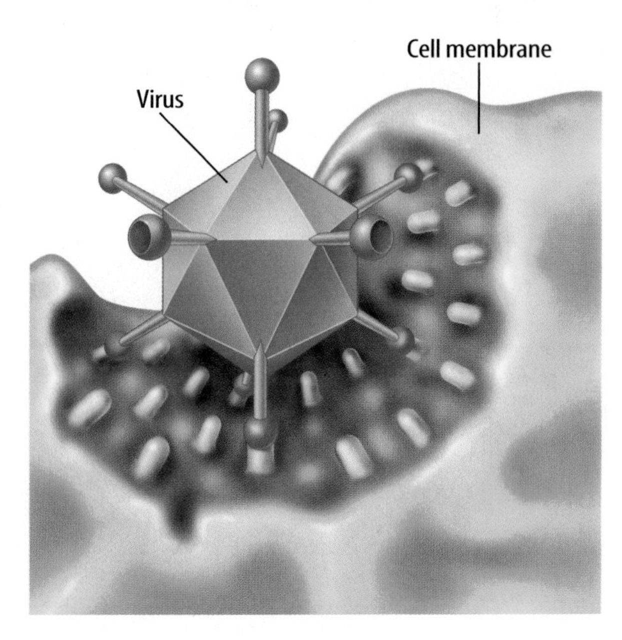

Treating and Preventing Viral Diseases

Antibiotics are used to treat bacterial infections, but they do not work against viral diseases. Antiviral drugs can be given to infected patients to help fight a virus; however, they are not widely used because of adverse side effects. Prevention is the best way to fight the diseases, because treatment is very difficult.

Public health measures for preventing viral diseases include vaccinating people, improving sanitary conditions, separating patients with diseases, and controlling animals that spread the disease. Yellow fever was wiped out completely in the United States through mosquito-control programs. Annual rabies vaccinations protect humans by keeping pets and farm animals free from infection.

Reading Check *What are some methods used to prevent viral infections?*

Natural Immunity One way your body can stop viral infections is by making interferons. Interferons are proteins that protect cells from viruses. These proteins are produced rapidly by infected cells and move to noninfected cells in the host. They cause the noninfected cells to produce protective substances.

Vaccines Vaccinations, like the one the child is receiving in **Figure 19,** are used to prevent disease. A vaccine is made from weakened virus particles that cause your body to produce interferons to fight the infection. Vaccines have been made to prevent many diseases, including measles, mumps, smallpox, chicken pox, polio, and rabies. Edward Jenner is credited with developing the first vaccine in 1796. He injected a weakened form of the cowpox virus into healthy people, which protected them from smallpox. Jenner did not know how his virus worked, or that smallpox and cowpox were related viruses.

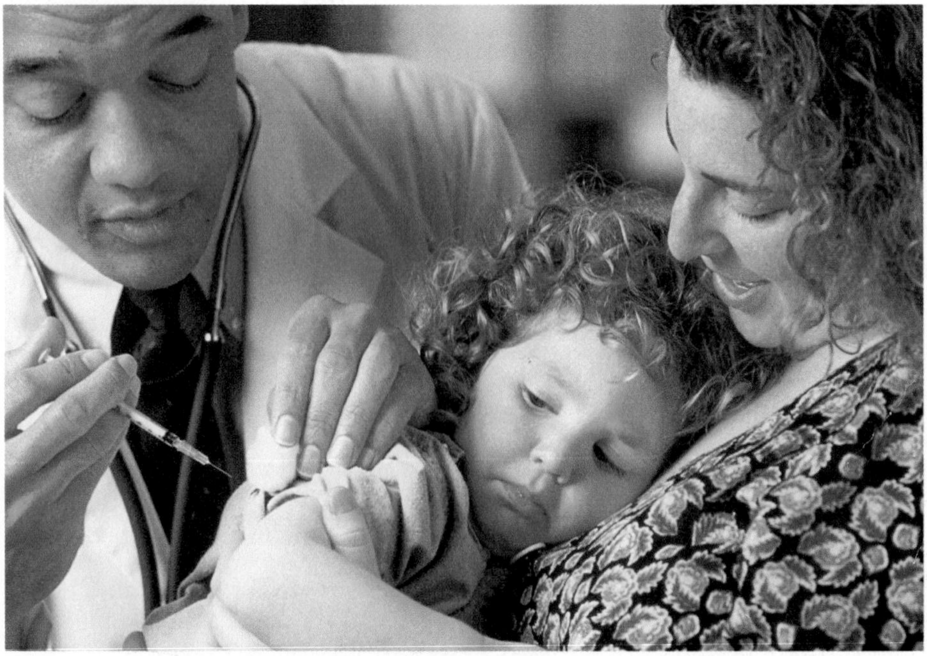

Figure 19 A child is getting vaccinated before entering school. **Describe** *how a vaccine works.*

Research with Viruses

You might think viruses are always harmful. However, scientists are discovering helpful uses for some viruses through research. One use, called gene therapy, is being tried on cells with defective hereditary material. Normal hereditary material is enclosed in viruses. The viruses then "infect" defective cells, taking the new hereditary material into the cells to replace the defective material. Using gene therapy, scientists hope to help people with genetic disorders and find a cure for cancer.

Reading Check *How does gene therapy work?*

Table 3 HIV/AIDS in the World

Adults age 15–49 with HIV/AIDS, 2001	37,100,000
New HIV infections, 2001	5,000,000
Adult HIV prevalence (%), 2001	1.20
Women age 15–49 with HIV/AIDS, 2001	18,500,000
Children with HIV/AIDS, 2001	3,000,000
AIDS deaths, 2001	3,000,000

Source: UNAIDS

An active area of viral research is HIV/AIDS research. HIV stands for human immuno-deficiency virus, a virus that attacks the immune system. The immune system is the system that protects your body from disease. Eventually, this virus leads to Acquired Immune Deficiency Syndrome, or AIDS. A weak immune system allows the body to be attacked by other diseases and infections, like pneumonia and certain types of cancer. AIDS occurs worldwide, with 95 percent of the cases in developing countries. **Table 3** shows some recent calculations of infected individuals. Currently, there is no known cure for AIDS. The research will hopefully lead to better treatments, a vaccine, and eventually a cure.

section 4 review

Summary

What are viruses?

- A virus is a strand of hereditary material surrounded by a protein coating, which multiplies by making copies of itself using a host cell.
- Almost every living thing can be infected by a virus.

Fighting Viruses

- Viruses can be prevented by vaccines, antiviral drugs, improving sanitary conditions, separating infected patients, and controlling animals that spread disease.
- Your body produces interferons to naturally protect itself.

Self Check

1. **List** the steps an active virus follows when attacking a cell.
2. **Compare and contrast** a latent virus and an active virus. Give an example of each.
3. **Explain** how vaccines prevent infection.
4. **Infer** how some viruses might be helpful.
5. **Think Critically** Explain why a doctor might not give you any medication if you have a viral disease.

Applying Skills

6. **Interpret Scientific Illustrations** Using **Figure 18,** explain why most viruses infect one species.
7. **Research Information** Find one precaution your community uses to prevent viral disease.

Design Your Own

Comparing Light Microscopes

Goals

- **Learn** how to correctly use a stereomicroscope and a compound light microscope.
- **Compare** the uses of the stereomicroscope and compound light microscope.

Possible Materials

compound light microscope
stereomicroscope
items from the classroom—include some living or once-living items (8)
microscope slides and coverslips
plastic petri dishes
distilled water
dropper

Safety Precautions

Real-World Question

You're a technician in a police forensic laboratory. You use a stereomicroscope, which uses two eyepieces to see larger objects in three dimensions, and a compound light microscope to see a smaller specimen. A detective just returned from a crime scene with bags of evidence. You must examine each piece of evidence under a microscope. How do you decide which microscope is the best tool to use? Will all of the evidence that you've collected be viewable through both microscopes?

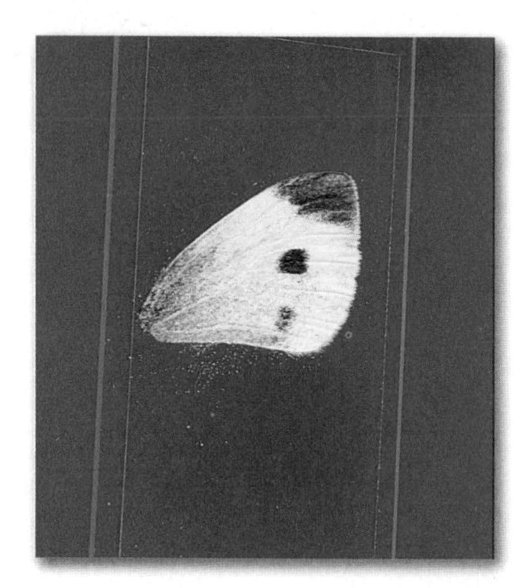

Form a Hypothesis

Form a hypothesis to predict which items in an actual police laboratory, such as blood and dirt samples, would be viewed under a compound light microscope. A stereomicroscope? What qualifications would the technician use to decide?

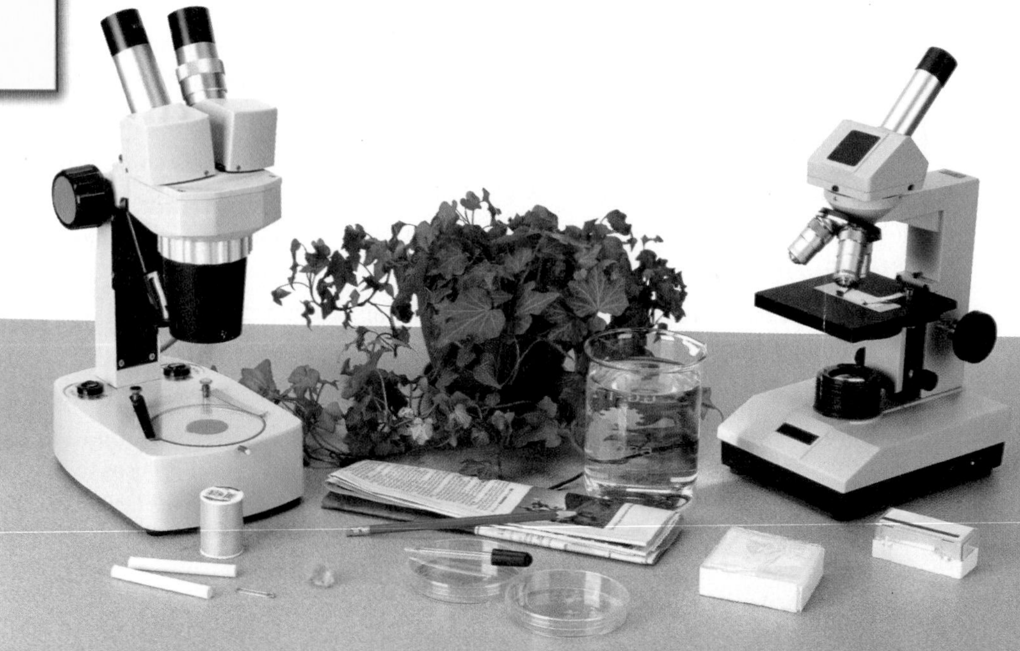

Test Your Hypothesis

Make a Plan

1. As a group, decide how you will test your hypothesis.
2. **Describe** how you will carry out this experiment using a series of specific steps. Make sure the steps are in a logical order. Remember that you must place an item in the bottom of a plastic petri dish to examine it under the stereomicroscope, and you must make a wet mount of any item to be examined under the compound light microscope. For more help, see the Reference Handbook.
3. If you need a data table or an observation table, design one in your Science Journal.

Follow Your Plan

1. Make sure your teacher approves the objects you'll examine, your plan, and your data table before you start.
2. Carry out the experiment.
3. While doing the experiment, record your observations and complete the data table.

Analyze Your Data

1. **Compare** the items you examined with those of your classmates.
2. Based on this experiment, classify the eight items you observed.

Conclude and Apply

1. Were you correct in your original hypothesis about the correct microscope to use? For which objects would you reconsider the microscope used?
2. **Infer** which microscope a scientist might use to examine a blood sample, fibers, and live snails.
3. **List** five careers that require people to use a stereomicroscope. List five careers that require people to use a compound light microscope. Enter the lists in your Science Journal.
4. **Describe** If you examined an item under a compound light microscope and a stereomicroscope, how would the images differ?
5. **Name** the microscope that was better for looking at large or possibly live items.

Communicating Your Data

In your Science Journal, write a short description of an imaginary crime scene and the evidence found there. Sort the evidence into two lists—items to be examined under a stereomicroscope and items to be examined under a compound light microscope.

Cobb Against Cancer

This colored scanning electron micrograph (SEM) shows two breast cancer cells in the final stage of cell division.

Jewel Plummer Cobb is a cell biologist who did important background research on the use of drugs against cancer. She removed cells from cancerous tumors and cultured them in the lab. Then, in a controlled study, she tried a series of different drugs against batches of the same cells. Her goal was to find the right drug to cure each patient's particular cancer. Cobb never met that goal, but her research laid the groundwork for modern chemotherapy—the use of chemicals to treat people with cancer.

Jewel Plummer Cobb also has influenced sciences as dean or president of several universities. She was able to promote equal opportunity for students of all backgrounds, especially in the sciences. She retired in 1990 from her post as president of California State University at Fullerton, and continues to be active in the sciences.

Light Up a Cure

Building on Cobb's work, Professor Julia Levy and her research team at the University of British Columbia actually go inside cells, and even organelles, to work against cancer. One technique they are pioneering is the use of light to guide cancer drugs to the right cells. First, the patient is given a chemotherapy drug that reacts to light. Next, a fiber optic tube is inserted into the tumor. Finally, laser light is passed through the tube. The light activates the light-sensitive drug—but only in the tumor itself. This technique keeps healthy cells healthy while killing sick cells.

Write Report on Cobb's experiments on cancer cells. What were her dependent and independent variables? What would she have used as a control? What sources of error did she have to guard against? Answer the same questions about Levy's work.

For more information, visit green.msscience.com/time

Reviewing Main Ideas

Section 1 Living Things

1. Organisms are made of cells, use energy, reproduce, respond, maintain homeostasis, grow, and develop.
2. Organisms need a food source, water, and a place to live.

Section 2 How are living things classified?

1. Scientists today use phylogeny to group organisms into six kingdoms.
2. All organisms are given a two-word scientific name using a system called binomial nomenclature.

Section 3 Cell Structure

1. The invention of the microscope led to the cell theory.
2. Cells without membrane-bound structures are prokaryotic cells. Cells with membrane-bound structures are eukaryotic cells.
3. Most many-celled organisms are organized into tissues, organs, and organ systems.

Section 4 Viruses

1. A virus is a nonliving structure containing hereditary material surrounded by a protein coating. It can only make copies of itself when inside a living host cell.
2. Viruses cause diseases in organisms.

Visualizing Main Ideas

Copy and complete the following concept map of the basic units of life.

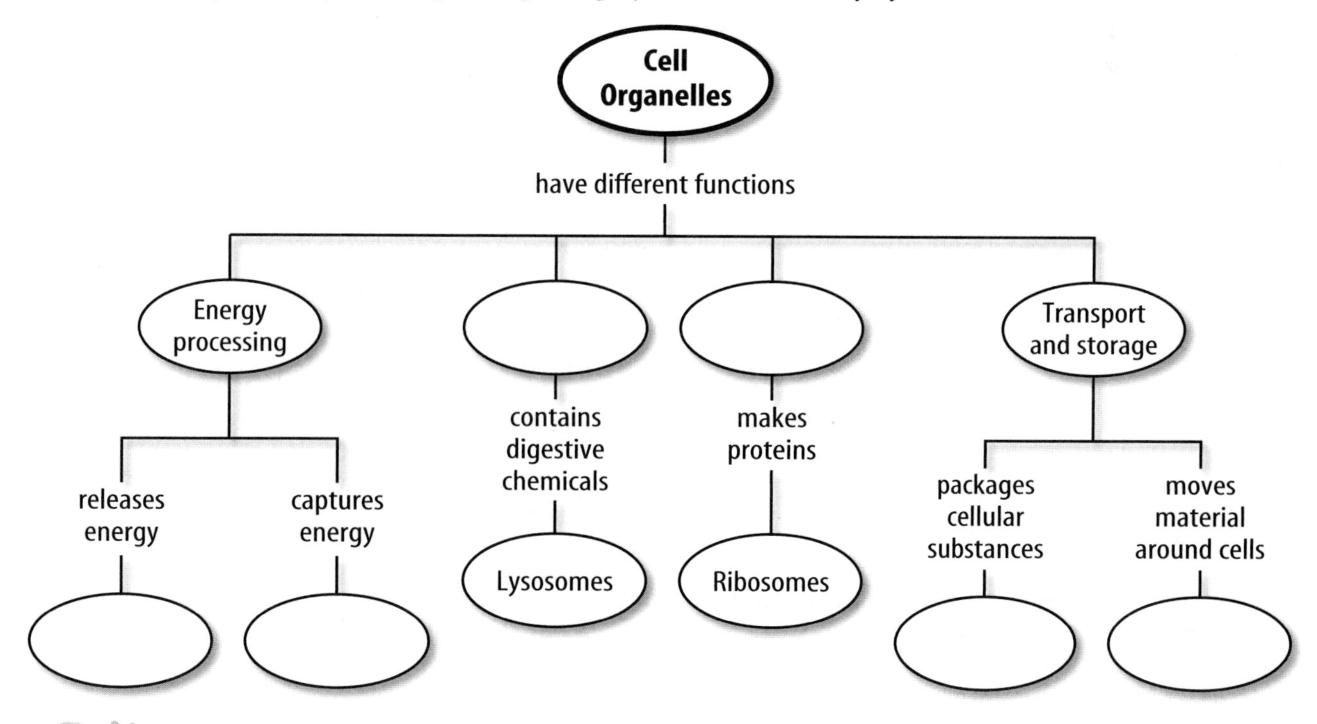

Using Vocabulary

Write the correct vocabulary word for each phrase.

1. the smallest unit of life
2. maintaining proper internal conditions
3. similar structures
4. *Homo sapiens*
5. directs all cell activities
6. plant-cell organelle that processes energy
7. powerhouse of a cell
8. a structure that surrounds every cell
9. made up of cells
10. contains only hereditary material

Checking Concepts

Choose the word or phrase that best answers the question.

11. What category of organisms can mate and produce fertile offspring?
    A) family **C)** genus
    B) class **D)** species

12. What is the closest relative of *Canis lupus?*
    A) *Quercus alba* **C)** *Felis tigris*
    B) *Equus zebra* **D)** *Canis familiaris*

13. What is the source of energy for plants?
    A) the Sun **C)** water
    B) carbon dioxide **D)** oxygen

14. What is the length of time an organism is expected to live?
    A) life span **C)** homeostasis
    B) stimulus **D)** theory

15. What is the first word in a two-word name of an organism?
    A) kingdom **C)** phylum
    B) species **D)** genus

Use the following photo to answer question 16.

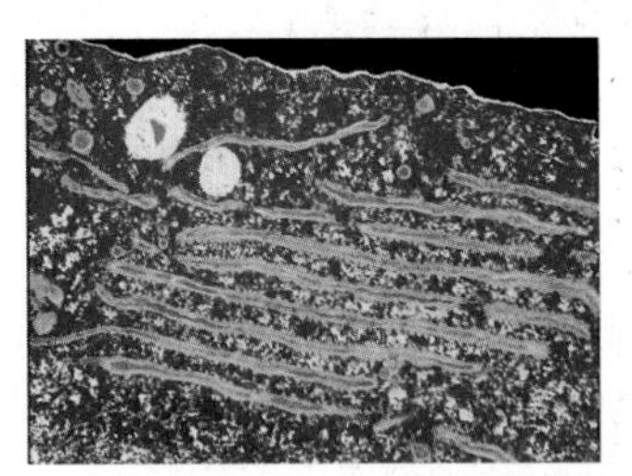

16. Identify the folded membranes that move materials around the cell, pictured above.
    A) nucleus
    B) cytoplasm
    C) Golgi body
    D) endoplasmic reticulum

17. What are the structures in the cytoplasm of a eukaryotic cell called?
    A) organs **C)** organ systems
    B) organelles **D)** tissues

18. Groups of different organ systems form which of the following?
    A) organ **C)** organ system
    B) organelle **D)** organism

19. According to the cell theory, what is the basic unit of life?
    A) the Sun **C)** a cell
    B) a tissue **D)** a chloroplast

20. Where does a virus multiply?
    A) in the ground **C)** a host cell
    B) in water **D)** in the air

Science Online green.msscience.com/vocabulary_puzzlemaker

Thinking Critically

21. **Draw Conclusions** Using a bird as an example, explain how it has the traits of living things.

22. **Explain** what binomial nomenclature is and why it is important.

23. **Infer** what *Lathyrus odoratus,* the name for a sweet pea, tells you about one of its characteristics.

24. **Determine** why it is difficult to treat a viral disease.

25. **Predict** what would happen to a plant cell that suddenly lost its chloroplasts.

26. **Concept Map** Make an events-chain concept map of the following from simple to complex: *small intestine, circular muscle cell, human,* and *digestive system.*

27. **Interpret Scientific Illustrations** Use the illustrations in **Figure 1** to describe how the shape of a cell is related to its function.

28. **Compare and Contrast** Copy and complete the following table to compare and contrast the structures of a prokaryotic cell to those of a eukaryotic cell.

Cell Structure

Structure	Prokaryotic Cell	Eukaryotic Cell
Cell Membrane		Yes
Cytoplasm	Yes	
Nucleus		Yes
Endoplasmic Reticulum		
Golgi Bodies		

29. **Describe** how you would decide if a cell was animal, plant, or prokaryotic.

Performance Activities

30. **Make a Time Line** Make and illustrate a time line to show the development of the cell theory. Begin with the development of the microscope and end with Virchow. Include the contributions of Leeuwenhoek, Hooke, Schleiden, and Schwann.

31. **Make a Model** Use materials that resemble cell parts or that represent their functions to make a model of a plant cell or an animal cell. Make a key to the cell parts to explain your model.

32. **Make a Brochure** Research the types of vaccinations required in your community and at what ages they are usually administered. Contact your local Health Department for current information. Prepare a brochure for new mothers.

Applying Math

Use the table below to answer questions 33 and 34.

Temperature v. Number of Viruses

Temperature	Number of Viruses
37°C	1.0 million
37.5°C	0.5 million
37.8°C	0.25 million
38.3°C	0.1 million
38.9°C	0.05 million

33. **Temperature Dependence** Use a computer to make a line graph of the above data.

34. **Number of Viruses** From the line graph you have just made, determine the number of viruses at 37.3°C

chapter 8 Standardized Test Practice

Part 1 Multiple Choice

Record your answers on the answer sheet provided by your teacher or on a sheet of paper.

Use the photo below to answer question 1.

1. The puffball above is releasing millions of spores. What characteristic of living things is represented by this picture?
   A. reproduction
   B. development
   C. organization
   D. use of energy

2. Which of the following is the best example of homeostasis?
   A. the ability of plants to use the Sun's energy, carbon dioxide and water to make food
   B. the movement of a cat toward food
   C. the ability of some organisms to keep their temperature within a certain range
   D. the production of eggs by most insects

3. What plant cell structure is made up mostly of cellulose?
   A. cell wall
   B. chloroplast
   C. cell membrane
   D. mitochondrion

Test-Taking Tip

Read Carefully Read each question carefully for full understanding.

Question 2 Reread this question so that you thoroughly understand it. Also, read all of the choices before deciding which one is the best answer to the question.

4. Which one of the following structures is found in prokaryotic cells?
   A. chloroplast
   B. Golgi bodies
   C. mitochondrion
   D. ribosomes

5. Which of the following is the evolutionary history of an organism?
   A. phylogeny
   B. classification
   C. nomenclature
   D. dichotomy

6. What is the two-word system used to name various species?
   A. binomial species
   B. binomial nomenclature
   C. dichotomous key
   D. modern classification

Use the illustration below to answer questions 7 and 8.

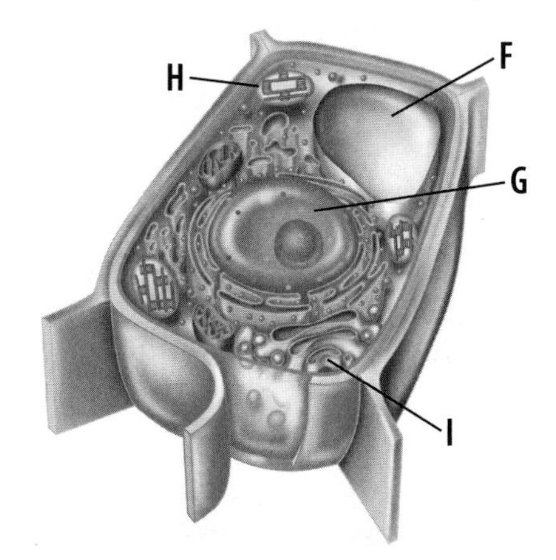

7. Which letter corresponds to the nucleus?
   A. F
   B. G
   C. H
   D. I

8. Which part is used by plant cells to make food?
   A. F
   B. G
   C. H
   D. I

9. What are the smallest units that make up your body called?
   A. cells
   B. tissues
   C. organisms
   D. organs

Part 2 | Short Response/Grid In

Record your answers on the answer sheet provided by your teacher or on a sheet of paper.

10. Explain why your heart would be called an organ and not a tissue.

Use the illustration below to answer questions 11 and 12.

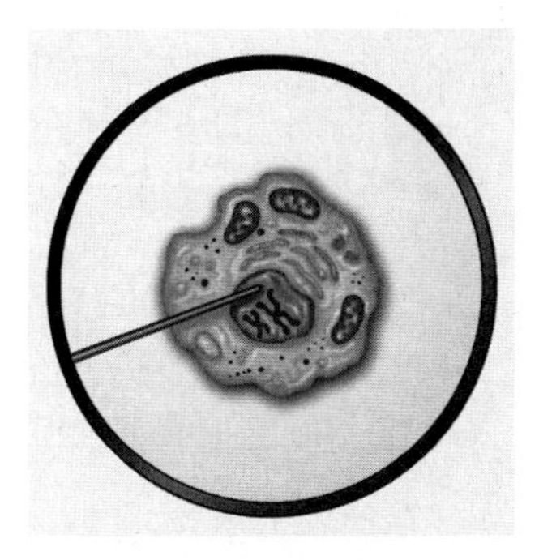

11. A scientist is studying living cells. Above is an image of one of the cells that is being studied. If the pointer shown above with the cell is 10 micrometers in length, then about how wide is this cell?

12. What tool was the scientist probably using to view the living cell?

13. List three different organisms whose cells are enclosed in cell walls.

14. Explain why water is important to living things.

15. A scientist observes a cell under a microscope and notices that the cell's cytoplasm is almost completely full of rough endoplasmic reticulum. What would you hypothesize these cells are making in large amounts? Explain your answer.

16. What is the smallest classification category?

17. What are four reasons scientific names are used to describe organisms rather than common names?

Part 3 | Open Ended

Record your answers on a sheet of paper.

18. Explain stimulus and response. How is response to a stimulus related to homeostasis? Give an example using the stimulus of heat.

19. Draw diagrams of a prokaryotic cell and of a eukaryotic cell. Label the parts to show the differences between the two cell types.

20. How might you go about writing a field guide to use to identify plants that might grow in a garden? What would you need to do if you wanted to include a dichotomous key in the field guide?

21. Explain how DNA is related to the chromosomes found in the nucleus of a cell. What is the importance of DNA in the nucleus?

22. The nucleus of a cell directs all the activities of a cell. But bacteria, which are prokaryotic cells, do not have nuclei. Explain how bacterial cells can function without a nucleus.

Use the photo below to answer questions 23 and 24.

23. Both of these organisms use energy. Describe the difference between their sources of energy. What are the common ways that these organisms use energy?

24. How are the needs of the two organisms alike? Explain why the plant fulfills a need for the beetle. When the beetle dies, how could it fill a need for the plant?

chapter 9

Cell Processes

chapter preview

sections

The Science of Gardening

Growing a garden is hard work for both you and the plants. Like you, plants need water and food for energy. How plants get food and water is different from you. Understanding how living things get the energy they need to survive will make a garden seem like much more than just plants and dirt.

Science Journal Describe two ways in which you think plants get food for energy.

Start-Up Activities

Why does water enter and leave plant cells?

If you forget to water a plant, it will wilt. After you water the plant, it probably will straighten up and look healthier. In the following lab, find out how water causes a plant to wilt and straighten.

1. Label a small bowl *Salt Water.* Pour 250 mL of water into the bowl. Then add 15 g of salt to the water and stir.
2. Pour 250 mL of water into another small bowl.
3. Place two carrot sticks into each bowl. Also, place two carrot sticks on the lab table.
4. After 30 min, remove the carrot sticks from the bowls and keep them next to the bowl they came from. Examine all six carrot sticks, then describe them in your Science Journal.
5. **Think Critically** Write a paragraph in your Science Journal that describes what would happen if you moved the carrot sticks from the plain water to the lab table, the ones from the salt water into the plain water, and the ones from the lab table into the salt water for 30 min. Now move the carrot sticks as described and write the results in your Science Journal.

FOLDABLES™ Study Organizer

How Living Things Survive Make the following vocabulary Foldable to help you understand the chemistry of living things and how energy is obtained for life.

STEP 1 **Fold** a vertical sheet of notebook paper from side to side.

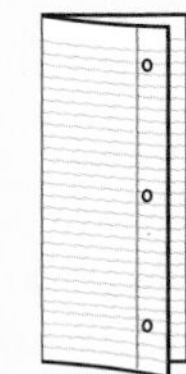

STEP 2 **Cut** along every third line of only the top layer to form tabs.

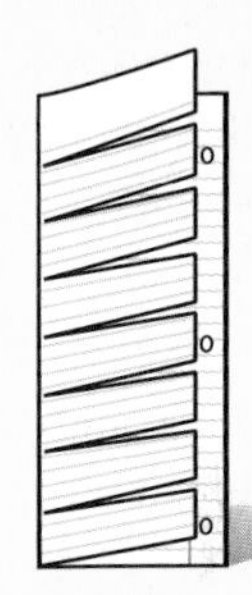

Build Vocabulary As you read this chapter, list the vocabulary words about cell processes on the tabs. As you learn the definitions, write them under the tab for each vocabulary word. Write a sentence about one of the cell processes using the vocabulary word on the tab.

Preview this chapter's content and activities at green.msscience.com

section 1

Chemistry of Life

as you read

What You'll Learn

- **List** the differences among atoms, elements, molecules, and compounds.
- **Explain** the relationship between chemistry and life science.
- **Discuss** how organic compounds are different from inorganic compounds.

Why It's Important

You grow because of chemical reactions in your body.

Review Vocabulary
cell: the smallest unit of a living thing that can perform the functions of life

New Vocabulary
- mixture
- organic compound
- enzyme
- inorganic compound

The Nature of Matter

Think about everything that surrounds you—chairs, books, clothing, other students, and air. What are all these things made up of? You're right if you answer "matter and energy." Matter is anything that has mass and takes up space. Energy is anything that brings about change. Everything in your environment, including you, is made of matter. Energy can hold matter together or break it apart. For example, the food you eat is matter that is held together by chemical energy. When food is cooked, energy in the form of heat can break some of the bonds holding the matter in food together.

Atoms Whether it is solid, liquid, or gas, matter is made of atoms. **Figure 1** shows a model of an oxygen atom. At the center of an atom is a nucleus that contains protons and neutrons. Although they have nearly equal masses, a proton has a positive charge and a neutron has no charge. Outside the nucleus are electrons, each of which has a negative charge. It takes about 1,837 electrons to equal the mass of one proton. Electrons are important because they are the part of the atom that is involved in chemical reactions. Look at **Figure 1** again and you will see that an atom is mostly empty space. Energy holds the parts of an atom together.

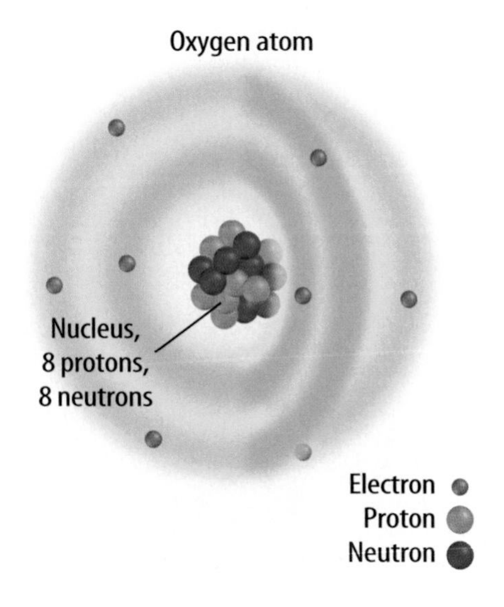

Figure 1 An oxygen atom model shows the placement of electrons, protons, and neutrons.

Table 1 Elements in the Human Body

Symbol	Element	Percent
O	Oxygen	65.0
C	Carbon	18.5
H	Hydrogen	9.5
N	Nitrogen	3.2
Ca	Calcium	1.5
P	Phosphorus	1.0
K	Potassium	0.4
S	Sulfur	0.3
Na	Sodium	0.2
Cl	Chlorine	0.2
Mg	Magnesium	0.1
	Other elements	0.1

Elements When something is made up of only one kind of atom, it is called an element. An element can't be broken down into a simpler form by chemical reactions. The element oxygen is made up of only oxygen atoms, and hydrogen is made up of only hydrogen atoms. Scientists have given each element its own one- or two-letter symbol.

All elements are arranged in a chart known as the periodic table of elements. You can find this table at the back of this book. The table provides information about each element including its mass, how many protons it has, and its symbol.

Everything is made up of elements. Most things, including all living things, are made up of a combination of elements. Few things exist as pure elements. **Table 1** lists elements that are in the human body. What two elements make up most of your body?

Six of the elements listed in the table are important because they make up about 99 percent of living matter. The symbols for these elements are S, P, O, N, C, and H. Use **Table 1** to find the names of these elements.

Reading Check *What types of things are made up of elements?*

Figure 2 The words *atoms, molecules,* and *compounds* are used to describe substances.
Explain *how these terms are related to each other.*

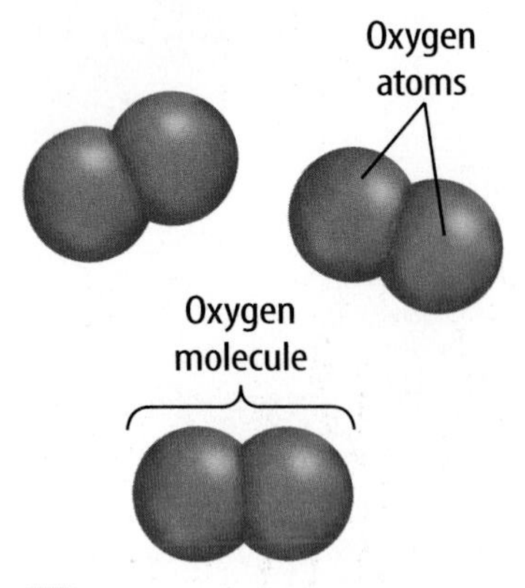

A Some elements, like oxygen, occur as molecules. These molecules contain atoms of the same element bonded together.

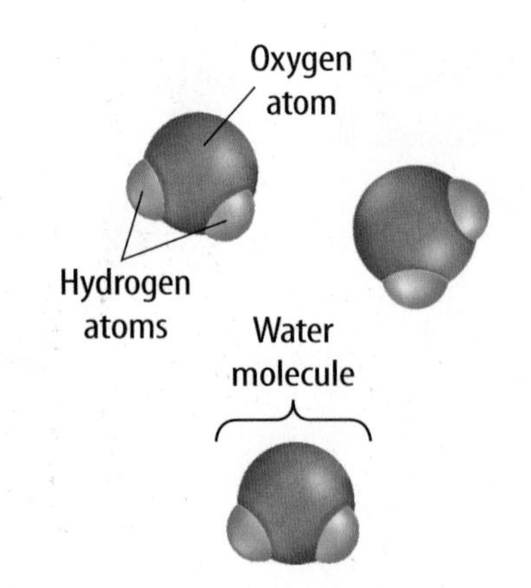

B Compounds also are composed of molecules. Molecules of compounds contain atoms of two or more different elements bonded together, as shown by these water molecules.

Compounds and Molecules

Suppose you make a pitcher of lemonade using a powdered mix and water. The water and the lemonade mix, which is mostly sugar, contain the elements oxygen and hydrogen. Yet, in one, they are part of a nearly tasteless liquid—water. In the other they are part of a sweet solid—sugar. How can the same elements be part of two materials that are so different? Water and sugar are compounds. Compounds are made up of two or more elements in exact proportions. For example, pure water, whether one milliliter of it or one million liters, is always made up of hydrogen atoms bonded to oxygen atoms in a ratio of two hydrogen atoms to one oxygen atom. Compounds have properties different from the elements they are made of. There are two types of compounds—molecular compounds and ionic compounds.

Molecular Compounds The smallest part of a molecular compound is a molecule. A molecule is a group of atoms held together by the energy of chemical bonds, as shown in **Figure 2.** When chemical reactions occur, chemical bonds break, atoms are rearranged, and new bonds form. The molecules produced are different from those that began the chemical reaction.

Molecular compounds form when different atoms share their outermost electrons. For example, two atoms of hydrogen each can share one electron on one atom of oxygen to form one molecule of water, as shown in **Figure 2B.** Water does not have the same properties as oxygen and hydrogen. Under normal conditions on Earth, oxygen and hydrogen are gases. Yet, water can be a liquid, a solid, or a gas. When hydrogen and oxygen combine, changes occur and a new substance forms.

Ions Atoms also combine because they've become positively or negatively charged. Atoms are usually neutral—they have no overall electric charge. When an atom loses an electron, it has more protons than electrons, so it becomes positively charged. When an atom gains an electron, it has more electrons than protons, so it becomes negatively charged. Electrically charged atoms—positive or negative—are called ions.

Ionic Compounds Ions of opposite charges attract one another to form electrically neutral compounds called ionic compounds. Table salt is made of sodium (Na) and chlorine (Cl) ions, as shown in **Figure 3B.** When they combine, a chlorine atom gains an electron from a sodium atom. The chlorine atom becomes a negatively charged ion, and the sodium atom becomes a positively charged ion. These oppositely charged ions then are attracted to each other and form the ionic compound sodium chloride, NaCl.

Ions are important in many life processes that take place in your body and in other organisms. For example, messages are sent along your nerves as potassium and sodium ions move in and out of nerve cells. Calcium ions are important in causing your muscles to contract. Ions also are involved in the transport of oxygen by your blood. The movement of some substances into and out of a cell would not be possible without ions.

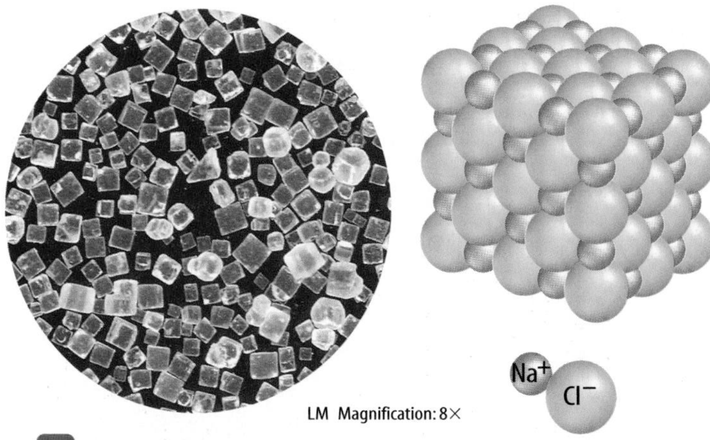

A Magnified crystals of salt look like this.

B A salt crystal is held together by the attractions between sodium ions and chlorine ions.

Figure 3 Table salt crystals are held together by ionic bonds.

Mixtures

Some substances, such as a combination of sugar and salt, can't change each other or combine chemically. A **mixture** is a combination of substances in which individual substances retain their own properties. Mixtures can be solids, liquids, gases, or any combination of them.

Reading Check *Why is a combination of sugar and salt said to be a mixture?*

Most chemical reactions in living organisms take place in mixtures called solutions. You've probably noticed the taste of salt when you perspire. Sweat is a solution of salt and water. In a solution, two or more substances are mixed evenly. A cell's cytoplasm is a solution of dissolved molecules and ions.

Living things also contain mixtures called suspensions. A suspension is formed when a liquid or a gas has another substance evenly spread throughout it. Unlike solutions, the substances in a suspension eventually sink to the bottom. If blood, shown in **Figure 4,** is left undisturbed, the red blood cells and white blood cells will sink gradually to the bottom. However, the pumping action of your heart constantly moves your blood and the blood cells remain suspended.

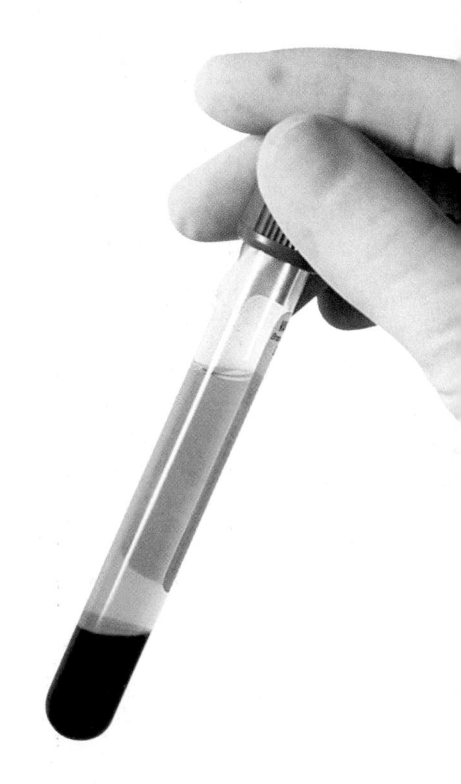

Figure 4 When a test tube of whole blood is left standing, the blood cells sink in the watery plasma.

Table 2 Organic Compounds Found in Living Things

	Carbohydrates	Lipids	Proteins	Nucleic Acids
Elements	carbon, hydrogen, and oxygen	carbon, oxygen, hydrogen, and phosphorus	carbon, oxygen, hydrogen, nitrogen, and sulfur	carbon, oxygen, hydrogen, nitrogen, and phosphorus
Examples	sugars, starch, and cellulose	fats, oils, waxes, phospholipids, and cholesterol	enzymes, skin, and hair	DNA and RNA
Function	supply energy for cell processes; form plant structures; short-term energy storage	store large amounts of energy long term; form boundaries around cells	regulate cell processes and build cell structures	carry hereditary information; used to make proteins

Organic Compounds

You and all living things are made up of compounds that are classified as organic or inorganic. Rocks and other nonliving things contain inorganic compounds, but most do not contain large amounts of organic compounds. **Organic compounds** always contain carbon and hydrogen and usually are associated with living things. One exception would be nonliving things that are products of living things. For example, coal contains organic compounds because it was formed from dead and decaying plants. Organic molecules can contain hundreds or even thousands of atoms that can be arranged in many ways. **Table 2** compares the four groups of organic compounds that make up all living things—carbohydrates, lipids, proteins, and nucleic acids.

Topic: Air Quality
Visit green.msscience.com for Web links to information about air quality.

Activity Organic compounds such as soot, smoke, and ash can affect air quality. Look up the air quality forecast for today. List three locations where the air quality forecast is good, and three locations where it is unhealthy.

Carbohydrates Carbohydrates are organic molecules that supply energy for cell processes. Sugars and starches are carbohydrates that cells use for energy. Some carbohydrates also are important parts of cell structures. For example, a carbohydrate called cellulose is an important part of plant cells.

Lipids Another type of organic compound found in living things is a lipid. Lipids do not mix with water. Lipids such as fats and oils store and release even larger amounts of energy than carbohydrates do. One type of lipid, the phospholipid, is a major part of cell membranes.

What are three types of lipids?

Proteins Organic compounds called proteins have many important functions in living organisms. They are made up of smaller molecules called amino acids. Proteins are the building blocks of many structures in organisms. Your muscles contain large amounts of protein. Proteins are scattered throughout cell membranes. Certain proteins called **enzymes** regulate nearly all chemical reactions in cells.

Nucleic Acids Large organic molecules that store important coded information in cells are called nucleic acids. One nucleic acid, deoxyribonucleic acid, or DNA—genetic material—is found in all cells at some point in their lives. It carries information that directs each cell's activities. Another nucleic acid, ribonucleic acid, or RNA, is needed to make enzymes and other proteins.

Inorganic Compounds

Most **inorganic compounds** are made from elements other than carbon. Generally, inorganic molecules contain fewer atoms than organic molecules. Inorganic compounds are the source for many elements needed by living things. For example, plants take up inorganic compounds from the soil. These inorganic compounds can contain the elements nitrogen, phosphorus, and sulfur. Many foods that you eat contain inorganic compounds. **Table 3** shows some of the inorganic compounds that are important to you. One of the most important inorganic compounds for living things is water.

Observing How Enzymes Work

Procedure

1. Get two small cups of **prepared gelatin** from your teacher. Do not eat or drink anything in lab.
2. On the gelatin in one of the cups, place a piece of **fresh pineapple.**
3. Let both cups stand undisturbed overnight.
4. Observe what happens to the gelatin.

Analysis

1. What effect did the piece of fresh pineapple have on the gelatin?
2. What does fresh pineapple contain that caused it to have the effect on the gelatin you observed?
3. Why do the preparation directions on a box of gelatin dessert tell you not to mix it with fresh pineapple?

Table 3 Some Inorganic Compounds Important in Humans

Compound	Use in Body
Water	makes up most of the blood; most chemical reactions occur in water
Calcium phosphate	gives strength to bones
Hydrochloric acid	breaks down foods in the stomach
Sodium bicarbonate	helps the digestion of food to occur
Salts containing sodium, chlorine, and potassium	important in sending messages along nerves

Importance of Water Some scientists hypothesize that life began in the water of Earth's ancient oceans. Chemical reactions might have occurred that produced organic molecules. Similar chemical reactions can take place in cells in your body.

Living things are composed of more than 50 percent water and depend on water to survive. You can live for weeks without food but only for a few days without water. **Figure 5** shows where water is found in your body. Although seeds and spores of plants, fungi, and bacteria can exist without water, they must have water if they are to grow and reproduce. All the chemical reactions in living things take place in water solutions, and most organisms use water to transport materials through their bodies. For example, many animals have blood that is mostly water and moves materials. Plants use water to move minerals and sugars between the roots and leaves.

Applying Math — Solve an Equation

CALCULATE THE IMPORTANCE OF WATER All life on Earth depends on water for survival. Water is the most vital part of humans and other animals. It is required for all of the chemical processes that keep us alive. At least 60 percent of an adult human body consists of water. If an adult man weighs 90 kg, how many kilograms of water does his body contain?

Solution

1. *This is what you know:*
   - adult human body = 60% water
   - man = 90 kg
2. *This is what you need to find:* How many kilograms of water does the adult man have?
3. *This is the procedure you need to use:*
   - Set up the ratio: $60/100 = x/90$.
   - Solve the equation for x: $(60 \times 90)/100$.
   - The adult man has 54 kg of water.
4. *Check your answer:* Divide your answer by 90, then multiply by 100. You should get 60%.

Practice Problems

1. A human body at birth consists of 78 percent water. This gradually decreases to 60 percent in an adult. Assume a baby weighed 3.2 kg at birth and grew into an adult weighing 95 kg. Calculate the approximate number of kilograms of water the human gained.
2. Assume an adult woman weighs 65 kg and an adult man weighs 90 kg. Calculate how much more water, in kilograms, the man has compared to the woman.

For more practice, visit green.msscience.com/math_practice

INTEGRATE Physics

Characteristics of Water The atoms of a water molecule are arranged in such a way that the molecule has areas with different charges. Water molecules are like magnets. The negative part of a water molecule is attracted to the positive part of another water molecule just like the north pole of a magnet is attracted to the south pole of another magnet. This attraction, or force, between water molecules is why a film forms on the surface of water. The film is strong enough to support small insects because the forces between water molecules are stronger than the force of gravity on the insect.

When heat is added to any substance, its molecules begin to move faster. Because water molecules are so strongly attracted to each other, the temperature of water changes slowly. The large percentage of water in living things acts like an insulator. The water in a cell helps keep its temperature constant, which allows life-sustaining chemical reactions to take place.

You've seen ice floating on water. When water freezes, ice crystals form. In the crystals, each water molecule is spaced at a certain distance from all the others. Because this distance is greater in frozen water than in liquid water, ice floats on water. Bodies of water freeze from the top down. The floating ice provides insulation from extremely cold temperatures and allows living things to survive in the cold water under the ice.

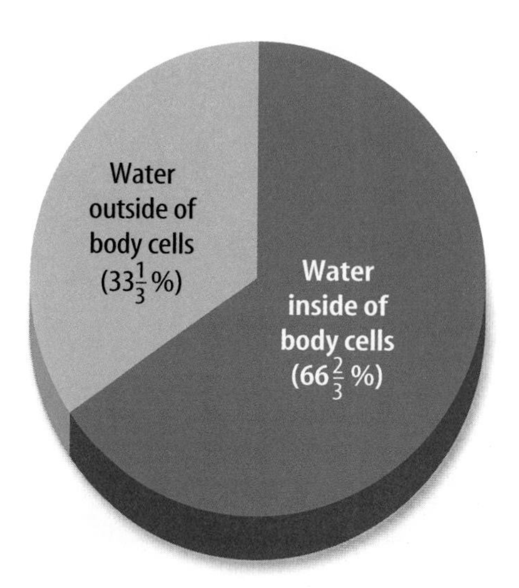

Figure 5 About two-thirds of your body's water is located within your body's cells. Water helps maintain the cells' shapes and sizes. One-third of your body's water is outside of your body's cells.

section 1 review

Summary

The Nature of Matter

- Atoms are made up of protons, neutrons, and electrons.
- Elements are made up of only one kind of atom.
- Compounds are made up of two or more elements.

Mixtures

- Solutions are made of two or more substances and are mixed evenly, whereas substances in suspension eventually will sink to the bottom.

Organic Compounds

- All living things contain organic compounds.

Inorganic Compounds

- Water is one of the most important inorganic compounds for living things.

Self Check

1. **Compare and contrast** atoms and molecules.
2. **Describe** the differences between an organic and an inorganic compound. Given an example of each type of compound.
3. **List** the four types of organic compounds found in all living things.
4. **Infer** why life as we know it depends on water.
5. **Think Critically** If you mix salt, sand, and sugar with water in a small jar, will the resulting mixture be a suspension, a solution, or both?

Applying Skills

6. **Interpret** Carefully observe **Figure 1** and determine how many protons, neutrons, and electrons an atom of oxygen has.

section 2

Moving Cellular Materials

as you read

What You'll Learn

- **Describe** the function of a selectively permeable membrane.
- **Explain** how the processes of diffusion and osmosis move molecules in living cells.
- **Explain** how passive transport and active transport differ.

Why It's Important

Cell membranes control the substances that enter and leave the cells in your body.

Review Vocabulary

cytoplasm: constantly moving gel-like mixture inside the cell membrane that contains hereditary material and is the location of most of a cell's life process

New Vocabulary

- passive transport
- diffusion
- equilibrium
- osmosis
- active transport
- endocytosis
- exocytosis

Passive Transport

"Close that window. Do you want to let in all the bugs and leaves?" How do you prevent unwanted things from coming through the window? As seen in **Figure 6,** a window screen provides the protection needed to keep unwanted things outside. It also allows some things to pass into or out of the room like air, unpleasant odors, or smoke.

Cells take in food, oxygen, and other substances from their environments. They also release waste materials into their environments. A cell has a membrane around it that works for a cell like a window screen does for a room. A cell's membrane is selectively permeable (PUR mee uh bul). It allows some things to enter or leave the cell while keeping other things outside or inside the cell. The window screen also is selectively permeable based on the size of its openings.

Things can move through a cell membrane in several ways. Which way things move depends on the size of the molecules or particles, the path taken through the membrane, and whether or not energy is used. The movement of substances through the cell membrane without the input of energy is called **passive transport.** Three types of passive transport can occur. The type depends on what is moving through the cell membrane.

Figure 6 A cell membrane, like a screen, will let some things through more easily than others. Air gets through a screen, but insects are kept out.

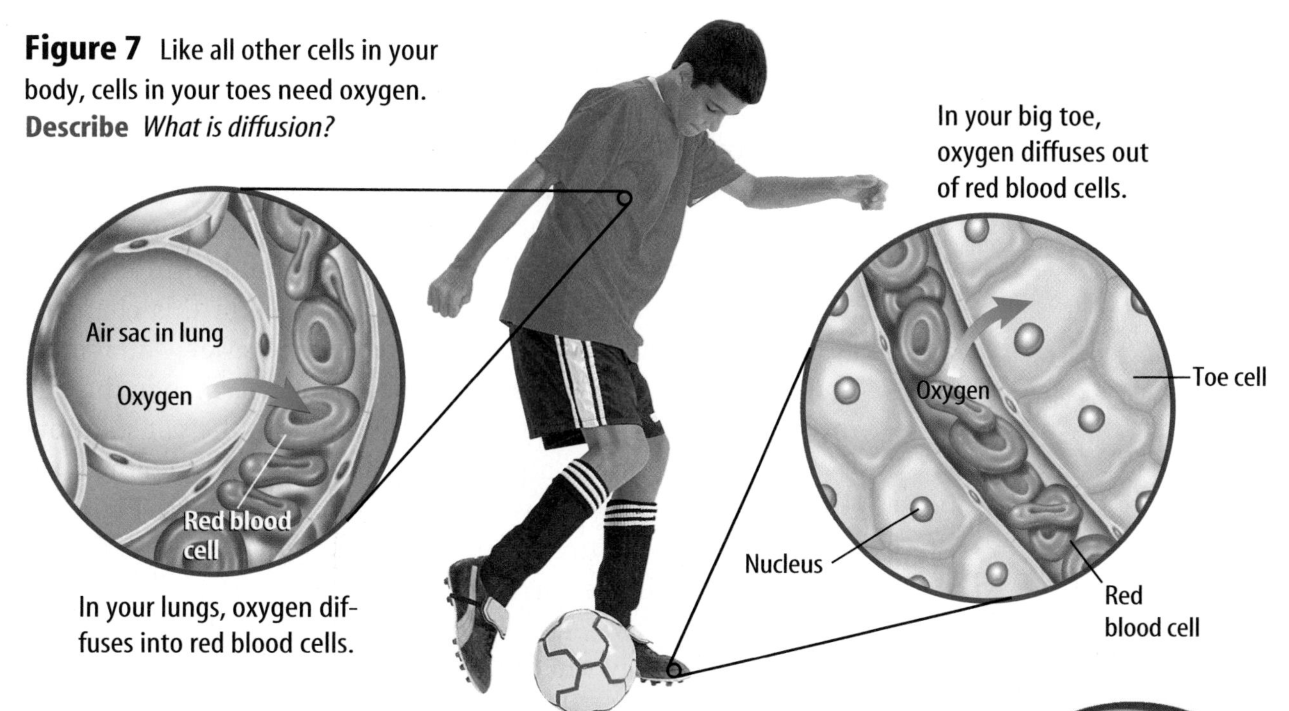

Figure 7 Like all other cells in your body, cells in your toes need oxygen.
Describe *What is diffusion?*

Diffusion Molecules in solids, liquids, and gases move constantly and randomly. You might smell perfume when you sit near or as you walk past someone who is wearing it. This is because perfume molecules randomly move throughout the air. This random movement of molecules from an area where there is relatively more of them into an area where there is relatively fewer of them is called **diffusion.** Diffusion is one type of cellular passive transport. Molecules of a substance will continue to move from one area into another until the relative number of these molecules is equal in the two areas. When this occurs, **equilibrium** is reached and diffusion stops. After equilibrium occurs, it is maintained because molecules continue to move.

Reading Check ***What is equilibrium?***

Every cell in your body uses oxygen. When you breathe, how does oxygen get from your lungs to cells in your big toe? Oxygen is carried throughout your body in your blood by the red blood cells. When your blood is pumped from your heart to your lungs, your red blood cells do not contain much oxygen. However, your lungs have more oxygen molecules than your red blood cells do, so the oxygen molecules diffuse into your red blood cells from your lungs, as shown in **Figure 7.** When the blood reaches your big toe, there are more oxygen molecules in your red blood cells than in your big toe cells. The oxygen diffuses from your red blood cells and into your big toe cells, as shown also in **Figure 7.**

Mini LAB

Observing Diffusion

Procedure

1. Use two clean glasses of equal size. Label one *Hot,* then fill it until half full with **very warm water.** Label the other *Cold,* then fill it until half full with **cold water. WARNING:** *Do not use boiling hot water.*
2. Add one drop of **food coloring** to each glass. Carefully release the drop just at the water's surface to avoid splashing the water.
3. Observe the water in the glasses. Record your observations immediately and again after 15 min.

Analysis

1. Describe what happens when food coloring is added to each glass.
2. How does temperature affect the rate of diffusion?

Try at Home

Osmosis—The Diffusion of Water Remember that water makes up a large part of living matter. Cells contain water and are surrounded by water. Water molecules move by diffusion into and out of cells. The diffusion of water through a cell membrane is called **osmosis.**

If cells weren't surrounded by water that contains few dissolved substances, water inside of cells would diffuse out of them. This is why water left the carrot cells in this chapter's Launch Lab. Because there were relatively fewer water molecules in the salt solution around the carrot cells than in the carrot cells, water moved out of the cells and into the salt solution.

Losing water from a plant cell causes its cell membrane to come away from its cell wall, as shown on the left in **Figure 8.** This reduces pressure against its cell wall, and a plant cell becomes limp. If the carrot sticks were taken out of salt water and put in pure water, the water around the cells would move into them and they would fill with water. Their cell membranes would press against their cell walls, as shown on the right in **Figure 8,** pressure would increase, and the cells would become firm. That is why the carrot sticks would be crisp again.

Reading Check *Why do carrots in salt water become limp?*

Osmosis also takes place in animal cells. If animal cells were placed in pure water, they too would swell up. However, animal cells are different from plant cells. Just like an overfilled water balloon, animal cells will burst if too much water enters the cell.

Figure 8 Cells respond to differences between the amount of water inside and outside the cell.
Define *What is osmosis?*

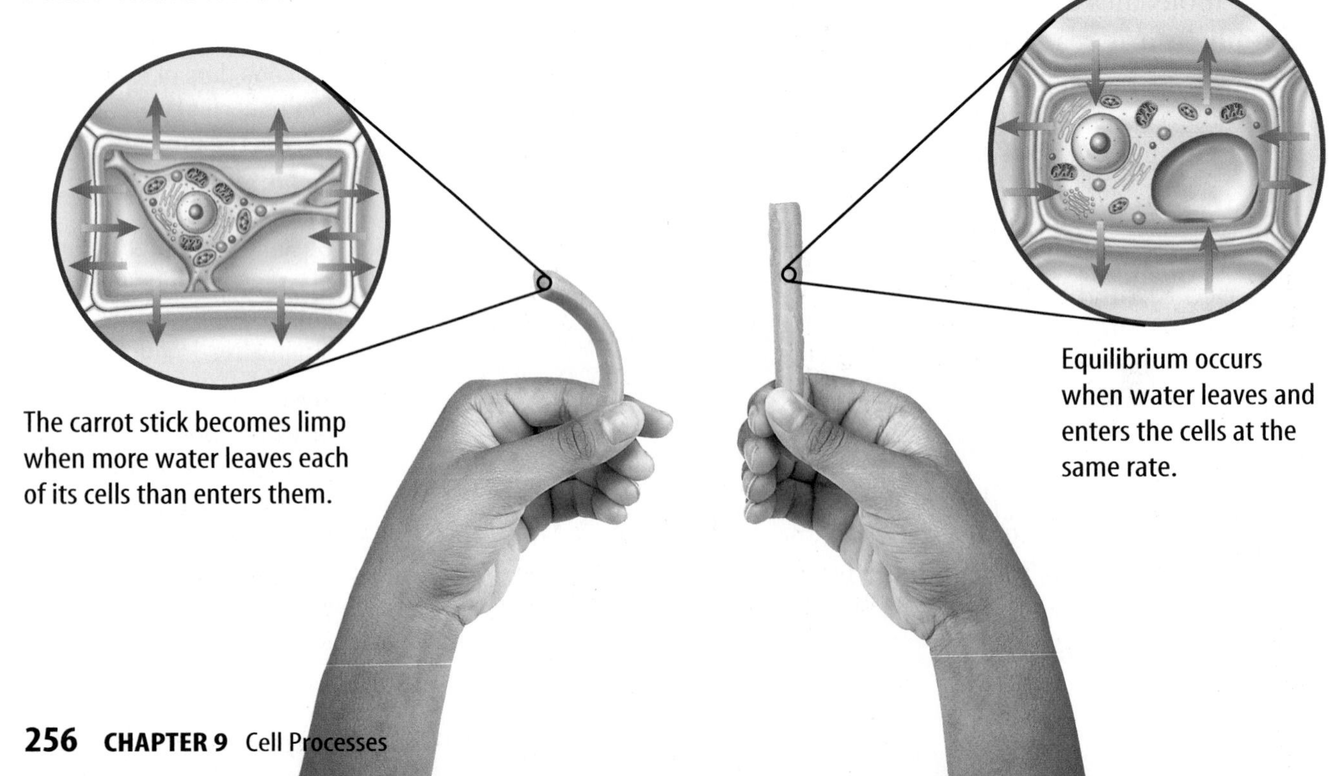

Facilitated Diffusion Cells take in many substances. Some substances pass easily through the cell membrane by diffusion. Other substances, such as glucose molecules, are so large that they can enter the cell only with the help of molecules in the cell membrane called transport proteins. This process, a type of passive transport, is known as facilitated diffusion. Have you ever used the drive through at a fast-food restaurant to get your meal? The transport proteins in the cell membrane are like the drive-through window at the restaurant. The window lets you get food out of the restaurant and put money into the restaurant. Similarly, transport proteins are used to move substances into and out of the cell.

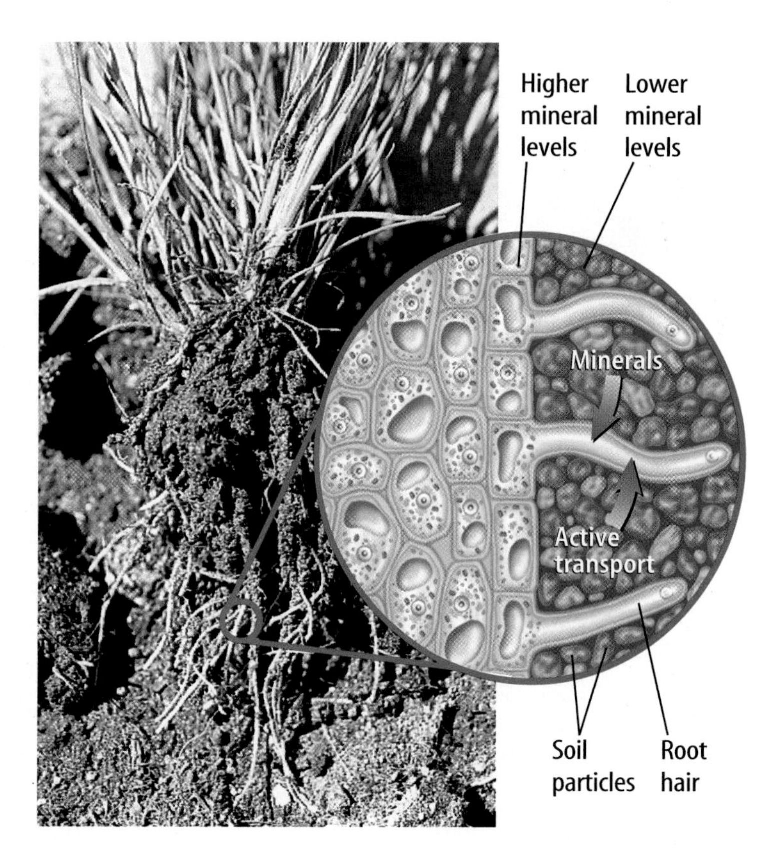

Figure 9 Some root cells have extensions called root hairs that may be 5 mm to 8 mm long. Minerals are taken in by active transport through the cell membranes of root hairs.

Active Transport

Imagine that a football game is over and you leave the stadium. As soon as you get outside of the stadium, you remember that you left your jacket on your seat. Now you have to move against the crowd coming out of the stadium to get back in to get your jacket. Which required more energy—leaving the stadium with the crowd or going back to get your jacket? Something similar to this happens in cells.

Sometimes, a substance is needed inside a cell even though the amount of that substance inside the cell is already greater than the amount outside the cell. For example, root cells require minerals from the soil. The roots of the plant in **Figure 9** already might contain more of those mineral molecules than the surrounding soil does. The tendency is for mineral molecules to move out of the root by diffusion or facilitated diffusion. But they need to move back across the cell membrane and into the cell just like you had to move back into the stadium. When an input of energy is required to move materials through a cell membrane, **active transport** takes place.

Active transport involves transport proteins, just as facilitated diffusion does. In active transport, a transport protein binds with the needed particle and cellular energy is used to move it through the cell membrane. When the particle is released, the transport protein can move another needed particle through the membrane.

Transport Proteins Your health depends on transport proteins. Sometimes transport proteins are missing or do not function correctly. What would happen if proteins that transport cholesterol across membranes were missing? Cholesterol is an important lipid used by your cells. Write your ideas in your Science Journal.

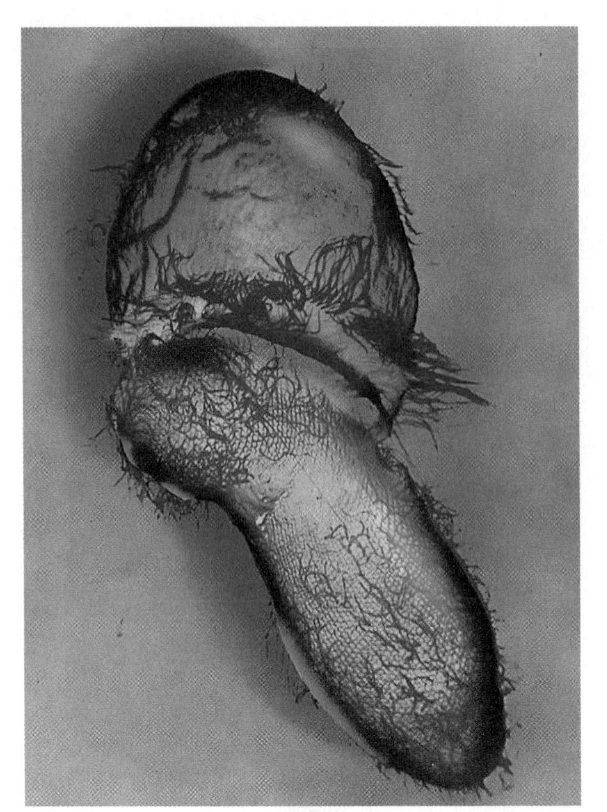
Color-enhanced TEM Magnification: 1,400×

Figure 10 One-celled organisms like this egg-shaped one can take in other one-celled organisms using endocytosis.

Endocytosis and Exocytosis

Some molecules and particles are too large to move by diffusion or to use the cell membrane's transport proteins. Large protein molecules and bacteria, for example, can enter a cell when they are surrounded by the cell membrane. The cell membrane folds in on itself, enclosing the item in a sphere called a vesicle. Vesicles are transport and storage structures in a cell's cytoplasm. The sphere pinches off, and the resulting vesicle enters the cytoplasm. A similar thing happens when you poke your finger into a partially inflated balloon. Your finger is surrounded by the balloon in much the same way that the protein molecule is surrounded by the cell membrane. This process of taking substances into a cell by surrounding it with the cell membrane is called **endocytosis** (en duh si TOH sus). Some one-celled organisms, as shown in **Figure 10,** take in food this way.

The contents of a vesicle can be released by a cell using the process called **exocytosis** (ek soh si TOH sus). Exocytosis occurs in the opposite way that endocytosis does. A vesicle's membrane fuses with a cell's membrane, and the vesicle's contents are released. Cells in your stomach use this process to release chemicals that help digest food. The ways that materials can enter or leave a cell are summarized in **Figure 11.**

section 2 review

Summary

Passive Transport

- Cells take in substances and release waste through their cell membranes.
- Facilitated diffusion and osmosis are types of passive transport.

Active Transport

- Transport proteins are involved in active transport.
- Transport proteins can be reused many times.

Endocytosis and Exocytosis

- Vesicles are formed when a cell takes in a substance by endocytosis.
- Contents of a vesicle are released to the outside of a cell by exocytosis.

Self Check

1. **Describe** how cell membranes are selectively permeable.
2. **Compare and contrast** the processes of osmosis and diffusion.
3. **Infer** why endocytosis and exocytosis are important processes to cells.
4. **Think Critically** Why are fresh fruits and vegetables sprinkled with water at produce markets?

Applying Skills

5. **Communicate** Seawater is saltier than tap water. Explain why drinking large amounts of seawater would be dangerous for humans.

green.msscience.com/self_check_quiz

NATIONAL GEOGRAPHIC

VISUALIZING CELL MEMBRANE TRANSPORT

Figure 11

A flexible yet strong layer, the cell membrane is built of two layers of lipids (gold) pierced by protein "passageways" (purple). Molecules can enter or exit the cell by slipping between the lipids or through the protein passageways. Substances that cannot enter or exit the cell in these ways may be surrounded by the membrane and drawn into or expelled from the cell.

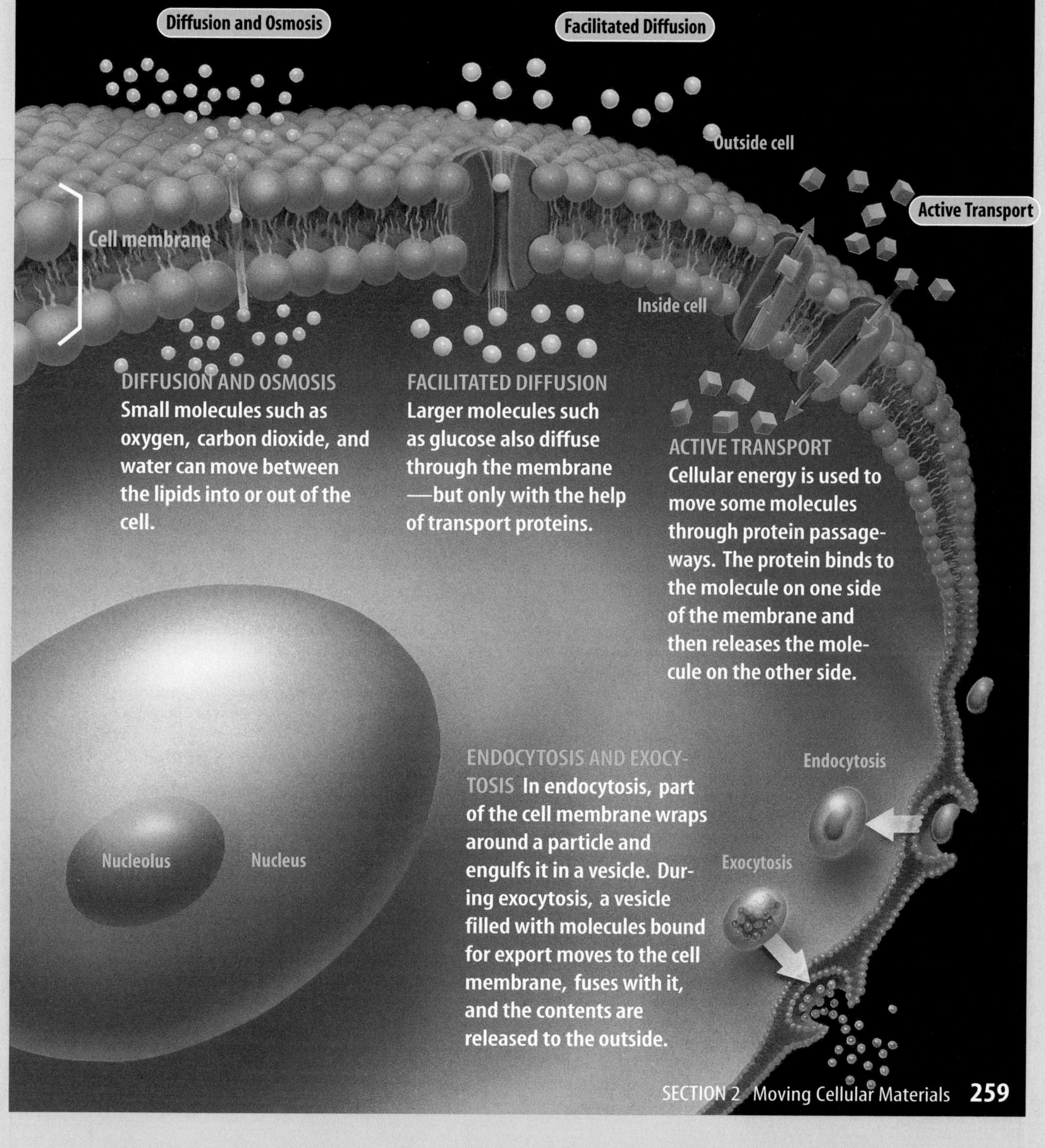

Diffusion and Osmosis

Facilitated Diffusion

Outside cell

Active Transport

Cell membrane

Inside cell

DIFFUSION AND OSMOSIS Small molecules such as oxygen, carbon dioxide, and water can move between the lipids into or out of the cell.

FACILITATED DIFFUSION Larger molecules such as glucose also diffuse through the membrane —but only with the help of transport proteins.

ACTIVE TRANSPORT Cellular energy is used to move some molecules through protein passageways. The protein binds to the molecule on one side of the membrane and then releases the molecule on the other side.

ENDOCYTOSIS AND EXOCYTOSIS In endocytosis, part of the cell membrane wraps around a particle and engulfs it in a vesicle. During exocytosis, a vesicle filled with molecules bound for export moves to the cell membrane, fuses with it, and the contents are released to the outside.

Endocytosis

Exocytosis

Nucleolus

Nucleus

Observing Osmosis

It is difficult to observe osmosis in cells because most cells are so small. However, a few cells can be seen without the aid of a microscope. Try this lab to observe osmosis.

Real-World Question

How does osmosis occur in an egg cell?

Materials

unshelled egg*
balance
spoon
distilled water (250 mL)
light corn syrup (250 mL)
500-mL container

an egg whose shell has been dissolved by vinegar

Goals

- **Observe** osmosis in an egg cell.
- **Determine** what affects osmosis.

Safety Precautions

WARNING: *Eggs may contain bacteria. Avoid touching your face.*

Procedure

1. Copy the table below into your Science Journal and use it to record your data.

Egg Mass Data

	Beginning Egg Mass	Egg Mass After Two Days
Distilled water	Do not write in this book.	
Corn syrup		

2. Obtain an unshelled egg from your teacher. Handle the egg gently. Use a balance to find the egg's mass and record it in the table.

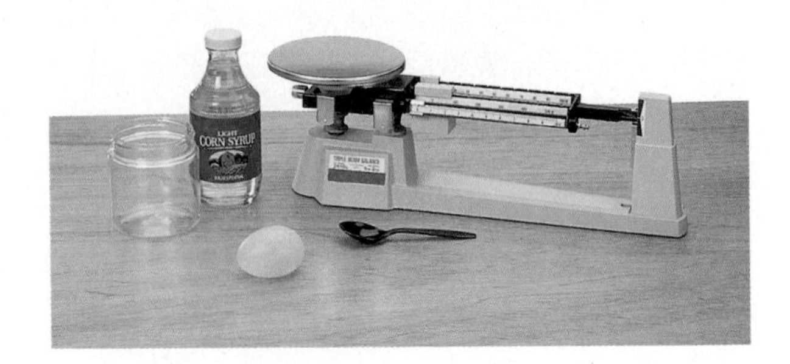

3. Place the egg in the container and add enough distilled water to cover it.
4. **Observe** the egg after 30 min, one day, and two days. After each observation, record the egg's appearance in your Science Journal.
5. After day two, remove the egg with a spoon and allow it to drain. Find the egg's mass and record it in the table.
6. Empty the container, then put the egg back in. Now add enough corn syrup to cover it. Repeat steps 4 and 5.

Conclude and Apply

1. **Explain** the difference between what happened to the egg in water and in corn syrup.
2. **Calculate** the mass of water that moved into and out of the egg.
3. **Hypothesize** why you used an unshelled egg for this investigation.
4. **Infer** what part of the egg controlled water's movement into and out of the egg.

Compare your conclusions with those of other students in your class. **For more help, refer to the** Science Skill Handbook.

section 3

Energy for Life

Trapping and Using Energy

Think of all the energy that players use in a basketball game. Where does the energy come from? The simplest answer is "from the food they eat." The chemical energy stored in food molecules is changed inside of cells into forms needed to perform all the activities necessary for life. In every cell, these changes involve chemical reactions. All of the activities of an organism involve chemical reactions in some way. The total of all chemical reactions in an organism is called **metabolism.**

The chemical reactions of metabolism need enzymes. What do enzymes do? Suppose you are hungry and decide to open a can of spaghetti. You use a can opener to open the can. Without a can opener, the spaghetti is unusable. The can of spaghetti changes because of the can opener, but the can opener does not change. The can opener can be used again later to open more cans of spaghetti. Enzymes in cells work something like can openers. The enzyme, like the can opener, causes a change, but the enzyme is not changed and can be used again, as shown in **Figure 12.** Unlike the can opener, which can only cause things to come apart, enzymes also can cause molecules to join. Without the right enzyme, a chemical reaction in a cell cannot take place. Each chemical reaction in a cell requires a specific enzyme.

as you read

What You'll Learn

- **List** the differences between producers and consumers.
- **Explain** how the processes of photosynthesis and respiration store and release energy.
- **Describe** how cells get energy from glucose through fermentation.

Why It's Important

Because of photosynthesis and respiration, you use the Sun's energy.

Review Vocabulary

mitochondrion: cell organelle that breaks down lipids and carbohydrates and releases energy

New Vocabulary

- metabolism
- photosynthesis
- respiration
- fermentation

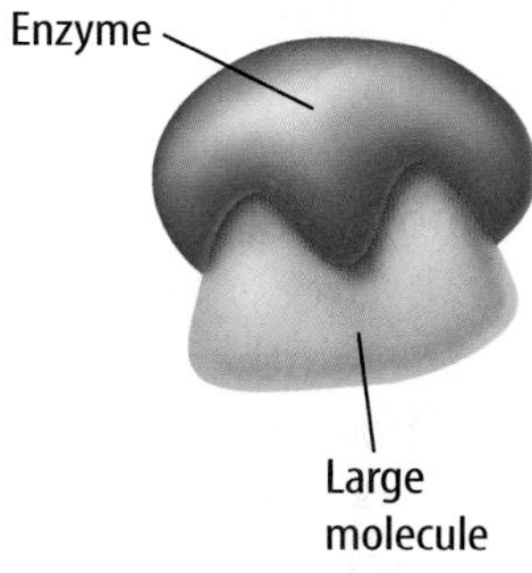

The enzyme attaches to the large molecule it will help change.

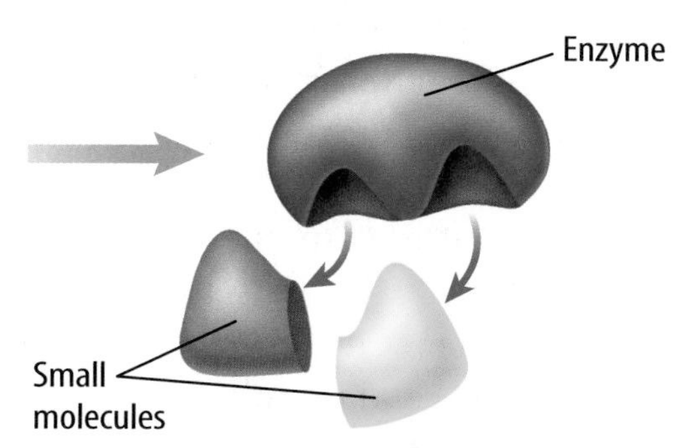

The enzyme causes the larger molecule to break down into two smaller molecules. The enzyme is not changed and can be used again.

Figure 12 Enzymes are needed for most chemical reactions that take place in cells.
Determine *What is the sum of all chemical reactions in an organism called?*

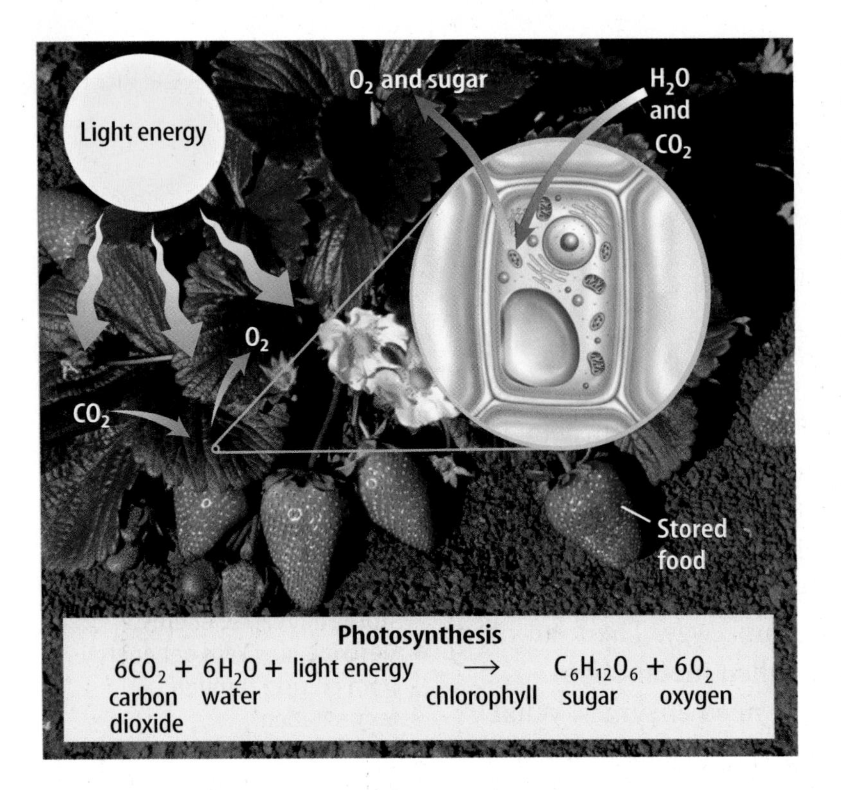

Figure 13 Plants use photosynthesis to make food.
Determine *According to the chemical equation, what raw materials would the plant in the photo need for photosynthesis?*

Photosynthesis Living things are divided into two groups—producers and consumers—based on how they obtain their food. Organisms that make their own food, such as plants, are called producers. Organisms that cannot make their own food are called consumers.

If you have ever walked barefoot across a sidewalk on a sunny summer day, you probably moved quickly because the sidewalk was hot. Sunlight energy was converted into thermal energy and heated the sidewalk. Plants and many other producers can convert light energy into another kind of energy—chemical energy. The process they use is called photosynthesis. During **photosynthesis,** producers use light energy to make sugars, which can be used as food.

Producing Carbohydrates Producers that use photosynthesis are usually green because they contain a green pigment called chlorophyll (KLOR uh fihl). Chlorophyll and other pigments are used in photosynthesis to capture light energy. In plant cells, these pigments are found in chloroplasts.

The captured light energy powers chemical reactions that produce sugar and oxygen from the raw materials, carbon dioxide and water. For plants, the raw materials come from air and soil. Some of the captured light energy is stored in the chemical bonds that hold the sugar molecules together. **Figure 13** shows what happens during photosynthesis in a plant. Enzymes also are needed before these reactions can occur.

Storing Carbohydrates Plants make more sugar during photosynthesis than they need for survival. Excess sugar is changed and stored as starches or used to make other carbohydrates. Plants use these carbohydrates as food for growth, maintenance, and reproduction.

Why is photosynthesis important to consumers? Do you eat apples? Apple trees use photosynthesis to produce apples. Do you like cheese? Some cheese comes from milk, which is produced by cows that eat plants. Consumers take in food by eating producers or other consumers. No matter what you eat, photosynthesis was involved directly or indirectly in its production.

Respiration Imagine that you get up late for school. You dress quickly, then run three blocks to school. When you get to school, you feel hot and are breathing fast. Why? Your muscle cells use a lot of energy when you run. To get this energy, muscle cells break down food. Some of the energy from the food is used when you move and some of it becomes thermal energy, which is why you feel warm or hot. Most cells also need oxygen to break down food. You were breathing fast because your body was working to get oxygen to your muscles. Your muscle cells were using the oxygen for the process of respiration. During **respiration,** chemical reactions occur that break down food molecules into simpler substances and release their stored energy. Just as in photosynthesis, enzymes are needed for the chemical reactions of respiration.

What must happen to food molecules for respiration to take place?

Breaking Down Carbohydrates The food molecules most easily broken down by cells are carbohydrates. Respiration of carbohydrates begins in the cytoplasm of the cell. The carbohydrates are broken down into glucose molecules. Each glucose molecule is broken down further into two simpler molecules. As the glucose molecules are broken down, energy is released.

The two simpler molecules are broken down again. This breakdown occurs in the mitochondria of the cells of plants, animals, fungi, and many other organisms. This process uses oxygen, releases much more energy, and produces carbon dioxide and water as wastes. When you exhale, you breathe out carbon dioxide and some of the water.

Respiration occurs in the cells of all living things. **Figure 14** shows how respiration occurs in one consumer. As you are reading this section of the chapter, millions of cells in your body are breaking down glucose, releasing energy, and producing carbon dioxide and water.

Microbiologist Dr. Harold Amos is a microbiologist who has studied cell processes in bacteria and mammals. He has a medical degree and a doctorate in bacteriology and immunology. He has also received many awards for his scientific work and his contributions to the careers of other scientists. Research microbiology careers, and write what you find in your Science Journal.

Figure 14 Producers and consumers carry on respiration that releases energy from foods.

Fermentation Remember imagining you were late and had to run to school? During your run, your muscle cells might not have received enough oxygen, even though you were breathing rapidly. When cells do not have enough oxygen for respiration, they use a process called **fermentation** to release some of the energy stored in glucose molecules.

Like respiration, fermentation begins in the cytoplasm. Again, as the glucose molecules are broken down, energy is released. But the simple molecules from the breakdown of glucose do not move into the mitochondria. Instead, more chemical reactions occur in the cytoplasm. These reactions release some energy and produce wastes. Depending on the type of cell, the wastes may be lactic acid or alcohol and carbon dioxide, as shown in **Figure 15.** Your muscle cells can use fermentation to change the simple molecules into lactic acid while releasing energy. The presence of lactic acid is why your muscle cells might feel stiff and sore after you run to school.

Science Online

Topic: Beneficial Microorganisms

Visit green.msscience.com for Web links to information about how microorganisms are used to produce many useful products.

Activity Find three other ways that microorganisms are beneficial.

Reading Check *Where in a cell does fermentation take place?*

Some microscopic organisms, such as bacteria, carry out fermentation and make lactic acid. Some of these organisms are used to produce yogurt and some cheeses. These organisms break down a sugar in milk and release energy. The lactic acid produced causes the milk to become more solid and gives these foods some of their flavor.

Have you ever used yeast to make bread? Yeasts are one-celled living organisms. Yeast cells use fermentation and break down sugar in bread dough. They produce alcohol and carbon dioxide as wastes. The carbon dioxide waste is a gas that makes bread dough rise before it is baked. The alcohol is lost as the bread bakes.

Figure 15 Organisms that use fermentation produce several different wastes.

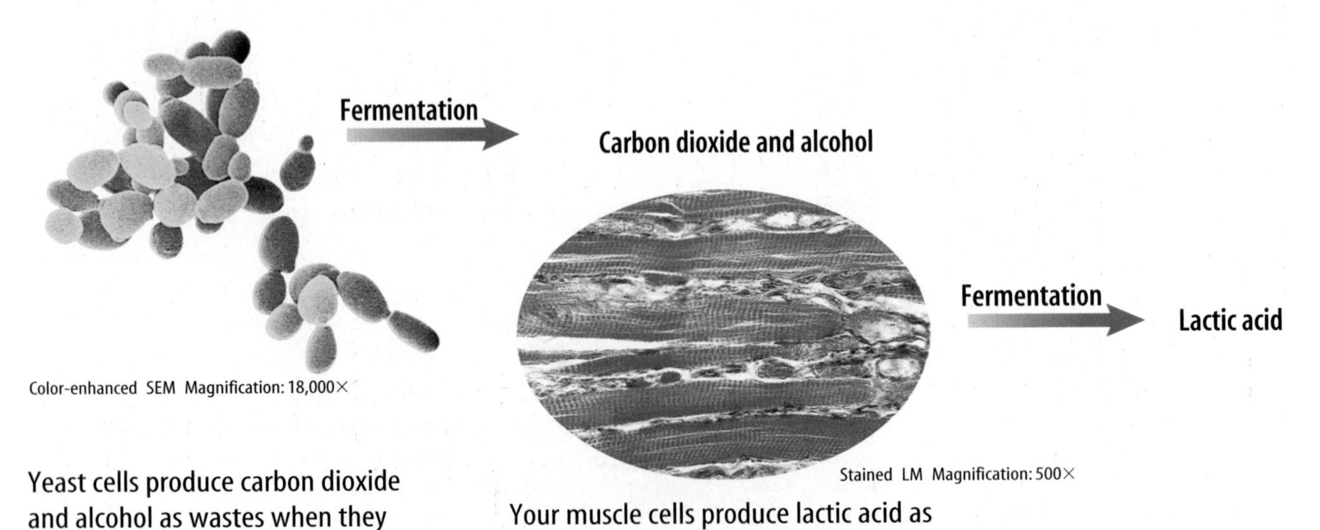

Yeast cells produce carbon dioxide and alcohol as wastes when they undergo fermentation.

Your muscle cells produce lactic acid as a waste when they undergo fermentation.

Figure 16 The chemical reactions of photosynthesis and respiration could not take place without each other.

Related Processes How are photosynthesis, respiration, and fermentation related? Some producers use photosynthesis to make food. All living things use respiration or fermentation to release energy stored in food. If you think carefully about what happens during photosynthesis and respiration, you will see that what is produced in one is used in the other, as shown in **Figure 16.** These two processes are almost the opposite of each other. Photosynthesis produces sugars and oxygen, and respiration uses these products. The carbon dioxide and water produced during respiration are used during photosynthesis. Most life would not be possible without these important chemical reactions.

section 3 review

Summary

Trapping and Using Energy

- Metabolism is the total of all chemical reactions in an organism.
- During photosynthesis, light energy is used to make sugars.
- Chlorophyll and other pigments capture light energy.
- Consumers take in energy by eating producers and other consumers.
- Living cells break down glucose and release energy. This is called respiration.
- Fermentation changes simple molecules and releases energy.
- Without photosynthesis and respiration, most life would not be possible.

Self Check

1. **Explain** the difference between producers and consumers and give three examples of each.
2. **Infer** how the energy used by many living things on Earth can be traced back to sunlight.
3. **Compare and contrast** respiration and fermentation.
4. **Think Critically** How can some indoor plants help to improve the quality of air in a room?

Applying Math

5. **Solve** Refer to the chemical equation for photosynthesis. Calculate then compare the number of carbon, hydrogen, and oxygen atoms before and after photosynthesis.

Photosynthesis and Respiration

LM Magnification: 225×

Goals

- **Observe** green water plants in the light and dark.
- **Determine** whether plants carry on photosynthesis and respiration.

Materials

16-mm test tubes (3)
150-mm test tubes with stoppers (4)
**small, clear-glass baby food jars with lids (4)*
test-tube rack
stirring rod
scissors
carbonated water (5 mL)
bromthymol blue solution in dropper bottle
aged tap water (20 mL)
**distilled water (20 mL)*
sprigs of *Elodea* (2)
**other water plants*
**Alternate materials*

Safety Precautions

WARNING: *Wear splash-proof safety goggles to protect eyes from hazardous chemicals.*

Real-World Question

Every living cell carries on many chemical processes. Two important chemical processes are respiration and photosynthesis. All cells, including the ones in your body, carry on respiration. However, some plant cells can carry on both processes. In this experiment you will investigate when these processes occur in plant cells. How could you find out when plants were using these processes? Are the products of photosynthesis and respiration the same? When do plants carry on photosynthesis and respiration?

Procedure

1. In your Science Journal, copy and complete the test-tube data table as you perform this lab.

Test-Tube Data

Test Tube	Color at Start	Color After 30 Minutes
1		
2	Do not write in this book.	
3		
4		

2. Label each test tube using the numbers *1, 2, 3,* and *4.* Pour 5 mL of aged tap water into each test tube.
3. Add 10 drops of carbonated water to test tubes *1* and *2.*
4. Add 10 drops of bromthymol blue to all of the test tubes. Bromthymol blue turns green to yellow in the presence of an acid.
5. Cut two 10-cm sprigs of *Elodea.* Place one sprig in test tube *1* and one sprig in test tube *3.* Stopper all test tubes.
6. Place test tubes *1* and *2* in bright light. Place tubes *3* and *4* in the dark. Observe the test tubes for 45 min or until the color changes. Record the color of each of the four test tubes.

Analyze Your Data

1. **Identify** what is indicated by the color of the water in all four test tubes at the start of the activity.
2. **Infer** what process occurred in the test tube or tubes that changed color after 30 min.

Conclude and Apply

1. **Describe** the purpose of test tubes *2* and *4* in this experiment.
2. **Explain** whether or not the results of this experiment show that photosynthesis and respiration occur in plants.

Communicating Your Data

Choose one of the following activities to **communicate** your data. Prepare an oral presentation that explains how the experiment showed the differences between products of photosynthesis and respiration. Draw a cartoon strip to **explain** what you did in this experiment. Use each panel to show a different step. **For more help, refer to the Science Skill Handbook.**

Science and Language Arts

from "Tulip"

by Penny Harter

I watched its first green push
through bare dirt, where the builders
had dropped boards, shingles,
plaster—
killing everything.
I could not recall what grew there,
what returned each spring,
but the leaves looked tulip,
and one morning it arrived,
a scarlet slash against the aluminum siding.
Mornings, on the way to my car,
I bow to the still bell
of its closed petals; evenings,
it greets me, light ringing
at the end of my driveway.
Sometimes I kneel
to stare into the yellow throat . . . .
It opens and closes my days.
It has made me weak with love . . . .

Understanding Literature

Personification Using human traits or emotions to describe an idea, animal, or inanimate object is called personification. When the poet writes that the tulip has a "yellow throat," she uses personification. Where else does the poet use personification?

Respond to the Reading

1. Why do you suppose the tulip survived the builders' abuse?
2. What is the yellow throat that the narrator is staring into?
3. **Linking Science and Writing** Keep a gardener's journal of a plant for a month, describing weekly the plant's condition, size, health, color, and other physical qualities.

Because most chemical reactions in plants take place in water, plants must have water in order to grow. The water carries nutrients and minerals from the soil into the plant. The process of active transport allows needed nutrients to enter the roots. The cell membranes of root cells contain proteins that bind with the needed nutrients. Cellular energy is used to move these nutrients through the cell membrane.

Reviewing Main Ideas

Section 1 Chemistry of Life

1. Matter is anything that has mass and takes up space.
2. Energy in matter is in the chemical bonds that hold matter together.
3. All organic compounds contain the elements hydrogen and carbon. The organic compounds in living things are carbohydrates, lipids, proteins, and nucleic acids.
4. Organic and inorganic compounds are important to living things.

Section 2 Moving Cellular Materials

1. The selectively permeable cell membrane controls which molecules can pass into and out of the cell.
2. In diffusion, molecules move from areas where there are relatively more of them to areas where there are relatively fewer of them.
3. Osmosis is the diffusion of water through a cell membrane.
4. Cells use energy to move molecules by active transport but do not use energy for passive transport.
5. Cells move large particles through cell membranes by endocytosis and exocytosis.

Section 3 Energy for Life

1. Photosynthesis is the process by which some producers change light energy into chemical energy.
2. Respiration that uses oxygen releases the energy in food molecules and produces waste carbon dioxide and water.
3. Some one-celled organisms and cells that lack oxygen use fermentation to release small amounts of energy from glucose. Wastes such as alcohol, carbon dioxide, and lactic acid are produced.

Visualizing Main Ideas

Copy and complete the following table on energy processes.

Energy Processes

	Photosynthesis	Respiration	Fermentation
Energy source		food (glucose)	food (glucose)
In plant and animal cells, occurs in			
Reactants are			
Products are			

Using Vocabulary

active transport p. 257
diffusion p. 255
endocytosis p. 258
enzyme p. 251
equilibrium p. 255
exocytosis p. 258
fermentation p. 264
inorganic compound p. 251
metabolism p. 261
mixture p. 249
organic compound p. 250
osmosis p. 256
passive transport p. 254
photosynthesis p. 262
respiration p. 263

Use what you know about the vocabulary words to answer the following questions.

1. What is the diffusion of water called?
2. What type of protein regulates nearly all chemical reactions in cells?
3. How do large food particles enter an amoeba?
4. What type of compound is water?
5. What process is used by some producers to convert light energy into chemical energy?
6. What type of compounds always contain carbon and hydrogen?
7. What process uses oxygen to break down glucose?
8. What is the total of all chemical reactions in an organism called?

Checking Concepts

Choose the word or phrase that best answers the question.

9. What is it called when cells use energy to move molecules?
   A) diffusion **C)** active transport
   B) osmosis **D)** passive transport

Use the photo below to answer question 10.

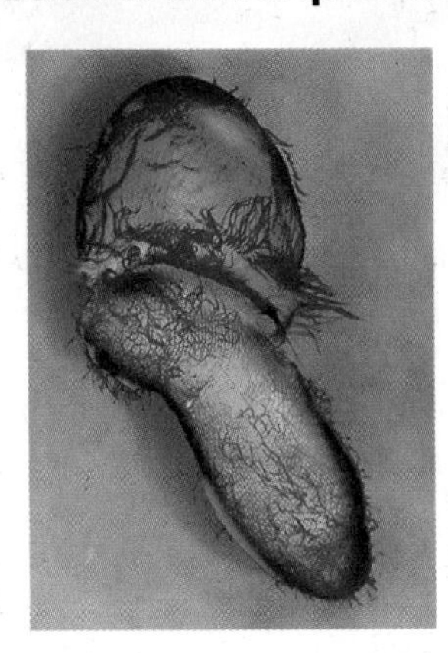

10. What cell process is occurring in the photo?
    A) osmosis **C)** exocytosis
    B) endocytosis **D)** diffusion
11. What occurs when the number of molecules of a substance is equal in two areas?
    A) equilibrium **C)** fermentation
    B) metabolism **D)** cellular respiration
12. Which of the following substances is an example of a carbohydrate?
    A) enzymes **C)** waxes
    B) sugars **D)** proteins
13. What is RNA an example of?
    A) carbon dioxide **C)** lipid
    B) water **D)** nucleic acid
14. What organic molecule stores the greatest amount of energy?
    A) carbohydrate **C)** lipid
    B) water **D)** nucleic acid
15. Which of these formulas is an example of an organic compound?
    A) $C_6H_{12}O_6$ **C)** H_2O
    B) NO_2 **D)** O_2
16. What are organisms that cannot make their own food called?
    A) biodegradables **C)** consumers
    B) producers **D)** enzymes

Science Online green.msscience.com/vocabulary_puzzlemaker

Thinking Critically

17. **Concept Map** Copy and complete the events-chain concept map to sequence the following parts of matter from smallest to largest: *atom, electron,* and *compound.*

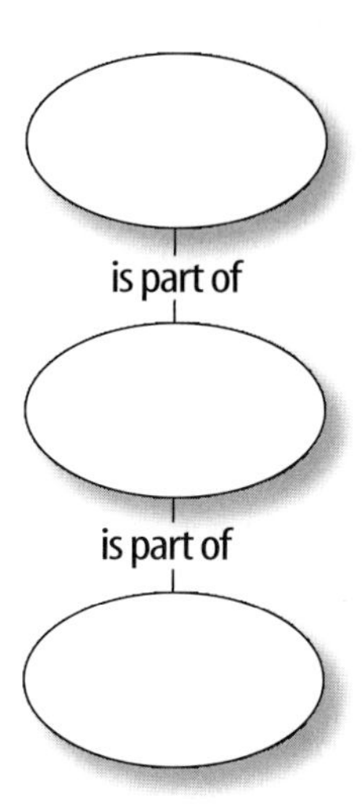

Use the table below to answer question 18.

Photosynthesis in Water Plants

Beaker Number	Distance from Light (cm)	Bubbles per Minute
1	10	45
2	30	30
3	50	19
4	70	6
5	100	1

18. **Interpret Data** Water plants were placed at different distances from a light source. Bubbles coming from the plants were counted to measure the rate of photosynthesis. What can you say about how the distance from the light affected the rate?

19. **Infer** why, in snowy places, salt is used to melt ice on the roads. Explain what could happen to many roadside plants as a result.

20. **Draw a conclusion** about why sugar dissolves faster in hot tea than in iced tea.

21. **Predict** what would happen to the consumers in a lake if all the producers died.

22. **Explain** how meat tenderizers affect meat.

23. **Form a hypothesis** about what will happen to wilted celery when placed in a glass of plain water.

Performance Activities

24. **Puzzle** Make a crossword puzzle with words describing ways substances are transported across cell membranes. Use the following words in your puzzle: *diffusion, osmosis, facilitated diffusion, active transport, endocytosis,* and *exocytosis.* Make sure your clues give good descriptions of each transport method.

Applying Math

25. **Light and Photosynthesis** Using the data from question 18, make a line graph that shows the relationship between the rate of photosynthesis and the distance from light.

26. **Importance of Water** Assume the brain is 70% water. If the average adult human brain weighs 1.4 kg, how many kilograms of water does it contain?

Use the equation below to answer question 27.

Photosynthesis

$$6CO_2 + 6H_2O + \text{light energy} \xrightarrow{\text{chlorophyll}} C_6H_{12}O_6 + 6O_2$$

carbon dioxide, water → sugar, oxygen

27. **Photosynthesis** Refer to the chemical equation above. If 18 CO_2 molecules and 18 H_2O molecules are used with light energy to make sugar, how many sugar molecules will be produced? How many oxygen molecules will be produced?

Part 1 Multiple Choice

Record your answers on the answer sheet provided by your teacher or on a sheet of paper.

1. An element cannot be broken down by chemical reactions and is made up of only one kind of
   A. electron. **C.** atom.
   B. carbohydrate. **D.** molecule.

Use the illustration below to answer questions 2 and 3.

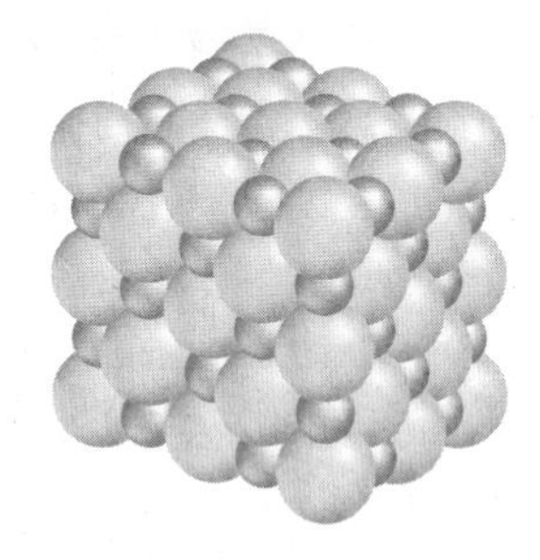

2. What kind of chemical compound do salt and water form?
   A. covalent
   B. ionic
   C. solution
   D. lipid

3. Salt is very important in the human body. What kind of compound is salt?
   A. organic
   B. carbohydrate
   C. protein
   D. inorganic

4. A cell that contains 40% water is placed in a solution that is 20% water. The cell and the solution will reach equilibrium when they both contain how much water?
   A. 30% **C.** 60%
   B. 40% **D.** 20%

5. All chemical reactions in living things take place in what kind of a solution?
   A. protein **C.** gas
   B. water **D.** solid

6. The sum of all the chemical reactions in an organism is
   A. respiration. **C.** fermentation.
   B. metabolism. **D.** endocytosis.

7. What is needed for all chemical reactions in cells?
   A. enzymes **C.** DNA
   B. lipids **D.** cell membrane

8. The carbon dioxide that you exhale is a product of
   A. osmosis.
   B. DNA synthesis.
   C. photosynthesis.
   D. respiration.

9. Matter cannot be held together or broken apart without
   A. gas.
   B. liquid.
   C. energy.
   D. temperature.

Use the table below to answer question 10.

Cell Substances		
Organic Compound	**Flexibility**	**Found in**
Keratin	Not very flexible	Hair and skin of mammals
Collagen	Not very flexible	Skin, bones, and tendons of mammals
Chitin	Very rigid	Tough outer shell of insects and crabs
Cellulose	Very flexible	Plant cell walls

10. According to this information, which organic compound is the least flexible?
    A. keratin
    B. collagen
    C. chitin
    D. cellulose

Part 2 | Short Response/Grid In

Record your answers on the answer sheet provided by your teacher or on a sheet of paper.

11. Explain the structure of an atom.

12. How does chewing food affect your body's ability to release the chemical energy of the food?

13. Ice fishing is a popular sport in the winter. What properties of water is this sport based on?

14. Explain where the starch in a potato comes from.

15. Does fermentation or respiration release more energy for an athlete's muscles? Which process would be responsible for making muscles sore?

Use the table below to answer question 16.

Classification of Compounds			
Compound	**Organic**	**Inorganic**	**Type of organic compound**
Salt			
Fat			
Skin			
DNA			
Sugar			
Water			
Potassium			

16. Copy and complete the table above. Identify each item as inorganic or organic. If the item is an organic compound further classify it as a protein, carbohydrate, lipid or nucleic acid.

17. Define selectively permeable and discuss why it is important for the cell membrane.

18. What is the source of energy for the photosynthesis reactions and where do they take place in a cell?

Part 3 | Open Ended

Record your answers on a sheet of paper.

19. Give examples of each of the four types of organic molecules and why they are needed in a plant cell.

20. Trace the path of how oxygen molecules are produced in a plant cell to how they are used in human cells.

21. Describe four ways a large or small molecule can cross the cell membrane.

22. Discuss how water is bonded together and the unique properties that result from the bonds.

Use the illustration below to answer question 23.

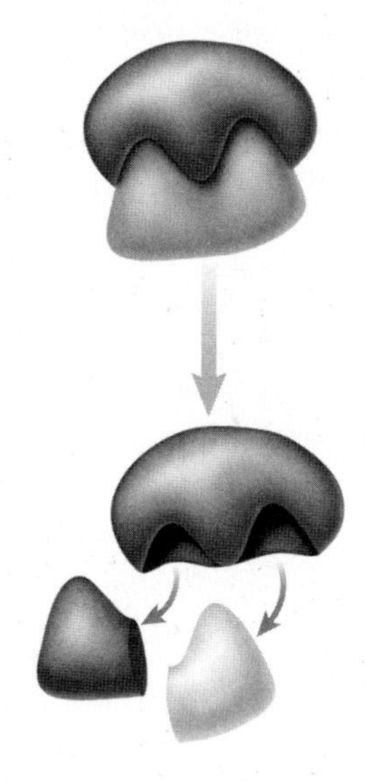

23. Describe in detail what process is taking place in this diagram and its significance for a cell.

24. How do plants use carbon dioxide? Why would plants need oxygen?

Test-Taking Tip

Diagrams Study a diagram carefully, being sure to read all labels and captions.

Cell Reproduction

chapter preview

sections

Why a turtle, not a chicken?

A sweet potato plant can be grown from just one potato, but turtles and most other animals need to have two parents. A cut on your finger heals. How do these things happen? In this chapter, you will find answers to these questions as you learn about cell reproduction.

Science Journal Write three things that you know about how and why cells reproduce.

Start-Up Activities

Infer About Seed Growth

Most flower and vegetable seeds sprout and grow into entire plants in just a few weeks. Although all of the cells in a seed have information and instructions to produce a new plant, only some of the cells in the seed use the information. Where are these cells in seeds? Do the following lab to find out.

1. Carefully split open two bean seeds that have soaked in water overnight.
2. Observe both halves and record your observations.
3. Wrap all four halves in a moist paper towel. Then put them into a self-sealing, plastic bag and seal the bag.
4. Make observations every day for a few days.
5. **Think Critically** Write a paragraph that describes what you observe. Hypothesize which cells in seeds use information about how plants grow.

Preview this chapter's content and activities at green.msscience.com

How and Why Cells Divide Make the following Foldable to help you organize information from the chapter about cell reproduction.

STEP 1 **Draw** a mark at the midpoint of a vertical sheet of paper along the side edge.

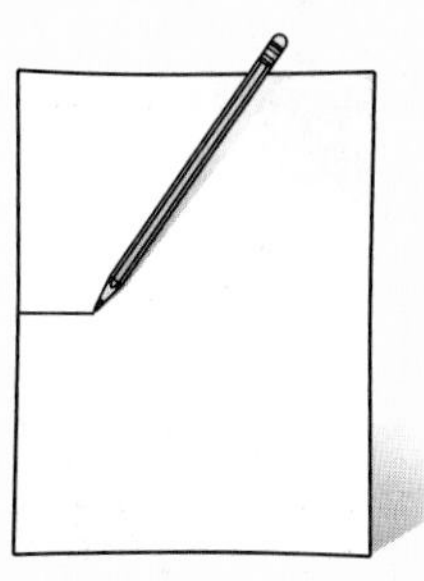

STEP 2 **Turn** the paper horizontally and **fold** the outside edges in to touch at the midpoint mark.

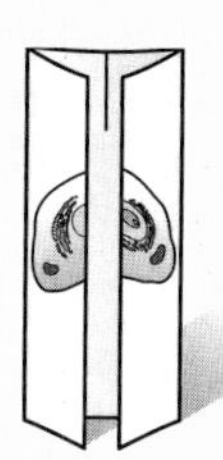

STEP 3 **Use** a pencil to draw a cell on the front of your Foldable as shown.

Analyze As you read the chapter, write under the flaps how cells divide. In the middle section, list why cells divide.

section 1

Cell Division and Mitosis

as you read

What You'll Learn

- **Explain** why mitosis is important.
- **Examine** the steps of mitosis.
- **Compare** mitosis in plant and animal cells.
- **List** two examples of asexual reproduction.

Why It's Important

Your growth, like that of many organisms, depends on cell division.

Review Vocabulary

nucleus: organelle that controls all the activities of a cell and contains hereditary material made of proteins and DNA

New Vocabulary

- mitosis
- chromosome
- asexual reproduction

Why is cell division important?

What do you, an octopus, and an oak tree have in common? You share many characteristics, but an important one is that you are all made of cells—trillions of cells. Where did all of those cells come from? As amazing as it might seem, many organisms start as just one cell. That cell divides and becomes two, two become four, four become eight, and so on. Many-celled organisms, including you, grow because cell division increases the total number of cells in an organism. Even after growth stops, cell division is still important. Every day, billions of red blood cells in your body wear out and are replaced. During the few seconds it takes you to read this sentence, your bone marrow produced about six million red blood cells. Cell division is important to one-celled organisms, too—it's how they reproduce themselves, as shown in **Figure 1.** Cell division isn't as simple as just cutting the cell in half, so how do cells divide?

The Cell Cycle

A living organism has a life cycle. A life cycle begins with the organism's formation, is followed by growth and development, and finally ends in death. Right now, you are in a stage of your life cycle called adolescence, which is a period of active growth and development. Individual cells also have life cycles.

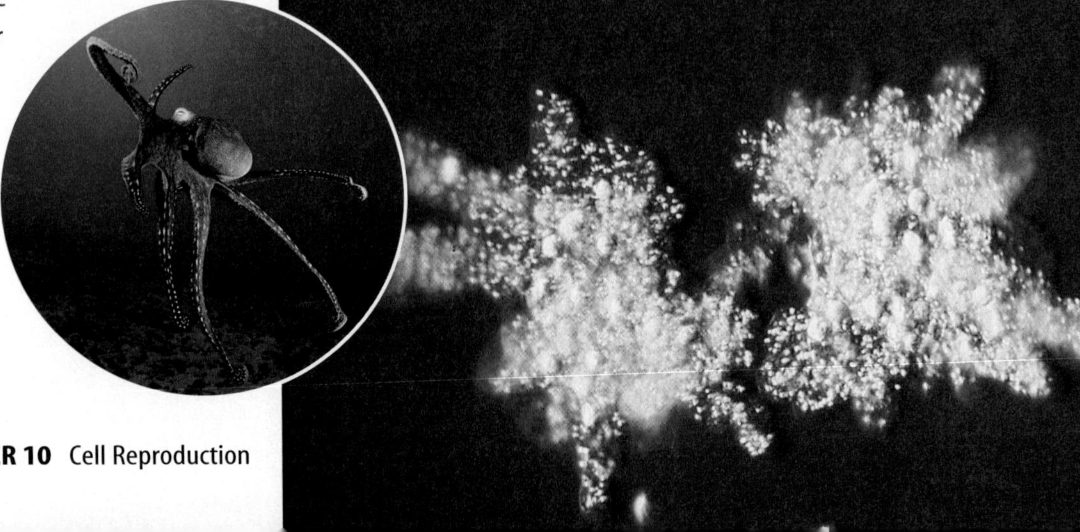

Figure 1 All organisms use cell division. Many-celled organisms, such as this octopus, grow by increasing the numbers of their cells.

Like this dividing amoeba, a one-celled organism reaches a certain size and then reproduces.

Figure 2 Interphase is the longest part of the cell cycle.
Identify *When do chromosomes duplicate?*

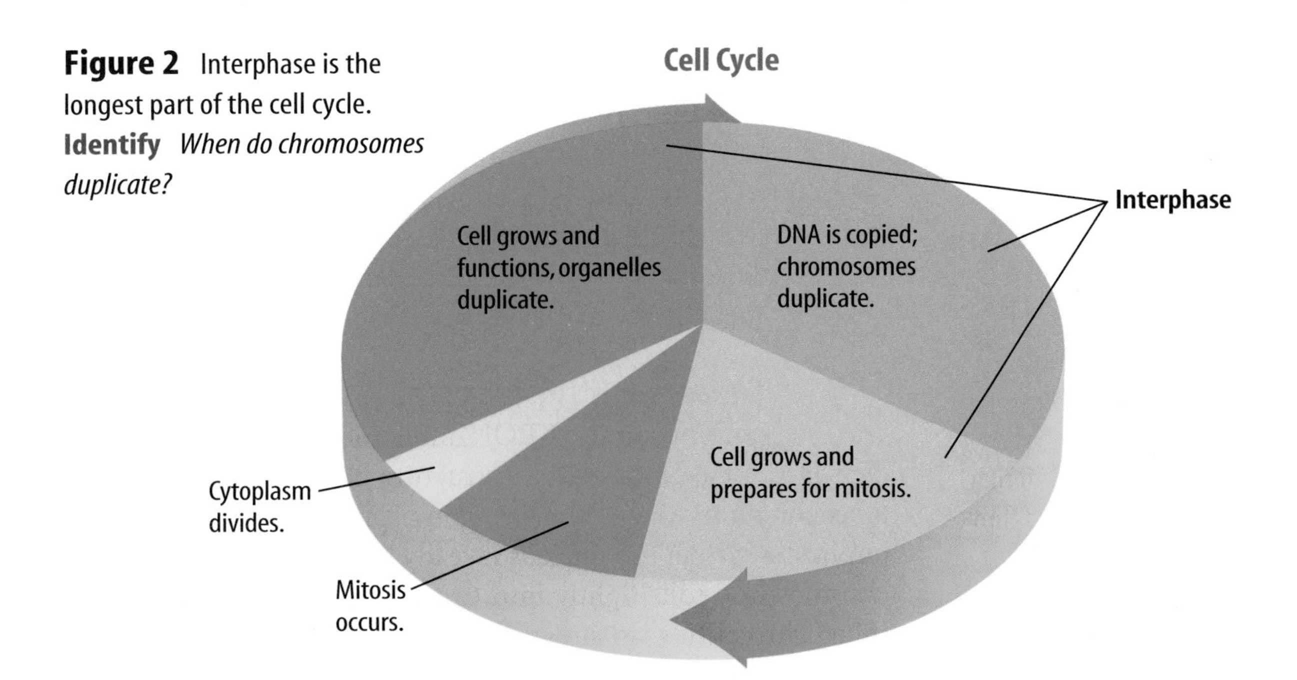

Length of Cycle The cell cycle, as shown in **Figure 2,** is a series of events that takes place from one cell division to the next. The time it takes to complete a cell cycle is not the same in all cells. For example, the cycle for cells in some bean plants takes about 19 h to complete. Cells in animal embryos divide rapidly and can complete their cycles in less than 20 min. In some human cells, the cell cycle takes about 16 h. Cells in humans that are needed for repair, growth, or replacement, like skin and bone cells, constantly repeat the cycle.

Interphase Most of the life of any eukaryotic cell—a cell with a nucleus—is spent in a period of growth and development called interphase. Cells in your body that no longer divide, such as nerve and muscle cells, are always in interphase. An actively dividing cell, such as a skin cell, copies its hereditary material and prepares for cell division during interphase.

Why is it important for a cell to copy its hereditary information before dividing? Imagine that you have a part in a play and the director has one complete copy of the script. If the director gave only one page to each person in the play, no one would have the entire script. Instead the director makes a complete, separate copy of the script for each member of the cast so that each one can learn his or her part. Before a cell divides, a copy of the hereditary material must be made so that each of the two new cells will get a complete copy. Just as the actors in the play need the entire script, each cell needs a complete set of hereditary material to carry out life functions.

After interphase, cell division begins. The nucleus divides, and then the cytoplasm separates to form two new cells.

Oncologist In most cells, the cell cycle is well controlled. Cancer cells, however, have uncontrolled cell division. Doctors who diagnose, study, and treat cancer are called oncologists. Someone wanting to become an oncologist must first complete medical school before training in oncology. Research the subspecialities of oncology. List and describe them in your Science Journal.

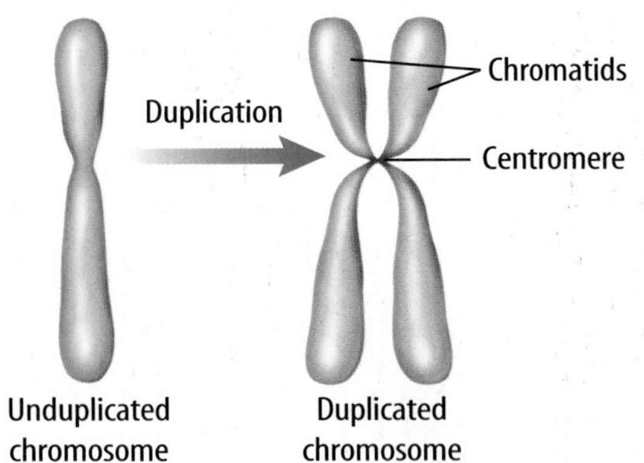

Figure 3 DNA is copied during interphase. An unduplicated chromosome has one strand of DNA. A duplicated chromosome has two identical DNA strands, called chromatids, that are held together at a region called the centromere.

Mitosis

Mitosis (mi TOH sus) is the process in which the nucleus divides to form two identical nuclei. Each new nucleus also is identical to the original nucleus. Mitosis is described as a series of phases, or steps. The steps of mitosis in order are named prophase, metaphase, anaphase, and telophase.

Steps of Mitosis When any nucleus divides, the chromosomes (KROH muh sohmz) play the important part. A **chromosome** is a structure in the nucleus that contains hereditary material. During interphase, each chromosome duplicates. When the nucleus is ready to divide, each duplicated chromosome coils tightly into two thickened, identical strands called chromatids, as shown in **Figure 3.**

Reading Check *How are chromosomes and chromatids related?*

During prophase, the pairs of chromatids are fully visible when viewed under a microscope. The nucleolus and the nuclear membrane disintegrate. Two small structures called centrioles (SEN tree olz) move to opposite ends of the cell. Between the centrioles, threadlike spindle fibers begin to stretch across the cell. Plant cells also form spindle fibers during mitosis but do not have centrioles.

In metaphase, the pairs of chromatids line up across the center of the cell. The centromere of each pair usually becomes attached to two spindle fibers—one from each side of the cell.

In anaphase, each centromere divides and the spindle fibers shorten. Each pair of chromatids separates, and chromatids begin to move to opposite ends of the cell. The separated chromatids are now called chromosomes. In the final step, telophase, spindle fibers start to disappear, the chromosomes start to uncoil, and a new nucleus forms.

Figure 4 The cell plate shown in this plant cell appears when the cytoplasm is being divided.
Identify *what phase of mitosis will be next.*

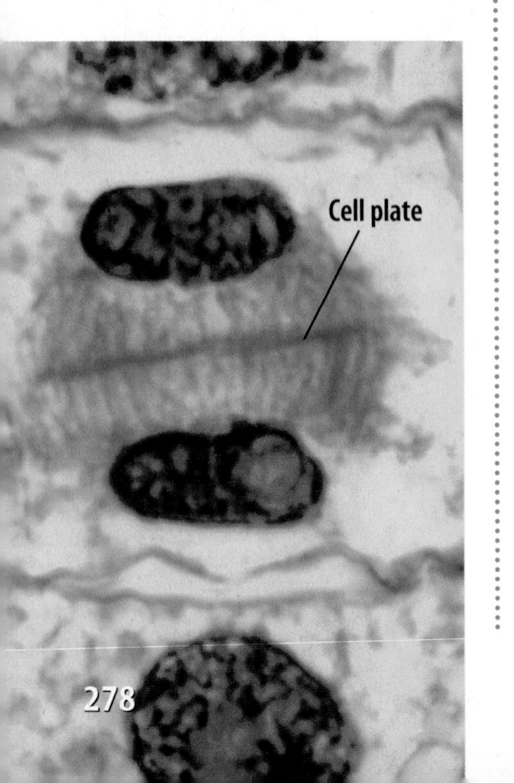

Division of the Cytoplasm For most cells, after the nucleus has divided, the cytoplasm separates and two new cells are formed. In animal cells, the cell membrane pinches in the middle, like a balloon with a string tightened around it, and the cytoplasm divides. In plant cells, the appearance of a cell plate, as shown in **Figure 4,** tells you that the cytoplasm is being divided. New cell walls form along the cell plate, and new cell membranes develop inside the cell walls. Following division of the cytoplasm, most new cells begin the period of growth, or interphase, again. Review cell division for an animal cell using the illustrations in **Figure 5.**

Figure 5 Cell division for an animal cell is shown here. Each micrograph shown in this figure is magnified 600 times.

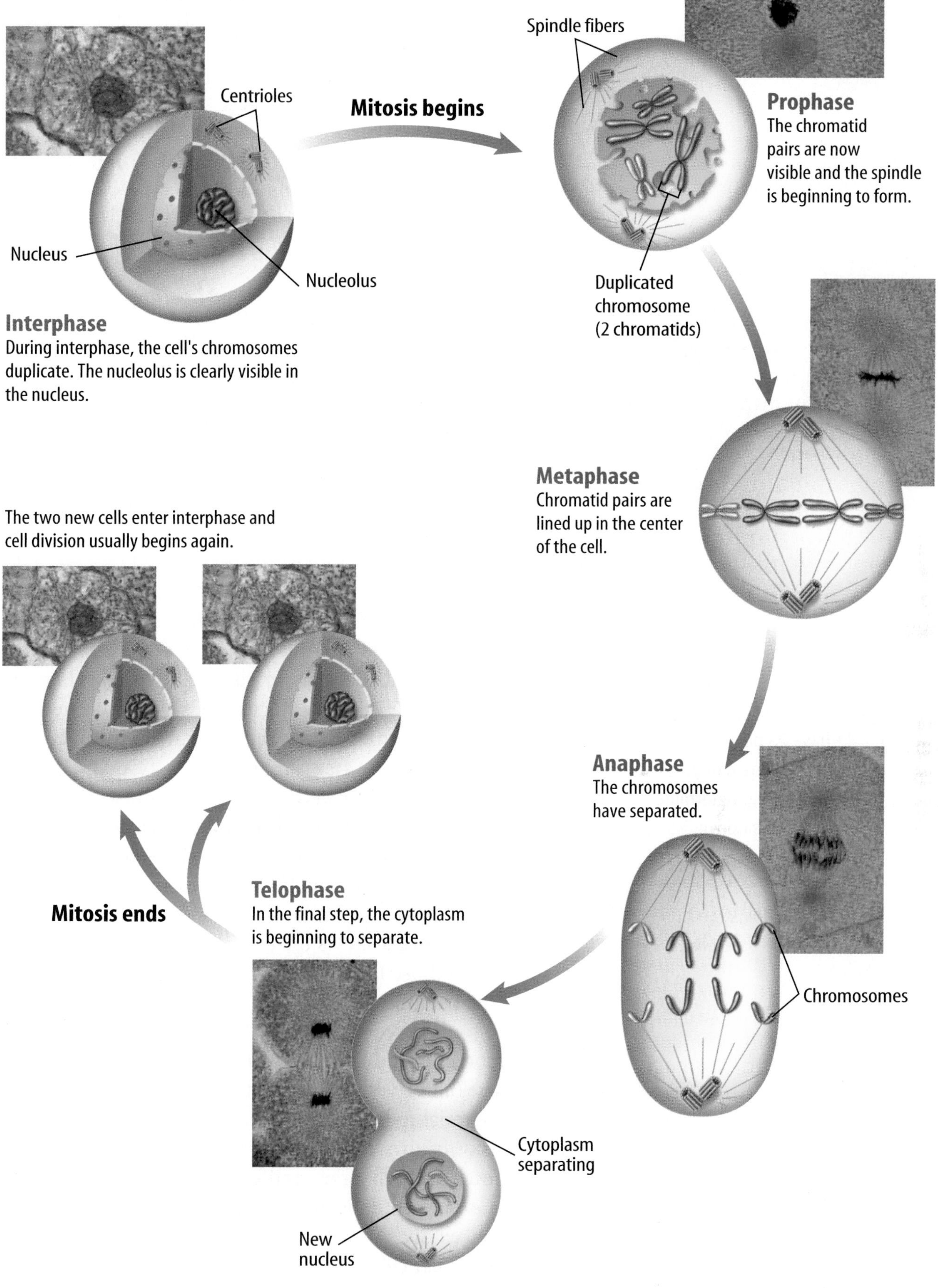

Centrioles

Nucleus

Nucleolus

Interphase
During interphase, the cell's chromosomes duplicate. The nucleolus is clearly visible in the nucleus.

Mitosis begins

Spindle fibers

Prophase
The chromatid pairs are now visible and the spindle is beginning to form.

Duplicated chromosome (2 chromatids)

Metaphase
Chromatid pairs are lined up in the center of the cell.

Anaphase
The chromosomes have separated.

Chromosomes

Telophase
In the final step, the cytoplasm is beginning to separate.

Cytoplasm separating

New nucleus

Mitosis ends

The two new cells enter interphase and cell division usually begins again.

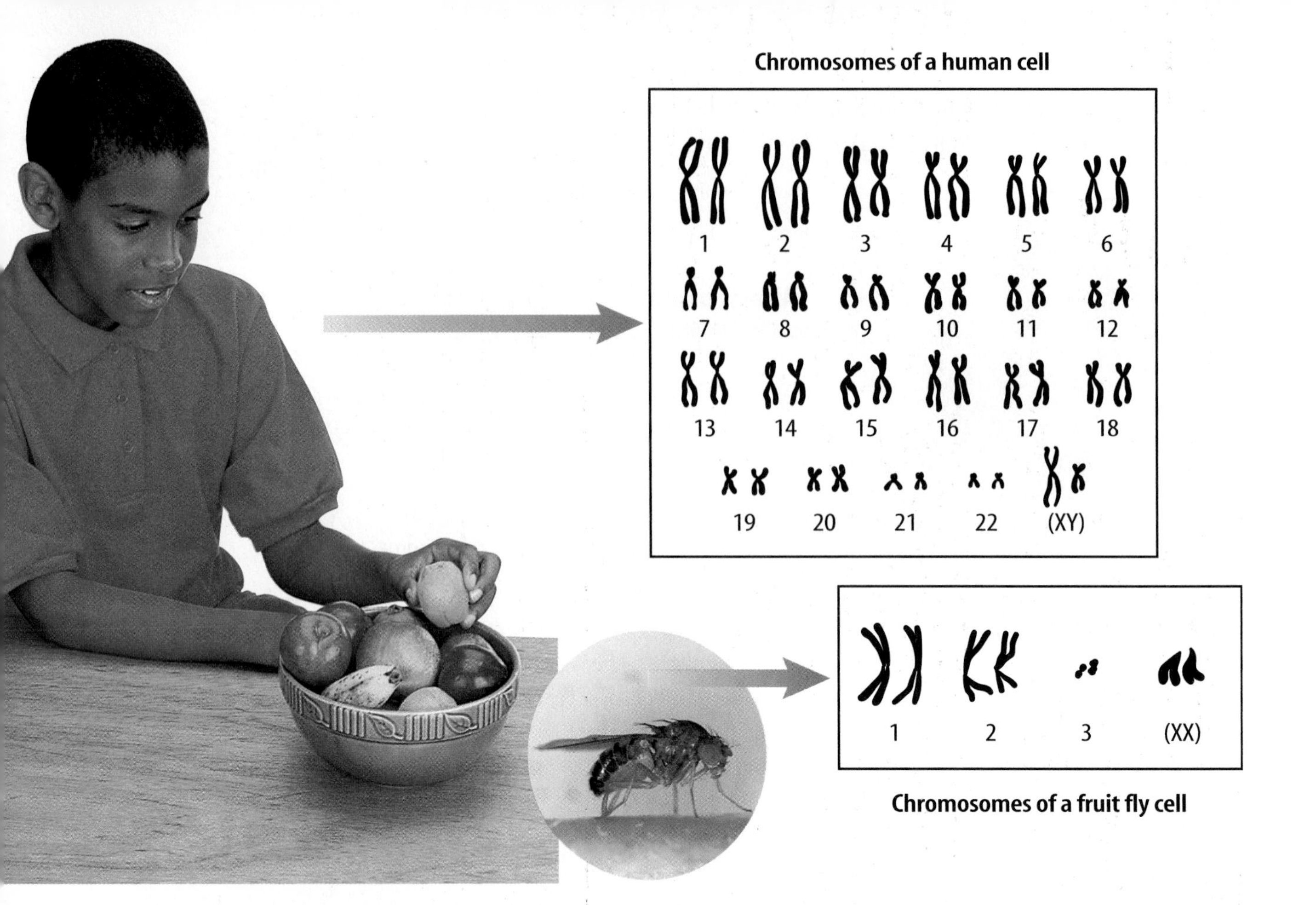

Figure 6 Pairs of chromosomes are found in the nucleus of most cells. All chromosomes shown here are in their duplicated form. Most human cells have 23 pairs of chromosomes including one pair of chromosomes that help determine sex such as the XY pair above. Most fruit fly cells have four pairs of chromosomes.
Infer *What do you think the XX pair in fruit flies helps determine?*

Results of Mitosis You should remember two important things about mitosis. First, it is the division of a nucleus. Second, it produces two new nuclei that are identical to each other and the original nucleus. Each new nucleus has the same number and type of chromosomes. Every cell in your body, except sex cells, has a nucleus with 46 chromosomes—23 pairs. This is because you began as one cell with 46 chromosomes in its nucleus. Skin cells, produced to replace or repair your skin, have the same 46 chromosomes as the original single cell you developed from. Each cell in a fruit fly has eight chromosomes, so each new cell produced by mitosis has a copy of those eight chromosomes. **Figure 6** shows the chromosomes found in most human cells and those found in most fruit fly cells.

Each of the trillions of cells in your body, except sex cells, has a copy of the same hereditary material. Even though all actors in a play have copies of the same script, they do not learn the same lines. Likewise, all of your cells use different parts of the same hereditary material to become different types of cells.

Cell division allows growth and replaces worn out or damaged cells. You are much larger and have more cells than a baby mainly because of cell division. If you cut yourself, the wound heals because cell division replaces damaged cells. Another way some organisms use cell division is to produce new organisms.

Asexual Reproduction

Reproduction is the process by which an organism produces others of its same kind. Among living organisms, there are two types of reproduction—sexual and asexual. Sexual reproduction usually requires two organisms. In **asexual reproduction,** a new organism (sometimes more than one) is produced from one organism. The new organism will have hereditary material identical to the hereditary material of the parent organism.

Reading Check *How many organisms are needed for asexual reproduction?*

Cellular Asexual Reproduction Organisms with eukaryotic cells asexually reproduce by cell division. A sweet potato growing in a jar of water is an example of asexual reproduction. All the stems, leaves, and roots that grow from the sweet potato have been produced by cell division and have the same hereditary material. New strawberry plants can be reproduced asexually from horizontal stems called runners. **Figure 7** shows asexual reproduction in a potato and a strawberry plant.

Recall that mitosis is the division of a nucleus. However, bacteria do not have a nucleus so they can't use mitosis. Instead, bacteria reproduce asexually by fission. During fission, an organism whose cells do not contain a nucleus copies its genetic material and then divides into two identical organisms.

Modeling Mitosis

Procedure

1. Make models of cell division using **materials supplied by your teacher.**
2. Use four chromosomes in your model.
3. When finished, arrange the models in the order in which mitosis occurs.

Analysis

1. In which steps is the nucleus visible?
2. How many cells does a dividing cell form?

Figure 7 Many plants can reproduce asexually.

A new potato plant can grow from each sprout on this potato.

Infer *how the genetic material in the small strawberry plant above compares to the genetic material in the large strawberry plant.*

Figure 8 Some organisms use cell division for budding and regeneration.

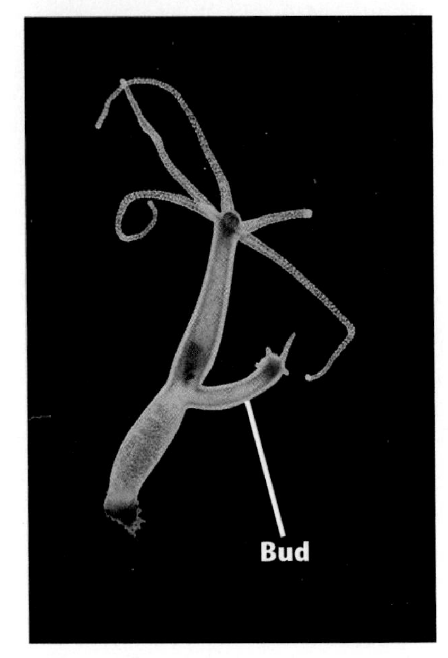

A Hydra, a freshwater animal, can reproduce asexually by budding. The bud is a small exact copy of the adult.

B This sea star is regenerating four new arms.

Budding and Regeneration Look at **Figure 8A.** A new organism is growing from the body of the parent organism. This organism, called a hydra, is reproducing by budding. Budding is a type of asexual reproduction made possible because of cell division. When the bud on the adult becomes large enough, it breaks away to live on its own.

Could you grow a new finger? Some organisms can regrow damaged or lost body parts, as shown in **Figure 8B.** Regeneration is the process that uses cell division to regrow body parts. Sponges, planaria, sea stars, and some other organisms can use regeneration for asexual reproduction. If these organisms break into pieces, a whole new organism will grow from each piece. Because sea stars eat oysters, oyster farmers dislike them. What would happen if an oyster farmer collected sea stars, cut them into pieces, and threw them back into the ocean?

section 1 review

Summary

The Cell Cycle

- The cell cycle is a series of events from one cell division to the next.
- Most of a eukaryotic cell's life is interphase.

Mitosis

- Mitosis is a series of four phases or steps.
- Each new nucleus formed by mitosis has the same number and type of chromosomes.

Asexual Reproduction

- In asexual reproduction, a new organism is produced from one organism.
- Cellular, budding, and regeneration are forms of asexual reproduction.

Self Check

1. **Define** mitosis. How does it differ in plants and animals?
2. **Identify** two examples of asexual reproduction in many-celled organisms.
3. **Describe** what happens to chromosomes before mitosis.
4. **Compare and contrast** the two new cells formed after mitosis and cell division.
5. **Think Critically** Why is it important for the nuclear membrane to disintegrate during mitosis?

Applying Math

6. **Solve One-Step Equations** If a cell undergoes cell division every 5 min, how many cells will there be after 1 h?

Science Online green.msscience.com/self_check_quiz

Mitosis in Plant Cells

Reproduction of most cells in plants and animals uses mitosis and cell division. In this lab, you will study mitosis in plant cells by examining prepared slides of onion root-tip cells.

Real-World Question

How can plant cells in different stages of mitosis be distinguished from each other?

Goals

- **Compare** cells in different stages of mitosis and observe the location of their chromosomes.
- **Observe** what stage of mitosis is most common in onion root tips.

Materials

prepared slide of an onion root tip
microscope

Safety Precautions

Procedure

1. Copy the data table in your Science Journal.

Number of Root-Tip Cells Observed

Stage of Mitosis	Number of Cells Observed	Percent of Cells Observed
Prophase		
Metaphase		
Anaphase	Do not write in this book.	
Telophase		
Total		

2. **Obtain** a prepared slide of cells from an onion root tip.

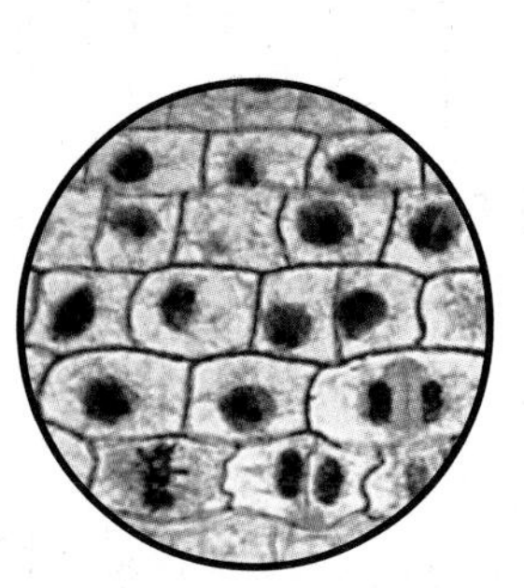

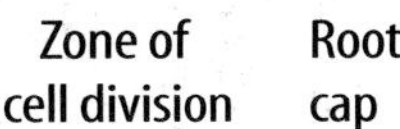

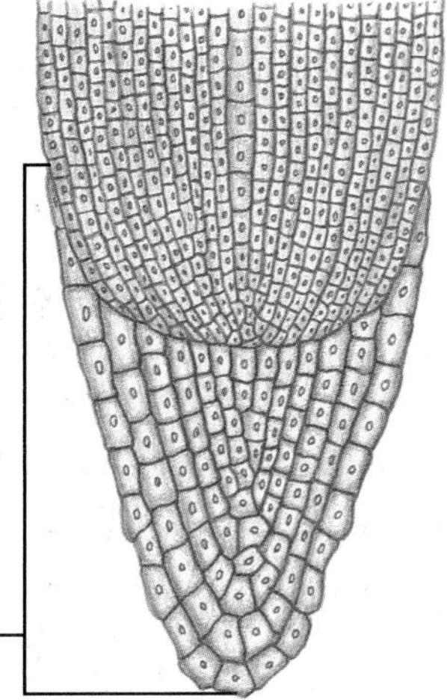

3. Set your microscope on low power and examine the slide. The large, round cells at the root tip are called the root cap. Move the slide until you see the cells just behind the root cap. Turn to the high-power objective.
4. Find an area where you can see the most stages of mitosis. Count and record how many cells you see in each stage.
5. Return the nosepiece to low power. Remove the onion root-tip slide.

Conclude and Apply

1. **Compare** the cells in the region behind the root cap to those in the root cap.
2. **Calculate** the percent of cells found in each stage of mitosis. Infer which stage of mitosis takes the longest period of time.

Communicating Your Data

Write and illustrate a story as if you were a cell undergoing mitosis. Share your story with your class. **For more help, refer to the** Science Skill Handbook.

section 2

Sexual Reproduction and Meiosis

as you read

What You'll Learn

- **Describe** the stages of meiosis and how sex cells are produced.
- **Explain** why meiosis is needed for sexual reproduction.
- **Name** the cells that are involved in fertilization.
- **Explain** how fertilization occurs in sexual reproduction.

Why It's Important

Meiosis and sexual reproduction are the reasons why no one else is exactly like you.

Review Vocabulary

organism: any living thing; uses energy, is made of cells, reproduces, responds, grows, and develops

New Vocabulary

- sexual reproduction
- sperm
- egg
- fertilization
- zygote
- diploid
- haploid
- meiosis

Sexual Reproduction

Sexual reproduction is another way that a new organism can be produced. During **sexual reproduction,** two sex cells, sometimes called an egg and a sperm, come together. Sex cells, like those in **Figure 9,** are formed from cells in reproductive organs. **Sperm** are formed in the male reproductive organs. **Eggs** are formed in the female reproductive organs. The joining of an egg and a sperm is called **fertilization,** and the cell that forms is called a **zygote** (ZI goht). Generally, the egg and the sperm come from two different organisms of the same species. Following fertilization, cell division begins. A new organism with a unique identity develops.

Diploid Cells Your body forms two types of cells—body cells and sex cells. Body cells far outnumber sex cells. Your brain, skin, bones, and other tissues and organs are formed from body cells. A typical human body cell has 46 chromosomes. Each chromosome has a mate that is similar to it in size and shape and has similar DNA. Human body cells have 23 pairs of chromosomes. When cells have pairs of similar chromosomes, they are said to be **diploid** (DIH ployd).

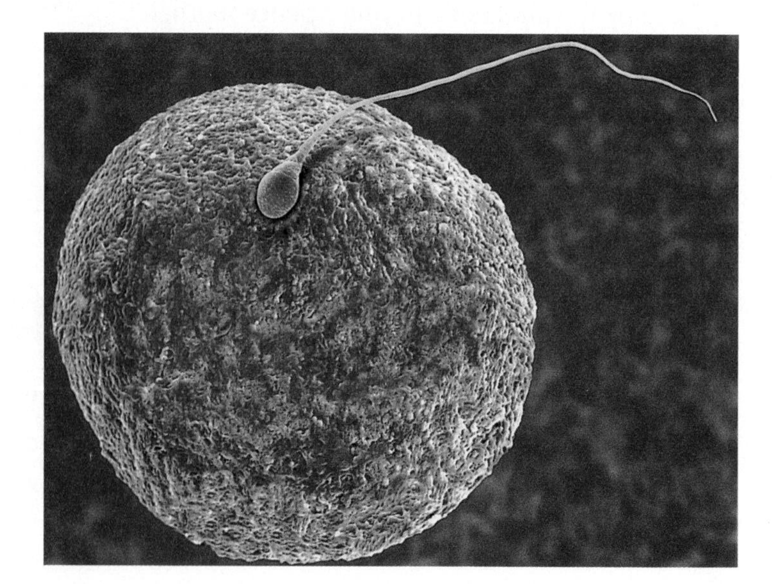

Figure 9 A human egg and a human sperm at fertilization.

Haploid Cells Because sex cells do not have pairs of chromosomes, they are said to be **haploid** (HA ployd). They have only half the number of chromosomes as body cells. *Haploid* means "single form." Human sex cells have only 23 chromosomes—one from each of the 23 pairs of similar chromosomes. Compare the chromosomes found in a sex cell, as shown in **Figure 9,** to the full set of human chromosomes seen in **Figure 6.**

Reading Check *How many chromosomes are usually in each human sperm?*

Meiosis and Sex Cells

A process called **meiosis** (mi OH sus) produces haploid sex cells. What would happen in sexual reproduction if two diploid cells combined? The offspring would have twice as many chromosomes as its parent. Although plants with twice the number of chromosomes as the parent plants are often produced, most animals do not survive with a double number of chromosomes. Meiosis ensures that the offspring will have the same diploid number as its parent, as shown in **Figure 10.** After two haploid sex cells combine, a diploid zygote is produced that develops into a new diploid organism.

During meiosis, two divisions of the nucleus occur. These divisions are called meiosis I and meiosis II. The steps of each division have names like those in mitosis and are numbered for the division in which they occur.

Diploid Zygote The human egg releases a chemical into the surrounding fluid that attracts sperm. Usually, only one sperm fertilizes the egg. After the sperm nucleus enters the egg, the cell membrane of the egg changes in a way that prevents other sperm from entering. What adaptation in this process guarantees that the zygote will be diploid? Write a paragraph describing your ideas in your Science Journal.

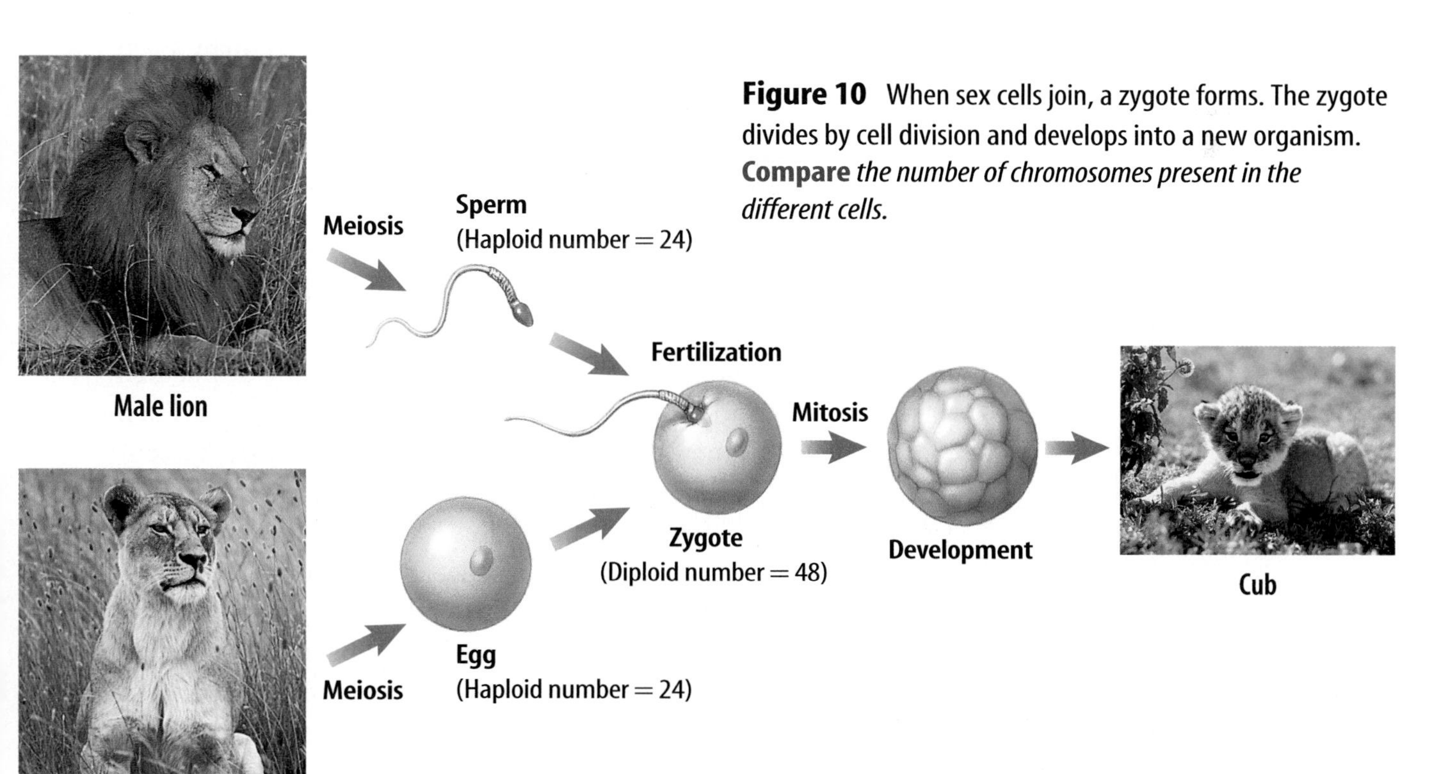

Figure 10 When sex cells join, a zygote forms. The zygote divides by cell division and develops into a new organism. **Compare** *the number of chromosomes present in the different cells.*

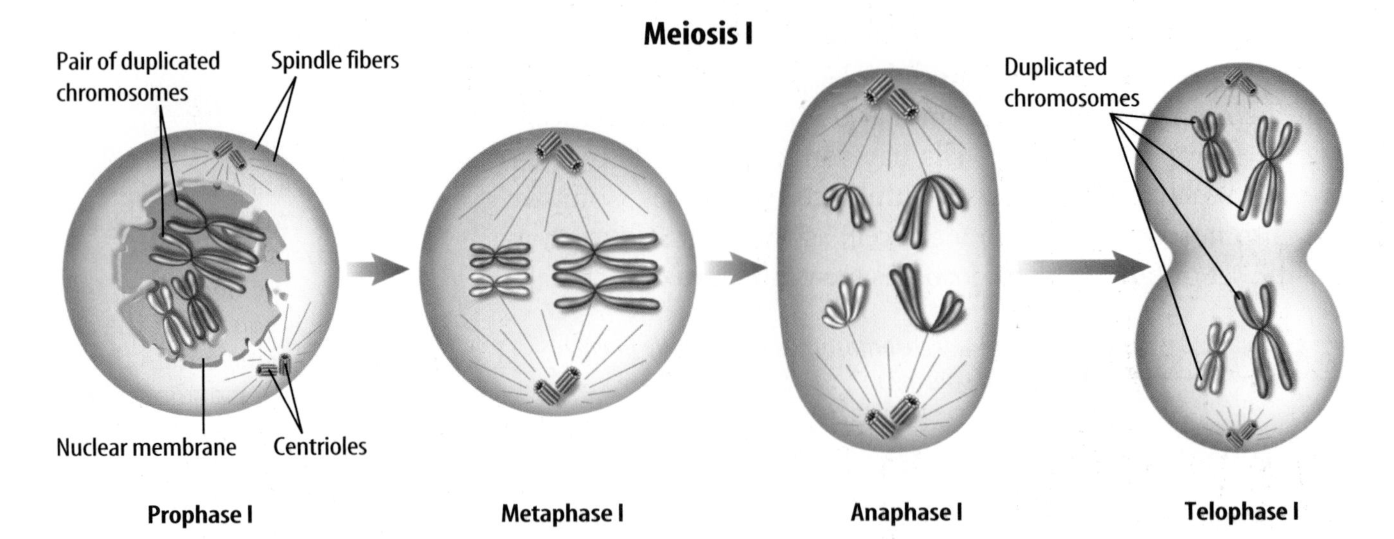

Figure 11 Meiosis has two divisions of the nucleus—meiosis I and meiosis II.
Determine *how many sex cells are finally formed after both divisions are completed.*

Meiosis I Before meiosis begins, each chromosome is duplicated, just as in mitosis. When the cell is ready for meiosis, each duplicated chromosome is visible under the microscope as two chromatids. As shown in **Figure 11,** the events of prophase I are similar to those of prophase in mitosis. In meiosis, each duplicated chromosome comes near its similar duplicated mate. In mitosis they do not come near each other.

In metaphase I, the pairs of duplicated chromosomes line up in the center of the cell. The centromere of each chromatid pair becomes attached to one spindle fiber, so the chromatids do not separate in anaphase I. The two pairs of chromatids of each similar pair move away from each other to opposite ends of the cell. Each duplicated chromosome still has two chromatids. Then, in telophase I, the cytoplasm divides, and two new cells form. Each new cell has one duplicated chromosome from each similar pair.

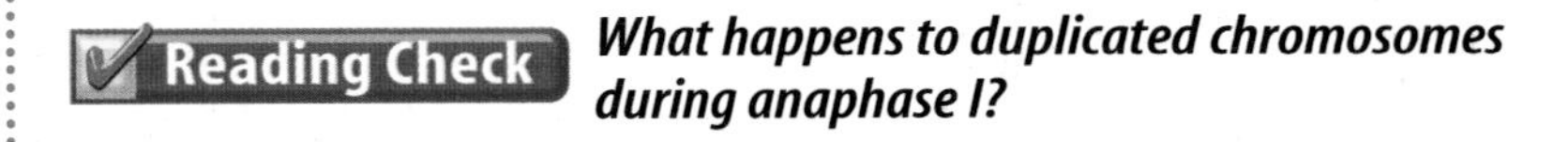

What happens to duplicated chromosomes during anaphase I?

Meiosis II The two cells formed during meiosis I now begin meiosis II. The chromatids of each duplicated chromosome will be separated during this division. In prophase II, the duplicated chromosomes and spindle fibers reappear in each new cell. Then in metaphase II, the duplicated chromosomes move to the center of the cell. Unlike what occurs in metaphase I, each centromere now attaches to two spindle fibers instead of one. The centromere divides during anaphase II, and the chromatids separate and move to opposite ends of the cell. Each chromatid now is an individual chromosome. As telophase II begins, the spindle fibers disappear, and a nuclear membrane forms around the chromosomes at each end of the cell. When meiosis II is finished, the cytoplasm divides.

Meiosis II

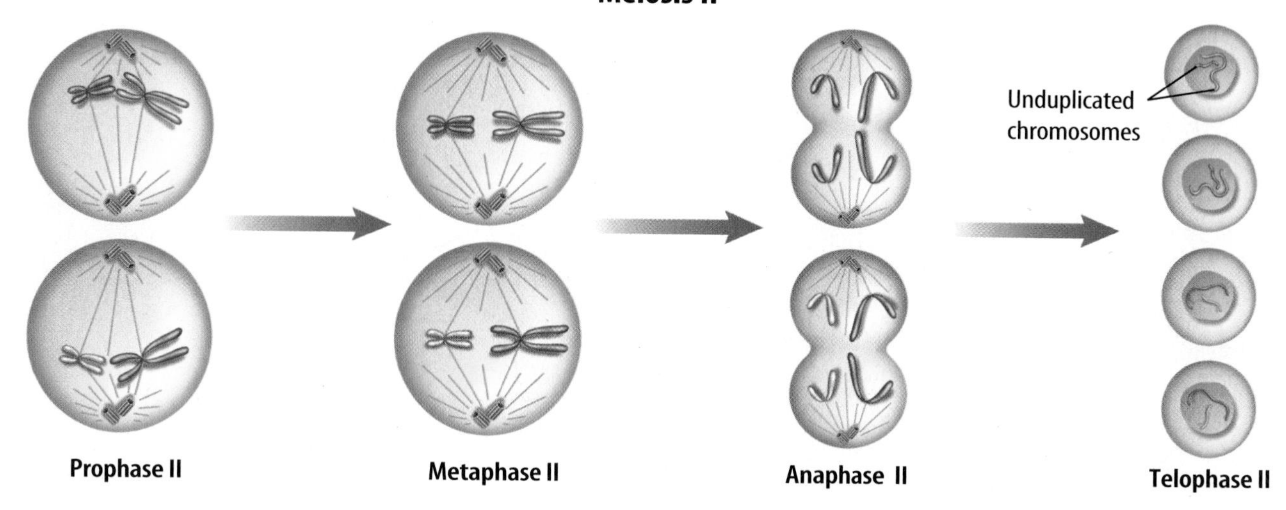

Summary of Meiosis Two cells form during meiosis I. In meiosis II, both of these cells form two cells. The two divisions of the nucleus result in four sex cells. Each has one-half the number of chromosomes in its nucleus that was in the original nucleus. From a human cell with 46 paired chromosomes, meiosis produces four sex cells each with 23 unpaired chromosomes.

Applying Science

How can chromosome numbers be predicted?

Offspring get half of their chromosomes from one parent and half from the other. What happens if each parent has a different diploid number of chromosomes?

Identifying the Problem

A zebra and a donkey can mate to produce a zonkey. Zebras have a diploid number of 46. Donkeys have a diploid number of 62.

Solving the Problem

1. How many chromosomes would the zonkey receive from each parent?
2. What is the chromosome number of the zonkey?
3. What would happen when meiosis occurs in the zonkey's reproductive organs?
4. Predict why zonkeys are usually sterile.

Donkey
62 chromosomes

Zonkey

Zebra
46 chromosomes

NATIONAL GEOGRAPHIC VISUALIZING POLYPLOIDY IN PLANTS

Figure 12

You received a haploid (n) set of chromosomes from each of your parents, making you a diploid (2n) organism. In nature, however, many plants are polyploid—they have three (3n), four (4n), or more sets of chromosomes. We depend on some of these plants for food.

▲ TRIPLOID Bright yellow bananas typically come from triploid (3n) banana plants. Plants with an odd number of chromosome sets usually cannot reproduce sexually and have very small seeds or none at all.

▲ TETRAPLOID Polyploidy occurs naturally in many plants—including peanuts and daylilies—due to mistakes in mitosis or meiosis.

▼ HEXAPLOID Modern cultivated strains of oats have six sets of chromosomes, making them hexaploid (6n) plants.

▲ OCTAPLOID Polyploid plants often are bigger than nonpolyploid plants and may have especially large leaves, flowers, or fruits. Strawberries are an example of octaploid (8n) plants.

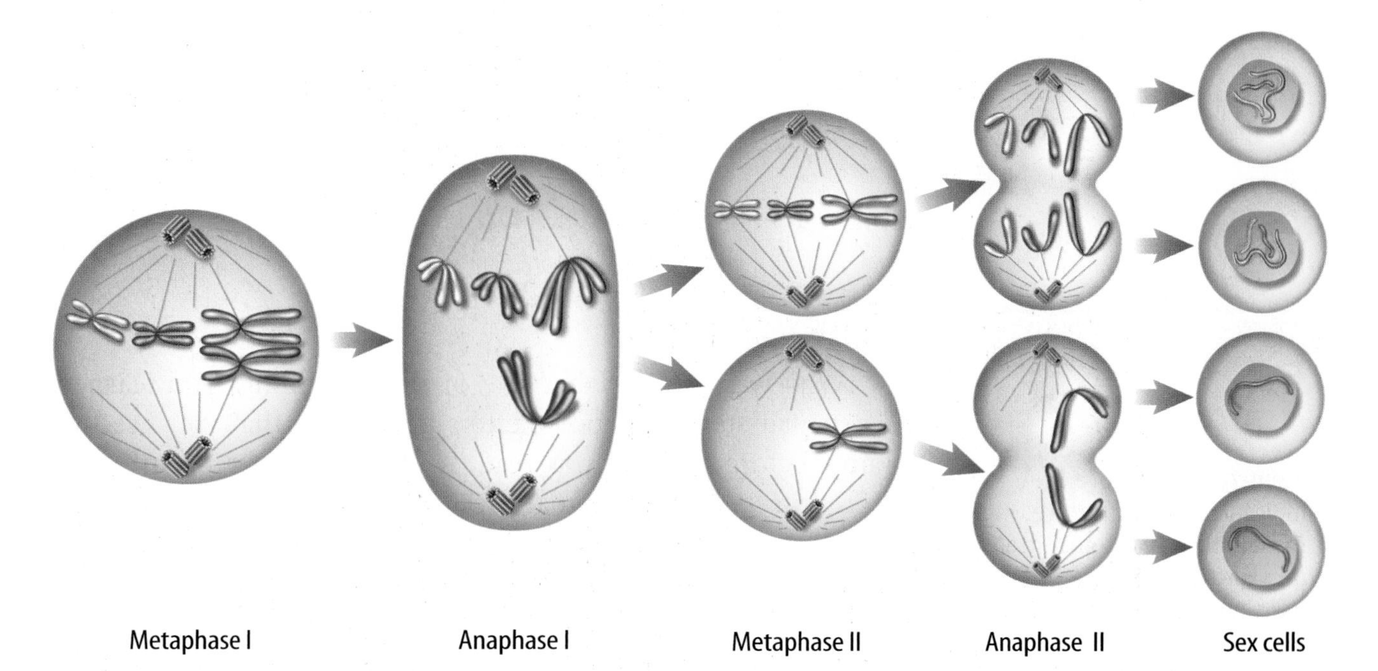

Figure 13 This diploid cell has four chromosomes. During anaphase I, one pair of duplicated chromosomes did not separate. **Infer** *how many chromosomes each sex cell usually has.*

Mistakes in Meiosis Meiosis occurs many times in reproductive organs. Although mistakes in plants, as shown in **Figure 12,** are common, mistakes are less common in animals. These mistakes can produce sex cells with too many or too few chromosomes, as shown in **Figure 13.** Sometimes, zygotes produced from these sex cells die. If the zygote lives, every cell in the organism that grows from that zygote usually will have the wrong number of chromosomes. Organisms with the wrong number of chromosomes may not grow normally.

section 2 review

Summary

Sexual Reproduction

- During sexual reproduction, two sex cells come together.
- Cell division begins after fertilization.
- A typical human body cell has 46 chromosomes, and a human sex cell has 23 chromosomes.

Meiosis and Sex Cells

- Each chromosome is duplicated before meiosis, then two divisions of the nucleus occur.
- During meiosis I, duplicated chromosomes are separated into new cells.
- Chromatids separate during meiosis II.
- Meiosis I and meiosis II result in four sex cells.

Self Check

1. **Describe** a zygote and how it is formed.
2. **Explain** where sex cells form.
3. **Compare** what happens to chromosomes during anaphase I and anaphase II.
4. **Think Critically** Plants grown from runners and leaf cuttings have the same traits as the parent plant. Plants grown from seeds can vary from the parent plants in many ways. Why can this happen?

Applying Skills

5. **Make and use a table** to compare mitosis and meiosis in humans. Vertical headings should include: *What Type of Cell (Body or Sex), Beginning Cell (Haploid or Diploid), Number of Cells Produced, End-Product Cell (Haploid or Diploid),* and *Number of Chromosomes in New Cells.*

section 3

DNA

as you read

What You'll Learn

- **Identify** the parts of a DNA molecule and its structure.
- **Explain** how DNA copies itself.
- **Describe** the structure and function of each kind of RNA.

Why It's Important

DNA helps determine nearly everything your body is and does.

Review Vocabulary
heredity: the passing of traits from parents to offspring

New Vocabulary
- DNA
- gene
- RNA
- mutation

What is DNA?

Why was the alphabet one of the first things you learned when you started school? Letters are a code that you need to know before you learn to read. A cell also uses a code that is stored in its hereditary material. The code is a chemical called deoxyribonucleic (dee AHK sih ri boh noo klay ihk) acid, or **DNA.** It contains information for an organism's growth and function. **Figure 14** shows how DNA is stored in cells that have a nucleus. When a cell divides, the DNA code is copied and passed to the new cells. In this way, new cells receive the same coded information that was in the original cell. Every cell that has ever been formed in your body or in any other organism contains DNA.

INTEGRATE Chemistry

Discovering DNA Since the mid-1800s, scientists have known that the nuclei of cells contain large molecules called nucleic acids. By 1950, chemists had learned what the nucleic acid DNA was made of, but they didn't understand how the parts of DNA were arranged.

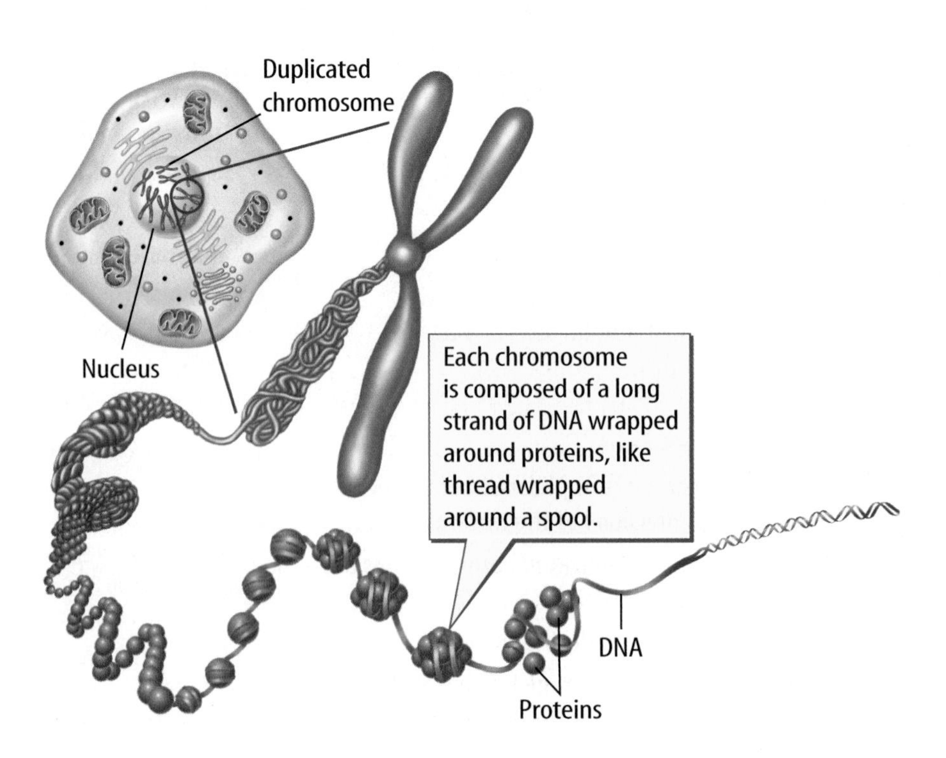

Figure 14 DNA is part of the chromosomes found in a cell's nucleus.

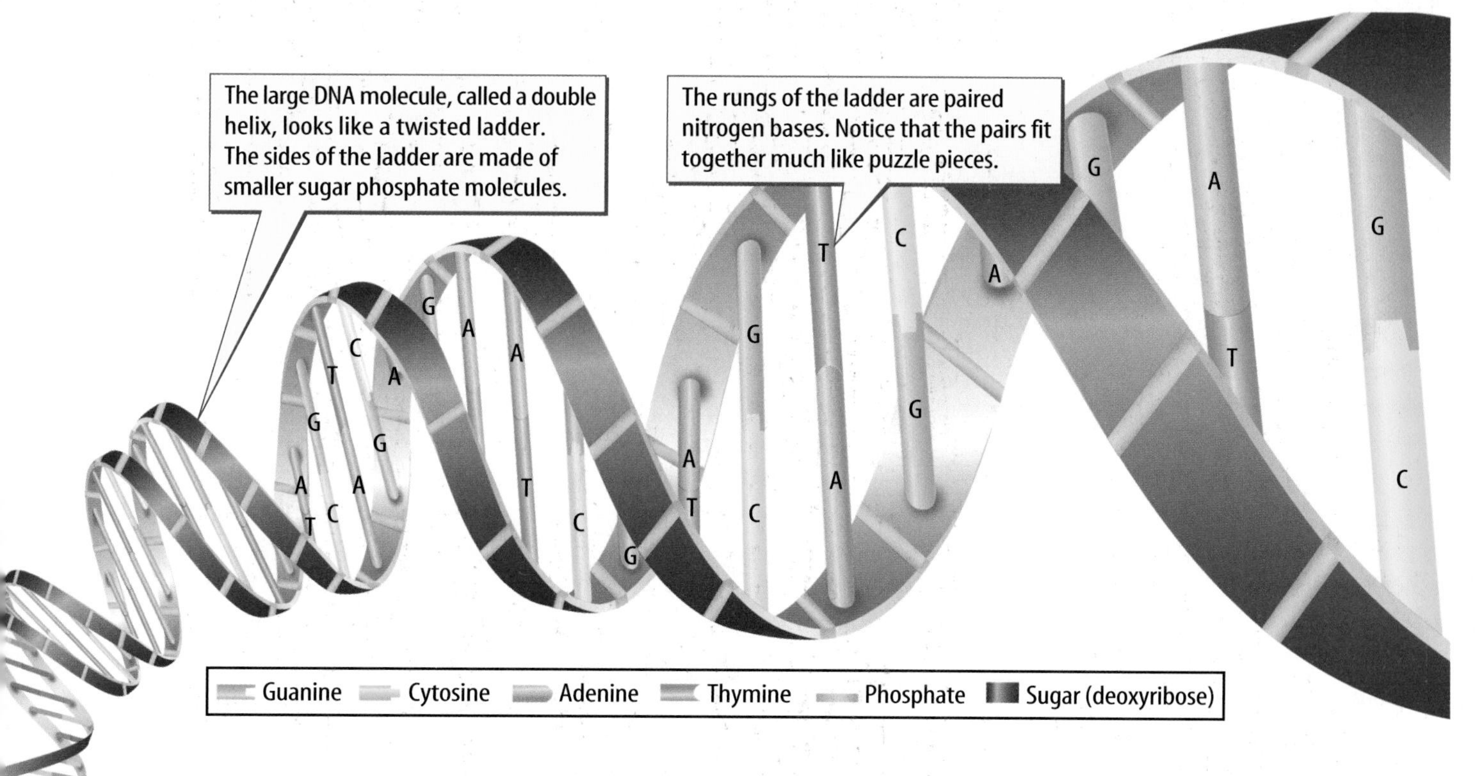

DNA's Structure In 1952, scientist Rosalind Franklin discovered that DNA is two chains of molecules in a spiral form. By using an X-ray technique, Dr. Franklin showed that the large spiral was probably made up of two spirals. As it turned out, the structure of DNA is similar to a twisted ladder. In 1953, using the work of Franklin and others, scientists James Watson and Francis Crick made a model of a DNA molecule.

A DNA Model What does DNA look like? According to the Watson and Crick DNA model, each side of the ladder is made up of sugar-phosphate molecules. Each molecule consists of the sugar called deoxyribose (dee AHK sih ri bohs) and a phosphate group. The rungs of the ladder are made up of other molecules called nitrogen bases. Four kinds of nitrogen bases are found in DNA—adenine (A duh neen), guanine (GWAH neen), cytosine (SI tuh seen), and thymine (THI meen). The bases are represented by the letters A, G, C, and T. The amount of cytosine in cells always equals the amount of guanine, and the amount of adenine always equals the amount of thymine. This led to the hypothesis that these bases occur as pairs in DNA. **Figure 14** shows that adenine always pairs with thymine, and guanine always pairs with cytosine. Like interlocking pieces of a puzzle, each base bonds only with its correct partner.

What are the nitrogen base pairs in a DNA molecule?

Modeling DNA Replication

Procedure

1. Suppose you have a segment of DNA that is six nitrogen base pairs in length. On **paper,** using the letters A, T, C, and G, write a combination of six pairs, remembering that A and T are always a pair and C and G are always a pair.
2. Duplicate your segment of DNA. On paper, diagram how this happens and show the new DNA segments.

Analysis

Compare the order of bases of the original DNA to the new DNA molecules.

Try at Home

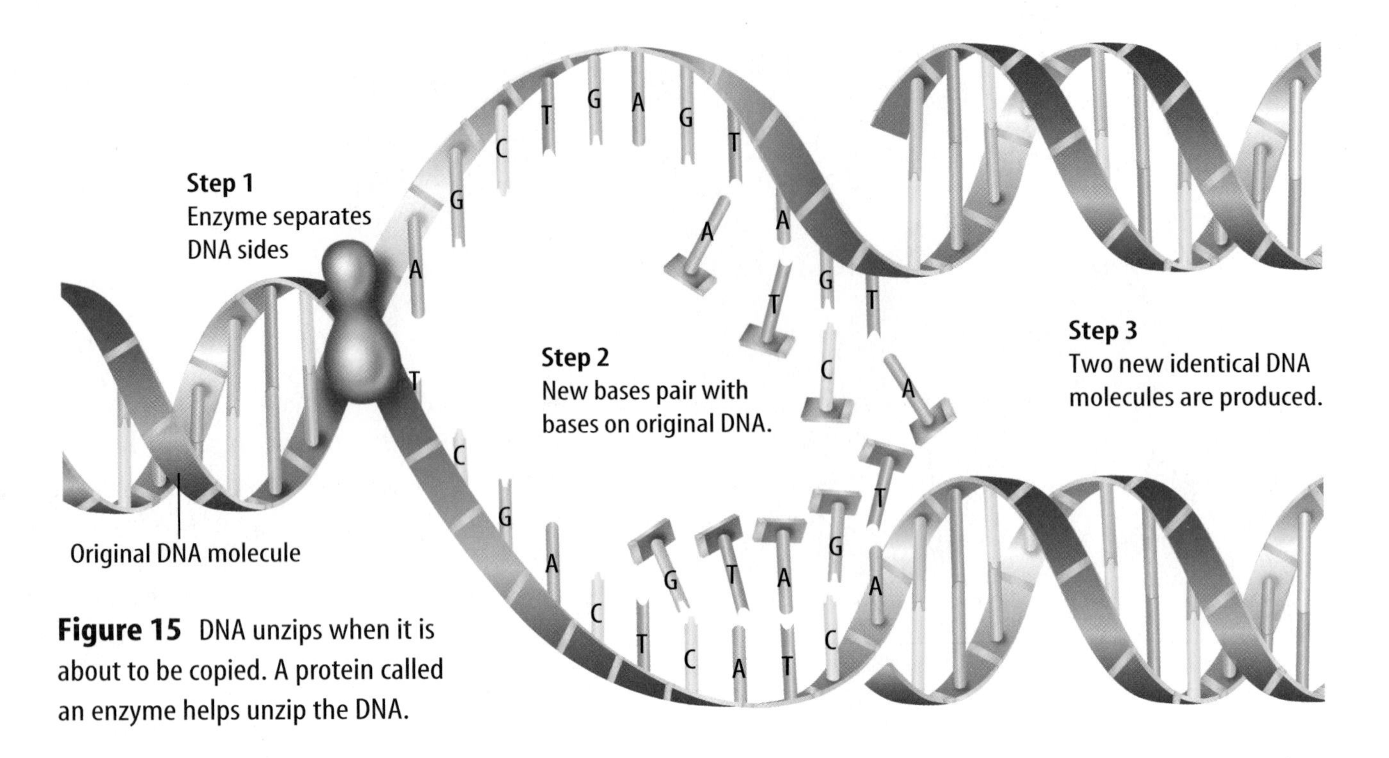

Figure 15 DNA unzips when it is about to be copied. A protein called an enzyme helps unzip the DNA.

Copying DNA When chromosomes are duplicated before mitosis or meiosis, the amount of DNA in the nucleus is doubled. The Watson and Crick model shows how this takes place. The two sides of DNA unwind and separate. Each side then becomes a pattern on which a new side forms, as shown in **Figure 15.** The new DNA has bases that are identical to those of the original DNA and are in the same order.

Figure 16 This diagram shows just a few of the genes that have been identified on human chromosome 7. The bold print is the name that has been given to each gene.

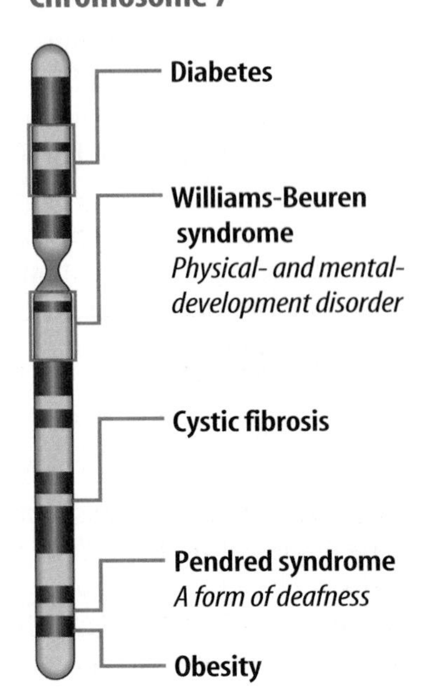

Genes

Most of your characteristics, such as the color of your hair, your height, and even how things taste to you, depend on the kinds of proteins your cells make. DNA in your cells stores the instructions for making these proteins.

Proteins build cells and tissues or work as enzymes. The instructions for making a specific protein are found in a **gene** which is a section of DNA on a chromosome. As shown in **Figure 16,** each chromosome contains hundreds of genes. Proteins are made of chains of hundreds or thousands of amino acids. The gene determines the order of amino acids in a protein. Changing the order of the amino acids makes a different protein. What might occur if an important protein couldn't be made or if the wrong protein was made in your cells?

Making Proteins Genes are found in the nucleus, but proteins are made on ribosomes in cytoplasm. The codes for making proteins are carried from the nucleus to the ribosomes by another type of nucleic acid called ribonucleic acid, or **RNA.**

Ribonucleic Acid RNA is made in the nucleus on a DNA pattern. However, RNA is different from DNA. If DNA is like a ladder, RNA is like a ladder that has all its rungs sawed in half. Compare the DNA molecule in **Figure 14** to the RNA molecule in **Figure 17.** RNA has the bases A, G, and C like DNA but has the base uracil (U) instead of thymine (T). The sugar-phosphate molecules in RNA contain the sugar ribose, not deoxyribose.

The three main kinds of RNA made from DNA in a cell's nucleus are messenger RNA (mRNA), ribosomal RNA (rRNA), and transfer RNA (tRNA). Protein production begins when mRNA moves into the cytoplasm. There, ribosomes attach to it. Ribosomes are made of rRNA. Transfer RNA molecules in the cytoplasm bring amino acids to these ribosomes. Inside the ribosomes, three nitrogen bases on the mRNA temporarily match with three nitrogen bases on the tRNA. The same thing happens for the mRNA and another tRNA molecule, as shown in **Figure 17.** The amino acids that are attached to the two tRNA molecules bond. This is the beginning of a protein.

The code carried on the mRNA directs the order in which the amino acids bond. After a tRNA molecule has lost its amino acid, it can move about the cytoplasm and pick up another amino acid just like the first one. The ribosome moves along the mRNA. New tRNA molecules with amino acids match up and add amino acids to the protein molecule.

Topic: The Human Genome Project

Visit green.msscience.com for Web links to information about the Human Genome Project.

Activity Find out when chromosomes 5, 16, 29, 21, and 22 were completely sequenced. Write about what scientists learned about each of these chromosomes.

Figure 17 Cells need DNA, RNA, and amino acids to make proteins.

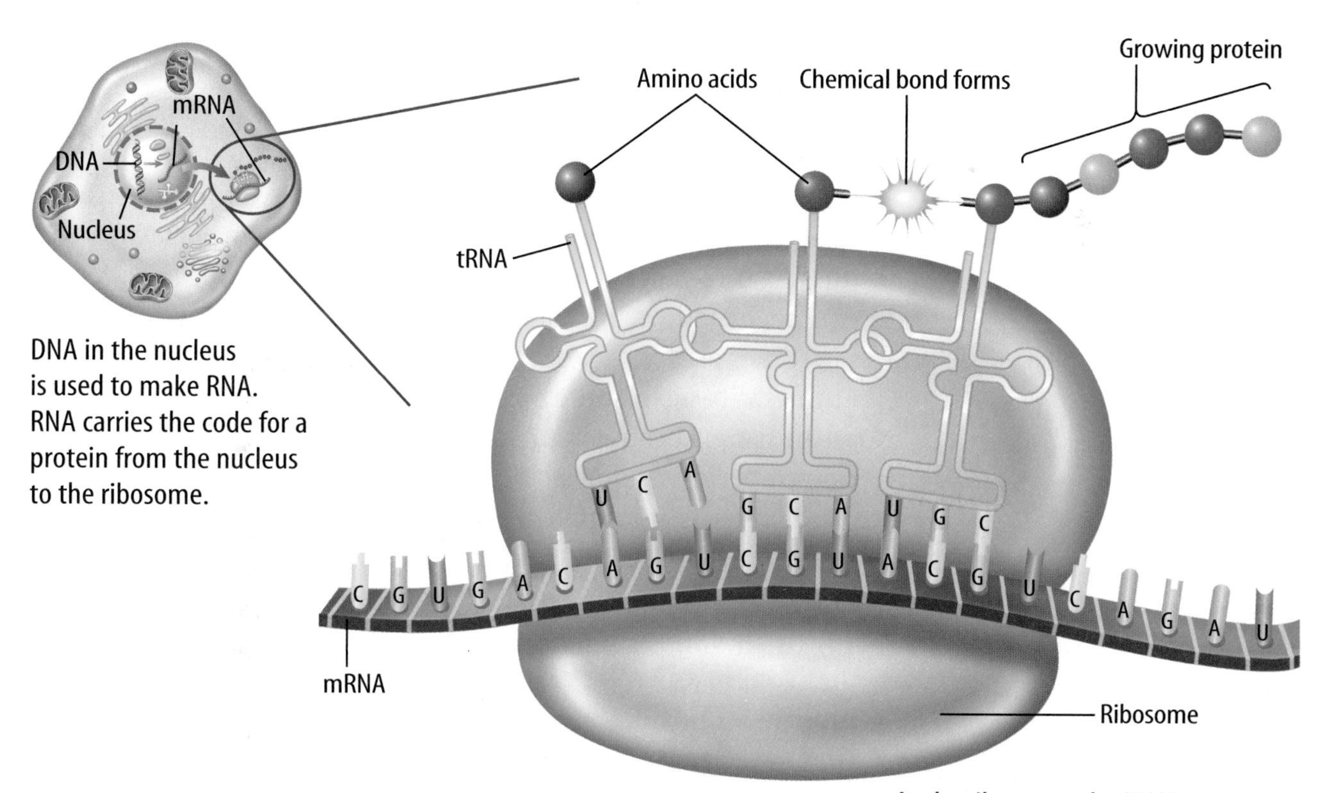

DNA in the nucleus is used to make RNA. RNA carries the code for a protein from the nucleus to the ribosome.

At the ribosome, the RNA's message is translated into a specific protein.

Figure 18 Each cell in the body produces only the proteins that are necessary to do its job.

Controlling Genes You might think that because most cells in an organism have exactly the same chromosomes and the same genes, they would make the same proteins, but they don't. In many-celled organisms like you, each cell uses only some of the thousands of genes that it has to make proteins. Just as each actor uses only the lines from the script for his or her role, each cell uses only the genes that direct the making of proteins that it needs. For example, muscle proteins are made in muscle cells, as represented in **Figure 18,** but not in nerve cells.

Cells must be able to control genes by turning some genes off and turning other genes on. They do this in many different ways. Sometimes the DNA is twisted so tightly that no RNA can be made. Other times, chemicals bind to the DNA so that it cannot be used. If the incorrect proteins are produced, the organism cannot function properly.

Mutations

Sometimes mistakes happen when DNA is being copied. Imagine that the copy of the script the director gave you was missing three pages. You use your copy to learn your lines. When you begin rehearsing for the play, everyone is ready for one of the scenes except for you. What happened? You check your copy of the script against the original and find that three of the pages are missing. Because your script is different from the others, you cannot perform your part correctly.

If DNA is not copied exactly, the proteins made from the instructions might not be made correctly. These mistakes, called **mutations,** are any permanent change in the DNA sequence of a gene or chromosome of a cell. Some mutations include cells that receive an entire extra chromosome or are missing a chromosome. Outside factors such as X rays, sunlight, and some chemicals have been known to cause mutations.

Reading Check *When are mutations likely to occur?*

Figure 19 Because of a defect on chromosome 2, the mutant fruit fly has short wings and cannot fly.
Predict *Could this defect be transferred to the mutant's offspring? Explain.*

Results of a Mutation Genes control the traits you inherit. Without correctly coded proteins, an organism can't grow, repair, or maintain itself. A change in a gene or chromosome can change the traits of an organism, as illustrated in **Figure 19.**

If the mutation occurs in a body cell, it might or might not be life threatening to the organism. However, if a mutation occurs in a sex cell, then all the cells that are formed from that sex cell will have that mutation. Mutations add variety to a species when the organism reproduces. Many mutations are harmful to organisms, often causing their death. Some mutations do not appear to have any effect on the organism, and some can even be beneficial. For example, a mutation to a plant might cause it to produce a chemical that certain insects avoid. If these insects normally eat the plant, the mutation will help the plant survive.

Topic: Fruit Fly Genes
Visit green.msscience.com for Web links to information about what genes are present on the chromosomes of a fruit fly.

Activity Draw a picture of one of the chromosomes of a fruit fly and label some of its genes.

section 3 review

Summary

What is DNA?

- Each side of the DNA ladder is made up of sugar-phosphate molecules, and the rungs of the ladder are made up of nitrogenous bases.
- When DNA is copied, the new DNA has bases that are identical to those of the original DNA.

Genes

- The instructions for making a specific protein are found in genes in the cell nucleus. Proteins are made on ribosomes in the cytoplasm.
- There are three main kinds of RNA—mRNA, rRNA, and tRNA.

Mutations

- If DNA is not copied exactly, the resulting mutations may cause proteins to be made incorrectly.

Self Check

1. **Describe** how DNA makes a copy of itself.
2. **Explain** how the codes for proteins are carried from the nucleus to the ribosomes.
3. **Apply** A strand of DNA has the bases AGTAAC. Using letters, show a matching DNA strand.
4. **Determine** how tRNA is used when cells build proteins.
5. **Think Critically** You begin as one cell. Compare the DNA in your brain cells to the DNA in your heart cells.

Applying Skills

6. **Concept Map** Using a Venn diagram, compare and contrast DNA and RNA.
7. **Use a word processor** to make an outline of the events that led up to the discovery of DNA. Use library resources to find this information.

Science Online green.msscience.com/self_check_quiz

Use the Internet

Mutations

Real-World Question

Mutations can result in dominant or recessive genes. A recessive characteristic can appear only if an organism has two recessive genes for that characteristic. However, a dominant characteristic can appear if an organism has one or two dominant genes for that characteristic. Why do some mutations result in more common traits while others do not? Form a hypothesis about how a mutation can become a common trait.

Fantail pigeon

Make a Plan

1. **Observe** common traits in various animals, such as household pets or animals you might see in a zoo.
2. **Learn** what genes carry these traits in each animal.
3. **Research** the traits to discover which ones are results of mutations. Are all mutations dominant? Are any of these mutations beneficial?

Goals

- **Observe** traits of various animals.
- **Research** how mutations become traits.
- Gather data about mutations.
- Make a frequency table of your findings and communicate them to other students.

Data Source

Science Online

Visit green.msscience.com/internet_lab for more information on common genetic traits in different animals, recessive and dominant genes, and data from other students.

White tiger

Follow Your Plan

1. Make sure your teacher approves your plan before you start.
2. Visit the link shown below to access different Web sites for information about mutations and genetics.
3. **Decide** if a mutation is beneficial, harmful, or neither. Record your data in your Science Journal.

Analyze Your Data

1. **Record** in your Science Journal a list of traits that are results of mutations.
2. **Describe** an animal, such as a pet or an animal you've seen in the zoo. Point out which traits are known to be the result of a mutation.
3. **Make** a chart that compares recessive mutations to dominant mutations. Which are more common?
4. **Share** your data with other students by posting it at the link shown below.

Siberian Husky's eyes

Conclude and Apply

1. **Compare** your findings to those of your classmates and other data at the link shown below. What were some of the traits your classmates found that you did not? Which were the most common?
2. Look at your chart of mutations. Are all mutations beneficial? When might a mutation be harmful to an organism?
3. **Predict** how your data would be affected if you had performed this lab when one of these common mutations first appeared. Do you think you would see more or less animals with this trait?
4. Mutations occur every day but we only see a few of them. Infer how many mutations over millions of years can lead to a new species.

Communicating Your Data

Find this lab using the link below. **Post** your data in the table provided. Combine your data with that of other students and make a chart that shows all of the data.

Science Online

green.msscience.com/internet_lab

A Tangled Tale

How did a scientist get chromosomes to separate?

Thanks to chromosomes, each of us is unique!

Viewed under the microscope, chromosomes in cells sometimes look a lot like spaghetti. That's why scientists had such a hard time figuring out how many chromosomes are in each human cell. Imagine, then, how Dr. Tao-Chiuh Hsu (dow shew•SEW) must have felt when he looked into a microscope and saw "beautifully scattered chromosomes." The problem was, Hsu didn't know what he had done to separate the chromosomes into countable strands.

"I tried to study those slides and set up some more cultures to repeat the miracle," Hsu explained. "But nothing happened."

These chromosomes are magnified 500 times.

For three months Hsu tried changing every variable he could think of to make the chromosomes separate again. In April 1952, his efforts were finally rewarded. Hsu quickly realized that the chromosomes separated because of osmosis.

Osmosis is the movement of water molecules through cell membranes. This movement occurs in predictable ways. The water molecules move from areas with higher concentrations of water to areas with lower concentrations of water. In Hsu's case, the solution he used to prepare the cells had a higher concentration of water then the cell did. So water moved from the solution into the cell and the cell swelled until it finally exploded. The chromosomes suddenly were visible as separate strands.

What made the cells swell the first time? Apparently a technician had mixed the solution incorrectly. "Since nearly four months had elapsed, there was no way to trace who actually had prepared that particular [solution]," Hsu noted. "Therefore, this heroine must remain anonymous."

Research What developments led scientists to conclude that the human cell has 46 chromosomes? Visit the link shown to the right to get started.

For more information, visit green.msscience.com/oops

Reviewing Main Ideas

Section 1 Cell Division and Mitosis

1. The life cycle of a cell has two parts—growth and development, and cell division.
2. In mitosis, the nucleus divides to form two identical nuclei. Mitosis occurs in four continuous steps, or phases—prophase, metaphase, anaphase, and telophase.
3. Cell division in animal cells and plant cells is similar, but plant cells do not have centrioles and animal cells do not form cell walls.
4. Organisms use cell division to grow, to replace cells, and for asexual reproduction. Asexual reproduction produces organisms with DNA identical to the parent's DNA. Fission, budding, and regeneration can be used for asexual reproduction.

Section 2 Sexual Reproduction and Meiosis

1. Sexual reproduction results when an egg and sperm join. This event is called fertilization, and the cell that forms is called the zygote.
2. Meiosis occurs in the reproductive organs, producing four haploid sex cells.
3. During meiosis, two divisions of the nucleus occur.
4. Meiosis ensures that offspring produced by fertilization have the same number of chromosomes as their parents.

Section 3 DNA

1. DNA is a large molecule made up of two twisted strands of sugar-phosphate molecules and nitrogen bases.
2. All cells contain DNA. The section of DNA on a chromosome that directs the making of a specific protein is a gene.
3. DNA can copy itself and is the pattern from which RNA is made. Messenger RNA, ribosomal RNA, and transfer RNA are used to make proteins.
4. Permanent changes in DNA are called mutations.

Visualizing Main Ideas

Think of four ways that organisms can use mitosis. Copy and complete the spider diagram below.

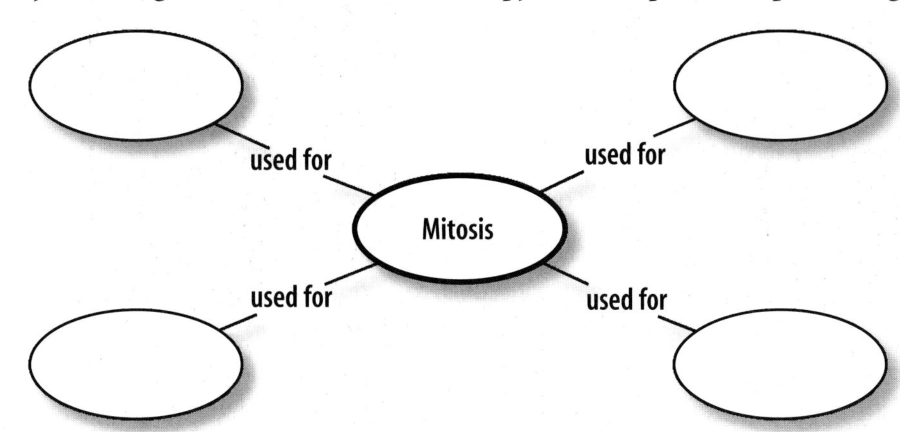

Using Vocabulary

asexual reproduction p. 281
chromosome p. 278
diploid p. 284
DNA p. 290
egg p. 284
fertilization p. 284
gene p. 292
haploid p. 285
meiosis p. 285
mitosis p. 278
mutation p. 294
RNA p. 292
sexual reproduction p. 284
sperm p. 284
zygote p. 284

Fill in the blanks with the correct vocabulary word or words.

1. _________ and _________ cells are sex cells.

2. _________ produces two identical cells.

3. An example of a nucleic acid is _________.

4. A(n) _________ is the code for a protein.

5. A(n) _________ sperm is formed during meiosis.

6. Budding is a type of _________.

7. A(n) _________ is a structure in the nucleus that contains hereditary material.

8. _________ produces four sex cells.

9. As a result of _________, a new organism develops that has its own unique identity.

10. An error made during the copying of DNA is called a(n) _________.

Checking Concepts

Choose the word or phrase that best answers the question.

11. Which of the following is a double spiral molecule with pairs of nitrogen bases?
A) RNA
B) amino acid
C) protein
D) DNA

12. What is in RNA but not in DNA?
A) thymine
B) thyroid
C) adenine
D) uracil

13. If a diploid tomato cell has 24 chromosomes, how many chromosomes will the tomato's sex cells have?
A) 6
B) 12
C) 24
D) 48

14. During a cell's life cycle, when do chromosomes duplicate?
A) anaphase
B) metaphase
C) interphase
D) telophase

15. When do chromatids separate during mitosis?
A) anaphase
B) prophase
C) metaphase
D) telophase

16. How is the hydra shown in the picture reproducing?
A) asexually, by budding
B) sexually, by budding
C) asexually, by fission
D) sexually, by fission

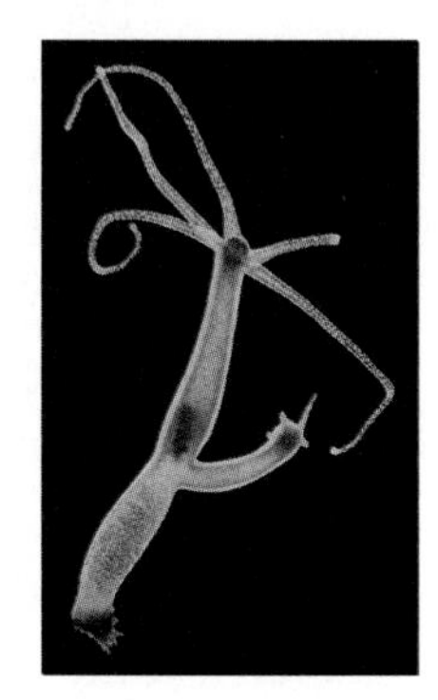

17. What is any permanent change in a gene or a chromosome called?
A) fission
B) reproduction
C) replication
D) mutation

18. What does meiosis produce?
A) cells with the diploid chromosome number
B) cells with identical chromosomes
C) sex cells
D) a zygote

19. What type of nucleic acid carries the codes for making proteins from the nucleus to the ribosome?
A) DNA
B) RNA
C) protein
D) genes

Science Online green.msscience.com/vocabulary_puzzlemaker

Thinking Critically

20. **List** the base sequence of a strand of RNA made using the DNA pattern ATCCGTC. Look at **Figure 14** for a hint.

21. **Predict** whether a mutation in a human skin cell can be passed on to the person's offspring. Explain.

22. **Explain** how a zygote could end up with an extra chromosome.

23. **Classify** Copy and complete this table about DNA and RNA.

DNA and RNA

	DNA	RNA
Number of strands		
Type of sugar	Do not write in this book.	
Letter names of bases		
Where found		

24. **Concept Map** Make an events-chain concept map of what occurs from interphase in the parent cell to the formation of the zygote. Tell whether the chromosome's number at each stage is haploid or diploid.

25. **Concept Map** Copy and complete the events-chain concept map of DNA synthesis.

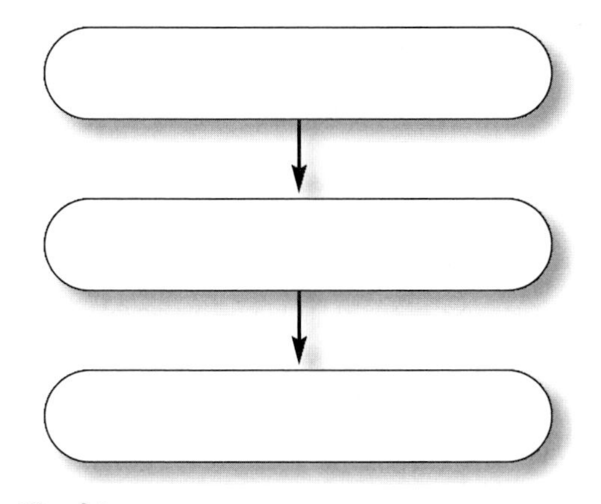

26. **Compare and Contrast** Meiosis is two divisions of a reproductive cell's nucleus. It occurs in a continuous series of steps. Compare and contrast the steps of meiosis I to the steps of meiosis II.

27. **Describe** what occurs in mitosis that gives the new cells identical DNA.

28. **Form a hypothesis** about the effect of an incorrect mitotic division on the new cells produced.

29. **Determine** how many chromosomes are in the original cell compared to those in the new cells formed by cell division. Explain.

Performance Activities

30. **Flash Cards** Make a set of 11 flash cards with drawings of a cell that show the different stages of meiosis. Shuffle your cards and then put them in the correct order. Give them to another student in the class to try.

Applying Math

31. **Cell Cycle** Assume an average human cell has a cell cycle of 20 hours. Calculate how many cells there would be after 80 hours.

Use the diagram below to answer question 32.

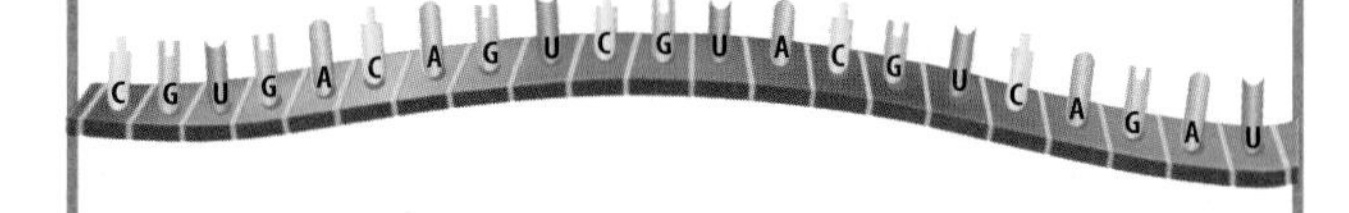

32. **Amino Acids** Sets of three nitrogen bases code for an amino acid. How many amino acids will make up the protein molecule that is coded for by the mRNA molecule above?

Part 1 Multiple Choice

Record your answers on the answer sheet provided by your teacher or on a sheet of paper.

1. What stage of the cell cycle involves growth and function?

A. prophase **C.** mitosis
B. interphase **D.** cytoplasmic division

2. During interphase, which structure of a cell is duplicated?

A. cell plate
B. mitochondrion
C. chromosome
D. chloroplast

Use the figure below to answer questions 3 and 4.

3. What form of asexual reproduction is shown here?

A. regeneration **C.** sprouting
B. cell division **D.** meiosis

4. How does the genetic material of the new organism above compare to that of the parent organism?

A. It is exactly the same.
B. It is a little different.
C. It is completely different.
D. It is haploid.

5. Organisms with three or more sets of chromosomes are called

A. monoploid. **C.** haploid.
B. diploid. **D.** polyploid.

6. If a sex cell has eight chromosomes, how many chromosomes will there be after fertilization?

A. 8 **C.** 32
B. 16 **D.** 64

Use the diagram below to answer questions 7 and 8.

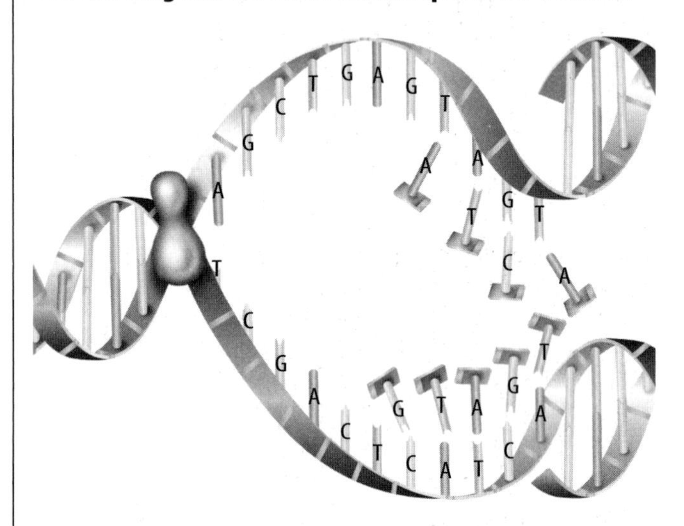

7. What does this diagram illustrate?

A. DNA duplication
B. RNA
C. cell reproduction
D. RNA synthesis

8. When does the process shown occur in the cell cycle?

A. prophase **C.** interphase
B. metaphase **D.** anaphase

9. Proteins are made of

A. genes **C.** amino acids
B. bases **D.** chromosomes

Test-Taking Tip

Prepare Avoid rushing on test day. Prepare your clothes and test supplies the night before. Wake up early and arrive at school on time on test day.

Part 2 Short Response/Grid In

Record your answers on the answer sheet provided by your teacher or on a sheet of paper.

10. In the human body, which cells are constantly dividing? Why is this important? How can this be potentially harmful?

11. Arrange the following terms in the correct order: *fertilization, sex cells, meiosis, zygote, mitosis.*

12. What are the three types of RNA used during protein synthesis? What is the function of each type of RNA?

13. Describe the relationship between gene, protein, DNA and chromosome.

Use the table below to answer question 14.

Phase of Cell Cycle	Action within the Cell
	Chromosomes duplicate
Prophase	
Metaphase	
	Chromosomes have separated
Telophase	

14. Fill in the blanks in the table with the appropriate term or definition.

15. What types of cells would constantly be in interphase?

16. Why is regeneration important for some organisms? In what way could regeneration of nerve cells be beneficial for humans?

17. What types of organisms are polyploidy? Why are they important?

18. What happens to chromosomes in meiosis I and meiosis II?

19. Describe several different ways that organisms can reproduce.

Part 3 Open Ended

Record your answers on a sheet of paper.

Use the photo below to answer question 20.

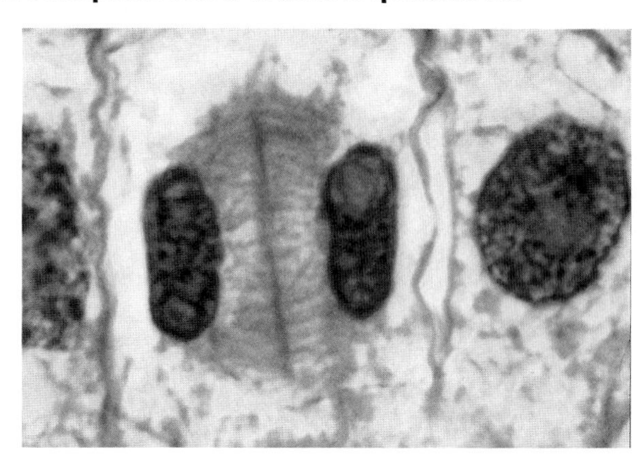

20. Is this a plant or an animal cell? Compare and contrast animal and plant cell division.

21. Describe in detail the structure of DNA.

Use the diagram below to answer question 22.

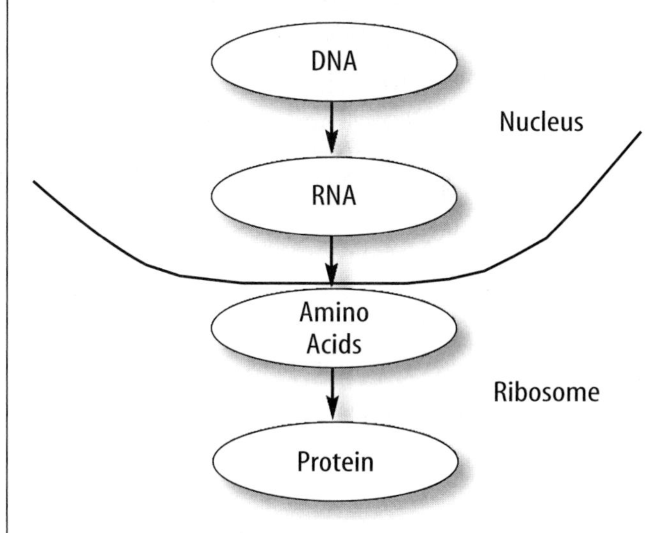

22. Discuss in detail what is taking place at each step of protein synthesis diagrammed above.

23. If a skin cell and a stomach cell have the same DNA then why are they so different?

24. What is mutation? Give examples where mutations could be harmful, beneficial or neutral.

chapter 11

Heredity

chapter preview

sections

1 Genetics
Lab *Predicting Results*

2 Genetics Since Mendel

3 Advances in Genetics
Lab *Tests for Color Blindness*

Virtual Lab *How are traits passed from parents to offspring?*

Why do people look different?

People have different skin colors, different kinds of hair, and different heights. Knowing how these differences are determined will help you predict when certain traits might appear. This will help you understand what causes hereditary disorders and how these are passed from generation to generation.

Science Journal Write three traits that you have and how you would determine how those traits were passed to you.

Start-Up Activities

Who around you has dimples?

You and your best friend enjoy the same sports, like the same food, and even have similar haircuts. But, there are noticeable differences between your appearances. Most of these differences are controlled by the genes you inherited from your parents. In the following lab, you will observe one of these differences.

1. Notice the two students in the photographs. One student has dimples when she smiles, and the other student doesn't have dimples.
2. Ask your classmates to smile naturally. In your Science Journal, record the name of each classmate and whether each one has dimples.
3. **Think Critically** In your Science Journal, calculate the percentage of students who have dimples. Are facial dimples a common feature among your classmates?

Classify Characteristics As you read this chapter about heredity, you can use the following Foldable to help you classify characteristics as inherited or not inherited.

STEP 1 **Fold** the top of a vertical piece of paper down and the bottom up to divide the paper into thirds.

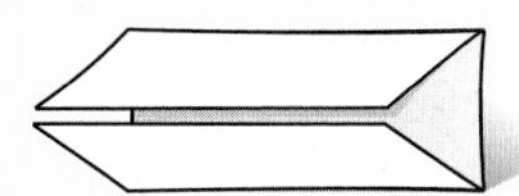

STEP 2 **Turn** the paper horizontally; **unfold and label** the three columns as shown.

Personal Characteristics | Inherited | Not Inherited
eyes
hair
dimples

Read for Main Ideas Before you read the chapter, list personal characteristics and predict which are inherited or not inherited. As you read the chapter, check and change your list.

Preview this chapter's content and activities at green.msscience.com

section 1

Genetics

as you read

What You'll Learn

- **Explain** how traits are inherited.
- **Identify** Mendel's role in the history of genetics.
- **Use** a Punnett square to predict the results of crosses.
- **Compare and contrast** the difference between an individual's genotype and phenotype.

Why It's Important

Heredity and genetics help explain why people are different.

Review Vocabulary

meiosis: reproductive process that produces four haploid sex cells from one diploid cell

New Vocabulary

- heredity
- allele
- genetics
- hybrid
- dominant
- recessive
- Punnett square
- genotype
- phenotype
- homozygous
- heterozygous

Inheriting Traits

Do you look more like one parent or grandparent? Do you have your father's eyes? What about Aunt Isabella's cheekbones? Eye color, nose shape, and many other physical features are some of the traits that are inherited from parents, as **Figure 1** shows. An organism is a collection of traits, all inherited from its parents. **Heredity** (huh REH duh tee) is the passing of traits from parent to offspring. What controls these traits?

What is genetics? Generally, genes on chromosomes control an organism's form and function. The different forms of a trait that a gene may have are called **alleles** (uh LEELZ). When a pair of chromosomes separates during meiosis (mi OH sus), alleles for each trait also separate into different sex cells. As a result, every sex cell has one allele for each trait, as shown in **Figure 2.** The allele in one sex cell may control one form of the trait, such as having facial dimples. The allele in the other sex cell may control a different form of the trait—not having dimples. The study of how traits are inherited through the interactions of alleles is the science of **genetics** (juh NE tihks).

Figure 1 Note the strong family resemblance among these four generations.

Figure 2 An allele is one form of a gene. Alleles separate into separate sex cells during meiosis. In this example, the alleles that control the trait for dimples include *D*, the presence of dimples, and *d*, the absence of dimples.

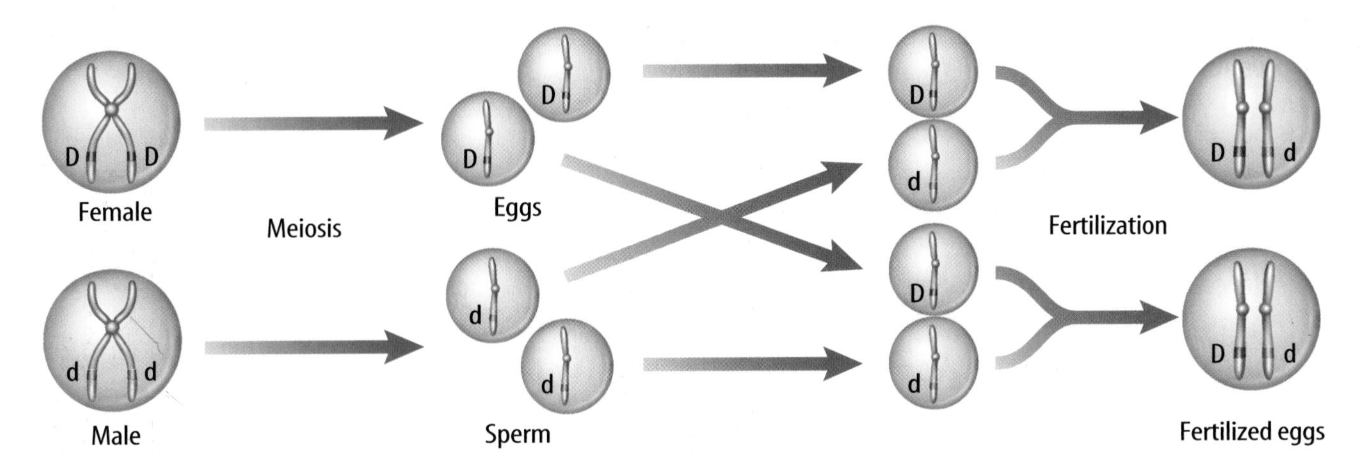

The alleles that control a trait are located on each duplicated chromosome.

During meiosis, duplicated chromosomes separate.

During fertilization, each parent donates one chromosome. This results in two alleles for the trait of dimples in the new individual formed.

Mendel—The Father of Genetics

Did you know that an experiment with pea plants helped scientists understand why your eyes are the color that they are? Gregor Mendel was an Austrian monk who studied mathematics and science but became a gardener in a monastery. His interest in plants began as a boy in his father's orchard where he could predict the possible types of flowers and fruits that would result from crossbreeding two plants. Curiosity about the connection between the color of a pea flower and the type of seed that same plant produced inspired him to begin experimenting with garden peas in 1856. Mendel made careful use of scientific methods, which resulted in the first recorded study of how traits pass from one generation to the next. After eight years, Mendel presented his results with pea plants to scientists.

Before Mendel, scientists mostly relied on observation and description, and often studied many traits at one time. Mendel was the first to trace one trait through several generations. He was also the first to use the mathematics of probability to explain heredity. The use of math in plant science was a new concept and not widely accepted then. Mendel's work was forgotten for a long time. In 1900, three plant scientists, working separately, reached the same conclusions as Mendel. Each plant scientist had discovered Mendel's writings while doing his own research. Since then, Mendel has been known as the father of genetics.

Topic: Genetics
Visit green.msscience.com for Web links to information about early genetics experiments.

Activity List two other scientists who studied genetics, and what organism they used in their research.

Table 1 Traits Compared by Mendel

Traits	Shape of Seeds	Color of Seeds	Color of Pods	Shape of Pods	Plant Height	Position of Flowers	Flower Color
Dominant trait	Round	Yellow	Green	Full	Tall	At leaf junctions	Purple
Recessive trait	Wrinkled	Green	Yellow	Flat, constricted	short	At tips of branches	White

Genetics in a Garden

Each time Mendel studied a trait, he crossed two plants with different expressions of the trait and found that the new plants all looked like one of the two parents. He called these new plants **hybrids** (HI brudz) because they received different genetic information, or different alleles, for a trait from each parent. The results of these studies made Mendel even more curious about how traits are inherited.

Garden peas are easy to breed for pure traits. An organism that always produces the same traits generation after generation is called a purebred. For example, tall plants that always produce seeds that produce tall plants are purebred for the trait of tall height. **Table 1** shows other pea plant traits that Mendel studied.

Why might farmers plant purebred crop seeds?

Dominant and Recessive Factors In nature, insects randomly pollinate plants as they move from flower to flower. In his experiments, Mendel used pollen from the flowers of purebred tall plants to pollinate by hand the flowers of purebred short plants. This process is called cross-pollination. He found that tall plants crossed with short plants produced seeds that produced all tall plants. Whatever caused the plants to be short had disappeared. Mendel called the tall form the **dominant** (DAH muh nunt) factor because it dominated, or covered up, the short form. He called the form that seemed to disappear the **recessive** (rih SE sihv) factor. Today, these are called dominant alleles and recessive alleles. What happened to the recessive form? **Figure 3** answers this question.

Comparing Common Traits

Procedure

1. Safely survey as many **dogs** in your neighborhood as you can for the presence of a solid color or spotted coat, short or long hair, and floppy or upright ears.
2. Make a data table that lists each of the traits. Record your data in the data table.

Analysis

1. Compare the number of dogs that have one form of a trait with those that have the other form.
2. What can you conclude about the variations you noticed in the dogs?

Try at Home

NATIONAL GEOGRAPHIC

VISUALIZING MENDEL'S EXPERIMENTS

Figure 3

Gregor Mendel discovered that the experiments he carried out on garden plants provided an understanding of heredity. For eight years he crossed plants that had different characteristics and recorded how those characteristics were passed from generation to generation. One such characteristic, or trait, was the color of pea pods. The results of Mendel's experiment on pea pod color are shown below.

Parents

1st Generation

2nd Generation

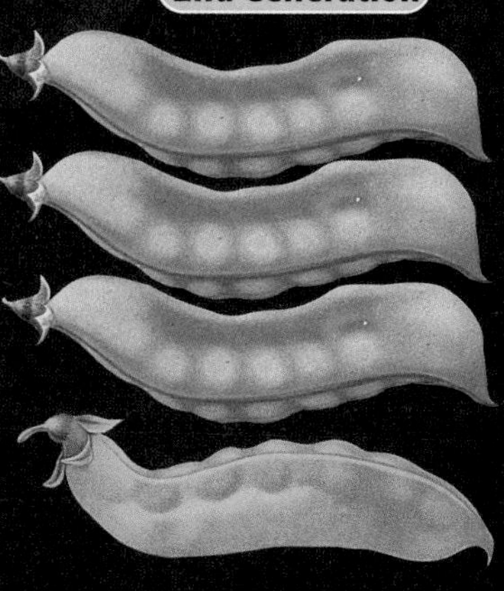

A One of the so-called "parent plants" in Mendel's experiment had pods that were green, a dominant trait. The other parent plant had pods that were yellow, a recessive trait.

B Mendel discovered that the two "parents" produced a generation of plants with green pods. The recessive color—yellow—did not appear in any of the pods.

C Next, Mendel collected seeds from the first-generation plants and raised a second generation. He discovered that these second-generation plants produced plants with either green or yellow pods in a ratio of about three plants with green pods for every one plant with yellow pods. The recessive trait had reappeared. This 3:1 ratio proved remarkably consistent in hundreds of similar crosses, allowing Mendel to accurately predict the ratio of pod color in second-generation plants.

Using Probability to Make Predictions If you and your sister can't agree on what movie to see, you could solve the problem by tossing a coin. When you toss a coin, you're dealing with probabilities. Probability is a branch of mathematics that helps you predict the chance that something will happen. If your sister chooses tails while the coin is in the air, what is the probability that the coin will land tail-side up? Because a coin has two sides, there are two possible outcomes, heads or tails. One outcome is tails. Therefore, the probability of one side of a coin showing is one out of two, or 50 percent.

Mendel also dealt with probabilities. One of the things that made his predictions accurate was that he worked with large numbers of plants. He studied almost 30,000 pea plants over a period of eight years. By doing so, Mendel increased his chances of seeing a repeatable pattern. Valid scientific conclusions need to be based on results that can be duplicated.

Figure 4 This snapdragon's phenotype is red.
Determine *Can you tell what the flower's genotype for color is? Explain your answer.*

Punnett Squares Suppose you wanted to know what colors of pea plant flowers you would get if you pollinated white flowers on one pea plant with pollen from purple flowers on a different plant. How could you predict what the offspring would look like without making the cross? A handy tool used to predict results in Mendelian genetics is the **Punnett** (PUH nut) **square.** In a Punnett square, letters represent dominant and recessive alleles. An uppercase letter stands for a dominant allele. A lowercase letter stands for a recessive allele. The letters are a form of code. They show the **genotype** (JEE nuh tipe), or genetic makeup, of an organism. Once you understand what the letters mean, you can tell a lot about the inheritance of a trait in an organism.

The way an organism looks and behaves as a result of its genotype is its **phenotype** (FEE nuh tipe), as shown in **Figure 4.** If you have brown hair, then the phenotype for your hair color is brown.

Alleles Determine Traits Most cells in your body have two alleles for every trait. These alleles are located on chromosomes within the nucleus of cells. An organism with two alleles that are the same is called **homozygous** (hoh muh ZI gus). For Mendel's peas, this would be written as *TT* (homozygous for the tall-dominant trait) or *tt* (homozygous for the short-recessive trait). An organism that has two different alleles for a trait is called **heterozygous** (he tuh roh ZI gus). The hybrid plants Mendel produced were all heterozygous for height, *Tt*.

What is the difference between homozygous and heterozygous organisms?

Making a Punnett Square In a Punnett square for predicting one trait, the letters representing the two alleles from one parent are written along the top of the grid, one letter per section. Those of the second parent are placed down the side of the grid, one letter per section. Each square of the grid is filled in with one allele donated by each parent. The letters that you use to fill in each of the squares represent the genotypes of possible offspring that the parents could produce.

Applying Math — Calculate Percentages

PUNNET SQUARE One dog carries heterozygous, black-fur traits (*Bb*), and its mate carries homogeneous, blond-fur traits (*bb*). Use a Punnett square to determine the probability of one of their puppies having black fur.

Solution

1 *This is what you know:*

- dominant allele is represented by *B*
- recessive allele is represented by *b*

2 *This is what you need to find out:*

What is the probability of a puppy's fur color being black?

3 *This is the procedure you need to use:*

- Complete the Punnett square.
- There are two *Bb* genotypes and four possible outcomes.
- %(black fur) = $\frac{\text{number of ways to get black fur}}{\text{number of possible outcomes}}$ $= \frac{2}{4} = \frac{1}{2} = 50\%$

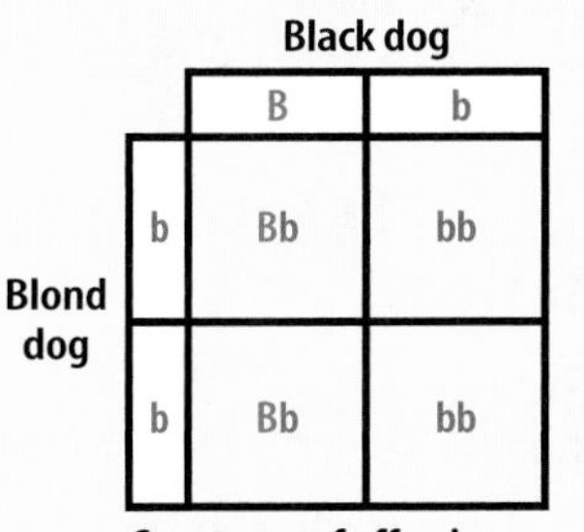

Genotypes of offspring: 2Bb, 2bb
Phenotypes of offspring: 2 black, 2 blond

4 *Check your answer:*

$\frac{1}{2}$ of 4 is 2, which is the number of black dogs.

Practice Problems

1. In peas, the color yellow (*Y*) is dominant to the color green (*y*). According to the Punnett square, what is the probability of an offspring being yellow?
2. What is the probability of an offspring having the *yy* genotype?

Parent (Yy)	Y	y
Y	YY	Yy
y	Yy	yy

Parent (Yy) (side)

Science Online — For more practice, visit green.msscience.com/math_practice

Principles of Heredity Even though Gregor Mendel didn't know anything about DNA, genes, or chromosomes, he succeeded in beginning to describe and mathematically represent how inherited traits are passed from parents to offspring. He realized that some factor in the pea plant produced certain traits. Mendel also concluded that these factors separated when the pea plant reproduced. Mendel arrived at his conclusions after years of detailed observation, careful analysis, and repeated experimentation. **Table 2** summarizes Mendel's principles of heredity.

Table 2 Principles of Heredity

1	Traits are controlled by alleles on chromosomes.
2	An allele's effect is dominant or recessive.
3	When a pair of chromosomes separates during meiosis, the different alleles for a trait move into separate sex cells.

section 1 review

Summary

Inheriting Traits

- Heredity is the passing of traits from parent to offspring.
- Genetics is the study of how traits are inherited through the interactions of alleles.

Mendel—The Father of Genetics

- In 1856, Mendel began experimenting with garden peas, using careful scientific methods.
- Mendel was the first to trace one trait through several generations.
- In 1900, three plant scientists separately reached the same conclusions as Mendel.

Genetics in a Garden

- Hybrids receive different genetic information for a trait from each parent.
- Genetics involves dominant and recessive factors.
- Punnett squares can be used to predict the results of a cross.
- Mendel's conclusions led to the principles of heredity.

Self Check

1. **Contrast** Alleles are described as being dominant or recessive. What is the difference between a dominant and a recessive allele?
2. **Describe** how dominant and recessive alleles are represented in a Punnett square.
3. **Explain** the difference between genotype and phenotype. Give examples.
4. **Infer** Gregor Mendel, an Austrian monk who lived in the 1800s, is known as the father of genetics. Explain why Mendel has been given this title.
5. **Think Critically** If an organism expresses a recessive phenotype, can you tell the genotype? Explain your answer by giving an example.

Applying Math

6. **Use Percentages** One fruit fly is heterozygous for long wings, and another fruit fly is homozygous for short wings. Long wings are dominant to short wings. Use a Punnett square to find the expected percent of offspring with short wings.

Science Online green.msscience.com/self_check_quiz

Predicting Results

Could you predict how many brown rabbits would result from crossing two heterozygous black rabbits? Try this investigation to find out. Brown color is a recessive trait for hair color in rabbits.

Real-World Question

How does chance affect combinations of genes?

Goals

- **Model** chance events in heredity.
- **Compare and contrast** predicted and actual results.

Materials

paper bags (2)
red beans (100)
white beans (100)

Safety Precautions

WARNING: *Do not taste, eat, or drink any materials used in the lab.*

Procedure

1. Use a Punnett square to predict how many red/red, red/white, and white/white bean combinations are possible. The combinations represent the coat colors in rabbit offspring.
2. Place 50 red beans and 50 white beans in a paper bag. Place 50 red beans and 50 white beans in a second bag. Red beans represent black alleles and white beans represent brown alleles.
3. Label one of the bags *Female* for the female parent. Label the other bag *Male* for the male parent.
4. Use a data table to record the combination each time you remove two beans. Your table will need to accommodate 100 picks.
5. Without looking, remove one bean from each bag. The two beans represent the alleles that combine when sperm and egg join. After recording, return the beans to their bags.
6. **Count** and record the total numbers for each of the three combinations in your data table.
7. **Compile and record** the class totals.

Conclude and Apply

1. **Name** the combination that occurred most often.
2. **Calculate** the ratio of red/red to red/white to white/white. What hair color in rabbits do these combinations represent?
3. **Compare** your predicted (expected) results with your observed (actual) results.
4. **Hypothesize** how you could get predicted results to be closer to actual results.

Gene Combinations

Rabbits	Red/ Red	Red/ White	White/ White
Your total	Do not write in this book.		
Class total			

Communicating Your Data

Write a paragraph that clearly describes your results. Have another student read your paragraph. Ask if he or she could understand what happened. If not, rewrite your paragraph and have the other student read it again. **For more help, refer to the** Science Skill Handbook.

section 2

Genetics Since Mendel

as you read

What You'll Learn

- **Explain** how traits are inherited by incomplete dominance.
- **Compare** multiple alleles and polygenic inheritance, and give examples of each.
- **Describe** two human genetic disorders and how they are inherited.
- **Explain** how sex-linked traits are passed to offspring.

Why It's Important

Most of your inherited traits involve more complex patterns of inheritance than Mendel discovered.

Review Vocabulary

gene: section of DNA on a chromosome that contains instructions for making specific proteins

New Vocabulary

- incomplete dominance
- polygenic inheritance
- sex-linked gene

Incomplete Dominance

Not even in science do things remain the same. After Mendel's work was rediscovered in 1900, scientists repeated his experiments. For some plants, such as peas, Mendel's results proved true. However, when different plants were crossed, the results were sometimes different. One scientist crossed purebred red four-o'clock plants with purebred white four-o'clock plants. He expected to get all red flowers, but they were pink. Neither allele for flower color seemed dominant. Had the colors become blended like paint colors? He crossed the pink-flowered plants with each other, and red, pink, and white flowers were produced. The red and white alleles had not become blended. Instead, when the allele for white flowers and the allele for red flowers combined, the result was an intermediate phenotype—a pink flower.

When the offspring of two homozygous parents show an intermediate phenotype, this inheritance is called **incomplete dominance.** Other examples of incomplete dominance include the flower color of some plant breeds and the coat color of some horse breeds, as shown in **Figure 5.**

Chestnut horse

Cremello horse

Figure 5 When a chestnut horse is bred with a cremello horse, all offspring will be palomino. The Punnett square shown on the opposite page can be used to predict this result. **Explain** *how the color of the palomino horse shows that the coat color of horses may be inherited by incomplete dominance.*

Multiple Alleles Mendel studied traits in peas that were controlled by just two alleles. However, many traits are controlled by more than two alleles. A trait that is controlled by more than two alleles is said to be controlled by multiple alleles. Traits controlled by multiple alleles produce more than three phenotypes of that trait.

Imagine that only three types of coins are made—nickels, dimes, and quarters. If every person can have only two coins, six different combinations are possible. In this problem, the coins represent alleles of a trait. The sum of each two-coin combination represents the phenotype. Can you name the six different phenotypes possible with two coins?

Blood type in humans is an example of multiple alleles that produce only four phenotypes. The alleles for blood types are called A, B, and O. The O allele is recessive to both the A and B alleles. When a person inherits one A allele and one B allele for blood type, both are expressed—phenotype AB. A person with phenotype A blood has the genetic makeup, or genotype—AA or AO. Someone with phenotype B blood has the genotype BB or BO. Finally, a person with phenotype O blood has the genotype OO.

What are the six different blood type genotypes?

Topic: Blood Types
Visit green.msscience.com for Web links to information about the importance of blood types in blood transfusions.

Activity Make a chart showing which blood types can be used for transfusions into people with A, B, AB, or O blood phenotypes.

Palomino horse

Punnett square

Cremello horse (C'C') \ Chestnut horse (CC)	C	C
C'	CC'	CC'
C'	CC'	CC'

Genotypes: All CC'
Phenotypes: All palomino horses

Interpreting Polygenic Inheritance

Procedure

1. Measure the hand spans of your classmates.
2. Using a **ruler,** measure from the tip of the thumb to the tip of the little finger when the hand is stretched out. Read the measurement to the nearest centimeter.
3. Record the name and hand-span measurement of each person in a data table.

Analysis

1. What range of hand spans did you find?
2. Are hand spans inherited as a simple Mendelian pattern or as a polygenic or incomplete dominance pattern? Explain.

Polygenic Inheritance

Eye color is an example of a trait that is produced by a combination of many genes. **Polygenic** (pah lih JEH nihk) **inheritance** occurs when a group of gene pairs acts together to produce a trait. The effects of many alleles produces a wide variety of phenotypes. For this reason, it may be hard to classify all the different shades of eye color.

Your height and the color of your eyes and skin are just some of the many human traits controlled by polygenic inheritance. It is estimated that three to six gene pairs control your skin color. Even more gene pairs might control the color of your hair and eyes. The environment also plays an important role in the expression of traits controlled by polygenic inheritance. Polygenic inheritance is common and includes such traits as grain color in wheat and milk production in cows. Egg production in chickens is also a polygenic trait.

Impact of the Environment

Your environment plays a role in how some of your genes are expressed or whether they are expressed at all, as shown in **Figure 6.** Environmental influences can be internal or external. For example, most male birds are more brightly colored than females. Chemicals in their bodies determine whether the gene for brightly colored feathers is expressed.

Although genes determine many of your traits, you might be able to influence their expression by the decisions you make. Some people have genes that make them at risk for developing certain cancers. Whether they get cancer might depend on external environmental factors. For instance, if some people at risk for skin cancer limit their exposure to the Sun and take care of their skin, they might never develop cancer.

Reading Check *What environmental factors might affect the size of leaves on a tree?*

Figure 6 Himalayan rabbits have alleles for dark-colored fur. However, this allele is able to express itself only at lower temperatures. Only the areas located farthest from the rabbit's main body heat (ears, nose, feet, tail) have dark-colored fur.

Human Genes and Mutations

Sometimes a gene undergoes a change that results in a trait that is expressed differently. Occasionally errors occur in the DNA when it is copied inside of a cell. Such changes and errors are called mutations. Not all mutations are harmful. They might be helpful or have no effect on an organism.

Certain chemicals are known to produce mutations in plants or animals, including humans. X rays and radioactive substances are other causes of some mutations. Mutations are changes in genes.

Genetic Counselor Testing for genetic disorders may allow many affected individuals to seek treatment and cope with their diseases. Genetic counselors are trained to analyze a family's history to determine a person's health risk. Research what a genetic counselor does and how to become a genetic counselor. Record what you learn in your Science Journal.

Chromosome Disorders In addition to individual mutations, problems can occur if the incorrect number of chromosomes is inherited. Every organism has a specific number of chromosomes. However, mistakes in the process of meiosis can result in a new organism with more or fewer chromosomes than normal. A change in the total number of human chromosomes is usually fatal to the unborn embryo or fetus, or the baby may die soon after birth.

Look at the human chromosomes in **Figure 7.** If three copies of chromosome 21 are produced in the fertilized human egg, Down's syndrome results. Individuals with Down's syndrome can be short, exhibit learning disabilities, and have heart problems. Such individuals can lead normal lives if they have no severe health complications.

Figure 7 Humans usually have 23 pairs of chromosomes. Notice that three copies of chromosome 21 are present in this photo, rather than the usual two chromosomes. This change in chromosome number results in Down's syndrome. Chris Burke, a well-known actor, has this syndrome.

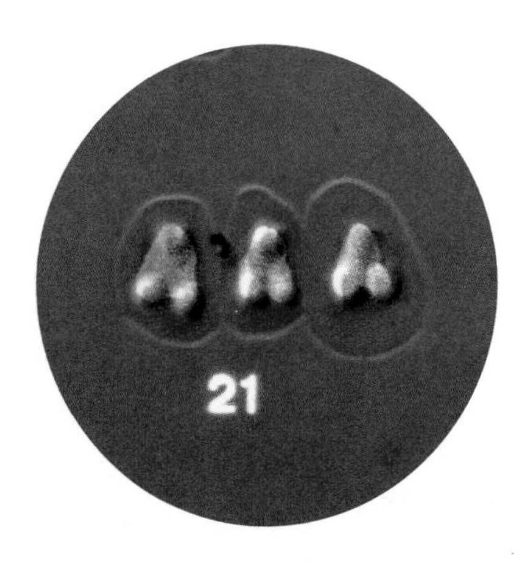

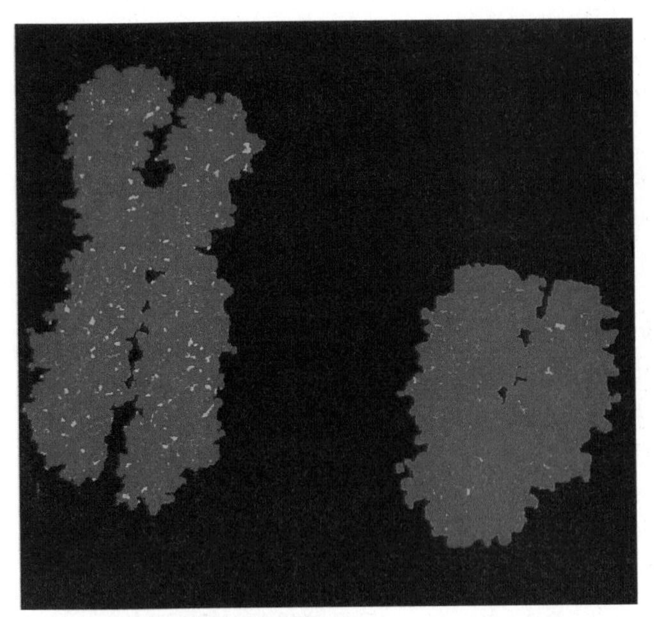

Color-enhanced SEM Magnification: 16000×

Figure 8 Sex in many organisms is determined by X and Y chromosomes.
Observe *How do the X (left) and Y (right) chromosomes differ from one another in shape and size?*

Recessive Genetic Disorders

Many human genetic disorders, such as cystic fibrosis, are caused by recessive genes. Some recessive genes are the result of a mutation within the gene. Many of these alleles are rare. Such genetic disorders occur when both parents have a recessive allele responsible for this disorder. Because the parents are heterozygous, they don't show any symptoms. However, if each parent passes the recessive allele to the child, the child inherits both recessive alleles and will have a recessive genetic disorder.

Reading Check *How is cystic fibrosis inherited?*

Cystic fibrosis is a homozygous recessive disorder. It is the most common genetic disorder that can lead to death among Caucasian Americans. In most people, a thin fluid is produced that lubricates the lungs and intestinal tract. People with cystic fibrosis produce thick mucus instead of this thin fluid. The thick mucus builds up in the lungs and makes it hard to breathe. This buildup often results in repeated bacterial respiratory infections. The thick mucus also reduces or prevents the flow of substances necessary for digesting food. Physical therapy, special diets, and new drug therapies have increased the life spans of patients with cystic fibrosis.

Sex Determination

What determines the sex of an individual? Much information on sex inheritance came from studies of fruit flies. Fruit flies have only four pairs of chromosomes. Because the chromosomes are large and few in number, they are easy to study. Scientists identified one pair that contains genes that determine the sex of the organism. They labeled the pair XX in females and XY in males. Geneticists use these labels when studying organisms, including humans. You can see human X and Y chromosomes in **Figure 8.**

Each egg produced by a female normally contains one X chromosome. Males produce sperm that normally have either an X or a Y chromosome. When a sperm with an X chromosome fertilizes an egg, the offspring is a female, XX. A male offspring, XY, is the result of a Y-containing sperm fertilizing an egg. What pair of sex chromosomes is in each of your cells? Sometimes chromosomes do not separate during meiosis. When this occurs, an individual can inherit an abnormal number of sex chromosomes.

Sex-Linked Disorders

Some inherited conditions are linked with the X and Y chromosomes. An allele inherited on a sex chromosome is called a **sex-linked gene.** Color blindness is a sex-linked disorder in which people cannot distinguish between certain colors, particularly red and green. This trait is a recessive allele on the X chromosome. Because males have only one X chromosome, a male with this allele on his X chromo-some is color-blind. However, a color-blind female occurs only when both of her X chromosomes have the allele for this trait.

The allele for the distinct patches of three different colors found in calico cats is recessive and carried on the X chromosome. As shown in **Figure 9,** calico cats have inherited two X chromosomes with this recessive allele—one from both parents.

Female carrier of calico gene (X^CX)

Male carrier of calico gene (X^CY)

	X^C	X
X^C	X^CX^C	XX^C
Y	X^CY	XY

Genotypes: X^CX^C, X^CX, X^CY, XY
Phenotypes: one calico female, one carrier female, one carrier male, one normal male

Figure 9 Calico cat fur is a homozygous recessive sex-linked trait. Female cats that are heterozygous are not calico but are only carriers. Two recessive alleles must be present for this allele to be expressed.
Determine *Why aren't all the females calico?*

Pedigrees Trace Traits

How can you trace a trait through a family? A pedigree is a visual tool for following a trait through generations of a family. Males are represented by squares and females by circles. A completely filled circle or square shows that the trait is seen in that person. Half-colored circles or squares indicate carriers. A carrier is heterozygous for the trait, and it is not seen. People represented by empty circles or squares do not have the trait and are not carriers. The pedigree in **Figure 10** shows how the trait for color blindness is carried through a family.

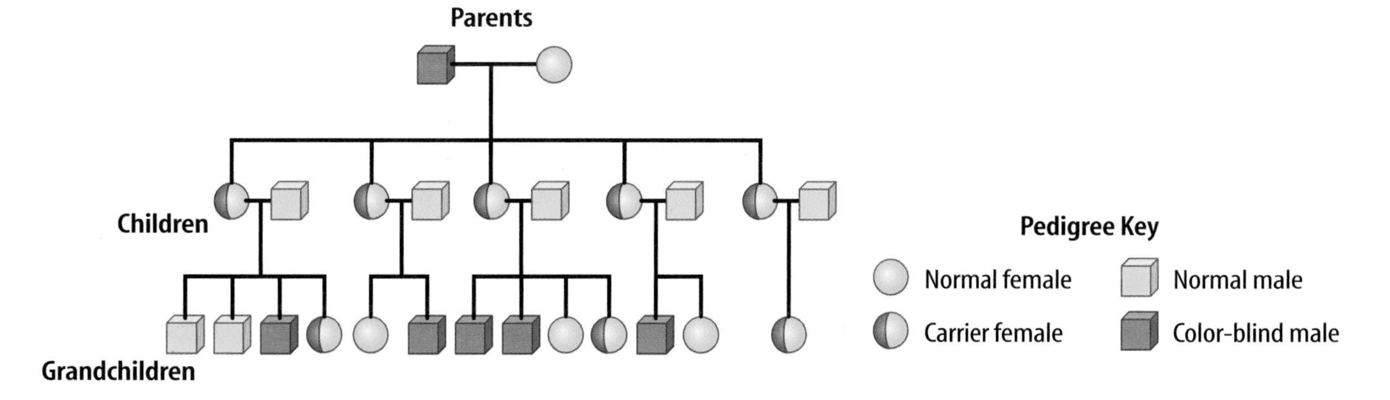

Figure 10 The symbols in this pedigree's key mean the same thing on all pedigree charts. The grandfather in this family was color-blind and married to a woman who was not a carrier of the color-blind allele.
Infer *why no women in this family are color-blind.*

Shih Tzu

Black Labrador

Figure 11
A variety of traits are considered when breeding dogs.

Using Pedigrees A pedigree is a useful tool for a geneticist. Sometimes a geneticist needs to understand who has had a trait in a family over several generations to determine its pattern of inheritance. A geneticist determines if a trait is recessive, dominant, sex-linked, or has some other pattern of inheritance. When geneticists understand how a trait is inherited, they can predict the probability that a baby will be born with a specific trait.

Pedigrees also are important in breeding animals or plants. Because livestock and plant crops are used as sources of food, these organisms are bred to increase their yield and nutritional content. Breeders of pets and show animals, like the dogs pictured in **Figure 11,** also examine pedigrees carefully for possible desirable physical and ability traits. Issues concerning health also are considered when researching pedigrees.

section 2 review

Summary

Incomplete Dominance

- Incomplete dominance is when a dominant and recessive allele for a trait show an intermediate phenotype.
- Many traits are controlled by more than two alleles.
- A wide variety of phenotypes is produced by polygenic inheritance.

Human Genes and Mutations

- Errors can occur when DNA is copied.
- Mistakes in meiosis can result in an unequal number of chromosomes in sex cells.
- Recessive genes control many human genetic disorders.

Sex Determination

- An allele inherited on a sex chromosome is called a sex-linked gene.
- Pedigrees are visual tools to trace a trait through generations of a family.

Self Check

1. **Compare** how inheritance by multiple alleles and polygenic inheritance are similar.
2. **Explain** why a trait inherited by incomplete dominance is not a blend of two alleles.
3. **Discuss** Choose two genetic disorders and discuss how they are inherited.
4. **Apply** Using a Punnett square, explain why males are affected more often than females by sex-linked genetic disorders.
5. **Think Critically** Calico male cats are rare. Explain how such a cat can exist.

Applying Skills

6. **Predict** A man with blood type B marries a woman with blood type A. Their first child has blood type O. Use a Punnett square to predict what other blood types are possible for their offspring.
7. **Communicate** In your Science Journal, explain why offspring may or may not resemble either parent.

green.msscience.com/self_check_quiz

section 3

Advances in Genetics

Why is genetics important?

If Mendel were to pick up a daily newspaper in any country today, he'd probably be surprised. News articles about developments in genetic research appear almost daily. The term *gene* has become a common word. The principles of heredity are being used to change the world.

Genetic Engineering

You may know that chromosomes are made of DNA and are in the nucleus of a cell. Sections of DNA in chromosomes that direct cell activities are called genes. Through **genetic engineering,** scientists are experimenting with biological and chemical methods to change the arrangement of DNA that makes up a gene. Genetic engineering already is used to help produce large volumes of medicine. Genes also can be inserted into cells to change how those cells perform their normal functions, as shown in **Figure 12.** Other research is being done to find new ways to improve crop production and quality, including the development of plants that are resistant to disease.

as you read

What You'll Learn

- **Evaluate** the importance of advances in genetics.
- **Sequence** the steps in making genetically engineered organisms.

Why It's Important

Advances in genetics can affect your health, the foods that you eat, and your environment.

Review Vocabulary
DNA: deoxyribonucleic acid; the genetic material of all organisms

New Vocabulary
- genetic engineering

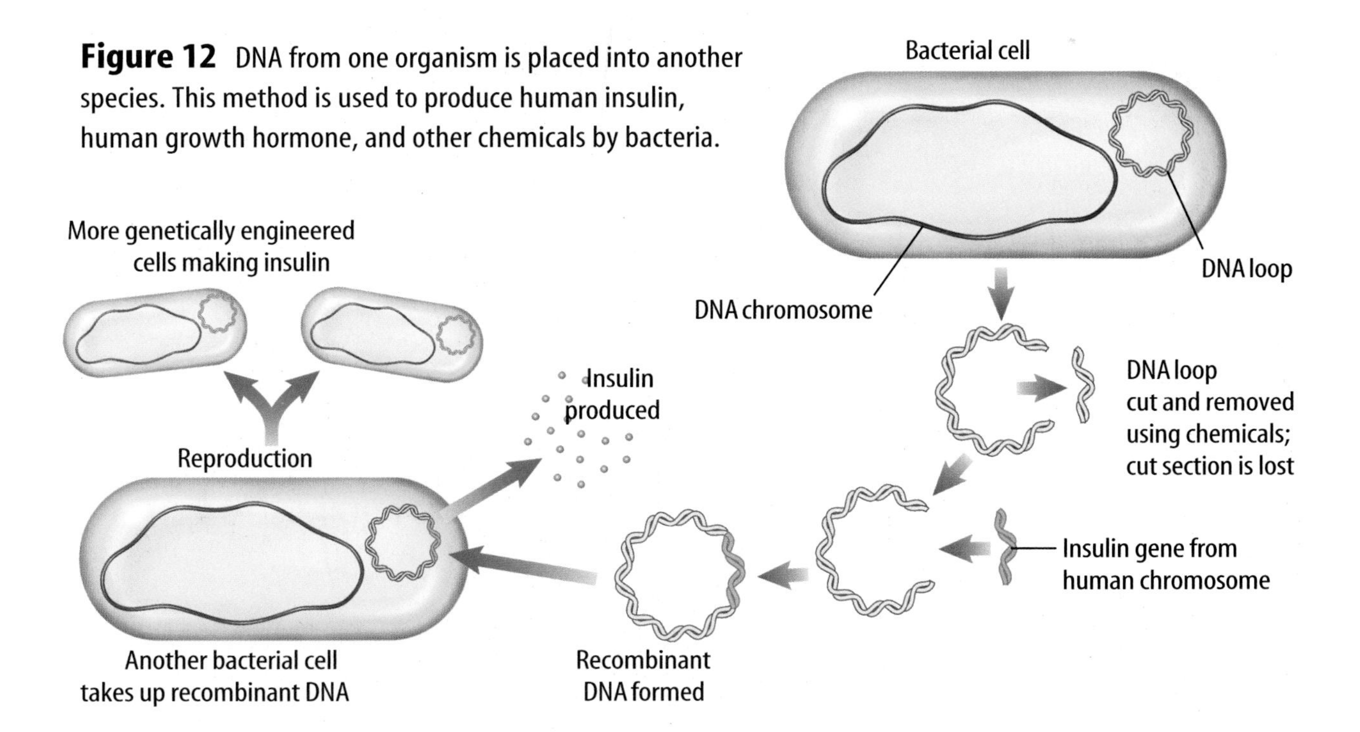

Figure 12 DNA from one organism is placed into another species. This method is used to produce human insulin, human growth hormone, and other chemicals by bacteria.

Genetically Engineered Crops Crop plants are now being genetically engineered to produce chemicals that kill specific pests that feed on them. Some of the pollen from pesticide-resistant canola crops is capable of spreading up to 8 km from the plant, while corn and potato pollen can spread up to 1 km. What might be the effects of pollen landing on other plants?

Recombinant DNA Making recombinant DNA is one method of genetic engineering. Recombinant DNA is made by inserting a useful segment of DNA from one organism into a bacterium, as illustrated in **Figure 12.** Large quantities of human insulin are made by some genetically engineered organisms. People with Type 1 diabetes need this insulin because their pancreases produce too little or no insulin. Other uses include the production of growth hormone to treat dwarfism and chemicals to treat cancer.

Gene Therapy Gene therapy is a kind of genetic engineering. In gene therapy, a normal allele is placed in a virus, as shown in **Figure 13.** The virus then delivers the normal allele when it infects its target cell. The normal allele replaces the defective one. Scientists are conducting experiments that use this method to test ways of controlling cystic fibrosis and some kinds of cancer. More than 2,000 people already have taken part in gene therapy experiments. Gene therapy might be a method of curing several other genetic disorders in the future.

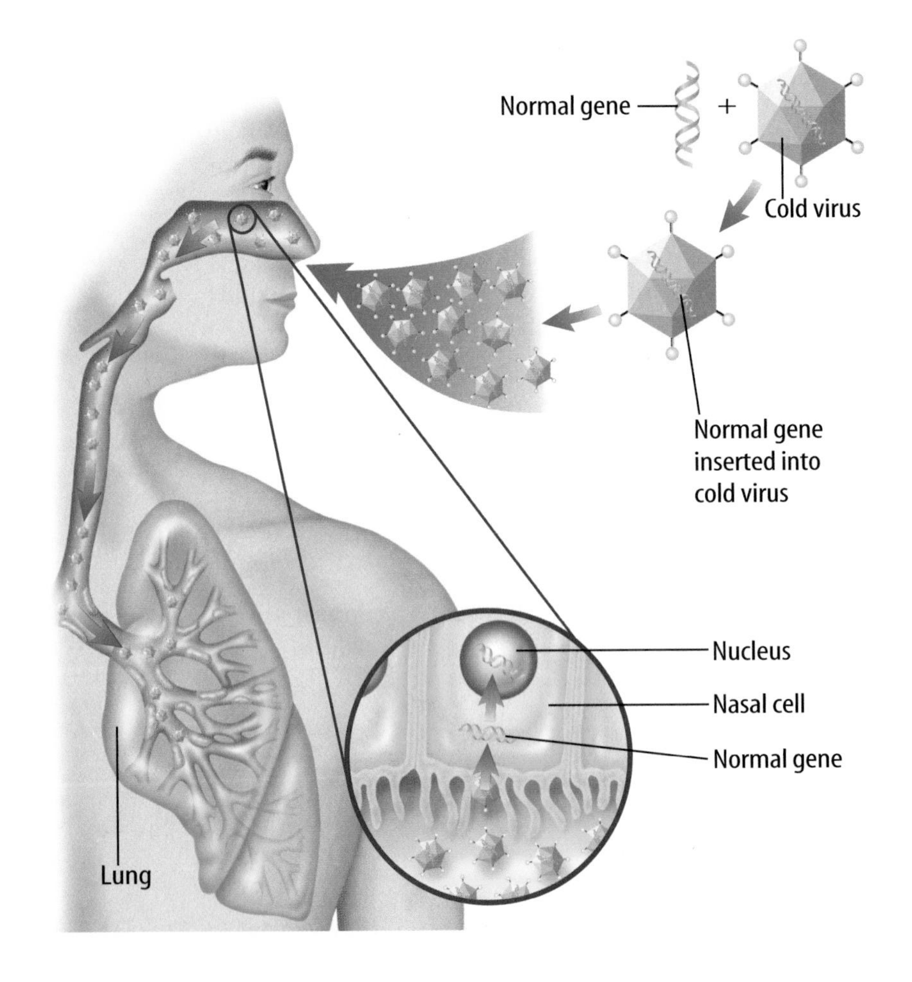

Figure 13 Gene therapy involves placing a normal allele into a cell that has a mutation. When the normal allele begins to function, a genetic disorder such as cystic fibrosis (CF) may be corrected.

Genetically Engineered Plants For thousands of years people have improved the plants they use for food and clothing even without the knowledge of genotypes. Until recently, these improvements were the results of selecting plants with the most desired traits to breed for the next generation. This process is called selective breeding. Recent advances in genetics have not replaced selective breeding. Although a plant can be bred for a particular phenotype, the genotype and pedigree of the plants also are considered.

Genetic engineering can produce improvements in crop plants, such as corn, wheat, and rice. One type of genetic engineering involves finding the genes that produce desired traits in one plant and then inserting those genes into a different plant. Scientists recently have made genetically engineered tomatoes with a gene that allows tomatoes to be picked green and transported great distances before they ripen completely. Ripe, firm tomatoes are then available in the local market. In the future, additional food crops may be genetically engineered so that they are not desirable food for insects.

What other types of traits would be considered desirable in plants?

Because some people might prefer foods that are not changed genetically, some stores label such produce, as shown in **Figure 14.** The long-term effects of consuming genetically engineered plants are unknown.

Figure 14 Genetically engineered produce is sometimes labeled. This allows consumers to make informed choices about their foods.

section  review

Summary

Why is genetics important?

- Developments in genetic research appear in newspapers almost daily.
- The world is being changed by the principles of heredity.

Genetic Engineering

- Scientists work with biological and chemical methods to change the arrangement of DNA that makes up a gene.
- One method of genetic engineering is making recombinant DNA.
- A normal allele is replaced in a virus and then delivers the normal allele when it infects its target cell.

Self Check

1. **Apply** Give examples of areas in which advances in genetics are important.
2. **Compare and contrast** the technologies of using recombinant DNA and gene therapy.
3. **Infer** What are some benefits of genetically engineered crops?
4. **Describe** how selective breeding differs from genetic engineering.
5. **Think Critically** Why might some people be opposed to genetically engineered plants?

Applying Skills

6. **Concept Map** Make an events-chain concept map of the steps used in making recombinant DNA.

Design Your Own

Tests for Color Blindness

Goals

- **Design** an experiment that tests for a specific type of color blindness in males and females.
- **Calculate** the percentage of males and females with the disorder.

Possible Materials

white paper or poster board
colored markers: red, orange, yellow, bright green, dark green, blue
*computer and color printer

*Alternate materials

Real-World Question

What do color-blind people see? People who have inherited color blindness can see most colors, but they have difficulty telling the difference between two specific colors. You have three genes that help you see color. One gene lets you see red, another blue, and the third gene allows you to see green. In the most common type of color blindness, red-green color blindness, the green gene does not work properly. What percentages of males and females in your school are color-blind?

Form a Hypothesis

Based on your reading and your own experiences, form a hypothesis about how common color blindness is among males and females.

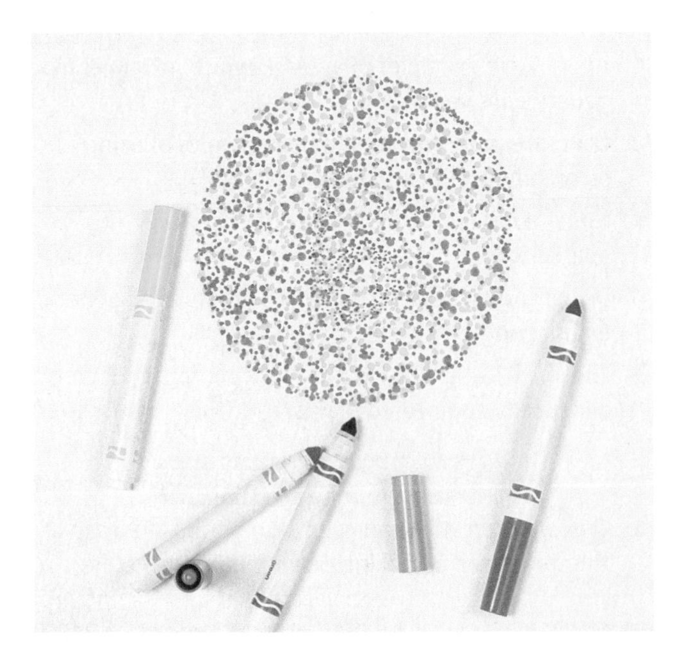

Test Your Hypothesis

Make a Plan

1. Decide what type of color blindness you will test for—the common green-red color blindness or the more rare green-blue color blindness.
2. **List** the materials you will need and describe how you will create test pictures. Tests for color blindness use many circles of red, orange, and yellow as a background, with circles of dark and light green to make a picture or number. List the steps you will take to test your hypothesis.
3. Prepare a data table in your Science Journal to record your test results.

4. **Examine** your experiment to make sure all steps are in logical order.
5. **Identify** which pictures you will use as a control and which pictures you will use as variables.

Follow Your Plan

1. Make sure your teacher approves your plan before you start.
2. **Draw** the pictures that you will use to test for color blindness.
3. Carry out your experiment as planned and record your results in your data table.

Analyze Your Data

1. **Calculate** the percentage of males and females that tested positive for color blindness.
2. **Compare** the frequency of color blindness in males with the frequency of color blindness in females.

Conclude and Apply

1. **Explain** whether or not the results supported your hypothesis.
2. **Explain** why color blindness is called a sex-linked disorder.
3. **Infer** how common the color-blind disorder is in the general population.
4. **Predict** your results if you were to test a larger number of people.

Communicating Your Data

Using a word processor, write a short article for the advice column of a fashion magazine about how a color-blind person can avoid wearing outfits with clashing colors. **For more help, refer to the Science Skill Handbook.**

SCIENCE Stats

The Human Genome

Did you know...

. . . The biggest advance in genetics in years took place in February 2001. Scientists successfully mapped the human genome. There are 30,000 to 40,000 genes in the human genome. Genes are in the nucleus of each of the several trillion cells in your body.

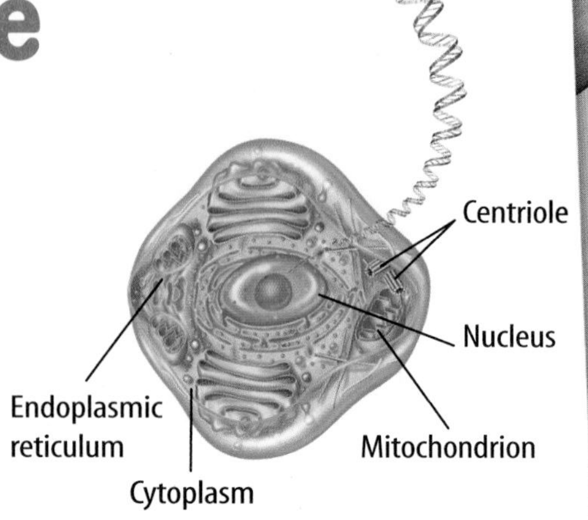

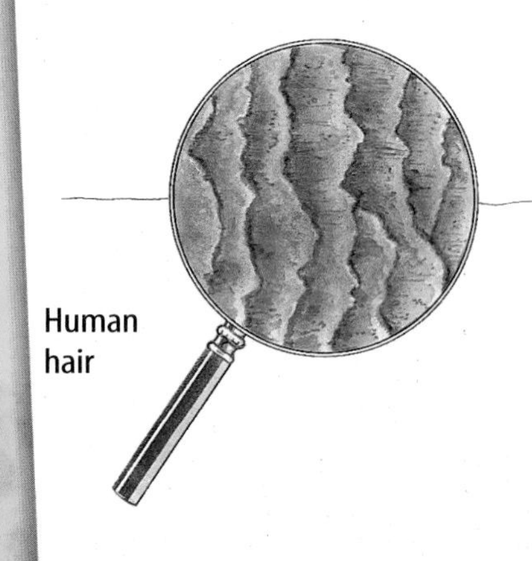

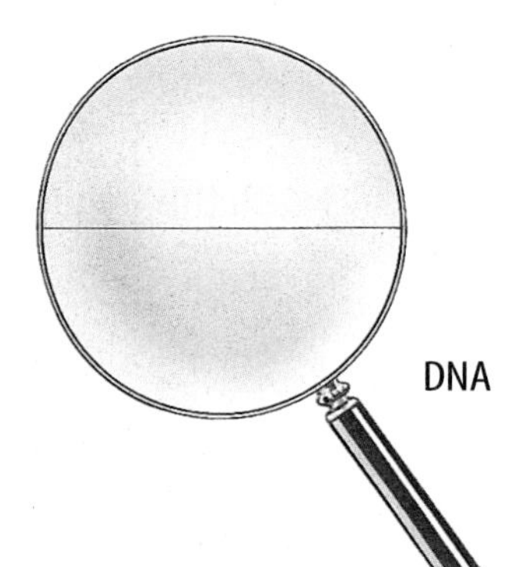

. . . The strands of DNA in the human genome, if unwound and connected end to end, would be more than 1.5 m long—but only about 130 trillionths of a centimeter wide. Even an average human hair is as much as 200,000 times wider than that.

. . . It would take about nine and one-half years to read aloud without stopping the 3 billion bits of instructions (called base pairs) in your genome.

Applying Math If one million base pairs of DNA take up 1 megabyte of storage space on a computer, how many gigabytes (1,024 megabytes) would the whole genome fill?

Find Out About It

Human genome scientists hope to identify the location of disease-causing genes. Visit green.msscience.com/science_stats to research a genetic disease and share your results with your class.

Reviewing Main Ideas

Section 1 Genetics

1. Genetics is the study of how traits are inherited. Gregor Mendel determined the basic laws of genetics.
2. Traits are controlled by alleles on chromosomes.
3. Some alleles can be dominant or recessive.
4. When a pair of chromosomes separates during meiosis, the different alleles move into separate sex cells. Mendel found that he could predict the outcome of genetic crosses.

Section 2 Genetics Since Mendel

1. Inheritance patterns studied since Mendel include incomplete dominance, multiple alleles, and polygenic inheritance.
2. These inheritance patterns allow a variety of phenotypes to be produced.
3. Some disorders are the results of inheritance and can be harmful and even deadly.
4. Pedigree charts help reveal patterns of the inheritance of a trait in a family. Pedigrees show that sex-linked traits are expressed more often in males than in females.

Section 3 Advances in Genetics

1. Genetic engineering uses biological and chemical methods to change genes.
2. Recombinant DNA is one method of genetic engineering to make useful chemicals, including hormones.
3. Gene therapy shows promise for correcting many human genetic disorders by inserting normal alleles into cells.
4. Breakthroughs in the field of genetic engineering are allowing scientists to do many things, such as producing plants that are resistant to disease.

Visualizing Main Ideas

Examine the following pedigree for diabetes and explain the inheritance pattern.

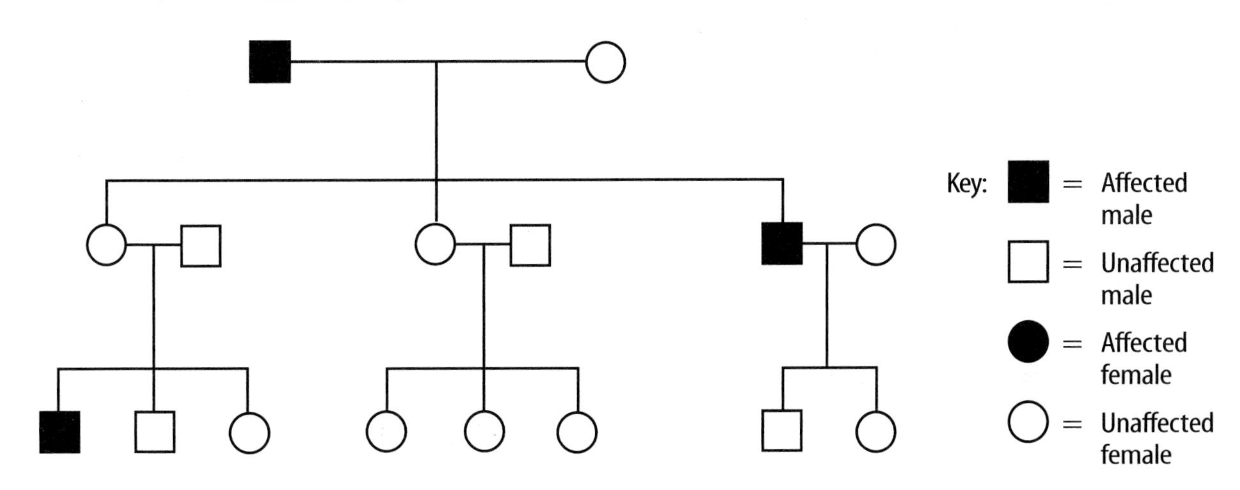

chapter 11 Review

Using Vocabulary

allele p. 306	hybrid p. 308
dominant p. 308	incomplete dominance p. 314
genetic engineering p. 321	phenotype p. 310
genetics p. 306	polygenic inheritance p. 316
genotype p. 310	Punnett square p. 310
heredity p. 306	recessive p. 308
heterozygous p. 310	sex-linked gene p. 319
homozygous p. 310	

Fill in the blanks with the correct word.

1. Alternate forms of a gene are called ________.
2. The outward appearance of a trait is a(n) ________.
3. Human height, eye color, and skin color are all traits controlled by ________.
4. An allele that produces a trait in the heterozygous condition is ________.
5. ________ is the branch of biology that deals with the study of heredity.
6. The actual combination of alleles of an organism is its ________.
7. ________ is moving fragments of DNA from one organism and inserting them into another organism.
8. A(n) ________ is a helpful device for predicting the proportions of possible genotypes.
9. ________ is the passing of traits from parents to offspring.
10. Red-green color blindness and hemophilia are two human genetic disorders that are caused by a(n) ________.

Checking Concepts

Choose the word or phrase that best answers the question.

11. Which of the following describes the allele that causes color blindness?
    A) dominant
    B) carried on the Y chromosome
    C) carried on the X chromosome
    D) present only in males

12. What is it called when the presence of two different alleles results in an intermediate phenotype?
    A) incomplete dominance
    B) polygenic inheritance
    C) multiple alleles
    D) sex-linked genes

13. What separates during meiosis?
    A) proteins **C)** alleles
    B) phenotypes **D)** pedigrees

14. What controls traits in organisms?
    A) cell membrane **C)** genes
    B) cell wall **D)** Punnett squares

15. What term describes the inheritance of cystic fibrosis?
    A) polygenic inheritance
    B) multiple alleles
    C) incomplete dominance
    D) recessive genes

16. What phenotype will the offspring represented in the Punnett square have?
    A) all recessive
    B) all dominant
    C) half recessive, half dominant
    D) Each will have a different phenotype.

	F	f
F	FF	Ff
F	FF	Ff

Thinking Critically

17. **Explain** the relationship between DNA, genes, alleles, and chromosomes.

18. **Classify** the inheritance pattern for each of the following:
    a. many different phenotypes produced by one pair of alleles
    b. many phenotypes produced by more than one pair of alleles; two phenotypes from two alleles; three phenotypes from two alleles.

Use the illustration below to answer question 19.

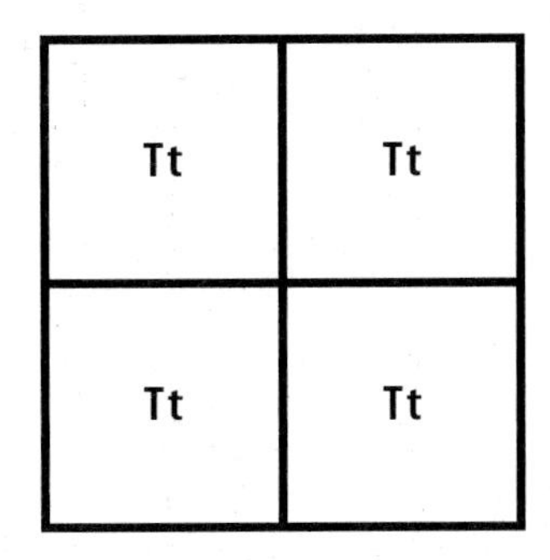

19. **Interpret Scientific Illustrations** What were the genotypes of the parents that produced the Punnett Square shown above?

20. **Explain** why two rabbits with the same genes might not be colored the same if one is raised in northern Maine and one is raised in southern Texas.

21. **Apply** Why would a person who receives genetic therapy for a disorder still be able to pass the disorder to his or her children?

22. **Predict** Two organisms were found to have different genotypes but the same phenotype. Predict what these phenotypes might be. Explain.

23. **Compare and contrast** Mendelian inheritance with incomplete dominance.

Performance Activities

24. **Newspaper Article** Write a newspaper article to announce a new, genetically engineered plant. Include the method of developing the plant, the characteristic changed, and the terms that you would expect to see. Read your article to the class.

25. **Predict** In humans, the widow's peak allele is dominant, and the straight hairline allele is recessive. Predict how both parents with widow's peaks could have a child without a widow's peak hairline.

26. **Use a word processor** or program to write predictions about how advances in genetics might affect your life in the next ten years.

Applying Math

27. **Human Genome** If you wrote the genetic information for each gene in the human genome on a separate sheet of 0.2-mm-thick paper and stacked the sheets, how tall would the stack be?

Use the table below to answer question 28.

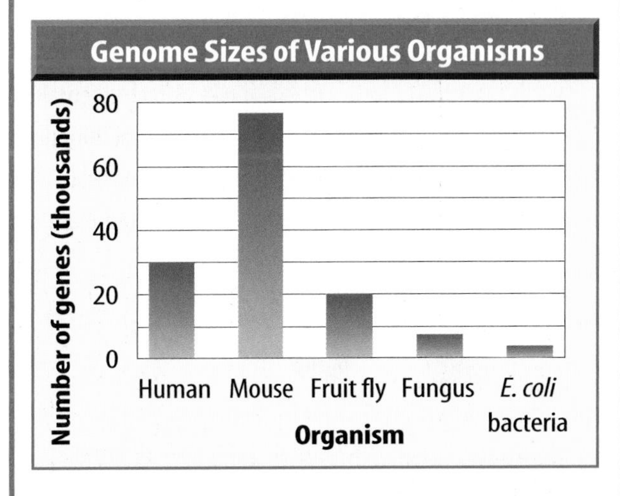

28. **Genes** Consult the graph above. How many more genes are in the human genome than the genome of the fruit fly?

chapter 11 Standardized Test Practice

Part 1 Multiple Choice

Record your answers on the answer sheet provided by your teacher or on a sheet of paper.

1. Heredity includes all of the following except
   - **A.** traits.
   - **B.** chromosomes.
   - **C.** nutrients.
   - **D.** phenotype.

2. What is a mutation?
   - **A.** A change in a gene which is harmful, beneficial, or has no effect at all.
   - **B.** A change in a gene which is only beneficial.
   - **C.** A change in a gene which is only harmful.
   - **D.** No change in a gene.

3. Sex of the offspring is determined by
   - **A.** only the mother, because she has two X chromosomes.
   - **B.** only the father, because he has one X and one Y chromosome.
   - **C.** an X chromosome from the mother and either an X or Y chromosome from the father.
   - **D.** mutations.

Use the pedigree below to answer questions 4–6.

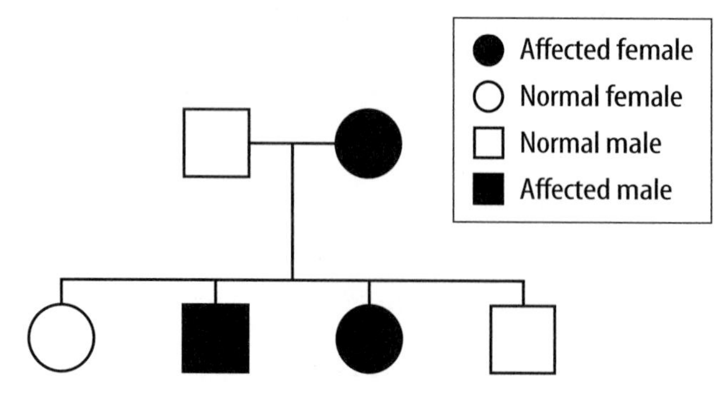

Huntington disease has a dominant (DD or Dd) inheritance pattern.

4. What is the genotype of the father?
   - **A.** DD
   - **B.** Dd
   - **C.** dd
   - **D.** D

5. What is the genotype of the mother?
   - **A.** DD
   - **B.** Dd
   - **C.** dd
   - **D.** D

6. The genotype of the unaffected children is
   - **A.** DD.
   - **B.** Dd.
   - **C.** dd.
   - **D.** D.

7. Manipulating the arrangement of DNA that makes up a gene is called
   - **A.** genetic engineering.
   - **B.** chromosomal migration.
   - **C.** viral reproduction.
   - **D.** cross breeding.

Use the Punnett square below to answer question 8.

	A	O
A	AA	AO
B	AB	BO

8. How many phenotypes would result from the following Punnett square?
   - **A.** 1
   - **B.** 2
   - **C.** 3
   - **D.** 4

9. Down's Syndrome is an example of
   - **A.** incomplete dominance.
   - **B.** genetic engineering.
   - **C.** a chromosome disorder.
   - **D.** a sex linked disorder.

Test-Taking Tip

Complete Charts Write directly on complex charts such as a Punnett square.

Question 10 Draw a Punnett square to answer all parts of the question.

Part 2 Short Response/Grid In

Record your answers on the answer sheet provided by your teacher or on a sheet of paper.

Use the table below to answer questions 10–11.

Some Traits Compared by Mendel			
Traits	Shape of Seeds	Shape of Pods	Flower Color
Dominant Trait	Round	Full	Purple
Recessive Trait	Wrinkled	Flat, constricted	White

10. Create a Punnett square using the *Shape of Pods* trait crossing heterozygous parents. What percentage of the offspring will be heterozygous? What percentage of the offspring will be homozygous? What percentage of the offspring will have the same phenotype as the parents?

11. Gregor Mendel studied traits in pea plants that were controlled by single genes. Explain what would have happened if the alleles for flower color were an example of incomplete dominance. What phenotypes would he have observed?

12. Why are heterozygous individuals called carriers for non-sex-linked and X-linked recessive patterns of inheritance?

13. How many alleles does a body cell have for each trait? What happens to the alleles during meiosis?

Part 3 Open Ended

Record your answers on a sheet of paper.

14. Genetic counseling helps individuals determine the genetic risk or probability a disorder will be passed to offspring. Why would a pedigree be a very important tool for the counselors? Which patterns of inheritance (dominant, recessive, x-linked) would be the easiest to detect?

15. Explain the process of gene therapy. What types of disorders would this therapy be best suited? How has this therapy helped patients with cystic fibrosis?

Refer to the figure below to answer question 16.

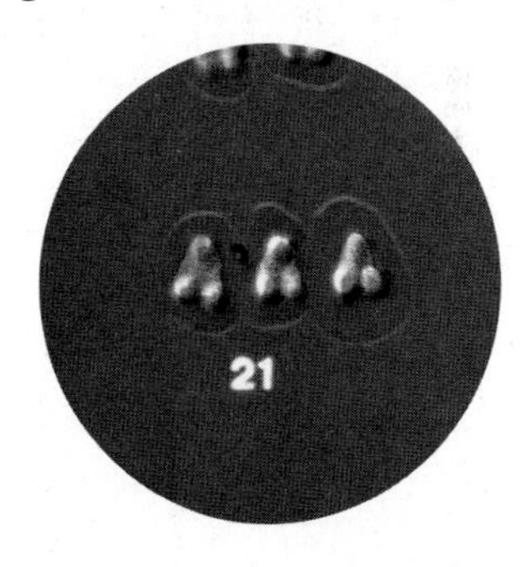

16. What is the disorder associated with the karyotype shown above? How does this condition occur? What are the characteristics of someone with this disorder?

17. Explain why the parents of someone with cystic fibrosis do not show any symptoms. How are the alleles for cystic fibrosis passed from parents to offspring?

18. What is recombinant DNA and how is it used to help someone with Type I diabetes?

19. If each kernel on an ear of corn represents a separate genetic cross, would corn be a good plant to use to study genetics? Why or why not? What process could be used to control pollination?

chapter 12

Adaptations over Time

chapter preview

sections

Adaptation? No problem.

Cockroaches have existed for millions of years, yet they are still adapted to their environment. Since they first appeared, many species have disappeared, and other well-adapted species have evolved.

Science Journal Pick a favorite plant or animal and list in your Science Journal all the ways it is well-suited to its environment.

Start-Up Activities

Adaptation for a Hunter

The cheetah is nature's fastest hunter, but it can run swiftly for only short distances. Its fur blends in with tall grass, making it almost invisible as it hides and waits for prey. Then the cheetah pounces, capturing the prey before it can run away.

1. Spread a sheet of newspaper classified ads on the floor.
2. Using a hole puncher, make 100 circles from each of the following types of paper: white paper, black paper, and classified ads.
3. Scatter all the circles on the newspaper on the floor. For 10 s, pick up as many circles as possible, one at a time. Have a partner time you.
4. Count the number of each kind of paper circle that you picked up. Record your results in your Science Journal.
5. **Think Critically** Which paper circles were most difficult to find? What can you infer about a cheetah's coloring from this activity? Enter your responses to these questions in your Science Journal.

Principles of Natural Selection Make the following Foldable to help you understand the process of natural selection.

STEP 1 **Fold** a sheet of paper in half lengthwise.

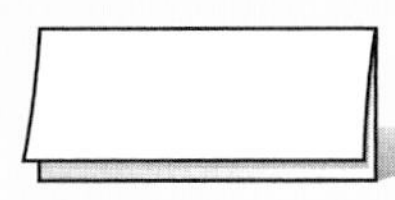

STEP 2 **Fold** paper down 2.5 cm from the top. (Hint: From the tip of your index finger to your middle knuckle is about 2.5 cm.)

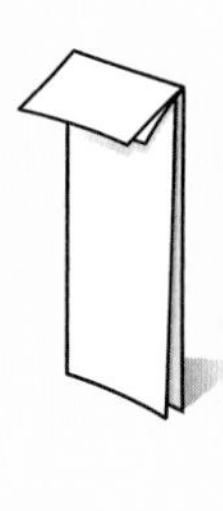

STEP 3 **Open and draw** lines along the 2.5-cm fold and the center fold. **Label** as shown.

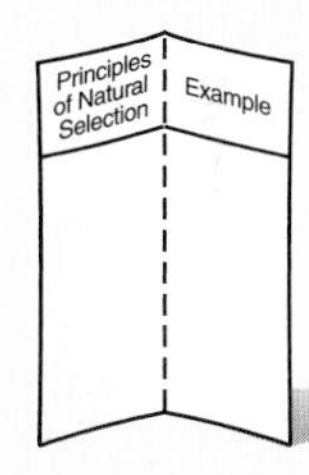

Summarize in a Table As you read, list the five principles of natural selection in the left-hand column. In the right-hand column, briefly write an example for each principle.

Preview this chapter's content and activities at green.msscience.com

section 1

Ideas About Evolution

as you read

What You'll Learn

- **Describe** Lamarck's hypothesis of acquired characteristics and Darwin's theory of natural selection.
- **Identify** why variations in organisms are important.
- **Compare and contrast** gradualism and punctuated equilibrium.

Why It's Important

The theory of evolution suggests why there are so many different living things.

Review Vocabulary

gene: a section of DNA that contains instructions for making specific proteins

New Vocabulary

- species
- evolution
- natural selection
- variation
- adaptation
- gradualism
- punctuated equilibrium

Early Models of Evolution

Millions of species of plants, animals, and other organisms live on Earth today. Do you suppose they are exactly the same as they were when they first appeared—or have any of them changed? A **species** is a group of organisms that share similar characteristics and can reproduce among themselves to produce fertile offspring. Many characteristics of a species are inherited when they pass from parent to offspring. Change in these inherited characteristics over time is **evolution. Figure 1** shows how the characteristics of the camel have changed over time.

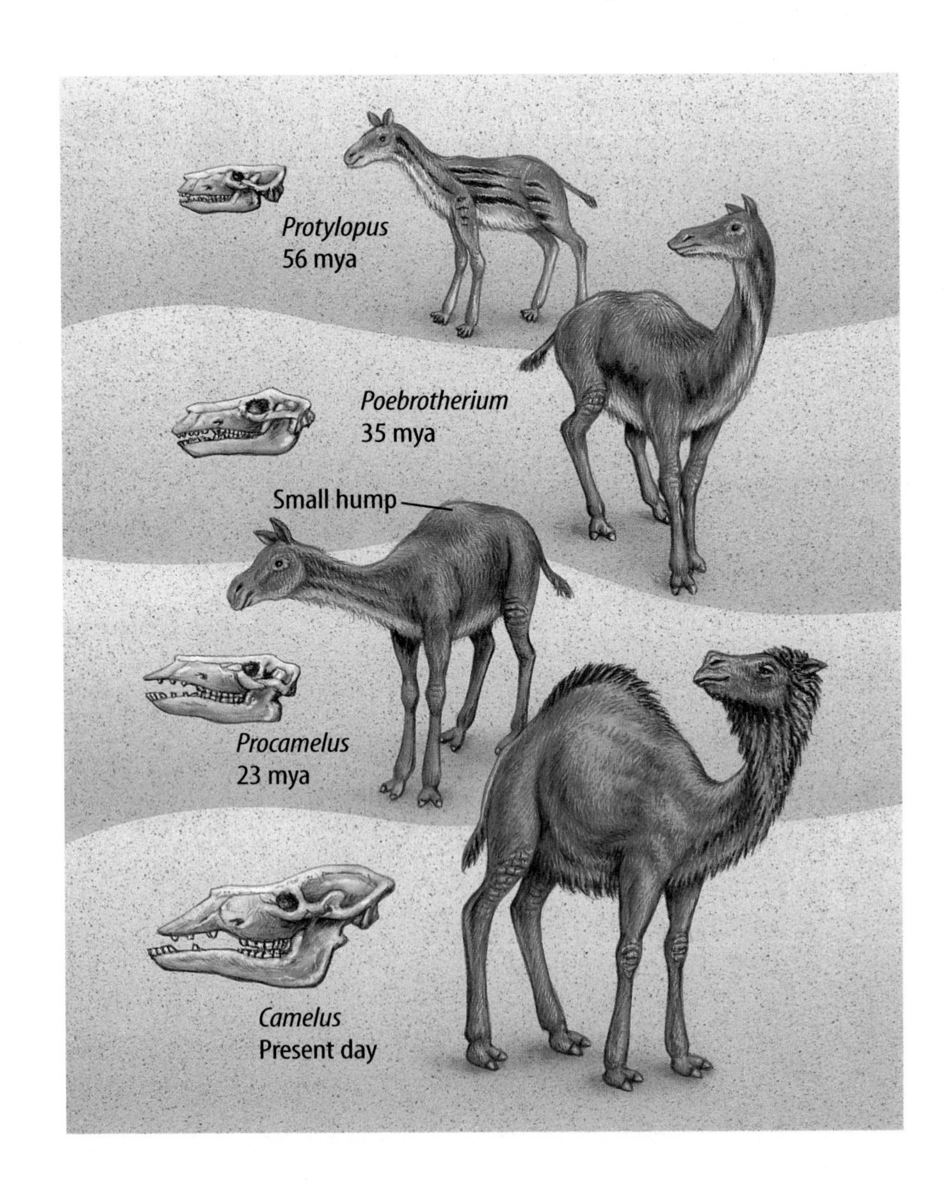

Figure 1 By studying fossils, scientists have traced the hypothesized evolution of the camel.
Discuss *the changes you observe in camels over time.*

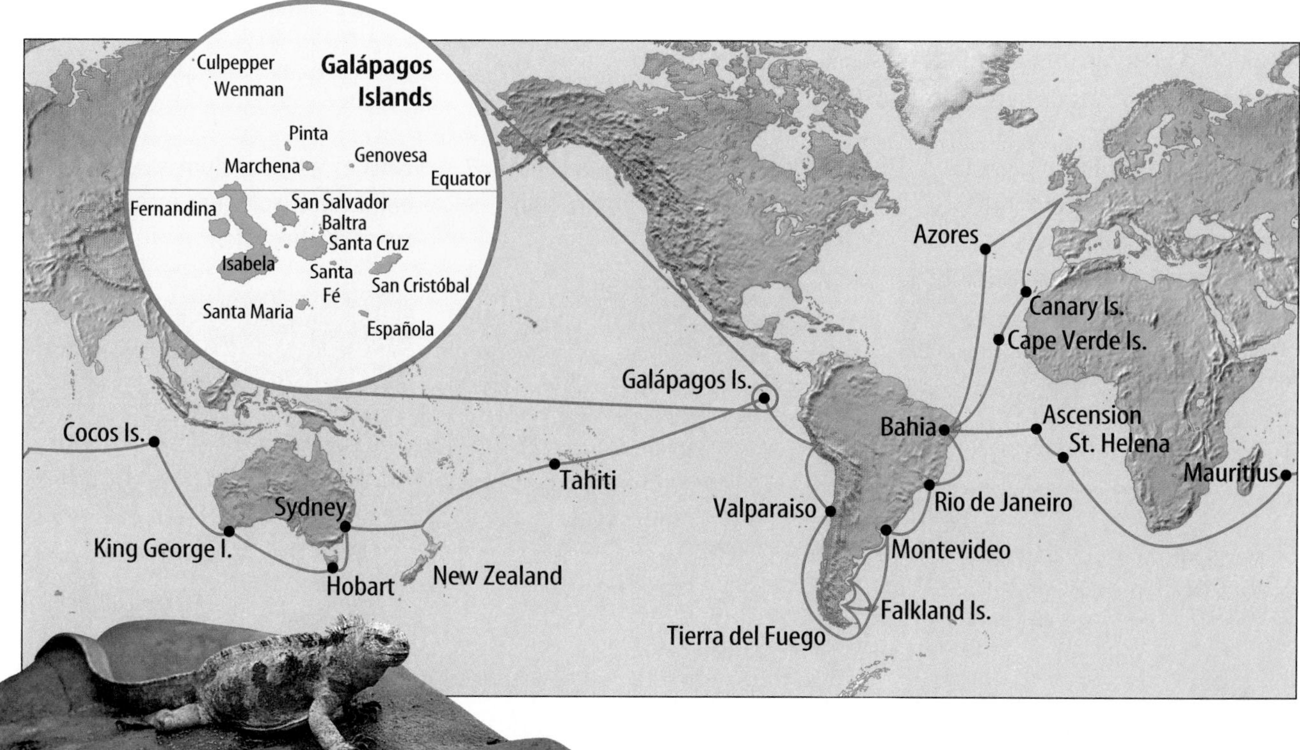

Figure 2 This map shows the route of Darwin's voyage on the HMS *Beagle*. Darwin noticed many species on the Galápagos Islands that he had not seen along the coast of South America, including the marine iguana. This species is the only lizard in the world known to enter the ocean and feed on seaweed.

Hypothesis of Acquired Characteristics In 1809, Jean Baptiste de Lamarck proposed a hypothesis to explain how species change over time. He suggested that characteristics, or traits, developed during a parent organism's lifetime are inherited by its offspring. His hypothesis is called the inheritance of acquired characteristics. Scientists collected data on traits that are passed from parents to offspring. The data showed that traits developed during a parent's lifetime, such as large muscles built by hard work or exercise, are not passed on to offspring. The evidence did not support Lamarck's hypothesis.

Reading Check *What was Lamarck's explanation of evolution?*

Darwin's Model of Evolution

In December 1831, the HMS *Beagle* sailed from England on a journey to explore the South American coast. On board was a young naturalist named Charles Darwin. During the journey, Darwin recorded observations about the plants and animals he saw. He was amazed by the variety of life on the Galápagos Islands, which are about 1,000 km from the coast of Ecuador. Darwin hypothesized that the plants and animals on the Galápagos Islands originally must have come from Central and South America. But the islands were home to many species he had not seen in South America, including giant cactus trees, huge land tortoises, and the iguana shown in **Figure 2.**

Figure 3 Darwin observed that the beak shape of each species of Galápagos finch is related to its eating habits.

Finches that eat nuts and seeds have short, strong beaks for breaking hard shells.

Finches that feed on insects have long, slender beaks for probing beneath tree bark.

Finches with medium-sized beaks eat a variety of foods including seeds and insects.

Topic: Darwin's Finches
Visit green.msscience.com for Web links to information about the finches Darwin observed.

Activity In your Science Journal, describe the similarities and differences of any two species of Galápagos finches.

Darwin's Observations Darwin observed 13 species of finches on the Galápagos Islands. He noticed that all 13 species were similar, except for differences in body size, beak shape, and eating habits, as shown in **Figure 3.** He also noticed that all the Galápagos finch species were similar to one finch species he had seen on the South American coast.

Darwin reasoned that the Galápagos finches must have had to compete for food. Finches with beak shapes that allowed them to eat available food survived longer and produced more offspring than finches without those beak shapes. After many generations, these groups of finches became separate species.

How did Darwin explain the evolution of the different species of Galápagos finches?

Natural Selection

After the voyage, Charles Darwin returned to England and continued to think about his observations. He collected more evidence on inherited traits by breeding racing pigeons. He also studied breeds of dogs and varieties of flowers. In the mid 1800s, Darwin developed a theory of evolution that is accepted by most scientists today. He described his ideas in a book called *On the Origin of Species,* which was published in 1859.

Darwin's Theory Darwin's observations led many other scientists to conduct experiments on inherited characteristics. After many years, Darwin's ideas became known as the theory of evolution by natural selection. **Natural selection** means that organisms with traits best suited to their environment are more likely to survive and reproduce. Their traits are passed to more offspring. All living organisms produce more offspring than survive. Galápagos finches lay several eggs every few months. Darwin realized that in just a few years, several pairs of finches could produce a large population. A population is all of the individuals of a species living in the same area. Members of a large population compete for living space, food, and other resources. Those that are best able to survive are more likely to reproduce and pass on their traits to the next generation.

The principles that describe how natural selection works are listed in **Table 1.** Over time, as new data was gathered and reported, changes were made to Darwin's original ideas about evolution by natural selection. His theory remains one of the most important ideas in the study of life science.

Table 1 The Principles of Natural Selection
1. Organisms produce more offspring than can survive.
2. Differences, or variations, occur among individuals of a species.
3. Some variations are passed to offspring.
4. Some variations are helpful. Individuals with helpful variations survive and reproduce better than those without these variations.
5. Over time, the offspring of individuals with helpful variations make up more of a population and eventually may become a separate species.

Applying Science

Does natural selection take place in a fish tank?

Alejandro raises tropical fish as a hobby. Could the observations that he makes over several weeks illustrate the principles of natural selection?

Identifying the Problem

Alejandro keeps a detailed journal of his observations, some of which are given in the table to the right.

Fish Tank Observations

Date	Observation
June 6	6 fish are placed in aquarium tank.
July 22	16 new young appear.
July 24	3 young have short or missing tail fins. 13 young have normal tail fins.
July 28	Young with short or missing tail fins die.
August 1	2 normal fish die—from overcrowding?
August 12	30 new young appear.
August 15	5 young have short or missing tail fins. 25 young have normal tail fins.
August 18	Young with short or missing tail fins die.
August 20	Tank is overcrowded. Fish are divided equally into two tanks.

Solving the Problem

Refer to **Table 1** and match each of Alejandro's journal entries with the principle(s) it demonstrates. Here's a hint: *Some entries may not match any of the principles of natural selection. Some entries may match more than one principle.*

INTEGRATE Language Arts

Evolution of English
If someone from Shakespeare's time were to speak to you today, you probably would not understand her. Languages, like species, change over time. In your Science Journal, discuss some words or phrases that you use that your parents or teachers do not use correctly.

Variation and Adaptation

Darwin's theory of evolution by natural selection emphasizes the differences among individuals of a species. These differences are called variations. A **variation** is an inherited trait that makes an individual different from other members of its species. Variations result from permanent changes, or mutations, in an organism's genes. Some gene changes produce small variations, such as differences in the shape of human hairlines. Other gene changes produce large variations, such as an albino squirrel in a population of gray squirrels or fruit without seeds. Over time, more and more individuals of the species might inherit these variations. If individuals with these variations continue to survive and reproduce over many generations, a new species can evolve. It might take hundreds, thousands, or millions of generations for a new species to evolve.

Some variations are more helpful than others. An **adaptation** is any variation that makes an organism better suited to its environment. The variations that result in an adaptation can involve an organism's color, shape, behavior, or chemical makeup. Camouflage (KA muh flahj) is an adaptation. A camouflaged organism, like the one shown in **Figure 4,** blends into its environment and is more likely to survive and reproduce.

Figure 4 Variations that provide an advantage tend to increase in a population over time. Variations that result in a disadvantage tend to decrease in a population over time.

Albinism can prevent an organism from blending into its environment. **Infer** *what might happen to an albino lemur in its natural environment.*

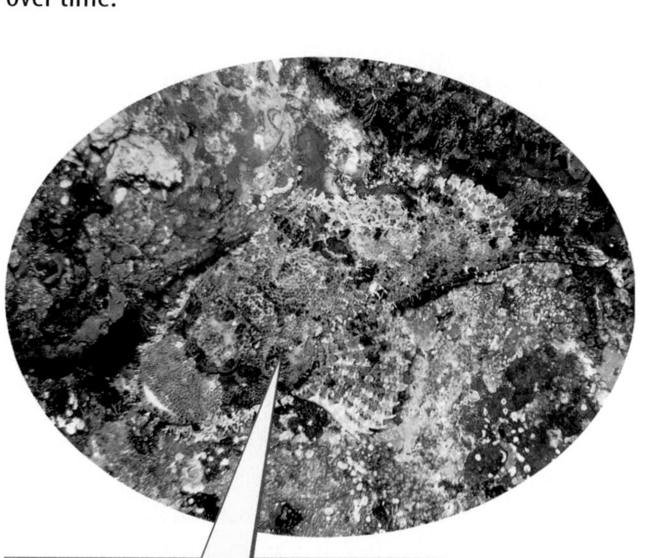

Camouflage allows organisms to blend into their environments. **Infer** *how its coloration gives this scorpion fish a survival advantage.*

Figure 5 About 600 years ago, European rabbits were introduced to the Canary Islands from a visiting Portuguese ship. The Canary Islands are in the Atlantic Ocean off the northwest coast of Africa. Over time, the Canary Island rabbits became a separate species.

European rabbits, like the one above, feed during the day and are fairly large.

Canary Island rabbits feed during the night. **Explain** *why large eyes might be considered a helpful adaptation in Canary Island rabbits.*

Changes in the Sources of Genes Over time, the genetic makeup of a species might change its appearance. For example, as the genetic makeup of a species of seed-eating Galápagos finch changed, so did the size and shape of its beak. Many kinds of environmental factors help bring about changes. When individuals of the same species move into or out of an area, they might bring in or remove genes and variations. Suppose a family from another country moves to your neighborhood. They might bring different foods, customs, and ways of speaking with them. In a similar way, when new individuals enter an existing population, they can bring in different genes and variations.

Geographic Isolation Sometimes mountains, lakes, or other geologic features isolate a small number of individuals from the rest of a population. Over several generations, variations that do not exist in the larger population might begin to be more common in the isolated population. Also, gene mutations can occur that add variations to populations. Over time, the two populations can become so different that they no longer can breed with each other. The two populations of rabbits shown in **Figure 5** have been geographically isolated from each other for thousands of generations.

Relating Evolution to Species

Procedure
1. On a piece of **paper,** print the word *train.*
2. Add, subtract, or change one letter to make a new word.
3. Repeat step 2 with the new word.
4. Repeat steps 2 and 3 two more times.
5. Make a "family tree" that shows how your first word changed over time.

Analysis
1. Compare your tree to those of other people. Did you produce the same words?
2. How is this process similar to evolution by natural selection?

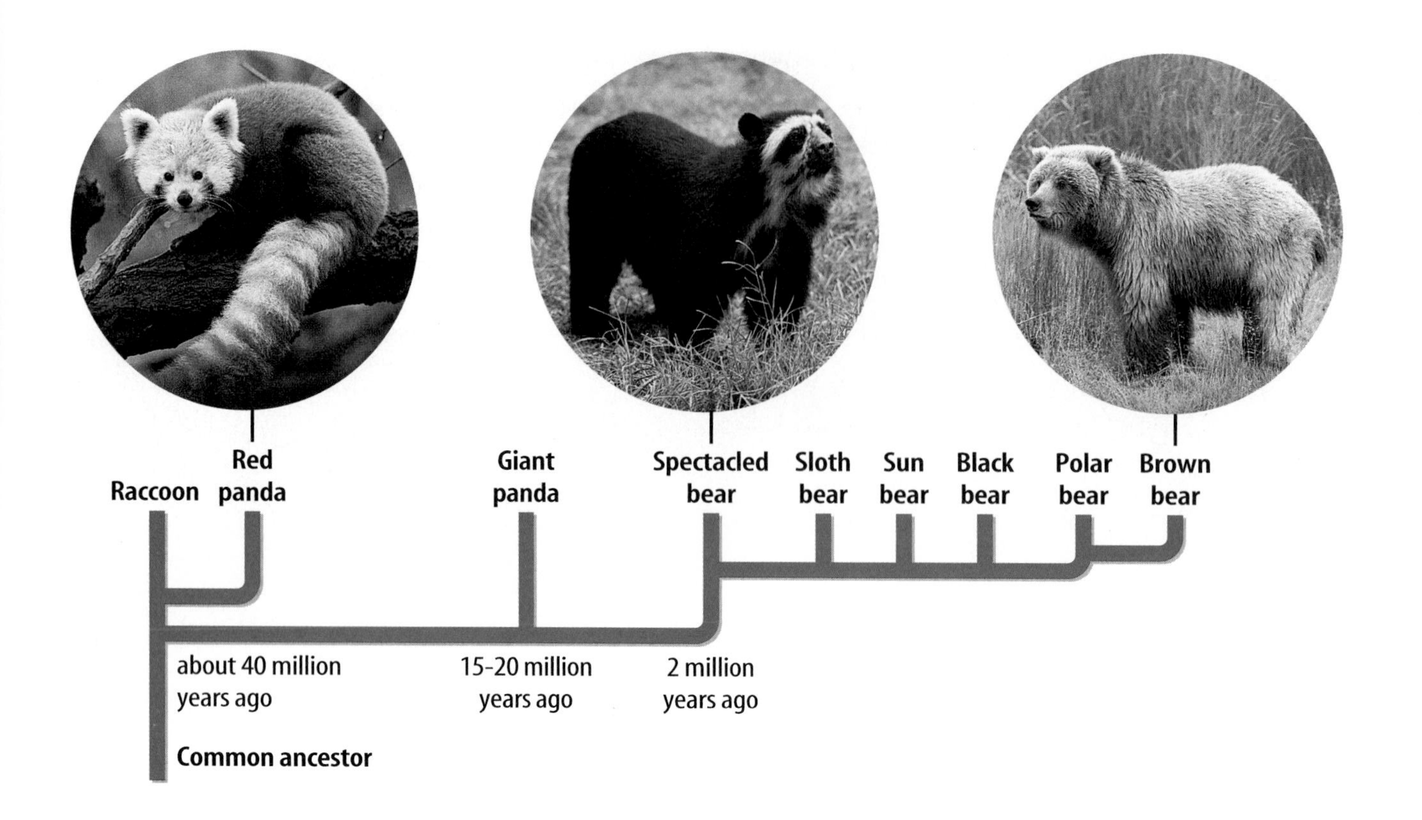

Figure 6 The hypothesized evolution of bears illustrates the punctuated equilibrium model of evolution.
Discuss *how the six species on the far right are explained better by punctuated equlibrium.*

The Speed of Evolution

Scientists do not agree on how quickly evolution occurs. Many scientists hypothesize that evolution occurs slowly, perhaps over tens or hundreds of millions of years. Other scientists hypothesize that evolution can occur quickly. Most scientists agree that evidence supports both of these models.

Gradualism Darwin hypothesized that evolution takes place slowly. The model that describes evolution as a slow, ongoing process by which one species changes to a new species is known as **gradualism.** According to the gradualism model, a continuing series of mutations and variations over time will result in a new species. Look back at **Figure 1,** which shows the evolution of the camel over tens of millions of years. Fossil evidence shows a series of intermediate forms that indicate a gradual change from the earliest camel species to today's species.

Punctuated Equilibrium Gradualism doesn't explain the evolution of all species. For some species, the fossil record shows few intermediate forms—one species suddenly changes to another. According to the **punctuated equilibrium** model, rapid evolution comes about when the mutation of a few genes results in the appearance of a new species over a relatively short period of time. The fossil record gives examples of this type of evolution, as you can see in **Figure 6.**

Punctuated Equilibrium Today Evolution by the punctuated equilibrium model can occur over a few thousand or million years, and sometimes even faster. For example, many bacteria have changed in a few decades. The antibiotic penicillin originally came from the fungus shown in **Figure 7.** But many bacteria species that were once easily killed by penicillin no longer are harmed by it. These bacteria have developed resistance to the drug. Penicillin has been in use since 1943. Just four years later, in 1947, a species of bacteria that causes pneumonia and other infections already had developed resistance to the drug. By the 1990s, several disease-producing bacteria had become resistant to penicillin and many other antibiotics.

How did penicillin-resistant bacteria evolve so quickly? As in any population, some organisms have variations that allow them to survive unfavorable living conditions when other organisms cannot. When penicillin was used to kill bacteria, those with the penicillin-resistant variation survived, reproduced, and passed this trait to their offspring. Over a period of time, this bacteria population became penicillin-resistant.

Figure 7 The fungus growing in this petri dish is *Penicillium,* the original source of penicillin. It produces an antibiotic substance that prevents the growth of certain bacteria.

section 1 review

Summary

Early Models of Evolution

- Evolution is change in the characteristics of a species over time.
- Lamarck proposed the hypothesis of inherited acquired characteristics.

Natural Selection

- Darwin proposed evolution by natural selection, a process by which organisms best suited to their environments are most likely to survive and reproduce.
- Organisms have more offspring than can survive, individuals of a species vary, and many of these variations are passed to offspring.

Variation and Adaptation

- Adaptations are variations that help an organism survive or reproduce in its environment.
- Mutations are the source of new variations.

The Speed of Evolution

- Evolution may be a slow or fast process depending on the species under study.

Self Check

1. **Compare** Lamarck's and Darwin's ideas about how evolution takes place.
2. **Explain** why variations are important to understanding change in a population over time.
3. **Discuss** how the gradualism model of evolution differs from the punctuated equilibrium model of evolution.
4. **Describe** how geographic isolation contributes to evolution.
5. **Think Critically** What adaptations would be helpful for an animal species that was moved to the Arctic?
6. **Concept Map** Use information given in **Figure 6** to make a map that shows how raccoons, red pandas, giant pandas, polar bears, and black bears are related to a common ancestor.

Applying Math

7. **Use Percentages** The evolution of the camel can be traced back at least 56 million years. Use **Figure 1** to estimate the percent of this time that the modern camel has existed.

Hidden Frogs

Through natural selection, animals become adapted for survival in their environment. Adaptations include shapes, colors, and even textures that help an animal blend into its surroundings. These adaptations are called camouflage. The red-eyed tree frog's mint green body blends in with tropical forest vegetation as shown in the photo on the right. Could you design camouflage for a desert frog? A temperate forest frog?

Real-World Question

What type of camouflage would best suit a frog living in a particular habitat?

Goals

- **Create** a frog model camouflaged to blend in with its surroundings.

Materials (for each group)

cardboard form of a frog
colored markers
crayons
colored pencils
glue
beads
sequins
modeling clay

Safety Precautions

Procedure

1. Choose one of the following habitats for your frog model: muddy shore of a pond, orchid flowers in a tropical rain forest, multicolored clay in a desert, or the leaves and branches of trees in a temperate forest.
2. **List** the features of your chosen habitat that will determine the camouflage your frog model will need.
3. **Brainstorm** with your group the body shape, coloring, and skin texture that would make the best camouflage for your model. Record your ideas in your Science Journal.
4. **Draw** in your Science Journal samples of colors, patterns, texture, and other features your frog model might have.
5. Show your design ideas to your teacher and ask for further input.
6. **Construct** your frog model.

Conclude and Apply

1. **Explain** how the characteristics of the habitat helped you decide on the specific frog features you chose.
2. **Infer** how the color patterns and other physical features of real frogs develop in nature.
3. **Explain** why it might be harmful to release a frog into a habitat for which it is not adapted.

Communicating Your Data

Create a poster or other visual display that represents the habitat you chose for this activity. Use your display to show classmates how your design helps camouflage your frog model. **For more help, refer to the** Science Skill Handbook.

section 2

Clues About Evolution

Clues from Fossils

Imagine going on a fossil hunt in Wyoming. Your companions are paleontologists—scientists who study the past by collecting and examining fossils. As you climb a low hill, you notice a curved piece of stone jutting out of the sandy soil. One of the paleontologists carefully brushes the soil away and congratulates you on your find. You've discovered part of the fossilized shell of a turtle like the one shown in **Figure 8.**

The Green River Formation covers parts of Wyoming, Utah, and Colorado. On your fossil hunt, you learn that about 50 million years ago, during the Eocene Epoch, this region was covered by lakes. The water was home to fish, crocodiles, lizards, and turtles. Palms, fig trees, willows, and cattails grew on the lakeshores. Insects and birds flew through the air. How do scientists know all this? After many of the plants and animals of that time died, they were covered with silt and mud. Over millions of years, they became the fossils that have made the Green River Formation one of the richest fossil deposits in the world.

Figure 8 The desert of the Green River Formation is home to pronghorn antelope, elks, coyotes, and eagles. Fossil evidence shows that about 50 million years ago the environment was much warmer and wetter than it is today.

as you read

What You'll Learn

- **Identify** the importance of fossils as evidence of evolution.
- **Explain** how relative and radiometric dating are used to estimate the age of fossils.
- **List** examples of five types of evidence for evolution.

Why It's Important

The scientific evidence for evolution helps you understand why this theory is so important to the study of biology.

Review Vocabulary

epoch: next-smaller division of geological time after a period; is characterized by differences in life-forms that may vary regionally

New Vocabulary

- sedimentary rock
- radioactive element
- embryology
- homologous
- vestigial structure

Figure 9 Examples of several different types of fossils are shown here.
Infer *which of these would most likely be found in a layer of sedimentary rock.*

Mineralized fossils Minerals can replace wood or bone to create a piece of petrified wood as shown to the left or a mineralized bone fossil.

Imprint fossils A leaf, feather, bones, or even the entire body of an organism can leave an imprint on sediment that later hardens to become rock.

Frozen fossils The remains of organisms like this mammoth can be trapped in ice that remains frozen for thousands of years.

Cast fossils Minerals can fill in the hollows of animal tracks, as shown to the right, a mollusk shell, or other parts of an organism to create a cast.

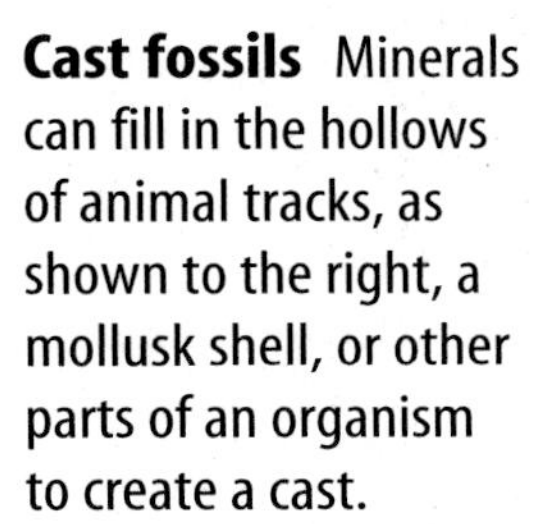

Fossils in amber When the sticky resin of certain cone-bearing plants hardens over time, amber forms. It can contain the remains of trapped insects.

Types of Fossils

Most of the evidence for evolution comes from fossils. A fossil is the remains, an imprint, or a trace of a prehistoric organism. Several types of fossils are shown in **Figure 9.** Most fossils are found in sedimentary rock. **Sedimentary rock** is formed when layers of sand, silt, clay, or mud are compacted and cemented together, or when minerals are deposited from a solution. Limestone, sandstone, and shale are all examples of sedimentary rock. Fossils are found more often in limestone than in any other kind of sedimentary rock. The fossil record provides evidence that living things have evolved.

Determining a Fossil's Age

Paleontologists use detective skills to determine the age of dinosaur fossils or the remains of other ancient organisms. They can use clues provided by unique rock layers and the fossils they contain. The clues provide information about the geology, weather, and life-forms that must have been present during each geologic time period. Two basic methods—relative dating and radiometric dating—can be used, alone or together, to estimate the ages of rocks and fossils.

Topic: Fossil Finds

Visit green.msscience.com for Web links to information about recent fossil discoveries.

Activity Prepare a newspaper article describing how one of these discoveries was made, what it reveals about past life on Earth, and how it has impacted our understanding of what the past environments of Earth were like.

Relative Dating One way to find the approximate age of fossils found within a rock layer is relative dating. Relative dating is based on the idea that in undisturbed areas, younger rock layers are deposited on top of older rock layers, as shown in **Figure 10.** Relative dating provides only an estimate of a fossil's age. The estimate is made by comparing the ages of rock layers found above and below the fossil layer. For example, suppose a 50-million-year-old rock layer lies below a fossil, and a 35-million-year-old layer lies above it. According to relative dating, the fossil is between 35 million and 50 million years old.

Why can relative dating be used only to estimate the age of a fossil?

Radiometric Dating Scientists can obtain a more accurate estimate of the age of a rock layer by using radioactive elements. A **radioactive element** gives off a steady amount of radiation as it slowly changes to a nonradioactive element. Each radioactive element gives off radiation at a different rate. Scientists can estimate the age of the rock by comparing the amount of radioactive element with the amount of nonradioactive element in the rock. This method of dating does not always produce exact results, because the original amount of radioactive element in the rock can never be determined for certain.

Figure 10 In Bryce Canyon, erosion by water and wind has cut through the sedimentary rock, exposing the layers.
Infer *the relative age of rocks in the lowest layers compared to the top layer.*

NATIONAL GEOGRAPHIC **VISUALIZING THE GEOLOGIC TIME SCALE**

Figure 11

Earth is roughly 4.5 billion years old. As shown here, the vast period of time from Earth's beginning to the present day has been organized into the geologic time scale. The scale is divided into eras and periods. Dates on this scale are given as millions of years ago (mya).

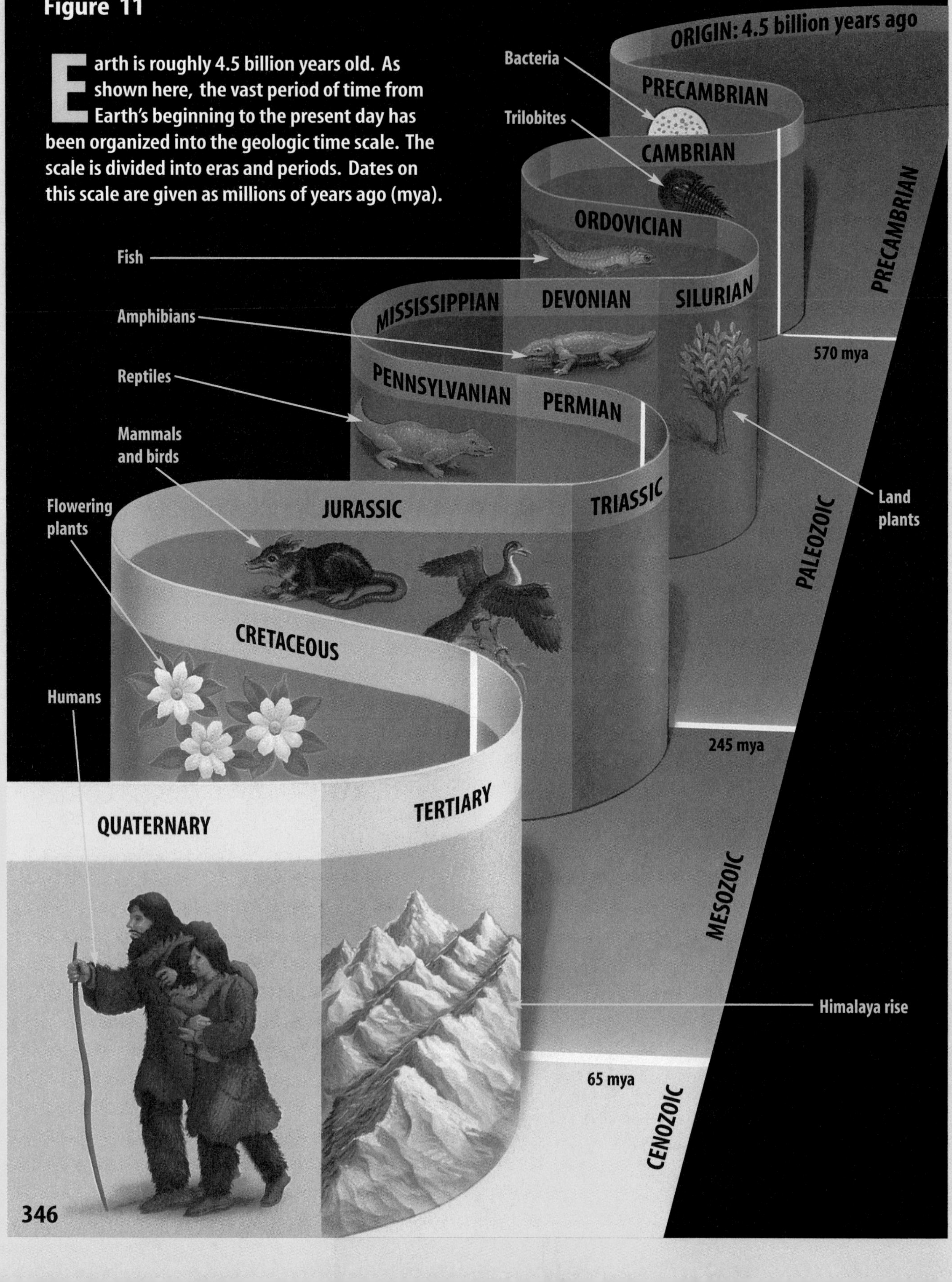

Fossils and Evolution

Fossils provide a record of organisms that lived in the past. However, the fossil record is incomplete, or has gaps, much like a book with missing pages. The gaps exist because most organisms do not become fossils. By looking at fossils, scientists conclude that many simpler forms of life existed earlier in Earth's history, and more complex forms of life appeared later, as shown in **Figure 11.** Fossils provide indirect evidence that evolution has occurred on Earth.

Almost every week, fossil discoveries are made somewhere in the world. When fossils are found, they are used to help scientists understand the past. Scientists can use fossils to make models that show what the organisms might have looked like. From fossils, scientists can sometimes determine whether the organisms lived in family groups or alone, what types of food they ate, what kind of environment they lived in, and many other things about them. Most fossils represent extinct organisms. From a study of the fossil record, scientists have concluded that more than 99 percent of all organisms that have ever existed on Earth are now extinct.

Evolution in Fossils Many organisms have a history that has been preserved in sedimentary rock. Fossils show that the bones of animals such as horses and whales have become reduced in size or number over geologic time, as the species has evolved. In your Science Journal, explain what information can be gathered from changes in structures that occur over time.

More Clues About Evolution

Besides fossils, what other clues do humans have about evolution? Sometimes, evolution can be observed directly. Plant breeders observe evolution when they use cross-breeding to produce genetic changes in plants. The development of antibiotic resistance in bacteria is another direct observation of evolution. Entomologists have noted similar rapid evolution of pesticide-resistant insect species. These observations provide direct evidence that evolution occurs. Also, many examples of indirect evidence for evolution exist. They include similarities in embryo structures, the chemical makeup of organisms including DNA, and the way organisms develop into adults. Indirect evidence does not provide proof of evolution, but it does support the idea that evolution takes place over time.

Embryology The study of embryos and their development is called **embryology** (em bree AH luh jee). An embryo is the earliest growth stage of an organism. A tail and pharyngeal pouches are found at some point in the embryos of fish, reptiles, birds, and mammals, as **Figure 12** shows. Fish develop gills, but the other organisms develop other structures as their development continues. Fish, birds, and reptiles keep their tails, but many mammals lose theirs. These similarities suggest an evolutionary relationship among all vertebrate species.

Figure 12 Similarities in the embryos of fish, chickens, and rabbits show evidence of evolution. **Evaluate** *these embryos as evidence for evolution.*

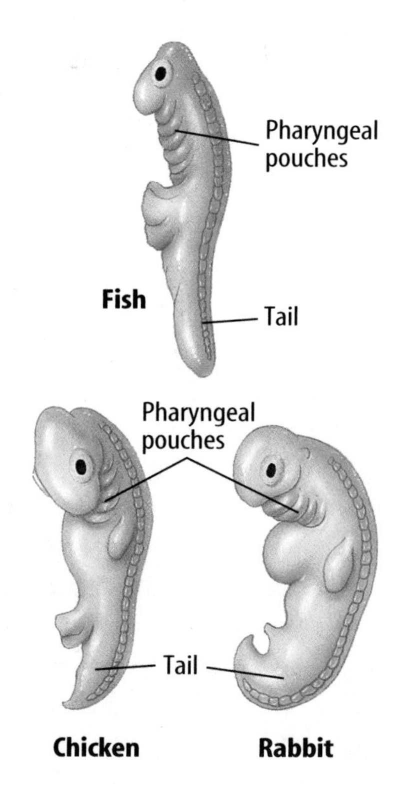

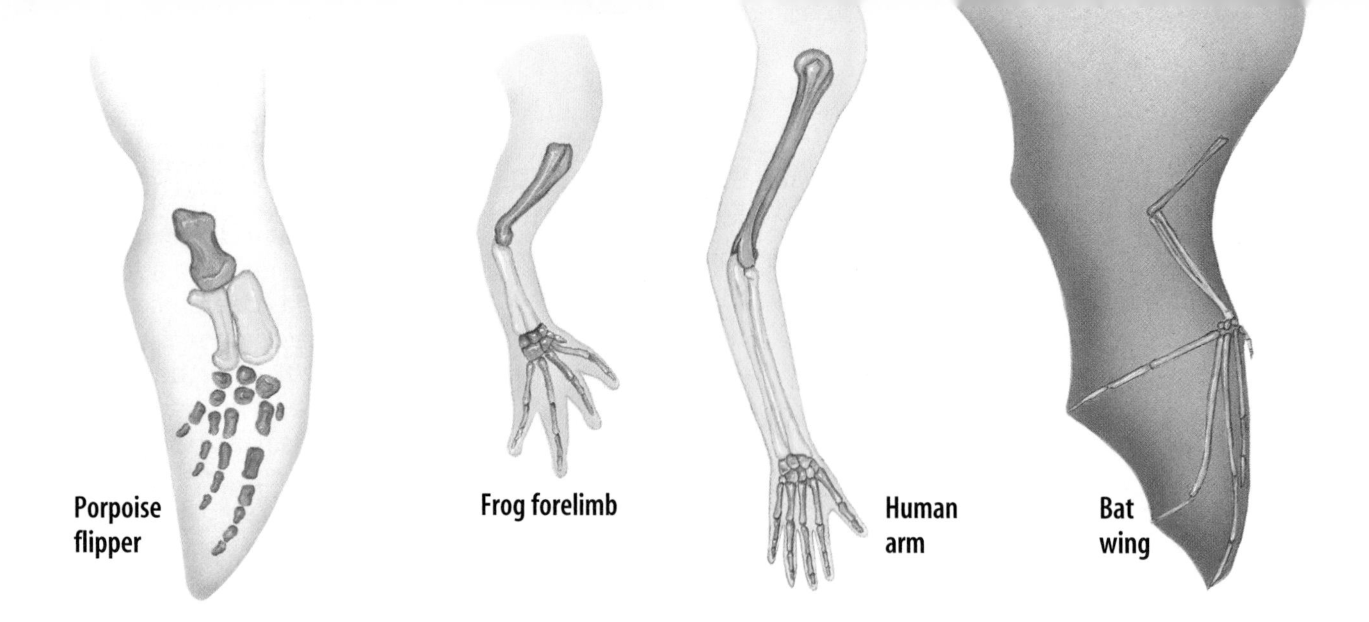

Figure 13 A porpoise flipper, frog forelimb, human arm, and bat wing are homologous. These structures show different arrangements and shapes of the bones of the forelimb. They have the same number of bones, muscles, and blood vessels, and they developed from similar tissues.

Homologous Structures What do the structures shown in **Figure 13** have in common? Although they have different functions, each of these structures is made up of the same kind of bones. Body parts that are similar in origin and structure are called **homologous** (hoh MAH luh gus). Homologous structures also can be similar in function. They often indicate that two or more species share common ancestors.

Reading Check *What do homologous structures indicate?*

Vestigial Structures The bodies of some organisms include **vestigial** (veh STIH jee ul) **structures**—structures that don't seem to have a function. Vestigial structures also provide evidence for evolution. For example, manatees, snakes, and whales no longer have back legs, but, like all animals with legs, they still have pelvic bones. The human appendix is a vestigial structure. The appendix appears to be a small version of the cecum, which is an important part of the digestive tract of many mammals. Scientists hypothesize that vestigial structures, like those shown in **Figure 14,** are body parts that once functioned in an ancestor.

Figure 14 Humans have three small muscles around each ear that are vestigial. In some mammals, such as horses, these muscles are large. They allow a horse to turn its ears toward the source of a sound. Humans cannot rotate their ears, but some people can wiggle their ears.

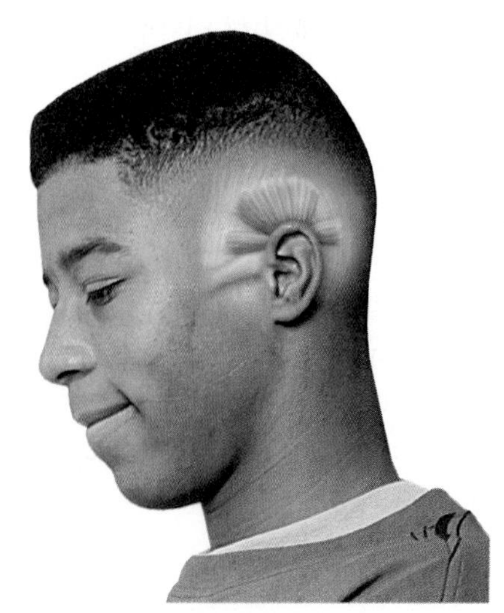

DNA If you enjoy science fiction, you probably have read books or seen movies in which scientists re-create dinosaurs and other extinct organisms from DNA taken from fossils. DNA is the molecule that controls heredity and directs the development of every organism. In a cell with a nucleus, DNA is found in genes that make up the chromosomes. Scientists compare DNA from living organisms to identify similarities among species. Examinations of ancient DNA often provide additional evidence of how some species evolved from their extinct ancestors. By looking at DNA, scientists also can determine how closely related organisms are. For example, DNA studies indicate that dogs are the closest relatives of bears.

Similar DNA also can suggest common ancestry. Apes such as the gorillas shown in **Figure 15,** chimpanzees, and orangutans have 24 pairs of chromosomes. Humans have 23 pairs. When two of an ape's chromosomes are laid end to end, a match for human chromosome number 2 is formed. Also, similar proteins such as hemoglobin—the oxygen-carrying protein in red blood cells—are found in many primates. This can be further evidence that primates have a common ancestor.

Figure 15 Gorillas have DNA and proteins that are similar to humans and other primates.

section 2 review

Summary

Clues from Fossils

- Scientists learn about past life by studying fossils.

Determining a Fossil's Age

- The relative date of a fossil can be estimated from the ages of rocks in nearby layers.
- Radiometric dating using radioactive elements gives more accurate dates for fossils.

Fossils and Evolution

- The fossil record has gaps which may yet be filled with later discoveries.

More Clues About Evolution

- Homologous structures, similar embryos, or vestigial structures can show evolutionary relationships.
- Evolutionary relationships among organisms can be inferred from DNA comparisons.

Self Check

1. **Compare and contrast** relative dating and radiometric dating.
2. **Discuss** the importance of fossils as evidence of evolution and describe five different kinds of fossils.
3. **Explain** how DNA can provide some evidence of evolution.
4. **List** three examples of direct evidence for evolution.
5. **Interpret Scientific Illustrations** According to data in **Figure 11,** what was the longest geologic era? What was the shortest era? In what period did mammals appear?
6. **Think Critically** Compare and contrast the five types of evidence that support the theory of evolution.

Applying Math

7. **Use Percentages** The Cenozoic Era represents about 65 million years. Approximately what percent of Earth's 4.5-billion-year history does this era represent?

section 3

The Evolution of Primates

as you read

What You'll Learn

- **Describe** the differences among living primates.
- **Identify** the adaptations of primates.
- **Discuss** the evolutionary history of modern primates.

Why It's Important

Studying primate evolution will help you appreciate the differences among primates.

Review Vocabulary
opposable: can be placed against another digit of a hand or foot

New Vocabulary
- primate
- hominid
- *Homo sapiens*

Primates

Humans, monkeys, and apes belong to the group of mammals known as the **primates.** All primates have opposable thumbs, binocular vision, and flexible shoulders that allow the arms to rotate. These shared characteristics indicate that all primates may have evolved from a common ancestor.

Having an opposable thumb allows you to cross your thumb over your palm and touch your fingers. This means that you can grasp and hold things with your hands. An opposable thumb allows tree-dwelling primates to hold on to branches.

Binocular vision permits you to judge depth or distance with your eyes. In a similar way, it allows tree-dwelling primates to judge the distances as they move between branches. Flexible shoulders and rotating forelimbs also help tree-dwelling primates move from branch to branch. They also allow humans to do the backstroke, as shown in **Figure 16.**

Primates are divided into two major groups. The first group, the strepsirhines (STREP suh rines), includes lemurs and tarsiers like those shown in **Figure 17.** The second group, haplorhines (HAP luh rines), includes monkeys, apes, and humans.

Figure 16 The ability to rotate the shoulder in a complete circle allows humans to swim through water and tree-dwelling primates to travel through treetops.

Figure 17 Tarsiers and lemurs are active at night. Tarsiers are commonly found in the rain forests of Southeast Asia. Lemurs live on Madagascar and other nearby islands.
List *the traits that distinguish these animals as primates.*

Hominids About 4 million to 6 million years ago, humanlike primates appeared that were different from the other primates. These ancestors, called **hominids,** ate both meat and plants and walked upright on two legs. Hominids shared some characteristics with gorillas, orangutans, and chimpanzees, but a larger brain separated them from the apes.

African Origins In the early 1920s, a fossil skull was discovered in a quarry in South Africa. The skull had a small space for the brain, but it had a humanlike jaw and teeth. The fossil, named *Australopithecus*, was one of the oldest hominids discovered. An almost-complete skeleton of *Australopithecus* was found in northern Africa in 1974. This hominid fossil, shown in **Figure 18,** was called Lucy and had a small brain but is thought to have walked upright. This fossil indicates that modern hominids might have evolved from similar ancestors.

Figure 18 The fossil remains of Lucy are estimated to be 2.9 million to 3.4 million years old.

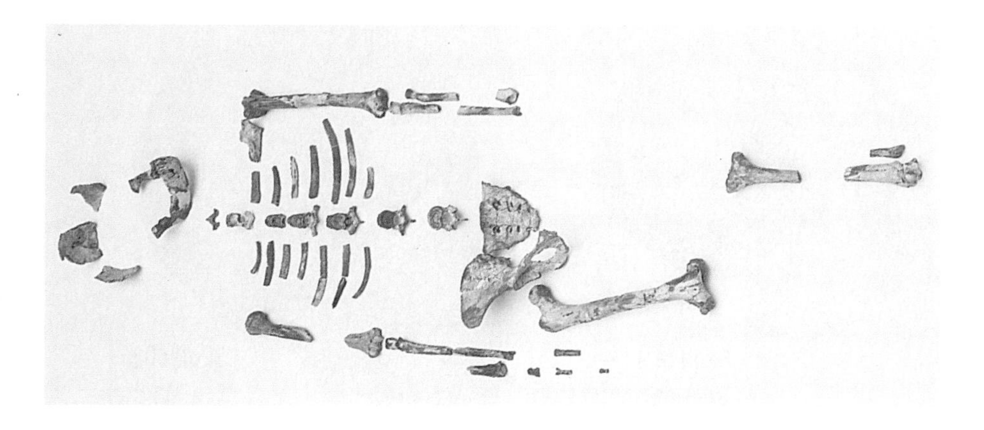

Living Without Thumbs

Procedure

1. Using **tape,** fasten down each of your thumbs next to the palm of each hand.
2. Leave your thumbs taped down for at least 1 h. During this time, do the following activities: eat a meal, change clothes, and brush your teeth. Be careful not to try anything that could be dangerous.
3. Untape your thumbs, then write about your experiences in your **Science Journal.**

Analysis

1. Did not having use of your thumbs significantly affect the way you did anything? Explain.
2. Infer how having opposable thumbs could have influenced primate evolution.

Try at Home

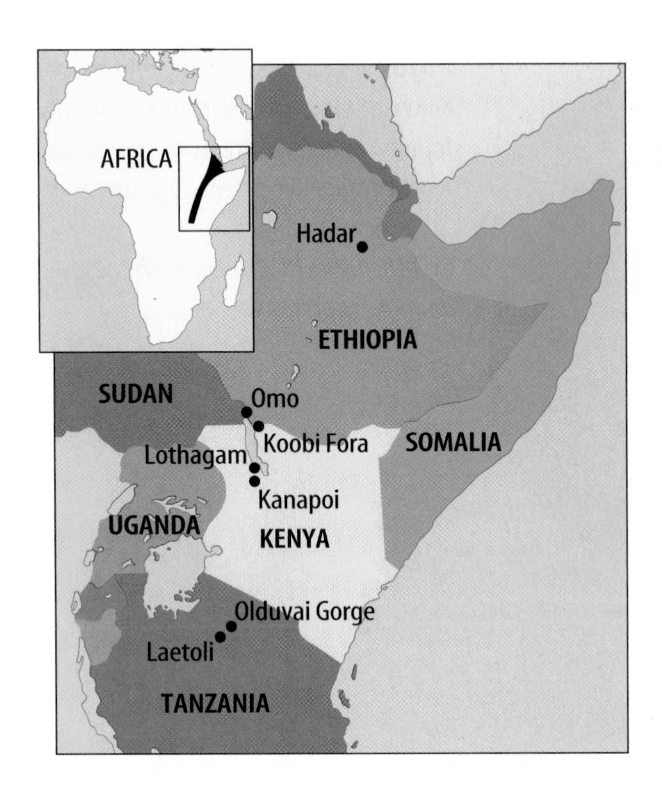

Figure 19 Many of the oldest humanlike skeletons have been found in this area of east Africa.

Early Humans In the 1960s in the region of Africa shown in **Figure 19,** a hominid fossil, which was more like present-day humans than *Australopithecus,* was discovered. The hominid was named *Homo habilis,* meaning "handy man," because simple stone tools were found near him. *Homo habilis* is estimated to be 1.5 million to 2 million years old. Based upon many fossil comparisons, scientists have suggested that *Homo habilis* gave rise to another species, *Homo erectus,* about 1.6 million years ago. This hominid had a larger brain than *Homo habilis. Homo erectus* traveled from Africa to Southeast Asia, China, and possibly Europe. *Homo habilis* and *Homo erectus* are thought to be ancestors of humans because they had larger brains and more humanlike features than *Australopithecus.*

Why was Homo habilis *given that name?*

Humans

The fossil record indicates that ***Homo sapiens*** evolved about 400,000 years ago. By about 125,000 years ago, two early human groups, Neanderthals (nee AN dur tawlz) and Cro-Magnon humans, as shown in **Figure 20,** probably lived at the same time in parts of Africa and Europe.

Neanderthals Short, heavy bodies with thick bones, small chins, and heavy browridges were physical characteristics of Neanderthals. Family groups lived in caves and used well-made stone tools to hunt large animals. Neanderthals disappeared from the fossil record about 30,000 years ago. They probably are not direct ancestors of modern humans, but represent a side branch of human evolution.

Figure 20 Compare the skull of a Neanderthal with the skull of a Cro-Magnon. **Describe** *what differences you can see between these two skulls.*

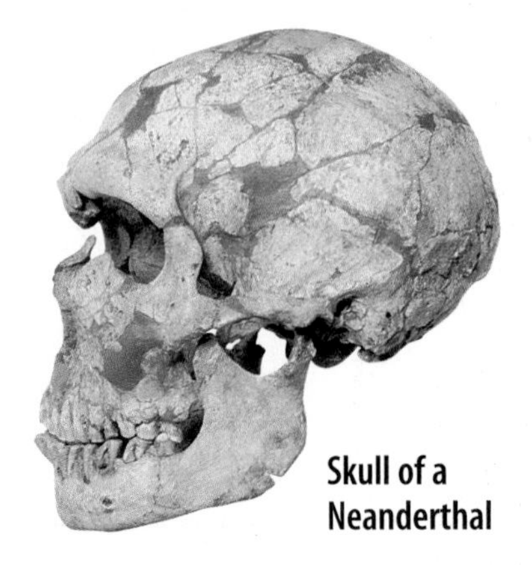

Skull of a Neanderthal

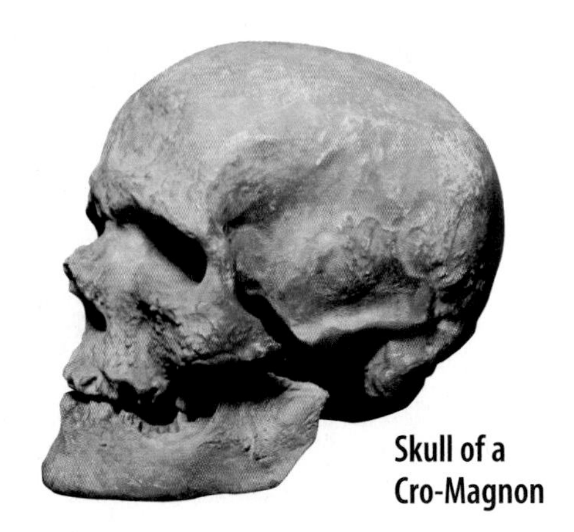

Skull of a Cro-Magnon

Figure 21 Paintings on cave walls have led scientists to hypothesize that Cro-Magnon humans had a well-developed culture.

Cro-Magnon Humans Cro-Magnon fossils have been found in Europe, Asia, and Australia and date from 10,000 to about 40,000 years in age. Standing about 1.6 m to 1.7 m tall, the physical appearance of Cro-Magnon people was almost the same as that of modern humans. They lived in caves, made stone carvings, and buried their dead. As shown in **Figure 21,** the oldest recorded art has been found on the walls of caves in France, where Cro-Magnon humans first painted bison, horses, and people carrying spears. Cro-Magnon humans are thought to be direct ancestors of early humans, *Homo sapiens,* which means "wise human." Evidence indicates that modern humans, *Homo sapiens sapiens,* evolved from *Homo sapiens.*

section 3 review

Summary

Primates

- Primates are an order of mammals characterized by opposable thumbs, binocular vision, and flexible shoulder joints.
- Primates are divided into strepsirrhines and haplorhines.
- Hominids are human ancestors that first appeared in Africa 4–6 million years ago.
- Hominids in the genus *Homo* first used tools and had larger brains than previous primates.

Humans

- *Homo sapiens* first appeared about 400,000 years ago.
- Cro-Magnon humans and Neanderthals coexisted in many places until Neanderthals disappeared about 30,000 years ago.
- *Homo sapiens* looked like modern humans and are believed to be our direct ancestors.

Self Check

1. **Describe** three kinds of evidence suggesting that all primates might have shared a common ancestor.
2. **Discuss** the importance of *Australopithecus.*
3. **Compare and contrast** Neanderthals, Cro-Magnon humans, and early humans.
4. **Identify** three groups most scientists consider to be direct ancestors of modern humans.
5. **Think Critically** Propose a hypothesis to explain why teeth are the most abundant fossil of hominids.

Applying Skills

6. **Concept Map** Make a concept map to show in what sequence hominids appeared. Use the following: *Homo sapiens sapiens,* Neanderthal, *Homo habilis, Australopithecus, Homo sapiens,* and Cro-Magnon human.
7. **Write** a story in your Science Journal about what life might have been like when both Neanderthals and Cro-Magnon humans were alive.

Design Your Own

Recognizing Variation in a Population

Goals

- **Design** an experiment that will allow you to collect data about variation in a population.
- **Observe, measure, and analyze** variations in a population.

Possible Materials

fruit and seeds from one plant species
metric ruler
magnifying lens
graph paper

Safety Precautions

WARNING: *Do not put any fruit or seeds in your mouth.*

Real-World Question

When you first observe a flock of pigeons, you might think all the birds look alike. However, if you look closer, you will notice minor differences, or variations, among the individuals. Different pigeons might have different color markings, or some might be smaller or larger than others. Individuals of the same species—whether they're birds, plants, or worms—might look alike at first, but some variations undoubtedly exist. According to the principles of natural selection, evolution could not occur without variations. What kinds of variations have you noticed among species of plants or animals? How can you measure variation in a plant or animal population?

Form a Hypothesis

Make a hypothesis about the amount of variation in the fruit and seeds of one species of plant.

Test Your Hypothesis

Make a Plan

1. As a group, agree upon and write out the prediction.
2. **List** the steps you need to take to test your prediction. Be specific. Describe exactly what you will do at each step. List your materials.
3. **Decide** what characteristic of fruit and seeds you will study. For example, you could measure the length of fruit and seeds or count the number of seeds per fruit.
4. **Design** a data table in your Science Journal to collect data about two variations. Use the table to record the data your group collects.
5. **Identify** any constants, variables, and controls of the experiment.
6. How many fruit and seeds will you examine? Will your data be more accurate if you examine larger numbers?
7. **Summarize** your data in a graph or chart.

Follow Your Plan

1. Make sure your teacher approves your plan before you start.
2. Carry out the experiment as planned.
3. While the experiment is going on, write down any observations you make and complete the data table in your Science Journal.

Analyze Your Data

1. **Calculate** the mean and range of variation in your experiment. The range is the difference between the largest and the smallest measurements. The mean is the sum of all the data divided by the sample size.
2. **Graph** your group's results by making a line graph for the variations you measured. Place the range of variation on the *x*-axis and the number of organisms that had that measurement on the *y*-axis.

Conclude and Apply

1. **Explain** your results in terms of natural selection.
2. **Discuss** the factors you used to determine the amount of variation present.
3. **Infer** why one or more of the variations you observed in this activity might be helpful to the survival of the individual.

Create a poster or other exhibit that illustrates the variations you and your classmates observed.

TIME SCIENCE AND HISTORY

SCIENCE CAN CHANGE THE COURSE OF HISTORY!

Fighting HIV

The first cases of AIDS, or acquired immune deficiency syndrome, in humans were reported in the early 1980s. AIDS is caused by the human immunodeficiency virus, or HIV.

A major problem in AIDS research is the rapid evolution of HIV. When HIV multiplies inside a host cell, new versions of the virus are produced as well as identical copies of the virus that invaded the cell. New versions of the virus soon can outnumber the original version. A treatment that works against today's HIV might not work against tomorrow's version.

These rapid changes in HIV also mean that different strains of the virus exist in different places around the world. Treatments developed in the United States work only for people who contracted the virus in the United States. This leaves people in some parts of the world without effective treatments. So, researchers such as geneticist Flossie Wong-Staal at the University of California in San Diego, must look for new ways to fight the evolving virus.

Working Backwards

Flossie Wong-Staal is taking a new approach. First, her team identifies the parts of a human cell that HIV depends on and the parts of the human cell that HIV needs but the human cell doesn't need. Then the team looks for a way to remove—or inactivate—those unneeded parts. This technique limits the virus's ability to multiply.

Wong-Staal's research combines three important aspects of science—a deep understanding of how cells and genes operate, great skill in the techniques of genetics, and great ideas. Understanding, skill, and great ideas are the best weapons so far in the fight to conquer HIV.

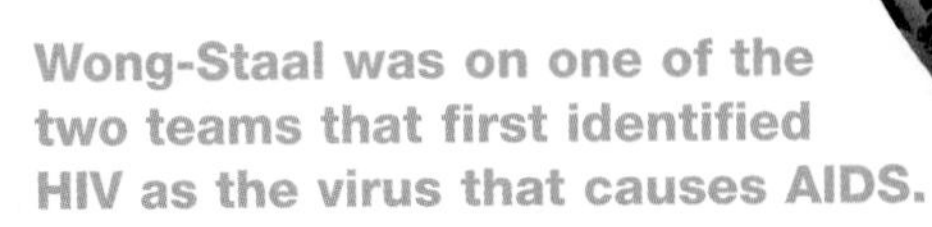

Wong-Staal was on one of the two teams that first identified HIV as the virus that causes AIDS.

Research Use the link to the right and other sources to determine which nations have the highest rates of HIV infection. Which nation has the highest rate? Where does the U.S. rank? Next, find data from ten years ago. Have the rankings changed?

For more information, visit green.msscience.com/time

Reviewing Main Ideas

Section 1 Ideas About Evolution

1. Evolution is one of the central ideas of biology. It explains how living things have changed in the past and is a basis for predicting how they might change in the future.
2. Charles Darwin developed the theory of evolution by natural selection to explain how evolutionary changes account for the diversity of organisms on Earth.
3. Natural selection includes concepts of variation, overproduction, and competition.
4. According to natural selection, organisms with traits best suited to their environment are more likely to survive and reproduce.

Section 2 Clues About Evolution

1. Fossils provide evidence for evolution.
2. Relative dating and radiometric dating can be used to estimate the age of fossils.
3. The evolution of antibiotic-resistant bacteria, pesticide-resistant insects, and rapid genetic changes in plant species provides direct evidence that evolution occurs.
4. Homologous structures, vestigial structures, comparative embryology, and similarities in DNA provide indirect evidence of evolution.

Section 3 The Evolution of Primates

1. Primates include monkeys, apes, and humans. Hominids are humanlike primates.
2. The earliest known hominid fossil is *Australopithecus*.
3. *Homo sapiens* are thought to have evolved from Cro-Magnon humans about 400,000 years ago.

Visualizing Main Ideas

Copy and complete the following spider map on evolution.

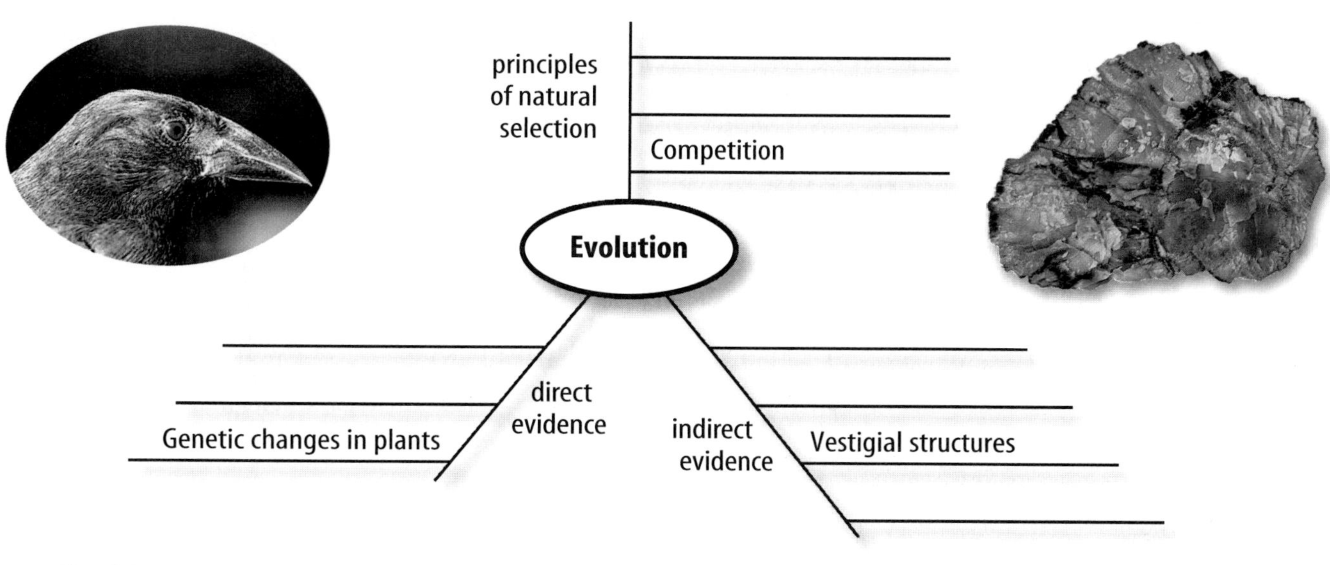

chapter 12 Review

Using Vocabulary

adaptation p. 338
embryology p. 347
evolution p. 334
gradualism p. 340
hominid p. 351
Homo sapiens p. 352
homologous p. 348
natural selection p. 337
primate p. 350
punctuated equilibrium p. 340
radioactive element p. 345
sedimentary rock p. 344
species p. 334
variation p. 338
vestigial structure p. 348

Fill in the blanks with the correct vocabulary word or words.

1. ________ contains many different kinds of fossils.
2. The muscles that move the human ear appear to be ________.
3. Forelimbs of bats, humans, and seals are ________.
4. Opposable thumbs are a characteristic of ________.
5. The study of ________ can provide evidence of evolution.
6. The principles of ________ include variation and competition.
7. ________ likely evolved directly from Cro-Magnons.

Checking Concepts

Choose the word or phrase that best answers the question.

8. What is an example of adaptation?
   A) a fossil
   B) gradualism
   C) camouflage
   D) embryo
9. What method provides the most accurate estimate of a fossil's age?
   A) natural selection
   B) radiometric dating
   C) relative dating
   D) camouflage
10. What do homologous structures, vestigial structures, and fossils provide evidence of?
    A) gradualism **C)** populations
    B) food choice **D)** evolution
11. Which model of evolution shows change over a relatively short period of time?
    A) embryology
    B) adaptation
    C) gradualism
    D) punctuated equilibrium
12. What might a series of helpful variations in a species result in?
    A) adaptation **C)** embryology
    B) fossils **D)** climate change

Use the following chart to answer question 13.

Homo habilis
↓
Homo erectus
↓
?
____
↓
Homo sapiens

13. Which of the following correctly fills the gap in the line of descent from *Homo habilis?*
    A) Neanderthal
    B) *Australopithecus*
    C) Cro-Magnon human
    D) chimpanzee
14. What is the study of an organism's early development called?
    A) adaptation **C)** natural selection
    B) relative dating **D)** embryology

Science Online green.msscience.com/vocabulary_puzzlemaker

Thinking Critically

15. **Predict** what type of bird the foot pictured at right would belong to. Explain your reasoning.

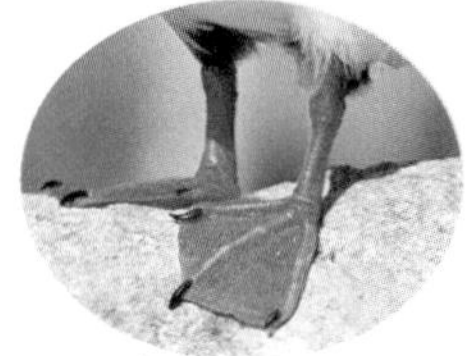

16. **Discuss** how Lamarck and Darwin would have explained the large eyes of an owl.
17. **Explain,** using an example, how a new species of organism could evolve.
18. **Identify** how the color-changing ability of chameleons is an adaptation.
19. **Form a hypothesis** as to why ponds are not overpopulated by frogs in summer. Use the concept of natural selection to help you.
20. **Sequence** Make an events-chain concept map of the events that led Charles Darwin to his theory of evolution by natural selection.

Use the table below to answer question 21.

Chemicals Present in Bacteria	
Species 1	A, G, T, C, L, E, S, H
Species 2	A, G, T, C, L, D, H
Species 3	A,G, T, C, L, D, P, U, S, R, I, V
Species 4	A, G, T, C, L, D, H

21. **Interpret Data** Each letter above represents a chemical found in a species of bacteria. Which species are most closely related?
22. **Discuss** the evidence you would use to determine whether the evolution of a group were best explained by gradualism. How would this differ from a group that followed a punctuated equilibrium model?
23. **Describe** the processes a scientist would use to figure out the age of a fossil.
24. **Evaluate** the possibility for each of the five types of fossils in **Figure 9** to yield a DNA sample. Remember that only biological tissue will contain DNA.

Performance Activities

25. **Collection** With permission, collect fossils from your area and identify them. Show your collection to your class.
26. **Brochure** Assume that you are head of an advertising company. Develop a brochure to explain Darwin's theory of evolution by natural selection.

Applying Math

27. **Relative Age** The rate of radioactive decay is measured in half-lives—the amount of time it takes for one half of a radioactive element to decay. Determine the relative age of a fossil given the following information:
    - Rock layers are undisturbed.
    - The layer below the fossil has potassium-40 with a half-life of 1 million years and only one half of the original potassium is left.
    - The layer above the fossil has carbon-14 with a half-life of 5,730 years and one-sixteenth of the carbon isotope remains.

Use the graph below to answer question 28.

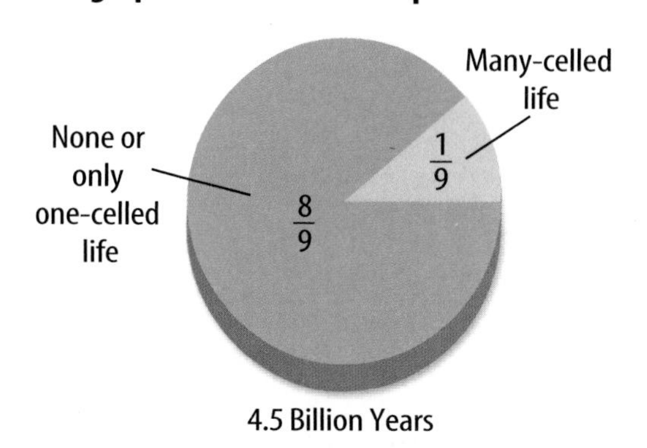

28. **First Appearances** If Earth is 4.5 billion years old, how long ago did the first many-celled life-forms appear?

Standardized Test Practice

Part 1 Multiple Choice

Record your answers on the answer sheet provided by your teacher or on a sheet of paper.

1. A species is a group of organisms
   A. that lives together with similar characteristics.
   B. that shares similar characteristics and can reproduce among themselves to produce fertile offspring.
   C. across a wide area that cannot reproduce.
   D. that chooses mates from among themselves.

2. Which of the following is considered an important factor in natural selection?
   A. limited reproduction
   B. competition for resources
   C. no variations within a population
   D. plentiful food and other resources

3. The marine iguana of the Galápagos Islands enters the ocean and feeds on seaweed. What is this an example of?
   A. adaptation
   B. gradualism
   C. survival of the fittest
   D. acquired characteristic

Use the illustration below to answer question 4.

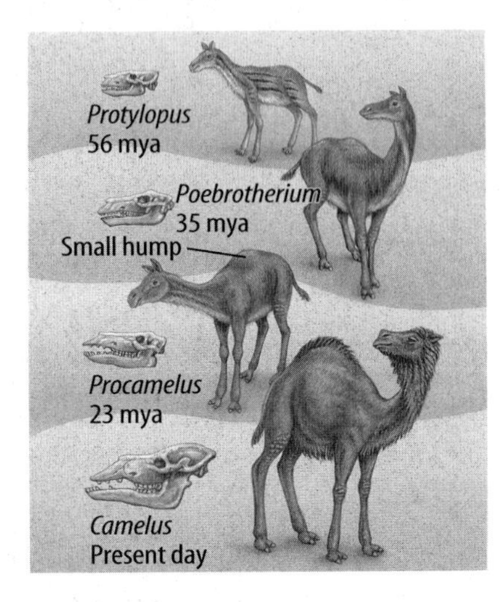

4. According to Lamarck's hypothesis of acquired characteristics, which statement best explains the changes in the camel over time?
   A. All characteristics developed during an individual's lifetime are passed on to offspring.
   B. Characteristics that do not help the animal survive are passed to offspring.
   C. Variation of the species leads to adaptation.
   D. Individuals moving from one area to another carry with them new characteristics.

Use the illustrations below to answer question 5.

5. What, besides competition for food, contributed to the evolution of the species of Darwin's finches?
   A. predation
   B. natural disaster
   C. DNA
   D. variation in beak shapes

6. Some harmless species imitate or mimic a poisonous species as a means for increased survival. What is this an example of?
   A. acquired characteristics
   B. adaptation
   C. variation
   D. geographic isolation

Part 2 Short Response/Grid In

Record your answers on the answer sheet provided by your teacher or on a sheet of paper.

7. How does camouflage benefit a species?

Use the photo below to answer question 8.

8. Describe an environment where the albino lemur would not be at a disadvantage.
9. Variation between members of a species plays an important role in Darwin's theory of evolution. What happens to variation in endangered species where the number of individuals is very low?
10. Describe what happens to an endangered species if a variation provides an advantage for the species. What would happen if the variation resulted in a disadvantage?
11. Using the theory of natural selection, hypothesize why the Cro-Magnon humans survived and the Neanderthals disappeared.

Test-Taking Tip

Never Leave Any Answer Blank Answer each question as best you can. You can receive partial credit for partially correct answers.

Question 16 If you cannot remember all primate characteristics, list as many as you can.

Part 3 Open Ended

Record your answers on a sheet of paper.

12. What are the two groups of early humans that lived about 125,000 years ago in Africa and Europe? Describe their general appearance and characteristics. Compare these characteristics to modern humans.
13. Explain how bacterial resistance to antibiotics is an example of punctuated equilibrium.
14. Why are radioactive elements useful in dating fossils? Does this method improve accuracy over relative dating?

Use the illustrations below to answer question 15.

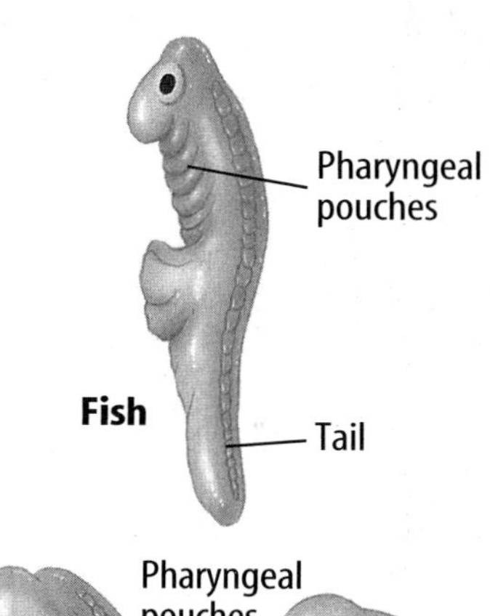

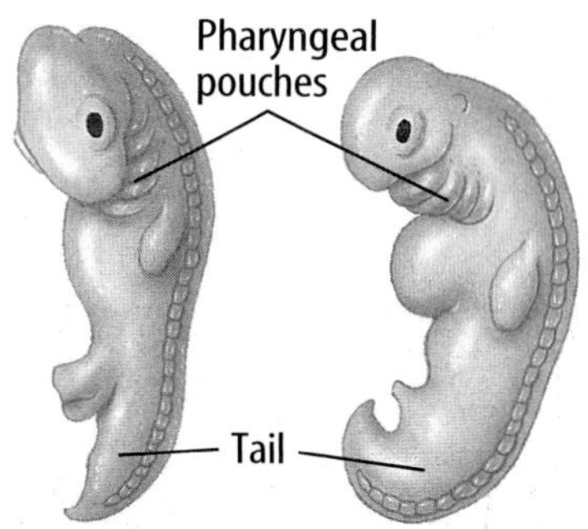

15. Why would scientists study embryos? What features of these three embryos support evolution?
16. How does DNA evidence provide support that primates have a common ancestor?

unit 4 Human Body Systems

How are Chickens & Rice Connected?

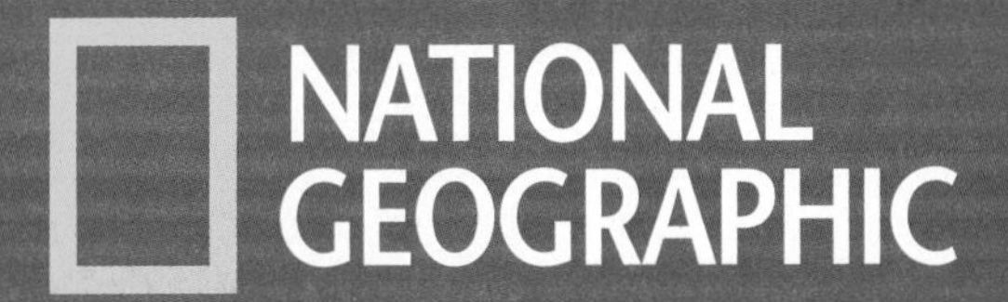

Back in the 1800s, a mysterious disease called beriberi affected people in certain parts of Asia. One day, a doctor in Indonesia noticed some chickens staggering around, a symptom often seen in people with beriberi. It turned out that the chickens had been eating white rice—the same kind of rice that was being eaten by human beriberi sufferers. White rice has had the outer layers, including the bran, removed. When the sick chickens were fed rice that still had its bran, they quickly recovered. It turned out that the same treatment worked for people with beriberi! Research eventually showed that rice bran contains a vitamin, B_1, which is essential for good health. Today, white rice usually is "vitamin-enriched" to replace B_1 and other nutrients lost in processing.

unit projects

Visit green.msscience.com/unit_project to find project ideas and resources.
Projects include:

- **History** Contribute to a class "remedy journal" with interesting, out-dated medical treatments, and how techniques have improved.
- **Technology** Investigate rare and interesting medical conditions, including their history, characteristics, and treatments. Present a colorful poster with photos and information for class display.
- **Model** Research and create a menu that includes vitamin-rich foods. Prepare a sample and a recipe card for a class food fair.

Understand the *History of Disease Prevention,* and how science has progressed through history. Become acquainted with famous scientists and learn how healthy lifestyles prevent disease.

chapter 13

Circulation and Immunity

chapter preview

sections

1 Blood
2 Circulation
3 Immunity
4 Diseases
Lab *Microorganisms and Disease*
Lab *Blood Type Reactions*

Virtual Lab *How does the body protect itself against foreign substances?*

The Flow of Traffic

This highway interchange is simple compared to how blood travels within your body. In this chapter, you will discover how complex your circulatory system is—from parts of your blood to how it travels through your body and fights disease.

Science Journal Write three questions that you have about blood, circulation, or how diseases are spread.

Start-Up Activities

Transportation by Road and Vessel

Your circulatory system is like a road system. Just as roads are used to transport goods to homes and factories, your blood vessels transport substances throughout your body. You'll find out how similar roads and blood vessels are in this lab.

1. Observe a map of your city, county, or state.
2. Identify roads that are interstates, as well as state and county roads, using the map key.
3. Plan a route to a destination that your teacher describes. Then plan a different return trip.
4. Draw a diagram in your Science Journal showing your routes to and from the destination.
5. **Think Critically** If the destination represents your heart, what do the routes represent? In your Science Journal, draw a comparison between a blocked road on your map and a clogged artery in your body.

Circulation Make the following Foldable to help you organize information and diagram ideas about circulation.

STEP 1 **Fold** a sheet of paper in half lengthwise. Make the back edge about 5 cm longer than the front edge.

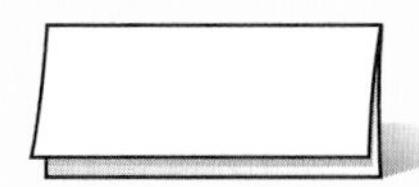

STEP 2 **Turn** the paper so the fold is on the bottom. Then **fold** it into thirds.

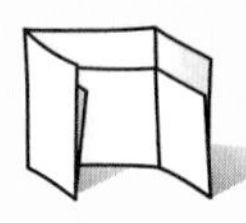

STEP 3 **Unfold and cut** only the top layer along both folds to make three tabs.

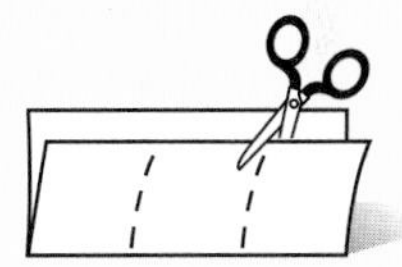

STEP 4 **Label** the Foldable as shown.

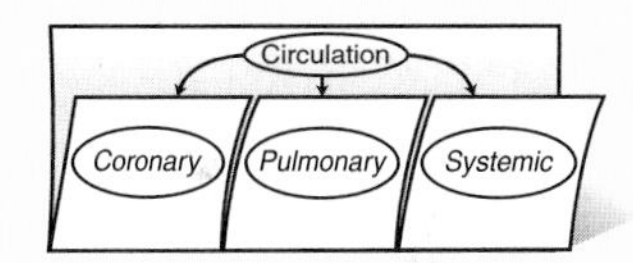

Read and Write As you read the chapter, write information about each circulatory system under the appropriate tab.

Preview this chapter's content and activities at green.msscience.com

section 1

Blood

as you read

What You'll Learn

- **Identify** the parts and functions of blood.
- **Explain** why blood types are checked before a transfusion.
- **Give** examples of diseases of blood.

Why It's Important

Blood plays a part in every major activity of your body.

Review Vocabulary
diffusion: a type of passive transport within cells in which molecules move from areas where there are more of them to areas where there are fewer of them

New Vocabulary
- plasma
- platelet
- hemoglobin

Functions of Blood

You take a last, deep, calming breath before plunging into a dark, vessel-like tube. Water is everywhere. You take a hard right turn, then left as you streak through a narrow tunnel of twists and turns. The water transports you down the slide much like the way blood carries substances to all parts of your body. Blood has four important functions.

1. Blood carries oxygen from your lungs to all your body cells. Carbon dioxide diffuses from your body cells into your blood. Your blood carries carbon dioxide to your lungs to be exhaled.
2. Blood carries waste products from your cells to your kidneys to be removed.
3. Blood transports nutrients and other substances to your body cells.
4. Cells and molecules in blood fight infections and help heal wounds.

Anything that disrupts or changes these functions affects all the tissues of your body. Can you understand why blood is sometimes called the tissue of life?

Parts of Blood

A close look at blood tells you that blood is not just a red-colored liquid. Blood is a tissue made of plasma (PLAZ muh), red and white blood cells, and platelets (PLAYT luts), shown in **Figure 1.** Blood makes up about eight percent of your body's total mass. If you weigh 45 kg, you have about 3.6 kg of blood moving through your body.

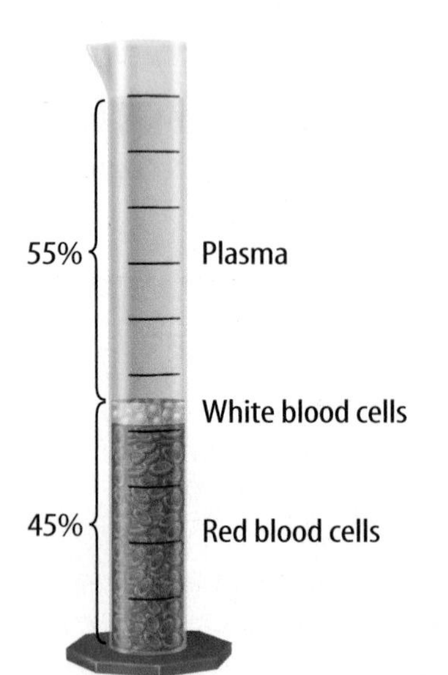

Figure 1 The blood in this graduated cylinder has separated into its parts. Each part plays a key role in body functions.

Plasma The liquid part of blood, which is made mostly of water, is called **plasma.** It makes up more than half the volume of blood. Nutrients, minerals, and oxygen are dissolved in plasma so that they can be carried to body cells. Wastes from body cells also are carried in plasma.

Blood Cells Disk-shaped red blood cells, shown in **Figure 2,** are different from other cells in your body because they have no nuclei when they mature. They contain **hemoglobin** (HEE muh gloh bun), which is a molecule that carries oxygen and carbon dioxide. Hemoglobin carries oxygen from your lungs to your body cells. Then it carries some of the carbon dioxide from your body cells back to your lungs. The rest of the carbon dioxide is carried in the cytoplasm of red blood cells and in plasma.

Red blood cells have a life span of about 120 days. They are made at a rate of 2 million to 3 million per second in the center of long bones, like the femur in your thigh. Red blood cells wear out and are destroyed at about the same rate.

A cubic millimeter of blood, about the size of a grain of rice, has about 5 million red blood cells. In contrast, a cubic millimeter of blood has about 5,000 to 10,000 white blood cells. White blood cells fight bacteria, viruses, and other invaders of your body. Your body reacts to invaders by increasing the number of white blood cells. These cells leave the blood through capillary walls and go into the tissues that have been invaded. Here, they destroy bacteria and viruses and absorb dead cells. The life span of white blood cells varies from a few days to many months.

Circulating with the red and white blood cells are platelets. **Platelets** are irregularly shaped cell fragments that help clot blood. A cubic millimeter of blood can contain as many as 400,000 platelets. Platelets have a life span of five to nine days.

Topic: Human White Blood Cells

Visit green.msscience.com for Web links to information about the types of human white blood cells and their functions.

Activity Make a table showing the functions of the various types of white blood cells.

Figure 2 Red blood cells supply your body with oxygen, and white blood cells and platelets have protective roles.

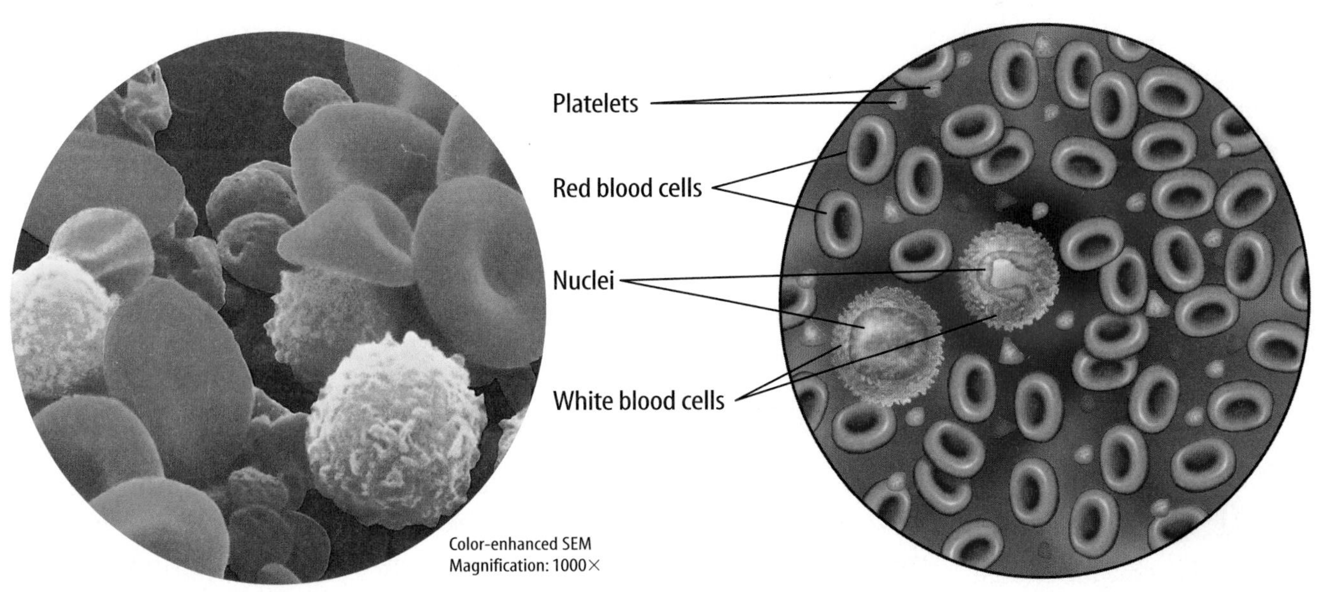

Platelets help stop bleeding. Platelets not only plug holes in small vessels, they also release chemicals that help form filaments of fibrin.

Several types, sizes, and shapes of white blood cells exist. These cells destroy bacteria, viruses, and foreign substances.

Figure 3 When the skin is damaged, a sticky blood clot seals the leaking blood vessel. Eventually, a scab forms to protect the wound from further damage and allow it to heal.

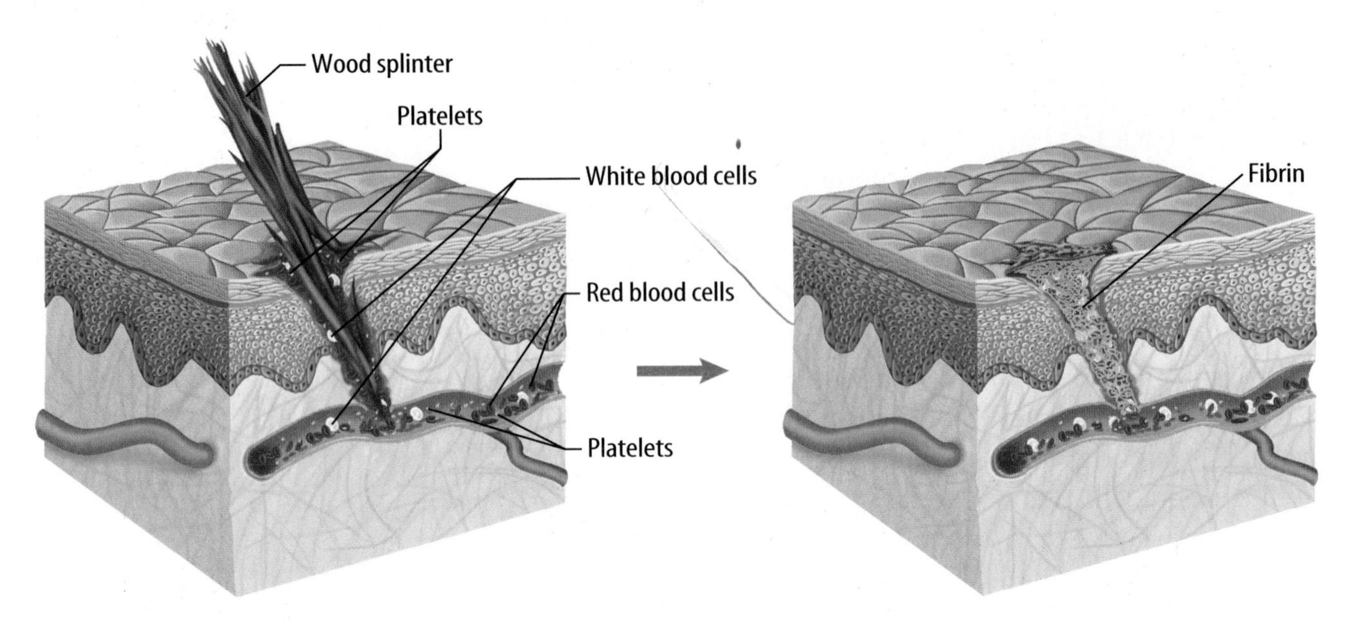

Modeling Scab Formation

Procedure

1. Place a 5-cm × 5-cm square of **gauze** on a piece of **aluminum foil.**
2. Place several drops of a **liquid bandage solution** onto the gauze and let it dry. Keep the liquid bandage away from eyes and mouth.
3. Use a **dropper** to place one drop of **water** onto the area of the liquid bandage. Place another drop of water in another area of the gauze.

Analysis

1. Compare the drops of water in both areas.
2. Describe how the treated area of the gauze is like a scab.

Blood Clotting

You're running with your dog in a park, when suddenly you trip and fall down. Your knee starts to bleed, but the bleeding stops quickly. Already the wounded area has begun to heal. Bleeding stops because platelets and clotting factors in your blood make a blood clot that plugs the wounded blood vessels.

A blood clot also acts somewhat like a bandage. When you cut yourself, platelets stick to the wound and release chemicals. Then substances, called clotting factors, carry out a series of chemical reactions. These reactions cause threadlike fibers called fibrin (FI brun) to form a sticky net, as shown in **Figure 3.** This net traps escaping blood cells and plasma and forms a clot. The clot helps stop more blood from escaping. After the clot is in place and becomes hard, skin cells begin the repair process under the scab. Eventually, the scab is lifted off. Bacteria that get into the wound during the healing process usually are destroyed by white blood cells.

Reading Check *What blood components help form blood clots?*

Most people will not bleed to death from a minor wound, such as a cut or scrape. However, some people have a genetic condition called hemophilia (hee muh FIH lee uh). Their plasma lacks one of the clotting factors that begins the clotting process. A minor injury can be a life-threatening problem for a person with hemophilia.

Blood Types

Blood clots stop blood loss quickly in a minor wound, but with a serious wound a person might lose a lot of blood. A blood transfusion might be necessary. During a blood transfusion, a person receives donated blood or parts of blood. The medical provider must be sure that the right type of blood is given. If the wrong type is given, the red blood cells will clump together. Then, clots form in the blood vessels and the person could die.

Blood Transfusions In 1665, the first successful blood transfusion was performed between two dogs. The first successful human-to-human blood transfusion was performed in 1818. However, many failures followed. The different blood types and the problems that result when they are mixed were unknown at that time. Research the discovery of the four types of blood and write a summary in your Science Journal.

The ABO Identification System People can inherit one of four types of blood: A, B, AB, or O. Types A, B, and AB have chemical identification tags called antigens (AN tih junz) on their red blood cells. Type O red blood cells have no antigens.

Each blood type also has specific antibodies in its plasma. Antibodies are proteins that destroy or neutralize substances that do not belong in or are not part of your body. Because of these antibodies, certain blood types cannot be mixed. This limits blood transfusion possibilities, as shown in **Table 1.** If type A blood is mixed with type B blood, the antibodies in type A blood determine that type B blood does not belong there. The antibodies in type A blood cause the type B red blood cells to clump. In the same way, type B blood antibodies cause type A blood to clump. Type AB blood has no antibodies, so people with this blood type can receive blood from A, B, AB, and O types. Type O blood has both A and B antibodies.

Reading Check *Why are people with type O blood called universal donors?*

The Rh Factor Another inherited chemical identification tag in blood is the Rh factor. If the Rh factor is on red blood cells, the person has Rh-positive (Rh+) blood. If it is not present, the person has Rh-negative (Rh−) blood. If an Rh− person receives a blood transfusion from an Rh+ person, he or she will produce antibodies against the Rh factor. These antibodies can cause Rh+ cells to clump. Clots then form in the blood vessels and the person could die. In the same way, an Rh− mother can make antibodies against her Rh+ baby during pregnancy. If the antibodies pass into the baby's blood, they can destroy the baby's red blood cells. To prevent deadly results, blood groups and Rh factor are checked before transfusions and during pregnancies.

Table 1 Blood Transfusion Possibilities

Type	Can Receive	Can Donate To
A	0, A	A, AB
B	0, B	B, AB
AB	all	AB
0	0	all

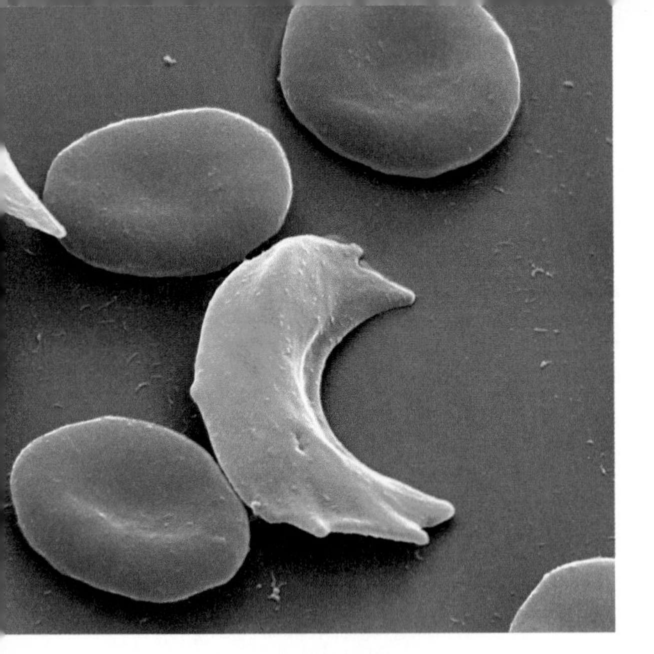

Figure 4 Persons with sickle-cell disease have misshapened red blood cells. The sickle-shaped cells clog the capillaries of a person with this disease. Oxygen cannot reach tissues served by the capillaries, and wastes cannot be removed. **Explain** *how this damages the affected tissues.*

Diseases of Blood

Because blood circulates to all parts of your body and performs so many important functions, any disease of the blood is a cause for concern. One common disease of the blood is anemia (uh NEE mee uh). In this disease of red blood cells, body tissues can't get enough oxygen and are unable to carry on their usual activities. Anemia has many causes. Sometimes, anemia is caused by the loss of large amounts of blood. A diet lacking iron or certain vitamins also might cause anemia. Still other types of anemia are inherited problems related to the structure of the red blood cells. Cells from one such type of anemia, sickle-cell disease, are shown in **Figure 4.**

Leukemia (lew KEE mee uh) is a disease in which one or more types of white blood cells are made in excessive numbers. These cells are immature and do not fight infections well. These immature cells fill the bone marrow and crowd out the normal, mature cells. Then not enough red blood cells, normal white blood cells, and platelets can be made. Some types of leukemia affect children. Other kinds are more common in adults. Medicines, blood transfusions, and bone marrow transplants are used to treat this disease. If the treatments are not successful, the person will eventually die from related complications.

section 1 review

Summary

Functions and Parts of Blood

- Blood carries oxygen, carbon dioxide, wastes, and nutrients.
- Blood contains cells that help fight infections and heal wounds.
- Blood is a tissue made of plasma, red and white blood cells, and platelets.

Blood Clotting and Blood Types

- Platelets and clotting factors form blood clots to stop bleeding from a wound.
- Blood type—A, B, AB, or O—must be identified before a person receives a transfusion.

Diseases of Blood

- Anemia affects red blood cells, while leukemia affects white blood cells.

Self Check

1. **List** the four functions of blood in the body.
2. **Compare and contrast** red blood cells, white blood cells, and platelets.
3. **Describe** how anemia and leukemia affect the blood.
4. **Explain** why blood type and Rh factor are checked before a transfusion.
5. **Think Critically** Think about the main job of your red blood cells. If red blood cells couldn't deliver oxygen to your cells, what would be the condition of your body tissues?

Applying Skills

6. **Interpret Data** Look at the data in **Table 1** about blood group interactions. To which group(s) can people with blood type AB donate blood?

green.msscience.com/self_check_quiz

section 2

Circulation

The Body's Delivery System

It's time to get ready for school, but your younger sister is taking a long time in the shower. "Don't use up all the water," you shout. Water is carried throughout your house in pipes that are part of the plumbing system. The plumbing system supplies water for your needs and carries away wastes. Just as you expect water to flow when you turn on the faucet, your body needs a continuous supply of oxygen and nutrients and a way to remove wastes. In a similar way, materials are moved throughout your body by your cardiovascular (kar dee oh VAS kyuh lur) system. It includes your heart, kilometers of blood vessels, and blood. Blood vessels carry the blood to every part of your body, as shown in **Figure 5.** Recall that blood moves oxygen and nutrients to cells and carries carbon dioxide and other wastes away from the cells.

as you read

What You'll Learn

- **Compare and contrast** arteries, veins, and capillaries.
- **Explain** how blood moves through the heart.
- **Identify** the functions of the pulmonary and systemic circulation systems.
- **Describe** functions of the lymphatic system.

Why It's Important

Your body's cells depend on blood vessels to deliver nutrients and remove wastes. The lymphatic system helps protect you from infections and disease.

Review Vocabulary

tissue: group of similar cells that work together to do one job

New Vocabulary

- capillary
- artery
- vein
- lymph

The Heart

Your heart is an organ made of cardiac muscle tissue. It is located behind your breastbone, called the sternum, and between your lungs. Your heart has four compartments called chambers. The two upper chambers are called the right and left atriums (AY tree umz). The two lower chambers are called the right and left ventricles (VEN trih kulz). A one-way valve separates each atrium from the ventricle below it. The blood flows from an atrium to a ventricle, then from a ventricle into a blood vessel. A wall between the two atriums or the two ventricles keeps blood rich in oxygen separate from blood low in oxygen.

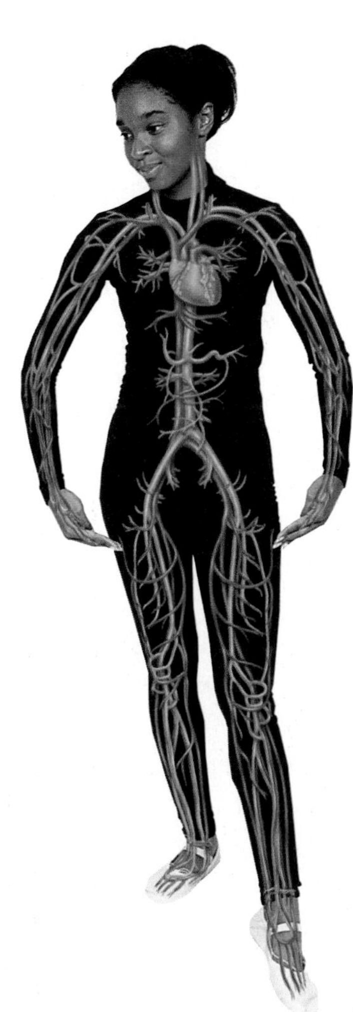

Figure 5 The blood is pumped by the heart to all the cells of the body and then back to the heart through a network of blood vessels.

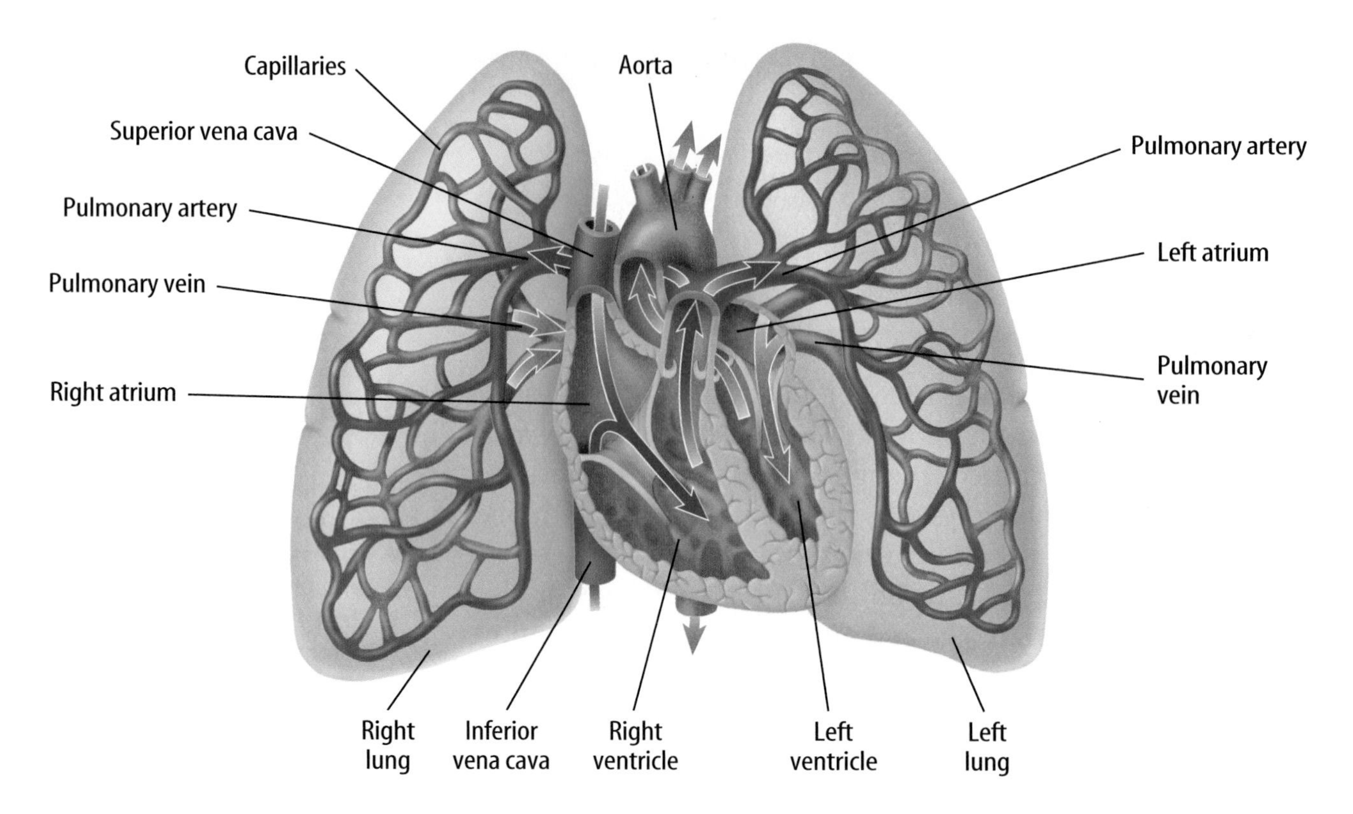

Figure 6 Pulmonary circulation moves blood between the heart and lungs.

Types of Circulation

Scientists have divided the circulatory system into three sections—coronary (KOR uh ner ee) circulation, pulmonary (PUL muh ner ee) circulation, and systemic circulation. The beating of your heart controls blood flow through each section.

Coronary Circulation Your heart has its own blood vessels that supply it with nutrients and oxygen and remove wastes. Coronary circulation is the flow of blood to and from the tissues of the heart. When the coronary circulation is blocked, oxygen and nutrients cannot reach all the cells of the heart. This can result in a heart attack.

Pulmonary Circulation The flow of blood through the heart to the lungs and back to the heart is called pulmonary circulation. Use **Figure 6** to trace the path blood takes through this part of the circulatory system. The blood returning from the body through the right side of the heart and to the lungs contains cellular wastes. The wastes include molecules of carbon dioxide and other substances. In the lungs, gaseous wastes diffuse out of the blood, and oxygen diffuses into the blood. Then the blood returns to the left side of the heart. In the final step of pulmonary circulation, the oxygen-rich blood is pumped from the left ventricle into the aorta (ay OR tuh), the largest artery in your body. From there, the oxygen-rich blood flows to all parts of your body.

Systemic Circulation Oxygen-rich blood moves to all of your organs and body tissues, except the heart and lungs, and oxygen-poor blood returns to the heart by a process called systemic circulation. Systemic circulation is the largest of the three sections of your circulatory system. Oxygen-rich blood flows from your heart in the arteries of this system. Then nutrients and oxygen are delivered by blood to your body cells and exchanged for carbon dioxide and wastes. Finally, the blood returns to your heart in the veins of the systemic circulation system.

Blood Vessels

In the middle 1600s, scientists discovered that blood moves by the pumping of the heart and flows in one direction from arteries to veins. But they couldn't explain how blood gets from arteries to veins. Using a new invention of that time, the microscope, scientists discovered **capillaries** (KA puh ler eez), the blood vessels that connect arteries and veins.

Arteries As blood is pumped out of the heart, it travels through arteries, capillaries, and then veins, shown in **Figure 7.** **Arteries** are blood vessels that carry blood away from the heart. Arteries have thick, elastic walls made of connective tissue and smooth muscle tissue.

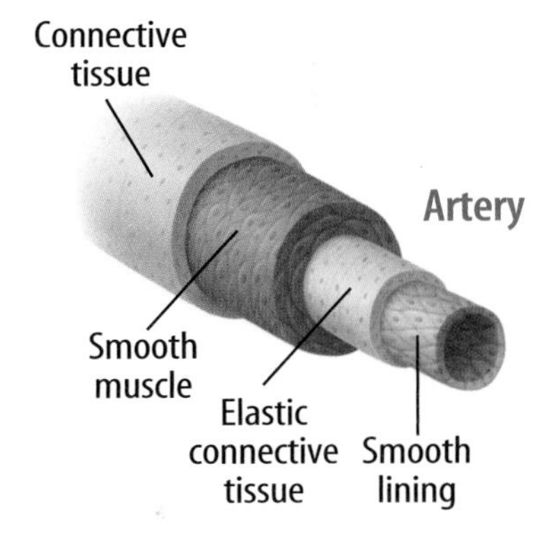

Veins The blood vessels that carry blood back to the heart are called **veins.** Veins have one-way valves that keep blood moving toward the heart. If blood flows backward, the pressure of the blood against the valves causes them to close. Blood flow in veins also is helped by your skeletal muscles. When skeletal muscles contract, this action squeezes veins and helps blood move toward the heart.

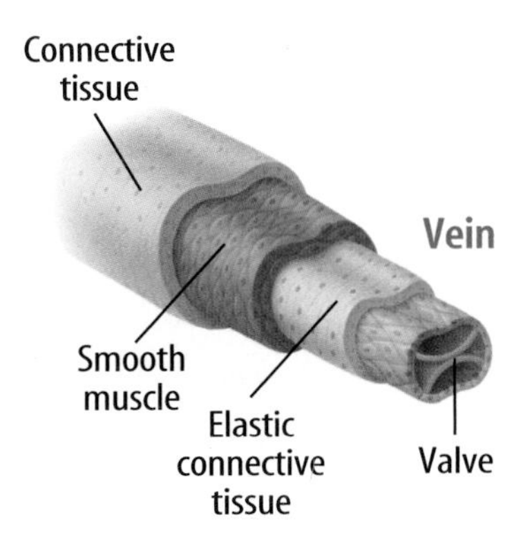

Reading Check *What are the similarities and differences between arteries and veins?*

Capillaries The walls of capillaries are only one cell thick. Nutrients and oxygen diffuse into body cells from capillaries. Waste materials and carbon dioxide diffuse from body cells into the capillaries.

Capillary

Figure 7 The structures of arteries, veins, and capillaries are different. Valves in veins prevent blood from flowing backward. Capillaries are much smaller. Capillary walls are only one cell thick.

Blood Pressure

INTEGRATE Physics

If you fill a balloon with water and then push on it, the pressure moves through the water in all directions, as shown in **Figure 8.** Your circulatory system is like the water balloon. When your heart pumps blood through the circulatory system, the pressure of the push moves through the blood. The force of the blood on the walls of the blood vessels is called blood pressure. This pressure is highest in arteries and lowest in veins. When you take your pulse, you can feel the waves of pressure. This rise and fall of pressure occurs with each heartbeat.

Controlling Blood Pressure Special nerve cells in the walls of some arteries sense changes in blood pressure. When pressure is higher or lower than normal, messages are sent to your brain. Then the brain sends messages that speed up or slow the heart rate. This helps keep blood pressure constant within your arteries so that enough blood reaches all organs and tissues in your body and delivers needed nutrients to every cell.

Cardiovascular Disease

Any disease that affects the cardiovascular system—the heart, blood vessels, and blood—can seriously affect the health of your entire body. Heart disease is the leading cause of death in humans.

Atherosclerosis One leading cause of heart disease is called atherosclerosis (ah thur oh skluh ROH sus). In this condition, fatty deposits build up on arterial walls. Atherosclerosis can occur in any artery in the body, but fatty deposits in coronary arteries are especially serious. If a coronary artery is blocked, a heart attack can occur. Open-heart surgery then may be needed to correct the problem.

Figure 8 When pressure is exerted on a fluid in a closed container, the pressure is transmitted through the liquid in all directions. Your circulatory system is like a closed container. Blood pressure is measured using a blood pressure cuff and a stethoscope.

Hypertension Another condition of the cardiovascular system is called hypertension (hi pur TEN chun), or high blood pressure. When blood pressure is higher than normal most of the time, the heart must work harder to keep blood flowing. One cause of hypertension is atherosclerosis. A clogged artery can increase pressure within the vessel, causing the walls to become stiff and hard. The artery walls no longer contract and dilate easily because they have lost their elasticity.

Preventing Cardiovascular Disease Having a healthy lifestyle is important for the health of your cardiovascular system. The choices you make now to maintain good health may reduce your risk of future serious illness. Regular checkups, a healthful diet, and exercise are all part of a heart-healthy lifestyle.

Another way to prevent cardiovascular disease is to not smoke. Smoking causes blood vessels to contract and makes the heart beat faster and harder. Smoking also increases carbon monoxide levels in the blood. Not smoking helps prevent heart disease and a number of respiratory system problems.

Functions of the Lymphatic System

You turn on the water faucet and fill a glass with water. The excess water runs down the drain. In a similar way, your body's tissue fluid is removed by the lymphatic (lihm FA tihk) system, shown in **Figure 9.** The nutrient, water, and oxygen molecules in blood diffuse through capillary walls to nearby cells. Water and other substances become part of the tissue fluid that is found between cells. This fluid is collected and returned to the blood by the lymphatic system.

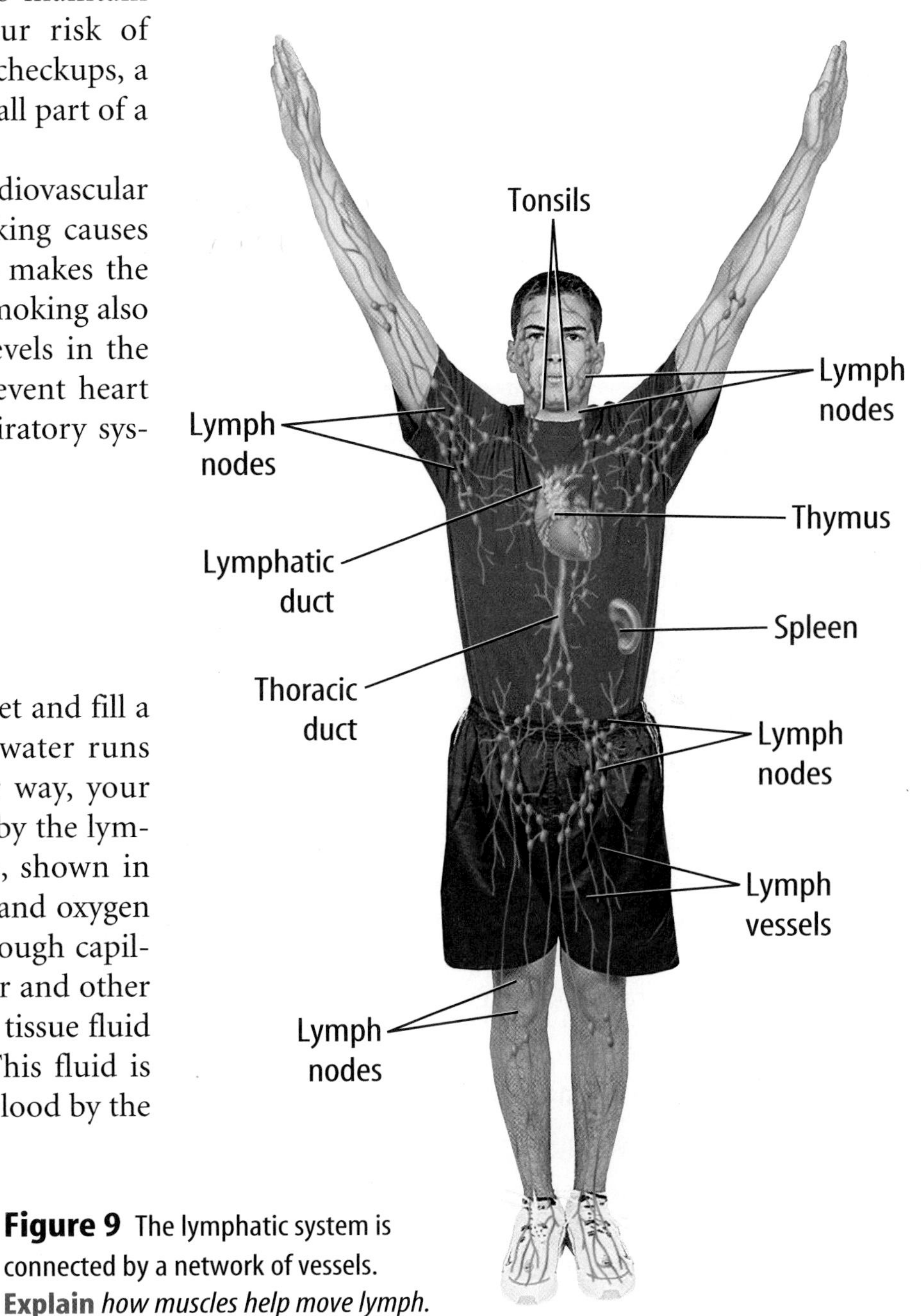

Figure 9 The lymphatic system is connected by a network of vessels.
Explain *how muscles help move lymph.*

Topic: Hodgkin's Disease
Visit green.msscience.com for Web links to information about Hodgkin's Disease.

Activity Create a brochure about Hodgkin's Disease, including what it is, its symptoms, risk factors, and treatment.

Lymph After tissue fluid diffuses into the lymphatic capillaries, it is called **lymph** (LIHMF). In addition to water and dissolved substances, lymph contains lymphocytes (LIHM fuh sites), a type of white blood cell. Lymphocytes help your body defend itself against disease-causing organisms. If the lymphatic system is not working properly, severe swelling occurs because the tissue fluid cannot get back to the blood.

Reading Check *What is lymph?*

Your lymphatic system carries lymph through a network of lymph capillaries and larger lymph vessels. Then, the lymph passes through lymph nodes, which are bean-shaped organs found throughout the body. Lymph nodes filter out microorganisms and foreign materials that have been taken up by lymphocytes. After it is filtered, lymph enters the bloodstream through large veins near the neck. No heartlike structure pumps the lymph through the lymphatic system. The movement of lymph depends on the contraction of smooth muscles in lymph vessels and skeletal muscles. Lymphatic vessels, like veins, have valves that keep lymph from flowing backward.

section 2 review

Summary

The Body's Delivery System

- Blood vessels carry blood to the body.

The Heart and Types of Circulation

- Your heart controls blood flow through the circulatory system.
- In the lungs, carbon dioxide leaves the blood and oxygen diffuses into the blood.

Blood Vessels and Blood Pressure

- The three types of blood vessels are arteries, veins, and capillaries.
- The force of the blood on the walls of the blood vessels is called blood pressure.

Cardiovascular Disease

- Heart disease is a leading cause of death.

Functions of the Lymphatic System

- Lymph is tissue fluid from cells that has entered the lymph vessels.
- Lymphocytes help fight disease.

Self Check

1. **Compare and contrast** veins, arteries, and capillaries.
2. **Identify** the vessels in the pulmonary and systemic circulation systems that carry oxygen-rich blood.
3. **Describe** the functions of the lymphatic system.
4. **Explain** how blood moves through the heart.
5. **Explain** why blood type and Rh factor are checked before a transfusion.
6. **Think Critically** What waste product builds up in blood and cells when the heart is unable to pump blood efficiently?

Applying Skills

7. **Use a Database** Research diseases of the circulatory system. Make a database showing what part of the circulatory system is affected by each disease. Categories should include the organs and vessels of the circulatory system.
8. **Concept Map** Make an events-chain concept map to show pulmonary circulation beginning at the right atrium and ending at the aorta.

section 3

Immunity

Lines of Defense

Your body has many ways to defend itself. Its first-line defenses work against harmful substances and all types of disease-causing organisms, called pathogens (PA thuh junz). Your second-line defenses are specific and work against specific pathogens. This complex group of defenses is called your immune system. Tonsils are one of the organs in the immune system that protect your body.

Reading Check *What types of defenses does your body have?*

First-Line Defenses Your skin and respiratory, digestive, and circulatory systems are first-line defenses against pathogens, like those in **Figure 10.** The skin is a barrier that prevents many pathogens from entering your body. However, pathogens can get into your body easily through a cut or through your mouth and the membranes in your nose and eyes. The conditions on the skin can affect pathogens. Perspiration contains substances that can slow the growth of some pathogens. At times, secretions from the skin's oil glands and perspiration are acidic. Some pathogens cannot grow in this acidic environment.

Internal First-Line Defenses Your respiratory system traps pathogens with hairlike structures, called cilia (SIH lee uh), and mucus. Mucus contains an enzyme that weakens the cell walls of some pathogens. When you cough or sneeze, you get rid of some of these trapped pathogens.

Your digestive system has several defenses against pathogens—saliva, enzymes, hydrochloric acid solution, and mucus. Saliva in your mouth contains substances that kill bacteria. Also, enzymes (EN zimez) in your stomach, pancreas, and liver help destroy pathogens. Hydrochloric acid solution in your stomach helps digest your food. It also kills some bacteria and stops the activity of some viruses that enter your body on the food that you eat. The mucus found on the walls of your digestive tract contains a chemical that coats bacteria and prevents them from binding to the inner lining of your digestive organs.

as you read

What You'll Learn

- **Explain** the difference between an antigen and an antibody.
- **Compare and contrast** active and passive immunity.

Why It's Important

Your body's defenses fight the pathogens that you are exposed to every day.

Review Vocabulary
enzyme: a type of protein that speeds up the rate of a chemical reaction in your body

New Vocabulary
- antigen
- antibody
- active immunity
- passive immunity

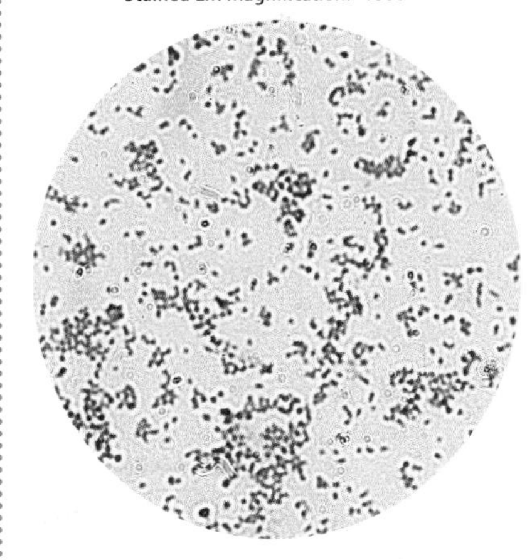

Figure 10 Most pathogens, such as the staphylococci bacteria shown below, cannot get through unbroken skin.

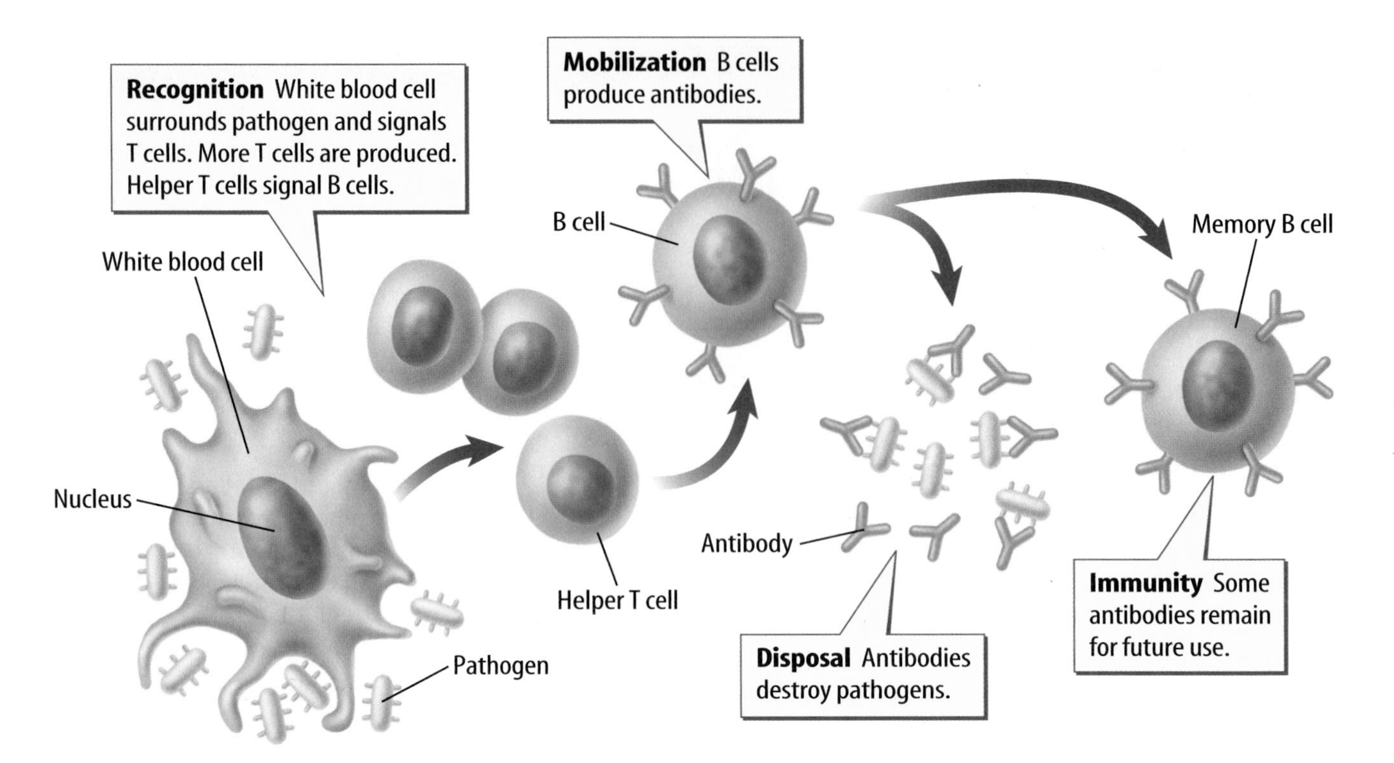

Figure 11 The response of your immune system to disease-causing organisms can be divided into four steps—recognition, mobilization, disposal, and immunity.
Describe *the function of B cells.*

White Blood Cells Your circulatory system contains white blood cells that surround and digest foreign organisms and chemicals. These white blood cells constantly patrol your body, sweeping up and digesting bacteria that invade.

Inflammation When tissue is damaged or infected by pathogens, it can become inflamed—becomes red, feels warm, swells, and hurts. Chemicals released by damaged cells expand capillary walls, allowing more blood to flow into the area. Other chemicals released by damaged tissue attract certain white blood cells that surround and take in pathogenic bacteria. If pathogens get past these first-line defenses, your body uses another line of defense called specific immunity.

Specific Immunity When your body fights disease, it is battling complex molecules called **antigens** that don't belong there. Antigens can be separate molecules or they can be found on the surface of a pathogen.

When your immune system recognizes foreign molecules, as in **Figure 11,** special lymphocytes called T cells respond. One type of T cells, called killer T cells, releases enzymes that help destroy invading foreign matter. Another type of T cells, called helper T cells, turns on the immune system. They stimulate other lymphocytes, known as B cells, to form antibodies. An **antibody** is a protein made in response to a specific antigen. The antibody attaches to the antigen and makes it useless.

Memory B Cells Another type of lymphocyte, called memory B cells, also has antibodies for the specific pathogen. Memory B cells remain in the blood, ready to defend against an invasion by that same pathogen at another time.

Active Immunity Antibodies help your body build defenses in two ways—actively and passively. In **active immunity** your body makes its own antibodies in response to an antigen. **Passive immunity** results when antibodies that have been produced in another animal are introduced into your body.

When a pathogen invades your body, the pathogen quickly multiplies and you get sick. Your body immediately starts to make antibodies to attack the pathogen. After enough antibodies form, you usually get better. Some antibodies stay on duty in your blood, and more are produced rapidly if the pathogen enters your body again. Because of this defense system, you usually don't get certain diseases, such as chicken pox, more than once.

How does active immunity differ from passive immunity?

Vaccination Another way to develop active immunity to a disease is to be inoculated with a vaccine, as shown in **Figure 12.** The process of giving a vaccine by injection or by mouth is called vaccination. A vaccine is a form of the antigen that gives you active immunity against a disease.

A vaccine can prevent a disease, but it is not a cure. As you grow older, you will be exposed to many more types of pathogens and will build a separate immunity to each one.

Determining Reproduction Rates

Procedure

1. Place **one penny** on a table. Imagine that the penny is a bacterium that can divide every 10 min.
2. Place **two pennies** below to form a triangle with the first penny. These indicate the two new bacteria present after a bacterium divides.
3. Repeat three more divisions, placing two pennies under each penny in the row above.
4. Calculate how many bacteria you would have after 5 h of reproduction. Graph your data.

Analysis

1. How many bacteria are present after 5 h?
2. Why is it important to take antibiotics promptly if you have an infection?

Try at Home

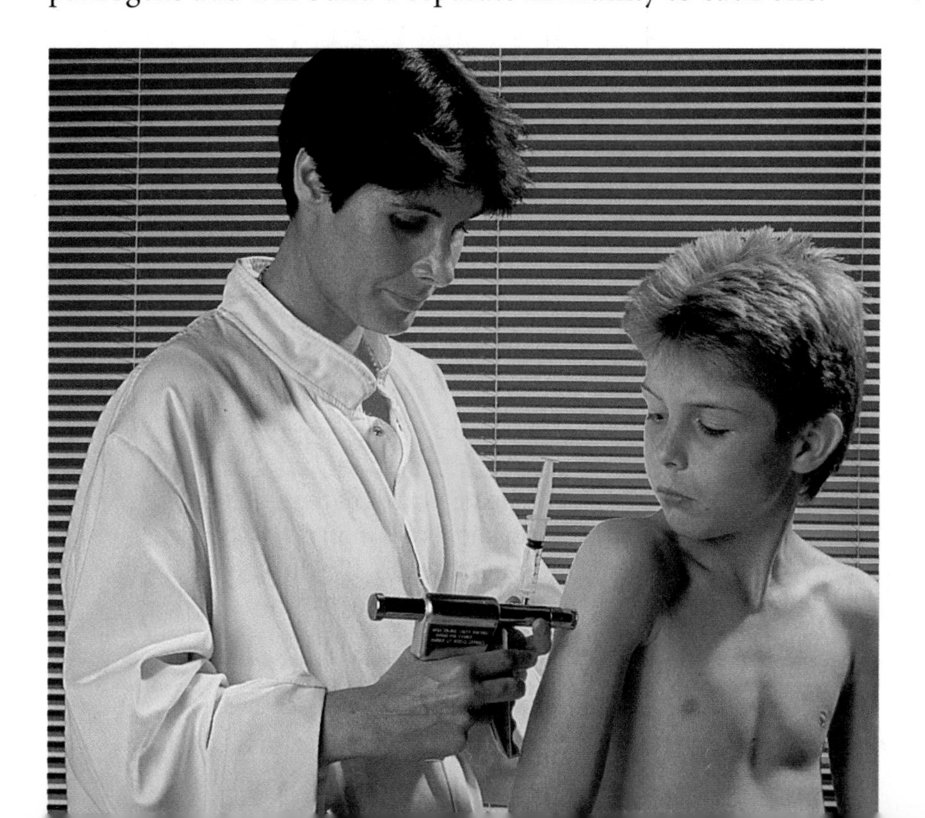

Figure 12 The Td vaccine protects against tetanus and diphtheria, an infectious disease of the respiratory system, and usually is injected into the arm.

Table 2 Cases of Disease Before and After Vaccine Availability in the U.S.

Disease	Average Number of Cases per Year Before Vaccine Available	Cases in 1998 After Vaccine Available
Measles	503,282	89
Diphtheria	175,885	1
Tetanus	1,314	34
Mumps	152,209	606
Rubella	47,745	345
Pertussis (whooping cough)	147,271	6,279

Data from the National Immunization Program, CDC

Passive Immunity Passive immunity does not last as long as active immunity does. For example, you were born with all the antibodies that your mother had in her blood. However, these antibodies stayed with you for only a few months. Because newborn babies lose their passive immunity in a few months, they need to be vaccinated to develop their own immunity. Vaccines have helped reduce the number of cases of many childhood diseases, as shown in **Table 2.**

section 3 review

Summary

Lines of Defense

- The purpose of the immune system is to fight disease.
- Your skin and your respiratory, digestive, and circulatory systems are first-line defenses against pathogens.
- Your body's second line of defense is called specific immunity.
- In active immunity, your body makes its own antibodies in response to an antigen.
- Vaccinations can give you active immunity against a disease.
- Passive immunity results when antibodies that have been produced in another animal are introduced into your body.

Self Check

1. **Describe** how harmful bacteria can cause infections in your body.
2. **List** the natural defenses that your body has against harmful substances and disease.
3. **Explain** how an active vaccine works to protect the human body.
4. **Think Critically** Several diseases have symptoms similar to those of measles. Why doesn't the measles vaccine protect you from all of these diseases?

Applying Skills

5. **Make Models** Create models of the different types of T cells, antigens, and B cells from clay, construction paper, or other art materials. Use them to explain how T cells function in the immune system.

green.msscience.com/self_check_quiz

section 4

Diseases

Disease in History

Throughout time, the plague, smallpox, and influenza have killed millions of people worldwide. Today, the causes of these diseases are known, and treatments can prevent or cure them. But even today, some diseases cannot be cured, and outbreaks of new diseases, such as severe acute respiratory syndrome (SARS), occur.

Discovering Disease Organisms With the invention of the microscope in the latter part of the seventeenth century, bacteria, yeast, and mold spores were seen for the first time. However, scientists did not make a connection between microorganisms and disease transmission until the late 1800s and early 1900s.

The French chemist Louis Pasteur learned that microorganisms cause disease in humans. Many scientists of his time did not believe that microorganisms could harm larger organisms, such as humans. However, Pasteur discovered that microorganisms could spoil wine and milk. He then realized that microorganisms could attack the human body in the same way. Pasteur invented **pasteurization** (pas chuh ruh ZAY shun), which is the process of heating a liquid to a temperature that kills most bacteria.

Disease Organisms **Table 3** lists some of the diseases caused by various groups of pathogens. Bacteria and viruses cause many common diseases.

Table 3 Human Diseases and Their Agents

Agent	Diseases
Bacteria	Tetanus, tuberculosis, typhoid fever, strep throat, bacterial pneumonia, plague
Protists	Malaria, sleeping sickness
Fungi	Athlete's foot, ringworm
Viruses	Colds, influenza, AIDS, measles, mumps, polio, smallpox, SARS

as you read

What You'll Learn

- **Describe** the work of Pasteur, Koch, and Lister in the discovery and prevention of disease.
- **Identify** diseases caused by viruses and bacteria.
- **Explain** how HIV affects the immune system.
- **Define** noninfectious diseases and list their causes.
- **Explain** what happens during an allergic reaction.

Why It's Important

You can help prevent certain illnesses if you know what causes disease and how disease spreads.

Review Vocabulary
virus: tiny piece of genetic material surrounded by a protein coating that infects and multiplies in host cells

New Vocabulary
- pasteurization
- infectious disease
- noninfectious disease
- allergen

Antibiotics Soil contains many microorganisms—some that are harmful, such as tetanus bacteria, and some that are helpful. Some infections are treated with antibiotics made from bacteria and molds found in the soil. One such antibiotic is streptomycin. In your Science Journal, write a brief report about the drug streptomycin.

Pathogens The conditions in your body, such as temperature and available nutrients, help harmful bacteria that enter your body grow and multiply. Bacteria can slow down the normal growth and metabolic activities of body cells and tissues. Some bacteria even produce toxins that kill cells on contact.

A virus infects and multiplies in host cells. The host cells die when the viruses break out of them. These new viruses infect other cells, leading to the destruction of tissues or the interruption of vital body activities.

Reading Check *What is the relationship between a virus and a host cell?*

Pathogenic protists, such as the organisms that cause malaria, can destroy tissues and blood cells or interfere with normal body functions. In a similar manner, fungus infections can cause athlete's foot, nonhealing wounds, chronic lung disease, or inflammation of the membranes of the brain.

Koch's Rules Many diseases caused by pathogens can be treated with medicines. In many cases, these organisms need to be identified before specific treatment can begin. Today, a method developed in the nineteenth century by Robert Koch still is used to identify organisms, as shown in **Figure 13.**

Infectious Diseases

A disease that is caused by a virus, bacterium, protist, or fungus and is spread from an infected organism or the environment to another organism is called an **infectious disease.** Infectious diseases are spread by direct contact with the infected organism, through water and air, on food, by contact with contaminated objects, and by disease-carrying organisms called biological vectors. Examples of vectors that have been sources of disease are rats, birds, cats, dogs, mosquitoes, fleas, and flies, as shown in **Figure 14.**

Figure 14 When flies land on food, they can transport pathogens from one location to another.

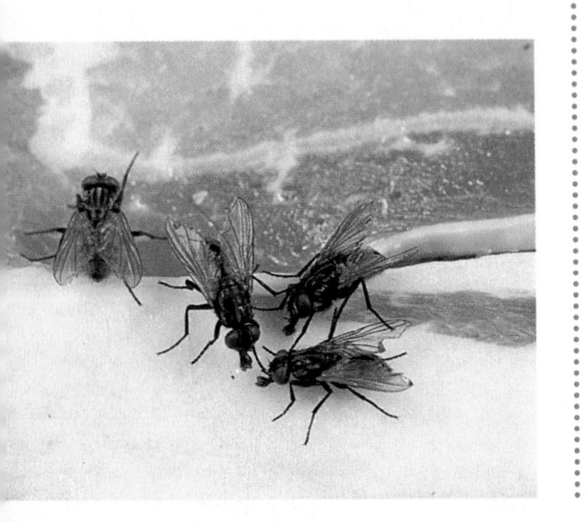

Human Vectors People also can be carriers of disease. Colds and many other diseases are spread through contact. Each time you turn a doorknob or use a telephone, your skin comes in contact with bacteria and viruses, which is why washing your hands frequently should be part of your daily routine.

Joseph Lister, an English surgeon, recognized the relationship between infections and cleanliness. Lister dramatically reduced the number of deaths among his patients by washing their skin and his hands with carbolic (kar BAH lihk) acid, which is a liquid that kills pathogens.

NATIONAL GEOGRAPHIC **VISUALIZING KOCH'S RULES**

Figure 13

In the 1880s, German doctor Robert Koch developed a series of methods for identifying which organism was the cause of a particular disease. Koch's Rules are still in use today. Developed mainly for determining the cause of particular diseases in humans and other animals, these rules have been used for identifying diseases in plants as well.

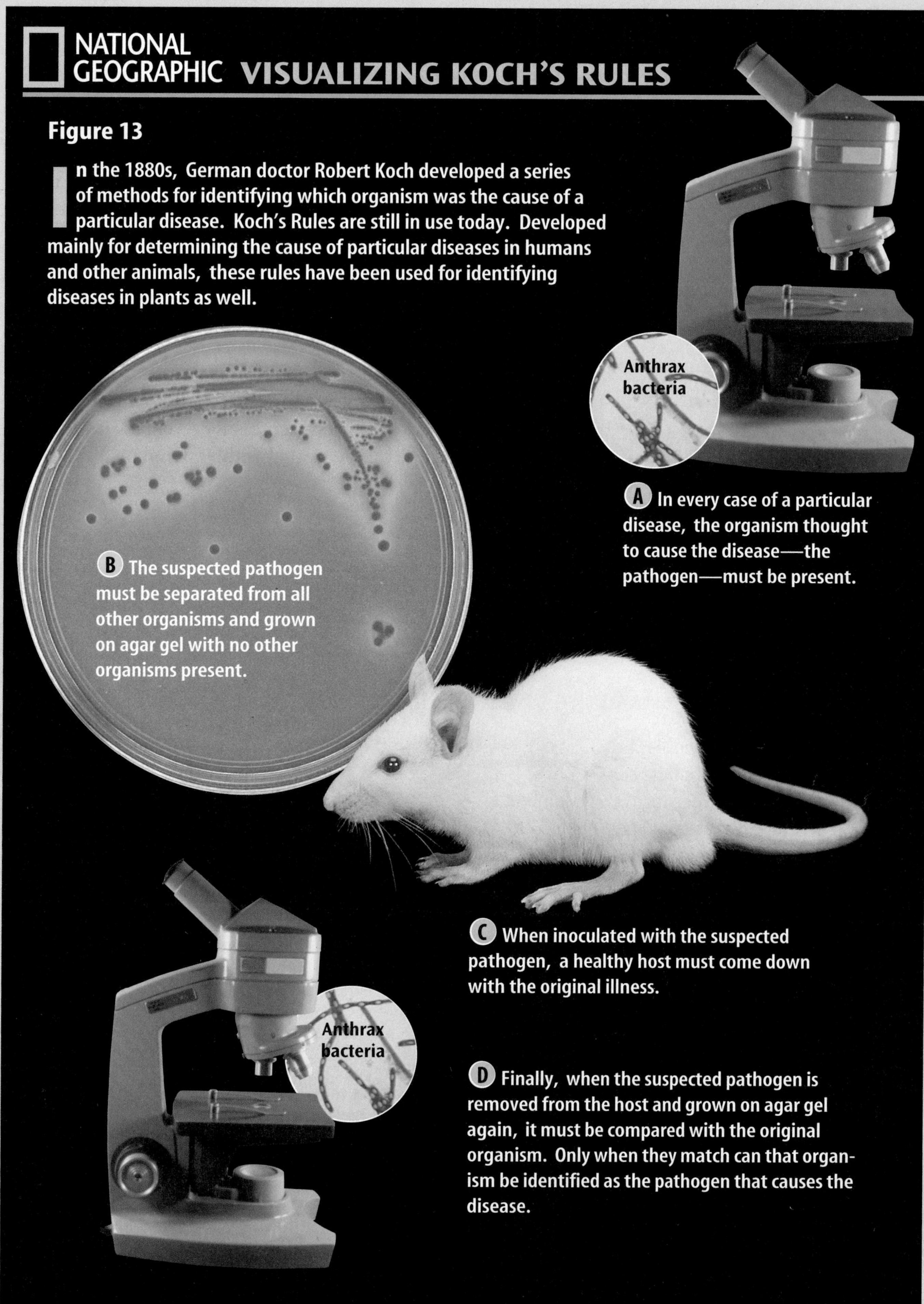

A In every case of a particular disease, the organism thought to cause the disease—the pathogen—must be present.

B The suspected pathogen must be separated from all other organisms and grown on agar gel with no other organisms present.

C When inoculated with the suspected pathogen, a healthy host must come down with the original illness.

D Finally, when the suspected pathogen is removed from the host and grown on agar gel again, it must be compared with the original organism. Only when they match can that organism be identified as the pathogen that causes the disease.

Sexually Transmitted Diseases Infectious diseases that are passed from person to person during sexual contact are called sexually transmitted diseases (STDs). STDs are caused by bacteria or viruses.

Gonorrhea (gah nuh REE uh), chlamydia (kluh MIH dee uh), and syphilis (SIH fuh lus) are STDs caused by bacteria. Antibiotics are used to treat these diseases. If they are untreated, gonorrhea and chlamydia can leave a person sterile because the reproductive organs can be damaged permanently. Untreated syphilis may infect cardiovascular and nervous systems, resulting in damage to body organs that cannot be reversed.

Genital herpes, a lifelong viral disease, causes painful blisters on the sex organs. This type of herpes can be transmitted during sexual contact or from an infected mother to her child during birth. Herpes has no cure, and no vaccine can prevent it. However, the symptoms of herpes can be treated with antiviral medicines.

Reading Check *Why should STDs be treated in the early stages?*

Applying Science

Has the annual percentage of deaths from major diseases changed?

Each year, many people die from diseases. Medical science has found numerous ways to treat and cure disease. Have new medicines, improved surgery techniques, and healthier lifestyles helped decrease the number of deaths from disease? By using your ability to interpret data tables, you can find out.

Percentage of Deaths Due to Major Diseases

Disease	Year			
	1950	1980	1990	2000
Heart	37.1	38.3	33.5	29.6
Cancer	14.6	20.9	23.5	23.0
Stroke	10.8	8.6	6.7	7.0
Diabetes	1.7	1.8	2.2	2.9
Pneumonia and flu	3.3	2.7	3.7	2.7

Identifying the Problem

The table above shows the percentage of total deaths due to six major diseases for a 50-year time period. Study the data for each disease. Can you see any trends in the percentage of deaths?

Solving the Problem

1. Has the percentage increased for any disease that is listed?
2. What factors could have contributed to this increase?

HIV and Your Immune System

Human immunodeficiency virus (HIV) can exist in blood and body fluids. This virus can hide in body cells, sometimes for years. You can become infected with HIV by having sex with an HIV-infected person or by reusing an HIV-contaminated hypodermic needle for an injection. However, a freshly unwrapped sterile needle cannot transmit infection. The risk of getting HIV through blood transfusion is small because all donated blood is tested for the presence of HIV. A pregnant woman with HIV can infect her child when the virus passes through the placenta. The child also may become infected from contacts with blood during the birth process or when nursing after birth.

HIV cannot multiply outside the body, and it does not survive long in the environment. The virus cannot be transmitted by touching an infected person, by handling objects used by the person unless they are contaminated with body fluids, or from contact with a toilet seat.

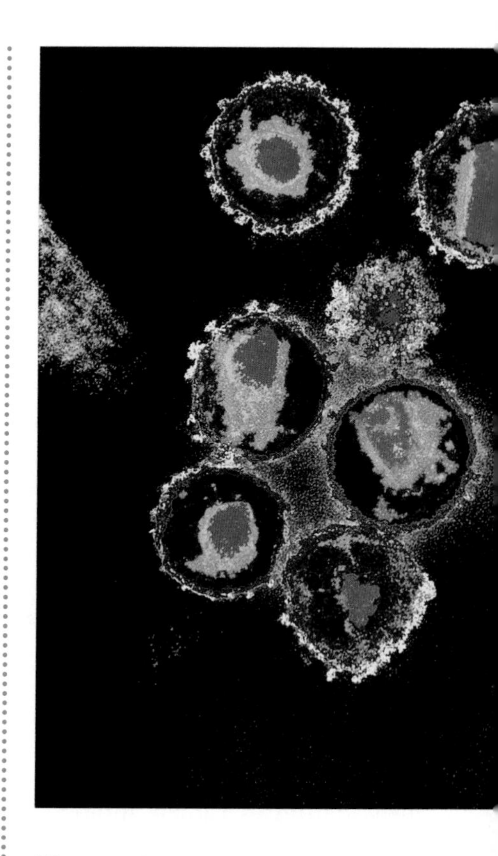

Figure 15 A person can be infected with HIV and not show any symptoms of the infection for several years.
Explain *why this characteristic makes the spread of AIDS more likely.*

AIDS An HIV infection can lead to Acquired Immune Deficiency Syndrome (AIDS), which is a disease that attacks the body's immune system. HIV, as shown in **Figure 15,** is different from other viruses. It attacks the helper T cells in the immune system. The virus enters the T cell and multiplies. When the infected cell bursts open, it releases more HIV. These infect other T cells. Soon, so many T cells are destroyed that not enough B cells are stimulated to produce antibodies. The body no longer has an effective way to fight invading antigens. The immune system then is unable to fight HIV or any other pathogen.

In December 2003, it was estimated that nearly 40 million people worldwide have HIV/AIDS. At this time the disease has no known cure. However, several medications help treat AIDS in some patients.

Fighting Disease

Washing a small wound with soap and water is the first step in preventing an infection. Cleaning the wound with an antiseptic and covering it with a bandage are other steps. Is it necessary to wash your body to help prevent diseases? Yes! In addition to reducing body odor, washing your body removes and destroys some surface microorganisms.

In your mouth, microorganisms are responsible for mouth odor and tooth decay. Using dental floss and routine tooth brushing keep these organisms under control.

Figure 16 Dust mites are smaller than a period at the end of a sentence. They can live in pillows, mattresses, carpets, furniture, and other places.

Healthy Choices Exercise and good nutrition help the circulatory and respiratory systems work more effectively. Good health habits, including getting enough rest and eating well-balanced meals, can make you less susceptible to the actions of disease organisms such as those that cause colds and flu. Keeping up with recommended immunizations and having annual health checkups also can help you stay healthy.

Chronic Disease

Not all diseases are caused by pathogens. Diseases and disorders such as diabetes, allergies, asthma, cancer, and heart disease are **noninfectious diseases.** They are not spread from one person to another. Many are chronic (KRAH nihk). This means that they can last for a long time. Although some chronic diseases can be cured, others cannot.

Some infectious diseases can be chronic too. For example, deer ticks carry a bacterium that causes Lyme disease. This bacterium can affect the nervous system, heart, and joints for weeks to years. It can become chronic if not treated. Antibiotics will kill the bacteria, but some damage cannot be reversed.

Allergies Many people have allergies. Some people react to cosmetics, shellfish, strawberries, peanuts, or insect stings. An allergy is an overly strong reaction of the immune system to a foreign substance. Most allergic reactions are minor. However, severe allergic reactions can occur, causing shock and even death if they aren't treated promptly.

Substances that cause an allergic response are called **allergens.** Some chemicals, certain foods, pollen, molds, some antibiotics, and dust are allergens for some people. Dust can contain cat and dog dander and dust mites, shown in **Figure 16.**

When you come in contact with an allergen, your immune system usually forms antibodies. Your body reacts by releasing chemicals called histamines (HIHS tuh meenz) that promote red, swollen tissues. Antihistamines are medications that can be used to treat allergic reactions and asthma, a lung disorder associated with reactions to allergens. Some severe allergies are treated with repeated injections of small doses of the allergen. This allows your body to become less sensitive to the allergen.

Diabetes A chronic disease associated with the levels of insulin produced by the pancreas is diabetes. Insulin is a hormone that enables glucose to pass from the bloodstream into your cells. Doctors recognize two types of diabetes—Type 1 and Type 2. Type 1 diabetes is the result of too little or no insulin production. In Type 2 diabetes, your body cannot properly process insulin. Symptoms of diabetes include fatigue, excessive thirst, frequent urination, and tingling sensations in the hands and feet.

If glucose levels in the blood remain high for a long time, other health problems can develop. These problems can include blurred vision, kidney failure, heart attack, stroke, loss of feeling in the feet, and the loss of consciousness (diabetic coma).

Cancer

Cancer is the name given to a group of closely related diseases that result from uncontrolled cell growth. It is a complicated disease, and no one fully understands how cancers form. Characteristics of cancer cells are shown in **Table 4.** Tumors can occur anywhere in your body. Cancerous cells can leave a tumor, spread throughout the body via blood and lymph vessels, and then invade other tissues.

Reading Check *How do cancers spread?*

Causes In the latter part of the eighteenth century, a British physician recognized the association of soot to cancer in chimney sweeps. Since that time, scientists have learned more about causes of cancer. Research done in the 1940s and 1950s first related genes to cancer.

Although not all the causes of cancer are known, many causes have been identified. Smoking has been linked to lung cancer—the leading cause of cancer deaths for males in the United States. Exposure to certain chemicals also can increase your chances of developing cancer. These substances, called carcinogens (kar SIH nuh junz), include asbestos, various solvents, heavy metals, alcohol, and home and garden chemicals. Exposure to X rays, nuclear radiation, and ultraviolet radiation of the Sun also increases your risk of cancer.

Table 4 Characteristics of Cancer Cells
Cell growth is uncontrolled.
These cells do not function as part of your body.
The cells take up space and interfere with normal body functions.
The cells travel throughout the body.
The cells produce tumors and abnormal growths anywhere in your body.

Table 5 Early Warning Signs of Cancer
Changes in bowel or bladder habits
A sore that does not heal
Unusual bleeding or discharge
Thickening or lump in the breast or elsewhere
Indigestion or difficulty swallowing
Obvious change in a wart or mole
Nagging cough or hoarseness

from the National Cancer Institute

Prevention Knowing some causes of cancer might help you prevent it. The first step is to know the early warning signs, shown in **Table 5.** Medical attention and treatments such as chemotherapy or surgery in the early stages of some cancers can cure or keep them inactive.

A second step in cancer prevention concerns lifestyle choices. Choosing not to use tobacco and alcohol products can help prevent mouth and lung cancers and the other associated respiratory and circulatory system diseases. Selecting a healthy diet without many foods that are high in fats, salt, and sugar also might reduce your chances of developing cancer. Using sunscreen and limiting the amount of time that you expose your skin to direct sunlight are good preventive measures against skin cancer. Careful handling of harmful home and garden chemicals will help you avoid the dangers connected with these substances.

section 4 review

Summary

Disease in History

- Pasteur, Koch, and Lister made important discoveries about the causes and how to prevent the spread of diseases.

Infectious Diseases and HIV

- Bacteria, fungi, protists, and viruses can cause infectious disease.
- STDs are passed during sexual contact and are caused by bacteria or viruses.
- HIV infection can lead to AIDS, a disease that attacks the immune system.

Fighting Disease

- Good health habits can help prevent the spread of disease.

Chronic Disease and Cancer

- Allergies, diabetes, and cancer are chronic noninfectious diseases.
- Early detection and lifestyle choices can help treat or prevent some cancers.

Self Check

1. **Name** an infectious disease caused by each of the following: a virus, a bacterium, a protist, and a fungus.
2. **Compare and contrast** how HIV and other viruses affect the immune system.
3. **Explain** why diabetes is classified as a noninfectious disease.
4. **Recognize** how poor hygiene is related to the spread of disease.
5. **Describe** how your body might respond to an allergen.
6. **Think Critically** In what ways does Koch's procedure demonstrate the use of scientific methods?

Applying Math

7. **Make and Use Graphs** Make a bar graph using the following data about the number of deaths from AIDS-related diseases for children younger than 13 years old: 1995, 536; 1996, 420; 1997, 209; 1998, 115; and 1999, 76.

green.msscience.com/self_check_quiz

Microorganisms and Disease

Microorganisms are everywhere. Washing your hands and disinfecting items you use helps remove some of these organisms.

Real-World Question

How do microorganisms cause infection?

Goals

- **Observe** the transmission of microorganisms.
- **Relate** microorganisms to infections.

Materials

fresh apples (6)
rotting apple
rubbing alcohol (5 mL)
self-sealing plastic bags (6)
labels and pencil
gloves
paper towels
sandpaper
cotton ball
soap and water
newspaper

Safety Precautions

WARNING: *Do not eat the apples. Do not remove goggles until the lab and cleanup are completed.* When you complete the experiment, give all bags to your teacher for disposal.

Procedure

1. **Label** the plastic bags 1 through 6. Put on gloves. Place a fresh apple in bag 1.
2. Rub the rotting apple over the other five apples. This is your source of microorganisms. **WARNING:** *Don't touch your face.*
3. Put one apple in bag 2.
4. Hold one apple 1.5 m above the floor and drop it on a newspaper. Put it in bag 3.
5. Rub one apple with sandpaper. Place this apple in bag 4.
6. Wash one apple with soap and water. Dry it well. Put this apple in bag 5.
7. Use a cotton ball to spread alcohol over the last apple. Let it air-dry. Place it in bag 6.
8. Seal all bags and put them in a dark place.
9. On day 3 and day 7, compare all of the apples without removing them from the bags. Record your observations in a data table.

Apple Observations

Condition	Day 3	Day 7
1. Fresh		
2. Untreated		
3. Dropped		
4. Rubbed with sandpaper	Do not write	in this book.
5. Washed with soap and water		
6. Covered with alcohol		

Conclude and Apply

1. **Infer** How does this experiment relate to infections on your skin?
2. **Explain** why it is important to clean a wound.

Communicating Your Data

Prepare a poster illustrating the advantages of washing hands to avoid the spread of disease. Get permission to put the poster near a school rest room. **For more help, refer to the** Science Skill Handbook.

Design Your Own

Blood Type Reactions

Goals

- **Design** an experiment that simulates the reactions between different blood types.
- **Identify** which blood types can donate to which other blood types.

Possible Materials

simulated blood (10 mL low-fat milk and 10 mL water plus red food coloring)
lemon juice as antigen A (for blood types B and O)
water as antigen A (for blood types A and AB)
droppers
small paper cups
marking pen
10-mL graduated cylinder

Safety Precautions

WARNING: *Do not taste, eat, or drink any materials used in the lab.*

Real-World Question

Human blood can be classified into four main blood types—A, B, AB, and O. These types are determined by the presence or absence of antigens on the red blood cells. After blood is collected into a transfusion bag, it is tested to determine the blood type. The type is labeled clearly on the bag. Blood is refrigerated to keep it fresh and available for transfusion. What happens when two different blood types are mixed?

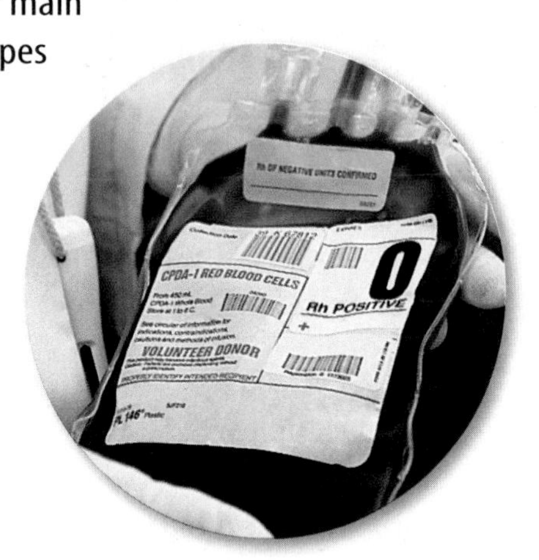

Form a Hypothesis

Based on your reading and observations, form a hypothesis to explain how different blood types will react to each other.

Test Your Hypothesis

Make a Plan

1. As a group, agree upon a hypothesis and decide how you will test it. Identify the results that will confirm the hypothesis.
2. **List** the steps you must take and the materials you will need to test your hypothesis. Be specific. Describe exactly what you will do in each step.
3. **Prepare** a data table like the one at the right in your Science Journal to record your observations.

Blood Type Reactions

Blood Type	Clumping (Yes or No)
A	
B	Do not write in this book.
AB	
O	

4. Reread the entire experiment to make sure all steps are in logical order.
5. **Identify** constants and variables. Blood type O will be the control.

Follow Your Plan

1. Make sure your teacher approves your plan before you start.
2. Carry out the experiment according to the approved plan.
3. While doing the experiment, record your observations and complete the data table in your Science Journal.

Analyze Your Data

1. **Compare** the reactions of each blood type (A, B, AB, and O) when antigen A was added to the blood.
2. **Observe** where clumping took place.
3. **Compare** your results with those of other groups.
4. What was the control factor in this experiment?
5. What were your variables?

Conclude and Apply

1. Did the results support your hypothesis? Explain.
2. **Predict** what might happen to a person if other antigens are not matched properly.
3. What would happen in an investigation with antigen B added to each blood type?

Communicating Your Data

Write a brief report on how blood is tested to determine blood type. Describe why this is important to know before receiving a blood transfusion. **For more help, refer to the** Science Skill Handbook.

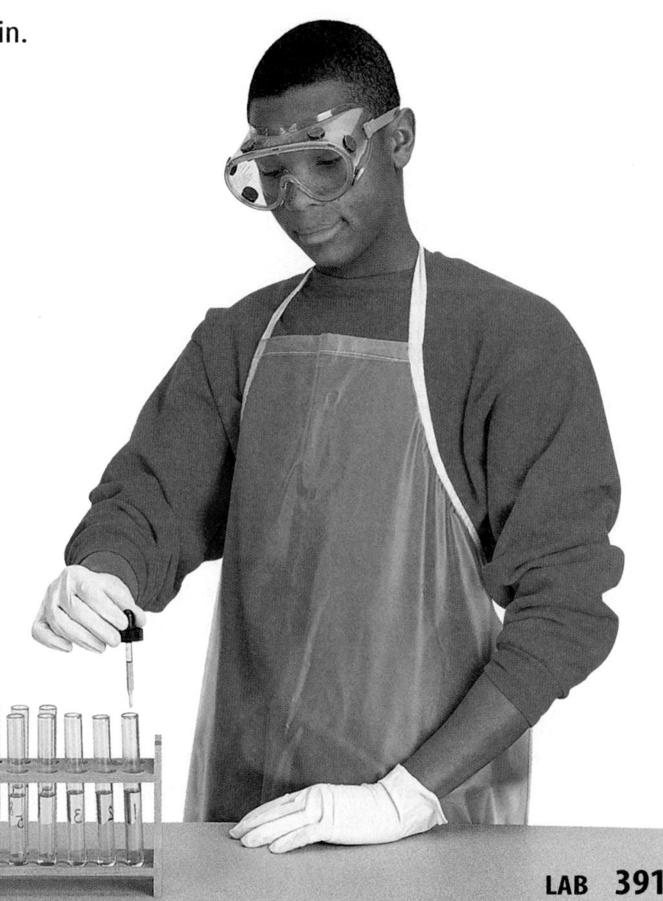

TIME SCIENCE AND HISTORY

SCIENCE CAN CHANGE THE COURSE OF HISTORY!

Dr. Daniel Hale Williams was a pioneer in open-heart surgery.

Have a Heart

People didn't always know where blood came from or how it moved through the body.

You prick your finger, and when blood starts to flow out of the cut, you put on a bandage. But if you were a scientist living long ago, you might have also asked yourself some questions: How did your blood get to the tip of your finger? And why and how does it flow through (and sometimes out of!) your body?

As early as the 1500s, a Spanish scientist named Miguel Serveto (mee GEL • ser VEH toh) asked that question. His studies led him to the theory that blood circulated throughout the human body, but he didn't know how or why.

About 100 years later, William Harvey, an English doctor, explored Serveto's idea. Harvey studied animals to develop a theory about how the heart and the circulatory system work. Blood was pumped from the heart throughout the body, Harvey hypothesized. Then it returned to the heart and recirculated. He published his ideas in 1628 in his famous book, *On the Motion of the Heart and Blood in Animals.* His theories were correct, and Harvey's book became the basis for all modern research on heart and blood vessels.

Medical Pioneer

More than two centuries later, another pioneer, Dr. Daniel Hale Williams, stepped forward and used Harvey's ideas to change the science frontier again. He performed the first open-heart surgery by removing a knife from the heart of a stabbing victim. He stitched the wound in the fluid sac surrounding the heart, and the patient lived for several years afterward.

Report **Identify a pioneer in science or medicine who has changed our lives for the better. Find out how this person started in the field, and how they came to make an important discovery. Give a presentation to the class.**

For more information, visit green.msscience.com/time

Reviewing Main Ideas

Section 1 Blood

1. Red blood cells carry oxygen and carbon dioxide, platelets form clots, and white blood cells fight infection.
2. A, B, AB, and O blood types are determined by the presence or absence of antigens on red blood cells.

Section 2 Circulation

1. Arteries carry blood away from the heart and veins return blood to the heart. Capillaries connect arteries to veins.
2. The circulatory system can be divided into three sections—coronary, pulmonary, and systemic circulation.
3. Lymph structures filter blood, produce white blood cells, and destroy worn out blood cells.

Section 3 Immunity

1. Your body is protected against most pathogens by the immune system.
2. Active immunity is long lasting, but passive immunity is not.

Section 4 Diseases

1. Pasteur and Koch discovered that microorganisms cause diseases. Lister learned that cleanliness helps control microorganisms.
2. Bacteria, viruses, fungi, and protists can cause infectious diseases.
3. HIV damages your body's immune system, which can cause AIDS.
4. Causes of noninfectious diseases, such as diabetes and cancer, include genetics, a poor diet, chemicals, and uncontrolled cell growth.

Visualizing Main Ideas

Copy and complete this concept map on the functions of the parts of the blood.

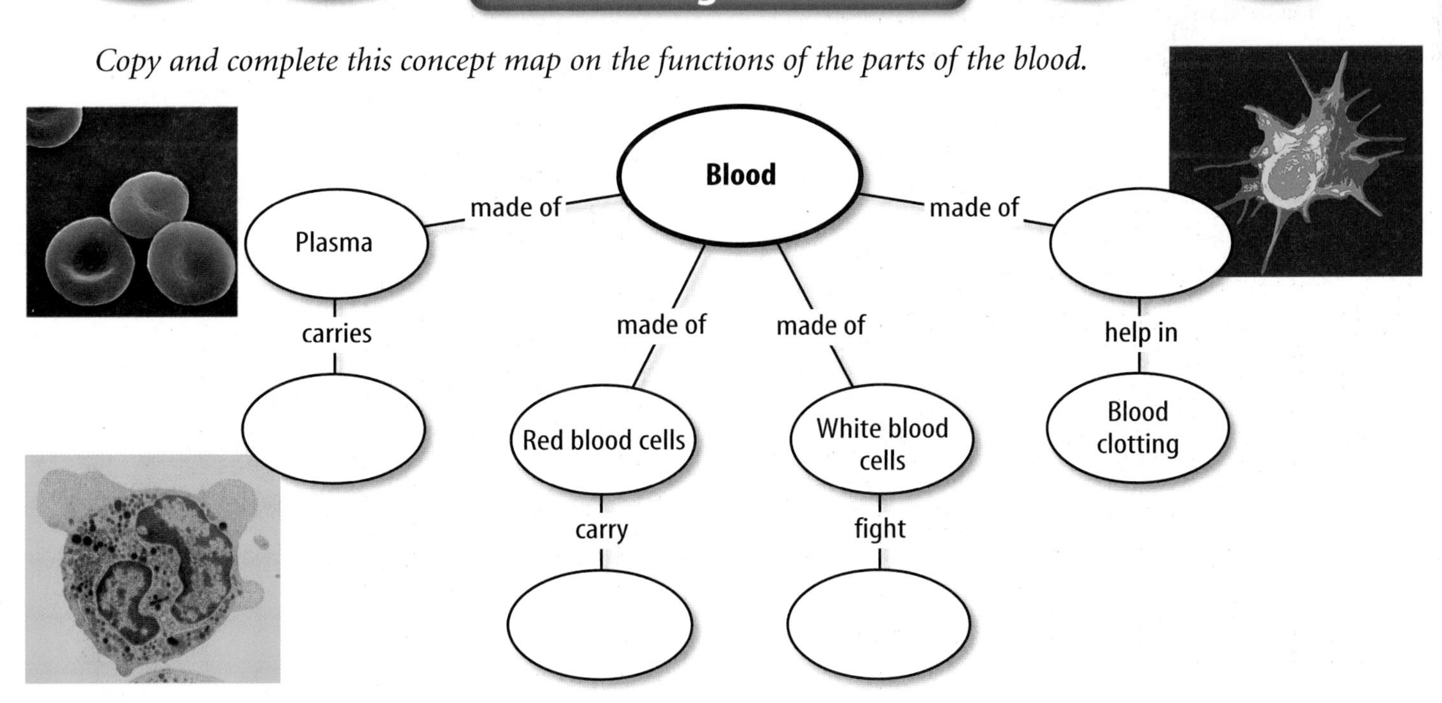

chapter 13 Review

Using Vocabulary

active immunity p. 379
allergen p. 386
antibody p. 378
antigen p. 378
artery p. 373
capillary p. 373
hemoglobin p. 367
infectious disease p. 382
lymph p. 376
noninfectious disease p. 386
passive immunity p. 379
pasteurization p. 381
plasma p. 366
platelet p. 367
vein p. 373

Fill in the blanks with the correct vocabulary word or words.

1. _________ is the chemical in red blood cells.
2. _________ are cell fragments that help clot blood.
3. _________ occurs when your body makes its own antibodies.
4. A(n) _________ stimulates histamine release.
5. Heating a liquid to kill harmful bacteria is called _________.

Checking Concepts

Choose the word or phrase that best answers the question.

6. Where does the exchange of food, oxygen, and wastes occur?
   A) arteries **C)** veins
   B) capillaries **D)** lymph vessels
7. How can infectious diseases be caused?
   A) heredity **C)** chemicals
   B) allergies **D)** organisms
8. Where is blood under greatest pressure?
   A) arteries **C)** veins
   B) capillaries **D)** lymph vessels
9. Which cells fight off infection?
   A) red blood **C)** white blood
   B) bone **D)** nerve
10. Of the following, which carries oxygen in blood?
    A) red blood cells **C)** white blood cells
    B) platelets **D)** lymph
11. What is required to clot blood?
    A) plasma **C)** platelets
    B) oxygen **D)** carbon dioxide

Use the table below to answer question 12.

Table 1 Blood Types

Blood Type	Antigen	Antibody
A	A	Anti-B
B	B	Anti-A
AB	A, B	None
O	None	Anti-A, Anti-B

12. Using the table above, what kind of antigen does type O blood have?
    A) A **C)** A and B
    B) B **D)** no antigen
13. Where does oxygen-rich blood enter first?
    A) right atrium
    B) left atrium
    C) left ventricle
    D) right ventricle
14. What is formed in the blood to fight invading antigens?
    A) hormones **C)** pathogens
    B) allergens **D)** antibodies
15. Which disease is caused by a virus that attacks white blood cells?
    A) AIDS **C)** flu
    B) measles **D)** polio

Science Online green.msscience.com/vocabulary_puzzlemaker

Thinking Critically

16. **Compare and contrast** the life spans of red blood cells, white blood cells, and platelets.

17. **Sequence** blood clotting from the wound to forming a scab.

18. **Compare and contrast** the functions of arteries, veins, and capillaries.

19. **Analyze** how antibodies, antigens, and antibiotics differ.

20. **Recognize Cause and Effect** Use library references to identify the cause—bacteria, virus, fungus, or protist—of each of these diseases: athlete's foot, AIDS, cold, dysentery, flu, pinkeye, acne, and strep throat.

21. **Classify** Using word processing software, make a table to classify the following diseases as infectious or noninfectious: diabetes, gonorrhea, herpes, strep throat, syphilis, cancer, and flu.

Use the graph below to answer question 22.

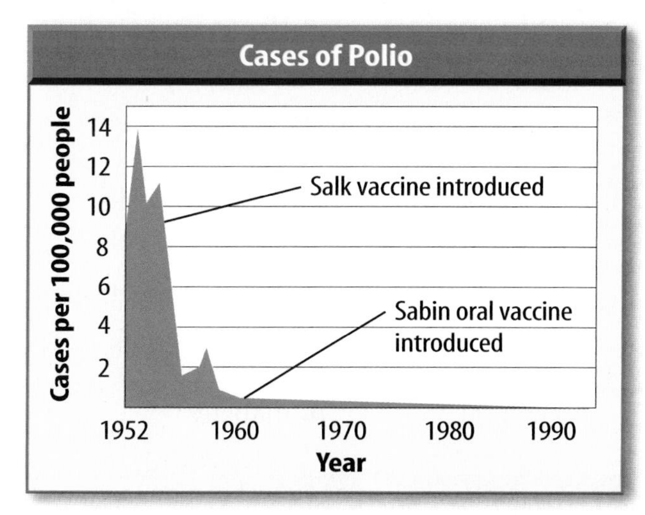

22. **Explain** the rate of polio cases between 1952 and 1965. What conclusions can you draw about the effectiveness of the polio vaccine?

Performance Activities

23. **Scientific Drawing** Prepare a drawing of the human heart and label its parts. Use arrows to show the flow of blood through the heart.

24. **Poster** Design and construct a poster to illustrate how a person with the flu could spread the disease to family members, classmates, and others.

25. **Pamphlet** Prepare a pamphlet describing heart transplants. Include an explanation of why the patient is given drugs that suppress the immune system and describe the patient's life after the operation.

Applying Math

26. **Percentages of Blood Cells** A cubic millimeter of blood has about five million red blood cells, 7,500 white blood cells, and 400,000 platelets. Find the total number of red blood cells, white blood cells, and platelets in 1 mm^3 of blood. Calculate what percentage of the total each type is.

Use the table below to answer question 27.

Gender and Heart Rate

Sex	Pulse/Minute
Male 1	72
Male 2	64
Male 3	65
Female 1	67
Female 2	84
Female 3	74

27. **Heart Rates** Interpret the data listed in the table above. Find the average heart rate of the three males and the three females and compare the two averages.

Standardized Test Practice

Part 1 Multiple Choice

Record your answers on the answer sheet provided by your teacher or on a sheet of paper.

1. Which of the following is a cause of cardiovascular disease?
   A. smoking
   B. jogging
   C. asbestos exposure
   D. ultraviolet radiation

Use the graph below to answer questions 2 and 3.

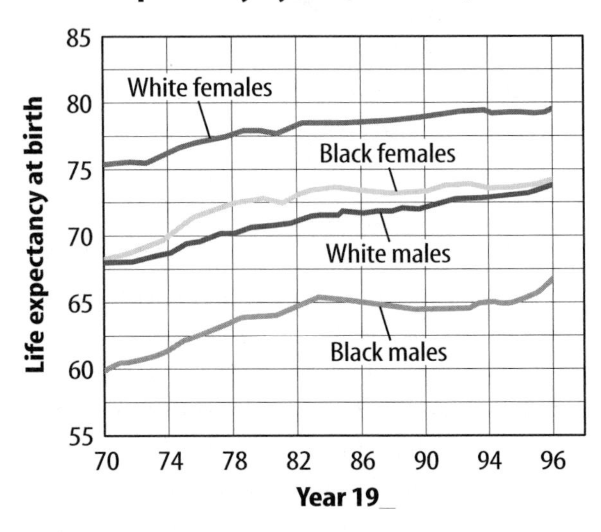

2. According to the information in the graph, which group had the lowest life expectancy in both 1975 and 1994?
   A. white males
   B. black females
   C. white females
   D. black males

3. A reasonable hypothesis based on the information in the graph is that
   A. life expectancy has decreased for black males between 1970 and 1984.
   B. life expectancy is longer for females than for males.
   C. life expectancy has decreased for white males between 1970 and 1980.
   D. life expectancy is longer for males than for females.

Use the table below to answer questions 4 and 5.

Results from Ashley's Activities

Activity	Pulse Rate (beats/min)	Body Temperature	Degree of Sweating
1	80	98.6°F	None
2	90	98.8°F	Minimal
3	100	98.9°F	Little
4	120	99.1°F	Moderate
5	150	99.5°F	Considerable

4. According to the information in the table, which of the following activities caused Ashley's pulse to be less than 100 beats per minute?
   A. Activity 2
   B. Activity 3
   C. Activity 4
   D. Activity 5

5. A reasonable hypothesis based on these data is that during Activity 2, Ashley was probably
   A. sprinting.
   B. marching.
   C. sitting down.
   D. walking slowly.

6. Which of the following is a function of blood?
   A. It carries saliva to the mouth.
   B. It excretes salts from the body.
   C. It transports nutrients to body cells.
   D. It removes lymph from around cells.

7. Which of the following is a noninfectious disease?
   A. tetanus
   B. malaria
   C. influenza
   D. diabetes

Test-Taking Tip

Answer Bubbles For each question, double check that you are filling in the correct answer bubble for the question number you are working on.

Part 2 | Short Response/Grid In

Record your answers on the answer sheet provided by your teacher or on a sheet of paper.

8. If red blood cells are made at the rate of 2 million per second in the center of long bones, how many red blood cells are made in one hour?

9. If a cubic milliliter of blood has 7,500 white blood cells and 400,000 platelets, how many times more platelets than white blood cells are present in a cubic milliliter of blood?

10. What would happen if type A blood was given to a person with type O blood?

Use the illustration below to answer questions 11 and 12.

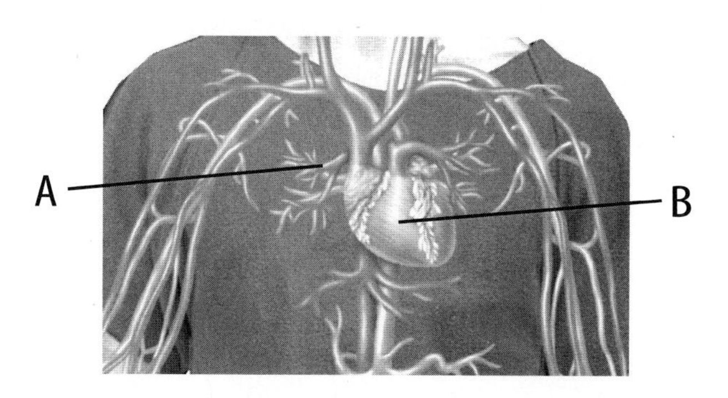

11. What might happen if there was a blood clot blocking vessel "A" in the illustration?

12. What might happen if there was a blood clot blocking vessel "B" in the illustration?

13. Explain why capillaries do not have thick elastic walls.

14. How does your skin help defend your body from diseases?

15. Describe some health practices that can help protect you from infectious diseases.

Part 3 | Open Ended

Record your answers on a sheet of paper.

16. How do the lymphatic and circulatory systems work together?

Use the illustration below to answer questions 17 and 18.

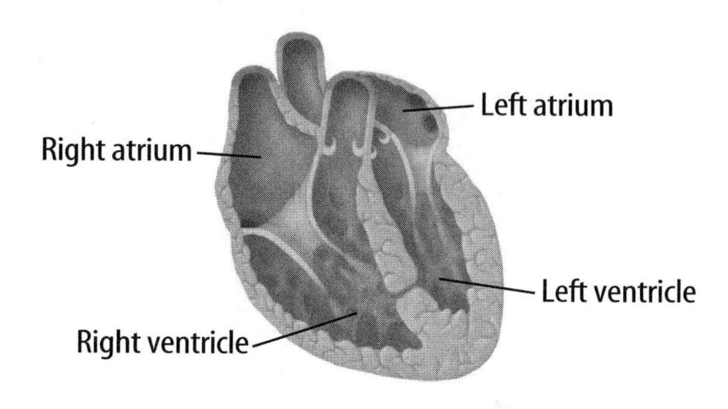

17. What is wrong with the heart in the illustration above? Explain your answer.

18. The left ventricle pumps blood under higher pressure than the right ventricle does. In which direction would you predict blood would flow through the hole in the heart? Compare the circulation in this heart with that of a normal heart.

19. About 950,000 Americans die from cardiovascular disease each year. What are some ways to prevent cardiovascular disease?

20. Which is longer lasting, active immunity or passive immunity? Explain.

21. Dr. Cavazos has isolated a bacterium that she thinks causes a recently discovered disease. How can she prove her hypothesis? What steps should she follow?

22. Compare and contrast infectious and noninfectious diseases.

chapter 14

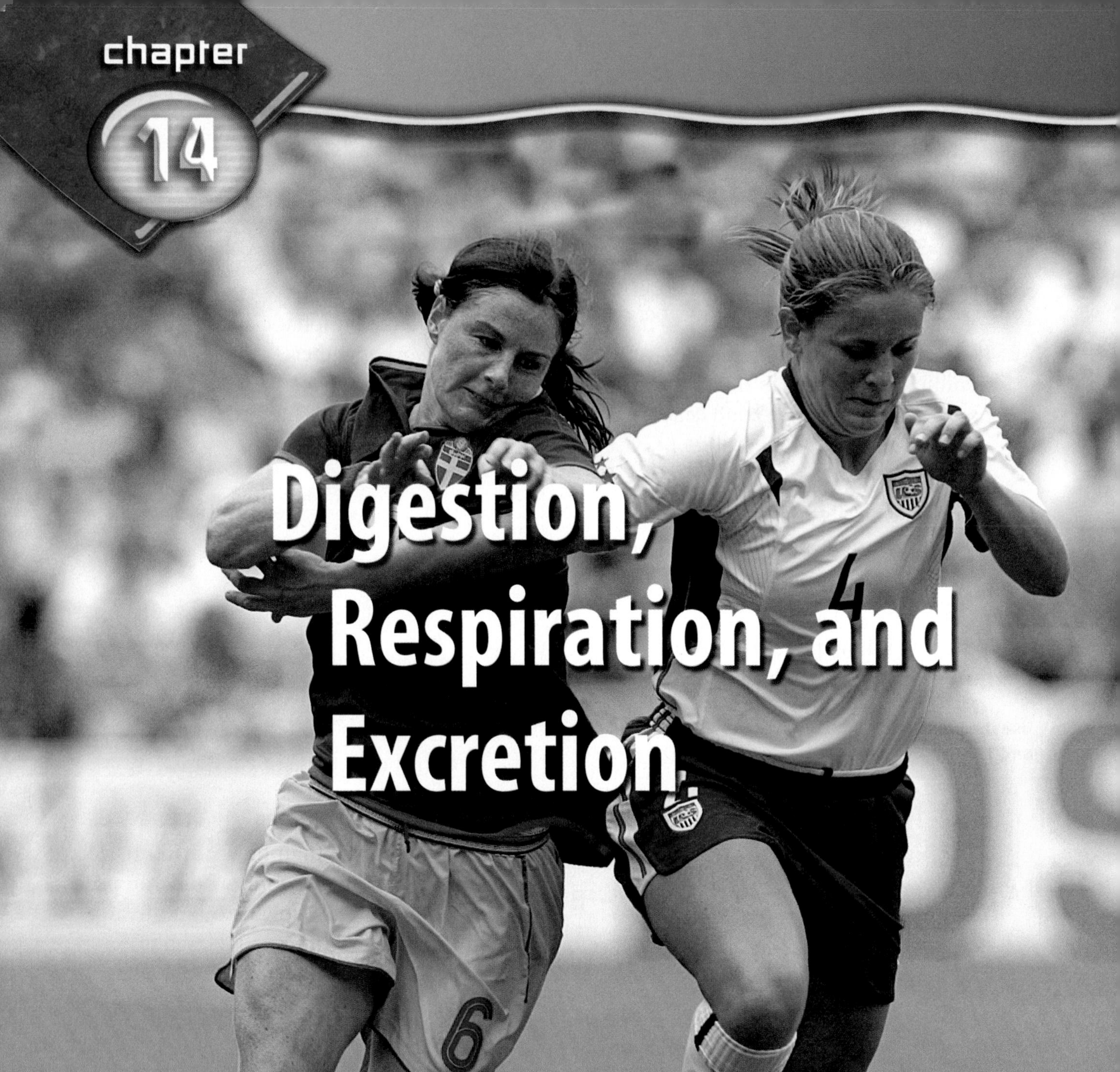

Digestion, Respiration, and Excretion

chapter preview

sections

Playing soccer is hard work.

If you're like most people, when you play an active game like soccer you probably breathe hard and perspire. You need a constant supply of oxygen and energy to keep your body cells functioning. Your body is adapted to meet that need.

Science Journal Write a paragraph describing what you do to help your body recover after an active game.

Start-Up Activities

Breathing Rate

Your body can store food and water, but it cannot store much oxygen. Breathing brings oxygen into your body. In the following lab, find out about one factor that can change your breathing rate.

1. Put your hand on the side of your rib cage. Using a watch or clock with a second hand, count the number of breaths you take for 15 s. Multiply this number by four to calculate your normal breathing rate for one minute.
2. Repeat step 1 two more times, then calculate your average breathing rate.
3. Do a physical activity described by your teacher for one minute and repeat step 1 to determine your breathing rate now.
4. Time how long it takes for your breathing rate to return to normal.
5. **Think Critically** In your Science Journal, write a paragraph explaining how breathing rate appears to be related to physical activity.

Respiration Make the following Foldable to help identify what you already know, what you want to know, and what you learn about respiration.

STEP 1 **Fold** a vertical sheet of paper from side to side. Make the front edge about 1.25 cm shorter than the back edge.

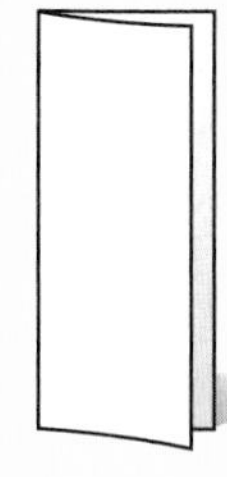

STEP 2 **Turn** lengthwise and **fold** into thirds.

STEP 3 **Unfold and cut** only the top layer along both folds to make three tabs. **Label** each tab.

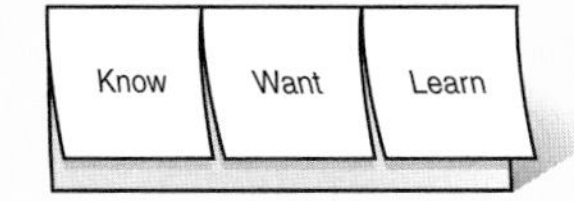

Identify Questions Before you read the chapter, write *I breathe* under the left tab, and write *Why do I breathe?* under the center tab. As you read the chapter, write the answer you learn under the right tab.

Preview this chapter's content and activities at green.msscience.com

section 1

The Digestive System

as you read

What You'll Learn

- **Distinguish** the differences between mechanical digestion and chemical digestion.
- **Identify** the organs of the digestive system and what takes place in each.
- **Explain** how homeostasis is maintained in digestion.

Why It's Important

The processes of the digestive system make the food you eat available to your cells.

Review Vocabulary

bacteria: one-celled organisms without membrane-bound organelles

New Vocabulary

- nutrient
- enzyme
- peristalsis
- chyme
- villi

Functions of the Digestive System

Food is processed in your body in four stages—ingestion, digestion, absorption, and elimination. Whether it is a piece of fruit or an entire meal, all the food you eat is treated to the same processes in your body. As soon as food enters your mouth, or is ingested, digestion begins. Digestion breaks down food so that nutrients (NEW tree unts) can be absorbed and moved into the blood. **Nutrients** are substances in food that provide energy and materials for cell development, growth, and repair. From the blood, these nutrients are transported across the cell membrane to be used by the cell. Unused substances pass out of your body as wastes.

Digestion is mechanical and chemical. Mechanical digestion takes place when food is chewed, mixed, and churned. Chemical digestion occurs when chemical reactions break down food.

Enzymes

Chemical digestion is possible only because of enzymes (EN zimez). An **enzyme** is a type of protein that speeds up the rate of a chemical reaction in your body. One way enzymes speed up reactions is by reducing the amount of energy necessary for a chemical reaction to begin. If enzymes weren't there to help, the rate of chemical reactions would be too slow. Some reactions might not even happen at all. As shown in **Figure 1,** enzymes work without being changed or used up.

Figure 1 Enzymes speed up the rate of certain body reactions.
Explain *what happens to the enzyme after it separates from the new molecule.*

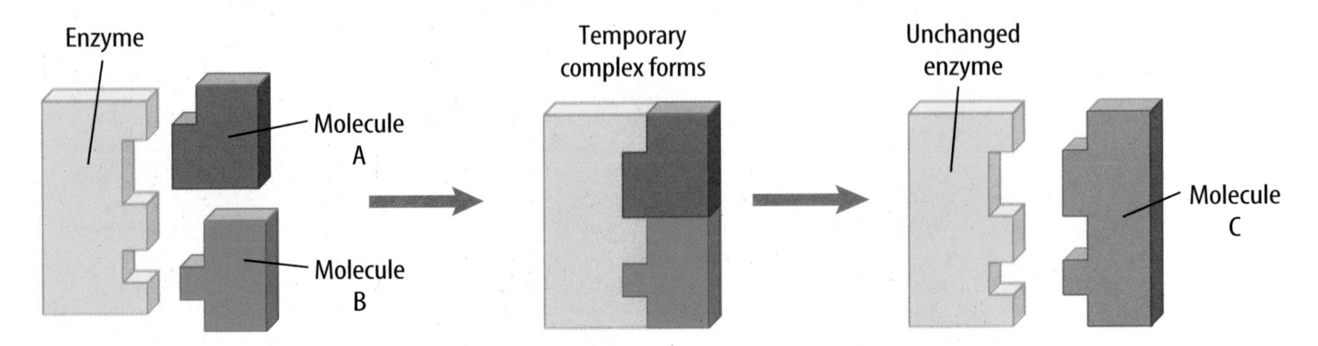

Enzymes in Digestion Many enzymes help you digest carbohydrates, proteins, and fats. These enzymes are produced in the salivary glands, stomach, small intestine, and pancreas.

Reading Check *What is the role of enzymes in the chemical digestion of food?*

Other Enzyme Actions Enzyme-aided reactions are not limited to the digestive process. Enzymes also help speed up chemical reactions responsible for building your body. They are involved in the energy-releasing activities of your muscle and nerve cells. Enzymes also aid in the blood-clotting process. Without enzymes, the chemical reactions in your body would happen too slowly for you to exist.

Organs of the Digestive System

Your digestive system has two parts—the digestive tract and the accessory organs. The major organs of your digestive tract—mouth, esophagus (ih SAH fuh gus), stomach, small intestine, large intestine, rectum, and anus—are shown in **Figure 2.** Food passes through all of these organs. The tongue, teeth, salivary glands, liver, gallbladder, and pancreas, also shown in **Figure 2,** are the accessory organs. Although food doesn't pass through them, they are important in mechanical and chemical digestion. Your liver, gallbladder, and pancreas produce or store enzymes and other chemicals that help break down food as it passes through the digestive tract.

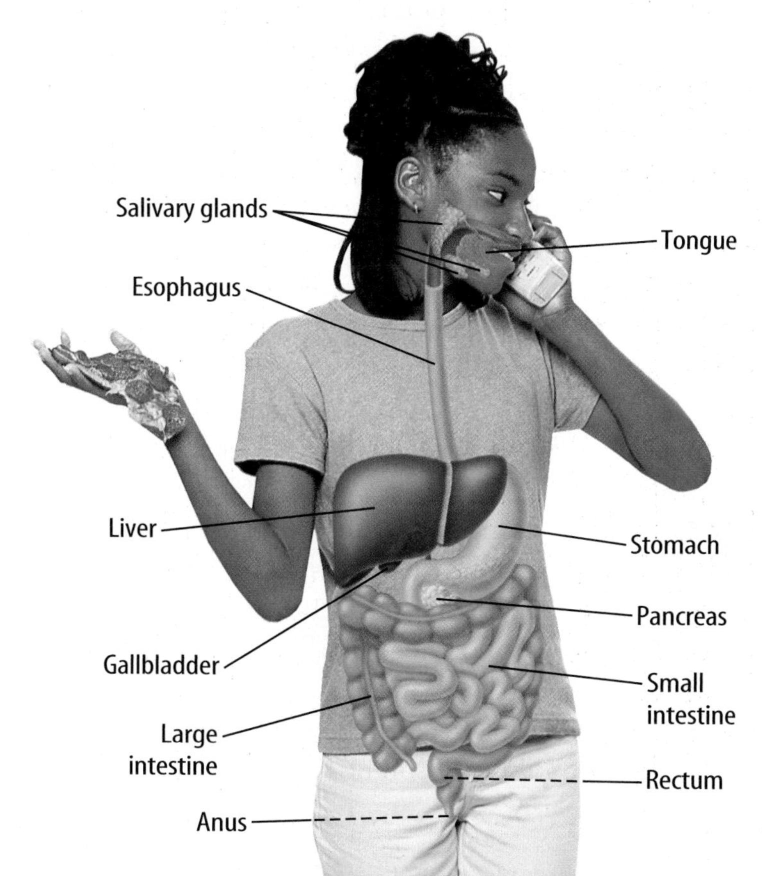

Figure 2 The human digestive system can be described as a tube divided into several specialized sections. If stretched out, an adult's digestive system is 6 m to 9 m long.

Figure 3 About 1.5 L of saliva are produced each day by salivary glands in your mouth.
Describe *what happens in your mouth when you think about a food you like.*

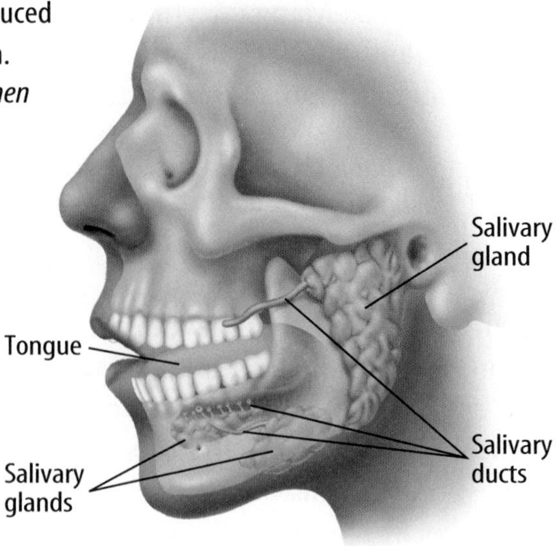

The Mouth Mechanical and chemical digestion begin in your mouth. Mechanical digestion happens when you chew your food with your teeth and mix it with your tongue. Chemical digestion begins with the addition of a watery substance called saliva (suh LI vuh), which contains water, mucus, and an enzyme that aids in the breakdown of starch into sugar. Saliva is produced by three sets of glands near your mouth, shown in **Figure 3.** Food mixed with saliva becomes a soft mass and is moved to the back of your mouth by your tongue. It is swallowed and passes into your esophagus. Now ingestion is complete, but the process of digestion continues.

The Esophagus Food moving into the esophagus passes over a flap of tissue called the epiglottis (eh puh GLAH tus). This structure automatically covers the opening to the windpipe to prevent food from entering it, otherwise you would choke. Your esophagus is a muscular tube about 25 cm long. No digestion takes place in the esophagus. Smooth muscles in the wall of the esophagus move food downward with a squeezing action. These waves of muscle contractions, called **peristalsis** (per uh STAHL sus), move food through the entire digestive tract. Secretions from the mucous glands in the wall of the esophagus keep food moist.

The Stomach The stomach is a muscular bag. When empty, it is somewhat sausage shaped with folds on the inside. As food enters from the esophagus, the stomach expands and the folds smooth out. Mechanical and chemical digestion take place here. Mechanically, food is mixed in the stomach by peristalsis. Chemically, food is mixed with enzymes and strong digestive solutions, such as hydrochloric acid solution, to help break it down.

Specialized cells in the stomach's walls release about two liters of hydrochloric acid solution each day. This solution works with the enzyme pepsin to digest protein and destroys bacteria that are present in food. The stomach also produces mucus, which makes food more slippery and protects the stomach from the strong, digestive solutions. Food is changed in the stomach into a thin, watery liquid called **chyme** (KIME). Slowly, chyme moves out of your stomach and into your small intestine.

Reading Check *Why isn't your stomach digested by the acidic digestive solution?*

The Small Intestine Your small intestine, shown in **Figure 4,** is small in diameter, but it measures 4 m to 7 m in length. As chyme leaves your stomach, it enters the first part of your small intestine, called the duodenum (doo AH duh num). Most digestion takes place in your duodenum. Here, bile—a greenish fluid from the liver—is added. The acidic solution from the stomach makes large fat particles float to the top of the chyme. Bile breaks up the large fat particles, similar to the way detergent breaks up grease.

Chemical digestion of carbohydrates, proteins, and fats occurs when a digestive solution from the pancreas is mixed in. This solution contains bicarbonate ions and enzymes. The bicarbonate ions help neutralize the stomach acid that is mixed with chyme. Your pancreas also makes insulin, a hormone that allows glucose to pass from the bloodstream into your cells.

Absorption of broken down food takes place in the small intestine. The wall of the small intestine, shown in **Figure 4,** has many ridges and folds. These folds are covered with fingerlike projections called **villi** (VIH li). Villi increase the surface area of the small intestine, which allows more places for nutrients to be absorbed. Nutrients move into blood vessels within the villi. From here, blood transports the nutrients to all cells of your body. Peristalsis continues to force the remaining undigested and unabsorbed materials slowly into the large intestine.

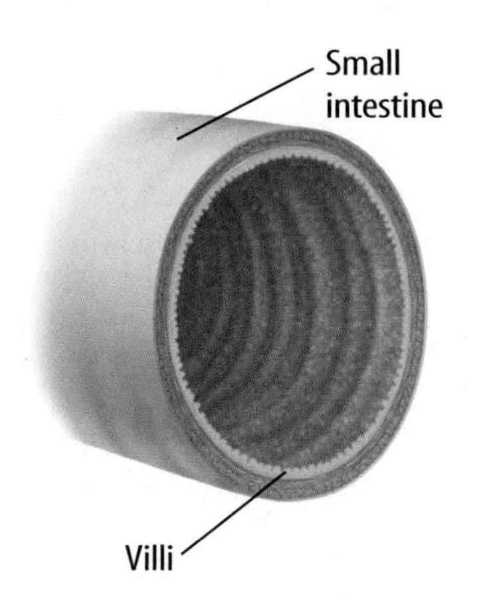

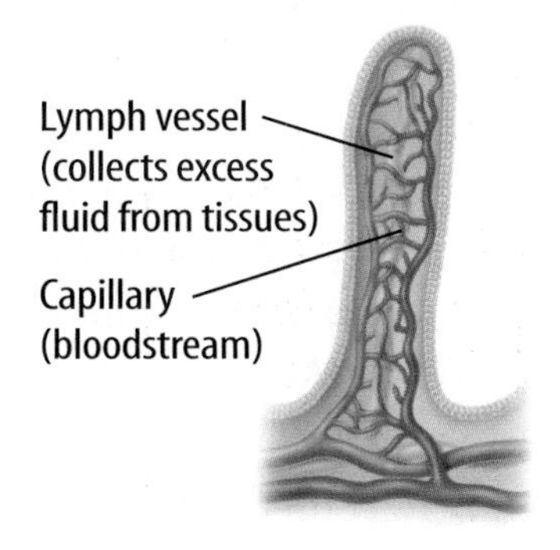

Figure 4 Hundreds of thousands of densely packed villi give the impression of a velvet cloth surface. If the surface area of your villi could be stretched out, it would cover an area the size of a tennis court.
Infer *what would happen to a person's weight if the number of villi were drastically reduced. Why?*

Large Intestine Bacteria The species of bacteria that live in your large intestine are adapted to their habitat. What do you think would happen to the bacteria if their environment were to change? How would this affect your large intestine? Discuss your ideas with a classmate and write your answers in your Science Journal.

The Large Intestine When the chyme enters the large intestine, it is still a thin, watery mixture. The large intestine absorbs water from the undigested mass, which helps maintain homeostasis (hoh mee oh STAY sus). Peristalsis usually slows down in the large intestine. After the excess water is absorbed, the remaining undigested materials become more solid. Muscles in the rectum, which is the last section of the large intestine, and the anus control the release of semisolid wastes from the body in the form of feces (FEE seez).

Bacteria Are Important

Many types of bacteria live in your body. Some bacteria live in many of the organs of your digestive tract including your mouth and large intestine. Some of these bacteria live in a relationship that is beneficial to the bacteria and to your body. The bacteria in your large intestine feed on undigested material like cellulose and make vitamins you need—vitamin K and two B vitamins. Vitamin K is needed for blood clotting. The two B vitamins, niacin and thiamine, are important for your nervous system and for other body functions. Bacterial action also converts bile pigments into new compounds. The breakdown of intestinal materials by bacteria produces gas.

section 1 review

Summary

Functions of the Digestive System

- Food is processed in four stages—ingestion, digestion, absorption, and elimination.

Enzymes

- Enzymes make chemical digestion possible.
- Enzymes are used in other chemical reactions, including blood clotting.

Organs of the Digestive System

- In the digestive system, food passes through the mouth, esophagus, stomach, small intestine, large intestine, rectum, and anus.
- Accessory digestive organs help in mechanical and chemical digestion.

Bacteria Are Important

- Some bacteria that live in the organs of the digestive tract are helpful to your body.

Self Check

1. **Compare and contrast** mechanical digestion and chemical digestion.
2. **Describe** the function of each organ through which food passes as it moves through the digestive tract.
3. **Explain** how activities in the large intestine help maintain homeostasis.
4. **Describe** how the accessory organs aid digestion.
5. **Think Critically** Crackers contain starch. Explain why a cracker begins to taste sweet after it is in your mouth for five minutes without being chewed.

Applying Skills

6. **Recognize Cause and Effect** What would happen to some of the nutrients in chyme if the pancreas did not secrete its solution into the small intestine?
7. **Communicate** Write a paragraph in your Science Journal explaining what would happen to the mechanical and chemical digestion in a person missing a large portion of his or her stomach.

Science Online green.msscience.com/self_check_quiz

section 2

Nutrition

Why do you eat?

You might choose a food because of its taste, because it's readily available, or quickly prepared. However, as much as you don't want to admit it, the nutritional value of and Calories in foods are more important. A Calorie is a measurement of the amount of energy available in food. The amount of food energy a person requires varies with activity level, body weight, age, sex, and natural body efficiency. A chocolate donut might be tasty, quick to eat, and provide plenty of Calories, but it has only some of the nutrients that your body needs.

Classes of Nutrients

Six kinds of nutrients are available in food—proteins, carbohydrates, fats, vitamins, minerals, and water. Proteins, carbohydrates, vitamins, and fats all contain carbon and are called organic nutrients. Inorganic nutrients, such as water and minerals, do not contain carbon. Foods containing carbohydrates, fats, and proteins need to be digested or broken down before your body can use them. Water, vitamins, and minerals don't require digestion and are absorbed directly into your bloodstream.

as you read

***What* You'll Learn**

- **Distinguish** among the six classes of nutrients.
- **Identify** the importance of each type of nutrient.
- **Explain** the relationship between diet and health.

***Why* It's Important**

You can make healthful food choices if you know what nutrients your body uses daily.

Review Vocabulary

molecule: the smallest particle of a substance that retains the properties of the substance and is composed of one or more atoms

New Vocabulary

- amino acid
- carbohydrate
- vitamin
- mineral

Proteins Your body uses proteins for replacement and repair of body cells and for growth. Proteins are large molecules that contain carbon, hydrogen, oxygen, nitrogen, and sometimes sulfur. A molecule of protein is made up of a large number of smaller units, or building blocks, called **amino acids.** You can see some sources of proteins in **Figure 5.**

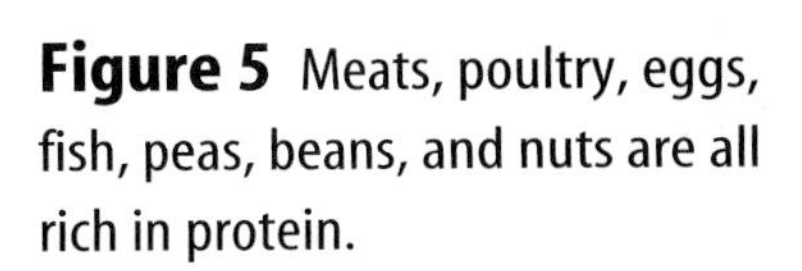

Figure 5 Meats, poultry, eggs, fish, peas, beans, and nuts are all rich in protein.

Topic: Fiber

Visit green.msscience.com for Web links to recent news or magazine articles about the importance of fiber in your diet.

Activity In your Science Journal, classify your favorite foods into two groups—*Good source of fiber* and *Little or no fiber.*

Protein Building Blocks Your body needs only 20 amino acids in various combinations to make the thousands of proteins used in your cells. Most of these amino acids can be made in your body's cells, but eight of them cannot. These eight are called essential amino acids. They have to be supplied by the foods you eat. Complete proteins provide all of the essential amino acids. Eggs, milk, cheese, and meat contain complete proteins. Incomplete proteins are missing one or more of the essential amino acids. If you are a vegetarian, you can get all of the essential amino acids by eating a wide variety of protein-rich vegetables, fruits, and grains.

Carbohydrates Study the nutrition label on several boxes of cereal. You'll notice that the number of grams of carbohydrates found in a typical serving of cereal is higher than the amounts of the other nutrients. **Carbohydrates** (kar boh HI drayts) usually are the main sources of energy for your body.

Three types of carbohydrates are sugar, starch, and fiber, shown in **Figure 6.** Sugars are called simple carbohydrates. You're probably most familiar with table sugar. However, fruits, honey, and milk also contain forms of sugar. Your cells break down glucose, a simple sugar.

The other two types of carbohydrates—starch and fiber—are called complex carbohydrates. Starch is found in potatoes and foods made from grains such as pasta. Starches are made up of many simple sugars. Fiber, such as cellulose, is found in the cell walls of plant cells. Foods like whole-grain breads and cereals, beans, peas, and other vegetables and fruits are good sources of fiber. Because different types of fiber are found in foods, you should eat a variety of fiber-rich plant foods. You cannot digest fiber, but it is needed to keep your digestive system running smoothly.

Figure 6 These foods contain carbohydrates that provide energy for all the things that you do. **Describe** *the role of carbohydrates in your body.*

Fats The term *fat* has developed a negative meaning for some people. However, fats, also called lipids, are necessary because they provide energy and help your body absorb vitamins. Fat tissue cushions your internal organs. A major part of every cell membrane is made up of a type of fat.

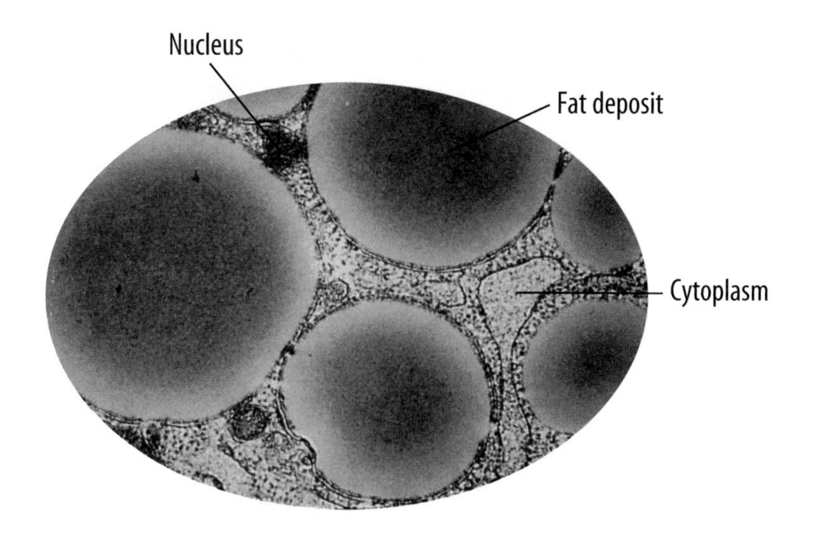

Figure 7 Fat is stored in certain cells in your body. Fat deposits push the cytoplasm and nucleus to the edge of the cell.

A gram of fat can release more than twice as much energy as a gram of carbohydrate can. Because fat is a good storage unit for energy, excess energy from the foods you eat is converted to fat and stored for later use, as shown in **Figure 7.**

Reading Check *Why is fat a good storage unit for energy?*

Fats are classified as unsaturated or saturated based on their chemical structure. Unsaturated fats are usually liquid at room temperature. Vegetable oils as well as fats found in seeds are unsaturated fats. Saturated fats are found in meats, animal products, and some plants and are usually solid at room temperature. Saturated fats have been associated with high levels of blood cholesterol. Your body makes cholesterol in your liver. Cholesterol is part of the cell membrane in all of your cells. However, a diet high in cholesterol may result in deposits forming on the inside walls of blood vessels. These deposits can block the blood supply to organs and increase blood pressure. This can lead to heart disease and strokes.

Vitamins Your bone cells need vitamin D to use calcium, and your blood needs vitamin K in order to clot. **Vitamins** are organic nutrients needed in small quantities for growth, regulating body functions, and preventing some diseases.

Vitamins are classified into two groups. Some vitamins dissolve easily in water and are called water-soluble vitamins. They are not stored by your body so you have to consume them daily. Other vitamins dissolve only in fat and are called fat-soluble vitamins. These vitamins are stored by your body. Although you eat or drink most vitamins, some are made by your body. Vitamin D is made when your skin is exposed to sunlight. Recall that vitamin K and two of the B vitamins are made in your large intestine with the help of bacteria that live there.

Mini LAB

Comparing the Fat Content of Foods

Procedure

1. Collect three pieces of each of the following foods: **potato chips; pretzels; peanuts;** and **small cubes of fruits, cheese, vegetables,** and **meat.**
2. Place the food items on a piece of **brown grocery bag.** Label the paper with the name of each food. Do not taste the foods.
3. Allow foods to sit for 30 min.
4. Remove the items, properly dispose of them, and observe the paper.

Analysis

1. Which items left a translucent (greasy) mark? Which left a wet mark?
2. How are the foods that left a greasy mark on the paper alike?
3. Use this test to determine which other foods contain fats. A greasy mark means the food contains fat. A wet mark means the food contains a lot of water.

Table 1 Minerals

Mineral	Health Effect	Food Sources
Calcium	strong bones and teeth, blood clotting, muscle and nerve activity	dairy products, eggs, green leafy vegetables, soy
Phosphorus	strong bones and teeth, muscle contraction, stores energy	cheese, meat, cereal
Potassium	balance of water in cells, nerve impulse conduction, muscle contraction	bananas, potatoes, nuts, meat, oranges
Sodium	fluid balance in tissues, nerve impulse conduction	meat, milk, cheese, salt, beets, carrots, nearly all foods
Iron	oxygen is transported in hemoglobin by red blood cells	red meat, raisins, beans, spinach, eggs
Iodine (trace)	thyroid activity, metabolic stimulation	seafood, iodized salt

Minerals Inorganic nutrients—nutrients that lack carbon and regulate many chemical reactions in your body—are called **minerals.** Of about 14 minerals that your body uses, calcium and phosphorus are used in the largest amounts for a variety of body functions. One of these functions is the formation and maintenance of bone. Some minerals, called trace minerals, are required only in small amounts. Copper and iodine usually are listed as trace minerals. Minerals are not used by the body as a source of energy. However, they do serve many different functions. Several minerals, their health effects, and some food sources for them are listed in **Table 1.**

Reading Check *Why is copper considered a trace mineral?*

Water Next to oxygen, water is the most important factor for survival. Different organisms need different amounts of water to survive. You could live for a few weeks without food but for only a few days without water because your cells need water to carry out their work. Most of the nutrients you have studied in this chapter can't be used by your body unless they are carried in a solution. This means that they have to be dissolved in water. In cells, chemical reactions take place in solutions.

Salt Mines The mineral halite is processed to make table salt. In the United States, most salt comes from underground mines. Research to find the location of these mines, then label them on a map.

The human body is about 60 percent water by mass. About two-thirds of your body water is located in your body cells. Water also is found around cells and in body fluids such as blood. **Table 2** shows how your body loses water every day. To replace water lost each day, you need to drink about 2 L of liquids. However, drinking liquids isn't the only way to supply cells with water. Most foods have more water than you realize. An apple is about 80 percent water, and many meats are 90 percent water.

Table 2 Water Loss

Method of Loss	Amount (mL/day)
Exhaled air	350
Feces	150
Skin (mostly as sweat)	500
Urine	1,500

Why do you get thirsty? Your body is made up of systems that operate together. When your body needs to replace lost water, messages are sent to your brain that result in a feeling of thirst. Drinking water satisfies your thirst and usually restores the body's homeostasis. When homeostasis is restored, the signal to the brain stops and you no longer feel thirsty.

Food Groups

Because no naturally occurring food has every nutrient, you need to eat a variety of foods. Nutritionists have developed a simple system, called the food pyramid, shown in **Figure 8,** to help people select foods that supply all the nutrients needed for energy and growth. The recommended daily amount for each food group will supply your body with the nutrients it needs for good health.

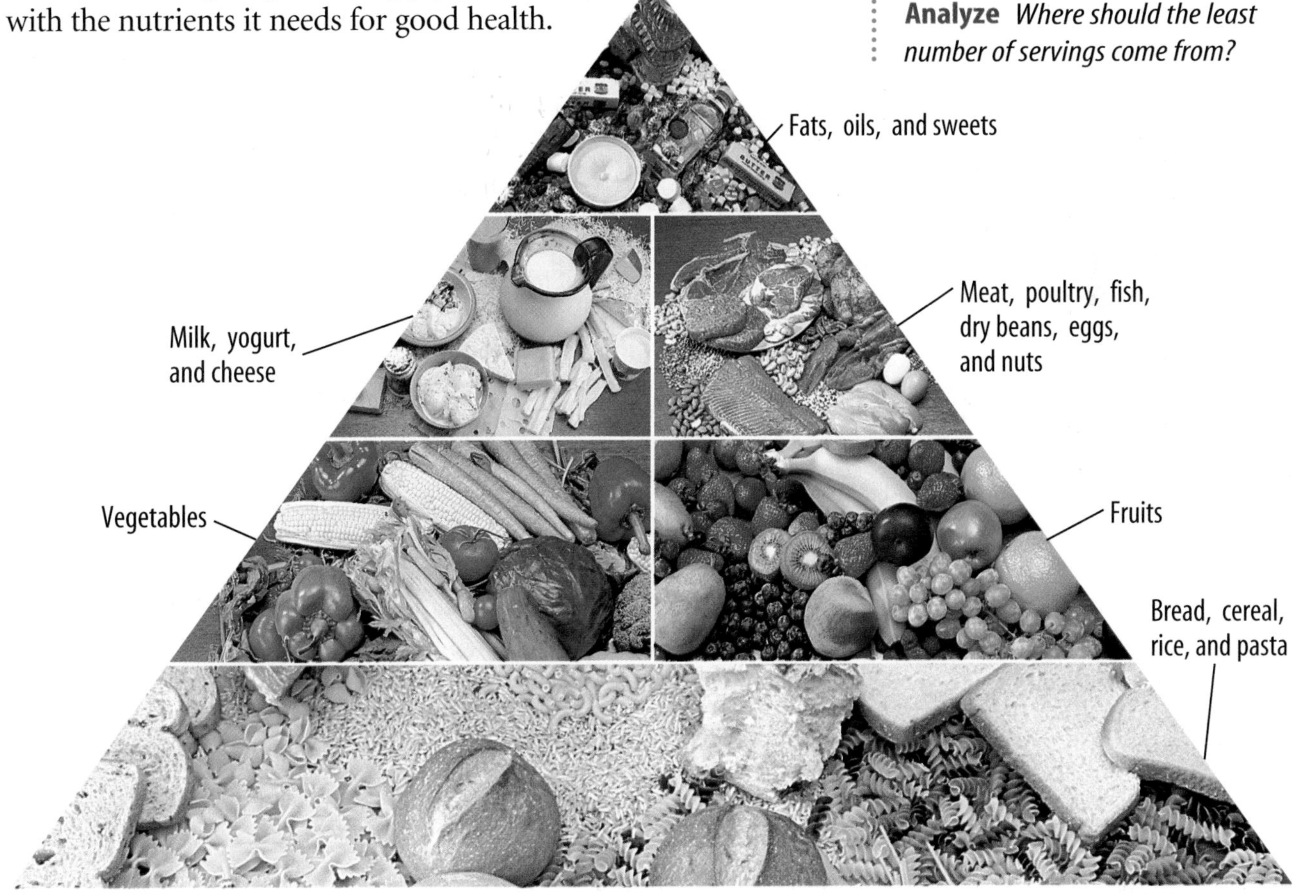

Figure 8 The pyramid shape reminds you that you should consume more servings from the bread and cereal group than from other groups.
Analyze *Where should the least number of servings come from?*

Figure 9 The information on a food label can help you decide what to eat.

Nutrition Facts
Serving Size 1 Meal

Amount Per Serving	
Calories 330	Calories from Fat 60
	% Daily Value*
Total Fat 7g	**10%**
Saturated Fat 3.5g	**17%**
Polyunsaturated Fat 1g	
Monounsaturated Fat 2.5g	
Cholesterol 35mg	**12%**
Sodium 460mg	**19%**
Total Carbohydrate 52g	**18%**
Dietary Fiber 6g	**24%**
Sugars 17g	
Protein 15g	
Vitamin A 15% •	Vitamin C 70%
Calcium 4% •	Iron 10%

* Percent Daily Values are based on a 2,000 calorie diet. Your daily values may be higher or lower depending on your calorie needs.

	Calories	2,000	2,500
Total Fat	Less than	65g	80g
Sat Fat	Less than	20g	25g
Cholesterol	Less than	300mg	300mg
Sodium	Less than	2,400mg	2,400mg
Total Carbohydrate		300g	375g
Dietary Fiber		25g	30g

Daily Servings Each day you should eat six to eleven servings from the bread and cereal group, three to five servings from the vegetable group, two to four servings from the fruit group, two to three servings from the milk group, and two to three servings from the meat and beans group. Only small amounts of fats, oils, and sweets should be consumed.

The size of a serving is different for different foods. For example, a slice of bread or one ounce of ready-to-eat cereal is a bread and cereal group serving. One cup of raw leafy vegetables or one-half cup of cooked or chopped raw vegetables make a serving from the vegetable group. One medium apple, banana, or orange, one-half cup of canned fruit, or three-quarter cup of fruit juice is a fruit serving. A serving from the milk group can be one cup of milk or yogurt. Two ounces of cooked lean meat, one-half cup of cooked dry beans, one egg, or two tablespoons of peanut butter counts as a serving from the meat and beans group.

Food Labels The nutritional facts found on all packaged foods make it easier to make healthful food choices. These labels, as shown in **Figure 9,** can help you plan meals that supply the daily recommended amounts of nutrients and meet special dietary requirements (for example, a low-fat diet).

section 2 review

Summary

Why do you eat?

- Nutrients in food provide energy and materials for cell development, growth, and repair.

Classes of Nutrients

- Six kinds of nutrients are found in food—proteins, carbohydrates, fats, vitamins, minerals, and water.
- Proteins are used for growth and repair, carbohydrates provide energy, and fats store energy and cushion organs.
- Vitamins and minerals regulate body functions.
- Next to oxygen, water is the most important factor for your survival.

Food Groups

- The food pyramid and nutritional labels can help you choose foods that supply all the nutrients you need for energy and growth.

Self Check

1. **List** one example of a food source for each of the six classes of nutrients.
2. **Explain** how your body uses each class of nutrients.
3. **Discuss** how food choices can positively and negatively affect your health.
4. **Explain** the importance of water in the body.
5. **Think Critically** What foods from each food group would provide a balanced breakfast? Explain.

Applying Skills

6. **Interpret Data** Nutritional information can be found on the labels of most foods. Interpret the labels found on three different types of food products.
7. **Use a Spreadsheet** Make a spreadsheet of the minerals listed in **Table 1.** Use reference books to gather information about minerals and add these to the table: *sulfur, magnesium, copper, manganese, cobalt,* and *zinc.*

Science Online green.msscience.com/self_check_quiz

Identifying Vitamin C Content

Vitamin C is found in many fruits and vegetables. Oranges have a high vitamin C content. Try this lab to test the vitamin C content in different orange juices.

Real-World Question

Which orange juice contains the most vitamin C?

Goals

- **Observe** the vitamin C content of different orange juices.

Materials

test tube (4)
paper cups
test-tube rack
masking tape
wooden stirrer (13)
graduated cylinder
graduated container
2% tincture of iodine
dropper
cornstarch
triple-beam balance
weighing paper
water (50 mL)
glass-marking pencil
dropper bottles (4) containing orange juice that is:
(1) freshly squeezed
(2) from frozen concentrate
(3) canned
(4) in a carton

** Alternate materials*

Safety Precautions

WARNING: *Do not taste any of the juices. Iodine is poisonous and can stain skin and clothing. It is an irritant and can cause damage if it comes in contact with your eyes. Notify your teacher if a spill occurs.*

Procedure

1. Make a data table like the example shown to record your observations.
2. Label four test tubes 1 through 4 and place them in the test-tube rack.

Drops of Iodine Needed to Change Color

Juice	Trial 1	Trial 2	Trial 3	Average
1 Fresh juice				
2 Frozen juice	Do not write in this book.			
3 Canned juice				
4 Carton juice				

3. Measure and pour 5 mL of juice from bottle 1 into test tube 1, 5 mL from bottle 2 into test tube 2, 5 mL from bottle 3 into test tube 3, and 5 mL from bottle 4 into test tube 4.
4. Measure 0.3 g of cornstarch, then put it in a container. Slowly mix in 50 mL of water until the cornstarch completely dissolves.
5. Add 5 mL of the cornstarch solution to each of the four test tubes. Stir well.
6. Add iodine to test tube 1, one drop at a time. Stir after each drop. Record the number of drops it takes for the juice to change to a purple color. The more vitamin C that is present, the more drops it takes to change color.
7. Repeat step 6 with test tubes 2, 3, and 4.
8. Empty and clean the test tubes. Repeat steps 3 through 7 two more times, then average your results.
9. Dispose of all materials as directed by your teacher. Wash your hands thoroughly.

Conclude and Apply

1. **Compare and contrast** the amount of vitamin C in the orange juices tested.
2. **Infer** why the amount of vitamin C varies in the orange juices.

section 3

The Respiratory System

as you read

What You'll Learn

- **Describe** the functions of the respiratory system.
- **Explain** how oxygen and carbon dioxide are exchanged in the lungs and in tissues.
- **Identify** the pathway of air in and out of the lungs.
- **Explain** the effects of smoking on the respiratory system.

Why It's Important

Your body's cells depend on your respiratory system to supply oxygen and remove carbon dioxide.

Review Vocabulary

diaphragm: muscle beneath the lungs that contracts and relaxes to move gases in and out of the body

New Vocabulary

- larynx
- trachea
- bronchi
- alveoli

Functions of the Respiratory System

Can you imagine an astronaut walking on the Moon without a space suit or a diver exploring the ocean without scuba gear? Of course not. They couldn't survive in either location under those conditions because humans need to breathe air.

People often confuse the terms *breathing* and *respiration.* Breathing is the movement of the chest that brings air into the lungs and removes waste gases. The air entering the lungs contains oxygen. It passes from the lungs into the circulatory system because there is less oxygen in blood when it enters the lungs than in cells of the lungs.

Blood carries oxygen and glucose from digested food to individual cells. In cells, they are raw materials for a series of chemical reactions called cellular respiration. Without oxygen, cellular respiration cannot occur. Cellular respiration results in the release of energy from glucose. Water and carbon dioxide are waste products of cellular respiration. Blood carries them back to the lungs. As shown in **Figure 10,** exhaling, or breathing out, eliminates waste carbon dioxide and some water molecules.

Reading Check *What is cellular respiration?*

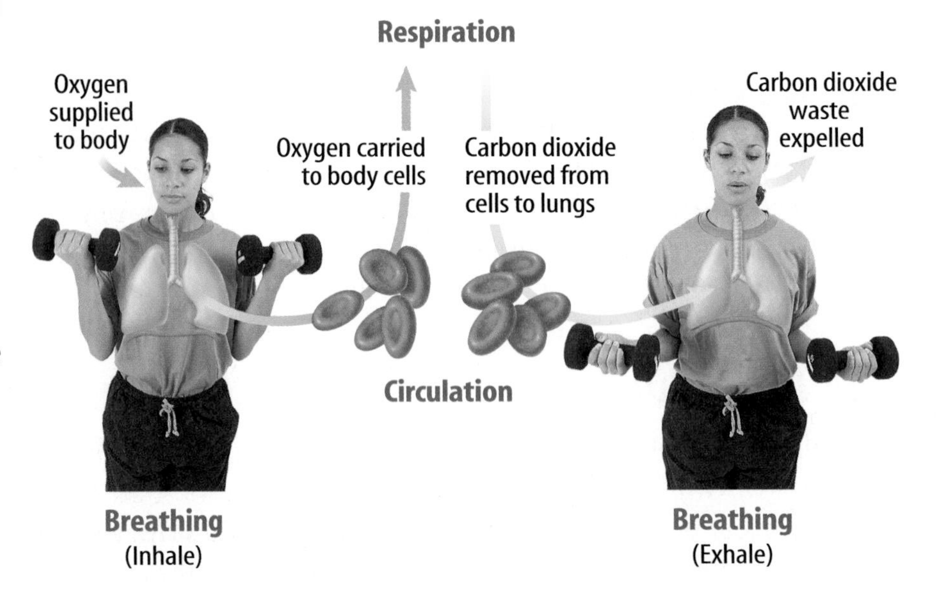

Figure 10 Several processes are involved in how the body obtains, transports, and uses oxygen.

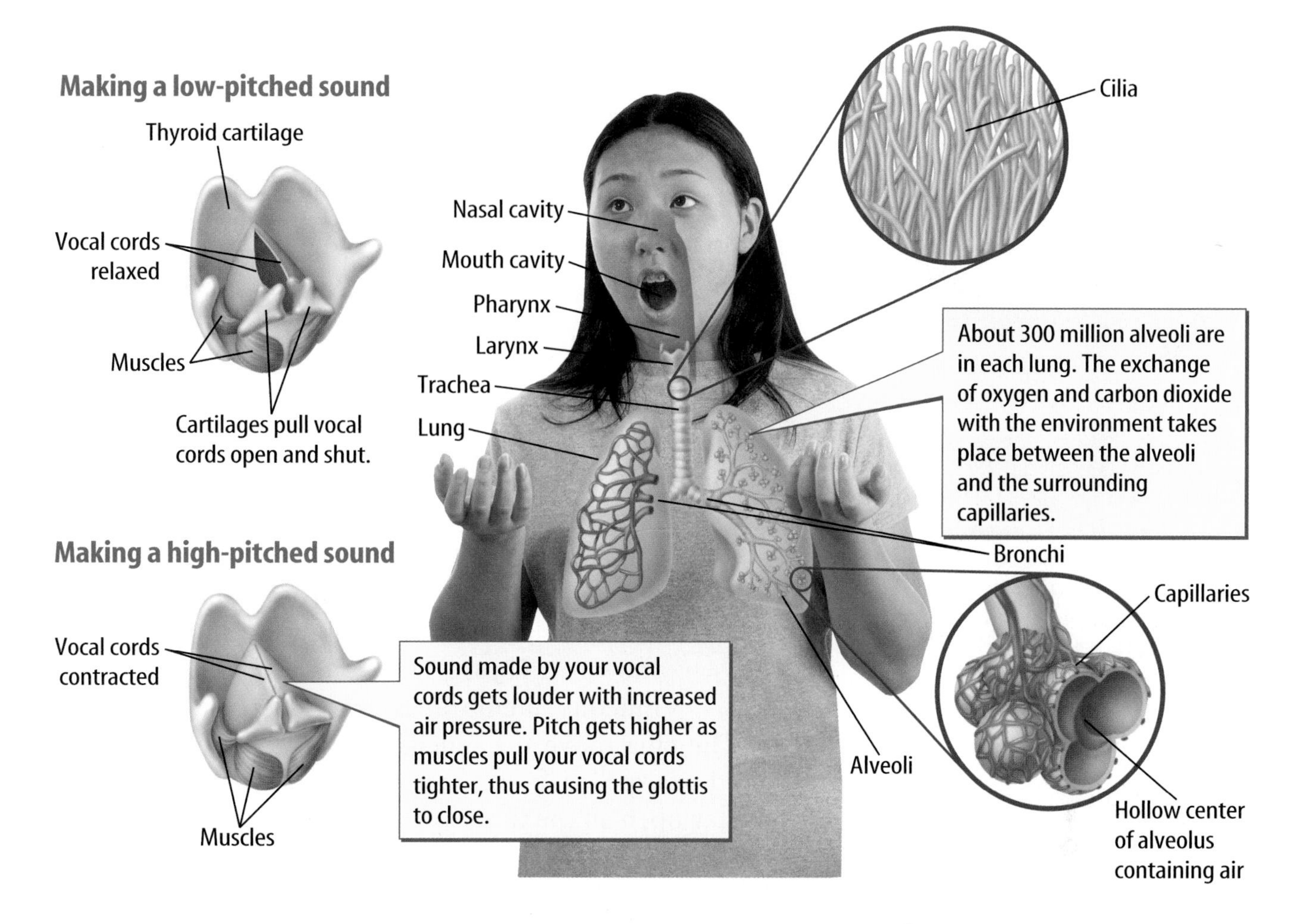

Figure 11 Air can enter the body through the nostrils and the mouth.
Explain *an advantage of having air enter through the nostrils.*

Organs of the Respiratory System

The respiratory system, shown in **Figure 11,** is made up of structures and organs that help move oxygen into the body and waste gases out of the body. Air enters your body through two openings in your nose called nostrils or through the mouth. Fine hairs inside the nostrils trap particles from the air. Air then passes through the nasal cavity, where it gets moistened and warmed by the body's heat. Glands that produce sticky mucus line the nasal cavity. The mucus traps particles that were not trapped by nasal hairs. This process helps filter and clean the air you breathe. Tiny, hairlike structures, called cilia (SIH lee uh), sweep mucus and trapped material to the back of the throat where it can be swallowed.

Pharynx Warmed, moist air then enters the pharynx (FER ingks), which is a tubelike passageway for food, liquids, and air. At the lower end of the pharynx is the epiglottis. When you swallow, your epiglottis folds down, which allows food or liquids to enter your esophagus instead of your airway. What do you think has happened if you begin to choke?

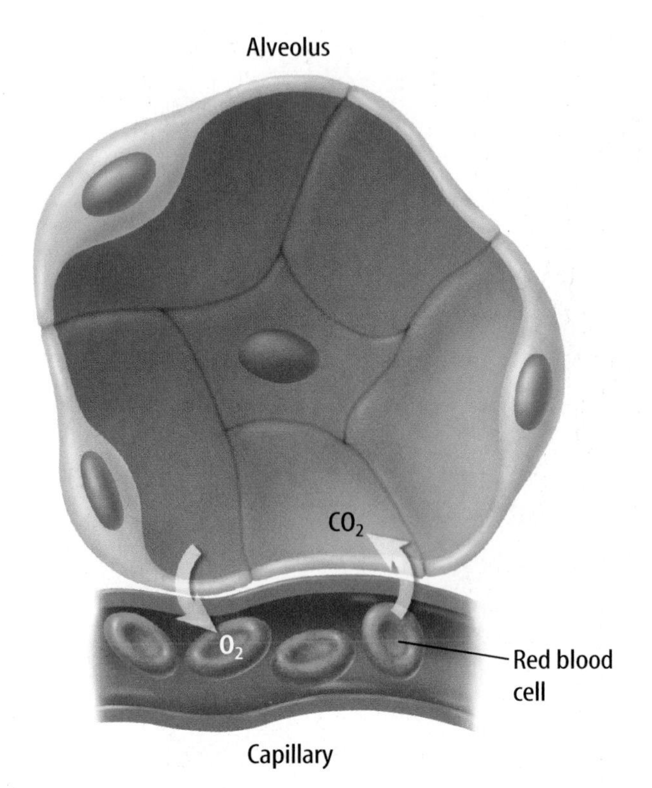

Figure 12 The thin capillary walls allow gases to be exchanged easily between the alveoli and the capillaries.
Name *the two gases that are exchanged by the capillaries and alveoli.*

Larynx and Trachea Next, the air moves into your larynx (LER ingks). The **larynx** is the airway to which two pairs of horizontal folds of tissue, called vocal cords, are attached, as shown in **Figure 11** on the previous page. Forcing air between the cords causes them to vibrate and produce sounds. When you speak, muscles tighten or loosen your vocal cords, resulting in different sounds. Your brain coordinates the movement of the muscles in your throat, tongue, cheeks, and lips when you talk, sing, or just make noise. Your teeth also are involved in forming letter sounds and words.

From the larynx, air moves into the **trachea** (TRAY kee uh). Strong, C-shaped rings of cartilage prevent the trachea from collapsing. It is lined with mucous membranes and cilia, also shown in **Figure 11** on the previous page. The mucous membranes trap dust, bacteria, and pollen. The cilia move the mucus upward, where it is either swallowed or expelled from the nose or mouth. Why must the trachea stay open all the time?

Bronchi and the Lungs Air is carried into your lungs by two short tubes called **bronchi** (BRAHN ki) (singular, *bronchus*) at the lower end of the trachea. Within the lungs, the bronchi branch into smaller and smaller tubes. The smallest tubes are called bronchioles (BRAHN kee ohlz). At the end of each bronchiole are clusters of tiny, thin-walled sacs called **alveoli** (al VEE uh li) (singular, *alveolus*). Air passes into the bronchi, then into the bronchioles, and finally into the alveoli. Lungs are masses of alveoli, like the one shown in **Figure 12,** arranged in grapelike clusters. The capillaries surround the alveoli like a net.

The exchange of oxygen and carbon dioxide takes place between the alveoli and capillaries. The walls of the alveoli and capillaries are only one cell thick, as shown in **Figure 12.** Oxygen moves through the cell membranes of alveoli and through cell membranes of the capillaries into the blood. In blood, oxygen is picked up by hemoglobin (HEE muh gloh bun), a molecule in red blood cells, and carried to all body cells. At the same time, carbon dioxide and other cellular wastes leave the body cells and move into capillaries. Then they are carried by the blood to the lungs. In the lungs, waste gases move through cell membranes from capillaries into alveoli. Then waste gases leave the body when you exhale.

Why do you breathe?

Signals from your brain tell the muscles in your chest and abdomen to contract and relax. You don't have to think about breathing to breathe, just like your heart beats without you telling it to beat. Your brain can change your breathing rate depending on the amount of carbon dioxide present in your blood. If a lot of carbon dioxide is present, your breathing rate increases. It decreases if less carbon dioxide is in your blood. You do have some control over your breathing—you can hold your breath if you want to. Eventually, your brain will respond to the buildup of carbon dioxide in your blood and signal your chest and abdomen muscles to work automatically. You will breathe whether you want to or not.

Inhaling and Exhaling Breathing is partly the result of changes in volume and resulting air pressure. Under normal conditions, a gas moves from an area of higher pressure to an area of lower pressure. When you squeeze an empty, soft-plastic bottle, air is pushed out. This happens because air pressure outside the top of the bottle is less than the pressure you create inside the bottle when you changed its volume. As you release your grip on the bottle, the air pressure inside the bottle becomes less than it is outside the bottle because the bottle's volume changed. Air rushes back in, and the bottle returns to its original shape.

Your lungs work in a way similar to the squeezed bottle. Your diaphragm (DI uh fram) contracts and relaxes, changing the volume of the chest, which helps move gases into and out of your lungs. **Figure 13** illustrates breathing.

Reading Check *How does your diaphragm help you breathe?*

When a person's airway is blocked, a rescuer can use abdominal thrusts, as shown in **Figure 14,** to save the life of the choking victim.

Comparing Surface Area

Procedure

1. Stand a **bathroom-tissue cardboard tube** in an **empty bowl.**
2. Drop **marbles** into the tube, filling it to the top.
3. Count the number of marbles used.
4. Repeat steps 2 and 3 two more times. Calculate the average number of marbles needed to fill the tube.
5. The tube's inside surface area is approximately 161.29 cm^2. Each marble has a surface area of approximately 8.06 cm^2. Calculate the surface area of the average number of marbles.

Analysis

1. Compare the inside surface area of the tube with the surface area of the average number of marbles needed to fill the tube.
2. If the tube represents a bronchus, what do the marbles represent?
3. Using this model, explain what makes gas exchange in the lungs efficient.

Try at Home

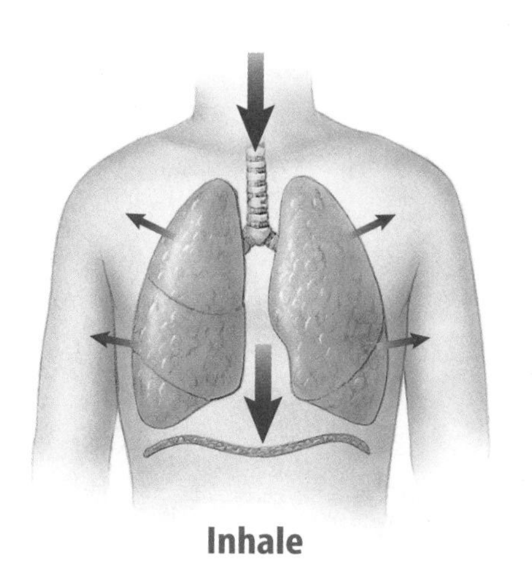

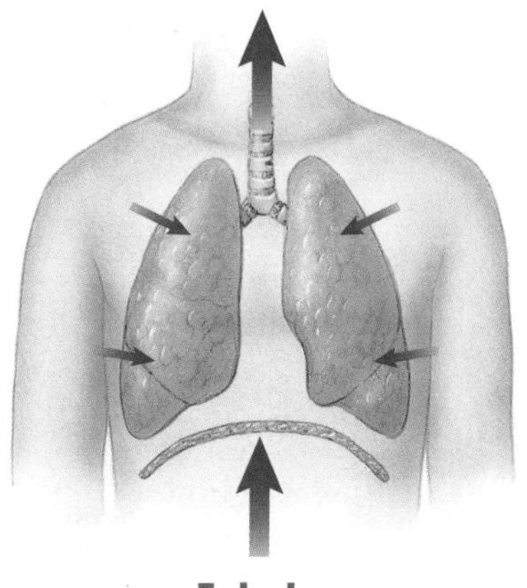

Figure 13 Your lungs inhale and exhale about 500 mL of air with an average breath. This can increase to 2,000 mL of air per breath when you do strenuous activity.

NATIONAL GEOGRAPHIC

VISUALIZING ABDOMINAL THRUSTS

Figure 14

When food or other objects become lodged in the trachea, airflow between the lungs and the mouth and nasal cavity is blocked. Death can occur in minutes. However, prompt action by someone can save the life of a choking victim. The rescuer uses abdominal thrusts to force the victim's diaphragm up. This decreases the volume of the chest cavity and forces air up in the trachea. The result is a rush of air that dislodges and expels the food or other object. The victim can breathe again. This technique is shown at right and should only be performed in emergency situations.

Food is lodged in the victim's trachea.

The rescuer places her fist against the victim's stomach.

The rescuer's second hand adds force to the fist.

A The rescuer stands behind the choking victim and wraps her arms around the victim's upper abdomen. She places a fist (thumb side in) against the victim's stomach. The fist should be below the ribs and above the navel.

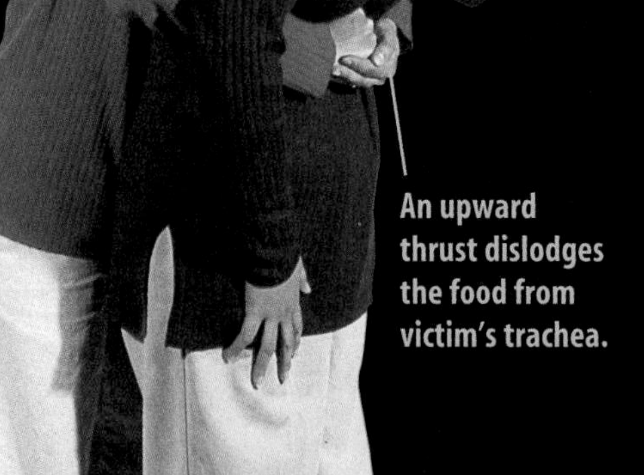

An upward thrust dislodges the food from victim's trachea.

B With a violent, sharp movement, the rescuer thrusts her fist up into the area below the ribs. This action should be repeated as many times as necessary.

Diseases and Disorders of the Respiratory System

INTEGRATE Health

If you were asked to make a list of some things that can harm your respiratory system, you probably would put smoking at the top. As you can see in **Table 3,** many serious diseases are related to smoking. The chemical substances in tobacco—nicotine and tars—are poisons and can destroy cells. The high temperatures, smoke, and carbon monoxide produced when tobacco burns also can injure a smoker's cells. Even if you are a nonsmoker, inhaling smoke from tobacco products—called secondhand smoke—is unhealthy and has the potential to harm your respiratory system. Smoking, polluted air, coal dust, and asbestos (as BES tus) have been related to respiratory problems such as asthma (AZ muh), bronchitis (brahn KI tus), emphysema (em fuh SEE muh), and cancer.

Table 3 Smokers' Risk of Death from Disease

Disease	Smokers' Risk Compared to Nonsmokers' Risk
Lung cancer	23 times higher for males; 11 times higher for females
Chronic bronchitis and emphysema	5 times higher
Heart disease	2 times higher

Respiratory Infections Bacteria, viruses, and other microorganisms can cause infections that affect any of the organs of the respiratory system. The common cold usually affects the upper part of the respiratory system—from the nose to the pharynx. The cold virus also can cause irritation and swelling in the larynx, trachea, and bronchi. The cilia that line the trachea and bronchi can be damaged. However, cilia usually heal rapidly.

Chronic Bronchitis When bronchial tubes are irritated and swell and too much mucus is produced, a disease called bronchitis develops. Many cases of bronchitis clear up within a few weeks, but the disease sometimes lasts for a long time. When this happens, it is called chronic (KRAH nihk) bronchitis.

Emphysema A disease in which the alveoli in the lungs enlarge is called emphysema. When cells in the alveoli are reddened and swollen, an enzyme is released that causes the walls of the alveoli to break down. As a result, alveoli can't push air out of the lungs, so less oxygen moves into the bloodstream from the alveoli. When blood becomes low in oxygen and high in carbon dioxide, shortness of breath occurs.

Topic: Secondhand Smoke
Visit green.msscience.com for Web links to information about the health aspects of secondhand smoke.

Activity Write a paragraph in your Science Journal summarizing the possible effects of secondhand smoke on your health.

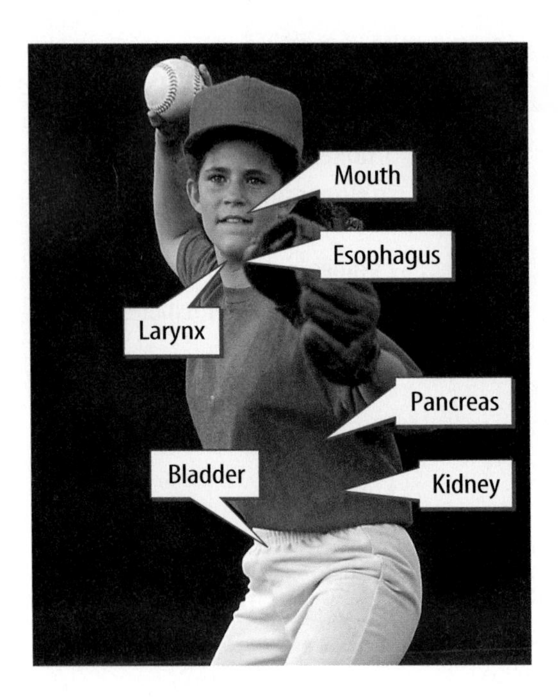

Figure 15 More than 85 percent of all lung cancer is related to smoking. Smoking also can play a part in the development of cancer in other body organs indicated above.

Lung Cancer The third leading cause of death in men and women in the United States is lung cancer. Inhaling the tar in cigarette smoke is the greatest contributing factor to lung cancer. In the body, tar and other ingredients found in smoke act as carcinogens (kar SIH nuh junz). Carcinogens are substances that can cause an uncontrolled growth of cells. In the lungs, this is called lung cancer. Lung cancer is not easy to detect in its early stages. Smoking also has been linked to the development of cancers of the mouth, esophagus, larynx, pancreas, kidney, and bladder. See **Figure 15.**

Asthma Shortness of breath, wheezing, or coughing can occur in a lung disorder called asthma. When a person has an asthma attack, the bronchial tubes contract quickly. Inhaling medicine that relaxes the bronchial tubes is the usual treatment for an asthma attack. Asthma can be an allergic reaction. An allergic reaction occurs when the body overreacts to a foreign substance. An asthma attack can result from breathing certain substances such as cigarette smoke or certain plant pollen, eating certain foods, or stress in a person's life.

section 3 review

Summary

Functions of the Respiratory System

- Breathing moves the chest to bring air into and remove wastes from the lungs.
- Cellular respiration uses oxygen to release energy from glucose.

Organs of the Respiratory System

- Air flows from your nostrils or mouth through the pharynx, larynx, trachea, bronchi, and into the alveoli of your lungs.
- Alveoli and capillaries exchange oxygen and carbon dioxide.

Why do you breathe?

- Your brain sends signals to your chest and abdominal muscles to contract and relax, which controls breathing rate.

Diseases of the Respiratory System

- Problems of the respiratory system include chronic bronchitis, emphysema, and lung cancer.

Self Check

1. **State** the main function of the respiratory system.
2. **Describe** the exchange of oxygen, carbon dioxide, and other waste gases in the lungs and body tissues.
3. **Explain** how air moves into and out of the lungs.
4. **Describe** the effects of smoking on the respiratory and circulatory systems.
5. **Think Critically** How is the work of the digestive and circulatory systems related to the respiratory system?

Applying Skills

6. **Research Information** Nicotine in tobacco is a poison. Using library references, find out how nicotine affects the body.
7. **Communicate** Use references to find out about lung disease common among coal miners, stonecutters, and sandblasters. Find out what safety measures are required now for these trades. In your Science Journal, write a paragraph about these safety measures.

green.msscience.com/self_check_quiz

The Excretory System

Functions of the Excretory System

It's your turn to take out the trash. You carry the bag outside and put it in the trash can. The next day, you bring out another bag of trash, but the trash can is full. When trash isn't collected, it piles up. Just as trash needs to be removed from your home to keep it livable, your body must eliminate wastes to remain healthy. Undigested material is eliminated by your large intestine. Waste gases are eliminated through the combined efforts of your circulatory and respiratory systems. Some salts are eliminated when you sweat. These systems function together as parts of your excretory system. If wastes aren't eliminated, toxic substances build up and damage organs. If not corrected, serious illness or death occurs.

The Urinary System

Figure 16 shows how the urinary system functions as a part of the excretory system. The urinary system rids the blood of wastes produced by the cells. It controls blood volume by removing excess water produced by body cells during cellular respiration. The urinary system also balances the amounts of certain salts and water that must be present for all cellular activities.

as you read

***What* You'll Learn**

- **Distinguish** between the excretory and urinary systems.
- **Describe** how the kidneys work.
- **Explain** what happens when urinary organs don't work.

***Why* It's Important**

The urinary system helps clean your blood of cellular wastes.

Review Vocabulary
capillary: blood vessel that connects arteries and veins

New Vocabulary
- nephron
- ureter
- bladder

Figure 16 The urinary, digestive, and respiratory systems, and the skin, make up the excretory system.

Digestive System
Food and liquid in
Water and undigested food out

Respiratory System
Oxygen in
Carbon dioxide and water out

Skin
Salt and some organic substances out

Urinary System
Water and salts in
Excess water, metabolic wastes, and salts out

Excretion

Regulating Fluid Levels To stay in good health, the fluid levels within the body must be balanced and normal blood pressure must be maintained. An area in the brain, the hypothalamus (hi poh THA luh mus), constantly monitors the amount of water in the blood. When the brain detects too much water in the blood, the hypothalamus releases a lesser amount of a specific hormone. This signals the kidneys to return less water to the blood and increase the amount of urine that is excreted.

Reading Check *How does the urinary system control the volume of water in the blood?*

Organs of the Urinary System Excretory organs is another name for the organs of the urinary system. The main organs of the urinary system are two bean-shaped kidneys. Kidneys are located on the back wall of the abdomen at about waist level. The kidneys filter blood that contains wastes collected from cells. In approximately 5 min, all of the blood in your body passes through the kidneys. The red-brown color of the kidneys is due to their enormous blood supply. In **Figure 17,** you can see that blood enters the kidneys through a large artery and leaves through a large vein.

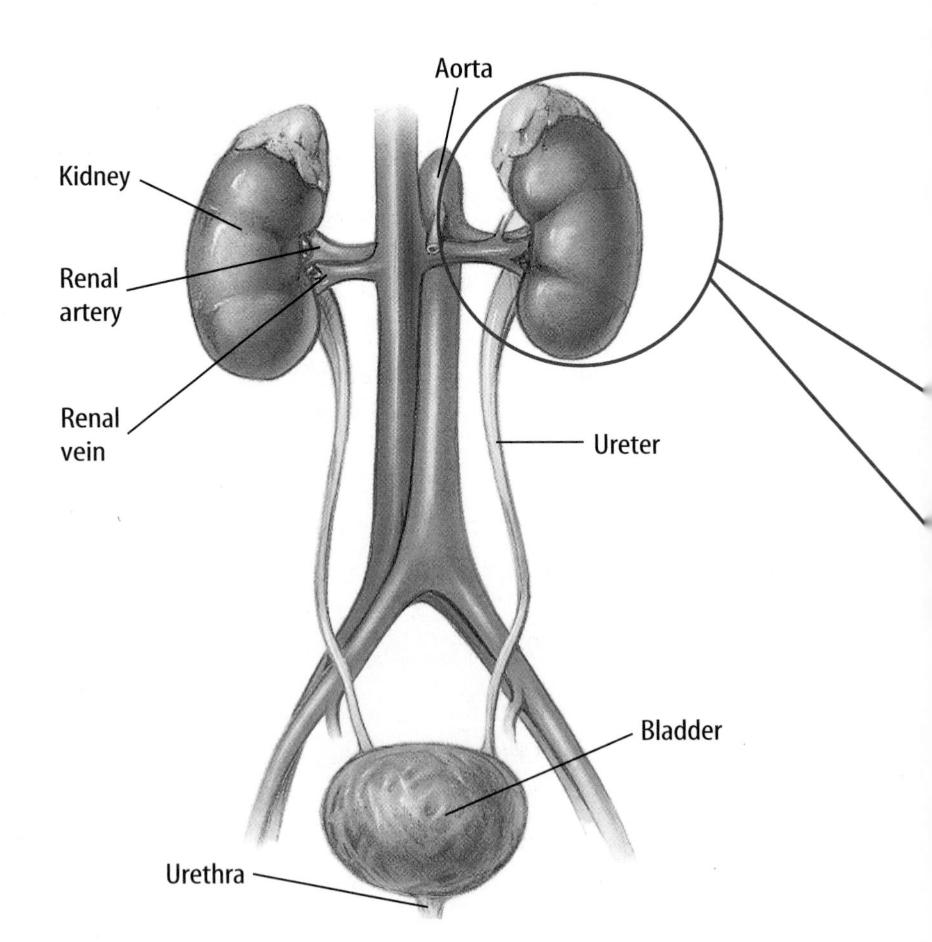

Figure 17 The urinary system removes wastes from the blood. The urinary system includes the kidneys, the bladder, and the connecting tubes.
Explain *how the kidneys help the body balance its fluid levels.*

Filtration in the Kidney A two-stage filtration system is an accurate description of a kidney, shown in **Figure 18.** It is made up of about one million tiny filtering units called **nephrons** (NE frahnz), also shown in **Figure 18.** Each nephron has a cuplike structure and a tubelike structure called a duct. Blood moves from a renal artery to capillaries in the cuplike structure. The first filtration occurs when water, sugar, salt, and wastes from the blood pass into the cuplike structure. Left behind in the blood are the red blood cells and proteins. Next, liquid in the cuplike structure is squeezed into a narrow tubule.

Capillaries that surround the tubule perform the second filtration. Most of the water, sugar, and salt are reabsorbed and returned to the blood. These collection capillaries merge to form small veins, which merge to form a renal vein in each kidney. Purified blood is returned to the main circulatory system. The liquid left behind flows into collecting tubules in each kidney. This wastewater, or urine, contains excess water, salts, and other wastes that are not reabsorbed by the body. An average-sized person produces about 1 L of urine per day.

Urine Collection and Release The urine in each collecting tubule drains into a funnel-shaped area of each kidney that leads to the **ureter** (YOO ruh tur). Ureters are tubes that lead from each kidney to the bladder. The **bladder** is an elastic, muscular organ that holds urine until it leaves the body. The elastic walls of the bladder can stretch to hold up to 0.5 L of urine. When empty, the bladder looks wrinkled and the cells lining the bladder are thick. When full, the bladder looks like an inflated balloon and the cells lining the bladder are stretched and thin. A tube called the urethra (yoo REE thruh) carries urine from the bladder to the outside of the body.

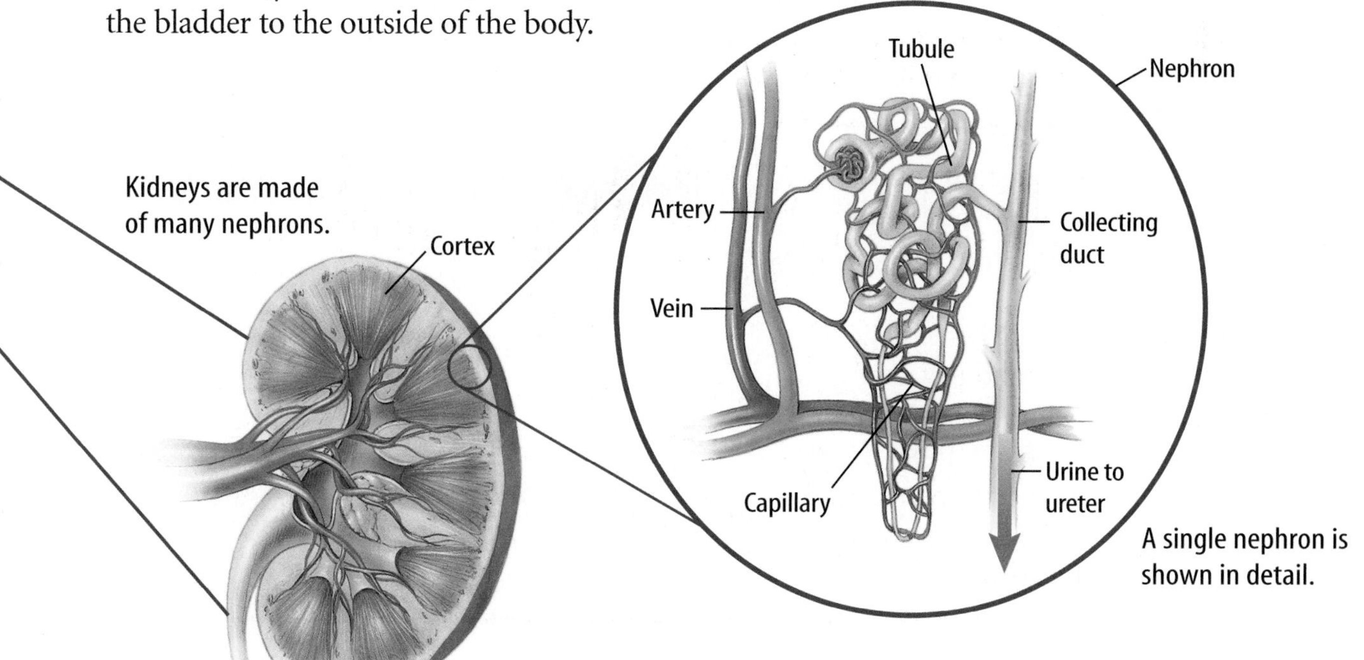

Figure 18 A single nephron is a complex structure.
Describe *the main function of a nephron.*

Urinary Diseases and Disorders

What happens when someone's kidneys don't work properly or stop working? Waste products that are not removed build up and act as poisons in body cells. Without excretion, an imbalance of salts occurs. The body responds by trying to restore this balance. If the balance isn't restored, the kidneys and other organs can be damaged. Kidney failure occurs when the kidneys don't work as they should. This is always a serious problem because the kidneys' job is so important to the rest of the body.

Applying Science

How does your body gain and lose water?

Your body depends on water. Without water, your cells could not carry out their activities and body systems could not function. Water is so important to your body that your brain and other body systems are involved in balancing water gain and water loss.

Identifying the Problem

Table A shows the major sources by which your body gains water. Oxidation of nutrients occurs when energy is released from nutrients by your body's cells. Water is a waste product of these reactions. **Table B** lists the major sources by which your body loses water. The data show you how daily gain and loss of water are related.

Table A Major Sources by Which Body Water Is Gained

Source	Amount (mL)	Percent
Oxidation of nutrients	250	10
Foods	750	30
Liquids	1,500	60
Total	**2,500**	**100**

Solving the Problem

1. What is the greatest source of water gained by your body? What is the greatest source of water lost by your body?
2. How would the percentages of water gained and lost change in a person who was working in extremely warm temperatures? In this case, what organ of the body would be the greatest contributor to water loss?

Table B Major Sources by Which Body Water Is Lost

Source	Amount (mL)	Percent
Urine	1,500	60
Skin	500	20
Lungs	350	14
Feces	150	6
Total	**2,500**	**100**

Because the ureters and urethra are narrow tubes, they can be blocked easily in some disorders. A blockage can cause serious problems because urine cannot flow out of the body properly. If the blockage is not corrected, the kidneys can be damaged.

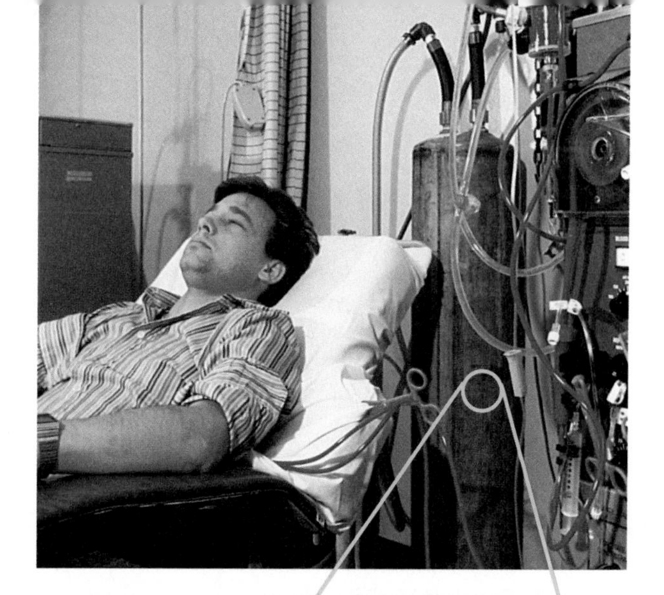

Why is a blocked ureter or urethra a serious problem?

Dialysis A person who has only one kidney still can live normally. The remaining kidney increases in size and works harder to make up for the loss of the other kidney. However, if both kidneys fail, the person will need to have his or her blood filtered by an artificial kidney machine in a process called dialysis (di AH luh sus), as shown in **Figure 19.**

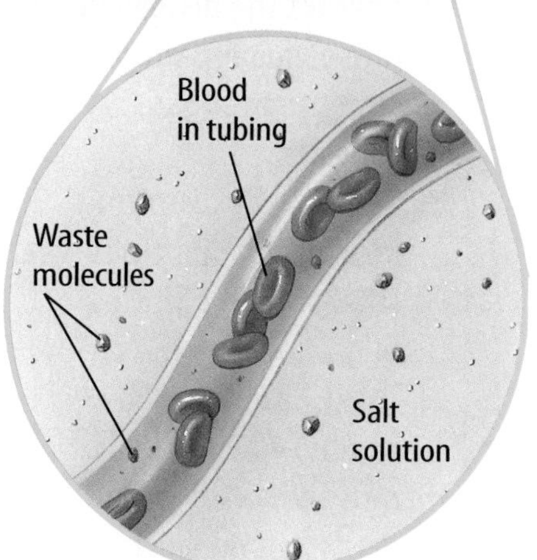

Figure 19 A dialysis machine can replace or help with some of the activities of the kidneys in a person with kidney failure. Like the kidney, the dialysis machine removes wastes from the blood.

section 4 review

Summary

Functions of the Excretory System

- The excretory system removes wastes from your body.
- The digestive, respiratory, and urinary systems and skin make up your excretory system.

The Urinary System

- The kidneys, which filter wastes from the blood, are the major organs of the urinary system.
- Urine moves from the kidneys through the ureters, into the bladder, then leaves the body through the urethra.

Urinary Diseases and Disorders

- Kidney failure can lead to a buildup of waste products in the body.
- An artificial kidney can be used to filter the blood in a process called dialysis.

Self Check

1. **List** the functions of a person's urinary system.
2. **Explain** how the kidneys remove wastes and keep fluids and salts in balance.
3. **Describe** what happens when the urinary system does not function properly.
4. **Compare** the excretory system and urinary system.
5. **Think Critically** Explain why reabsorption of certain materials in the kidneys is important to your health.

Applying Math

6. **Make and Use Graphs** Make a circle graph of major sources by which body water is gained. Use the data in **Table A** of the Applying Science activity.
7. **Concept Map** Using a network-tree concept map, compare the excretory functions of the kidneys and the lungs.

Particle Size and Absorption

Goals

- **Compare and contrast** the dissolving rates of different sized particles.
- **Predict** the dissolving rate of sugar particles larger than sugar cubes.
- **Predict** the dissolving rate of sugar particles smaller than particles of ground sugar.
- **Infer,** using the lab results, why the body must break down and dissolve food particles.

Materials

beakers or jars (3)
thermometers (3)
sugar granules
mortar and pestle
triple-beam balance
stirring rod
sugar cubes
weighing paper
warm water
stopwatch

Safety Precautions

WARNING: *Never taste, eat, or drink any materials used in the lab.*

Real-World Question

Before food reaches the small intestine, it is digested mechanically in the mouth and the stomach. The food mass is reduced to small particles. You can chew an apple into small pieces, but you would feed applesauce to a small child who didn't have teeth. What is the advantage of reducing the size of the food material? Does reducing the size of food particles aid the process of digestion?

Procedure

1. Copy the data table below into your Science Journal.

Dissolving Times of Sugar Particles		
Size of Sugar Particles	Mass	Time Until Dissolved
Sugar cube		
Sugar granules	Do not write in this book.	
Ground sugar particles		

2. Place a sugar cube into your mortar and grind up the cube with the pestle until the sugar becomes powder.
3. Using the triple-beam balance and weighing paper, measure the mass of the powdered sugar from your mortar. Using separate sheets of weighing paper, measure the mass of a sugar cube and the mass of a sample of the granular sugar. The masses of the powdered sugar, sugar cube, and granular sugar should be approximately equal to each other. Record the three masses in your data table.
4. Place warm water into the three beakers. Use the thermometers to be certain the water in each beaker is the same temperature.

5. Place the sugar cube in a beaker, the powdered sugar in a second beaker, and the granular sugar in a third beaker. Place all the sugar samples in the beakers at the same time and start the stopwatch when you put the sugar samples in the beaker.
6. Stir each sample equally.
7. Measure the time it takes each sugar sample to dissolve and record the times in your data table.

Analyze Your Data

1. **Identify** the experiment's constants and variables.
2. **Compare** the rates at which the sugar samples dissolved. What type of sugar dissolved most rapidly? Which was the slowest to dissolve?

Conclude and Apply

1. **Predict** how long it would take sugar particles larger than the sugar cubes to dissolve. Predict how long it would take sugar particles smaller than the powdered sugar to dissolve.
2. **Infer** and explain the reason why small particles dissolve more rapidly than large particles.
3. **Infer** why you should thoroughly chew your food.
4. **Explain** how reducing the size of food particles aids the process of digestion.

Communicating Your Data

Write a news column for a health magazine explaining to health-conscious people what they can do to digest their food better.

Does the same diet work for everyone?

Growing up in India in the first half of the twentieth century, R. Rajalakshmi (RAH jah lok shmee) saw many people around her who did not get enough food. Breakfast for a poor child might have been a cup of tea. Lunch might have consisted of a slice of bread. For dinner, a child might have eaten a serving of rice with a small piece of fish. This type of diet, low in calories and nutrients, produced children who were often sick and died young.

In the 1960s, R. Rajalakshmi was asked to help manage a program to improve nutrition in her country. North American and European nutritionists suggested foods that were common and worked well for people who lived in these nations. But Rajalakshmi knew this advice was useless in a country such as India.

The Proper Diet for India

Rajalakshmi knew that for a nutrition program to work, it had to fit Indian culture. First, she found out what healthy middle-class people in India ate. She took note of the nutrients available in those foods. Then she looked for cheap, easy-to-find foods that would provide the same nutrients. Rajalakshmi created a balanced diet of locally grown fruits, vegetables, and grains.

Rajalakshmi's ideas were thought unusual in the 1960s. For example, she insisted that a diet without meat could provide all major nutrients. It took persistence to get others to accept her ideas. Because of Rajalakshmi's program, Indian children almost doubled their food intake. Many children who would have been hungry and ill, grew healthy and strong.

Thanks to R. Rajalakshmi and other nutritionists, many children in India are eating well and staying healthy.

Report **Choose a continent and research what foods are native to that area. As a class, compile a list of the foods and where they originated. Using the class list, create a world map on a bulletin board that shows the origins of the different foods.**

For more information, visit green.msscience.com/time

Reviewing Main Ideas

Section 1 The Digestive System

1. Mechanical digestion breaks down food through chewing and churning. Enzymes and other chemicals aid chemical digestion.
2. Food passes through the mouth, esophagus, stomach, small intestine, large intestine, and rectum and then out the anus.
3. The large intestine absorbs water, which helps the body maintain homeostasis.

Section 2 Nutrition

1. Proteins, carbohydrates, fats, vitamins, minerals, and water are the six nutrients found in foods.
2. Health is affected by the combination of foods that make up a diet.

Section 3 The Respiratory System

1. The respiratory system brings oxygen into the body and removes carbon dioxide.
2. Breathing is the movement of the chest that allows air to move into the lungs and waste gases to leave the lungs.
3. The chemical reaction in cells that needs oxygen to release energy and produces carbon dioxide and water as wastes is called cellular respiration.
4. Smoking causes many respiratory problems, including chronic bronchitis, emphysema, and lung cancer.

Section 4 The Excretory System

1. The urinary system is part of the excretory system. The skin, lungs, liver, and large intestine are also excretory organs.
2. The kidneys are the major organs of the urinary system and have a two-stage filtration system that removes wastes.
3. When kidneys fail to work, an artificial kidney can be used to filter the blood in a process called dialysis.

Visualizing Main Ideas

Copy and complete the following table on the respiratory and excretory systems.

Human Body Systems

	Respiratory System	Excretory System
Major Organs		
Wastes Eliminated	Do not write	in this book.
Disorders		

chapter 14 Review

Using Vocabulary

alveoli p. 414	mineral p. 408
amino acid p. 405	nephron p. 421
bladder p. 421	nutrient p. 400
bronchi p. 414	peristalsis p. 402
carbohydrate p. 406	trachea p. 414
chyme p. 403	ureter p. 421
enzyme p. 400	villi p. 403
larynx p. 414	vitamin p. 407

Fill in the blanks with the correct vocabulary word or words.

1. ________ is the muscular contractions of the esophagus.
2. The building blocks of proteins are ________.
3. The liquid product of digestion is called ________.
4. ________ are inorganic nutrients.
5. ________ are the filtering units of the kidney.
6. ________ are thin-walled sacs in the lungs.
7. The ________ is an elastic muscular organ that holds urine.

Checking Concepts

Choose the word or phrase that best answers the question.

8. Where in humans does most chemical digestion occur?
   A) duodenum **C)** liver
   B) stomach **D)** large intestine

9. In which organ is water absorbed?
   A) liver **C)** small intestine
   B) esophagus **D)** large intestine

10. Which of these organs is an accessory organ?
    A) mouth **C)** small intestine
    B) stomach **D)** liver

11. What beneficial substances are produced by bacteria in the large intestine?
    A) fats **C)** vitamins
    B) minerals **D)** proteins

12. Which food group contains yogurt and cheese?
    A) dairy **C)** meat
    B) grain **D)** fruit

13. When you inhale, which of the following contracts and moves down?
    A) bronchioles **C)** nephrons
    B) diaphragm **D)** kidneys

14. Exchange of gases occurs between capillaries and which of the following structures?
    A) alveoli **C)** bronchioles
    B) bronchi **D)** trachea

15. Which of the following conditions does smoking worsen?
    A) arthritis **C)** excretion
    B) respiration **D)** emphysema

16. Urine is held temporarily in which of the following structures?

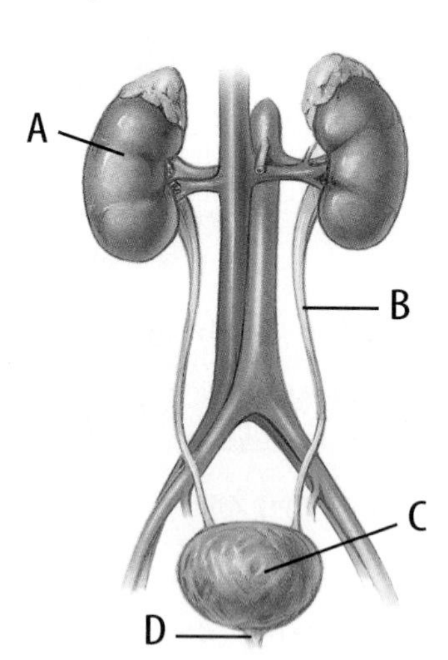

17. Which of the following substances is not reabsorbed by blood after it passes through the kidneys?
    A) salt **C)** wastes
    B) sugar **D)** water

Science Online green.msscience.com/vocabulary_puzzlemaker

Thinking Critically

18. Make and use a table to sequence the order of organs in the digestive system through which food passes. Indicate whether ingestion, digestion, absorption, or elimination takes place in each.

19. Compare and contrast the three types of carbohydrates—sugar, starch, and fiber.

20. Classify the parts of your favorite sandwich into three of the nutrient categories—carbohydrates, proteins, and fats.

21. Recognize cause and effect by discussing how lack of oxygen is related to lack of energy.

22. Form a hypothesis about the number of breaths a person might take per minute in each of these situations: asleep, exercising, and on top of Mount Everest. Give a reason for each hypothesis.

23. Concept Map Make an events-chain concept map showing how urine forms in the kidneys. Begin with, "In the nephron …"

Use the table below to answer question 24.

Materials Filtered by the Kidneys

Substance Filtered in Urine	Amount Moving Through Kidney	Amount Excreted
Water	125 L	1 L
Salt	350 g	10 g
Urea	1 g	1 g
Glucose	50 g	0 g

24. Interpret Data Study the data above. How much of each substance is reabsorbed into the blood in the kidneys? What substance is excreted completely in the urine?

25. Describe how bile aids the diegestive process.

26. Explain how the bacteria that live in your large intestine help your body.

Performance Activities

27. Questionnaire and Interview Prepare a questionnaire that can be used to interview a health specialist who works with lung cancer patients. Include questions on reasons for choosing the career, new methods of treatment, and the most encouraging or discouraging part of the job.

Applying Math

28. Kidney Blood Flow In approximately 5 min, all 5 L of blood in the body pass through the kidneys. Calculate the average rate of flow through the kidneys in liters per minute.

Use the graph below to answer question 29.

Total Lung Capacity

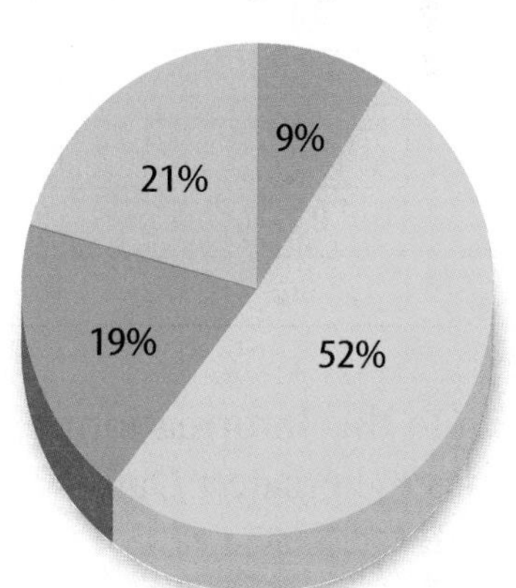

Total Lung Capacity = 5800 mL

- Volume of air normally inhaled or exhaled
- Volume of additional air that can be inhaled forcefully
- Volume of additional air that can be exhaled forcefully
- Volume of air left in lungs after forcefully exhaling

29. Total Lung Capacity What volume of air (mL) is left in the lungs after forcefully exhaling?

Part 1 Multiple Choice

Record your answers on the answer sheet provided by your teacher or on a sheet of paper.

Use the table below to answer questions 1 and 2.

Nutrition Facts of Vanilla Ice Cream		
Item	**Amount**	**DV (Daily Values)**
Serving Size	112 g	0
Calories	208	0
Total Fat	19 g	29%
Saturated Fat	11 g	55%
Cholesterol	0.125 g	42%
Sodium	0.90 g	4%
Total Carbohydrates	22 g	7%
Fiber	0 g	0%
Sugars	22 g	n/a
Protein	5 g	n/a
Calcium	0.117 g	15%
Iron	n/a	0%

1. According to this information, which mineral has the greatest Daily Value (DV) percentage?
A. sodium
B. cholesterol
C. iron
D. calcium

2. How many grams of saturated fat and Daily Value (DV) percentage in two servings of this ice cream?
A. 11 g, 110%
B. 22 g, 110%
C. 21 g, 55%
D. 5.5 g, 110%

Test-Taking Tip

Answer All Questions Never skip a question. If you are unsure of an answer, mark your best guess on another sheet of paper and mark the question in your test booklet to remind you to come back to it at the end of the test.

Use the illustration below to answer question 3.

3. What is the structure shown above and to what body system does it belong?
A. capillary—circulatory
B. alveolus—respiratory
C. nephron—urinary
D. ureter—excretory

4. If all of the blood in your body passes through the kidneys in 5 minutes, how many times does all of your blood pass through the kidneys in one hour?
A. 12 times
B. 6 times
C. 5 times
D. 20 times

5. Which of the following is the correct sequence of the organs of the digestive tract?
A. mouth, stomach, esophagus, small intestine, large intestine
B. esophagus, mouth, stomach, small intestine, large intestine
C. mouth, small intestine, stomach, large intestine
D. mouth, esophagus, stomach, small intestine, large intestine

6. Which of the following diseases may be caused by smoking?
A. lung cancer
B. diabetes
C. influenza
D. bladder infection

Part 2 Short Response/Grid In

Record your answers on the answer sheet provided by your teacher or on a sheet of paper.

7. Explain the difference between organic and inorganic nutrients. Name a class of nutrients for each.

8. Enzymes play an important role in the digestive process. But enzyme-aided reactions are also involved in other body systems. Give an example of how enzymes are used by the body in a way that does not involve the digestive system.

Use the table and paragraph below to answer questions 9–12.

For one week, research scientists collected and accurately measured the amount of body water lost and gained per day for four different patients. The following table lists results from their investigation.

Body Water Gained (+) and Lost (−)				
Patient	Day 1 (L)	Day 2 (L)	Day 3 (L)	Day 4 (L)
Mr. Stoler	+0.15	+0.15	−0.35	+0.12
Mr. Jemma	−0.01	0.00	−0.20	−0.01
Mr. Lowe	0.00	+0.20	−0.28	+0.01
Mr. Cheng	−0.50	−0.50	−0.55	−0.32

9. What was Mr. Cheng's average daily body water loss for the 4 days shown in the table?

10. Which patient had the greatest amount of body water gained on days 1 and 2?

11. According to the data in the table, on which day was the temperature in each patient's hospital room probably the hottest?

12. Which patient had the highest total gain in body water over the 4-day period?

Part 3 Open Ended

Record your answers on a sheet of paper.

13. Explain the role of cilia in the respiratory system. In chronic bronchitis, cilia are damaged. What effects does this damage have on the respiratory system?

14. Antibiotics may be given to help a person fight off a bacterial infection. If a person is taking antibiotics, what might happen to the normal bacteria living in the large intestine? How would this affect the body?

Use the illustration below to answer questions 15 and 16.

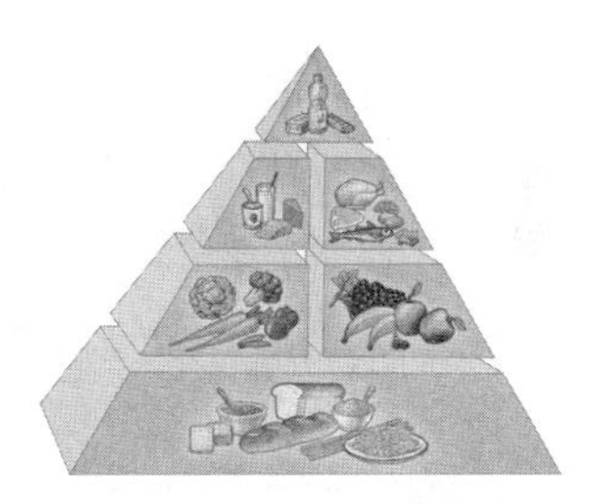

15. Identify the food group shown at the base of the food pyramid. Explain why the greatest number of servings should come from this group.

16. Identify the food group at the top of the pyramid. Explain why the least number of servings should come from this group.

17. Sometimes the esophagus can be affected by a disease in which the smooth muscle in the wall of the esophagus does not work properly. What do you think would happen to food that the person swallowed? Why?

18. Compare and contrast the roles of mucus in the digestive and respiratory systems.

19. Urine can be tested for any signs of a urinary tract disease. Mrs. Chavez had a urine test that showed protein in the urine. What might the results of this urine test mean?

chapter 15

Support, Movement, and Responses

chapter preview

sections

How are you like a building?

Buildings are supported and protected by internal and external structures. Your body is supported by your skeleton and protected by your skin. In this chapter, you also will learn how your body senses and responds to the world around you.

Science Journal Imagine for a moment that your body does not have a support system. How will you perform your daily activities? Explain your reasoning.

Start-Up Activities

Effect of Muscles on Movement

The expression "Many hands make light work" is also true when it comes to muscles in your body. In fact, hundreds of muscles and bones work together to bring about smooth, easy movement. Muscle interactions enable you to pick up a penny or lift a 10-kg weight.

1. Sit on a chair at an empty table and place the palm of one hand under the edge of the table.
2. Push your hand up against the table. Do not push too hard.
3. Use your other hand to feel the muscles located on both sides of your upper arm, as shown in the photo.
4. Next, place your palm on top of the table and push down. Again, feel the muscles in your upper arm.
5. **Think Critically** Describe in your Science Journal how the different muscles in your upper arm were working during each movement.

Support, Movement, and Responses Make the following Foldable to help you understand the functions of skin, muscles, bones, and nerves.

STEP 1 **Fold** a sheet of paper in half lengthwise. Make the back edge about 1.25 cm longer than the front edge.

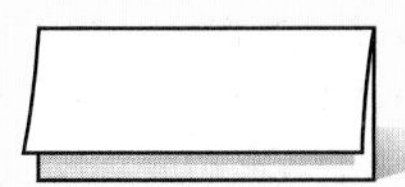

STEP 2 **Fold** the paper in half widthwise, twice.

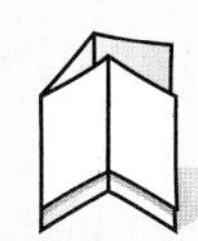

STEP 3 **Unfold and cut** only the top layer along the three folds to make four tabs. **Label** the tabs as shown.

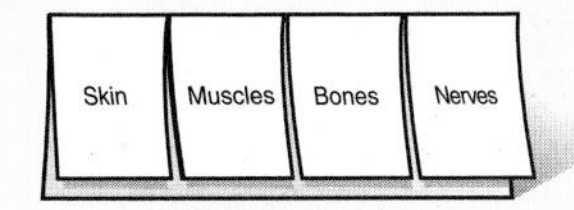

Read and Write As you read this chapter, list the functions that skin, muscles, bones, and nerves have in support, movement, and responses.

Preview this chapter's content and activities at green.msscience.com

section 1

The Skin

as you read

What You'll Learn

- **Distinguish** between the epidermis and dermis of the skin.
- **Identify** the functions of the skin.
- **Explain** how skin protects the body from disease and how it heals itself.

Why It's Important

Skin plays a vital role in protecting your body against injury and disease.

Review Vocabulary
organ: a structure, such as the heart, made up of different types of tissues that work together

New Vocabulary
- epidermis
- melanin
- dermis

Skin Structures

Your skin is the largest organ of your body. Much of the information you receive about your environment comes through your skin. You can think of your skin as your largest sense organ.

Skin is made up of three layers of tissue—the epidermis, the dermis, and a fatty layer—as shown in **Figure 1.** Each layer is made of different cell types. The **epidermis** is the outer, thinnest layer. The epidermis's outermost cells are dead and water repellent. Thousands of epidermal cells rub off every time you take a shower, shake hands, or scratch your elbow. New cells are produced constantly at the base of the epidermis. These new cells move up and eventually replace those that are rubbed off. The **dermis** is the layer of cells directly below the epidermis. This layer is thicker than the epidermis and contains blood vessels, nerves, muscles, oil and sweat glands, and other structures. Below the dermis is a fatty region that insulates the body. This is where much of the fat is deposited when a person gains weight.

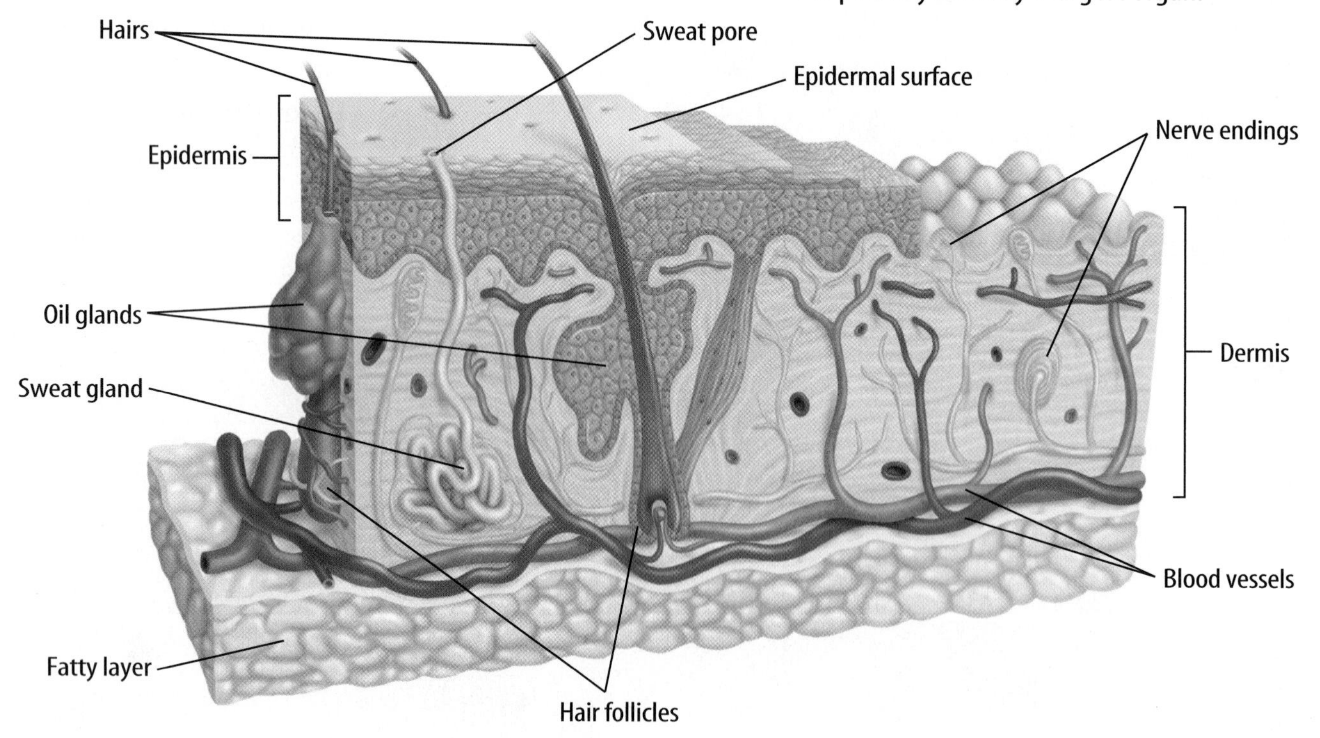

Figure 1 Hair, sweat glands, and oil glands are part of your body's largest organ.

Melanin Cells in the epidermis produce the chemical **melanin** (ME luh nun), a pigment that protects your skin and gives it color. The different amounts of melanin produced by cells result in differences in skin color, as shown in **Figure 2.** When your skin is exposed to ultraviolet rays, melanin production increases and your skin becomes darker. Lighter skin tones have less protection. Such skin burns more easily and can be more susceptible to skin cancer.

High Altitude and Skin There have been many books written about mountain climbing. Search the library for such a book and find references to the effects of sunlight and weather on the skin at high altitudes. In your Science Journal, list the book title and author and summarize the passage that describes the effects of sunlight and weather on the skin.

Skin Functions

Your skin carries out several major functions, including protection, sensory response, formation of vitamin D, regulation of body temperature, and ridding the body of wastes. The most important function of the skin is protection. The skin forms a protective covering over the body that prevents physical and chemical injury. Some bacteria and other disease-causing organisms cannot pass through skin as long as it is unbroken. Glands in the skin secrete fluids that can damage or destroy some bacteria. The skin also slows water loss from body tissues.

Specialized nerve cells in the skin detect and relay information to the brain. Because of these cells, you are able to sense the softness of a cat, the sharpness of a pin, or the heat of a frying pan.

Another important function of skin is the formation of vitamin D. Small amounts of this vitamin are produced in the presence of ultraviolet light from a fatlike molecule in your epidermis. Vitamin D is essential for good health because it helps your body absorb calcium into your blood from food in your digestive tract.

Figure 2 Melanin gives skin and eyes their color. The more melanin that is present, the darker the color is. This pigment provides protection from damage caused by harmful light energy.

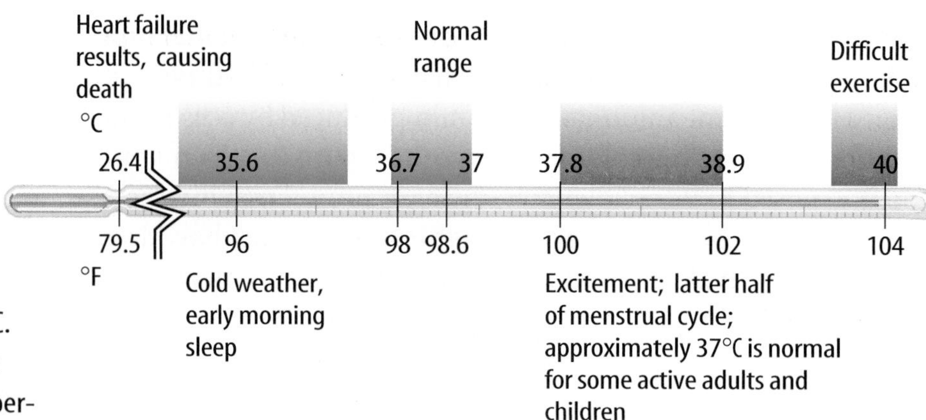

Figure 3 Normal human body temperature is about 37°C. Temperature varies throughout the day. The highest body temperature is reached at about 11 A.M. and the lowest at around 4 A.M. At 43°C (109.5°F), fatal bleeding results, causing death.

Heat and Waste Exchange Humans can withstand a limited range of body temperatures, as shown in **Figure 3.** Your skin plays an important role in regulating your body temperature. Blood vessels in the skin can help release or hold heat. If the blood vessels expand, or dilate, blood flow increases and heat is released. In contrast, less heat is released when the blood vessels constrict. Think of yourself after running—are you flushed red or pale and shivering?

An adult human's dermis has about 3 million sweat glands that help regulate the body's temperature and excrete wastes. When blood vessels dilate, pores open in the skin that lead to the sweat glands. Perspiration, or sweat, moves out onto the skin. Heat transfers from the body to the sweat on the skin. Eventually, this sweat evaporates, removing the heat and cooling the skin. This process eliminates excess heat produced by muscle contractions.

Reading Check *What are two functions of sweat glands?*

Wastes are produced when nutrients are broken down in cells. Such wastes, if not removed, can act as poisons. In addition to helping regulate your body's temperature, sweat glands release waste products, such as water and salt. If too much water and salt are released during periods of extreme heat or physical exertion, you might feel light-headed or even faint.

Recognizing Why You Sweat

Procedure

1. Examine the epidermis and the pores of your skin using a **magnifying lens.**
2. Place a **clear-plastic sandwich bag** on your hand. Use **tape** to seal the bag around your wrist. **WARNING:** *Do not wrap the tape too tightly.*
3. Quietly study your textbook for 10 min, then look at your hand. Remove the bag.
4. Describe what happened to your hand while it was inside the bag.

Analysis

1. Identify what formed inside the bag. Where did this substance come from?
2. Why does this substance form even when you are not active?

Skin Injuries and Repair

Your skin often can be bruised, scratched, burned, ripped, or exposed to harsh conditions like cold, dry air. In response, the epidermis produces new cells and the dermis repairs tears. When the skin is injured, disease-causing organisms can enter the body rapidly. An infection often results.

Bruises When tiny blood vessels burst under unbroken skin, a bruise results. Red blood cells from these broken blood vessels leak into the surrounding tissue. These blood cells then break down, releasing a chemical called hemoglobin that gradually breaks down into its components, called pigments. The colors of these pigments cause the bruised area to turn shades of blue, red, and purple, as shown in **Figure 4.** Swelling also may occur. As the injury heals, the bruise eventually turns yellow as the pigment in the red blood cells is broken down even more and reenters the bloodstream. After all of the pigment is absorbed into the bloodstream, the bruise disappears and the skin appears normal again.

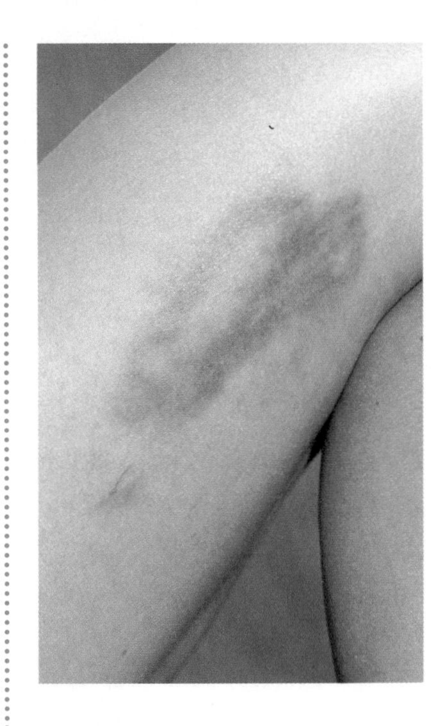

Figure 4 Bruising occurs when tiny blood vessels beneath the skin burst.
Infer *whether this bruise is new or is already healing.*

Reading Check *What is the source of the yellow color of a bruise that is healing?*

The body usually can repair bruises and small cuts. What happens when severe burns, some diseases, and surgeries result in injury to large areas of skin? Sometimes there are not enough skin cells left to produce new skin. If not treated, this can lead to rapid water loss from skin and muscle tissues, leading to infection and possibly death. Skin grafts can prevent this. Pieces of skin are cut from one part of a person's body and moved to the injured or burned area where there is no skin. This new graft is kept alive by nearby blood vessels and soon becomes part of the surrounding skin.

section 1 review

Summary

Skin Structures

- Skin is the largest organ in your body.
- There are three layers of tissue in skin and each layer is made of different cell types.
- Melanin protects your skin and gives it color.

Skin Functions

- The most important function of skin is protection.
- Specialized nerve cells in the skin detect and relay information to the brain.

Skin Injuries and Repair

- When skin is injured, disease-causing organisms can enter the body rapidly.
- When skin is damaged, the epidermis produces new cells and the dermis repairs tears.

Self Check

1. **Compare and contrast** the epidermis and dermis.
2. **Identify** the major functions of the skin.
3. **Describe** the role that your skin plays in regulating body temperature.
4. **Explain** how skin helps prevent disease in the body.
5. **Describe** one way doctors are able to repair severe skin damage from burns, injuries, or surgeries.
6. **Think Critically** Why is a person who has been severely burned in danger of dying from loss of water?

Applying Math

7. **Solve One-Step Equations** The skin of eyelids is 0.5 mm thick. On the soles of your feet, skin is up to 0.4 cm thick. How many times thicker is the skin on the soles of your feet compared to your eyelids?

Measuring Skin Surface

Skin covers the entire surface of your body and is your body's largest organ. Skin cells make up a layer of skin about 2 mm thick. How big is this organ?

Real-World Question

How much skin covers your body?

Goals

- **Estimate** the surface area of skin that covers the body of a middle-school student.

Materials

10 large sheets of newspaper
scissors
tape
meterstick or ruler

Safety Precautions

Procedure

1. Form groups of three or four, either all female or all male. Select one person from your group to measure the surface area of his or her skin.
2. **Estimate** how much skin covers the average student in your classroom. In your Science Journal, record your estimation.
3. Wrap newspaper snugly around each part of your classmate's body. Overlapping the sheets of paper, use tape to secure the paper. Small body parts, such as fingers and toes, do not need to be wrapped individually. Cover entire hands and feet. Do not cover the face.
4. After your classmate is completely covered with paper, carefully cut the newspaper off his or her body. **WARNING:** *Do not cut any clothing or skin.*
5. Lay all of the overlapping sheets of newspaper on the floor. Using scissors and more tape, cut and piece the paper suit together to form a rectangle.
6. Using a meterstick, measure the length and width of the resulting rectangle. Multiply these two measurements for an estimate of the surface area of your classmate's skin.

Conclude and Apply

1. Was your estimation correct? Explain.
2. How accurate are your measurements of your classmate's skin surface area? How could your measurements be improved?
3. **Calculate** the volume of your classmate's skin, using 2 mm as the average thickness and your calculated surface area from this activity.

Communicating Your Data

Using a table, record the estimated skin surface area from all the groups in your class. Find the average surface areas for both males and females. Discuss any differences in these two averages.

The Muscular System

Movement of the Human Body

Muscles help make all of your daily movements possible. In the process of relaxing, contracting, and providing the force for movements, energy is used and work is done. Imagine how much energy the more than 600 muscles in your body use each day. No matter how still you might try to be, some muscles in your body are always moving. You're breathing, your heart is beating, and your digestive system is working.

Muscle Control Your hand, arm, and leg muscles are voluntary. So are the muscles of your face, as shown in **Figure 5.** You can choose to move them or not to move them. Muscles that you are able to control are called **voluntary muscles.** Muscles that you can't control consciously are **involuntary muscles.** They work all day long, all your life. Blood is pumped through blood vessels, and food is moved through your digestive system by the action of involuntary muscles.

Reading Check *What is another body activity that is controlled by involuntary muscles?*

as you read

What You'll Learn

- **Identify** the major function of the muscular system.
- **Compare** and contrast the three types of muscles.
- **Explain** how muscle action results in the movement of body parts.

Why It's Important

The muscular system is responsible for how you move and the production of heat in your body. Muscles also give your body its shape.

Review Vocabulary

muscle: an organ that can relax, contract, and provide the force to move bones and body parts

New Vocabulary

- voluntary muscle
- involuntary muscle
- tendon

Figure 5 Facial expressions generally are controlled by voluntary muscles. It takes only 13 muscles to smile, but 43 muscles to frown.

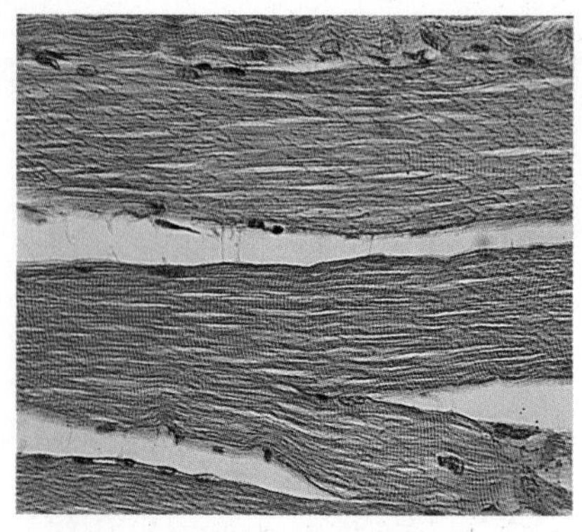

Skeletal muscles move bones. The muscle tissue appears striped, or striated, and is attached to bone.

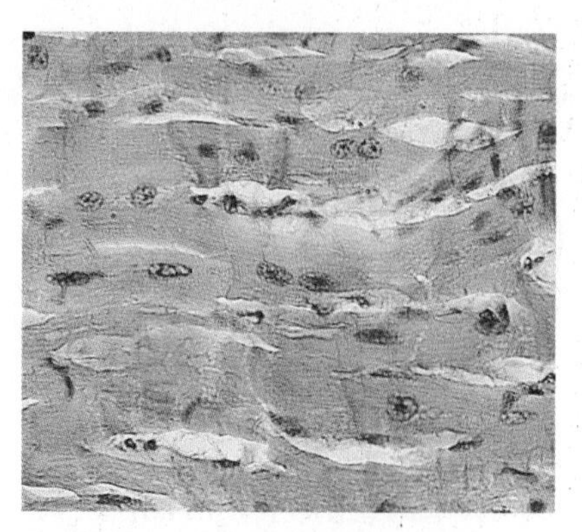

Cardiac muscle is found only in the heart. The muscle tissue has striations.

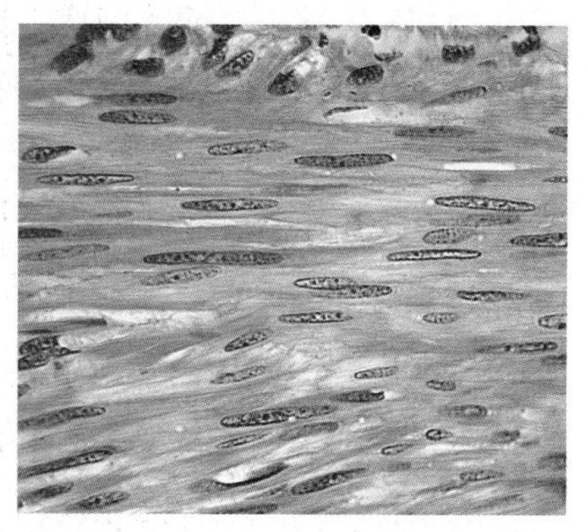

Smooth muscle is found in many of your internal organs, such as the digestive tract. This muscle tissue is nonstriated.

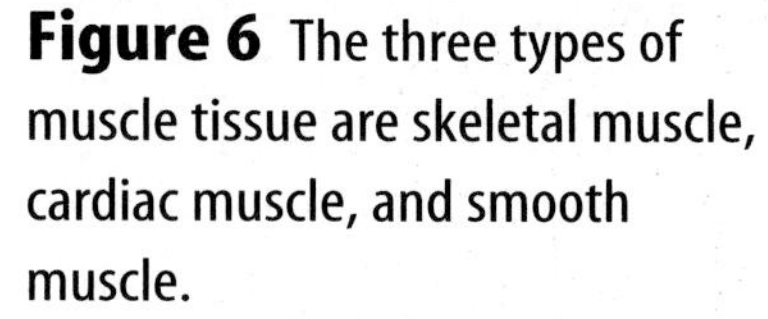

Figure 6 The three types of muscle tissue are skeletal muscle, cardiac muscle, and smooth muscle.
Infer *what type of muscle tissue makes up the walls of veins, which carry blood.*

Classification of Muscle Tissue

Humans have three types of muscle tissue: skeletal, smooth, and cardiac. Skeletal muscles are voluntary muscles that move bones. They are more common than other muscle types, and are attached to bones by thick bands of tissue called **tendons.** Skeletal muscle cells are striated (STRI ay tud), and when viewed under a microscope, appear striped. You can see the striations in **Figure 6.**

The remaining two types of muscles also are shown in **Figure 6.** Cardiac muscle is found only in the heart. Like skeletal muscle, cardiac muscle is striated. This type of muscle contracts about 70 times per minute every day of your life. Smooth muscles are nonstriated involuntary muscles and are found in your intestines, bladder, blood vessels, and other internal organs.

Your Body's Simple Machines—Levers

INTEGRATE Physics

Your skeletal system and muscular system work together when you move, like a machine. A machine, such as a bicycle, is any device that makes work easier. A simple machine does work with only one movement, like a hammer. The hammer is a type of simple machine called a lever, which is a rod or plank that pivots or turns about a point. This point is called a fulcrum. The action of muscles, bones, and joints working together is like a lever. In your body, bones are rods, joints are fulcrums, and contraction and relaxation of muscles provide the force to move body parts. Levers are classified into three types—first class, second class, and third class. Examples of the three types of levers that are found in the human body are shown in **Figure 7.**

NATIONAL GEOGRAPHIC **VISUALIZING HUMAN BODY LEVERS**

Figure 7

All three types of levers—first class, second class, and third class—are found in the human body. In the photo below, a tennis player prepares to serve a ball. As shown in the accompanying diagrams, the tennis player's stance demonstrates the operation of all three classes of levers in the human body.

▲ Fulcrum
Effort force
■ Load

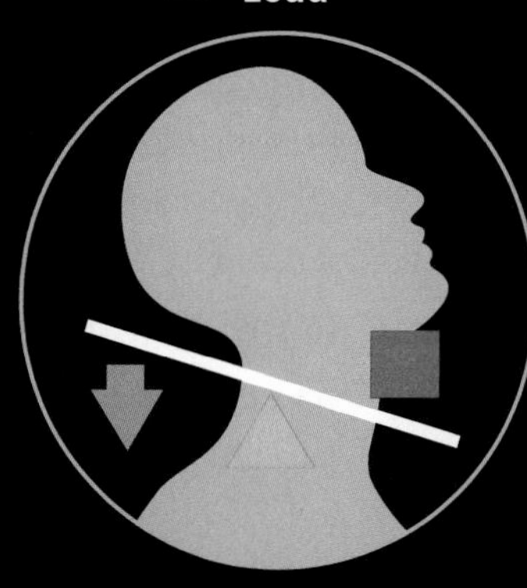

FIRST-CLASS LEVER
The fulcrum lies between the effort force and the load. This happens when the tennis player uses his neck muscles to tilt his head back.

THIRD-CLASS LEVER
The effort force is between the fulcrum and the load. This happens when the tennis player flexes the muscles in his arm and shoulder.

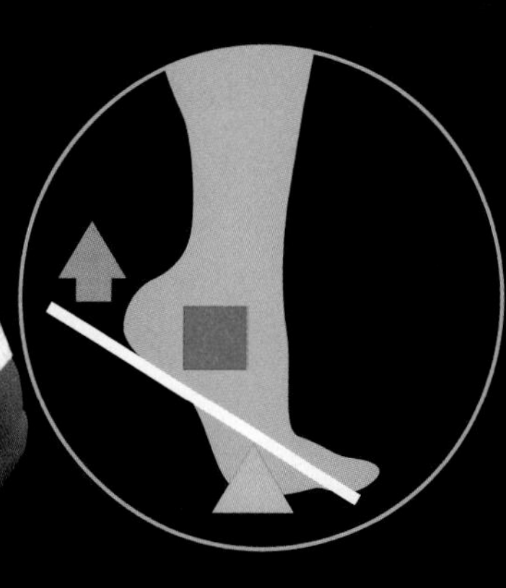

SECOND-CLASS LEVER
The load lies between the fulcrum and the effort force. This happens when the tennis player's calf muscles lift the weight of his body up on his toes.

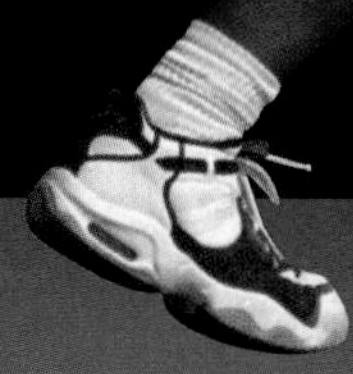

Working Muscles

How do muscles allow you to move your body? You move because pairs of skeletal muscles work together. When one muscle of a pair of muscles contracts, the other muscle relaxes or returns to its original length, as shown in **Figure 8.** Muscles always pull; they never push. When the muscles on the back of your upper leg contract, they shorten and pull your lower leg back and up. When you straighten your leg, the back muscles lengthen and relax, and the muscles on the front of your upper leg contract. Compare how your leg muscles work with how the muscles of your arms work.

Changes in Muscles Over time, muscles can become larger or smaller, depending on whether or not they are used. Also, muscles that are given regular exercise respond quickly to stimuli. Skeletal muscles that do a lot of work, such as those in your writing hand, can become stronger and larger. Some of this change in muscle size is because of an increase in the number of muscle cells. However, most of this change in muscle size is because individual muscle cells become larger. For example, many soccer and basketball players have noticeably larger, defined leg muscles. In contrast, someone who only participates in nonactive pastimes such as watching television or playing computer games, instead of participating in more active pastimes, will have smaller and weaker muscles. Muscles that aren't exercised become smaller in size. Paralyzed muscles also become smaller because they cannot be moved or have limited movement.

Reading Check *How do muscles increase their size?*

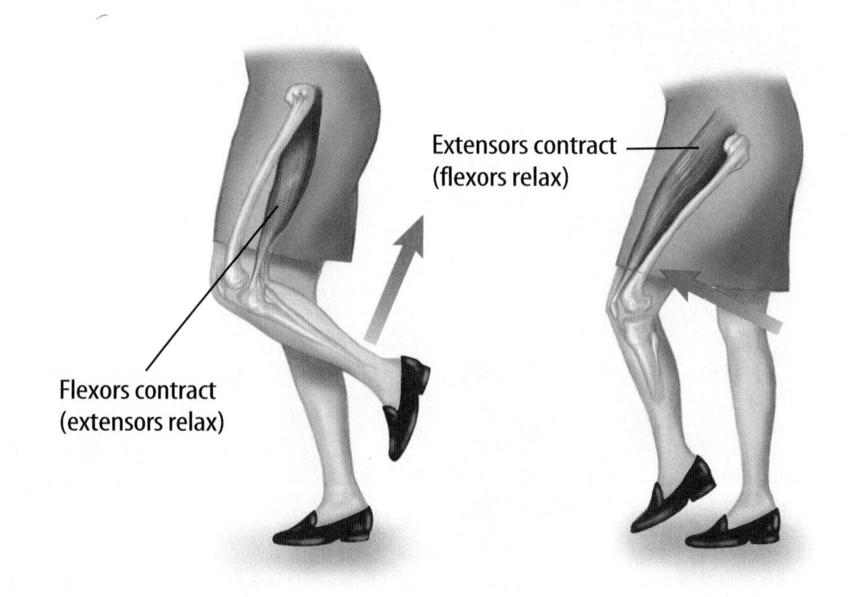

Figure 8 When the flexor (hamstring) muscles of your thigh contract, the lower leg is brought toward the thigh. When the extensor (quadriceps) muscles contract, the lower leg is straightened. **Describe** *the class of lever shown to the right.*

How Muscles Move

Your muscles need energy to contract and relax. Your blood carries energy-rich molecules to your muscle cells, where the chemical energy stored in these molecules is released. As the muscle contracts, this released energy changes to mechanical energy (movement) and thermal energy (heat), as shown in **Figure 9.** The heat produced by muscle contractions helps keep your body temperature constant. When the supply of energy-rich molecules in a muscle is used up, the muscle becomes tired and needs to rest. During this resting period, your blood supplies more energy-rich molecules to your muscle cells.

Figure 9 Chemical energy is needed for muscle activity. During activity, chemical energy supplied by food is changed into mechanical energy (movement) and thermal energy (heat).

How do muscles obtain energy to contract and relax?

section 2 review

Summary

Movement of the Human Body

- Muscles contract to move bones and body parts.
- You can control voluntary muscles, but you cannot consciously control involuntary muscles.

Classification of Muscle Tissue

- Skeletal muscles are voluntary, smooth muscles control movement of internal organs, and cardiac muscle is striated and involuntary.

Your Body's Simple Machines—Levers

- Your muscles, bones, and joints work together like levers to move your body.

Working Muscles

- Muscles always pull, and when one muscle of a pair contracts, the other relaxes.
- Chemical energy is needed for muscle activity.

Self Check

1. **Describe** the function of muscles.
2. **Compare and contrast** the three types of muscle tissue.
3. **Identify** and describe the appearance of the type of muscle tissue found in your heart.
4. **Explain** how your muscles, bones, and joints work together to move your body.
5. **Describe** how a muscle attaches to a bone.
6. **Think Critically** What happens to your upper arm muscles when you bend your arm at the elbow to eat your favorite sandwich?

Applying Skills

7. **Concept Map** Using a concept map, sequence the activities that take place when you bend your leg at the knee.
8. **Communicate** Write a paragraph in your Science Journal about the three forms of energy involved in a muscle contraction.

section 3

The Skeletal System

as you read

What You'll Learn

- **Identify** five functions of the skeletal system.
- **Compare and contrast** movable and immovable joints.

Why It's Important

You'll begin to understand how each of your body parts moves and what happens that allows you to move them.

Review Vocabulary
skeleton: a framework of living bones that supports your body

New Vocabulary
- periosteum
- cartilage
- joint
- ligament

Functions of Your Skeletal System

The skeletal system includes all the bones in your body and has five major functions.

1. The skeleton gives shape and support to your body.
2. Bones protect your internal organs.
3. Major muscles are attached to bones and help them move.
4. Blood cells form in the red marrow of many bones.
5. Major quantities of calcium and phosphorous compounds are stored in the skeleton for later use. Calcium and phosphorus make bones hard.

Bone Structure

Looking at bone through a magnifying glass will show you that it isn't smooth. Bones have bumps, edges, round ends, rough spots, and many pits and holes. Muscles and ligaments attach to some of the bumps and pits. In your body, blood vessels and nerves enter and leave through the holes in bones. How a bone looks from the inside and the outside is shown in **Figure 10.**

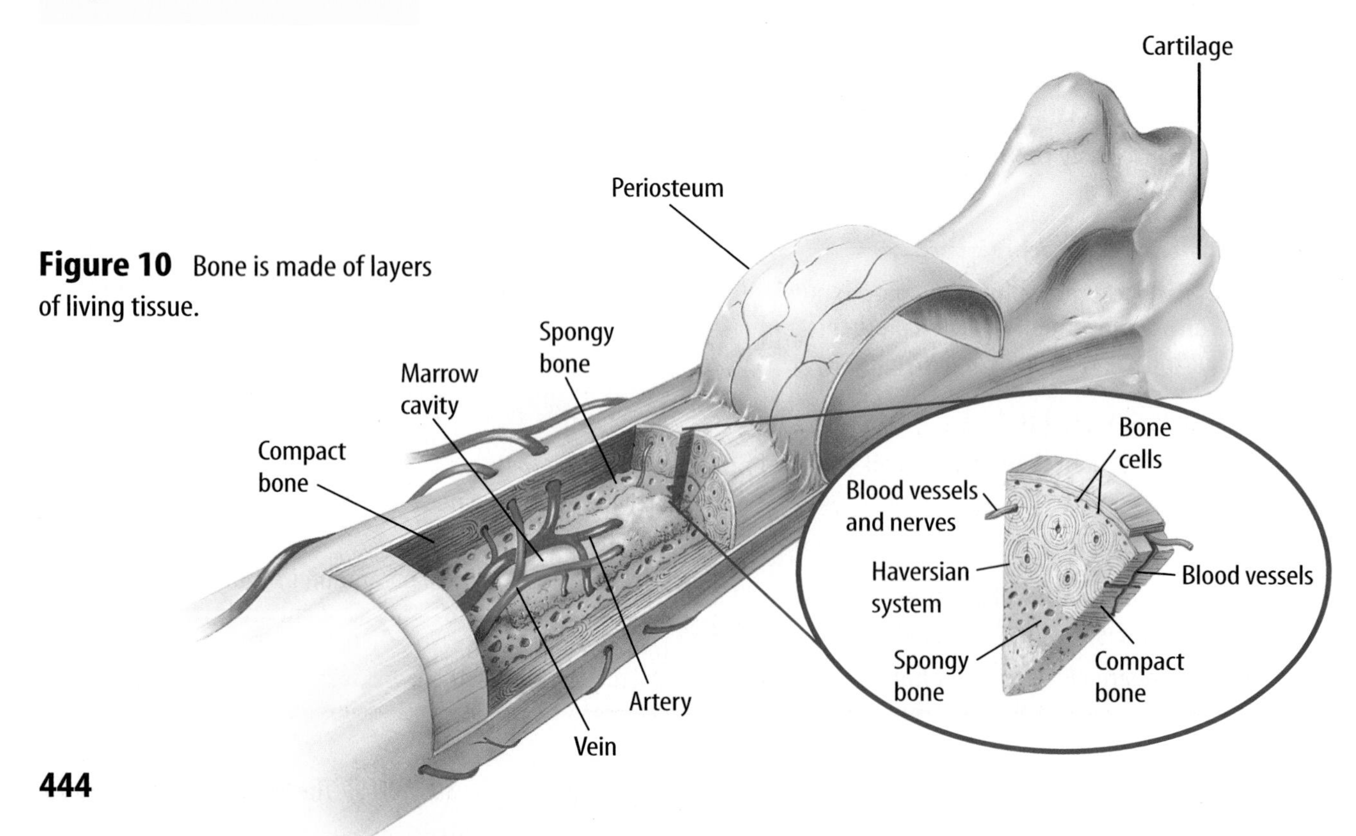

Figure 10 Bone is made of layers of living tissue.

Bone Tissue Living bone is an organ made of several different tissues. A living bone's surface is covered with a tough, tight-fitting membrane called the **periosteum** (pur ee AHS tee um). Small blood vessels in the periosteum carry nutrients into the bone and its nerves signal pain. Under the periosteum are compact bone and spongy bone.

Compact bone gives bones strength. It has a framework containing deposits of calcium phosphate that make the bone hard. Spongy bone is located toward the ends of long bones, such as those in your thigh and upper arm. Spongy bone has many small, open spaces that make bones lightweight.

In the centers of long bones are large openings called cavities. These cavities and the spaces in spongy bone are filled with a substance called marrow. Some marrow is yellow and is composed of fat cells. Red marrow produces red blood cells at a rate of 2 million to 3 million cells per second.

Topic: Bone Fractures
Visit green.msscience.com for Web links to information about new techniques for treating bone fractures.

Activity Using captions, illustrate one technique in your Science Journal.

Cartilage The ends of bones are covered with a smooth, slippery, thick layer of tissue called **cartilage.** Cartilage does not contain blood vessels or minerals. It is flexible and important in joints because it acts as a shock absorber. It also makes movement easier by reducing friction that would be caused by bones rubbing together.

Reading Check *What is cartilage?*

Bone Formation

Your bones have not always been as hard as they are now. Months before your birth, your skeleton was made of cartilage. Gradually, the cartilage broke down and was replaced by bone, as illustrated in **Figure 11.** Bone-forming cells called osteoblasts (AHS tee oh blasts) deposit calcium and phosphorus in bones, making the bone tissue hard. At birth, your skeleton was made up of more than 300 bones. As you developed, some bones fused, or grew together, so that now you have only 206 bones.

Healthy bone tissue is always being formed and re-formed. Osteoblasts build up bone. Another type of bone cell, called an osteoclast, breaks down bone tissue in other areas of the bone. This is a normal process in a healthy person. When osteoclasts break bone down, they release calcium and phosphorus into the bloodstream. These elements are necessary for the working of your body, including the movement of your muscles.

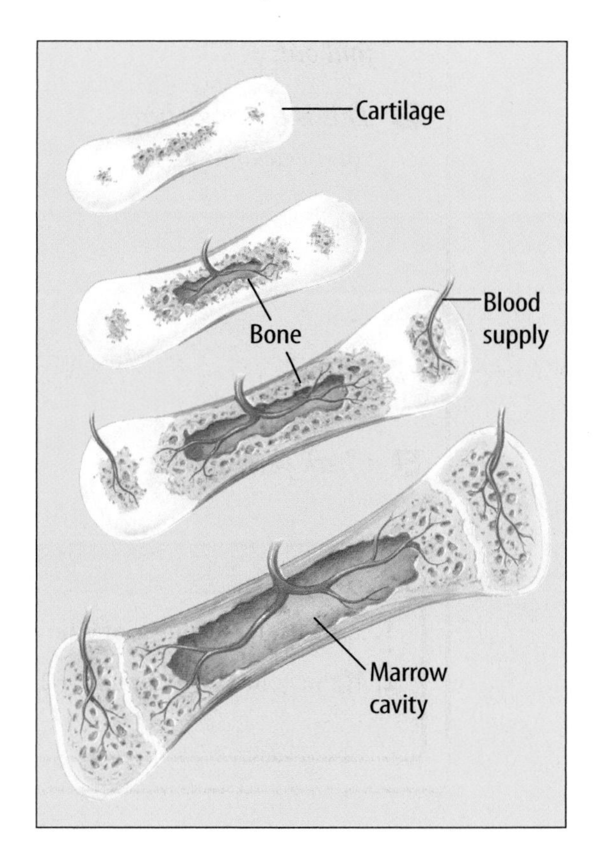

Figure 11 Cartilage is replaced slowly by bone as solid tissue grows outward. Over time, the bone reshapes to include blood vessels, nerves, and marrow.
Describe *the type of bone cell that builds up bone.*

Joints

What will you do at school today? You may sit at a table, chew and swallow your lunch, or walk to class. All of these motions are possible because your skeleton has joints.

Any place where two or more bones come together is a **joint.** Bones are held in place at joints by a tough band of tissue called a **ligament.** Many joints, such as your knee, are held together by more than one ligament. Muscles move bones by moving joints. The bones in healthy joints are separated by a thin layer of cartilage so that they do not rub against each other as they move.

Applying Math — Estimate

VOLUME OF BONES Although bones are not perfectly shaped, many of them are cylindrical. This cylindrical shape allows your bones to withstand great pressure. Estimate the volume of a bone that is 36 cm long and 7 cm in diameter.

Solution

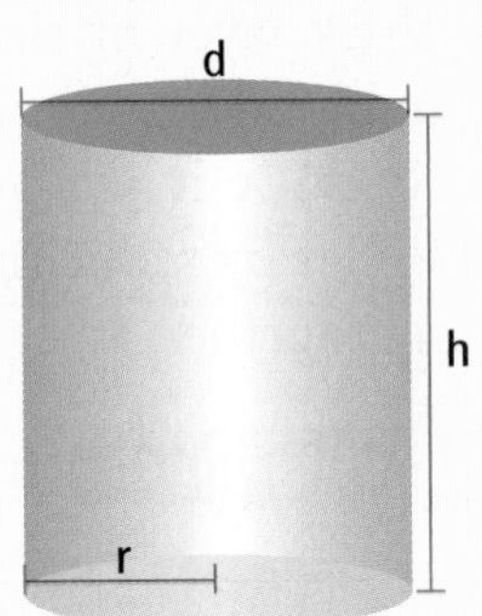

1 *This is what you know:* The bone has a shape of a cylinder whose height, h, measures 36 cm and whose diameter is 7.0 cm.

2 *This is what you want to find out:* Volume of the cylinder

3 *This is the procedure you need to use:*

- Use this equation:
  Volume = $\pi \times (\text{radius})^2 \times$ height, or $V = \pi \times r^2 \times h$
- A radius is one-half the diameter $\left(\frac{1}{2} \times 7 \text{ cm}\right)$, so $r = 3.5$ cm, $h = 36$ cm, and $\pi = 3.14$.
- $V = 3.14 \times (3.5 \text{ cm})^2 \times 36 \text{ cm}$
  $V = 1{,}384.74 \text{ cm}^3$
- The volume of the bone is approximately 1,384.74 cm^3.

4 *Check your answer:* Divide your answer by 3.14 and then divide that number by $(3.5)^2$. This number should be the height of the bone.

Practice Problems

1. Estimate the volume of a bone that has a height of 12 cm and a diameter of 2.4 cm.
2. If the volume of a bone is 62.8 cm^3 and its height is 20 cm, what is its diameter?

For more practice, visit green.msscience.com/math_practice

Immovable Joints Joints are broadly classified as immovable or movable. An immovable joint allows little or no movement. The joints of the bones in your skull and pelvis are classified as immovable joints.

Reading Check *How are bones held in place at joints?*

Movable Joints All movements, including somersaulting and working the controls of a video game, require movable joints, such as those in **Figure 12.** A movable joint allows the body to make a wide range of motions. There are several types of movable joints—pivot, ball and socket, hinge, and gliding.

In a pivot joint, one bone rotates in a ring of another bone that does not move. Turning your head is an example of a pivot movement. A ball-and-socket joint consists of a bone with a rounded end that fits into a cuplike cavity on another bone. A ball-and-socket joint provides a wider range of motion than a pivot joint does. That's why your legs and arms can swing in almost any direction.

A third type of joint is a hinge joint, which has a back-and-forth movement like hinges on a door. Elbows, knees, and fingers have hinge joints. Hinge joints have a smaller range of motion than the ball-and-socket joint. They are not dislocated, or pulled apart, as easily as a ball-and-socket joint can be.

A fourth type of joint is a gliding joint, in which one part of a bone slides over another bone. Gliding joints also move in a back-and-forth motion and are found in your wrists and ankles and between vertebrae. Gliding joints are used the most in your body. You can't write a word, use a joy stick, or take a step without using one of your gliding joints.

Figure 12 When a basketball player shoots a ball, several types of joints are in action.
Name *other activities that use several types of joints.*

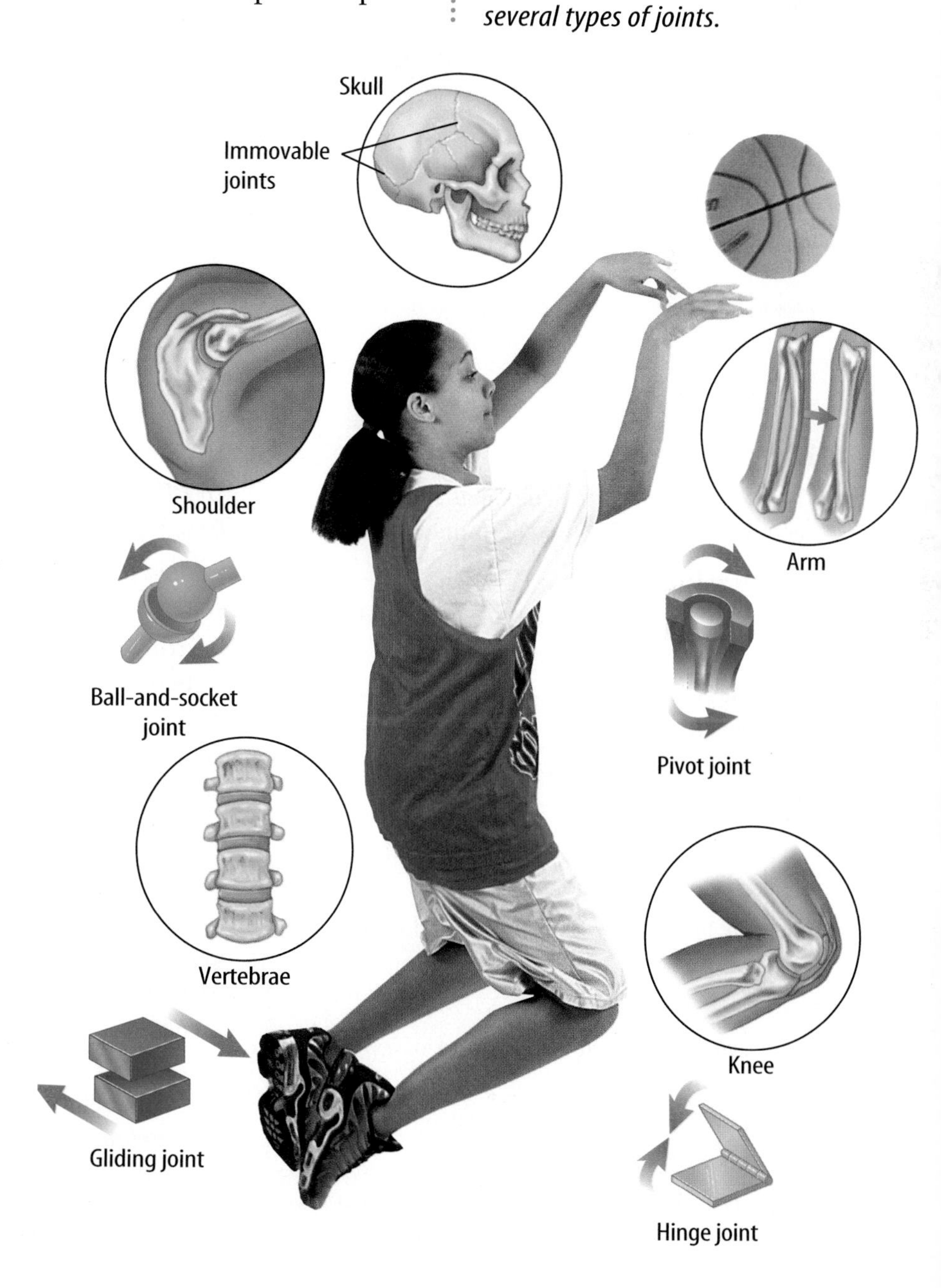

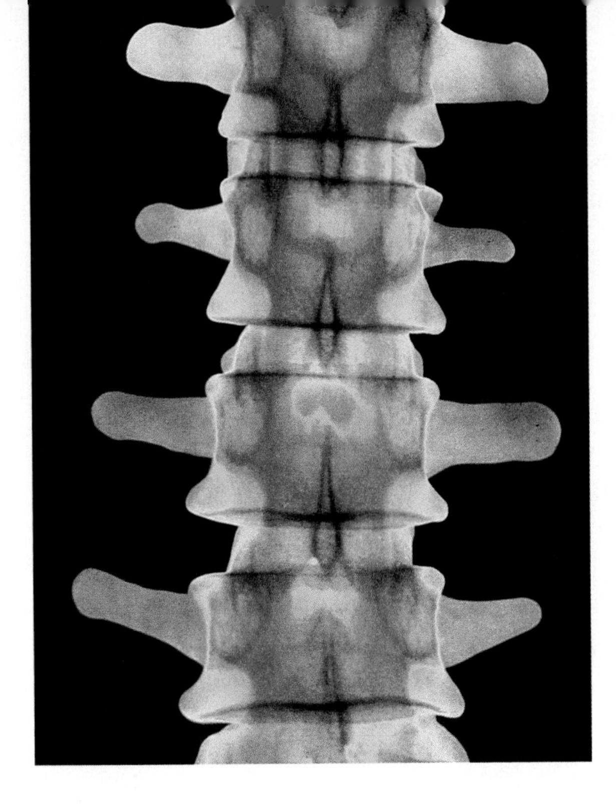

Figure 13 A colored X ray of the human backbone shows disks of cartilage between the vertebrae.

Moving Smoothly When you rub two pieces of chalk together, their surfaces begin to wear away, and they get reshaped. Without the protection of the cartilage at the end of your bones, they also would wear away at the joints. Cartilage helps make joint movement easier. It reduces friction and allows bones to slide more easily over each other. As shown in **Figure 13,** pads of cartilage, called disks, are located between the vertebrae in your back. They act as cushions and prevent injury to your spinal cord. A fluid that comes from nearby blood vessels also lubricates the joint.

Common Joint Problems Arthritis is the most common joint problem. The term *arthritis* describes more than 100 different diseases that can damage joints. About one out of every seven people in the United States suffers from arthritis. All forms of arthritis begin with the same symptoms: pain, stiffness, and swelling of the joints.

section 3 review

Summary

Functions of Your Skeletal System

- The skeletal system includes all the bones in your body and is your body's framework.

Bone Structure

- Bones are living organs that need nutrients.
- Compact bone is hard and strong, and spongy bone has many open spaces to make it lightweight.
- Cartilage covers the ends of bones.

Bone Formation

- Bone-forming cells deposit calcium and phosphorus to make the bone tissue hard.
- Healthy bone tissue is always being formed and reformed.

Joints

- Immovable joints allow little or no movement.
- Movable joints include pivot, ball and socket, hinge, and gliding joints.
- Cartilage helps make joint movement easier.

Self Check

1. **List** the five major functions of the human skeletal system.
2. **Describe** and give an example of an immovable joint.
3. **Explain** the functions of cartilage in your skeletal system.
4. **Describe** ligaments and their function in the skeletal system.
5. **Think Critically** A thick band of bone forms around a broken bone as it heals. In time, the thickened band disappears. Explain how this extra bone can disappear over time.

Applying Skills

6. **Make and Use Tables** Use a table to classify the bones of the human body as follows: *long, short, flat,* and *irregular.*
7. **Use graphics software** to make a graph that shows how an adult's bones are distributed: *29 skull bones, 26 vertebrae, 25 ribs, four shoulder bones, 60 arm and hand bones, two hip bones, and 60 leg and feet bones.*

section 4

The Nervous System

as you read

What You'll Learn

- **Describe** the basic structure of a neuron and how an impulse moves across a synapse.
- **Compare and contrast** the central and peripheral nervous systems.
- **List** the sensory receptors in each sense organ.
- **Explain** what type of stimulus each sense organ responds to and how.
- **Explain** how drugs affect the body.

Why It's Important

Your body reacts to your environment because of your nervous system.

Review Vocabulary

homeostasis: regulation of an organism's internal, life-maintaining conditions despite changes in its environment

New Vocabulary

- neuron
- synapse
- central nervous system
- peripheral nervous system

How the Nervous System Works

After doing the dishes and finishing your homework, you settle down in your favorite chair and pick up that mystery novel you've been trying to finish. Only three pages to go. . . Who did it? Why did she do it? Crash! You scream. What made that unearthly noise? You turn around to find that your dog's wagging tail has just swept the lamp off the table. Suddenly, you're aware that your heart is racing and your hands are shaking. After a few minutes, your breathing returns to normal and your heartbeat is back to its regular rate. What's going on?

Responding to Stimuli The scene described above is an example of how your body responds to changes in its environment. Any internal or external change that brings about a response is called a stimulus (STIHM yuh lus). Each day, you're bombarded by thousands of stimuli, as shown in **Figure 14.** Noise, light, the smell of food, and the temperature of the air are all stimuli from outside your body. Chemical substances such as hormones are examples of stimuli from inside your body. Your body adjusts to changing stimuli with the help of your nervous system.

Figure 14 Stimuli are found everywhere and all the time, even when you're enjoying being with your friends.
List *the types of stimuli that are present at this party.*

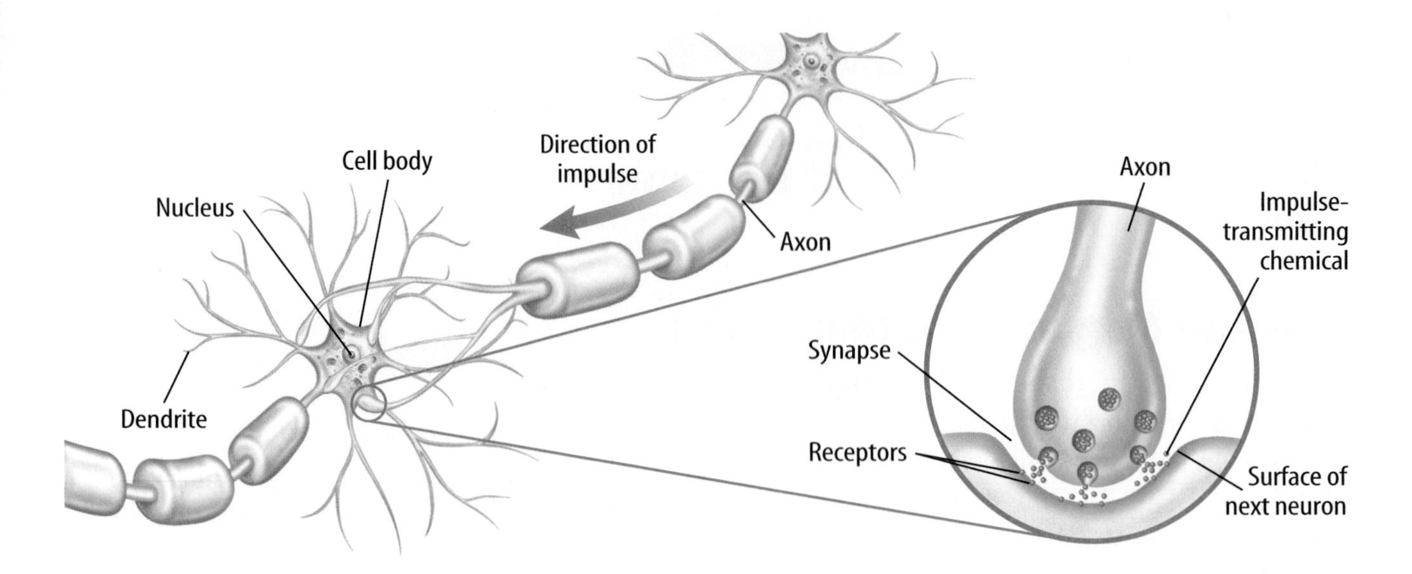

Figure 15 A neuron is made up of a cell body, dendrites, and an axon. An impulse moves in only one direction across a synapse—from an axon to the dendrites or cell body of another neuron.

Homeostasis It's amazing how your body handles all these stimuli. Control systems maintain homeostasis. They keep steady, life-maintaining conditions inside your body, despite changes around you. Examples of homeostasis are the regulation of your breathing, heartbeat, and digestion. Your nervous system is one of several control systems used by your body to maintain homeostasis.

Nerve Cells

The basic functioning units of the nervous system are nerve cells, or **neurons** (NOO rahnz). As shown in **Figure 15,** a neuron is made up of a cell body, branches called dendrites, and axons (AK sahns). Any message carried by a neuron is called an impulse. Your neurons are adapted in such a way that impulses move in only one direction. Dendrites receive impulses from other neurons and send them to the cell body. Axons carry impulses away from the cell body. The end of the axon branches. This allows the impulses to move to many other muscles, neurons, or glands.

Three types of neurons—sensory neurons, motor neurons, and interneurons—transport impulses. Sensory neurons receive information and send impulses to the brain or spinal cord, where interneurons relay these impulses to motor neurons. Motor neurons then conduct impulses from the brain or spinal cord to muscles or glands throughout your body.

Synapses Neurons don't touch each other. As an impulse moves from one neuron to another it crosses a small space called a **synapse** (SIH naps). In **Figure 15,** note that when an impulse reaches the end of an axon, the axon releases a chemical. This chemical flows across the synapse and stimulates the impulse in the dendrite of the next neuron.

The Divisions of the Nervous System

Figure 16 shows how organs of the nervous system are grouped into two major divisions—the central nervous system (CNS) and the peripheral (puh RIH fuh rul) nervous system (PNS). The **central nervous system** includes the brain and spinal cord. The brain is the control center for all activities in the body. It is made of billions of neurons. The spinal cord is made up of bundles of neurons. An adult's spinal cord is about the width of a thumb and about 43 cm long. Sensory neurons send impulses to the brain or spinal cord.

Topic: Nervous System
Visit green.msscience.com for Web links to information about the nervous system.

Activity Make a brochure outlining recent medical advances.

The Peripheral Nervous System

All the nerves outside the CNS that connect the brain and spinal cord to other body parts are part of the **peripheral nervous system.** The PNS includes 12 pairs of nerves from your brain called cranial nerves, and 31 pairs of nerves from your spinal cord called spinal nerves. Spinal nerves are made up of bundles of sensory and motor neurons bound together by connective tissue. They carry impulses from all parts of the body to the brain and from the brain to all parts of your body. A single spinal nerve can have impulses going to and from the brain at the same time. Some nerves contain only sensory neurons, and some contain only motor neurons, but most nerves contain both types of neurons.

Somatic and Autonomic Systems

The peripheral nervous system has two major divisions. The somatic system controls voluntary actions. It is made up of the cranial and spinal nerves that go from the central nervous system to your skeletal muscles. The autonomic system controls involuntary actions—those not under conscious control—such as your heart rate, breathing, digestion, and glandular functions.

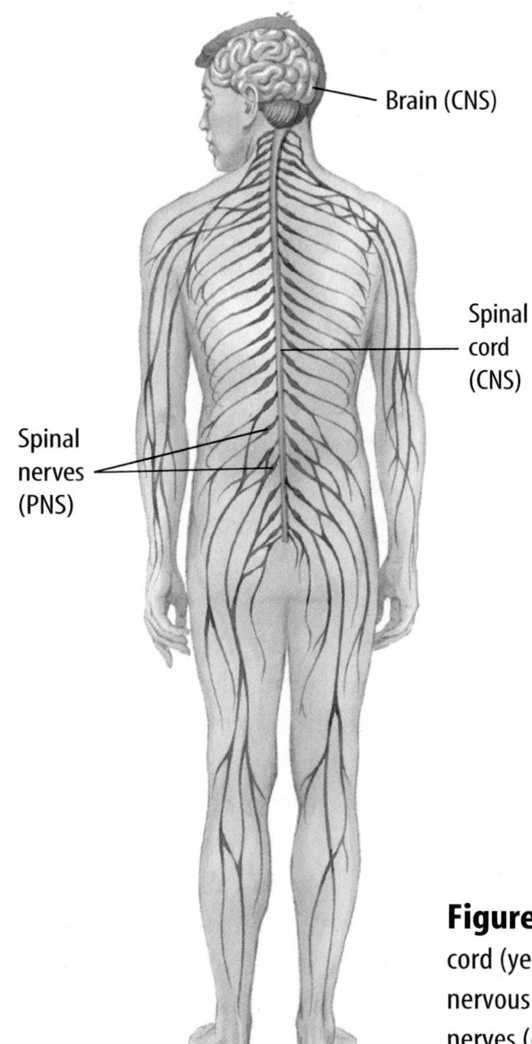

Figure 16 The brain and spinal cord (yellow) form the central nervous system (CNS). All other nerves (green) are part of the peripheral nervous system (PNS).

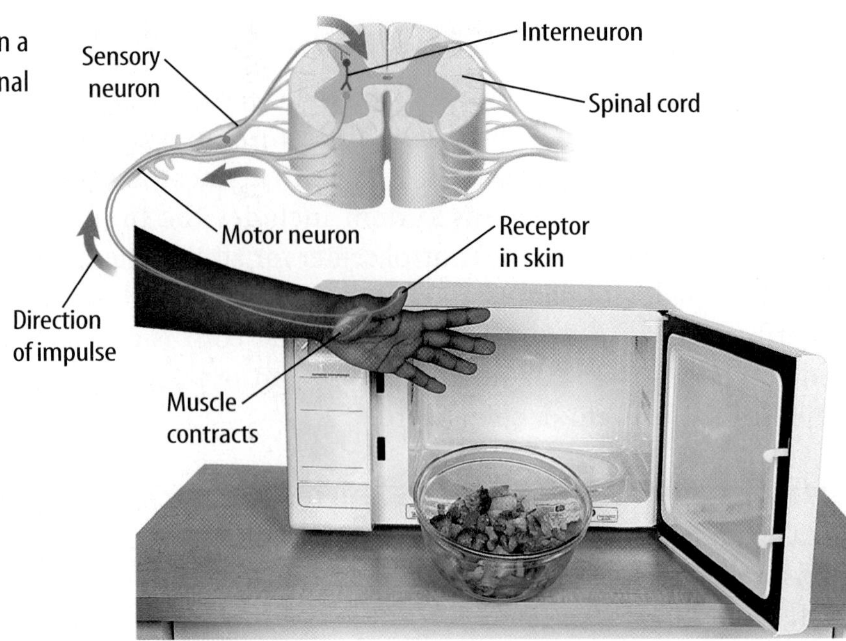

Figure 17 Your response in a reflex is controlled in your spinal cord, not your brain.

Safety and the Nervous System

Every mental process and physical action of the body involves structures of the central and peripheral nervous systems. Therefore, any injury to them can be serious. A severe blow to the head can bruise the brain and cause temporary or permanent loss of mental and physical abilities. For example, an injury to the back of the brain could result in the loss of vision.

Although the spinal cord is surrounded by the vertebrae of your spine, spinal cord injuries do occur. They can be just as dangerous as a brain injury. Injury to the spine can bring about damage to nerve pathways and result in paralysis (puh RAH luh suhs), which is the loss of muscle movement. Major causes of head and spinal injuries include automobile, motorcycle, and bicycle accidents, as well as sports injuries. Just like wearing seat belts in automobiles, it is important to wear the appropriate safety gear while playing sports and riding on bicycles and skateboards.

Neuron Chemical
Acetylcholine (uh see tul KOH leen) is a chemical produced by neurons that carries an impulse across a synapse to the next neuron. After the impulse is started, acetylcholine breaks down rapidly. In your Science Journal, make an inference about why the rapid breakdown of acetylcholine is important.

Reflexes You experience a reflex if you accidentally touch something sharp, something extremely hot or cold, or when you cough or vomit. A reflex is an involuntary, automatic response to a stimulus. You can't control reflexes because they occur before you know what has happened. A reflex involves a simple nerve pathway called a reflex arc, as illustrated in **Figure 17.**

A reflex allows the body to respond without having to think about what action to take. Reflex responses are controlled in your spinal cord, not in your brain. Your brain acts after the reflex to help you figure out what to do to make the pain stop.

Why are reflexes important?

The Senses

Sense organs are adapted for intercepting stimuli, such as light rays, sound waves, heat, chemicals, or pressure, and converting them into impulses for the nervous system. Your internal organs have several kinds of sensory receptors that respond to touch, pressure, pain, and temperature and transmit impulses to the brain or spinal cord. In turn, your body responds to this new information. All of your body's senses work together to maintain homeostasis.

Sensory receptors also are located throughout your skin. Your lips are sensitive to heat and can prevent you from drinking something so hot that it would burn you. Pressure-sensitive skin cells warn you of danger and enable you to move to avoid injury.

Vision Think about the different kinds of objects you might look at every day. The eye, shown in **Figure 18,** is a sense organ. Your eyes have unique adaptations that usually enable you to see shapes of objects, shadows, and color. It's amazing that at one glance you might see the words on this page, the color illustrations, and your classmate sitting next to you.

How do you see? Light travels in a straight line unless something causes it to refract or change direction. Your eyes have structures that refract light. Two of these structures are the cornea and the lens. As light enters the eye, it passes through the cornea—the transparent section at the front of the eye—and is refracted. Then light passes through a lens and is refracted again. The lens directs the light onto the retina (RET nuh), which is a tissue at the back of the eye that is sensitive to light energy. Two types of cells called rods and cones are found in the retina. Cones respond to bright light and color. Rods respond to dim light. They are used to help you detect shape and movement.

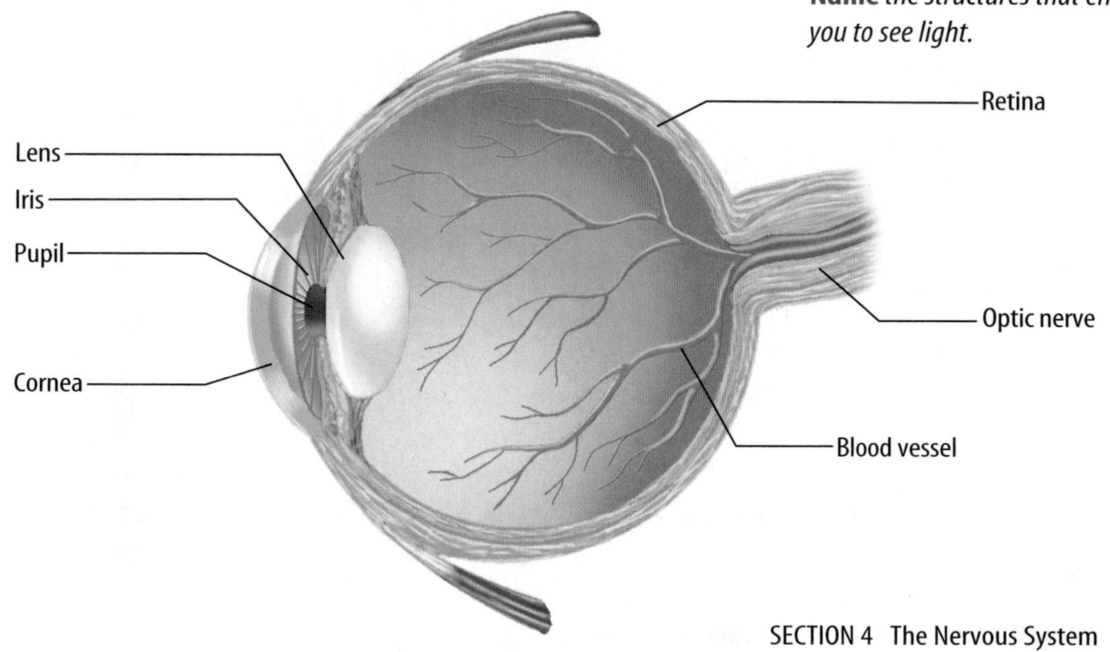

Figure 18 Light moves through the cornea and the lens—before striking the retina.
Name *the structures that enable you to see light.*

Images Light energy stimulates impulses in rods and cones that pass to the optic nerve. This nerve carries the impulses to the vision area of the brain. The image transmitted from the retina to the brain is upside down and reversed. The brain interprets the image correctly, and you see what you are looking at. The brain also interprets the images received by both eyes. It blends them into one image that gives you a sense of distance. This allows you to tell how close or how far away something is.

Hearing Whether it's the roar of a rocket launch, the cheers at a football game, or the distant song of a robin in a tree, sound waves are necessary for hearing. Sound is to hearing as light is to vision. When an object vibrates, sound waves are produced. Sound waves can travel through solids, liquids, and gases. When sound waves reach your ear, they usually stimulate nerve cells deep within your ear. Impulses from these cells are sent to the brain. When the sound impulse reaches the hearing area of the brain, it responds and you hear a sound.

Figure 19 shows that your ear is divided into three sections—the outer ear, middle ear, and inner ear. Your outer ear intercepts sound waves and funnels them down the ear canal to the middle ear. The sound waves cause the eardrum to vibrate much like the membrane on a musical drum vibrates when you tap it. These vibrations then move through three tiny bones called the hammer, anvil, and stirrup. The stirrup bone rests against a second membrane on an opening to the inner ear.

The inner ear includes the cochlea (KOH klee uh) and the semicircular canals. The cochlea is a fluid-filled structure shaped like a snail's shell. When the stirrup vibrates, fluids in the cochlea begin to vibrate. These vibrations bend sensory hair cells in the cochlea, which cause electrical impulses to be sent to the brain by a nerve. Depending on how the nerve endings are stimulated, you hear a different type of sound.

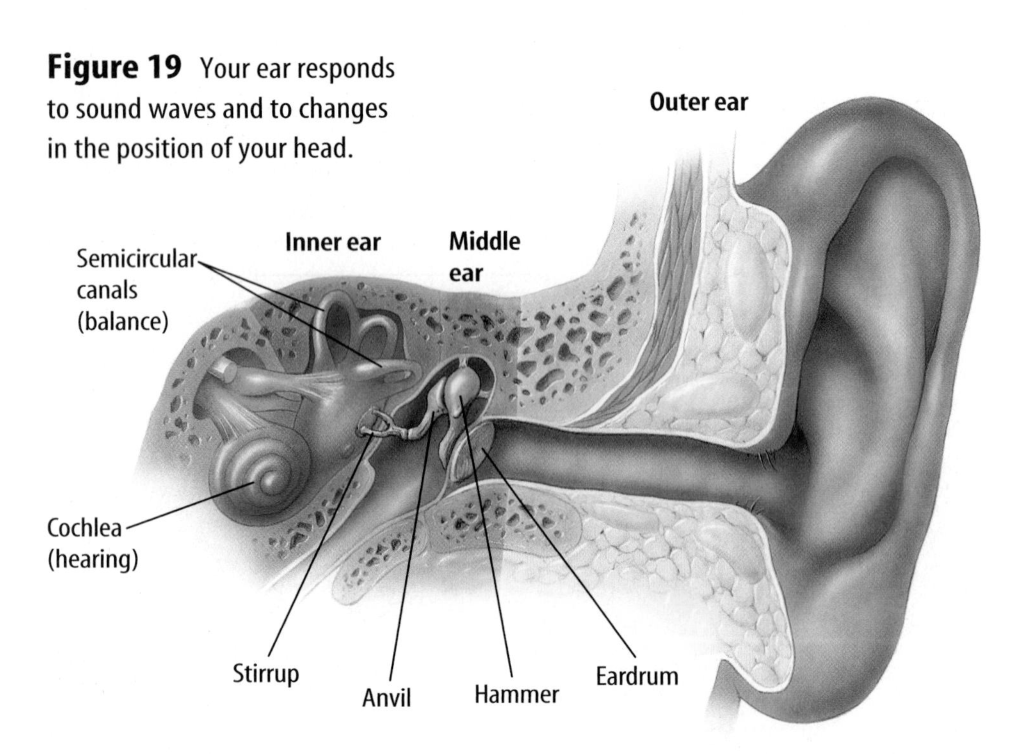

Figure 19 Your ear responds to sound waves and to changes in the position of your head.

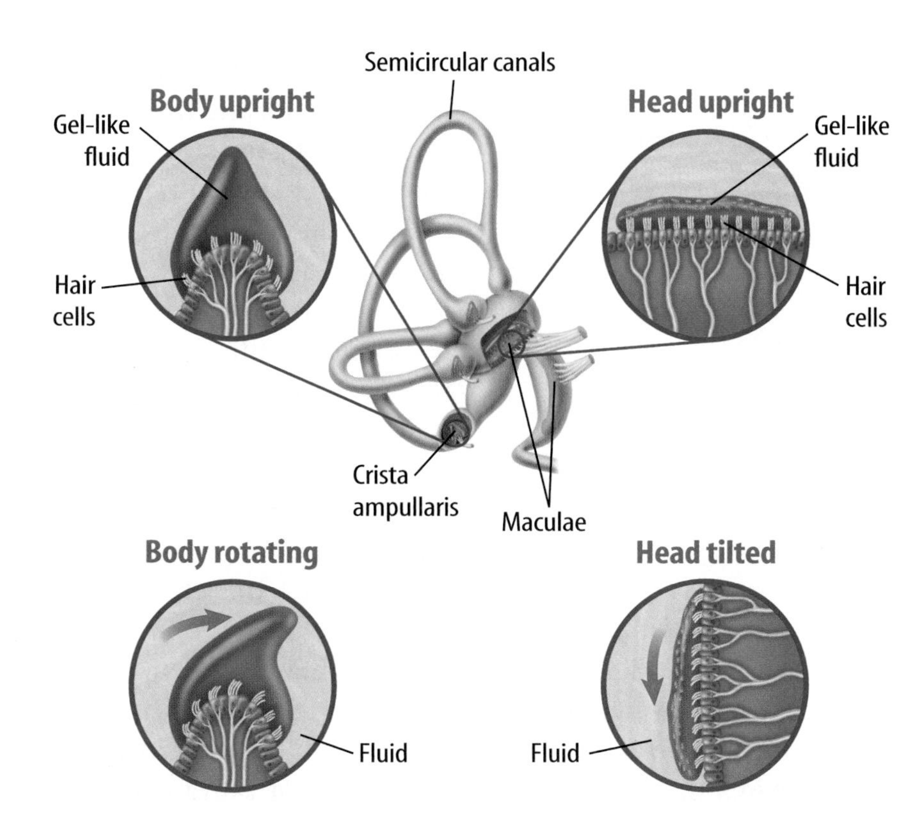

Figure 20 In your inner ear the cristae ampullaris react to rotating movements of your body, and the maculae check the position of your head with respect to the ground. **Explain** *why spinning around makes you dizzy.*

Balance Structures in your inner ear also control your body's balance. Structures called the cristae ampullaris (KRIHS tee • am pyew LEER ihs) and the maculae (MA kyah lee), illustrated in **Figure 20,** sense body movement. The cristae ampullaris react to rotating body movements and the maculae responds to the tilt of your head. Both structures contain tiny hair cells. As your body moves, gel-like fluid surrounding the hair cells moves and stimulates the nerve cells at the base of the hair cells. This produces nerve impulses that are sent to the brain, which interprets the body movements. The brain, in turn, sends impulses to skeletal muscles, resulting in other body movements that maintain balance.

Smell How can you smell your favorite food? You can smell food because molecules from the food move into the air. If they enter your nasal passages, these molecules stimulate sensitive nerve cells, called olfactory (ohl FAK tree) cells. Olfactory cells are kept moist by mucus. When molecules in the air dissolve in this moisture, the cells become stimulated. If enough molecules are present, an impulse starts in these cells, then travels to the brain where the stimulus is interpreted. If the stimulus is recognized from a previous experience, you can identify the odor. If you don't recognize a particular odor, it is remembered and may be identified the next time you encounter it.

What produces nerve impulses that interpret body movement?

Observing Balance Control

Procedure

1. Place two narrow strips of **paper** on the wall to form two parallel, vertical lines 20–25 cm apart. Have a person stand between them for 3 min, without leaning on the wall.
2. Observe how well balance is maintained.
3. Have the person close his or her eyes, then stand within the lines for 3 min.

Analysis

1. When was balance more difficult to maintain? Why?
2. What other factors might cause a person to lose his or her sense of balance?

Try at Home

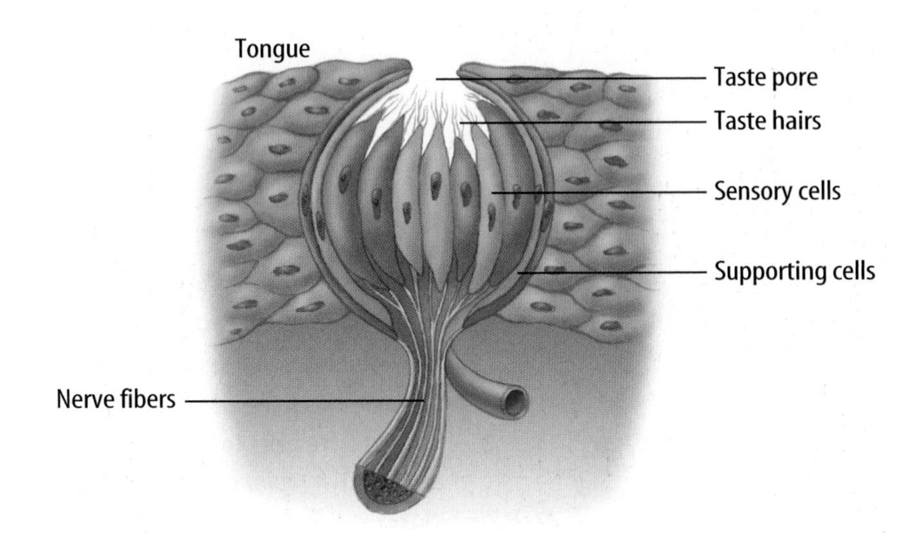

Figure 21 Taste buds are made up of a group of sensory cells with tiny taste hairs projecting from them. When food is taken into the mouth, it is dissolved in saliva. This mixture then stimulates receptor sites on the taste hairs, and an impulse is sent to the brain.

Taste Sometimes you taste a new food with the tip of your tongue, it tastes sweet. Then when you chew it, it tastes bitter. Taste buds on your tongue are the major sensory receptors for taste. About 10,000 taste buds all over your tongue enable you to tell one taste from another. Most taste buds respond to several taste sensations. However, certain areas of the tongue are more receptive to one taste than another. The five taste sensations are sweet, salty, sour, bitter, and the taste of MSG (monosodium glutamate).

A taste bud, shown in **Figure 21,** responds to chemical stimuli. In order to taste something, it has to be dissolved in water. Saliva begins this process. When a solution of saliva and food washes over taste buds, impulses are sent to your brain. The brain interprets the impulses, and you identify the tastes.

What needs to happen to food before you are able to taste it?

Smell and Taste The sense of smell is needed to identify some foods such as chocolate. When saliva in your mouth mixes with chocolate, odors travel up the nasal passage in the back of your throat. Olfactory cells in the nose are stimulated, and the taste and smell of chocolate are sensed. So when you have a stuffy nose and some foods seem tasteless, it may be because the food's molecules are blocked from contacting the olfactory cells in your nasal passages.

Drugs Affect the Nervous System

Many drugs, such as alcohol and caffeine, directly affect your nervous system. When swallowed, alcohol directly passes into cells of the stomach and small intestine then into the circulatory system. After it is in the circulatory system, it can travel throughout your body. Upon reaching neurons, alcohol moves through their cell membranes and disrupts their normal cell functions. As a result, this drug slows the activities of the central nervous system and is classified as a depressant. Muscle control, judgment, reasoning, memory, and concentration also are impaired. Heavy alcohol use destroys brain and liver cells.

Stimulants Any substance that speeds up the activity of the central nervous system is called a stimulant. Caffeine is a stimulant found in coffee, tea, cocoa, and many soft drinks, as shown in **Figure 22.** Too much caffeine can increase heart rate and aggravates restlessness, tremors, and insomnia in some people. It also can stimulate the kidneys to produce more urine.

Do you remember reading at the beginning of this section about being frightened after a lamp was broken? Think again about that scare. The organs of your nervous system control and coordinate responses to maintain homeostasis within your body. This task might be more difficult when your body must cope with the effects of drugs.

Figure 22 Caffeine, a substance found in colas, coffee, chocolate, and some teas, can cause excitability and sleeplessness.

section 4 review

Summary

How the Nervous System Works

- The nervous system responds to stimuli to maintain homeostasis.

Nerve Cells

- Neurons are the basic functioning units of the nervous system.
- To move from one neuron to another, an impulse crosses a synapse.

The Divisions of the Nervous System

- The autonomic system controls involuntary actions like heart rate and breathing.
- The somatic system controls voluntary actions.

Safety and the Nervous System

- Reflex responses are automatic and are controlled by the spinal cord.

The Senses

- Sense organs respond to stimuli and work together to maintain homeostasis.

Drugs Affect the Nervous System

- Drugs can slow or stimulate your nervous system.

Self Check

1. **Draw and label** the parts of a neuron and describe the function of each part.
2. **Name** the sensory receptors for the eyes, ears, and nose.
3. **Compare and contrast** the central and peripheral nervous systems.
4. **Explain** why you have trouble falling asleep after drinking several cups of hot cocoa.
5. **Identify** the role of saliva in tasting.
6. **Explain** why is it important to have sensory receptors for pain and pressure in your internal organs.
7. **Think Critically** Explain why many medications caution the consumer not to operate heavy machinery.

Applying Skills

8. **Communicate** Write a paragraph in your Science Journal that describes what each of the following objects would feel like: ice cube, snake, silk blouse, sandpaper, jelly, and smooth rock.
9. **Make and Use Tables** Organize the information on senses in a table that names the sense organs and which stimuli they respond to.

Design Your Own

Skin Sensitivity

Real-World Question

Your body responds to touch, pressure, temperature, and other stimuli. Not all parts of your body are equally sensitive to stimuli. Some areas are more sensitive than others are. For example, your lips are sensitive to heat. This protects you from burning your mouth and tongue. Now think about touch. How sensitive to touch is the skin on various parts of your body ? Which areas can distinguish the smallest amount of distance between stimuli?

Goals

- **Observe** the sensitivity to touch on specific areas of the body.
- **Design** an experiment that tests the effects of a variable, such as how close the contact points are, to determine which body areas can distinguish which stimuli are closest to one another.

Possible Materials

3-in × 5-in index card
toothpicks
tape
glue
metric ruler

**Alternate materials*

Safety Precautions

WARNING: *Do not apply heavy pressure when touching the toothpicks to the skin of your classmates.*

Form a Hypothesis

Based on your experiences, state a hypothesis about which of the following five areas of the body—fingertip, forearm, back of the neck, palm, and back of the hand—you believe to be most sensitive. Rank the areas from 5 (the most sensitive) to 1 (the least sensitive).

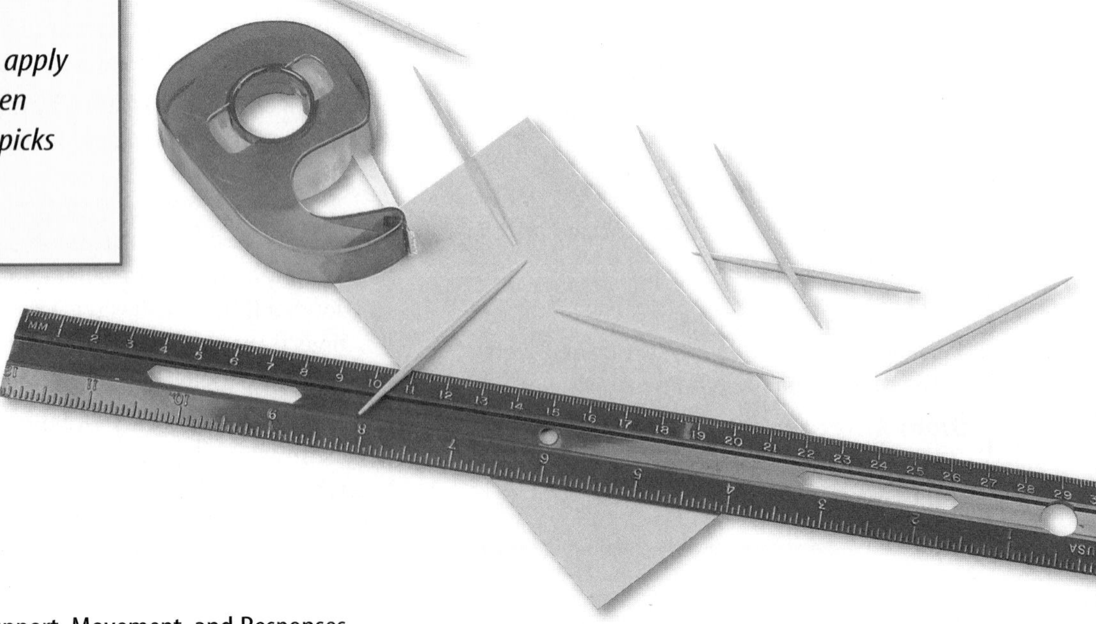

Using Scientific Methods

Test Your Hypothesis

Make a Plan

1. As a group, agree upon and write the hypothesis statement.
2. As a group, list the steps you need to test your hypothesis. Describe exactly what you will do at each step. Consider the following as you list the steps. How will you know that sight is not a factor? How will you use the card shown on the right to determine sensitivity to touch? How will you determine that one or both points are sensed?
3. Design a data table in your Science Journal to record your observations.
4. Reread your entire experiment to make sure that all steps are in the correct order.
5. Identify constants, variables, and controls of the experiment.

Follow Your Plan

1. Make sure your teacher approves your plan before you start.
2. Carry out the experiment as planned.
3. While the experiment is going on, write down any observations that you make and complete the data table in your Science Journal.

Analyze Your Data

1. **Identify** which part of the body tested can distinguish between the closest stimuli.
2. **Compare** your results with those of other groups.
3. Rank body parts tested from most to least sensitive. Did your results from this investigation support your hypothesis? Explain.

Conclude and Apply

1. **Infer** Based on the results of your investigation, what can you infer about the distribution of touch receptors on the skin?
2. **Predict** what other parts of your body would be less sensitive? Explain your predictions.

Write a report to share with your class about body parts of animals that are sensitive to touch.

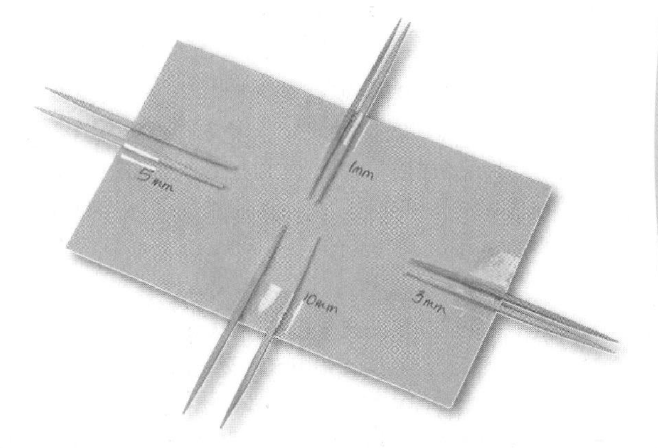

A fashion doll is doing her part for medical science! It turns out that the plastic joints that make it possible for one type of doll's legs to bend make good joints in prosthetic (artificial) fingers for humans.

Jane Bahor works at Duke University Medical Center in Durham, North Carolina. She makes lifelike body parts for people who have lost legs, arms, or fingers. A few years ago, she met a patient named Jennifer Jordan, an engineering student who'd lost a finger. The artificial finger that Bahor made looked real, but it couldn't bend. She and Jordan began to discuss the problem.

The engineer went home and borrowed one of her sister's dolls. Returning with it to Bahor's office, she and Bahor operated on the fashion doll's legs and removed the knee joints.

"It turns out that the doll's knee joints flexed the same way that human finger joints do," says Bahor. "We could see that using these joints would allow patients more use and flexibility with their 'new' fingers." Because these new prosthetic fingers can bend, the wearers can hold a pen, pick up a cup, or grab a steering wheel.

Bahor called the company that makes the fashion doll and shared the surprising discovery. The toymaker was so impressed that Bahor now has a ten-year supply of plastic knee joints—free of charge! But supplies come from other sources, too. "A Girl Scout troop in New Jersey just sent me a big box of donated dolls for the cause," reports Bahor. "It's really great to have kids' support in this effort."

Invent Choose a "problem" you can solve. Need a better place to store your notebooks in your locker, for instance? Use what Bahor calls "commonly found materials" to solve the problem. Then, make a model or a drawing of the problem-solving device.

For more information, visit green.msscience.com/oops

Reviewing Main Ideas

Section 1 The Skin

1. The epidermis produces melanin. Cells at the base of the epidermis produce new skin cells. The dermis contains nerves, sweat and oil glands, and blood vessels.
2. The skin protects the body, reduces water loss, produces vitamin D, and helps to maintain body temperature.
3. Severe skin damage can lead to infection and death if left untreated.

Section 2 The Muscular System

1. Skeletal muscle is voluntary and moves bones. Smooth muscle is involuntary and controls movement of internal organs. Cardiac muscle is involuntary and located only in the heart.
2. Muscles only can contract. When one skeletal muscle contracts, the other relaxes.

Section 3 The Skeletal System

1. Bones are living structures that protect, support, make blood, store minerals, and provide for muscle attachment.
2. Joints are either immovable or moveable.

Section 4 The Nervous System

1. The nervous system responds to stimuli to maintain homeostasis.
2. A neuron is the basic unit of structure and function of the nervous system.
3. A reflex is an automatic response.
4. The central nervous system is the brain and spinal cord. The peripheral nervous system includes cranial and spinal nerves.
5. Your senses enable you to react to your environment.
6. Many drugs affect your nervous system.

Visualizing Main Ideas

Copy and complete the following concept map on body movement.

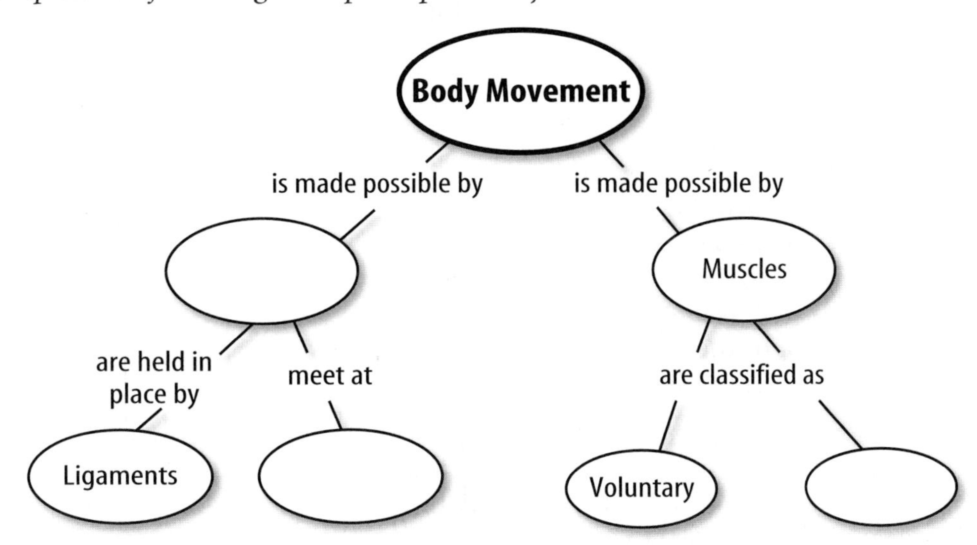

Using Vocabulary

cartilage p. 445	melanin p. 435
central nervous system p. 451	neuron p. 450
dermis p. 434	periosteum p. 445
epidermis p.434	peripheral nervous system p. 451
involuntary muscle p. 439	synapse p. 450
joint p. 446	tendon p. 440
ligament p. 446	voluntary muscle p. 439

Match the definitions with the correct vocabulary word.

1. outer layer of skin
2. thick band of tissue that attaches muscle to a bone
3. a muscle that you control
4. basic functioning unit of the nervous system
5. small space across which an impulse moves
6. tough outer covering of bone
7. a tough band of tissue that holds two bones together

Checking Concepts

Choose the word or phrase that best answers the question.

8. Where are blood cells made?
   A) compact bone **C)** cartilage
   B) periosteum **D)** marrow
9. What are the ends of bones covered with?
   A) cartilage **C)** ligaments
   B) tendons **D)** muscle
10. Where are human immovable joints found?
   A) at the elbow **C)** in the wrist
   B) at the neck **D)** in the skull
11. Which vitamin is made in the skin?
   A) A **C)** D
   B) B **D)** K
12. Which of the following structures helps retain fluids in the body?
   A) bone **C)** skin
   B) muscle **D)** joint
13. How do impulses cross synapses between neurons?
   A) by osmosis
   B) through interneurons
   C) through a cell body
   D) by a chemical
14. What are the neurons called that detect stimuli in the skin and eyes?
   A) interneurons **C)** motor neurons
   B) synapses **D)** sensory neurons
15. What does the somatic system of the PNS control?
   A) gland **C)** skeletal muscles
   B) heart **D)** salivary glands
16. What part of the eye is light finally focused on?
   A) lens **C)** pupil
   B) retina **D)** cornea
17. Which of the following is in the inner ear?
   A) anvil **C)** eardrum
   B) hammer **D)** cochlea

Use the illustration below to answer question 18.

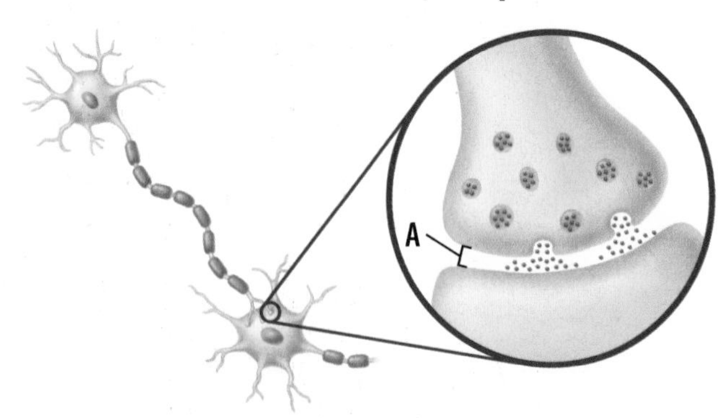

18. What is the name given to A?
   A) axon **C)** synapse
   B) dendrite **D)** nucleus

Science Online green.msscience.com/vocabulary_puzzlemaker

Thinking Critically

19. **Infer** why an infant's skull joints are flexible, but those of a teenager have fused together and are immovable.
20. **Predict** what would happen if a person's sweat glands didn't produce sweat.
21. **Compare and contrast** the functions of ligaments and tendons.
22. **Form a Hypothesis** Your body has about 3 million sweat glands. Make a hypothesis about where these sweat glands are on your body. Are they distributed evenly throughout your body?
23. **Draw Conclusions** If an impulse traveled down one neuron but failed to move on to the next neuron, what might you conclude about the first neuron?
24. **Concept Map** Copy and complete this events-chain concept map to show the correct sequence of the structures through which light passes in the eye.

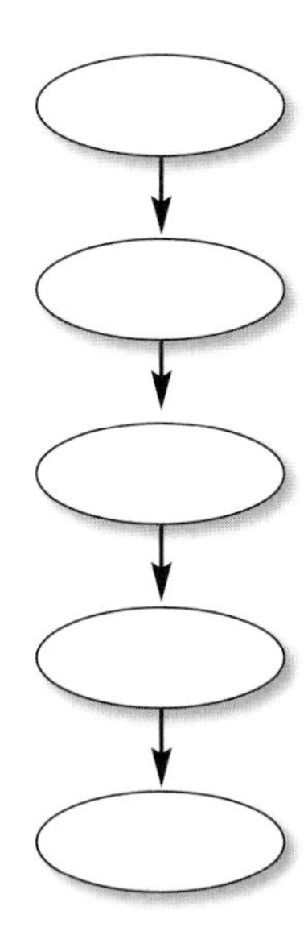

25. **List** what factors a doctor might consider before choosing a method of skin repair for a severe burn victim.
26. **Explain** why skin might not be able to produce enough vitamin D.

Performance Activities

27. **Illustrate** While walking on a sandy beach, a pain suddenly shoots through your foot. You look down and see that you stepped on the sharp edge of a broken shell. Draw and label the reflex arc that results from this stimulus.

Applying Math

Use the graph below to answer question 28.

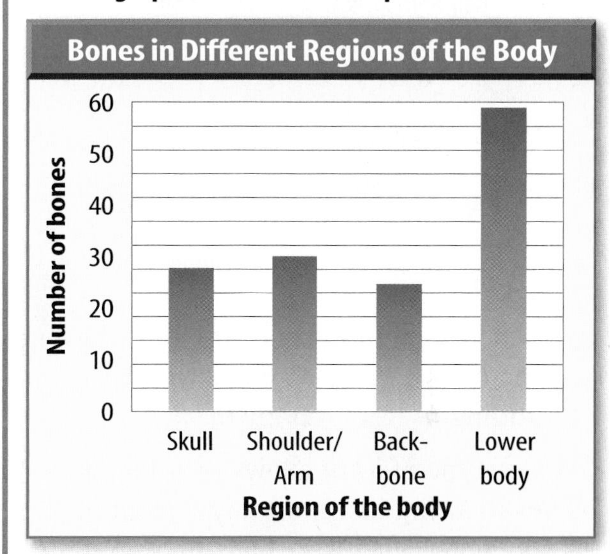

28. **Bone Tally** The total number of bones in the human body is 206. Approximately what percentage of bones is located in the backbone?

    A) 2%
    B) 12%
    C) 50%
    D) 75%

29. **Fireworks** You see the flash of fireworks and then four seconds later, you hear the boom. Light travels so fast that you see far away things instantaneously. Sound, on the other hand, travels at 340 m/s. How far away are you from the fireworks?

Standardized Test Practice

Part 1 Multiple Choice

Record your answers on the answer sheet provided by your teacher or on a sheet of paper.

1. Which of the following is NOT released by sweat glands?

A. water
B. salt
C. waste products
D. oil

Use the illustration below to answer questions 2 and 3.

Ball-and-socket joint

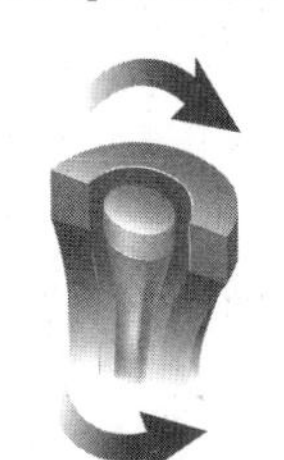

Pivot joint

Gliding joint

Hinge joint

2. Which type of joint do your elbows have?

A. hinge
B. gliding
C. ball and socket
D. pivot

3. Which type of joint allows your legs and arms to swing in almost any direction?

A. hinge
B. gliding
C. ball and socket
D. pivot

4. An internal or external change that brings about a response is called a

A. reflex
B. stimulus
C. receptor
D. heartbeat

Test-Taking Tip

Rest to be Alert Get plenty of sleep—at least eight hours every night—the week before the test and during test week.

Use the table below to answer questions 5 and 6.

Approximate Blood Alcohol Percentage for Men

Drinks	Body Weight in Kilograms							
	45.4	54.4	63.5	72.6	81.6	90.7	99.8	108.9
1	0.04	0.03	0.03	0.02	0.02	0.02	0.02	0.02
2	0.08	0.06	0.05	0.05	0.04	0.04	0.03	0.03
3	0.11	0.09	0.08	0.07	0.06	0.06	0.05	0.05
4	0.15	0.12	0.11	0.09	0.08	0.08	0.07	0.06
5	0.19	0.16	0.13	0.12	0.11	0.09	0.09	0.08

Subtract 0.01% for each 40 min of drinking. One drink is 40 mL of 80-proof liquor, 355 mL of beer, or 148 mL of table wine.

5. In Michigan underage drivers can be arrested for drinking and driving if their blood alcohol percentage is more than 0.02 percent. According to the table above, how many drinks would it take for a 72-kg man to exceed this limit?

A. three
B. two
C. one
D. zero

6. In some states, the legal blood alcohol percentage limit for driving while under the influence of alcohol is 0.08 percent. According to the table above, how many drinks would it take for a 54-kg man to exceed this limit?

A. four
B. three
C. two
D. one

7. A 90-kg man has been tested for blood alcohol content. His blood alcohol percentage is 0.08. Based upon the information in the table, about how much has he had to drink?

A. three drinks
B. 120 mL of 80-proof liquor
C. 396 mL of table wine
D. 1,420 mL of beer

Part 2 | Short Response/Grid In

Record your answers on the answer sheet provided by your teacher or on a sheet of paper.

8. One in seven people in the United States suffers from arthritis. Calculate the percentage of people that suffer from arthritis.

9. Explain the difference between voluntary and involuntary muscles.

Use the table below to answer questions 10–12.

Number of Bicycle Deaths per Year		
Year	Male	Female
1996	654	107
1997	712	99
1998	658	99
1999	656	94
2000	605	76

Data from Insurance Institute for Highway Safety

10. Head injuries are the most serious injuries that are found in people who died in bicycle accidents. Ninety percent of the deaths were in people who were not wearing bicycle helmets. Using the data in the table, approximately how many of the people (male and female) who died in bicycle accidents in 1998 were wearing bicycle helmets?

11. In 2000, what percentage of the people who died were women?

12. Which of the years from 1996 to 2000 had the greatest total number of bicycle deaths?

13. Explain why alcohol is classified as a depressant.

14. The brain is made up of approximately 100 billion neurons, which is about 10% of all neurons in the body. How many neurons are there in the human body?

Part 3 | Open Ended

Record your answers on a sheet of paper.

15. Explain how bone cells help maintain homeostasis.

16. Describe the changes that occur in muscles that do a lot of work. Compare these muscles to the muscles of a person who only does inactive pastimes.

17. Inez and Maria went to the ice cream parlor. They both ordered strawberry sundaes. Marie thought that the sundae was made with fresh strawberries, because it tasted so great. Inez thought that her sundae didn't have much flavor. What could be the reason that Inez's sundae was tasteless? Explain.

Use the illustration below to answer question 18.

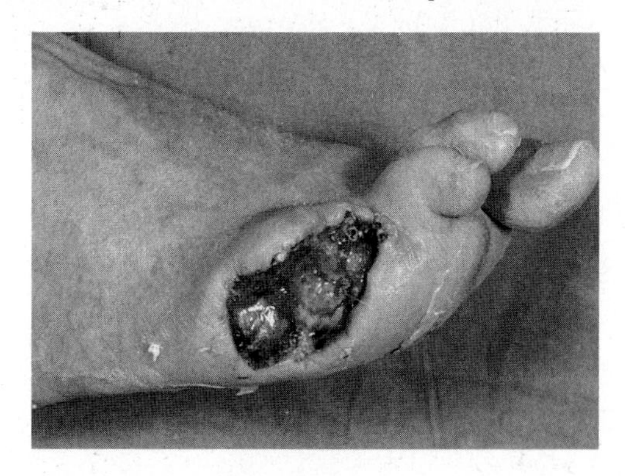

18. The person with this foot sore has diabetes. People with diabetes often lose sensation in their feet. Explain why a sore like the one in the photograph might develop if skin sensory receptors were not working properly.

19. What might happen to your body temperature if blood vessels in the skin did not contain smooth muscle?

20. Sam bumped into another player during soccer practice and bruised his leg. Describe the sequence of events from the time of injury until the injury disappears.

chapter 16

Regulation and Reproduction

chapter preview

sections

Where's the emergency?

The fire station control room blinks with panels of buttons and monitors. Dispatchers can access and relay emergency information quickly using this complex monitoring system. In a similar way, your body's endocrine system monitors and controls the actions of many of your body's functions.

Science Journal Write a paragraph describing how an emergency call might be handled at a fire station.

Start-Up Activities

Model a Chemical Message

Your body has systems that work together to coordinate your body's activities. One of these systems sends chemical messages through your blood to certain tissues, which, in turn, respond. Do the lab below to see how a chemical signal can be sent.

1. Cut a 10-cm-tall *Y* shape from filter paper and place it on a plastic, ceramic, or glass plate.
2. Sprinkle baking soda on one arm of the *Y* and salt on the other arm.
3. Using a dropper, place five or six drops of vinegar halfway up the leg of the *Y*.
4. **Think Critically** Describe in your Science Journal how the chemical moves along the paper and the reaction(s) it causes.

Preview this chapter's content and activities at green.msscience.com

Stages of Life Make the following Foldable to help you predict the stages of life.

STEP 1 **Fold** a vertical sheet of paper in half from top to bottom. Then fold it in half again top to bottom two more times. Unfold all the folds.

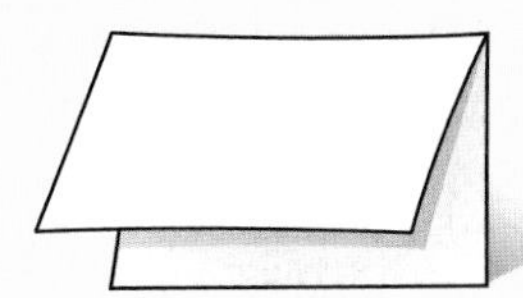

STEP 2 **Refold** the paper into a fan, using the folds as a guide. Unfold all the folds again.

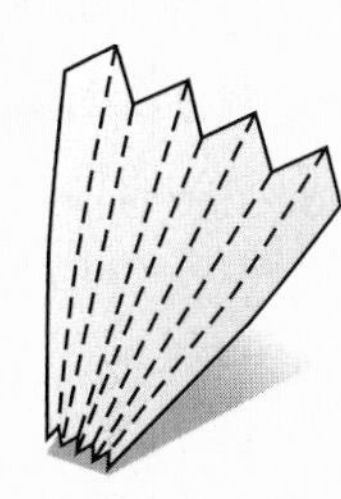

STEP 3 **Label** as shown.

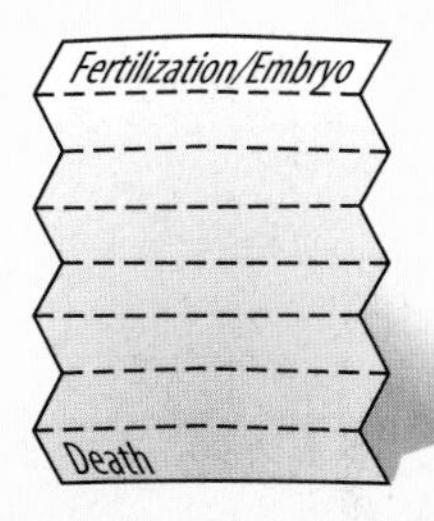

Read and Write Before you read the chapter, list as many stages of life as you can on your Foldable. Add to your list as you read the chapter.

section 1

The Endocrine System

as you read

What You'll Learn

- **Define** how hormones function.
- **Identify** different endocrine glands and the effects of the hormones they produce.
- **Describe** how a feedback system works in your body.

Why It's Important

The endocrine system uses chemicals to control many systems in your body.

Review Vocabulary
tissue: groups of cells that work together to perform a specific function

New Vocabulary
- hormone

Functions of the Endocrine System

You go through the dark hallways of a haunted house. You can't see a thing. Your heart is pounding. Suddenly, a monster steps out in front of you. You scream and jump backwards. Your body is prepared to defend itself or get away. Preparing the body for fight or flight in times of emergency, as shown in **Figure 1,** is one of the functions of the body's control systems.

Control Systems All of your body's systems work together, but the endocrine (EN duh krun) and the nervous systems are your body's control systems. The endocrine system sends chemical messages in your blood that affect specific tissues called target tissues. The nervous system sends rapid impulses to and from your brain, then throughout your body. Your body does not respond as quickly to chemical messages as it does to impulses.

Endocrine Glands

Tissues found throughout your body called endocrine glands produce the chemical messages called **hormones** (HOR mohnz). Hormones can speed up or slow down certain cellular processes. Some glands in your body release their products through small tubes called ducts. Endocrine glands are ductless and each endocrine gland releases its hormone directly into the blood. Then, the blood transports the hormone to the target tissue. A target tissue usually is located in the body far from the location of the endocrine gland that produced the hormone to which it responds.

Reading Check *What is the function of hormones?*

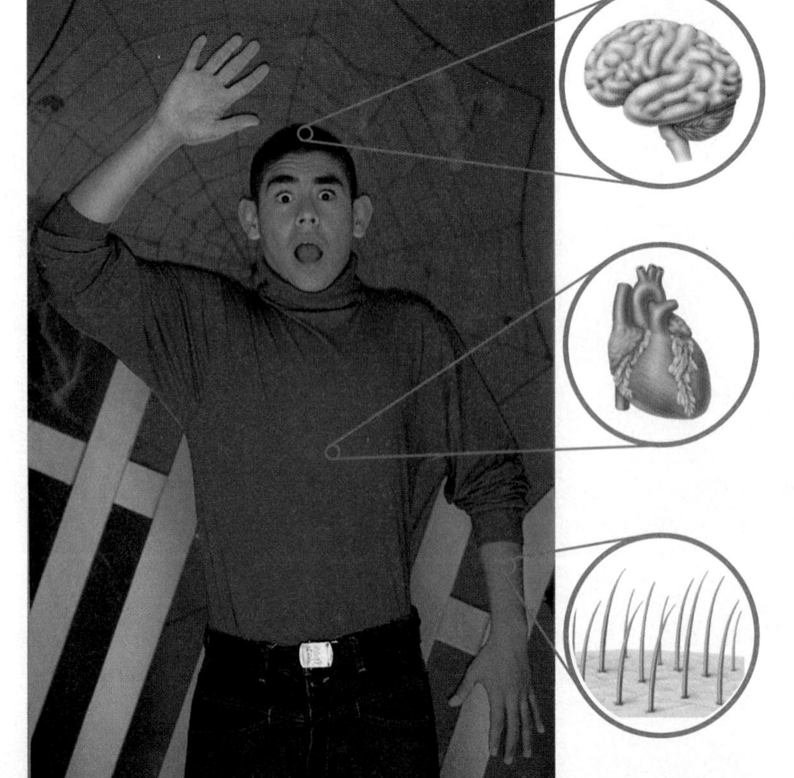

Figure 1 Your endocrine system enables many parts of your body to respond immediately in a fearful situation.

Gland Functions Endocrine glands have many functions in the body. The functions include the regulation of its internal environment, adaptation to stressful situations, promotion of growth and development, and the coordination of circulation, digestion, and the absorption of food. **Figure 2** on the next two pages shows some of the body's endocrine glands.

Applying Math Use Percentages

GLUCOSE LEVELS Calculate how much higher the blood sugar (glucose) level of a diabetic is before breakfast when compared to a nondiabetic before breakfast. Express this number as a percentage of the nondiabetic sugar level before breakfast.

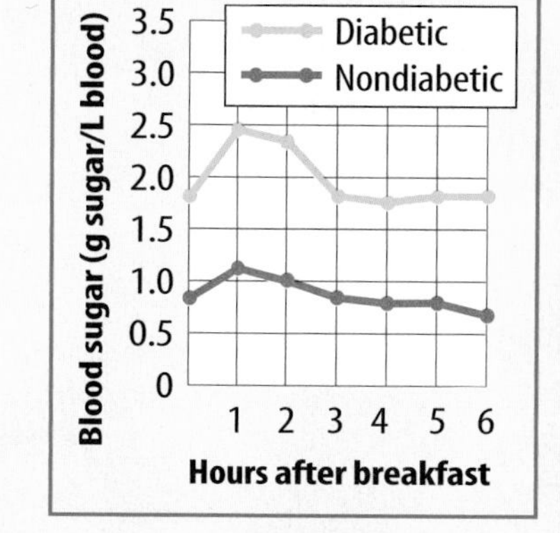

Solution

1 *This is what you know:*

- nondiabetic blood sugar at 0 h = 0.85 g sugar/L blood
- diabetic blood sugar at 0 h = 1.8 g sugar/L blood

2 *This is what you need to find out:*

How much higher is the glucose level of a diabetic person than that of a nondiabetic person before breakfast?

3 *This is the procedure you need to use:*

- Find the difference in glucose levels: 1.8 g/L − 0.85 g/L = 0.95 g/L
- Use this equation:

$$\frac{\text{difference between values}}{\text{nondiabetic value}} \times 100\% = \text{percent difference}$$

- Substitute in the known values:

$$\frac{0.95}{0.85} \times 100\% = 112\%$$

- Before breakfast, a diabetic's blood sugar is about 112 percent higher than that of a nondiabetic.

4 *Check your answer:*

Change 112% to a decimal then multiply it by 0.85. You should get 0.95.

Practice Problems

1. Express as a percentage how much higher the blood sugar value is for a diabetic person compared to a nondiabetic person 1 h after breakfast.
2. Express as a percentage how much higher the blood sugar value is for a diabetic person compared to a nondiabetic person 3 h and 6 h after breakfast.

For more practice, visit green.msscience.com/math_practice

NATIONAL GEOGRAPHIC

VISUALIZING THE ENDOCRINE SYSTEM

Figure 2

Your endocrine system is involved in regulating and coordinating many body functions, from growth and development to reproduction. This complex system consists of many diverse glands and organs, including the nine shown here. Endocrine glands produce chemical messenger molecules, called hormones, that circulate in the bloodstream. Hormones exert their influence only on the specific target cells to which they bind.

PINEAL GLAND Shaped like a tiny pinecone, the pineal gland lies deep in the brain. It produces melatonin, a hormone that may function as a sort of body clock by regulating wake/sleep patterns.

PITUITARY GLAND A pea-size structure attached to the hypothalamus of the brain, the pituitary gland produces hormones that affect a wide range of body activities, from growth to reproduction.

THYMUS The thymus is located in the upper chest, just behind the sternum. Hormones produced by this organ stimulate the production of certain infection-fighting cells.

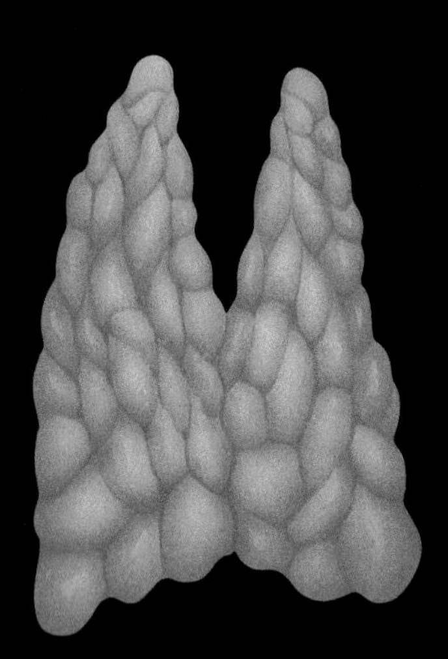

TESTES These paired male reproductive organs primarily produce testosterone, a hormone that controls the development and maintenance of male sexual traits. Testosterone also plays an important role in the production of sperm.

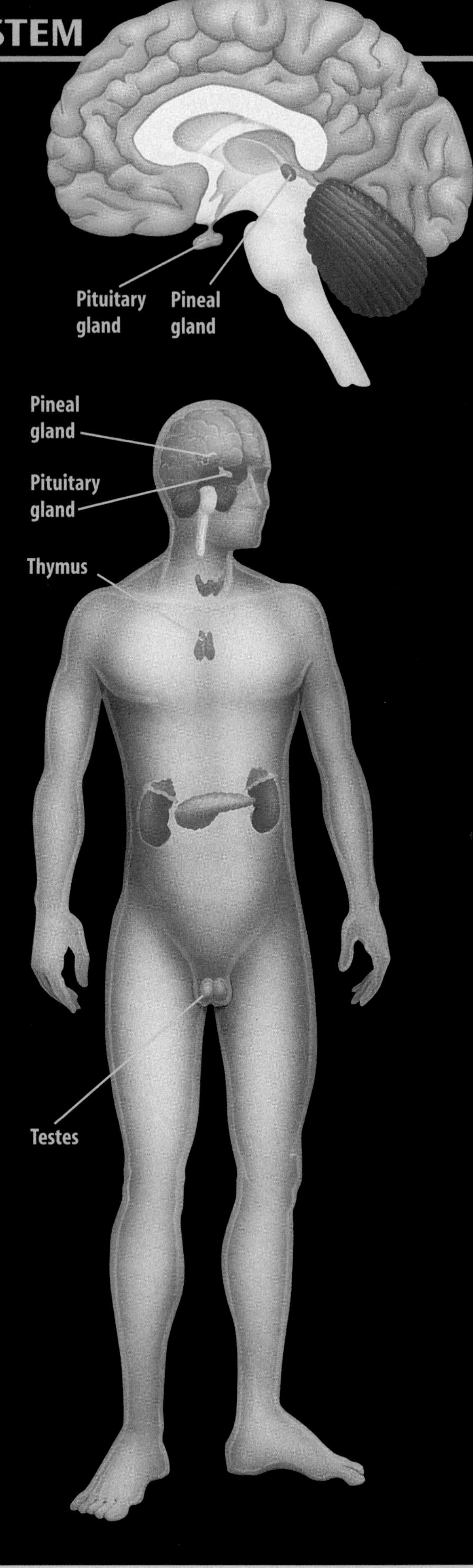

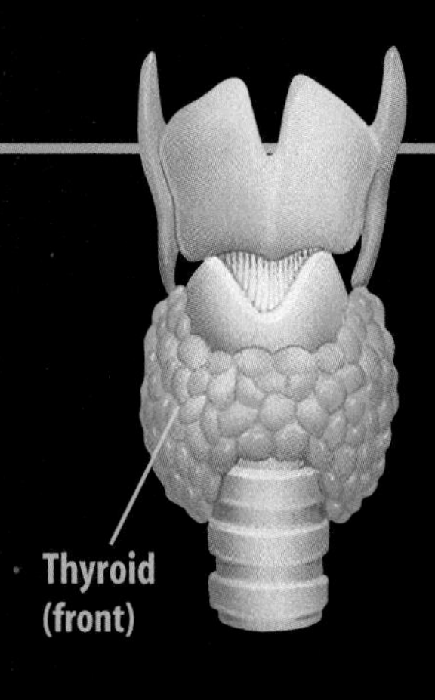

THYROID GLAND Located below the larynx, the bilobed thyroid gland is richly supplied with blood vessels. It produces hormones that regulate metabolic rate, control the uptake of calcium by bones, and promote normal nervous system development.

PARATHYROID GLANDS Attached to the back surface of the thyroid are tiny parathyroids, which help regulate calcium levels in the body. Calcium is important for bone growth and maintenance, as well as for muscle contraction and nerve impulse transmission.

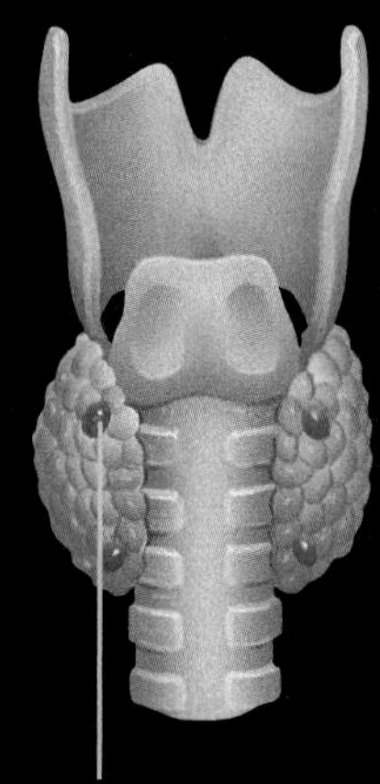

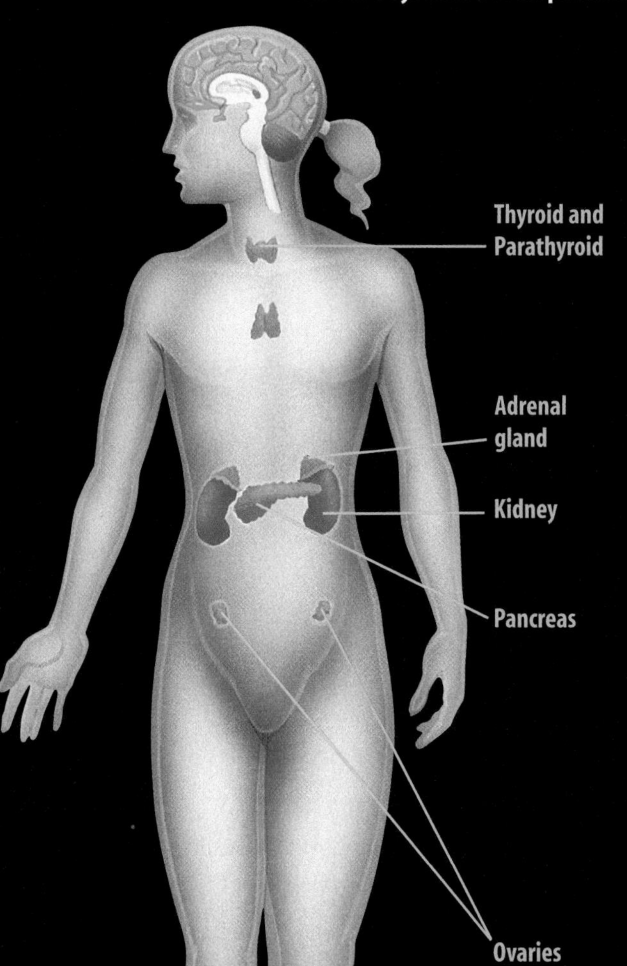

ADRENAL GLANDS On top of each of your kidneys is an adrenal gland. This complex endocrine gland produces a variety of hormones. Some play a critical role in helping your body adapt to physical and emotional stress. Others help stabilize blood sugar levels.

PANCREAS Scattered throughout the pancreas are millions of tiny clusters of endocrine tissue called the islets of Langerhans. Cells that make up the islets produce hormones that help control sugar levels in the bloodstream.

OVARIES Found deep in the pelvic cavity, ovaries produce female sex hormones known as estrogen and progesterone. These hormones regulate the female reproductive cycle and are responsible for producing and maintaining female sex characteristics.

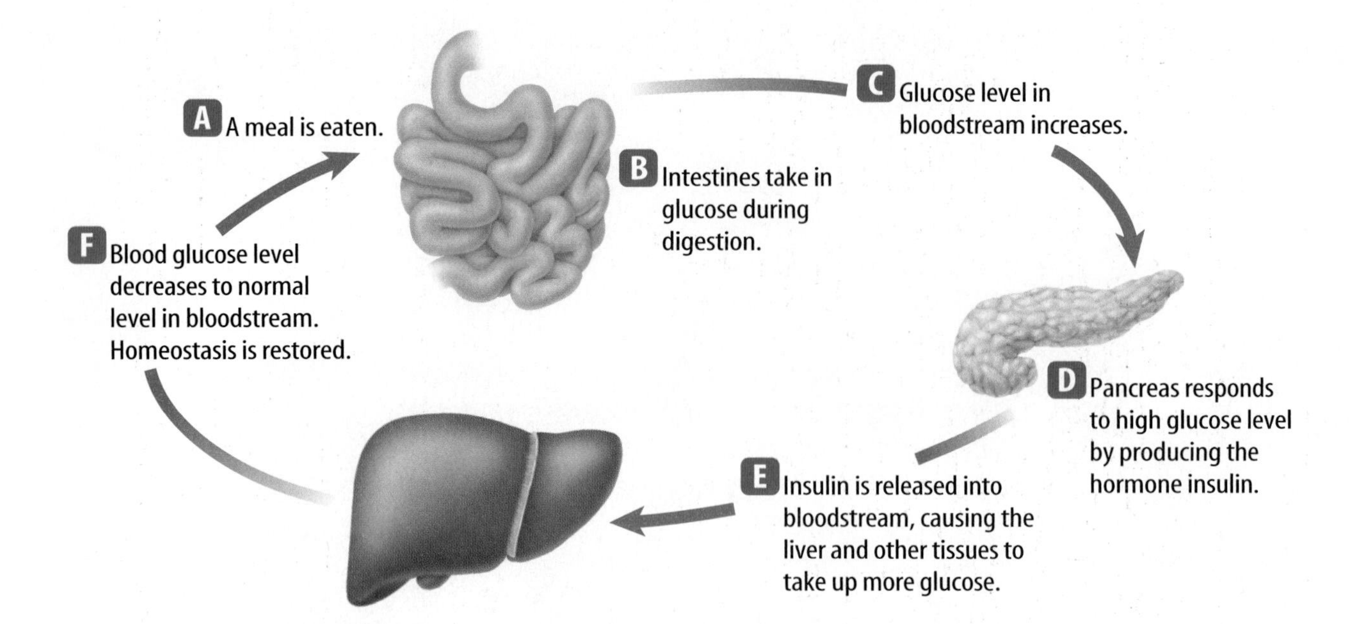

Figure 3 Many internal body conditions, such as hormone level, blood sugar level, and body temperature, are controlled by negative-feedback systems.

A Negative-Feedback System

To control the amount of hormones that are in your body, the endocrine system sends chemical messages back and forth within itself. This is called a negative-feedback system. It works much the way a thermostat works. When the temperature in a room drops below a set level, the thermostat signals the furnace to turn on. Once the furnace has raised the temperature in the room to the set level, the thermostat signals the furnace to shut off. It will continue to stay off until the thermostat signals that the temperature has dropped again. **Figure 3** shows how a negative-feedback system controls the level of glucose in your bloodstream.

section 1 review

Summary

Functions of the Endocrine System

- The nervous system and the endocrine system are the control systems of your body.
- The endocrine system uses hormones to deliver messages to the body.

Endocrine Glands

- Endocrine glands release hormones directly into the bloodstream.

A Negative-Feedback System

- The endocrine system uses a negative-feedback system to control the amount of hormones in your body.

Self Check

1. **Explain** the function of hormones.
2. **Choose** one endocrine gland. How does it work?
3. **Describe** a negative-feedback system.
4. **Think Critically** Glucose is required for cellular respiration, the process that releases energy within cells. How would a lack of insulin affect this process?

Applying Skills

5. **Predict** why the circulatory system is a good mechanism for delivering hormones throughout the body.
6. **Research** recent treatments for growth disorders involving the pituitary gland. Write a brief paragraph of your results in your Science Journal.

Science Online green.msscience.com/self_check_quiz

section 2

The Reproductive System

Reproduction and the Endocrine System

Reproduction is the process that continues life on Earth. Most human body systems, such as the digestive system and the nervous system, are the same in males and females, but this is not true for the reproductive system. Males and females each have structures specialized for their roles in reproduction. Although structurally different, both the male and female reproductive systems are adapted to allow for a series of events that can lead to the birth of a baby.

Hormones are the key to how the human reproductive system functions, as shown in **Figure 4.** Sex hormones are necessary for the development of sexual characteristics, such as breast development in females and facial hair growth in males. Hormones from the pituitary gland also begin the production of eggs in females and sperm in males. Eggs and sperm transfer hereditary information from one generation to the next.

as you read

What You'll Learn

- **Identify** the function of the reproductive system.
- **Compare and contrast** the major structures of the male and female reproductive systems.
- **Sequence** the stages of the menstrual cycle.

Why It's Important

Human reproductive systems help ensure that human life continues on Earth.

Review Vocabulary
cilia: short, hairlike structures that extend from a cell

New Vocabulary
- testes
- sperm
- semen
- ovary
- ovulation
- uterus
- vagina
- menstrual cycle
- menstruation

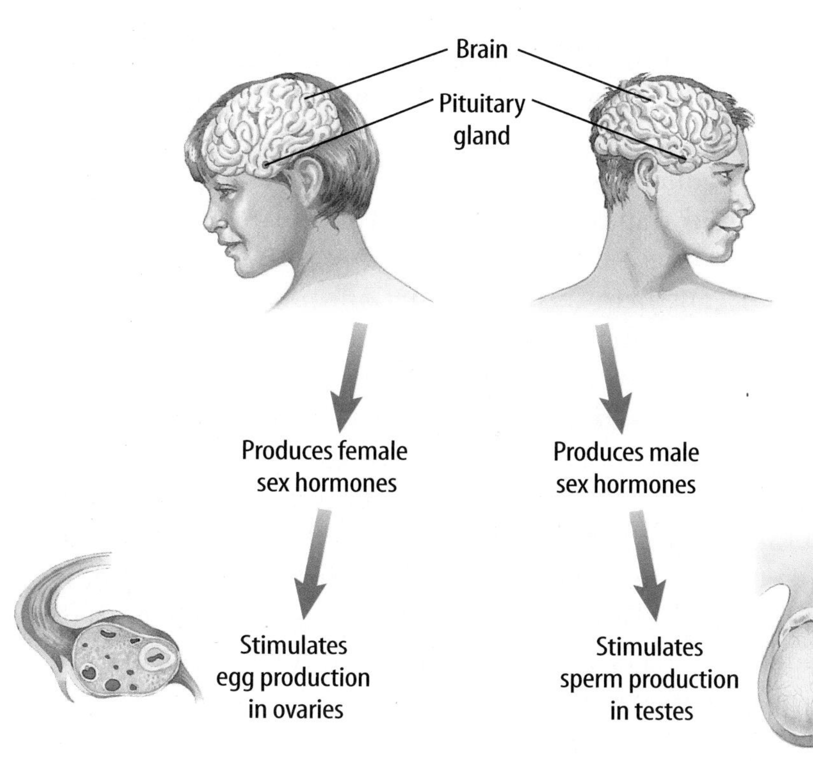

Figure 4 The pituitary gland produces hormones that control the male and female reproductive systems.

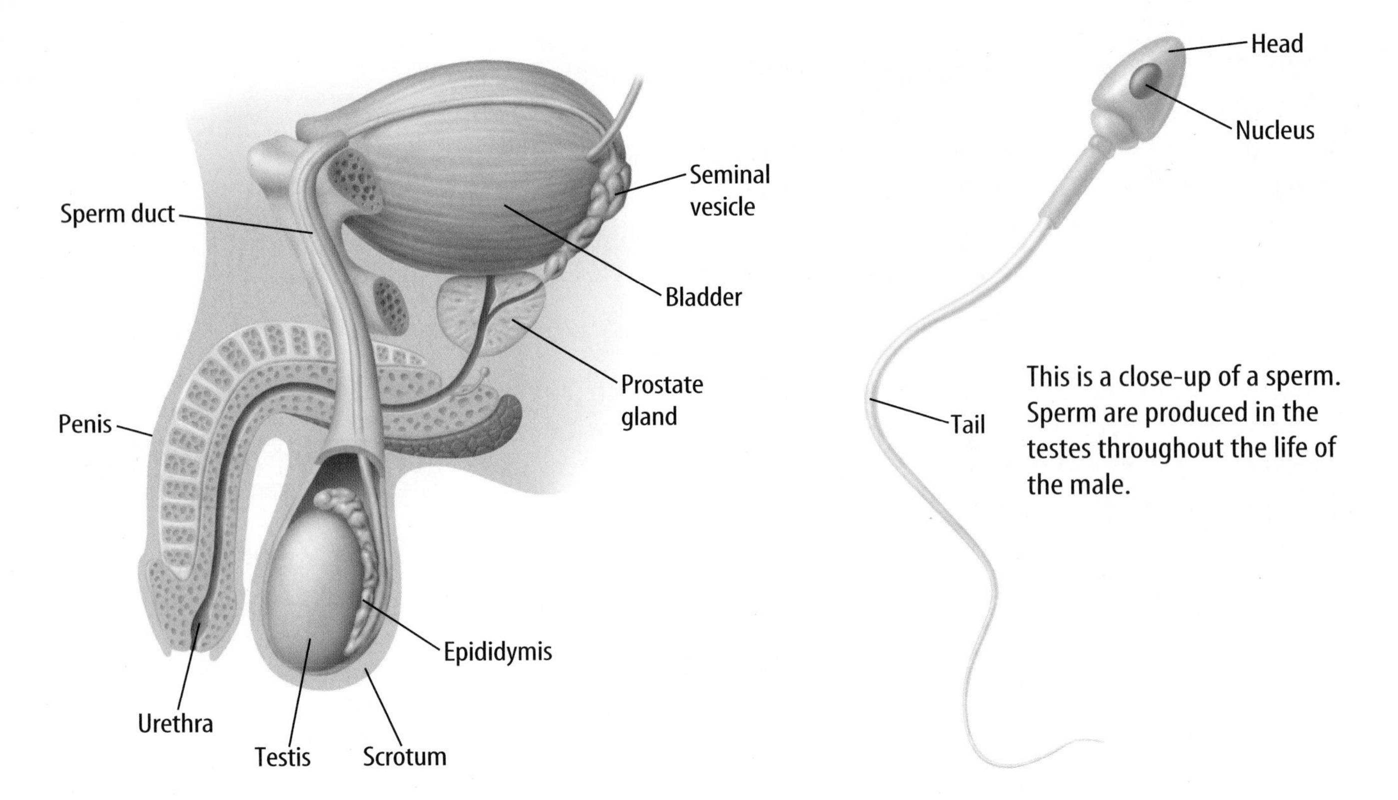

Figure 5 The structures of the male reproductive system are shown from the side of the body.

The Male Reproductive System

The male reproductive system is made up of external and internal organs. The external organs of the male reproductive system are the penis and scrotum, shown in **Figure 5.** The scrotum contains two organs called testes (TES teez). As males mature sexually, the **testes** begin to produce testosterone, the male hormone, and **sperm,** which are male reproductive cells.

Sperm Each sperm cell has a head and tail. The head contains hereditary information, and the tail moves the sperm. Because the scrotum is located outside the body cavity, the testes, where sperm are produced, are kept at a lower temperature than the rest of the body. Sperm are produced in greater numbers at lower temperatures.

Many organs help in the production, transportation, and storage of sperm. After sperm are produced, they travel from the testes through sperm ducts that circle the bladder. Behind the bladder, a gland called the seminal vesicle provides sperm with a fluid. This fluid supplies the sperm with an energy source and helps them move. This mixture of sperm and fluid is called **semen** (SEE mun). Semen leaves the body through the urethra, which is the same tube that carries urine from the body. However, semen and urine never mix. A muscle at the back of the bladder contracts to prevent urine from entering the urethra as sperm leave the body.

The Female Reproductive System

Unlike male reproductive organs, most of the reproductive organs of the female are inside the body. The **ovaries**—the female sex organs—are located in the lower part of the body cavity. Each of the two ovaries is about the size and shape of an almond. **Figure 6** shows the different organs of the female reproductive system.

The Egg When a female is born, she already has all of the cells in her ovaries that eventually will develop into eggs—the female reproductive cells. At puberty, eggs start to develop in her ovaries because of specific sex hormones.

About once a month, an egg is released from an ovary in a hormone-controlled process called **ovulation** (ahv yuh LAY shun). The two ovaries release eggs on alternating months. One month, an egg is released from an ovary. The next month, the other ovary releases an egg, and so on. After the egg is released, it enters the oviduct. If a sperm fertilizes the egg, it usually happens in an oviduct. Short, hairlike structures called cilia help sweep the egg through the oviduct toward the uterus (YEW tuh rus).

Reading Check *When are eggs released by the ovaries?*

The **uterus** is a hollow, pear-shaped, muscular organ with thick walls in which a fertilized egg develops. The lower end of the uterus, the cervix, narrows and is connected to the outside of the body by a muscular tube called the **vagina** (vuh JI nuh). The vagina also is called the birth canal because during birth, a baby travels through this tube from the uterus to the outside of the mother's body.

Topic: Ovarian Cysts
Visit green.msscience.com for Web links to information about ovarian cysts.

Activity Make a small pamphlet explaining what cysts are and how they can be treated.

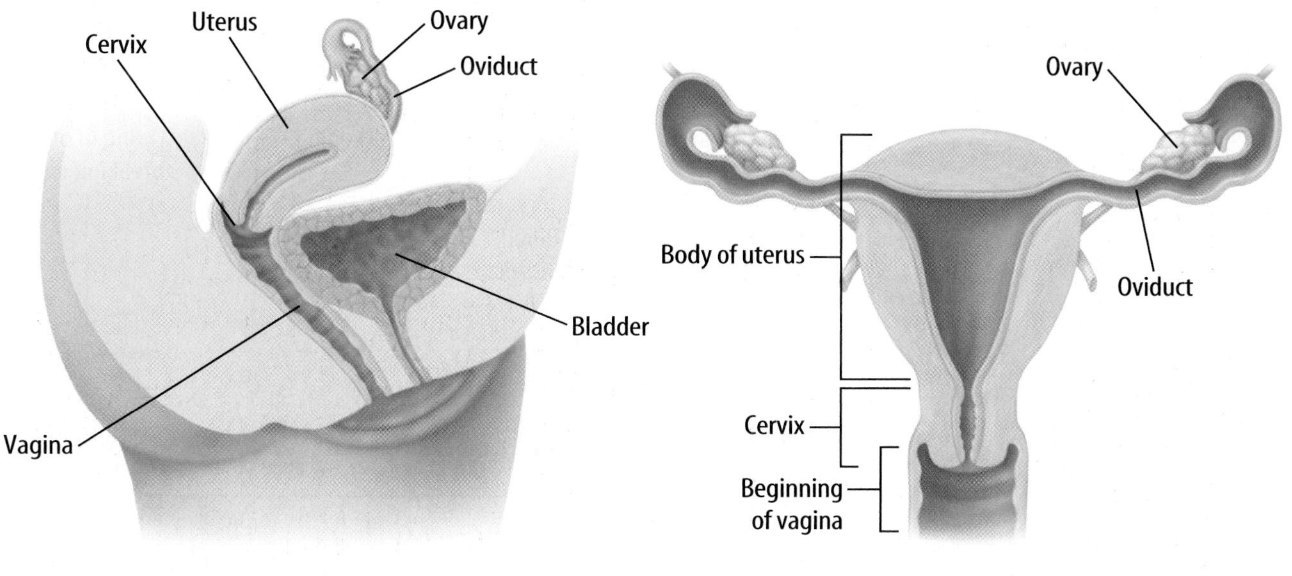

Figure 6 The structures of the female reproductive system are internal.
Name *where eggs develop in the female reproductive system.*

Graphing Hormone Levels

Procedure
Make a line graph of this table.

Hormone Changes

Day	Level of Hormone
1	12
5	14
9	15
13	70
17	13
21	12
25	8

Analysis
1. On what day is the highest level of hormone present?
2. What event takes place around the time of the highest hormone level?

The Menstrual Cycle

How is the female body prepared for having a baby? The **menstrual cycle** is the monthly cycle of changes in the female reproductive system. Before and after an egg is released from an ovary, the uterus undergoes changes. The menstrual cycle of a human female averages 28 days. However, the cycle can vary in some individuals from 20 to 40 days. Changes include the maturing of an egg, the production of female sex hormones, the preparation of the uterus to receive a fertilized egg, and menstrual flow.

Reading Check *What is the menstrual cycle?*

Endocrine Control Hormones control the entire menstrual cycle. The pituitary gland responds to chemical messages from the hypothalamus by releasing several hormones. These hormones start the development of eggs in the ovary. They also start the production of other hormones in the ovary, including estrogen (ES truh jun) and progesterone (proh JES tuh rohn). The interaction of all these hormones results in the physical processes of the menstrual cycle.

Phase One As shown in **Figure 7,** the first day of phase 1 starts when menstrual flow begins. Menstrual flow consists of blood and tissue cells released from the thickened lining of the uterus. This flow usually continues for four to six days and is called **menstruation** (men STRAY shun).

Figure 7 The three phases of the menstrual cycle make up the monthly changes in the female reproductive system.
Explain *why the uterine lining thickens.*

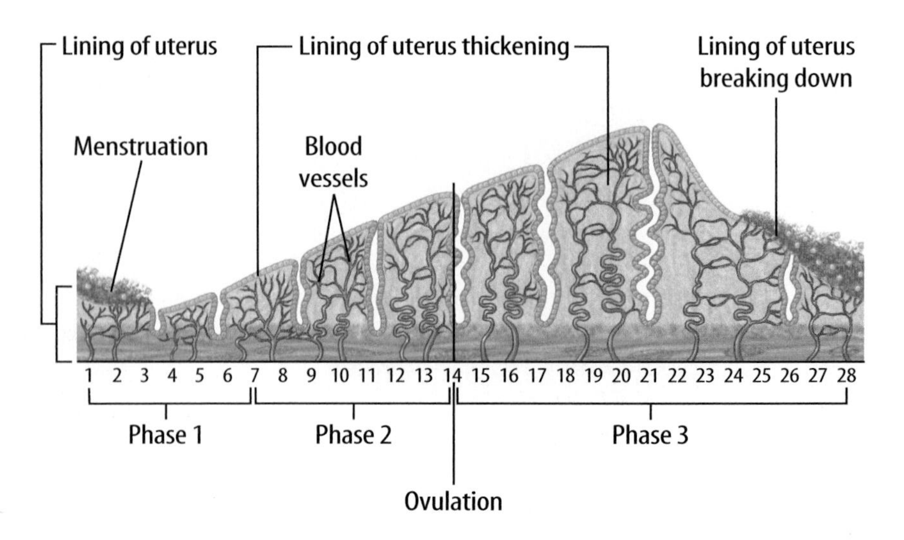

Phase Two Hormones cause the lining of the uterus to thicken in phase 2. Hormones also control the development of an egg in the ovary. Ovulation occurs about 14 days before menstruation begins. Once the egg is released, it must be fertilized within 24 h or it usually begins to break down. Because sperm can survive in a female's body for up to three days, fertilization can occur soon after ovulation.

Phase Three Hormones produced by the ovaries continue to cause an increase in the thickness of the uterine lining during phase 3. If a fertilized egg does arrive, the uterus is ready to support and nourish the developing embryo. If the egg is not fertilized, the lining of the uterus breaks down as the hormone levels decrease. Menstruation begins and the cycle repeats itself.

Menopause For most females, the first menstrual period happens between ages nine years and 13 years and continues until 45 years of age to 60 years of age. Then, a gradual reduction of menstruation takes place as hormone production by the ovaries begins to shut down. Menopause occurs when both ovulation and menstrual periods end. It can take several years for the completion of menopause. As **Figure 8** indicates, menopause does not inhibit a woman's ability to enjoy an active life.

Figure 8 This older woman enjoys exercising with her granddaughter.

section 2 review

Summary

Reproduction and the Endocrine System

- Reproduction is the process that continues life.
- The human reproductive system needs hormones to function.

The Male Reproductive System

- Sperm are produced in the testes and leave the male through the penis.

The Female Reproductive System

- Eggs are produced in the ovaries and, if fertilized, can develop in the uterus.

The Menstrual Cycle

- A female's menstrual cycle occurs approximately every 28 days.
- If an egg is not fertilized, the lining of the uterus breaks down and is shed in a process called menstruation.

Self Check

1. **Identify** the major function of male and female reproductive systems in humans.
2. **Explain** the movement of sperm through the male reproductive system.
3. **Compare and contrast** the major organs and structures of the male and female reproductive systems.
4. **Sequence** the stages of the menstrual cycle in a human female using diagrams and captions.
5. **Think Critically** Adolescent females often require additional amounts of iron in their diet. Explain.

Applying Math

6. **Order of Operations** Usually, one egg is released each month during a female's reproductive years. If menstruation begins at 12 years of age and ends at 50 years of age, calculate the number of eggs her body can release during her reproductive years.

Interpreting Diagrams

Starting in adolescence, hormones cause the development of eggs in the ovary and changes in the uterus. These changes prepare the uterus to accept a fertilized egg that can attach itself in the wall of the uterus. What happens to an unfertilized egg?

Real-World Question

What changes occur to the uterus during a female's monthly menstrual cycle?

Goals

- **Observe** the stages of the menstrual cycle in the diagram.
- **Relate** the process of ovulation to the cycle.

Materials

paper
pencil

Procedure

1. The diagrams below illustrate the menstrual cycle.
2. Copy and complete the data table using information in this chapter and diagrams below.
3. On approximately what day in a 28-day cycle is the egg released from the ovary?

Menstruation Cycle

Days	Condition of Uterus	What Happens
1–6		
7–12	Do not write in this book.	
13–14		
15–18		

Conclude and Apply

1. **Infer** how many days the average menstrual cycle lasts.
2. **State** on what days the lining of the uterus builds up.
3. **Infer** why this process is called a cycle.
4. **Calculate** how many days before menstruation ovulation usually occurs.

Communicating Your Data

Compare your data table with those of other students in your class. **For more help, refer to the** Science Skill Handbook.

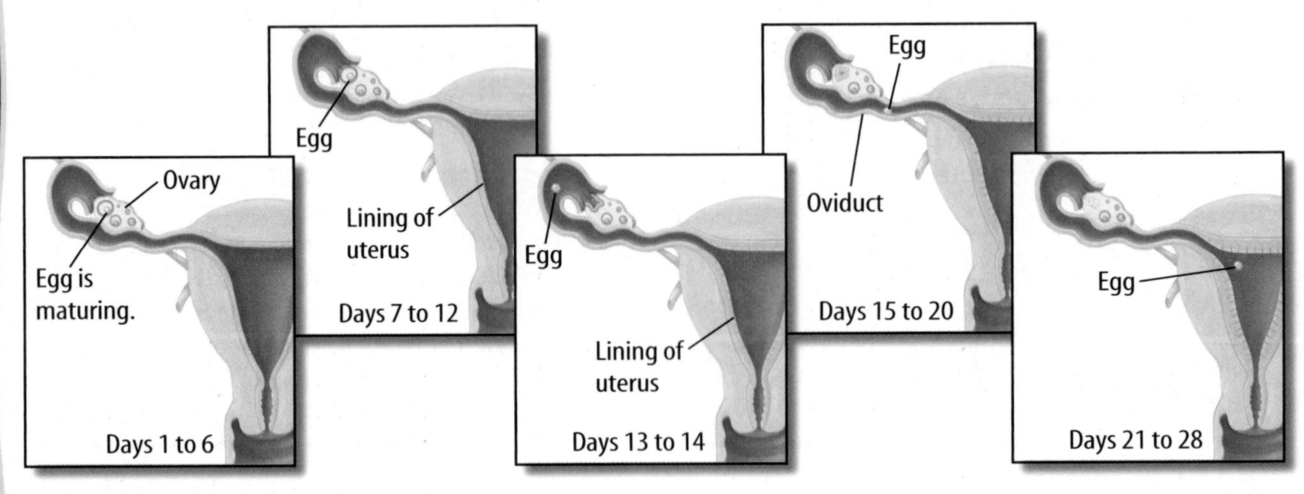

section 3

Human Life Stages

The Function of the Reproductive System

Before the invention of powerful microscopes, some people imagined an egg or a sperm to be a tiny person that grew inside a female. In the latter part of the 1700s, experiments using amphibians showed that contact between an egg and sperm is necessary for the development of life. With the development of the cell theory in the 1800s, scientists recognized that a human develops from an egg that has been fertilized by a sperm. The uniting of a sperm and an egg is known as fertilization. Fertilization, as shown in **Figure 9,** usually takes place in the oviduct.

as you read

What You'll Learn

- **Describe** the fertilization of a human egg.
- **List** the major events in the development of an embryo and fetus.
- **Describe** the developmental stages of infancy, childhood, adolescence, and adulthood.

Why It's Important

Fertilization begins the entire process of human growth and development.

Review Vocabulary

nutrient: substance in food that provides energy and materials for cell development, growth, and repair

New Vocabulary

- pregnancy
- embryo
- amniotic sac
- fetus
- fetal stress

Fertilization

Although 200 million to 300 million sperm can be deposited in the vagina, only several thousand reach an egg in the oviduct. As they enter the female, the sperm come into contact with chemical secretions in the vagina. It appears that this contact causes a change in the membrane of the sperm. The sperm then become capable of fertilizing the egg. The one sperm that makes successful contact with the egg releases an enzyme from the saclike structure on its head. Enzymes help speed up chemical reactions that have a direct effect on the protective membranes on the egg's surface. The structure of the egg's membrane is disrupted, and the sperm head can enter the egg.

Zygote Formation Once a sperm has entered the egg, changes in the electric charge of the egg's membrane prevent other sperm from entering the egg. At this point, the nucleus of the successful sperm joins with the nucleus of the egg. This joining of nuclei creates a fertilized cell called the zygote. It begins to undergo many cell divisions.

Figure 9 After the sperm releases enzymes that disrupt the egg's membrane, it penetrates the egg.

Color-enhanced SEM Magnification: 340×

Figure 10 The development of fraternal and identical twins is different.

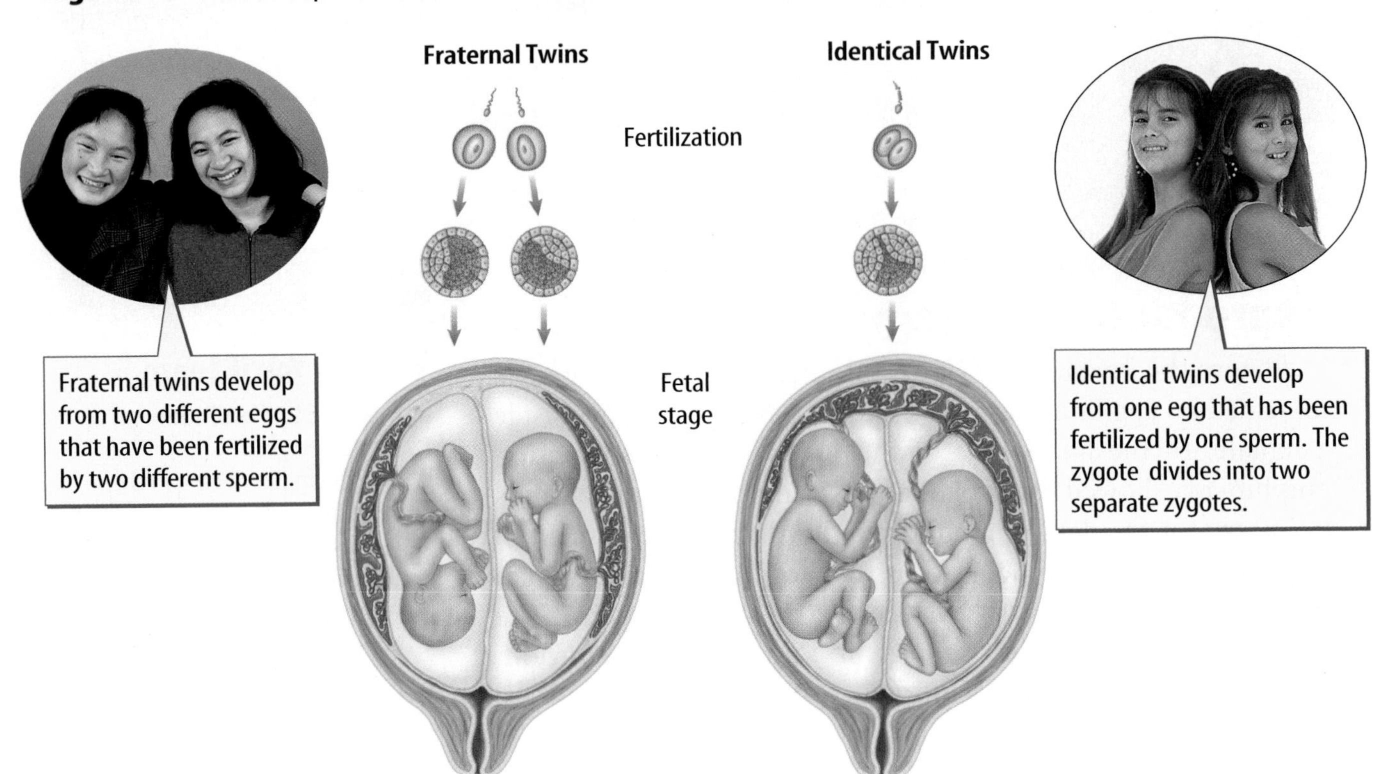

Multiple Births

Sometimes two eggs leave the ovary at the same time. If both eggs are fertilized and both develop, fraternal twins are born. Fraternal twins, as shown in **Figure 10,** can be two girls, two boys, or a boy and a girl. Because fraternal twins come from two eggs, they only resemble each other.

Because identical twin zygotes develop from the same egg and sperm, as explained in **Figure 10,** they have the same hereditary information. These identical zygotes develop into identical twins, which are either two girls or two boys. Multiple births also can occur when three or more eggs are produced at one time or when the zygote separates into three or more parts.

Midwives Some women choose to deliver their babies at home rather than at a hospital. An at-home birth can be attended by a certified nurse-midwife. Research to find the educational and skill requirements of a nurse-midwife.

Development Before Birth

After fertilization, the zygote moves along the oviduct to the uterus. During this time, the zygote is dividing and forming into a ball of cells. After about seven days, the zygote attaches to the wall of the uterus, which has been thickening in preparation to receive a zygote, as shown in **Figure 11.** If attached to the wall of the uterus, the zygote will develop into a baby in about nine months. This period of development from fertilized egg to birth is known as **pregnancy.**

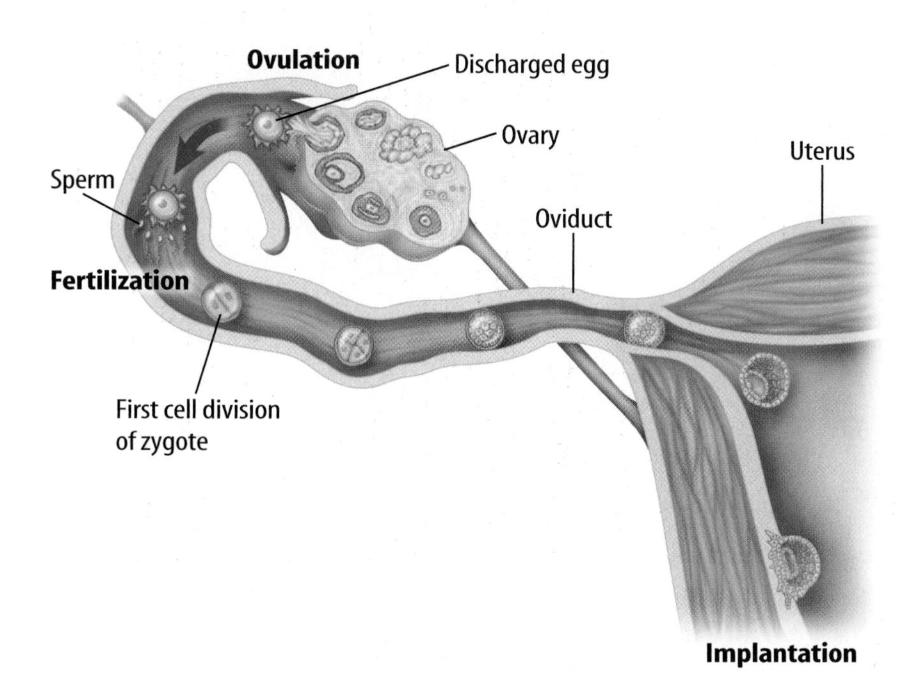

Figure 11 After a few days of rapid cell division, the zygote, now a ball of cells, reaches the lining of the uterus, where it attaches itself to the lining for development.

The Embryo After the zygote attaches to the wall of the uterus, it is known as an **embryo,** illustrated in **Figure 12.** It receives nutrients from fluids in the uterus until the placenta (plu SEN tuh) develops from tissues of the uterus and the embryo. An umbilical cord develops that connects the embryo to the placenta. In the placenta, materials diffuse between the mother's blood and the embryo's blood, but their bloods do not mix. Blood vessels in the umbilical cord carry nutrients and oxygen from the mother's blood through the placenta to the embryo. Other substances in the mother's blood can move into the embryo, including drugs, toxins, and disease organisms. Wastes from the embryo are carried in other blood vessels in the umbilical cord through the placenta to the mother's blood.

Reading Check ***Why must a pregnant woman avoid alcohol, tobacco, and harmful drugs?***

Pregnancy in humans lasts about 38 to 39 weeks. During the third week, a thin membrane called the **amniotic** (am nee AH tihk) **sac** begins to form around the embryo. The amniotic sac is filled with a clear liquid called amniotic fluid, which acts as a cushion for the embryo and stores nutrients and wastes.

During the first two months of development, the embryo's major organs form and the heart structure begins to beat. At five weeks, the embryo has a head with eyes, nose, and mouth features. During the sixth and seventh weeks, fingers and toes develop.

Figure 12 By two months, the developing embryo is about 2.5 cm long and is beginning to develop recognizable features.

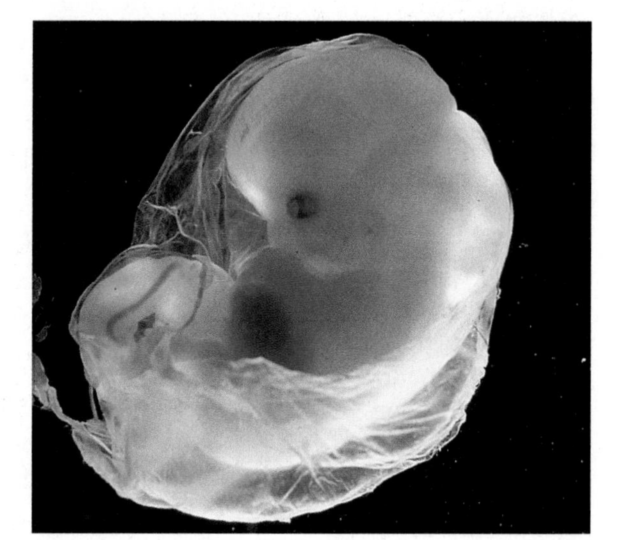

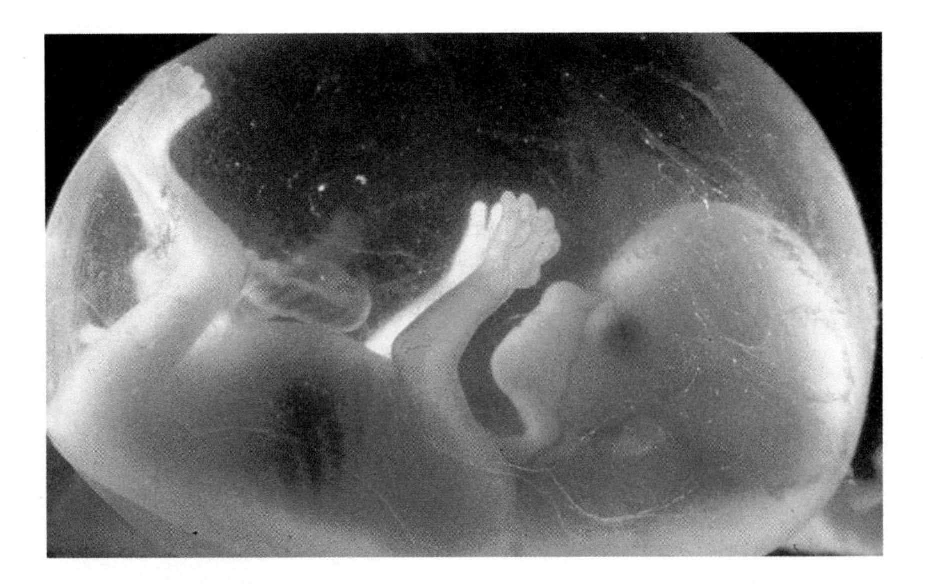

Figure 13 A fetus at about 16 weeks is approximately 15 cm long and weighs 140 g.
Describe *the changes that take place in a fetus by the end of the seventh month.*

The Fetus After the first two months of pregnancy, the developing embryo is called a **fetus,** shown in **Figure 13.** At this time, body organs are present. Around the third month, the fetus is 8 cm to 9 cm long. The mother may feel the fetus move. The fetus can even suck its thumb. By the fourth month, an ultrasound test can determine the sex of the fetus. The fetus is 30 cm to 38 cm in length by the end of the seventh month of pregnancy. Fatty tissue builds up under the skin, and the fetus looks less wrinkled. By the ninth month, the fetus usually has shifted to a head-down position within the uterus, a position beneficial for delivery. The head usually is in contact with the opening of the uterus to the vagina. The fetus is about 50 cm in length and weighs from 2.5 kg to 3.5 kg.

Interpreting Fetal Development

Procedure
Make a bar graph of the following data.

Fetal Development

End of Month	Length (cm)
3	8
4	15
5	25
6	30
7	35
8	40
9	51

Analysis
1. During which month does the greatest increase in length occur?
2. On average, how many centimeters does the baby grow per month?

Try at Home

The Birthing Process

The process of childbirth, as shown in **Figure 14,** begins with labor, the muscular contractions of the uterus. As the contractions increase in strength and number, the amniotic sac usually breaks and releases its fluid. Over a period of hours, the contractions cause the opening of the uterus to widen. More powerful and more frequent contractions push the baby out through the vagina into its new environment.

Delivery Often a mother is given assistance by a doctor during the delivery of the baby. As the baby emerges from the birth canal, a check is made to determine if the umbilical cord is wrapped around the baby's neck or any body part. When the head is free, any fluid in the baby's nose and mouth is removed by suction. After the head and shoulders appear, contractions force the baby out completely. Up to an hour after delivery, contractions occur that push the placenta out of the mother's body.

Cesarean Section Sometimes a baby must be delivered before labor begins or before it is completed. At other times, a baby cannot be delivered through the birth canal because the mother's pelvis might be too small or the baby might be in the wrong birthing position. In cases like these, surgery called a cesarean (suh SEER ee uhn) section is performed. An incision is made through the mother's abdominal wall, then through the wall of the uterus. The baby is delivered through this opening.

Reading Check *What is a cesarean section?*

Topic: Cesarean Sections
Visit green.msscience.com for Web links to information about cesarean section delivery.

Activity Make a chart listing the advantages and disadvantages of a cesarean section delivery.

After Birth When the baby is born, it is attached to the umbilical cord. The person assisting with the birth clamps the cord in two places and cuts it between the clamps. The baby does not feel any pain from this procedure. The baby might cry, which is the result of air being forced into its lungs. The scar that forms where the cord was attached is called the navel.

Figure 14 Childbirth begins with labor. The opening to the uterus widens, and the baby passes through.

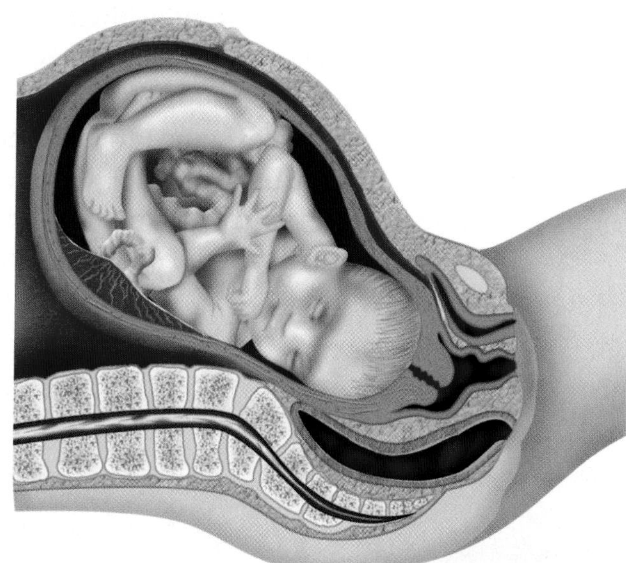

The fetus moves into the opening of the birth canal, and the uterus begins to widen.

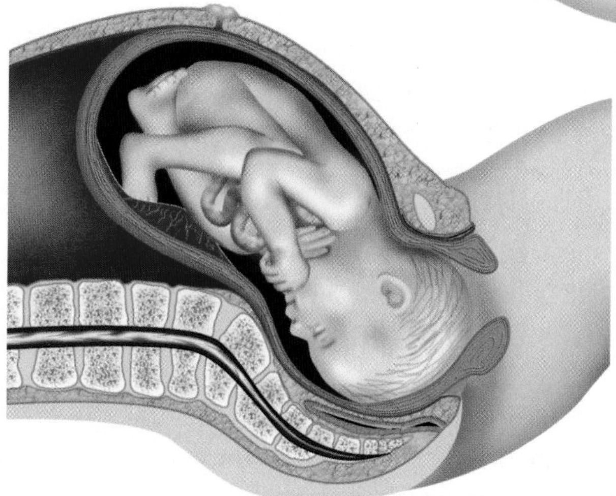

The base of the uterus is completely dilated.

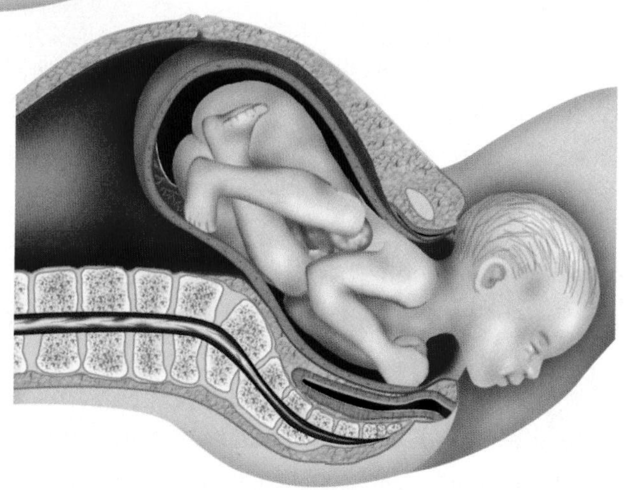

The fetus is pushed out through the birth canal.

Stages After Birth

Defined stages of development occur after birth, based on the major developments that take place during those specific years. Infancy lasts from birth to around 18 months of age. Childhood extends from the end of infancy to sexual maturity, or puberty. The years of adolescence vary, but they usually are considered to be the teen years. Adulthood covers the years of age from the early 20s until life ends, with older adulthood considered to be over 60. The age spans of these different stages are not set, and scientists differ in their opinions regarding them.

Infancy What type of environment must the infant adjust to after birth? The experiences the fetus goes through during birth cause **fetal stress.** The fetus has emerged from an environment that was dark, watery, a constant temperature, and nearly soundless. In addition, the fetus might have been forced through the constricted birth canal. However, in a short period of time, the infant's body becomes adapted to its new world.

The first four weeks after birth are known as the neonatal (nee oh NAY tul) period. The term *neonatal* means "newborn." During this time, the baby's body begins to function normally. Unlike the newborn of some other animals, human babies, such as the one shown in **Figure 15,** depend on other humans for their survival. In contrast, many other animals, such as the young horse also shown in **Figure 15,** begin walking a few hours after they are born.

Figure 15 Human babies are more dependent upon their caregivers than many other mammals are.

Infants and toddlers are completely dependent upon caregivers for all their needs.

Other young mammals are more self-sufficient. This colt is able to stand within an hour after birth.

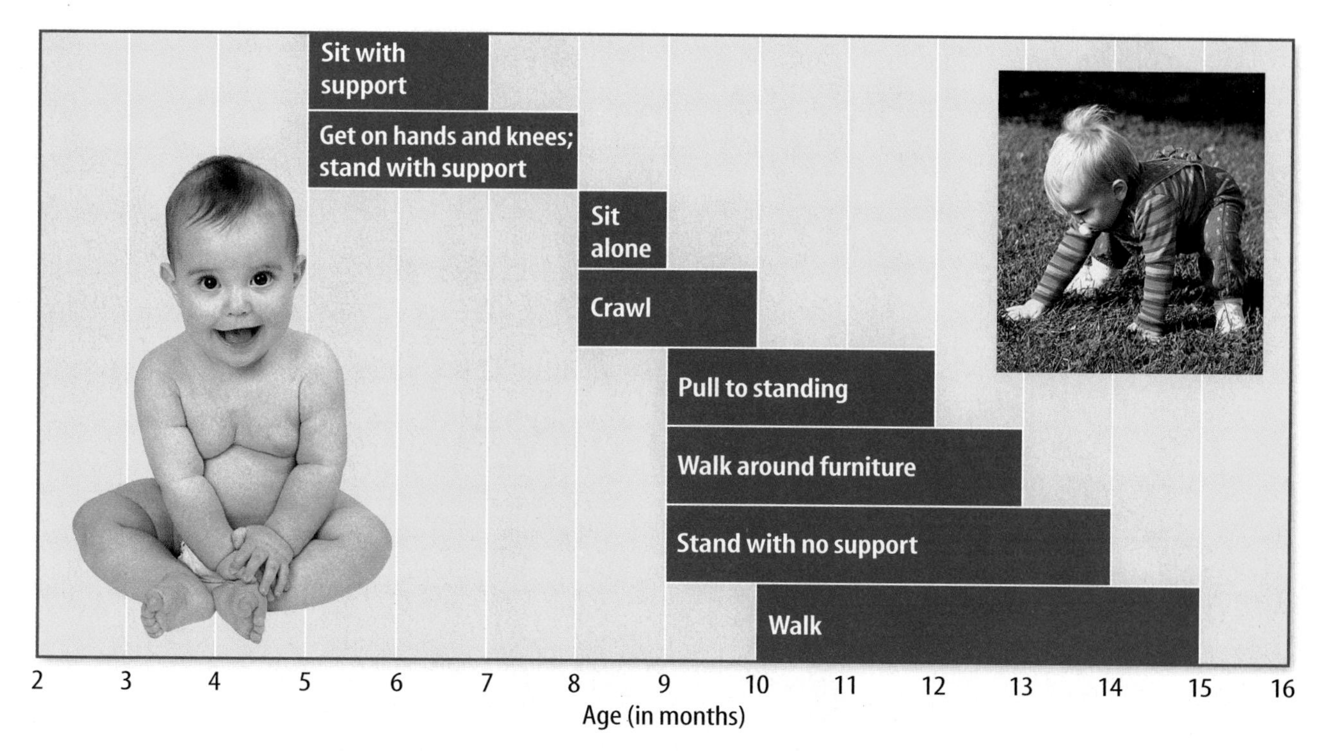

During these first 18 months, infants show increased physical coordination, mental development, and rapid growth. Many infants will triple their weight in the first year. **Figure 16** shows the extremely rapid development of the nervous and muscular systems during this stage, which enables infants to start interacting with the world around them.

Figure 16 Infants show rapid development in their nervous and muscular systems through 18 months of age.

Childhood After infancy is childhood, which lasts until about puberty, or sexual maturity. Sexual maturity occurs around 12 years of age. Overall, growth during early childhood is rather rapid, although the physical growth rate for height and weight is not as rapid as it is in infancy. Between two and three years of age, the child learns to control his or her bladder and bowels. At age two to three, most children can speak in simple sentences. Around age four, the child is able to get dressed and undressed with some help. By age five, many children can read a limited number of words. By age six, children usually have lost their chubby baby appearance, as seen in **Figure 17.** However, muscular coordination and mental abilities continue to develop. Throughout this stage, children develop their abilities to speak, read, write, and reason. These ages of development are only guidelines because each child develops at a different rate.

Figure 17 Children, like these kindergartners, grow and develop at different rates.

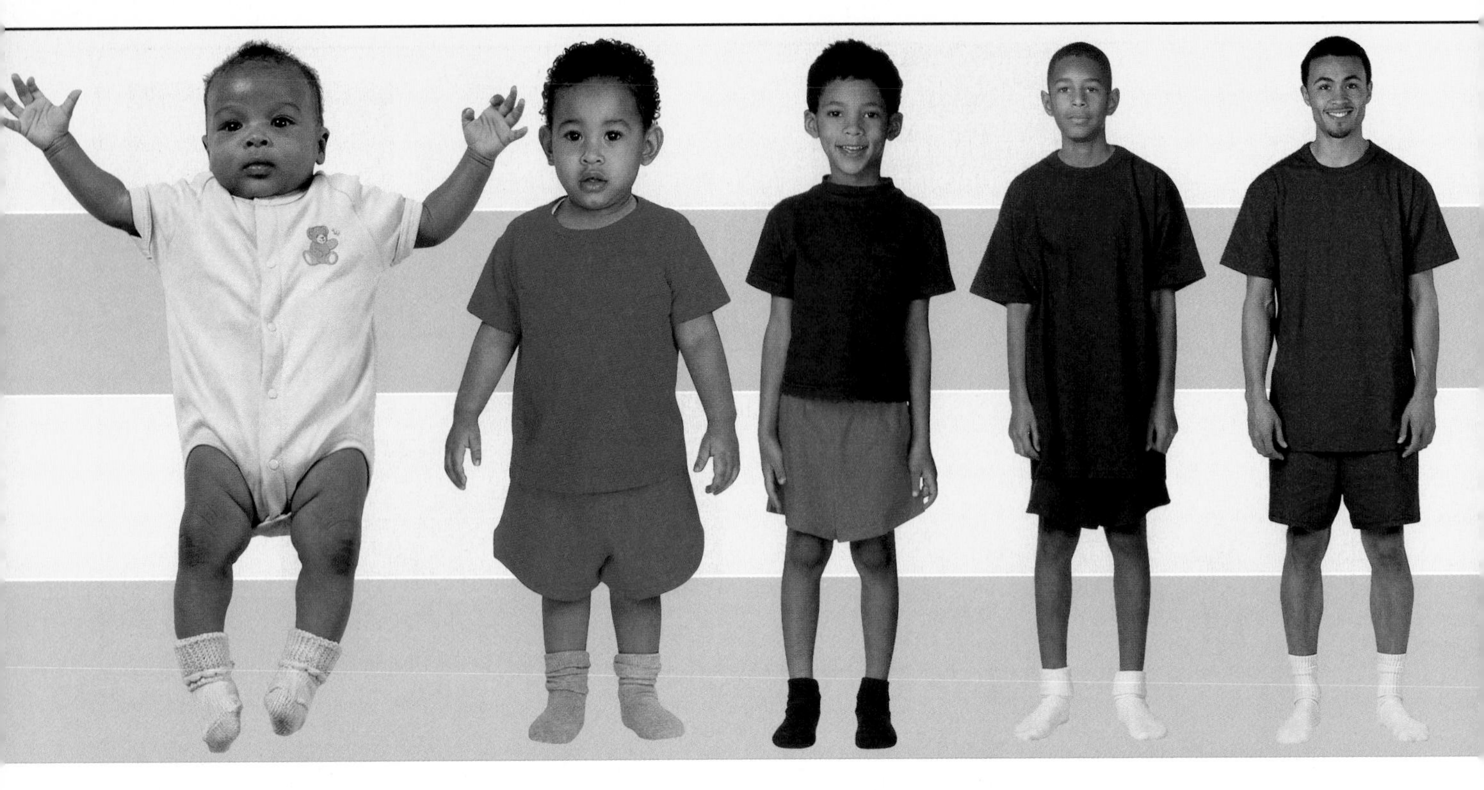

Figure 18 The proportions of body parts change over time as the body develops.
Describe *how the head changes proportion.*

Adolescence Adolescence usually begins around age 12 or 13. A part of adolescence is puberty—the time of development when a person becomes physically able to reproduce. For girls, puberty occurs between ages nine and 13. For boys, puberty occurs between ages 13 and 16. During puberty, hormones produced by the pituitary gland cause changes in the body. These hormones produce reproductive cells and sex hormones. Secondary sex characteristics also develop. In females, the breasts develop, pubic and underarm hair appears, and fatty tissue is added to the buttocks and thighs. In males, the hormones cause a deepened voice, an increase in muscle size, and the growth of facial, pubic, and underarm hair.

Adolescence usually is when the final growth spurt occurs. Because the time when hormones begin working varies among individuals and between males and females, growth rates differ. Girls often begin their final growth phase at about age 11 and end around age 16. Boys usually start their growth spurt at age 13 and end around 18 years of age.

Adolescent Growth During adolescence, body parts do not all grow at the same rate. Legs grow longer before the upper body lengthens. This changes the body's center of gravity, the point at which the body maintains its balance. This is one cause of teenager clumsiness. In your Science Journal, write a paragraph about how this might affect playing sports.

Adulthood The final stage of development, adulthood, begins with the end of adolescence and continues through old age. This is when the growth of the muscular and skeletal system stops. **Figure 18** shows how body proportions change as you age.

People from age 45 to age 60 are sometimes considered middle-aged adults. During these years, physical strength begins to decline. Blood circulation and respiration become less efficient. Bones become more brittle, and the skin becomes wrinkled.

Older Adulthood People over the age of 60 may experience an overall decline in their physical body systems. The cells that make up these systems no longer function as well as they did at a younger age. Connective tissues lose their elasticity, causing muscles and joints to be less flexible. Bones become thinner and more brittle. Hearing and vision are less sensitive. The lungs and heart work less efficiently. However, exercise and eating well over a lifetime can help extend the health of one's body systems. Many healthy older adults enjoy full lives and embrace challenges, as shown in **Figure 19.**

Figure 19 Astronaut and Senator John Glenn traveled into space twice. In 1962, at age 40, he was the first U.S. citizen to orbit Earth. He was part of the space shuttle crew in 1998 at age 77. Senator Glenn has helped change people's views of what many older adults are capable of doing.

Reading Check ***What physical changes occur during late adulthood?***

Human Life Spans Seventy-seven years is the average life span—from birth to death—of humans in the United States, although an increasing number of people live much longer. However, body systems break down with age, resulting in eventual death. Death can occur earlier than old age for many reasons, including diseases, accidents, and bad health choices.

section 3 review

Summary

Fertilization

- Fertilization is the uniting of a sperm and an egg.

Development Before Birth

- Pregnancy begins when an egg is fertilized and lasts until birth.

The Birthing Process

- Birth begins with labor. Contractions force the baby out of the mother's body.

Stages After Birth

- Infancy (birth to 18 months) and childhood (until age 12) are periods of physical and mental growth.
- A person becomes physically able to reproduce during adolescence. Adulthood is the final stage of development.

Self Check

1. **Describe** what happens when an egg is fertilized in a female.
2. **Explain** what happens to an embryo during the first two months of pregnancy.
3. **Describe** the major events that occur during childbirth.
4. **Name** the stage of development that you are in. What physical changes have occurred or will occur during this stage of human development?
5. **Think Critically** Why is it hard to compare the growth and development of different adolescents?

Applying Skills

6. **Use a Spreadsheet** Using your text and other resources, make a spreadsheet for the stages of human development from a zygote to a fetus. Title one column *Zygote,* another *Embryo,* and a third *Fetus.* Complete the spreadsheet.

Changing Body Proportions

Goals
- **Measure** specific body proportions of adolescents.
- **Infer** how body proportions differ between adolescent males and females.

Materials
tape measure
erasable pencil
graph paper

Real-World Question

The ancient Greeks believed that the perfect body was completely balanced. Arms and legs should not be too long or short. A person's head should not be too large or small. The extra-large muscles of a body builder would have been ugly to the Greeks. How do you think they viewed the bodies of infants and children? Infants and young children have much different body proportions than adults, and teenagers often go through growth spurts that quickly change their body proportions. How do the body proportions differ between adolescent males and females?

Procedure

1. Copy the data table in your Science Journal and record the gender of each person that you measure.
2. Measure each person's head circumference by starting in the middle of the forehead and wrapping the tape measure once around the head. Record these measurements.

3. Measure each person's arm length from the top of the shoulder to the tip of the middle finger while the arm is held straight out to the side of the body. Record these measurements.
4. Ask each person to remove his or her shoes and stand next to a wall. Mark their height with an erasable pencil and measure their height from the floor to the mark. Record these measurements in the data table.

Age and Body Measurements

Gender of Person	Head Circumference (cm)	Arm Length (cm)	Height (cm)
	Do not write in this book.		

5. **Combine** your data with that of your classmates. Find the averages of head circumference, arm length, and height. Then, find these averages for males and females.
6. Make a bar graph of your calculations in step 5. Plot the measurements on the *y*-axis and plot all of the averages along the *x*-axis.
7. **Calculate** the proportion of average head circumference to average height for everyone in your class by dividing the average head circumference by the average height. Repeat this calculation for males and females.
8. **Calculate** the proportion of average arm length to average height for everyone in your class by dividing the average arm length by the average height. Repeat this calculation for males and females.

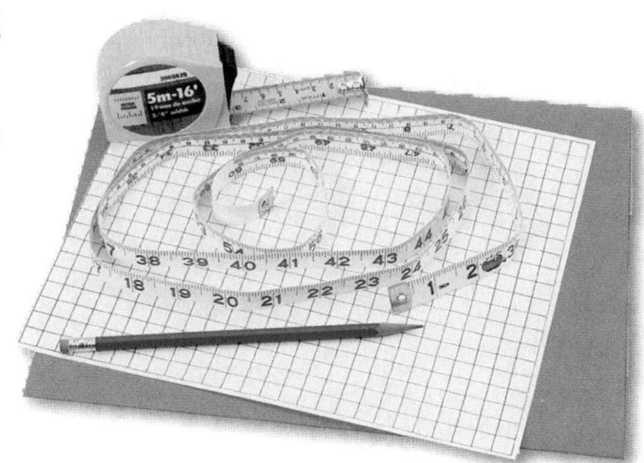

Analyze Your Data

Analyze whether adolescent males or females have larger head circumferences or longer arms. Which group has the larger proportion of head circumference or arm length to height?

Conclude and Apply

Explain if this lab supports the information in this chapter about the differences between growth rates of adolescent males and females.

Communicating Your Data

Construct data tables on poster board showing your results and those of your classmates. Discuss with your classmates why these results might be different.

SCIENCE Stats

Facts About Infants

Did you know...

...Humans and chimpanzees share about 99 percent of their genes. Although humans look different than chimps, reproduction is similar and gestation is the same—about nine months. Youngsters of both species lose their baby teeth at about six years of age.

Mammal Facts

Mammal	Average Gestation	Average Birth Weight	Average Adult Weight	Average Life Span (years)
African elephant	22 months	136 kg	4,989.5 kg	35
Blue whale	12 months	1,800 kg	135,000 kg	60
Human	**9 months**	**3.3 kg**	**59–76 kg**	**77***
Brown bear	7 months	0.23–0.5 kg	350 kg	22.5
Cat	2 months	99 g	2.7–7 kg	13.5
Kangaroo	1 month	0.75 –1.0 g	45 kg	5
Golden hamster	2.5 weeks	0.3 g	112 g	2

*In the United States

Applying Math Assume that a female of each mammal listed in the table above is pregnant once during her life. Which mammal is pregnant for the greatest proportion of her life?

Echidna

...Of about 4,000 species of mammals, only three lay eggs: the platypus, the short-beaked echidna (ih KIHD nuh), and the long-beaked echidna.

Find Out About It

Visit green.msscience.com/science_stats to research which species of vertebrate animals has the longest life span and which has the shortest. Present your findings in a table that also shows the life span of humans.

Reviewing Main Ideas

Section 1 The Endocrine System

1. Endocrine glands secrete hormones directly into the bloodstream. They affect specific tissues in the body.
2. A change in the body causes an endocrine gland to function. Hormone production slows or stops when homeostasis is reached.

Section 2 The Reproductive System

1. Reproductive systems allow new organisms to be formed.
2. The testes produce sperm, which leave the male body through the penis.
3. The female ovaries produce eggs. If fertilized, an egg develops into a fetus within the uterus.
4. An unfertilized egg and the built-up lining of the uterus are shed in menstruation.

Section 3 Human Life Stages

1. After fertilization, the zygote becomes an embryo, then a fetus. Twins occur when two eggs are fertilized or when a zygote divides after fertilization.
2. Birth begins with labor. The amniotic sac breaks. Then, usually after several hours, contractions force the baby out of the mother's body.
3. Infancy, from birth to 18 months of age, is a period of rapid growth of mental and physical skills. Childhood lasts until age 12 and involves further physical and mental development.
4. Adolescence is when a person becomes physically able to reproduce. In adulthood, physical development is complete and body systems become less efficient. Death occurs at the end of life.

Visualizing Main Ideas

Copy and complete the following table on life stages.

Human Development

Stages of Life	Age Range	Physical Development
Infant		sits, stands, words spoken
		walks, speaks, writes, reads
Adolescent		
		end of muscular and skeletal growth

Using Vocabulary

amniotic sac p. 481
embryo p. 481
fetal stress p. 484
fetus p. 482
hormone p. 468
menstrual cycle p. 476
menstruation p. 476
ovary p. 475
ovulation p. 475
pregnancy p. 480
semen p. 474
sperm p. 474
testes p. 474
uterus p. 475
vagina p. 475

Fill in the blank with the correct vocabulary word or words.

1. _________ is a mixture of sperm and fluid.
2. The time of the development until the birth of a baby is known as _________.
3. During the first two months of pregnancy, the unborn child is known as a(n) _________.
4. The _________ is a hollow, pear-shaped muscular organ.
5. The _________ is the membrane that protects the unborn child.
6. The _________ is the organ that produces eggs.

Checking Concepts

Choose the word or phrase that best answers the question.

7. Where is the egg usually fertilized?
   A) oviduct **C)** vagina
   B) uterus **D)** ovary
8. What are the chemicals produced by the endocrine system?
   A) enzymes **C)** hormones
   B) target tissues **D)** saliva
9. Which gland produces melatonin?
   A) adrenal **C)** pancreas
   B) thyroid **D)** pineal
10. Where does the embryo develop?
    A) oviduct **C)** uterus
    B) ovary **D)** vagina

Use the figure below to answer question 11.

Prevalence of Diabetes per 100 Adults, United States, 2001

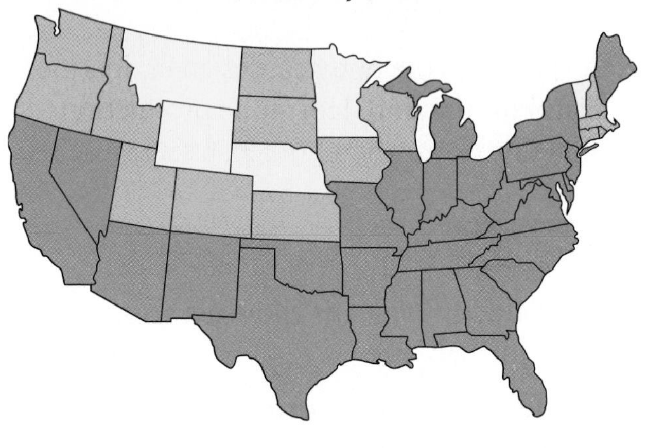

11. Using the figure above, which state has the lowest incidence of diabetes?
    A) Wyoming **C)** Michigan
    B) Florida **D)** Washington
12. What is the monthly process that releases an egg called?
    A) fertilization **C)** menstruation
    B) ovulation **D)** puberty
13. What is the union of an egg and a sperm?
    A) fertilization **C)** menstruation
    B) ovulation **D)** puberty
14. During what stage of development does the amniotic sac form?
    A) zygote **C)** fetus
    B) embryo **D)** newborn
15. When does puberty occur?
    A) childhood **C)** adolescence
    B) adulthood **D)** infancy
16. During which period does growth stop?
    A) childhood **C)** adolescence
    B) adulthood **D)** infancy

Science Online green.msscience.com/vocabulary_puzzlemaker

Thinking Critically

17. **List** the effects that adrenal gland hormones can have on your body as you prepare to run a race.

18. **Explain** the similar functions of the ovaries and testes.

Use the diagram below to answer question 19.

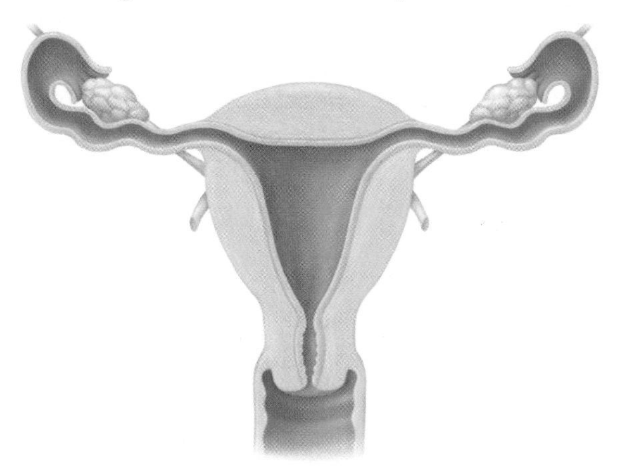

19. **Identify** the structure in the above diagram in which each process occurs: ovulation, fertilization, and implantation.

20. **Compare and contrast** your endocrine system with the thermostat in your home.

21. **Explain** if quadruplets—four babies born at one birth—are always identical or always fraternal, or if they can be either.

22. **Predict** During the ninth month of pregnancy, the fetus develops a white, greasy coating. Predict what the function of this coating might be.

23. **Form a hypothesis** about the effect of raising identical twins apart from each other.

24. **Classify** each of the following structures as female or male and internal or external: ovary, penis, scrotum, testes, uterus, and vagina.

Performance Activities

25. **Letter** Find newspaper or magazine articles on the effects of smoking on the health of the developing embryo and newborn. Write a letter to the editor about why a mother's smoking is damaging to her unborn baby's health.

Applying Math

26. **Blood Sugar Levels** Carol is diabetic and has a fasting blood sugar level of 180 mg/dL. Luisa does not have diabetes and has a fasting blood sugar level of 90 mg/dL. Express as a percentage how much higher the fasting blood sugar level is for Carol as compared to that for Luisa.

Use the graph below to answer questions 27 and 28.

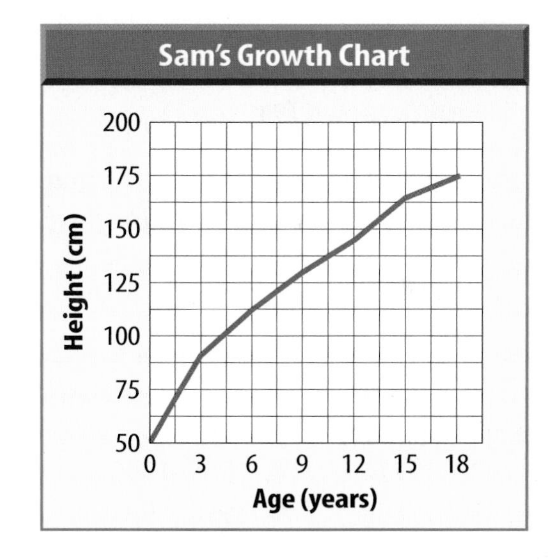

27. **Early Childhood Growth** The graph above charts Sam's growth from birth to 18 years of age. According to the graph, how much taller was Sam at 12 years of age than he was at 3 years of age?

28. **Adolescent Growth** According to the graph, how much did Sam grow between 12 and 18 years of age?

Part 1 Multiple Choice

Record your answers on the answer sheet provided by your teacher or on a sheet of paper.

1. When do eggs start to develop in the ovaries?
   A. before birth **C.** during childhood
   B. at puberty **D.** during infancy

Use the graph below to answers questions 2 and 3.

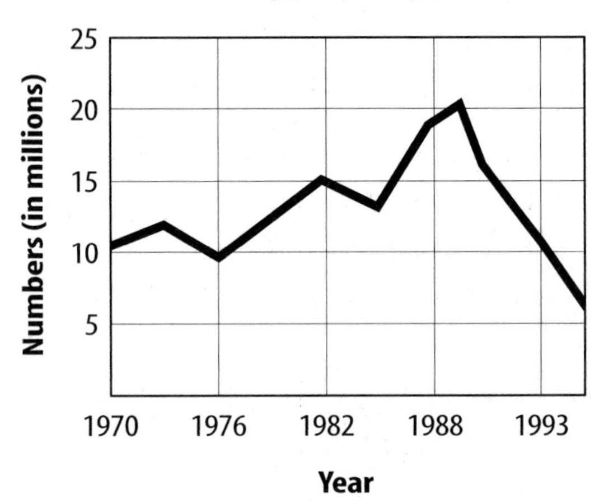

2. According to the information in the graph, in which year was the syphilis rate the lowest?
   A. 1976 **C.** 1988
   B. 1982 **D.** 1993

3. According to the information in the graph, during which years was there a decrease in the syphilis rate?
   A. 1970–1972 **C.** 1988–1990
   B. 1976–1982 **D.** 1990–1993

4. Which of the following glands is found in the neck?
   A. pineal **C.** thyroid
   B. adrenal **D.** pancreas

Test-Taking Tip

Bar Graphs On a bar graph, line up each bar with its corresponding value by laying your pencil between the two points.

5. What is the mixture of sperm and fluid called?
   A. semen **C.** seminal vesicle
   B. testes **D.** epididymis

Use the table below to answer questions 6–8.

Results of Folic Acid on Development of Neural Tube Defect		
Group	**Babies with Neural Tube Defect**	**Babies without Neural Tube Defect**
Group I—Folic Acid	6	497
Group II—No Folic Acid	21	581

(From CDC)

6. Researchers have found that the B-vitamin folic acid can prevent neural tube defects. In a study done in Europe in 1991, one group of pregnant women was given extra folic acid, and the other group did not receive extra folic acid. What percentage of babies were born with a neural tube defect in Group II?
   A. 1.0% **C.** 3.0%
   B. 2.5% **D.** 4.0%

7. What percentage of babies were born with a neural tube defect in Group I?
   A. 1.0% **C.** 3.0%
   B. 2.5% **D.** 4.0%

8. Which of the following statements is true regarding the data in this table?
   A. Folic acid had no effect on the percentage of babies with a neural tube defect.
   B. Extra folic acid decreased the percentage of babies with a neural tube defect.
   C. Extra folic acid increased the percentage of babies with a neural tube defect.
   D. Group I and Group II had the same percentage of babies born with a neural tube defect.

Part 2 Short Response/Grid In

Record your answers on the answer sheet provided by your teacher or on a sheet of paper.

9. How are endocrine glands different from salivary glands?
10. What does parathyroid hormone do for the body?
11. What is the function of the cilia in the oviduct?

Use the illustration below to answer questions 12 and 13.

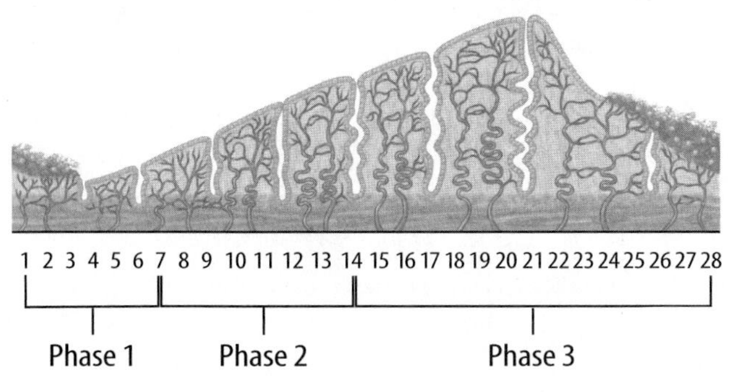

12. According to the illustration, what percentage of the menstrual cycle is phase 3?
13. According to the illustration, what percentage of the menstrual cycle is phase 2?
14. According to the illustration, on which day does ovulation occur?
15. During which stage of development before birth does amniotic fluid develop? What is the purpose of amniotic fluid?
16. During which stage of development after birth is physical growth and development the most rapid?
17. Rubella, also know as German measles, is caused by a virus. If a pregnant woman is infected with rubella, the virus can affect the formation of major organs, such as the heart, in the fetus. During which stage of development before birth would a rubella infection be most dangerous?

Part 3 Open Ended

Record your answers on a sheet of paper.

18. Predict how each of the following factors may affect sperm production: hot environment, illness with fever, testes located inside the body cavity, and injury to the testes. Explain your answer.
19. Sexually transmitted diseases can cause infection of the female reproductive organs, including the oviduct. Infection of the oviduct can result in scarring. What might happen to an egg that enters a scarred oviduct?

Use the table below to answer question 20.

Pre-eclampsia Risk in Pregnancy	
Risk Factors	**Risk Ratio**
First pregnancy	3:1
Over 40 years of age	3:1
Family history	5:1
Chronic hypertension	10:1
Chronic renal disease	20:1
Antiphospholipid syndrome	10:1
Diabetes mellitus	2:1
Twin birth	4:1
Angiotensinogen gene T235	
Homozygous	20:1
Heterozygous	4:1

20. Pre-eclampsia is a condition that can develop in a woman after 20 weeks of pregnancy. It involves the development of hypertension or high blood pressure, an abnormal amount of protein in urine, and swelling. Infer why a woman with chronic hypertension has a higher risk of developing pre-eclampsia than a woman without hypertension.

unit 5 The Interdependence of Life

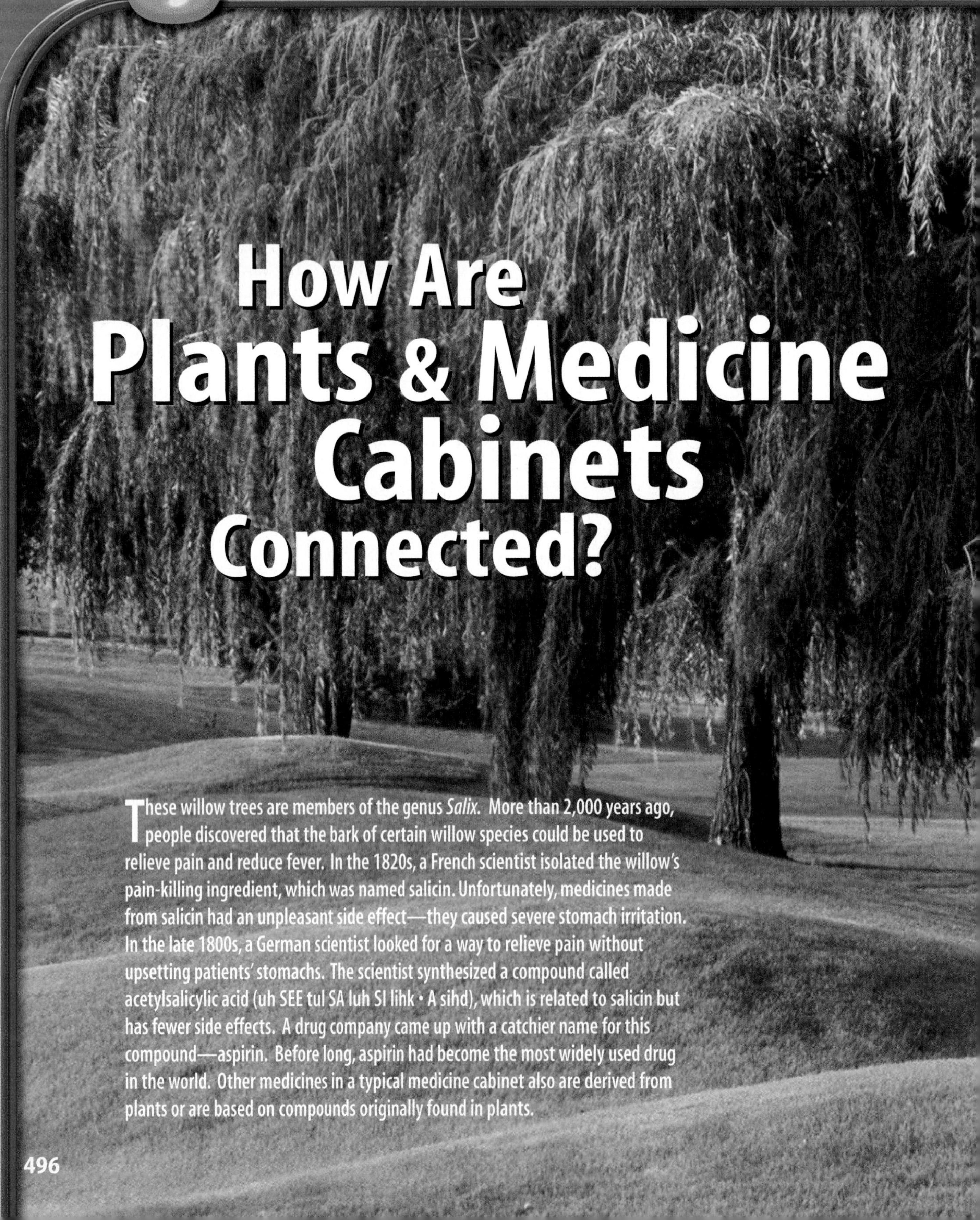

How Are Plants & Medicine Cabinets Connected?

These willow trees are members of the genus *Salix*. More than 2,000 years ago, people discovered that the bark of certain willow species could be used to relieve pain and reduce fever. In the 1820s, a French scientist isolated the willow's pain-killing ingredient, which was named salicin. Unfortunately, medicines made from salicin had an unpleasant side effect—they caused severe stomach irritation. In the late 1800s, a German scientist looked for a way to relieve pain without upsetting patients' stomachs. The scientist synthesized a compound called acetylsalicylic acid (uh SEE tul SA luh SI lihk • A sihd), which is related to salicin but has fewer side effects. A drug company came up with a catchier name for this compound—aspirin. Before long, aspirin had become the most widely used drug in the world. Other medicines in a typical medicine cabinet also are derived from plants or are based on compounds originally found in plants.

unit projects

Visit green.msscience.com/unit_project to find project ideas and resources.
Projects include:

- **History** Design a slide show to present information on medicines derived from plants and where these plants grow.
- **Technology** Make your own giant jigsaw puzzle illustrating the five systems of a seed plant, including labels and functions of each plant part.
- **Model** Construct a review game demonstrating knowledge of nitrogen and oxygen cycles. The game and instructions should be assembled in an eco-friendly box.

WebQuest Discover *Phytochemicals and a Healthy Diet.* Compare your diet with the suggested diet that helps prevent cancer and heart disease.

chapter 17

Plants

chapter preview

How are all plants alike?

Plants are found nearly everywhere on Earth. A tropical rain forest like this one is crowded with lush, green plants. When you look at a plant, what do you expect to see? Do all plants have green leaves? Do all plants produce flowers and seeds?

Science Journal Write three characteristics that you think all plants have in common.

Start-Up Activities

How do you use plants?

Plants are just about everywhere—in parks and gardens, by streams, on rocks, in houses, and even on dinner plates. Do you use plants for things other than food?

1. Brainstorm with two other classmates and make a list of everything that you use in a day that comes from plants.
2. Compare your list with those of other groups in your class.
3. Search through old magazines for images of the items on your list.
4. As a class, build a bulletin board display of the magazine images.
5. **Think Critically** In your Science Journal, list things that were made from plants 100 years or more ago but today are made from plastics, steel, or some other material.

Preview this chapter's content and activities at green.msscience.com

Plants Make the following Foldable to help identify what you already know, what you want to know, and what you learned about plants.

STEP 1 **Fold** a vertical sheet of paper from side to side. Make the front edge 1.25 cm shorter than the back edge.

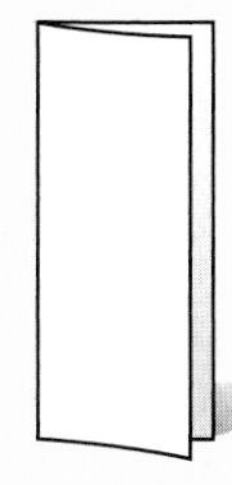

STEP 2 **Turn** lengthwise and fold into thirds.

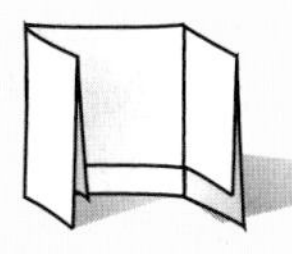

STEP 3 **Unfold and cut** only the top layer along both folds to make three tabs.

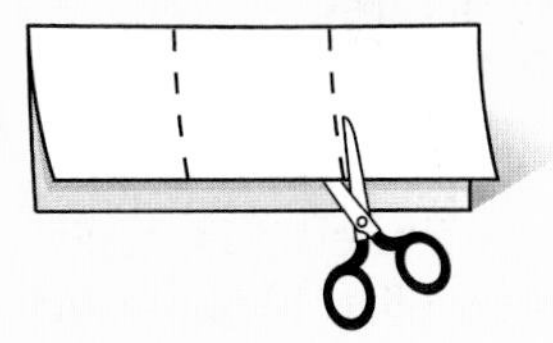

STEP 4 **Label** each tab as shown.

Identify Questions Before you read the chapter, write what you already know about plants under the left tab of your Foldable, and write questions about what you'd like to know under the center tab. After you read the chapter, list what you learned under the right tab.

section 1

An Overview of Plants

as you read

What You'll Learn

- **Identify** characteristics common to all plants.
- **Explain** which plant adaptations make it possible for plants to survive on land.
- **Compare and contrast** vascular and nonvascular plants.

Why It's Important

Plants produce food and oxygen, which are required for life by most organisms on Earth.

Review Vocabulary
species: closely related organisms that share similar characteristics and can reproduce among themselves

New Vocabulary
- cuticle
- cellulose
- vascular plant
- nonvascular plant

What is a plant?

What is the most common sight you see when you walk along nature trails in parks like the one shown in **Figure 1?** Maybe you've taken off your shoes and walked barefoot on soft, cool grass. Perhaps you've climbed a tree to see what things look like from high in its branches. In each instance, plants surrounded you.

If you named all the plants that you know, you probably would include trees, flowers, vegetables, fruits, and field crops like wheat, rice, or corn. Between 260,000 and 300,000 plant species have been discovered and identified. Scientists think many more species are still to be found, mainly in tropical rain forests. Plants are important food sources to humans and other consumers. Without plants, most life on Earth as we know it would not be possible.

Plant Characteristics Plants range in size from microscopic water ferns to giant sequoia trees that are sometimes more than 100 m in height. Most have roots or rootlike structures that hold them in the ground or onto some other object like a rock or another plant. Plants are adapted to nearly every environment on Earth. Some grow in frigid, ice-bound polar regions and others grow in hot, dry deserts. All plants need water, but some plants cannot live unless they are submerged in either freshwater or salt water.

Figure 1 All plants are many-celled and nearly all contain chlorophyll. Grasses, trees, shrubs, mosses, and ferns are all plants.

Plant Cells Like other living things, plants are made of cells. A plant cell has a cell membrane, a nucleus, and other cellular structures. In addition, plant cells have cell walls that provide structure and protection. Animal cells do not have cell walls.

Many plant cells contain the green pigment chlorophyll (KLOR uh fihl) so most plants are green. Plants need chlorophyll to make food using a process called photosynthesis. Chlorophyll is found in a cell structure called a chloroplast. Plant cells from green parts of the plant usually contain many chloroplasts.

Most plant cells have a large, membrane-bound structure called the central vacuole that takes up most of the space inside of the cell. This structure plays an important role in regulating the water content of the cell. Many substances are stored in the vacuole, including the pigments that make some flowers red, blue, or purple.

Origin and Evolution of Plants

Have plants always existed on land? The first plants that lived on land probably could survive only in damp areas. Their ancestors were probably ancient green algae that lived in the sea. Green algae are one-celled or many-celled organisms that use photosynthesis to make food. Today, plants and green algae have the same types of chlorophyll and carotenoids (kuh RAH tun oydz) in their cells. Carotenoids are red, yellow, or orange pigments that also are used for photosynthesis. These facts lead scientists to think that plants and green algae have a common ancestor.

Reading Check *How are plants and green algae alike?*

Fossil Record The fossil record for plants is not like that for animals. Most animals have bones or other hard parts that can fossilize. Plants usually decay before they become fossilized. The oldest fossil plants are about 420 million years old. **Figure 2** shows *Cooksonia,* a fossil of one of these plants. Other fossils of early plants are similar to the ancient green algae. Scientists hypothesize that some of these early plants evolved into the plants that exist today.

Cone-bearing plants, such as pines, probably evolved from a group of plants that grew about 350 million years ago. Fossils of these plants have been dated to about 300 million years ago. It is estimated that flowering plants did not exist until about 120 million years ago. However, the exact origin of flowering plants is not known.

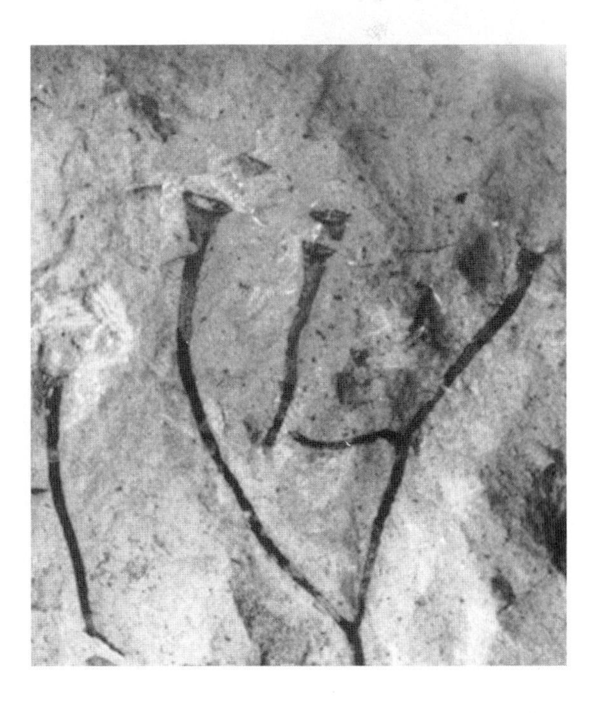

Figure 2 This is a fossil of a plant named *Cooksonia.* These plants grew about 420 million years ago and were about 2.5 cm tall.

Cellulose Plant cell walls are made mostly of cellulose. Anselme Payen, a French scientist, first isolated and identified the chemical composition of cellulose in 1838, while analyzing the chemical makeup of wood. Choose a type of wood and research to learn the uses of that wood. Make a classroom display of research results.

Life on Land

Life on land has some advantages for plants. More sunlight and carbon dioxide—needed for photosynthesis—are available on land than in water. During photosynthesis, plants give off oxygen. Long ago, as more and more plants adapted to life on land, the amount of oxygen in Earth's atmosphere increased. This was the beginning for organisms that depend on oxygen.

Adaptations to Land

What is life like for green algae, shown in **Figure 3,** as they float in a shallow pool? The water in the pool surrounds and supports them as the algae make their own food through the process of photosynthesis. Because materials can enter and leave through their cell membranes and cell walls, the algae cells have everything they need to survive as long as they have water.

If the pool begins to dry up, the algae are on damp mud and are no longer supported by water. As the soil becomes drier and drier, the algae will lose water too because water moves through their cell membranes and cell walls from where there is more water to where there is less water. Without enough water in their environment, the algae will die. Plants that live on land have adaptations that allow them to conserve water, as well as other differences that make it possible for survival.

Figure 3 The alga *Spirogyra*, like all algae, must have water to survive. If the pool where it lives dries up, it will die.

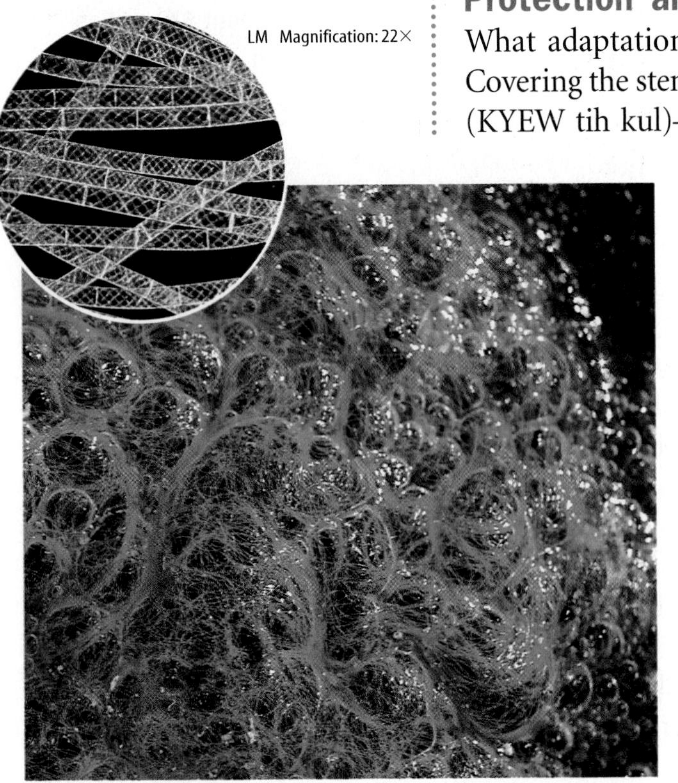

Protection and Support Water is important for plants. What adaptations would help a plant conserve water on land? Covering the stems, leaves, and flowers of many plants is a **cuticle** (KYEW tih kul)—a waxy, protective layer secreted by cells onto the surface of the plant. The cuticle slows the loss of water. The cuticle and other adaptations shown in **Figure 4** enable plants to survive on land.

What is the function of a plant's cuticle?

Supporting itself is another problem for a plant on land. Like all cells, plant cells have cell membranes, but they also have rigid cell walls outside the membrane. Cell walls contain **cellulose** (SEL yuh lohs), which is a chemical compound that plants can make out of sugar. Long chains of cellulose molecules form tangled fibers in plant cell walls. These fibers provide structure and support.

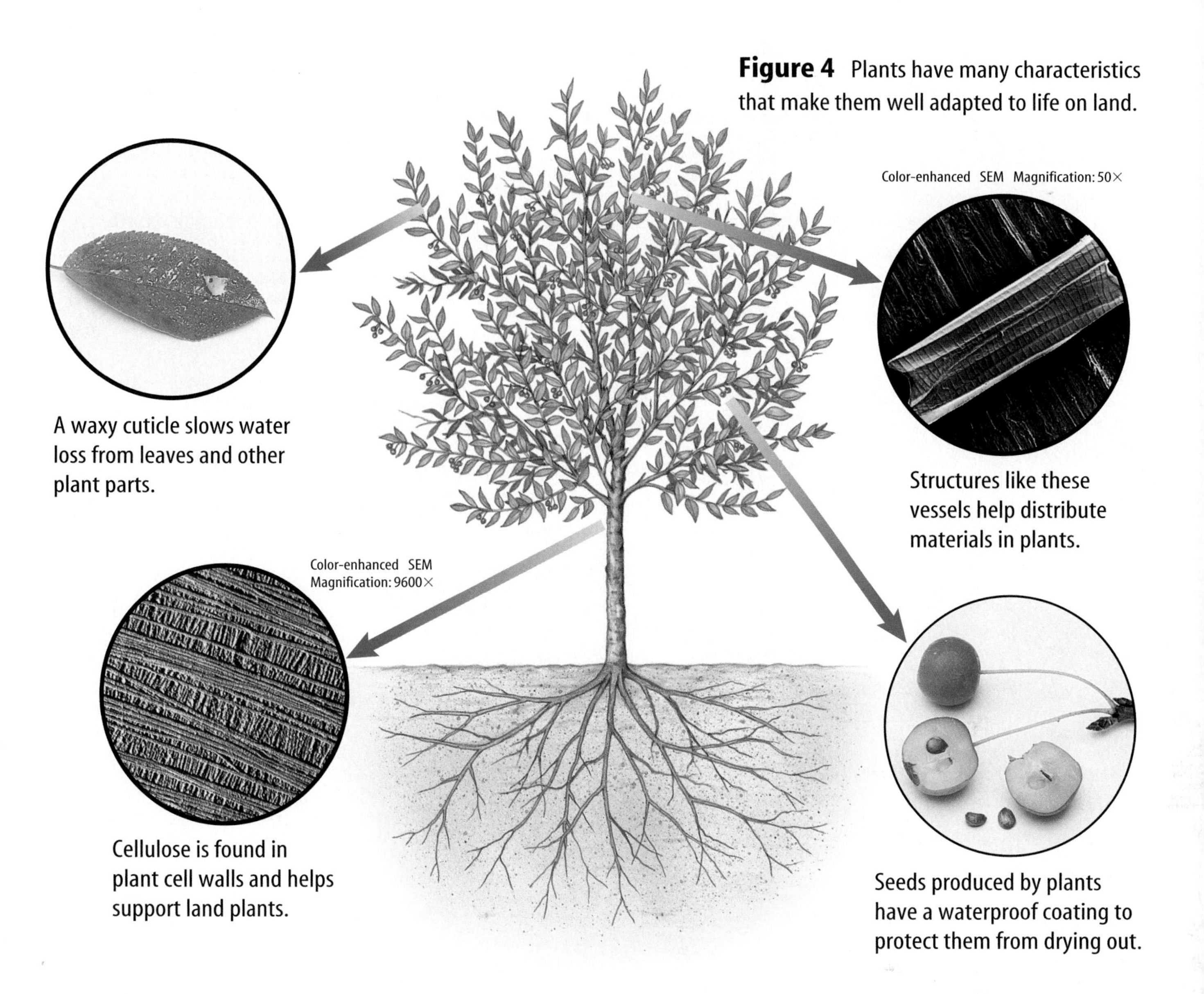

Figure 4 Plants have many characteristics that make them well adapted to life on land.

A waxy cuticle slows water loss from leaves and other plant parts.

Structures like these vessels help distribute materials in plants.

Cellulose is found in plant cell walls and helps support land plants.

Seeds produced by plants have a waterproof coating to protect them from drying out.

Other Cell Wall Substances Cells of some plants secrete other substances into the cellulose that make the cell wall even stronger. Trees, such as oaks and pines, could not grow without these strong cell walls. Wood from trees can be used for construction mostly because of strong cell walls.

Life on land means that each plant cell is not surrounded by water and dissolved nutrients that can move into the cell. Through adaptations, structures developed in many plants that distribute water, nutrients, and food to all plant cells. These structures also help provide support for the plant.

Reproduction Changes in reproduction were necessary if plants were to survive on land. The presence of water-resistant spores helped some plants reproduce successfully. Other plants adapted by producing water-resistant seeds in cones or in flowers that developed into fruits.

NATIONAL GEOGRAPHIC
VISUALIZING PLANT CLASSIFICATION
Figure 5
Scientists group plants as either vascular—those with water- and food-conducting cells in their stems—or nonvascular. Vascular plants are further divided into those that produce spores and those that make seeds.
Sunflower
Joint fir
Cycad
Douglas fir
Ginkgo
Horsetail
Fern
Club moss
Moss
Liverwort
Hornwort
Vascular
Nonvascular
Seed vascular
Seedless vascular
Flowering
Joint firs
Cycads
Conifers
Ginkgoes
Horsetails
Ferns
Club mosses
Mosses
Liverworts
Hornworts

Classification of Plants

The plant kingdom is classified into major groups called divisions. A division is the same as a phylum in other kingdoms. Another way to group plants is as vascular (VAS kyuh lur) or nonvascular plants, as illustrated in **Figure 5.** **Vascular plants** have tubelike structures that carry water, nutrients, and other substances throughout the plant. **Nonvascular plants** do not have these tubelike structures and use other ways to move water and substances.

Figure 6 Although these two plants are both called daisies, they are not the same species of plant. Using their binomial names helps eliminate the confusion that might come from using their common names.

Naming Plants Why do biologists call a pecan tree *Carya illinoiensis* and a white oak *Quercus alba?* They are using words that accurately name the plant. In the third century B.C., most plants were grouped as trees, shrubs, or herbs and placed into smaller groups by leaf characteristics. This simple system survived until late in the eighteenth century when a Swedish botanist, Carolus Linnaeus, developed a new system. His new system used many characteristics to classify a plant. He also developed a way to name plants called binomial nomenclature (bi NOH mee ul • NOH mun klay chur). Under this system, every plant species is given a unique two-word name like the names above for the pecan tree and white oak and for the two daisies in **Figure 6.**

section 1 review

Summary

What is a plant?

- All plant cells are surrounded by a cell wall.
- Many plant cells contain chlorophyll.

Origin and Evolution of Plants

- Ancestors of land plants were probably ancient green algae.

Adaptations to Land

- A waxy cuticle helps conserve water.
- Cellulose strengthens cell walls.

Classification of Plants

- The plant kingdom is divided into two groups—nonvascular plants and vascular plants.
- Vascular tissues transport nutrients.

Self Check

1. **List** the characteristics of plants.
2. **Compare and contrast** the characteristics of vascular and nonvascular plants.
3. **Identify** three adaptations that allow plants to survive on land.
4. **Explain** why binomial nomenclature is used to name plants.
5. **Thinking Critically** If you left a board lying on the grass for a few days, what would happen to the grass underneath the board? Why?

Applying Skills

6. **Form a hypothesis** about adaptations a land plant might undergo if it lived submerged in water.

Seedless Plants

as you read

What You'll Learn

- **Distinguish** between characteristics of seedless nonvascular plants and seedless vascular plants.
- **Identify** the importance of some nonvascular and vascular plants.

Why It's Important

Seedless plants are among the first to grow in damaged or disturbed environments and help build soil for the growth of other plants.

Review Vocabulary
spore: waterproof reproductive cell

New Vocabulary
- rhizoid
- pioneer species

Seedless Nonvascular Plants

If you were asked to name the parts of a plant, you probably would list roots, stems, leaves, and flowers. You also might know that many plants grow from seeds. However, some plants, called nonvascular plants, don't grow from seeds and they do not have all of these parts. **Figure 7** shows some common types of nonvascular plants.

Nonvascular plants are usually just a few cells thick and only 2 cm to 5 cm in height. Most have stalks that look like stems and green, leaflike growths. Instead of roots, threadlike structures called **rhizoids** (RI zoydz) anchor them where they grow. Most nonvascular plants grow in places that are damp. Water is absorbed and distributed directly through their cell membranes and cell walls. Nonvascular plants also do not have flowers or cones that produce seeds. They reproduce by spores. Mosses, liverworts, and hornworts are examples of nonvascular plants.

Mosses Most nonvascular plants are classified as mosses, like the ones in **Figure 7.** They have green, leaflike growths arranged around a central stalk. Their rhizoids are made of many cells. Sometimes stalks with caps grow from moss plants. Reproductive cells called spores are produced in the caps of these stalks. Mosses often grow on tree trunks and rocks or the ground. Although they commonly are found in damp areas, some are adapted to living in deserts.

Figure 7 The seedless nonvascular plants include mosses, liverworts, and hornworts.

Close-up of moss plants

Close-up of a liverwort

Close-up of a hornwort

Figure 8 Mosses can grow in the thin layer of soil that covers these rocks.

Liverworts In the ninth century, liverworts were thought to be useful in treating diseases of the liver. The suffix *-wort* means "herb," so the word *liverwort* means "herb for the liver." Liverworts are rootless plants with flattened, leaflike bodies, as shown in **Figure 7.** They usually have one-celled rhizoids.

Hornworts Most hornworts are less than 2.5 cm in diameter and have a flattened body like liverworts, as shown in **Figure 7.** Unlike other nonvascular plants, almost all hornworts have only one chloroplast in each of their cells. Hornworts get their name from their spore-producing structures, which look like tiny horns of cattle.

Nonvascular Plants and the Environment Mosses and liverworts are important in the ecology of many areas. Although they require moist conditions to grow and reproduce, many of them can withstand long, dry periods. They can grow in thin soil and in soils where other plants could not grow, as shown in **Figure 8.**

Spores of mosses and liverworts are carried by the wind. They will grow into plants if growing conditions are right. Mosses often are among the first plants to grow in new or disturbed environments, such as lava fields or after a forest fire. Organisms that are the first to grow in new or disturbed areas are called **pioneer species.** As pioneer plants grow and die, decaying material builds up. This, along with the slow breakdown of rocks, builds soil. When enough soil has formed, other organisms can move into the area.

Why are pioneer plant species important in disturbed environments?

Mini LAB

Measuring Water Absorption by a Moss

Procedure

1. Place a few teaspoons of ***Sphagnum* moss** on a piece of **cheesecloth.** Gather the corners of the cloth and twist, then tie them securely to form a ball.
2. Weigh the ball.
3. Put 200 mL of **water** in a **container** and add the ball.
4. After 15 min, remove the ball and drain the excess water into the container.
5. Weigh the ball and measure the amount of water left in the container.
6. Wash your hands after handling the moss.

Analysis

In your **Science Journal,** calculate how much water was absorbed by the *Sphagnum* moss.

Topic: Medicinal Plants

Visit green.msscience.com for Web links to information about plants used as medicines.

Activity In your Science Journal, list four medicinal plants and their uses.

Seedless Vascular Plants

The fern in **Figure 9** is growing next to some moss plants. Ferns and mosses are alike in one way. Both reproduce by spores instead of seeds. However, ferns are different from mosses because they have vascular tissue. The vascular tissue in seedless vascular plants, like ferns, is made up of long, tubelike cells. These cells carry water, minerals, and food to cells throughout the plant. Why is vascular tissue an advantage to a plant? Nonvascular plants like the moss are usually only a few cells thick. Each cell absorbs water directly from its environment. As a result, these plants cannot grow large. Vascular plants, on the other hand, can grow bigger and thicker because the vascular tissue distributes water and nutrients to all plant cells.

Applying Science

What is the value of rain forests?

Throughout history, cultures have used plants for medicines. Some cultures used willow bark to cure headaches. Willow bark contains salicylates (suh LIH suh layts), the main ingredient in aspirin. Heart problems were treated with foxglove, which is the main source of digitalis (dih juh TAH lus), a drug prescribed for heart problems. Have all medicinal plants been identified?

Identifying the Problem

Tropical rain forests have the largest variety of organisms on Earth. Many plant species are still unknown. These forests are being destroyed rapidly. The map below shows the rate of destruction of the rain forests. Some scientists estimate that most tropical rain forests will be destroyed in 30 years.

Solving the Problem

1. What country has the most rain forest destroyed each year?
2. Where can scientists go to study rain forest plants before the plants are destroyed?
3. Predict how the destruction of rain forests might affect research on new drugs from plants.

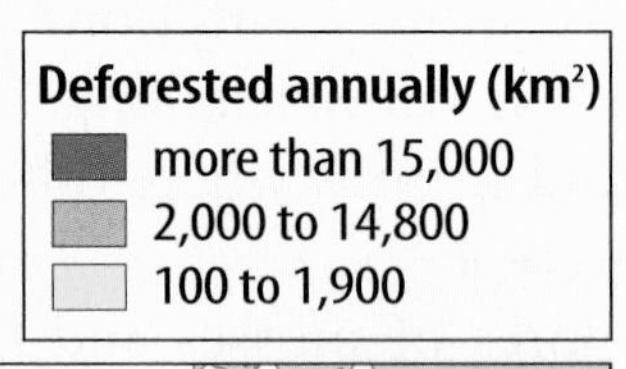

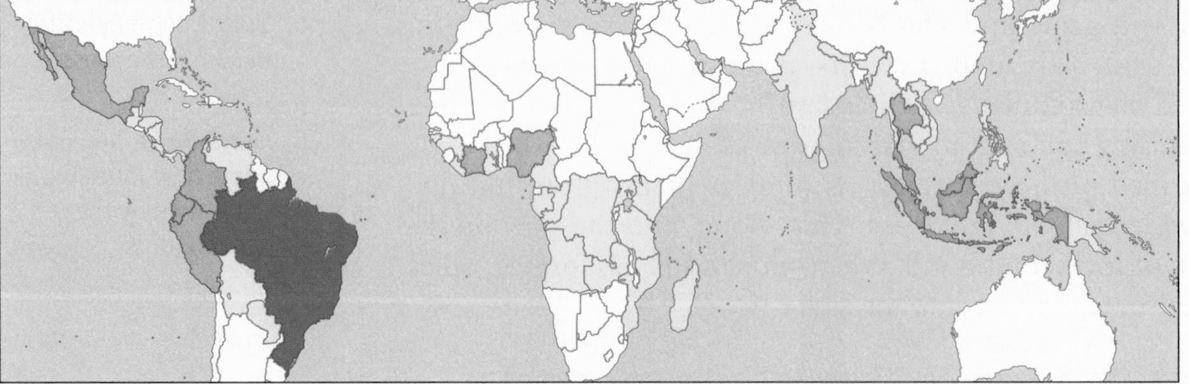

Types of Seedless Vascular Plants

Besides ferns, seedless vascular plants include ground pines, spike mosses, and horsetails. About 1,000 species of ground pines, spike mosses, and horsetails are known to exist. Ferns are more abundant, with at least 12,000 known species. Many species of seedless vascular plants are known only from fossils. They flourished during the warm, moist period 360 million to 286 million years ago. Fossil records show that some horsetails grew 15 m tall, unlike modern species, which grow only 1 m to 2 m tall.

Figure 9 The mosses and ferns pictured here are seedless plants. **Explain** *why the fern can grow taller than the moss.*

Ferns The largest group of seedless vascular plants is the ferns. They include many different forms, as shown in **Figure 10.** They have stems, leaves, and roots. Fern leaves are called fronds. Ferns produce spores in structures that usually are found on the underside of their fronds. Thousands of species of ferns now grow on Earth, but many more existed long ago. From clues left in rock layers, scientists infer that about 360 million years ago much of Earth was tropical. Steamy swamps covered large areas. The tallest plants were species of ferns. The ancient ferns grew as tall as 25 m—as tall as the tallest fern species alive today. Most modern tree ferns are about 3 m to 5 m in height and grow in tropical regions of the world.

Figure 10 Ferns come in many different shapes and sizes.

The sword fern has a typical fern shape. Spores are produced in structures on the back of the frond.

This fern grows on other plants, not in the soil.
Infer *why it's called the staghorn fern.*

Tree ferns, like this one in Hawaii, grow in tropical areas.

Figure 11 Photographers once used the dry, flammable spores of club mosses as flash powder. It burned rapidly and produced the light that was needed to take photographs.

Club Mosses Ground pines and spike mosses are groups of plants that often are called club mosses. They are related more closely to ferns than to mosses. These seedless vascular plants have needle-like leaves. Spores are produced at the end of the stems in structures that look like tiny pine cones. Ground pines, shown in **Figure 11,** are found from arctic regions to the tropics, but rarely in large numbers. In some areas, they are endangered because they have been over collected to make wreaths and other decorations.

Reading Check *Where are spores in club mosses produced?*

Spike mosses resemble ground pines. One species of spike moss, the resurrection plant, is adapted to desert conditions. When water is scarce, the plant curls up and seems dead. When water becomes available, the resurrection plant unfurls its green leaves and begins making food again. The plant can repeat this process whenever necessary.

Horsetails The stem structure of horsetails is unique among the vascular plants. The stem is jointed and has a hollow center surrounded by a ring of vascular tissue. At each joint, leaves grow out from around the stem. In **Figure 12,** you can see these joints. If you pull on a horsetail stem, it will pop apart in sections. Like the club mosses, spores from horsetails are produced in a conelike structure at the tips of some stems. The stems of the horsetails contain silica, a gritty substance found in sand. For centuries, horsetails have been used for polishing objects, sharpening tools, and scouring cooking utensils. Another common name for horsetails is scouring rush.

Figure 12 Most horsetails grow in damp areas and are less than 1 m tall.
Identify *where spores would be produced on this plant.*

Importance of Seedless Plants

When many ancient seedless plants died, they became submerged in water and mud before they decomposed. As this plant material built up, it became compacted and compressed and eventually turned into coal—a process that took millions of years.

Today, a similar process is taking place in bogs, which are poorly drained areas of land that contain decaying plants. The plants in bogs are mostly seedless plants like mosses and ferns.

Peat When bog plants die, the waterlogged soil slows the decay process. Over time, these decaying plants are compressed into a substance called peat. Peat, which forms from the remains of sphagnum moss, is mined from bogs to use as a low-cost fuel in places such as Ireland and Russia, as shown in **Figure 13.** Peat supplies about one-third of Ireland's energy requirements. Scientists hypothesize that over time, if additional layers of soil bury, compact, and compress the peat, it will become coal.

Figure 13 Peat is cut from bogs and used for a fuel in some parts of Europe.

Uses of Seedless Vascular Plants Many people keep ferns as houseplants. Ferns also are sold widely as landscape plants for shady areas. Peat and sphagnum mosses also are used for gardening. Peat is an excellent soil conditioner, and sphagnum moss often is used to line hanging baskets. Ferns also are used as weaving material for basketry.

Although most mosses are not used for food, parts of many other seedless vascular plants can be eaten. The rhizomes and young fronds of some ferns are edible. The dried stems of one type of horsetail can be ground into flour. Seedless plants have been used as folk medicines for hundreds of years. For example, ferns have been used to treat bee stings, burns, fevers, and even dandruff.

section 2 review

Summary

Seedless Nonvascular Plants

- Seedless nonvascular plants include mosses, liverworts, and hornworts.
- They are usually only a few cells thick and no more than a few centimeters tall.
- They produce spores rather than seeds.

Seedless Vascular Plants

- Seedless vascular plants include ferns, club mosses, and horsetails.
- Vascular plants grow taller and can live farther from water than nonvascular plants.

Importance of Seedless Plants

- Nonvascular plants help build new soil.
- Coal deposits formed from ancient seedless plants that were buried in water and mud before they began to decay.

Self Check

1. **Compare and contrast** the characteristics of mosses and ferns.
2. **Explain** what fossil records tell about seedless plants that lived on Earth long ago.
3. **Identify** growing conditions in which you would expect to find pioneer plants such as mosses and liverworts.
4. **Summarize** the functions of vascular tissues.
5. **Think Critically** The electricity that you use every day might be produced by burning coal. What is the connection between electricity production and seedless vascular plants?

Applying Math

6. **Use Fractions** Approximately 8,000 species of liverworts and 9,000 species of mosses exist today. Estimate what fraction of these seedless nonvascular plants are mosses.

section 3

Seed Plants

as you read

What You'll Learn

- **Identify** the characteristics of seed plants.
- **Explain** the structures and functions of roots, stems, and leaves.
- **Describe** the main characteristics and importance of gymnosperms and angiosperms.
- **Compare** similarities and differences between monocots and dicots.

Why It's Important

Humans depend on seed plants for food, clothing, and shelter.

Review Vocabulary
seed: plant embryo and food supply in a protective coating

New Vocabulary

- stomata
- guard cell
- xylem
- phloem
- cambium
- gymnosperm
- angiosperm
- monocot
- dicot

Characteristics of Seed Plants

What foods from plants have you eaten today? Apples? Potatoes? Carrots? Peanut butter and jelly sandwiches? All of these foods and more come from seed plants.

Most of the plants you are familiar with are seed plants. Most seed plants have leaves, stems, roots, and vascular tissue. They also produce seeds, which usually contain an embryo and stored food. The stored food is the source of energy for the embryo's early growth as it develops into a plant. Most of the plant species that have been identified in the world today are seed plants. The seed plants generally are classified into two major groups—gymnosperms (JIHM nuh spurmz) and angiosperms (AN jee uh spurmz).

Leaves Most seed plants have leaves. Leaves are the organs of the plant where the food-making process—photosynthesis—usually occurs. Leaves come in many shapes, sizes, and colors. Examine the structure of a typical leaf, shown in **Figure 14.**

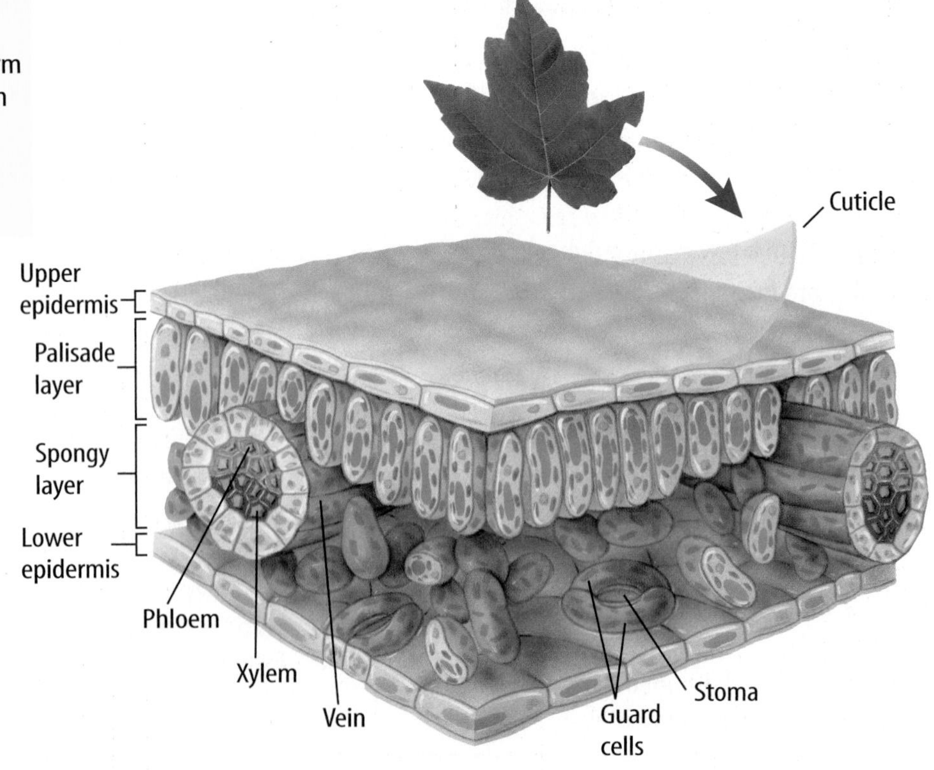

Figure 14 The structure of a typical leaf is adapted for photosynthesis.
Explain *why cells in the palisade layer have more chloroplasts than cells in the spongy layer.*

Leaf Cell Layers A typical leaf is made of several different layers of cells. On the upper and lower surfaces of a leaf is a thin layer of cells called the epidermis, which covers and protects the leaf. A waxy cuticle coats the epidermis of some leaves. Most leaves have small openings in the epidermis called **stomata** (STOH muh tuh) (singular, *stoma*). Stomata allow carbon dioxide, water, and oxygen to enter into and exit from a leaf. Each stoma is surrounded by two **guard cells** that open and close it.

Just below the upper epidermis is the palisade layer. It consists of closely packed, long, narrow cells that usually contain many chloroplasts. Most of the food produced by plants is made in the palisade cells. Between the palisade layer and the lower epidermis is the spongy layer. It is a layer of loosely arranged cells separated by air spaces. In a leaf, veins containing vascular tissue are found in the spongy layer.

Stems The trunk of a tree is really the stem of the tree. Stems usually are located above ground and support the branches, leaves, and reproductive structures. Materials move between leaves and roots through the vascular tissue in the stem. Stems also can have other functions, as shown in **Figure 15.**

Plant stems are either herbaceous (hur BAY shus) or woody. Herbaceous stems usually are soft and green, like the stems of a tulip, while trees and shrubs have hard, rigid, woody stems. Lumber comes from woody stems.

Observing Water Moving in a Plant

Procedure

1. Into a **clear container** pour **water** to a depth of 1.5 cm. Add 25 drops of **red food coloring** to the water.
2. Put the root end of a **green onion** into the container. Do not cut the onion in any way. Wash your hands.
3. The next day, examine the outside of the onion. Peel off the onion's layers and examine them. **WARNING:** *Do not eat the onion.*

Analysis

In your **Science Journal,** infer how the location of red color inside the onion might be related to vascular tissue.

Try at Home

Figure 15 Some plants have stems with special functions.

These potatoes are stems that grow underground and store food for the plant.

The stems of this cactus store water and can carry on photosynthesis.

Some stems of this grape plant help it climb on other plants.

Figure 16 The root system of a tree can be as long as the tree can be tall.
Infer *why the root system of a tree would need to be so large.*

Roots Imagine a lone tree growing on top of a hill. What is the largest part of this plant? Maybe you guessed the trunk or the branches. Did you consider the roots, like those shown in **Figure 16?** The root systems of most plants are as large or larger than the aboveground stems and leaves.

Roots are important to plants. Water and other substances enter a plant through its roots. Roots have vascular tissue in which water and dissolved substances move from the soil through the stems to the leaves. Roots also act as anchors, preventing plants from being blown away by wind or washed away by moving water. Underground root systems support other plant parts that are aboveground—the stem, branches, and leaves of a tree. Sometimes, part of or all of the roots are aboveground, too.

Roots can store food. When you eat carrots or beets, you eat roots that contain stored food. Plants that continue growing from one year to the next use this stored food to begin new growth in the spring. Plants that grow in dry areas often have roots that store water.

Root tissues also can perform functions such as absorbing oxygen that is used in the process of respiration. Because water does not contain as much oxygen as air does, plants that grow with their roots in water might not be able to absorb enough oxygen. Some swamp plants have roots that grow partially out of the water and take in oxygen from the air. In order to perform all these functions, the root systems of plants must be large.

Reading Check *What are several functions of roots in plants?*

Vascular Tissue Three tissues usually make up the vascular system in a seed plant. **Xylem** (ZI lum) tissue is made up of hollow, tubular cells that are stacked one on top of the other to form a structure called a vessel. These vessels transport water and dissolved substances from the roots throughout the plant. The thick cell walls of xylem are also important because they help support the plant.

Phloem (FLOH em) is a plant tissue also made up of tubular cells that are stacked to form structures called tubes. Tubes are different from vessels. Phloem tubes move food from where it is made to other parts of the plant where it is used or stored.

In some plants, a cambium is between xylem and phloem. **Cambium** (KAM bee um) is a tissue that produces most of the new xylem and phloem cells. The growth of this new xylem and phloem increases the thickness of stems and roots. All three tissues are illustrated in **Figure 17.**

Vascular Systems Plants have vascular tissue, and you have a vascular system. Your vascular system transports oxygen, food, and wastes through blood vessels. Instead of xylem and phloem, your blood vessels include veins and arteries. In your Science Journal write a paragraph describing the difference between veins and arteries.

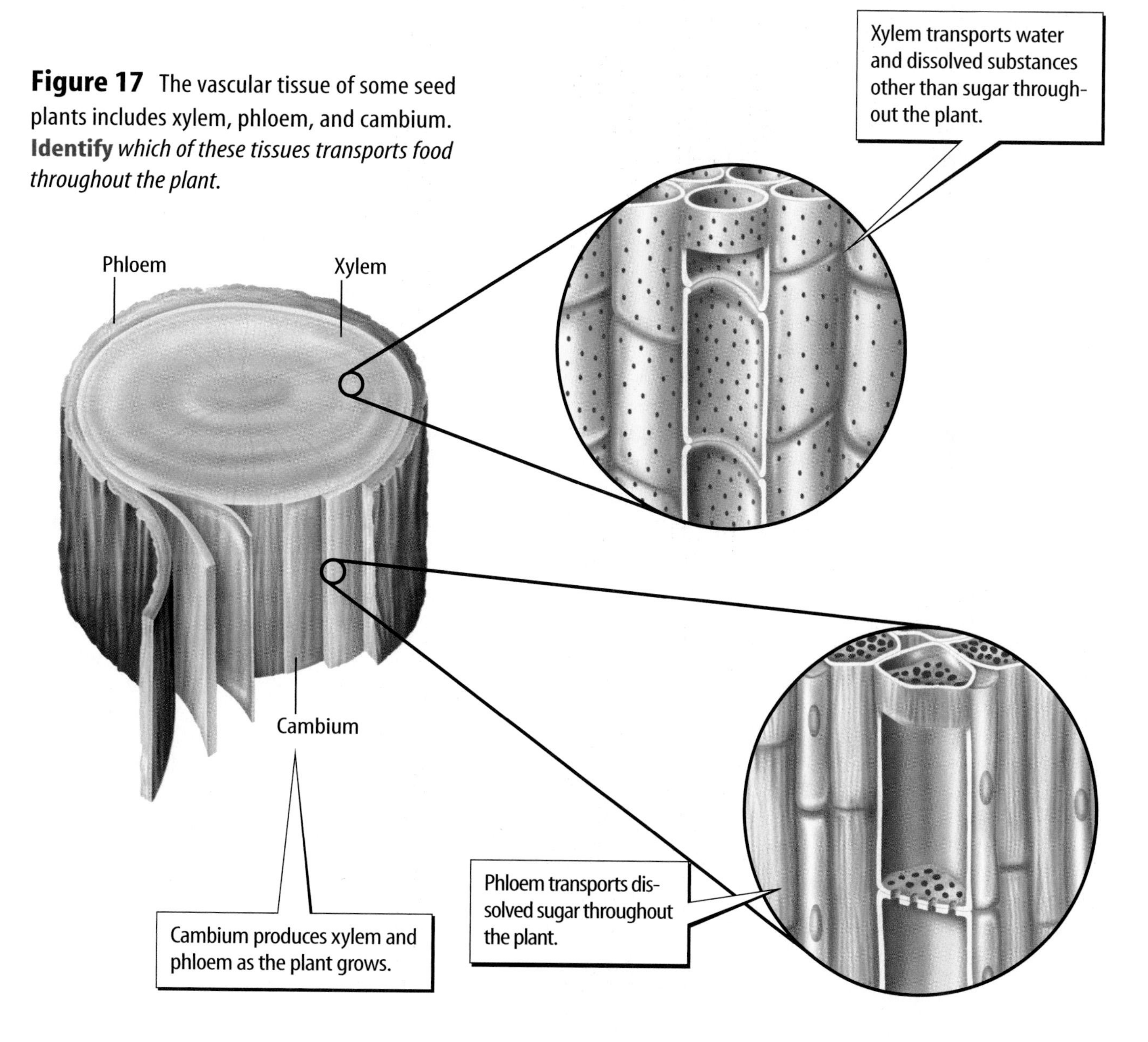

Figure 17 The vascular tissue of some seed plants includes xylem, phloem, and cambium.
Identify *which of these tissues transports food throughout the plant.*

Figure 18 The gymnosperms include four divisions of plants.

About 100 species of cycads exist today. Only one genus is native to the United States.

The ginkgoes are represented by one living species. Ginkgoes lose their leaves in the fall.
Explain *how this is different from most gymnosperms.*

Conifers are the largest, most diverse division. Most conifers are evergreen plants, such as this ponderosa pine (above).

More than half of the 70 species of gnetophytes, such as this joint fir, are in one genus.

Gymnosperms

The oldest trees alive are gymnosperms. A bristlecone pine tree in the White Mountains of eastern California is estimated to be 4,900 years old. **Gymnosperms** are vascular plants that produce seeds that are not protected by fruit. The word *gymnosperm* comes from the Greek language and means "naked seed." Another characteristic of gymnosperms is that they do not have flowers. Leaves of most gymnosperms are needlelike or scalelike. Many gymnosperms are called evergreens because some green leaves always remain on their branches.

Four divisions of plants—conifers, cycads, ginkgoes, and gnetophytes (NE tuh fites)—are classified as gymnosperms. **Figure 18** shows examples of the four divisions. You are probably most familiar with the division Coniferophyta (kuh NIH fur uh fi tuh), the conifers. Pines, firs, spruces, redwoods, and junipers belong to this division. It contains the greatest number of gymnosperm species. All conifers produce two types of cones—male and female. Both types usually are found on the same plant. Cones are the reproductive structures of conifers. Seeds develop on the female cone but not on the male cone.

What is the importance of cones to gymnosperms?

Angiosperms

When people are asked to name a plant, most name an angiosperm. An **angiosperm** is a vascular plant that flowers and produces fruits with one or more seeds, such as the peaches shown in **Figure 19.** The fruit develops from a part or parts of one or more flowers. Angiosperms are familiar plants no matter where you live. They grow in parks, fields, forests, jungles, deserts, freshwater, salt water, and in the cracks of sidewalks. You might see them dangling from wires or other plants, and one species of orchid even grows underground. Angiosperms make up the plant division Anthophyta (AN thoh fi tuh). More than half of the known plant species belong to this division.

Flowers The flowers of angiosperms vary in size, shape, and color. Duckweed, an aquatic plant, has a flower that is only 0.1 mm long. A plant in Indonesia has a flower that is nearly 1 m in diameter and can weigh 9 kg. Nearly every color can be found in some flower, although some people would not include black. Multicolored flowers are common. Some plants have flowers that are not recognized easily as flowers, such as the flowers of ash trees, shown below.

Some flower parts develop into a fruit. Most fruits contain seeds, like an apple, or have seeds on their surface, like a strawberry. If you think all fruits are juicy and sweet, there are some that are not. The fruit of the vanilla orchid, as shown to the right, contains seeds and is dry.

Angiosperms are divided into two groups—the monocots and the dicots—shortened forms of the words *monocotyledon* (mah nuh kah tuh LEE dun) and *dicotyledon* (di kah tuh LEE dun).

Figure 19 Angiosperms have a wide variety of flowers and fruits.

The fruit of the vanilla orchid is the source of vanilla flavoring.

The flowers and fruit of a peach tree are typical of many angiosperms.

Ash flowers are not large and colorful. Their fruits are small and dry.

Monocots and Dicots A cotyledon is part of a seed often used for food storage. The prefix *mono* means "one," and *di* means "two." Therefore, **monocots** have one cotyledon inside their seeds and **dicots** have two. The flowers, leaves, and stems of monocots and dicots are shown in **Figure 20.**

Many important foods come from monocots, including corn, rice, wheat, and barley. If you eat bananas, pineapple, or dates, you are eating fruit from monocots. Lilies and orchids also are monocots.

Dicots also produce familiar foods such as peanuts, green beans, peas, apples, and oranges. You might have rested in the shade of a dicot tree. Most shade trees, such as maple, oak, and elm, are dicots.

Figure 20 By observing a monocot and a dicot, you can determine their plant characteristics.

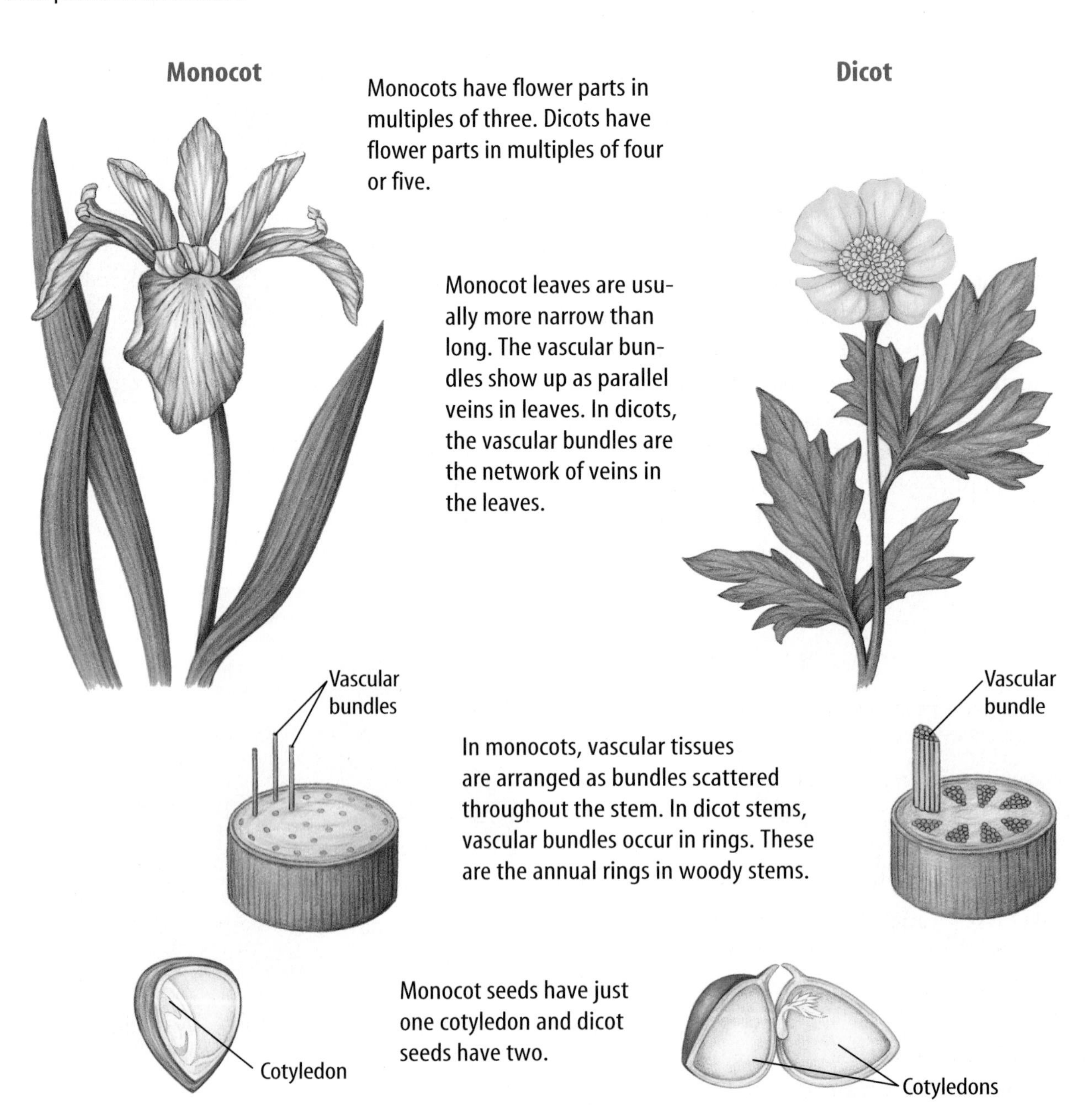

Petunias

Parsley

Pecan tree

Figure 21 Life cycles of angiosperms include annuals, biennials, and perennials. Petunias, which are annuals, complete their life cycle in one year. Parsley plants, which are biennials, do not produce flowers and seeds the first year. Perennials, such as the pecan tree, flower and produce fruits year after year.

Life Cycles of Angiosperms Flowering plants vary greatly in appearance. Their life cycles are as varied as the kinds of plants, as shown in **Figure 21.** Some angiosperms grow from seeds to mature plants with their own seeds in less than a month. The life cycles of other plants can take as long as a century. If a plant's life cycle is completed within one year, it is called an annual. These plants must be grown from seeds each year.

Plants called biennials (bi EH nee ulz) complete their life cycles within two years. Biennials such as parsley store a large amount of food in an underground root or stem for growth in the second year. Biennials produce flowers and seeds only during the second year of growth. Angiosperms that take more than two years to grow to maturity are called perennials. Herbaceous perennials such as peonies appear to die each winter but grow and produce flowers each spring. Woody perennials such as fruit trees produce flowers and fruits on stems that survive for many years.

Importance of Seed Plants

What would a day at school be like without seed plants? One of the first things you'd notice is the lack of paper and books. Paper is made from wood pulp that comes from trees, which are seed plants. Are the desks and chairs at your school made of wood? They would need to be made of something else if no seed plants existed. Clothing that is made from cotton would not exist because cotton comes from seed plants. At lunchtime, you would have trouble finding something to eat. Bread, fruits, and potato chips all come from seed plants. Milk, hamburgers, and hot dogs all come from animals that eat seed plants. Unless you like to eat plants such as mosses and ferns, you'd go hungry. Without seed plants, your day at school would be different.

Topic: Renewable Resources
Visit green.msscience.com for Web links to information and recent news or magazine articles about the timber industry's efforts to replant trees.

Activity List in your Science Journal the species of trees that are planted and some of their uses.

Table 1 Some Products of Seed Plants			
From Gymnosperms		**From Angiosperms**	
lumber, paper, soap, varnish, paints, waxes, perfumes, edible pine nuts, medicines		foods, sugar, chocolate, cotton cloth, linen, rubber, vegetable oils, perfumes, medicines, cinnamon, flavorings, dyes, lumber	

Products of Seed Plants Conifers are the most economically important gymnosperms. Most wood used for construction and for paper production comes from conifers. Resin, a waxy substance secreted by conifers, is used to make chemicals found in soap, paint, varnish, and some medicines.

The most economically important plants on Earth are the angiosperms. They form the basis of the diets of most animals. Angiosperms were the first plants that humans grew. They included grains, such as barley and wheat, and legumes, such as peas and lentils. Angiosperms are also the source of many of the fibers used in clothing. Besides cotton, linen fabrics come from plant fibers. **Table 1** shows just a few of the products of angiosperms and gymnosperms.

section 3 review

Summary

Characteristics of Seed Plants

- Leaves are organs in which photosynthesis takes place.
- Stems support leaves and branches and contain vascular tissues.
- Roots absorb water and nutrients from soil.

Gymnosperms

- Gymnosperms do not have flowers and produce seeds that are not protected by a fruit.

Angiosperms

- Angiosperms produce flowers that develop into a fruit with seeds.

Importance of Seed Plants

- The diets of most animals are based on angiosperms.

Self Check

1. **List** four characteristics common to all seed plants.
2. **Compare and contrast** the characteristics of gymnosperms and angiosperms.
3. **Classify** a flower with five petals as a monocot or a dicot.
4. **Explain** why the root system might be the largest part of a plant.
5. **Think Critically** The cuticle and epidermis of leaves are transparent. If they weren't, what might be the result?

Applying Skills

6. **Form a hypothesis** about what substance or substances are produced in palisade cells but not in xylem cells.

Science Online green.msscience.com/self_check_quiz

Identifying Conifers

How can you tell a pine from a spruce or a cedar from a juniper? One way is to observe their leaves. The leaves of most conifers are either needlelike—shaped like needles—or scale-like—shaped like the scales on a fish. Examine and identify some conifer branches using the key to the right.

Real-World Question

How can leaves be used to classify conifers?

Goals

- **Identify** the difference between needlelike and scalelike leaves.
- **Classify** conifers according to their leaves.

Materials

short branches of the following conifers:

pine	Douglas fir	redwood
cedar	hemlock	arborvitae
spruce	fir	juniper

**illllustrations of the conifers above*

**Alternate materials*

Safety Precautions

Wash your hands after handling leaves.

Procedure

1. **Observe** the leaves or illustrations of each conifer, then use the key to identify it.
2. **Write** the number and name of each conifer you identify in your Science Journal.

Conclude and Apply

1. **Name** two traits of hemlock leaves.
2. **Compare and contrast** pine and cedar leaves.

Key to Classifying Conifer Leaves

1. All leaves are needlelike.
   a. yes, go to 2
   b. no, go to 8
2. Needles are in clusters.
   a. yes, go to 3
   b. no, go to 4
3. Clusters contain two, three, or five needles.
   a. yes, pine
   b. no, cedar
4. Needles grow on all sides of the stem.
   a. yes, go to 5
   b. no, go to 7
5. Needles grow from a woody peg.
   a. yes, spruce
   b. no, go to 6
6. Needles appear to grow from the branch.
   a. yes, Douglas fir
   b. no, hemlock
7. Most of the needles grow upward.
   a. yes, fir
   b. no, redwood
8. All the leaves are scalelike but not prickly.
   a. yes, arborvitae
   b. no, juniper

Communicating Your Data

Use the key above to identify conifers growing on your school grounds. Draw and label a map that locates these conifers. Post the map in your school. **For more help, refer to the Science Skill Handbook.**

Use the Internet

Plants as Medicine

Goals

- **Identify** two plants that can be used as a treatment for illness or as a supplement to support good health.
- **Research** the cultural and historical use of each of the two selected plants as medical treatments.
- **Review** multiple sources to understand the effectiveness of each of the two selected plants as a medical treatment.
- **Compare and contrast** the research and form a hypothesis about the medicinal effectiveness of each of the two plants.

Data Source

Science Online

Visit green.msscience.com/internet_lab for more information about plants that can be used for maintaining good health and for data collected by other students.

Real-World Question

You may have read about using peppermint to relieve an upset stomach, or taking *Echinacea* to boost your immune system and fight off illness. But did you know that pioneers brewed a cough medicine from lemon mint? In this lab, you will explore plants and their historical use in treating illness, and the benefits and risks associated with using plants as medicine. How are plants used in maintaining good health?

Echinacea

Make a Plan

1. Search for information about plants that are used as medicine and identify two plants to investigate.
2. **Research** how these plants are currently recommended for use as medicine or to promote good health. Find out how each has been used historically.
3. **Explore** how other cultures used these plants as a medicine.

Mentha

Follow Your Plan

1. Make sure your teacher approves your plan before you start.
2. **Record** data you collect about each plant in your Science Journal.

Analyze Your Data

1. **Write** a description of how different cultures have used each plant as medicine.
2. How have the plants you investigated been used as medicine historically?
3. **Record** all the uses suggested by different sources for each plant.
4. **Record** the side effects of using each plant as a treatment.

Conclude and Apply

1. After conducting your research, what do you think are the benefits and drawbacks of using these plants as alternative medicines?
2. **Describe** any conflicting information about using each of these plants as medicine.
3. Based on your analysis, would you recommend the use of each of these two plants to treat illness or promote good health? Why or why not?
4. What would you say to someone who was thinking about using any plant-based, over-the-counter, herbal supplement?

Communicating Your Data

Find this lab using the link below. Post your data for the two plants you investigated in the tables provided. **Compare** your data to those of other students. Review data that other students have entered about other plants that can be used as medicine.

Science Online

green.msscience.com/internet_lab

Monarda

Oops! Accidents in SCIENCE

SOMETIMES GREAT DISCOVERIES HAPPEN BY ACCIDENT!

A LOOPY Idea Inspires a "Fastenating" Invention

A wild cocklebur plant inspired the hook-and-loop fastener.

Scientists often spend countless hours in the laboratory dreaming up useful inventions. Sometimes, however, the best ideas hit them in unexpected places at unexpected times. That's why scientists are constantly on the lookout for things that spark their curiosity.

One day in 1948, a Swiss inventor named George deMestral strolled through a field with his dog. When they returned home, deMestral discovered that the dog's fur was covered with cocklebursts, parts of a prickly plant. These burs were also stuck to deMestral's jacket and pants. Curious about what made the burs so sticky, the inventor examined one under a microscope.

DeMestral noticed that the cocklebur was covered with lots of tiny hooks. By clinging to animal fur and fabric, this plant is carried to other places. While studying these burs, he got the idea to invent a new kind of fastener that could do the work of buttons, snaps, zippers, and laces—but better!

After years of experimentation, deMestral came up with a strong, durable hook-and-loop fastener made of two strips of nylon fabric. One strip has thousands of small, stiff hooks; the other strip is covered with soft, tiny loops. Today, this hook-and-loop fastening tape is used on shoes and sneakers, watchbands, hospital equipment, space suits, clothing, book bags, and more. You may have one of those hook-and-loop fasteners somewhere on you right now. They're the ones that go rrrrrrrrip when you open them.

So, if you ever get a fresh idea that clings to your mind like a hook to a loop, stick with it and experiment! Who knows? It may lead to a fabulous invention that changes the world!

This photo provides a close-up view of a hook-and-loop fastener.

List Make a list of ten ways hook-and-loop tape is used today. Think of three new uses for it. Since you can buy strips of hook-and-loop fastening tape in most hardware and fabric stores, try out some of your favorite ideas.

For more information, visit green.msscience.com/oops

Reviewing Main Ideas

Section 1 An Overview of Plants

1. Plants are made up of eukaryotic cells and vary greatly in size and shape.
2. Plants usually have some form of leaves, stems, and roots.
3. As plants evolved from aquatic to land environments, changes occurred in how they reproduced, supported themselves, and moved substances from one part of the plant to another.
4. The plant kingdom is classified into groups called divisions.

Section 2 Seedless Plants

1. Seedless plants include nonvascular and vascular types.
2. Most seedless nonvascular plants have no true leaves, stems, or roots. Reproduction usually is by spores.
3. Seedless vascular plants have vascular tissues that move substances throughout the plant. These plants may reproduce by spores.
4. Many ancient forms of these plants underwent a process that resulted in the formation of coal.

Section 3 Seed Plants

1. Seed plants are adapted to survive in nearly every environment on Earth.
2. Seed plants produce seeds and have vascular tissue, stems, roots, and leaves.
3. The two major groups of seed plants are gymnosperms and angiosperms. Gymnosperms generally have needlelike leaves and some type of cone. Angiosperms are plants that flower and are classified as monocots or dicots.
4. Seed plants are the most economically important plants on Earth.

Visualizing Main Ideas

Copy and complete the following concept map about the seed plants.

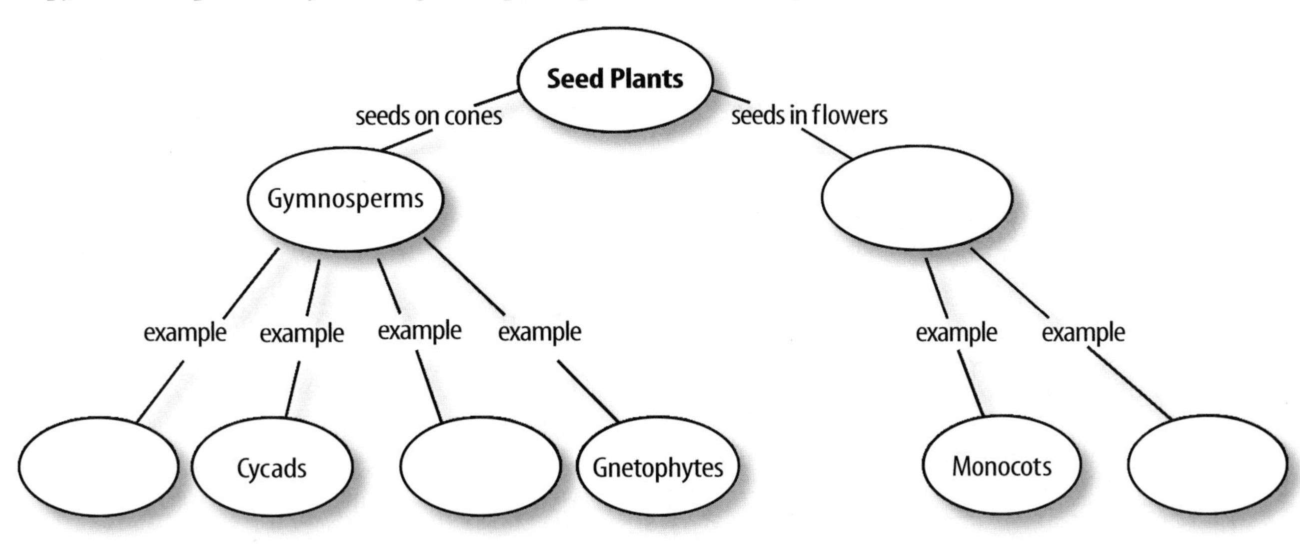

chapter 17 Review

Using Vocabulary

angiosperm p. 517	nonvascular plant p. 505
cambium p. 515	phloem p. 515
cellulose p. 502	pioneer species p. 507
cuticle p. 502	rhizoid p. 506
dicot p. 518	stomata p. 513
guard cell p. 513	vascular plant p. 505
gymnosperm p. 516	xylem p. 515
monocot p. 518	

Complete each analogy by providing the missing vocabulary word.

1. Angiosperm is to flower as _________ is to cone.
2. Dicot is to two seed leaves as _________ is to one seed leaf.
3. Root is to fern as _________ is to moss.
4. Phloem is to food transport as _________ is to water transport.
5. Vascular plant is to horsetail as _________ is to liverwort.
6. Cellulose is to support as _________ is to protect.
7. Fuel is to ferns as _________ is to bryophytes.
8. Cuticle is to wax as _________ is to fibers.

Checking Concepts

Choose the word or phrase that best answers the question.

9. Which of the following is a seedless vascular plant?
   A) moss **C)** horsetail
   B) liverwort **D)** pine
10. What are the small openings in the surface of a leaf surrounded by guard cells called?
    A) stomata **C)** rhizoids
    B) cuticles **D)** angiosperms
11. What are the plant structures that anchor the plant called?
    A) stems **C)** roots
    B) leaves **D)** guard cells
12. Where is most of a plant's new xylem and phloem produced?
    A) guard cell **C)** stomata
    B) cambium **D)** cuticle
13. What group has plants that are only a few cells thick?
    A) gymnosperms **C)** ferns
    B) cycads **D)** mosses
14. The oval plant parts shown to the right are found only in which plant group?

    A) nonvascular **C)** gymnosperms
    B) seedless **D)** angiosperms
15. What kinds of plants have structures that move water and other substances?
    A) vascular **C)** nonvascular
    B) protist **D)** bacterial
16. In what part of a leaf does most photosynthesis occur?
    A) epidermis **C)** stomata
    B) cuticle **D)** palisade layer
17. Which one of the following do ferns have?
    A) cones **C)** spores
    B) rhizoids **D)** seeds
18. Which of these is an advantage to life on land for plants?
    A) more direct sunlight
    B) less carbon dioxide
    C) greater space to grow
    D) less competition for food

Science Online green.msscience.com/vocabulary_puzzlemaker

Thinking Critically

19. Predict what might happen if a land plant's waxy cuticle was destroyed.

20. Draw Conclusions On a walk through the woods with a friend, you find a plant neither of you has seen before. The plant has green leaves and yellow flowers. Your friend says it is a vascular plant. How does your friend know this?

21. Infer Plants called succulents store large amounts of water in their leaves, stems, and roots. In what environments would you expect to find succulents growing naturally?

22. Explain why mosses usually are found in moist areas.

23. Recognize Cause and Effect How do pioneer species change environments so that other plants can grow there?

24. Concept Map Copy and complete this map for the seedless plants of the plant kingdom.

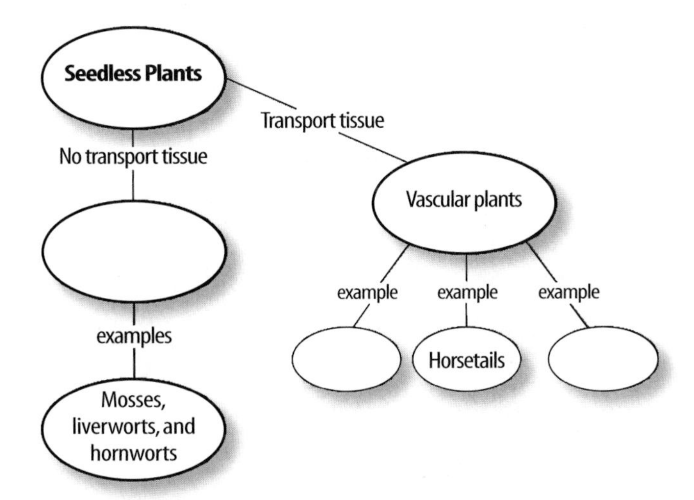

25. Interpret Scientific Illustrations Using **Figure 20** in this chapter, compare and contrast the number of cotyledons, bundle arrangement in the stem, veins in leaves, and number of flower parts for monocots and dicots.

26. Sequence Put the following events in order to show how coal is formed from plants: *living seedless plants, coal is formed, dead seedless plants decay,* and *peat is formed.*

27. Predict what would happen if a ring of bark and camium layer were removed from around the trunk of a tree.

Performance Activities

28. Poem Choose a topic in this chapter that interests you. Look it up in a reference book, in an encyclopedia, or on a CD-ROM. Write a poem to share what you learn.

29. Display Use dried plant material, photos, drawings, or other materials to make a poster describing the form and function of roots, stems, and leaves.

Applying Math

Use the table below to answer questions 30–32.

Number of Stomata (per mm^2)

Plant	Upper Surface	Lower Surface
Pine	50	71
Bean	40	281
Fir	0	228
Tomato	12	13

30. Gas Exchange What do the data in this table tell you about where gas exchange occurs in the leaf of each plant species?

31. Compare Leaf Surfaces Make two circle graphs—upper surface and lower surface—using the table above.

32. Guard Cells On average, how many guard cells are found on the lower surface of a bean leaf?

chapter 17 Standardized Test Practice

Part 1 Multiple Choice

Record your answers on the answer sheet provided by your teacher or on a sheet of paper.

1. Which of the following do plants use to photosynthesize?
   A. blood
   B. iron
   C. chlorophyll
   D. cellulose

2. Which of the following describes the function of the central vacuole in plant cells?
   A. It helps in reproduction.
   B. It helps regulate water content.
   C. It plays a key role in photosynthesis.
   D. It stores food.

Use the illustration below to answer questions 3 and 4.

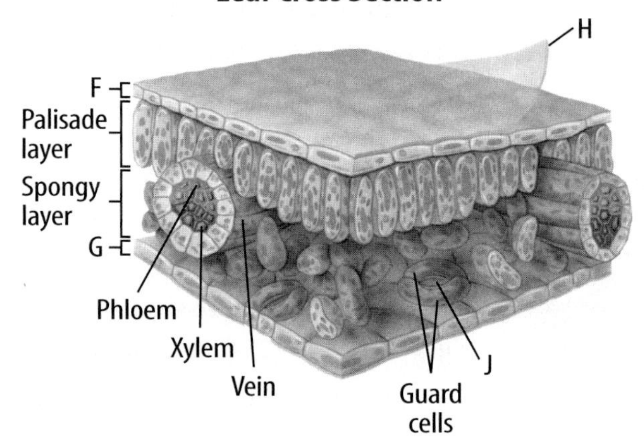

3. In the leaf cross section, what is indicated by H?
   A. upper epidermis
   B. cuticle
   C. stoma
   D. lower epidermis

4. What flows through the structure indicated by J?
   A. water only
   B. carbon dioxide and water only
   C. oxygen and carbon dioxide only
   D. water, carbon dioxide, and oxygen

5. In seed plants, vascular tissue refers to which of the following?
   A. xylem and phloem only
   B. xylem only
   C. phloem only
   D. xylem, phloem, and cambium

Use the illustration below to answer questions 6 and 7.

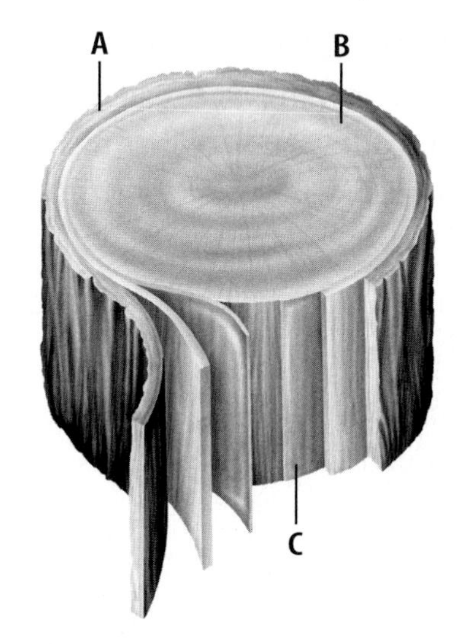

6. What is the function of the structure labeled C?
   A. It transports nutrients throughout the plant.
   B. It produces new xylem and phloem.
   C. It transports water from the roots to other parts of the plant.
   D. It absorbs water from outside the plant.

7. What type of vascular tissue is indicated by B?
   A. xylem
   B. cambium
   C. phloem
   D. cellulose

Test-Taking Tip

Eliminate Answer Choices If you don't know the answer to a multiple-choice question, eliminate as many incorrect choices as possible. Mark your best guess from the remaining answers before moving on to the next question.

Part 2 | Short Response/Grid In

Record your answers on the answer sheet provided by your teacher or on a sheet of paper.

Use the two illustrations below to answer questions 8–10.

8. Identify the flowers shown above as a monocot or a dicot. Explain the differences between the flowers of monocots and dicots.

9. Give three examples of plants represented by Plant A.

10. Give three examples of plants represented by Plant B.

11. How are plants that live on land able to conserve water?

12. Explain why reproductive adaptations were necessary in order for plants to survive on land.

13. You are hiking through a dense forest area and notice some unusual plants growing on the trunk of a tall tree. The plants are no taller than about 3 cm and have delicate stalks. They do not appear to have flowers. Based on this information, what type of plants would you say you found?

14. What is a conifer? To which major group of plants does it belong?

Part 3 | Open Ended

Record your answers on a sheet of paper.

Use the two diagrams below to answer questions 15–16.

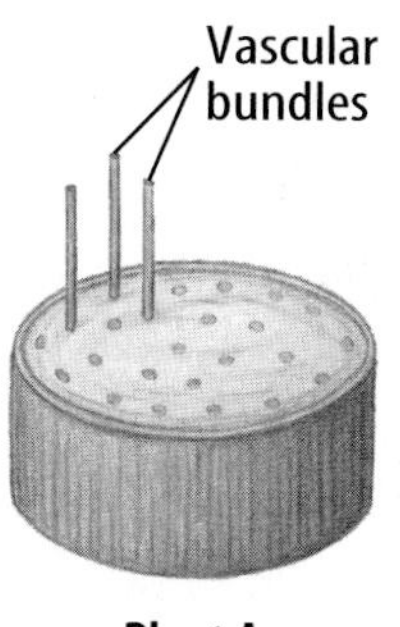

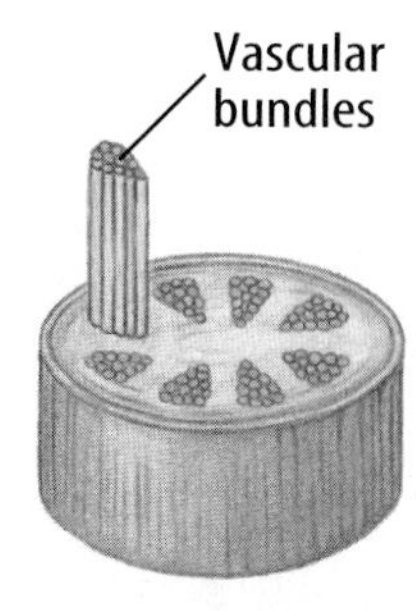

15. Two plants, A and B, have stem cross sections as shown in the diagrams above. What does the different vascular bundle arrangement tell you about each plant?

16. Draw what the seed from each plant would look like.

17. Create a diagram that describes the life cycle of an annual angiosperm.

18. Discuss the importance of plants in your daily life. Give examples of plants or plant products that you use or consume regularly.

19. Compare and contrast vascular and nonvascular plants. Include examples of each type of plant.

20. Describe the group of plants known as the seedless vascular plants. How do these plants reproduce without seeds?

21. Explain what peat is and how it is formed. How is peat used today?

22. How would our knowledge of ancient plants be different if the fossil record for plants was as plentiful as it is for animals?

chapter 18

Interactions of Living Things

chapter preview

sections

1. The Environment
   Lab Delicately Balanced Ecosystems
2. Interactions Among Living Organisms
3. Matter and Energy
   Lab Identifying a Limiting Factor

Virtual Lab How is energy transferred through a community of organisms?

Interactions at a Waterhole

How many different kinds of animals can you see in the photo? How are the animals interacting with each other? Animals and other organisms in an area not only interact with each other, but with the nonliving factors of the area as well. What non-living factors can you identify?

Science Journal Write a list of things you interact with each day.

Start-Up Activities

Space and Interactions

Imagine that you are in a crowded elevator. Everyone jostles and bumps each other. The temperature increases and ordinary noises seem louder. Like people in an elevator, plants and animals in an area interact. How does the amount of space available to each organism affect its interaction with other organisms?

1. Use a meterstick to measure the length and width of the classroom.
2. Multiply the length by the width to find the area of the room in square meters.
3. Count the number of individuals in your class. Divide the area of the classroom by the number of individuals. In your Science Journal, record how much space each person has.
4. **Think Critically** Write a prediction in your Science Journal about what might happen if the number of students in your classroom doubled.

Preview this chapter's content and activities at green.msscience.com

Biotic and Abiotic Make the following Foldable to help you understand the cause and effect relationship of biotic and abiotic things.

STEP 1 **Fold** a vertical sheet of paper in half from top to bottom.

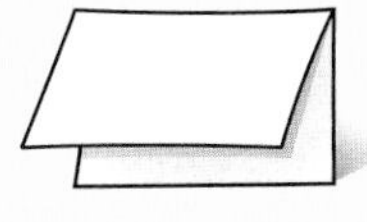

STEP 2 **Fold** in half from side to side with the fold at the top.

STEP 3 **Unfold** the paper once. **Cut** only the fold of the top flap to make two tabs.

STEP 4 **Turn** the paper vertically and **label** the front tabs as shown.

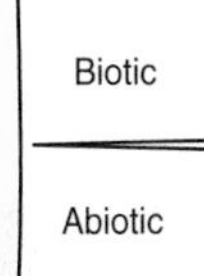

Illustrate and Label Before you read the chapter, list examples of biotic and abiotic things around you on the tabs. As you read, write about each.

section 1

The Environment

as you read

What You'll Learn

- **Identify** biotic and abiotic factors in an ecosystem.
- **Describe** the different levels of biological organization.
- **Explain** how ecology and the environment are related.

Why It's Important

Abiotic and biotic factors interact to make up your ecosystem. The quality of your ecosystem can affect your health. Your actions can affect the health of the ecosystem.

Review Vocabulary
climate: the average weather conditions of an area over time

New Vocabulary
- ecology
- abiotic factor
- biotic factor
- population
- community
- ecosystem
- biosphere

Ecology

All organisms, from the smallest bacteria to a blue whale, interact with their environment. **Ecology** is the study of the interactions among organisms and their environment. Ecologists, such as the one in **Figure 1,** are scientists who study these relationships. Ecologists organize the environmental factors that influence organisms into two groups—nonliving and living or once-living. **Abiotic** (ay bi AH tihk) **factors** are the nonliving parts of the environment. Living or once-living organisms in the environment are called **biotic** (bi AH tihk) **factors.**

Reading Check *Why is a rotting log considered a biotic factor in the environment?*

Abiotic Factors

In any environment, birds, insects, and other living things, including humans, depend on one another for food and shelter. They also depend on the abiotic factors that surround them, such as water, sunlight, temperature, air, and soil. All of these factors and others are important in determining which organisms are able to live in a particular environment.

Figure 1 Ecologists study biotic and abiotic factors in an environment and the relationships among them. Many times, ecologists must travel to specific environments to examine the organisms that live there.

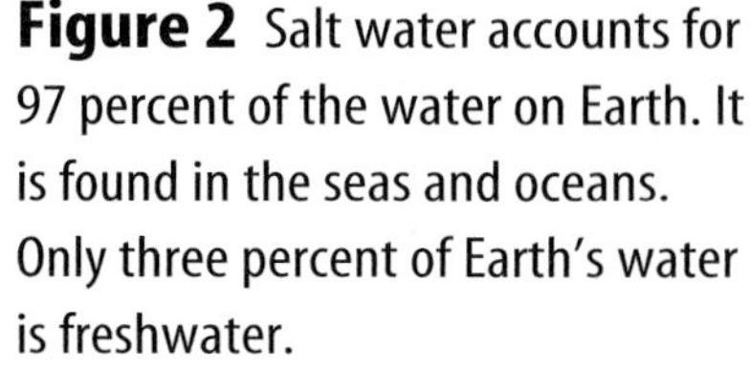

Figure 2 Salt water accounts for 97 percent of the water on Earth. It is found in the seas and oceans. Only three percent of Earth's water is freshwater.

Water All living organisms need water to survive. The bodies of most organisms are 50 percent to 95 percent water. Water is an important part of the cytoplasm in cells and the fluid that surrounds cells. Respiration, photosynthesis, digestion, and other important life processes can only occur in the presence of water.

More than 95 percent of Earth's surface water is found in the oceans. The saltwater environment in the oceans is home to a vast number of species. Freshwater environments, like the one in **Figure 2,** also support thousands of types of organisms.

Light and Temperature The abiotic factors of light and temperature also affect the environment. The availability of sunlight is a major factor in determining where green plants and other photosynthetic organisms live, as shown in **Figure 3.** By the process of photosynthesis, energy from the Sun is changed into chemical energy that is used for life processes. Most green algae live near the water's surface where sunlight can penetrate. In dense forests where little sunlight penetrates through to the forest floor, very few photosynthetic plants grow.

The temperature of a region also determines which plants and animals can live there. Some areas of the world have a fairly consistent temperature year round, but other areas have seasons during which temperatures vary. Water environments throughout the world also have widely varied temperatures. Living organisms are found in the freezing cold Arctic, in the extremely hot water near ocean vents, and at almost every temperature in between.

Figure 3 Flowers that grow on the forest floor, such as these bluebells, grow during the spring when they receive the most sunlight. **Infer** *why there is little sunlight on the forest floor during the summer.*

Figure 4 Air pollution can come from many different sources. Air quality in an area affects the health and survival of the species that live there.

Air Pollution Engineer Have you ever wondered who monitors the air you breathe? Air pollution engineers are people who make sure air quality standards are being met. They also design new technologies to reduce air pollution, such as improved machinery, filters, and ventilation systems, to try and solve problems like "sick building syndrome".

Air Although you can't see the air that surrounds you, it has an impact on the lives of most species. Air is composed of a mixture of gases including nitrogen, oxygen, and carbon dioxide. Most plants and animals depend on the gases in air for respiration. The atmosphere is the layer of gases and airborne particles that surrounds Earth. Polluted air, like the air in **Figure 4,** can cause the species in an area to change, move, or die off.

Clouds and weather occur in the bottom 8 km to 16 km of the atmosphere. All species are affected by the weather in the area where they live. The ozone layer is 20 km to 50 km above Earth's surface and protects organisms from harmful radiation from the Sun. Air pressure, which is the weight of air pressing down on Earth, changes depending on altitude. Higher altitudes have less air pressure. Few organisms live at extreme air pressures.

How does pollution in the atmosphere affect the species in an area?

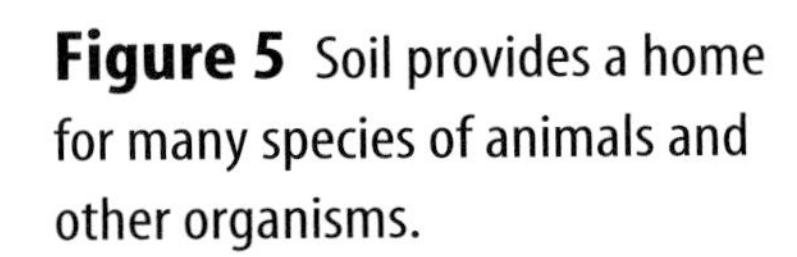

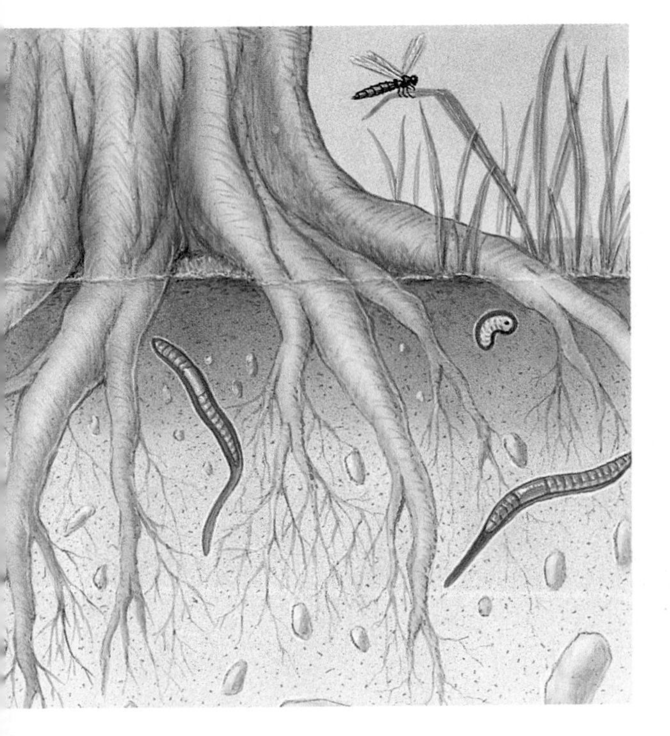

Figure 5 Soil provides a home for many species of animals and other organisms.

Soil From one enviroment to another, soil, as shown in **Figure 5,** can vary greatly. Soil type is determined by the amounts of sand, silt, and clay it contains. Various kinds of soil contain different amounts of nutrients, minerals, and moisture. Different plants need different kinds of soil. Because the types of plants in an area help determine which other organisms can survive in that area, soil affects every organism in an environment.

Biotic Factors

Abiotic factors do not provide everything an organism needs for survival. Organisms depend on other organisms for food, shelter, protection, and reproduction. How organisms interact with one another and with abiotic factors can be described in an organized way.

Levels of Organization The living world is highly organized. Atoms are arranged into molecules, which in turn might be organized into cells. Cells form tissues, tissues form organs, and organs form organ systems. Together, organ systems form organisms. Biotic and abiotic factors also can be arranged into levels of biological organization, as shown in **Figure 6.**

Figure 6 The living world is organized in levels.

An organism is one individual from a population.

Population

All of the individuals of one species that live in the same area at the same time make up a population.

The populations of different species that interact in some way are called a community.

All of the communities in an area and the abiotic factors they interact with make up an ecosystem.

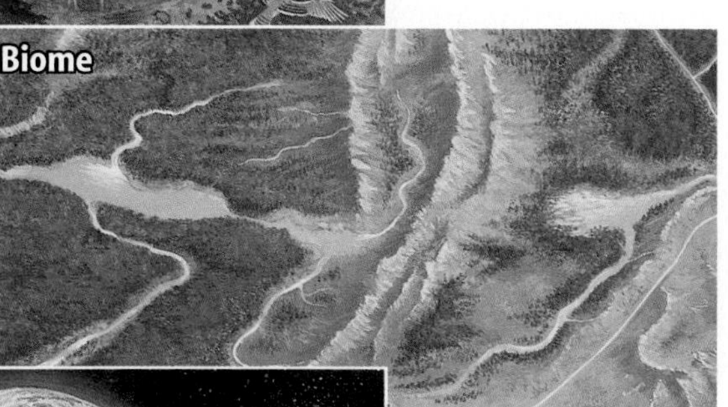

A biome is a large region with plants and animals well adapted to the soil and climate of the region.

The level of biological organization that is made up of all the ecosystems on Earth is the biosphere.

Figure 7 Members of a penguin population compete for resources. **Infer** *what resources these penguins might be using.*

Populations All the members of one species that live together make up a **population.** For example, all of the humans living on Earth at the same time make up a population. Part of a population of penguins is shown in **Figure 7.** Members of a population compete for food, water, mates, and space. The resources of the environment and the ways the organisms use these resources determine how large a population can become.

Communities Most populations of organisms do not live alone. They live and interact with populations of other types of organisms. Groups of populations that interact with each other in a given area form a **community.** For example, a population of penguins and all of the species that they interact with form a community. Populations of organisms in a community depend on each other for food, shelter, and other needs.

Ecosystems In addition to interactions among populations, ecologists also study interactions among populations and their physical surroundings. An **ecosystem** is made up of a biotic community and the abiotic factors that affect it. Examples of ecosystems include coral reefs, forests, and ponds. You will learn more about the interactions that occur in ecosystems later in this chapter.

Biomes Scientists divide Earth into different regions called biomes. A biome (BI ohm) is a large region with plant and animal groups that are well adapted to the soil and climate of the region. Many different ecosystems are found in a biome. Examples of biomes include tundra, as shown in **Figure 8,** tropical rain forests, and grasslands.

Topic: Earth's Biomes
Visit green.msscience.com for Web links to information about Earth's different biomes.

Activity Select one of Earth's biomes and research what plants, animals, and other organisms live there. Prepare a display that includes pictures and text about your selected biome.

Figure 8 Biomes contain many different ecosystems. This mountaintop ecosystem is part of the alpine tundra biome.

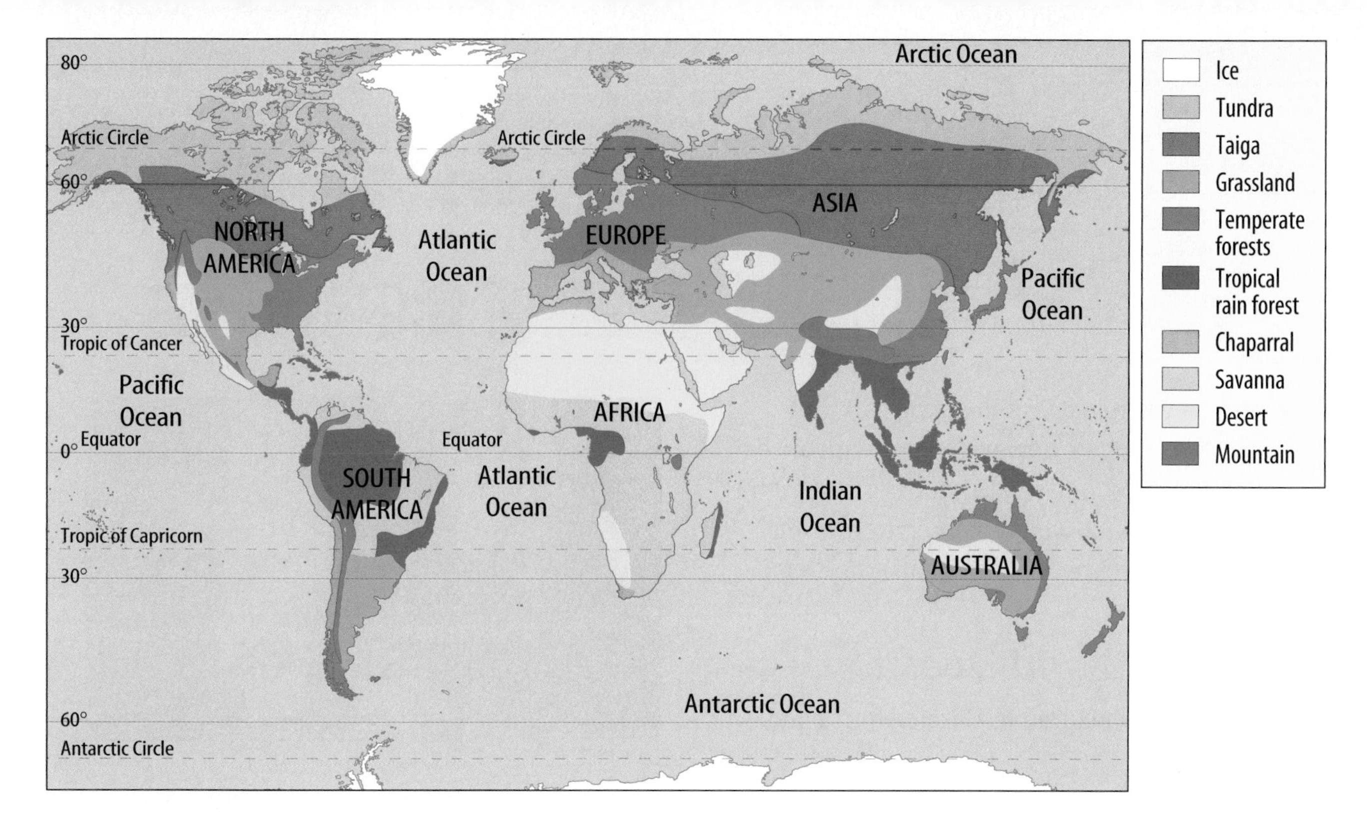

The Biosphere Where do all of Earth's organisms live? Living things can be found 11,000 m deep in the ocean, 9,000 m high on mountains, and 4.5 km high in Earth's atmosphere. The part of Earth that supports life is the **biosphere** (BI uh sfihr). It includes the top part of Earth's crust, all the waters that cover Earth's surface, the surrounding atmosphere, and all biomes, including those in **Figure 9.**

Figure 9 This map shows some of the major biomes of the world. **Determine** *what biome you live in.*

section 1 review

Summary

Abiotic Factors

- Organisms interact with and depend on factors in their environments.
- More than 95 percent of Earth's surface is water.
- The amount of sunlight determines where green plants can grow.
- Temperature determines which organisms can live in a region.
- Air is needed by most organisms. Polluted air can harm organisms.
- Soil can determine organisms in an area.

Biotic Factors

- Organisms depend on other organisms for food, shelter, protection, and reproduction.
- The living world is organized into levels.

Self Check

1. **Compare and contrast** abiotic factors and biotic factors. Give five examples of each that are in your ecosystem.
2. **Describe** a population and a community.
3. **Define** the term *ecosystem.*
4. **Explain** how the terms *ecology* and *environment* are related.
5. **Think Critically** Explain how biotic factors change in an ecosystem that has flooded.

Applying Skills

6. **Record Observations** Each person lives in a population as part of a community. Describe your population and community.
7. **Use a database** to research biomes. Find the name of the biome that best describes where you live.

Science Online green.msscience.com/self_check_quiz

Delicately Balanced Ecosystems

Each year you might visit the same park, but notice little change. However, ecosystems are delicately balanced, and small changes can upset this balance. In this lab, you will observe how small amounts of fertilizer can disrupt an ecosystem.

Real-World Question

How do manufactured fertilizers affect pond systems?

Goals

- **Observe** the effects of manufactured fertilizer on water plants.
- **Predict** the effects of fertilizers on pond and stream ecosystems.

Materials

large glass jars of equal size (4)
clear plastic wrap
stalks of *Elodea* (8)
**another aquatic plant*
garden fertilizer
**houseplant fertilizer*
rubber bands (4)
pond water
triple-beam balance
**electronic scale*
weighing paper
spoon
metric ruler

**Alternate materials*

Safety Precautions

Procedure

1. Working in a group, label four jars *A, B, C,* and *D.*
2. **Measure** eight *Elodea* stalks to be certain that they are all about equal in length.
3. Fill the jars with equal volumes of pond water and place two stalks of *Elodea* in each jar.
4. Add 5 g of fertilizer to jar B, 10 g to jar C, and 30 g to jar D. Put no fertilizer in jar A.

5. Cover each jar with plastic wrap and secure it with a rubber band. Use your pencil to punch three small holes through the plastic wrap.
6. Place all jars in a well-lit area.
7. **Observe** the jars daily for three weeks. Record your observations in your Science Journal.
8. **Measure and record** the length of each *Elodea* stalk in your Science Journal.

Conclude and Apply

1. **List** the control and variables you used in this experiment.
2. **Compare** the growth of *Elodea* in each jar.
3. **Predict** what might happen to jar A if you added 5 g of fertilizer to it each week.
4. **Infer** what effects manufactured fertilizers might have on pond and stream ecosystems.

Communicating Your Data

Compare your results with the results of other students. Research how fertilizer runoff from farms and lawns has affected aquatic ecosystems in your area.

section 2

Interactions Among Living Organisms

Characteristics of Populations

You, the person sitting next to you, everyone in your class, and every other organism on Earth is a member of a specific population. Populations can be described by their characteristics such as spacing and density.

Population Size The number of individuals in the population is the population's size, as shown in **Figure 10.** Population size can be difficult to measure. If a population is small and made up of organisms that do not move, the size can be determined by counting the individuals. Usually individuals are too widespread or move around too much to be counted. The population size then is estimated. The number of organisms of one species in a small section is counted and this value is used to estimate the population of the larger area.

Suppose you spent several months observing a population of field mice that live in a pasture. You probably would observe changes in the size of the population. Older mice die. Mice are born. Some are eaten by predators, and some mice move away to new nests. The size of a population is always changing. The rate of change in population size varies from population to population. In contrast to a mouse population, the number of pine trees in a mature forest changes slowly, but a forest fire or disease could reduce the pine tree population quickly.

as you read

What You'll Learn

- **Identify** the characteristics that describe populations.
- **Examine** the different types of relationships that occur among populations in a community.
- **Determine** the habitat and niche of a species in a community.

Why It's Important

You must interact with other organisms to survive.

Review Vocabulary

coexistence: living together in the same place at the same time

New Vocabulary

- population density
- limiting factor
- symbiosis
- niche
- habitat

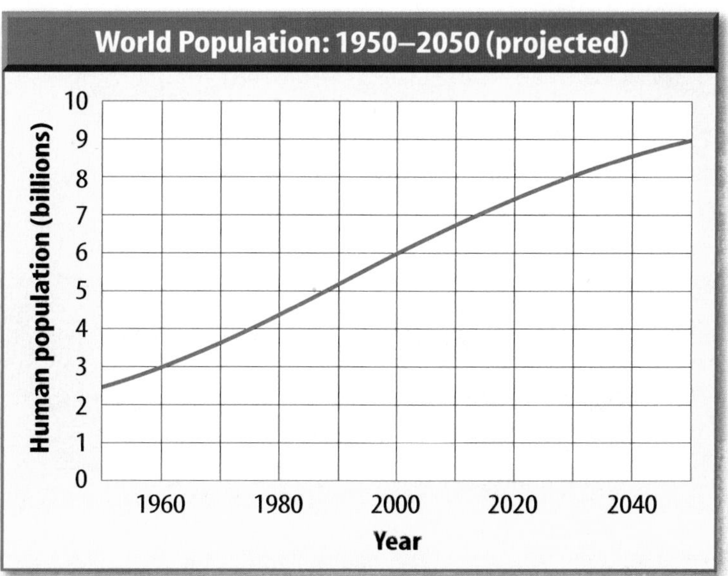

Figure 10 The size of the human population is increasing each year. By the year 2050, the human population is projected to be more than 9 billion.

Figure 11 Population density can be shown on a map. This map uses different colors to show varying densities of a population of northern bobwhites, a type of bird.

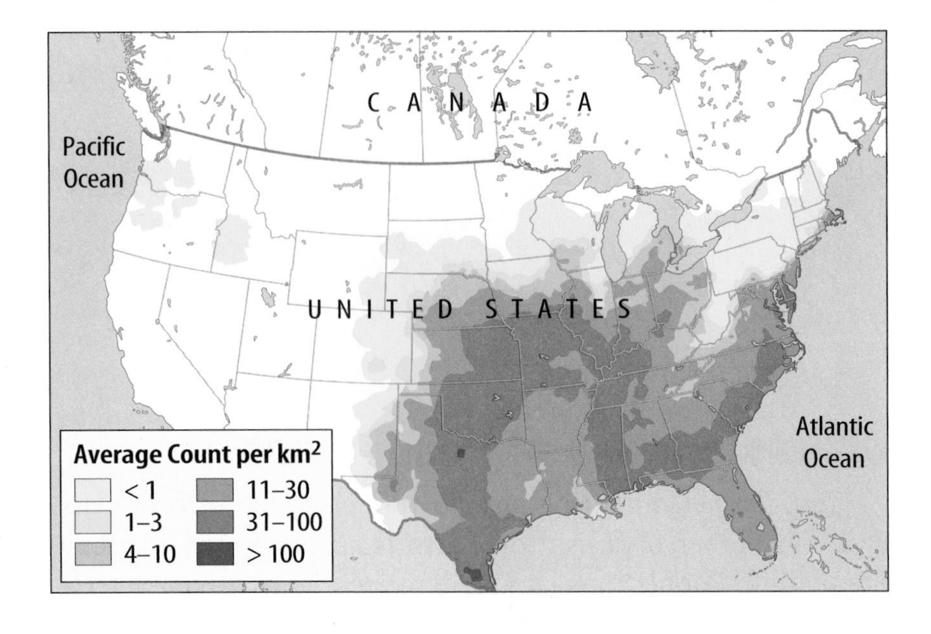

Topic: Human Population
Visit green.msscience.com for Web links to information about human population and densities.

Activity Select at least three different areas of the world and prepare a bar graph to compare population density of each area. Compare the population density of where you live to the three areas of the world you select.

Population Density At the beginning of this chapter, when you figured out how much space is available to each student in your classroom, you were measuring another population characteristic. The number of individuals in a population that occupy a definite area is called **population density.** For example, if 100 mice live in an area of one square kilometer, the population density is 100 mice per square kilometer. When more individuals live in a given amount of space, as seen in **Figure 11,** the population is more dense.

Population Spacing Another characteristic of populations is spacing, or how the organisms are arranged in a given area. They can be evenly spaced, randomly spaced, or clumped together. If organisms have a fairly consistent distance between them, as shown in **Figure 12,** they are evenly spaced. In random spacing, each organism's location is independent of the locations of other organisms in the population. Random spacing of plants usually results when wind or birds disperse seeds. Clumped spacing occurs when resources such as food or living space are clumped. Clumping results when animals gather in groups or plants grow near each other in groups.

Figure 12 Some populations, such as creosote bushes in the desert, are evenly spaced throughout an area.

Limiting Factors Populations, such as the antelopes in **Figure 13,** cannot continue to grow larger forever. All ecosystems have a limited amount of food, water, living space, mates, nesting sites, and other resources. A **limiting factor** is any biotic or abiotic factor that limits the number of individuals in a population. A limiting factor also can affect other populations in the community indirectly. For example, a drought might reduce the number of seed-producing plants in a forest clearing. Fewer plants means that food can become a limiting factor for deer that eat the plants and for a songbird population that feeds on the seeds of these plants. Food also could become a limiting factor for animals that feed on the songbirds.

Figure 13 These antelope and zebra populations live in the grasslands of Africa.
Infer *what limiting factors might affect the plant and animal populations shown here.*

Reading Check *What is an example of a limiting factor?*

Competition is the struggle among organisms to obtain the same resources needed to survive and reproduce, as shown in **Figure 14.** As population density increases, so does competition among individuals for the resources in their environment.

Carrying Capacity Suppose a population increases in size year after year. At some point, food, nesting space, or other resources become so scarce that some individuals are not able to survive or reproduce. When this happens, the environment has reached its carrying capacity. Carrying capacity is the largest number of individuals of a species that an environment can support and maintain for a long period of time. If a population gets bigger than the carrying capacity of the environment, some individuals are left without adequate resources. They will die or be forced to move elsewhere.

Figure 14 During dry summers, the populations of animals at existing watering holes increase because some watering holes have dried up. This creates competition for water, a valuable resource.

Observing Symbiosis

Procedure

1. Carefully wash and examine the roots of a **legume plant** and a **nonlegume plant.**
2. Use a **magnifying lens** to examine the roots of the legume plant.

Analysis

1. What differences do you observe in the roots of the two plants?
2. Bacteria and legume plants help one another thrive. What type of symbiotic relationship is this?

Biotic Potential What would happen if a population's environment had no limiting factors? The size of the population would continue to increase. The maximum rate at which a population increases when plenty of food and water are available, the weather is ideal, and no diseases or enemies exist, is its biotic potential. Most populations never reach their biotic potential, or they do so for only a short period of time. Eventually, the carrying capacity of the environment is reached and the population stops increasing.

Symbiosis and Other Interactions

In ecosystems, many species of organisms have close relationships that are necessary for their survival. **Symbiosis** (sihm bee OH sus) is any close interaction between two or more different species. Symbiotic relationships can be identified by the type of interaction between organisms. Mutualism is a symbiotic relationship in which two different species of organisms cooperate and both benefit. **Figure 15** shows one example of mutualism.

Commensalism is a form of symbiosis that benefits one organism without affecting the other organism. For example, a species of flatworm benefits by living in the gills of horseshoe crabs, eating scraps of the horseshoe crab's meals. The horseshoe crab is unaffected by the flatworms.

Parasitism is a symbiotic relationship between two species in which one species benefits and the other species is harmed. Some species of mistletoe are parasites because their roots grow into a tree's tissue and take nutrients from the tree.

Reading Check ***What form of symbiosis exists between a bee and a flower?***

Figure 15 The partnership between the desert yucca plant and the yucca moth is an example of mutualism.

Predation One way that population size is regulated is by predation (prih DAY shun). Predation is the act of one organism hunting, killing, and feeding on another organism. Owls are predators of mice, as shown in **Figure 16.** Mice are their prey. Predators are biotic factors that limit the size of the prey population. Availability of prey is a biotic factor that can limit the size of the predator population. Because predators are more likely to capture old, ill, or young prey, the strongest individuals in the prey population are the ones that manage to reproduce. This improves the prey population over several generations.

Figure 16 Owls use their keen senses of sight and hearing to hunt for mice in the dark.

Habitats and Niches In a community, every species plays a particular role. For example, some are producers and some are consumers. Each also has a particular place to live. The role, or job, of an organism in the ecosystem is called its **niche** (NICH). What a species eats, how it gets its food, and how it interacts with other organisms are all parts of its niche. The place where an organism lives is called its **habitat.** For example, an earthworm's habitat is soil. An earthworm's niche includes loosening, aerating, and enriching the soil.

section 2 review

Summary

Characteristics of Populations

- Populations can be described by size, density, and spacing.
- Limiting factors affect population size.
- The number of individuals an environment can support and maintain over time is called the carrying capacity.
- The biotic potential is the rate a population would increase without limiting factors.

Symbiosis and Other Interactions

- A close interaction between two or more different species is called symbiosis.
- Mutualism, commensalism, and parasitism are types of symbiotic relationships that can exist between organisms.
- Predators are biotic limiting factors of prey.
- The role an organism plays is called its niche.

Self Check

1. **Determine** the population of students in your classroom.
2. **Describe** how limiting factors can affect a population.
3. **Explain** the difference between a habitat and a niche.
4. **Describe** and give an example of two symbiotic relationships that occur among populations in a community.
5. **Explain** how sound could be used to relate the size of the cricket population in one field to the cricket population in another field.
6. **Think Critically** A parasite obtains food from its host. Most parasites weaken but do not kill their hosts. Why?

Applying Math

7. **Solve One-Step Equations** A 15-m^2 wooded area has the following: 30 ferns, 150 grass plants, and 6 oak trees. What is the population density per m^2 of each of the above species?

section 3

Matter and Energy

as you read

What You'll Learn

- **Explain** the difference between a food chain and a food web.
- **Describe** how energy flows through ecosystems.
- **Examine** how materials such as water, carbon, and nitrogen are used repeatedly.

Why It's Important

You are dependent upon the recycling of matter and the transfer of energy for survival.

Review Vocabulary

consumer: organism that obtains energy by eating other organisms

New Vocabulary

- food chain
- food web
- water cycle

Energy Flow Through Ecosystems

Life on Earth is not simply a collection of independent organisms. Even organisms that seem to spend most of their time alone interact with other members of their species. They also interact with members of other species. Most of the interactions among members of different species occur when one organism feeds on another. Food contains nutrients and energy needed for survival. When one organism is food for another organism, some of the energy in the first organism (the food) is transferred to the second organism (the eater).

Producers are organisms that take in and use energy from the Sun or some other source to produce food. Some use the Sun's energy for photosynthesis to produce carbohydrates. For example, plants, algae, and some one-celled, photosynthetic organisms are producers. Consumers are organisms that take in energy when they feed on producers or other consumers. The transfer of energy does not end there. When organisms die, other organisms called decomposers, as shown in **Figure 17,** take in energy as they break down the remains of organisms. This movement of energy through a community can be diagrammed as a food chain or a food web.

Food Chains A **food chain,** as shown in **Figure 18,** is a model, a simple way of showing how energy, in the form of food, passes from one organism to another. When drawing a food chain, arrows between organisms indicate the direction of energy transfer. An example of a pond food chain follows.

aquatic plants → insects → bluegill → bass → humans

Food chains usually have only three or four links. This is because the available energy decreases from one link to the next link. At each transfer of energy, a portion of the energy is lost as heat due to the activities of the organisms. In a food chain, the amount of energy left for the last link is only a small portion of the energy in the first link.

Figure 17 These mushrooms are decomposers. They obtain needed energy for life when they break down organic material.

NATIONAL GEOGRAPHIC

VISUALIZING A FOOD CHAIN

Figure 18

In nature, energy in food passes from one organism to another in a sequence known as a food chain. All living things are linked in food chains, and there are millions of different chains in the world. Each chain is made up of organisms in a community. The photographs here show a food chain in a North American meadow community.

E The last link in many food chains is a top carnivore, an animal that feeds on other animals, including other carnivores. This great horned owl is a top carnivore.

D The fourth link of this food chain is a garter snake, which feeds on toads.

A The first link in any food chain is a producer—in this case, grass. Grass gets its energy from sunlight.

B The second link of a food chain is usually an herbivore like this grasshopper. Herbivores are animals that feed only on producers.

C The third link of this food chain is a carnivore, an animal that feeds on other animals. This woodhouse toad feeds on grasshoppers.

Food Webs Food chains are too simple to describe the many interactions among organisms in an ecosystem. A **food web** is a series of overlapping food chains that exist in an ecosystem. A food web provides a more complete model of the way energy moves through an ecosystem. They also are more accurate models because food webs show how many organisms, including humans, are part of more than one food chain in an ecosystem.

Humans are a part of many land and aquatic food webs. Most people eat foods from several different levels of a food chain. Every time you eat a hamburger, an apple, or other food, you have become a link in a food web. Can you picture the steps in the food web that led to the food in your lunch?

Applying Science

How do changes in Antarctic food webs affect populations?

The food webs in the icy Antarctic Ocean are based on phytoplankton, which are microscopic algae that float near the water's surface. The algae are eaten by tiny, shrimplike krill, which are consumed by baleen whales, squid, and fish. Toothed whales, seals, and penguins eat the fish and squid. How would changes in any of these populations affect the other populations?

Identifying the Problem

Worldwide, the hunting of most baleen whales has been illegal since 1986. It is hoped that the baleen whale population will increase. How will an increase in the whale population affect the food web illustrated below?

Solving the Problem

1. Populations of seals, penguins, and krill-eating fish increased in size as populations of baleen whales declined. Explain why this occurred.
2. What might happen if the number of baleen whales increases but the amount of krill does not?

Ecological Pyramids Most of the energy in the biosphere comes from the Sun. Producers take in and transform only a small part of the energy that reaches Earth's surface. When an herbivore eats a plant, some of the energy in the plant passes to the herbivore. However, most of it is given off into the atmosphere as heat. The same thing happens when a carnivore eats an herbivore. An ecological pyramid models the number of organisms at each level of a food chain. The bottom of an ecological pyramid represents the producers of an ecosystem. The rest of the levels represent successive consumers.

Reading Check *What is an ecological pyramid?*

Chemosynthesis Certain bacteria take in energy through a process called chemosynthesis. In chemosynthesis, the bacteria produce food using the energy in chemical compounds instead of light energy. In your Science Journal, predict where these bacteria are found.

Energy Pyramid The flow of energy from grass to the hawk in **Figure 19** can be illustrated by an energy pyramid. An energy pyramid compares the energy available at each level of the food chain in an ecosystem. Just as most food chains have three or four links, a pyramid of energy usually has three or four levels. Only about ten percent of the energy at each level of the pyramid is available to the next level. By the time the top level is reached, the amount of energy available is greatly reduced.

Figure 19 An energy pyramid illustrates that available energy decreases at each successive feeding step.
Determine *why an energy pyramid doesn't have more levels.*

Modeling the Water Cycle

Procedure

1. With a **marker,** make a line halfway up on a **plastic cup.** Fill the cup to the mark with water.
2. Cover the top with **plastic wrap** and secure it with a **rubber band or tape.**
3. Put the cup in direct sunlight. Observe the cup for three days. Record your observations.
4. Remove the plastic wrap and observe the cup for seven more days.

Analysis

1. What parts of the water cycle did you observe during this activity?
2. How did the water level in the cup change after the plastic wrap was removed?

Try at Home

The Cycles of Matter

The energy available as food is constantly renewed by plants using sunlight. However, think about the matter that makes up the bodies of living organisms. The law of conservation of mass states that matter on Earth is never lost or gained. It is used over and over again. In other words, it is recycled. The carbon atoms in your body might have been on Earth since the planet formed billions of years ago. They have been recycled billions of times. Many important materials that make up your body cycle through the environment. Some of these materials are water, carbon, and nitrogen.

Water Cycle Water molecules on Earth constantly rise into the atmosphere, fall to Earth, and soak into the ground or flow into rivers and oceans. The **water cycle** involves the processes of evaporation, condensation, and precipitation.

Heat from the Sun causes water on Earth's surface to evaporate, or change from a liquid to a gas, and rise into the atmosphere as water vapor. As the water vapor rises, it encounters colder and colder air and the molecules of water vapor slow down. Eventually, the water vapor changes back into tiny droplets of water. It condenses, or changes from a gas to a liquid. These water droplets clump together to form clouds. When the droplets become large and heavy enough, they fall back to Earth as rain or other precipitation. This process is illustrated in **Figure 20.**

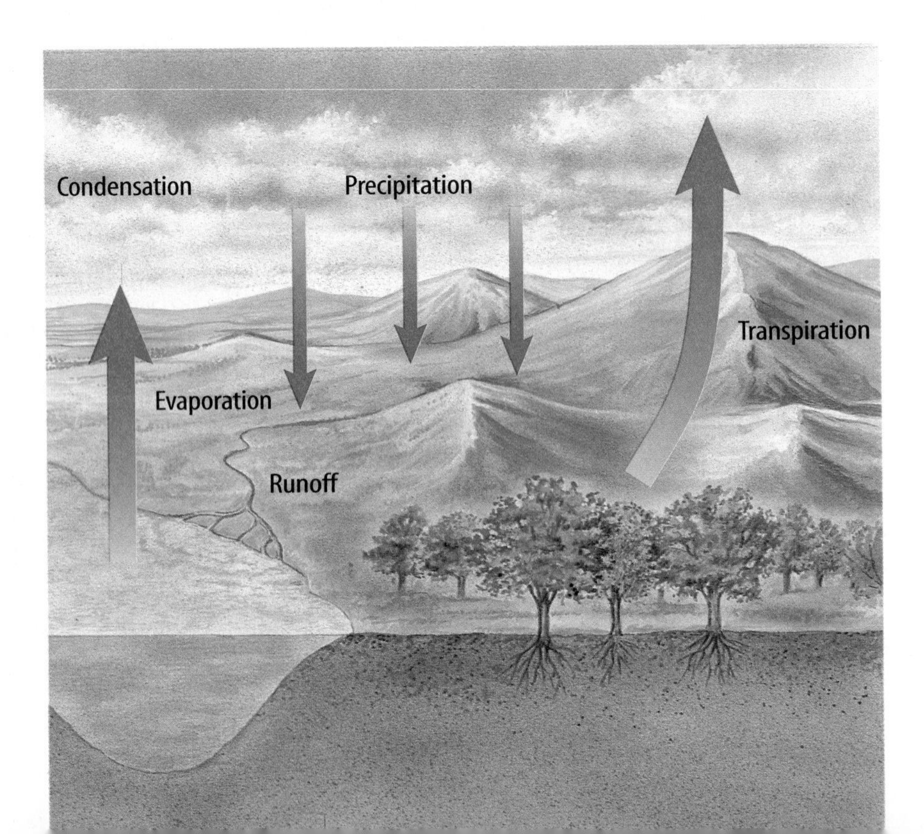

Figure 20 A water molecule that falls as rain can follow several paths through the water cycle. **Identify** *these paths in this diagram.*

Other Cycles in Nature You and all organisms contain carbon. Earth's atmosphere contains about 0.03 percent carbon in the form of carbon dioxide gas. The movement of carbon through Earth's biosphere is called the carbon cycle, as shown in **Figure 21.**

Nitrogen is an element found in proteins and nucleic acids. The nitrogen cycle begins with the transfer of nitrogen from the atmosphere to producers then to consumers. The nitrogen then moves back to the atmosphere or directly into producers again.

Phosphorus, sulfur, and other elements needed by living organisms also are used and returned to the environment. Just as you recycle aluminum, glass, and paper products, the matter that organisms need to live is recycled continuously in the biosphere.

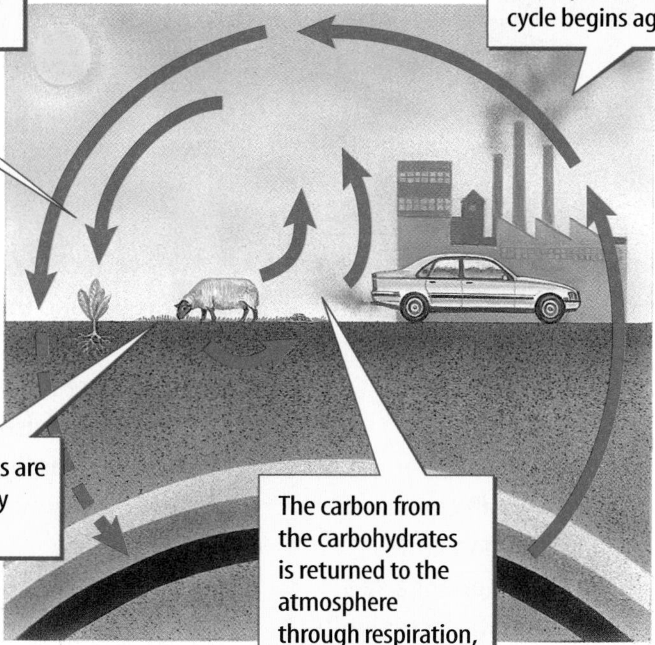

Plants remove carbon dioxide from the air and use it to make carbohydrates.

After the carbon is returned to the atmosphere, the cycle begins again.

The carbohydrates are eaten and used by other organisms.

The carbon from the carbohydrates is returned to the atmosphere through respiration, combustion, and decay.

Figure 21 Carbon can follow several different paths through the carbon cycle. Some carbon is stored in Earth's biomass.

section 3 review

Summary

Energy Flow Through Ecosystems

- A food chain models one pathway of energy through an ecosystem, and a food web is made of many food chains.
- Humans are part of different food webs.
- Ecological pyramids model the number of organisms at each level of a food chain.
- Energy pyramids illustrate the available energy at each level of a food chain.

The Cycles of Matter

- Energy is constantly renewed by the Sun, but matter must be recycled.
- The water cycle involves evaporation, condensation, and precipitation.
- Other matter that cycles includes carbon, nitrogen, phosphorus, and sulfur.

Self Check

1. **Draw and label** a food web that includes you and what you've eaten today.
2. **Compare and contrast** producers, consumers, and decomposers.
3. **Explain** how carbon flows through ecosystems.
4. **Think Critically** Use your knowledge of food chains and the energy pyramid to explain why fewer lions than gazelles live on the African plains.

Applying Skills

5. **Classify** Look at the food chain in **Figure 18.** Classify each organism as a producer or a consumer.
6. **Communicate** In your Science Journal, write a short essay about how the water cycle, carbon cycle, and nitrogen cycle are important to living organisms.

Design Your Own

Identifying a Limiting Factor

Real-World Question

Organisms depend upon many biotic and abiotic factors in their environment to survive. When these factors are limited or are not available, it can affect an organism's survival. How do abiotic factors such as light, water, and temperature affect the germination of seeds?

Form a Hypothesis

Based on what you have learned about limiting factors, make a hypothesis about how one specific abiotic factor might affect the germination of a bean seed. Be sure to consider factors that you can change easily.

Goals

- **Observe** the effects of an abiotic factor on the germination and growth of bean seedlings.
- **Design** an experiment that demonstrates whether or not a specific abiotic factor limits the germination of bean seeds.

Possible Materials

bean seeds
small planting containers
soil
water
label
trowel
**spoon*
aluminum foil
sunny window
**other light source*
refrigerator or oven
**Alternate materials*

Safety Precautions

Test Your Hypothesis

Make a Plan

1. As a group, agree upon and write out a hypothesis statement.
2. **Decide** on a way to test your group's hypothesis. Keep available materials in mind as you plan your procedure. List your materials.
3. **Design** a data table in your Science Journal for recording data.
4. Remember to test only one variable at a time and use suitable controls.

5. Read over your entire experiment to make sure that all steps are in logical order.
6. **Identify** any constants, variables, and controls in your experiment.
7. Be sure the factor that you will test is measurable.

Follow Your Plan

1. Make sure your teacher approves your plan before you start.
2. Carry out your approved plan.
3. While the experiment is going on, record any observations that you make and complete the data table in your Science Journal.

Analyze Your Data

1. **Compare** the results of this experiment with those of other groups in your class.
2. **Infer** how the abiotic factor you tested affected the germination of bean seeds.
3. **Graph** your results in a bar graph that compares the number of bean seeds that germinated in the experimental container with the number of seeds that germinated in the control container.

Conclude and Apply

1. **Identify** which factor had the greatest effect on the germination of the seeds.
2. **Determine** whether or not you could change more than one factor in this experiment and still have germination of seeds.

Communicating Your Data

Write a set of instructions that could be included on a packet of this type of seeds. Describe the best conditions for seed germination.

Science and Language Arts

The Solace of Open Spaces

a novel by Gretel Ehrlich

Animals give us their constant, unjaded[1] faces and we burden them with our bodies and civilized ordeals. We're both humbled by and imperious[2] with them. We're comrades who save each other's lives. The horse we pulled from a boghole this morning bucked someone off later in the day; one stock dog refuses to work sheep, while another brings back a calf we had overlooked. . . . What's stubborn, secretive, dumb, and keen[3] in us bumps up against those same qualities in them. . . .

Living with animals makes us redefine our ideas about intelligence. Horses are as mischievous as they are dependable. Stupid enough to let us use them, they are cunning enough to catch us off guard. . . .

We pay for their loyalty; They can be willful, hard to catch, dangerous to shoe and buck on frosty mornings. In turn, they'll work themselves into a lather cutting cows, not for the praise they'll get but for the simple glory of outdodging a calf or catching up with an errant steer. . . .

1 *Jaded* means "to be weary with fatigue," so *unjaded* means "not to be weary with fatigue."

2 domineering or overbearing

3 intellectually smart or sharp

Understanding Literature

Informative Writing This passage is informative because it describes the real relationship between people and animals on a ranch in Wyoming. The author speaks from her own point of view, not from the point of view of a disinterested party. How might this story have been different if it had been told from the point of view of a visiting journalist?

Respond to the Reading

1. Describe the relationship between people and animals in this passage.
2. What words does the author use to indicate that horses are intelligent?
3. **Linking Science and Writing** Write a short passage about an experience you have had with a pet. Put yourself in the passage without overusing the word "I".

Animals and ranchers are clearly dependent on each other. Ranchers provide nutrition and shelter for animals on the ranch and, in turn, animals provide food, companionship, and perform work for the ranchers. You might consider the relationship between horses and ranchers to be a symbiotic one. Symbiosis (sihm bee OH sus) is any close interaction among two or more different species.

Reviewing Main Ideas

Section 1 The Environment

1. Ecology is the study of interactions among organisms and their environment.
2. The nonliving features of the environment are abiotic factors, and the organisms in the environment are biotic factors.
3. Ecosystems include biotic and abiotic factors.
4. The region of Earth and its atmosphere in which all organisms live is the biosphere.

Section 2 Interactions Among Living Organisms

1. Characteristics that can describe populations include size, spacing, and density.
2. Any biotic or abiotic factor that limits the number of individuals in a population is a limiting factor.
3. A close relationship between two or more species is a symbiotic relationship.
4. The place where an organism lives is its habitat, and its role in the environment is its niche.

Section 3 Matter and Energy

1. Food chains and food webs are models that describe the flow of energy.
2. At each level of a food chain, organisms lose energy as heat. Energy on Earth is renewed constantly by sunlight.
3. Matter on Earth is never lost or gained. It is used over and over again, or recycled.

Visualizing Main Ideas

Copy and complete the following concept map on the biosphere.

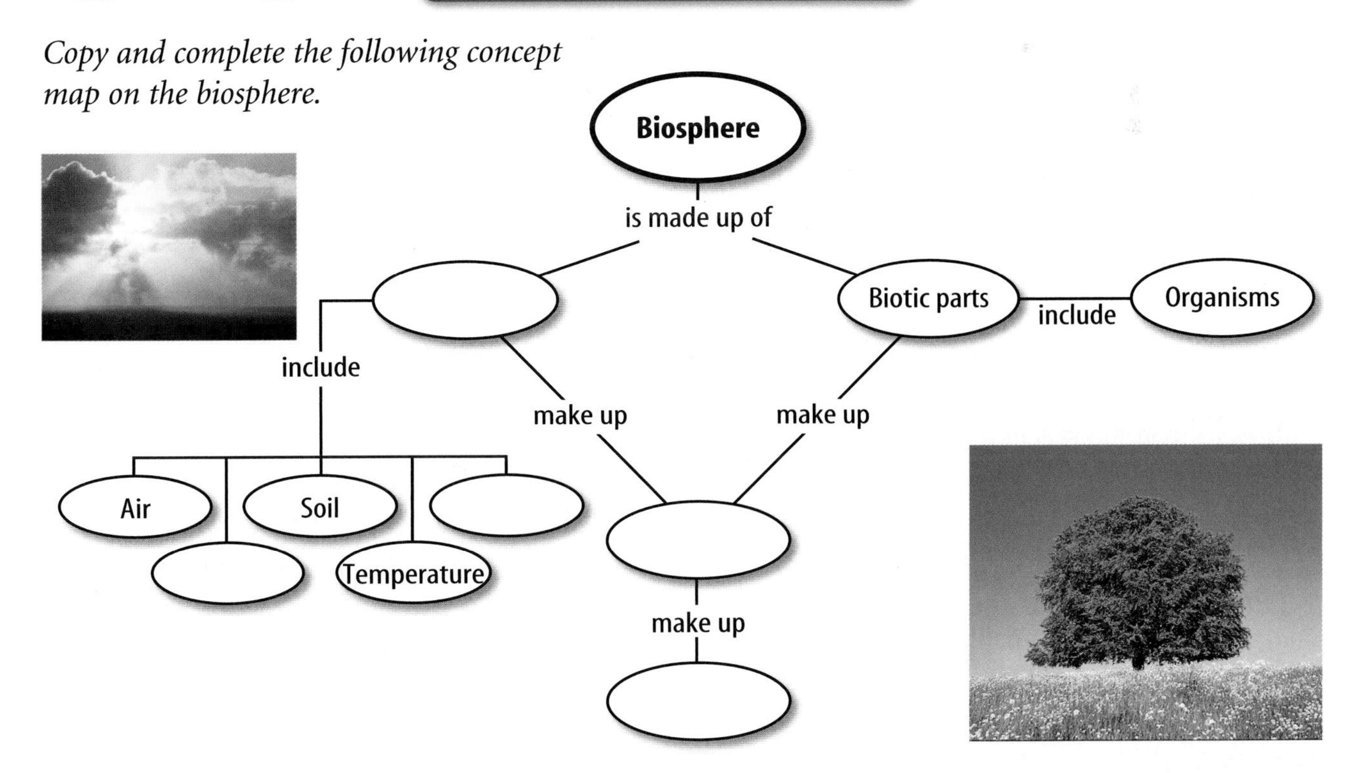

Using Vocabulary

abiotic factor p. 532	habitat p. 543
biosphere p. 537	limiting factor p. 541
biotic factor p. 532	niche p. 543
community p. 536	population p. 536
ecology p. 532	population density p. 540
ecosystem p. 536	symbiosis p. 542
food chain p. 544	water cycle p. 548
food web p. 546	

Fill in the blanks with the correct vocabulary word or words.

1. A(n) ________ is any living thing in the environment.
2. A series of overlapping food chains makes up a(n) ________.
3. The size of a population that occupies an area of definite size is its ________.
4. Where an organism lives in an ecosystem is its ________.
5. The part of Earth that supports life is the ________.
6. Any close relationship between two or more species is ________.

Checking Concepts

Choose the word or phrase that best answers the question.

7. Which of the following is a model that shows the amount of energy available as it flows through an ecosystem?
   A) niche
   B) energy pyramid
   C) carrying capacity
   D) food chain
8. Which of the following is a biotic factor?
   A) animals **C)** sunlight
   B) air **D)** soil
9. What is made up of all populations in an area?
   A) niche **C)** community
   B) habitat **D)** ecosystem
10. What is the term for the total number of individuals in a population occupying a certain area?
   A) clumping **C)** spacing
   B) size **D)** density
11. What is the tree to the right an example of?

   A) prey
   B) consumer
   C) producer
   D) predator
12. Which level of the food chain has the most energy?
   A) consumer **C)** decomposers
   B) herbivores **D)** producers
13. What is the symbitotc relationship called in which one organism is helped and the other organism is harmed?
   A) mutualism
   B) parasitism
   C) commensalism
   D) consumer
14. Which of the following is NOT cycled in the biosphere?
   A) nitrogen **C)** water
   B) soil **D)** carbon
15. What are coral reefs, forests, and ponds examples of?
   A) niches **C)** populations
   B) habitats **D)** ecosystems
16. What are all of the individuals of one species that live in the same area at the same time called?
   A) community **C)** biosphere
   B) population **D)** organism

Thinking Critically

Use the illustration below to answer question 17.

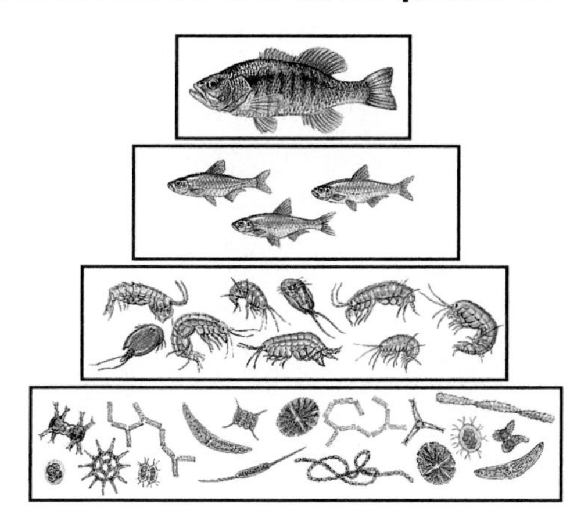

17. **Infer** why each level of the energy pyramid shown above is smaller than the one below it.
18. **Compare and contrast** the role of producers, consumers, and decomposers in an ecosystem.
19. **Explain** what carrying capacity has to do with whether or not a population reaches its biotic potential.
20. **Infer** why decomposers are vital to the cycling of matter in an ecosystem.
21. **Write** a paragraph that describes your own habitat and niche.
22. **Classify** the following as the result of either evaporation or condensation.
    - **a.** A puddle disappears after a rainstorm.
    - **b.** Rain falls.
    - **c.** A lake becomes shallower.
    - **d.** Clouds form.
23. **Concept Map** Use the following information to draw a food web of organisms living in a goldenrod field. *Aphids eat goldenrod sap, bees eat goldenrod nectar, beetles eat goldenrod pollen and goldenrod leaves, stinkbugs eat beetles, spiders eat aphids,* and *assassin bugs eat bees.*
24. **Record Observations** A home aquarium contains water, an air pump, a light, algae, a goldfish, and algae-eating snails. What are the abiotic factors in this environment?
25. **Determine** why viruses are considered parasites.

Performance Activities

26. **Poster** Use your own observations or the results of library research to develop a food web for a nearby park, pond, or other ecosystem. Make a poster display illustrating the food web.
27. **Oral Presentation** Research the steps in the phosphorous cycle. Find out what role phosphorus plays in the growth of algae in ponds and lakes. Present your findings to the class.

Applying Math

Use the table below to answer questions 28 and 29.

Arizona Deer Population

Year	Deer Per 400 Hectares
1905	5.7
1915	35.7
1920	142.9
1925	85.7
1935	25.7

28. **Deer Population** Use the data above to graph the population density of a deer population over the years. Plot the number of deer on the *y*-axis and years on the *x*-axis. Predict what might have happened to cause the changes in the size of the population.
29. **Population Trend** What might the population of deer be in 1940 if the trend continued?

Part 1 Multiple Choice

Record your answers on the answer sheet provided by your teacher or on a sheet of paper.

1. Many of the processes necessary for life can only occur in the presence of water. Which of the following is not one of these necessary life processes?
   - **A.** photosynthesis
   - **B.** respiration
   - **C.** digestion
   - **D.** dehydration

Use the illustration below to answer questions 2 and 3.

2. Little light reaches the plants on the floor of this deciduous forest. Which season would let the bluebells pictured grow the best?
   - **A.** spring
   - **B.** summer
   - **C.** fall
   - **D.** winter

3. What process do these bluebells use to transform energy from the Sun into stored chemical energy for their life processes?
   - **A.** respiration
   - **B.** radiation
   - **C.** photosynthesis
   - **D.** desertification

4. Soil that receives little rain can be changed and a desert can form. What is this process known as?
   - **A.** pollution
   - **B.** radiation
   - **C.** desertification
   - **D.** respiration

5. Which of the following is NOT considered when determining soil type?
   - **A.** amount of clay
   - **B.** amount of silt
   - **C.** amount of sand
   - **D.** amount of plant

Use the illustration below to answer questions 6 and 7.

6. Antelope and zebras interacting is an example of a(n)
   - **A.** community.
   - **B.** population.
   - **C.** ecosystem.
   - **D.** biome.

7. These two groups both eat plants, biotic factors that help determine the number of individuals that will survive in that area. This is known as a
   - **A.** carrying capacity.
   - **B.** limiting factor.
   - **C.** biotic potential.
   - **D.** population spacing.

8. The maximum rate at which a population increases when plenty of food and water are available is known as
   - **A.** carrying capacity.
   - **B.** limiting factor.
   - **C.** biotic potential.
   - **D.** population density.

9. A close interaction between two or more different species is known as
   - **A.** symbiosis.
   - **B.** mutualism.
   - **C.** commensalism.
   - **D.** botulism.

10. The job of an organism in the ecosystem is called its
    - **A.** habitat.
    - **B.** ecosystem.
    - **C.** niche.
    - **D.** community.

Part 2 Short Response/Grid In

Record your answers on the answer sheet provided by your teacher or on a sheet of paper.

11. How do producers and consumers interact?

12. Give an example of an abiotic factor and explain how living organisms interact with it.

13. What is a limiting factor? Give a non-living limiting factor and show how it affects living organisms.

14. What is symbiosis? Name three types of symbiosis.

Use the illustration below to answer questions 15 and 16.

15. What is competition and how does it relate to population density? Use the illustration above as an example of competition. Explain the environment and how it impacts their survival.

16. Use the illustration to explain populations and communities.

17. What is biotic potential and how does it relate to limiting factors?

18. How is population size controlled by predators? Use the owl/mouse relationship as an example.

Part 3 Open Ended

Record your answers on a sheet of paper.

19. How do humans interact with land and water food webs? What would happen to humans if these webs were destroyed?

20. Describe the carbon cycle.

21. Describe how better food production, sanitation, and disease prevention have contributed to the yearly increase in the human population.

Use the illustration below to answer questions 22 and 23.

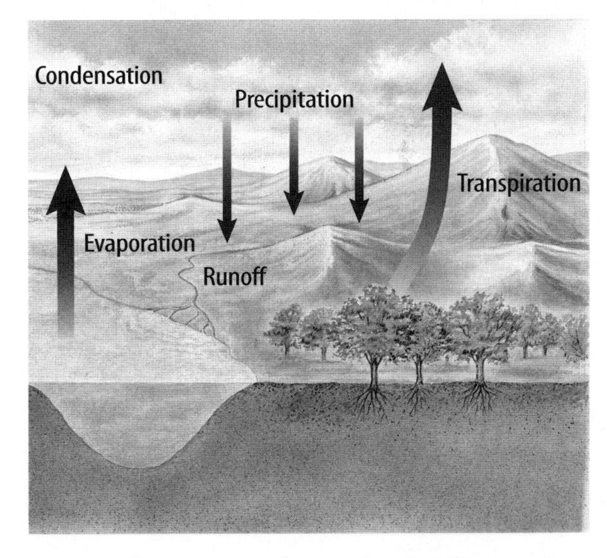

22. Name the process shown in the illustration above.

23. Explain why this process is necessary for living organisms.

Test-Taking Tip

Answer All Parts of the Question Make sure each part of the question is answered when listing discussion points. For example, if the question asks you to compare and contrast, make sure you list both similarities and differences.

Question 21 Notice that this question asks you to describe how three things contribute to human population growth.

chapter 19

Conserving Resources

chapter preview

sections

1. Resources
2. Pollution
   Lab The Greenhouse Effect
3. The Three Rs of Conservation
   Lab Solar Cooking

Virtual Lab When is water safe to drink?

Resources Fuel Our Lives

Resources, such as clean water and air, are commonly taken for granted. We depend on water and air to survive. Fossil fuels are another type of resource, and we depend on them for energy. However, fossil fuels can pollute our air and water.

Science Journal List some other resources that we depend on and describe how we use them.

Start-Up Activities

What happens when topsoil is left unprotected?

Plants grow in the top, nutrient-rich layer, called topsoil. Plants help keep topsoil in place by protecting it from wind and rain. Try the following experiment to find out what happens when topsoil is left unprotected.

1. Use a mixture of moist sand and potting soil to create a miniature landscape in a plastic basin or aluminum-foil baking pan. Form hills and valleys in your landscape.
2. Use clumps of moss to cover areas of your landscape. Leave some sloping portions without plant cover.
3. Simulate a rainstorm over your landscape by spraying water on it from a spray bottle or by pouring a slow stream of water on it from a beaker.
4. **Think Critically** In your Science Journal, record your observations and describe what happened to the land that was not protected by plant cover.

Preview this chapter's content and activities at green.msscience.com

Resources Make the following Foldable to help you organize information and diagram ideas about renewable and nonrenewable resources.

STEP 1 **Fold** a sheet of paper in half lengthwise. Make the back edge about 5 cm longer than the front edge.

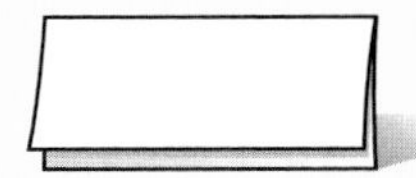

STEP 2 **Turn** the paper so the fold is on the bottom. Then **fold** in half.

STEP 3 **Unfold and cut** only the top layer along the fold to make two tabs. **Label** the Foldable as shown.

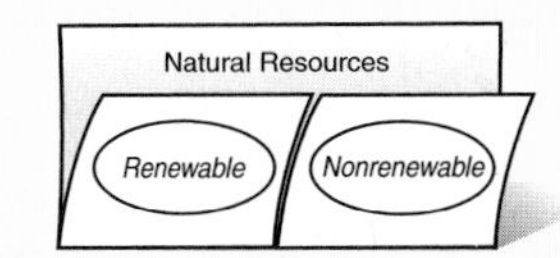

Make a Concept Map Before you read the chapter, list examples of each type of natural resource you already know on the back of the appropriate tabs. As you read the chapter, add to your lists.

section 1

Resources

as you read

What You'll Learn

- **Compare** renewable and nonrenewable resources.
- **List** uses of fossil fuels.
- **Identify** alternatives to fossil fuel use.

Why It's Important

Wise use of natural resources is important for the health of all life on Earth.

Review Vocabulary

geyser: a spring that emits intermittent jets of heated water and steam

New Vocabulary

- natural resource
- renewable resource
- nonrenewable resource
- petroleum
- fossil fuel
- hydroelectric power
- nuclear energy
- geothermal energy

Natural Resources

An earthworm burrowing in moist soil eats decaying plant material. A robin catches the worm and flies to a tree. The leaves of the tree use sunlight during photosynthesis. Leaves fall to the ground, decay, and perhaps become an earthworm's meal. What do these living things have in common? They rely on Earth's **natural resources**—the parts of the environment that are useful or necessary for the survival of living organisms.

What kinds of natural resources do you use? Like other organisms, you need food, air, and water. You also use resources that are needed to make everything from clothes to cars. Natural resources supply energy for automobiles and power plants. Although some natural resources are plentiful, others are not.

Renewable Resources The Sun, an inexhaustible resource, provides a constant supply of heat and light. Rain fills lakes and streams with water. When plants carry out photosynthesis, they add oxygen to the air. Sunlight, water, air, and the crops shown in **Figure 1** are examples of renewable resources. A **renewable resource** is any natural resource that is recycled or replaced constantly by nature.

Figure 1 Cotton and wood are renewable resources. Cotton cloth is used for rugs, curtains, and clothing. A new crop of cotton can be grown every year. Wood is used for furniture, building materials, and paper. It will take 20 years for these young trees to grow large enough to harvest.

Supply and Demand Even though renewable resources are recycled or replaced, they are sometimes in short supply. Rain and melted snow replace the water in streams, lakes, and reservoirs. Sometimes, there may not be enough rain or snowmelt to meet all the needs of people, plants, and animals. In some parts of the world, especially desert regions, water and other resources usually are scarce. Other resources can be used instead, as shown in **Figure 2.**

Figure 2 In parts of Africa, firewood has become scarce. People in this village now use solar energy instead of wood for cooking.

Nonrenewable Resources Natural resources that are used up more quickly than they can be replaced by natural processes are **nonrenewable resources.** Earth's supply of nonrenewable resources is limited. You use nonrenewable resources when you take home groceries in a plastic bag, paint a wall, or travel by car. Plastics, paints, and gasoline are made from an important nonrenewable resource called petroleum, or oil. **Petroleum** is formed mostly from the remains of microscopic marine organisms buried in Earth's crust. It is nonrenewable because it takes hundreds of millions of years for it to form.

Reading Check *What are nonrenewable resources?*

Minerals and metals found in Earth's crust are nonrenewable resources. Petroleum is a mineral. So are diamonds and the graphite in pencil lead. The aluminum used to make soft-drink cans is a metal. Iron, copper, tin, gold, silver, tungsten, and uranium also are metals. Many manufactured items, like the car shown in **Figure 3,** are made from nonrenewable resources.

Figure 3 Iron, a nonrenewable resource, is the main ingredient in steel. Steel is used to make cars, trucks, appliances, buildings, bridges, and even tires.
Infer *what other nonrenewable resources are used to build a car.*

Observing Mineral Mining Effects

Procedure

1. Place a **chocolate-chip cookie** on a **paper plate.** Pretend the chips are mineral deposits and the rest of the cookie is Earth's crust.
2. Use a **toothpick** to locate and dig up the mineral deposits. Try to disturb the land as little as possible.
3. When mining is completed, try to restore the land to its original condition.

Analysis

1. How well were you able to restore the land?
2. Compare the difficulty of digging for mineral deposits found close to the surface with digging for those found deep in Earth's crust.
3. Describe environmental changes that might result from a mining operation.

Try at Home

Fossil Fuels

Coal, oil, and natural gas are nonrenewable resources that supply energy. Most of the energy you use comes from these fossil fuels, as the graph in **Figure 4** shows. **Fossil fuels** are fuels formed in Earth's crust over hundreds of millions of years. Cars, buses, trains, and airplanes are powered by gasoline, diesel fuel, and jet fuel, which are made from oil. Coal is used in many power plants to produce electricity. Natural gas is used in manufacturing, for heating and cooking, and sometimes as a vehicle fuel.

Fossil Fuel Conservation Billions of people all over the world use fossil fuels every day. Because fossil fuels are nonrenewable, Earth's supply of them is limited. In the future, they may become more expensive and difficult to obtain. Also, the use of fossil fuels can lead to environmental problems. For example, mining coal can require stripping away thick layers of soil and rock, as shown in **Figure 4,** which destroys ecosystems. Another problem is that fossil fuels must be burned to release the energy stored in them. The burning of fossil fuels produces waste gases that cause air pollution, including smog and acid rain. For these reasons, many people suggest reducing the use of fossil fuels and finding other sources of energy.

You can use simple conservation measures to help reduce fossil fuel use. Switch off the light when you leave a room and turn off the television when you're not watching it. These actions reduce your use of electricity, which often is produced in power plants that burn fossil fuels. Hundreds of millions of automobiles are in use in the United States. Riding in a car pool or taking public transportation uses fewer liters of gasoline than driving alone in a car. Walking or riding a bicycle uses even less fossil fuel. Reducing fossil fuel use has an added benefit—the less you use, the more money you save.

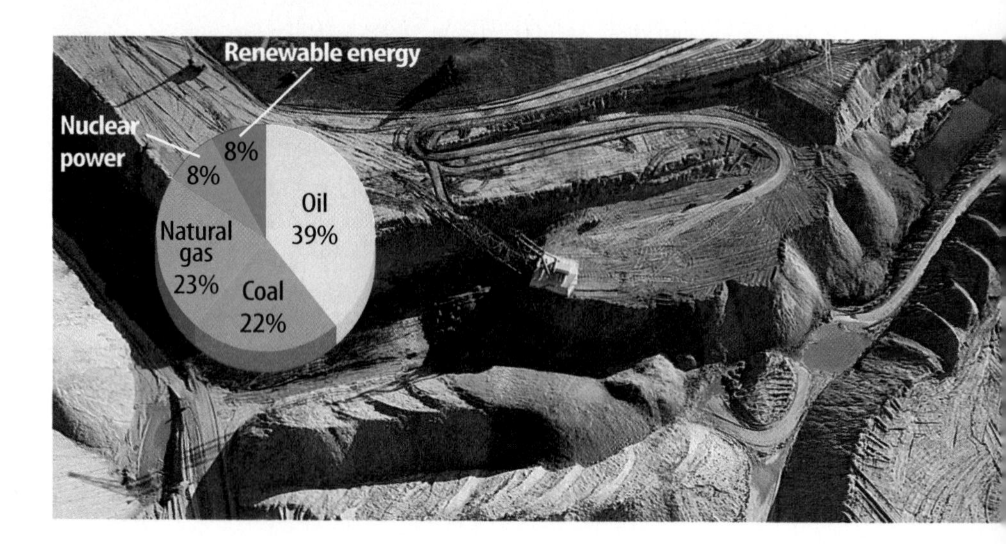

Figure 4 Coal is a fossil fuel. It often is obtained by strip mining, which removes all the soil above the coal deposit. The soil is replaced, but it takes many years for the ecosystem to recover.
Identify *the resource that provided 84 percent of the energy used in the United States in 1999.*

Figure 5 Most power plants use turbine generators to produce electricity. In fossil fuel plants, burning fuel boils water and produces steam that turns the turbine.

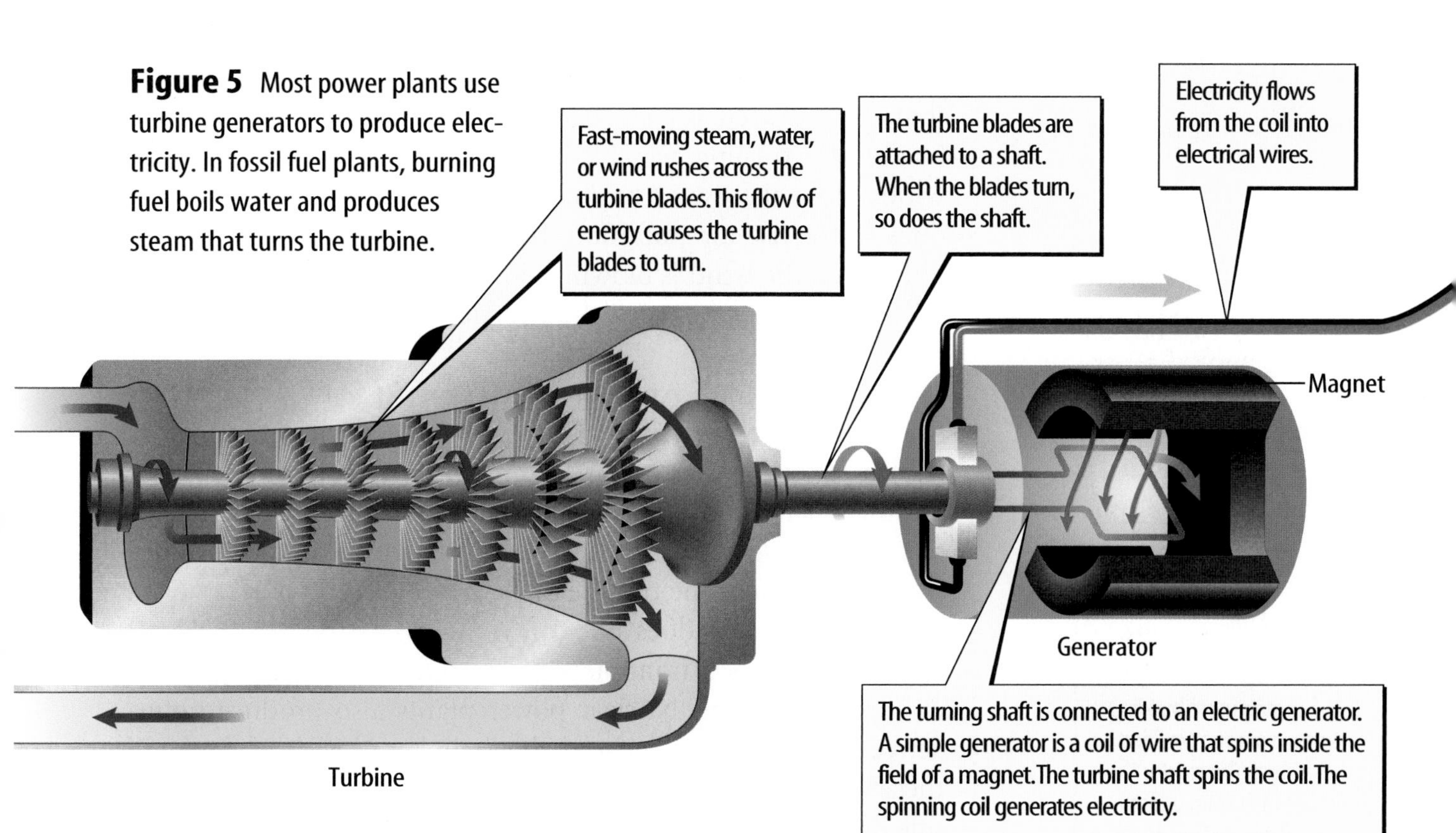

Alternatives to Fossil Fuels

Another approach to reducing fossil fuel use is to develop other sources of energy. Much of the electricity used today comes from power plants that burn fossil fuels. As **Figure 5** shows, electricity is generated when a rotating turbine turns a coil of wires in the magnetic field of an electric generator. Fossil-fuel power plants boil water to produce steam that turns the turbine. Alternative energy sources, including water, wind, and atomic energy can be used instead of fossil fuels to turn turbines. Also, solar cells can produce electricity using only sunlight, with no turbines at all. Some of these alternative energy sources—particularly wind and solar energy—are so plentiful they could be considered inexhaustible resources.

Water Power Water is a renewable energy source that can be used to generate electricity. **Hydroelectric power** is electricity that is produced when the energy of falling water is used to turn the turbines of an electric generator. Hydroelectric power does not contribute to air pollution because no fuel is burned. However, it does present environmental concerns. Building a hydroelectric plant usually involves constructing a dam across a river. The dam raises the water level high enough to produce the energy required for electricity generation. Many acres behind the dam are flooded, destroying land habitats and changing part of the river into a lake.

Energy Oil and natural gas are used to produce over 60 percent of the energy supply in the United States. Over half of the oil used is imported from other countries. Many scientists suggest that emissions from the burning of fossil fuels are principally responsible for global warming. In your Science Journal, write what you might do to persuade utility companies to increase their use of water, wind, and solar power.

Wind Power Wind power is another renewable energy source that can be used for electricity production. Wind turns the blades of a turbine, which powers an electric generator. When winds blow at least 32 km/h, energy is produced. Wind power does not cause air pollution, but electricity can be produced only when the wind is blowing. So far, wind power accounts for only a small percentage of the electricity used worldwide.

Nuclear Power Another alternative to fossil fuels makes use of the huge amounts of energy in the nuclei of atoms, as shown in **Figure 6. Nuclear energy** is released when billions of atomic nuclei from uranium, a radioactive element, are split apart in a nuclear fission reaction. This energy is used to produce steam that rotates the turbine blades of an electric generator.

Nuclear power does not contribute to air pollution. However, uranium is a nonrenewable resource, and mining it can disrupt ecosystems. Nuclear power plants also produce radioactive wastes that can seriously harm living organisms. Some of these wastes remain radioactive for thousands of years, and their safe disposal is a problem that has not yet been solved. Accidents also are a danger.

Figure 6 Nuclear power plants are designed to withstand the high energy produced by nuclear reactions.
Describe *how heat is produced in a nuclear reactor.*

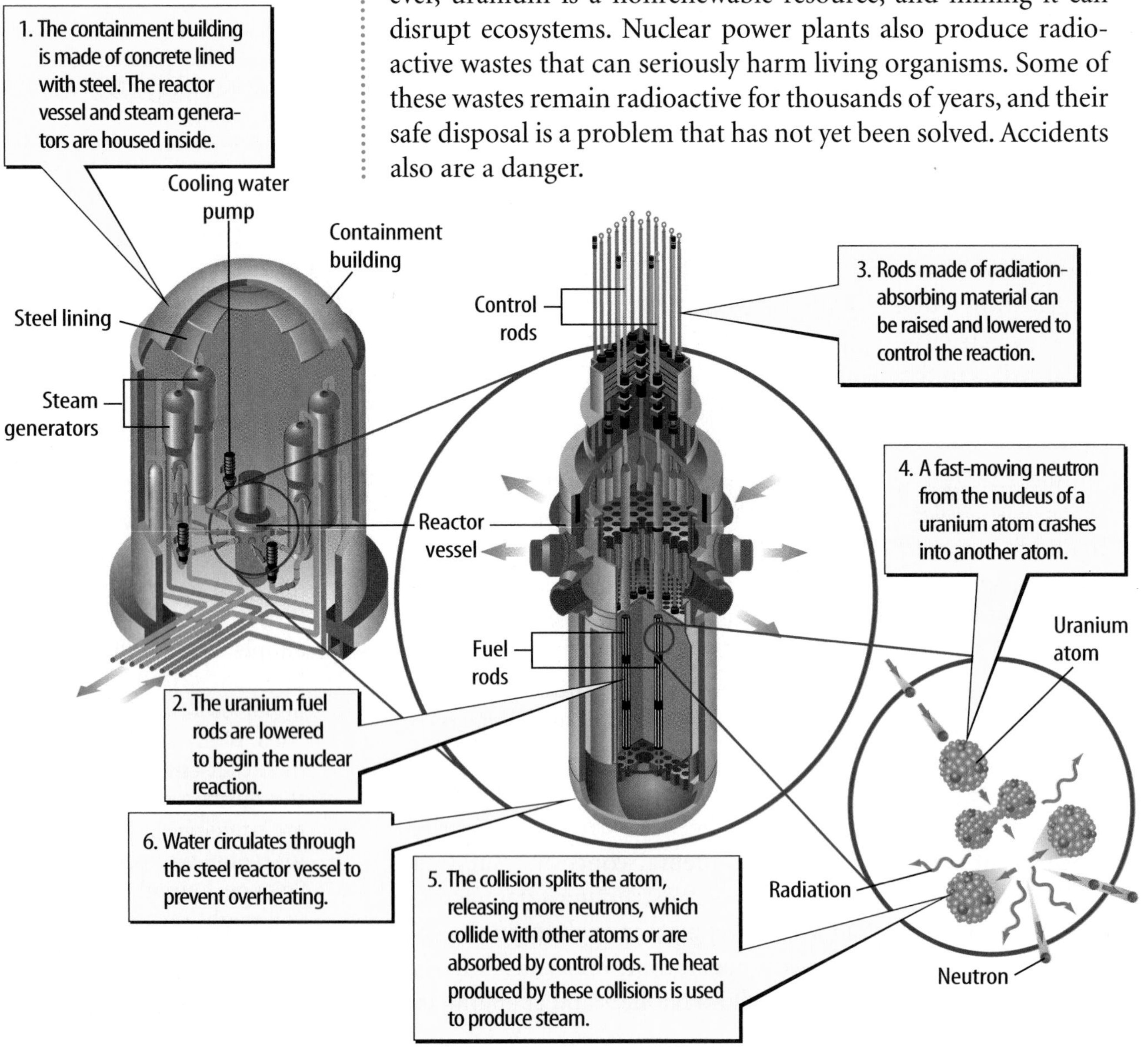

1. The containment building is made of concrete lined with steel. The reactor vessel and steam generators are housed inside.
Cooling water pump
Containment building
Steel lining
Steam generators
Control rods
3. Rods made of radiation-absorbing material can be raised and lowered to control the reaction.
4. A fast-moving neutron from the nucleus of a uranium atom crashes into another atom.
Reactor vessel
Uranium atom
Fuel rods
2. The uranium fuel rods are lowered to begin the nuclear reaction.
6. Water circulates through the steel reactor vessel to prevent overheating.
5. The collision splits the atom, releasing more neutrons, which collide with other atoms or are absorbed by control rods. The heat produced by these collisions is used to produce steam.
Radiation
Neutron

Geothermal Energy The hot, molten rock that lies deep beneath Earth's surface is also a source of energy. You see the effects of this energy when lava and hot gases escape from an erupting volcano or when hot water spews from a geyser. The heat energy contained in Earth's crust is called **geothermal energy.** Most geothermal power plants use this energy to produce steam to generate electricity.

Geothermal energy for power plants is available only where natural geysers or volcanoes are found. A geothermal power plant in California uses steam produced by geysers. The island nation of Iceland was formed by volcanoes, and geothermal energy is plentiful there. Geothermal power plants supply heat and electricity to about 90 percent of the homes in Iceland. Outdoor swimming areas also are heated with geothermal energy, as shown in **Figure 7.**

Figure 7 In Iceland, a geothermal power plant pumps hot water out of the ground to heat buildings and generate electricity. Leftover hot water goes into this lake, making it warm enough for swimming even when the ground is covered with snow.

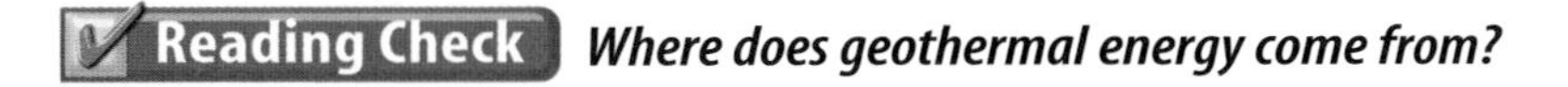

Reading Check *Where does geothermal energy come from?*

Solar Energy The most inexhaustible source of energy for all life on Earth is the Sun. Solar energy is an alternative to fossil fuels. One use of solar energy is in solar-heated buildings. During winter in the northern hemisphere, the parts of a building that face south receive the most sunlight. Large windows placed on the south side of a building help heat it by allowing warm sunshine into the building during the day. Floors and walls of most solar-heated buildings are made of materials that absorb heat during the day. During the night, the stored heat is released slowly, keeping the building warm. **Figure 8** shows how solar energy can be used.

Figure 8 The Zion National Park Visitor Center in Utah is a solar-heated building designed to save energy. The roof holds solar panels that are used to generate electricity. High windows can be opened to circulate air and help cool the building on hot days. The overhanging roof shades the windows during summer.

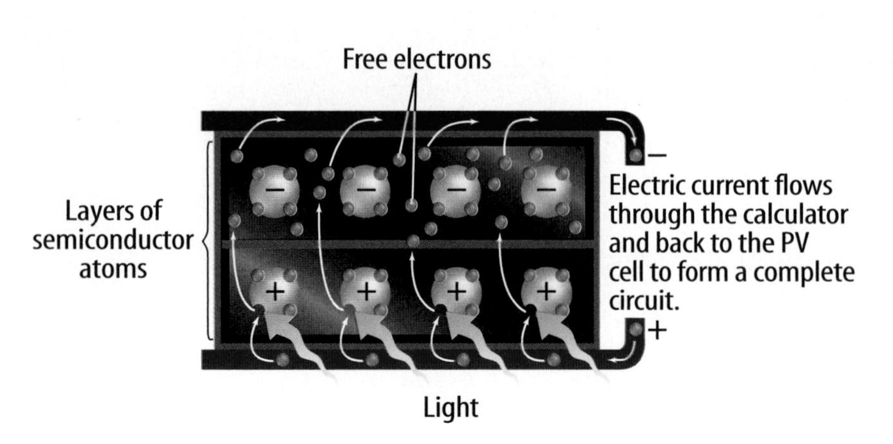

Figure 9 Light energy from the Sun travels in tiny packets of energy called photons. Photons crash into the atoms of PV cells, knocking electrons loose. These electrons create an electric current.

Solar Cells Do you know how a solar-powered calculator works? How do spacecraft use sunlight to generate electricity? These devices use photovoltaic (foh toh vohl TAY ihk) cells to turn sunlight into electric current, as shown in **Figure 9.** Photovoltaic (PV) cells are small and easy to use. However, they produce electricity only in sunlight, so batteries are needed to store electricity for use at night or on cloudy days. Also, PV cells presently are too expensive to use for generating large amounts of electricity. Improvements in this technology continue to be made, and prices probably will go down in the future. As **Figure 10** shows, solar buildings and PV cells are just two of the many ways solar energy can be used to replace fossil fuels.

section 1 review

Summary

Natural Resources

- All living things depend on natural resources to survive.
- Some resources are renewable, while other resources, such as petroleum, are nonrenewable.

Fossil Fuels

- Most of the energy that humans use comes from fossil fuels.
- Fossil fuels must be burned to release the energy stored in them, which causes air pollution.

Alternatives to Fossil Fuels

- Alternatives to fossil fuels include water power, wind power, nuclear power, geothermal energy, and solar energy.
- The Sun provides the most inexhaustible supply of energy for all life on Earth.

Self Check

1. **Summarize** What are natural resources?
2. **Compare and contrast** renewable and nonrenewable resources. Give five examples of each.
3. **Describe** the advantages and disadvantages of using nuclear power.
4. **Describe** two ways solar energy can be used to reduce fossil fuel use.
5. **Think Critically** Explain why the water that is used to cool the reactor vessel of a nuclear power plant is kept separate from the water that is heated to produce steam for the turbine generators.

Applying Math

6. **Solve One-Step Equations** Most cars in the U.S. are driven about 10,000 miles each year. If a car can travel 30 miles on one gallon of gasoline, how many gallons will it use in a year?

NATIONAL GEOGRAPHIC

VISUALIZING SOLAR ENERGY

Figure 10

Sunlight is a renewable energy source that provides an alternative to fossil fuels. Solar technologies use the Sun's energy in many ways—from heating to electricity generation.

▲ POWER PLANTS In California's Mojave Desert, an experimental solar power plant used hundreds of mirrors to focus sunlight on a water-filled tower. The steam produced by this system generates enough electricity to power 2,400 homes.

▼ ELECTRICITY Photovoltaic (PV) cells turn sunlight into electric current. They are commonly used to power small devices, such as calculators. Panels that combine many PV cells provide enough electricity for a home—or an orbiting satellite, such as the International Space Station, below.

▼ COOKING In hot, sunny weather, a solar oven or panel cooker can be used to cook a pot of rice or heat water. The powerful solar cooker shown below reaches even higher temperatures. It is being used to fry food.

▼ INDOOR HEATING South-facing windows and heat-absorbing construction materials turn a room into a solar collector that can help heat an entire building, such as this Connecticut home.

▲ WATER HEATING Water is heated as it flows through small pipes in this roof-mounted solar heat collector. The hot water then flows into an insulated tank for storage.

section 2

Pollution

as you read

What You'll Learn

- **Describe** types of air pollution.
- **Identify** causes of water pollution.
- **Explain** methods that can be used to prevent erosion.

Why It's Important

By understanding the causes of pollution, you can help solve pollution problems.

Review Vocabulary
atmosphere: the whole mass of air surrounding Earth

New Vocabulary
- pollutant
- acid precipitation
- greenhouse effect
- ozone depletion
- erosion
- hazardous waste

Keeping the Environment Healthy

More than six billion people live on Earth. This large human population puts a strain on the environment, but each person can make a difference. You can help safeguard the environment by paying attention to how your use of natural resources affects air, land, and water.

Air Pollution

On a still, sunny day in almost any large city, you might see a dark haze in the air, like that in **Figure 11.** The haze comes from pollutants that form when wood or fuels are burned. A **pollutant** is a substance that contaminates the environment. Air pollutants include soot, smoke, ash, and gases such as carbon dioxide, carbon monoxide, nitrogen oxides, and sulfur oxides. Wherever cars, trucks, airplanes, factories, homes, or power plants are found, air pollution is likely. Air pollution also can be caused by volcanic eruptions, wind-blown dust and sand, forest fires, and the evaporation of paints and other chemicals.

Smog is a form of air pollution created when sunlight reacts with pollutants produced by burning fuels. It can irritate the eyes and make breathing difficult for people with asthma or other lung diseases. Smog can be reduced if people take buses or trains instead of driving or if they use vehicles, such as electric cars, that produce fewer pollutants than gasoline-powered vehicles.

Figure 11 The term *smog* was used for the first time in the early 1900s to describe the mixture of smoke and fog that often covers large cities in the industrial world. **Infer** *how smog can be reduced in large cities.*

Figure 12 Compare these two photographs of the same statue. The photo on the left was taken before acid rain became a problem. The photo on the right shows acid rain damage. The pH scale, shown below, indicates whether a solution is acidic or basic.

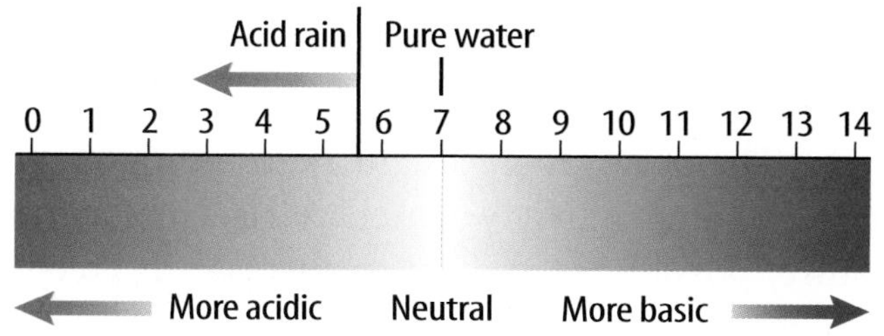

Acid Precipitation

INTEGRATE Chemistry

Water vapor condenses on dust particles in the air to form droplets that combine to create clouds. Eventually, the droplets become large enough to fall to the ground as precipitation—mist, rain, snow, sleet, or hail. Air pollutants from the burning of fossil fuels can react with water in the atmosphere to form strong acids. Acidity is measured by a value called pH, as shown in **Figure 12. Acid precipitation** has a pH below 5.6.

Effects of Acid Rain Acid precipitation washes nutrients from the soil, which can lead to the death of trees and other plants. Runoff from acid rain that flows into a lake or pond can lower the pH of the water. If algae and microscopic organisms cannot survive in the acidic water, fish and other organisms that depend on them for food also die.

Preventing Acid Rain Sulfur from burning coal and nitrogen oxides from vehicle exhaust are the pollutants primarily responsible for acid rain. Using low-sulfur fuels, such as natural gas or low-sulfur coal, can help reduce acid precipitation. However, these fuels are less plentiful and more expensive than high-sulfur coal. Smokestacks that remove sulfur dioxide before it enters the atmosphere also help. Reducing automobile use and keeping car engines properly tuned can reduce acid rain caused by nitrogen oxide pollution. The use of electric cars, or hybrid-fuel cars that can run on electricity as well as gasoline, also could help.

Measuring Acid Rain

Procedure

1. Collect **rainwater** by placing a clean **cup** outdoors. Do not collect rainwater that has been in contact with any object or organism.
2. Dip a piece of **pH indicator paper** into the sample.
3. Compare the color of the paper to the pH chart provided. Record the pH of the rainwater.
4. Use separate pieces of pH paper to test the pH of **tap water** and **distilled water**. Record these results.

Analysis

1. Is the rainwater acidic, basic, or neutral?
2. How does the pH of the rainwater compare with the pH of tap water? With the pH of distilled water?

Topic: Global Warming

Visit green.msscience.com for Web links to information about global warming.

Activity Describe three possible impacts of global warming. Provide one fact that supports global warming and one fact that does not.

Greenhouse Effect

Sunlight travels through the atmosphere to Earth's surface. Some of this sunlight normally is reflected back into space. The rest is trapped by certain atmospheric gases, as shown in **Figure 13.** This heat-trapping feature of the atmosphere is the **greenhouse effect.** Without it, temperatures on Earth probably would be too cold to support life.

Atmospheric gases that trap heat are called greenhouse gases. One of the most important greenhouse gases is carbon dioxide (CO_2). CO_2 is a normal part of the atmosphere. It is also a waste product that forms when fossil fuels are burned. Over the past century, more fossil fuels have been burned than ever before, which is increasing the percentage of CO_2 in the atmosphere. The atmosphere might be trapping more of the Sun's heat, making Earth warmer. A rise in Earth's average temperature, possibly caused by an increase in greenhouse gases, is known as global warming.

Global Warming Temperature data collected from 1895 through 1995 indicate that Earth's average temperature increased about 1°C during that 100-year period. No one is certain whether this rise was caused by human activities or is a natural part of Earth's weather cycle. What kinds of changes might be caused by global warming? Changing rainfall patterns could alter ecosystems and affect the kinds of crops that can be grown in different parts of the world. The number of storms and hurricanes might increase. The polar ice caps might begin to melt, raising sea levels and flooding coastal areas. Warmer weather might allow tropical diseases, such as malaria, to become more widespread. Many people feel that the possibility of global warming is a good reason to reduce fossil fuel use.

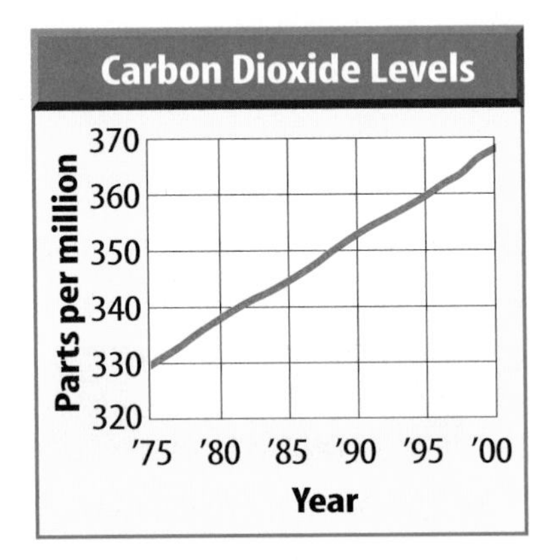

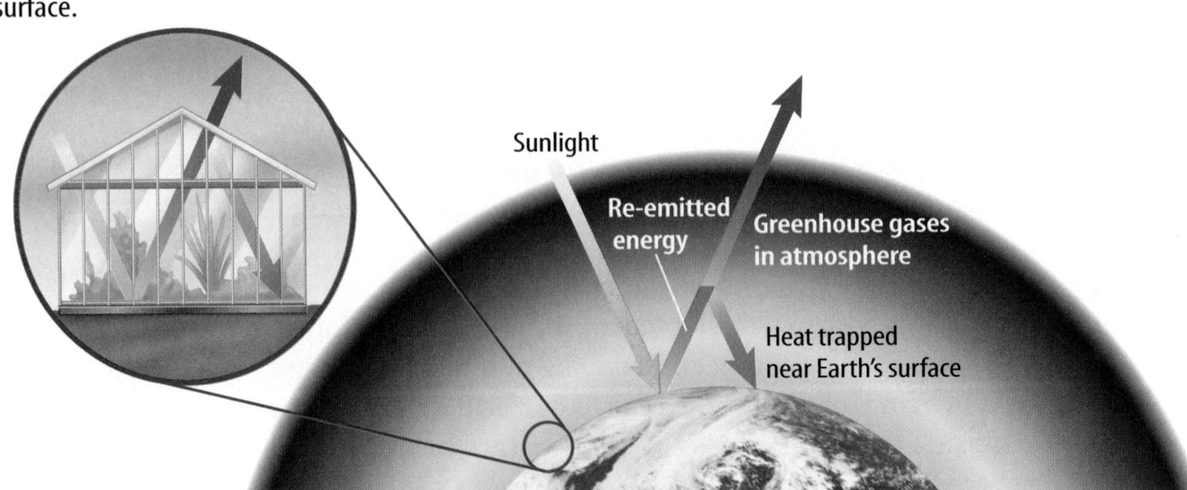

Figure 13 The moment you step inside a greenhouse, you feel the results of the greenhouse effect. Heat trapped by the glass walls warms the air inside. In a similar way, atmospheric greenhouse gases trap heat close to Earth's surface.

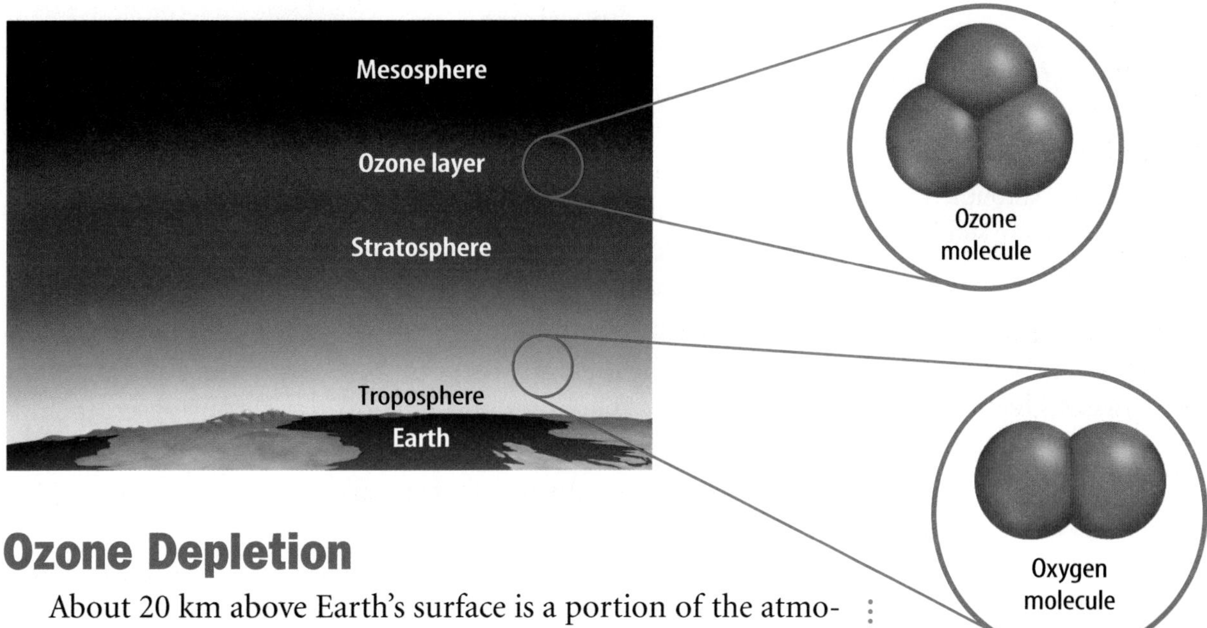

Figure 14 The atmosphere's ozone layer absorbs large amounts of UV radiation, preventing it from reaching Earth's surface. Ozone molecules are made of three oxygen atoms. They are formed in a chemical reaction between sunlight and oxygen. The oxygen you breathe has two oxygen atoms in each molecule.
Infer *what will happen if the ozone layer continues to thin.*

Ozone Depletion

About 20 km above Earth's surface is a portion of the atmosphere known as the ozone (OH zohn) layer. Ozone is a form of oxygen, as shown in **Figure 14.** The ozone layer absorbs some of the Sun's harmful ultraviolet (UV) radiation. UV radiation can damage living cells.

Every year, the ozone layer temporarily becomes thinner over each polar region during its spring season. The thinning of the ozone layer is called **ozone depletion.** This problem is caused by certain pollutant gases, especially chlorofluorocarbons (klor oh FLOR oh kar bunz) (CFCs). CFCs are used in the cooling systems of refrigerators, freezers, and air conditioners. When CFCs leak into the air, they slowly rise into the atmosphere until they arrive at the ozone layer. CFCs react chemically with ozone, breaking apart the ozone molecules.

UV Radiation Because of ozone depletion, the amount of UV radiation that reaches Earth's surface could be increasing. UV radiation could be causing a rise in the number of skin cancer cases in humans. It also might be harming other organisms. The ozone layer is so important to the survival of life on Earth that world governments and industries have agreed to stop making and using CFCs.

Ozone that is high in the upper atmosphere protects life on Earth. Near Earth's surface though, it can be harmful. Ozone is produced when fossil fuels are burned. This ozone stays in the lower atmosphere, where it pollutes the air. Ozone damages the lungs and other sensitive tissues of animals and plants. For example, it can cause the needles of a Ponderosa pine to drop, harming growth.

Reading Check *What is the difference between ozone in the upper atmosphere and ozone in the lower atmosphere?*

Air Quality Carbon monoxide enters the body through the lungs. It attaches to red blood cells, preventing the cells from absorbing oxygen. In your Science Journal, explain why heaters and barbecues designed for outdoor use never should be used indoors.

Indoor Air Pollution

Air pollution can occur indoors. Today's buildings are better insulated to conserve energy. However, better insulation reduces the flow of air into and out of a building, so air pollutants can build up indoors. For example, burning cigarettes release hazardous particles and gases into the air. Even nonsmokers can suffer ill effects from secondhand cigarette smoke. As a result, smoking no longer is allowed in many public and private buildings. Paints, carpets, glues and adhesives, printers, and photocopy machines also give off dangerous gases, including formaldehyde. Like cigarette smoke, formaldehyde is a carcinogen, which means it can cause cancer.

Carbon Monoxide Carbon monoxide (CO) is a poisonous gas that is produced whenever charcoal, natural gas, kerosene, or other fuels are burned. CO poisoning can cause serious illness or death. Fuel-burning stoves and heaters must be designed to prevent CO from building up indoors. CO is colorless and odorless, so it is difficult to detect. Alarms that provide warning of a dangerous buildup of CO are being used in more and more homes.

Radon Radon is a naturally occurring, radioactive gas that is given off by some types of rock and soil, as shown in **Figure 15.** Radon has no color or odor. It can seep into basements and the lower floors of buildings. Radon exposure is the second leading cause of lung cancer in this country. A radon detector sounds an alarm when levels of the gas in indoor air become too high. If radon is present, increasing a building's ventilation can eliminate any damaging effects.

Figure 15 The map shows the potential for radon exposure in different parts of the United States. **Identify** *the area of the country with soils that produce the most radon gas.*

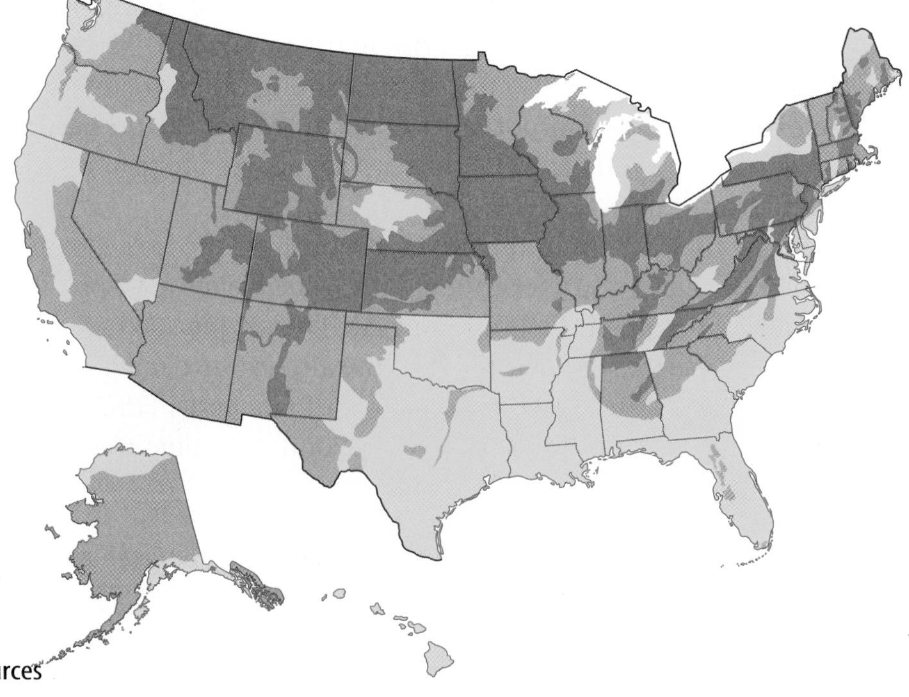

Geologic radon potential
Low Moderate High

When rain falls on roads and parking lots, it can wash oil and grease onto the soil and into nearby streams.

Rain can wash agricultural pesticides and fertilizers into lakes, streams, or oceans.

Industrial wastes are sometimes released directly into surface waters.

Figure 16 Pollution of surface waters can occur in several ways, as shown above.

Water Pollution

Pollutants enter water, too. Air pollutants can drift into water or be washed out of the sky by rain. Rain can wash land pollutants into waterways, as shown in **Figure 16.** Wastewater from factories and sewage-treatment plants often is released into waterways. In the United States and many other countries, laws require that wastewater be treated to remove pollutants before it is released. But, in many parts of the world, wastewater treatment is not always possible. Pollution also enters water when people dump litter or waste materials into rivers, lakes, and oceans.

Surface Water Some water pollutants poison fish and other wildlife, and can be harmful to people who swim in or drink the water. For example, chemical pesticides sprayed on farmland can wash into lakes and streams. These chemicals can harm the insects that fish, turtles, or frogs rely on for food. Shortages of food can lead to deaths among water-dwelling animals. Some pollutants, especially those containing mercury and other metals, can build up in the tissues of fish. Eating contaminated fish and shellfish can transfer these metals to people, birds, and other animals. In some areas, people are advised not to eat fish or shellfish taken from polluted waterways.

Algal blooms are another water pollution problem. Raw sewage and excess fertilizer contain large amounts of nitrogen. If they are washed into a lake or pond, they can cause the rapid growth of algae. When the algae die, they are decomposed by huge numbers of bacteria that use up much of the oxygen in the water. Fish and other organisms can die from a lack of oxygen in the water.

Figure 17 In 1996, the oil tanker *Sea Empress* spilled more than 72 million kg of oil into the sea along the coast of Wales. More than $40 million was spent on the cleanup effort, but thousands of ocean organisms were destroyed, including birds, fish, and shellfish.

Ocean Water Rivers and streams eventually flow into oceans, bringing their pollutants along. Also, polluted water can enter the ocean in coastal areas where factories, sewage-treatment plants, or shipping activities are located. Oil spills are a well-known ocean pollution problem. About 4 billion kg of oil are spilled into ocean waters every year. Much of that oil comes from ships that use ocean water to wash out their fuel tanks. Oil also can come from oil tanker wrecks, as shown in **Figure 17.**

Groundwater Pollution can affect water that seeps underground, as shown in **Figure 18.** Groundwater is water that collects between particles of soil and rock. It comes from precipitation and runoff that soaks into the soil. This water can flow slowly through permeable layers of rock called aquifers. If this water comes into contact with pollutants as it moves through the soil and into an aquifer, the aquifer could become polluted. Polluted groundwater is difficult—and sometimes impossible—to clean. In some parts of the country, chemicals leaking from underground storage tanks have created groundwater pollution problems.

Figure 18 Water from rainfall slowly filters through sand or soil until it is trapped in underground aquifers. Pollutants picked up by the water as it filters through the soil can contaminate water wells.

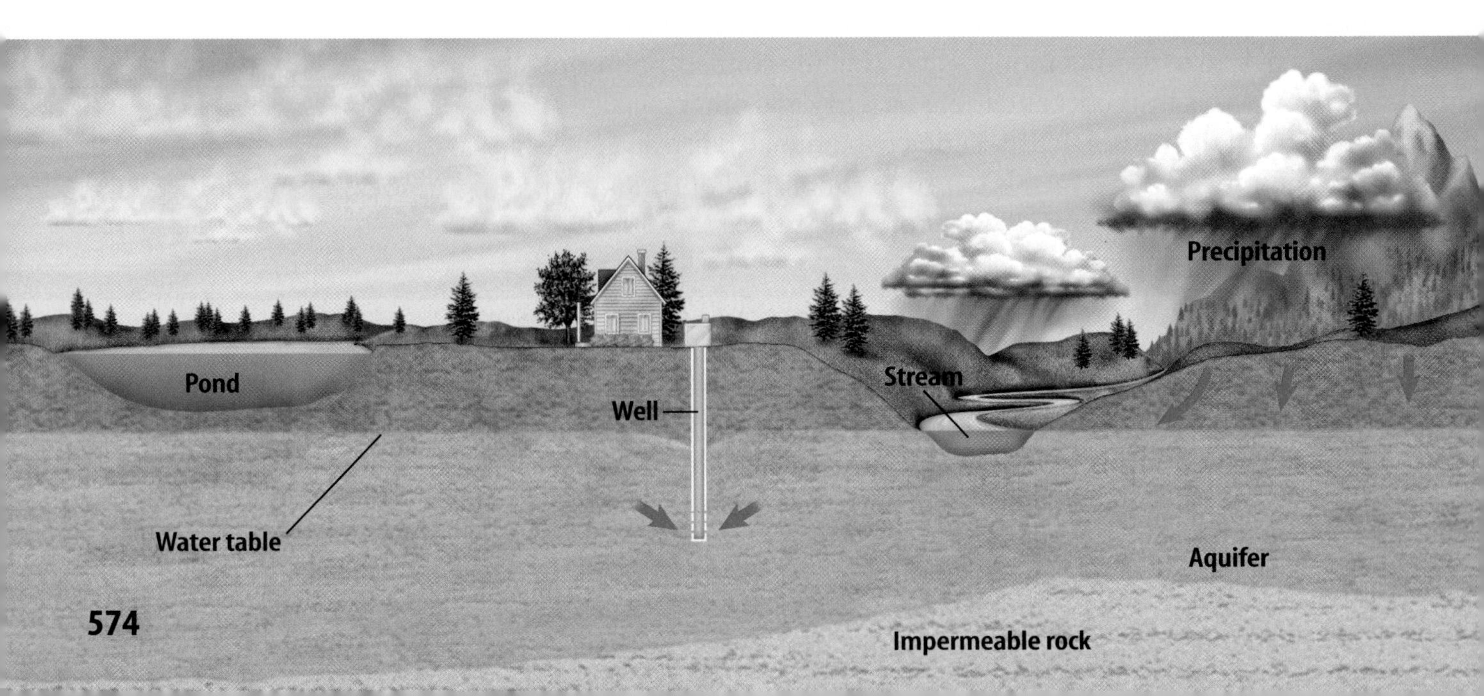

Contour plowing reduces the downhill flow of water.

Figure 19 The farming methods shown here help prevent soil erosion.
Infer *why soil erosion is a concern for farmers.*

On steep hillsides, flat areas called terraces reduce downhill flow.

In strip cropping, cover crops are planted between rows to reduce wind erosion.

In no-till farming, soil is never left bare.

Soil Loss

Fertile topsoil is important to plant growth. New topsoil takes hundreds or thousands of years to form. The Launch Lab at the beginning of this chapter shows that rain washes away loose topsoil. Wind also blows it away. The movement of soil from one place to another is called **erosion** (ih ROH zhun). Eroded soil that washes into a river or stream can block sunlight and slow photosynthesis. It also can harm fish, clams, and other organisms. Erosion is a natural process, but human activities increase it. When a farmer plows a field or a forest is cut down, soil is left bare. Bare soil is more easily carried away by rain and wind. **Figure 19** shows some methods farmers use to reduce soil erosion.

Soil Pollution

Soil can become polluted when air pollutants drift to the ground or when water leaves pollutants behind as it flows through the soil. Soil also can be polluted when people toss litter on the ground or dispose of trash in landfills.

Solid Wastes What happens to the trash you throw out every week? What do people do with old refrigerators, TVs, and toys? Most of this solid waste is dumped in landfills. Most landfills are designed to seal out air and water. This helps prevent pollutants from seeping into surrounding soil, but it slows normal decay processes. Even food scraps and paper, which usually break down quickly, can last for decades in a landfill. In populated areas, landfills fill up quickly. Reducing the amount of trash people generate can reduce the need for new landfills.

Figure 20 Leftover paints, batteries, pesticides, drain cleaners, and medicines are hazardous wastes that should not be discarded in the trash. They should never be poured down a drain, onto the ground, or into a storm sewer. Most communities have collection facilities where people can dispose of hazardous materials like these.

Hazardous Wastes Waste materials that are harmful to human health or poisonous to living organisms are **hazardous wastes.** They include dangerous chemicals, such as pesticides, oil, and petroleum-based solvents used in industry. They also include radioactive wastes from nuclear power plants, from hospitals that use radioactive materials to treat disease, and from nuclear weapons production. Many household items also are considered hazardous, such as those shown in **Figure 20.** If these materials are dumped into landfills, they could seep into the soil, surface water, or groundwater over time. Hazardous wastes usually are handled separately from trash. They are treated in ways that prevent environmental pollution.

Reading Check *What are hazardous wastes?*

section 2 review

Summary

Air Pollution and Acid Precipitation

- Vehicles, volcanoes, forest fires, and even wind-blown dust and sand can cause air pollution.
- Acid rain washes nutrients from the soil, which can harm plants.

Greenhouse Effect and Ozone Depletion

- CO_2 is a greenhouse gas that helps warm Earth.
- The ozone layer protects life on Earth.

Indoor Air Pollution, Water Pollution, Soil Loss, and Soil Pollution

- Pollutants can build up inside of buildings.
- There are many sources of water pollutants.
- Wind and rain can erode bare soil.
- Pollutants in soil decay more slowly than in air.

Self Check

1. **List** four ways that air pollution affects the environment.
2. **Explain** how an algal bloom can affect other pond organisms.
3. **Describe** possible causes and effects of ozone depletion.
4. **Think Critically** How could hazardous wastes in landfills eventually affect groundwater?

Applying Math

5. **Solve a One-Step Equation** A solution of pH 4 is 10 times more acidic than one of pH 5, and it is 10 times more acidic than a solution of pH 6. How many times more acidic is the solution of pH 4 than the one of pH 6?

green.msscience.com/self_check_quiz

The Greenhouse Effect

You can create models of Earth with and without heat-reflecting green-house gases. Then, experiment with the models to observe the greenhouse effect.

Real-World Question

How does the greenhouse effect influence temperatures on Earth?

Goals

- **Observe** the greenhouse effect.
- **Describe** the effect that a heat source has on an environment.

Materials

1-L clear-plastic, soft-drink bottles with tops cut off and labels removed (2)
thermometers (2)
**temperature probe*
potting soil
masking tape
plastic wrap
rubber band
lamp with 100-W lightbulb
watch or clock with second hand
**Alternate materials*

Safety Precautions

Procedure

1. Copy the data table and use it to record your temperature measurements.
2. Put an equal volume of potting soil in the bottom of each container.
3. Use masking tape to attach a thermometer to the inside of each container. Place each thermometer at the same height above the soil. Shield each thermometer bulb by putting a double layer of masking tape over it.

Changes in Temperature

Time (min)	Open Container Temperature (°C)	Closed Container Temperature (°C)
0		
2	Do not write in this book.	
4		
6		

4. Seal the top of one container with plastic wrap held in place with a rubber band.
5. Place the lamp with the exposed 100-W lightbulb between the two containers and exactly 1 cm away from each. Do not turn on the light.
6. Let the setup sit for 5 min, then record the temperature in each container.
7. Turn on the light. Record the temperature in each container every 2 min for 15 min to 20 min. Graph the results.

Conclude and Apply

1. **Compare and contrast** temperatures in each container at the end of the experiment.
2. **Infer** What does the lightbulb represent in this experimental model? What does the plastic wrap represent?

Communicating Your Data

Average the data obtained in the experiments conducted by all the groups in your class. Prepare a line graph of these data. **For more help, refer to the** Science Skill Handbook.

section 3

The Three Rs of Conservation

as you read

What You'll Learn

- **Recognize** ways you can reduce your use of natural resources.
- **Explain** how you can reuse resources to promote conservation.
- **Describe** how many materials can be recycled.

Why It's Important

Conservation preserves resources and reduces pollution.

Review Vocabulary
reprocessing: to subject to a special process or treatment in preparation for reuse

New Vocabulary
- recycling

Conservation

A teacher travels to school in a car pool. In the school cafeteria, students place glass bottles and cans in separate containers from the rest of the garbage. Conservation efforts like these can help prevent shortages of natural resources, slow growth of landfills, reduce pollution levels, and save people money. Every time a new landfill is created, an ecosystem is disturbed. Reducing the need for landfills is a major benefit of conservation. The three Rs of conservation are reduce, reuse, and recycle.

Reduce

You contribute to conservation whenever you reduce your use of natural resources. You use less fossil fuel when you walk or ride a bicycle instead of taking the bus or riding in a car. If you buy a carton of milk, reduce your use of petroleum by telling the clerk you don't need a plastic bag to carry it in.

You also can avoid buying things you don't need. For example, most of the paper, plastic, and cardboard used to package items for display on store shelves is thrown away as soon as the product is brought home. You can look for products with less packaging or with packaging made from recycled materials. What are some other ways you can reduce your use of natural resources?

Figure 21 Worn-out automobile tires can have other useful purposes.

Reuse

Another way to help conserve natural resources is to use items more than once. Reusing an item means using it again without changing it or reprocessing it, as shown in **Figure 21.** Bring reusable canvas bags to the grocery store to carry home your purchases. Donate clothes you've outgrown to charity so that others can reuse them. Take reusable plates and utensils on picnics instead of disposable paper items.

Recycle

If you can't avoid using an item, and if you can't reuse it, the next best thing is to recycle it. **Recycling** is a form of reuse that requires changing or reprocessing an item or natural resource. If your city or town has a curbside recycling program, you already separate recyclables from the rest of your garbage. Materials that can be recycled include glass, metals, paper, plastics, and yard and kitchen waste.

Reading Check *How is recycling different from reusing?*

Plastics Plastic is more difficult to recycle than other materials, mainly because several types of plastic are in use. A recycle code marked on every plastic container indicates the type of plastic it is made of. Plastic soft-drink bottles, like the one shown in **Figure 22,** are made of type 1 plastic and are the easiest to recycle. Most plastic bags are made of type 2 or type 4 plastic; they can be reused as well as recycled. Types 6 and 7 can't be recycled at all because they are made of a mixture of different plastics. Each type of plastic must be separated carefully before it is recycled because a single piece of a different type of plastic can ruin an entire batch.

Figure 22 Many soft-drink bottles are made of PETE, which is the most common type of recyclable plastic. It can be melted down and spun into fibers to make carpets, paintbrushes, rope, and clothing. **Identify** *other products made out of recycled materials.*

Topic: Recycling

Visit green.msscience.com for Web links to information about recycling bottles and cans.

Activity Write one argument in support of a money deposit for bottles and cans and one argument against it. Provide data to support one of your arguments.

Metals The manufacturing industry has been recycling all kinds of metals, especially steel, for decades. At least 25 percent of the steel in cans, appliances, and automobiles is recycled steel. Up to 100 percent of the steel in plates and beams used to build skyscrapers is made from reprocessed steel. About one metric ton of recycled steel saves about 1.1 metric tons of iron ore and 0.5 metric ton of coal. Using recycled steel to make new steel products reduces energy use by 75 percent. Other metals, including iron, copper, aluminum, and lead also can be recycled.

You can conserve metals by recycling food cans, which are mostly steel, and aluminum cans. It takes less energy to make a can from recycled aluminum than from raw materials. Also, remember that recycled cans do not take up space in landfills.

Glass When sterilized, glass bottles and jars can be reused. They also can be melted and re-formed into new bottles, especially those made of clear glass. Most glass bottles already contain at least 25 percent recycled glass. Glass can be recycled again and again. It never needs to be thrown away. Recycling about one metric ton of glass saves more than one metric ton of mineral resources and reduces the energy used to make new glass by 25 percent or more.

Applying Science

What items are you recycling at home?

Many communities have recycling programs. Recyclable items may be picked up at the curbside, taken to a collection site, or the resident may hire a licensed recycling handler to pick them up. What do you recycle in your home?

Recycling Rates of Key Household Items

Percent: 0, 10, 20, 30, 40, 50, 60, 70

1990 | 1995 | 2000

Aluminum cans | Yard waste | Old newsprint | Steel cans | Plastic soda bottles | Glass containers

Source: U.S. EPA, 2003

Identifying the Problem

This bar graph shows the recycling rates in the U. S. of six types of household items for the years 1990, 1995, and 2000. What are you and your classmates' recycling rates?

Solving the Problem

For one week, list each glass, plastic, and aluminum item you use. Note which items you throw away and which ones you recycle. Calculate the percentage of glass, plastic, and aluminum you recycled. How do your percentages compare with those on the graph?

Paper Used paper is recycled into paper towels, insulation, newsprint, cardboard, and stationery. Ranchers and dairy farmers sometimes use shredded paper instead of straw for bedding in barns and stables. Used paper can be made into compost. Recycling about one metric ton of paper saves 17 trees, more than 26,000 L of water, close to 1,900 L of oil, and more than 4,000 kW of electric energy. You can do your part by recycling newspapers, notebook and printer paper, cardboard, and junk mail.

Figure 23 Composting is a way of turning plant material you would otherwise throw away into rich garden soil. Dry leaves and weeds, grass clippings, vegetable trimmings, and nonmeat food scraps can be composted.

Reading Check *What nonrenewable resource(s) do you conserve by recycling paper?*

Compost Grass clippings, leaves, and fruit and vegetable scraps that are discarded in a landfill can remain there for decades without breaking down. The same items can be turned into soil-enriching compost in just a few weeks, as shown in **Figure 23.** Many communities distribute compost bins to encourage residents to recycle fruit and vegetable scraps and yard waste.

Buy Recycled People have become so good at recycling that recyclable materials are piling up, just waiting to be put to use. You can help by reading labels when you shop and choosing products that contain recycled materials. What other ways of recycling natural resources can you think of?

section 3 review

Summary

Conservation

- The three Rs of conservation are reduce, reuse, and recycle.

Reduce

- You can contribute to conservation by reducing your use of natural resources.

Reuse

- Some items can be used more than once, such as reusable canvas bags for groceries.

Recycle

- Some items can be recycled, including some plastics, metal, glass, and paper.
- Grass clippings, leaves, and fruit and vegetable scraps can be composted into rich garden soil.

Self Check

1. **Describe** at least three actions you could take to reduce your use of natural resources.
2. **Describe** how you could reuse three items people usually throw away.
3. **Think Critically** Why is reusing something better than recycling it?

Applying Skills

4. **Make and Use Tables** Make a table of data of the number of aluminum cans thrown away in the United States: 2.7 billion in 1970; 11.1 billion in 1974; 21.3 billion in 1978; 22.7 billion in 1982; 35.0 billion in 1986; 33.8 billion in 1990; 38.5 billion in 1994; 45.5 billion in 1998; 50.7 billion in 2001.

Model and Invent

Solar Cooking

Goals

- **Research** designs for solar panel cookers or box cookers.
- **Design** a solar cooker that can be used to cook food.
- **Plan** an experiment to measure the effectiveness of your solar cooker.

Possible Materials

poster board
cardboard boxes
aluminum foil
string
wire coat hangers
clear plastic sheets
**oven bags*
black cookware
thermometer
stopwatch
**timer*
glue
tape
scissors
**Alternate materials*

Safety Precautions

WARNING: *Be careful when cutting materials. Your solar cooker will get hot. Use insulated gloves or tongs to handle hot objects.*

Real-World Question

The disappearance of forests in some places on Earth has made firewood extremely difficult and expensive to obtain. People living in these regions often have to travel long distances or sell some of their food to get firewood. This can be a serious problem for people who may not have much food to begin with. Is there a way they could cook food without using firewood? How would you design and build a cooking device that uses the Sun's energy?

Make the Model

1. **Design** a solar cooker. In your Science Journal, explain why you chose this design and draw a picture of it.
2. **Write** a summary explaining how you will measure the effectiveness of your solar cooker. What will you measure? How will you collect and organize your data? How will you present your results?

3. **Compare** your solar cooker design to those of other students.
4. Share your experimental plan with students in your class. Discuss the reasoning behind your plan. Be specific about what you intend to test and how you are going to test it.
5. Make sure your teacher approves your plan before you start working on your model.
6. Using all of the information you have gathered, construct a solar cooker that follows your design.

Test the Model

1. **Test** your design to determine how well it works. Try out a classmate's design. How do the two compare?

Analyze Your Data

1. Combine the results for your entire class and decide which type of solar cooker was most effective. How could you design a more effective solar cooker, based on what you learned from this activity?
2. **Infer** Do you think your results might have been different if you tested your solar cooker on a different day? Explain. Why might a solar cooker be more useful in some regions of the world than in others?

Conclude and Apply

1. **Infer** Based on what you've read and the results obtained by you and your classmates, do you think that your solar cooker could boil water? Explain.
2. **Compare** the amount of time needed to cook food with a solar cooker and with more traditional cooking methods. Assuming plenty of sunlight is available, would you prefer to use a solar cooker or a traditional oven? Explain.

Prepare a demonstration showing how to use a solar cooker. Present your demonstration to another class of students or to a group of friends or relatives. **For more help, refer to the** Science Skill Handbook.

Beauty Plagiarized

by Amitabha Mukerjee

I wandered lonely as a cloud –
Except for a motorboat,
Nary a soul in sight.
Beside the lake beneath the trees,
Next to the barbed wire fence,
There was a picnic table
And beer bottle caps from many years.
A boat ramp to the left,
And the chimney from a power station on the
other side,
A summer haze hung in the air,
And the lazy drone of traffic far away.

Crimson autumn of mists and mellow fruitfulness
Blue plastic covers the swimming pools
The leaves fall so I can see
Dark glass reflections in the building
That came up
where the pine cones crunched underfoot . . .

And then it is snow
White lining on trees and rooftops . . .
And through my windshield wipers
The snow is piled dark and grey . . .
Next to my driveway where I check my mail
Little footprints on fresh snow —
A visiting rabbit.

I knew a bank where the wild thyme blew
Over-canopied with luscious woodbine
It is now a landfill —
Fermentation of civilization
Flowers on TV
Hyacinth rose tulip chrysanthemum
Acres of colour
Wind up wrapped in decorous plastic,
In this landfill where oxlips grew. . .

Understanding Literature

Cause and Effect Recognizing cause-and-effect relationships can help you make sense out of what you read. One event causes another event. The second event is the effect of the first event. In the poem, the author describes the causes and effects of pollution and waste. What effects do pollution and the use of nonrenewable resources have on nature in the poem?

Respond to the Reading

1. To plagiarize is to copy without giving credit to the source. In this poem, who or what has plagiarized beauty?
2. What do the four verses in the poem correspond to?
3. **Linking Science and Writing** Write a poem that shows how conservation methods could restore the beauty in nature.

The poet makes a connection between the four seasons of the year and the pollution and waste products created by human activity, or civilization. For example, in the spring, a landfill for dumping garbage replaces a field of wildflowers. Describing four seasons instead of one reinforces the poet's message that the beauty of nature has been stolen, or plagiarized.

Reviewing Main Ideas

Section 1 Resources

1. Natural resources are the parts of the environment that supply materials needed for the survival of living organisms.
2. Renewable resources are being replaced continually by natural processes.
3. Nonrenewable resources cannot be replaced or are replaced very slowly.
4. Energy sources include fossil fuels, wind, solar energy, geothermal energy, hydroelectric power, and nuclear power.

Section 2 Pollution

1. Most air pollution is made up of waste products from the burning of fossil fuels.
2. The greenhouse effect is the warming of Earth by a blanket of heat-reflecting gases in the atmosphere.
3. Water can be polluted by acid rain and by the spilling of oil or other wastes into waterways.
4. Solid wastes and hazardous wastes dumped on land or disposed of in landfills can pollute the soil. Erosion can cause the loss of fertile topsoil.

Section 3 The Three Rs of Conservation

1. You can reduce your use of natural resources in many ways.
2. Reusing items is an excellent way to practice conservation.
3. In recycling, materials are changed in some way so that they can be used again.
4. Materials that can be recycled include paper, metals, glass, plastics, yard waste, and nonmeat kitchen scraps.

Visualizing Main Ideas

Copy and complete the following concept map using the terms smog, acid precipitation, *and* ozone depletion.

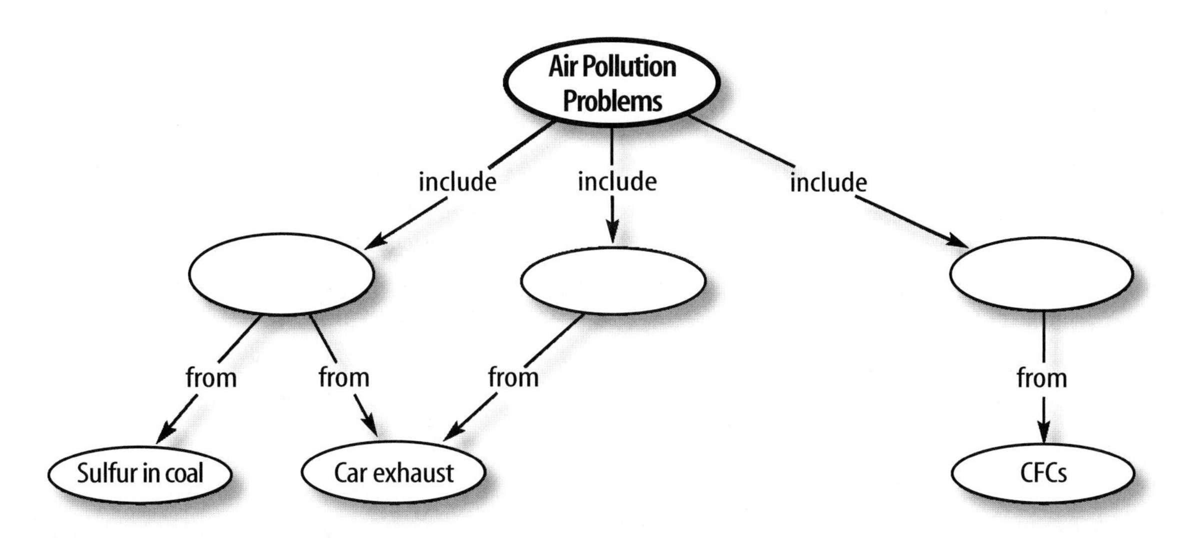

Using Vocabulary

acid precipitation p. 569
erosion p. 575
fossil fuel p. 562
geothermal energy p. 565
greenhouse effect p. 570
hazardous waste p. 576
hydroelectric power p. 563
natural resource p. 560
nonrenewable resource p. 561
nuclear energy p. 564
ozone depletion p. 571
petroleum p. 561
pollutant p. 568
recycling p. 579
renewable resource p. 560

Explain the differences in the vocabulary words given below. Then explain how the words are related. Use complete sentences in your answers.

1. fossil fuel—petroleum
2. erosion—pollutant
3. ozone depletion—acid precipitation
4. greenhouse effect—fossil fuels
5. hazardous wastes—nuclear energy
6. hydroelectric power—fossil fuels
7. acid precipitation—fossil fuels
8. ozone depletion—pollutant
9. recycle—nonrenewable resources
10. geothermal energy—fossil fuels

Checking Concepts

Choose the word or phrase that best answers the question.

11. An architect wants to design a solar house in the northern hemisphere. For maximum warmth, which side of the house should have the most windows?
    A) north **C)** east
    B) south **D)** west

12. Of the following, which is considered a renewable resource?
    A) coal **C)** sunlight
    B) oil **D)** aluminum

Use the photo below to answer question 13.

13. Which energy resource is shown in the photo?
    A) solar energy
    B) geothermal energy
    C) hydroelectric energy
    D) photovoltaic energy

14. Which of the following is a fossil fuel?
    A) wood **C)** nuclear power
    B) oil **D)** photovoltaic cell

15. Which of the following contributes to ozone depletion?
    A) carbon dioxide **C)** CFCs
    B) radon **D)** carbon monoxide

16. What is a substance that contaminates the environment called?
    A) acid rain **C)** pollutant
    B) pollution **D)** ozone

17. If there were no greenhouse effect in Earth's atmosphere, which of the following statements would be true?
    A) Earth would be much hotter.
    B) Earth would be much colder.
    C) The temperature of Earth would be the same.
    D) The polar ice caps would melt.

18. Which of the following can change solar energy into electricity?
    A) photovoltaic cells
    B) smog
    C) nuclear power plants
    D) geothermal power plants

Science Online green.msscience.com/vocabulary_puzzlemaker

Thinking Critically

19. **Explain** how geothermal energy is used to produce electricity.
20. **Infer** why burning wood and burning fossil fuels produce similar pollutants.

Use the photos below to answer question 21.

21. **Draw a Conclusion** Which would make a better location for a solar power plant—a polar region (left) or a desert region (right)? Why?
22. **Explain** why it is beneficial to grow a different crop on soil after the major crop has been harvested.
23. **Infer** Is garbage a renewable resource? Why or why not?
24. **Summarize** Solar, nuclear, wind, water, and geothermal energy are alternatives to fossil fuels. Are they all renewable? Why or why not?
25. **Draw Conclusions** Would you save more energy by recycling or reusing a plastic bag?
26. **Recognize Cause and Effect** Forests use large amounts of carbon dioxide during photosynthesis. How might cutting down a large percentage of Earth's forests affect the greenhouse effect?
27. **Form a hypothesis** about why Americans throw away more aluminum cans each year.
28. **Compare and contrast** contour farming, terracing, strip cropping, and no-till farming.

Performance Activities

29. **Poster** Create a poster to illustrate and describe three things students at your school can do to conserve natural resources.

Applying Math

Use the table below to answer questions 30 and 31.

Estimated Recycling Rates

Item	Percent Recycled
Aluminum cans	60
Glass beverage bottles	31
Plastic soft-drink containers	37
Newsprint	56
Magazines	23

30. **Recycling Rates** Make a bar graph of the data above.
31. **Bottle Recycling** For every 1,000 glass beverage bottles that are produced, how many are recycled?
32. **Nonrenewable Resources** 45.8 billion (45,800,000,000) cans were thrown away in 2000. If it takes 33.79 cans to equal one pound and the average scrap value is \$0.58/lb, then what was the total dollar value of the discarded cans?
33. **Ozone Depletion** The thin ozone layer called the "ozone hole" over Antarctica reached nearly 27,000,000 km^2 in 1998. To conceptualize this, the United States has a geographical area of 9,363,130 km^2. How much larger is the "ozone hole" in comparison to the United States?
34. **Increased CO_2 Levels** To determine the effects of increased CO_2 levels in the atmosphere, scientists increased the CO_2 concentration by 70 percent in an enclosed rain forest environment. If the initial CO_2 concentration was 430 parts per million, what was it after the increase?

Standardized Test Practice

Part 1 Multiple Choice

Record your answers on the answer sheet provided by your teacher or on a sheet of paper.

1. From what natural resource are plastics, paints, and gasoline made?
   A. coal **C.** iron ore
   B. petroleum **D.** natural gas

Use the illustration below to answer questions 2–4.

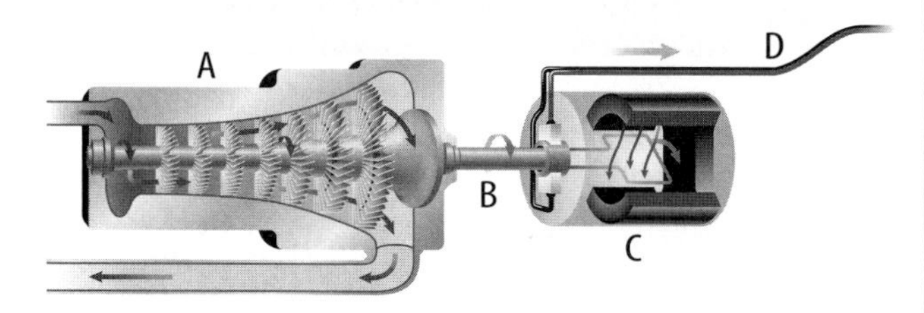

2. What is produced by the mechanism shown in the illustration?
   A. electricity **C.** petroleum
   B. coal **D.** plastic

3. In which section are the turbine blades found?
   A. A **C.** C
   B. B **D.** D

4. Which section represents the generator?
   A. A **C.** C
   B. B **D.** D

5. Which of the following is necessary for the production of hydroelectric power?
   A. wind
   B. access to a river
   C. exposure to sunlight
   D. heat from below Earth's crust

6. With which type of alternative energy are photovoltaic cells used?
   A. hydroelectric power
   B. geothermal energy
   C. nuclear energy
   D. solar energy

7. Which of the following is a type of air pollution that results when sunlight reacts with pollutants produced by burning fuels?
   A. ozones **C.** smog
   B. acid rain **D.** UV radiation

Use the photograph below to answer questions 8 and 9.

8. What is the name of the method of farming illustrated above?
   A. contour plowing **C.** terracing
   B. strip cropping **D.** no-till farming

9. What is the purpose of the method shown in the illustration?
   A. to decrease soil erosion from wind
   B. to decrease soil erosion from water flow
   C. to decrease acid rain production
   D. to increase the return of nutrients to the soil

Test-Taking Tip

Qualifying Terms Look for qualifiers in a question. Such questions are not looking for absolute answers. Qualifiers could be words such as most likely, most common, or least common.

Question 18 The qualifier in this question is *possible*. This indicates that there is uncertainty about the effects of global warming.

Part 2 | Short Response/Grid In

Record your answers on the answer sheet provided by your teacher or on a sheet of paper.

10. Give one example of a renewable source of energy and one example of a nonrenewable source of energy.

Use the illustration below to answer questions 11 and 12.

11. What type of alternative energy is the girl using in the diagram?

12. Name one benefit and one drawback to using this type of energy for cooking.

13. What are two ways that smog can be reduced?

14. A group of students collects rain outside their classroom, then tests the pH of the collected rain. The pH of the rain is 7.2. Can the students say that their rain is acid rain? Why or why not?

15. Why do we depend on the greenhouse effect for survival?

16. What is the cause of algal blooms in lakes and ponds?

Part 3 | Open Ended

Record your answers on a sheet of paper.

17. Are renewable resources always readily available? Explain.

18. What are the possible worldwide effects of global warming? What causes global warming? Why do some people think that using fossil fuels less will decrease global warming?

19. A family lives in a house that uses solar panels to heat the hot water, a wood-burning stove to heat the house, and a windmill for pumping water from a well into a tower where it is stored and then piped into the house as needed. What would be the result if there was no sunlight for two weeks?

Use the illustration below to answer question 20.

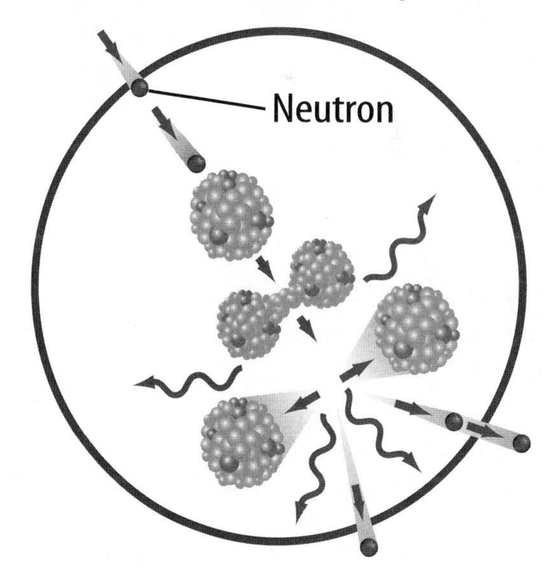

20. What does the diagram represent?

21. Explain how different kinds of plastics are recycled.

How Are Refrigerators & Frying Pans Connected?

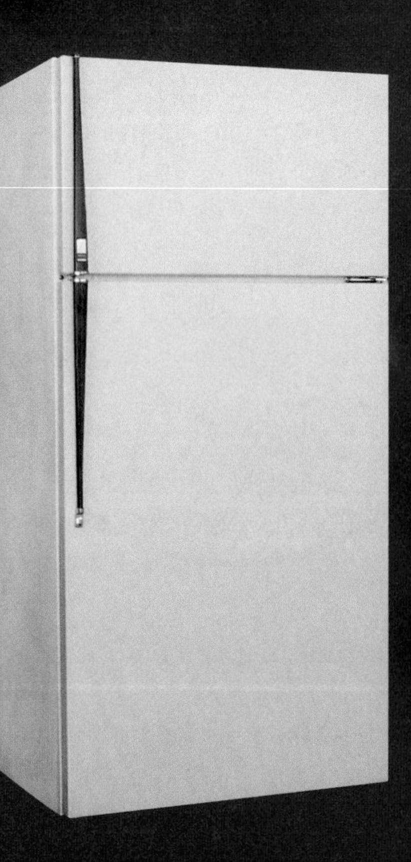

In the late 1930s, scientists were experimenting with a gas that they hoped would work as a new coolant in refrigerators. They filled several metal canisters with the gas and stored the canisters on dry ice. Later, when they opened the canisters, they were surprised to find that the gas had disappeared and that the inside of each canister was coated with a slick, powdery, white solid. The gas had undergone a chemical change. That is, the chemical bonds in its molecules had broken and new bonds had formed, turning one kind of matter into a completely different kind of matter. Strangely, the mysterious white powder proved to be just about the slipperiest substance that anyone had ever encountered. Years later, a creative Frenchman obtained some of the slippery stuff and tried applying it to his fishing tackle to keep the lines from tangling. His wife noticed what he was doing and suggested putting the substance on the inside of a frying pan to keep food from sticking. He did, and nonstick cookware was born!

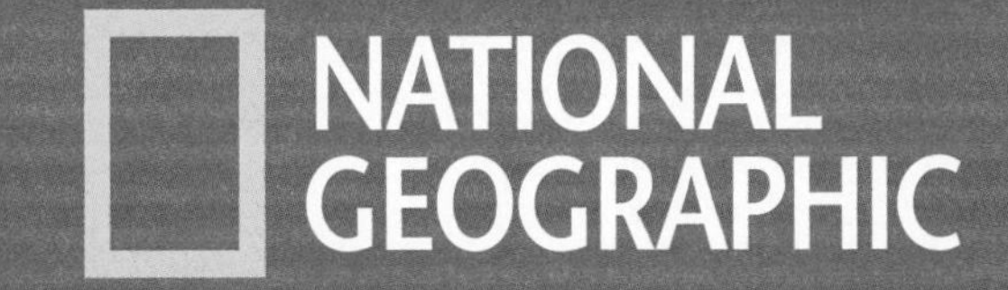

unit projects

Visit green.msscience.com/unit_project to find project ideas and resources.
Projects include:

- **History** Research the French chemist Antoine-Laurent Lavoisier. Design a time line with 20 of his contributions to chemistry.
- **Technology** Design a classroom periodic table wall mural. Use the information as a learning tool and review game.
- **Model** Demonstrate your knowledge of the characteristics of physical and chemical change by preparing a simple snack to share.

Investigating the Sun is an exploration into the composition of our nearest star, the energy it produces, and the possibilities of harnessing that energy for everyday use.

chapter 20

Properties and Changes of Matter

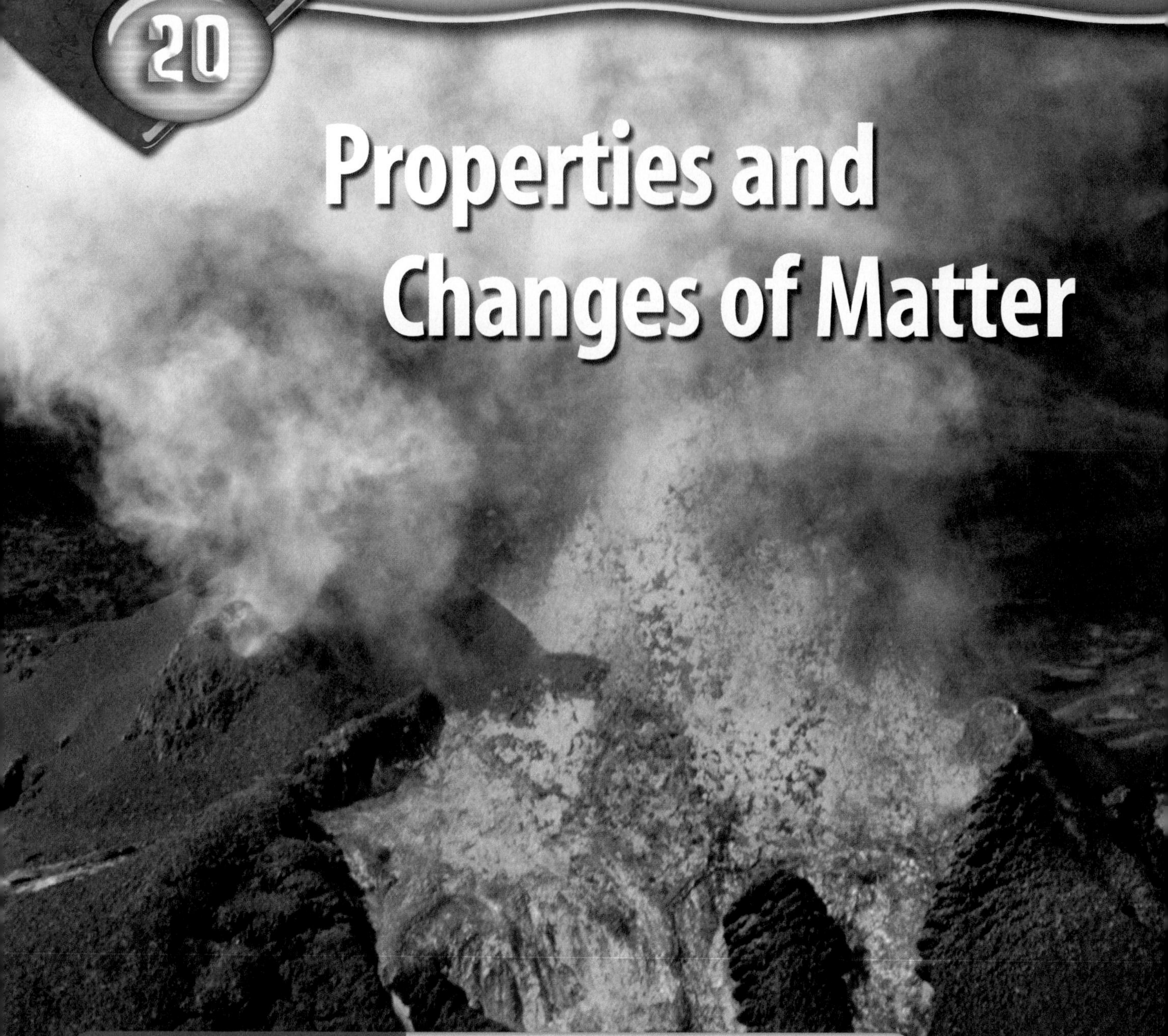

chapter preview

sections

Volcanic Eruptions

At very high temperatures deep within Earth, solid rock melts. One of the properties of rock is its state—solid, liquid, or gas. As lava changes from a liquid to a solid, what happens to its properties? In this chapter, you will learn about physical and chemical properties and changes of matter.

Science Journal Think about what happens when you crack a glow stick. What types of changes are you observing?

Start-Up Activities

The Changing Face of a Volcano

When a volcano erupts, it spews lava and gases. Lava is hot, melted rock from deep within the Earth. After it reaches the Earth's surface, the lava cools and hardens into solid rock. The minerals and gases within the lava, as well as the rate at which it cools, determine the characteristics of the resulting rocks. In this lab, you will compare two types of volcanic rock.

1. Obtain similar-sized samples of the rocks obsidian (ub SIH dee un) and pumice (PUH mus) from your teacher.
2. Compare the colors of the two rocks.
3. Decide which sample is heavier.
4. Look at the surfaces of the two rocks. How are the surfaces different?
5. Place each rock in water and observe.
6. **Think Critically** What characteristics are different about these rocks? In your Science Journal, make a table that compares your observations.

Preview this chapter's content and activities at green.msscience.com

Changes of Matter Make the following Foldable to help you organize your thoughts about properties and changes.

STEP 1 **Fold** a sheet of paper in half lengthwise. Make the back edge about 1.25 cm longer than the front edge.

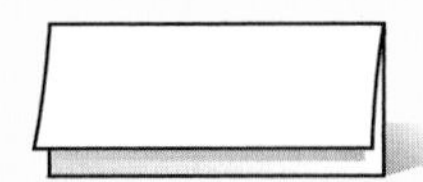

STEP 2 **Fold** in half, then fold in half again to make three folds.

STEP 3 **Unfold and cut** only the top layer along the three folds to make four tabs.

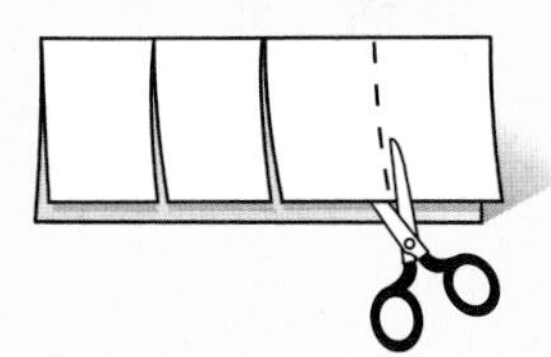

STEP 4 **Label** the tabs as shown.

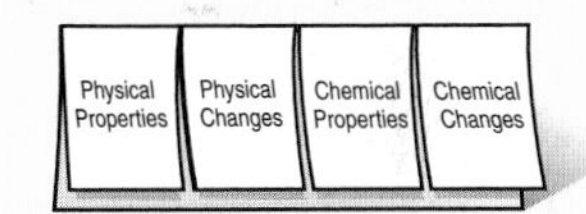

Find Main Ideas As you read the chapter, write information about matter's physical and chemical properties and changes.

section 1

Physical and Chemical Properties

as you read

What You'll Learn

- **Identify** physical and chemical properties of matter.
- **Classify** objects based on physical properties.

Why It's Important

Understanding the different properties of matter will help you to better describe the world around you.

Review Vocabulary
matter: anything that has mass and takes up space

New Vocabulary
- physical property
- chemical property

Physical Properties

It's a busy day at the state fair as you and your classmates navigate your way through the crowd. While you follow your teacher, you can't help but notice the many sights and sounds that surround you. Eventually, you fall behind the group as you spot the most amazing ride you have ever seen. You inspect it from one end to the other. How will you describe it to the group when you catch up to them? What features will you use in your description?

Perhaps you will mention that the ride is large, blue, and made of wood. These features are all physical properties, or characteristics, of the ride. A **physical property** is a characteristic that you can observe without changing or trying to change the composition of the substance. How something looks, smells, sounds, or tastes are all examples of physical properties. In **Figure 1** you can describe and differentiate all types of matter by observing their properties.

Reading Check *What is a physical property of matter?*

Figure 1 All matter can be described by physical properties that can be observed using the five senses.
Identify *the types of matter you think you could see, hear, taste, touch, and smell at the fair.*

Using Your Senses Some physical properties describe the appearance of matter. You can detect many of these properties with your senses. For example, you can see the color and shape of the ride at the fair. You can also touch it to feel its texture. You can smell the odor or taste the flavor of some matter. (You should never taste anything in the laboratory.) Consider the physical properties of the items in **Figure 2.**

Figure 2 Some matter has a characteristic color, such as this sulfur pile. You can use a characteristic smell or taste to identify these fruits. Even if you didn't see it, you could probably identify this sponge by feeling its texture.

State To describe a sample of matter, you need to identify its state. Is the ride a solid, a liquid, or a gas? This property, known as the state of matter, is another physical property that you can observe. The ride, your chair, a book, and a pen are examples of matter in the solid state. Milk, gasoline, and vegetable oil are examples of matter in the liquid state. The helium in a balloon, air in a tire, and neon in a sign are examples of matter in the gas state. You can see examples of solids, liquids, and gases in **Figure 3.**

Perhaps you are most familiar with the three states of water. You can drink or swim in liquid water. You use the solid state of water, which is ice, when you put ice cubes in a drink or skate on a frozen lake. Although you can't see it, water in the gas state is all around you in the air.

Figure 3 The state of a sample of matter is an important physical property.

This rock formation is in the solid state.

The oil flowing out of a bottle is in the liquid state.

This colorful sign uses the element neon, which is generally found in the gaseous state.

Measuring Properties

Procedure

1. Measure the mass of a **10-mL graduated cylinder.**
2. Fill the graduated cylinder with **water** to the 10-mL mark and remeasure the mass of the graduated cylinder with the water.
3. Determine the mass of the water by subtracting the mass of the graduated cylinder from the mass of the graduated cylinder and water.
4. Determine the density of water by dividing the mass of the water by the volume of the water.

Analysis

1. Why did you need to measure the mass of the empty graduated cylinder?
2. How would your calculated density be affected if you added more than 10 mL of water?

Size-Dependent Properties Some physical properties depend on the size of the object. Suppose you need to move a box. The size of the box would be important in deciding if you need to use your backpack or a truck. You begin by measuring the width, height, and depth of the box. If you multiply them together, you calculate the box's volume. The volume of an object is the amount of space it occupies.

Another physical property that depends on size is mass. Recall that the mass of an object is a measurement of how much matter it contains. A bowling ball has more mass than a basketball. Weight is a measurement of force. Weight depends on the mass of the object and on gravity. If you were to travel to other planets, your weight would change but your size and mass would not. Weight is measured using a spring scale like the one in **Figure 4.**

Size-Independent Properties Another physical property, density, does not depend on the size of an object. Density measures the amount of mass in a given volume. To calculate the density of an object, divide its mass by its volume. The density of water is the same in a glass as it is in a tub. The density of an object will change, however, if the mass changes and the volume remains the same. Another property, solubility, also does not depend on size. Solubility is the number of grams of one substance that will dissolve in 100 g of another substance at a given temperature. The amount of drink mix that can be dissolved in 100 g of water is the same in a pitcher as it is when it is poured into a glass. Size-dependent and independent properties are shown in **Table 1.**

Melting and Boiling Point Melting and boiling point also do not depend upon an object's size. The temperature at which a solid changes into a liquid is called its melting point. The temperature at which a liquid changes into a gas is called its boiling point. The melting and boiling points of several substances, along with some of their other physical properties, are shown in **Table 2.**

Figure 4 A spring scale is used to measure an object's weight.

Table 1 Properties of Matter

	Physical Properties
Dependent on sample size	mass, weight, volume
Independent of sample size	density, melting/boiling point, solubility, ability to attract a magnet, state of matter, color

Table 2 Physical Properties of Several Substances

Substance	State	Density (g/cm^3)	Melting point (°C)	Boiling point (°C)	Solubility in cold water (g/100 mL)
Ammonia	gas	0.7710	-78	-33	89.9
Bromine	liquid	3.12	-7	59	4.17
Calcium carbonate	solid	2.71	1,339	898	0.0014
Iodine	solid	4.93	113.5	184	0.029
Potassium hydroxide	solid	2.044	360	1,322	107
Sodium chloride	solid	2.17	801	1,413	35.7
Water	liquid	1	0	100	—

Magnetic Properties Some matter can be described by the specific way in which it behaves. For example, some materials pull iron toward them. These materials are said to be magnetic. The lodestone in **Figure 5** is a rock that is naturally magnetic.

Other materials can be made into magnets. You might have magnets on your refrigerator or locker at school. The door of your refrigerator also has a magnet within it that holds the door shut tightly.

Reading Check *What are some examples of physical properties of matter?*

Figure 5 This lodestone attracts certain metals to it. Lodestone is a natural magnet.

Identifying an Unknown Substance

Procedure

1. Obtain data from your teacher (mass, volume, solubility, melting or boiling point) for an unknown substance(s).
2. Calculate density and solubility in units of g/100 mL for your unknown substance(s).
3. Using Table 2 and the information you have, identify your unknown substance(s).

Analysis

1. Describe the procedure used to determine the density of your unknown substance(s).
2. Identify three characteristics of your substance(s).
3. Explain how the solubility of your substance would be affected if the water was hot.

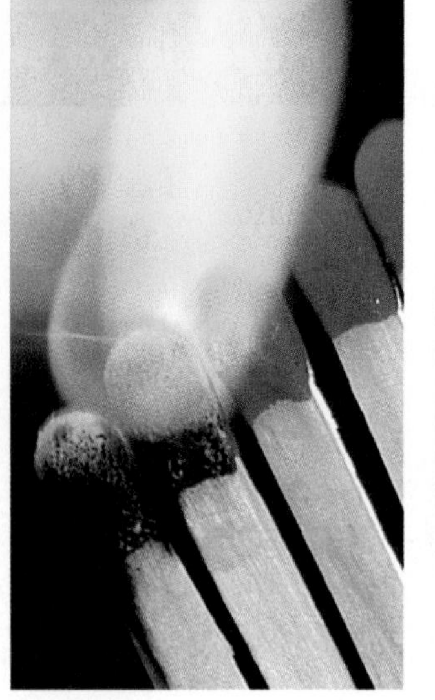
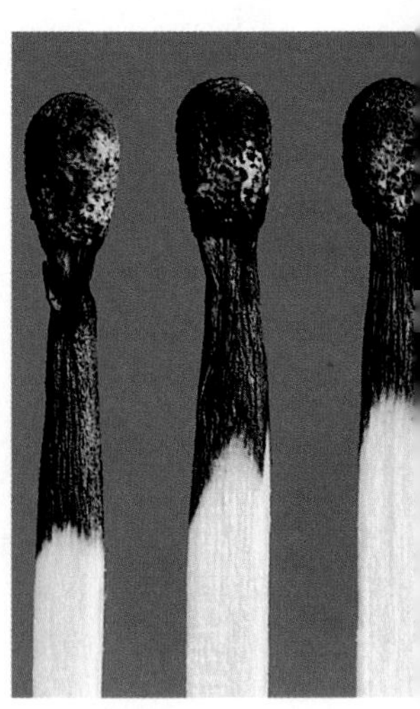

Figure 6 Notice the difference between the new matches and the matches that have been burned. The ability to burn is a chemical property of matter.

Topic: Measuring Matter
Visit green.msscience.com for Web links to information about methods of measuring matter.

Activity Find an object around the house. Use two methods of measuring matter to describe it.

Chemical Properties

Some properties of matter cannot be identified just by looking at a sample. For example, nothing happens if you look at the matches in the first picture. But if someone strikes the matches on a hard, rough surface they will burn, as shown in the second picture. The ability to burn is a chemical property. A **chemical property** is a characteristic that cannot be observed without altering the substance. As you can see in the last picture, the matches are permanently changed after they are burned. Therefore this property can be observed only by changing the composition of the match. Another way to define a chemical property, then, is the ability of a substance to undergo a change that alters its identity. You will learn more about changes in matter in the following section.

section 1 review

Summary

Physical Properties

- Matter exists in solid, liquid, and gaseous states.
- Volume, mass, and weight are size-dependent properties.
- Properties such as density, solubility, boiling and melting points, and ability to attract a magnet are size-independent.
- Density relates the mass of an object to its volume.

Chemical Properties

- Chemical properties have characteristics that cannot be observed without altering the identity of the substance.

Self Check

1. **Infer** How are your senses important for identifying physical properties of matter?
2. **Describe** the physical properties of a baseball.
3. **Think Critically** Explain why solubility is a size-independent physical property.
4. **Compare and Contrast** How do chemical and physical properties differ?

Applying Math

5. **Solve One-Step Equations** The volume of a bucket is 5 L and you are using a cup with a volume of 50 mL. How many cupfuls will you need to fill the bucket? Hint: 1 L = 1,000 mL

Finding the Difference

Real-World Question

You can identify an unknown object by comparing its physical and chemical properties to the properties of identified objects.

Goals

- **Identify** the physical properties of objects.
- **Compare and contrast** the properties.
- **Categorize** the objects based on their properties.

Materials

meterstick
spring scale
block of wood
metal bar or metal ruler
plastic bin
drinking glass
water
rubber ball
paper
carpet
magnet
rock
plant or flower
soil
sand
apple (or other fruit)
vegetable
slice of bread
dry cereal
egg
feather

Safety Precautions

Procedure

1. List at least six properties that you will observe, measure, or calculate for each object. Describe how to determine each property.
2. In your Science Journal, create a data table with a column for each property and rows for the objects.
3. Complete your table by determining the properties for each object.

Conclude and Apply

1. **Describe** Which properties were you able to observe easily? Which required making measurements? Which required calculations?
2. **Compare and contrast** the objects based on the information in your table.
3. **Draw Conclusions** Choose a set of categories and group your objects into those categories. Some examples of categories are large/medium/small, heavy/moderate/light, bright/moderate/dull, solid/liquid/gas, etc. Were the categories you chose useful for grouping your objects? Why or why not?

Communicating Your Data

Compare your results with those of other students in your class. **Discuss** the properties of objects that different groups included on their tables. Make a large table including all of the objects that students in the class studied.

section 2

Physical and Chemical Changes

as you read

What You'll Learn

- **Compare** several physical and chemical changes.
- **Identify** examples of physical and chemical changes.

Why It's Important

From modeling clay to watching the leaves turn colors, physical and chemical changes are all around us.

Review Vocabulary
solubility: the amount of a substance that will dissolve in a given amount of another substance

New Vocabulary
- physical change
- vaporization
- condensation
- sublimation
- deposition
- chemical change
- law of conservation of mass

Physical Changes

What happens when the artist turns the lump of clay shown in **Figure 7** into bowls and other shapes? The composition of the clay does not change. Its appearance, however, changes dramatically. The change from a lump of clay to different shapes is a physical change. A **physical change** is one in which the form or appearance of matter changes, but not its composition. The lake in **Figure 7** also experiences a physical change. Although the water changes state due to a change in temperature, it is still made of the elements hydrogen and oxygen.

Changing Shape Have you ever crumpled a sheet of paper into a ball? If so, you caused physical change. Whether it exists as one flat sheet or a crumpled ball, the matter is still paper. Similarly, if you cut fruit into pieces to make a fruit salad, you do not change the composition of the fruit. You change only its form. Generally, whenever you cut, tear, grind, or bend matter, you are causing a physical change.

Figure 7 Although each sample looks quite different after it experiences a change, the composition of the matter remains the same. These changes are examples of physical changes.

Dissolving What type of change occurs when you add sugar to iced tea, as shown in **Figure 8?** Although the sugar seems to disappear, it does not. Instead, the sugar dissolves. When this happens, the particles of sugar spread out in the liquid. The composition of the sugar stays the same, which is why the iced tea tastes sweet. Only the form of the sugar has changed.

Figure 8 Physical changes are occurring constantly. The sugar blending into the iced tea is an example of a physical change. **Define** *What is a physical change?*

Changing State Another common physical change occurs when matter changes from one state to another. When an ice cube melts, for example, it becomes liquid water. The solid ice and the liquid water have the same composition. The only difference is the form.

Matter can change from any state to another. Freezing is the opposite of melting. During freezing, a liquid changes into a solid. A liquid also can change into a gas. This process is known as **vaporization.** During the reverse process, called **condensation,** a gas changes into a liquid. **Figure 9** summarizes these changes.

In some cases, matter changes between the solid and gas states without ever becoming a liquid. The process in which a solid changes directly into a gas is called **sublimation.** The opposite process, in which a gas changes into a solid, is called **deposition.**

Figure 9 Look at the photographs below to identify the different physical changes that bromine undergoes as it changes from one state to another.

Chemical Changes

It's the Fourth of July in New York City. Brilliant fireworks are exploding in the night sky. When you look at fireworks, such as these in **Figure 10,** you see dazzling sparkles of red and white trickle down in all directions. The explosion of fireworks is an example of a chemical change. During a **chemical change,** substances are changed into different substances. In other words, the composition of the substance changes.

You are familiar with another chemical change if you have ever left your bicycle out in the rain. After awhile, a small chip in the paint leads to an area of a reddish, powdery substance. This substance is rust. When iron in steel is exposed to oxygen and water in air, iron and oxygen atoms combine to form the principle component in rust. In a similar way, silver coins tarnish when exposed to air. These chemical changes are shown in **Figure 11.**

Figure 10 These brilliant fireworks result from chemical changes.
Define *What is a chemical change?*

How is a chemical change different from a physical change?

Figure 11 Each of these examples shows the results of a chemical change. In each case, the substances that are present after the change are different from those that were present before the change.

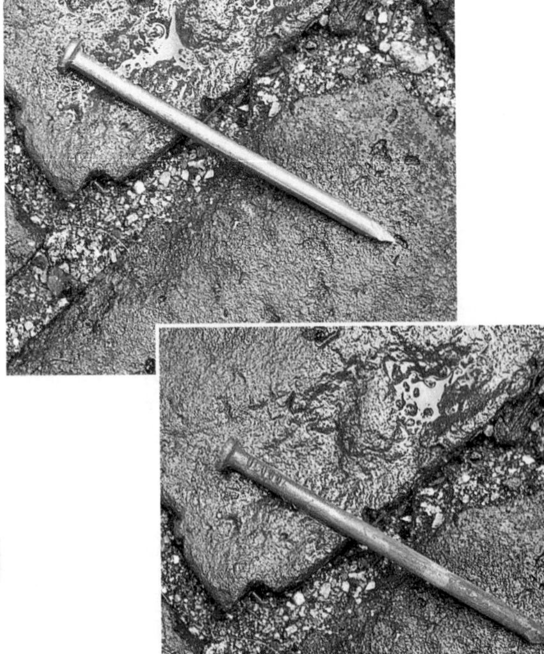

Figure 12 In the fall, the chlorophyll in this tree's leaves undergoes a chemical change into colorless chemicals. This allows the red pigment to be seen.

Topic: Recognizing Chemical Changes

Visit green.msscience.com for Web links to information about how chemical equations can be used to model chemical changes.

Activity Describe the chemical reactions that are involved in making and baking a yeast bread.

Signs of Chemical Changes

Physical changes are relatively easy to identify. If only the form of a substance changes, you have observed a physical change. How can you tell whether a change is a chemical change? If you think you are unfamiliar with chemical changes, think again.

INTEGRATE Life Science You have witnessed a spectacular chemical change if you have seen the leaves on a tree change from green to bright yellow, red, or orange. But, it is not a change from a green pigment to a red pigment, as you might think. Pigments are chemicals that give leaves their color. In **Figure 12,** the green pigment that you see during the summer is chlorophyll (KLOHR uh fihl). In autumn, however, changes in temperature and rainfall amounts cause trees to stop producing chlorophyll. The chlorophyll already in the leaves undergoes a chemical change into colorless chemicals. Where do the bright fall colors come from? The pigments that produce fall colors have been present in the leaves all along. However, in the summer, chlorophyll is present in large enough amounts to mask these pigments. In the fall, when chlorophyll production stops, the bright pigments become visible.

Color Perhaps you have found that a half-eaten apple turns brown. The reason is that a chemical change occurs when the apple is exposed to air. Maybe you have toasted a marshmalow or a slice of bread and watched them turn black. In each case, the color of the food changes as it is cooked because a chemical change occurs.

Comparing Changes

Procedure

1. Separate a piece of **fine steel wool** into two halves.
2. Dip one half in **tap water.**
3. Place each piece of steel wool on a separate **paper plate** and let them sit overnight.

Analysis

1. Did you observe any changes in the steel wool? If so, describe them.
2. If you observed changes, were they physical or chemical? How do you know?

Try at Home

Energy Another sign of a chemical change is the release or gain of energy by an object. Many substances must absorb energy in order to undergo a chemical change. For example, energy is absorbed during the chemical changes involved in cooking. When you bake a cake or make pancakes, energy is absorbed by the batter as it changes from a runny mix into what you see in **Figure 13.**

Another chemical change in which a substance absorbs energy occurs during the production of cement. This process begins with the heating of limestone. Ordinarily, limestone will remain unchanged for centuries. But when it absorbs energy during heating, it undergoes a chemical change in which it turns into lime and carbon dioxide.

Figure 13 Cake batter undergoes a chemical change as it absorbs energy during cooking.

Energy also can be released during a chemical change. The fireworks you read about earlier released energy in the form of light that you can see. As shown in **Figure 14,** a chemical change within a firefly releases energy in the form of light. Fuel burned in the camping stove releases energy you see as light and feel as heat. You also can see that energy is released when sodium and chlorine are combined and ignited in the last picture. During this chemical change, the original substances change into sodium chloride, which is ordinary table salt.

Figure 14 Energy is released when a firefly glows, when fuel is burned in a camping stove, and when sodium and chlorine undergo a chemical change to form table salt.

Odor It takes only one experience with a rotten egg to learn that they smell much different than fresh eggs. When eggs and other foods spoil, they undergo chemical change. The change in odor is a clue to the chemical change. This clue can save lives. When you smell an odd odor in foods, such as chicken, pork, or mayonnaise, you know that the food has undergone a chemical change. You can use this clue to avoid eating spoiled food and protect yourself from becoming ill.

Gases or Solids Look at the antacid tablet in **Figure 15.** You can produce similar bubbles if you pour vinegar on baking soda. The formation of a gas is a clue to a chemical change. What other products undergo chemical changes and produce bubbles?

Figure 15 also shows another clue to a chemical change—the formation of a solid. A solid that separates out of a solution during a chemical change is called a precipitate. The precipitate in the photograph forms when a solution containing sodium iodide is mixed with a solution containing lead nitrate.

Meteoroid A meteoroid is a chunk of metal or stone in space. Every day, meteoroids enter Earth's atmosphere. When this happens, the meteoroid burns as a result of friction with gases in the atmosphere. It is then referred to as a meteor, or shooting star. The burning produces streaks of light. The burning is an example of a chemical change. In your Science Journal, infer why most meteoroids never reach Earth's surface.

Figure 15 The bubbles of gas formed when this antacid tablet is dropped into water indicate a chemical change. The solid forming from two liquids is another sign that a chemical change has taken place.

Figure 16 As wood burns, it turns into a pile of ashes and gases that rise into the air.
Determine *Can you turn ashes back into wood?*

Not Easily Reversed How do physical and chemical changes differ from one another? Think about ice for a moment. After solid ice melts into liquid water, it can refreeze into solid ice if the temperature drops enough. Freezing and melting are physical changes. The substances produced during a chemical change cannot be changed back into the original substances by physical means. For example, the wood in **Figure 16** changes into ashes and gases that are released into the air. After wood is burned, it cannot be restored to its original form as a log.

Think about a few of the chemical changes you just read about to see if this holds true. An antacid tablet cannot be restored to its original form after being dropped in water. Rotten eggs cannot be made fresh again, and pancakes cannot be turned back into batter. The substances that existed before the chemical change no longer exist.

Reading Check *What signs indicate a chemical change?*

Applying Math — Solve for an Unknown

CONVERTING TEMPERATURES Fahrenheit is a non-SI temperature scale. Because it is used so often, it is useful to be able to convert from Fahrenheit to Celsius. The equation that relates Celsius degrees to Fahrenheit degrees is: $(°C \times 1.8) + 32 = °F$. What is 15°F on the Celsius scale?

Solution

1 *This is what you know:*
- temperature = 15°F
- $(°C \times 1.8) + 32 = °F$

2 *This is what you need to find out:*
- temperature in degrees Celsius

3 *This is the procedure you need to use:*
- $(°C \times 1.8) + 32 = °F$
- $°C = (°F - 32)/1.8$
- $°C = (15 - 32)/1.8 = -9.4°C$

4 *Check your answer:*
Substitute the Celsius temperature into the original equation. Did you calculate the Fahrenheit temperature that was given?

Practice Problems

1. Water is being heated on the stove at 156°F. What is this temperature on the Celsius scale?
2. The boiling point of ethylene glycol is 199°C. What is the temperature on the Fahrenheit scale?

Science Online — For more practice, visit green.msscience.com/math_practice

Chemical Versus Physical Change

Now you have learned about many different physical and chemical changes. You have read about several characteristics that you can use to distinguish between physical and chemical changes. The most important point for you to remember is that in a physical change, the composition of a substance does not change and in a chemical change, the composition of a substance does change. When a substance undergoes a physical change, only its form changes. In a chemical change, both form and composition change.

When the wood and copper in **Figure 17** undergo physical changes, the original wood and copper still remain after the change. When a substance undergoes a chemical change, however, the original substance is no longer present after the change. Instead, different substances are produced during the chemical change. When the wood and copper in **Figure 17** undergo chemical changes, wood and copper have changed into new substances with new physical and chemical properties.

Physical and chemical changes are used to recycle or reuse certain materials. **Figure 18** discusses the importance of some of these changes in recycling.

Figure 17 When a substance undergoes a physical change, its composition stays the same. When a substance undergoes a chemical change, it is changed into different substances.

NATIONAL GEOGRAPHIC VISUALIZING RECYCLING

Figure 18

Recycling is a way to separate wastes into their component parts and then reuse those components in new products. In order to be recycled, wastes need to be physically—and sometimes chemically—changed. The average junked automobile contains about 62 percent iron and steel, 28 percent other materials such as aluminum, copper, and lead, and 10 percent rubber, plastics, and various materials.

◀ Rubber tires can be shredded and added to asphalt pavement and playground surfaces. New recycling processes make it possible to supercool tires to a temperature at which the rubber is shattered like glass. A magnet can then draw out steel from the tires and other parts of the car.

▼ After being crushed and flattened, car bodies are chopped into small pieces. Metals are separated from other materials using physical processes. Some metals are separated using powerful magnets. Others are separated by hand.

◀ Glass can be pulverized and used in asphalt pavement, new glass, and even artwork. This sculpture, named *Groundswell,* was created by artist Maya Lin using windshield glass.

▲ Some plastics can be melted and formed into new products. Others are ground up or shredded and used as fillers or insulating materials.

Conservation of Mass

During a chemical change, the form or the composition of the matter changes. The particles within the matter rearrange to form new substances, but they are not destroyed and new particles are not created. The number and type of particles remains the same. As a result, the total mass of the matter is the same before and after a physical or chemical change. This is known as the **law of conservation of mass.**

This law can sometimes be difficult to believe, especially when the materials remaining after a chemical change might look quite different from those before it. In many chemical changes in which mass seems to be gained or lost, the difference is often due to a gas being given off or taken in. When the candle burns in **Figure 19,** gases in the air combine with the candle wax. New gases are formed that go into the air. The mass of the wax, which is burned and the gases that combine with the wax equal the mass of the gases produced by burning

The scientist who first performed the careful experiments necessary to prove that mass is conserved was Antoine Lavoisier (AN twan • luh VWAH see ay) in the eighteenth century. It was Lavoisier who recognized that the mass of gases that are given off or taken from the air during chemical changes account for any differences in mass.

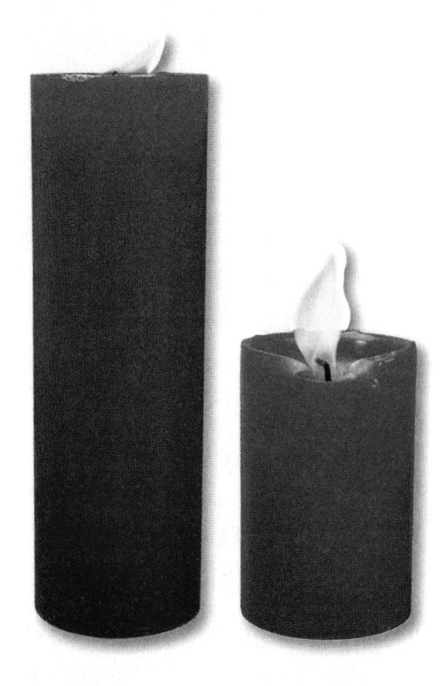

Figure 19 The candle lost mass when it was burned. The mass lost by the candle combined with the gases in the air to form new substances. As a result, mass was not created or destroyed.

section 2 review

Summary

Physical Changes

- The form of matter, its shape or state, is altered during a physical change.
- The composition of matter remains the same.

Chemical Changes

- Both form and composition of matter are altered during a chemical change.
- Some signs of a chemical change are altered color, energy, odor, and formation of a gas or solid.
- Chemical changes are not easily reversed.

Conservation of Mass

- The total mass of the matter is the same before and after a physical or chemical change.

Self Check

1. **List** five physical changes that you can observe in your home.
2. **Determine** what kind of change occurs on the surface of bread when it is toasted.
3. **Infer** How is mass conserved during a chemical change?
4. **Think Critically** A log is reduced to a small pile of ash when it burns. Explain the difference in mass between the log and the ash.

Applying Math

5. **Solve One-Step Equations** Magnesium and oxygen undergo a chemical change to form magnesium oxide. How many grams of magnesium oxide will be produced when 0.486 g of oxygen completely react with 0.738 g of magnesium?

Design Your Own

Battle of the Toothpastes

Goals
- **Observe** how toothpaste helps prevent tooth decay.
- **Design** an experiment to test the effectiveness of various types and brands of toothpaste.

Possible Materials
3 or 4 different brands and types of toothpaste
drinking glasses or bowls
hard-boiled eggs
concentrated lemon juice
apple juice
water
artist's paint brush

Safety Precautions

Real-World Question

Your teeth are made of a compound called hydroxyapatite (hi DRAHK see A puh tite). The sodium fluoride in toothpaste undergoes a chemical reaction with hydroxyapatite to form a new compound on the surface of your teeth. This compound resists food acids that cause tooth decay, another chemical change. In this lab, you will design an experiment to test the effectiveness of different toothpaste brands. The compound found in your teeth is similar to the mineral compound found in eggshells. Treating hard-boiled eggs with toothpaste is similar to brushing your teeth with toothpaste. Soaking the eggs in food acids such as vinegar for several days will produce similar conditions as eating foods, which contain acids that will produce a chemical change in your teeth, for several months.

Form a Hypothesis

Form a hypothesis about the effectiveness of different brands of toothpaste.

Test Your Hypothesis

Make a Plan

1. **Describe** how you will use the materials to test the toothpaste.
2. **List** the steps you will follow to test your hypothesis.
3. **Decide** on the length of time that you will conduct your experiment.

4. **Identify** the control and variables you will use in your experiment.
5. **Create** a data table in your Science Journal to record your observations, measurements, and results.
6. **Describe** how you will measure the amount of protection each toothpaste brand provides.

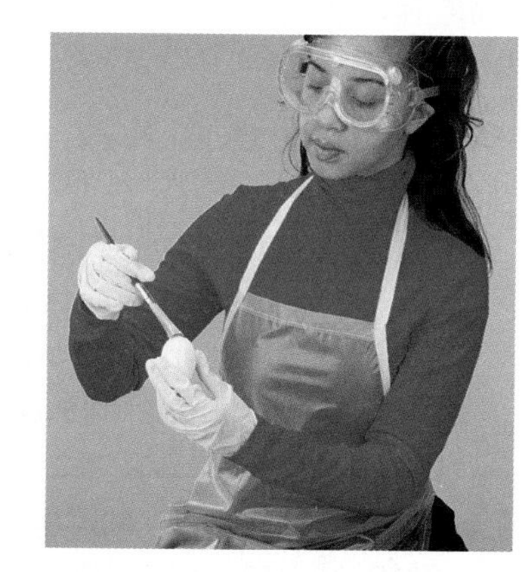

Follow Your Plan

1. Make sure your teacher approves your plan before you start.
2. **Conduct** your experiment as planned. Be sure to follow all proper safety precautions.
3. **Record** your observations in your data table.

Analyze Your Data

1. **Compare** the untreated eggshells with the shells you treated with toothpaste.
2. **Compare** the condition of the eggshells you treated with different brands of toothpaste.
3. **Compare** the condition of the eggshells soaked in lemon juice and in apple juice.
4. **Identify** unintended variables you discovered in your experiment that might have influenced the results.

Conclude and Apply

1. **Identify** Did the results support your hypothesis? Describe the strengths and weaknesses of your hypothesis.
2. **Explain** why the eggshells treated with toothpaste were better-protected than the untreated eggshells.
3. **Identify** which brands of toothpaste, if any, best protected the eggshells from decay.
4. **Evaluate** the scientific explanation for why adding fluoride to toothpaste and drinking water prevents tooth decay.
5. **Predict** what would happen to your protected eggs if you left them in the food acids for several weeks.
6. **Infer** why it is a good idea to brush with fluoride toothpaste.

Communicating Your Data

Compare your results with the results of your classmates. **Create** a poster advertising the benefits of fluoride toothpaste.

SCIENCE Stats

Strange Changes

Did you know...

... A hair colorist is also a chemist! Colorists use hydrogen peroxide and ammonia to swell and open the cuticle-like shafts on your hair. Once these are open, the chemicals in hair dye can get into your natural pigment molecules and chemically change your hair color. The first safe commercial hair color was created in 1909 in France.

... Americans consume about 175 million kg of sauerkraut each year. During the production of sauerkraut, bacteria produce lactic acid. The acid chemically breaks down the material in the cabbage, making it translucent and tangy.

Applying Math There are 275 million people in the United States. Calculate the average amount of sauerkraut consumed by each person in the United States in one year.

... More than 450,000 metric tons of plastic packaging are recycled each year in the U.S. Discarded plastics undergo physical changes including melting and shredding. They are then converted into flakes or pellets, which are used to make new products. Recycled plastic is used to make clothes, furniture, carpets, and even lumber.

Projected Recycling Rates by Material, 2000

Material	1995 Recycling	Proj. Recycling
Paper/Paperboard	40.0%	43 to 46%
Glass	24.5%	27 to 36%
Ferrous metal	36.5%	42 to 55%
Aluminum	34.6%	46 to 48%
Plastics	5.3%	7 to 10%
Yard waste	30.3%	40 to 50%
Total Materials	**27.0%**	**30 to 35%**

Find Out About It

Every time you cook, you make physical and chemical changes to food. Visit green.msscience.com/science_stats or to your local or school library to find out what chemical or physical changes take place when cooking ingredients are heated or cooled.

Reviewing Main Ideas

Section 1 Physical and Chemical Properties

1. Matter can be described by its characteristics, or properties, and can exist in different states—solid, liquid, or gas.
2. A physical property is a characteristic that can be observed without altering the composition of the sample.
3. Physical properties include color, shape, smell, taste, and texture, as well as measurable quantities such as mass, volume, density, melting point, and boiling point.
4. A chemical property is a characteristic that cannot be observed without changing what the sample is made of.

Section 2 Physical and Chemical Changes

1. During a physical change, the composition of matter stays the same but the appearance changes in some way.
2. Physical changes occur when matter changes from one state to another.
3. A chemical change occurs when the composition of matter changes.
4. Signs of chemical change include changes in energy, color, odor, or the production of gases or solids.
5. According to the law of conservation of mass, mass cannot be created or destroyed.

Visualizing Main Ideas

Copy and complete the following concept map on matter.

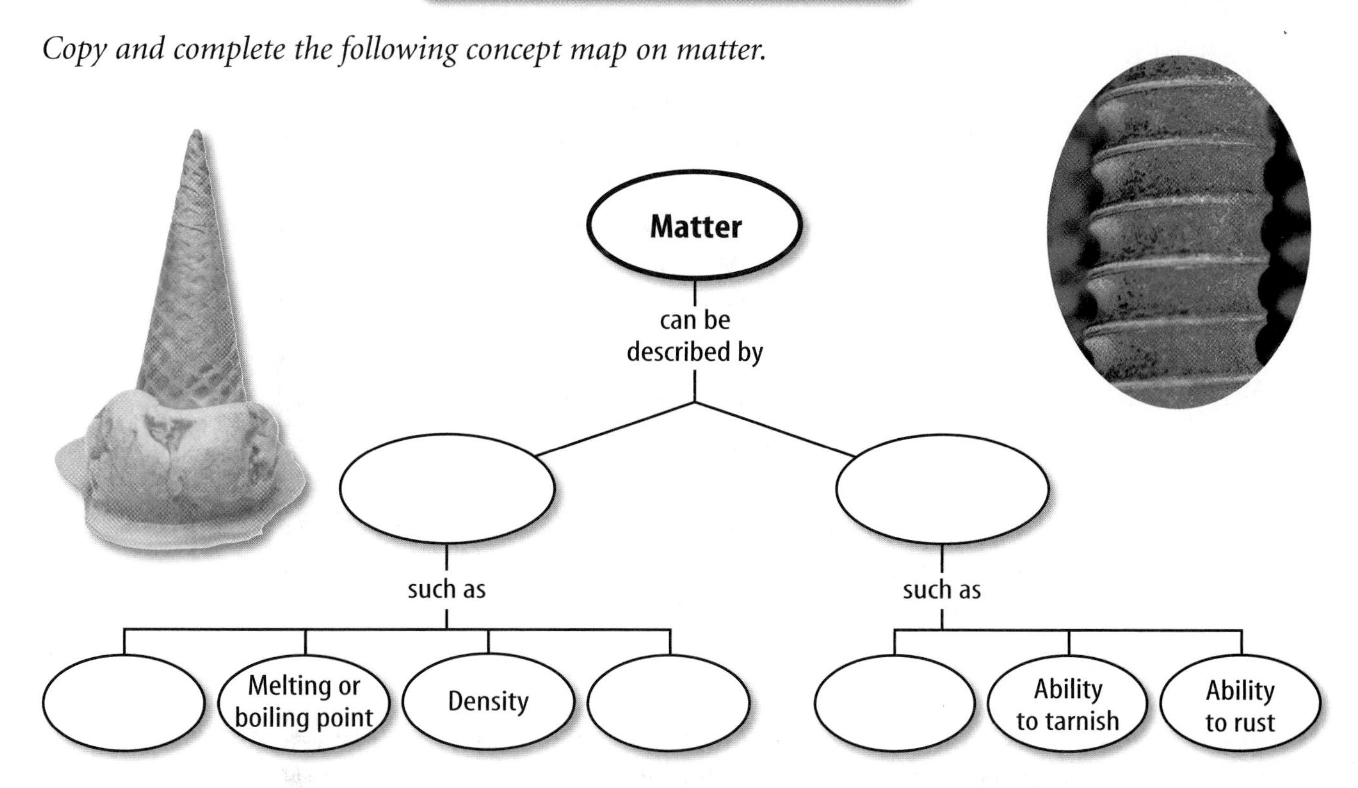

Using Vocabulary

chemical change p. 602	physical change p. 600
chemical property p. 598	physical property p. 594
condensation p. 601	sublimation p. 601
deposition p. 601	vaporization p. 601
law of conservation of mass p. 609	

Use what you know about the vocabulary words to answer the following questions. Use complete sentences.

1. Why is color a physical property?
2. What is a physical property that does not change with the amount of matter?
3. What happens during a physical change?
4. What type of change is a change of state?
5. What happens during a chemical change?
6. What are three clues that a chemical change has occurred?
7. What is an example of a chemical change?
8. What is the law of conservation of mass?

Checking Concepts

Choose the word or phrase that best answers the question.

9. What changes when the mass of an object increases while volume stays the same?
   A) color **C)** density
   B) length **D)** height
10. What word best describes the type of materials that attract iron?
   A) magnetic **C)** mass
   B) chemical **D)** physical
11. Which is an example of a chemical property?
   A) color **C)** density
   B) mass **D)** ability to burn
12. Which is an example of a physical change?
   A) metal rusting **C)** water boiling
   B) silver tarnishing **D)** paper burning
13. What characteristic best describes what happens during a physical change?
   A) composition changes
   B) composition stays the same
   C) form stays the same
   D) mass is lost
14. Which is an example of a chemical change?
   A) water freezes **C)** bread is baked
   B) wood is carved **D)** wire is bent
15. Which is NOT a clue that could indicate a chemical change?
   A) change in color
   B) change in shape
   C) change in energy
   D) change in odor
16. What property stays the same during physical and chemical changes?
   A) density
   B) shape
   C) mass
   D) arrangement of particles

Use the illustration below to answer question 17.

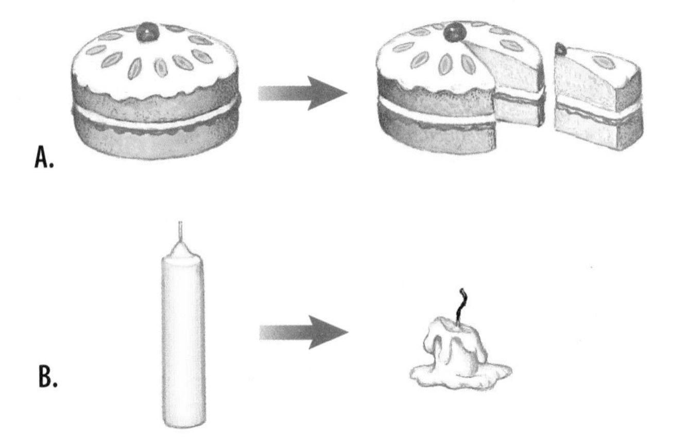

17. Which is an example of a physical change and which is a chemical change?

Science Online green.msscience.com/vocabulary_puzzlemaker

Thinking Critically

18. **Draw Conclusions** When asked to give the physical properties of a painting, your friend says the painting is beautiful. Why isn't this description a true scientific property?

19. **Draw Conclusions** You are told that a sample of matter gives off energy as it changes. Can you conclude which type of change occurred? Why or why not?

20. **Describe** what happens to mass during chemical and physical changes. Explain.

21. **Classify** Decide whether the following properties are physical or chemical.
    - **a.** Sugar can change into alcohol.
    - **b.** Iron can rust.
    - **c.** Alcohol can vaporize.
    - **d.** Paper can burn.
    - **e.** Sugar can dissolve.

Use the table below to answer question 22 and 23.

Physical Properties

Substance	Melting Point (°C)	Density (g/cm^3)
Benzoic acid	122.1	1.075
Sucrose	185.0	1.581
Methane	−182.0	0.466
Urea	135.0	1.323

22. **Determine** A scientist has a sample of a substance with a mass of 1.4 g and a volume of 3.0 mL. According to the table above, which substance might it be?

23. **Conclude** Using the table above, which substance would take the longest time to melt? Explain your reasoning.

24. **Determine** A jeweler bends gold into a beautiful ring. What type of change is this? Explain.

25. **Compare and Contrast** Relate such human characteristics as hair and eye color and height and weight to physical properties of matter. Relate human behavior to chemical properties. Think about how you observe these properties.

Performance Activities

26. **Write a Story** Write a story describing an event that you have experienced. Then go back through the story and circle any physical or chemical properties you mentioned. Underline any physical or chemical changes you included.

Applying Math

27. **Brick Volume** What is the volume of a brick that is 20 cm long, 10 cm wide, and 3 cm high?

28. **Density of an Object** What is the density of an object with a mass of 50 g and a volume of 5 cm^3?

Use the table below to answer question 29.

Mineral Samples

Sample	Mass	Volume
A	96.5 g	5 cm^3
B	38.6 g	4 cm^3

29. **Density of Gold** The density of gold is 19.3 g/cm^3. Which sample is the gold?

30. **Ammonia Solubility** 89.9 g of ammonia will dissolve in 100 mL of cold water. How much ammonia is needed to dissolve in 1.5 L of water?

Part 1 Multiple Choice

Record your answers on the answer sheet provided by your teacher or on a sheet of paper.

Use the photograph below to answer questions 1 and 2.

1. Which of the following could you do to the ball in the photograph above to cause a chemical change?
   A. cut in half
   B. paint
   C. flatten
   D. burn

2. Which of the following physical properties of the ball is size independent?
   A. density
   B. mass
   C. volume
   D. weight

3. Each of the following procedures results in the formation of bubbles. Which of these is a physical change?
   A. pouring an acid onto calcium carbonate
   B. dropping an antacid tablet into water
   C. heating water to its boiling point
   D. pouring vinegar onto baking soda

4. Which of the following occurs as you heat a liquid to its boiling point?
   A. condensation
   B. melting
   C. vaporization
   D. freezing

Test-Taking Tip

Essay Questions Spend a few minutes listing and organizing the main points that you plan to discuss. Make sure to do all of this work on your scratch paper, not on the answer sheet.

5. During an experiment, you find that you can dissolve 4.2 g of a substance in 250 mL of water at 25°C. How much of the substance would you predict that you could dissolve in 500 mL of water at the same temperature?
   A. 2.1 g
   B. 4.2 g
   C. 6.3 g
   D. 8.4 g

6. Which of the following is a chemical reaction?
   A. making ice cubes
   B. toasting bread
   C. slicing a carrot
   D. boiling water

7. When you make and eat scrambled eggs, many changes occur to the eggs. Which of the following best describes a chemical change?
   A. crack the eggs
   B. scramble the eggs
   C. cook the eggs
   D. chew the eggs

Use the table below to answer questions 8 and 9.

Physical Properties of Bromide	
Density	3.12 g/cm^3
Boiling point	59°C
Melting point	–7°C

8. According to the table above, what is the mass of 4.34 cm^3 of bromine?
   A. 0.719 g
   B. 1.39 g
   C. 7.46 g
   D. 13.5 g

9. At which of the following temperatures is bromine a solid?
   A. –10°C
   B. 10°C
   C. 40°C
   D. 80°C

Part 2 Short Response/Grid In

Record your answers on the answer sheet provided by your teacher or on a sheet of paper.

10. A precipitate is one clue that a chemical change has occurred. What is a precipitate and when is it observed?

Use the photo below to answer questions 11 and 12.

11. The photograph above shows a rusted chain. Explain why rusting is a physical or a chemical change.

12. What are some physical properties of the rusty chain that you can see? What are some physical properties that you can't see?

13. You measure the density of a 12.3-g sample of limestone as 2.72 g/cm^3. What is the density of a 36.9 g sample?

14. A scientist measures the masses of two chemicals. He then combines the chemicals and measures their total mass. The total mass is less than the sum of each individual mass. Has this violated the law of conservation of mass? Explain what might have happened when the chemicals were combined.

15. A scientist measures 275 mL of water into a beaker. She then adds 51.0 g of lead into the beaker. After the addition of the lead, the volume of water in the beaker increases by 4.50 mL. What is the density of the lead?

Part 3 Open Ended

Record your answers on a sheet of paper.

16. Suppose you have a gas in a closed container. Explain what would happen to the mass and density of the gas if you compressed it into half the volume.

17. Color change is an indication that a chemical change may have occurred. Mixing yellow and blue modeling clay makes green modeling clay. Is this a chemical reaction? Explain why or why not.

18. At a temperature of 40°C, you find that 40 g of ammonium chloride easily dissolves in 100 mL of water. When you stir 40 g of potassium chloride into a beaker containing 100 mL of water at 40°C, you find that some of the potassium chloride remains in the bottom of the beaker. Explain why this occurs and how to make to make the remaining potassium chloride dissolve.

Use the photo below to answer questions 19 and 20.

19. What would happen if you left the glass of cold water shown in the photograph above in the hot Sun for several hours? Describe how some physical properties of the water would change.

20. What properties of the water would not change? Explain why the density of the water would or would not change.

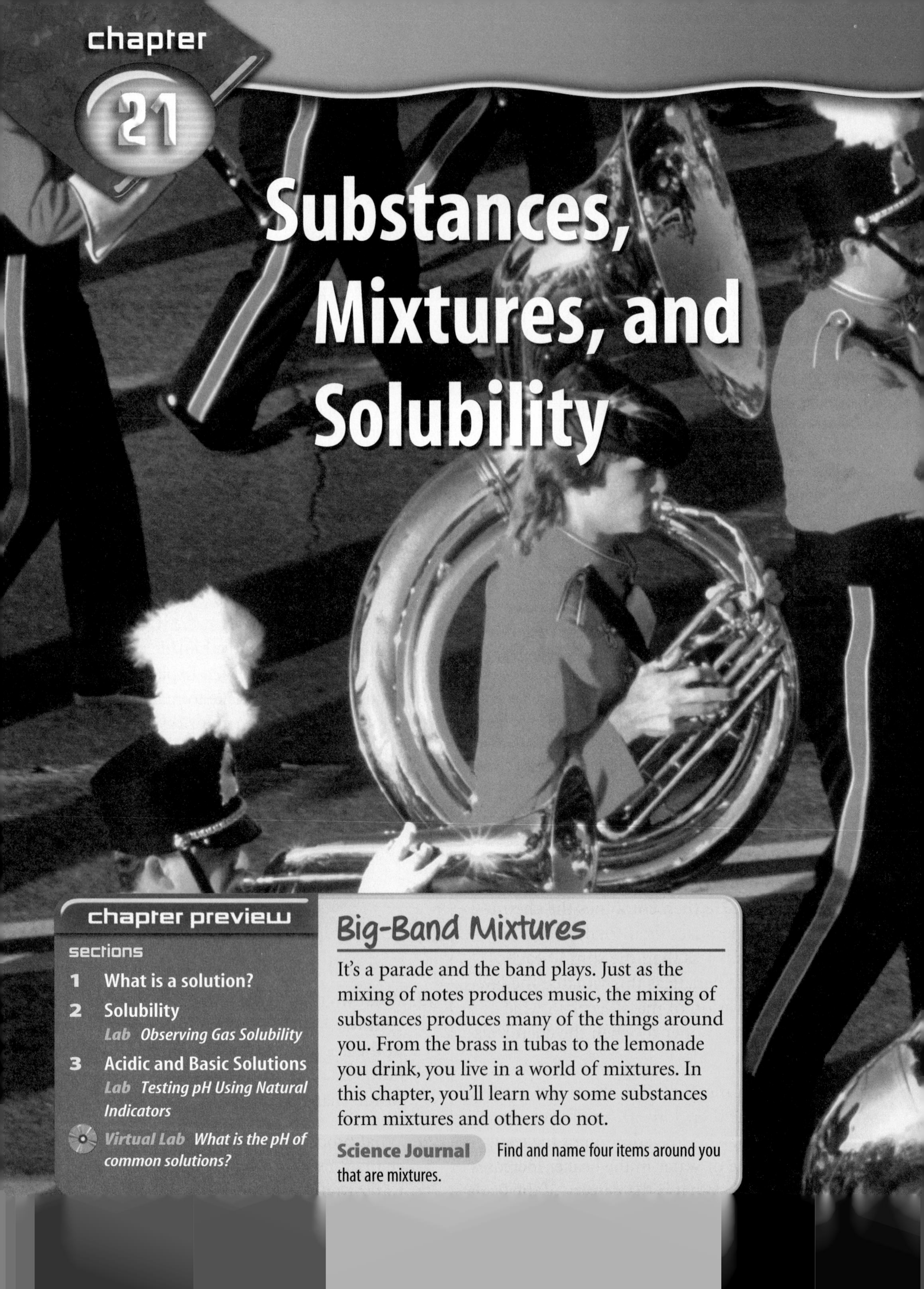

chapter 21

Substances, Mixtures, and Solubility

chapter preview

sections

1 What is a solution?

2 Solubility
Lab Observing Gas Solubility

3 Acidic and Basic Solutions
Lab Testing pH Using Natural Indicators

Virtual Lab What is the pH of common solutions?

Big-Band Mixtures

It's a parade and the band plays. Just as the mixing of notes produces music, the mixing of substances produces many of the things around you. From the brass in tubas to the lemonade you drink, you live in a world of mixtures. In this chapter, you'll learn why some substances form mixtures and others do not.

Science Journal Find and name four items around you that are mixtures.

Start-Up Activities

Particle Size and Dissolving Rates

Why do drink mixes come in powder form? What would happen if you dropped a big chunk of drink mix into the water? Would it dissolve quickly? Powdered drink mix dissolves faster in water than chunks do because it is divided into smaller particles, exposing more of the mix to the water. See for yourself how particle size affects the rate at which a substance dissolves.

1. Pour 400 mL of water into each of two 600-mL beakers.
2. Carefully grind a bouillon cube into powder using a mortar and pestle.
3. Place the bouillon powder into one beaker and drop a whole bouillon cube into the second beaker.
4. Stir the water in each beaker for 10 s and observe.
5. **Think Critically** Write a paragraph in your Science Journal comparing the color of the two liquids and the amount of undissolved bouillon at the bottom of each beaker. How does the particle size affect the rate at which a substance dissolves?

FOLDABLES™ Study Organizer

Solutions Make the following Foldable to help classify solutions based on their common features.

STEP 1 **Fold** a vertical sheet of paper from side to side. Make the front edge about 1.25 cm shorter than the back edge.

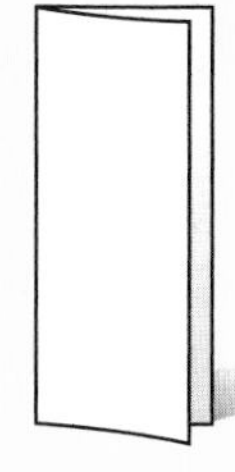

STEP 2 **Turn** lengthwise and **fold** into thirds.

STEP 3 **Unfold and cut** only the top layer along both folds to make three tabs.

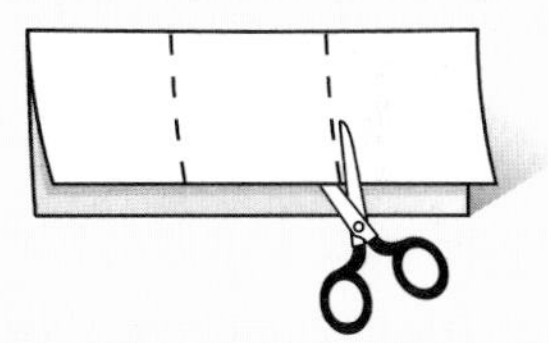

STEP 4 **Label** each tab as shown.

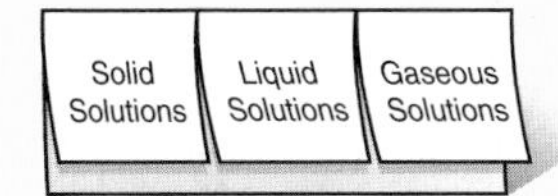

Find Main Ideas As you read the chapter, classify solutions based on their states and list them under the appropriate tabs. On your Foldable, circle the solutions that are acids and underline the solutions that are bases.

Preview this chapter's content and activities at green.msscience.com

section 1

What is a solution?

as you read

What You'll Learn

- **Distinguish** between substances and mixtures.
- **Describe** two different types of mixtures.
- **Explain** how solutions form.
- **Describe** different types of solutions.

Why It's Important

The air you breathe, the water you drink, and even parts of your body are all solutions.

Review Vocabulary
proton: positively charged particle located in the nucleus of an atom

New Vocabulary
- substance
- heterogeneous mixture
- homogeneous mixture
- solution
- solute
- solvent
- precipitate

Substances

Water, salt water, and pulpy orange juice have some obvious differences. These differences can be explained by chemistry. Think about pure water. No matter what you do to it physically—freeze it, boil it, stir it, or strain it—it still is water. On the other hand, if you boil salt water, the water turns to gas and leaves the salt behind. If you strain pulpy orange juice, it loses its pulp. How does chemistry explain these differences? The answer has to do with the chemical compositions of the materials.

Atoms and Elements Recall that atoms are the basic building blocks of matter. Each atom has unique chemical and physical properties which are determined by the number of protons it has. For example, all atoms that have eight protons are oxygen atoms. A **substance** is matter that has the same fixed composition and properties. It can't be broken down into simpler parts by ordinary physical processes, such as boiling, grinding, or filtering. Only a chemical process can change a substance into one or more new substances. **Table 1** lists some examples of physical and chemical processes. An element is an example of a pure substance; it cannot be broken down into simpler substances. The number of protons in an element, like oxygen, are fixed—it cannot change unless the element changes.

Table 1 Examples of Physical and Chemical Processes

Physical Processes	Chemical Processes
Boiling	Burning
Changing pressure	Reacting with other chemicals
Cooling	Reacting with light
Sorting	

Compounds Water is another example of a substance. It is always water even when you boil it or freeze it. Water, however, is not an element. It is an example of a compound which is made of two or more elements that are chemically combined. Compounds also have fixed compositions. The ratio of the atoms in a compound is always the same. For example, when two hydrogen atoms combine with one oxygen atom, water is formed. All water—whether it's in the form of ice, liquid, or steam—has the same ratio of hydrogen atoms to oxygen atoms.

Separation by magnetism

Separation by straining

Figure 1 Mixtures can be separated by physical processes. **Explain** *why the iron-sand mixture and the pulpy lemonade are not pure substances.*

Mixtures

Imagine drinking a glass of salt water. You would know right away that you weren't drinking pure water. Like salt water, many things are not pure substances. Salt water is a mixture of salt and water. Mixtures are combinations of substances that are not bonded together and can be separated by physical processes. For example, you can boil salt water to separate the salt from the water. If you had a mixture of iron filings and sand, you could separate the iron filings from the sand with a magnet. **Figure 1** shows some mixtures being separated.

Unlike compounds, mixtures do not always contain the same proportions of the substances that they are composed of. Lemonade is a mixture that can be strong tasting or weak tasting, depending on the amounts of water and lemon juice that are added. It also can be sweet or sour, depending on how much sugar is added. But whether it is strong, weak, sweet, or sour, it is still lemonade.

Topic: Desalination
Visit green.msscience.com for Web links to information about how salt is removed from salt water to provide drinking water.

Activity Compare and contrast the two most common methods used for desalination.

Heterogeneous Mixtures It is easy to tell that some things are mixtures just by looking at them. A watermelon is a mixture of fruit and seeds. The seeds are not evenly spaced through the whole melon—one bite you take might not have any seeds in it and another bite might have several seeds. A type of mixture where the substances are not mixed evenly is called a **heterogeneous** (he tuh ruh JEE nee us) **mixture.** The different areas of a heterogeneous mixture have different compositions. The substances in a heterogeneous mixture are usually easy to tell apart, like the seeds from the fruit of a watermelon. Other examples of heterogeneous mixtures include a bowl of cold cereal with milk and the mixture of pens, pencils, and books in your backpack.

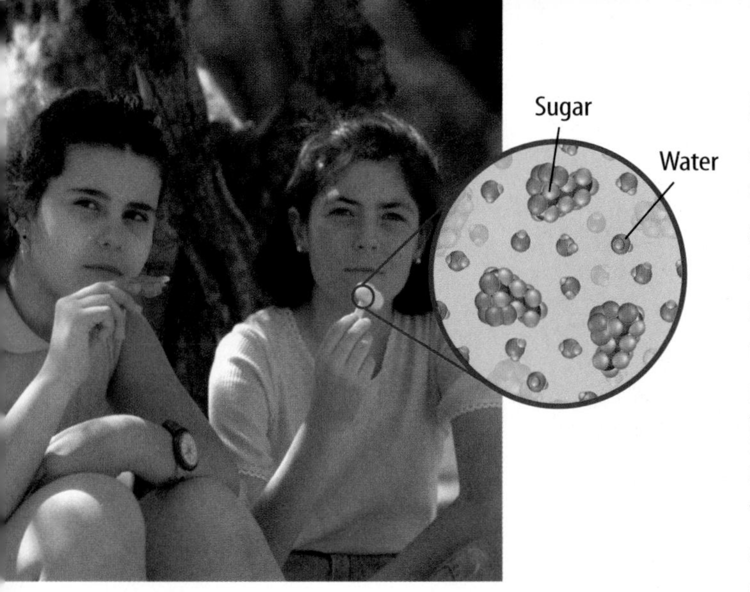

Figure 2 Molecules of sugar and water are evenly mixed in frozen pops.

Homogeneous Mixtures Your shampoo contains many ingredients, but you can't see them when you look at the shampoo. It is the same color and texture throughout. Shampoo is an example of a homogeneous (hoh muh JEE nee us) mixture. A **homogeneous mixture** contains two or more substances that are evenly mixed on a molecular level but still are not bonded together. Another name for a homogeneous mixture is a **solution.** The sugar and water in the frozen pops shown in **Figure 2,** are a solution—the sugar is evenly distributed in the water, and you can't see the sugar.

Reading Check *What is another name for a homogeneous mixture?*

How Solutions Form

How do you make sugar water for a hummingbird feeder? You might add sugar to water and heat the mixture until the sugar disappears. The sugar molecules would spread out until they were evenly spaced throughout the water, forming a solution. This is called dissolving. The substance that dissolves—or seems to disappear—is called the **solute.** The substance that dissolves the solute is called the **solvent.** In the hummingbird feeder solution, the solute is the sugar and the solvent is water. The substance that is present in the greatest quantity is the solvent.

Forming Solids from Solutions Under certain conditions, a solute can come back out of its solution and form a solid. This process is called crystallization. Sometimes this occurs when the solution is cooled or when some of the solvent evaporates. Crystallization is the result of a physical change. When some solutions are mixed, a chemical reaction occurs, forming a solid. This solid is called a **precipitate** (prih SIH puh tayt). A precipitate is the result of a chemical change. Precipitates probably have formed in your sink or shower because of chemical reactions. Minerals that are dissolved in tap water react chemically with soap. The product of this reaction leaves the water as a precipitate called soap scum, shown in **Figure 3.**

Figure 3 Minerals and soap react to form soap scum, which comes out of the water solution and coats the tiles of a shower.

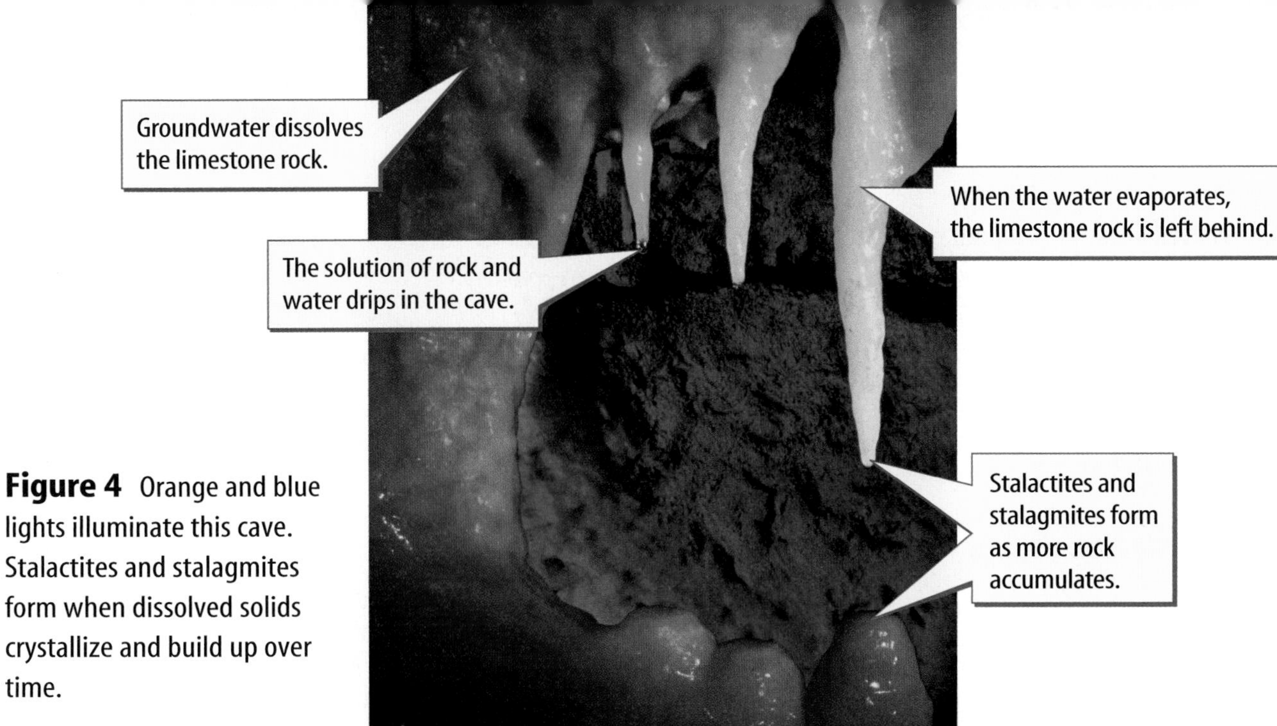

Figure 4 Orange and blue lights illuminate this cave. Stalactites and stalagmites form when dissolved solids crystallize and build up over time.

INTEGRATE Environment Stalactites and stalagmites in caves are formed from solutions, as shown in **Figure 4.** First, minerals dissolve in water as it flows through rocks at the top of the cave. This solution of water and dissolved minerals drips from the ceiling of the cave. When drops of the solution evaporate from the roof of the cave, the minerals are left behind. They create the hanging rock formations called stalactites. When drops of the solution fall onto the floor of the cave and evaporate, they form stalagmites. Very often, a stalactite develops downward while a stalagmite develops upward until the two meet. One continuous column of minerals is formed. This process will be discussed later.

Types of Solutions

So far, you've learned about types of solutions in which a solid solute dissolves in a liquid solvent. But solutions can be made up of different combinations of solids, liquids, and gases, as shown in **Table 2.**

Table 2 Examples of Common Solutions

	Solvent/ State	Solute/ State	State of Solution
Earth's atmosphere	nitrogen/gas	oxygen/gas carbon dioxide/gas argon/gas	gas
Ocean water	water/liquid	salt/solid oxygen/gas carbon dioxide/gas	liquid
Carbonated beverage	water/liquid	carbon dioxide/gas	liquid
Brass	copper/solid	zinc/solid	solid

Figure 5 Acetic acid (a liquid), carbon dioxide (a gas), and drink-mix crystals (a solid) can be dissolved in water (a liquid). **Determine** *whether one liquid solution could contain all three different kinds of solute.*

Liquid Solutions

You're probably most familiar with liquid solutions like the ones shown in **Figure 5,** in which the solvent is a liquid. The solute can be another liquid, a solid, or even a gas. You've already learned about liquid-solid solutions such as sugar water and salt water. When discussing solutions, the state of the solvent usually determines the state of the solution.

Liquid-Gas Solutions Carbonated beverages are liquid-gas solutions—carbon dioxide is the gaseous solute, and water is the liquid solvent. The carbon dioxide gas gives the beverage its fizz and some of its tartness. The beverage also might contain other solutes, such as the compounds that give it its flavor and color.

Reading Check *What are the solutes in a carbonated beverage?*

Liquid-Liquid Solutions In a liquid-liquid solution, both the solvent and the solute are liquids. Vinegar, which you might use to make salad dressing, is a liquid-liquid solution made of 95 percent water (the solvent) and 5 percent acetic acid (the solute).

Gaseous Solutions

In gaseous solutions, a smaller amount of one gas is dissolved in a larger amount of another gas. This is called a gas-gas solution because both the solvent and solute are gases. The air you breathe is a gaseous solution. Nitrogen makes up about 78 percent of dry air and is the solvent. The other gases are the solutes.

Figure 6 Metal alloys can contain either metal or nonmetal solutes dissolved in a metal solvent.

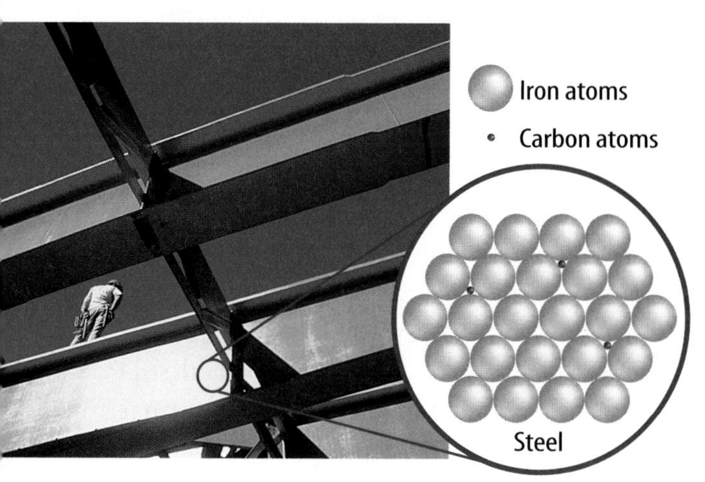

Steel is a solid solution of the metal iron and the nonmetal carbon.

Brass is a solid solution made of copper and zinc.

Solid Solutions In solid solutions, the solvent is a solid. The solute can be a solid, liquid, or gas. The most common solid solutions are solid-solid solutions—ones in which the solvent and the solute are solids. A solid-solid solution made from two or more metals is called an alloy. It's also possible to include elements that are not metals in alloys. For example, steel is an alloy that has carbon dissolved in iron. The carbon makes steel much stronger and yet more flexible than iron. Two alloys are shown in **Figure 6.**

section 1 review

Summary

Substances

- Elements are substances that cannot be broken down into simpler substances.
- A compound is made up of two or more elements bonded together.

Mixtures and Solutions

- Mixtures are either heterogeneous or homogeneous.
- Solutions have two parts—solute and solvent.
- Crystallization and precipitation are two ways that solids are formed from solutions.

Types of Solutions

- The solutes and solvents can be solids, liquids, or gases.

Self Check

1. **Compare and contrast** substances and mixtures. Give two examples of each.
2. **Describe** how heterogeneous and homogeneous mixtures differ.
3. **Explain** how a solution forms.
4. **Identify** the common name for a solid-solid solution of metals.
5. **Think Critically** The tops of carbonated-beverage cans usually are made with a different aluminum alloy than the pull tabs are made with. Explain.

Applying Skills

6. **Compare and contrast** the following solutions: a helium-neon laser, bronze (a copper-tin alloy), cloudy ice cubes, and ginger ale.

section 2

Solubility

as you read

What You'll Learn

- **Explain** why water is a good general solvent.
- **Describe** how the structure of a compound affects which solvents it dissolves in.
- **Identify** factors that affect how much of a substance will dissolve in a solvent.
- **Describe** how temperature affects reaction rate.
- **Explain** how solute particles affect physical properties of water.

Why It's Important

How you wash your hands, clothes, and dishes depends on which substances can dissolve in other substances.

Review Vocabulary

polar bond: a bond resulting from the unequal sharing of electrons

New Vocabulary

- aqueous
- saturated
- solubility
- concentration

Water—The Universal Solvent

In many solutions, including fruit juice and vinegar, water is the solvent. A solution in which water is the solvent is called an **aqueous** (A kwee us) solution. Because water can dissolve so many different solutes, chemists often call it the universal solvent. To understand why water is such a great solvent, you must first know a few things about atoms and bonding.

Molecular Compounds When certain atoms form compounds, they share electrons. Sharing electrons is called covalent bonding. Compounds that contain covalent bonds are called molecular compounds, or molecules.

If a molecule has an even distribution of electrons, like the one in **Figure 7,** it is called nonpolar. The atoms in some molecules do not have an even distribution of electrons. For example, in a water molecule, two hydrogen atoms share electrons with a single oxygen atom. However, as **Figure 7** shows, the electrons spend more time around the oxygen atom than they spend around the hydrogen atoms. As a result, the oxygen portion of the water molecule has a partial negative charge and the hydrogen portions have a partial positive charge. The overall charge of the water molecule is neutral. Such a molecule is said to be polar, and the bonds between its atoms are called polar covalent bonds.

Figure 7 Some atoms share electrons to form covalent bonds.

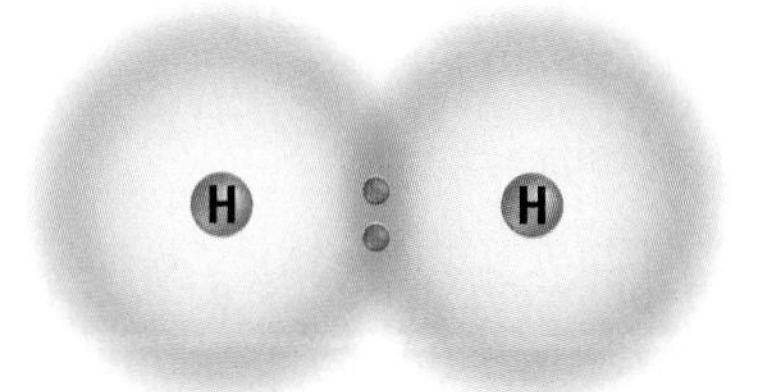

Two atoms of hydrogen share their electrons equally. Such a molecule is nonpolar.

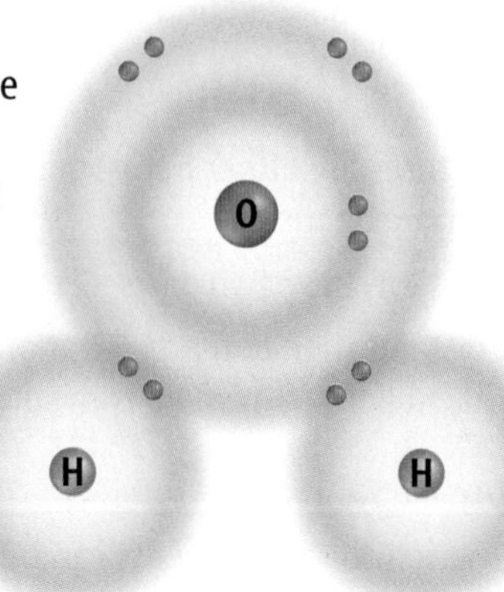

The electrons spend more time around the oxygen atom than the hydrogen atoms. Such a molecule is polar.

Ionic Bonds Some atoms do not share electrons when they join with other atoms to form compounds. Instead, these atoms lose or gain electrons. When they do, the number of protons and electrons within an atom are no longer equal, and the atom becomes positively or negatively charged. Atoms with a charge are called ions. Bonds between ions that are formed by the transfer of electrons are called ionic bonds, and the compound that is formed is called an ionic compound. Table salt is an ionic compound that is made of sodium ions and chloride ions. Each sodium atom loses one electron to a chlorine atom and becomes a positively charged sodium ion. Each chlorine atom gains one electron from a sodium atom, becoming a negatively charged chloride ion.

How does an ionic compound differ from a molecular compound?

Solutions Seawater is a solution that contains nearly every element found on Earth. Most elements are present in tiny quantities. Sodium and chloride ions are the most common ions in seawater. Several gases, including oxygen, nitrogen, and carbon dioxide, also are dissolved in seawater.

How Water Dissolves Ionic Compounds Now think about the properties of water and the properties of ionic compounds as you visualize how an ionic compound dissolves in water. Because water molecules are polar, they attract positive and negative ions. The more positive part of a water molecule—where the hydrogen atoms are—is attracted to negatively charged ions. The more negative part of a water molecule—where the oxygen atom is—attracts positive ions. When an ionic compound is mixed with water, the different ions of the compound are pulled apart by the water molecules. **Figure 8** shows how sodium chloride dissolves in water.

Figure 8 Water dissolves table salt because its partial charges are attracted to the charged ions in the salt.

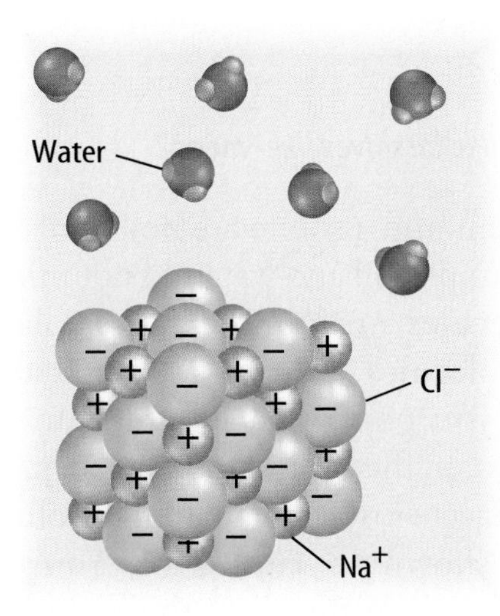

The partially negative oxygen in the water molecule is attracted to a positive sodium ion.

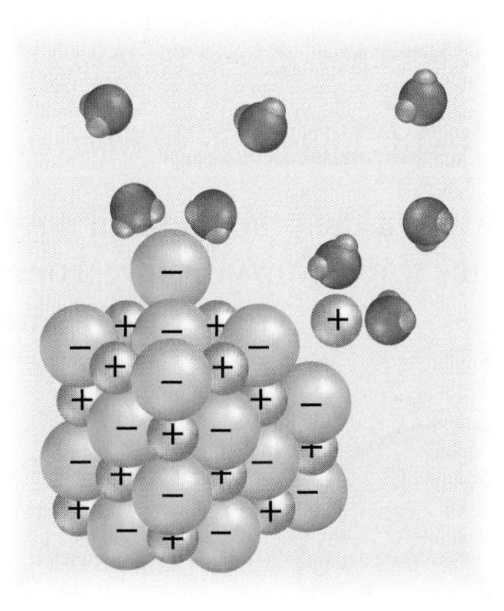

The partially positive hydrogen atoms in another water molecule are attracted to a negative chloride ion.

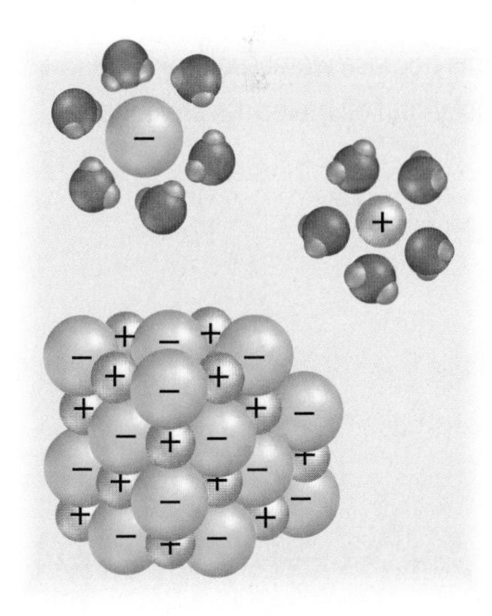

The sodium and chloride ions are pulled apart from each other, and more water molecules are attracted to them.

How Water Dissolves Molecular Compounds

Can water also dissolve molecular compounds that are not made of ions? Water does dissolve molecular compounds, such as sugar, although it doesn't break each sugar molecule apart. Water simply moves between different molecules of sugar, separating them. Like water, a sugar molecule is polar. Polar water molecules are attracted to the positive and negative portions of the polar sugar molecules. When the sugar molecules are separated by the water and spread throughout it, as **Figure 9** shows, they have dissolved.

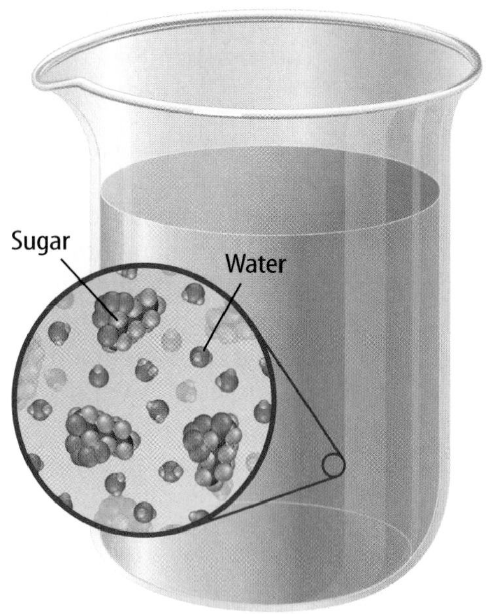

Figure 9 Sugar molecules that are dissolved in water spread out until they are spaced evenly in the water.

What will dissolve?

When you stir a spoonful of sugar into iced tea, all of the sugar dissolves but none of the metal in the spoon does. Why does sugar dissolve in water, but metal does not? A substance that dissolves in another is said to be soluble in that substance. You would say that the sugar is soluble in water but the metal of the spoon is insoluble in water, because it does not dissolve readily.

Like Dissolves Like When trying to predict which solvents can dissolve which solutes, chemists use the rule of "like dissolves like." This means that polar solvents dissolve polar solutes and nonpolar solvents dissolve nonpolar solutes. In the case of sugar and water, both are made up of polar molecules, so sugar is soluble in water. In the case of salt and water, the sodium and chloride ion pair is like the water molecule because it has a positive charge at one end and a negative charge at the other end.

Reading Check *What does "like dissolves like" mean?*

On the other hand, if a solvent and a solute are not similar, the solute won't dissolve. For example, oil and water do not mix. Oil molecules are nonpolar, so polar water molecules are not attracted to them. If you pour vegetable oil into a glass of water, the oil and the water separate into layers instead of forming a solution, as shown in **Figure 10.** You've probably noticed the same thing about the oil-and-water mixtures that make up some salad dressings. The oil stays on the top. Oils generally dissolve better in solvents that have nonpolar molecules.

Figure 10 Water and oil do not mix because water molecules are polar and oil molecules are nonpolar.

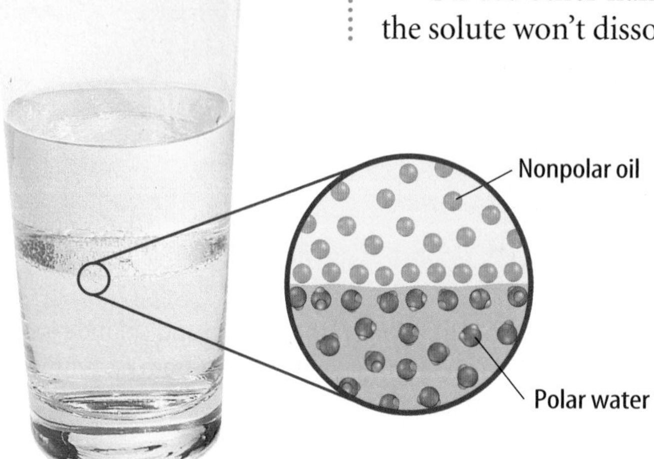

How much will dissolve?

Even though sugar is soluble in water, if you tried to dissolve 1 kg of sugar into one small glass of water, not all of the sugar would dissolve. **Solubility** (sahl yuh BIH luh tee) is a measurement that describes how much solute dissolves in a given amount of solvent. The solubility of a material has been described as the amount of the material that can dissolve in 100 g of solvent at a given temperature. Some solutes are highly soluble, meaning that a large amount of solute can be dissolved in 100 g of solvent. For example, 63 g of potassium chromate can be dissolved in 100 g of water at 25°C. On the other hand, some solutes are not very soluble. For example, only 0.00025 g of barium sulfate will dissolve in 100 g of water at 25°C. When a substance has an extremely low solubility, like barium sulfate does in water, it usually is considered insoluble.

Reading Check ***What is an example of a substance that is considered to be insoluble in water?***

Figure 11 The solubility of some solutes changes as the temperature of the solvent increases.
Use a Graph *According to the graph, is it likely that warm ocean water contains any more sodium chloride than cold ocean water does?*

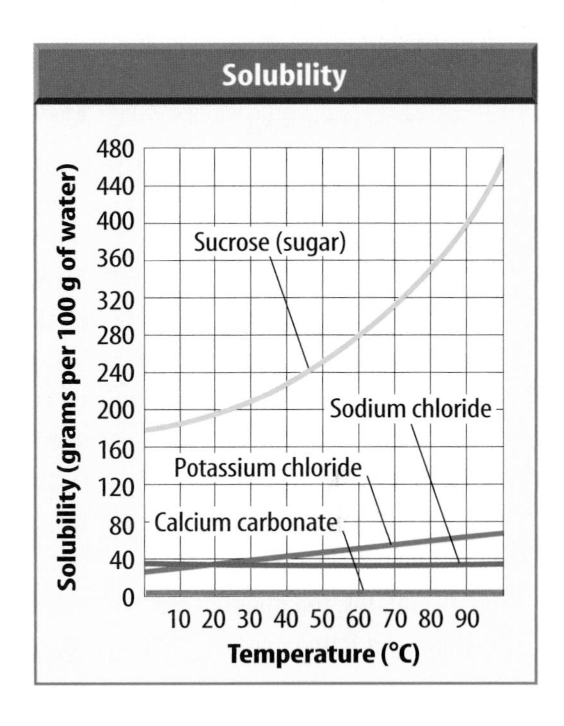

Solubility in Liquid-Solid Solutions

Did you notice that the temperature was included in the explanation about the amount of solute that dissolves in a quantity of solvent? The solubility of many solutes changes if you change the temperature of the solvent. For example, if you heat water, not only does the sugar dissolve at a faster rate, but more sugar can dissolve in it. However, some solutes, like sodium chloride and calcium carbonate, do not become more soluble when the temperature of water increases. The graph in **Figure 11** shows how the temperature of the solvent affects the solubility of some solutes.

Solubility in Liquid-Gas Solutions

Unlike liquid-solid solutions, an increase in temperature decreases the solubility of a gas in a liquid-gas solution. You might notice this if you have ever opened a warm carbonated beverage and it bubbled up out of control while a chilled one barely fizzed. Carbon dioxide is less soluble in a warm solution. What keeps the carbon dioxide from bubbling out when it is sitting at room temperature on a supermarket shelf? When a bottle is filled, extra carbon dioxide gas is squeezed into the space above the liquid, increasing the pressure in the bottle. This increased pressure increases the solubility of gas and forces most of it into the solution. When you open the cap, the pressure is released and the solubility of the carbon dioxide decreases.

Why does a bottle of carbonated beverage go "flat" after it has been opened for a few days?

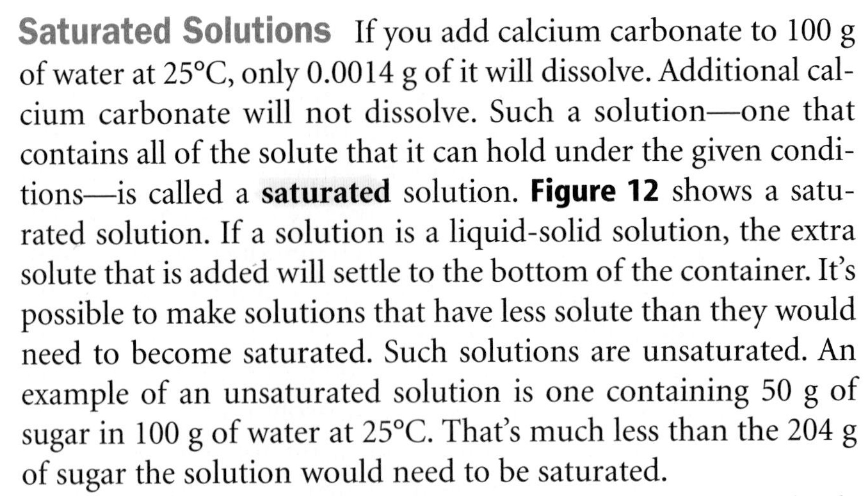

Saturated Solutions If you add calcium carbonate to 100 g of water at 25°C, only 0.0014 g of it will dissolve. Additional calcium carbonate will not dissolve. Such a solution—one that contains all of the solute that it can hold under the given conditions—is called a **saturated** solution. **Figure 12** shows a saturated solution. If a solution is a liquid-solid solution, the extra solute that is added will settle to the bottom of the container. It's possible to make solutions that have less solute than they would need to become saturated. Such solutions are unsaturated. An example of an unsaturated solution is one containing 50 g of sugar in 100 g of water at 25°C. That's much less than the 204 g of sugar the solution would need to be saturated.

A hot solvent usually can hold more solute than a cool solvent can. When a saturated solution cools, some of the solute usually falls out of the solution. But if a saturated solution is cooled slowly, sometimes the excess solute remains dissolved for a period of time. Such a solution is said to be supersaturated, because it contains more than the normal amount of solute.

Rate of Dissolving

Solubility does not tell you how fast a solute will dissolve—it tells you only how much of a solute will dissolve at a given temperature. Some solutes dissolve quickly, but others take a long time to dissolve. A solute dissolves faster when the solution is stirred or shaken or when the temperature of the solution is increased. These methods increase the rate at which the surfaces of the solute come into contact with the solvent. Increasing the area of contact between the solute and the solvent can also increase the rate of dissolving. This can be done by breaking up the solute into smaller pieces, which increases the surface area of the solute that is exposed to the solvent.

Molecules are always moving and colliding. The collisions must take place for chemical processes to occur. The chemical processes take place at a given rate of reaction. Temperature has a large effect on that rate. The higher the temperature, the more collisions occur and the higher the rate of reaction. The opposite is also true. The lower the temperature, the less collisions occur and the lower the rate of reaction. Refrigerators are an example of slowing the reaction rate—and therefore the chemical process—down to prevent food spoilage.

Observing Chemical Processes

Procedure

1. Pour **two small glasses of milk.**
2. Place one glass of milk in the **refrigerator.** Leave the second glass on the counter.
3. Allow the milk to sit overnight. **WARNING:** *Do not drink the milk that sat out overnight.*
4. On the following day, smell both glasses of milk. Record your observations.

Analysis

1. Compare and contrast the smell of the refrigerated milk to the non-refrigerated milk.
2. Explain why refrigeration is needed.

Try at Home

Figure 12 The Dead Sea has an extremely high concentration of dissolved minerals. When the water evaporates, the minerals are left behind and form pillars.

Concentration

What makes strong lemonade strong and weak lemonade weak? The difference between the two drinks is the amount of water in each one compared to the amount of lemon. The lemon is present in different concentrations in the solution. The **concentration** of a solution tells you how much solute is present compared to the amount of solvent. You can give a simple description of a solution's concentration by calling it either concentrated or dilute. These terms are used when comparing the concentrations of two solutions with the same type of solute and solvent. A concentrated solution has more solute per given amount of solvent than a dilute solution.

Measuring Concentration Can you imagine a doctor ordering a dilute intravenous, or IV, solution for a patient? Because dilute is not an exact measurement, the IV could be made with a variety of amounts of medicine. The doctor would need to specify the exact concentration of the IV solution to make sure that the patient is treated correctly.

Pharmacist Doctors rely on pharmacists to formulate IV solutions. Pharmacists begin with a concentrated form of the drug, which is supplied by pharmaceutical companies. This is the solute of the IV solution. The pharmacist adds the correct amount of solvent to a small amount of the solute to achieve the concentration requested by the doctor. There may be more than one solute per IV solution in varying concentrations.

Applying Science

How can you compare concentrations?

A solute is a substance that can be dissolved in another substance called a solvent. Solutions vary in concentration, or strength, depending on the amount of solute and solvent being used. Fruit drinks are examples of such a solution. Stronger fruit drinks appear darker in color and are the result of more drink mix being dissolved in a given amount of water. What would happen if more water were added to the solution?

Glucose Solutions (g/100 mL)

Solute Glucose (g)	Solvent Water (mL)	Solution Concentration of Glucose (%)
2	100	2
4	100	4
10	100	10
20	100	20

Identifying the Problem

The table on the right lists different concentration levels of glucose solutions, a type of carbohydrate your body uses as a source of energy. The glucose is measured in grams, and the water is measured in milliliters.

Solving the Problem

A physician writes a prescription for a patient to receive 1,000 mL of a 20 percent solution of glucose. How many grams of glucose must the pharmacist add to 1,000 mL of water to prepare this 20 percent concentration level?

Figure 13 Concentrations can be stated in percentages.
Identify *the percentage of this fruit drink that is water, assuming there are no other dissolved substances.*

One way of giving the exact concentration is to state the percentage of the volume of the solution that is made up of solute. Labels on fruit drinks show their concentration like the one in **Figure 13.** When a fruit drink contains 15 percent fruit juice, the remaining 85 percent of the drink is water and other substances such as sweeteners and flavorings. This drink is more concentrated than another brand that contains 10 percent fruit juice, but it's more dilute than pure juice, which is 100 percent juice. Another way to describe the concentration of a solution is to give the percentage of the total mass that is made up of solute.

Effects of Solute Particles All solute particles affect the physical properties of the solvent, such as its boiling point and freezing point. The effect that a solute has on the freezing or boiling point of a solvent depends on the number of solute particles.

When a solvent such as water begins to freeze, its molecules arrange themselves in a particular pattern. Adding a solute such as sodium chloride to this solvent changes the way the molecules arrange themselves. To overcome this interference of the solute, a lower temperature is needed to freeze the solvent.

When a solvent such as water begins to boil, the solvent molecules are gaining enough energy to move from the liquid state to the gaseous state. When a solute such as sodium chloride is added to the solvent, the solute particles interfere with the evaporation of the solvent particles. More energy is needed for the solvent particles to escape from the liquid, and the boiling point of the solution will be higher.

section 2 review

Summary

The Universal Solvent

- Water is known as the universal solvent.
- A molecule that has an even distribution of electrons is a nonpolar molecule.
- A molecule that has an uneven distribution of electrons is a polar molecule.
- A compound that loses or gains electrons is an ionic compound.

Dissolving a Substance

- Chemists use the rule "like dissolves like."

Concentration

- Concentration is the quantity of solute present compared to the amount of solvent.

Self Check

1. **Identify** the property of water that makes it the universal solvent.
2. **Describe** the two methods to increase the rate at which a substance dissolves.
3. **Infer** why it is important to add sodium chloride to water when making homemade ice cream.
4. **Think Critically** Why can the fluids used to dry-clean clothing remove grease even when water cannot?

Applying Skills

5. **Recognize Cause and Effect** Why is it more important in terms of reaction rate to take groceries straight home from the store when it is 25°C than when it is 2°C?

 Science Online green.msscience.com/self_check_quiz

Observing Gas Solubility

On a hot day, a carbonated beverage will cool you off. If you leave the beverage uncovered at room temperature, it quickly loses its fizz. However, if you cap the beverage and place it in the refrigerator, it will still have its fizz hours later. In this lab you will explore why this happens.

Real-World Question

What effect does temperature have on the fizz, or carbon dioxide, in your carbonated beverage?

Goals

- **Observe** the effect that temperature has on solubility of a gas in a liquid.
- **Compare** the amount of carbon dioxide released at room temperature and in hot tap water.

Materials

carbonated beverages in plastic bottles, thoroughly chilled (2)
balloons (2)
tape
fabric tape measure
string
**ruler*
container
hot tap water

**Alternative materials*

Safety Precautions

WARNING: *DO NOT point the bottles at anyone at any time during the lab.*

Procedure

1. Carefully remove the caps from the thoroughly chilled plastic bottles one at a time. Create as little agitation as possible.
2. Quickly cover the opening of each bottle with an uninflated balloon.
3. Use tape to secure and tightly seal the balloons to the top of the bottles.
4. Gently agitate one bottle from side to side for two minutes. Measure the circumference of the balloon.

WARNING: *Contents under pressure can cause serious accidents. Be sure to wear safety goggles, and DO NOT point the bottles at anyone.*

5. Gently agitate the second bottle in the same manner as in step 4. Then, place the bottle in a container of hot tap water for ten minutes. Measure the circumference of the balloon.

Conclude and Apply

1. **Contrast** the relative amounts of carbon dioxide gas released from the cold and the warm carbonated beverages.
2. **Infer** Why does the warmed carbonated beverage release a different amount of carbon dioxide than the chilled one?

Communicating Your Data

Compare the circumferences of your balloons with those of members of your class. **For more help, refer to the Science Skill Handbook.**

section 3

Acidic and Basic Solutions

as you read

What You'll Learn

- **Compare** acids and bases and their properties.
- **Describe** practical uses of acids and bases.
- **Explain** how pH is used to describe the strength of an acid or base.
- **Describe** how acids and bases react when they are brought together.

Why It's Important

Many common products, such as batteries and bleach, work because of acids or bases.

Review Vocabulary
physical property: any characteristic of a material that can be seen or measured without changing the material

New Vocabulary
- acid
- hydronium ion
- base
- pH
- indicator
- neutralization

Acids

What makes orange juice, vinegar, dill pickles, and grapefruit tangy? Acids cause the sour taste of these and other foods. **Acids** are substances that release positively charged hydrogen ions, H^+, in water. When an acid mixes with water, the acid dissolves, releasing a hydrogen ion. The hydrogen ion then combines with a water molecule to form a hydronium ion, as shown in **Figure 14. Hydronium ions** are positively charged and have the formula H_3O^+.

Properties of Acidic Solutions Sour taste is one of the properties of acidic solutions. The taste allows you to detect the presence of acids in your food. However, even though you can identify acidic solutions by their sour taste, you should never taste anything in the laboratory, and you should never use taste to test for the presence of acids in an unknown substance. Many acids can cause serious burns to body tissues.

Another property of acidic solutions is that they can conduct electricity. The hydronium ions in an acidic solution can carry the electric charges in a current. This is why some batteries contain an acid. Acidic solutions also are corrosive, which means they can break down certain substances. Many acids can corrode fabric, skin, and paper. The solutions of some acids also react strongly with certain metals. The acid-metal reaction forms metallic compounds and hydrogen gas, leaving holes in the metal in the process.

$$H^+ + H_2O \longrightarrow H_3O^+$$

Hydrogen ion + Water molecule → Hydronium ion

Figure 14 One hydrogen ion can combine with one water molecule to form one positively charged hydronium ion.
Identify *what kinds of substances are sources of hydrogen ions.*

Figure 15 Each of these products contains an acid or is made with the help of an acid.
Describe *how your life would be different if acids were not available to make these products.*

Uses of Acids You're probably familiar with many acids. Vinegar, which is used in salad dressing, contains acetic acid. Lemons, limes, and oranges have a sour taste because they contain citric acid. Your body needs ascorbic acid, which is vitamin C. Ants that sting inject formic acid into their victims.

Figure 15 shows other products that are made with acids. Sulfuric acid is used in the production of fertilizers, steel, paints, and plastics. Acids often are used in batteries because their solutions conduct electricity. For this reason, it sometimes is referred to as battery acid. Hydrochloric acid, which is known commercially as muriatic acid, is used in a process called pickling. Pickling is a process that removes impurities from the surfaces of metals. Hydrochloric acid also can be used to clean mortar from brick walls. Nitric acid is used in the production of fertilizers, dyes, and plastics.

Acid in the Environment Carbonic acid plays a key role in the formation of caves and of stalactites and stalagmites. Carbonic acid is formed when carbon dioxide in soil is dissolved in water. When this acidic solution comes in contact with calcium carbonate—or limestone rock—it can dissolve it, eventually carving out a cave in the rock. A similar process occurs when acid rain falls on statues and eats away at the stone, as shown in **Figure 16.** When this acidic solution drips from the ceiling of the cave, water evaporates and carbon dioxide becomes less soluble, forcing it out of solution. The solution becomes less acidic and the limestone becomes less soluble, causing it to come out of solution. These solids form stalactites and stalagmites.

Observing a Nail in a Carbonated Drink

Procedure

1. Observe the initial appearance of an **iron nail.**
2. Pour enough **carbonated soft drink** into a **cup or beaker** to cover the nail.
3. Drop the nail into the soft drink and observe what happens.
4. Leave the nail in the soft drink overnight and observe it again the next day.

Analysis

1. Describe what happened when you first dropped the nail into the soft drink and the appearance of the nail the following day.
2. Based upon the fact that the soft drink was carbonated, explain why you think the drink reacted with the nail as you observed.

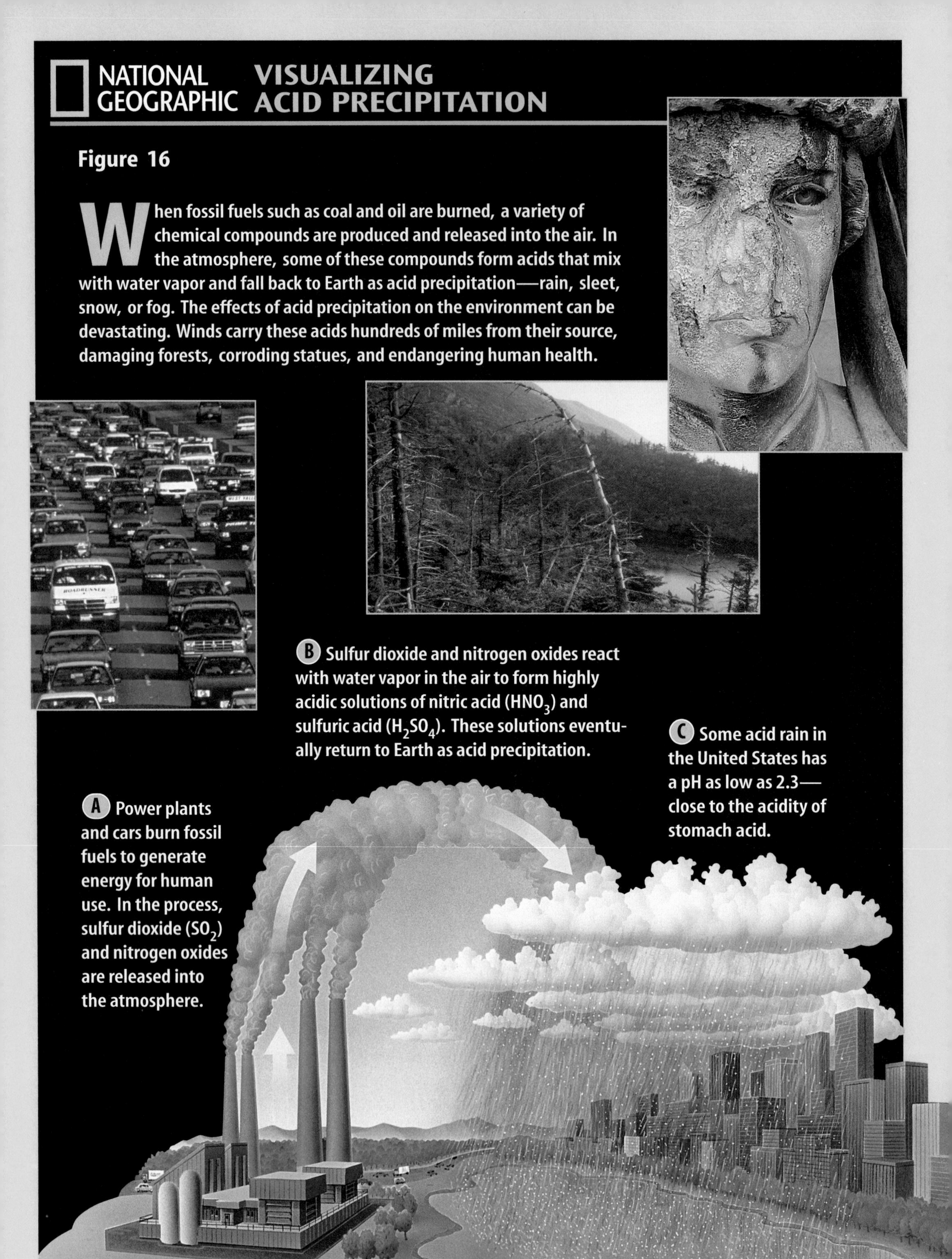
NATIONAL GEOGRAPHIC
VISUALIZING ACID PRECIPITATION
Figure 16
When fossil fuels such as coal and oil are burned, a variety of chemical compounds are produced and released into the air. In the atmosphere, some of these compounds form acids that mix with water vapor and fall back to Earth as acid precipitation—rain, sleet, snow, or fog. The effects of acid precipitation on the environment can be devastating. Winds carry these acids hundreds of miles from their source, damaging forests, corroding statues, and endangering human health.
B Sulfur dioxide and nitrogen oxides react with water vapor in the air to form highly acidic solutions of nitric acid (HNO_3) and sulfuric acid (H_2SO_4). These solutions eventually return to Earth as acid precipitation.
C Some acid rain in the United States has a pH as low as 2.3—close to the acidity of stomach acid.
A Power plants and cars burn fossil fuels to generate energy for human use. In the process, sulfur dioxide (SO_2) and nitrogen oxides are released into the atmosphere.

Bases

People often use ammonia solutions to clean windows and floors. These solutions have different properties from those of acidic solutions. Ammonia is called a base. **Bases** are substances that can accept hydrogen ions. When bases dissolve in water, some hydrogen atoms from the water molecules are attracted to the base. A hydrogen atom in the water molecule leaves behind the other hydrogen atom and oxygen atom. This pair of atoms is a negatively charged ion called a hydroxide ion. A hydroxide ion has the formula OH^-. Most bases contain a hydroxide ion, which is released when the base dissolves in water. For example, sodium hydroxide is a base with the formula NaOH. When NaOH dissolves in water, a sodium ion and the hydroxide ion separate.

Topic: Calcium Hydroxide
Visit green.msscience.com for Web links to information about the uses for calcium hydroxide.

Activity Describe the chemical reaction that converts limestone (calcium carbonate) to calcium hydroxide.

Properties of Basic Solutions Most soaps are bases, so if you think about how soap feels, you can figure out some of the properties of basic solutions. Basic solutions feel slippery. Acids in water solution taste sour, but bases taste bitter—as you know if you have ever accidentally gotten soap in your mouth.

Like acids, bases are corrosive. Bases can cause burns and damage tissue. You should never touch or taste a substance to find out whether it is a base. Basic solutions contain ions and can conduct electricity. Basic solutions are not as reactive with metals as acidic solutions are.

Uses of Bases Many uses for bases are shown in **Figure 17.** Bases give soaps, ammonia, and many other cleaning products some of their useful properties. The hydroxide ions produced by bases can interact strongly with certain substances, such as dirt and grease.

Chalk and oven cleaner are examples of familiar products that contain bases. Your blood is a basic solution. Calcium hydroxide, often called lime, is used to mark the lines on athletic fields. It also can be used to treat lawns and gardens that have acidic soil. Sodium hydroxide, known as lye, is a strong base that can cause burns and other health problems. Lye is used to make soap, clean ovens, and unclog drains.

Figure 17 Many products, including soaps, cleaners, and plaster contain bases or are made with the help of bases.

pH Levels Most life-forms can't exist at extremely low pH levels. However, some bacteria thrive in acidic environments. Acidophils are bacteria that exist at low pH levels. These bacteria have been found in the Hot Springs of Yellowstone National Park in areas with pH levels ranging from 1 to 3.

What is pH?

You've probably heard of pH-balanced shampoo or deodorant, and you might have seen someone test the pH of the water in a swimming pool. **pH** is a measure of how acidic or basic a solution is. The pH scale ranges from 0 to 14. Acidic solutions have pH values below 7. A solution with a pH of 0 is very acidic. Hydrochloric acid can have a pH of 0. A solution with a pH of 7 is neutral, meaning it is neither acidic nor basic. Pure water is neutral. Basic solutions have pH values above 7. A solution with a pH of 14 is very basic. Sodium hydroxide can have a pH of 14. **Figure 18** shows where various common substances fall on the pH scale.

The pH of a solution is related directly to its concentrations of hydronium ions (H_3O^+) and hydroxide ions (OH^-). Acidic solutions have more hydronium ions than hydroxide ions. Neutral solutions have equal numbers of the two ions. Basic solutions have more hydroxide ions than hydronium ions.

Reading Check *In a neutral solution, how do the numbers of hydronium ions and hydroxide ions compare?*

pH Scale The pH scale is not a simple linear scale like mass or volume. For example, if one book has a mass of 2 kg and a second book has a mass of 1 kg, the mass of the first book is twice that of the second. However, a change of 1 pH unit represents a tenfold change in the acidity of the solution. For example, if one solution has a pH of 1 and a second solution has a pH of 2, the first solution is not twice as acidic as the second—it is ten times more acidic. To determine the difference in pH strength, use the following calculation: 10^n, where n = the difference between pHs. For example: pH3 − pH1 = 2, $10^2 = 100$ times more acidic.

Figure 18 The pH scale classifies a solution as acidic, basic, or neutral.

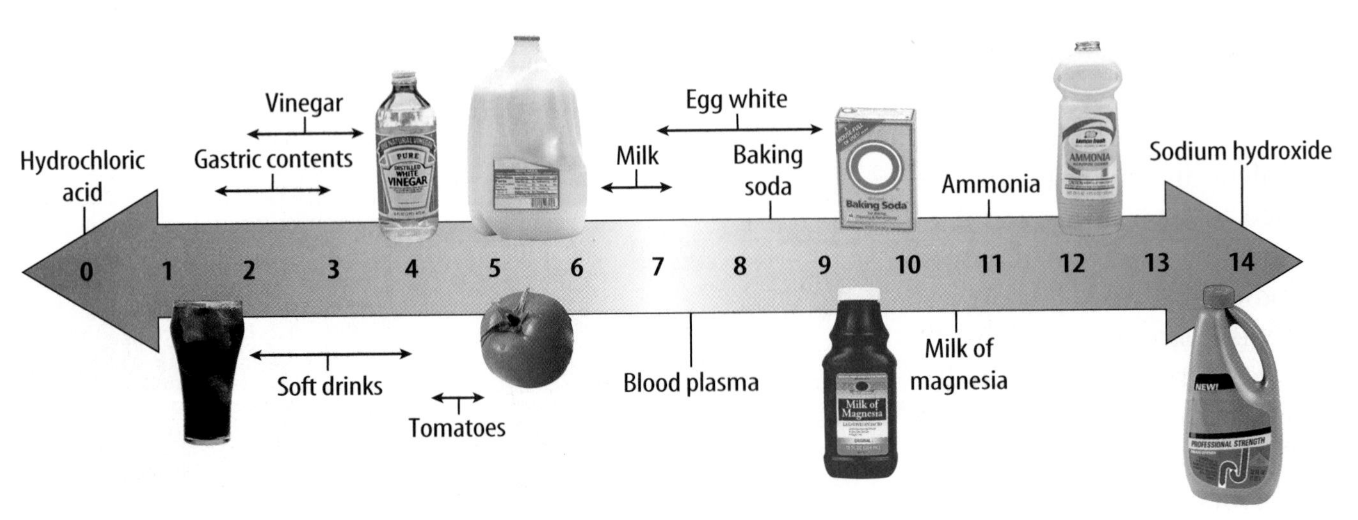

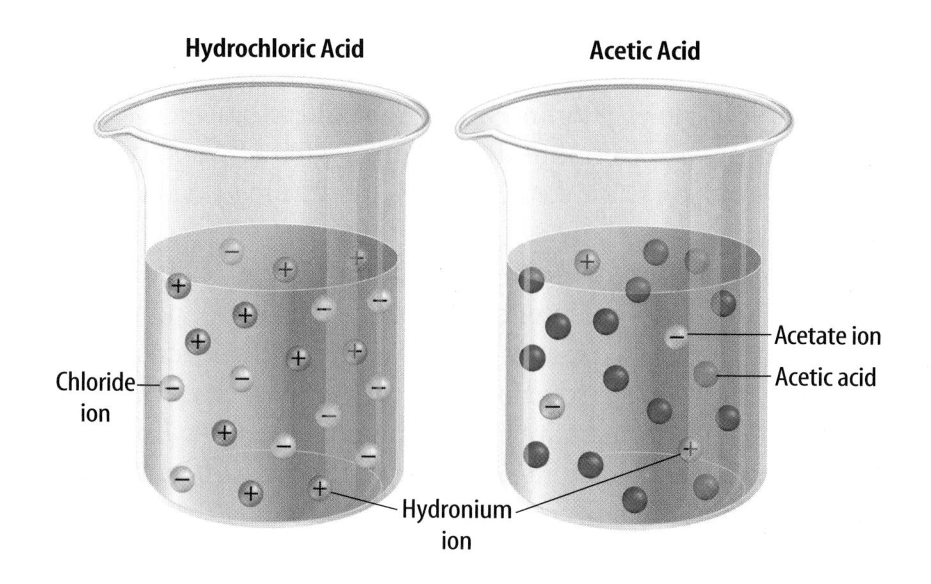

Figure 19 Hydrochloric acid separates into ions more readily than acetic acid does when it dissolves in water. Therefore, hydrochloric acid exists in water as separated ions. Acetic acid exists in water almost entirely as molecules.

Strengths of Acids and Bases You've learned that acids give foods a sour taste but also can cause burns and damage tissue. The difference between food acids and the acids that can burn you is that they have different strengths. The acids in food are fairly weak acids, while the dangerous acids are strong acids. The strength of an acid is related to how easily the acid separates into ions, or how easily a hydrogen ion is released, when the acid dissolves in water. Look at **Figure 19.** In the same concentration, a strong acid—like hydrochloric acid—forms more hydronium ions in solution than a weak acid does—like acetic acid. More hydronium ions means the strong-acid solution has a lower pH than the weak-acid solution. Similarly, the strength of a base is related to how easily the base separates into ions, or how easily a hydroxide ion is released, when the base dissolves in water. The relative strengths of some common acids and bases are shown in **Table 3.**

Reading Check *What determines the strength of an acid or a base?*

An acid containing more hydrogen atoms, such as carbonic acid, H_2CO_3, is not necessarily stronger than an acid containing fewer hydrogen atoms, such as nitric acid, HNO_3. An acid's strength is related to how easily a hydrogen ion separates—not to how many hydrogen atoms it has. For this reason, nitric acid is stronger than carbonic acid.

Table 3 Strengths of Some Acids and Bases

	Acid	Base
Strong	hydrochloric (HCl) sulfuric (H_2SO_4) nitric (HNO_3)	sodium hydroxide ($NaOH$) potassium hydroxide (KOH)
Weak	acetic (CH_3COOH) carbonic (H_2CO_3) ascorbic ($H_2C_6H_6O_6$)	ammonia (NH_3) aluminum hydroxide ($Al(OH)_3$) iron (III) hydroxide ($Fe(OH)_3$)

Indicators

What is a safe way to find out how acidic or basic a solution is? **Indicators** are compounds that react with acidic and basic solutions and produce certain colors, depending on the solution's pH.

Because they are different colors at different pHs, indicators can help you determine the pH of a solution. Some indicators, such as litmus, are soaked into paper strips. When litmus paper is placed in an acidic solution, it turns red. When placed in a basic solution, litmus paper turns blue. Some indicators can change through a wide range of colors, with each different color appearing at a different pH value.

Topic: Indicators
Visit green.msscience.com for Web links to information about the types of pH indicators.

Activity Describe how plants can act as indicators in acidic and basic solutions.

Neutralization

Perhaps you've heard someone complain about heartburn or an upset stomach after eating spicy food. To feel better, the person might have taken an antacid. Think about the word *antacid* for a minute. How do antacids work?

Heartburn or stomach discomfort is caused by excess hydrochloric acid in the stomach. Hydrochloric acid helps break down the food you eat, but too much of it can irritate your stomach or digestive tract. An antacid product, often made from the base magnesium hydroxide, $Mg(OH)_2$, neutralizes the excess acid. **Neutralization** (new truh luh ZAY shun) is the reaction of an acid with a base. It is called this because the properties of both the acid and base are diminished, or neutralized. In most cases, the reaction produces a water and a salt. **Figure 20** illustrates the relative amounts of hydronium and hydroxide ions between pH 0 and pH 14.

Reading Check *What are the products of neutralization?*

Figure 20 The pH of a solution is more acidic when greater amounts of hydronium ions are present.
Define *what makes a pH 7 solution neutral.*

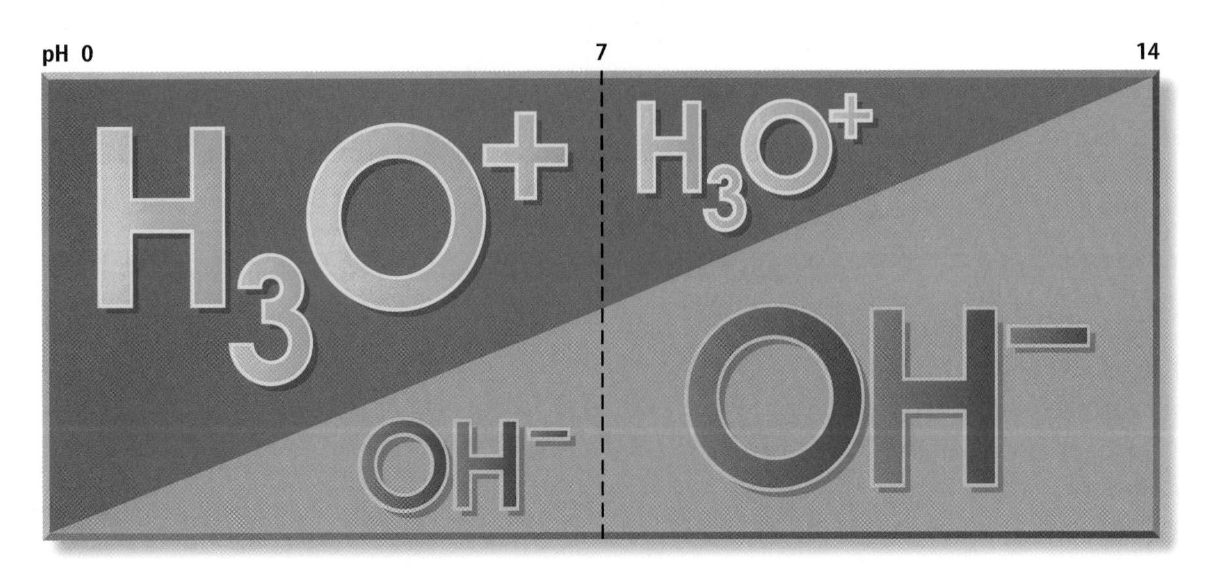

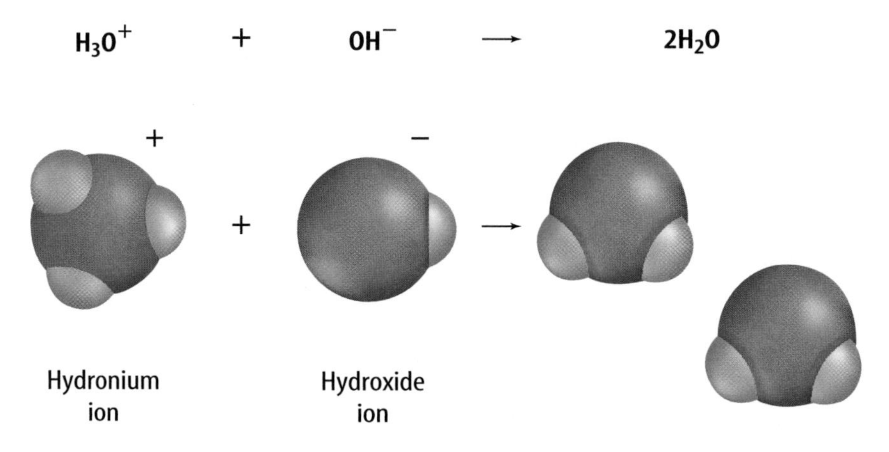

Figure 21 When acidic and basic solutions react, hydronium and hydroxide ions react to form water.
Determine *why the pH of the solution changes.*

How does neutralization occur? Recall that every water molecule contains two hydrogen atoms and one oxygen atom. As **Figure 21** shows, when one hydronium ion reacts with one hydroxide ion, the product is two water molecules. This reaction occurs during acid-base neutralization. Equal numbers of hydronium ions from the acidic solution and hydroxide ions from the basic solution react to produce water. Pure water has a pH of 7, which means that it's neutral.

What happens to acids and bases during neutralization?

section 3 review

Summary

Acids and Bases

- Acids are substances that release positively charged hydrogen ions in water.
- Substances that accept hydrogen ions in water are bases.
- Acidic and basic solutions can conduct electricity.

pH

- pH measures how acidic or basic a solution is.
- The scale ranges from 0 to 14.

Neutralization

- Neutralization is the interaction between an acid and a base to form water and a salt.

Self Check

1. **Identify** what ions are produced by acids in water and bases in water. Give two properties each of acids and bases.
2. **Name** three acids and three bases and list an industrial or household use of each.
3. **Explain** how the concentration of hydronium ions and hydroxide ions are related to pH.
4. **Think Critically** In what ways might a company that uses a strong acid handle an acid spill on the factory floor?

Applying Math

5. **Solve One-Step Equations** How much more acidic is a solution with a pH of 2 than one with a pH of 6? How much more basic is a solution with a pH of 13 than one with a pH of 10?

Testing pH Using Natural Indicators

Goals
- **Determine** the relative acidity or basicity of several common solutions.
- **Compare** the strengths of several common acids and bases.

Materials
small test tubes (9)
test-tube rack
concentrated red cabbage juice in a dropper bottle
labeled bottles containing: household ammonia, baking soda solution, soap solution, 0.1*M* hydrochloric acid solution, white vinegar, colorless carbonated soft drink, borax soap solution, distilled water
grease pencil
droppers (9)

Safety Precautions

WARNING: *Many acids and bases are poisonous, can damage your eyes, and can burn your skin. Wear goggles and gloves AT ALL TIMES. Tell your teacher immediately if a substance spills. Wash your hands after you finish but before removing your goggles.*

Real-World Question

You have learned that certain substances, called indicators, change color when the pH of a solution changes. The juice from red cabbage is a natural indicator. How do the pH values of various solutions compare to each other? How can you use red cabbage juice to determine the relative pH of several solutions?

Procedure

1. **Design** a data table to record the names of the solutions to be tested, the colors caused by the added cabbage juice indicator, and the relative strengths of the solutions.
2. Mark each test tube with the identity of the acid or base solution it will contain.
3. Half-fill each test tube with the solution to be tested.
   WARNING: *If you spill any liquids on your skin, rinse the area immediately with water. Alert your teacher if any liquid spills in the work area or on your skin.*

4. Add ten drops of the cabbage juice indicator to each of the solutions to be tested. Gently agitate or wiggle each test tube to mix the cabbage juice with the solution.
5. **Observe** and record the color of each solution in your data table.

Determining pH Values	
Cabbage Juice Color	Relative Strength of Acid or Base
	strong acid
	medium acid
	weak acid
	neutral
	weak base
	medium base
	strong base

Analyze Your Data

1. **Compare** your observations with the table above. Record in your data table the relative acid or base strength of each solution you tested.
2. **List** the solutions by pH value starting with the most acidic and finishing with the most basic.

Conclude and Apply

1. **Classify** which solutions were acidic and which were basic.
2. **Identify** which solution was the weakest acid. The strongest base? The closest to neutral?
3. **Predict** what ion might be involved in the cleaning process based upon your data for the ammonia, soap, and borax soap solutions.

Form a Hypothesis

Form a hypothesis that explains why the borax soap solution was less basic than an ammonia solution of approximately the same concentration.

Communicating Your Data

Use your data to create labels for the solutions you tested. Include the relative strength of each solution and any other safety information you think is important on each label. **For more help, refer to the Science Skill Handbook.**

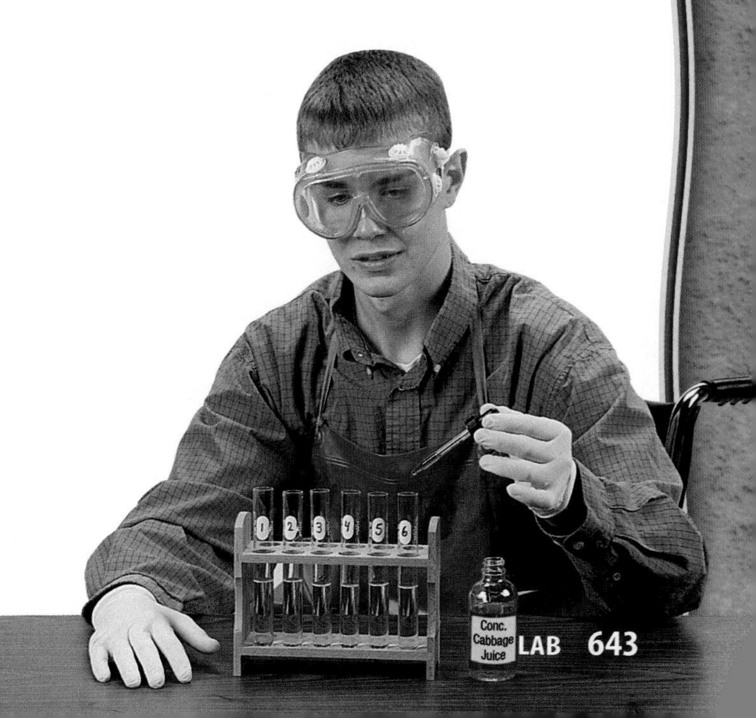

SCIENCE Stats

Salty Solutions

Did you know...

...Seawater is certainly a salty solution. Ninety-nine percent of all salt ions in the sea are sodium, chlorine, sulfate, magnesium, calcium, and potassium. The major gases in the sea are nitrogen, oxygen, carbon dioxide, argon, neon, and helium.

...Tears and saliva have a lot in common. Both are salty solutions that protect you from harmful bacteria, keep tissues moist, and help spread nutrients. Bland-tasting saliva, however, is 99 percent water. The remaining one percent is a combination of many ions, including sodium and several proteins.

...The largest salt lake in the United States is the Great Salt Lake. It covers more than 4,000 km^2 in Utah and is up to 13.4 m deep. The Great Salt Lake and the Salt Lake Desert were once part of the enormous, prehistoric Lake Bonneville, which was 305 m deep at some points.

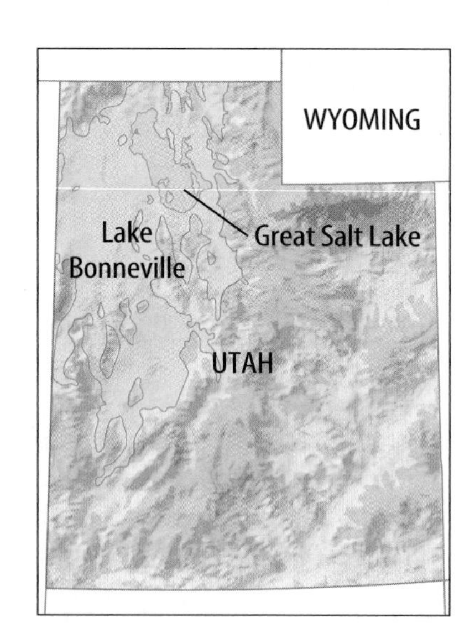

Applying Math At its largest, Lake Bonneville covered about 32,000 km^2. What percentage of that area does the Great Salt Lake now cover?

...Salt can reduce pain. Gargled salt water is a disinfectant; it fights the bacteria that cause some sore throats.

Graph It

Visit green.msscience.com/science_stats to research and learn about other elements in seawater. Create a graph that shows the amounts of the ten most common elements in 1 L of seawater.

Reviewing Main Ideas

Section 1 What is a solution?

1. Elements and compounds are pure substances, because their compositions are fixed. Mixtures are not pure substances.
2. Heterogeneous mixtures are not mixed evenly. Homogeneous mixtures, also called solutions, are mixed evenly on a molecular level.
3. Solutes and solvents can be gases, liquids, or solids, combined in many different ways.

Section 2 Solubility

1. Because water molecules are polar, they can dissolve many different solutes. Like dissolves like.
2. Temperature and pressure can affect solubility.
3. Solutions can be unsaturated, saturated, or supersaturated, depending on how much solute is dissolved compared to the solubility of the solute in the solvent.
4. The concentration of a solution is the amount of solute in a particular volume of solvent.

Section 3 Acidic and Basic Solutions

1. Acids release H+ ions and produce hydronium ions when they are dissolved in water. Bases accept H+ ions and produce hydroxide ions when dissolved in water.
2. pH expresses the concentrations of hydronium ions and hydroxide ions in aqueous solutions.
3. In a neutralization reaction, an acid reacts with a base to form water and a salt.

Visualizing Main Ideas

Copy and complete the concept map on the classification of matter.

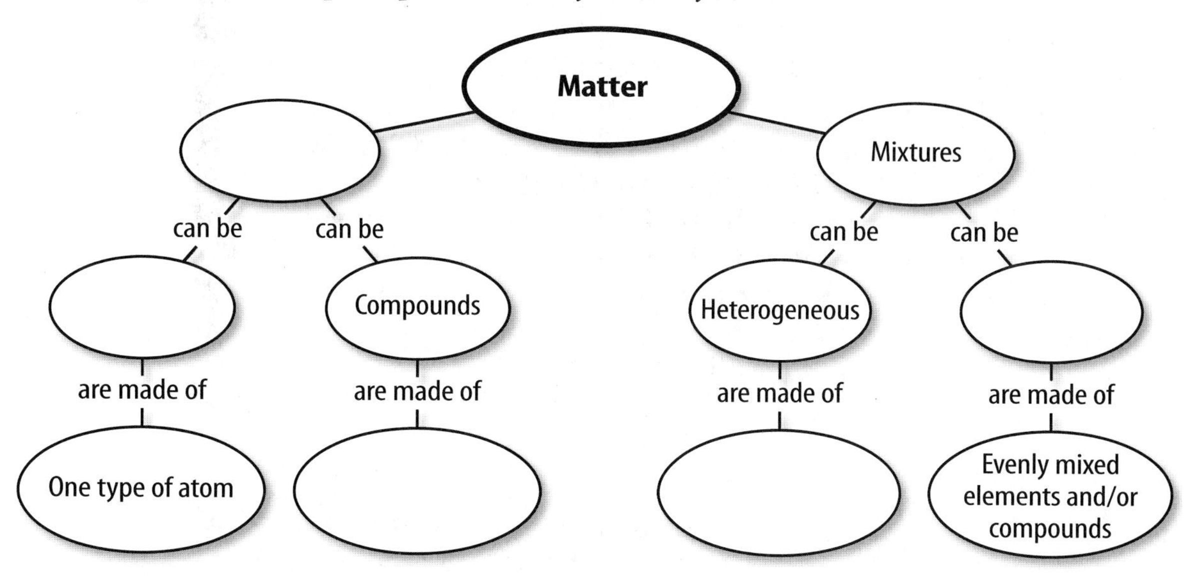

Using Vocabulary

acid p. 634	neutralization p. 640
aqueous p. 626	pH p. 638
base p. 637	precipitate p. 622
concentration p. 631	saturated p. 630
heterogeneous mixture p. 621	solubility p. 629
homogeneous mixture p. 622	solute p. 622
hydronium ion p. 634	solution p. 622
indicator p. 640	solvent p. 622
	substance p. 620

Fill in the blanks with the correct vocabulary word.

1. A base has a(n) ________ value above 7.
2. A measure of how much solute is in a solution is its ________.
3. The amount of a solute that can dissolve in 100 g of solvent is its ________.
4. The ________ is the substance that is dissolved to form a solution.
5. The reaction between an acidic and basic solution is called ________.
6. A(n) ________ has a fixed composition.

Checking Concepts

Choose the word or phrase that best answers the question.

7. Which of the following is a solution?
   A) pure water
   B) an oatmeal-raisin cookie
   C) copper
   D) vinegar
8. What type of compounds will not dissolve in water?
   A) polar **C)** nonpolar
   B) ionic **D)** charged
9. What type of molecule is water?
   A) polar **C)** nonpolar
   B) ionic **D)** precipitate
10. When chlorine compounds are dissolved in pool water, what is the water?
   A) the alloy
   B) the solvent
   C) the solution
   D) the solute

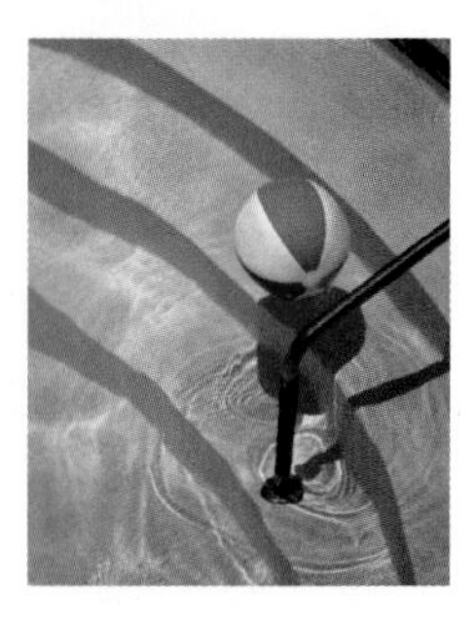

11. A solid might become less soluble in a liquid when you decrease what?
   A) particle size **C)** temperature
   B) pressure **D)** container size
12. Which acid is used in the industrial process known as pickling?
   A) hydrochloric **C)** sulfuric
   B) carbonic **D)** nitric
13. A solution is prepared by adding 100 g of solid sodium hydroxide, NaOH, to 1,000 mL of water. What is the solid NaOH called?
   A) solution **C)** solvent
   B) solute **D)** mixture
14. Given equal concentrations, which of the following will produce the most hydronium ions in an aqueous solution?
   A) a strong base **C)** a strong acid
   B) a weak base **D)** a weak acid
15. Bile, an acidic body fluid used in digestion, has a high concentration of hydronium ions. Predict its pH.
   A) 11 **C)** less than 7
   B) 7 **D)** greater than 7
16. When you swallow an antacid, what happens to your stomach acid?
   A) It is more acidic.
   B) It is concentrated.
   C) It is diluted.
   D) It is neutralized.

Science Online green.msscience.com/vocabulary_puzzlemaker

Thinking Critically

17. **Infer** why deposits form in the steam vents of irons in some parts of the country.

18. **Explain** if it is possible to have a dilute solution of a strong acid.

19. **Draw Conclusions** Antifreeze is added to water in a car's radiator to prevent freezing in cold months. It also prevents overheating or boiling. Explain how antifreeze does both.

Use the illustration below to answer question 20.

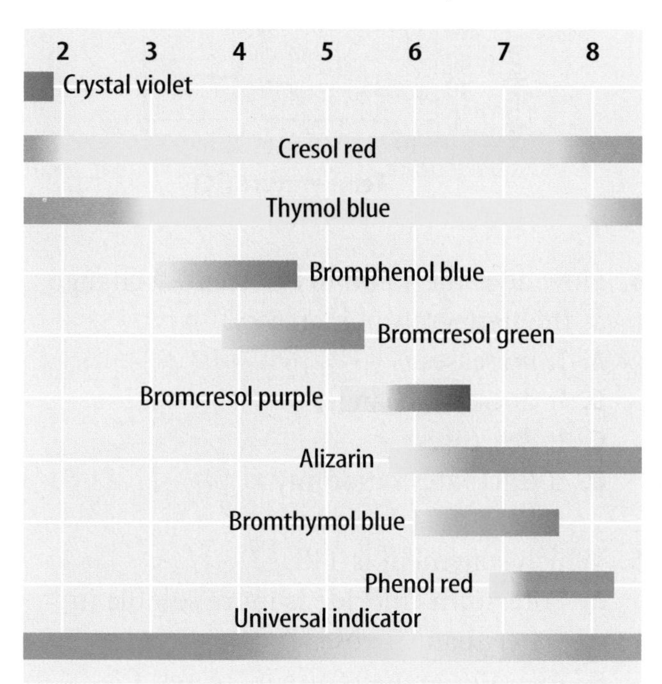

20. **Interpret** Chemists use a variety of indicators. Using the correct indicator is important. The color change must occur at the proper pH or the results could be misleading. Looking at the indicator chart, what indicators could be used to produce a color change at both pH 2 and pH 8?

21. **Explain** Water molecules can break apart to form H^+ ions and OH^- ions. Water is known as an amphoteric substance, which is something that can act as an acid or a base. Explain how this can be so.

22. **Describe** how a liquid-solid solution forms. How is this different from a liquid-gas solution? How are these two types of solutions different from a liquid-liquid solution? Give an example of each with your description.

23. **Compare and contrast** examples of heterogeneous and homogeneous mixtures from your daily life.

24. **Form a Hypotheses** A warm carbonated beverage seems to fizz more than a cold one when it is opened. Explain this based on the solubility of carbon dioxide in water.

Performance Activities

25. **Poem** Write a poem that explains the difference between a substance and a mixture.

Applying Math

Use the graph below to answer question 26.

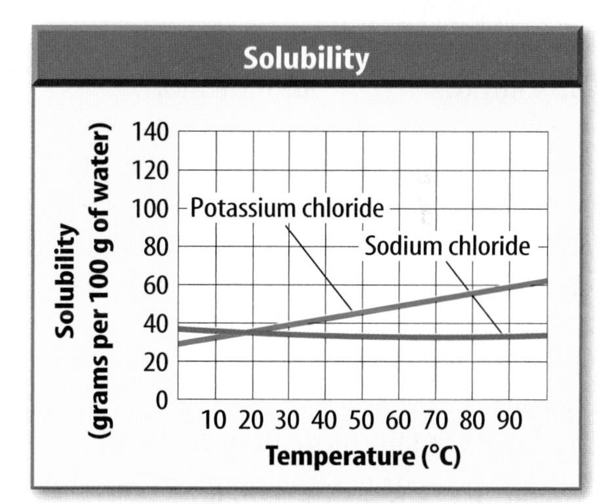

26. **Solubility** Using the solubility graph above, estimate the solubilities of potassium chloride and sodium chloride in grams per 100 g of water at 80°C.

27. **Juice Concentration** You made a one-liter (1,000 mL) container of juice. How much concentrate, in mL, did you add to make a concentration of 18 percent?

Standardized Test Practice

Part 1 Multiple Choice

Record your answers on the answer sheet provided by your teacher or on a sheet of paper.

Use the illustration below to answer questions 1 and 2.

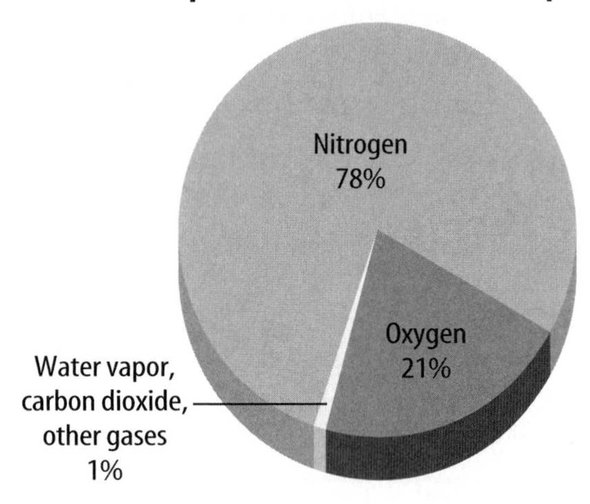

1. Which term best describes Earth's atmosphere?

A. saturated
B. solution
C. precipitate
D. indicator

2. Which of these is the solvent in Earth's atmosphere?

A. nitrogen
B. oxygen
C. water vapor
D. carbon dioxide

3. What characteristic do aqueous solutions share?

A. They contain more than three solutes.
B. No solids or gases are present as solutes in them.
C. All are extremely concentrated.
D. Water is the solvent in them.

Test-Taking Tip

Start the Day Right The morning of the test, eat a healthy breakfast with a balanced amount of protein and carbohydrates.

Use the illustration below to answer questions 4 and 5.

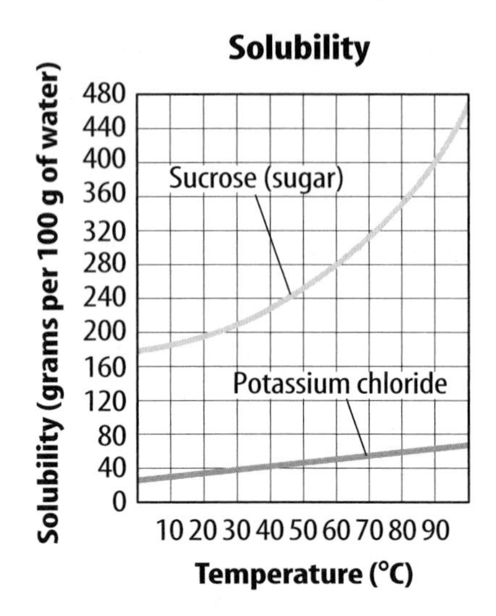

4. How does the solubility of sucrose change as the temperature increases?

A. It increases.
B. It does not change.
C. It decreases.
D. It fluctuates randomly.

5. Which statement is TRUE?

A. Potassium chloride is more soluble in water than sucrose.
B. As water temperature increases, the solubility of potassium chloride decreases.
C. Sucrose is more soluble in water than potassium chloride.
D. Water temperature has no effect on the solubility of these two chemicals.

6. Which of these is a property of acidic solutions?

A. They taste sour.
B. They feel slippery.
C. They are in many cleaning products.
D. They taste bitter.

Part 2 Short Response/Grid In

Record your answers on the answer sheet provided by your teacher or on a sheet of paper.

7. Identify elements present in the alloy steel. Compare the flexibility and strength of steel and iron.

Use the illustration below to answer questions 8 and 9.

8. How can you tell that the matter in this bowl is a mixture?

9. What kind of mixture is this? Define this type of mixture, and give three additional examples.

10. Explain why a solute broken into small pieces will dissolve more quickly than the same type and amount of solute in large chunks.

11. Compare the concentration of two solutions: Solution A is composed of 5 grams of sodium chloride dissolved in 100 grams of water. Solution B is composed of 27 grams of sodium chloride dissolved in 100 grams of water.

12. Give the pH of the solutions vinegar, blood plasma, and ammonia. Compare the acidities of soft drinks, tomatoes, and milk.

13. Describe how litmus paper is used to determine the pH of a solution.

Part 3 Open Ended

Record your answers on a sheet of paper.

14. Compare and contrast crystallization and a precipitation reaction.

15. Why is a carbonated beverage defined as a liquid-gas solution? In an open container, the ratio of liquid solvent to gas solute changes over time. Explain.

Use the illustration below to answer questions 16 and 17.

(Partial negative charge)

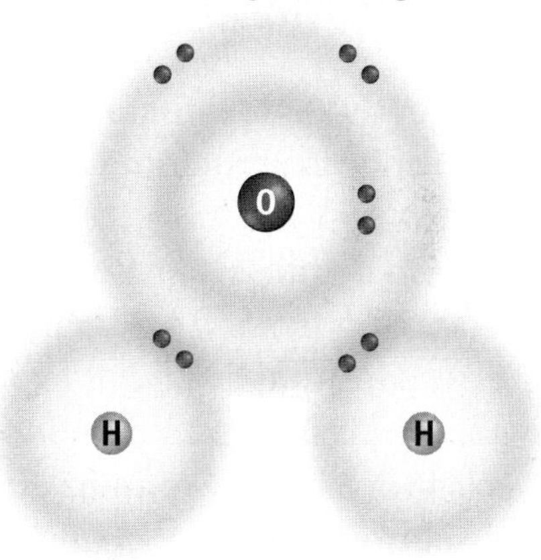

(Partial positive charge)

16. The diagram shows a water molecule. Use the distribution of electrons to describe this molecule's polarity.

17. Explain how the polarity of water molecules makes water effective in dissolving ionic compounds.

18. Marble statues and building facades in many of the world's cities weather more quickly today than when first constructed. Explain how the pH of water plays a role in this process.

19. Acetic acid, CH_3COOH, has more hydrogen atoms than the same concentration of hydrochloric acid, HCl. Hydrogen ions separate more easily from hydrochloric than acetic acid. Which acid is strongest? Why?

chapter 22

States of Matter

chapter preview

sections

Ahhh!

A long, hot soak on a snowy day! This Asian monkey called a macaque is experiencing the effects of heat—the transfer of thermal energy from a warmer object to a colder object. In this chapter, you will learn about heat and the three common states of matter on Earth.

Science Journal Write about what you think is the source of the warm water.

Start-Up Activities

Experiment with a Freezing Liquid

Have you ever thought about how and why you might be able to ice-skate on a pond in the winter but swim in the same pond in the summer? Many substances change form as temperature changes.

1. Make a table to record temperature and appearance. Obtain a test tube containing an unknown liquid from your teacher. Place the test tube in a rack.
2. Insert a thermometer into the liquid. **WARNING:** *Do not allow the thermometer to touch the bottom of the test tube.* Starting immediately, observe and record the substance's temperature and appearance every 30 s.
3. Continue making measurements and observations until you're told to stop.
4. **Think Critically** In your Science Journal, describe your investigation and observations. Did anything unusual happen while you were observing? If so, what?

Preview this chapter's content and activities at green.msscience.com

Changing States of Matter Make the following Foldable to help you study the changes in water.

STEP 1 **Fold** a vertical sheet of paper from left to right two times. Unfold.

STEP 2 **Fold** the paper in half from top to bottom two times.

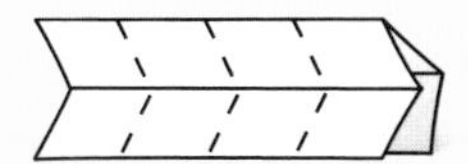

STEP 3 **Unfold** and draw lines along the folds.

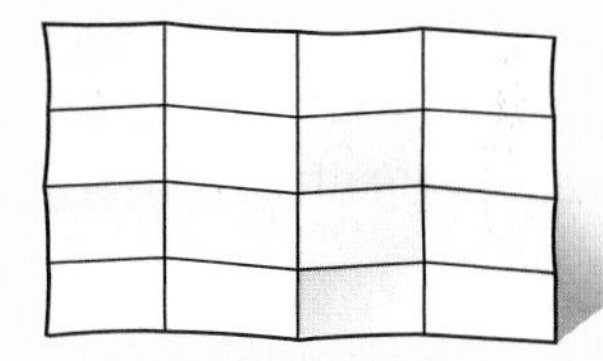

STEP 4 **Label** the top row and first column as shown below.

	Define States	+ Heat	– Heat
Liquid Water			
Water as a Gas			
Water as a Solid (Ice)			

Read and Write As you read the chapter, define the states of matter as listed on your Foldable in the *Define States* column. Write what happens when heat is added to or lost from the three states of matter.

section 1

Matter

as you read

What You'll Learn

- **Recognize** that matter is made of particles in constant motion.
- **Relate** the three states of matter to the arrangement of particles within them.

Why It's Important

Everything you can see, taste, and touch is matter.

Review Vocabulary
atom: a small particle that makes up most types of matter

New Vocabulary
- matter
- solid
- liquid
- viscosity
- surface tension
- gas

What is matter?

Take a look at the beautiful scene in **Figure 1.** What do you see? Perhaps you notice the water and ice. Maybe you are struck by the Sun in the background. All of these images show examples of matter. **Matter** is anything that takes up space and has mass. Matter doesn't have to be visible—even air is matter.

States of Matter All matter is made up of tiny particles, such as atoms, molecules, or ions. Each particle attracts other particles. In other words, each particle pulls other particles toward itself. These particles also are constantly moving. The motion of the particles and the strength of attraction between the particles determine a material's state of matter.

Reading Check *What determines a material's state of matter?*

There are three familiar states of matter—solid, liquid, and gas. A fourth state of matter known as plasma occurs at extremely high temperatures. Plasma is found in stars, lightning, and neon lights. Although plasma is common in the universe, it is not common on Earth. For that reason, this chapter will focus only on the three states of matter that are common on Earth.

Figure 1 Matter exists in all four states in this scene.
Identify *the solid, liquid, gas, and plasma in this photograph.*

Solids

What makes a substance a solid? Think about some familiar solids. Chairs, floors, rocks, and ice cubes are a few examples of matter in the solid state. What properties do all solids share? A **solid** is matter with a definite shape and volume. For example, when you pick up a rock from the ground and place it in a bucket, it doesn't change shape or size. A solid does not take the shape of a container in which it is placed. This is because the particles of a solid are packed closely together, as shown in **Figure 2.**

Figure 2 The particles in a solid vibrate in place while maintaining a constant shape and volume.

Particles in Motion The particles that make up all types of matter are in constant motion. Does this mean that the particles in a solid are moving too? Although you can't see them, a solid's particles are vibrating in place. The particles do not have enough energy to move out of their fixed positions.

Reading Check *What motion do solid particles have?*

Crystalline Solids In some solids, the particles are arranged in a repeating, three-dimensional pattern called a crystal. These solids are called crystalline solids. In **Figure 3** you can see the arrangement of particles in a crystal of sodium chloride, which is table salt. The particles in the crystal are arranged in the shape of a cube. Diamond, another crystalline solid, is made entirely of carbon atoms that form crystals that look more like pyramids. Sugar, sand, and snow are other crystalline solids.

Figure 3 The particles in a crystal of sodium chloride (NaCl) are arranged in an orderly pattern.

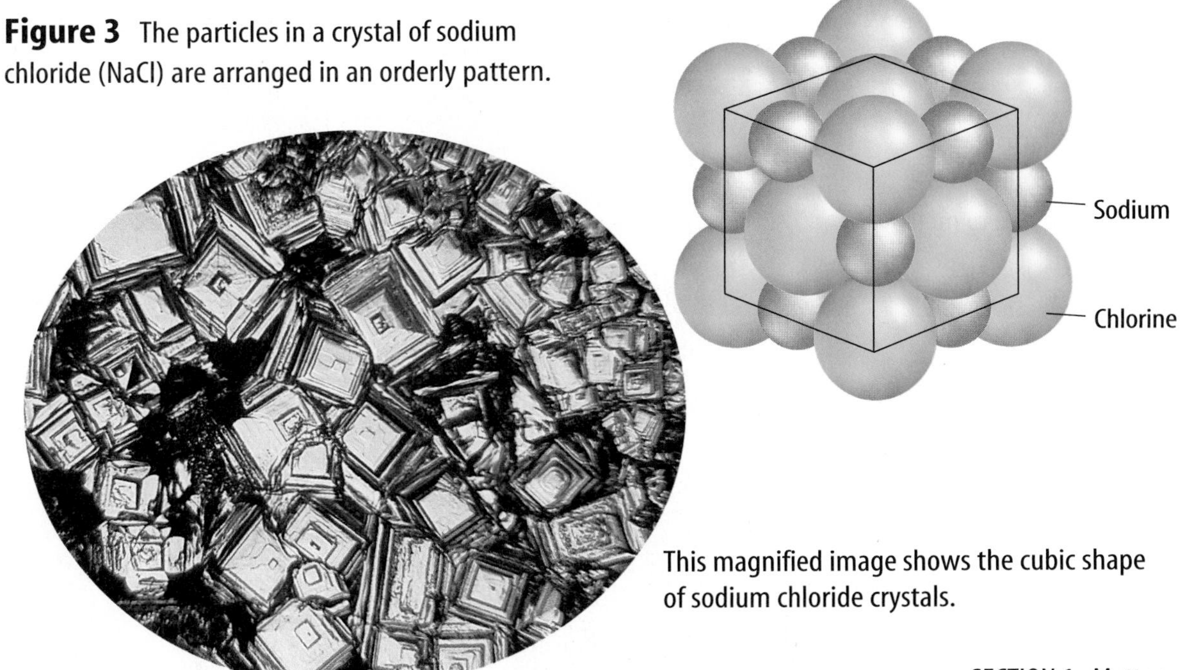

This magnified image shows the cubic shape of sodium chloride crystals.

Amorphous Solids Some solids come together without forming crystal structures. These solids often consist of large particles that are not arranged in a repeating pattern. Instead, the particles are found in a random arrangement. These solids are called amorphous (uh MOR fuhs) solids. Rubber, plastic, and glass are examples of amorphous solids.

Fresh Water Early settlers have always decided to build their homes near water. The rivers provided ways for people to travel, drinking water for themselves and their animals, and irrigation for farming. Over time, small communities became larger communities with industry building along the same water.

Reading Check *How is a crystalline solid different from an amorphous solid?*

Liquids

From the orange juice you drink with breakfast to the water you use to brush your teeth at night, matter in the liquid state is familiar to you. How would you describe the characteristics of a liquid? Is it hard like a solid? Does it keep its shape? A **liquid** is matter that has a definite volume but no definite shape. When you pour a liquid from one container to another, the liquid takes the shape of the container. The volume of a liquid, however, is the same no matter what the shape of the container. If you pour 50 mL of juice from a carton into a pitcher, the pitcher will contain 50 mL of juice. If you then pour that same juice into a glass, its shape will change again but its volume will not.

Free to Move The reason that a liquid can have different shapes is because the particles in a liquid move more freely, as shown in **Figure 4,** than the particles in a solid. The particles in a liquid have enough energy to move out of their fixed positions but not enough energy to move far apart.

Figure 4 The particles in a liquid stay close together, although they are free to move past one another.

Viscosity Do all liquids flow the way water flows? You know that honey flows more slowly than water and you've probably heard the phrase "slow as molasses." Some liquids flow more easily than others. A liquid's resistance to flow is known as the liquid's **viscosity.** Honey has a high viscosity. Water has a lower viscosity. The slower a liquid flows, the higher its viscosity is. The viscosity results from the strength of the attraction between the particles of the liquid. For many liquids, viscosity increases as the liquid becomes colder.

Topic: Plasma

Visit green.msscience.com for Web links to information about the states of matter.

Activity List four ways that plasma differs from the other three states of matter

Surface Tension If you're careful, you can float a needle on the surface of water. This is because attractive forces cause the particles on the surface of a liquid to pull themselves together and resist being pushed apart. You can see in **Figure 5** that particles beneath the surface of a liquid are pulled in all directions. Particles at the surface of a liquid are pulled toward the center of the liquid and sideways along the surface. No liquid particles are located above to pull on them. The uneven forces acting on the particles on the surface of a liquid are called **surface tension.** Surface tension causes the liquid to act as if a thin film were stretched across its surface. As a result you can float a needle on the surface of water. For the same reason, the water strider can move around on the surface of a pond or lake. When a liquid is present in small amounts, surface tension causes the liquid to form small droplets.

Figure 5 Surface tension exists because the particles at the surface experience different forces than those at the center of the liquid.

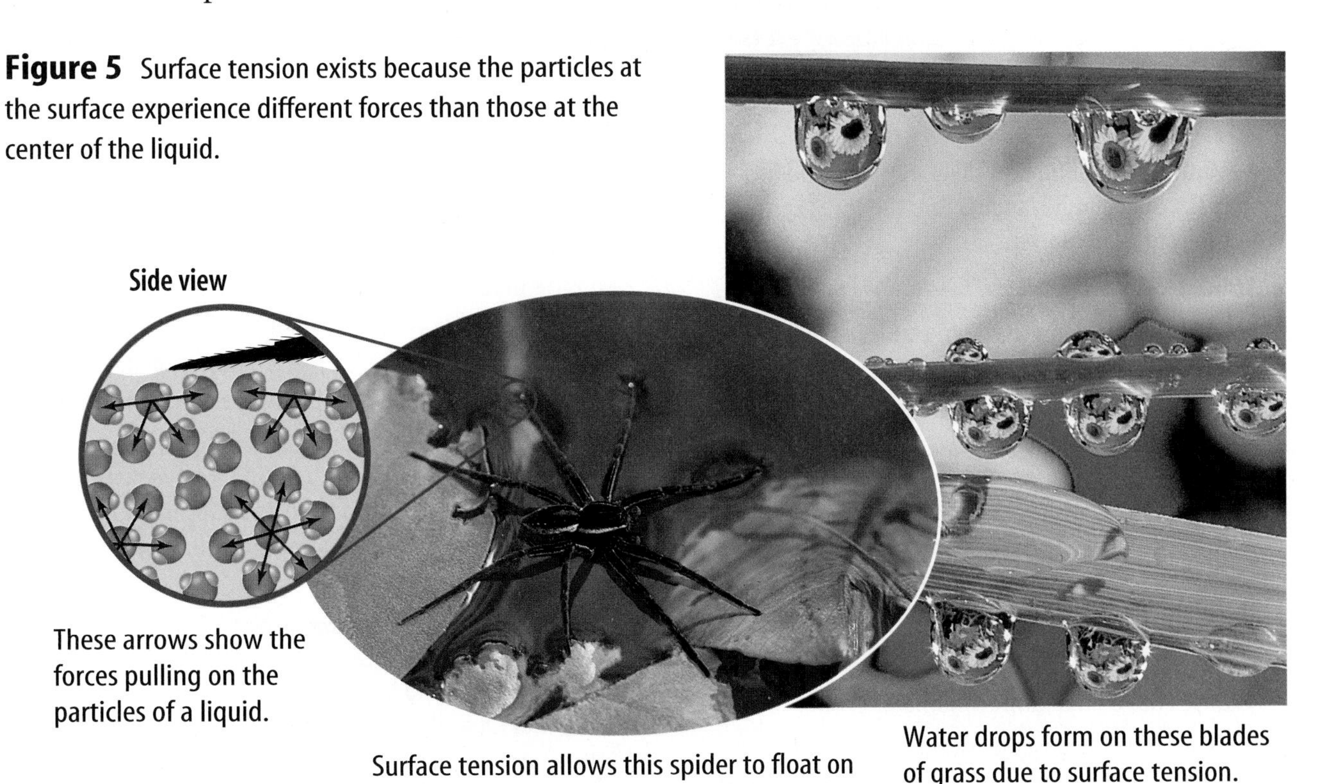

These arrows show the forces pulling on the particles of a liquid.

Surface tension allows this spider to float on water as if the water had a thin film.

Water drops form on these blades of grass due to surface tension.

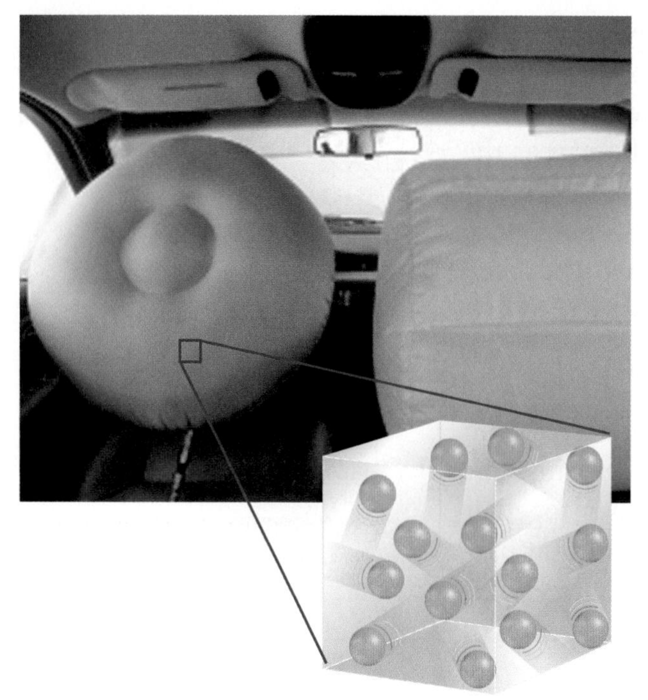

Figure 6 The particles in gas move at high speeds in all directions. The gas inside these air bags spreads out to fill the entire volume of the bag.

Gases

Unlike solids and liquids, most gases are invisible. The air you breathe is a mixture of gases. The gas in the air bags in **Figure 6** and the helium in some balloons are examples of gases. **Gas** is matter that does not have a definite shape or volume. The particles in gas are much farther apart than those in a liquid or solid. Gas particles move at high speeds in all directions. They will spread out evenly, as far apart as possible. If you poured a small volume of a liquid into a container, the liquid would stay in the bottom of the container. However, if you poured the same volume of a gas into a container, the gas would fill the container completely. A gas can expand or be compressed. Decreasing the volume of the container squeezes the gas particles closer together.

Vapor Matter that exists in the gas state but is generally a liquid or solid at room temperature is called vapor. Water, for example, is a liquid at room temperature. Thus, water vapor is the term for the gas state of water.

section 1 review

Summary

What is matter?

- Matter is anything that takes up space and has mass. Solid, liquid, and gas are the three common states of matter.

Solids

- Solids have a definite volume and shape.
- Solids with particles arranged in order are called crystalline solids. The particles in amorphous solids are not in any order.

Liquids

- Liquids have definite volume but no defined shape.
- Viscosity is a measure of how easily liquids flow.

Gases

- Gases have no definite volume or shape.
- Vapor refers to gaseous substances that are normally liquids or solids at room temperature.

Self Check

1. **Define** the two properties of matter that determine its state.
2. **Describe** the movement of particles within solids, liquids, and gases.
3. **Name** the property that liquids and solids share. What property do liquids and gases share?
4. **Infer** A scientist places 25 mL of a yellow substance into a 50-mL container. The substance quickly fills the entire container. Is it a solid, liquid, or gas?
5. **Think Critically** The particles in liquid A have a stronger attraction to each other than the particles in liquid B. If both liquids are at the same temperature, which liquid has a higher viscosity? Explain.

Applying Skills

6. **Concept Map** Draw a Venn diagram in your Science Journal and fill in the characteristics of the states of matter.

green.msscience.com/self_check_quiz

Changes of State

Thermal Energy and Heat

Shards of ice fly from the sculptor's chisel. As the crowd looks on, a swan slowly emerges from a massive block of ice. As the day wears on, however, drops of water begin to fall from the sculpture. Drip by drip, the sculpture is transformed into a puddle of liquid water. What makes matter change from one state to another? To answer this question, you need to think about the particles that make up matter.

Energy Simply stated, energy is the ability to do work or cause change. The energy of motion is called kinetic energy. Particles within matter are in constant motion. The amount of motion of these particles depends on the kinetic energy they possess. Particles with more kinetic energy move faster and farther apart. Particles with less energy move more slowly and stay closer together.

The total kinetic and potential energy of all the particles in a sample of matter is called **thermal energy.** Thermal energy, an extensive property, depends on the number of particles in a substance as well as the amount of energy each particle has. If either the number of particles or the amount of energy in each particle changes, the thermal energy of the sample changes. With identically sized samples, the warmer substance has the greater thermal energy. In **Figure 7,** the particles of hot water from the hot spring have more thermal energy than the particles of snow on the surrounding ground.

as you read

What You'll Learn

- **Define and compare** thermal energy and temperature.
- **Relate** changes in thermal energy to changes of state.
- **Explore** energy and temperature changes on a graph.

Why It's Important

Matter changes state as it heats up or cools down.

Review Vocabulary
energy: the ability to do work or cause change

New Vocabulary
- thermal energy
- temperature
- heat
- melting
- freezing
- vaporization
- condensation

Figure 7 These girls are enjoying the water from the hot spring. **Infer** *why the girls appear to be comfortable in the hot spring while there is snow on the ground.*

Figure 8 The particles in hot tea move faster than those in iced tea. The temperature of hot tea is higher than the temperature of iced tea.
Identify *which tea has the higher kinetic energy.*

Temperature Not all of the particles in a sample of matter have the same amount of energy. Some have more energy than others. The average kinetic energy of the individual particles is the **temperature,** an intensive property, of the substance. You can find an average by adding up a group of numbers and dividing the total by the number of items in the group. For example, the average of the numbers 2, 4, 8, and 10 is $(2 + 4 + 8 + 10) \div 4 = 6$. Temperature is different from thermal energy because thermal energy is a total and temperature is an average.

You know that the iced tea is colder than the hot tea, as shown in **Figure 8.** Stated differently, the temperature of iced tea is lower than the temperature of hot tea. You also could say that the average kinetic energy of the particles in the iced tea is less than the average kinetic energy of the particles in the hot tea.

Types of Energy Thermal energy is one of several different forms of energy. Other forms include the chemical energy in chemical compounds, the electrical energy used in appliances, the electromagnetic energy of light, and the nuclear energy stored in the nucleus of an atom. Make a list of examples of energy that you are familiar with.

Heat When a warm object is brought near a cooler object, thermal energy will be transferred from the warmer object to the cooler one. The movement of thermal energy from a substance at a higher temperature to one at a lower temperature is called **heat.** When a substance is heated, it gains thermal energy. Therefore, its particles move faster and its temperature rises. When a substance is cooled, it loses thermal energy, which causes its particles to move more slowly and its temperature to drop.

Reading Check *How is heat related to temperature?*

Specific Heat

As you study more science, you will discover that water has many unique properties. One of those is the amount of heat required to increase the temperature of water as compared to most other substances. The specific heat of a substance is the amount of heat required to raise the temperature of 1 g of a substance 1°C.

Figure 9 The specific heat of water is greater than that of sand. The energy provided by the Sun raises the temperature of the sand much faster than the water.

Substances that have a low specific heat, such as most metals and the sand in **Figure 9,** heat up and cool down quickly because they require only small amounts of heat to cause their temperatures to rise. A substance with a high specific heat, such as the water in **Figure 9,** heats up and cools down slowly because a much larger quantity of heat is required to cause its temperature to rise or fall by the same amount.

Changes Between the Solid and Liquid States

Matter can change from one state to another when thermal energy is absorbed or released. This change is known as change of state. The graph in **Figure 11** shows the changes in temperature as thermal energy is gradually added to a container of ice.

Melting As the ice in **Figure 11** is heated, it absorbs thermal energy and its temperature rises. At some point, the temperature stops rising and the ice begins to change into liquid water. The change from the solid state to the liquid state is called **melting.** The temperature at which a substance changes from a solid to a liquid is called the melting point. The melting point of water is 0°C.

Amorphous solids, such as rubber and glass, don't melt in the same way as crystalline solids. Because they don't have crystal structures to break down, these solids get softer and softer as they are heated, as you can see in **Figure 10.**

Figure 10 Rather than melting into a liquid, glass gradually softens. Glass blowers use this characteristic to shape glass into beautiful vases while it is hot.

NATIONAL GEOGRAPHIC VISUALIZING STATES OF MATTER

Figure 11

Like most substances, water can exist in three distinct states—solid, liquid, or gas. At certain temperatures, water changes from one state to another. This diagram shows what changes occur as water is heated or cooled.

MELTING When ice melts, its temperature remains constant until all the ice turns to water. Continued heating of liquid water causes the molecules to vibrate even faster, steadily raising the temperature.

FREEZING When liquid water freezes, it releases thermal energy and turns into the solid state, ice.

VAPORIZATION When water reaches its boiling point of 100°C, water molecules are moving so fast that they break free of the attractions that hold them together in the liquid state. The result is vaporization—the liquid becomes a gas. The temperature of boiling water remains constant until all of the liquid turns to steam.

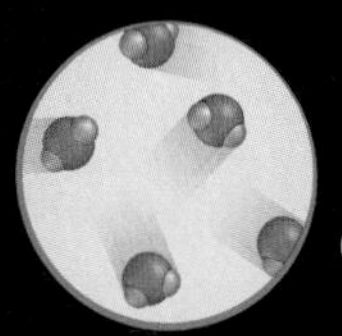

CONDENSATION When steam is cooled, it releases thermal energy and turns into its liquid state. This process is called condensation.

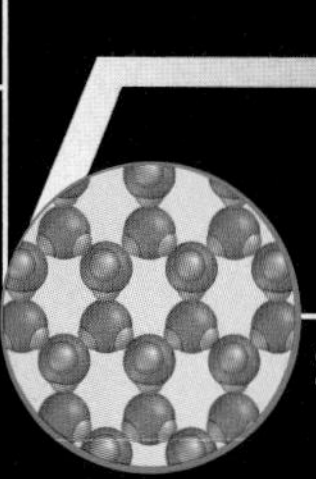

Solid state: ice

Liquid state: water

Gaseous state: steam

Freezing The process of melting a crystalline solid can be reversed if the liquid is cooled. The change from the liquid state to the solid state is called **freezing.** As the liquid cools, it loses thermal energy. As a result, its particles slow down and come closer together. Attractive forces begin to trap particles, and the crystals of a solid begin to form. As you can see in **Figure 11,** freezing and melting are opposite processes.

The temperature at which a substance changes from the liquid state to the solid state is called the freezing point. The freezing point of the liquid state of a substance is the same temperature as the melting point of the solid state. For example, solid water melts at 0°C and liquid water freezes at 0°C.

During freezing, the temperature of a substance remains constant while the particles in the liquid form a crystalline solid. Because particles in a liquid have more energy than particles in a solid, energy is released during freezing. This energy is released into the surroundings. After all of the liquid has become a solid, the temperature begins to decrease again.

Topic: Freezing Point Study

Visit green.msscience.com for Web links to information about freezing.

Activity Make a list of several substances and the temperatures at which they freeze. Find out how the freezing point affects how the substance is used.

Applying Science

How can ice save oranges?

During the spring, Florida citrus farmers carefully watch the fruit when temperatures drop close to freezing. When the temperatures fall below 0°C, the liquid in the cells of oranges can freeze and expand. This causes the cells to break, making the oranges mushy and the crop useless for sale. To prevent this, farmers spray the oranges with water just before the temperature reaches 0°C. How does spraying oranges with water protect them?

Identifying the Problem

Using the diagram in **Figure 11,** consider what is happening to the water at 0°C. Two things occur. What are they?

Solving the Problem

1. What change of state and what energy changes occur when water freezes?
2. How does the formation of ice on the orange help the orange?

Observing Vaporization

Procedure

1. Use a **dropper** to place one drop of **rubbing alcohol** on the back of your hand.
2. Describe how your hand feels during the next 2 min.
3. Wash your hands.

Analysis

1. What changes in the appearance of the rubbing alcohol did you notice?
2. What sensation did you feel during the 2 min? How can you explain this sensation?
3. Infer how sweating cools the body.

Changes Between the Liquid and Gas States

After an early morning rain, you and your friends enjoy stomping through the puddles left behind. But later that afternoon when you head out to run through the puddles once more, the puddles are gone. The liquid water in the puddles changed into a gas. Matter changes between the liquid and gas states through vaporization and condensation.

Vaporization As liquid water is heated, its temperature rises until it reaches 100°C. At this point, liquid water changes into water vapor. The change from a liquid to a gas is known as **vaporization** (vay puh ruh ZAY shun). You can see in **Figure 11** that the temperature of the substance does not change during vaporization. However, the substance absorbs thermal energy. The additional energy causes the particles to move faster until they have enough energy to escape the liquid as gas particles.

Two forms of vaporization exist. Vaporization that takes place below the surface of a liquid is called boiling. When a liquid boils, bubbles form within the liquid and rise to the surface, as shown in **Figure 12.** The temperature at which a liquid boils is called the boiling point. The boiling point of water is 100°C.

Vaporization that takes place at the surface of a liquid is called evaporation. Evaporation, which occurs at temperatures below the boiling point, explains how puddles dry up. Imagine that you could watch individual water molecules in a puddle. You would notice that the molecules move at different speeds. Although the temperature of the water is constant, remember that temperature is a measure of the average kinetic energy of the molecules. Some of the fastest-moving molecules overcome the attractive forces of other molecules and escape from the surface of the water.

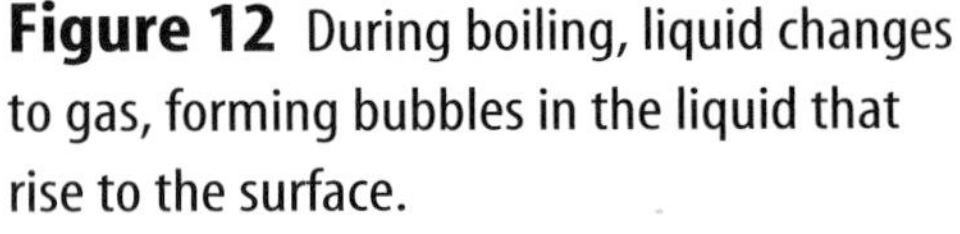

Figure 12 During boiling, liquid changes to gas, forming bubbles in the liquid that rise to the surface.
Define *the word that describes a liquid changing to the gas.*

Figure 13 The drops of water on these glasses and pitcher of lemonade were formed when water vapor in the air lost enough energy to return to the liquid state. This process is called condensation.

Location of Molecules It takes more than speed for water molecules to escape the liquid state. During evaporation, these faster molecules also must be near the surface, heading in the right direction, and they must avoid hitting other water molecules as they leave. With the faster particles evaporating from the surface of a liquid, the particles that remain are the slower, cooler ones. Evaporation cools the liquid and anything near the liquid. You experience this cooling effect when perspiration evaporates from your skin.

Topic: Condensation
Visit green.msscience.com for Web links to information about how condensation is involved in weather.

Activity Find out how condensation is affected by the temperature as well as the amount of water in the air.

Condensation Pour a nice, cold glass of lemonade and place it on the table for a half hour on a warm day. When you come back to take a drink, the outside of the glass will be covered by drops of water, as shown in **Figure 13.** What happened? As a gas cools, its particles slow down. When particles move slowly enough for their attractions to bring them together, droplets of liquid form. This process, which is the opposite of vaporization, is called **condensation.** As a gas condenses to a liquid, it releases the thermal energy it absorbed to become a gas. During this process, the temperature of the substance does not change. The decrease in energy changes the arrangement of particles. After the change of state is complete, the temperature continues to drop, as you saw in **Figure 11.**

What energy change occurs during condensation?

Condensation formed the droplets of water on the outside of your glass of lemonade. In the same way, water vapor in the atmosphere condenses to form the liquid water droplets in clouds. When the droplets become large enough, they can fall to the ground as rain.

Figure 14 The solid carbon dioxide (dry ice) at the bottom of this beaker of water is changing directly into gaseous carbon dioxide. This process is called sublimation.

Changes Between the Solid and Gas States

Some substances can change from the solid state to the gas state without ever becoming a liquid. During this process, known as sublimation, the surface particles of the solid gain enough energy to become a gas. One example of a substance that undergoes sublimation is dry ice. Dry ice is the solid form of carbon dioxide. It often is used to keep materials cold and dry. At room temperature and pressure, carbon dioxide does not exist as a liquid. Therefore, as dry ice absorbs thermal energy from the objects around it, it changes directly into a gas. When dry ice becomes a gas, it absorbs thermal energy from water vapor in the air. As a result, the water vapor cools and condenses into liquid water droplets, forming the fog you see in **Figure 14.**

section 2 review

Summary

Thermal Energy and Heat

- Thermal energy depends on the amount of the substance and the kinetic energy of particles in the substance.
- Heat is the movement of thermal energy from a warmer substance to a cooler one.

Specific Heat

- Specific heat is a measure of the amount of energy required to raise 1 g of a substance 1°C.

Changes Between Solid and Liquid States

- During all changes of state, the temperature of a substance stays the same.

Changes Between Liquid and Gas States

- Vaporization is the change from the liquid state to a gaseous state.
- Condensation is the change from the gaseous state to the liquid state.

Changes Between Solid and Gas States

- Sublimation is the process of a substance going from the solid state to the gas state without ever being in the liquid state.

Self Check

1. **Describe** how thermal energy and temperature are similar. How are they different?
2. **Explain** how a change in thermal energy causes matter to change from one state to another. Give two examples.
3. **List** the three changes of state during which energy is absorbed.
4. **Describe** the two types of vaporization.
5. **Think Critically** How can the temperature of a substance remain the same even if the substance is absorbing thermal energy?
6. **Write** a paragraph in your Science Journal that explains why you can step out of the shower into a warm bathroom and begin to shiver.

Applying Math

7. **Make and Use Graphs** Use the data you collected in the Launch Lab to plot a temperature-time graph. Describe your graph. At what temperature does the graph level off? What was the liquid doing during this time period?
8. **Use Numbers** If sample A requires 10 calories to raise the temperature of a 1-g sample 1°C, how many calories does it take to raise a 5-g sample 10°C?

green.msscience.com/self_check_quiz

The Water Cycle

Water is all around us and you've used water in all three of its common states. This lab will give you the opportunity to observe the three states of matter and to discover for yourself if ice really melts at 0°C and if water boils at 100°C.

Real-World Question

How does the temperature of water change as it is heated from a solid to a gas?

Goals

- **Measure** the temperature of water as it heats.
- **Observe** what happens as the water changes from one state to another.
- **Graph** the temperature and time data.

Materials

hot plate
ice cubes (100 mL)
Celsius thermometer
**electronic temperature probe*
wall clock
**watch with second hand*
stirring rod
250-mL beaker
**Alternate materials*

Safety Precautions

Procedure

1. Make a data table similar to the table shown.
2. Put 150 mL of water and 100 mL of ice into the beaker and place the beaker on the hot plate. Do not touch the hot plate.
3. Put the thermometer into the ice/water mixture. Do not stir with the thermometer or allow it to rest on the bottom of the beaker. After 30 s, read and record the temperature in your data table.

Characteristics of Water Sample

Time (min)	Temperature (°C)	Physical State
	Do not write in this book.	

4. Plug in the hot plate and turn the temperature knob to the medium setting.
5. Every 30 s, read and record the temperature and physical state of the water until it begins to boil. Use the stirring rod to stir the contents of the beaker before making each temperature measurement. Stop recording. Allow the water to cool.

Analyze Your Data

Use your data to make a graph plotting time on the x-axis and temperature on the y-axis. Draw a smooth curve through the data points.

Conclude and Apply

1. **Describe** how the temperature of the ice/water mixture changed as you heated the beaker.
2. **Describe** the shape of the graph during any changes of state.

Communicating Your Data

Add labels to your graph. Use the detailed graph to explain to your class how water changes state. **For more help, refer to the** Science Skill Handbook.

section 3

Behavior of Fluids

as you read

What You'll Learn

- **Explain** why some things float but others sink.
- **Describe** how pressure is transmitted through fluids.

Why It's Important

Pressure enables you to squeeze toothpaste from a tube, and buoyant force helps you float in water.

Review Vocabulary

force: a push or pull

New Vocabulary

- pressure
- buoyant force
- Archimedes' principle
- density
- Pascal's principle

Pressure

It's a beautiful summer day when you and your friends go outside to play volleyball, much like the kids in **Figure 15.** There's only one problem—the ball is flat. You pump air into the ball until it is firm. The firmness of the ball is the result of the motion of the air particles in the ball. As the air particles in the ball move, they collide with one another and with the inside walls of the ball. As each particle collides with the inside walls, it exerts a force, pushing the surface of the ball outward. A force is a push or a pull. The forces of all the individual particles add together to make up the pressure of the air.

Pressure is equal to the force exerted on a surface divided by the total area over which the force is exerted.

$$\text{pressure} = \frac{\text{force}}{\text{area}}$$

When force is measured in newtons (N) and area is measured in square meters (m^2), pressure is measured in newtons per square meter (N/m^2). This unit of pressure is called a pascal (Pa). A more useful unit when discussing atmospheric pressure is the kilopascal (kPa), which is 1,000 pascals.

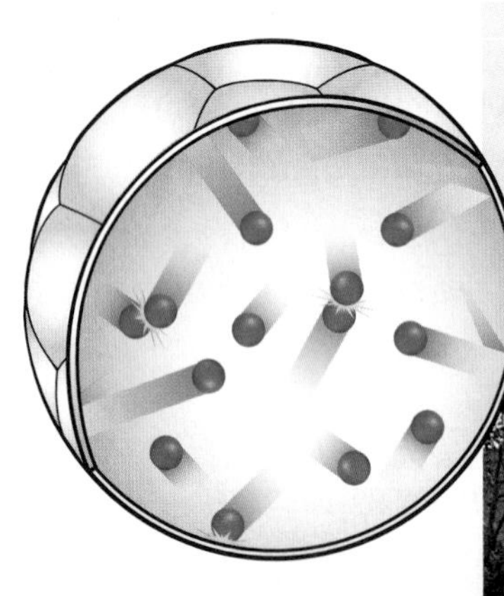

Figure 15 Without the pressure of air inside this volleyball, the ball would be flat.

Figure 16 The force of the dancer's weight on pointed toes results in a higher pressure than the same force on flat feet. **Explain** *why the pressure is higher.*

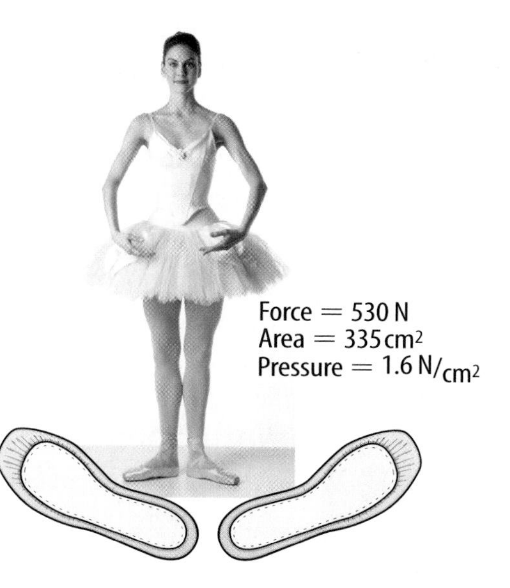

Force and Area You can see from the equation on the opposite page that pressure depends on the quantity of force exerted and the area over which the force is exerted. As the force increases over a given area, pressure increases. If the force decreases, the pressure will decrease. However, if the area changes, the same amount of force can result in different pressure. **Figure 16** shows that if the force of the ballerina's weight is exerted over a smaller area, the pressure increases. If that same force is exerted over a larger area, the pressure will decrease.

Reading Check ***What variables does pressure depend on?***

Atmospheric Pressure You can't see it and you usually can't feel it, but the air around you presses on you with tremendous force. The pressure of air also is known as atmospheric pressure because air makes up the atmosphere around Earth. Atmospheric pressure is 101.3 kPa at sea level. This means that air exerts a force of about 101,000 N on every square meter it touches. This is approximately equal to the weight of a large truck.

It might be difficult to think of air as having pressure when you don't notice it. However, you often take advantage of air pressure without even realizing it. Air pressure, for example, enables you to drink from a straw. When you first suck on a straw, you remove the air from it. As you can see in **Figure 17,** air pressure pushes down on the liquid in your glass then forces liquid up into the straw. If you tried to drink through a straw inserted into a sealed, airtight container, you would not have any success because the air would not be able to push down on the surface of the drink.

Figure 17 The downward pressure of air pushes the juice up into the straw.

Figure 18 Atmospheric pressure exerts a force on all surfaces of this dancer's body.
Explain *why she can't feel this pressure.*

Balanced Pressure If air is so forceful, why don't you feel it? The reason is that the pressure exerted outward by the fluids in your body balances the pressure exerted by the atmosphere on the surface of your body. Look at **Figure 18.** The atmosphere exerts a pressure on all surfaces of the dancer's body. She is not crushed by this pressure because the fluids in her body exert a pressure that balances atmospheric pressure.

Variations in Atmospheric Pressure Atmospheric pressure changes with altitude. Altitude is the height above sea level. As altitude increases atmospheric pressure decreases. This is because fewer air particles are found in a given volume. Fewer particles have fewer collisions, and therefore exert less pressure. This idea was tested in the seventeenth century by a French physician named Blaise Pascal. He designed an experiment in which he filled a balloon only partially with air. He then had the balloon carried to the top of a mountain. **Figure 19** shows that as Pascal predicted, the balloon expanded while being carried up the mountain. Although the amount of air inside the balloon stayed the same, the air pressure pushing in on it from the outside decreased. Consequently, the particles of air inside the balloon were able to spread out further.

Figure 19 Notice how the balloon expands as it is carried up the mountain. The reason is that atmospheric pressure decreases with altitude. With less pressure pushing in on the balloon, the gas particles within the balloon are free to expand.

Air Travel If you travel to higher altitudes, perhaps flying in an airplane or driving up a mountain, you might feel a popping sensation in your ears. As the air pressure drops, the air pressure in your ears becomes greater than the air pressure outside your body. The release of some of the air trapped inside your ears is heard as a pop. Airplanes are pressurized so that the air pressure within the cabin does not change dramatically throughout the course of a flight.

Changes in Gas Pressure

In the same way that atmospheric pressure can vary as conditions change, the pressure of gases in confined containers also can change. The pressure of a gas in a closed container changes with volume and temperature.

Pressure and Volume If you squeeze a portion of a filled balloon, the remaining portion of the balloon becomes more firm. By squeezing it, you decrease the volume of the balloon, forcing the same number of gas particles into a smaller space. As a result, the particles collide with the walls more often, thereby producing greater pressure. This is true as long as the temperature of the gas remains the same. You can see the change in the motion of the particles in **Figure 20.** What will happen if the volume of a gas increases? If you make a container larger without changing its temperature, the gas particles will collide less often and thereby produce a lower pressure.

Predicting a Waterfall

Procedure

1. Fill a **plastic cup** to the brim with **water.**
2. Cover the top of the cup with an **index card.**
3. Predict what will happen if you turn the cup upside down.
4. While holding the index card in place, turn the cup upside down over a sink. Then let go of the card.

Analysis

1. What happened to the water when you turned the cup?
2. How can you explain your observation in terms of the concept of fluid pressure?

Try at Home

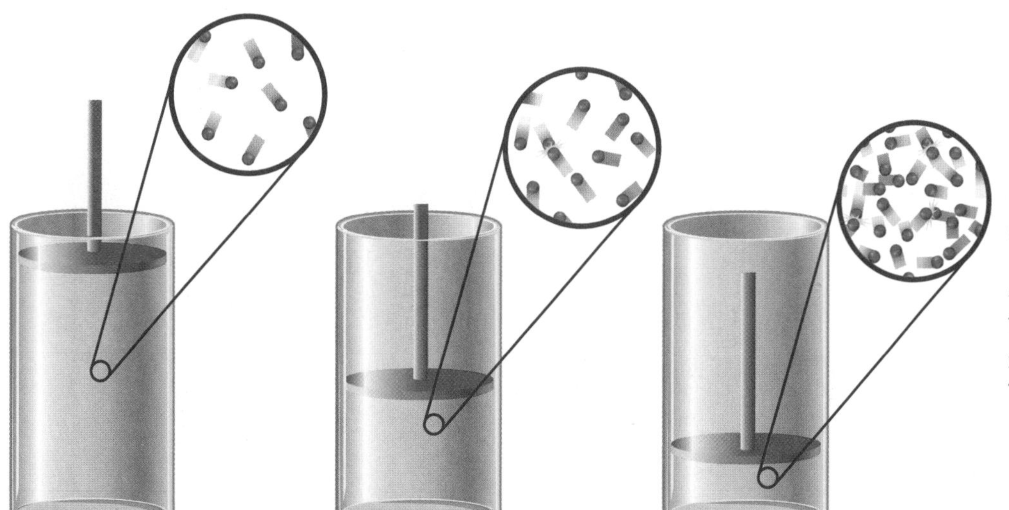

Figure 20 As volume decreases, pressure increases.

As the piston is moved down, the gas particles have less space and collide more often. The pressure increases.

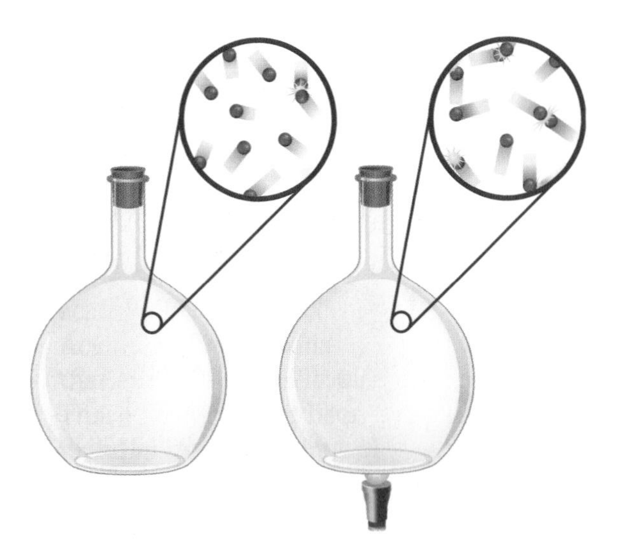

Pressure and Temperature When the volume of a confined gas remains the same, the pressure can change as the temperature of the gas changes. You have learned that temperature rises as the kinetic energy of the particles in a substance increases. The greater the kinetic energy is, the faster the particles move. The faster the speed of the particles is, the more they collide and the greater the pressure is. If the temperature of a confined gas increases, the pressure of the gas will increase, as shown in **Figure 21.**

Reading Check *Why would a sealed container of air be crushed after being frozen?*

Figure 21 Even though the volume of this container does not change, the pressure increases as the substance is heated. **Describe** *what will happen if the substance is heated too much.*

Float or Sink

You may have noticed that you feel lighter in water than you do when you climb out of it. While you are under water, you experience water pressure pushing on you in all directions. Just as air pressure increases as you walk down a mountain, water pressure increases as you swim deeper in water. Water pressure increases with depth. As a result, the pressure pushing up on the bottom of an object is greater than the pressure pushing down on it because the bottom of the object is deeper than the top.

The difference in pressure results in an upward force on an object immersed in a fluid, as shown in **Figure 22.** This force is known as the **buoyant force.** If the buoyant force is equal to the weight of an object, the object will float. If the buoyant force is less than the weight of an object, the object will sink.

Figure 22 The pressure pushing up on an immersed object is greater than the pressure pushing down on it. This difference results in the buoyant force.

Weight is a force in the downward direction. The buoyant force is in the upward direction. An object will float if the upward force is equal to the downward force.

Archimedes' Principle What determines the buoyant force? According to **Archimedes'** (ar kuh MEE deez) **principle,** the buoyant force on an object is equal to the weight of the fluid displaced by the object. In other words, if you place an object in a beaker that already is filled to the brim with water, some water will spill out of the beaker, as in **Figure 23.** If you weigh the spilled water, you will find the buoyant force on the object.

Density Understanding density can help you predict whether an object will float or sink. **Density** is mass divided by volume.

$$\text{density} = \frac{\text{mass}}{\text{volume}}$$

An object will float in a fluid that is more dense than itself and sink in a fluid that is less dense than itself. If an object has the same density, the object will neither sink nor float but instead stay at the same level in the fluid.

Figure 23 When the golf ball was dropped in the large beaker, it displaced some of the water, which was collected and placed into the smaller beaker. **Communicate** *what you know about the weight and the volume of the displaced water.*

Applying Math — Find an Unknown

CALCULATING DENSITY You are given a sample of a solid that has a mass of 10.0 g and a volume of 4.60 cm^3. Will it float in liquid water, which has a density of 1.00 g/cm^3?

Solution

1 *This is what you know:*
- mass = 10.0 g
- volume = 4.60 cm^3
- density of water = 1.00 g/cm^3

2 *This is what you need to find:* the density of the sample

3 *This is the procedure you need to use:*
- density = mass/volume
- density = 10.0 g/4.60 cm^3 = 2.17 g/cm^3
- The density of the sample is greater than the density of water. The sample will sink.

4 *Check your answer:*
- Find the mass of your sample by multiplying the density and the volume.

Practice Problems

1. A 7.40-cm^3 sample of mercury has a mass of 102 g. Will it float in water?
2. A 5.0-cm^3 sample of aluminum has a mass of 13.5 g. Will it float in water?

For more practice, visit green.msscience.com/math_practice

Pascal's Principle

What happens if you squeeze a plastic container filled with water? If the container is closed, the water has nowhere to go. As a result, the pressure in the water increases by the same amount everywhere in the container—not just where you squeeze or near the top of the container. When a force is applied to a confined fluid, an increase in pressure is transmitted equally to all parts of the fluid. This relationship is known as **Pascal's principle.**

Hydraulic Systems You witness Pascal's principle when a car is lifted up to have its oil changed or if you are in a dentist's chair as it is raised or lowered, as shown in **Figure 24.** These devices, known as hydraulic (hi DRAW lihk) systems, use Pascal's principle to increase force. Look at the tube in **Figure 25.** The force applied to the piston on the left increases the pressure within the fluid. That increase in pressure is transmitted to the piston on the right. Recall that pressure is equal to force divided by area. You can solve for force by multiplying pressure by area.

Figure 24 A hydraulic lift utilizes Pascal's principle to help lift this car and this dentist's chair.

$$\text{pressure} = \frac{\text{force}}{\text{area}} \quad \text{or} \quad \text{force} = \text{pressure} \times \text{area}$$

If the two pistons on the tube have the same area, the force will be the same on both pistons. If, however, the piston on the right has a greater surface area than the piston on the left, the resulting force will be greater. The same pressure multiplied by a larger area equals a greater force. Hydraulic systems enable people to lift heavy objects using relatively small forces.

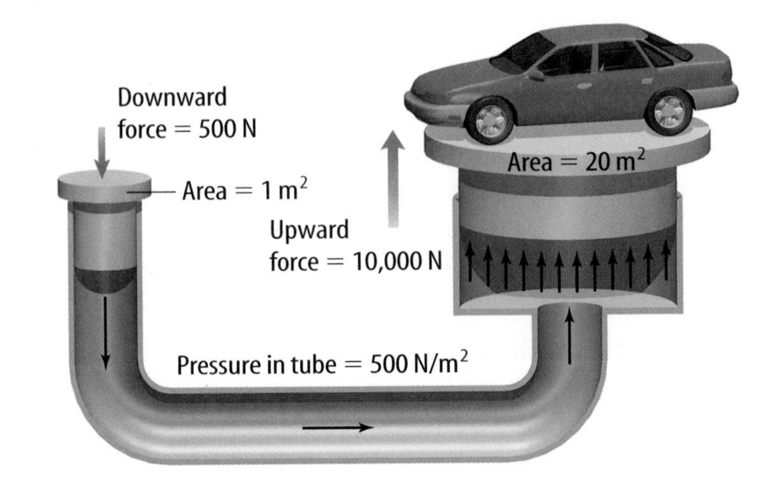

Figure 25 By increasing the area of the piston on the right side of the tube, you can increase the force exerted on the piston. In this way a small force pushing down on the left piston can result in a large force pushing up on the right piston. The force can be great enough to lift a car.

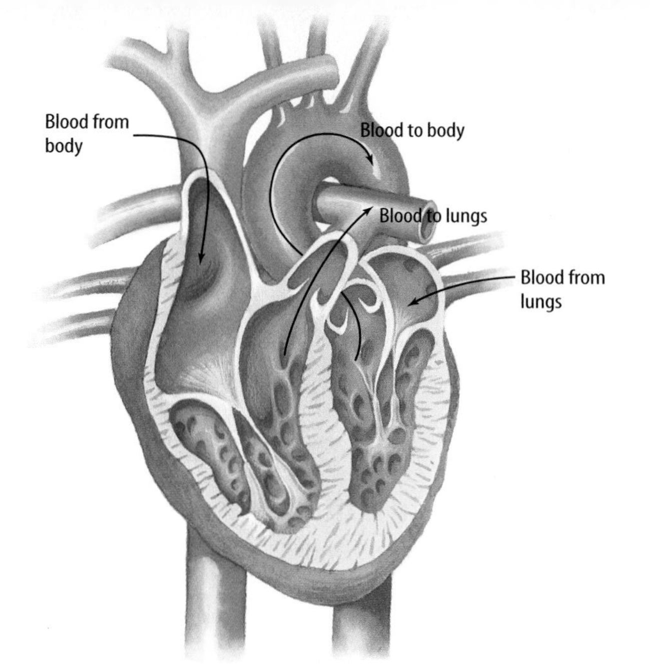

Figure 26 The heart is responsible for moving blood throughout the body. Two force pumps work together to move blood to and from the lungs and to the rest of the body.

Force Pumps If an otherwise closed container has a hole in it, any fluid in the container will be pushed out the opening when you squeeze it. This arrangement, known as a force pump, makes it possible for you to squeeze toothpaste out of a tube or mustard from a plastic container.

Your heart has two force pumps. One pump pushes blood to the lungs, where it picks up oxygen. The other force pump pushes the oxygen-rich blood to the rest of your body. These pumps are shown in **Figure 26.**

Topic: Blood Pressure
Visit green.msscience.com for Web links to information about blood pressure. Find out what the term means, how it changes throughout the human body, and why it is unhealthy to have high blood pressure.

Activity Write a paragraph in your Science Journal that explains why high blood pressure is dangerous.

section 3 review

Summary

Pressure

- Pressure depends on force and area.
- The air around you exerts a pressure.
- The pressure inside your body matches the pressure exerted by air.

Changes in Gas Pressure

- The pressure exerted by a gas depends on its volume and its temperature.

Float or Sink

- Whether an object floats or sinks depends on its density relative to the density of the fluid it's in.

Pascal's Principle

- This principle relates pressure and area to force.

Self Check

1. **Describe** what happens to pressure as the force exerted on a given area increases.
2. **Describe** how atmospheric pressure changes as altitude increases.
3. **State** Pascal's principle in your own words.
4. **Infer** An object floats in a fluid. What can you say about the buoyant force on the object?
5. **Think Critically** All the air is removed from a sealed metal can. After the air has been removed, the can looks as if it were crushed. Why?

Applying Math

6. **Simple Equations** What pressure is created when 5.0 N of force are applied to an area of 2.0 m^2? How does the pressure change if the force is increased to 10.0 N? What about if instead the area is decreased to 1.0 m^2?

Design Your Own

Design Your Own Ship

Goals

- **Design** an experiment that uses Archimedes' principle to determine the size of ship needed to carry a given amount of cargo in such a way that the top of the ship is even with the surface of the water.

Possible Materials

balance
small plastic cups (2)
graduated cylinder
metric ruler
scissors
marbles (cupful)
sink
**basin, pan, or bucket*
**Alternate materials*

Safety Precautions

Real-World Question

It is amazing to watch ships that are taller than buildings float easily on water. Passengers and cargo are carried on these ships in addition to the tremendous weight of the ship itself. How can you determine the size of a ship needed to keep a certain mass of cargo afloat?

Cargo ship

Form a Hypothesis

Think about Archimedes' principle and how it relates to buoyant force. Form a hypothesis to explain how the volume of water displaced by a ship relates to the mass of cargo the ship can carry.

Test Your Hypothesis

Make a Plan

1. Obtain a set of marbles or other items from your teacher. This is the cargo that your ship must carry. Think about the type of ship

you will design. Consider the types of materials you will use. Decide how your group is going to test your hypothesis.

2. **List** the steps you need to follow to test your hypothesis. Include in your plan how you will measure the mass of your ship and cargo, calculate the volume of water your ship must displace in order to float with its cargo, and measure the volume and mass of the displaced water. Also, explain how you will design your ship so that it will float with the top of the ship even with the surface of the water. Make the ship.
3. **Prepare** a data table in your Science Journal to use as your group collects data. Think about what data you need to collect.

Follow Your Plan

1. Make sure your teacher approves your plan before you start.
2. Perform your experiment as planned. Be sure to follow all proper safety procedures. In particular, clean up any spilled water immediately.
3. Record your observations carefully and complete the data table in your Science Journal.

Analyze Your Data

1. **Write** your calculations showing how you determined the volume of displaced water needed to make your ship and cargo float.
2. Did your ship float at the water's surface, sink, or float above the water's surface? Draw a diagram of your ship in the water.
3. **Explain** how your experimental results agreed or failed to agree with your hypothesis.

Conclude and Apply

1. If your ship sank, how would you change your experiment or calculations to correct the problem? What changes would you make if your ship floated too high in the water?
2. What does the density of a ship's cargo have to do with the volume of cargo the ship can carry? What about the density of the water?

Compare your results with other students' data. Prepare a combined data table or summary showing how the calculations affect the success of the ship. **For more help, refer to the Science Skill Handbook.**

SOMETIMES GREAT DISCOVERIES HAPPEN BY ACCIDENT!

The Incredible Stretching Goo

A serious search turns up a toy

During World War II, when natural resources were scarce and needed for the war effort, the U.S. government asked an engineer to come up with an inexpensive alternative to synthetic rubber. While researching the problem and looking for solutions, the engineer dropped boric acid into silicone oil. The result of these two substances mixing together was—a goo!

Because of its molecular structure, the goo could bounce and stretch in all directions. The engineer also discovered the goo could break into pieces. When strong pressure is applied to the substance, it reacts like a solid and breaks apart. Even though the combination was versatile—and quite amusing, the U.S. government decided the new substance wasn't a good substitute for synthetic rubber.

A few years later, the recipe for the stretch material fell into the hands of a businessperson, who saw the goo's potential—as a toy. The toymaker paid $147 for rights to the boric acid and silicone oil mixture. And in 1949 it was sold at toy stores for the first time. The material was packaged in a plastic egg and it took the U.S. by storm. Today, the acid and oil mixture comes in a multitude of colors and almost every child has played with it at some time.

The substance can be used for more than child's play. Its sticky consistency makes it good for cleaning computer keyboards and removing small specks of lint from fabrics.

People use it to make impressions of newspaper print or comics. Athletes strengthen their grips by grasping it over and over. Astronauts use it to anchor tools on spacecraft in zero gravity. All in all, a most *eggs-cellent* idea!

Research **As a group, examine a sample of the colorful, sticky, stretch toy made of boric acid and silicone oil. Then brainstorm some practical—and impractical—uses for the substance.**

For more information, visit green.msscience.com/oops

Reviewing Main Ideas

Section 1 Matter

1. All matter is composed of tiny particles that are in constant motion.
2. In the solid state, the attractive force between particles holds them in place to vibrate.
3. Particles in the liquid state have defined volumes and are free to move about within the liquid.

Section 2 Changes of State

1. Thermal energy is the total energy of the particles in a sample of matter. Temperature is the average kinetic energy of the particles in a sample.
2. An object gains thermal energy when it changes from a solid to a liquid, or when it changes from a liquid to a gas.
3. An object loses thermal energy when it changes from a gas to a liquid, or when it changes from a liquid to a solid.

Section 3 Behavior of Fluids

1. Pressure is force divided by area.
2. Fluids exert a buoyant force in the upward direction on objects immersed in them.
3. An object will float in a fluid that is more dense than itself.
4. Pascal's principle states that pressure applied to a liquid is transmitted evenly throughout the liquid.

Visualizing Main Ideas

Copy and complete the following concept map on matter.

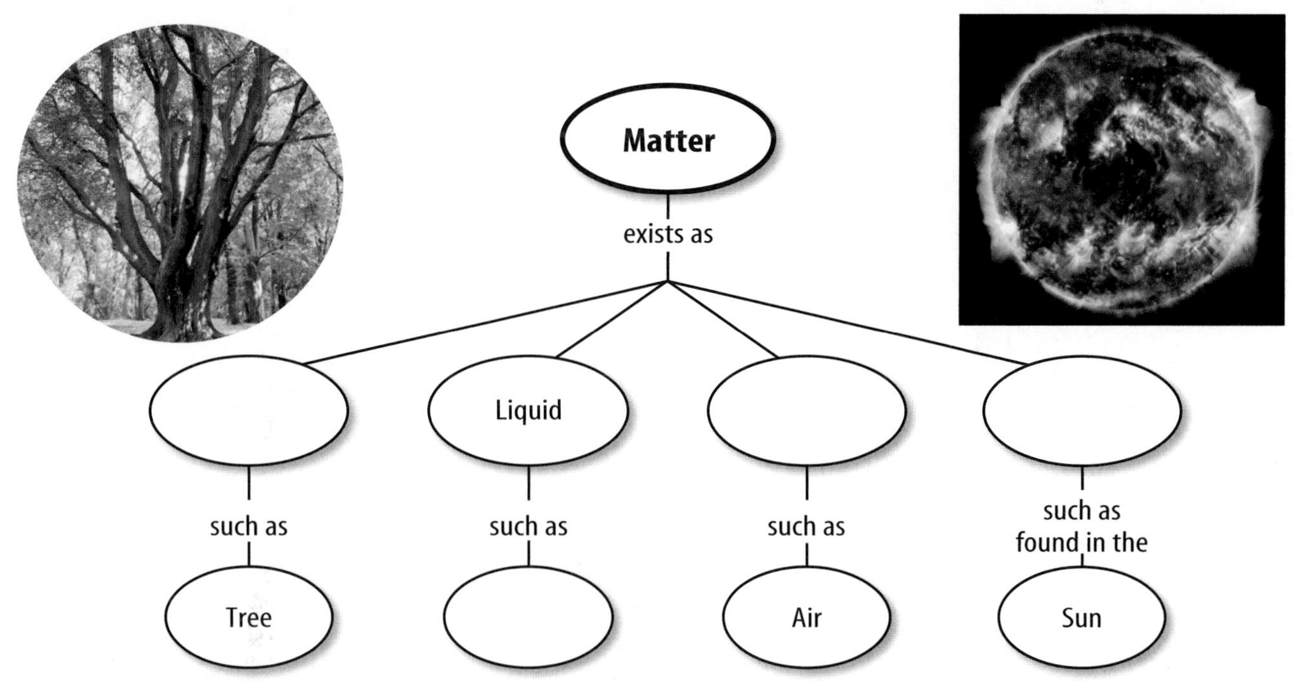

Using Vocabulary

Archimedes' principle p. 671
buoyant force p. 670
condensation p. 663
density p. 671
freezing p. 661
gas p. 656
heat p. 658
liquid p. 654
matter p. 652
melting p. 659
Pascal's principle p. 672
pressure p. 666
solid p. 653
surface tension p. 655
temperature p. 658
thermal energy p. 657
vaporization p. 662
viscosity p. 655

Fill in the blanks with the correct vocabulary word.

1. A(n) _________ can change shape and volume.
2. A(n) _________ has a different shape but the same volume in any container.
3. _________ is thermal energy moving from one substance to another.
4. _________ is a measure of the average kinetic energy of the particles of a substance.
5. A substance changes from a gas to a liquid during the process of _________.
6. A liquid becomes a gas during _________.
7. _________ is mass divided by volume.
8. _________ is force divided by area.
9. _________ explains what happens when force is applied to a confined fluid.

Checking Concepts

Choose the word or phrase that best answers the question.

10. Which of these is a crystalline solid?
    A) glass **C)** rubber
    B) sugar **D)** plastic

11. Which description best describes a solid?
    A) It has a definite shape and volume.
    B) It has a definite shape but not a definite volume.
    C) It adjusts to the shape of its container.
    D) It can flow.

12. What property enables you to float a needle on water?
    A) viscosity **C)** surface tension
    B) temperature **D)** crystal structure

13. What happens to an object as its kinetic energy increases?
    A) It holds more tightly to nearby objects.
    B) Its mass increases.
    C) Its particles move more slowly.
    D) Its particles move faster.

14. During which process do particles of matter release energy?
    A) melting **C)** sublimation
    B) freezing **D)** boiling

15. How does water vapor in air form clouds?
    A) melting **C)** condensation
    B) evaporation **D)** sublimation

16. Which is a unit of pressure?
    A) N **C)** g/cm^3
    B) kg **D)** N/m^2

17. Which change results in an increase in gas pressure in a balloon?
    A) decrease in temperature
    B) decrease in volume
    C) increase in volume
    D) increase in altitude

18. In which case will an object float on a fluid?
    A) Buoyant force is greater than weight.
    B) Buoyant force is less than weight.
    C) Buoyant force equals weight.
    D) Buoyant force equals zero.

Science Online green.msscience.com/vocabulary_puzzlemaker

Use the photo below to answer question 19.

19. In the photo above, the water in the small beaker was displaced when the golf ball was added to the large beaker. What principle does this show?
A) Pascal's principle
B) the principle of surface tension
C) Archimedes' principle
D) the principle of viscosity

20. Which is equal to the buoyant force on an object?
A) volume of the object
B) weight of the displaced fluid
C) weight of object
D) volume of fluid

Thinking Critically

21. Explain why steam causes more severe burns than boiling water.

22. Explain why a bathroom mirror becomes fogged while you take a shower.

23. Form Operational Definitions Write operational definitions that explain the properties of and differences among solids, liquids, and gases.

24. Determine A king's crown has a volume of 110 cm^3 and a mass of 1,800 g. The density of gold is 19.3 g/cm^3. Is the crown pure gold?

25. Infer Why do some balloons pop when they are left in sunlight for too long?

Performance Activities

26. Storyboard Create a visual-aid storyboard to show ice changing to steam. There should be a minimum of five frames.

Applying Math

Use the graph below to answer question 27.

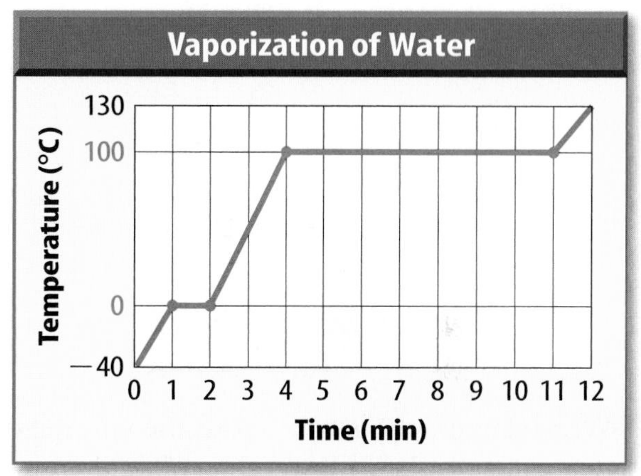

27. Explain how this graph would change if a greater volume of water were heated. How would it stay the same?

Use the table below to answer question 28.

Water Pressure

Depth (m)	Pressure (atm)	Depth (m)	Pressure (atm)
0	1.0	100	11.0
25	3.5	125	13.5
50	6.0	150	16.0
75	8.5	175	18.5

28. Make and Use Graphs In July of 2001, Yasemin Dalkilic of Turkey dove to a depth of 105 m without any scuba equipment. Make a depth-pressure graph for the data above. Based on your graph, how does water pressure vary with depth? Note: The pressure at sea level, 101.3 kPa, is called one atmosphere (atm).

Part 1 Multiple Choice

Record your answers on the answer sheet provided by your teacher or on a sheet of paper.

1. In which state of matter do particles stay close together, yet are able to move past one another?

A. solid
B. gas
C. liquid
D. plasma

Use the illustration below to answer questions 2 and 3.

2. Which statement is true about the volume of the water displaced when the golf ball was dropped into the large beaker?

A. It is equal to the volume of the golf ball.
B. It is greater than the volume of the golf ball.
C. It is less than the volume of the golf ball.
D. It is twice the volume of a golf ball.

3. What do you know about the buoyant force on the golf ball?

A. It is equal to the density of the water displaced.
B. It is equal to the volume of the water displaced.
C. It is less than the weight of the water displaced.
D. It is equal to the weight of the water displaced.

4. What is the process called when a gas cools to form a liquid?

A. condensation
B. sublimation
C. boiling
D. freezing

5. Which of the following is an amorphous solid?

A. diamond
B. sugar
C. glass
D. sand

6. Which description best describes a liquid?

A. It has a definite shape and volume.
B. It has a definite volume but not a definite shape.
C. It expands to fill the shape and volume of its container.
D. It cannot flow.

7. During which processes do particles of matter absorb energy?

A. freezing and boiling
B. condensation and melting
C. melting and vaporization
D. sublimation and freezing

Use the illustration below to answer questions 8 and 9.

8. What happens as the piston moves down?

A. The volume of the gas increases.
B. The volume of the gas decreases.
C. The gas particles collide less often.
D. The pressure of the gas decreases.

9. What relationship between the volume and pressure of a gas does this illustrate?

A. As volume decreases, pressure decreases.
B. As volume decreases, pressure increases.
C. As volume decreases, pressure remains the same.
D. As the volume increases, pressure remains the same.

Part 2 | Short Response/Grid In

Record your answers on the answer sheet provided by your teacher or on a sheet of paper.

10. A balloon filled with helium bursts in a closed room. What space will the helium occupy?

Use the illustration below to answer questions 11 and 12.

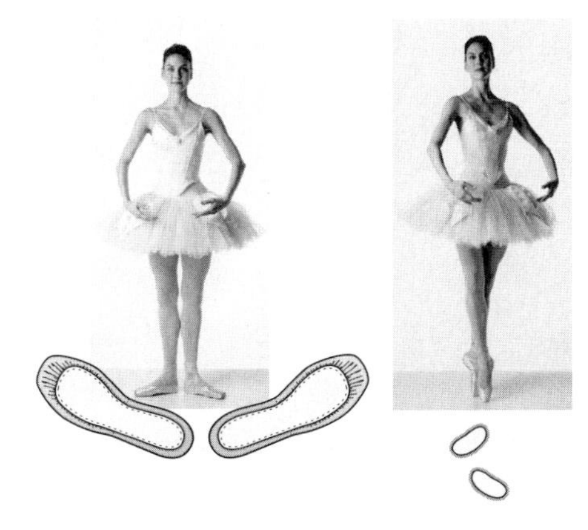

11. If the force exerted by the dancer is 510 N, what is the pressure she exerts if the area is 335 cm^2 on the left and 37 cm^2 on the right?

12. Compare the pressure the dancer would exert on the floor if she were wearing large clown shoes to the photo on the left.

13. If a balloon is blown up and tied closed, air is held inside it. What will happen to the balloon if it is then pushed into hot water or held over a heater? Why does this happen?

14. What is the relationship of heat and thermal energy?

15. Why are some insects able to move around on the surface of a lake or pond?

16. How does the weight of a floating object compare with the buoyant force acting on the object?

17. What is the mass of an object that has a density of 0.23 g/cm^3 and whose volume is 52 cm^3?

Part 3 | Open Ended

Record your answer on a sheet of paper.

18. Compare and contrast evaporation and boiling.

Use the illustration below to answer questions 19 and 20.

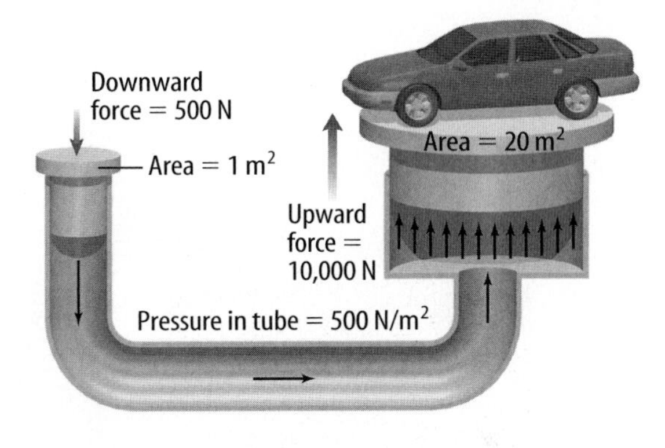

19. Name and explain the principle that is used in lifting the car.

20. Explain what would happen if you doubled the area of the piston on the right side of the hydraulic system.

21. Explain why a woman might put dents in a wood floor when walking across it in high-heeled shoes, but not when wearing flat sandals.

22. Explain why the tires on a car might become flattened on the bottom after sitting outside in very cold weather.

23. Compare the arrangement and movement of the particles in a solid, a liquid, and a gas.

24. Explain why the water in a lake is much cooler than the sand on beach around it on a sunny summer day.

Test-Taking Tip

Show Your Work For open-ended questions, show all of your work and any calculations on your answer sheet.

Hint: In question 20, the pressure in the tube does not change.

chapter 23

Newton's Laws of Motion

chapter preview

sections

How do they fly?

Humans have been flying in airplanes since 1903 and spacecraft since the 1960s. Although both planes and space vehicles are incredibly complex machines, their motion is governed by the same principles that explain the motion of a person walking along a sidewalk.

Science Journal List three questions that you would ask an astronaut about space flight.

Start-Up Activities

Observe Motion Along Varying Slopes

Billions of stars orbit around the center of the galaxy. The wind rattles the leaves in a grove of trees creating a cool breeze. A roller coaster plummets down a steep hill. Just three laws of motion explain why objects move the way they do and how different types of forces cause these movements.

1. Lean one end of a ruler on top of three books to form a ramp. Use a plastic ruler with a groove down the middle.
2. Tap a marble so it rolls up the ramp. Measure how far it rolls up the ramp before stopping and rolling back down.
3. Repeat step 2 using two books, one book, and no books. The same person should tap the marble each time, trying to keep the force constant.
4. **Think Critically** What would happen if you tapped the marble on a flat surface with no friction? Write your predictions and observations in your Science Journal.

Newton's Laws of Motion Make the following Foldable to help you better understand Newton's laws of motion as you read the chapter.

STEP 1 **Draw** a mark at the midpoint of a sheet of paper along the side edge. Then **fold** the top and bottom edges in to touch the midpoint.

STEP 2 **Fold** in half from side to side.

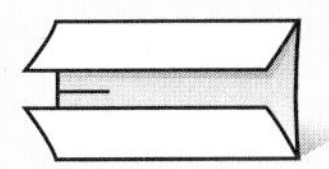

STEP 3 **Turn** the paper vertically. **Open and cut** along the inside fold lines to form four tabs.

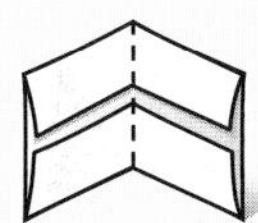

STEP 4 **Label** the tabs *Motion, First Law, Second Law,* and *Third Law.*

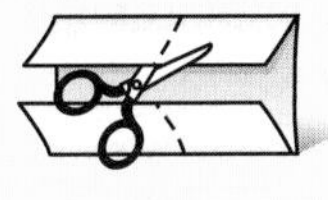

Find Main Ideas As you read the chapter, list the main ideas in each section under the appropriate tab.

Preview this chapter's content and activities at green.msscience.com

section 1

Motion

as you read

What **You'll Learn**

- **Contrast** distance and displacement.
- **Define** speed, velocity, and acceleration.
- **Calculate** speed, velocity, and acceleration.

Why **It's Important**

You, and many of the things around you, are in motion every day.

Review Vocabulary
meter: SI unit of distance equal to approximately 39.37 in

New Vocabulary
- displacement
- speed
- velocity
- acceleration

What is motion?

Cars drive past on streets and freeways. Inside you, your heart pumps and sends blood throughout your body. When walking, skating, jumping, or dancing, you are in motion. In fact, motion is all around you. Even massive pieces of Earth's surface move—if only a few centimeters per year.

A person wants to know the distance between home and another place. Is it close enough to walk? The coach at your school wants to analyze various styles of running. Who is fastest? Who can change direction easily in a sprint? To answer these questions, you must be able to describe motion.

Distance and Displacement One way you can describe the motion of an object is by how it changes position. There are two ways to describe how something changes position. One way is to describe the entire path the object travels. The other way is to give only the starting and stopping points. Picture yourself on a hiking trip through the mountains. The path you follow is shown in **Figure 1.** By following the trail, you walk 22 km but end up only 12 km from where you started. The total distance you travel is 22 km. However, you end up only 12 km northeast of where you started. **Displacement** is the distance and direction between starting and ending positions. Your displacement is 12 km northeast.

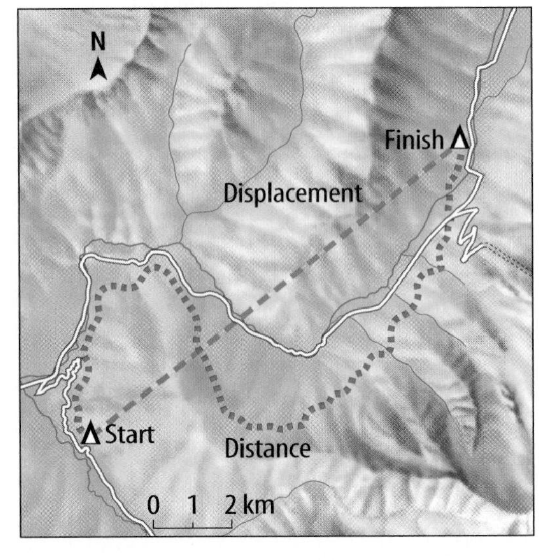

Figure 1 A hiker might be interested in the total distance traveled along a hike or in the displacement. **Infer** *whether the displacement can be greater than the total distance.*

Figure 2 The student has been in motion if he changes position relative to a reference point, such as the plant.
Explain *how the description of the motion depends on the reference point chosen.*

Relative Motion Something that is in motion changes its position. The position of an object is described relative to another object, which is the reference point. Suppose you look out the window in the morning and see a truck parked next to a tree. When you look out the window later, the truck is parked further down the street. If you choose the tree as your reference point, then the truck has been in motion because it has changed position relative to the tree. Has the student in **Figure 2** been in motion?

Reading Check ***How can you tell whether an object has changed position?***

Speed

When running, you are interested not only in the distance you travel, but also in how fast you are moving. Your speed describes how quickly or slowly you are moving. **Speed** is the distance traveled divided by the time needed to travel the distance.

Speed Equation

$$\textbf{speed} \text{ (in meters/second)} = \frac{\textbf{distance} \text{ (in meters)}}{\textbf{time} \text{ (in seconds)}}$$

$$s = \frac{d}{t}$$

The unit for speed is a distance unit divided by a time unit. The SI unit for speed is meters per second, which is abbreviated as m/s. However, sometimes it is more convenient to measure distance and time in other units. For example, a speedometer shows a car's speed in units of km/h.

Figure 3 When you read a speedometer, you are finding your instantaneous speed.
Identify *when your instantaneous speed would be the same as your average speed.*

Constant Speed If you are riding in a car with the cruise control on, the speed doesn't change. In other words, the car moves at a constant speed. If you are running at a constant speed of 5 m/s, you run 5 m each second. When you are traveling at a constant speed, the speed at any instant of time is the same.

Changing Speed For some motion, speed is not constant. If you are riding your bike, you must slow down for intersections and turns and then pedal faster to resume your speed. Your average speed for your trip would be the total distance traveled divided by the time you were riding. If you wanted to know how fast you were going at just one instant, you would be interested in your instantaneous speed. A car speedometer like the one in **Figure 3** shows instantaneous speed—how fast the car is moving at any moment. When speed is constant, average speed and instantaneous speed are the same.

Applying Math — Solve a Simple Equation

A SWIMMER'S AVERAGE SPEED It takes a swimmer 57.2 s to swim a distance of 100 m. What is the swimmer's average speed?

Solution

1. *This is what you know:*
   - distance: $d = 100.0$ m
   - time: $t = 57.2$ s
2. *This is what you need to find:* speed: s
3. *This is the procedure you need to use:* Substitute the known values for distance and time into the speed equation and calculate the speed:

$$s = \frac{d}{t} = \frac{(100.0 \text{ m})}{(57.2 \text{ s})} = 1.75 \text{ m/s}$$

4. *Check your answer:* Multiply your answer by the time, 57.2 s. The result should be the given distance, 100.0 m.

Practice Problems

1. A bicycle coasting downhill travels 170.5 m in 21.0 s. What is the bicycle's average speed?
2. A car travels a distance of 870 km with an average speed of 91.0 km/h. How many hours were needed to make the trip?

Science Online — For more practice, visit green.msscience.com/math_practice

Velocity

Sometimes you might be interested not only in how fast you are going, but also in the direction. When direction is important, you want to know your velocity. **Velocity** is the displacement divided by time. For example, if you were to travel 1 km east in 0.5 h, you would calculate your velocity as follows.

$$\text{velocity} = \frac{\text{displacement}}{\text{time}}$$

$$\text{velocity} = \frac{(1 \text{ km east})}{(0.5 \text{ h})}$$

$$\text{velocity} = 2 \text{ km/h east}$$

Figure 4 The speedometers and compasses on an airplane instrument panel tell the pilot the airplane's velocity.

Like displacement, velocity includes a direction. Velocity is important to pilots flying airplanes. They rely on control panels like the one in **Figure 4** because they need to know not only how fast they are flying, but also in what direction.

Acceleration

Displacement and velocity describe how far, how fast, and where something is moving. You also might want to know how motion is changing. Is a car speeding up, or is it slowing down? Is it moving in a straight line or changing direction? **Acceleration** is the change in velocity divided by the amount of time required for the change to occur. Because velocity includes speed and direction, so does acceleration. If an object changes its speed, its direction, or both, it is accelerating.

Speeding Up and Slowing Down When you think of an object accelerating, you might think of it moving faster and faster. If someone says a car accelerates, you think of it moving forward and increasing its velocity.

However, when an object slows down, it also is accelerating. Why? Recall that an object is accelerating when its velocity changes. Velocity can change if the speed of an object changes, whether the speed increases or decreases, or if it changes direction. If an object slows down, its speed changes. Therefore, if an object is speeding up or slowing down, it is accelerating.

Measuring Motion

Procedure

1. Measure a fixed distance such as the length of your driveway.
2. Use a **watch** to measure the time it takes you to stroll, rapidly walk, and run this distance.

Analysis

1. Calculate your speed in each case.
2. Use your results to predict how long it would take you to go 100 m by each method.

Try at Home

Calculating Acceleration If an object changes speed but not direction, you can use the following equation to calculate the object's acceleration.

Acceleration Equation

acceleration (in m/s^2) =

$$\frac{\textbf{final speed (in m/s)} - \textbf{initial speed (in m/s)}}{\textbf{time (in s)}}$$

$$a = \frac{s_f - s_i}{t}$$

In the acceleration equation, the symbol s_f stands for the final speed and s_i stands for the initial speed. The SI unit for acceleration is m/s^2, which means meters/(seconds × seconds). The unit m/s^2 is the result when the unit m/s is divided by the unit s.

Applying Math — Solve a Simple Equation

SKATEBOARD ACCELERATION It takes a skateboarder 12.0 s to speed up from 2.0 m/s to 8.0 m/s. What is the skateboarder's acceleration?

Solution

1. *This is what you know:*
   - initial speed: s_f = 2.0 m/s
   - final speed: s_i = 8.0 m/s
   - time: t = 12.0 s
2. *This is what you need to find:* acceleration: a
3. *This is the procedure you need to use:* Substitute the known values for initial speed and time into the acceleration equation, and calculate the acceleration:

$$a = \frac{(s_f - s_i)}{t} = \frac{(8.0\ \text{m/s} - 2.0\ \text{m/s})}{(12.0\ \text{s})} = 0.5\ \text{m/s}^2$$

4. *Check your answer:* Multiply your answer by the time 12.0 s and then add the initial speed. The result should be the given final speed, 8.0 m/s.

Practice Problems

1. A horse speeds up from a speed of 11 m/s to 17 m/s in 3 s. What is the horse's acceleration?
2. What is the acceleration of a sports car initially at rest that reaches a speed of 30.0 m/s in 5.0 s?

For more practice, visit green.msscience.com/math_practice

When the race starts, the runner accelerates by speeding up.

When rounding the corners of the track, the runner accelerates by turning.

After crossing the finish line, the runner accelerates by slowing down.

Figure 5 During a race, a runner accelerates in different ways.

Turning When an object turns or changes direction, its velocity changes. This means that any object that changes direction is accelerating. To help understand acceleration, picture running a race, as shown in **Figure 5.** As soon as the race begins, you start from rest and speed up. When you follow the turn in the track, you are changing your direction. After you pass the finish line, you slow down to a steady walk. In each case, you are accelerating.

section 1 review

Summary

What is motion?

- The position of an object is measured relative to a reference point.
- The motion of an object can be described by how the position of an object changes.

Speed

- The speed of an object can be calculated from:

$$s = \frac{d}{t}$$

- Average speed is the total distance traveled divided by the total time; instantaneous speed is the speed at an instant of time.

Acceleration

- Acceleration occurs when the speed of an object or its direction of motion changes.
- The acceleration of an object moving in a straight line can be calculated from:

$$a = \frac{s_f - s_i}{t}$$

Self Check

1. **Evaluate** whether the velocity of a jogger can be determined from the information that the jogger travels 2 km in 10 min.
2. **Determine** your distance traveled and displacement if you walk 100 m forward and then 35 m backward.
3. **Describe** how a speedometer needle moves when a car is moving with constant velocity, speeding up, and slowing down.
4. **Think Critically** How could two observers measure a different speed for the same moving object?

Applying Math

5. **Evaluate Motion** Two people swim for 60 s. Each travels a distance of 40 m swimming out from shore, and 40 m swimming back. One swimmer returns to the starting point and the other lands 6 m north of the starting point. Determine the total distance, displacement, and average speed of each swimmer.
6. **Calculate Acceleration** What is the acceleration of a car moving at 12.0 m/s that comes to a stop in 4 s?

section 2

Newton's First Law

as you read

***What* You'll Learn**

- **Define** force.
- **Describe** Newton's first law of motion.
- **Contrast** balanced and unbalanced forces.

***Why* It's Important**

Newton's first law explains why the motion of objects change.

Review Vocabulary

mass: a measure of the amount of matter in an object

New Vocabulary

- force
- first law of motion
- balanced forces
- unbalanced forces

Laws of Motion

A picture might tell a thousand words, but sometimes it takes a thousand pictures to show a motion. Look at the strobe photo in **Figure 6.** The smoothness of the overall motion is the result of many individual motions—hands move and legs flex. Some trainers who work with athletes use such films to help them understand and improve performance.

What causes such complex motions to occur? Each individual motion can be explained and predicted by a set of principles that were first stated by Isaac Newton. These principles are called Newton's laws of motion.

Force An object's motion changes in response to a force. A **force** is a push or a pull. A force has a size and a direction—both are important in determining an object's motion. For example, pushing this book on one side will slide it across the desk in the same direction. Pushing it harder will make it move faster in the direction of your push. However, if you push the book straight down, it will not move. The force you exert on the book when you push is called a contact force. A contact force is a force that is exerted when two objects are touching each other.

Reading Check *Name two contact forces that act on you when you sit at a table. Include the directions.*

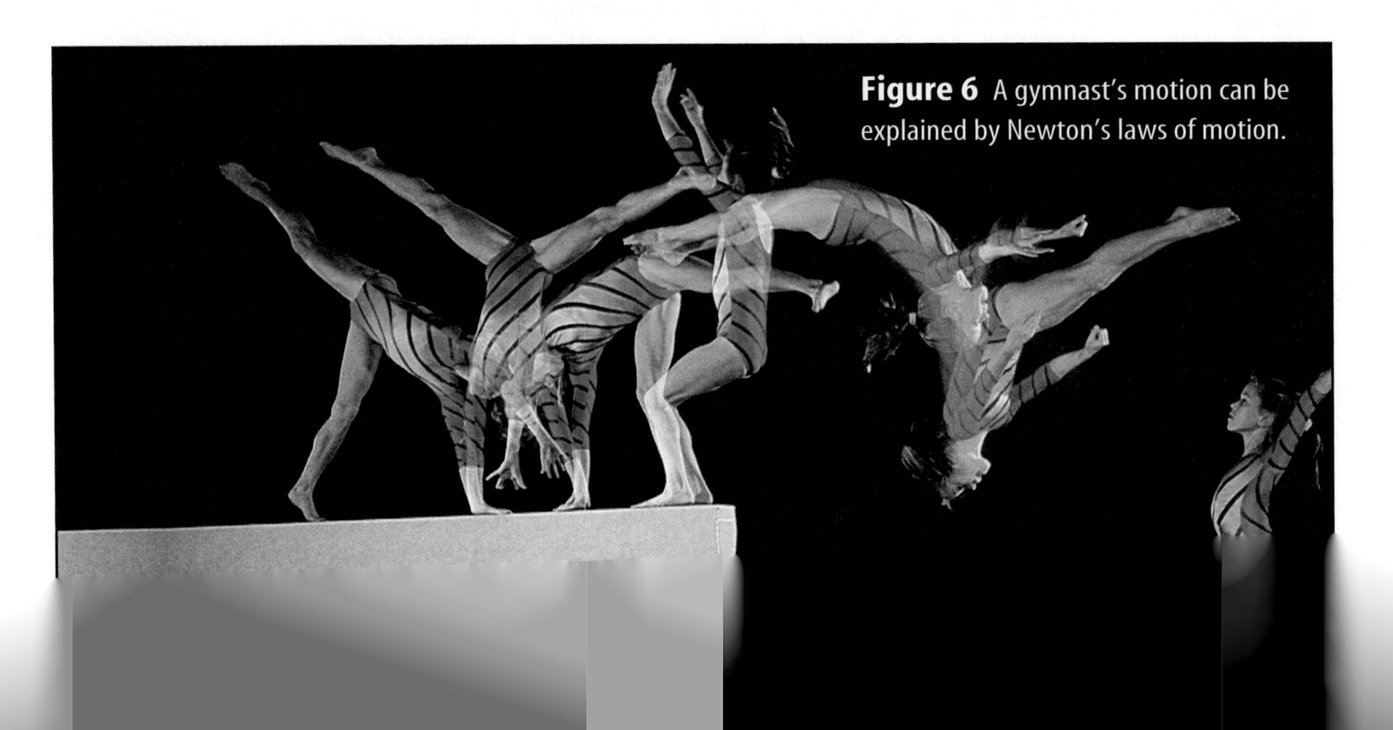

Figure 6 A gymnast's motion can be explained by Newton's laws of motion.

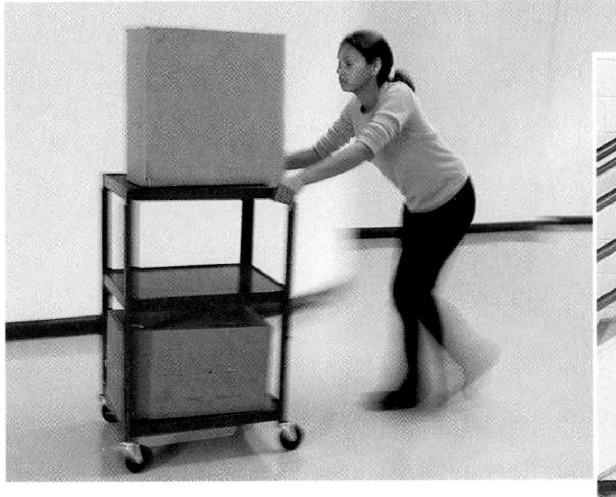

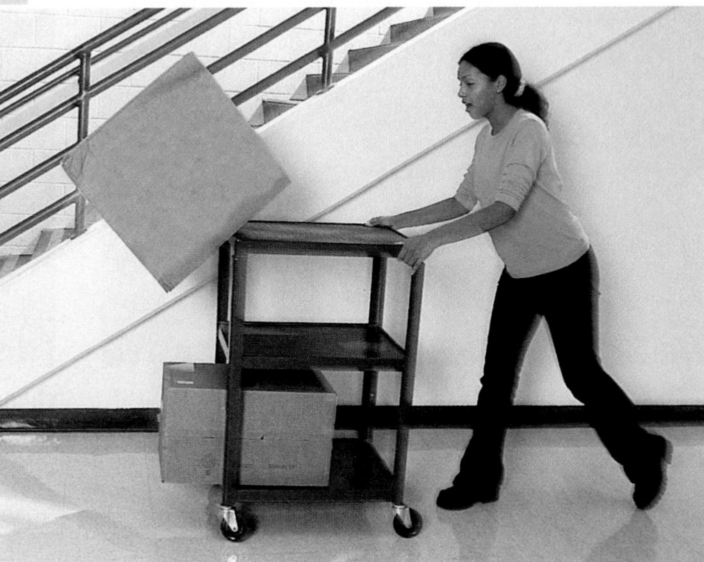

Figure 7 If you were to stop the cart suddenly, the boxes would keep moving.
Infer *why the boxes move downward after they slide off the cart.*

Long-Range Forces However, a force can be exerted if two objects are not in contact. If you bring a magnet close to a paper clip, the paper clip moves toward the magnet, so a force must be acting on the paper clip. The same is true if you drop a ball—the ball will fall downward, even though nothing appears to be touching it. The forces acting on the paper clip and the ball are long-range forces. These forces can cause an object's motion to change without direct contact. Long-range forces include gravity, magnetism, and electricity.

In SI units the unit of force is the newton, which is abbreviated N and named for Isaac Newton. One newton is about the amount of force needed to lift a half cup of water.

Newton's First Law of Motion

When you are riding in a car and it comes to a sudden stop, you feel yourself continuing to move forward. Your seat belt slows you down and pulls you back into your seat. This can be explained by the **first law of motion**—An object will remain at rest or move in a straight line with constant speed unless it is acted upon by a force. For a long time it was thought that all objects come to rest naturally. It seemed that a force had to be applied continually to keep an object moving. Newton and others theorized that if an object already is moving, it will continue to move in a straight line with constant speed. For the object to slow down, a force has to act on it.

Think about what keeps you in your seat when a car comes to a sudden stop. You can feel yourself moving forward, but the seat belt applies a force to pull you back and stop your motion. If you were to push a cart with boxes on it, as in **Figure 7,** and suddenly stop the cart, the boxes would continue moving and slide off the cart.

Car Safety Features In a car crash, a car comes to a sudden stop. According to Newton's first law, passengers inside the car will continue to move unless they are held in place. Research the various safety features that have been designed to protect passengers during car crashes. Write a paragraph on what you've learned in your Science Journal.

It is easy to speed up or slow down a small child in a wagon.

Changing the motion of a wagon carrying two children is harder.

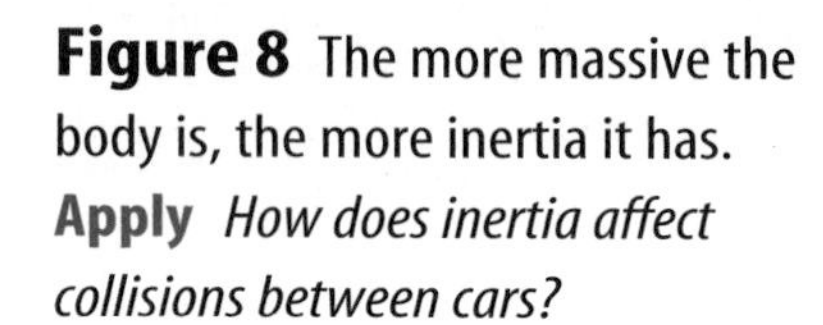

Figure 8 The more massive the body is, the more inertia it has.
Apply *How does inertia affect collisions between cars?*

Inertia and Mass The first law of motion is sometimes called the law of inertia. Inertia measures an object's tendency to remain at rest or keep moving with constant velocity. Inertia depends on the mass of the object. The more mass an object has, the more inertia it has and the harder it is to change the motion of the object. Look at **Figure 8.** It is easy to get a wagon to start moving if only one small child is sitting in it. You must exert more force to start the wagon moving if it holds more mass. The same is true if you try to stop the wagon after it is moving. You need to exert a larger force to stop the wagon that has more mass.

Reading Check *What is inertia?*

Science Online

Topic: Aircraft Carriers

Visit green.msscience.com for Web links to information about how airplanes land on aircraft carriers. Because these airplanes have a large mass, their inertia is also large. This makes the planes hard to stop on the short runway.

Activity Investigate the effects of the large forces encountered in aircraft and spacecraft acceleration on the human body.

Adding Forces

According to Newton's first law, the motion of an object changes only if a force is acting on the object. Sometimes more than one force acts on an object, as when several people push a stalled car to the side of the road. Motion depends upon the size and direction of all the forces.

If two people push in opposite directions on a box with an equal amount of force, the box will not move. Because the forces are equal but in opposite directions, they will cancel each other out and are called **balanced forces.** When forces on an object are balanced, no change will occur in the object's motion because the total force on the object is zero.

If one force pushing on the box is greater than the other, the forces do not cancel. Instead, the box will move in the direction of the larger force. Forces acting on an object that do not cancel are **unbalanced forces.** The motion of an object changes only if the forces acting on it are unbalanced. The change in motion is in the direction of the unbalanced force.

Changes in Motion and Forces

How do unbalanced forces affect the motion of an object? Look at the dancer in **Figure 9.** When she jumps, she pushes off the stage with a force greater than the force of gravity pulling her down. This creates an unbalanced force upward, and the dancer moves upward. After she has left the ground and her feet are no longer in contact with the floor, gravity becomes the only force acting on her. The forces on her are unbalanced and her motion changes—she slows down as she rises into the air and speeds up as she returns to the ground.

The motion of an object changes only when unbalanced forces act on the object. Recall that if the motion of an object changes, the object is accelerating. The object can speed up, slow down, or turn. In all cases, an object acted on by an unbalanced force changes velocity.

Figure 9 Because the motion of this dancer is changing, the forces on her must be unbalanced.

section 2 review

Summary

Forces

- A force is a push or a pull. An object's motion changes in response to a force.
- A contact force is a force exerted when two objects are touching each other.
- Long-range forces are exerted between objects that are not touching. Gravity and electric and magnetic forces are long-range forces.

Newton's First Law of Motion

- Newton's first law of motion states that an object will remain at rest or move with a constant velocity unless it is acted upon by an unbalanced force.
- The tendency of an object to resist a change in motion is inertia, which increases as the mass of the object increases.
- Balanced forces cancel and do not cause a change in motion. Unbalanced forces cause a change in motion.

Self Check

1. **Apply** the first law of motion to the motion of an ice skater sliding across the ice at a constant velocity.
2. **Describe** the information that must be given to specify a force.
3. **Explain** When you sit in a chair, the force of gravity is pulling you downward. Is this a balanced or an unbalanced force?
4. **Explain** why a greater force is needed to move a refrigerator than is needed to move a book.
5. **Think Critically** Explain whether the following statement is true: if an object is moving, there must be a force acting on it.

Applying Skills

6. **Apply** It was once thought that the reason objects slowed down and stopped was that there was no force acting on them. A force was necessary to keep an object moving. According to the first law of motion, why do objects slow down and stop?

section 3

Newton's Second Law

as you read

What You'll Learn

- **Predict** changes in motion using Newton's second law.
- **Describe** the gravitational force between objects.
- **Contrast** different types of friction.

Why It's Important

Newton's second law explains how motion changes.

Review Vocabulary
solar system: the Sun, together with the planets and other objects that revolve around the Sun

New Vocabulary
- second law of motion
- gravitational force
- friction

The Second Law of Motion

Newton's first law of motion can help predict when an object's motion will change. Can you predict what the change in motion will be? You know that when you kick a ball, as in **Figure 10,** your foot exerts a force on the ball, causing it to move forward and upward. The force of gravity then pulls the ball downward. The motion of the ball can be explained by Newton's second law of motion. According to the **second law of motion,** an object acted on by an unbalanced force will accelerate in the direction of the force with an acceleration given by the following equation.

Newton's Second Law of Motion

$$\textbf{acceleration}\ (\text{m/s}^2) = \frac{\textbf{force}\ (\text{in N})}{\textbf{mass}\ (\text{in kg})}$$

$$a = \frac{F}{m}$$

If more than one force acts on the object, the force in this formula is the combination of all the forces, or the total force that acts on the object. What is the acceleration if the total force is zero?

Reading Check *In what direction does an object accelerate when acted on by an unbalanced force?*

Figure 10 When you kick a ball, the forces on the ball are unbalanced. The ball moves in the direction of the unbalanced force.

Using the Second Law

The second law of motion enables the acceleration of an object to be calculated. Knowing the acceleration helps determine the speed or velocity of an object at any time. For motion in a straight line an acceleration of 5 m/s^2 means that every second the speed is increasing by 5 m/s.

If you know the mass and acceleration of an object, you can use Newton's second law to find the force. The equation for Newton's second law of motion can be solved for force if both sides of the equation are multiplied by mass as shown below:

$$\frac{F}{m} = a$$

$$\cancel{m} \times \frac{F}{\cancel{m}} = m \times a$$

$$F = ma$$

Applying Math — Solve a Simple Equation

THE FORCE ON A BIKE AND RIDER A bike and a rider together have a mass of 60.0 kg. If the bike and rider have an acceleration of 2.0 m/s^2, what is the force on the bike and rider?

Solution

1. *This is what you know:*
   - mass: $m = 60.0$ kg
   - acceleration: $a = 2.0\ m/s^2$
2. *This is what you need to find:* force: F
3. *This is the procedure you need to use:* Substitute the known values for mass and acceleration into the equation for Newton's second law of motion, and calculate force:

   $$F = ma = (60.0\ kg)(2.0\ m/s^2) = 120\ N$$

4. *Check your answer:* Divide your answer by the mass 60.0 kg. The result should be the given acceleration 2.0 m/s^2.

Practice Problems

1. You are lifting a backpack with a mass of 12.0 kg so it has an acceleration of 0.5 m/s^2. What force are you exerting on the backpack?
2. The space shuttle lifts off with an acceleration of 16.0 m/s^2. If the shuttle's mass is about 2,000,000 kg, what force is being exerted on the shuttle at liftoff?

For more practice visit green.msscience.com/math_practice

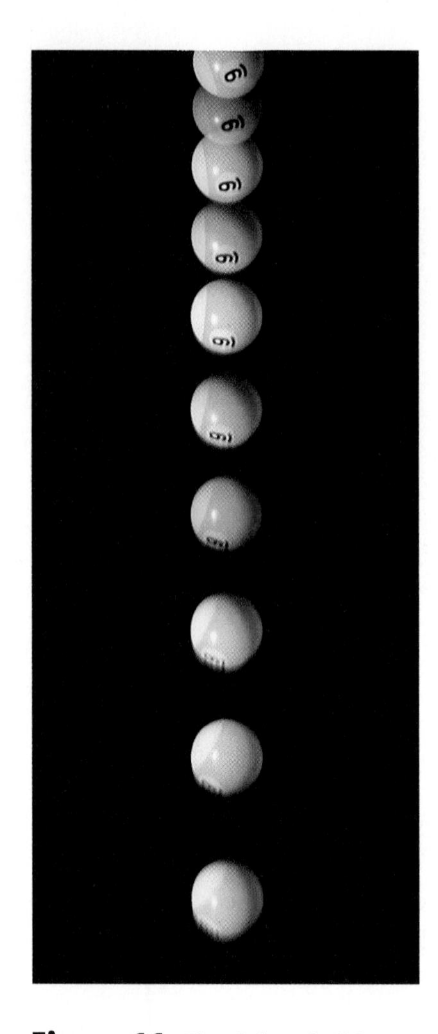

Figure 11 This falling ball has been photographed using a strobe light that flashes on and off at a steady rate. Because the ball is accelerating, its speed is increasing. As a result, the distance it travels between each strobe flash also increases.

The Force of Gravity Every object exerts an attractive gravitational force on every other object. The **gravitational force** between two objects depends on the masses of the objects and the distance between them. This force increases as the objects get closer together or if their masses increase.

You don't notice the gravitational force between you and ordinary objects. Your mass and the masses of these objects are too small for the gravitational force produced to be large enough to feel. However, objects with much greater mass, such as planets and stars, can exert much greater gravitational forces. For example, the gravitational forces between the Sun, planets, and other nearby space objects are large enough to hold the solar system together.

You can feel the gravitational force exerted by Earth because Earth has much more mass than any other nearby object. Gravity causes all objects at or near Earth's surface to be pulled downward toward Earth's center with the same acceleration, 9.8 m/s^2. The motion of a falling object is shown in **Figure 11.** The gravitational force on a falling object with mass, m, is

$$F = m \times (9.8\ m/s^2)$$

Even if you are standing on the ground, Earth still is pulling on you with this force. You are at rest because the ground exerts an upward force that balances the force of gravity.

Mass and Weight The force of Earth's gravity on an object near Earth is the weight of that object. The weight of an object is not the same as the object's mass. Weight is a force, just like a push of your hand is a force. Weight changes when the force of gravity changes. For example, your weight on the Moon is less than your weight on Earth because the force of gravity at the Moon's surface is only about 16 percent as large as the force of Earth's gravity. Mass, on the other hand, is a measure of the amount of matter in an object. It is the same no matter where you are. An astronaut would have the same mass on Earth as on the Moon.

Reading Check *How are mass and weight different?*

Recall that mass is a measure of inertia, which is how hard it is to change the motion of an object. Think of how hard it would be to stop a 130-kg football player rushing at you. Now suppose you were in outer space. If you were far enough from Earth, the force of Earth's gravity would be weak. Your weight and the football player's weight would be small. However, if he were moving toward you, he would be just as hard to stop as on Earth. He still has the same mass and the same inertia, even though he is nearly weightless.

Friction

Rub your hands together. You can feel a force between your hands, slowing their motion. If your hands are covered in lotion, making them slippery, the force is weaker. If anything gritty or sticky is on your hands, this force increases. **Friction** is a force that resists sliding motion between surfaces that are touching.

Friction always is present when two surfaces of two objects slide past each other. To keep the objects moving, a force has to be applied to overcome the force of friction. However, friction is also essential to everyday motion. How hard would it be to move without friction? When you walk up a ramp, friction prevents you from sliding to the bottom. When you stand on a skateboard, friction keeps you from sliding off. When you ride a bike, friction enables the wheels to propel the bike forward. Friction sometimes can be reduced by adding oil or grease to the surfaces, but it always is present. There are several types of friction, and any form of motion will include one or more of them.

Topic: Gravity on Other Planets

Visit green.msscience.com for Web links to information about the force of gravity on other planets and moons in the solar system.

Activity Make a table showing your weight in newtons on the other planets of the solar system.

Static Friction Gently push horizontally on this book. You should be able to push against it without moving it. This is because of the static friction between the cover of the book and the top of the desk. Static friction keeps an object at rest from moving on a surface when a force is applied to the object. When you stand on a slight incline, you don't slide down because of the static friction between your shoes and the ground.

When you carry a tray of food to a table and stop to sit down, Newton's first law predicts that the food on the tray will keep moving and slide off the tray. Happily, static friction keeps the food from sliding off the tray, allowing you to sit down and enjoy your meal. Another example of static friction is shown in **Figure 12.**

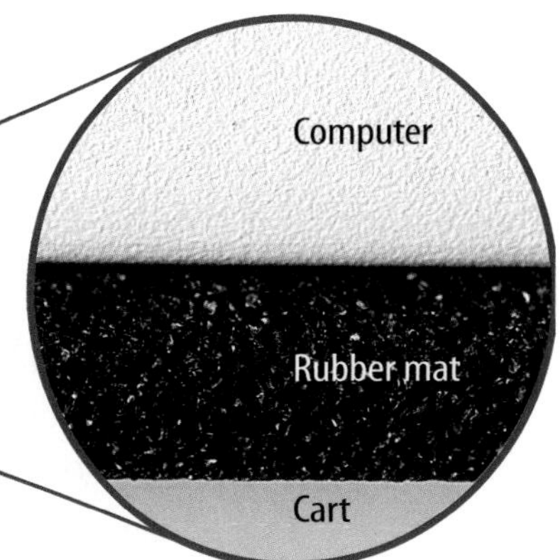

Figure 12 The rubber mat increases the static friction acting on the computer. This static friction keeps the computer from sliding.
Explain *what would happen when the cart stopped if there were no static friction.*

Figure 13 Sliding friction between the rider's shoe and the sidewalk slows the skateboarder down.
Compare and contrast *static and sliding friction.*

Sliding Friction Push on your book again so that it slides. Notice that once the book is in motion and leaves your hand, it comes to a stop. Sliding friction is the force that slows the book down. Sliding friction occurs when two surfaces slide past each other. To keep the book moving, you have to keep applying a force to overcome sliding friction.

The friction between a skier and the snow, a sliding baseball player and the ground, and your shoes and the skin where a blister is forming are examples of sliding friction. When you apply the brakes to a bike, a car, or the skateboard shown in **Figure 13,** you use sliding friction to slow down.

Rolling Friction A car stuck in snow or mud spins its wheels but doesn't move. Rolling friction makes a wheel roll forward or backward. If the rolling friction is large enough, a wheel will roll without slipping. The car that is stuck doesn't move because mud or snow makes the ground too slippery. Then there is not enough rolling friction to keep the wheels from slipping.

Because rolling friction is the force that enables a wheel to roll on a surface, the force of rolling friction is in the same direction as the wheel is rolling. If the wheel is rolling forward, the rolling friction force also points forward, as shown in **Figure 14.** Some of the ways that static, sliding, and rolling friction are useful to a bike rider are shown in **Figure 15.**

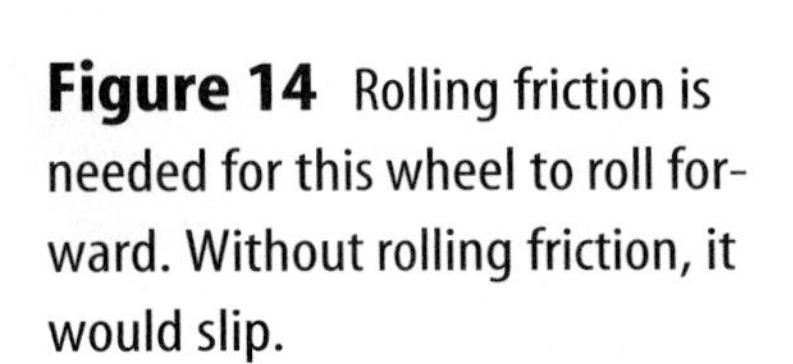
Figure 14 Rolling friction is needed for this wheel to roll forward. Without rolling friction, it would slip.

NATIONAL GEOGRAPHIC

VISUALIZING FRICTION

Figure 15

Friction is a force that opposes motion. But sometimes friction can be used to your advantage. For example, you could not ride a bicycle without friction. A: The rolling friction between the tires and the pavement pushes the bottom of the tires forward; this helps rotate the tires and propels the cyclist forward. B: Brake pads push against the rim of the tire; this creates sliding friction, which slows the wheel. C: Bicycle pedals have a rough surface that increases static friction and keeps the cyclist's feet from slipping off.

Air Resistance Do you know why it is harder to walk in deep water than in shallow water? It's because the deeper the water is, the more of it you have to push out of your way to move forward. The same is true for walking on dry land. To move forward, you must push air out of your way.

Molecules in air collide with the forward-moving surface of an object, slowing its motion. This is called air resistance. Air resistance is less for a narrow, pointed object than for a large, flat object. Air resistance increases as the speed of an object increases. Because air resistance is a type of friction, it acts in the direction opposite to an object's motion.

Figure 16 Gravity and air resistance are the forces acting on a parachutist. When the forces balance each other, the parachutist falls at a constant velocity.

Before the sky diver in **Figure 16** opens the parachute, his air resistance is small. The force of air resistance is upward, but it is not large enough to balance the downward force of gravity. As a result, the sky diver falls rapidly. When he opens his parachute, the air resistance is much greater because the parachute has a large surface area. Then the force of air resistance is large enough to slow his fall and balance the force of gravity.

section 3 review

Summary

Newton's Second Law

- The second law states that an object's acceleration can be calculated from this equation:

$$a = \frac{F}{m}$$

Gravitational Force and Friction

- The gravitational force between two objects depends on the masses of the objects and the distance between them.
- The weight of an object on Earth is the gravitational force exerted by Earth on the object and is given by:

$$F = m(9.8 \text{ m/s}^2)$$

- Friction is a force that resists motion between surfaces that are in contact.
- Static friction acts between surfaces that are not sliding, sliding friction acts between surfaces that are sliding past each other.

Self Check

1. **Identify** the force that keeps a box from sliding down an angled conveyor belt that slopes upward.
2. **Identify** the force that causes a book to slow down and stop as it slides across a table top.
3. **Explain** why you feel Earth's gravitational force, but not the gravitational force exerted by this book.
4. **Determine** how the acceleration of an object changes when the total force on the object increases.
5. **Think Critically** A 1-kg book is at rest on a desk. Determine the force the desk exerts on the book.

Applying Math

6. **Calculate Weight** A student has a mass of 60 kg. What is the student's weight?
7. **Calculate Acceleration** A sky diver has a mass of 70 kg. If the total force on the sky diver is 105 N, what is the sky diver's acceleration?

Science Online green.msscience.com/self_check_quiz

Static and Sliding Friction

Static friction can hold an object in place when you try to push or pull it. Sliding friction explains why you must continually push on something to keep it sliding across a horizontal surface.

Real-World Question

How do the forces of static friction and sliding friction compare?

Goals

- **Observe** static and sliding friction.
- **Measure** static and sliding frictional forces.
- **Compare and contrast** static and sliding friction.

Materials

spring scale
block of wood or other material
tape

Safety Precautions

Procedure

1. Attach a spring scale to the block and set it on the table. Experiment with pulling the block with the scale so you have an idea of how hard you need to pull to start it in motion and continue the motion.
2. **Measure** the force needed just to start the block moving. This is the force of static friction.
3. **Measure** the force needed to keep the block moving at a steady speed. This is the force of sliding friction on the block.
4. Repeat steps 2 and 3 on a different surface, such as carpet. Record your measurements in your Science Journal.

Analyze Your Data

1. **Compare** the forces of static friction and sliding friction on both horizontal surfaces. Which force is greater?
2. On which horizontal surface is the force of static friction greater?
3. On which surface is the force of sliding friction greater?

Conclude and Apply

1. **Draw Conclusions** Which surface is rougher? How do static and sliding friction depend on the roughness of the surface?
2. **Explain** how different materials affect the static and sliding friction between two objects.

Communicating Your Data

Compare your conclusions with those of other students in your class.

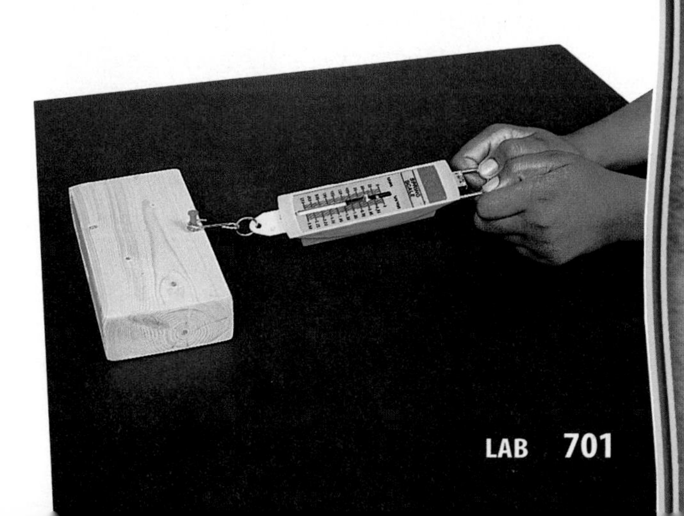

Newton's Third Law

as you read

What You'll Learn

- **Interpret** motion using Newton's third law.
- **Analyze** motion using all three laws.

Why It's Important

The third law of motion explains how objects exert forces on each other.

Review Vocabulary
contact force: force exerted when two objects are touching each other

New Vocabulary
- third law of motion

The Third Law of Motion

When you push off the side of a pool, you accelerate. By Newton's laws of motion, when you accelerate, a force must be exerted on you. Where does this force come from?

Newton's first two laws explain how forces acting on a single object affect its motion. The third law describes the connection between the object receiving a force and the object supplying that force. According to the **third law of motion,** forces always act in equal but opposite pairs. This idea often is restated—for every action, there is an equal but opposite reaction. For example, if object A exerts a force on object B, then object B exerts a force of the same size on object A.

Reading Check *What is Newton's third law of motion?*

Action and Reaction Forces Newton's third law means that when you lift your book bag, your book bag pulls back. When you jump, you push down on the ground, which pushes up on you. When you walk, you push back on the ground, and the ground pushes forward on you. When you throw a ball, you push forward on the ball, and it pushes back on your hand. **Figure 17** shows that when you exert a force on the floor, the floor exerts an equal force on you in the opposite direction.

The same is true for any two objects, regardless of whether the force between the objects is a contact force or a long-range force. For example, if you place two bar magnets with opposite poles facing one another, they will move toward each other. Each magnet applies a force to the other, even though they are not in direct contact. No matter how one object exerts a force on another, the other object always exerts an equal force on the first object in the opposite direction.

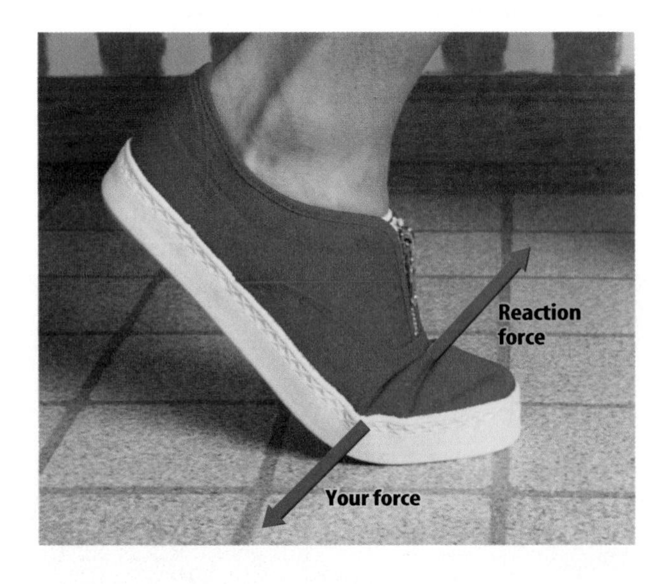

Figure 17 When you walk, you exert a backward force on the ground, yet you move forward. This is because the ground is exerting an equal force forward on you.

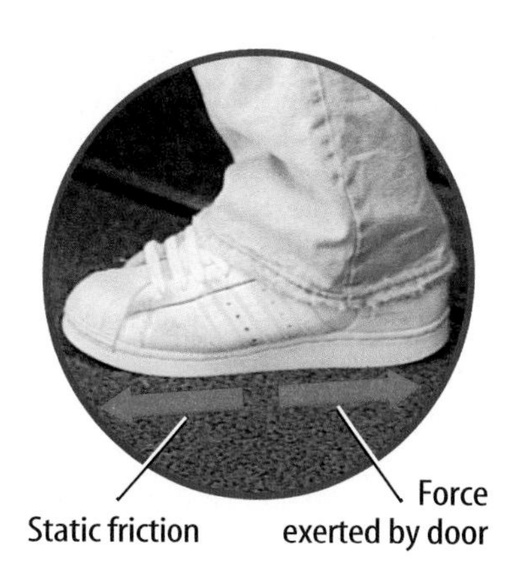

Figure 18 Static friction between your shoes and the ground makes it possible to push a heavy door forward. Without static friction, pushing the door would make you slide backwards.

Applying the Third Law

Action and reaction forces are not the same as balanced forces. Recall that balanced forces are forces that act on the same object and cancel each other. Action and reaction forces act on different objects. When you kick a soccer ball, your force on the ball equals the ball's force on you. The harder you kick, the greater the force the ball exerts on your foot. A hard kick can hurt because the ball exerts a large force on your foot. Unlike balanced forces, action and reaction forces can cause the motion of objects to change.

How do balanced forces differ from action and reaction forces?

Using Friction When you push a heavy door, as in **Figure 18,** you exert a force on the door to move it. The door exerts an equal force back on you. Why don't you move? When you push on the door, your feet are touching Earth, and static friction keeps you from sliding. The reaction force is exerted on you and Earth together. You don't move because the door doesn't exert a large enough reaction force to move both you and Earth.

However, if you wear slippery shoes, or if the floor is very smooth, your feet might slide when you push on the door. Because static friction is smaller when the surfaces are smooth, the static friction force might not be large enough to keep you attached to Earth. Then the reaction force exerted by the door acts only on you, and not on you and Earth together.

Observing the Laws of Motion

Procedure

1. Incline one end of a **board** to a height of 20 cm.
2. Lay a **meterstick** on the floor with one end flush to the board and place a **softball** where the board and meterstick meet.
3. Hold a **baseball** 10 cm up the slope and roll it into the softball. Measure the distance the softball is pushed.
4. Repeat step 3 from different heights.

Analysis

1. Compare the distances the baseball pushes the softball.
2. Use the laws of motion to explain your results.

Motion Caused by Force Pairs Although the action and reaction forces in a force pair are the same size, they can have different effects on the objects they act upon. Suppose a 50-kg student and a 20-kg box are in the middle of an ice-skating rink. The student pushes on the box with a force of 10 N, and the box slides on the ice. The reaction force is the box pushing on the student with a force of 10 N, and the student slides in the opposite direction. These forces are exerted only while the student and the box are in contact.

Although the same size force is acting on the student and the box, they will have different accelerations because their masses are different. The acceleration of each can be calculated using Newton's second law. The acceleration of the box is:

$$\text{acceleration} = \frac{\text{force}}{\text{mass}} = \frac{10\ \text{N}}{20\ \text{kg}} = 0.5\ \text{m/s}^2$$

The acceleration of the student can be calculated by replacing 20 kg with 50 kg in the above formula and is only 0.2 m/s^2. The student and the box accelerate only while they are in contact. As a result, the student moves more slowly than the box moves.

Gravity and the Third Law Gravity is pulling you down to the ground. According to the third law, you aren't just pulled toward Earth—Earth also is pulled toward you. When you jump into a swimming pool, how far does Earth move? The force you exert on Earth is the same as the force Earth exerts on you. However, Earth is trillions of times more massive than you are. Because Earth has such a large mass, the force you exert on it doesn't have a noticeable effect.

Newton's laws of motion apply to all objects, even the distant galaxies shown in **Figure 19.** The Sun exerts a gravitational force on Earth, so according to the third law of motion Earth exerts an equal force on the Sun. This force has a small effect on the motion of the Sun. Planets that might be orbiting stars other than the Sun are too far away to be seen from Earth. But they also affect the motion of the stars they orbit. Astronomers look for variations in the motions of stars that might be caused by an orbiting planet. More than 100 planets have been detected around stars other than the Sun using this method.

Figure 19 Newton's laws of motion apply to the billions of stars in these distant galaxies.

Combining the Laws

The laws of motion describe how any object moves when forces act on it. Consider what happens during a jump, as shown in **Figure 20.**

First, when you push on the ground, the ground pushes up on you with an equal and opposite force. Therefore, two forces are acting on you—gravity pulls you down, and the force from the ground pushes you up. The overall force is upward, so as the second law predicts, you accelerate upward as your foot pushes against the ground.

When your feet leave the ground, gravity is the only force acting on you. Again according to the second law, you accelerate in the direction of this unbalanced force. This downward acceleration slows you until you stop at the top of your jump and then causes you to increase your speed downward until you reach the ground.

When your feet hit the ground, the ground exerts an upward force on you. The force must be greater than the downward force of gravity to slow you down. When you stop moving, all of the forces on you are balanced. As the first law predicts, you remain at rest.

Figure 20 All of Newton's laws apply to a simple jump.
Identify *the forces that are acting in each part of this jump.*

section 4 review

Summary

The Third Law of Motion

- Forces always occur in equal and opposite pairs called action and reaction forces.
- Newton's third law of motion states that if object A exerts a force on object B, object B exerts an equal force in the opposite direction on object A.

Applying the Third Law

- Action and reaction forces are not balanced forces because they act on different objects.
- Even though the forces in an action-reaction force pair are equal in size, the motion of the objects they act on depends on the masses of the objects.
- The gravitational force you exert on Earth has no noticeable effect because Earth's mass is so large.

Self Check

1. **Identify** the action and reaction forces acting on this book when it rests on your hands.
2. **Explain** why, when you jump from a boat, the boat moves back as you move forward.
3. **Infer** You and a child with half your mass are standing on ice. If the child pushes you, who will have the larger acceleration?
4. **Compare and contrast** the first two laws of motion with the third law of motion.
5. **Think Critically** Identify the force that causes you to move forward when you walk on a floor.

Applying Math

6. **Calculate Acceleration** A skater standing on ice skates pushes on the side of a skating rink with a force of 100 N. If the skater's mass is 60 kg, what is the skater's resulting acceleration?

Design Your Own

Balanced and Unbalanced Forces

Goals

- **Describe** how to create balanced and unbalanced forces on an object.
- **Demonstrate** forces that change the speed and the direction of an object's motion.

Possible Materials

block
book
pieces of string (2)
spring scales (2)

Safety Precautions

Real-World Question

Newton's laws tell you that to change the velocity of an object, there must be an unbalanced force acting on the object. Changing the velocity can involve changing the speed of the object, changing the direction of motion, or changing both. How can you apply an unbalanced force to an object? How does the motion change when you exert a force in different ways?

Form a Hypothesis

Predict how the motion of a block will change when different forces are applied to it. Consider both speed and direction.

Test Your Hypothesis

Make a Plan

1. **Describe** how you are going to exert forces on the block using the available materials.
2. **List** several different ways to exert forces or combinations of forces on the block. Think about how strong each force will need to be to change the motion of the block. Include at least one force or combination of forces that you think will not change the object's motion.
3. **Predict** which forces will change the object's direction, its speed, both, or neither. Are the forces balanced or unbalanced?

Check Your Plan

1. Make sure that your teacher approves your plans before going any further.
2. **Compare** your plans for exerting forces with those of others in your class. Discuss why each of you chose the forces you chose.

Follow Your Plan

1. Set up your model so that you can exert each of the forces that you listed.
2. **Collect data** by exerting each of the forces in turn and recording how each one affects the object's motion.

Analyze Your Data

1. **Identify Variables** For each of the forces or combinations of forces that you applied to the object, list all of the forces acting on the object. Was the number of forces acting always the same? Was there a situation when only a single force was being applied? Explain.
2. **Record Observations** What happened when you exerted balanced forces on the object? Were the results for unbalanced forces the same for different combinations of forces? Why or why not?

Conclude and Apply

1. Were your predictions correct? Explain how you were able to predict the motion of the block and any mistaken predictions you might have made.
2. **Summarize** Which of Newton's laws of motion did you demonstrate in this lab?
3. **Apply** Suppose you see a pole that is supposed to be vertical, but is starting to tip over. What could you do to prevent the pole from falling over? Describe the forces acting on the pole as it starts to tip and after you do something. Are the forces balanced or unbalanced?

Compare your results with those of other students in your class. **Discuss** how different combinations of forces affect the motion of the objects.

TIME SCIENCE AND Society

SCIENCE ISSUES THAT AFFECT YOU!

Should there be a limit to the size and speed of roller coasters?

Bigger, HIGHER, Faster

If you've been to an amusement park lately, you know that roller coasters are taller and faster than ever. The thrill of their curves and corkscrews makes them incredibly popular. However, the increasingly daring designs have raised concerns about safety.

Dangerous Coasters

The 1990s saw a sharp rise in amusement park injuries, with 4,500 injuries (many on coasters) in 1998 alone. A new 30-story-high roller coaster will drop you downhill at speeds nearing 160 km/h. The excitement of such a high-velocity coaster is undeniable, but skeptics argue that, even with safety measures, accidents on supercoasters will be more frequent and more severe. "Technology and ride design are outstripping our understanding of the health effects of high forces on riders," said one lawmaker.

Safe As Can Be

Supporters of new rides say that injuries and deaths are rare when you consider the hundreds of millions of annual riders. They also note that most accidents or deaths result from breakdowns or foolish rider behavior, not bad design.

Designers emphasize that rides are governed by Newton's laws of motion. Factors such as the bank and tightness of a curve are carefully calculated according to these laws to safely balance the forces on riders.

The designs can't account for riders who don't follow instructions, however. The forces on a standing rider might be quite different from those on a seated rider that is strapped in properly, and might cause the standing rider to be ejected.

Supercoasters are here to stay, but with accidents increasing, designers and riders of roller coasters must consider both safety and thrills.

Research Choose an amusement park ride. Visit the link at the right to research how forces act on you while you're on the ride to give you thrills but still keep you safe. Write a report with diagrams showing how the forces work.

For more information, visit green.msscience.com/time

Reviewing Main Ideas

Section 1 Motion

1. Motion occurs when an object changes its position relative to a reference point.
2. Distance is the path length an object travels. Displacement is the distance and direction between start and end points.
3. Speed is the distance divided by the time. Velocity is the displacement divided by the time. Acceleration is the change in velocity divided by the time.

Section 2 Newton's First Law

1. A force is a push or a pull. Forces are balanced if they cancel.
2. The first law of motion states that the motion of an object does not change unless it is acted upon by an unbalanced force.

Section 3 Newton's Second Law

1. The second law of motion states acceleration is in the direction of the unbalanced force and equals the force divided by the mass.
2. Friction resists the sliding motion between two surfaces in contact.

Section 4 Newton's Third Law

1. The third law states that forces always act in equal but opposite pairs.
2. Action and reaction forces do not cancel because they act on different objects.

Visualizing Main Ideas

Copy and complete the following table on the laws of motion.

Laws of Motion

	First Law	Second Law	Third Law
Statement	An object will remain at rest or in motion at constant velocity until it is acted upon by an unbalanced force.		
Describes Relation Between What?		motion of object and force on object	

chapter 23 Review

Using Vocabulary

acceleration p. 687
balanced forces p. 692
displacement p. 684
first law of motion p. 691
force p. 690
friction p. 697
gravitational force p. 696
second law of motion p. 694
speed p. 685
third law of motion p. 702
unbalanced forces p. 692
velocity p. 687

For each set of vocabulary words below, explain the relationship that exists.

1. displacement and velocity
2. force and third law of motion
3. speed and velocity
4. force and friction
5. unbalanced forces and second law of motion
6. balanced forces and first law of motion
7. acceleration and second law of motion
8. acceleration and force

Checking Concepts

Choose the word or phrase that best answers the question.

9. Which does NOT change when an unbalanced force acts on an object?
   A) displacement **C)** mass
   B) velocity **D)** motion
10. What is the force that keeps you from sliding off a sled when it starts moving?
    A) sliding friction **C)** air resistance
    B) static friction **D)** rolling friction
11. Which of the following indicates that the forces on an object are balanced?
    A) The object speeds up.
    B) The object slows down.
    C) The object moves at a constant velocity.
    D) The object turns.
12. Which of the following includes direction?
    A) mass **C)** velocity
    B) speed **D)** distance
13. The inertia of an object increases when which of the following increases?
    A) speed **C)** mass
    B) force **D)** acceleration
14. The unbalanced force on a football is 5 N downward. Which of the following best describes its acceleration?
    A) Its acceleration is upward.
    B) Its acceleration is downward.
    C) Its acceleration is zero.
    D) Its acceleration depends on its motion.
15. Which of the following exerts the force that pushes you forward when you walk?
    A) static friction **C)** gravity
    B) sliding friction **D)** air resistance
16. If the action force on an object is 3 N to the left, what is the reaction force?
    A) 6 N left **C)** 3 N right
    B) 3 N left **D)** 6 N right
17. Which of the following forces is not a long-range force?
    A) gravitational force
    B) friction
    C) electric force
    D) magnetic force
18. The gravitational force between two objects depends on which of the following?
    A) their masses and their velocities
    B) their masses and their weights
    C) their masses and their inertia
    D) their masses and their separation
19. Which of the following depends on the force of gravity on an object?
    A) inertia **C)** mass
    B) weight **D)** friction

Science Online green.msscience.com/vocabulary_puzzlemaker

Thinking Critically

20. Apply You are skiing down a hill at constant speed. What do Newton's first two laws say about your motion?

21. Explain how an object can be moving if there is no unbalanced force acting on it.

22. Compare and Contrast When you come to school, what distance do you travel? Is the distance you travel greater than your displacement from home?

23. Draw Conclusions If the gravitational force between Earth and an object always causes the same acceleration, why does a feather fall more slowly than a hammer does?

24. Recognize Cause and Effect Why does a baseball thrown from the outfield to home plate follow a curved path rather than a straight line?

25. Concept Map Copy and complete the concept maps below on Newton's three laws of motion.

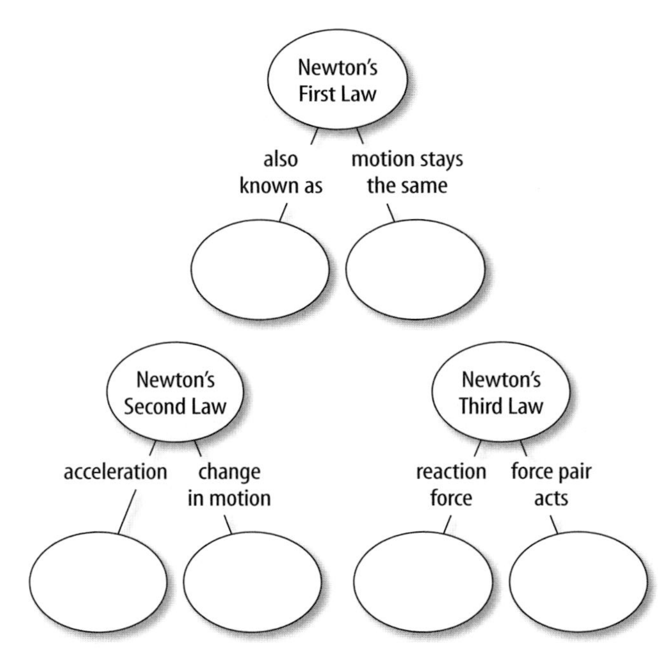

Performance Activities

26. Make a poster identifying all the forces acting in each of the following cases. When is there an unbalanced force? How do Newton's laws apply?

- **a.** You push against a box; it doesn't move.
- **b.** You push harder, and the box starts to move.
- **c.** You push the box across the floor at a constant speed.
- **d.** You don't touch the box. It sits on the floor.

Applying Math

27. Acceleration of a Ball A bowling ball has a mass of 6.4 kg. If a bowler exerts an unbalanced force of 16.0 N on the ball, what is its acceleration?

28. Distance Walked A person walks at a speed of 1.5 m/s. How far does the person walk in 0.5 h?

29. Force on a Ball A tennis ball has a mass of 57 g and has an acceleration of 300 m/s^2 when it hits the tennis racket. What is the force exerted on the ball by the tennis racket?

30. Acceleration of a Car A car has a speed of 20.0 m/s. If the speed of the car increases to 30.0 m/s in 5.0 s, what is the car's acceleration?

Use the table below to answer question 31.

Speed of Sound in Different Materials

Material	Speed (m/s)
Air	335
Water	1,554
Wood	3,828
Iron	5,103

31. The Speed of Sound Sound travels a distance of 1.00 m through a material in 0.003 s. According to the table above, what is the material?

Part 1 Multiple Choice

Record your answers on the answer sheet provided by your teacher or on a sheet of paper.

1. What is the difference in the force of gravity on an object with a mass of 23.5 kg and on an object with a mass of 14.7 kg?
   A. 86.2 N **C.** 1.1 N
   B. 8.8 N **D.** no difference

Use the illustration below to answer questions 2 and 3.

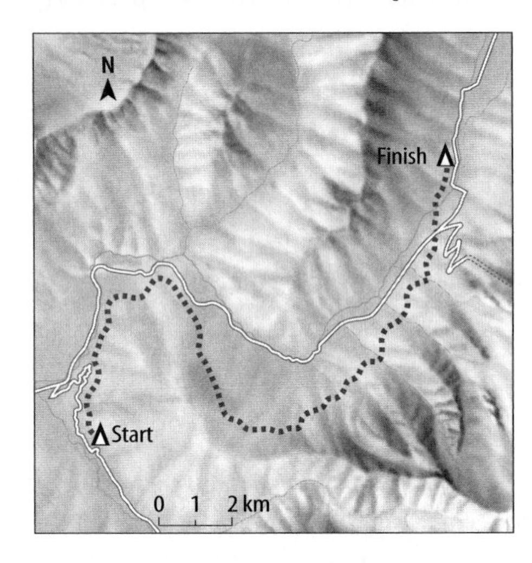

2. The illustration above shows a mountain hiking path. What is the approximate displacement of a hiker who travels from Start to Finish?
   A. 22 km southwest
   B. 22 km northeast
   C. 12 km southwest
   D. 12 km northeast

3. About how long would it take a hiker to travel along the path from Start to Finish if the hiker's average speed were 2.6 km/h?
   A. 2.3 h **C.** 4.6 h
   B. 4.3 h **D.** 8.5 h

4. What force makes it difficult to start a box moving across the floor?
   A. inertia **C.** rolling friction
   B. static friction **D.** sliding friction

5. A person pushes against a 8.3-kg box with a force of 12 N. The frictional force on the box is 2.5 N. What is the acceleration of the box?
   A. 0.87 m/s^2 **C.** 1.4 m/s^2
   B. 1.1 m/s^2 **D.** 1.7 m/s^2

6. What is the acceleration of a 4-kg ball tossed vertically upward when the ball is at its highest point?
   A. 39.2 m/s^2 **C.** 9.8 m/s^2
   B. 19.6 m/s^2 **D.** no acceleration

Use the table below to answer questions 7 and 8.

Distance (m)	Time (m)
0	0
4	2
8	4
14	6
20	8
26	10
32	12

7. The table above shows the distances traveled by an object every 2 s for 12 s. What is the average speed of the object over the time interval 6 s to 12 s?
   A. 2.0 m/s **C.** 3.0 m/s
   B. 2.7 m/s **D.** 5.3 m/s

8. Over which time interval did the object accelerate?
   A. 2 to 6 s **C.** 6 to 10 s
   B. 4 to 8 s **D.** 8 to 12 s

9. A train moves along a straight section of track at a velocity of 95 km/h north. A person walks toward the back of the train with a velocity of 3 km/h south. What is the velocity of the person relative to the ground?
   A. 92 km north **C.** 98 km north
   B. 92 km south **D.** 98 km south

Part 2 Short Response/Grid In

Record your answers on the answer sheet provided by your teacher or on a sheet of paper.

10. Earth exerts a gravitational force on an apple falling from a tree. Use Newton's third law of motion of describe the gravitational pull that the apple exerts on Earth.

11. Do astronauts in orbit around Earth experience the same gravitational force that they do when they are on the ground? Explain why or why not.

Use the photograph below to answer questions 12 and 13.

12. If the runners in the photograph above are moving at a constant speed, are they accelerating? Explain.

13. What force prevents the runners from sliding as they go around the curve?

14. What is the acceleration of a 12-kg object if the net force on it is 32 N?

15. According to Newton's third law of motion, for every action there is an equal but opposite reaction. Explain why the action and reaction forces don't cancel and result in no movement of an object.

Part 3 Open Ended

Record your answers on a sheet of paper.

16. Use Newton's laws to describe the forces that cause a ball thrown against the ground to bounce.

17. Three objects, all with the same mass, fall from the same height. One object is narrow and pointed, another is spherical, and the third is shaped like a box. Compare the speeds of the objects as they fall.

18. A 70-kg adult and a 35-kg child stand on ice facing each other. They push against each other with a force of 9.5 N. Assuming there is no friction, describe and compare the acceleration they each experience.

Use the photograph below to answer questions 19 and 20.

19. Name and describe the forces that enable the boy in the photograph to ride his bicycle.

20. Explain how Newton's first law of motion influences the movement of the bike.

Test-Taking Tip

Answer all Parts of the Question Make sure each part of the question is answered when listing discussion points. For example, if the question asks you to compare and contrast, make sure you list both similarities and differences.

Question 18 Be sure to describe and compare the acceleration of each person.

chapter 24

Energy and Energy Resources

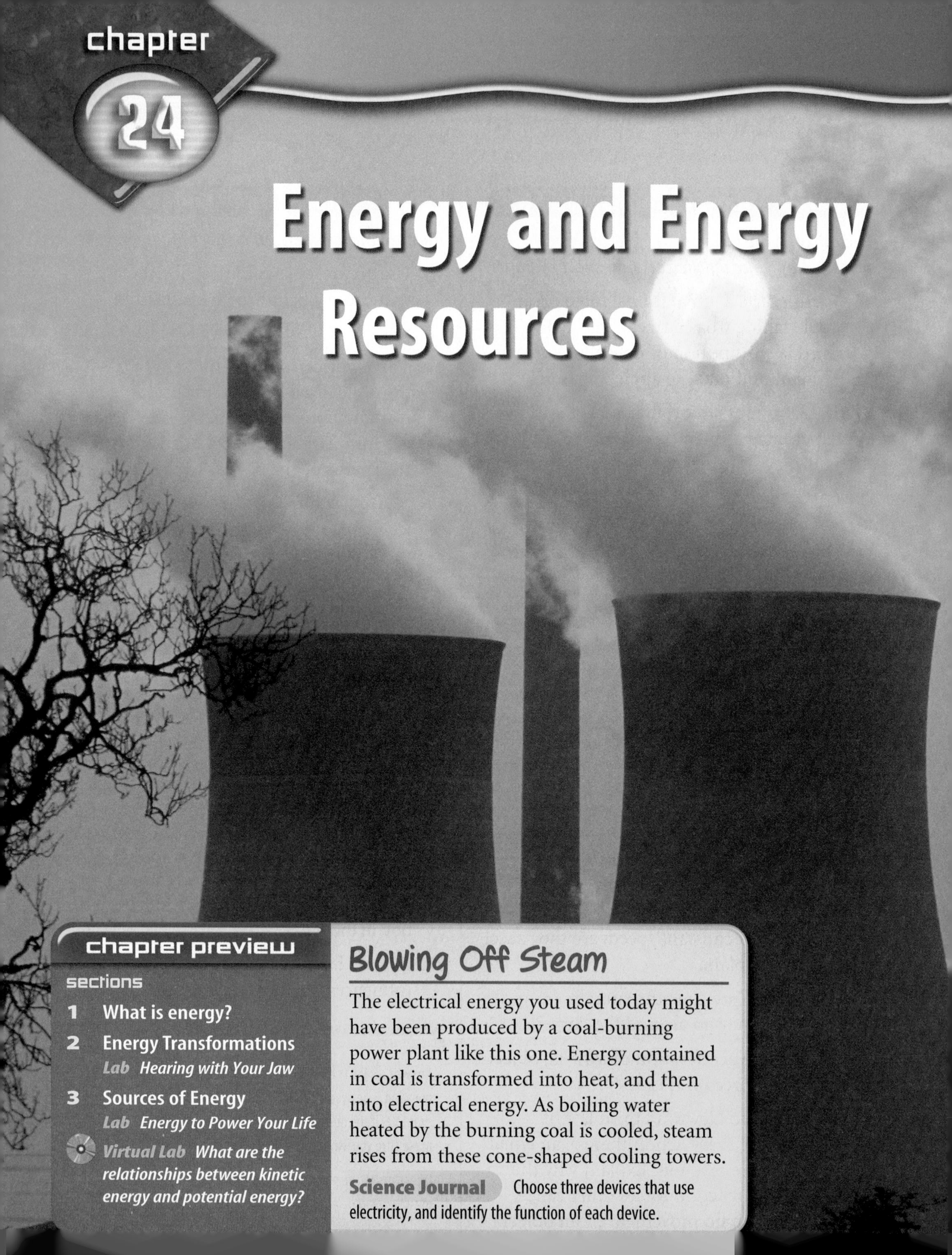

chapter preview

sections

1. What is energy?
2. Energy Transformations
   Lab *Hearing with Your Jaw*
3. Sources of Energy
   Lab *Energy to Power Your Life*

Virtual Lab *What are the relationships between kinetic energy and potential energy?*

Blowing Off Steam

The electrical energy you used today might have been produced by a coal-burning power plant like this one. Energy contained in coal is transformed into heat, and then into electrical energy. As boiling water heated by the burning coal is cooled, steam rises from these cone-shaped cooling towers.

Science Journal Choose three devices that use electricity, and identify the function of each device.

Start-Up Activities

Marbles and Energy

What's the difference between a moving marble and one at rest? A moving marble can hit something and cause a change to occur. How can a marble acquire energy—the ability to cause change?

1. Make a track on a table by slightly separating two metersticks placed side by side.
2. Using a book, raise one end of the track slightly and measure the height.
3. Roll a marble down the track. Measure the distance from its starting point to where it hits the floor. Repeat. Calculate the average of the two measurements.
4. Repeat steps 2 and 3 for three different heights. Predict what will happen if you use a heavier marble. Test your prediction and record your observations.
5. **Think Critically** In your Science Journal, describe how the distance traveled by the marble is related to the height of the ramp. How is the motion of the marble related to the ramp height?

Energy Make the following Foldable to help identify what you already know, what you want to know, and what you learned about energy.

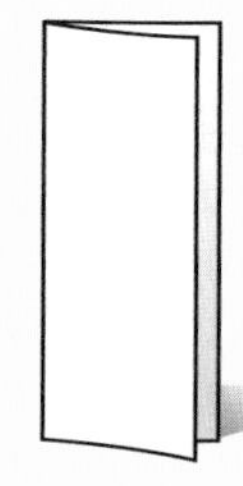

STEP 1 **Fold** a vertical sheet of paper from side to side. Make the front edge about 1 cm shorter than the back edge.

STEP 2 **Turn** lengthwise and **fold** into thirds

STEP 3 **Unfold, cut, and label** each tab for only the top layer along both folds to make three tabs.

Identify Questions Before you read the chapter, write what you know and what you want to know about the types, sources, and transformation of energy under the appropriate tabs. As you read the chapter, correct what you have written and add more questions under the *Learned* tab.

Preview this chapter's content and activities at green.msscience.com

section 1

What is energy?

as you read

What You'll Learn

- **Explain** what energy is.
- **Distinguish** between kinetic energy and potential energy.
- **Identify** the various forms of energy.

Why It's Important

Energy is involved whenever a change occurs.

Review Vocabulary
mass: a measure of the amount of matter in an object

New Vocabulary
- energy
- kinetic energy
- potential energy
- thermal energy
- chemical energy
- radiant energy
- electrical energy
- nuclear energy

The Nature of Energy

What comes to mind when you hear the word *energy?* Do you picture running, leaping, and spinning like a dancer or a gymnast? How would you define energy? When an object has energy, it can make things happen. In other words, **energy** is the ability to cause change. What do the items shown in **Figure 1** have in common?

Look around and notice the changes that are occurring—someone walking by or a ray of sunshine that is streaming through the window and warming your desk. Maybe you can see the wind moving the leaves on a tree. What changes are occurring?

Transferring Energy You might not realize it, but you have a large amount of energy. In fact, everything around you has energy, but you notice it only when a change takes place. Anytime a change occurs, energy is transferred from one object to another. You hear a footstep because energy is transferred from a foot hitting the ground to your ears. Leaves are put into motion when energy in the moving wind is transferred to them. The spot on the desktop becomes warmer when energy is transferred to it from the sunlight. In fact, all objects, including leaves and desktops, have energy.

Figure 1 Energy is the ability to cause change.
Explain *how these objects cause change.*

Energy of Motion

Things that move can cause change. A bowling ball rolls down the alley and knocks down some pins, as in **Figure 2A.** Is energy involved? A change occurs when the pins fall over. The bowling ball causes this change, so the bowling ball has energy. The energy in the motion of the bowling ball causes the pins to fall. As the ball moves, it has a form of energy called kinetic energy. **Kinetic energy** is the energy an object has due to its motion. If an object isn't moving, it doesn't have kinetic energy.

Figure 2 The kinetic energy of an object depends on the mass and speed of the object.

A This ball has kinetic energy because it is rolling down the alley.

B This ball has more kinetic energy because it has more speed.

C This ball has less kinetic energy because it has less mass.

Kinetic Energy and Speed

If you roll the bowling ball so it moves faster, what happens when it hits the pins? It might knock down more pins, or it might cause the pins to go flying farther. A faster ball causes more change to occur than a ball that is moving slowly. Look at **Figure 2B.** The professional bowler rolls a fast-moving bowling ball. When her ball hits the pins, pins go flying faster and farther than for a slower-moving ball. All that action signals that her ball has more energy. The faster the ball goes, the more kinetic energy it has. This is true for all moving objects. Kinetic energy increases as an object moves faster.

Reading Check *How does kinetic energy depend on speed?*

Kinetic Energy and Mass

Suppose, as shown in **Figure 2C,** you roll a volleyball down the alley instead of a bowling ball. If the volleyball travels at the same speed as a bowling ball, do you think it will send pins flying as far? The answer is no. The volleyball might not knock down any pins. Does the volleyball have less energy than the bowling ball even though they are traveling at the same speed?

An important difference between the volleyball and the bowling ball is that the volleyball has less mass. Even though the volleyball is moving at the same speed as the bowling ball, the volleyball has less kinetic energy because it has less mass. Kinetic energy also depends on the mass of a moving object. Kinetic energy increases as the mass of the object increases.

Figure 3 The potential energy of an object depends on its mass and height above the ground.
Determine *which vase has more potential energy, the red one or the blue one.*

Energy of Position

An object can have energy even though it is not moving. For example, a glass of water sitting on the kitchen table doesn't have any kinetic energy because it isn't moving. If you accidentally nudge the glass and it falls on the floor, changes occur. Gravity pulls the glass downward, and the glass has energy of motion as it falls. Where did this energy come from?

When the glass was sitting on the table, it had potential (puh TEN chul) energy. **Potential energy** is the energy stored in an object because of its position. In this case, the position is the height of the glass above the floor. The potential energy of the glass changes to kinetic energy as the glass falls. The potential energy of the glass is greater if it is higher above the floor. Potential energy also depends on mass. The more mass an object has, the more potential energy it has. Which object in **Figure 3** has the most potential energy?

Forms of Energy

Food, sunlight, and wind have energy, yet they seem different because they contain different forms of energy. Food and sunlight contain forms of energy different from the kinetic energy in the motion of the wind. The warmth you feel from sunlight is another type of energy that is different from the energy of motion or position.

Thermal Energy The feeling of warmth from sunlight signals that your body is acquiring more thermal energy. All objects have **thermal energy** that increases as its temperature increases. A cup of hot chocolate has more thermal energy than a cup of cold water, as shown in **Figure 4.** Similarly, the cup of water has more thermal energy than a block of ice of the same mass. Your body continually produces thermal energy. Many chemical reactions that take place inside your cells produce thermal energy. Where does this energy come from? Thermal energy released by chemical reactions comes from another form of energy called chemical energy.

Figure 4 The hotter an object is, the more thermal energy it has. A cup of hot chocolate has more thermal energy than a cup of cold water, which has more thermal energy than a block of ice with the same mass.

Figure 5 Complex chemicals in food store chemical energy. During activity, the chemical energy transforms into kinetic energy.

Chemical Energy When you eat a meal, you are putting a source of energy inside your body. Food contains chemical energy that your body uses to provide energy for your brain, to power your movements, and to fuel your growth. As in **Figure 5,** food contains chemicals, such as sugar, which can be broken down in your body. These chemicals are made of atoms that are bonded together, and energy is stored in the bonds between atoms. **Chemical energy** is the energy stored in chemical bonds. When chemicals are broken apart and new chemicals are formed, some of this energy is released. The flame of a candle is the result of chemical energy stored in the wax. When the wax burns, chemical energy is transformed into thermal energy and light energy.

Reading Check *When is chemical energy released?*

Light Energy Light from the candle flame travels through the air at an incredibly fast speed of 300,000 km/s. This is fast enough to circle Earth almost eight times in 1 s. When light strikes something, it can be absorbed, transmitted, or reflected. When the light is absorbed by an object, the object can become warmer. The object absorbs energy from the light and this energy is transformed into thermal energy. Then energy carried by light is called **radiant energy. Figure 6** shows a coil of wire that produces radiant energy when it is heated. To heat the metal, another type of energy can be used—electrical energy.

Figure 6 Electrical energy is transformed into thermal energy in the metal heating coil. As the metal becomes hotter, it emits more radiant energy.

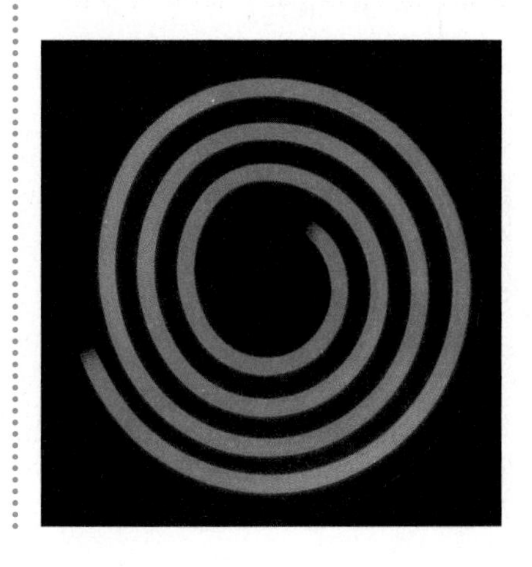

Electrical Energy Electrical lighting is one of the many ways electrical energy is used. Look around at all the devices that use electricity. Electric current flows in these devices when they are connected to batteries or plugged into an electric outlet. **Electrical energy** is the energy that is carried by an electric current. An electric device uses the electrical energy provided by the current flowing in the device. Large electric power plants generate the enormous amounts of electrical energy used each day. About 20 percent of the electrical energy used in the United States is generated by nuclear power plants.

Figure 7 Complex power plants are required to obtain useful energy from the nucleus of an atom.

Nuclear Energy Nuclear power plants use the energy stored in the nucleus of an atom to generate electricity. Every atomic nucleus contains energy—**nuclear energy**—that can be transformed into other forms of energy. However, releasing the nuclear energy is a difficult process. It involves the construction of complex power plants, shown in **Figure 7.** In contrast, all that is needed to release chemical energy from wood is a lighted match.

section 1 review

Summary

The Nature of Energy

- Energy is the ability to cause change.
- Kinetic energy is the energy an object has due to its motion. Kinetic energy depends on an object's speed and mass.
- Potential energy is the energy an object has due to its position. Potential energy depends on an object's height and mass.

Forces of Energy

- Thermal energy increases as temperature increases.
- Chemical energy is the energy stored in chemical bonds in molecules.
- Light energy, also called radiant energy, is the energy contained in light.
- Electrical energy is the energy carried by electric current.
- Nuclear energy is the energy contained in the nucleus of an atom.

Self Check

1. **Explain** why a high-speed collision between two cars would cause more damage than a low-speed collision between the same two cars.
2. **Describe** the energy transformations that occur when a piece of wood is burned.
3. **Identify** the form of energy that is converted into thermal energy by your body.
4. **Explain** how, if two vases are side by side on a shelf, one could have more potential energy.
5. **Think Critically** A golf ball and a bowling ball are moving and both have the same kinetic energy. Which one is moving faster? If they move at the same speed, which one has more kinetic energy?

Applying Skills

6. **Communicate** In your Science Journal, record different ways the word *energy* is used. Which ways of using the word *energy* are closest to the definition of energy given in this section?

green.msscience.com/self_check_quiz

section 2

Energy Transformations

as you read

What You'll Learn

- **Apply** the law of conservation of energy to energy transformations.
- **Identify** how energy changes form.
- **Describe** how electric power plants produce energy.

Why It's Important

Changing energy from one form to another is what makes cars run, furnaces heat, telephones work, and plants grow.

Review Vocabulary
transformation: a change in composition or structure

New Vocabulary
- law of conservation of energy
- generator
- turbine

Changing Forms of Energy

Chemical, thermal, radiant, and electrical are some of the forms that energy can have. In the world around you, energy is transforming continually between one form and another. You observe some of these transformations by noticing a change in your environment. Forest fires are a dramatic example of an environmental change that can occur naturally as a result of lightning strikes. A number of changes occur that involve energy as the mountain biker in **Figure 8** pedals up a hill. What energy transformations cause these changes to occur?

Tracking Energy Transformations As the mountain biker pedals, his leg muscles transform chemical energy into kinetic energy. The kinetic energy of his leg muscles transforms into kinetic energy of the bicycle as he pedals. Some of this energy transforms into potential energy as he moves up the hill. Also, some energy is transformed into thermal energy. His body is warmer because chemical energy is being released. Because of friction, the mechanical parts of the bicycle are warmer, too. Energy in the form of heat is almost always one of the products of an energy transformation. The energy transformations that occur when people exercise, when cars run, when living things grow and even when stars explode, all produce heat.

Figure 8 The ability to transform energy allows the biker to climb the hill.
Identify *all the forms of energy that are represented in the photograph.*

The Law of Conservation of Energy

It can be a challenge to track energy as it moves from object to object. However, one extremely important principle can serve as a guide as you trace the flow of energy. According to the **law of conservation of energy,** energy is never created or destroyed. The only thing that changes is the form in which energy appears. When the biker is resting at the summit, all his original energy is still around. Some of the energy is in the form of potential energy, which he will use as he coasts down the hill. Some of this energy was changed to thermal energy by friction in the bike. Chemical energy was also changed to thermal energy in the biker's muscles, making him feel hot. As he rests, this thermal energy moves from his body to the air around him. No energy is missing—it can all be accounted for.

Topic: Energy Transformations
Visit green.msscience.com for Web links to information about energy transformations that occur during different activities and processes.

Activity Choose an activity or process and make a graph showing how the kinetic and potential energy change during it.

Reading Check *Can energy ever be lost? Why or why not?*

Changing Kinetic and Potential Energy

The law of conservation of energy can be used to identify the energy changes in a system. For example, tossing a ball into the air and catching it is a simple system. As shown in **Figure 9,** as the ball leaves your hand, most of its energy is kinetic. As the ball rises, it slows and its kinetic energy decreases. But, the total energy of the ball hasn't changed. The decrease in kinetic energy equals the increase in potential energy as the ball flies higher in the air. The total amount of energy remains constant. Energy moves from place to place and changes form, but it never is created or destroyed.

Figure 9 During the flight of the baseball, energy is transforming between kinetic and potential energy.
Determine *where the ball has the most kinetic energy. Where does the ball have the most total energy?*

Figure 10 Hybrid cars that use an electric motor and a gasoline engine for power are now available. Hybrid cars make energy transformations more efficient.

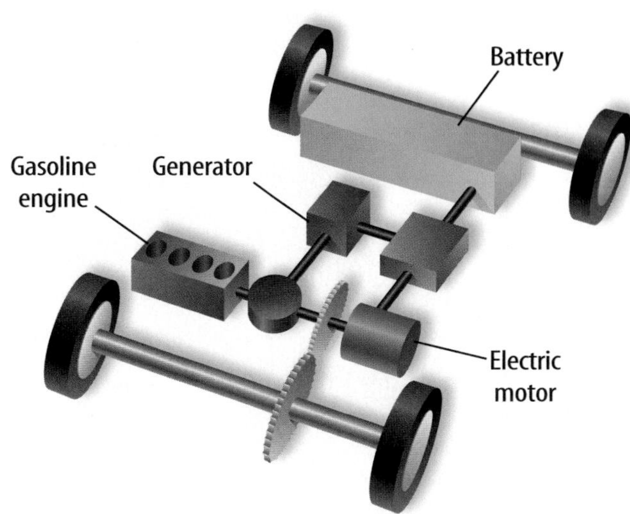

Energy Changes Form

Energy transformations occur constantly all around you. Many machines are devices that transform energy from one form to another. For example, an automobile engine transforms the chemical energy in gasoline into energy of motion. However, not all of the chemical energy is converted into kinetic energy. Instead, some of the chemical energy is converted into thermal energy, and the engine becomes hot. An engine that converts chemical energy into more kinetic energy is a more efficient engine. New types of cars, like the one shown in **Figure 10,** use an electric motor along with a gasoline engine. These engines are more efficient so the car can travel farther on a gallon of gas.

INTEGRATE Life Science

Transforming Chemical Energy

Inside your body, chemical energy also is transformed into kinetic energy. Look at **Figure 11.** The transformation of chemical to kinetic energy occurs in muscle cells. There, chemical reactions take place that cause certain molecules to change shape. Your muscle contracts when many of these changes occur, and a part of your body moves.

The matter contained in living organisms, also called biomass, contains chemical energy. When organisms die, chemical compounds in their biomass break down. Bacteria, fungi, and other organisms help convert these chemical compounds to simpler chemicals that can be used by other living things.

Thermal energy also is released as these changes occur. For example, a compost pile can contain plant matter, such as grass clippings and leaves. As the compost pile decomposes, chemical energy is converted into thermal energy. This can cause the temperature of a compost pile to reach 60°C.

Analyzing Energy Transformations

Procedure

1. Place soft **clay** on the floor and smooth out its surface.
2. Hold a **marble** 1.5 m above the clay and drop it. Measure the depth of the crater made by the marble.
3. Repeat this procedure using a **golf ball** and a **plastic golf ball.**

Analysis

1. Compare the depths of the craters to determine which ball had the most kinetic energy as it hit the clay.
2. Explain how potential energy was transformed into kinetic energy during your activity.

NATIONAL GEOGRAPHIC **VISUALIZING ENERGY TRANSFORMATIONS**

Figure 11

Paddling a raft, throwing a baseball, playing the violin — your skeletal muscles make these and countless other body movements possible. Muscles work by pulling, or contracting. At the cellular level, muscle contractions are powered by reactions that transform chemical energy into kinetic energy.

▶ Energy transformations taking place in your muscles provide the power to move.

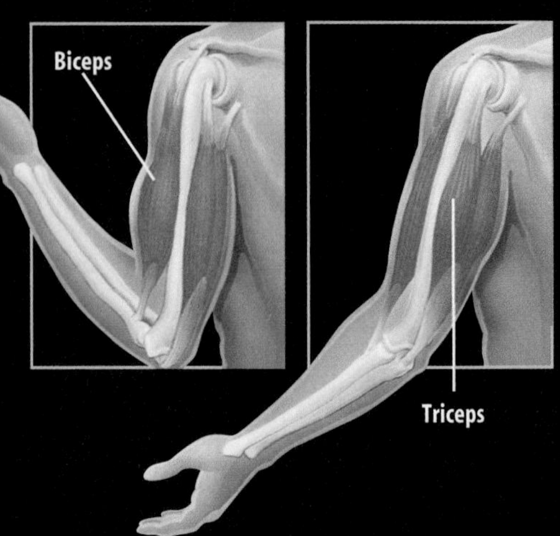

▲ Many skeletal muscles are arranged in pairs that work in opposition to each other. When you bend your arm, the biceps muscle contracts, while the triceps relaxes. When you extend your arm the triceps contracts, and the biceps relaxes.

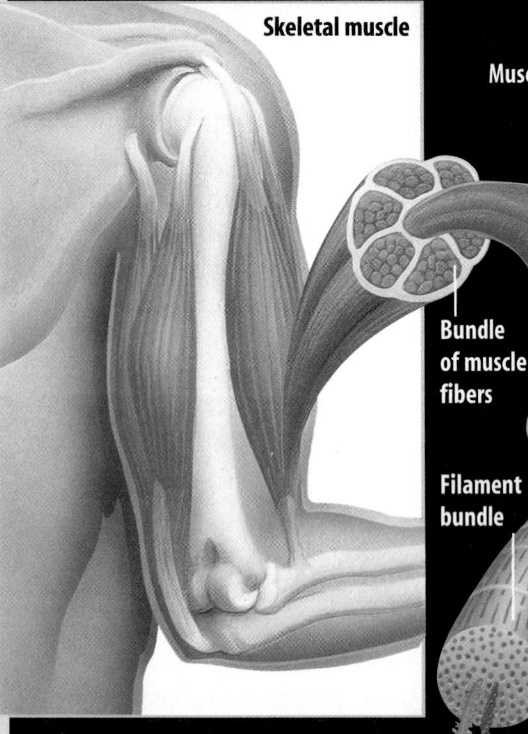

▲ Skeletal muscles are made up of bundles of muscle cells, or fibers. Each fiber is composed of many bundles of muscle filaments.

▲ A signal from a nerve fiber starts a chemical reaction in the muscle filament. This causes molecules in the muscle filament to gain energy and move. Many filaments moving together cause the muscle to contract.

Figure 12 The simple act of listening to a radio involves many energy transformations. A few are diagrammed here.

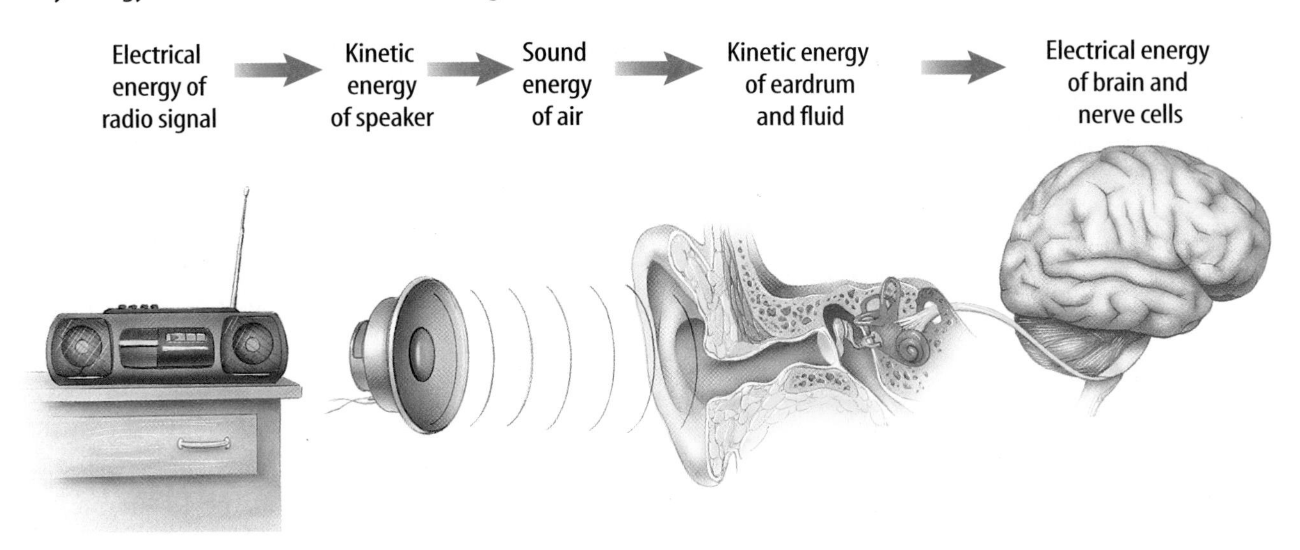

Transforming Electrical Energy Every day you use electrical energy. When you flip a light switch, or turn on a radio or television, or use a hair drier, you are transforming electrical energy to other forms of energy. Every time you plug something into a wall outlet, or use a battery, you are using electrical energy. **Figure 12** shows how electrical energy is transformed into other forms of energy when you listen to a radio. A loudspeaker in the radio converts electrical energy into sound waves that travel to your ear—energy in motion. The energy that is carried by the sound waves causes parts of the ear to move also. This energy of motion is transformed again into chemical and electrical energy in nerve cells, which send the energy to your brain. After your brain interprets this energy as a voice or music, where does the energy go? The energy finally is transformed into thermal energy.

Transforming Thermal Energy Different forms of energy can be transformed into thermal energy. For example, chemical energy changes into thermal energy when something burns. Electrical energy changes into thermal energy when a wire that is carrying an electric current gets hot. Thermal energy can be used to heat buildings and keep you warm. Thermal energy also can be used to heat water. If water is heated to its boiling point, it changes to steam. This steam can be used to produce kinetic energy by steam engines, like the steam locomotives that used to pull trains. Thermal energy also can be transformed into radiant energy. For example, when a bar of metal is heated to a high temperature, it glows and gives off light.

Controlling Body Temperature Most organisms have some adaptation for controlling the amount of thermal energy in their bodies. Some living in cooler climates have thick fur coats that help prevent thermal energy from escaping, and some living in desert regions have skin that helps keep thermal energy out. Research some of the adaptations different organisms have for controlling the thermal energy in their bodies.

How Thermal Energy Moves Thermal energy can move from one place to another. Look at **Figure 13.** The hot chocolate has thermal energy that moves from the cup to the cooler air around it, and to the cooler spoon. Thermal energy only moves from something at a higher temperature to something at a lower temperature.

Generating Electrical Energy

The enormous amount of electrical energy that is used every day is too large to be stored in batteries. The electrical energy that is available for use at any wall socket must be generated continually by power plants. Every power plant works on the same principle—energy is used to turn a large generator. A **generator** is a device that transforms kinetic energy into electrical energy. In fossil fuel power plants, coal, oil, or natural gas is burned to boil water. As the hot water boils, the steam rushes through a **turbine,** which contains a set of narrowly spaced fan blades. The steam pushes on the blades and turns the turbine, which in turn rotates a shaft in the generator to produce the electrical energy, as shown in **Figure 14.**

Reading Check *What does a generator do?*

Figure 13 Thermal energy moves from the hot chocolate to the cooler surroundings. **Explain** *what happens to the hot chocolate as it loses thermal energy.*

Figure 14 A coal-burning power plant transforms the chemical energy in coal into electrical energy. **List** *some of the other energy sources that power plants use.*

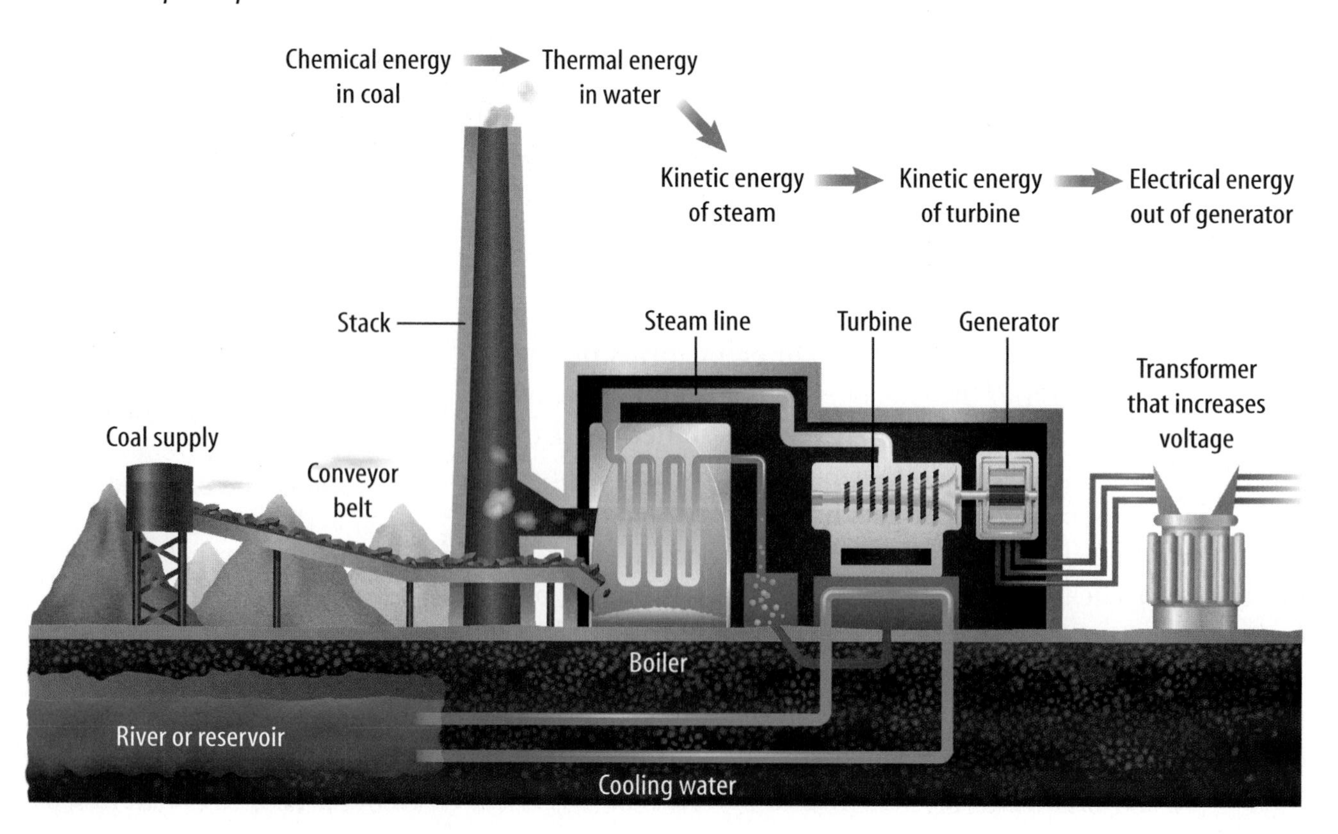

Power Plants Almost 90 percent of the electrical energy generated in the United States is produced by nuclear and fossil fuel power plants, as shown in **Figure 15.** Other types of power plants include hydroelectric (hi droh ih LEK trihk) and wind. Hydroelectric power plants transform the kinetic energy of moving water into electrical energy. Wind power plants transform the kinetic energy of moving air into electrical energy. In these power plants, a generator converts the kinetic energy of moving water or wind to electrical energy.

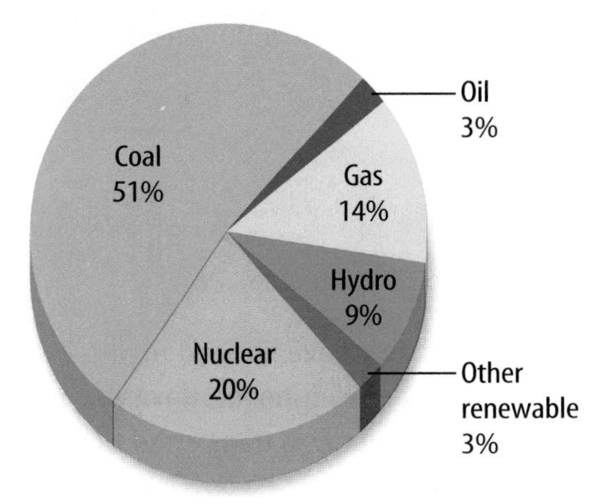

Figure 15 The graph shows sources of electrical energy in the United States.
Name *the energy source that you think is being used to provide the electricity for the lights overhead.*

To analyze the energy transformations in a power plant, you can diagram the energy changes using arrows. A coal-burning power plant generates electrical energy through the following series of energy transformations.

chemical energy of coal → thermal energy of water → kinetic energy of steam → kinetic energy of turbine → electrical energy out of generator

Nuclear power plants use a similar series of transformations. Hydroelectric plants, however, skip the steps that change water into steam because the water strikes the turbine directly.

section 2 review

Summary

Changing Forms of Energy

- Heat usually is one of the forms of energy produced in energy transformations.
- The law of conservation of energy states that energy cannot be created or destroyed; it can only change form.
- The total energy doesn't change when an energy transformation occurs.
- As an object rises and falls, kinetic and potential energy are transformed into each other, but the total energy doesn't change.

Generating Electrical Energy

- A generator converts kinetic energy into electrical energy.
- Burning fossil fuels produces thermal energy that is used to boil water and produce steam.
- In a power plant, steam is used to spin a turbine which then spins an electric generator.

Self Check

1. **Describe** the conversions between potential and kinetic energy that occur when you shoot a basketball at a basket.
2. **Explain** whether your body gains or loses thermal energy if your body temperature is 37°C and the temperature around you is 25°C.
3. **Describe** a process that converts chemical energy to thermal energy.
4. **Think Critically** A lightbulb converts 10 percent of the electrical energy it uses into radiant energy. Make a hypothesis about the other form of energy produced.

Applying Math

5. **Use a Ratio** How many times greater is the amount of electrical energy produced in the United States by coal-burning power plants than the amount produced by nuclear power plants?

Hearing with Your Jaw

You probably have listened to music using speakers or headphones. Have you ever considered how energy is transferred to get the energy from the radio or CD player to your brain? What type of energy is needed to power the radio or CD player? Where does this energy come from? How does that energy become sound? How does the sound get to you? In this activity, the sound from a radio or CD player is going to travel through a motor before entering your body through your jaw instead of your ears.

Real-World Question

How can energy be transferred from a radio or CD player to your brain?

Goals

- **Identify** energy transfers and transformations.
- **Explain** your observations using the law of conservation of energy.

Materials

radio or CD player
small electrical motor
headphone jack

Procedure

1. Go to one of the places in the room with a motor/radio assembly.
2. Turn on the radio or CD player so that you hear the music.
3. Push the headphone jack into the headphone plug on the radio or CD player.
4. Press the axle of the motor against the side of your jaw.

Conclude and Apply

1. **Describe** what you heard in your Science Journal.
2. **Identify** the form of energy produced by the radio or CD player.
3. **Draw** a diagram to show all of the energy transformations taking place.
4. **Evaluate** Did anything get hotter as a result of this activity? Explain.
5. **Explain** your observations using the law of conservation of energy.

Compare your conclusions with those of other students in your class. **For more help, refer to the** Science Skill Handbook.

section 3

Sources of Energy

Using Energy

Every day, energy is used to provide light and to heat and cool homes, schools, and workplaces. According to the law of conservation of energy, energy can't be created or destroyed. Energy only can change form. If a car or refrigerator can't create the energy they use, then where does this energy come from?

Energy Resources

Energy cannot be made, but must come from the natural world. As you can see in **Figure 16,** the surface of Earth receives energy from two sources—the Sun and radioactive atoms in Earth's interior. The amount of energy Earth receives from the Sun is far greater than the amount generated in Earth's interior. Nearly all the energy you used today can be traced to the Sun, even the gasoline used to power the car or school bus you came to school in.

as you read

What You'll Learn

- **Explain** what renewable, nonrenewable, and alternative resources are.
- **Describe** the advantages and disadvantages of using various energy sources.

Why It's Important

Energy is vital for survival and making life comfortable. Developing new energy sources will improve modern standards of living.

Review Vocabulary

resource: a natural feature or phenomenon that enhances the quality of life

New Vocabulary

- nonrenewable resource
- renewable resource
- alternative resource
- inexhaustible resource
- photovoltaic

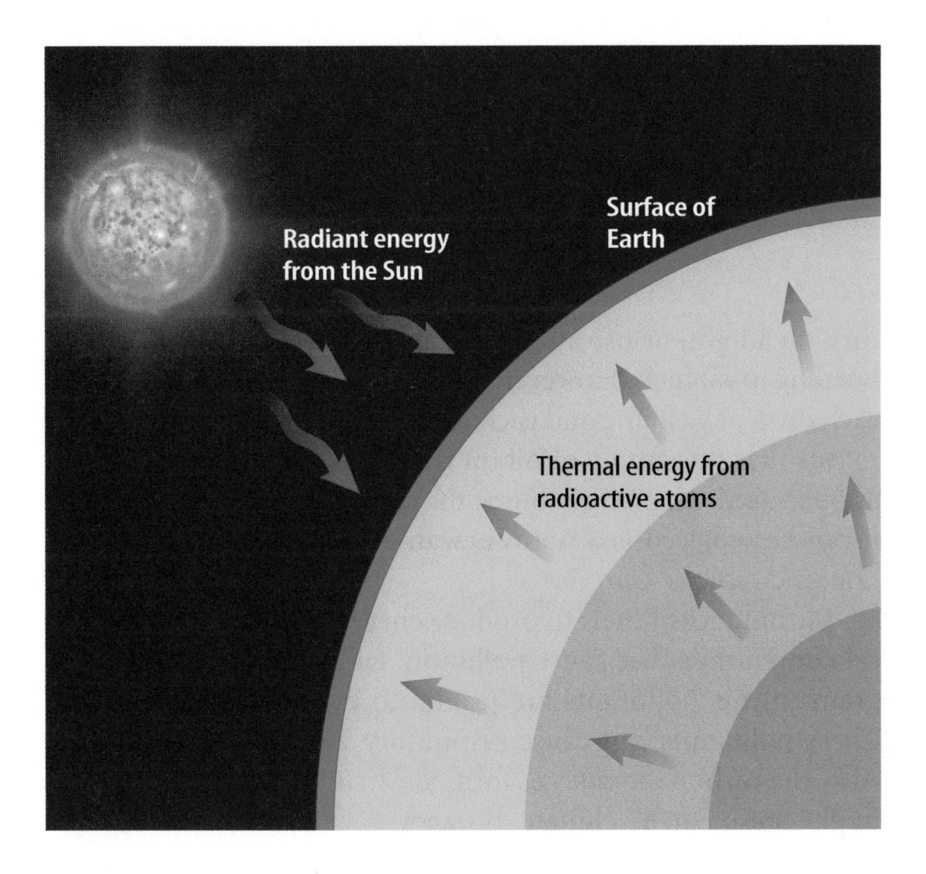

Figure 16 All the energy you use can be traced to one of two sources—the Sun or radioactive atoms in Earth's interior.

Figure 17 Coal is formed after the molecules in ancient plants are heated under pressure for millions of years. The energy stored by the molecules in coal originally came from the Sun.

Fossil Fuels

Fossil fuels are coal, oil, and natural gas. Oil and natural gas were made from the remains of microscopic organisms that lived in Earth's oceans millions of years ago. Heat and pressure gradually turned these ancient organisms into oil and natural gas. Coal was formed by a similar process from the remains of ancient plants that once lived on land, as shown in **Figure 17.**

Through the process of photosynthesis, ancient plants converted the radiant energy in sunlight to chemical energy stored in various types of molecules. Heat and pressure changed these molecules into other types of molecules as fossil fuels formed. Chemical energy stored in these molecules is released when fossil fuels are burned.

Using Fossil Fuels The energy used when you ride in a car, turn on a light, or use an electric appliance usually comes from burning fossil fuels. However, it takes millions of years to replace each drop of gasoline and each lump of coal that is burned. This means that the supply of oil on Earth will continue to decrease as oil is used. An energy source that is used up much faster than it can be replaced is a **nonrenewable resource.** Fossil fuels are nonrenewable resources.

Burning fossil fuels to produce energy also generates chemical compounds that cause pollution. Each year billions of kilograms of air pollutants are produced by burning fossil fuels. These pollutants can cause respiratory illnesses and acid rain. Also, the carbon dioxide gas formed when fossil fuels are burned might cause Earth's climate to warm.

Energy Source Origins
The kinds of fossil fuels found in the ground depend on the kinds of organisms (animal or plant) that died and were buried in that spot. Research coal, oil, and natural gas to find out what types of organisms were primarily responsible for producing each.

Nuclear Energy

Can you imagine running an automobile on 1 kg of fuel that releases almost 3 million times more energy than 1 L of gas? What could supply so much energy from so little mass? The answer is the nuclei of uranium atoms. Some of these nuclei are unstable and break apart, releasing enormous amounts of energy in the process. This energy can be used to generate electricity by heating water to produce steam that spins an electric generator, as shown in **Figure 18.** Because no fossil fuels are burned, generating electricity using nuclear energy helps make the supply of fossil fuels last longer. Also, unlike fossil fuel power plants, nuclear power plants produce almost no air pollution. In one year, a typical nuclear power plant generates enough energy to supply 600,000 homes with power and produces only 1 m^3 of waste.

Nuclear Wastes Like all energy sources, nuclear energy has its advantages and disadvantages. One disadvantage is the amount of uranium in Earth's crust is nonrenewable. Another is that the waste produced by nuclear power plants is radioactive and can be dangerous to living things. Some of the materials in the nuclear waste will remain radioactive for many thousands of years. As a result the waste must be stored so no radioactivity is released into the environment for a long time. One method is to seal the waste in a ceramic material, place the ceramic in protective containers, and then bury the containers far underground. However, the burial site would have to be chosen carefully so underground water supplies aren't contaminated. Also, the site would have to be safe from earthquakes and other natural disasters that might cause radioactive material to be released.

Figure 18 To obtain electrical energy from nuclear energy, a series of energy transformations must occur.

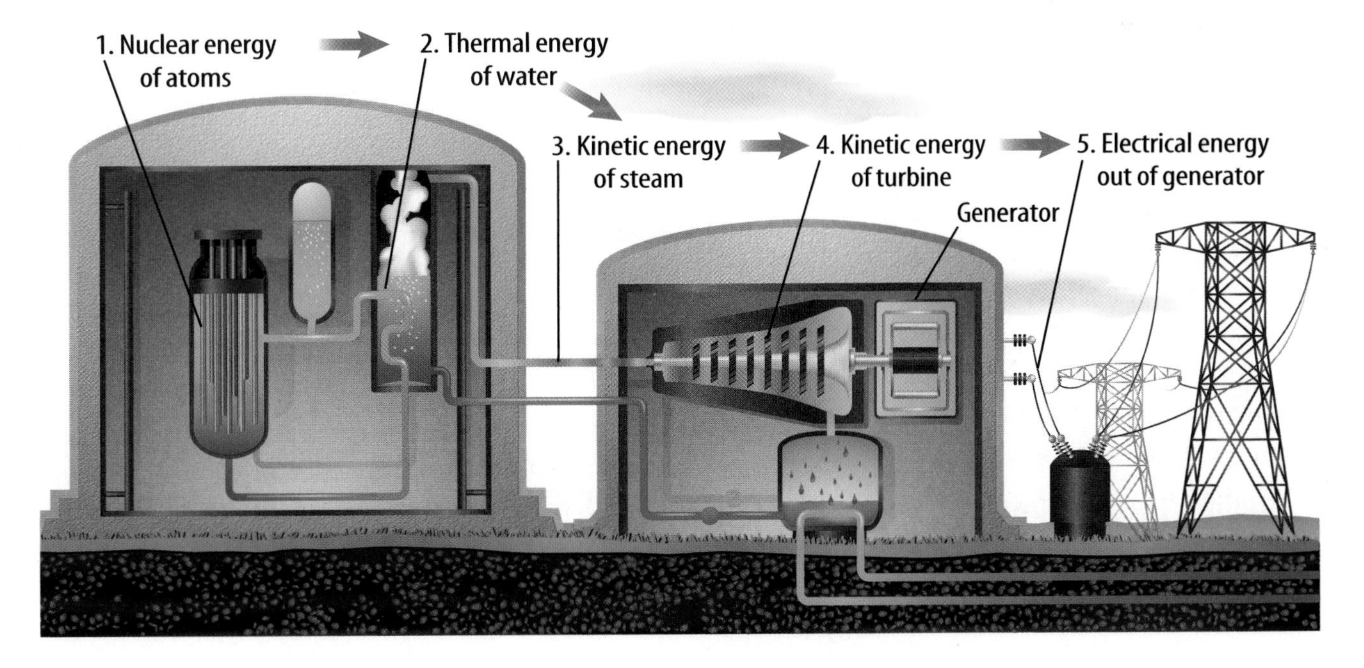

Hydroelectricity

Currently, transforming the potential energy of water that is trapped behind dams supplies the world with almost 20 percent of its electrical energy. Hydroelectricity is the largest renewable source of energy. A **renewable resource** is an energy source that is replenished continually. As long as enough rain and snow fall to keep rivers flowing, hydroelectric power plants can generate electrical energy, as shown in **Figure 19.**

Although production of hydroelectricity is largely pollution free, it has one major problem. It disrupts the life cycle of aquatic animals, especially fish. This is particularly true in the Northwest where salmon spawn and run. Because salmon return to the spot where they were hatched to lay their eggs, the development of dams has hindered a large fraction of salmon from reproducing. This has greatly reduced the salmon population. Efforts to correct the problem have resulted in plans to remove a number of dams. In an attempt to help fish bypass some dams, fish ladders are being installed. Like most energy sources, hydroelectricity has advantages and disadvantages.

Topic: Hydroelectricity
Visit green.msscience.com for Web links to information about the use of hydroelectricity in various parts of the world.

Activity On a map of the world, show where the use of hydroelectricity is the greatest.

Applying Science

Is energy consumption outpacing production?

You use energy every day—to get to school, to watch TV, and to heat or cool your home. The amount of energy consumed by an average person has increased over time. Consequently, more energy must be produced.

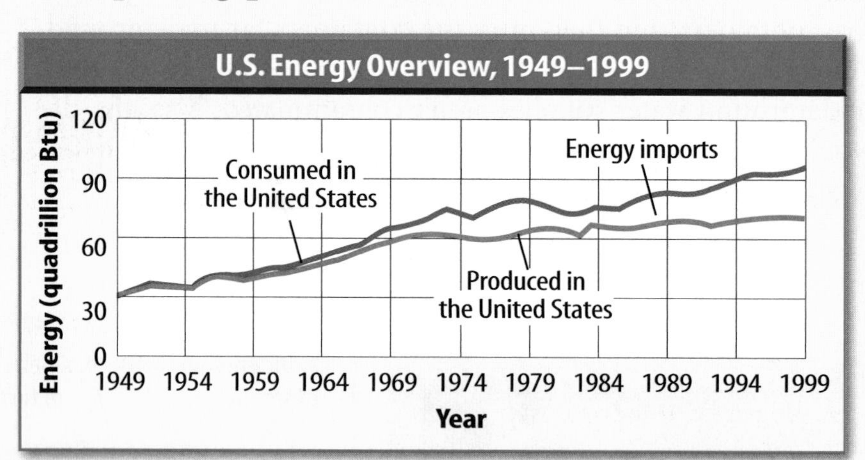

Identifying the Problem

The graph above shows the energy produced and consumed in the United States from 1949 to 1999. How does energy that is consumed by Americans compare with energy that is produced in the United States?

Solving the Problem

1. Determine the approximate amount of energy produced in 1949 and in 1999 and how much it has increased in 50 years. Has it doubled or tripled?
2. Do the same for consumption. Has it doubled or tripled?
3. Using your answers for steps 1 and 2 and the graph, where does the additional energy that is needed come from? Give some examples.

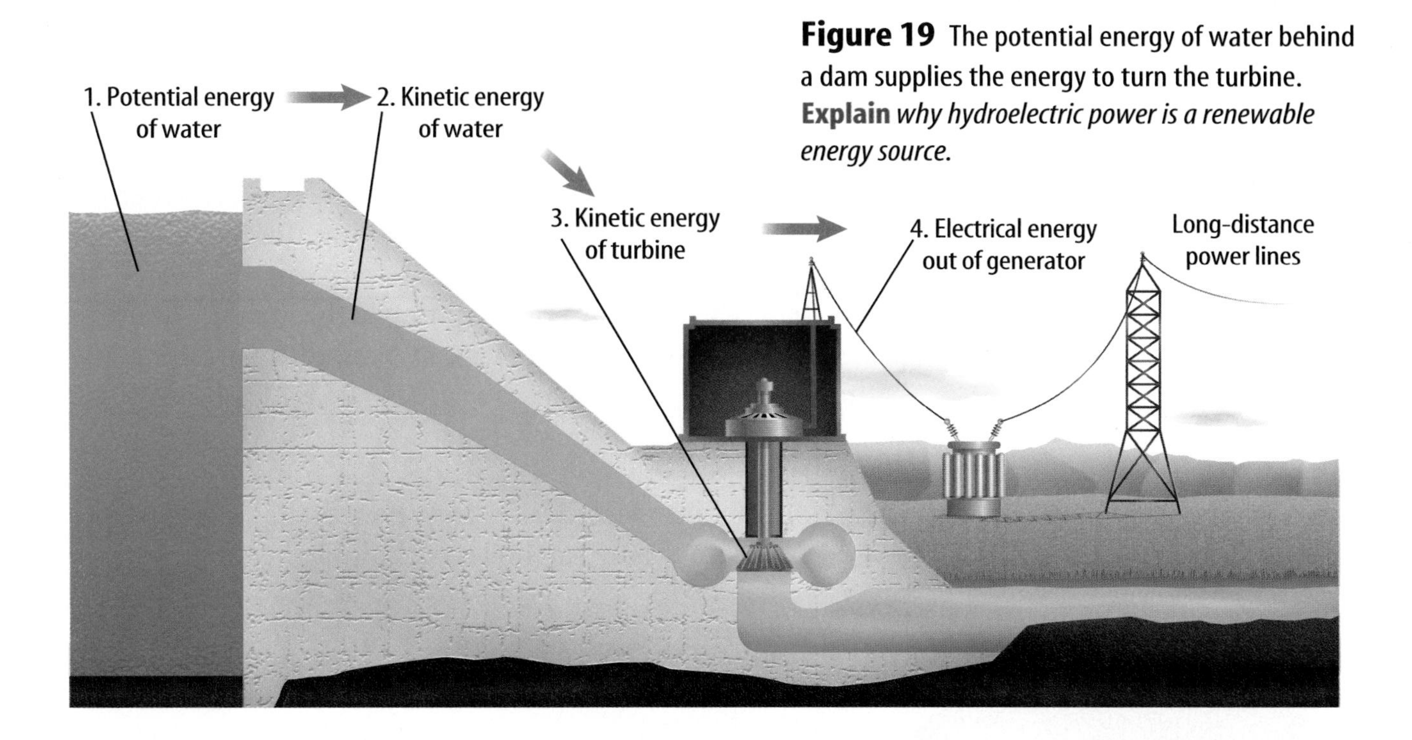

Figure 19 The potential energy of water behind a dam supplies the energy to turn the turbine. **Explain** *why hydroelectric power is a renewable energy source.*

Alternative Sources of Energy

Electrical energy can be generated in several ways. However, each has disadvantages that can affect the environment and the quality of life for humans. Research is being done to develop new sources of energy that are safer and cause less harm to the environment. These sources often are called **alternative resources.** These alternative resources include solar energy, wind, and geothermal energy.

Solar Energy

The Sun is the origin of almost all the energy that is used on Earth. Because the Sun will go on producing an enormous amount of energy for billions of years, the Sun is an inexhaustible source of energy. An **inexhaustible resource** is an energy source that can't be used up by humans.

Each day, on average, the amount of solar energy that strikes the United States is more than the total amount of energy used by the entire country in a year. However, less than 0.1 percent of the energy used in the United States comes directly from the Sun. One reason is that solar energy is more expensive to use than fossil fuels. However, as the supply of fossil fuels decreases, the cost of finding and mining these fuels might increase. Then, it may be cheaper to use solar energy or other energy sources to generate electricity and heat buildings than to use fossil fuels.

What is an inexhaustible energy source?

Building a Solar Collector

Procedure

1. Line a **large pot** with **black plastic** and fill with **water.**
2. Stretch **clear-plastic wrap** over the pot and tape it taut.
3. Make a slit in the top and slide a **thermometer** or a **computer probe** into the water.
4. Place your solar collector in direct sunlight and monitor the temperature change every 3 min for 15 min.
5. Repeat your experiment without using any black plastic.

Analysis

1. Graph the temperature changes in both setups.
2. Explain how your solar collector works.

Collecting the Sun's Energy Two types of collectors capture the Sun's rays. If you look around your neighborhood, you might see large, rectangular panels attached to the roofs of buildings or houses. If, as in **Figure 20,** pipes come out of the panel, it is a thermal collector. Using a black surface, a thermal collector heats water by directly absorbing the Sun's radiant energy. Water circulating in this system can be heated to about 70°C. The hot water can be pumped through the house to provide heat. Also, the hot water can be used for washing and bathing. If the panel has no pipes, it is a photovoltaic (foh toh vol TAY ihk) collector, like the one pictured in **Figure 20.** A **photovoltaic** is a device that transforms radiant energy directly into electrical energy. Photovoltaics are used to power calculators and satellites, including the *International Space Station.*

Figure 20 Solar energy can be collected and utilized by individuals using thermal collectors or photovoltaic collectors.

Reading Check *What does a photovoltaic do?*

Geothermal Energy

Imagine you could take a journey to the center of Earth—down to about 6,400 km below the surface. As you went deeper and deeper, you would find the temperature increasing. In fact, after going only about 3 km, the temperature could have increased enough to boil water. At a depth of 100 km, the temperature could be over 900°C. The heat generated inside Earth is called geothermal energy. Some of this heat is produced when unstable radioactive atoms inside Earth decay, converting nuclear energy to thermal energy.

At some places deep within Earth the temperature is hot enough to melt rock. This molten rock, or magma, can rise up close to the surface through cracks in the crust. During a volcanic eruption, magma reaches the surface. In other places, magma gets close to the surface and heats the rock around it.

Geothermal Reservoirs In some regions where magma is close to the surface, rainwater and water from melted snow can seep down to the hot rock through cracks and other openings in Earth's surface. The water then becomes hot and sometimes can form steam. The hot water and steam can be trapped under high pressure in cracks and pockets called geothermal reservoirs. In some places, the hot water and steam are close enough to the surface to form hot springs and geysers.

Geothermal Power Plants In places where the geothermal reservoirs are less than several kilometers deep, wells can be drilled to reach them. The hot water and steam produced by geothermal energy then can be used by geothermal power plants, like the one in **Figure 21,** to generate electricity.

Most geothermal reservoirs contain hot water under high pressure. **Figure 22** shows how these reservoirs can be used to generate electricity. While geothermal power is an inexhaustible source of energy, geothermal power plants can be built only in regions where geothermal reservoirs are close to the surface, such as in the western United States.

Figure 21 This geothermal power plant in Nevada produces enough electricity to power about 50,000 homes.

Heat Pumps Geothermal heat helps keep the temperature of the ground at a depth of several meters at a nearly constant temperature of about 10° to 20°C. This constant temperature can be used to cool and heat buildings by using a heat pump.

A heat pump contains a water-filled loop of pipe that is buried to a depth where the temperature is nearly constant. In summer the air is warmer than this underground temperature. Warm water from the building is pumped through the pipe down into the ground. The water cools and then is pumped back to the house where it absorbs more heat, and the cycle is repeated. During the winter, the air is cooler than the ground below. Then, cool water absorbs heat from the ground and releases it into the house.

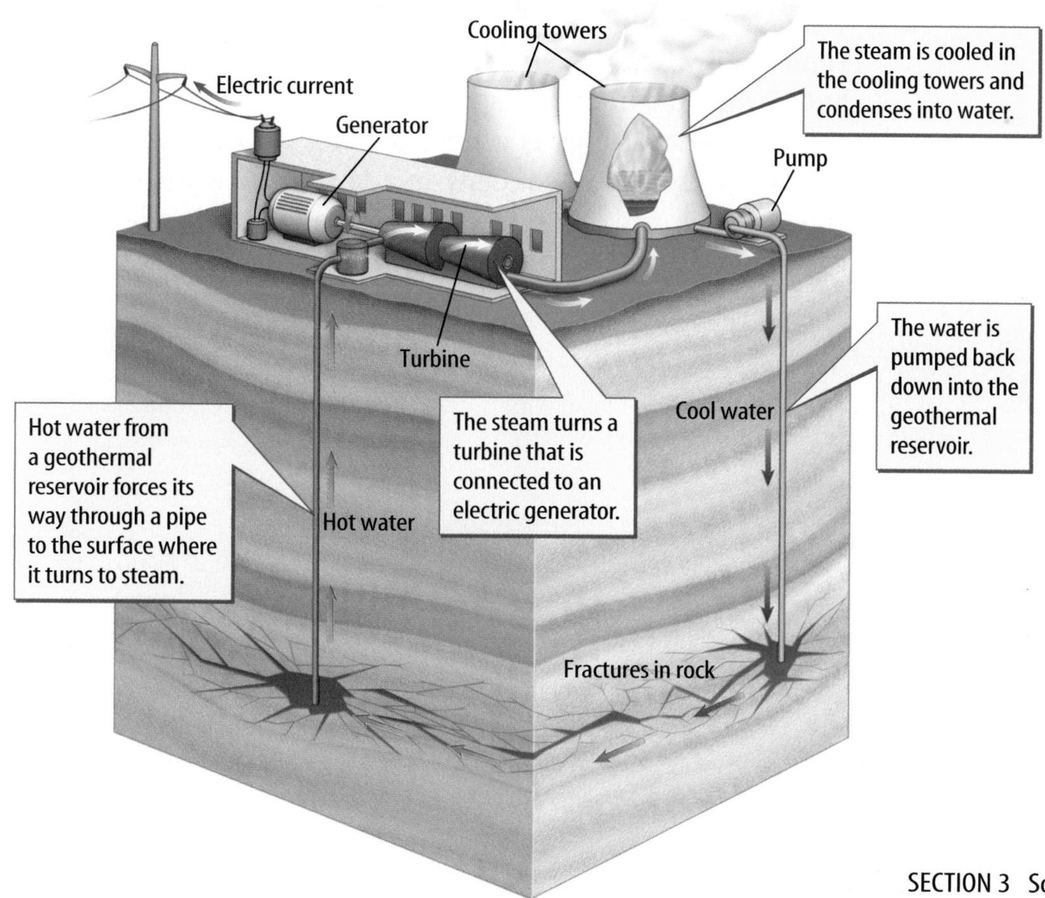

Figure 22 The hot water in a geothermal reservoir is used to generate electricity in a geothermal power plant.

Energy from the Oceans

The ocean is in constant motion. If you've been to the seashore you've seen waves roll in. You may have seen the level of the ocean rise and fall over a period of about a half day. This rise and fall in the ocean level is called a tide. The constant movement of the ocean is an inexhaustible source of mechanical energy that can be converted into electric energy. While methods are still being developed to convert the motion in ocean waves to electric energy, several electric power plants using tidal motion have been built.

Figure 23 This tidal power plant in Annapolis Royal, Nova Scotia, is the only operating tidal power plant in North America.

Using Tidal Energy A high tide and a low tide each occur about twice a day. In most places the level of the ocean changes by less than a few meters. However, in some places the change is much greater. In the Bay of Fundy in Eastern Canada, the ocean level changes by 16 m between high tide and low tide. Almost 14 trillion kg of water move into or out of the bay between high and low tide.

Figure 23 shows an electric power plant that has been built along the Bay of Fundy. This power plant generates enough electric energy to power about 12,000 homes. The power plant is constructed so that as the tide rises, water flows through a turbine that causes an electric generator to spin, as shown in **Figure 24A.** The water is then trapped behind a dam. When the tide goes out, the trapped water behind the dam is released through the turbine to generate more electricity, as shown in **Figure 24B.** Each day electric power is generated for about ten hours when the tide is rising and falling.

While tidal energy is a nonpolluting, inexhaustible energy source, its use is limited. Only in a few places is the difference between high and low tide large enough to enable a large electric power plant to be built.

Figure 24 A tidal power plant can generate electricity when the tide is coming in and going out.

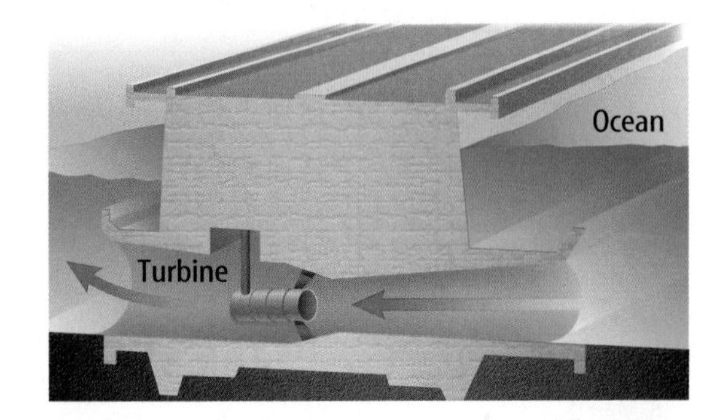

A As the tide comes in, it turns a turbine connected to a generator. When high tide occurs, gates are closed that trap water behind a dam.

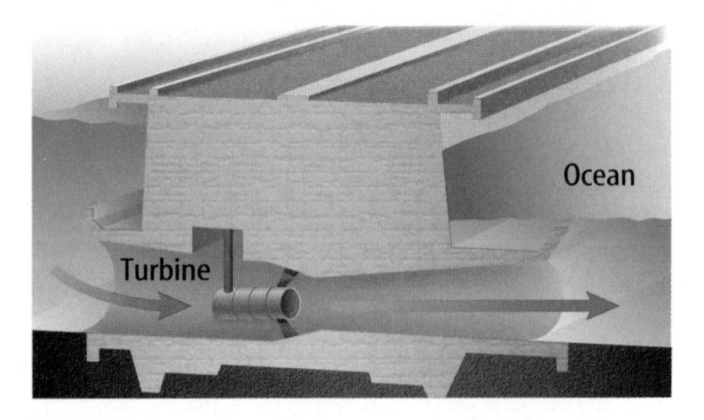

B As the tide goes out and the ocean level drops, the gates are opened and water from behind the dam flows through the turbine, causing it to spin and turn a generator.

Wind

Wind is another inexhaustible supply of energy. Modern windmills, like the ones in **Figure 25,** convert the kinetic energy of the wind to electrical energy. The propeller is connected to a generator so that electrical energy is generated when wind spins the propeller. These windmills produce almost no pollution. Some disadvantages are that windmills produce noise and that large areas of land are needed. Also, studies have shown that birds sometimes are killed by windmills.

Figure 25 Windmills work on the same basic principles as a power plant. Instead of steam turning a turbine, wind turns the rotors. **Describe** *some of the advantages and disadvantages of using windmills.*

Conserving Energy

Fossil fuels are a valuable resource. Not only are they burned to provide energy, but oil and coal also are used to make plastics and other materials. One way to make the supply of fossil fuels last longer is to use less energy. Reducing the use of energy is called conserving energy.

You can conserve energy and also save money by turning off lights and appliances such as televisions when you are not using them. Also keep doors and windows closed tightly when it's cold or hot to keep heat from leaking out of or into your house. Energy could also be conserved if buildings are properly insulated, especially around windows. The use of oil could be reduced if cars were used less and made more efficient, so they went farther on a liter of gas. Recycling materials such as aluminum cans and glass also helps conserve energy.

section 3 review

Summary

Nonrenewable Resources

- All energy resources have advantages and disadvantages.
- Nonrenewable energy resources are used faster than they are replaced.
- Fossil fuels include oil, coal, and natural gas and are nonrenewable resources. Nuclear energy is a nonrenewable resource.

Renewable and Alternative Resources

- Renewable energy resources, such as hydroelectricity, are resources that are replenished continually.
- Alternative energy sources include solar energy, wind energy, and geothermal energy.

Self Check

1. **Diagram** the energy conversions that occur when coal is formed, and then burned to produce thermal energy.
2. **Explain** why solar energy is considered an inexhaustible source of energy.
3. **Explain** how a heat pump is used to both heat and cool a building.
4. **Think Critically** Identify advantages and disadvantages of using fossil fuels, hydroelectricity, and solar energy as energy sources.

Applying Math

5. **Use a Ratio** Earth's temperature increases with depth. Suppose the temperature increase inside Earth is 500°C at a depth of 50 km. What is the temperature increase at a depth of 10 km?

Use the Internet

Energy to Power Your Life

Goals

- **Identify** how energy you use is produced and delivered.
- **Investigate** alternative sources for the energy you use.
- **Outline** a plan for how these alternative sources of energy could be used.

Data Source

Science Online

Visit green.msscience.com/internet_lab for more information about sources of energy and for data collected by other students.

Real-World Question

Over the past 100 years, the amount of energy used in the United States and elsewhere has greatly increased. Today, a number of energy sources are available, such as coal, oil, natural gas, nuclear energy, hydroelectric power, wind, and solar energy. Some of these energy sources are being used up and are nonrenewable, but others are replaced as fast as they are used and, therefore, are renewable. Some energy sources are so vast that human usage has almost no effect on the amount available. These energy sources are inexhaustible.

Think about the types of energy you use at home and school every day. In this lab, you will investigate how and where energy is produced, and how it gets to you. You will also investigate alternative ways energy can be produced, and whether these sources are renewable, nonrenewable, or inexhaustible. What are the sources of the energy you use every day?

Local Energy Information	
Energy Type	
Where is that energy produced?	
How is that energy produced?	Do not write in this book.
How is that energy delivered to you?	
Is the energy source renewable, nonrenewable, or inexhaustible?	
What type of alternative energy source could you use instead?	

Make a Plan

1. Think about the activities you do every day and the things you use. When you watch television, listen to the radio, ride in a car, use a hair drier, or turn on the air conditioning, you use energy. Select one activity or appliance that uses energy.
2. **Identify** the type of energy that is used.
3. **Investigate** how that energy is produced and delivered to you.
4. **Determine** if the energy source is renewable, nonrenewable, or inexhaustible.
5. If your energy source is nonrenewable, describe how the energy you use could be produced by renewable sources.

Follow Your Plan

1. Make sure your teacher approves your plan before you start.
2. Organize your findings in a data table, similar to the one that is shown.

Analyze Your Data

1. **Describe** the process for producing and delivering the energy source you researched. How is it created, and how does it get to you?
2. How much energy is produced by the energy source you investigated?
3. Is the energy source you researched renewable, nonrenewable, or inexhaustible? Why?

Conclude and Apply

1. **Describe** If the energy source you investigated is nonrenewable, how can the use of this energy source be reduced?
2. **Organize** What alternative sources of energy could you use for everyday energy needs? On the computer, create a plan for using renewable or inexhaustible sources.

Communicating Your Data

Find this lab using the link below. Post your data in the table that is provided. **Compare** your data to those of other students. **Combine** your data with those of other students and make inferences using the combined data.

green.msscience.com/internet_lab

SCIENCE Stats

Energy to Burn

Did you know...

... The energy released by the average hurricane is equal to about 200 times the total energy produced by all of the world's power plants. Almost all of this energy is released as heat when raindrops form.

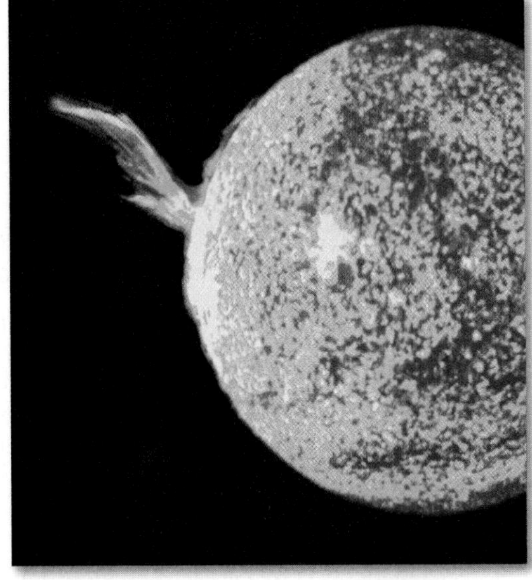

... The energy Earth gets each half hour from the Sun is enough to meet the world's demands for a year. Renewable and inexhaustible resources, including the Sun, account for only 18 percent of the energy that is used worldwide.

... The Calories in one medium apple will give you enough energy to walk for about 15 min, swim for about 10 min, or jog for about 9 min.

Applying Math If walking for 15 min requires 80 Calories of fuel (from food), how many Calories would someone need to consume to walk for 1 h?

Write About It

Where would you place solar collectors in the United States? Why? For more information on solar energy, go to green.msscience.com/science_stats.

Reviewing Main Ideas

Section 1 What is energy?

1. Energy is the ability to cause change.
2. A moving object has kinetic energy that depends on the object's mass and speed.
3. Potential energy is energy due to position and depends on an object's mass and height.
4. Light carries radiant energy, electric current carries electrical energy, and atomic nuclei contain nuclear energy.

Section 2 Energy Transformations

1. Energy can be transformed from one form to another. Thermal energy is usually produced when energy transformations occur.
2. The law of conservation of energy states that energy cannot be created or destroyed.
3. Electric power plants convert a source of energy into electrical energy. Steam spins a turbine which spins an electric generator.

Section 3 Sources of Energy

1. The use of an energy source has advantages and disadvantages.
2. Fossil fuels and nuclear energy are nonrenewable energy sources that are consumed faster than they can be replaced.
3. Hydroelectricity is a renewable energy source that is continually being replaced.
4. Alternative energy sources include solar, wind, and geothermal energy. Solar energy is an inexhaustible energy source.

Visualizing Main Ideas

Copy and complete the concept map using the following terms: fossil fuels, hydroelectric, solar, wind, oil, coal, photovoltaic, *and* nonrenewable resources.

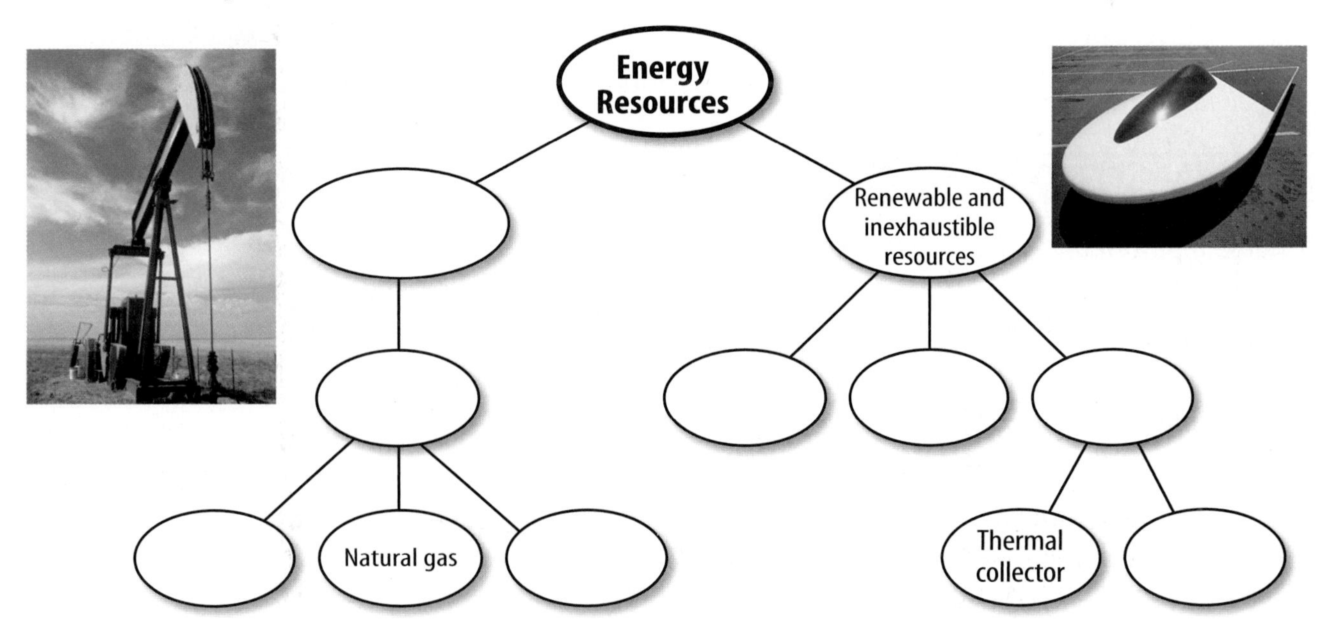

Using Vocabulary

alternative resource p. 733
chemical energy p. 719
electrical energy p. 720
energy p. 716
generator p. 726
inexhaustible resource p. 733
kinetic energy p. 717
law of conservation of energy p. 722
nonrenewable resource p. 730
nuclear energy p. 720
photovoltaic p. 734
potential energy p. 718
radiant energy p. 719
renewable resource p. 732
thermal energy p. 718
turbine p. 726

For each of the terms below, explain the relationship that exists.

1. electrical energy—nuclear energy
2. turbine—generator
3. photovoltaic—radiant energy—electrical energy
4. renewable resource—inexhaustible resource
5. potential energy—kinetic energy
6. kinetic energy—electrical energy—generator
7. thermal energy—radiant energy
8. law of conservation of energy—energy transformations
9. nonrenewable resource—chemical energy

Checking Concepts

Choose the word or phrase that best answers the question.

10. Objects that are able to fall have what type of energy?
    A) kinetic **C)** potential
    B) radiant **D)** electrical
11. Which form of energy does light have?
    A) electrical **C)** kinetic
    B) nuclear **D)** radiant
12. Muscles perform what type of energy transformation?
    A) kinetic to potential
    B) kinetic to electrical
    C) thermal to radiant
    D) chemical to kinetic
13. Photovoltaics perform what type of energy transformation?
    A) thermal to radiant
    B) kinetic to electrical
    C) radiant to electrical
    D) electrical to thermal
14. The form of energy that food contains is which of the following?
    A) chemical **C)** radiant
    B) potential **D)** electrical
15. Solar energy, wind, and geothermal are what type of energy resource?
    A) inexhaustible **C)** nonrenewable
    B) inexpensive **D)** chemical
16. Which of the following is a nonrenewable source of energy?
    A) hydroelectricity
    B) nuclear
    C) wind
    D) solar
17. A generator is NOT required to generate electrical energy when which of the following energy sources is used?
    A) solar **C)** hydroelectric
    B) wind **D)** nuclear
18. Which of the following are fossil fuels?
    A) gas **C)** oil
    B) coal **D)** all of these
19. Almost all of the energy that is used on Earth's surface comes from which of the following energy sources?
    A) radioactivity **C)** chemicals
    B) the Sun **D)** wind

Science Online green.msscience.com/vocabulary_puzzlemaker

Thinking Critically

20. **Explain** how the motion of a swing illustrates the transformation between potential and kinetic energy.

21. **Explain** what happens to the kinetic energy of a skateboard that is coasting along a flat surface, slows down, and comes to a stop.

22. **Describe** the energy transformations that occur in the process of toasting a bagel in an electric toaster.

23. **Compare and contrast** the formation of coal and the formation of oil and natural gas.

24. **Explain** the difference between the law of conservation of energy and conserving energy. How can conserving energy help prevent energy shortages?

25. **Make a Hypothesis** about how spacecraft that travel through the solar system obtain the energy they need to operate. Do research to verify your hypothesis.

26. **Concept Map** Copy and complete this concept map about energy.

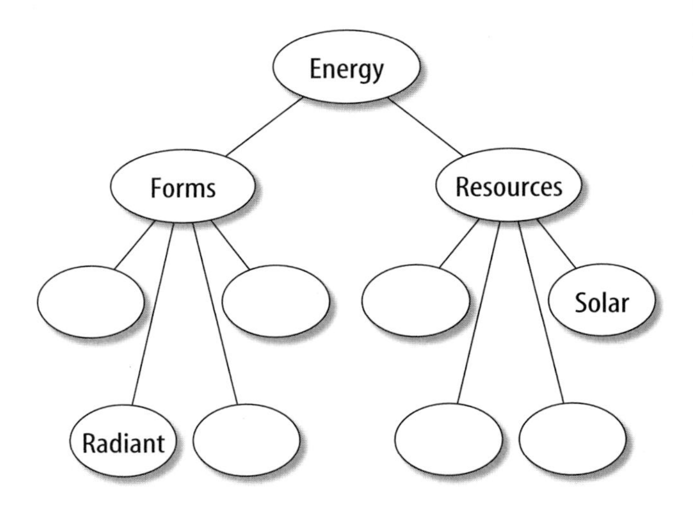

27. **Diagram** the energy transformations that occur when you rub sandpaper on a piece of wood and the wood becomes warm.

Performance Activities

28. **Multimedia Presentation** Alternative sources of energy that weren't discussed include biomass energy, wave energy, and hydrogen fuel cells. Research an alternative energy source and then prepare a digital slide show about the information you found. Use the concepts you learned from this chapter to inform your classmates about the future prospects of using such an energy source on a large scale.

Applying Math

29. **Calculate Number of Power Plants** A certain type of power plant is designed to provide energy for 10,000 homes. How many of these power plants would be needed to provide energy for 300,000 homes?

Use the table below to answer questions 30 and 31.

Energy Sources Used in the United States

Energy Source	Percent of Energy Used
Coal	23%
Oil	39%
Natural gas	23%
Nuclear	8%
Hydroelectric	4%
Other	3%

30. **Use Percentages** According to the data in the table above, what percentage of the energy used in the United States comes from fossil fuels?

31. **Calculate a Ratio** How many times greater is the amount of energy that comes from fossil fuels than the amount of energy from all other energy sources?

Part 1 Multiple Choice

Record your answers on the answer sheet provided by your teacher or on a sheet of paper.

1. The kinetic energy of a moving object increases if which of the following occurs?
   A. Its mass decreases.
   B. Its speed increases.
   C. Its height above the ground increases.
   D. Its temperature increases.

Use the graph below to answer questions 2–4.

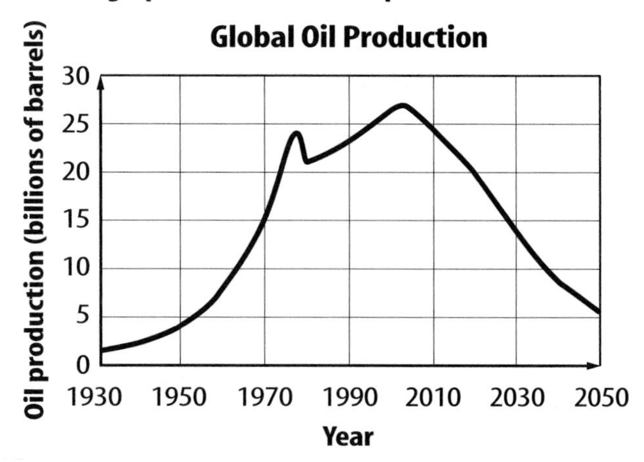

2. According to the graph above, in which year will global oil production be at a maximum?
   A. 1974 **C.** 2010
   B. 2002 **D.** 2050

3. Approximately how many times greater was oil production in 1970 than oil production in 1950?
   A. 2 times **C.** 6 times
   B. 10 times **D.** 3 times

4. In which year will the production of oil be equal to the oil production in 1970?
   A. 2010 **C.** 2022
   B. 2015 **D.** 2028

5. Which of the following energy sources is being used faster than it can be replaced?
   A. tidal **C.** fossil fuels
   B. wind **D.** hydroelectric

Use the circle graph below to answer question 6.

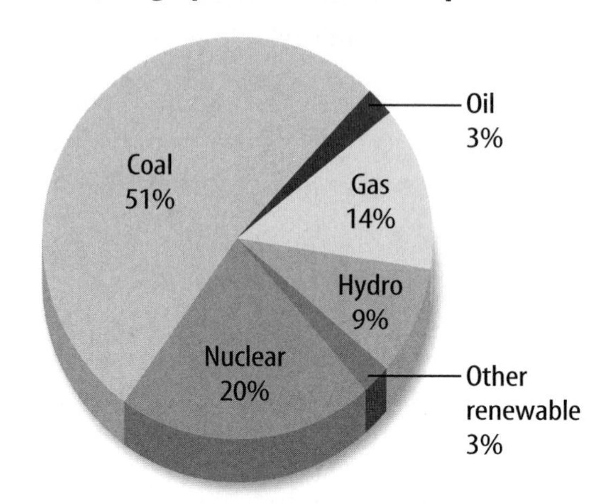

6. The circle graph shows the sources of electrical energy in the United States. In 2002, the total amount of electrical energy produced in the United States was 38.2 quads. How much electrical energy was produced by nuclear power plants?
   A. 3.0 quads **C.** 7.6 quads
   B. 3.8 quads **D.** 35.1 quads

7. When chemical energy is converted into thermal energy, which of the following must be true?
   A. The total amount of thermal energy plus chemical energy changes.
   B. Only the amount of chemical energy changes.
   C. Only the amount of thermal energy changes.
   D. The total amount of thermal energy plus chemical energy doesn't change.

8. A softball player hits a fly ball. Which of the following describes the energy conversion that occurs as it falls from its highest point?
   A. kinetic to potential
   B. potential to kinetic
   C. thermal to potential
   D. thermal to kinetic

Part 2 Short Response/Grid In

Record your answers on the answer sheet provided by your teacher or on a sheet of paper.

9. Why is it impossible to build a machine that produces more energy than it uses?

10. You toss a ball upward and then catch it on the way down. The height of the ball above the ground when it leaves your hand on the way up and when you catch it is the same. Compare the ball's kinetic energy when it leaves your hand and just before you catch it.

11. A basket ball is dropped from a height of 2 m and another identical basketball is dropped from a height of 4 m. Which ball has more kinetic energy just before it hits the ground?

Use the graph below to answer questions 12 and 13.

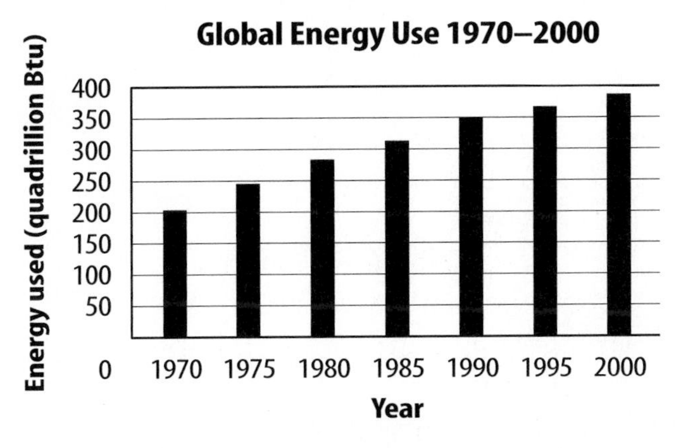

12. According to the graph above, by about how many times did the global use of energy increase from 1970 to 2000?

13. Over which five-year time period was the increase in global energy use the largest?

Test-Taking Tip

Do Your Studying Regularly Do not "cram" the night before the test. It can hamper your memory and make you tired.

Part 3 Open Ended

Record your answers on a sheet of paper.

14. When you drop a tennis ball, it hits the floor and bounces back up. But it does not reach the same height as released, and each successive upward bounce is smaller than the one previous. However, you notice the tennis ball is slightly warmer after it finishes bouncing. Explain how the law of conservation of energy is obeyed.

Use the graph below to answer questions 15–17.

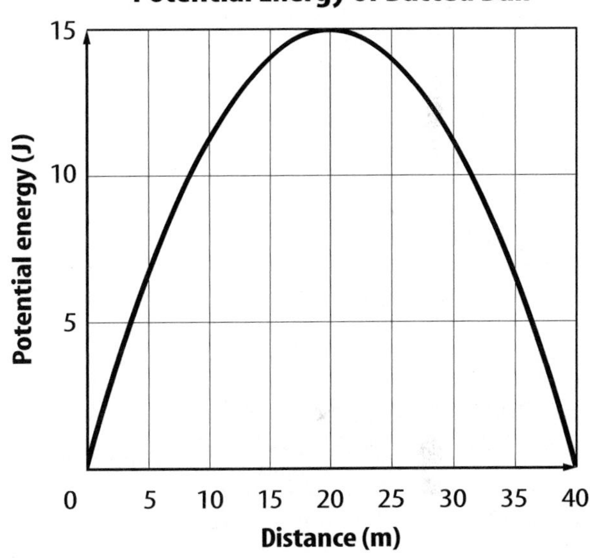

15. The graph shows how the potential energy of a batted ball depends on distance from the batter. At what distances is the kinetic energy of the ball the greatest?

16. At what distance from the batter is the height of the ball the greatest?

17. How much less is the kinetic energy of the ball at a distance of 20 m from the batter than at a distance of 0 m?

18. List advantages and disadvantages of the following energy sources: fossil fuels, nuclear energy, and geothermal energy.

Student Resources

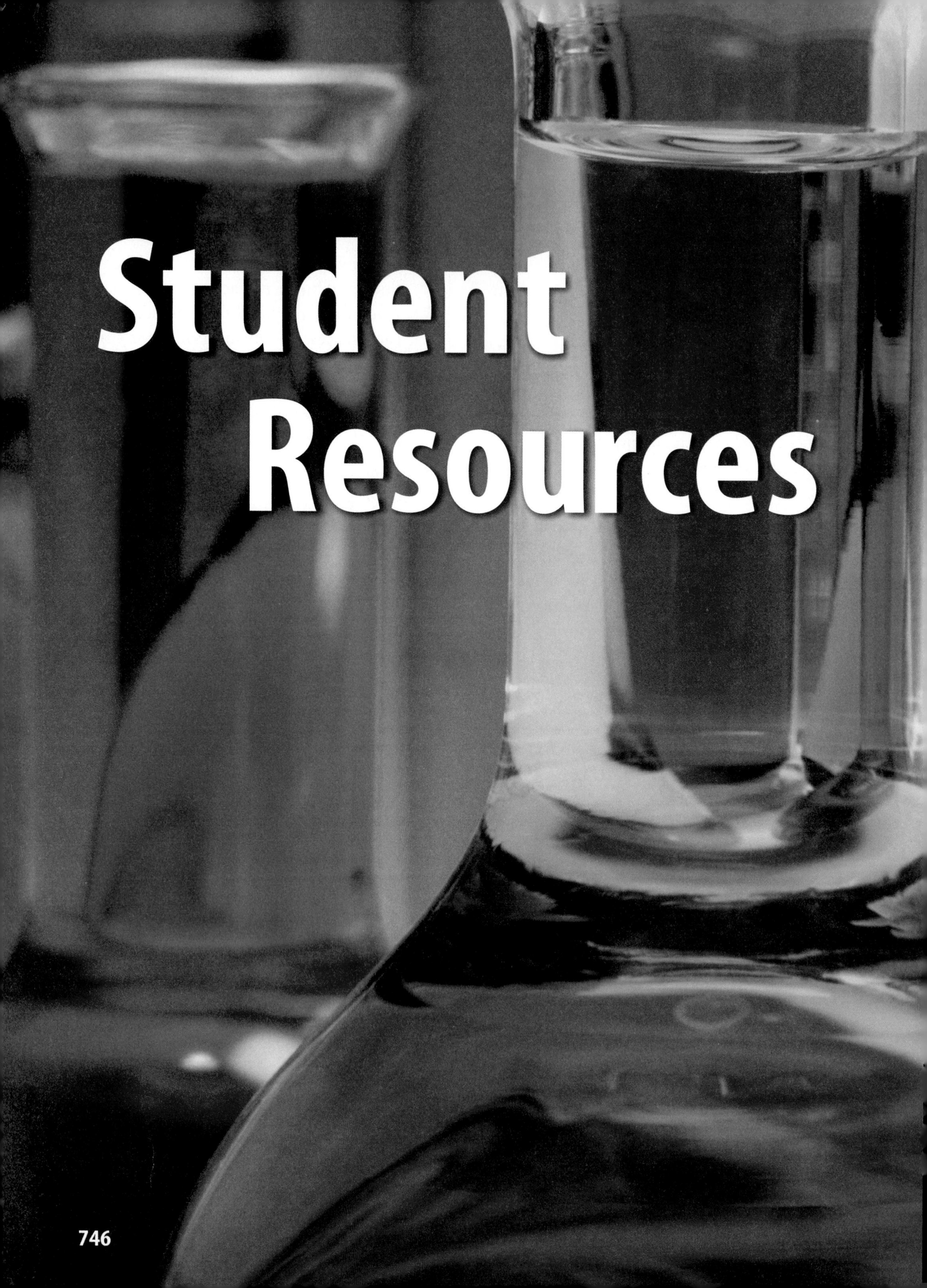

CONTENTS

Scientific Methods

Scientists use an orderly approach called the scientific method to solve problems. This includes organizing and recording data so others can understand them. Scientists use many variations in this method when they solve problems.

Identify a Question

The first step in a scientific investigation or experiment is to identify a question to be answered or a problem to be solved. For example, you might ask which gasoline is the most efficient.

Gather and Organize Information

After you have identified your question, begin gathering and organizing information. There are many ways to gather information, such as researching in a library, interviewing those knowledgeable about the subject, testing and working in the laboratory and field. Fieldwork is investigations and observations done outside of a laboratory.

Researching Information Before moving in a new direction, it is important to gather the information that already is known about the subject. Start by asking yourself questions to determine exactly what you need to know. Then you will look for the information in various reference sources, like the student is doing in **Figure 1.** Some sources may include textbooks, encyclopedias, government documents, professional journals, science magazines, and the Internet. Always list the sources of your information.

Figure 1 The Internet can be a valuable research tool.

Evaluate Sources of Information Not all sources of information are reliable. You should evaluate all of your sources of information, and use only those you know to be dependable. For example, if you are researching ways to make homes more energy efficient, a site written by the U.S. Department of Energy would be more reliable than a site written by a company that is trying to sell a new type of weatherproofing material. Also, remember that research always is changing. Consult the most current resources available to you. For example, a 1985 resource about saving energy would not reflect the most recent findings.

Sometimes scientists use data that they did not collect themselves, or conclusions drawn by other researchers. This data must be evaluated carefully. Ask questions about how the data were obtained, if the investigation was carried out properly, and if it has been duplicated exactly with the same results. Would you reach the same conclusion from the data? Only when you have confidence in the data can you believe it is true and feel comfortable using it.

Interpret Scientific Illustrations As you research a topic in science, you will see drawings, diagrams, and photographs to help you understand what you read. Some illustrations are included to help you understand an idea that you can't see easily by yourself, like the tiny particles in an atom in **Figure 2.** A drawing helps many people to remember details more easily and provides examples that clarify difficult concepts or give additional information about the topic you are studying. Most illustrations have labels or a caption to identify or to provide more information.

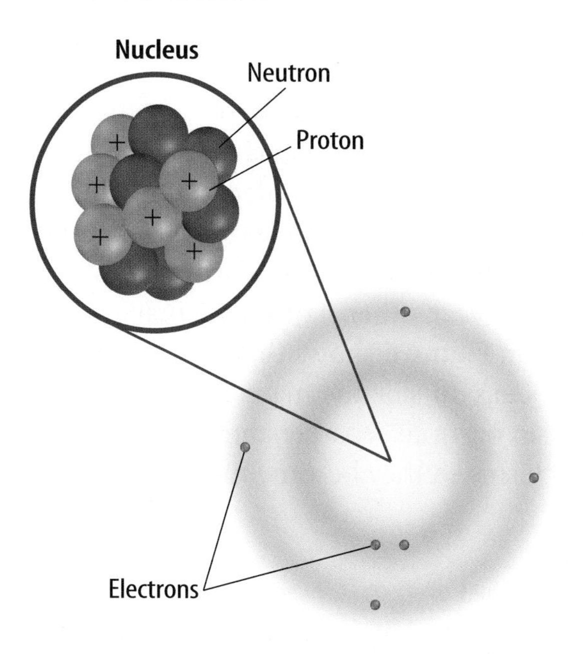

Figure 2 This drawing shows an atom of carbon with its six protons, six neutrons, and six electrons.

Concept Maps One way to organize data is to draw a diagram that shows relationships among ideas (or concepts). A concept map can help make the meanings of ideas and terms more clear, and help you understand and remember what you are studying. Concept maps are useful for breaking large concepts down into smaller parts, making learning easier.

Network Tree A type of concept map that not only shows a relationship, but how the concepts are related is a network tree, shown in **Figure 3.** In a network tree, the words are written in the ovals, while the description of the type of relationship is written across the connecting lines.

When constructing a network tree, write down the topic and all major topics on separate pieces of paper or notecards. Then arrange them in order from general to specific. Branch the related concepts from the major concept and describe the relationship on the connecting line. Continue to more specific concepts until finished.

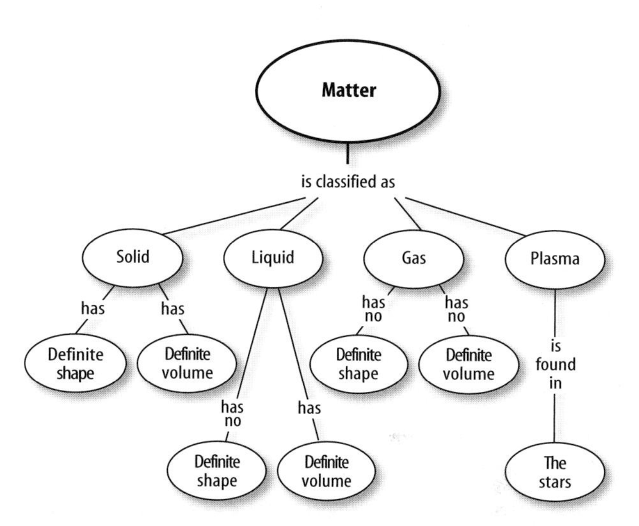

Figure 3 A network tree shows how concepts or objects are related.

Events Chain Another type of concept map is an events chain. Sometimes called a flow chart, it models the order or sequence of items. An events chain can be used to describe a sequence of events, the steps in a procedure, or the stages of a process.

When making an events chain, first find the one event that starts the chain. This event is called the initiating event. Then, find the next event and continue until the outcome is reached, as shown in **Figure 4.**

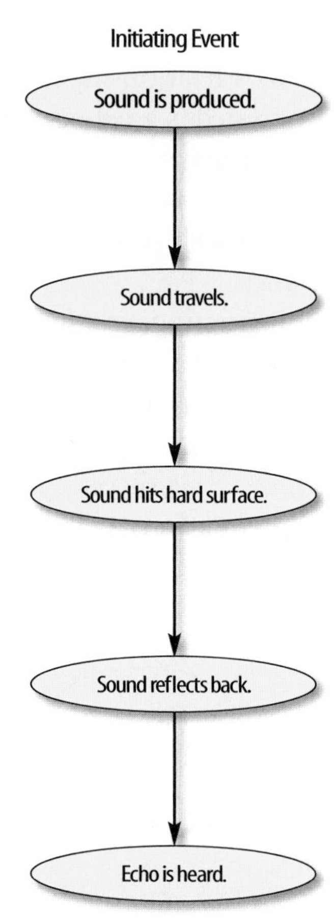

Figure 4 Events-chain concept maps show the order of steps in a process or event. This concept map shows how a sound makes an echo.

Cycle Map A specific type of events chain is a cycle map. It is used when the series of events do not produce a final outcome, but instead relate back to the beginning event, such as in **Figure 5.** Therefore, the cycle repeats itself.

To make a cycle map, first decide what event is the beginning event. This is also called the initiating event. Then list the next events in the order that they occur, with the last event relating back to the initiating event. Words can be written between the events that describe what happens from one event to the next. The number of events in a cycle map can vary, but usually contain three or more events.

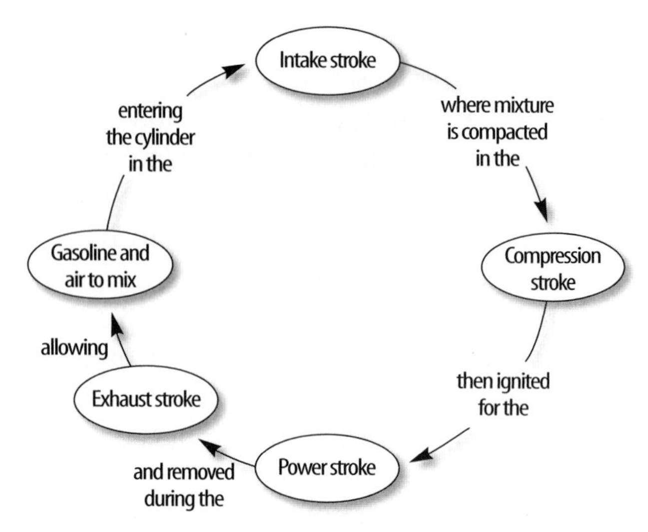

Figure 5 A cycle map shows events that occur in a cycle.

Spider Map A type of concept map that you can use for brainstorming is the spider map. When you have a central idea, you might find that you have a jumble of ideas that relate to it but are not necessarily clearly related to each other. The spider map on sound in **Figure 6** shows that if you write these ideas outside the main concept, then you can begin to separate and group unrelated terms so they become more useful.

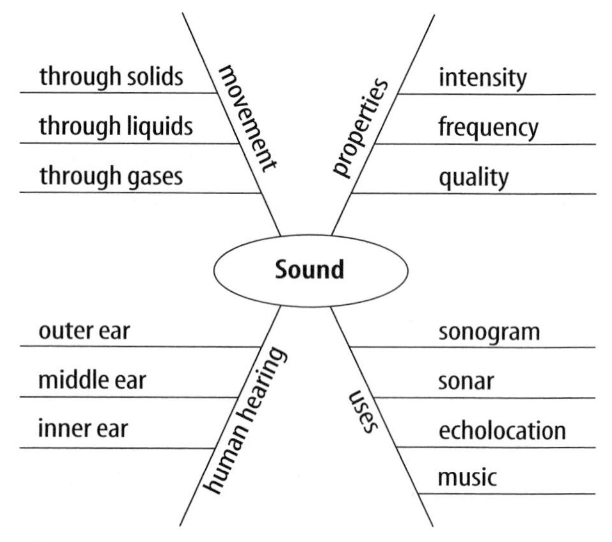

Figure 6 A spider map allows you to list ideas that relate to a central topic but not necessarily to one another.

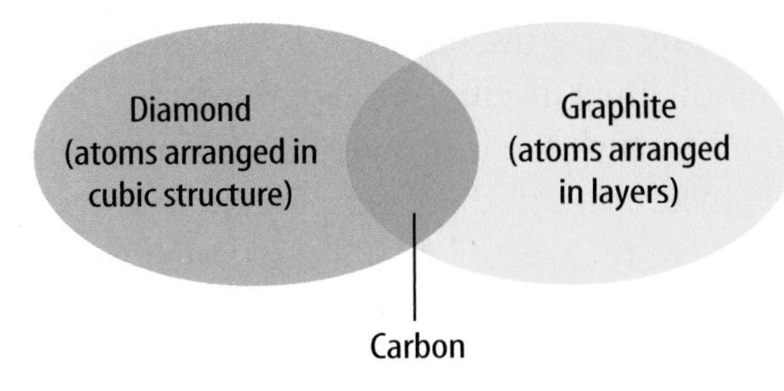

Figure 7 This Venn diagram compares and contrasts two substances made from carbon.

Venn Diagram To illustrate how two subjects compare and contrast you can use a Venn diagram. You can see the characteristics that the subjects have in common and those that they do not, shown in **Figure 7.**

To create a Venn diagram, draw two overlapping ovals that that are big enough to write in. List the characteristics unique to one subject in one oval, and the characteristics of the other subject in the other oval. The characteristics in common are listed in the overlapping section.

Make and Use Tables One way to organize information so it is easier to understand is to use a table. Tables can contain numbers, words, or both.

To make a table, list the items to be compared in the first column and the characteristics to be compared in the first row. The title should clearly indicate the content of the table, and the column or row heads should be clear. Notice that in **Table 1** the units are included.

Table 1 Recyclables Collected During Week

Day of Week	Paper (kg)	Aluminum (kg)	Glass (kg)
Monday	5.0	4.0	12.0
Wednesday	4.0	1.0	10.0
Friday	2.5	2.0	10.0

Make a Model One way to help you better understand the parts of a structure, the way a process works, or to show things too large or small for viewing is to make a model. For example, an atomic model made of a plastic-ball nucleus and pipe-cleaner electron shells can help you visualize how the parts of an atom relate to each other. Other types of models can by devised on a computer or represented by equations.

Form a Hypothesis

A possible explanation based on previous knowledge and observations is called a hypothesis. After researching gasoline types and recalling previous experiences in your family's car you form a hypothesis—our car runs more efficiently because we use premium gasoline. To be valid, a hypothesis has to be something you can test by using an investigation.

Predict When you apply a hypothesis to a specific situation, you predict something about that situation. A prediction makes a statement in advance, based on prior observation, experience, or scientific reasoning. People use predictions to make everyday decisions. Scientists test predictions by performing investigations. Based on previous observations and experiences, you might form a prediction that cars are more efficient with premium gasoline. The prediction can be tested in an investigation.

Design an Experiment A scientist needs to make many decisions before beginning an investigation. Some of these include: how to carry out the investigation, what steps to follow, how to record the data, and how the investigation will answer the question. It also is important to address any safety concerns.

Test the Hypothesis

Now that you have formed your hypothesis, you need to test it. Using an investigation, you will make observations and collect data, or information. This data might either support or not support your hypothesis. Scientists collect and organize data as numbers and descriptions.

Follow a Procedure In order to know what materials to use, as well as how and in what order to use them, you must follow a procedure. **Figure 8** shows a procedure you might follow to test your hypothesis.

Procedure

1. Use regular gasoline for two weeks.
2. Record the number of kilometers between fill-ups and the amount of gasoline used.
3. Switch to premium gasoline for two weeks.
4. Record the number of kilometers between fill-ups and the amount of gasoline used.

Figure 8 A procedure tells you what to do step by step.

Identify and Manipulate Variables and Controls In any experiment, it is important to keep everything the same except for the item you are testing. The one factor you change is called the independent variable. The change that results is the dependent variable. Make sure you have only one independent variable, to assure yourself of the cause of the changes you observe in the dependent variable. For example, in your gasoline experiment the type of fuel is the independent variable. The dependent variable is the efficiency.

Many experiments also have a control—an individual instance or experimental subject for which the independent variable is not changed. You can then compare the test results to the control results. To design a control you can have two cars of the same type. The control car uses regular gasoline for four weeks. After you are done with the test, you can compare the experimental results to the control results.

Collect Data

Whether you are carrying out an investigation or a short observational experiment, you will collect data, as shown in **Figure 9.** Scientists collect data as numbers and descriptions and organize it in specific ways.

Observe Scientists observe items and events, then record what they see. When they use only words to describe an observation, it is called qualitative data. Scientists' observations also can describe how much there is of something. These observations use numbers, as well as words, in the description and are called quantitative data. For example, if a sample of the element gold is described as being "shiny and very dense" the data are qualitative. Quantitative data on this sample of gold might include "a mass of 30 g and a density of 19.3 g/cm^3."

Figure 9 Collecting data is one way to gather information directly.

Figure 10 Record data neatly and clearly so it is easy to understand.

When you make observations you should examine the entire object or situation first, and then look carefully for details. It is important to record observations accurately and completely. Always record your notes immediately as you make them, so you do not miss details or make a mistake when recording results from memory. Never put unidentified observations on scraps of paper. Instead they should be recorded in a notebook, like the one in **Figure 10.** Write your data neatly so you can easily read it later. At each point in the experiment, record your observations and label them. That way, you will not have to determine what the figures mean when you look at your notes later. Set up any tables that you will need to use ahead of time, so you can record any observations right away. Remember to avoid bias when collecting data by not including personal thoughts when you record observations. Record only what you observe.

Estimate Scientific work also involves estimating. To estimate is to make a judgment about the size or the number of something without measuring or counting. This is important when the number or size of an object or population is too large or too difficult to accurately count or measure.

Sample Scientists may use a sample or a portion of the total number as a type of estimation. To sample is to take a small, representative portion of the objects or organisms of a population for research. By making careful observations or manipulating variables within that portion of the group, information is discovered and conclusions are drawn that might apply to the whole population. A poorly chosen sample can be unrepresentative of the whole. If you were trying to determine the rainfall in an area, it would not be best to take a rainfall sample from under a tree.

Measure You use measurements everyday. Scientists also take measurements when collecting data. When taking measurements, it is important to know how to use measuring tools properly. Accuracy also is important.

Length To measure length, the distance between two points, scientists use meters. Smaller measurements might be measured in centimeters or millimeters.

Length is measured using a metric ruler or meter stick. When using a metric ruler, line up the 0-cm mark with the end of the object being measured and read the number of the unit where the object ends. Look at the metric ruler shown in **Figure 11.** The centimeter lines are the long, numbered lines, and the shorter lines are millimeter lines. In this instance, the length would be 4.50 cm.

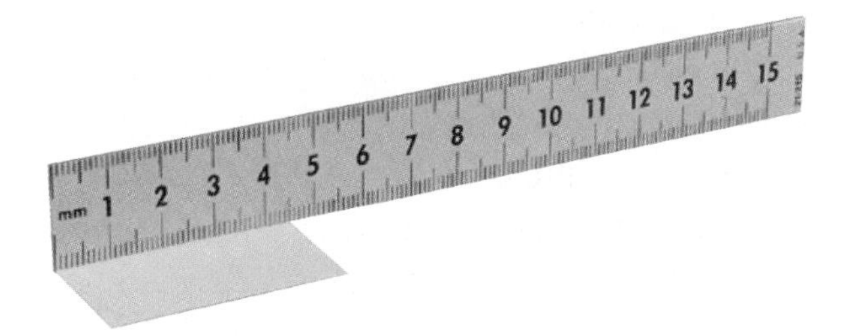

Figure 11 This metric ruler has centimeter and millimeter divisions.

Mass The SI unit for mass is the kilogram (kg). Scientists can measure mass using units formed by adding metric prefixes to the unit gram (g), such as milligram (mg). To measure mass, you might use a triple-beam balance similar to the one shown in **Figure 12.** The balance has a pan on one side and a set of beams on the other side. Each beam has a rider that slides on the beam.

When using a triple-beam balance, place an object on the pan. Slide the largest rider along its beam until the pointer drops below zero. Then move it back one notch. Repeat the process for each rider proceeding from the larger to smaller until the pointer swings an equal distance above and below the zero point. Sum the masses on each beam to find the mass of the object. Move all riders back to zero when finished.

Instead of putting materials directly on the balance, scientists often take a tare of a container. A tare is the mass of a container into which objects or substances are placed for measuring their masses. To mass objects or substances, find the mass of a clean container. Remove the container from the pan, and place the object or substances in the container. Find the mass of the container with the materials in it. Subtract the mass of the empty container from the mass of the filled container to find the mass of the materials you are using.

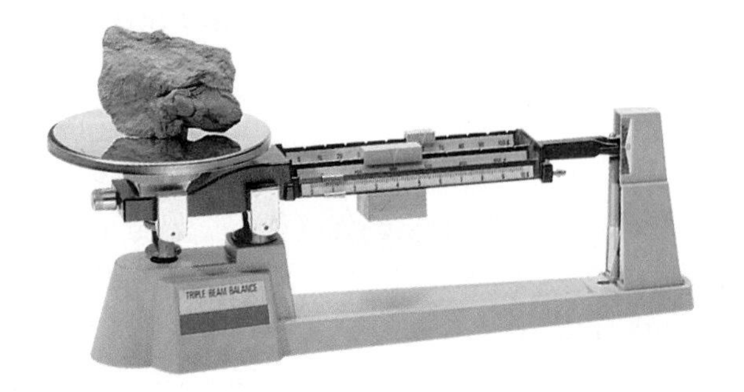

Figure 12 A triple-beam balance is used to determine the mass of an object.

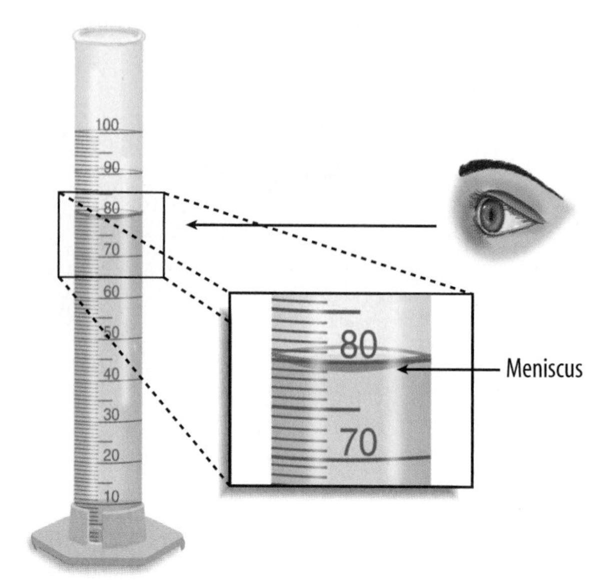

Figure 13 Graduated cylinders measure liquid volume.

Liquid Volume To measure liquids, the unit used is the liter. When a smaller unit is needed, scientists might use a milliliter. Because a milliliter takes up the volume of a cube measuring 1 cm on each side it also can be called a cubic centimeter ($cm^3 = cm \times cm \times cm$).

You can use beakers and graduated cylinders to measure liquid volume. A graduated cylinder, shown in **Figure 13,** is marked from bottom to top in milliliters. In lab, you might use a 10-mL graduated cylinder or a 100-mL graduated cylinder. When measuring liquids, notice that the liquid has a curved surface. Look at the surface at eye level, and measure the bottom of the curve. This is called the meniscus. The graduated cylinder in **Figure 13** contains 79.0 mL, or 79.0 cm^3, of a liquid.

Temperature Scientists often measure temperature using the Celsius scale. Pure water has a freezing point of 0°C and boiling point of 100°C. The unit of measurement is degrees Celsius. Two other scales often used are the Fahrenheit and Kelvin scales.

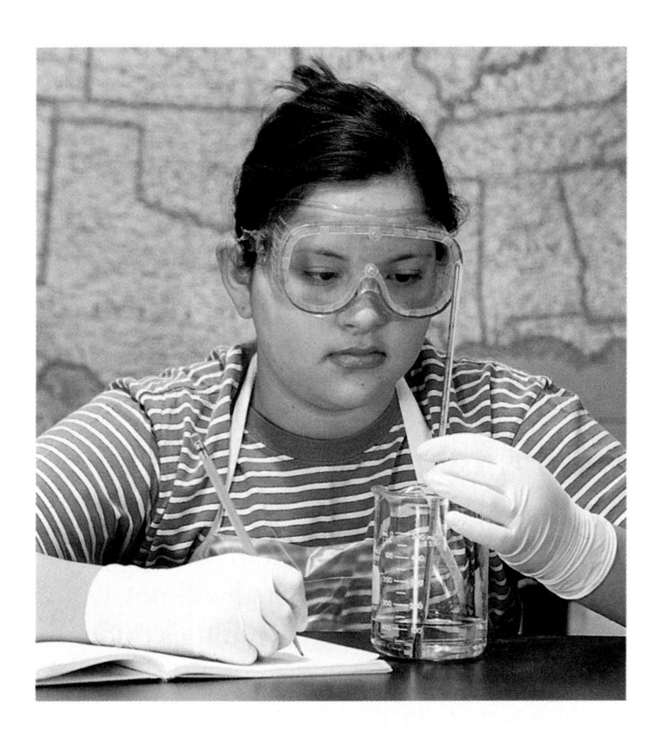

Figure 14 A thermometer measures the temperature of an object.

Scientists use a thermometer to measure temperature. Most thermometers in a laboratory are glass tubes with a bulb at the bottom end containing a liquid such as colored alcohol. The liquid rises or falls with a change in temperature. To read a glass thermometer like the thermometer in **Figure 14,** rotate it slowly until a red line appears. Read the temperature where the red line ends.

Form Operational Definitions An operational definition defines an object by how it functions, works, or behaves. For example, when you are playing hide and seek and a tree is home base, you have created an operational definition for a tree.

Objects can have more than one operational definition. For example, a ruler can be defined as a tool that measures the length of an object (how it is used). It can also be a tool with a series of marks used as a standard when measuring (how it works).

Analyze the Data

To determine the meaning of your observations and investigation results, you will need to look for patterns in the data. Then you must think critically to determine what the data mean. Scientists use several approaches when they analyze the data they have collected and recorded. Each approach is useful for identifying specific patterns.

Interpret Data The word *interpret* means "to explain the meaning of something." When analyzing data from an experiement, try to find out what the data show. Identify the control group and the test group to see whether or not changes in the independent variable have had an effect. Look for differences in the dependent variable between the control and test groups.

Classify Sorting objects or events into groups based on common features is called classifying. When classifying, first observe the objects or events to be classified. Then select one feature that is shared by some members in the group, but not by all. Place those members that share that feature in a subgroup. You can classify members into smaller and smaller subgroups based on characteristics. Remember that when you classify, you are grouping objects or events for a purpose. Keep your purpose in mind as you select the features to form groups and subgroups.

Compare and Contrast Observations can be analyzed by noting the similarities and differences between two more objects or events that you observe. When you look at objects or events to see how they are similar, you are comparing them. Contrasting is looking for differences in objects or events.

Recognize Cause and Effect A cause is a reason for an action or condition. The effect is that action or condition. When two events happen together, it is not necessarily true that one event caused the other. Scientists must design a controlled investigation to recognize the exact cause and effect.

Draw Conclusions

When scientists have analyzed the data they collected, they proceed to draw conclusions about the data. These conclusions are sometimes stated in words similar to the hypothesis that you formed earlier. They may confirm a hypothesis, or lead you to a new hypothesis.

Infer Scientists often make inferences based on their observations. An inference is an attempt to explain observations or to indicate a cause. An inference is not a fact, but a logical conclusion that needs further investigation. For example, you may infer that a fire has caused smoke. Until you investigate, however, you do not know for sure.

Apply When you draw a conclusion, you must apply those conclusions to determine whether the data supports the hypothesis. If your data do not support your hypothesis, it does not mean that the hypothesis is wrong. It means only that the result of the investigation did not support the hypothesis. Maybe the experiment needs to be redesigned, or some of the initial observations on which the hypothesis was based were incomplete or biased. Perhaps more observation or research is needed to refine your hypothesis. A successful investigation does not always come out the way you originally predicted.

Avoid Bias Sometimes a scientific investigation involves making judgments. When you make a judgment, you form an opinion. It is important to be honest and not to allow any expectations of results to bias your judgments. This is important throughout the entire investigation, from researching to collecting data to drawing conclusions.

Communicate

The communication of ideas is an important part of the work of scientists. A discovery that is not reported will not advance the scientific community's understanding or knowledge. Communication among scientists also is important as a way of improving their investigations.

Scientists communicate in many ways, from writing articles in journals and magazines that explain their investigations and experiments, to announcing important discoveries on television and radio. Scientists also share ideas with colleagues on the Internet or present them as lectures, like the student is doing in **Figure 15.**

Figure 15 A student communicates to his peers about his investigation.

SAFETY SYMBOLS	HAZARD	EXAMPLES	PRECAUTION	REMEDY
DISPOSAL	Special disposal procedures need to be followed.	certain chemicals, living organisms	Do not dispose of these materials in the sink or trash can.	Dispose of wastes as directed by your teacher.
BIOLOGICAL	Organisms or other biological materials that might be harmful to humans	bacteria, fungi, blood, unpreserved tissues, plant materials	Avoid skin contact with these materials. Wear mask or gloves.	Notify your teacher if you suspect contact with material. Wash hands thoroughly.
EXTREME TEMPERATURE	Objects that can burn skin by being too cold or too hot	boiling liquids, hot plates, dry ice, liquid nitrogen	Use proper protection when handling.	Go to your teacher for first aid.
SHARP OBJECT	Use of tools or glassware that can easily puncture or slice skin	razor blades, pins, scalpels, pointed tools, dissecting probes, broken glass	Practice common-sense behavior and follow guidelines for use of the tool.	Go to your teacher for first aid.
FUME	Possible danger to respiratory tract from fumes	ammonia, acetone, nail polish remover, heated sulfur, moth balls	Make sure there is good ventilation. Never smell fumes directly. Wear a mask.	Leave foul area and notify your teacher immediately.
ELECTRICAL	Possible danger from electrical shock or burn	improper grounding, liquid spills, short circuits, exposed wires	Double-check setup with teacher. Check condition of wires and apparatus.	Do not attempt to fix electrical problems. Notify your teacher immediately.
IRRITANT	Substances that can irritate the skin or mucous membranes of the respiratory tract	pollen, moth balls, steel wool, fiberglass, potassium permanganate	Wear dust mask and gloves. Practice extra care when handling these materials.	Go to your teacher for first aid.
CHEMICAL	Chemicals can react with and destroy tissue and other materials	bleaches such as hydrogen peroxide; acids such as sulfuric acid, hydrochloric acid; bases such as ammonia, sodium hydroxide	Wear goggles, gloves, and an apron.	Immediately flush the affected area with water and notify your teacher.
TOXIC	Substance may be poisonous if touched, inhaled, or swallowed.	mercury, many metal compounds, iodine, poinsettia plant parts	Follow your teacher's instructions.	Always wash hands thoroughly after use. Go to your teacher for first aid.
FLAMMABLE	Flammable chemicals may be ignited by open flame, spark, or exposed heat.	alcohol, kerosene, potassium permanganate	Avoid open flames and heat when using flammable chemicals.	Notify your teacher immediately. Use fire safety equipment if applicable.
OPEN FLAME	Open flame in use, may cause fire.	hair, clothing, paper, synthetic materials	Tie back hair and loose clothing. Follow teacher's instruction on lighting and extinguishing flames.	Notify your teacher immediately. Use fire safety equipment if applicable.

Eye Safety
Proper eye protection should be worn at all times by anyone performing or observing science activities.

Clothing Protection
This symbol appears when substances could stain or burn clothing.

Animal Safety
This symbol appears when safety of animals and students must be ensured.

Handwashing
After the lab, wash hands with soap and water before removing goggles.

Safety in the Science Laboratory

The science laboratory is a safe place to work if you follow standard safety procedures. Being responsible for your own safety helps to make the entire laboratory a safer place for everyone. When performing any lab, read and apply the caution statements and safety symbol listed at the beginning of the lab.

General Safety Rules

1. Obtain your teacher's permission to begin all investigations and use laboratory equipment.
2. Study the procedure. Ask your teacher any questions. Be sure you understand safety symbols shown on the page.
3. Notify your teacher about allergies or other health conditions which can affect your participation in a lab.
4. Learn and follow use and safety procedures for your equipment. If unsure, ask your teacher.
5. Never eat, drink, chew gum, apply cosmetics, or do any personal grooming in the lab. Never use lab glassware as food or drink containers. Keep your hands away from your face and mouth.
6. Know the location and proper use of the safety shower, eye wash, fire blanket, and fire alarm.

Prevent Accidents

1. Use the safety equipment provided to you. Goggles and a safety apron should be worn during investigations.
2. Do NOT use hair spray, mousse, or other flammable hair products. Tie back long hair and tie down loose clothing.
3. Do NOT wear sandals or other open-toed shoes in the lab.
4. Remove jewelry on hands and wrists. Loose jewelry, such as chains and long necklaces, should be removed to prevent them from getting caught in equipment.
5. Do not taste any substances or draw any material into a tube with your mouth.
6. Proper behavior is expected in the lab. Practical jokes and fooling around can lead to accidents and injury.
7. Keep your work area uncluttered.

Laboratory Work

1. Collect and carry all equipment and materials to your work area before beginning a lab.
2. Remain in your own work area unless given permission by your teacher to leave it.

3. Always slant test tubes away from yourself and others when heating them, adding substances to them, or rinsing them.
4. If instructed to smell a substance in a container, hold the container a short distance away and fan vapors towards your nose.
5. Do NOT substitute other chemicals/substances for those in the materials list unless instructed to do so by your teacher.
6. Do NOT take any materials or chemicals outside of the laboratory.
7. Stay out of storage areas unless instructed to be there and supervised by your teacher.

Laboratory Cleanup

1. Turn off all burners, water, and gas, and disconnect all electrical devices.
2. Clean all pieces of equipment and return all materials to their proper places.
3. Dispose of chemicals and other materials as directed by your teacher. Place broken glass and solid substances in the proper containers. Never discard materials in the sink.
4. Clean your work area.
5. Wash your hands with soap and water thoroughly BEFORE removing your goggles.

Emergencies

1. Report any fire, electrical shock, glassware breakage, spill, or injury, no matter how small, to your teacher immediately. Follow his or her instructions.
2. If your clothing should catch fire, STOP, DROP, and ROLL. If possible, smother it with the fire blanket or get under a safety shower. NEVER RUN.
3. If a fire should occur, turn off all gas and leave the room according to established procedures.
4. In most instances, your teacher will clean up spills. Do NOT attempt to clean up spills unless you are given permission and instructions to do so.
5. If chemicals come into contact with your eyes or skin, notify your teacher immediately. Use the eyewash or flush your skin or eyes with large quantities of water.
6. The fire extinguisher and first-aid kit should only be used by your teacher unless it is an extreme emergency and you have been given permission.
7. If someone is injured or becomes ill, only a professional medical provider or someone certified in first aid should perform first-aid procedures.

From Your Kitchen, Junk Drawer, or Yard

1 Testing Batteries

Real-World Question

Which type of battery is the best value?

Possible Materials

- long-lasting D batteries (4)
- less-expensive D batteries (4)
- identical flashlights (2)
- clock

Procedure

1. Put two of each type of battery into a flashlight. Record the price for each pair of batteries.
2. Turn on both flashlights. Record the time.
3. Examine the flashlights regularly. Record the approximate time when each set of batteries fails.
4. Switch battery types and flashlights and repeat the experiment.

Conclude and Apply

1. Identify the independent and dependent variables in this experiment.
2. Why did you repeat the experiment?
3. Which battery type lasted longest? Divide the price of the batteries by the amount of time each type lasted. Which battery was cheapest per hour of use?
4. Identify any advantages and disadvantages for each type of battery.

2 Panning Minerals

Real-World Question

How can minerals be separated from sand?

Possible Materials

- large, aluminum pie pan
- gallon jug filled with water
- empty gallon jug
- clean sand
- funnel
- coffee filter
- squirt bottle of water
- magnifying lens
- white paper
- hand magnet

Procedure

1. Conduct this lab outdoors.
2. Line the funnel with a coffee filter. Insert the funnel stem into an empty gallon jug.
3. Add a small amount of sand to the pie pan. Add some water and swirl the pan.
4. Continue to shake and swirl the pan until only black sand is left in the pan.

5. Use the squirt bottle to wash the black sand into the coffee filter. Repeat steps 3–5 until you have a good sample of black sand.
6. Let the black sand dry. Then observe it with a magnifying lens. Test the sand with a magnet.

Conclude and Apply

1. Why was black sand left in the gold pan after swirling it?
2. Describe how the sand looked under the lens. Did you see any well-shaped crystals?
3. What happened when you tested the sand with a magnet? Explain.

Adult supervision required for all labs.

3 Changing Rocks

Real-World Question
How can the change of metamorphic rock be modeled?

Possible Materials
- soil
- water
- measuring cup
- bowl
- spoon
- shale sample
- slate sample
- schist sample
- gneiss sample

Procedure
1. Mix equal parts of soil and water in a measuring cup or bowl. Stir the mixture until you make mud.
2. Place the bowl of mud on the table near the top edge.
3. Lay a sample of shale below the mud, a sample of slate below the shale, a sample of schist below the slate, a sample of gneiss below the schist.
4. Observe the different stages of sedimentary and metamorphic rocks that are formed by heat and pressure over long periods of time.

Conclude and Apply
1. Identify which rock sample(s) are sedimentary rock and which sample(s) are metamorphic rock.
2. Infer which type of rock is found at the greatest depth beneath the surface of Earth.

4 The Pressure's On

Real-World Question
How can atmospheric air pressure changes be modeled?

Possible Materials
- large pot
- stove or hotplate
- tongs
- oven mitt
- empty aluminum soda can
- water
- cold water
- small aquarium or large bowl
- large jar
- measuring cup

Procedure
1. Fill a small aquarium or large bowl with cold water.
2. Pour water into a large pot and boil it.
3. Pour 25 mL of water into an empty aluminum can.
4. Using an oven mitt and tongs, hold the bottom of the can in the boiling water for 1 min.
5. Remove the can from the pot and immediately submerge it upside-down in the cold water in the aquarium or large bowl.

Conclude and Apply
1. Describe what happened to the can in the cold water.
2. Infer why the can changed in the cold water.

5 Bottling a Tornado

Real-World Question
How can you model a tornado?

Possible Materials
- 2-L soda bottles (2)
- dish soap
- masking tape
- duct tape
- measuring cup
- towel

Procedure
1. Remove the labels from two 2-L soda bottles.
2. Fill one bottle with 1.5 L of water.
3. Add two drops of dish soap to the bottle with the water.
4. Invert the second bottle and connect the openings of the bottles.
5. Attach the two bottles together with duct tape.
6. Flip the bottles upside-down and quickly swirl the top bottle with a smooth motion. Observe the tornado pattern made in the water.

Conclude and Apply
1. Describe how you modeled a tornado.
2. Research how a real tornado forms.

6 Getting Warmer

Real-World Question
How do different surfaces affect temperature?

Possible Materials
- thermometer
- moist leaf litter
- large self-sealing bag
- stopwatch or watch

Procedure
1. Collect moist leaf litter in a large self-sealing bag.
2. Pile the leaf litter on a patch of grass that is exposed to direct sunlight.
3. Set the thermometer in the center of the leaf litter, wait 3 min, and measure the temperature.
4. Place the thermometer on the grass in direct sunlight, wait 3 min, and measure the temperature.
5. Place the thermometer on a cement surface in direct sunlight, wait 3 min, and measure the temperature.
6. Place the thermometer on an asphalt surface in direct sunlight, wait 3 min, and measure the temperature.

Conclude and Apply
1. Compare the temperatures of the different surfaces.
2. Explain how you measured the heat-island effect.

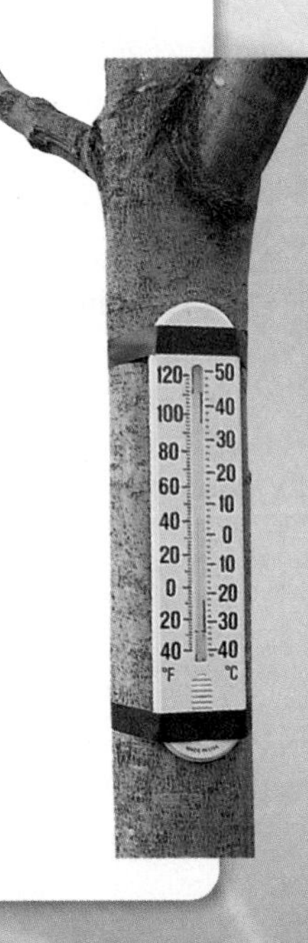

Adult supervision required for all labs.

7 Earth's Density

Real-World Question

How heavy would Earth feel if it were the size of a rock?

Possible Materials

- bathroom scale
- bucket
- water pitcher
- measuring cup
- large roasting pan
- turkey baster
- large chunk of wood, rock, or bundles of metal kitchen utensils held together with a rubber band

Procedure

1. Weigh yourself on a bathroom scale.
2. Pick up an object and stand on the scale. Calculate the mass of the object by subtracting your own mass.
3. Put the bucket in the roasting pan. Fill the bucket to the very brim with water. If you spill any water into the roasting pan, carefully mop it dry.
4. Put the object in the bucket.
5. Measure the water that spills into the roasting pan by collecting it with the turkey baster and putting it into the measuring cup. Depending on the size of your objects, you may need to empty the cup several times.
6. Use the formula: Density = Mass/Volume to calculate the density of each object.

Conclude and Apply

1. List the densities of the objects.
2. Look up Earth's density in this chapter. Which object has a density most like Earth's density?

8 Classification Poster

Real-World Question

How can you classify organisms?

Possible Materials

- old magazines or other resources with organism pictures
- poster board
- scissors
- glue
- colored pencils, markers, or crayons

Procedure

1. Search for pictures and photographs of a wide variety of organisms. Try to find organisms from all kingdoms.
2. Organize your pictures into a classification system you have researched or one you have made up.
3. Cut out the photographs and glue them to your poster.
4. Present your classification system to the class.

Conclude and Apply

1. Infer why you found more animal photographs than any other type of organism.
2. Infer why you found few protist photographs in the magazines.

9 Expanding Eggs

Real-World Question
How can you observe liquids passing through a cell membrane?

Possible Materials
- glass jar with lid
- white vinegar
- medium chicken egg
- tape measure or string and ruler
- tongs
- measuring cup

Procedure
1. Obtain a glass jar with a lid and a medium egg.
2. Make certain your egg easily fits into your jar.
3. Measure the circumference of your egg and then carefully place it in the jar.
4. Pour the white vinegar into the jar until the egg is completely submerged.
5. Place the lid on the jar but do not seal the jar.
6. Observe your egg each day for three days. Measure the circumference of the egg after three days.

Conclude and Apply
1. Describe the changes that happened to your egg.
2. Infer why the egg's circumference changed. *HINT: A hen's egg is a single cell.*

10 Putting Down Roots

Real-World Question
Can cells from a plant's stem produce root cells for a new plant?

Possible Materials
- houseplant
- scissors
- metric ruler
- glasses or jars (3)
- water
- magnifying lens

Procedure
1. Examine the stems of a houseplant, such as *Pothos*, and locate a node on three different stems. A node looks like a small bump.
2. Cut 3 stems off the plant at a 45° angle about 3–4 mm below the node.
3. Place the end of each stem into a separate glass of water and observe them for a week.

Conclude and Apply
1. Describe what happened to the ends of the stems.
2. Infer how plant stem cells can produce root cells.

Adult supervision required for all labs.

11 Do your ears hang low?

Real-World Question
Is ear lobe attachment a dominant or recessive trait?

Possible Materials
- Science Journal
- pencil
- pen
- calculator

Procedure
1. Ask your friends, family members, and other people you know if their ear lobes are attached or free.
2. Try to interview as many people as possible to collect a large sample of data.
3. Record the number of people who have attached ear lobes and the number with free ear lobes in your Science Journal.

Conclude and Apply
1. Calculate the percentage of people who have attached ear lobes and the percentage of people who have free ear lobes.
2. Infer whether or not attached ear lobes is a dominant or recessive trait. Do research to confirm your results.
3. Infer how many children in a family would have free ear lobes if their parents had attached ear lobes.

12 Frozen Fossils

Real-World Question
How can we model the formation of an amber fossil?

Possible Materials
- small glass jar with lid
- honey
- ruler
- dead insect or spider or small rubber insect or spider
- freezer

Procedure
1. Thoroughly wash and dry a small glass jar and its lid.
2. Pour 3 cm of honey into the jar. Do not pour honey down the sides of the jar.
3. Search for a dead insect or spider around your home or school and drop it into the center of the honey's surface.

4. Pour another 3 cm of honey into the jar to cover the organism.
5. Place the jar in the freezer overnight.

Conclude and Apply
1. Explain how you modeled the formation of an amber fossil.
2. Infer how amber fossils help scientists observe adaptations of organisms over time.

Adult supervision required for all labs.

13 Acid Defense

Real-World Question

How is stomach acid your internal first line of defense?

Possible Materials

- drinking glasses (2)
- milk
- cola or lemon juice
- masking tape
- marker
- measuring cup

Procedure

1. Pour 100 mL of milk into each glass.
2. Pour 20 mL of cola into the second glass.
3. Using the masking tape and marker, label the first glass *No Acid* and the second glass *Acid*.
4. Place the glasses in direct sunlight and observe the mixture each day for several days.

Conclude and Apply

1. Compare the odor of the mixture in both glasses after one or two days.
2. Infer how this experiment modeled one of your internal defenses against disease.

14 Vitamin Search

Real-World Question

How many vitamins and minerals are in the foods you eat?

Possible Materials

- labels from packaged foods and drinks
- nutrition guidebook or cookbook

Procedure

1. Create a data table to record the "% Daily Value" of important vitamins and minerals for a variety of foods.
2. Collect packages from a variety of packaged foods and check the Nutrition Facts chart for the "% Daily Value" of all the vitamins and minerals it contains. These values are listed at the bottom of the chart.
3. Use cookbooks or nutrition guidebooks to research the "% Daily Value" of vitamins and minerals found in several fresh fruits and vegetables such as strawberries, spinach, oranges, and lentils.

Conclude and Apply

1. Infer why a healthy diet includes fresh fruits and vegetables.
2. Infer why a healthy diet includes a wide variety of nutritious foods.

Nutrition Facts

Serving Size 1 Meal

Amount Per Serving	
Calories 330	Calories from Fat 60
	% Daily Value*
Total Fat 7g	**10%**
Saturated Fat 3.5g	**17%**
Polyunsaturated Fat 1g	
Monounsaturated Fat 2.5g	
Cholesterol 35mg	**12%**
Sodium 460mg	**19%**
Total Carbohydrate 52g	**18%**
Dietary Fiber 6g	**24%**
Sugars 17g	
Protein 15g	
Vitamin A 15%	Vitamin C 70%
Calcium 4%	Iron 10%

* Percent Daily Values are based on a 2,000 calorie diet. Your daily values may be higher or lower depending on your calorie needs.

	Calories	2,000	2,500
Total Fat	Less than	65g	80g
Sat Fat	Less than	20g	25g
Cholesterol	Less than	300mg	300mg
Sodium	Less than	2,400mg	2,400mg
Total Carbohydrate		300g	375g
Dietary Fiber		25g	30g

Adult supervision required for all labs.

15 Spinning Like a Top

Real-World Question
How can you observe your inner ear restoring your body's balance?

Possible Materials
- large pillows
- stopwatch or watch

Procedure
1. Choose a carpeted location several meters away from any furniture.
2. Lay large pillows on the floor around you.
3. Spin around in a circle once, stare straight ahead, and observe what the room looks like. Have a friend spot you to prevent you from falling.
4. Spin around in circles continuously for 5 s, stare straight ahead, and observe what the room looks like. Have a friend spot you.
5. After you stop, use a stopwatch to time how long it takes for the room to stop spinning.
6. With a friend spotting you, spin for 10 s and time how long it takes for the room to stop spinning.

Conclude and Apply
1. How long did it take for the room to stop spinning after you spun in circles for 5 s and for 10 s?
2. Infer why the room appeared to spin even after you stopped spinning.

16 Identifying Iodine

Real-World Question
How many iodine-rich foods are in your home?

Possible Materials
- Science Journal
- pen or pencil

Procedure
1. The element iodine is needed in a person's diet for the thyroid gland to work properly. Search your kitchen for the foods richest in iodine including fish, shellfish (clams, oysters, mussels, scallops), seaweed, and kelp.
2. Record all the iodine-rich foods you find in your Science Journal.
3. Search your kitchen for foods that may contain iodine if they are grown in rich soil such as onions, mushrooms, lettuce, spinach, green peppers, pineapple, peanuts, and whole wheat bread.
4. Record all of these foods that you find in your Science Journal.

Conclude and Apply
1. Infer whether or not your family eats a diet rich in iodine.
2. Infer why iodine is often added to table salt.
3. Research two diseases caused by a lack of iodine in the diet.

17 Prickly Plants

Real-World Question

Why does a cactus have spines?

Possible Materials

- toilet paper roll or paper towel roll (cut in half)
- transparent tape
- toothpicks (15)
- metric ruler
- oven mitt
- plastic bag or tissue paper

Procedure

1. Stuff the plastic bag or tissue paper into the toilet paper roll so that the bag or tissue is just inside the roll's rim.
2. Stand the roll on a table and hold it firmly with one hand. Place the oven mitt on your other hand and try to take the bag out of the roll.
3. If needed, place the bag back into the roll.
4. Securely tape toothpicks around the lip of the roll about 1 cm apart. About 4 cm of each toothpick should stick up above the rim.
5. Hold the roll on the table, put the oven mitt on, and try to take the bag out of the roll without breaking the toothpicks.

Conclude and Apply

1. Compare how easy it was to remove the plastic bag from the toilet paper roll with and without the toothpicks protecting it.
2. Describe the role of a cactus' spines.

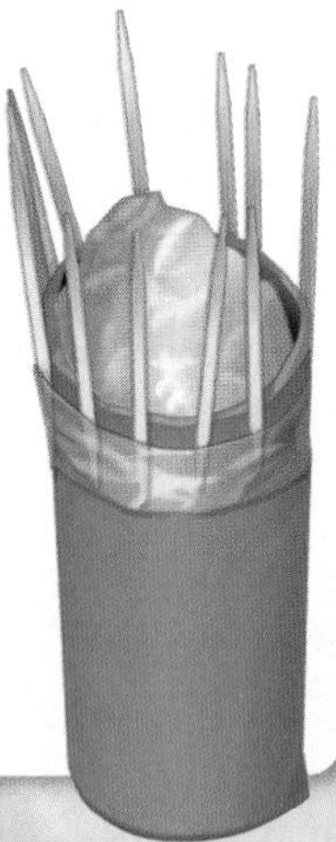

18 Feeding Frenzy

Real-World Question

How can you observe predation?

Possible Materials

- large jar with metal lid
- hammer and nail
- plant-eating insect
- insect net
- magnifying lens
- leaves, grass, flowers
- water

Procedure

1. Poke tiny holes in the metal lid of a jar with a nail so the insect has fresh air.
2. Observe a plant-eating insect in the wild. Collect one for indoor observation as well.
3. Put leaves, grass, and flowers in the bottom of the jar. Add a few drops of water.
4. Add the insect. Watch how it behaves.
5. When the experiment is finished, release the insect back into the wild.

Conclude and Apply

1. Describe the predation behaviors of the insect.
2. Compare the insect's behavior in the wild and in captivity. Did its diet change?

Adult supervision required for all labs.

19 UV Watch

Real-World Question
How can you find out about the risks of ultraviolet radiation each day?

Possible Materials
- daily newspaper with weekly weather forecasts
- graph paper

Procedure
1. Use the local newspaper or another resource to get the weather forecast for the day.
2. Check the UV (ultraviolet light) index for the day. If it provides an hourly UV index level, record the level for 1:00 P.M.
3. Find a legend or do research to discover what the numbers of the UV index mean.
4. Record the UV index everyday for ten days and graph your results on graph paper.

Conclude and Apply
1. Explain how the UV index system works.
2. Research several ways you can protect yourself from too much ultraviolet light exposure.

20 Good and Bad Apples

Real-World Question
How can the chemical reaction that turns apples brown be stopped?

Possible Materials
- apple
- concentrated lemon juice
- orange juice
- vitamin C tablet (1000 mg)
- water
- cola
- bowls (5)
- measuring cup
- kitchen knife
- paper plates (6)
- black marker

Procedure
1. Cut an apple into six equal slices.
2. Place one slice on a paper plate and label the plate *Untreated.*
3. Pour 100 mL of water into the first two bowls.
4. Dissolve a vitamin C tablet in the second bowl of water.
5. Pour 100 mL lemon juice, 100 mL of orange juice, and 100 mL of cola into the three remaining bowls.
6. Submerge an apple slice in each bowl for 10 min.
7. Label your other five plates *Water, Vitamin C Water, Lemon Juice, Orange Juice,* and *Cola.*
8. Take your apple wedges out of the bowls, place them on their correct plates. Observe the slices after one hour.

Conclude and Apply
1. Describe the results of your experiment.
2. Infer why some apple slices did not turn brown after being submerged.

Adult supervision required for all labs.

21 A Good Mix?

Real-World Question

What liquids will dissolve in water?

Possible Materials

- cooking oil
- water
- apple or grape juice
- rubbing alcohol
- spoon
- glass
- measuring cup

Procedure

1. Pour 100 mL of water into a large glass.
2. Pour 100 mL of apple juice into the glass and stir the water and juice together. Observe your mixture to determine whether juice is soluble in water.
3. Empty and rinse out your glass.
4. Pour 100 mL of water and 100 mL of cooking oil into the glass and stir them together. Observe your mixture to determine whether oil is soluble in water.
5. Empty and rinse out your glass.
6. Pour 100 mL of water and 100 mL of rubbing alcohol into the glass and stir them together. Observe your mixture to determine whether alcohol is soluble in water.

Conclude and Apply

1. List the liquid(s) that are soluble in water.
2. List the liquid(s) that are not soluble in water.
3. Infer why some liquids are soluble in water and others are not.

22 Microscopic Crystals

Real-World Question

What do crystalline and non-crystalline solids look like under a magnifying lens?

Possible Materials

- salt or sugar
- pepper
- magnifying lens
- paper
- bowl
- spoon
- measuring cup

Procedure

1. Pour 10 mL of salt into a bowl and grind the salt into small, powdery pieces with the back of the spoon.
2. Sprinkle a few grains of salt from the bowl onto a piece of paper and view the salt grains with the magnifying lens.
3. Clean out the bowl.
4. Pour 10 mL of pepper into the bowl and grind it into powder with the spoon.
5. Sprinkle a few grains of pepper from the bowl onto the paper and view the grains with the magnifying lens.

Conclude and Apply

1. Compare the difference between the salt and pepper grains under the magnifying lens.
2. Describe what a crystal is.

Adult supervision required for all labs.

23 Watch Out Below

Real-World Question

How does air resistance affect the velocity of falling objects?

Possible Materials

- tennis ball
- racquetball
- paper
- chair
- stopwatch

Procedure

1. Stand on a sturdy chair, hold a tennis ball above your head, and drop the ball to the floor.
2. Have a partner use a stopwatch to measure the time it takes for the ball to fall from your hand to the floor.
3. Drop a flat sheet of paper from the same height and measure the time it takes for the paper to fall to the floor.
4. Crumple the sheet of paper, drop it from the same height, and measure the time it takes for the paper ball to fall to the floor.

Conclude and Apply

1. List the amount of time it took for each of the objects to fall to the floor.
2. Infer how air resistance affects the velocity of falling objects.

24 The Heat is On

Real-World Question

How can different types of energy be transformed into thermal energy?

Possible Materials

- lamp
- incandescent light bulb
- black construction paper or cloth

Procedure

1. Feel the temperature of a black sheet of paper. Lay the paper in direct sunlight, wait 10 min, and observe how it feels.
2. Rub the palms of your hands together quickly for 10 s and observe how they feel.
3. Switch on a lamp that has a bare light bulb. *Without touching the lightbulb,* cup your hand 2 cm above the bulb for 30 s and observe what you feel.

Conclude and Apply

1. Infer the type of energy transforma tion that happened on the paper.
2. Infer the type of energy transformation that happened between the palms of your hands.
3. Infer the type of energy transformation that happened to the lightbulb.

Computer Skills

People who study science rely on computers, like the one in **Figure 16,** to record and store data and to analyze results from investigations. Whether you work in a laboratory or just need to write a lab report with tables, good computer skills are a necessity.

Using the computer comes with responsibility. Issues of ownership, security, and privacy can arise. Remember, if you did not author the information you are using, you must provide a source for your information. Also, anything on a computer can be accessed by others. Do not put anything on the computer that you would not want everyone to know. To add more security to your work, use a password.

Use a Word Processing Program

A computer program that allows you to type your information, change it as many times as you need to, and then print it out is called a word processing program. Word processing programs also can be used to make tables.

Figure 16 A computer will make reports neater and more professional looking.

Learn the Skill To start your word processing program, a blank document, sometimes called "Document 1," appears on the screen. To begin, start typing. To create a new document, click the *New* button on the standard tool bar. These tips will help you format the document.

- The program will automatically move to the next line; press *Enter* if you wish to start a new paragraph.
- Symbols, called non-printing characters, can be hidden by clicking the *Show/Hide* button on your toolbar.
- To insert text, move the cursor to the point where you want the insertion to go, click on the mouse once, and type the text.
- To move several lines of text, select the text and click the *Cut* button on your toolbar. Then position your cursor in the location that you want to move the cut text and click *Paste.* If you move to the wrong place, click *Undo.*
- The spell check feature does not catch words that are misspelled to look like other words, like "cold" instead of "gold." Always reread your document to catch all spelling mistakes.
- To learn about other word processing methods, read the user's manual or click on the *Help* button.
- You can integrate databases, graphics, and spreadsheets into documents by copying from another program and pasting it into your document, or by using desktop publishing (DTP). DTP software allows you to put text and graphics together to finish your document with a professional look. This software varies in how it is used and its capabilities.

Use a Database

A collection of facts stored in a computer and sorted into different fields is called a database. A database can be reorganized in any way that suits your needs.

Learn the Skill A computer program that allows you to create your own database is a database management system (DBMS). It allows you to add, delete, or change information. Take time to get to know the features of your database software.

- Determine what facts you would like to include and research to collect your information.
- Determine how you want to organize the information.
- Follow the instructions for your particular DBMS to set up fields. Then enter each item of data in the appropriate field.
- Follow the instructions to sort the information in order of importance.
- Evaluate the information in your database, and add, delete, or change as necessary.

Use the Internet

The Internet is a global network of computers where information is stored and shared. To use the Internet, like the students in **Figure 17,** you need a modem to connect your computer to a phone line and an Internet Service Provider account.

Learn the Skill To access internet sites and information, use a "Web browser," which lets you view and explore pages on the World Wide Web. Each page is its own site, and each site has its own address, called a URL. Once you have found a Web browser, follow these steps for a search (this also is how you search a database).

Figure 17 The Internet allows you to search a global network for a variety of information.

- Be as specific as possible. If you know you want to research "gold," don't type in "elements." Keep narrowing your search until you find what you want.
- Web sites that end in *.com* are commercial Web sites; *.org*, *.edu*, and *.gov* are non-profit, educational, or government Web sites.
- Electronic encyclopedias, almanacs, indexes, and catalogs will help locate and select relevant information.
- Develop a "home page" with relative ease. When developing a Web site, NEVER post pictures or disclose personal information such as location, names, or phone numbers. Your school or community usually can host your Web site. A basic understanding of HTML (hypertext mark-up language), the language of Web sites, is necessary. Software that creates HTML code is called authoring software, and can be downloaded free from many Web sites. This software allows text and pictures to be arranged as the software is writing the HTML code.

Use a Spreadsheet

A spreadsheet, shown in **Figure 18,** can perform mathematical functions with any data arranged in columns and rows. By entering a simple equation into a cell, the program can perform operations in specific cells, rows, or columns.

Learn the Skill Each column (vertical) is assigned a letter, and each row (horizontal) is assigned a number. Each point where a row and column intersect is called a cell, and is labeled according to where it is located—Column A, Row 1 (A1).

- Decide how to organize the data, and enter it in the correct row or column.
- Spreadsheets can use standard formulas or formulas can be customized to calculate cells.
- To make a change, click on a cell to make it activate, and enter the edited data or formula.
- Spreadsheets also can display your results in graphs. Choose the style of graph that best represents the data.

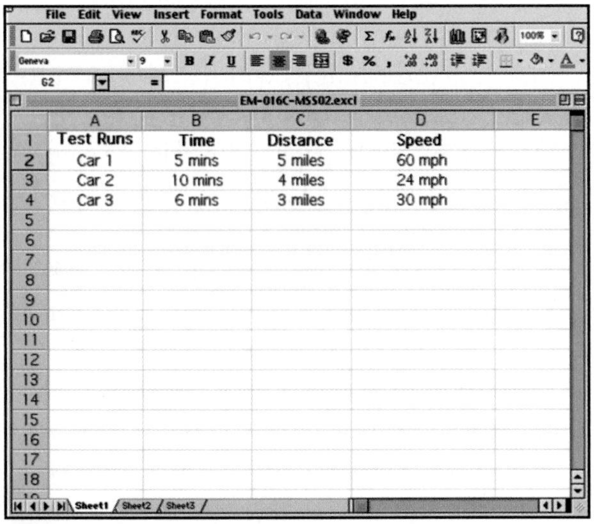

	A	B	C	D	E
1	Test Runs	Time	Distance	Speed	
2	Car 1	5 mins	5 miles	60 mph	
3	Car 2	10 mins	4 miles	24 mph	
4	Car 3	6 mins	3 miles	30 mph	

Figure 18 A spreadsheet allows you to perform mathematical operations on your data.

Use Graphics Software

Adding pictures, called graphics, to your documents is one way to make your documents more meaningful and exciting. This software adds, edits, and even constructs graphics. There is a variety of graphics software programs. The tools used for drawing can be a mouse, keyboard, or other specialized devices. Some graphics programs are simple. Others are complicated, called computer-aided design (CAD) software.

Learn the Skill It is important to have an understanding of the graphics software being used before starting. The better the software is understood, the better the results. The graphics can be placed in a word-processing document.

- Clip art can be found on a variety of internet sites, and on CDs. These images can be copied and pasted into your document.
- When beginning, try editing existing drawings, then work up to creating drawings.
- The images are made of tiny rectangles of color called pixels. Each pixel can be altered.
- Digital photography is another way to add images. The photographs in the memory of a digital camera can be downloaded into a computer, then edited and added to the document.
- Graphics software also can allow animation. The software allows drawings to have the appearance of movement by connecting basic drawings automatically. This is called in-betweening, or tweening.
- Remember to save often.

Presentation Skills

Develop Multimedia Presentations

Most presentations are more dynamic if they include diagrams, photographs, videos, or sound recordings, like the one shown in **Figure 19.** A multimedia presentation involves using stereos, overhead projectors, televisions, computers, and more.

Learn the Skill Decide the main points of your presentation, and what types of media would best illustrate those points.

- Make sure you know how to use the equipment you are working with.
- Practice the presentation using the equipment several times.
- Enlist the help of a classmate to push play or turn lights out for you. Be sure to practice your presentation with him or her.
- If possible, set up all of the equipment ahead of time, and make sure everything is working properly.

Figure 19 These students are engaging the audience using a variety of tools.

Computer Presentations

There are many different interactive computer programs that you can use to enhance your presentation. Most computers have a compact disc (CD) drive that can play both CDs and digital video discs (DVDs). Also, there is hardware to connect a regular CD, DVD, or VCR. These tools will enhance your presentation.

Another method of using the computer to aid in your presentation is to develop a slide show using a computer program. This can allow movement of visuals at the presenter's pace, and can allow for visuals to build on one another.

Learn the Skill In order to create multimedia presentations on a computer, you need to have certain tools. These may include traditional graphic tools and drawing programs, animation programs, and authoring systems that tie everything together. Your computer will tell you which tools it supports. The most important step is to learn about the tools that you will be using.

- Often, color and strong images will convey a point better than words alone. Use the best methods available to convey your point.
- As with other presentations, practice many times.
- Practice your presentation with the tools you and any assistants will be using.
- Maintain eye contact with the audience. The purpose of using the computer is not to prompt the presenter, but to help the audience understand the points of the presentation.

Math Review

Use Fractions

A fraction compares a part to a whole. In the fraction $\frac{2}{3}$, the 2 represents the part and is the numerator. The 3 represents the whole and is the denominator.

Reduce Fractions To reduce a fraction, you must find the largest factor that is common to both the numerator and the denominator, the greatest common factor (GCF). Divide both numbers by the GCF. The fraction has then been reduced, or it is in its simplest form.

Example Twelve of the 20 chemicals in the science lab are in powder form. What fraction of the chemicals used in the lab are in powder form?

Step 1 Write the fraction.

$$\frac{\text{part}}{\text{whole}} = \frac{12}{20}$$

Step 2 To find the GCF of the numerator and denominator, list all of the factors of each number.

Factors of 12: 1, 2, 3, 4, 6, 12 (the numbers that divide evenly into 12)

Factors of 20: 1, 2, 4, 5, 10, 20 (the numbers that divide evenly into 20)

Step 3 List the common factors.

1, 2, 4.

Step 4 Choose the greatest factor in the list.

The GCF of 12 and 20 is 4.

Step 5 Divide the numerator and denominator by the GCF.

$$\frac{12 \div 4}{20 \div 4} = \frac{3}{5}$$

In the lab, $\frac{3}{5}$ of the chemicals are in powder form.

Practice Problem At an amusement park, 66 of 90 rides have a height restriction. What fraction of the rides, in its simplest form, has a height restriction?

Add and Subtract Fractions To add or subtract fractions with the same denominator, add or subtract the numerators and write the sum or difference over the denominator. After finding the sum or difference, find the simplest form for your fraction.

Example 1 In the forest outside your house, $\frac{1}{8}$ of the animals are rabbits, $\frac{3}{8}$ are squirrels, and the remainder are birds and insects. How many are mammals?

Step 1 Add the numerators.

$$\frac{1}{8} + \frac{3}{8} = \frac{(1 + 3)}{8} = \frac{4}{8}$$

Step 2 Find the GCF.

$$\frac{4}{8} \text{ (GCF, 4)}$$

Step 3 Divide the numerator and denominator by the GCF.

$$\frac{4}{4} = 1, \ \frac{8}{4} = 2$$

$\frac{1}{2}$ of the animals are mammals.

Example 2 If $\frac{7}{16}$ of the Earth is covered by freshwater, and $\frac{1}{16}$ of that is in glaciers, how much freshwater is not frozen?

Step 1 Subtract the numerators.

$$\frac{7}{16} - \frac{1}{16} = \frac{(7 - 1)}{16} = \frac{6}{16}$$

Step 2 Find the GCF.

$$\frac{6}{16} \text{ (GCF, 2)}$$

Step 3 Divide the numerator and denominator by the GCF.

$$\frac{6}{2} = 3, \ \frac{16}{2} = 8$$

$\frac{3}{8}$ of the freshwater is not frozen.

Practice Problem A bicycle rider is going 15 km/h for $\frac{4}{9}$ of his ride, 10 km/h for $\frac{2}{9}$ of his ride, and 8 km/h for the remainder of the ride. How much of his ride is he going over 8 km/h?

Unlike Denominators To add or subtract fractions with unlike denominators, first find the least common denominator (LCD). This is the smallest number that is a common multiple of both denominators. Rename each fraction with the LCD, and then add or subtract. Find the simplest form if necessary.

Example 1 A chemist makes a paste that is $\frac{1}{2}$ table salt (NaCl), $\frac{1}{3}$ sugar ($C_6H_{12}O_6$), and the rest water (H_2O). How much of the paste is a solid?

Step 1 Find the LCD of the fractions.

$\frac{1}{2} + \frac{1}{3}$ (LCD, 6)

Step 2 Rename each numerator and each denominator with the LCD.

$1 \times 3 = 3,\ 2 \times 3 = 6$

$1 \times 2 = 2,\ 3 \times 2 = 6$

Step 3 Add the numerators.

$\frac{3}{6} + \frac{2}{6} = \frac{(3 + 2)}{6} = \frac{5}{6}$

$\frac{5}{6}$ of the paste is a solid.

Example 2 The average precipitation in Grand Junction, CO, is $\frac{7}{10}$ inch in November, and $\frac{3}{5}$ inch in December. What is the total average precipitation?

Step 1 Find the LCD of the fractions.

$\frac{7}{10} + \frac{3}{5}$ (LCD, 10)

Step 2 Rename each numerator and each denominator with the LCD.

$7 \times 1 = 7,\ 10 \times 1 = 10$

$3 \times 2 = 6,\ 5 \times 2 = 10$

Step 3 Add the numerators.

$\frac{7}{10} + \frac{6}{10} = \frac{(7 + 6)}{10} = \frac{13}{10}$

$\frac{13}{10}$ inches total precipitation, or $1\frac{3}{10}$ inches.

Practice Problem On an electric bill, about $\frac{1}{8}$ of the energy is from solar energy and about $\frac{1}{10}$ is from wind power. How much of the total bill is from solar energy and wind power combined?

Example 3 In your body, $\frac{7}{10}$ of your muscle contractions are involuntary (cardiac and smooth muscle tissue). Smooth muscle makes $\frac{3}{15}$ of your muscle contractions. How many of your muscle contractions are made by cardiac muscle?

Step 1 Find the LCD of the fractions.

$\frac{7}{10} - \frac{3}{15}$ (LCD, 30)

Step 2 Rename each numerator and each denominator with the LCD.

$7 \times 3 = 21,\ 10 \times 3 = 30$

$3 \times 2 = 6,\ 15 \times 2 = 30$

Step 3 Subtract the numerators.

$\frac{21}{30} - \frac{6}{30} = \frac{(21 - 6)}{30} = \frac{15}{30}$

Step 4 Find the GCF.

$\frac{15}{30}$ (GCF, 15)

$\frac{1}{2}$

$\frac{1}{2}$ of all muscle contractions are cardiac muscle.

Example 4 Tony wants to make cookies that call for $\frac{3}{4}$ of a cup of flour, but he only has $\frac{1}{3}$ of a cup. How much more flour does he need?

Step 1 Find the LCD of the fractions.

$\frac{3}{4} - \frac{1}{3}$ (LCD, 12)

Step 2 Rename each numerator and each denominator with the LCD.

$3 \times 3 = 9,\ 4 \times 3 = 12$

$1 \times 4 = 4,\ 3 \times 4 = 12$

Step 3 Subtract the numerators.

$\frac{9}{12} - \frac{4}{12} = \frac{(9 - 4)}{12} = \frac{5}{12}$

$\frac{5}{12}$ of a cup of flour.

Practice Problem Using the information provided to you in Example 3 above, determine how many muscle contractions are voluntary (skeletal muscle).

Multiply Fractions To multiply with fractions, multiply the numerators and multiply the denominators. Find the simplest form if necessary.

Example Multiply $\frac{3}{5}$ by $\frac{1}{3}$.

Step 1 Multiply the numerators and denominators.

$$\frac{3}{5} \times \frac{1}{3} = \frac{(3 \times 1)}{(5 \times 3)} = \frac{3}{15}$$

Step 2 Find the GCF.

$\frac{3}{15}$ (GCF, 3)

Step 3 Divide the numerator and denominator by the GCF.

$\frac{3}{3} = 1$, $\frac{15}{3} = 5$

$\frac{1}{5}$

$\frac{3}{5}$ multiplied by $\frac{1}{3}$ is $\frac{1}{5}$.

Practice Problem Multiply $\frac{3}{14}$ by $\frac{5}{16}$.

Find a Reciprocal Two numbers whose product is 1 are called multiplicative inverses, or reciprocals.

Example Find the reciprocal of $\frac{3}{8}$.

Step 1 Inverse the fraction by putting the denominator on top and the numerator on the bottom.

$\frac{8}{3}$

The reciprocal of $\frac{3}{8}$ is $\frac{8}{3}$.

Practice Problem Find the reciprocal of $\frac{4}{9}$.

Divide Fractions To divide one fraction by another fraction, multiply the dividend by the reciprocal of the divisor. Find the simplest form if necessary.

Example 1 Divide $\frac{1}{9}$ by $\frac{1}{3}$.

Step 1 Find the reciprocal of the divisor.

The reciprocal of $\frac{1}{3}$ is $\frac{3}{1}$.

Step 2 Multiply the dividend by the reciprocal of the divisor.

$$\frac{\frac{1}{9}}{\frac{1}{3}} = \frac{1}{9} \times \frac{3}{1} = \frac{(1 \times 3)}{(9 \times 1)} = \frac{3}{9}$$

Step 3 Find the GCF.

$\frac{3}{9}$ (GCF, 3)

Step 4 Divide the numerator and denominator by the GCF.

$\frac{3}{3} = 1$, $\frac{9}{3} = 3$

$\frac{1}{3}$

$\frac{1}{9}$ divided by $\frac{1}{3}$ is $\frac{1}{3}$.

Example 2 Divide $\frac{3}{5}$ by $\frac{1}{4}$.

Step 1 Find the reciprocal of the divisor.

The reciprocal of $\frac{1}{4}$ is $\frac{4}{1}$.

Step 2 Multiply the dividend by the reciprocal of the divisor.

$$\frac{\frac{3}{5}}{\frac{1}{4}} = \frac{3}{5} \times \frac{4}{1} = \frac{(3 \times 4)}{(5 \times 1)} = \frac{12}{5}$$

$\frac{3}{5}$ divided by $\frac{1}{4}$ is $\frac{12}{5}$ or $2\frac{2}{5}$.

Practice Problem Divide $\frac{3}{11}$ by $\frac{7}{10}$.

Use Ratios

When you compare two numbers by division, you are using a ratio. Ratios can be written 3 to 5, 3:5, or $\frac{3}{5}$. Ratios, like fractions, also can be written in simplest form.

Ratios can represent probabilities, also called odds. This is a ratio that compares the number of ways a certain outcome occurs to the number of outcomes. For example, if you flip a coin 100 times, what are the odds that it will come up heads? There are two possible outcomes, heads or tails, so the odds of coming up heads are 50:100. Another way to say this is that 50 out of 100 times the coin will come up heads. In its simplest form, the ratio is 1:2.

Example 1 A chemical solution contains 40 g of salt and 64 g of baking soda. What is the ratio of salt to baking soda as a fraction in simplest form?

Step 1 Write the ratio as a fraction.

$$\frac{\text{salt}}{\text{baking soda}} = \frac{40}{64}$$

Step 2 Express the fraction in simplest form.

The GCF of 40 and 64 is 8.

$$\frac{40}{64} = \frac{40 \div 8}{64 \div 8} = \frac{5}{8}$$

The ratio of salt to baking soda in the sample is 5:8.

Example 2 Sean rolls a 6-sided die 6 times. What are the odds that the side with a 3 will show?

Step 1 Write the ratio as a fraction.

$$\frac{\text{number of sides with a 3}}{\text{number of sides}} = \frac{1}{6}$$

Step 2 Multiply by the number of attempts.

$$\frac{1}{6} \times 6 \text{ attempts} = \frac{6}{6} \text{ attempts} = 1 \text{ attempt}$$

1 attempt out of 6 will show a 3.

Practice Problem Two metal rods measure 100 cm and 144 cm in length. What is the ratio of their lengths in simplest form?

Use Decimals

A fraction with a denominator that is a power of ten can be written as a decimal. For example, 0.27 means $\frac{27}{100}$. The decimal point separates the ones place from the tenths place.

Any fraction can be written as a decimal using division. For example, the fraction $\frac{5}{8}$ can be written as a decimal by dividing 5 by 8. Written as a decimal, it is 0.625.

Add or Subtract Decimals When adding and subtracting decimals, line up the decimal points before carrying out the operation.

Example 1 Find the sum of 47.68 and 7.80.

Step 1 Line up the decimal places when you write the numbers.

$$\begin{array}{r} 47.68 \\ +\ 7.80 \\ \hline \end{array}$$

Step 2 Add the decimals.

$$\begin{array}{r} 47.68 \\ +\ 7.80 \\ \hline 55.48 \end{array}$$

The sum of 47.68 and 7.80 is 55.48.

Example 2 Find the difference of 42.17 and 15.85.

Step 1 Line up the decimal places when you write the number.

$$\begin{array}{r} 42.17 \\ -15.85 \\ \hline \end{array}$$

Step 2 Subtract the decimals.

$$\begin{array}{r} 42.17 \\ -15.85 \\ \hline 26.32 \end{array}$$

The difference of 42.17 and 15.85 is 26.32.

Practice Problem Find the sum of 1.245 and 3.842.

Multiply Decimals To multiply decimals, multiply the numbers like any other number, ignoring the decimal point. Count the decimal places in each factor. The product will have the same number of decimal places as the sum of the decimal places in the factors.

Example Multiply 2.4 by 5.9.

Step 1 Multiply the factors like two whole numbers.
$24 \times 59 = 1416$

Step 2 Find the sum of the number of decimal places in the factors. Each factor has one decimal place, for a sum of two decimal places.

Step 3 The product will have two decimal places.
14.16

The product of 2.4 and 5.9 is 14.16.

Practice Problem Multiply 4.6 by 2.2.

Divide Decimals When dividing decimals, change the divisor to a whole number. To do this, multiply both the divisor and the dividend by the same power of ten. Then place the decimal point in the quotient directly above the decimal point in the dividend. Then divide as you do with whole numbers.

Example Divide 8.84 by 3.4.

Step 1 Multiply both factors by 10.
$3.4 \times 10 = 34$, $8.84 \times 10 = 88.4$

Step 2 Divide 88.4 by 34.

$$\begin{array}{r} 2.6 \\ 34\overline{)88.4} \\ -68 \\ \hline 20\,4 \\ -20\,4 \\ \hline 0 \end{array}$$

8.84 divided by 3.4 is 2.6.

Practice Problem Divide 75.6 by 3.6.

Use Proportions

An equation that shows that two ratios are equivalent is a proportion. The ratios $\frac{2}{4}$ and $\frac{5}{10}$ are equivalent, so they can be written as $\frac{2}{4} = \frac{5}{10}$. This equation is a proportion.

When two ratios form a proportion, the cross products are equal. To find the cross products in the proportion $\frac{2}{4} = \frac{5}{10}$, multiply the 2 and the 10, and the 4 and the 5. Therefore $2 \times 10 = 4 \times 5$, or $20 = 20$.

Because you know that both proportions are equal, you can use cross products to find a missing term in a proportion. This is known as solving the proportion.

Example The heights of a tree and a pole are proportional to the lengths of their shadows. The tree casts a shadow of 24 m when a 6-m pole casts a shadow of 4 m. What is the height of the tree?

Step 1 Write a proportion.
$$\frac{\text{height of tree}}{\text{height of pole}} = \frac{\text{length of tree's shadow}}{\text{length of pole's shadow}}$$

Step 2 Substitute the known values into the proportion. Let h represent the unknown value, the height of the tree.
$$\frac{h}{6} = \frac{24}{4}$$

Step 3 Find the cross products.
$h \times 4 = 6 \times 24$

Step 4 Simplify the equation.
$4h = 144$

Step 5 Divide each side by 4.
$$\frac{4h}{4} = \frac{144}{4}$$
$h = 36$

The height of the tree is 36 m.

Practice Problem The ratios of the weights of two objects on the Moon and on Earth are in proportion. A rock weighing 3 N on the Moon weighs 18 N on Earth. How much would a rock that weighs 5 N on the Moon weigh on Earth?

Use Percentages

The word *percent* means "out of one hundred." It is a ratio that compares a number to 100. Suppose you read that 77 percent of the Earth's surface is covered by water. That is the same as reading that the fraction of the Earth's surface covered by water is $\frac{77}{100}$. To express a fraction as a percent, first find the equivalent decimal for the fraction. Then, multiply the decimal by 100 and add the percent symbol.

Example Express $\frac{13}{20}$ as a percent.

Step 1 Find the equivalent decimal for the fraction.

$$\begin{array}{r} 0.65 \\ 20\overline{)13.00} \\ \underline{12\ 0} \\ 1\ 00 \\ \underline{1\ 00} \\ 0 \end{array}$$

Step 2 Rewrite the fraction $\frac{13}{20}$ as 0.65.

Step 3 Multiply 0.65 by 100 and add the % sign.

$0.65 \times 100 = 65 = 65\%$

So, $\frac{13}{20} = 65\%$.

This also can be solved as a proportion.

Example Express $\frac{13}{20}$ as a percent.

Step 1 Write a proportion.

$\frac{13}{20} = \frac{x}{100}$

Step 2 Find the cross products.

$1300 = 20x$

Step 3 Divide each side by 20.

$\frac{1300}{20} = \frac{20x}{20}$

$65\% = x$

Practice Problem In one year, 73 of 365 days were rainy in one city. What percent of the days in that city were rainy?

Solve One-Step Equations

A statement that two things are equal is an equation. For example, $A = B$ is an equation that states that A is equal to B.

An equation is solved when a variable is replaced with a value that makes both sides of the equation equal. To make both sides equal the inverse operation is used. Addition and subtraction are inverses, and multiplication and division are inverses.

Example 1 Solve the equation $x - 10 = 35$.

Step 1 Find the solution by adding 10 to each side of the equation.

$x - 10 = 35$

$x - 10 + 10 = 35 + 10$

$x = 45$

Step 2 Check the solution.

$x - 10 = 35$

$45 - 10 = 35$

$35 = 35$

Both sides of the equation are equal, so $x = 45$.

Example 2 In the formula $a = bc$, find the value of c if $a = 20$ and $b = 2$.

Step 1 Rearrange the formula so the unknown value is by itself on one side of the equation by dividing both sides by b.

$a = bc$

$\frac{a}{b} = \frac{bc}{b}$

$\frac{a}{b} = c$

Step 2 Replace the variables a and b with the values that are given.

$\frac{a}{b} = c$

$\frac{20}{2} = c$

$10 = c$

Step 3 Check the solution.

$a = bc$

$20 = 2 \times 10$

$20 = 20$

Both sides of the equation are equal, so $c = 10$ is the solution when $a = 20$ and $b = 2$.

Practice Problem In the formula $h = gd$, find the value of d if $g = 12.3$ and $h = 17.4$.

Use Statistics

The branch of mathematics that deals with collecting, analyzing, and presenting data is statistics. In statistics, there are three common ways to summarize data with a single number—the mean, the median, and the mode.

The **mean** of a set of data is the arithmetic average. It is found by adding the numbers in the data set and dividing by the number of items in the set.

The **median** is the middle number in a set of data when the data are arranged in numerical order. If there were an even number of data points, the median would be the mean of the two middle numbers.

The **mode** of a set of data is the number or item that appears most often.

Another number that often is used to describe a set of data is the range. The **range** is the difference between the largest number and the smallest number in a set of data.

A **frequency table** shows how many times each piece of data occurs, usually in a survey. **Table 2** below shows the results of a student survey on favorite color.

Table 2 Student Color Choice

Color	Tally	Frequency
red	||||	4
blue	~~||||~~	5
black	||	2
green	|||	3
purple	~~||||~~ ||	7
yellow	~~||||~~ |	6

Based on the frequency table data, which color is the favorite?

Example The speeds (in m/s) for a race car during five different time trials are 39, 37, 44, 36, and 44.

To find the mean:

Step 1 Find the sum of the numbers.
$39 + 37 + 44 + 36 + 44 = 200$

Step 2 Divide the sum by the number of items, which is 5.
$200 \div 5 = 40$

The mean is 40 m/s.

To find the median:

Step 1 Arrange the measures from least to greatest.
36, 37, 39, 44, 44

Step 2 Determine the middle measure.
36, 37, $\underline{39}$, 44, 44

The median is 39 m/s.

To find the mode:

Step 1 Group the numbers that are the same together.
44, 44, 36, 37, 39

Step 2 Determine the number that occurs most in the set.
$\underline{44, 44}$, 36, 37, 39

The mode is 44 m/s.

To find the range:

Step 1 Arrange the measures from largest to smallest.
44, 44, 39, 37, 36

Step 2 Determine the largest and smallest measures in the set.
$\underline{44}$, 44, 39, 37, $\underline{36}$

Step 3 Find the difference between the largest and smallest measures.
$44 - 36 = 8$

The range is 8 m/s.

Practice Problem Find the mean, median, mode, and range for the data set 8, 4, 12, 8, 11, 14, 16.

Use Geometry

The branch of mathematics that deals with the measurement, properties, and relationships of points, lines, angles, surfaces, and solids is called geometry.

Perimeter The **perimeter** (P) is the distance around a geometric figure. To find the perimeter of a rectangle, add the length and width and multiply that sum by two, or $2(l + w)$. To find perimeters of irregular figures, add the length of the sides.

Example 1 Find the perimeter of a rectangle that is 3 m long and 5 m wide.

Step 1 You know that the perimeter is 2 times the sum of the width and length.
$P = 2(3\text{ m} + 5\text{ m})$

Step 2 Find the sum of the width and length.
$P = 2(8\text{ m})$

Step 3 Multiply by 2.
$P = 16\text{ m}$

The perimeter is 16 m.

Example 2 Find the perimeter of a shape with sides measuring 2 cm, 5 cm, 6 cm, 3 cm.

Step 1 You know that the perimeter is the sum of all the sides.
$P = 2 + 5 + 6 + 3$

Step 2 Find the sum of the sides.
$P = 2 + 5 + 6 + 3$
$P = 16$

The perimeter is 16 cm.

Practice Problem Find the perimeter of a rectangle with a length of 18 m and a width of 7 m.

Practice Problem Find the perimeter of a triangle measuring 1.6 cm by 2.4 cm by 2.4 cm.

Area of a Rectangle The **area** (A) is the number of square units needed to cover a surface. To find the area of a rectangle, multiply the length times the width, or $l \times w$. When finding area, the units also are multiplied. Area is given in square units.

Example Find the area of a rectangle with a length of 1 cm and a width of 10 cm.

Step 1 You know that the area is the length multiplied by the width.
$A = (1\text{ cm} \times 10\text{ cm})$

Step 2 Multiply the length by the width. Also multiply the units.
$A = 10\text{ cm}^2$

The area is 10 cm^2.

Practice Problem Find the area of a square whose sides measure 4 m.

Area of a Triangle To find the area of a triangle, use the formula:

$$A = \frac{1}{2}(\text{base} \times \text{height})$$

The base of a triangle can be any of its sides. The height is the perpendicular distance from a base to the opposite endpoint, or vertex.

Example Find the area of a triangle with a base of 18 m and a height of 7 m.

Step 1 You know that the area is $\frac{1}{2}$ the base times the height.
$A = \frac{1}{2}(18\text{ m} \times 7\text{ m})$

Step 2 Multiply $\frac{1}{2}$ by the product of 18×7. Multiply the units.
$A = \frac{1}{2}(126\text{ m}^2)$
$A = 63\text{ m}^2$

The area is 63 m^2.

Practice Problem Find the area of a triangle with a base of 27 cm and a height of 17 cm.

Circumference of a Circle The **diameter** (d) of a circle is the distance across the circle through its center, and the **radius** (r) is the distance from the center to any point on the circle. The radius is half of the diameter. The distance around the circle is called the **circumference** (C). The formula for finding the circumference is:

$$C = 2\pi r \text{ or } C = \pi d$$

The circumference divided by the diameter is always equal to 3.1415926... This nonterminating and nonrepeating number is represented by the Greek letter π (pi). An approximation often used for π is 3.14.

Example 1 Find the circumference of a circle with a radius of 3 m.

Step 1 You know the formula for the circumference is 2 times the radius times π.
$C = 2\pi(3)$

Step 2 Multiply 2 times the radius.
$C = 6\pi$

Step 3 Multiply by π.
$C = 19$ m

The circumference is 19 m.

Example 2 Find the circumference of a circle with a diameter of 24.0 cm.

Step 1 You know the formula for the circumference is the diameter times π.
$C = \pi(24.0)$

Step 2 Multiply the diameter by π.
$C = 75.4$ cm

The circumference is 75.4 cm.

Practice Problem Find the circumference of a circle with a radius of 19 cm.

Area of a Circle The formula for the area of a circle is:
$A = \pi r^2$

Example 1 Find the area of a circle with a radius of 4.0 cm.

Step 1 $A = \pi(4.0)^2$

Step 2 Find the square of the radius.
$A = 16\pi$

Step 3 Multiply the square of the radius by π.
$A = 50 \text{ cm}^2$

The area of the circle is 50 cm^2.

Example 2 Find the area of a circle with a radius of 225 m.

Step 1 $A = \pi(225)^2$

Step 2 Find the square of the radius.
$A = 50625\pi$

Step 3 Multiply the square of the radius by π.
$A = 158962.5$

The area of the circle is 158,962 m^2.

Example 3 Find the area of a circle whose diameter is 20.0 mm.

Step 1 You know the formula for the area of a circle is the square of the radius times π, and that the radius is half of the diameter.
$A = \pi\left(\frac{20.0}{2}\right)^2$

Step 2 Find the radius.
$A = \pi(10.0)^2$

Step 3 Find the square of the radius.
$A = 100\pi$

Step 4 Multiply the square of the radius by π.
$A = 314 \text{ mm}^2$

The area is 314 mm^2.

Practice Problem Find the area of a circle with a radius of 16 m.

Volume The measure of space occupied by a solid is the **volume** (V). To find the volume of a rectangular solid multiply the length times width times height, or $V = l \times w \times h$. It is measured in cubic units, such as cubic centimeters (cm^3).

Example Find the volume of a rectangular solid with a length of 2.0 m, a width of 4.0 m, and a height of 3.0 m.

Step 1 You know the formula for volume is the length times the width times the height.

$V = 2.0\text{ m} \times 4.0\text{ m} \times 3.0\text{ m}$

Step 2 Multiply the length times the width times the height.

$V = 24\text{ m}^3$

The volume is 24 m^3.

Practice Problem Find the volume of a rectangular solid that is 8 m long, 4 m wide, and 4 m high.

To find the volume of other solids, multiply the area of the base times the height.

Example 1 Find the volume of a solid that has a triangular base with a length of 8.0 m and a height of 7.0 m. The height of the entire solid is 15.0 m.

Step 1 You know that the base is a triangle, and the area of a triangle is $\frac{1}{2}$ the base times the height, and the volume is the area of the base times the height.

$V = \left[\frac{1}{2}(b \times h)\right] \times 15$

Step 2 Find the area of the base.

$V = \left[\frac{1}{2}(8 \times 7)\right] \times 15$

$V = \left(\frac{1}{2} \times 56\right) \times 15$

Step 3 Multiply the area of the base by the height of the solid.

$V = 28 \times 15$

$V = 420\text{ m}^3$

The volume is 420 m^3.

Example 2 Find the volume of a cylinder that has a base with a radius of 12.0 cm, and a height of 21.0 cm.

Step 1 You know that the base is a circle, and the area of a circle is the square of the radius times π, and the volume is the area of the base times the height.

$V = (\pi r^2) \times 21$

$V = (\pi 12^2) \times 21$

Step 2 Find the area of the base.

$V = 144\pi \times 21$

$V = 452 \times 21$

Step 3 Multiply the area of the base by the height of the solid.

$V = 9490\text{ cm}^3$

The volume is 9490 cm^3.

Example 3 Find the volume of a cylinder that has a diameter of 15 mm and a height of 4.8 mm.

Step 1 You know that the base is a circle with an area equal to the square of the radius times π. The radius is one-half the diameter. The volume is the area of the base times the height.

$V = (\pi r^2) \times 4.8$

$V = \left[\pi\left(\frac{1}{2} \times 15\right)^2\right] \times 4.8$

$V = (\pi 7.5^2) \times 4.8$

Step 2 Find the area of the base.

$V = 56.25\pi \times 4.8$

$V = 176.63 \times 4.8$

Step 3 Multiply the area of the base by the height of the solid.

$V = 847.8$

The volume is 847.8 mm^3.

Practice Problem Find the volume of a cylinder with a diameter of 7 cm in the base and a height of 16 cm.

Science Applications

Measure in SI

The metric system of measurement was developed in 1795. A modern form of the metric system, called the International System (SI), was adopted in 1960 and provides the standard measurements that all scientists around the world can understand.

The SI system is convenient because unit sizes vary by powers of 10. Prefixes are used to name units. Look at **Table 3** for some common SI prefixes and their meanings.

Table 3 Common SI Prefixes

Prefix	Symbol	Meaning	
kilo-	k	1,000	thousand
hecto-	h	100	hundred
deka-	da	10	ten
deci-	d	0.1	tenth
centi-	c	0.01	hundredth
milli-	m	0.001	thousandth

Example How many grams equal one kilogram?

Step 1 Find the prefix *kilo* in **Table 3.**

Step 2 Using **Table 3,** determine the meaning of *kilo.* According to the table, it means 1,000. When the prefix *kilo* is added to a unit, it means that there are 1,000 of the units in a "*kilo*unit."

Step 3 Apply the prefix to the units in the question. The units in the question are grams. There are 1,000 grams in a kilogram.

Practice Problem Is a milligram larger or smaller than a gram? How many of the smaller units equal one larger unit? What fraction of the larger unit does one smaller unit represent?

Dimensional Analysis

Convert SI Units In science, quantities such as length, mass, and time sometimes are measured using different units. A process called dimensional analysis can be used to change one unit of measure to another. This process involves multiplying your starting quantity and units by one or more conversion factors. A conversion factor is a ratio equal to one and can be made from any two equal quantities with different units. If 1,000 mL equal 1 L then two ratios can be made.

$$\frac{1{,}000 \text{ mL}}{1 \text{ L}} = \frac{1 \text{ L}}{1{,}000 \text{ mL}} = 1$$

One can covert between units in the SI system by using the equivalents in **Table 3** to make conversion factors.

Example 1 How many cm are in 4 m?

Step 1 Write conversion factors for the units given. From **Table 3,** you know that 100 cm = 1 m. The conversion factors are

$$\frac{100 \text{ cm}}{1 \text{ m}} \text{ and } \frac{1 \text{ m}}{100 \text{ cm}}$$

Step 2 Decide which conversion factor to use. Select the factor that has the units you are converting from (m) in the denominator and the units you are converting to (cm) in the numerator.

$$\frac{100 \text{ cm}}{1 \text{ m}}$$

Step 3 Multiply the starting quantity and units by the conversion factor. Cancel the starting units with the units in the denominator. There are 400 cm in 4 m.

$$4 \cancel{\text{m}} \times \frac{100 \text{ cm}}{1 \cancel{\text{m}}} = 400 \text{ cm}$$

Practice Problem How many milligrams are in one kilogram? (Hint: You will need to use two conversion factors from **Table 3.**)

Table 4 Unit System Equivalents

Type of Measurement	Equivalent
Length	1 in = 2.54 cm 1 yd = 0.91 m 1 mi = 1.61 km
Mass and Weight*	1 oz = 28.35 g 1 lb = 0.45 kg 1 ton (short) = 0.91 tonnes (metric tons) 1 lb = 4.45 N
Volume	1 in^3 = 16.39 cm^3 1 qt = 0.95 L 1 gal = 3.78 L
Area	1 in^2 = 6.45 cm^2 1 yd^2 = 0.83 m^2 1 mi^2 = 2.59 km^2 1 acre = 0.40 hectares
Temperature	$°C = \frac{(°F - 32)}{1.8}$ K = °C + 273

*Weight is measured in standard Earth gravity.

Convert Between Unit Systems **Table 4** gives a list of equivalents that can be used to convert between English and SI units.

Example If a meterstick has a length of 100 cm, how long is the meterstick in inches?

Step 1 Write the conversion factors for the units given. From **Table 4,** 1 in = 2.54 cm.

$$\frac{1 \text{ in}}{2.54 \text{ cm}} \quad \textit{and} \quad \frac{2.54 \text{ cm}}{1 \text{ in}}$$

Step 2 Determine which conversion factor to use. You are converting from cm to in. Use the conversion factor with cm on the bottom.

$$\frac{1 \text{ in}}{2.54 \text{ cm}}$$

Step 3 Multiply the starting quantity and units by the conversion factor. Cancel the starting units with the units in the denominator. Round your answer based on the number of significant figures in the conversion factor.

$$100 \cancel{\text{cm}} \times \frac{1 \text{ in}}{2.54 \cancel{\text{cm}}} = 39.37 \text{ in}$$

The meterstick is 39.4 in long.

Practice Problem A book has a mass of 5 lbs. What is the mass of the book in kg?

Practice Problem Use the equivalent for in and cm (1 in = 2.54 cm) to show how 1 in^3 = 16.39 cm^3.

Math Skill Handbook

Precision and Significant Digits

When you make a measurement, the value you record depends on the precision of the measuring instrument. This precision is represented by the number of significant digits recorded in the measurement. When counting the number of significant digits, all digits are counted except zeros at the end of a number with no decimal point such as 2,050, and zeros at the beginning of a decimal such as 0.03020. When adding or subtracting numbers with different precision, round the answer to the smallest number of decimal places of any number in the sum or difference. When multiplying or dividing, the answer is rounded to the smallest number of significant digits of any number being multiplied or divided.

Example The lengths 5.28 and 5.2 are measured in meters. Find the sum of these lengths and record your answer using the correct number of significant digits.

Step 1 Find the sum.

5.28 m	2 digits after the decimal
+ 5.2 m	1 digit after the decimal
10.48 m	

Step 2 Round to one digit after the decimal because the least number of digits after the decimal of the numbers being added is 1.

The sum is 10.5 m.

Practice Problem How many significant digits are in the measurement 7,071,301 m? How many significant digits are in the measurement 0.003010 g?

Practice Problem Multiply 5.28 and 5.2 using the rule for multiplying and dividing. Record the answer using the correct number of significant digits.

Scientific Notation

Many times numbers used in science are very small or very large. Because these numbers are difficult to work with scientists use scientific notation. To write numbers in scientific notation, move the decimal point until only one non-zero digit remains on the left. Then count the number of places you moved the decimal point and use that number as a power of ten. For example, the average distance from the Sun to Mars is 227,800,000,000 m. In scientific notation, this distance is 2.278×10^{11} m. Because you moved the decimal point to the left, the number is a positive power of ten.

The mass of an electron is about 0.000 000 000 000 000 000 000 000 000 000 911 kg. Expressed in scientific notation, this mass is 9.11×10^{-31} kg. Because the decimal point was moved to the right, the number is a negative power of ten.

Example Earth is 149,600,000 km from the Sun. Express this in scientific notation.

Step 1 Move the decimal point until one non-zero digit remains on the left.

1.496 000 00

Step 2 Count the number of decimal places you have moved. In this case, eight.

Step 3 Show that number as a power of ten, 10^8.

The Earth is 1.496×10^8 km from the Sun.

Practice Problem How many significant digits are in 149,600,000 km? How many significant digits are in 1.496×10^8 km?

Practice Problem Parts used in a high performance car must be measured to 7×10^{-6} m. Express this number as a decimal.

Practice Problem A CD is spinning at 539 revolutions per minute. Express this number in scientific notation.

Make and Use Graphs

Data in tables can be displayed in a graph—a visual representation of data. Common graph types include line graphs, bar graphs, and circle graphs.

Line Graph A line graph shows a relationship between two variables that change continuously. The independent variable is changed and is plotted on the *x*-axis. The dependent variable is observed, and is plotted on the *y*-axis.

Example Draw a line graph of the data below from a cyclist in a long-distance race.

Table 5 Bicycle Race Data

Time (h)	Distance (km)
0	0
1	8
2	16
3	24
4	32
5	40

Step 1 Determine the *x*-axis and *y*-axis variables. Time varies independently of distance and is plotted on the *x*-axis. Distance is dependent on time and is plotted on the *y*-axis.

Step 2 Determine the scale of each axis. The *x*-axis data ranges from 0 to 5. The *y*-axis data ranges from 0 to 40.

Step 3 Using graph paper, draw and label the axes. Include units in the labels.

Step 4 Draw a point at the intersection of the time value on the *x*-axis and corresponding distance value on the *y*-axis. Connect the points and label the graph with a title, as shown in **Figure 20.**

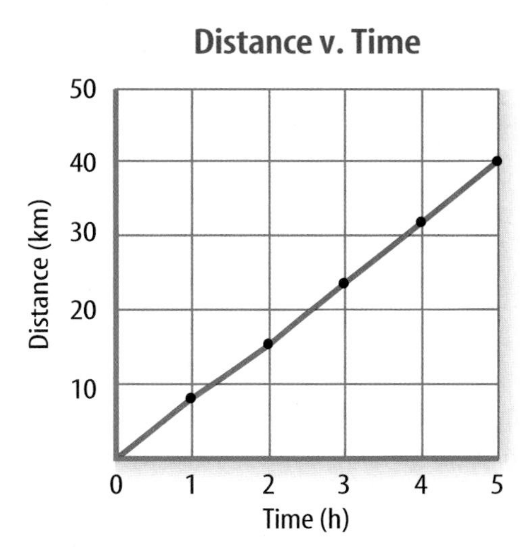

Figure 20 This line graph shows the relationship between distance and time during a bicycle ride.

Practice Problem A puppy's shoulder height is measured during the first year of her life. The following measurements were collected: (3 mo, 52 cm), (6 mo, 72 cm), (9 mo, 83 cm), (12 mo, 86 cm). Graph this data.

Find a Slope The slope of a straight line is the ratio of the vertical change, rise, to the horizontal change, run.

$$\text{Slope} = \frac{\text{vertical change (rise)}}{\text{horizontal change (run)}} = \frac{\text{change in } y}{\text{change in } x}$$

Example Find the slope of the graph in **Figure 20.**

Step 1 You know that the slope is the change in *y* divided by the change in *x*.

$$\text{Slope} = \frac{\text{change in } y}{\text{change in } x}$$

Step 2 Determine the data points you will be using. For a straight line, choose the two sets of points that are the farthest apart.

$$\text{Slope} = \frac{(40\text{–}0) \text{ km}}{(5\text{–}0) \text{ hr}}$$

Step 3 Find the change in *y* and *x*.

$$\text{Slope} = \frac{40 \text{ km}}{5\text{h}}$$

Step 4 Divide the change in *y* by the change in *x*.

$$\text{Slope} = \frac{8 \text{ km}}{\text{h}}$$

The slope of the graph is 8 km/h.

Bar Graph To compare data that does not change continuously you might choose a bar graph. A bar graph uses bars to show the relationships between variables. The x-axis variable is divided into parts. The parts can be numbers such as years, or a category such as a type of animal. The y-axis is a number and increases continuously along the axis.

Example A recycling center collects 4.0 kg of aluminum on Monday, 1.0 kg on Wednesday, and 2.0 kg on Friday. Create a bar graph of this data.

Step 1 Select the x-axis and y-axis variables. The measured numbers (the masses of aluminum) should be placed on the y-axis. The variable divided into parts (collection days) is placed on the x-axis.

Step 2 Create a graph grid like you would for a line graph. Include labels and units.

Step 3 For each measured number, draw a vertical bar above the x-axis value up to the y-axis value. For the first data point, draw a vertical bar above Monday up to 4.0 kg.

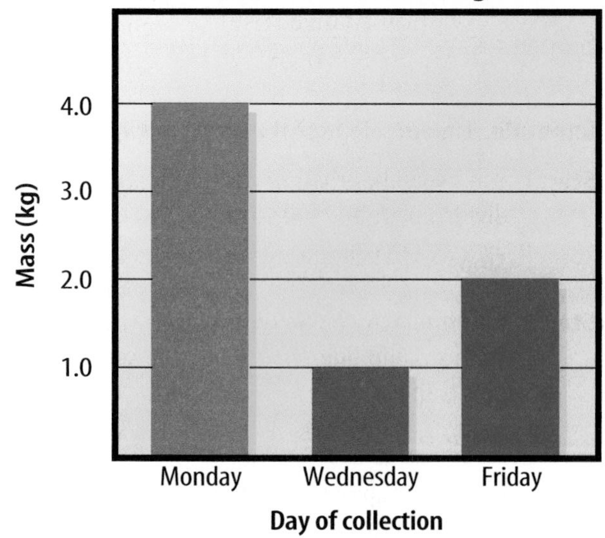

Practice Problem Draw a bar graph of the gases in air: 78% nitrogen, 21% oxygen, 1% other gases.

Circle Graph To display data as parts of a whole, you might use a circle graph. A circle graph is a circle divided into sections that represent the relative size of each piece of data. The entire circle represents 100%, half represents 50%, and so on.

Example Air is made up of 78% nitrogen, 21% oxygen, and 1% other gases. Display the composition of air in a circle graph.

Step 1 Multiply each percent by 360° and divide by 100 to find the angle of each section in the circle.

$$78\% \times \frac{360^\circ}{100} = 280.8^\circ$$

$$21\% \times \frac{360^\circ}{100} = 75.6^\circ$$

$$1\% \times \frac{360^\circ}{100} = 3.6^\circ$$

Step 2 Use a compass to draw a circle and to mark the center of the circle. Draw a straight line from the center to the edge of the circle.

Step 3 Use a protractor and the angles you calculated to divide the circle into parts. Place the center of the protractor over the center of the circle and line the base of the protractor over the straight line.

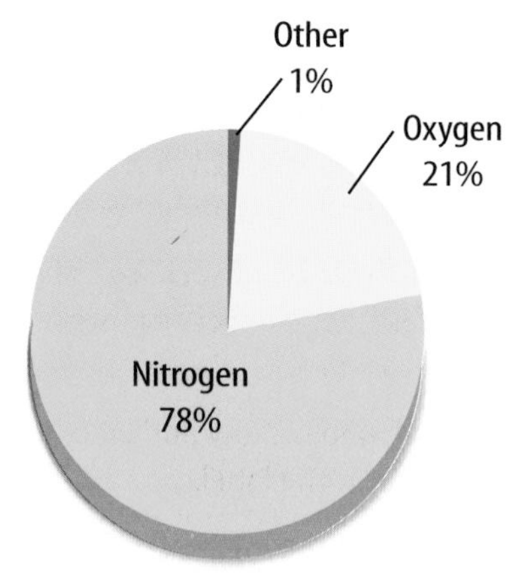

Practice Problem Draw a circle graph to represent the amount of aluminum collected during the week shown in the bar graph to the left.

Rocks

Rocks

Rock Type	Rock Name	Characteristics
Igneous (intrusive)	Granite	Large mineral grains of quartz, feldspar, hornblende, and mica. Usually light in color.
	Diorite	Large mineral grains of feldspar, hornblende, and mica. Less quartz than granite. Intermediate in color.
	Gabbro	Large mineral grains of feldspar, augite, and olivine. No quartz. Dark in color.
Igneous (extrusive)	Rhyolite	Small mineral grains of quartz, feldspar, hornblende, and mica, or no visible grains. Light in color.
	Andesite	Small mineral grains of feldspar, hornblende, and mica or no visible grains. Intermediate in color.
	Basalt	Small mineral grains of feldspar, augite, and possibly olivine or no visible grains. No quartz. Dark in color.
	Obsidian	Glassy texture. No visible grains. Volcanic glass. Fracture looks like broken glass.
	Pumice	Frothy texture. Floats in water. Usually light in color.
Sedimentary (detrital)	Conglomerate	Coarse grained. Gravel or pebble-size grains.
	Sandstone	Sand-sized grains 1/16 to 2 mm.
	Siltstone	Grains are smaller than sand but larger than clay.
	Shale	Smallest grains. Often dark in color. Usually platy.
Sedimentary (chemical or organic)	Limestone	Major mineral is calcite. Usually forms in oceans and lakes. Often contains fossils.
	Coal	Forms in swampy areas. Compacted layers of organic material, mainly plant remains.
Sedimentary (chemical)	Rock Salt	Commonly forms by the evaporation of seawater.
Metamorphic (foliated)	Gneiss	Banding due to alternate layers of different minerals, of different colors. Parent rock often is granite.
	Schist	Parallel arrangement of sheetlike minerals, mainly micas. Forms from different parent rocks.
	Phyllite	Shiny or silky appearance. May look wrinkled. Common parent rocks are shale and slate.
	Slate	Harder, denser, and shinier than shale. Common parent rock is shale.
Metamorphic (nonfoliated)	Marble	Calcite or dolomite. Common parent rock is limestone.
	Soapstone	Mainly of talc. Soft with greasy feel.
	Quartzite	Hard with interlocking quartz crystals. Common parent rock is sandstone.

Minerals

Minerals					
Mineral (formula)	**Color**	**Streak**	**Hardness**	**Breakage Pattern**	**Uses and Other Properties**
Graphite (C)	black to gray	black to gray	1–1.5	basal cleavage (scales)	pencil lead, lubricants for locks, rods to control some small nuclear reactions, battery poles
Galena (PbS)	gray	gray to black	2.5	cubic cleavage perfect	source of lead, used for pipes, shields for X rays, fishing equipment sinkers
Hematite (Fe_2O_3)	black or reddish-brown	reddish-brown	5.5–6.5	irregular fracture	source of iron; converted to pig iron, made into steel
Magnetite (Fe_3O_4)	black	black	6	conchoidal fracture	source of iron, attracts a magnet
Pyrite (FeS_2)	light, brassy, yellow	greenish-black	6–6.5	uneven fracture	fool's gold
Talc ($Mg_3Si_4O_{10}(OH)_2$)	white, greenish	white	1	cleavage in one direction	used for talcum powder, sculptures, paper, and tabletops
Gypsum ($CaSO_4 \cdot 2H_2O$)	colorless, gray, white, brown	white	2	basal cleavage	used in plaster of paris and dry wall for building construction
Sphalerite (ZnS)	brown, reddish-brown, brown, greenish	light to dark brown	3.5–4	cleavage in six directions	main ore of zinc; used in paints, dyes, and medicine
Muscovite ($KAl_3Si_3O_{10}(OH)_2$)	white, light gray, yellow, rose, green	colorless	2–2.5	basal cleavage	occurs in large, flexible plates; used as an insulator in electrical equipment, lubricant
Biotite ($K(Mg,Fe)_3(AlSi_3O_{10})(OH)_2$)	black to dark brown	colorless	2.5–3	basal cleavage	occurs in large, flexible plates
Halite (NaCl)	colorless, red, white, blue	colorless	2.5	cubic cleavage	salt; soluble in water; a preservative

Minerals

Minerals					
Mineral (formula)	**Color**	**Streak**	**Hardness**	**Breakage Pattern**	**Uses and Other Properties**
Calcite ($CaCO_3$)	colorless, white, pale blue	colorless, white	3	cleavage in three directions	fizzes when HCl is added; used in cements and other building materials
Dolomite ($CaMg(CO_3)_2$)	colorless, white, pink, green, gray, black	white	3.5–4	cleavage in three directions	concrete and cement; used as an ornamental building stone
Fluorite (CaF_2)	colorless, white, blue, green, red, yellow, purple	colorless	4	cleavage in four directions	used in the manufacture of optical equipment; glows under ultraviolet light
Hornblende ($(CaNa)_{2\text{-}3}(Mg,Al,Fe)_5\text{-}(Al,Si)_2Si_6O_{22}(OH)_2$)	green to black	gray to white	5–6	cleavage in two directions	will transmit light on thin edges; 6-sided cross section
Feldspar ($KAlSi_3O_8$) ($NaAlSi_3O_8$), ($CaAl_2Si_2O_8$)	colorless, white to gray, green	colorless	6	two cleavage planes meet at 90° angle	used in the manufacture of ceramics
Augite ($(Ca,Na)(Mg,Fe,Al)(Al,Si)_2O_6$)	black	colorless	6	cleavage in two directions	square or 8-sided cross section
Olivine ($(Mg,Fe)_2SiO_4$)	olive, green	none	6.5–7	conchoidal fracture	gemstones, refractory sand
Quartz (SiO_2)	colorless, various colors	none	7	conchoidal fracture	used in glass manufacture, electronic equipment, radios, computers, watches, gemstones

Weather Map Symbols

Sample Station Model

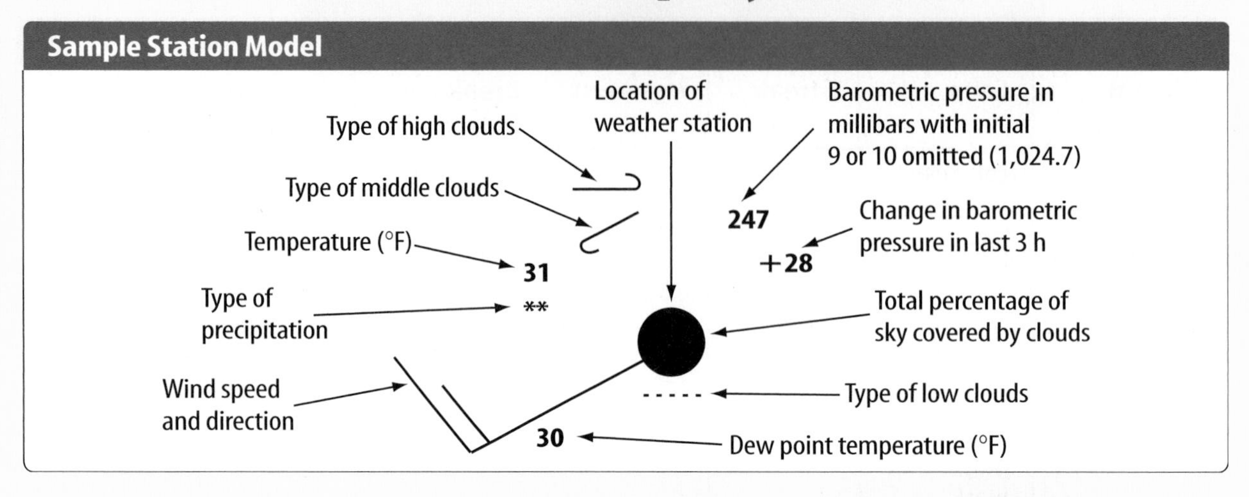

Sample Plotted Report at Each Station

Precipitation		Wind Speed and Direction		Sky Coverage		Some Types of High Clouds	
≡	Fog	○	0 calm	○	No cover		Scattered cirrus
★	Snow		1–2 knots		1/10 or less		Dense cirrus in patches
●	Rain		3–7 knots		2/10 to 3/10		Veil of cirrus covering entire sky
	Thunderstorm		8–12 knots		4/10		Cirrus not covering entire sky
,	Drizzle		13–17 knots		–		
▽	Showers		18–22 knots		6/10		
			23–27 knots		7/10		
			48–52 knots		Overcast with openings		
		1 knot = 1.852 km/h		●	Completely overcast		

Some Types of Middle Clouds		Some Types of Low Clouds		Fronts and Pressure Systems	
	Thin altostratus layer		Cumulus of fair weather	(H) or High (L) or Low	Center of high- or low-pressure system
	Thick altostratus layer		Stratocumulus		Cold front
	Thin altostratus in patches	-----	Fractocumulus of bad weather		Warm front
	Thin altostratus in bands	—	Stratus of fair weather		Occluded front
					Stationary front

Use and Care of a Microscope

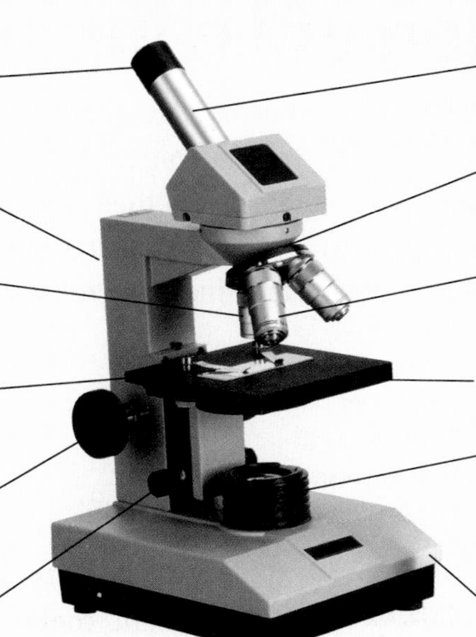

Eyepiece Contains magnifying lenses you look through.

Arm Supports the body tube.

Low-power objective Contains the lens with the lowest power magnification.

Stage clips Hold the microscope slide in place.

Coarse adjustment Focuses the image under low power.

Fine adjustment Sharpens the image under high magnification.

Body tube Connects the eyepiece to the revolving nosepiece.

Revolving nosepiece Holds and turns the objectives into viewing position.

High-power objective Contains the lens with the highest magnification.

Stage Supports the microscope slide.

Light source Provides light that passes upward through the diaphragm, the specimen, and the lenses.

Base Provides support for the microscope.

Caring for a Microscope

1. Always carry the microscope holding the arm with one hand and supporting the base with the other hand.
2. Don't touch the lenses with your fingers.
3. The coarse adjustment knob is used only when looking through the lowest-power objective lens. The fine adjustment knob is used when the high-power objective is in place.
4. Cover the microscope when you store it.

Using a Microscope

1. Place the microscope on a flat surface that is clear of objects. The arm should be toward you.
2. Look through the eyepiece. Adjust the diaphragm so light comes through the opening in the stage.
3. Place a slide on the stage so the specimen is in the field of view. Hold it firmly in place by using the stage clips.
4. Always focus with the coarse adjustment and the low-power objective lens first. After the object is in focus on low power, turn the nosepiece until the high-power objective is in place. Use ONLY the fine adjustment to focus with the high-power objective lens.

Making a Wet-Mount Slide

1. Carefully place the item you want to look at in the center of a clean, glass slide. Make sure the sample is thin enough for light to pass through.
2. Use a dropper to place one or two drops of water on the sample.
3. Hold a clean coverslip by the edges and place it at one edge of the water. Slowly lower the coverslip onto the water until it lies flat.
4. If you have too much water or a lot of air bubbles, touch the edge of a paper towel to the edge of the coverslip to draw off extra water and draw out unwanted air.

Diversity of Life: Classification of Living Organisms

A six-kingdom system of classification of organisms is used today. Two kingdoms—Kingdom Archaebacteria and Kingdom Eubacteria—contain organisms that do not have a nucleus and that lack membrane-bound structures in the cytoplasm of their cells. The members of the other four kingdoms have a cell or cells that contain a nucleus and structures in the cytoplasm, some of which are surrounded by membranes. These kingdoms are Kingdom Protista, Kingdom Fungi, Kingdom Plantae, and Kingdom Animalia.

Kingdom Archaebacteria

one-celled; some absorb food from their surroundings; some are photosynthetic; some are chemosynthetic; many are found in extremely harsh environments including salt ponds, hot springs, swamps, and deep-sea hydrothermal vents

Kingdom Eubacteria

one-celled; most absorb food from their surroundings; some are photosynthetic; some are chemosynthetic; many are parasites; many are round, spiral, or rod-shaped; some form colonies

Kingdom Protista

Phylum Euglenophyta one-celled; photosynthetic or take in food; most have one flagellum; euglenoids

Phylum Bacillariophyta one-celled; photosynthetic; have unique double shells made of silica; diatoms

Phylum Dinoflagellata one-celled; photosynthetic; contain red pigments; have two flagella; dinoflagellates

Phylum Chlorophyta one-celled, many-celled, or colonies; photosynthetic; contain chlorophyll; live on land, in freshwater, or salt water; green algae

Phylum Rhodophyta most are many-celled; photosynthetic; contain red pigments; most live in deep, saltwater environments; red algae

Phylum Phaeophyta most are many-celled; photosynthetic; contain brown pigments; most live in saltwater environments; brown algae

Phylum Rhizopoda one-celled; take in food; are free-living or parasitic; move by means of pseudopods; amoebas

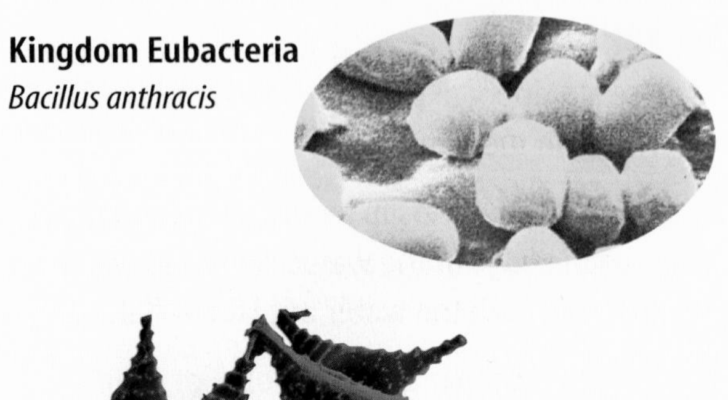

Kingdom Eubacteria
Bacillus anthracis

Phylum Chlorophyta
Desmids

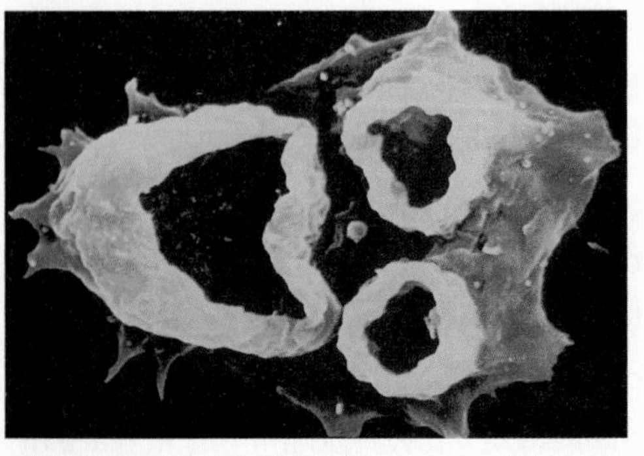

Amoeba

Phylum Zoomastigina one-celled; take in food; free-living or parasitic; have one or more flagella; zoomastigotes

Phylum Ciliophora one-celled; take in food; have large numbers of cilia; ciliates

Phylum Sporozoa one-celled; take in food; have no means of movement; are parasites in animals; sporozoans

Phylum Myxomycota
Slime mold

Phylum Oomycota
Phytophthora infestans

Phyla Myxomycota and Acrasiomycota one- or many-celled; absorb food; change form during life cycle; cellular and plasmodial slime molds

Phylum Oomycota many-celled; are either parasites or decomposers; live in freshwater or salt water; water molds, rusts and downy mildews

Kingdom Fungi

Phylum Zygomycota many-celled; absorb food; spores are produced in sporangia; zygote fungi; bread mold

Phylum Ascomycota one- and many-celled; absorb food; spores produced in asci; sac fungi; yeast

Phylum Basidiomycota many-celled; absorb food; spores produced in basidia; club fungi; mushrooms

Phylum Deuteromycota members with unknown reproductive structures; imperfect fungi; *Penicillium*

Phylum Mycophycota organisms formed by symbiotic relationship between an ascomycote or a basidiomycote and green alga or cyanobacterium; lichens

Lichens

Kingdom Plantae

Divisions Bryophyta (mosses), **Anthocerophyta** (hornworts), **Hepaticophyta** (liverworts), **Psilophyta** (whisk ferns) many-celled nonvascular plants; reproduce by spores produced in capsules; green; grow in moist, land environments

Division Lycophyta many-celled vascular plants; spores are produced in conelike structures; live on land; are photosynthetic; club mosses

Division Arthrophyta vascular plants; ribbed and jointed stems; scalelike leaves; spores produced in conelike structures; horsetails

Division Pterophyta vascular plants; leaves called fronds; spores produced in clusters of sporangia called sori; live on land or in water; ferns

Division Ginkgophyta deciduous trees; only one living species; have fan-shaped leaves with branching veins and fleshy cones with seeds; ginkgoes

Division Cycadophyta palmlike plants; have large, featherlike leaves; produces seeds in cones; cycads

Division Coniferophyta deciduous or evergreen; trees or shrubs; have needlelike or scale-like leaves; seeds produced in cones; conifers

Division Anthophyta
Tomato plant

Division Gnetophyta shrubs or woody vines; seeds are produced in cones; division contains only three genera; gnetum

Division Anthophyta dominant group of plants; flowering plants; have fruits with seeds

Kingdom Animalia

Phylum Porifera aquatic organisms that lack true tissues and organs; are asymmetrical and sessile; sponges

Phylum Cnidaria radially symmetrical organisms; have a digestive cavity with one opening; most have tentacles armed with stinging cells; live in aquatic environments singly or in colonies; includes jellyfish, corals, hydra, and sea anemones

Phylum Platyhelminthes bilaterally symmetrical worms; have flattened bodies; digestive system has one opening; parasitic and free-living species; flatworms

Division Bryophyta
Liverwort

Phylum Platyhelminthes
Flatworm

Phylum Chordata

Phylum Nematoda round, bilaterally symmetrical body; have digestive system with two openings; free-living forms and parasitic forms; roundworms

Phylum Mollusca soft-bodied animals, many with a hard shell and soft foot or footlike appendage; a mantle covers the soft body; aquatic and terrestrial species; includes clams, snails, squid, and octopuses

Phylum Annelida bilaterally symmetrical worms; have round, segmented bodies; terrestrial and aquatic species; includes earthworms, leeches, and marine polychaetes

Phylum Arthropoda largest animal group; have hard exoskeletons, segmented bodies, and pairs of jointed appendages; land and aquatic species; includes insects, crustaceans, and spiders

Phylum Echinodermata marine organisms; have spiny or leathery skin and a water-vascular system with tube feet; are radially symmetrical; includes sea stars, sand dollars, and sea urchins

Phylum Chordata organisms with internal skeletons and specialized body systems; most have paired appendages; all at some time have a notochord, nerve cord, gill slits, and a post-anal tail; include fish, amphibians, reptiles, birds, and mammals

PERIODIC TABLE OF THE ELEMENTS

Columns of elements are called groups. Elements in the same group have similar chemical properties.

Key: Element — Hydrogen; Atomic number — 1; Symbol — H; Atomic mass — 1.008; State of matter

Gas / Liquid / Solid / Synthetic

The first three symbols tell you the state of matter of the element at room temperature. The fourth symbol identifies elements that are not present in significant amounts on Earth. Useful amounts are made synthetically.

Period	1	2	3	4	5	6	7	8	9
1	Hydrogen 1 H 1.008 (Gas)								
2	Lithium 3 Li 6.941 (Solid)	Beryllium 4 Be 9.012 (Solid)							
3	Sodium 11 Na 22.990 (Solid)	Magnesium 12 Mg 24.305 (Solid)							
4	Potassium 19 K 39.098 (Solid)	Calcium 20 Ca 40.078 (Solid)	Scandium 21 Sc 44.956 (Solid)	Titanium 22 Ti 47.867 (Solid)	Vanadium 23 V 50.942 (Solid)	Chromium 24 Cr 51.996 (Solid)	Manganese 25 Mn 54.938 (Solid)	Iron 26 Fe 55.845 (Solid)	Cobalt 27 Co 58.933 (Solid)
5	Rubidium 37 Rb 85.468 (Solid)	Strontium 38 Sr 87.62 (Solid)	Yttrium 39 Y 88.906 (Solid)	Zirconium 40 Zr 91.224 (Solid)	Niobium 41 Nb 92.906 (Solid)	Molybdenum 42 Mo 95.94 (Solid)	Technetium 43 Tc (98) (Synthetic)	Ruthenium 44 Ru 101.07 (Solid)	Rhodium 45 Rh 102.906 (Solid)
6	Cesium 55 Cs 132.905 (Solid)	Barium 56 Ba 137.327 (Solid)	Lanthanum 57 La 138.906 (Solid)	Hafnium 72 Hf 178.49 (Solid)	Tantalum 73 Ta 180.948 (Solid)	Tungsten 74 W 183.84 (Solid)	Rhenium 75 Re 186.207 (Solid)	Osmium 76 Os 190.23 (Solid)	Iridium 77 Ir 192.217 (Solid)
7	Francium 87 Fr (223) (Solid)	Radium 88 Ra (226) (Solid)	Actinium 89 Ac (227) (Solid)	Rutherfordium 104 Rf (261) (Synthetic)	Dubnium 105 Db (262) (Synthetic)	Seaborgium 106 Sg (266) (Synthetic)	Bohrium 107 Bh (264) (Synthetic)	Hassium 108 Hs (277) (Synthetic)	Meitnerium 109 Mt (268) (Synthetic)

The number in parentheses is the mass number of the longest-lived isotope for that element.

Rows of elements are called periods. Atomic number increases across a period.

The arrow shows where these elements would fit into the periodic table. They are moved to the bottom of the table to save space.

Series					
Lanthanide series	Cerium 58 Ce 140.116 (Solid)	Praseodymium 59 Pr 140.908 (Solid)	Neodymium 60 Nd 144.24 (Solid)	Promethium 61 Pm (145) (Synthetic)	Samarium 62 Sm 150.36 (Solid)
Actinide series	Thorium 90 Th 232.038 (Solid)	Protactinium 91 Pa 231.036 (Solid)	Uranium 92 U 238.029 (Solid)	Neptunium 93 Np (237) (Synthetic)	Plutonium 94 Pu (244) (Synthetic)

Metal

Metalloid

Nonmetal

The color of an element's block tells you if the element is a metal, nonmetal, or metalloid.

ScienceOnline

Visit green.msscience.com for updates to the periodic table.

10	11	12	13	14	15	16	17	18
								Helium 2 **He** 4.003
			Boron 5 **B** 10.811	Carbon 6 **C** 12.011	Nitrogen 7 **N** 14.007	Oxygen 8 **O** 15.999	Fluorine 9 **F** 18.998	Neon 10 **Ne** 20.180
			Aluminum 13 **Al** 26.982	Silicon 14 **Si** 28.086	Phosphorus 15 **P** 30.974	Sulfur 16 **S** 32.065	Chlorine 17 **Cl** 35.453	Argon 18 **Ar** 39.948
Nickel 28 **Ni** 58.693	Copper 29 **Cu** 63.546	Zinc 30 **Zn** 65.409	Gallium 31 **Ga** 69.723	Germanium 32 **Ge** 72.64	Arsenic 33 **As** 74.922	Selenium 34 **Se** 78.96	Bromine 35 **Br** 79.904	Krypton 36 **Kr** 83.798
Palladium 46 **Pd** 106.42	Silver 47 **Ag** 107.868	Cadmium 48 **Cd** 112.411	Indium 49 **In** 114.818	Tin 50 **Sn** 118.710	Antimony 51 **Sb** 121.760	Tellurium 52 **Te** 127.60	Iodine 53 **I** 126.904	Xenon 54 **Xe** 131.293
Platinum 78 **Pt** 195.078	Gold 79 **Au** 196.967	Mercury 80 **Hg** 200.59	Thallium 81 **Tl** 204.383	Lead 82 **Pb** 207.2	Bismuth 83 **Bi** 208.980	Polonium 84 **Po** (209)	Astatine 85 **At** (210)	Radon 86 **Rn** (222)
Darmstadtium 110 **Ds** (281)	Roentgenium 111 **Rg** (272)	Ununbium * 112 **Uub** (285)		Ununquadium * 114 **Uuq** (289)				

* The names and symbols for elements 112 and 114 are temporary. Final names will be selected when the elements' discoveries are verified.

Europium 63 **Eu** 151.964	Gadolinium 64 **Gd** 157.25	Terbium 65 **Tb** 158.925	Dysprosium 66 **Dy** 162.500	Holmium 67 **Ho** 164.930	Erbium 68 **Er** 167.259	Thulium 69 **Tm** 168.934	Ytterbium 70 **Yb** 173.04	Lutetium 71 **Lu** 174.967
Americium 95 **Am** (243)	Curium 96 **Cm** (247)	Berkelium 97 **Bk** (247)	Californium 98 **Cf** (251)	Einsteinium 99 **Es** (252)	Fermium 100 **Fm** (257)	Mendelevium 101 **Md** (258)	Nobelium 102 **No** (259)	Lawrencium 103 **Lr** (262)

Cómo usar el glosario en español:
1. Busca el término en inglés que desees encontrar.
2. El término en español, junto con la definición, se encuentran en la columna de la derecha.

Pronunciation Key

Use the following key to help you sound out words in the glossary.

a	back (BAK)	**ew**	food (FEWD)
ay	day (DAY)	**yoo**	pure (PYOOR)
ah	father (FAH thur)	**yew**	few (FYEW)
ow	flower (FLOW ur)	**uh**	comma (CAH muh)
ar	car (CAR)	**u** (+ con)	rub (RUB)
e	less (LES)	**sh**	shelf (SHELF)
ee	leaf (LEEF)	**ch**	nature (NAY chur)
ih	trip (TRIHP)	**g**	gift (GIHFT)
i (i + con + e)	idea (i DEE uh)	**j**	gem (JEM)
oh	go (GOH)	**ing**	sing (SING)
aw	soft (SAWFT)	**zh**	vision (VIH zhun)
or	orbit (OR buht)	**k**	cake (KAYK)
oy	coin (COYN)	**s**	seed, cent (SEED, SENT)
oo	foot (FOOT)	**z**	zone, raise (ZOHN, RAYZ)

English — A — Español

abiotic (ay bi AHT ihk) factor: any nonliving part of the environment, such as water, sunlight, temperature, and air. (p. 532)

acceleration: change in velocity divided by the amount of time needed for that change to take place; occurs when an object speeds up, slows down, or changes direction. (p. 687)

acid: substance that releases H^+ ions and produces hydronium ions when dissolved in water. (p. 634)

acid precipitation: precipitation with a pH below 5.6—which occurs when air pollutants from the burning of fossil fuels react with water in the atmosphere to form strong acids—that can pollute water, kill fish and plants, and damage soils. (p. 569)

active immunity: long-lasting immunity that results when the body makes its own antibodies in response to a specific antigen. (p. 380)

active transport: energy-requiring process in which transport proteins bind with particles and move them through a cell membrane. (p. 257)

adaptation: any structural or behavioral change that helps an organism survive in its particular environment. (pp. 152, 338)

factor abiótico: cualquier parte no viva del medio ambiente, tal como el agua, la luz solar, la temperatura y el aire. (p. 532)

aceleración: cambio de velocidad dividido por la cantidad de tiempo necesario para que se realice dicho cambio; ocurre cuando un objeto viaja más rápido, más lento o cambia de dirección. (p. 687)

ácido: sustancia que libera iones H+ y produce iones de hidronio al ser disuelta en agua. (p. 634)

lluvia ácida: precipitación con un pH menor de 5.6—lo cual ocurre cuando los contaminantes del aire provenientes de la quema de combustibles fósiles reaccionan con el agua en la atmósfera para formar ácidos fuertes—que puede contaminar el agua, matar peces y plantas, y dañar los suelos. (p. 569)

inmunidad activa: inmunidad duradera que se presenta cuando el cuerpo crea sus propios anticuerpos como respuesta a un antígeno específico. (p. 380)

transporte activo: proceso que requiere energía y en el cual las proteínas de transporte se unen con partículas y las trasladan a través de la membrana celular. (p. 257)

adaptación: cualquier cambio de estructura o comportamiento que ayude a un organismo a sobrevivir en su medio ambiente particular. (pp. 152, 338)

air mass: large body of air that has the same characteristics of temperature and moisture content as the part of Earth's surface over which it formed. (p. 126)

masa de aire: gran cuerpo de aire que tiene las mismas características de temperatura y contenido de humedad que la parte de la superficie terrestre sobre la cual se formó. (p. 126)

allele (uh LEEL): an alternate form that a gene may have for a single trait; can be dominant or recessive. (p. 306)

alelo: forma alternativa que un gen puede tener para un rasgo único; puede ser dominante o recesivo. (p. 306)

allergen: substance that causes an allergic reaction. (p. 387)

alergeno: sustancia que produce una reacción alérgica. (p. 387)

alternative resource: new renewable or inexhaustible energy source; includes solar energy, wind, and geothermal energy. (p. 733)

recurso alternativo: nueva fuente de energía renovable o inagotable; incluye energía solar, eólica y geotérmica. (p. 733)

alveoli (al VEE uh li): tiny, thin-walled, grapelike clusters at the end of each bronchiole that are surrounded by capillaries; carbon dioxide and oxygen exchange takes place. (p. 414)

alvéolos: racimos parecidos a las uvas, pequeños y de paredes finas encontrados en el extremo de cada bronquiolo, los cuales están rodeados de capilares y en donde se realiza el intercambio de dióxido de carbono y oxígeno. (p. 414)

amino acid: building block of protein. (p. 405)

aminoácido: bloque formador de las proteínas. (p. 405)

amniotic (am nee AH tihk) sac: thin, liquid-filled, protective membrane that forms around the embryo. (p. 481)

saco amniótico: membrana protectora delgada y llena de líquido que se forma alrededor del embrión. (p. 481)

angiosperms: flowering vascular plants that produce fruits containing one or more seeds; monocots and dicots. (p. 517)

angiospermas: plantas vasculares que producen flores y frutos que contienen una o más semillas; pueden ser monocotiledóneas o dicotiledóneas. (p. 517)

antibody: a protein made in response to a specific antigen that can attach to the antigen and cause it to be useless. (p. 379)

anticuerpo: proteína creada como respuesta a un antígeno específico y que se puede adherir al antígeno inutilizándolo. (p. 379)

antigen (AN tih jun): complex molecule that is foreign to your body. (p. 379)

antígeno: molécula compleja extraña al cuerpo. (p. 379)

aqueous (A kwee us): solution in which water is the solvent. (p. 626)

acuoso: solución en la cual el agua es el solvente. (p. 626)

Archimedes' (ar kuh MEE deez) principle: states that the buoyant force on an object is equal to the weight of the fluid displaced by the object. (p. 671)

principio de Arquímedes: establece que la fuerza de empuje ejercida sobre un objeto es igual al peso del fluido desplazado por dicho objeto. (p. 671)

artery: blood vessel that carries blood away from the heart. (p. 374)

arteria: vaso sanguíneo que transporta la sangre desde el corazón. (p. 374)

asexual reproduction: a type of reproduction—fission, budding, and regeneration—in which a new organism is produced from one organism and has DNA identical to the parent organism. (p. 281)

reproducción asexual: tipo de reproducción—fisión, gemación y regeneración—en el que un organismo da origen a uno nuevo de ADN idéntico al organismo progenitor. (p. 281)

asteroid: small, rocky space object found in the asteroid belt between the orbits of Jupiter and Mars. (p. 200)

asteroide: pequeño objeto espacial rocoso encontrado en el cinturón de asteroides entre las órbitas de Júpiter y Marte. (p. 200)

astronomical unit: unit used to measure distances in the solar system; 1 AU equals 150,000,000 km. (p. 194)

unidad astronómica: unidad usada para medir distancias en el sistema solar; una unidad astronómica es igual a 150 millones de kilómetros. (p. 194)

atmosphere: Earth's air, which is made up of a thin layer of gases, solids, and liquids; forms a protective layer around the planet and is divided into five distinct layers. (p. 89)

atmósfera: el aire de la Tierra; está compuesta por una capa fina de gases, sólidos y líquidos, forma una capa protectora alrededor del planeta y está dividida en cinco capas distintas. (p. 89)

axis: imaginary line around which Earth spins; drawn from the north geographic pole through Earth to the south geographic pole. (p. 180)

eje terrestre: línea imaginaria alrededor de la cual gira la Tierra; se traza desde el polo norte geográfico al polo sur geográfico a través de la Tierra. (p. 180)

B

balanced forces: forces that are equal but opposite in direction; when they act on an object, they cancel each other out, and no change occurs in the object's motion. (p. 692)

fuerzas balanceadas: fuerzas que son iguales pero en direcciones opuestas; cuando actúan sobre un objeto se cancelan unas a otras pero no se produce ningún cambio en el movimiento del objeto. (p. 692)

basaltic: describes dense, dark-colored igneous rock formed from magma rich in magnesium and iron and poor in silica. (p. 65)

basáltica: describe roca ígnea densa de color oscuro que se forma a partir de magma rico en magnesio y hierro pero pobre en sílice. (p. 65)

base: substance that accepts H^+ ions and produces hydroxide ions when dissolved in water. (p. 637)

base: sustancia que acepta los iones H^+ y produce iones de hidróxido al ser disuelta en agua. (p. 637)

binomial nomenclature (bi NOH mee ul • NOH mun klay chur): two-word naming system that gives all organisms their scientific name. (p. 218)

nomenclatura binaria: sistema de dos palabras que da a todos los organismos su nombre científico. (p. 218)

biosphere (BI uh sfihr): part of Earth that supports life—the top part of Earth's crust, all the waters covering Earth's surface, and the surrounding atmosphere; includes all biomes, ecosystems, communities, and populations. (p. 537)

biosfera: parte de la Tierra que alberga la vida, es decir, la parte superior de la corteza terrestre, toda el agua que cubre la superficie terrestre y la atmósfera circundante; incluye todas las biomasas, ecosistemas, comunidades y poblaciones. (p. 537)

biotic (bi AH tihk) factor: any living or once-living organism in the environment. (p. 532)

factor biótico: cualquier organismo vivo, o que estuvo vivo, del medio ambiente. (p. 532)

bladder: elastic, muscular organ that holds urine until it leaves the body through the urethra. (p. 421)

vejiga: órgano muscular elástico que retiene la orina hasta que ésta sale del cuerpo a través de la uretra. (p. 421)

blizzard: winter storm that lasts at least three hours with temperatures of −12°C or below, poor visibility, and winds of at least 51 km/h. (p. 133)

nevasca: tormenta invernal que dura por lo menos tres horas con temperaturas de −12°C o menores, escasa visibilidad y vientos de por lo menos 51 km/h. (p. 133)

bronchi (BRAHN ki): two short tubes that branch off the lower end of the trachea and carry air into the lungs. (p. 414)

bronquios: dos tubos cortos que salen del extremo inferior de la tráquea y llevan el aire a los pulmones. (p. 414)

buoyant force: upward force exerted on an object immersed in a fluid. (p. 670)

fuerza de empuje: fuerza ascendente ejercida sobre un objeto inmerso en un fluido. (p. 670)

C

cambium (KAM bee um): vascular tissue that produces xylem and phloem cells as a plant grows. (p. 515)

cámbium: tejido vascular que produce las células del xilema y floema conforme crece la planta. (p. 515)

capillary (KAP uh ler ee): blood vessel that connects arteries and veins. (p. 374)

capilar: vaso sanguíneo que conecta las arterias y las venas. (p. 374)

carbohydrate (kar boh HI drayt): nutrient that usually is the body's main source of energy. (p. 406)

carbohidrato: nutriente que por lo general es la principal fuente de energía del cuerpo. (p. 406)

cartilage: thick, smooth, flexible and slippery tissue layer that covers the ends of bones, makes movement easier by reducing friction, and absorbing shocks. (p. 445)

cartílago: capa de tejido delgado, liso, flexible y resbaladizo que cubre los extremos de los huesos, facilita el movimiento al reducir la fricción y absorbe los golpes. (p. 445)

cell: smallest unit of an organism that can carry on life functions. (p. 214)

célula: la unidad más pequeña de un organismo que puede llevar a cabo funciones biológicas. (p. 214)

cell membrane: protective outer covering of all cells that regulates the interaction between the cell and the environment. (p. 224)

cell theory: states that all organisms are made up of one or more cells, the cell is the basic unit of life, and all cells come from other cells. (p. 221)

cell wall: rigid structure that encloses, supports, and protects the cells of plants, algae, fungi, and most bacteria. (p. 224)

cellulose (SEL yuh lohs): chemical compound made out of sugar; forms tangled fibers in the cell walls of many plants and provides structure and support. (p. 502)

cementation: sedimentary rock-forming process in which sediment grains are held together by natural cements that are produced when water moves through rock and soil. (p. 73)

central nervous system: division of the nervous system, made up of the brain and spinal cord. (p. 451)

chemical change: change in which the composition of a substance changes. (p. 602)

chemical energy: energy stored in chemical bonds. (p. 719)

chemical property: characteristic that cannot be observed without altering the sample. (p. 598)

chlorofluorocarbons (CFCs): group of chemical compounds used in refrigerators, air conditioners, foam packaging, and aerosol sprays that may enter the atmosphere and destroy ozone. (p. 96)

chloroplast: green, chlorophyll-containing, plant-cell organelle that captures light energy, which is used to make sugar. (p. 226)

chromosome: structure in a cell's nucleus that contains hereditary material. (p. 278)

chyme (KIME): liquid product of digestion. (p. 403)

cleavage: physical property of some minerals that causes them to break along smooth, flat surfaces. (p. 39)

climate: average weather pattern in an area over a long period of time; can be classified by temperature, humidity, precipitation, and vegetation. (p. 148)

comet: space object made of rocky particles and ice that forms a tail when orbiting near the Sun. (p. 200)

community: all of the populations of different species in a given area that interact in some way and depend on one another for food, shelter, and other needs. (p. 536)

compaction: process that forms sedimentary rocks when layers of sediments are compressed by the weight of the layers above them. (p. 72)

membrana celular: cubierta externa protectora de todas las células que regula la interacción entre la célula y su medio ambiente. (p. 224)

teoría celular: establece que todos los organismos están formados por una o más células, que las células son la unidad básica de la vida y que las células provienen de otras células. (p. 221)

pared celular: estructura rígida que rodea, mantiene y protege las células de las plantas, algas, hongos y la mayoría de las bacterias. (p. 224)

celulosa: compuesto químico formado por azúcares y que forma fibras intrincadas en la pared celular de muchas plantas proporcionando estructura y soporte. (p. 502)

cementación: proceso de formación de la roca sedimentaria en el que las partículas de sedimento están unidas por cementos naturales producidos cuando el agua se mueve a través de la roca y el suelo. (p. 73)

sistema nervioso central: división del sistema nervioso compuesto por el cerebro y la médula espinal. (p. 451)

cambio químico: cambio en el cual la composición de una sustancia es modificada. (p. 602)

energía química: energía almacenada en enlaces químicos. (p. 719)

propiedad química: característica que no puede ser observada sin alterar la muestra. (p. 598)

clorofluorocarbonos (CFCs): grupo de compuestos químicos usados en refrigeradores, acondicionadores de aire, espumas de empaque y aerosoles; pueden entrar en la atmósfera y destruir el ozono. (p. 96)

cloroplasto: organelo verde de las células vegetales que contiene clorofila y utiliza la energía de la luz para producir azúcar. (p. 226)

cromosoma: estructura en el núcleo celular que contiene el material hereditario. (p. 278)

quimo: líquido producido durante la digestión. (p. 403)

exfoliación: propiedad física de algunos minerales que causa que se rompan junto a superficies planas y lisas. (p. 39)

clima: modelo meteorológico en un área durante un periodo de tiempo largo; puede clasificarse por temperatura, humedad, precipitación y vegetación. (p. 148)

cometa: objeto espacial hecho de partículas rocosas y hielo, y que forma una cola cuando gira cerca del sol. (p. 200)

comunidad: todas las poblaciones de diferentes especies en una determinada área, que interactúan de alguna forma y que dependen unas de otras en cuanto a alimento, refugio y otras necesidades. (p. 536)

compactación: proceso que forma rocas sedimentarias cuando las capas de sedimento son comprimidas por el peso de las capas superiores. (p. 72)

Glossary/Glosario

concentration: describes how much solute is present in a solution compared to the amount of solvent. (p. 631)

condensation: change of matter from a gas to a liquid state. (pp. 101, 601, 659)

conduction: transfer of energy that occurs when molecules bump into each other. (p. 100)

constant: factor that stays the same through all phases of an experiment. (p. 16)

control: standard used for comparison in an experiment. (p. 16)

convection: transfer of heat by the flow of material. (p. 100)

Coriolis (kor ee OH lus) effect: causes moving air and water to turn left in the southern hemisphere and turn right in the northern hemisphere due to Earth's rotation. (p. 104)

crater: depression formed by impact of meteorites or comets; the more craters in a region, the older the surface. (p. 185)

crystal: solid in which the atoms are arranged in an orderly, repeating pattern. (p. 31)

cuticle (KYEW tih kul): waxy, protective layer that covers the stems, leaves, and flowers of many plants and helps prevent water loss. (p. 502)

cytoplasm: constantly moving gelatinlike mixture inside the cell membrane that contains heredity material and is the location of most of a cell's life processes. (p. 225)

concentración: describe la cantidad de soluto presente en una solución, comparada con la cantidad de solvente. (p. 631)

condensación: cambio de estado de la materia de gas a líquido. (pp. 101, 601, 659)

conducción: transferencia de energía que ocurre cuando las moléculas chocan unas con otras. (p. 100)

constante: factor que permanece constante durante todas las fases de un experimento. (p. 16)

control: estándar usado en un experimento para hacer comparaciones. (p. 16)

convección: transferencia de calor mediante flujo de material. (p. 100)

efecto de Coriolis: causa el movimiento del aire y agua hacia la izquierda en el hemisferio sur y hacia la derecha en el hemisferio norte; este efecto es debido a la rotación de la Tierra. (p. 104)

cráter: depresión formada por el impacto de meteoritos o cometas; entre más cráteres tenga una región, más vieja es la superficie. (p. 185)

cristal: sólido en el que los átomos están alineados en forma ordenada y repetitiva. (p. 31)

cutícula: capa cerosa protectora que recubre el tronco, hojas y flores de muchas plantas y ayuda a prevenir la pérdida de agua. (p. 502)

citoplasma: mezcla de apariencia gelatinosa que se mueve constantemente en el interior de las membranas de las células y contienen material hereditario; es donde se realizan la mayoría de los procesos biológicos de una célula. (p. 225)

D

deforestation: destruction and cutting down of forests—often to clear land for mining, roads, and grazing of cattle—resulting in increased atmospheric CO_2 levels. (p. 165)

density: mass of an object divided by its volume. (p. 671)

dependent variable: factor that will be measured in an experiment. (p. 16)

deposition: the process by which a gas changes into a solid. (p. 601)

dermis: skin layer below the epidermis that contains blood vessels, nerves, oil, and sweat glands, and other structures. (p. 434)

dew point: temperature at which air is saturated and condensation forms. (p. 121)

deforestación: destrucción y tala de los bosques—a menudo el despeje de la tierra para minería, carreteras y ganadería—resultando en el aumento de los niveles atmosféricos de dióxido de carbono. (p. 165)

densidad: masa de un objeto dividida por su volumen. (p. 671)

variable dependiente: factor que va a ser medido en un experimento. (p. 16)

deposición: el proceso mediante el cual un gas pasa a ser sólido. (p. 601)

dermis: capa de piel debajo de la epidermis que contiene vasos sanguíneos, nervios, grasa y glándulas sudoríparas, además de otra estructuras. (p. 434)

punto de condensación: temperatura a la que el aire se satura y se genera la condensación. (p. 121)

dicot: angiosperm with two cotyledons inside its seed, flower parts in multiples of four or five, and vascular bundles in rings. (p. 518)

dicotiledónea: angiosperma con dos cotiledones dentro de su semilla, partes florales en múltiplos de cuatro o cinco y haces vasculares distribuidos en anillos. (p. 518)

diffusion: a type of passive transport in cells in which molecules move from areas where there are more of them to areas where there are fewer of them. (p. 255)

difusión: tipo de transporte pasivo en las células en el que las moléculas se mueven de áreas de mayor concentración de éstas hacia áreas de menor concentración. (p. 255)

diploid (DIHP loyd): cell whose similar chromosomes occur in pairs. (p. 284)

diploide: célula cuyos cromosomas similares están en pares. (p. 284)

displacement: distance and direction from a starting point to an ending point. (p. 684)

desplazamiento: distancia y dirección desde un punto inicial hasta un punto final. (p. 684)

DNA: deoxyribonucleic acid; the genetic material of all organisms; made up of two twisted strands of sugar-phosphate molecules and nitrogen bases. (p. 290)

ADN: ácido desoxirribonucleico; material genético de todos los organismos constituido por dos cadenas trenzadas de moléculas de azúcar-fosfato y bases de nitrógeno (p. 290)

dominant (DAH muh nunt): describes a trait that covers over, or dominates, another form of that trait. (p. 308)

dominante: describe un rasgo que encubre o domina a otra forma de ese rasgo. (p. 308)

E

ecology: study of all the interactions among organisms and their environment. (p. 532)

ecología: estudio de todas las interacciones entre los organismos y su medio ambiente. (p. 532)

ecosystem: all of the communities in a given area and the abiotic factors that affect them. (p. 536)

ecosistema: todas las comunidades en un área determinada y los factores abióticos que las afectan. (p. 536)

egg: haploid sex cell formed in the female reproductive organs. (p. 284)

óvulo: célula sexual haploide que se forma en los órganos reproductivos femeninos. (p. 284)

El Niño (el NEEN yoh): climatic event that begins in the tropical Pacific Ocean; may occur when trade winds weaken or reverse, and can disrupt normal temperature and precipitation patterns around the world. (p. 157)

El Niño: evento climático que comienza en el Océano Pacífico tropical; puede ocurrir cuando los vientos alisios se debilitan o se invierten; puede desestabilizar los patrones normales de precipitación y temperatura del mundo. (p. 157)

electrical energy: energy carried by electric current. (p. 720)

energía eléctrica: energía transportada por corriente eléctrica. (p. 720)

embryo: fertilized egg that has attached to the wall of the uterus. (p. 481)

embrión: óvulo fertilizado que se ha adherido a la pared del útero. (p. 481)

embryology (em bree AH luh jee): study of embryos and their development. (p. 347)

embriología: el estudio de los embriones y su desarrollo. (p. 347)

endocytosis (en duh si TOH sus): process by which a cell takes in a substance by surrounding it with the cell membrane. (p. 258)

endocitosis: proceso mediante el cual una célula capta una sustancia rodeándola con su membrana celular. (p. 258)

endoplasmic reticulum (ER): cytoplasmic organelle that moves materials around in a cell and is made up of a complex series of folded membranes; can be rough (with attached ribosomes) or smooth (without attached ribosomes). (p. 228)

retículo endoplásmico (RE): organelo citoplásmico que mueve los materiales en el interior de una célula y está compuesto por una serie compleja de membranas plegadas; puede ser áspero (con ribosomas incorporados) o liso (sin ribosomas incorporados). (p. 228)

energy: the ability to cause change. (p. 716)

energía: capacidad de producir cambios. (p. 716)

enzyme: a type of protein that regulates chemical reactions in cells without being changed or used up itself. (pp. 251, 400)

epidermis: outer, thinnest skin layer, that constantly produces new cells to replace the dead cells that are rubbed off its surface. (p. 434)

equilibrium: occurs when molecules of one substance are spread evenly throughout another substance. (p. 255)

equinox: twice-yearly time when the Sun is directly above Earth's equator and there are equal hours of day and night. (p. 183)

erosion: movement of soil from one place to another. (p. 575)

evolution: change in inherited characteristics over time. (p. 334)

exocytosis (ek soh si TOH sus): process by which vesicles release their contents outside the cell. (p. 258)

extrusive: describes fine-grained igneous rock that forms when magma cools quickly at or near Earth's surface. (p. 63)

enzima: tipo de proteína que regula las reacciones químicas en las células sin que ésta sufra modificaciones o se agote. (pp. 251, 400)

epidermis: la capa más delgada y externa de la piel que constantemente produce células nuevas para reemplazar a las células muertas que han sido eliminadas de la superficie. (p. 434)

equilibrio: ocurre cuando las moléculas de una sustancia están diseminadas completa y uniformemente a lo largo de otra sustancia. (p. 255)

equinoccio: momento que ocurre dos veces al año cuando el sol está directamente por encima del ecuador de la Tierra, y el día y la noche tienen el mismo número de horas. (p. 183)

erosión: movimiento del suelo de un lugar a otro. (p. 575)

evolución: cambio en las características heredadas a través del tiempo. (p. 334)

exocitosis: proceso mediante el cual las vesículas liberan su contenido fuera de la célula. (p. 258)

extrusivo: describe rocas ígneas de grano fino que se forman cuando el magma se enfría rápidamente en o cerca de la superficie terrestre. (p. 63)

F

fermentation: process by which oxygen-lacking cells and some one-celled organisms release small amounts of energy from glucose molecules and produce wastes such as alcohol, carbon dioxide, and lactic acid. (p. 264)

fertilization: in sexual reproduction, the joining of a sperm and egg. (p. 284)

fetal stress: can occur during the birth process or after birth as an infant adjusts from a watery, dark, constant-temperature environment to its new environment. (p. 484)

fetus: in humans, a developing baby after the first two months of pregnancy until birth. (p. 482)

first law of motion: states that an object will remain at rest or move in a straight line at a constant speed unless it is acted upon by a force. (p. 691)

fog: a stratus cloud that forms when air is cooled to its dew point near the ground. (p. 123)

foliated: describes metamorphic rock, such as slate and gneiss, whose mineral grains line up in parallel layers. (p. 69)

fermentación: proceso mediante el cual las células carentes de oxígeno y algunos organismos unicelulares liberan pequeñas cantidades de energía a partir de moléculas de glucosa y producen desechos como alcohol, dióxido de carbono y ácido láctico. (p. 264)

fertilización: en la reproducción sexual, la unión de un óvulo y un espermatozoide. (p. 284)

estrés fetal: puede ocurrir durante el proceso del nacimiento o luego del mismo mientras un nuevo ser humano se adapta de un ambiente acuoso, oscuro y de temperatura constante a su nuevo ambiente. (p. 484)

feto: en los humanos, bebé en desarrollo desde los primeros dos meses de embarazo hasta el nacimiento. (p. 482)

primera ley del movimiento: establece que un objeto se mantiene en reposo o se mueve en línea recta a una velocidad constante a menos que actúe una fuerza sobre él. (p. 691)

niebla: nube de estrato que se forma cuando el aire se enfría a su punto de condensación cerca del suelo. (p. 123)

foliado: describe rocas metamórficas, como pizarra y gneis, cuyas vetas minerales se alinean en capas paralelas. (p. 69)

food chain: model that describes how energy passes from one organism to another. (p. 544)

food web: model that describes how energy from food moves through a community; a series of overlapping food chains. (p. 546)

force: push or a pull; has a size and direction. (p. 690)

fossil fuels: nonrenewable energy resources—coal, oil, and natural gas—that formed in Earth's crust over hundreds of millions of years. (p. 562)

fracture: physical property of some minerals that causes them to break with uneven, rough, or jagged surfaces. (p. 39)

freezing: change of matter from a liquid state to a solid state. (p. 661)

friction: force between two surfaces in contact that resists the sliding of the surfaces past each other. (p. 697)

front: boundary between two air masses with different temperatures, density, or moisture; can be cold, warm, occluded, and stationary. (p. 127)

cadena alimenticia: modelo que describe la forma como pasa la energía de un organismo a otro. (p. 544)

red alimenticia: modelo que describe cómo la energía proveniente de los alimentos pasa a una comunidad; es una serie de cadenas alimenticias entrelazadas. (p. 546)

fuerza: empujar o halar; tiene tamaño y dirección. (p. 690)

combustibles fósiles: recursos energéticos no renovables —carbón, petróleo y gas natural—que se formaron en la corteza terrestre durante cientos de millones de años. (p. 562)

fractura: propiedad física de algunos minerales que causa que se rompan formando superficies irregulares, ásperas o dentadas. (p. 39)

congelación: cambio de la materia de estado líquido a sólido. (p. 661)

fricción: fuerza entre dos superficies en contacto que opone resistencia al deslizamiento de las superficies al pasar una sobre la otra. (p. 697)

frente: límite entre dos masas de aire con temperatura, densidad o humedad diferentes; puede ser frío, caliente, ocluido o estacionario. (p. 127)

G

gas: matter that does not have a definite shape or volume; has particles that move at high speeds in all directions. (p. 656)

gem: beautiful, rare, highly prized mineral that can be worn in jewelry. (p. 41)

gene: section of DNA on a chromosome that contains instructions for making specific proteins. (p. 292)

generator: device that transforms kinetic energy into electrical energy. (p. 726)

genetic engineering: biological and chemical methods to change the arrangement of a gene's DNA to improve crop production, produce large volumes of medicine, and change how cells perform their normal functions. (p. 321)

genetics (juh NEH tihks): the study of how traits are inherited through the actions of alleles. (p. 306)

genotype (JEE nuh tipe): the genetic makeup of an organism. (p. 310)

genus: first word of the two-word scientific name used to identify a group of similar species. (p. 218)

gas: materia que no tiene ni forma ni volumen definidos; tiene partículas que se mueven a altas velocidades y en todas las direcciones. (p. 656)

gema: mineral hermoso, raro y altamente valorado que puede usarse como joya. (p. 41)

gen: sección de ADN en un cromosoma, el cual contiene instrucciones para la formación de proteínas específicas. (p. 292)

generador: dispositivo que transforma la energía cinética en energía eléctrica. (p. 726)

ingeniería genética: métodos biológicos y químicos para cambiar la disposición del ADN de un gen y así mejorar la producción de cosechas, producir grandes volúmenes de un medicamento, o cambiar la forma en que las células realizan sus funciones normales. (p. 321)

genética: estudio de la forma como se heredan los rasgos a través de las acciones de los alelos. (p. 306)

genotipo: composición genética de un organismo. (p. 310)

género: primera palabra del nombre científico de dos palabras usado para identificar un grupo de especies similares. (p. 218)

geothermal energy: heat energy within Earth's crust, available only where natural geysers or volcanoes are located. (p. 565)

global warming: increase in the average global temperature of Earth. (p. 164)

Golgi bodies: organelles that sort and package cellular materials and transport them within the cell or out of the cell. (p. 228)

gradualism: model describing evolution as a slow process by which one species changes into a new species through a continuing series of mutations and variations over time. (p. 340)

granitic: describes generally light-colored, silica-rich igneous rock that is less dense than basaltic rock. (p. 65)

gravitational force: force that every object exerts on every other object; depends on the masses of the objects and the distance between them. (p. 696)

greenhouse effect: heat-trapping feature of the atmosphere that occurs when certain gases in Earth's atmosphere, such as methane, CO_2, and water vapor, trap heat. (pp. 158, 575)

guard cells: pairs of cells that surround stomata and control their opening and closing. (p. 513)

gymnosperms: vascular plants that do not flower, generally have needlelike or scalelike leaves, and produce seeds that are not protected by fruit; conifers, cycads, ginkgoes, and gnetophytes. (p. 516)

energía geotérmica: energía calórica en el interior de la corteza terrestre disponible sólo donde existen géiseres o volcanes. (p. 565)

calentamiento global: incremento del promedio de la temperatura global. (p. 164)

aparato de Golgi: organelos que clasifican y recogen materiales celulares y los transportan hacia dentro o hacia afuera de la célula. (p. 228)

gradualismo: modelo que describe la evolución como un proceso lento mediante el cual una especie existente se convierte en una especie nueva a través de series continuas de mutaciones y variaciones a través del tiempo. (p. 340)

granítica: describe roca ígnea rica en sílice, generalmente de color claro y menos densa que la rocas basáltica. (p. 65)

fuerza gravitacional: fuerza que ejerce un objeto sobre otro, la cual depende de las masas de los objetos y de la distancia entre ellos. (p. 696)

efecto de invernadero: característica de la atmósfera que le permite atrapar calor y que ocurre cuando ciertos gases en la atmósfera terrestre, como el metano, el dióxido de carbono y el vapor de agua atrapan el calor. (pp. 158, 575)

células oclusoras: pares de células que rodean al estoma y que controlan su cierre y apertura. (p. 513)

gimnospermas: plantas vasculares que no florecen, generalmente tienen hojas en forma de aguja o de escama y producen semillas que no están protegidas por el fruto; se clasifican en coníferas, cicadáceas, ginkgoales y gnetofitas. (p. 516)

H

habitat: place where an organism lives. (p. 543)

haploid (HAP loyd): cell that has half the number of chromosomes as body cells. (p. 285)

hardness: measure of how easily a mineral can be scratched. (p. 37)

hazardous wastes: waste materials, such as pesticides and leftover paints, that are harmful to human health or poisonous to living organisms. (p. 576)

heat: movement of thermal energy from a substance at a higher temperature to a substance at a lower temperature. (p. 658)

hemoglobin (HEE muh gloh bun): a molecule in red blood cells that carries oxygen and carbon dioxide. (p. 367)

heredity (huh REH duh tee): the passing of traits from parent to offspring. (p. 306)

hábitat: lugar donde viven los organismos. (p. 543)

haploide: célula que posee la mitad del número de cromosomas que tienen las células somáticas. (p. 285)

dureza: medida de la facilidad con que un mineral puede ser rayado. (p. 37)

desperdicios peligrosos: materiales de desecho como los pesticidas y residuos de pintura nocivos para la salud humana o dañinos para los organismos vivos. (p. 576)

calor: movimiento de energía térmica de una sustancia que se encuentra a una alta temperatura hacia una sustancia a una baja temperatura. (p. 658)

hemoglobina: molécula presente en los glóbulos rojos que transporta oxígeno y dióxido de carbono. (p. 367)

herencia: transferencia de rasgos de un progenitor a su descendencia. (p. 306)

heterogeneous mixture: type of mixture where the substances are not evenly mixed. (p. 621)

heterozygous (heh tuh roh ZI gus): describes an organism with two different alleles for a trait. (p. 310)

hibernation: behavioral adaptation for winter survival in which an animal's activity is greatly reduced, its body temperature drops, and body processes slow down. (p. 154)

homeostasis: ability of an organism to keep proper internal conditions no matter what external stimuli are occurring. (p. 216)

hominid: humanlike primate that appeared about four million to six million years ago, ate both plants and meat, and walked upright on two legs. (p. 351)

Homo sapiens: early humans that likely evolved from Cro-Magnons. (p. 353)

homogeneous mixture: type of mixture where two or more substances are evenly mixed on a molecular level but are not bonded together. (p. 622)

homologous (huh MAH luh gus): body parts that are similar in structure and origin and can be similar in function. (p. 348)

homozygous (hoh muh ZI gus): describes an organism with two alleles that are the same for a trait. (p. 310)

hormone (HOR mohn): in humans, chemical produced by the endocrine system, released directly into the bloodstream by ductless glands; affects specific target tissues, and can speed up or slow down cellular activities. (p. 468)

host cell: living cell in which a virus can actively multiply or in which a virus can hide until activated by environmental stimuli. (p. 232)

humidity: amount of water vapor held in the air. (p. 120)

hurricane: large, severe storm that forms over tropical oceans, has winds of at least 120 km/h, and loses power when it reaches land. (p. 132)

hybrid (HI brud): an offspring that was given different genetic information for a trait from each parent. (p. 308)

hydroelectric power: electricity produced when the energy of falling water turns the blades of a generator turbine. (p. 563)

hydronium ion: hydrogen ion combines with a water molecule to form a hydronium ion, H_3O^+. (p. 634)

mezcla heterogénea: tipo de mezcla en la cual las sustancias no están mezcladas de manera uniforme. (p. 621)

heterocigoto: describe a un organismo con dos alelos diferentes para un rasgo. (p. 310)

hibernación: adaptación del comportamiento para sobrevivir durante el invierno en la cual la actividad del animal se ve fuertemente reducida, su temperatura corporal se reduce y los procesos corporales disminuyen su ritmo. (p. 154)

homeostasis: capacidad de un organismo para mantener las condiciones internas apropiadas, sin tener en cuenta los estímulos externos que ocurran. (p. 216)

homínido: primate con forma de humano que apareció entre cuatro y seis millones de años atrás, se alimentaba de plantas y carne, y caminaba erguido sobre sus dos pies. (p. 351)

Homo sapiens: humanos primitivos que probablemente evolucionaron a partir de los CroMagnon. (p. 353)

mezcla homogénea: tipo de mezcla en la cual dos o más sustancias están mezcladas en de manera uniforme a nivel molecular pero no están enlazadas. (p. 622)

homólogos: partes del cuerpo que son similares en estructura y origen y que pueden tener funciones similares. (p. 348)

homocigoto: describe a un organismo con dos alelos iguales para un rasgo. (p. 310)

hormona: en los humanos, sustancia química producida por el sistema endocrino, liberada directamente al torrente sanguíneo mediante glándulas sin conductos; afecta a tejidos que constituyen blancos específicos y puede acelerar o frenar actividades celulares. (p. 468)

célula huésped: célula viva en la cual un virus se puede multiplicar intensamente o en la que se puede ocultar hasta ser activado por un estímulo del medio ambiente. (p. 232)

humedad: cantidad de vapor de agua suspendido en el aire. (p. 120)

huracán: tormenta grande y severa que se forma sobre los océanos tropicales, tiene vientos de por lo menos 120 km/h y pierde su fuerza cuando alcanza la costa. (p. 132)

híbrido: un descendiente que recibe de cada progenitor información genética diferente para un rasgo. (p. 308)

energía hidroeléctrica: electricidad producida cuando la energía generada por la caída del agua hace girar las aspas de una turbina generadora. (p. 563)

ion de hidronio: ion de hidrógeno combinado con una molécula de agua para formar un ion de hidronio, H_3O^+. (p. 634)

hydrosphere: all the waters of Earth. (p. 101)

hidrosfera: toda el agua de la Tierra. (p. 101)

hypothesis (hi PAHTH uh sus): statement that can be tested. (p. 14)

hipótesis: enunciado que puede ser comprobado. (p. 14)

I

igneous rock: rock formed when magma or lava cools and hardens. (p. 62)

roca ígnea: roca formada cuando se enfría y endurece el magma o la lava. (p. 62)

incomplete dominance: production of a phenotype that is intermediate between the two homozygous parents. (p. 314)

dominancia incompleta: producción de un fenotipo intermedio entre dos progenitores homocigotos. (p. 314)

independent variable: single factor in an experiment that the experimenter changes. (p. 16)

variable independiente: único factor que el experimentador cambia en un experimento. (p. 16)

indicator: compound that changes color at different pH values when it reacts with acidic or basic solutions. (p. 640)

indicador: compuesto que cambia de color con diferentes valores de pH al reaccionar con soluciones ácidas o básicas. (p. 640)

inexhaustible resource: energy source that can't be used up by humans. (p. 733)

recurso inagotable: fuente de energía que no puede ser agotada por los seres humanos. (p. 733)

infectious disease: disease caused by a virus, bacterium, fungus, or protest that is spread from one person to another. (p. 383)

enfermedad infecciosa: enfermedad causada por un virus, bacteria, hongo o protista y que se propaga de una persona a otra. (p. 383)

inference: conclusion drawn from an observation. (p. 14)

deducción: conclusión hecha a partir de una observación. (p. 14)

inorganic compound: compound, such as H_2O, that is made from elements other than carbon and whose atoms usually can be arranged in only one structure. (p. 251)

compuesto inorgánico: compuesto, como H_2O, formado por elementos distintos al carbono y cuyos átomos generalmente pueden estar organizados en sólo una estructura. (p. 251)

intrusive: describes a type of igneous rock that generally contains large crystals and forms when magma cools slowly beneath Earth's surface. (p. 63)

intrusivo: describe un tipo de roca ígnea que generalmente contiene cristales grandes y se forma cuando el magma se enfría lentamente por debajo de la superficie terrestre. (p. 63)

involuntary muscle: muscle, such as heart muscle, that cannot be consciously controlled. (p. 439)

músculo involuntario: músculo, como el músculo cardíaco, que no puede ser controlado voluntariamente. (p. 439)

ionosphere: layer of electrically charged particles in the thermosphere that absorbs AM radio waves during the day and reflects them back at night. (p. 93)

ionosfera: capa de partículas con carga eléctrica presentes en la termosfera, la cual absorbe las ondas de radio AM durante el día y las refleja durante la noche. (p. 93)

isobars: lines drawn on a weather map that connect points having equal atmospheric pressure; also indicate the location of high- and low-pressure areas and can show wind speed. (p. 135)

isobaras: líneas dibujadas en un mapa meteorológico que conectan los puntos que tienen una presión atmosférica similar; también indican la ubicación de las áreas de baja y alta presión y pueden mostrar la velocidad del viento. (p. 135)

isotherm (I suh thurm): line drawn on a weather map that connects points having equal temperature. (p. 135)

isoterma: línea dibujada en un mapa meteorológico que conecta los puntos que tienen la misma temperatura. (p. 135)

J

jet stream: narrow belt of strong winds that blows near the top of the troposphere. (p. 106)

corriente de chorro: faja angosta de vientos fuertes que soplan cerca de la parte superior de la troposfera. (p. 106)

joint: any place where two or more bones come together; can be movable or immovable. (p. 446)

articulación: todo lugar en donde dos o más huesos se unen y la cual puede ser móvil o inmóvil. (p. 446)

K

kinetic energy: energy an object has due to its motion. (p. 717)

energía cinética: energía que posee un objeto debido a su movimiento. (p. 717)

kingdom: first and largest category used to classify organisms. (p. 219)

reino: la primera y más grande categoría usada para clasificar a los organismos. (p. 219)

L

land breeze: movement of air from land to sea at night, created when cooler, denser air from the land forces up warmer air over the sea. (p. 107)

brisa terrestre: movimiento de aire nocturno de la tierra al mar, generado cuando el aire denso y frío proveniente de la tierra empuja hacia arriba al aire caliente que está sobre el mar. (p. 107)

larynx: airway to which the vocal cords are attached. (p. 414)

laringe: pasaje aéreo al cual están adheridas las cuerdas vocales. (p. 414)

lava: molten rock that flows from volcanoes onto Earth's surface. (p. 62)

lava: roca derretida que fluye de los volcanes hacia la superficie terrestre. (p. 62)

law of conservation of energy: states that energy can change its form but is never created or destroyed. (p. 722)

ley de la conservación de la energía: establece que la energía puede cambiar de forma pero nunca puede ser creada ni destruida. (p. 722)

law of conservation of mass: states that mass is neither created nor destroyed—and as a result the mass of the substances before a physical or chemical change is equal to the mass of the substances present after the change. (p. 609)

ley de la conservación de masas: establece que la masa no puede ser creada ni destruida; como resultado, la masa de una sustancia antes de un cambio físico o químico es igual a la masa presente de la sustancia después del cambio. (p. 609)

ligament: tough band of tissue that holds bones together at joints. (p. 446)

ligamento: banda dura de tejido que mantiene los huesos unidos a las articulaciones. (p. 446)

limiting factor: any biotic or abiotic factor that limits the number of individuals in a population. (p. 541)

factor limitante: cualquier factor biótico o abiótico que limite el número de individuos en una población. (p. 541)

liquid: matter with a definite volume but no definite shape that can flow from one place to another. (p. 654)

líquido: materia con volumen definido pero no con forma definida que puede fluir de un sitio a otro. (p. 654)

lunar eclipse: occurs during a full moon, when the Sun, the Moon, and Earth line up in such a way that the Moon moves into Earth's shadow. (p. 190)

eclipse lunar: ocurre durante la luna llena cuando el sol, la luna y la Tierra se alinean de tal manera que la luna es cubierta por la sombra de la Tierra. (p. 190)

luster: describes the way a mineral reflects light from its surface; can be metallic or nonmetallic. (p. 38)

brillo: describe la forma en que un mineral refleja la luz desde su superficie; puede ser metálicos o no metálicos. (p. 38)

lymph: fluid that has diffused into the lymphatic capillaries. (p. 377)

linfa: fluido que se encuentra difundido en los capilares linfáticos. (p. 377)

M

magma: hot, melted rock material beneath Earth's surface. (p. 33)

matter: anything that takes up space and has mass. (p. 652)

meiosis (mi OH sus): reproductive process that produces four haploid sex cells from one diploid cell and ensures offspring will have the same number of chromosomes as the parent organisms. (p. 285)

melanin: pigment produced by the epidermis, that protects skin and gives skin and eyes their color. (p. 435)

melting: change of matter from a solid state to a liquid state. (p. 659)

menstrual cycle: hormone-controlled monthly cycle of changes in the female reproductive system that includes the maturation of an egg and preparation of the uterus for possible pregnancy. (p. 476)

menstruation (men STRAY shun): monthly flow of blood and tissue cells that occurs when the lining of the uterus breaks down and is shed. (p. 476)

metabolism: the total of all chemical reactions in an organism. (p. 261)

metamorphic rock: forms when heat, pressure, or fluids act on igneous, sedimentary, or other metamorphic rock to change its form or composition, or both. (p. 67)

meteorologist (meet ee uh RAHL uh just): studies weather and uses information from Doppler radar, weather satellites, computers and other instruments to make weather maps and provide forecasts. (p. 134)

mineral: inorganic nutrient that regulates many chemical reactions in the body. (p. 38); naturally occurring inorganic solid that has a definite chemical composition and an orderly internal atomic structure. (p. 400)

mitochondrion: cell organelle where food is broken down, which releases energy. (p. 226)

mitosis (mi TOH sus): cell process in which the nucleus divides to form two nuclei identical to each other, and identical to the original nucleus, in a series of steps (prophase, metaphase, anaphase, and telophase). (p. 278)

magma: material rocoso fundido y caliente que se encuentra por debajo de la superficie terrestre. (p. 33)

materia: cualquier cosa que ocupe espacio y tenga masa. (p. 652)

meiosis: proceso reproductivo que produce cuatro células sexuales haploides a partir de una célula diploide y asegura que la descendencia tendrá el mismo número de cromosomas que los organismos progenitores. (p. 285)

melanina: pigmento producido por la epidermis que protege la piel y da el color a los ojos y a la piel. (p. 435)

fusión: cambio de la materia de estado sólido a líquido. (p. 659)

ciclo menstrual: ciclo mensual de cambios en el sistema reproductor femenino, el cual es controlado por hormonas e incluye la maduración de un óvulo y la preparación del útero para un posible embarazo. (p. 476)

menstruación: flujo mensual de sangre y células tisulares que ocurre cuando el endometrio uterino se rompe y se desprende. (p. 476)

metabolismo: el conjunto de todas las reacciones químicas en un organismo. (p. 261)

roca metamórfica: se forma cuando el calor, la presión o los fluidos actúan sobre una roca ígnea, sedimentaria u otra roca metamórfica para cambiar su forma, composición o ambas. (p. 67)

meteorólogo: persona que estudia el clima y usa información del radar Doppler, satélites meteorológicos, computadoras y otros instrumentos para elaborar mapas del estado del tiempo y hacer pronósticos. (p. 134)

mineral: nutriente inorgánico que regula muchas reacciones químicas en el cuerpo. (p. 38); sólido inorgánico que se encuentra en la naturaleza, tiene una composición química definida y una estructura atómica ordenada. (p. 400)

mitocondria: organelo celular en donde se desdoblan los alimentos, los cuales liberan energía. (p. 226)

mitosis: proceso celular en el que el núcleo se divide para formar dos núcleos idénticos entre sí e idénticos al núcleo original, a través de varias etapas (profase, metafase, anafase y telofase). (p. 278)

mixture: a combination of substances in which the individual substances do not change or combine chemically but instead retain their own individual properties; can be gases, solids, liquids, or any combination of them. (p. 249)

mezcla: una combinación de sustancias en la que las sustancias individuales no cambian ni se combinan químicamente pero mantienen sus propiedades individuales; pueden ser gases, sólidos, líquidos o una combinación de ellos. (p. 249)

monocot: angiosperm with one cotyledon inside its seed, flower parts in multiples of three, and vascular tissues in bundles scattered throughout the stem. (p. 518)

monocotiledóneas: angiospermas con un solo cotiledón dentro de la semilla, partes florales dispuestas en múltiplos de tres y tejidos vasculares distribuidos en haces diseminados por todo el tallo. (p. 518)

Moon phases: changing views of the Moon as seen from Earth, which are caused by the Moon's revolution around Earth. (p. 187)

fases de la luna: cambios en la luna vistos desde la Tierra, producidos por la rotación de la luna alrededor de la Tierra. (p. 187)

mutation: any permanent change in a gene or chromosome of a cell; may be beneficial, harmful, or have little effect on an organism. (p. 294)

mutación: cualquier cambio permanente en un gen o cromosoma de una célula; puede ser benéfica, perjudicial o tener un pequeño efecto sobre un organismo. (p. 294)

N

natural resources: parts of Earth's environment that supply materials useful or necessary for the survival of living organisms. (p. 560)

recursos naturales: partes del medio ambiente terrestre que proporcionan materiales útiles o necesarios para la supervivencia de los organismos vivos. (p. 560)

natural selection: a process by which organisms with traits best suited to their environment are more likely to survive and reproduce; includes concepts of variation, overproduction, and competition. (p. 336)

selección natural: proceso mediante el cual los organismos con rasgos mejor adaptados a su ambiente tienen mayor probabilidad de sobrevivir y reproducirse; incluye los conceptos de variación, sobreproducción y competencia. (p. 336)

nebula: cloud of gas and dust particles in interstellar space. (p. 201)

nebulosa: nube de partículas de gas y polvo en el espacio interestelar. (p. 201)

nephron (NEF rahn): tiny filtering unit of the kidney. (p. 421)

nefrona: pequeña unidad de filtración en el riñón. (p. 421)

neuron (NOO rahn): basic functioning unit of the nervous system, made up of a cell body, dendrites, and axons. (p. 450)

neurona: unidad básica funcional del sistema nervioso compuesta por un cuerpo celular, dendritas y axones. (p. 450)

neutralization (new truh luh ZAY shun): reaction in which an acid reacts with a base and forms water and a salt. (p. 640)

neutralización: reacción en la cual un ácido reacciona con una base para formar agua y una sal. (p. 640)

niche (NICH): role of an organism in the ecosystem, including what it eats, how it interacts with other organisms, and how it gets its food. (p. 543)

nicho: papel que juega un organismo en un ecosistema, incluyendo lo que come, cómo interactúa con los otros organismos y cómo consigue su alimento. (p. 543)

nonfoliated: describes metamorphic rock, such as quartzite or marble, whose mineral grains grow and rearrange but generally do not form layers. (p. 70)

no foliado: describe rocas metamórficas, como la cuarcita o el mármol, cuyas vetas minerales se acumulan y reestructuran pero rara vez forman capas. (p. 70)

noninfectious disease: disease that is not caused by a pathogen. (p. 387)

enfermedad no infecciosa: enfermedad que no es causada por un patógeno. (p. 387)

nonrenewable resources: natural resources, such as petroleum, minerals, and metals, that are used more quickly than they can be replaced by natural processes. (pp. 561, 730)

recursos no renovables: recursos naturales, como el petróleo, los minerales y los metales, que son utilizados más rápidamente de lo que pueden ser reemplazados mediante procesos naturales. (pp. 561, 730)

nonvascular plant: plant that absorbs water and other substances directly through its cell walls instead of through tubelike structures. (p. 505)

nuclear energy: energy contained in atomic nuclei. (p. 564); energy produced from the splitting apart of billions of uranium nuclei by a nuclear fission reaction. (p. 720)

nucleus: organelle that controls all the activities of a cell and contains hereditary material made of DNA. (p. 226)

nutrients (NEW tree unts): substances in foods—proteins, carbohydrates, fats, vitamins, minerals, and water—that provide energy and materials for cell development, growth, and repair. (p. 400)

planta no vascular: planta que absorbe agua y otras sustancias directamente a través de sus paredes celulares en vez de utilizar estructuras tubulares. (p. 505)

energía nuclear: energía contenida en los núcleos de los atómos. (p. 564); a partir del fraccionamiento de billones de núcleos de uranio mediante una reacción de fisión nuclear. (p. 720)

núcleo: organelo que controla todas las actividades celulares y que contiene el material genético compuesto de ADN. (p. 226)

nutrientes: sustancias en los alimentos (proteínas, carbohidratos, grasas, vitaminas, minerales y agua) que suministran energía y materiales para el desarrollo, crecimiento y reparación de las células. (p. 400)

O

observation: bit of information gathered with the senses. (p. 14)

orbit: curved path followed by Earth as it moves around the Sun. (p. 181)

ore: deposit in which a mineral exists in large enough amounts to be mined at a profit. (p. 45)

organ: structure, such as the heart, made up of different types of tissues that all work together. (p. 230)

organ system: a group of organs working together to perform a certain function. (p. 230)

organelle: structure in the cytoplasm of a eukaryotic cell that can act as a storage site, process energy, move materials, or manufacture substances. (p. 226)

organic compounds: compounds that always contain hydrogen and carbon; carbohydrates, lipids, proteins, and nucleic acids are organic compounds found in living things. (p. 250)

organism: any living thing. (p. 214)

osmosis: a type of passive transport that occurs when water diffuses through a cell membrane. (p. 256)

ovary: in humans, female reproductive organ that produces eggs and is located in the lower part of the body. (p. 475)

ovulation (ahv yuh LAY shun): monthly process in which an egg is released from an ovary and enters the oviduct, where it can become fertilized by sperm. (p. 475)

observación: pequeña cantidad de información obtenida a través de los sentidos. (p. 14)

órbita: trayectoria curva que sigue la Tierra en su movimiento alrededor del sol. (p. 181)

mena: depósito en el que existe un mineral en cantidades suficientes para la explotación minera. (p. 45)

órgano: estructura, como el corazón, formada por diferentes tipos de tejidos que funcionan en conjunto. (p. 230)

sistema de órganos: grupo de órganos que funcionan conjuntamente para llevar a cabo una función determinada. (p. 230)

organelo: estructura citoplásmica de las células eucarióticas que puede servir para el almacenamiento, procesar energía, movilizar materiales o producir sustancias. (p. 226)

compuestos orgánicos: compuestos que siempre contienen hidrógeno y carbono; los carbohidratos, lípidos, proteínas y ácidos nucleicos son compuestos orgánicos que se encuentran en los seres vivos. (p. 250)

organismo: todo ser viviente. (p. 214)

ósmosis: tipo de transporte pasivo que ocurre cuando el agua se difunde a través de una membrana celular. (p. 256)

ovario: en los humanos, órgano reproductor femenino que produce óvulos y está localizado en la parte inferior del cuerpo. (p. 475)

ovulación: proceso mensual en el que un óvulo es liberado de un ovario y entra al oviducto, donde puede ser fertilizado por los espermatozoides. (p. 475)

ozone depletion: thinning of Earth's ozone layer caused by chlorofluorocarbons (CFCs) leaking into the air and reacting chemically with ozone, breaking the ozone molecules apart. (p. 571)

ozone layer: layer of the stratosphere with a high concentration of ozone; absorbs most of the Sun's harmful ultraviolet radiation. (p. 96)

agotamiento del ozono: adelgazamiento de la capa de ozono de la Tierra causado por los clorofluorocarbonos (CFC) que escapan al aire y reaccionan químicamente con el ozono rompiendo sus moléculas. (p. 571)

capa de ozono: capa de la estratosfera con una concentración alta de ozono y que absorbe la mayor parte de la radiación ultravioleta dañina del sol. (p. 96)

P

Pascal's principle: states that when a force is applied to a confined fluid, an increase in pressure is transmitted equally to all parts of the fluid. (p. 672)

passive immunity: immunity that results when antibodies produced in one animal are introduced into another's body; does not last as long as active immunity. (p. 380)

passive transport: movement of substances through a cell membrane without the use of cellular energy; includes diffusion, osmosis, and facilitated diffusion. (p. 254)

pasteurization (pas chur ruh ZAY shun): process in which a liquid is heated to a temperature that kills most bacteria. (p. 382)

periosteum (pur ee AHS tee um): tough, tight-fitting membrane, that covers a bone's surface and contains blood vessels that transport nutrients into the bone. (p. 445)

peripheral nervous system: division of the nervous system; includes of all the nerves outside the CNS; connects the brain and spinal cord to other body parts. (p. 451)

peristalsis (per uh STAHL sus): waves of muscular contractions that move food through the digestive tract. (p. 402)

petroleum: nonrenewable resource formed over hundreds of millions of years mostly from the remains of microscopic marine organisms buried in Earth's crust. (p. 561)

pH: measure of how acidic or basic a solution is, ranging in a scale from 0 to 14. (p. 638)

phenotype (FEE nuh tipe): outward physical appearance and behavior of an organism as a result of its genotype. (p. 310)

phloem (FLOH em): vascular tissue that forms tubes that transport dissolved sugar throughout a plant. (p. 515)

principio de Pascal: establece que cuando se ejerce una fuerza sobre un fluido encerrado, se transmite un incremento de presión uniforme a todas las partes del fluido. (p. 672)

inmunidad pasiva: inmunidad que se presenta cuando los anticuerpos producidos en un animal son introducidos en el cuerpo de otro animal, la cual no es tan duradera como la inmunidad activa. (p. 380)

transporte pasivo: movimiento de sustancias a través de la membrana celular sin usar energía celular; incluye difusión, ósmosis y difusión facilitada. (p. 254)

pasteurización: proceso mediante el cual un líquido es calentado a una temperatura que mata a la mayoría de las bacterias. (p. 382)

periostio: membrana fuerte y ajustada que cubre la superficie de los huesos y contiene vasos sanguíneos que transportan nutrientes a los huesos. (p. 445)

sistema nervioso periférico: parte del sistema nervioso compuesta por todos los nervios que no pertenecen al sistema nervioso central y que conecta el cerebro y la médula espinal con otras partes del cuerpo. (p. 451)

peristalsis: ondas de contracciones musculares que mueven al alimento a través del sistema digestivo. (p. 402)

petróleo: recurso no renovable formado durante cientos de millones de años, en su mayoría a partir de los restos de organismos marinos microscópicos sepultados en la corteza terrestre. (p. 561)

pH: medida para saber qué tan básica o ácida es una solución, en una escala de 0 a 14. (p. 638)

fenotipo: apariencia física externa y comportamiento de un organismo como resultado de su genotipo. (p. 310)

floema: tejido vascular que forma tubos que transportan azúcares disueltos a toda la planta. (p. 515)

photosynthesis: process by which plants and many other producers use light energy to produce a simple sugar from carbon dioxide and water and give off oxygen. (p. 262)

photovoltaic: device that transforms radiant energy directly into electrical energy. (p. 734)

phylogeny (fi LAH juh nee): evolutionary history of an organism; used today to group organisms into six kingdoms. (p. 219)

physical change: change in which the form or appearance of matter changes, but not its composition. (p. 600)

physical property: characteristic that can be observed, using the five senses, without changing or trying to change the composition of a substance. (p. 594)

pioneer species: species that break down rock and build up decaying plant material so that other plants can grow; first organisms to grow in new or disturbed areas. (p. 507)

plasma: the liquid part of blood, which is made mostly of water. (p. 366)

platelet: irregularly shaped cell fragments that help clot blood. (p. 367)

polar zones: climate zones that receive solar radiation at a low angle, extend from 66°N and S latitude to the poles, and are never warm. (p. 148)

pollutant: substance that contaminates any part of the environment. (p. 568)

polygenic (pah lih JEH nihk) inheritance: occurs when a group of gene pairs acts together and produces a specific trait, such as human eye color, skin color, or height. (p. 316)

population: all of the individuals of one species that live in the same space at the same time. (p. 536)

population density: number of individuals in a population that occupies an area of limited size. (p. 540)

potential energy: energy stored in an object due to its position. (p. 718)

precipitate: solid that comes back out of its solution because of a chemical reaction or physical change. (p. 622)

precipitation: water falling from clouds—including rain, snow, sleet, and hail—whose form is determined by air temperature. (p. 124)

pregnancy: period of development—usually about 38 or 39 weeks in female humans—from fertilized egg until birth. (p. 480)

fotosíntesis: proceso mediante el cual las plantas y muchos otros organismos productores usan la energía solar para producir azúcares simples a partir de dióxido de carbono y agua y desprender oxígeno. (p. 262)

fotovoltaico: dispositivo que transforma la energía radiante directamente en energía eléctrica. (p. 734)

filogenia: historia evolutiva de los organismos utilizada en la actualidad para agruparlos en seis reinos. (p. 219)

cambio físico: cambio en el cual varía la forma o apariencia de la materia pero no su composición. (p. 600)

propiedad física: característica que puede ser observada usando los cinco sentidos sin cambiar o tratar de cambiar la composición de una sustancia. (p. 594)

especies pioneras: especies que descomponen la roca y acumulan material vegetal en descomposición para que otras plantas puedan crecer; los primeros organismos que crecen en áreas nuevas o alteradas. (p. 507)

plasma: parte líquida de la sangre compuesto principalmente por agua. (p. 366)

plaqueta: fragmentos de célula de forma irregular que ayudan a coagular la sangre. (p. 367)

zonas polares: zonas climáticas que reciben radiación solar a un ángulo reducido, se extienden desde los 66° de latitud norte y sur hasta los polos y nunca son cálidas. (p. 148)

contaminante: sustancia que contamina cualquier parte del medio ambiente. (p. 568)

herencia poligénica: ocurre cuando un grupo de pares de genes actúa conjuntamente y produce un rasgo específico, tal como el color de los ojos, el color de la piel, o la estatura en los humanos. (p. 316)

población: todos los individuos de una especie que viven en el mismo espacio al mismo tiempo. (p. 536)

densidad de población: número de individuos en una población que ocupa un área de tamaño limitado. (p. 540)

energía potencial: energía almacenada en un objeto debido a su posición. (p. 718)

precipitado: sólido que se aísla de su solución mediante una reacción química o un cambio físico. (p. 622)

precipitación: agua que cae de las nubes—incluyendo lluvia, nieve, aguanieve y granizo—cuya forma está determinada por la temperatura del aire. (p. 124)

embarazo: período del desarrollo—generalmente unas 38 o 39 semanas en las hembras humanas—que va desde el óvulo fertilizado hasta el nacimiento. (p. 480)

pressure: force exerted on a surface divided by the total area over which the force is exerted. (p. 666)

presión: fuerza ejercida sobre una superficie dividida por el área total sobre la cual se ejerce dicha fuerza. (p. 666)

primates: group of mammals including humans, monkeys, and apes that share characteristics such as opposable thumbs, binocular vision, and flexible shoulders. (p. 350)

primates: grupo de mamíferos que incluye a los humanos, monos y simios, los cuales comparten características como pulgares opuestos, visión binocular y hombros flexibles. (p. 350)

punctuated equilibrium: model describing the rapid evolution that occurs when mutation of a few genes results in a species suddenly changing into a new species. (p. 340)

equilibrio punteado: modelo que describe la evolución rápida que ocurre cuando la mutación de unos pocos genes resulta en que una especie cambie rápidamente para convertirse en otra especie. (p. 340)

Punnett (PUN ut) square: a tool to predict the probability of certain traits in offspring that shows the different ways alleles can combine. (p. 310)

Cuadrado de Punnett: herramienta para predecir la probabilidad de ciertos rasgos en la descendencia mostrando las diferentes formas en que los alelos pueden combinarse. (p. 310)

R

radiant energy: energy carried by light. (p. 719)

energía radiante: energía transportada por la luz. (p. 719)

radiation: energy transferred by waves or rays. (p. 100)

radiación: energía transmitida por ondas o rayos. (p. 100)

radioactive element: element that gives off a steady amount of radiation as it slowly changes to a nonradioactive element. (p. 345)

elemento radiactivo: elemento que emite una cantidad estable de radiación mientras se convierte lentamente en un elemento no radiactivo. (p. 345)

recessive (rih SE sihv): describes a trait that is covered over, or dominated, by another form of that trait and seems to disappear. (p. 308)

recesivo: describe un rasgo que está encubierto, o que es dominado, por otra forma del mismo rasgo y que parece no estar presente. (p. 308)

recycling: conservation method that is a form of reuse and requires changing or reprocessing an item or natural resource. (p. 579)

reciclaje: método de conservación como una forma de reutilización y que requiere del cambio o reprocesamiento del producto o recurso natural. (p. 579)

relative humidity: measure of the amount of moisture held in the air compared with the amount it can hold at a given temperature; can range from 0 percent to 100 percent. (p. 120)

humedad relativa: medida de la cantidad de humedad suspendida en el aire en comparación con la cantidad que puede contener a una temperatura determinada; puede variar del cero al cien por ciento. (p. 120)

renewable resource: energy resource that is replenished continually. (p. 561); natural resources, such as water, sunlight, and biomass, that are constantly being recycled or replaced in nature. (p. 717)

recurso renovable: recurso energético regenerado continuamente. (p. 561); recursos naturales, como el agua, la luz solar y la biomasa, que son reciclados o reemplazados constantemente por la naturaleza. (p. 717)

respiration: process by which producers and consumers release stored energy from food molecules. (p. 263)

respiración: proceso mediante el cual los organismos productores y consumidores liberan la energía almacenada en las moléculas de los alimentos. (p. 263)

revolution: the motion of Earth around the Sun, which takes about 365 1/4 days, or one year, to complete. (p. 181)

traslación: movimiento de la Tierra alrededor del sol, el cual dura más o menos 365 1/4 días o un año en completarse. (p. 181)

rhizoids (RI zoydz): threadlike structures that anchor nonvascular plants to the ground. (p. 506)

rizoides: estructuras en forma de hilos que anclan las plantas no vasculares al suelo. (p. 506)

ribosome: small structure on which cells make their own proteins. (p. 225)

ribosoma: pequeña estructura en la que las células producen sus propias proteínas. (p. 225)

RNA: ribonucleic acid, a type of nucleic acid that carries codes for making proteins from the nucleus to the ribosomes. (p. 292)

rock: mixture of one or more minerals, rock fragments, volcanic glass, organic matter, or other natural materials; can be igneous, metamorphic, or sedimentary. (p. 57)

rock cycle: model that describes how rocks slowly change from one form to another through time. (p. 59)

rotation: spinning of Earth on its axis, which causes day and night; it takes 24 hours for Earth to complete one rotation. (p. 180)

ARN (Ácido ribonucleico): tipo de ácido nucleico que transporta los códigos para la formación de proteínas del núcleo a los ribosomas. (p. 292)

roca: mezcla de uno o más minerales, fragmentos de roca, obsidiana, materia orgánica u otros materiales naturales; puede ser ígnea, metamórfica o sedimentaria. (p. 57)

ciclo de la roca: modelo que describe cómo cambian lentamente las rocas de una forma a otra a través del tiempo. (p. 59)

rotación: giro de la Tierra sobre su propio eje, dando lugar al día y la noche; la Tierra dura 24 horas en completar una rotación. (p. 180)

S

saturated: describes a solution that holds the total amount of solute that it can hold under given conditions. (p. 630)

science: process of trying to understand the world. (p. 7)

scientific methods: step-by-step procedures of scientific problem solving, which can include identifying the problem, forming and testing a hypothesis, analyzing the test results, and drawing conclusions. (p. 12)

sea breeze: movement of air from sea to land during the day when cooler air from above the water moves over the land, forcing the heated, less dense air above the land to rise. (p. 107)

season: short period of climate change in an area caused by the tilt of Earth's axis as Earth revolves around the Sun. (p. 156)

second law of motion: states that an object acted on by an unbalanced force will accelerate in the direction of the force with an acceleration equal to the force divided by the object's mass. (p. 694)

sedimentary rock: a type of rock, such as limestone, that is most likely to contain fossils and is formed when layers of sand, silt, clay, or mud are cemented and compacted together or when minerals are deposited from a solution. (p. 68)

sediments: loose materials, such as rock fragments, mineral grains, and the remains of once-living plants and animals, that have been moved by wind, water, ice, or gravity. (p. 71)

saturado: describe a una solución que retiene toda la cantidad de soluto que puede retener bajo determinadas condiciones. (p. 630)

ciencia: proceso mediante el cual se trata de comprender el mundo. (p. 7)

métodos científicos: procedimientos paso por paso para resolver un problema científico, el cual puede incluir la identificación del problema, la formación y prueba de una hipótesis, el análisis y prueba de resultados y la obtención de conclusiones. (p. 12)

brisa marina: movimiento de aire del mar a la tierra durante el día, cuando el aire frío que está sobre el mar empuja al aire caliente y menos denso que está sobre la tierra. (p. 107)

estación: periodo corto de cambio climático en un área, causado por la inclinación del eje de la Tierra conforme gira alrededor del sol. (p. 156)

segunda ley del movimiento: establece que un objeto al que se le aplica una fuerza no balanceada se acelerará en la dirección de la fuerza con una aceleración igual a dicha fuerza dividida por la masa del objeto. (p. 694)

roca sedimentaria: tipo de roca, como la piedra caliza, con alta probabilidad de contener fósiles y que se forma cuando las capas de arena, sedimento, arcilla o lodo son cementadas y compactadas o cuando los minerales de una solución son depositados. (p. 68)

sedimentos: materiales sueltos, como fragmentos de roca, granos minerales y restos de animales y plantas, que han sido arrastrados por el viento, el agua, el hielo o la gravedad. (p. 71)

semen (SEE mun): mixture of sperm and a fluid that helps sperm move and supplies them with an energy source. (p. 474)

semen: mezcla de espermatozoides y un fluido que ayuda a la movilización de los espermatozoides y les suministra una fuente de energía. (p. 474)

sex-linked gene: an allele inherited on a sex chromosome and that can cause human genetic disorders such as color blindness and hemophilia. (p. 319)

gen ligado al sexo: un alelo heredado en un cromosoma sexual y que puede causar desórdenes genéticos humanos como daltonismo y hemofilia. (p. 319)

sexual reproduction: a type of reproduction in which two sex cells, usually an egg and a sperm, join to form a zygote, which will develop into a new organism with a unique identity. (p. 284)

reproducción sexual: tipo de reproducción en la que dos células sexuales, generalmente un óvulo y un espermatozoide, se unen para formar un zigoto, el cual se desarrollará para formar un nuevo organismo con identidad única. (p. 284)

silicate: describes a mineral that contains silicon and oxygen and usually one or more other elements. (p. 34)

silicato: describe mineral que contiene sílice y oxígeno y generalmente uno o varios elementos distintos. (p. 34)

solar eclipse: occurs during a new moon, when the Sun, the Moon, and Earth are lined up in a specific way and Earth moves into the Moon's shadow. (p. 189)

eclipse solar: ocurre durante la luna nueva, cuando el sol, la luna y la Tierra se alinean de una forma específica y la Tierra es cubierta por la sombra de la luna. (p. 189)

solar system: system that includes the Sun, planets, comets, meteoroids and other objects that orbit the Sun. (p. 194)

sistema solar: sistema que incluye el sol, los planetas, los cometas, los meteoritos y otros objetos que giran alrededor del sol. (p. 194)

solid: matter with a definite shape and volume; has tightly packed particles that move mainly by vibrating. (p. 653)

sólido: materia con forma y volumen definidos; tiene partículas fuertemente compactadas que se mueven principalmente por vibración. (p. 653)

solstice: time when the Sun reaches its greatest distance north or south of the equator. (p. 182)

solsticio: momento en el que el sol alcanza su mayor distancia al norte o sur del ecuador. (p. 182)

solubility (sahl yuh BIH luh tee): measure of how much solute can be dissolved in a certain amount of solvent. (p. 629)

solubilidad: medida de la cantidad de soluto que puede disolverse en cierta cantidad de solvente. (p. 629)

solute: substance that dissolves and seems to disappear into another substance. (p. 622)

soluto: sustancia que se disuelve y parece desaparecer en otra sustancia. (p. 622)

solution: homogeneous mixture whose elements and/or compounds are evenly mixed at the molecular level but are not bonded together. (p. 622)

solución: mezcla homogénea cuyos elementos o compuestos están mezclados de manera uniforme a nivel molecular pero no se enlazan. (p. 622)

solvent: substance that dissolves the solute. (p. 622)

solvente: sustancia que disuelve al soluto. (p. 622)

species: group of organisms that share similar characteristics and can reproduce among themselves producing fertile offspring. (p. 334)

especie: grupo de organismos que comparten características similares entre sí y que pueden reproducirse entre ellos dando lugar a una descendencia fértil. (p. 334)

specific gravity: ratio of a mineral's weight compared with the weight of an equal volume of water. (p. 38)

gravedad específica: cociente del peso de un mineral comparado con el peso de un volumen igual de agua. (p. 38)

speed: equals the distance traveled divided by the time needed to travel that distance. (p. 685)

rapidez: es igual a la distancia recorrida dividida por el tiempo necesario para recorrer dicha distancia. (p. 685)

sperm: haploid sex cell formed in the male reproductive organs. (p. 284); in humans, male reproductive cells produced in the testes. (p. 474)

espermatozoides: células sexuales haploides que se forman en los órganos reproductores masculinos. (p. 284); en los humanos, células reproductoras masculinas producidas por los testículos. (p. 474)

station model: indicates weather conditions at a specific location, using a combination of symbols on a map. (p. 135)

stomata (STOH muh tuh): tiny openings in a plant's epidermis through which carbon dioxide, water vapor, and oxygen enter and exit. (p. 513)

streak: color of a mineral when it is in powdered form. (p. 39)

sublimation: the process by which a solid changes directly into a gas. (p. 601)

substance: matter with a fixed composition whose identity can be changed by chemical processes but not by ordinary physical processes. (p. 620)

surface tension: the uneven forces acting on the particles on the surface of a liquid. (p. 655)

symbiosis (sihm bee OH sus): any close interaction among two or more different species, including mutualism, commensalism, and parasitism. (p. 542)

synapse (SIHN aps): small space across, which an impulse moves from an axon to the dendrites or cell body of another neuron. (p. 450)

modelo estacional: indica las condiciones del estado del tiempo en una ubicación específica, utilizando una combinación de símbolos en un mapa. (p. 135)

estomas: aperturas pequeñas en la superficie de la mayoría de las hojas de las plantas, las cuales permiten que entre y salga dióxido de carbono, agua y oxígeno. (p. 513)

veta: color de un mineral en forma de polvo. (p. 39)

sublimación: proceso mediante el cual un sólido se convierte directamente en gas. (p. 601)

sustancia: materia que tiene una composición fija cuya identidad puede ser cambiada mediante procesos químicos pero no mediante procesos físicos corrientes. (p. 620)

tensión superficial: fuerzas desiguales que actúan sobre las partículas que se encuentran en la superficie de un líquido. (p. 655)

simbiosis: cualquier interacción cercana entre dos o más especies diferentes, incluyendo mutualismo, asociaciones y parasitismo. (p. 542)

sinapsis: pequeño espacio a través del cual un impulso se mueve desde el axón hasta las dendritas o los cuerpos celulares de otra neurona. (p. 450)

T

technology: use of knowledge gained through scientific thinking and problem solving to make new products or tools. (p. 9)

temperate zones: climate zones with moderate temperatures that are located between the tropics and the polar zones. (p. 148)

temperature: measure of the average kinetic energy of the individual particles of a substance. (p. 658)

tendon: thick band of tissue that attaches bones to muscles. (p. 440)

testis: male organ that produces sperm and testosterone. (p. 474)

thermal energy: energy that all objects have that increases as the object's temperature increases. (p. 718)

third law of motion: states that forces act in equal but opposite pairs. (p. 702)

tissue: group of similar cells that work together to do one job. (p. 230)

tornado: violent, whirling windstorm that crosses land in a narrow path and can result from wind shears inside a thunderhead. (p. 130)

tecnología: uso de los conocimientos adquiridos a través del pensamiento científico y la solución de problemas para crear nuevos productos y herramientas. (p. 9)

zonas templadas: zonas climáticas con temperaturas moderadas que están localizadas entre los trópicos y las zonas polares. (p. 148)

temperatura: medida de la energía cinética promedio de las partículas individuales de una sustancia. (p. 658)

tendón: banda de tejido grueso que une los huesos y los músculos. (p. 440)

testículos: órganos masculinos que producen espermatozoides y testosterona. (p. 474)

energía térmica: energía que poseen todos los objetos y que aumenta al aumentar la temperatura de éstos. (p. 718)

tercera ley del movimiento: establece que las fuerzas actúan en pares iguales pero opuestos. (p. 702)

tejido: grupo de células similares que funcionan conjuntamente para llevar a cabo una función. (p. 230)

tornado: tormenta de viento en forma de remolino que cruza la tierra en un curso estrecho y puede resultar de vientos que se entrecruzan en direcciones opuestas dentro del frente de una tormenta. (p. 130)

trachea (TRAY kee uh): air-conducting tube that connects the larynx with the bronchi; is lined with mucous membranes and cilia, and contains strong cartilage rings. (p. 414)

tráquea: tubo conductor de aire que conecta a la laringe con los bronquios; está forrada con membranas mucosas y cilios y contiene fuertes anillos de cartílagos. (p. 414)

tropics: climate zone that receives the most solar radiation, is located between latitudes 23°N and 23°S, and is always hot, except at high elevations. (p. 148)

trópicos: zonas climáticas que reciben la mayor parte de la radiación solar, están localizadas entre los 23° de latitud norte y 23° de latitud sur y siempre son cálidas excepto a grandes alturas. (p. 148)

troposphere: layer of Earth's atmosphere that is closest to the ground, contains 99 percent of the water vapor and 75 percent of the atmospheric gases, and is where clouds and weather occur. (p. 92)

troposfera: capa de la atmósfera terrestre que se encuentra cerca del suelo, contiene el 99 por ciento del vapor de agua y el 75 por ciento de los gases atmosféricos; es donde se forman las nubes y las condiciones meteorológicas. (p. 92)

turbine: set of steam-powered fan blades that spins a generator at a power plant. (p. 726)

turbina: conjunto de aspas de ventilador impulsadas por vapor que hacen girar a un generador en una planta de energía eléctrica. (p. 726)

U

ultraviolet radiation: a type of energy that comes to Earth from the Sun, can damage skin and cause cancer, and is mostly absorbed by the ozone layer. (p. 96)

radiación ultravioleta: tipo de energía que llega a la Tierra desde el sol y que puede dañar la piel y causar cáncer; la mayor parte de esta radiación es absorbida por la capa de ozono. (p. 96)

unbalanced forces: unequal forces that do not cancel when acting on an object and cause a change in the object's motion. (p. 692)

fuerzas no balanceadas: fuerzas desiguales que no se cancelan al actuar sobre un objeto y producen un cambio en el movimiento de dicho objeto. (p. 692)

ureter: tube that carries urine from each kidney to the bladder. (p. 421)

uréter: tubo que transporta la orina desde cada uno de los riñones hasta la vejiga. (p. 421)

uterus: in female humans, hollow, muscular, pear-shaped organ where a fertilized egg develops into a baby. (p. 475)

útero: en seres hermanos femeninos, órgano en forma de pera, hueco y musculoso, en el que un óvulo fertilizado se desarrolla en bebé. (p. 475)

V

vagina (vuh JI nuh): muscular tube that connects the lower end of the uterus to the outside of the body; the birth canal through which a baby travels when being born. (p. 475)

vagina: tubo musculoso que conecta el extremo inferior del útero de una hembra con el exterior del cuerpo; el canal del nacimiento a través del cual sale un bebé al nacer. (p. 475)

vaporization: the process by which matter changes from a liquid state to a gas. (pp. 604, 659)

vaporización: proceso mediante el cual la materia cambia del estado líquido a gas. (pp. 604, 659)

variation: inherited trait that makes an individual different from other members of the same species and results from a mutation in the organism's genes. (p. 338)

variación: rasgo heredado que hace que un individuo sea diferente a otros miembros de su misma especie como resultado de una mutación de sus genes. (p. 338)

vascular plant: plant with tubelike structures that move minerals, water, and other substances throughout the plant. (p. 505)

planta vascular: planta con estructuras semejantes a tubos, las cuales sirven para movilizar minerales, agua y otras sustancias a toda la planta. (p. 505)

vein: blood vessel that carries blood to the heart. (p. 374)

vena: vasos sanguíneos que transportan la sangre al corazón. (p. 374)

velocity: speed and direction of a moving body; average velocity equals the displacement divided by the time. (p. 687)

vestigial (veh STIHJ ee ul) structure: structure, such as the human appendix, that doesn't seem to have a function and may once have functioned in the body of an ancestor. (p. 348)

villi (VIH li): fingerlike projections covering the wall of the small intestine that increase the surface area for food absorption. (p. 403)

virus: a strand of hereditary material surrounded by a protein coating. (p. 232)

viscosity: a liquid's resistance to flow. (p. 655)

vitamin: water-soluble or fat-soluble organic nutrient needed in small quantities for growth, for preventing some diseases, and for regulating body functions. (p. 407)

voluntary muscle: muscle, such as a leg or arm muscle, that can be consciously controlled. (p. 439)

velocidad: rapidez y dirección de un cuerpo en movimiento; la velocidad promedio es igual al desplazamiento dividido por el tiempo. (p. 687)

estructura vestigial: estructura, como el apéndice humano, que no parece tener alguna función pero que pudo haber funcionado en el cuerpo de un antepasado. (p. 348)

vellosidades: proyecciones en forma de dedo que cubren las paredes del intestino delgado y aumentan el área de superficie para la absorción de los alimentos. (p. 403)

virus: cadena de material genético rodeada de una capa proteica. (p. 232)

viscosidad: resistencia de un líquido al flujo. (p. 655)

vitamina: nutriente orgánico soluble en agua o en grasa, necesario en pequeñas cantidades para el crecimiento, para prevenir algunas enfermedades y para regular las funciones biológicas. (p. 407)

músculo voluntario: músculo, como el músculo de una pierna o de un brazo, que puede ser controlado voluntariamente. (p. 439)

W

water cycle: continuous cycle of water molecules on Earth as they rise into the atmosphere, fall back to Earth as rain or other precipitation, and flow into rivers and oceans through the processes of evaporation, condensation, and precipitation. (p. 548)

weather: state of the atmosphere at a specific time and place, determined by factors including air pressure, amount of moisture in the air, temperature, wind, and precipitation. (p. 118)

ciclo del agua: ciclo continuo de las moléculas de agua en la Tierra en su proceso de subir a la atmósfera, regresar a la Tierra en forma de lluvia u otra forma de precipitación y fluir hacia los ríos y océanos a través de la evaporación, condensación y precipitación. (p. 548)

estado del tiempo: estado de la atmósfera en un momento y lugar específicos, determinado por factores que incluyen la presión del aire, cantidad de humedad en el aire, temperatura, viento y precipitación. (p. 118)

X

xylem (ZI lum): vascular tissue that forms hollow vessels that transport substances, other than sugar, throughout a plant. (p. 516)

xilema: tejido vascular que forma vasos ahuecados que trasportan todo tipo de sustancias, excepto azúcares, en toda la planta. (p. 516)

Z

zygote: new diploid cell formed when a sperm fertilizes an egg; will divide by mitosis and develop into a new organism. (p. 284)

zigoto: célula diploide nueva formada cuando un espermatozoide fertiliza a un óvulo; se dividirá por mitosis y se desarrollará para formar un nuevo organismo. (p. 284)

Glossary/Glosario

Italic numbers = illustration/photo **Bold numbers = vocabulary term**
lab = a page on which the entry is used in a lab
act = a page on which the entry is used in an activity

A

Index

Index

Index

D

Index

T

U

V

Index

Magnification Key: Magnifications listed are the magnifications at which images were originally photographed.
LM–Light Microscope
SEM–Scanning Electron Microscope
TEM–Transmission Electron Microscope

Acknowledgments: Glencoe would like to acknowledge the artists and agencies who participated in illustrating this program: Absolute Science Illustration; Andrew Evansen; Argosy; Articulate Graphics; Craig Attebery represented by Frank & Jeff Lavaty; CHK America; John Edwards and Associates; Gagliano Graphics; Pedro Julio Gonzalez represented by Melissa Turk & The Artist Network; Robert Hynes represented by Mendola Ltd.; Morgan Cain & Associates; JTH Illustration; Laurie O'Keefe; Matthew Pippin represented by Beranbaum Artist's Representative; Precision Graphics; Publisher's Art; Rolin Graphics, Inc.; Wendy Smith represented by Melissa Turk & The Artist Network; Kevin Torline represented by Berendsen and Associates, Inc.; WILDlife ART; Phil Wilson represented by Cliff Knecht Artist Representative; Zoo Botanica.

Photo Credits

Cover PhotoDisc; **i ii** (bkgd)Will & Deni McIntyre/Getty Images, (b)Jeffrey Coolidge/Getty Images, (c)Daryl Benson/Masterfile; **vii** Aaron Haupt; **viii** John Evans; **ix** (t)PhotoDisc, (b)John Evans; **x** (l)John Evans, (r)Geoff Butler; **xi** (l)John Evans, (r)PhotoDisc; **xii** PhotoDisc; **xiii** CORBIS; **xiv** Bob Rowan/CORBIS; **xv** NASA; **xvi** Dave B. Fleetham/Tom Stack & Assoc.; **xvii** National Cancer Institute/Science Photo Library/Photo Researchers; **xviii** (t)Amanita Pictures, (b)Tom Stack & Assoc.; **xix** David Schultz/Stone/Getty Images; **xx** FPG/Getty Images; **xxi** Matt Meadows; **xxii** (l)NASA, (r)CORBIS; **xxiii** (t)Ron Kimball Photography, (b)SuperStock; **xxiv** Geoff Butler; **xxvii** Bob Daemmrich; **1** Roger Ressmeyer/CORBIS; **2–3** Coco McCoy from Rainbow/PictureQuest; **3** (inset)Mary Evans Picture Library; **4–5** Pascal Goetgheluck/Science Photo Library/Photo Researchers; **6** Charles Gupton/Stock Boston/PictureQuest; **7** (l)Coco McCoy from Rainbow, (r)Stephen J. Krasemann/Photo Researchers; **9** (t)Geoff Butler, (b)Courtesy Sensors & Software, Inc.; **10** (l)Alexander Nesbitt/Danita Delimont, Agent, (r)Georg Gerster/Photo Researchers; **11** Bob Daemmrich/Stock Boston; **13** (r)Doug Martin, (others) Dominic Oldershaw; **15** (tl)Lawrence Migdale, (tr)Holly Payne, (cl b)Katharine Payne; **17** Thomas Veneklasen; **18** Brent Turner/BLT Productions; **20** KS Studios; **21** John Evans; **22** file photo; **23** (l)Heinz Plenge/Peter Arnold, Inc., (r)Martha Cooper/Peter Arnold, Inc.; **26** (l)Alexander Nesbitt/Danita Delimont, Agent, (r)Bob Daemmrich/Stock Boston; **27** Nigel Cattlin/Photo Researchers, Inc.; **28–29** SuperStock; **30** Matt Meadows; **31** (inset)John R. Foster/Photo Researchers, (l tr)Mark A. Schneider/Visuals Unlimited; **32** (cl)A.J. Copley/Visuals Unlimited, (cr bl)Harry Taylor/DK Images, (others)Mark A. Schneider/Photo Researchers; **33** (inset)Patricia K. Armstrong/Visuals Unlimited, (r)Dennis Flaherty Photography/Photo Researchers; **35** KS Studios; **36** (l)Mark Burnett/Photo Researchers, (c)Dan Suzio/Photo Researchers, (r)Breck P. Kent/Earth Scenes; **37** (t)Bud Roberts/Visuals Unlimited, (inset)Icon Images, (b)Charles D. Winters/Photo Researchers; **38** (l)Andrew McClenaghan/Science Photo Library/Photo Researchers, (r)Charles D. Winters/Photo Researchers; **39** (t)Goeff Butler, (bl)Doug Martin, (br)Photo Researchers; **40** Matt Meadows; **41** Reuters NewMedia, Inc./CORBIS; **42** (Beryl, Spinel)Biophoto Associates/Photo Researchers, (Emerald, Topaz)H. Stern/Photo Researchers, (Ruby Spinel, Tanzanite)A.J. Copley/Visuals Unlimited, (Zoisite)Visuals Unlimited, (Uncut Topaz)Mark A. Schneider/Visuals Unlimited; **43** (Olivine)University of Houston, (Peridot) Charles D. Winters/Photo Researchers, (Garnet)Arthur R. Hill/Visuals Unlimited, (Almandine)David Lees/CORBIS, (Quartz, Corundum)Doug Martin, (Amethyst)A.J. Copley/Visuals Unlimited, (Blue Sapphire)Vaughan Fleming/Science Photo Library/Photo Researchers; **44** (l)Francis G. Mayer/CORBIS, (r)National Museum of Natural History/©Smithsonian Institution; **45** (l)Fred Whitehead/Earth Scenes, (inset)Doug Martin; **46** (t)Matt Meadows, (bl)Paul Silverman/Fundamental Photographs, (br)Biophoto Associates/Photo Researchers; **47** Jim Cummins/Getty Images; **48** Matt Meadows; **49** (t)Doug Martin, (bl)Andrew J. Martinez/Photo Researchers, (br)Charles D. Winter/Photo Researchers, (inset)José Manuel Sanchis Calvete/CORBIS; **50** (bkgd)Science Photo Library/Custom Medical Stock Photo, (bl)Bettmann/CORBIS; **51** José Manuel Sanchis Calvete/CORBIS; **52** R. Weller/Cochise College; **54** José Manuel Sanchis Calvete/CORBIS; **55** Breck P. Kent/Earth Scenes; **56–57** Michael T. Sedam/CORBIS; **58** (inset)Doug Martin, (l)CORBIS; **59** (tl)Steve Hoffman, (tr cr)Breck P. Kent/Earth Scenes, (cl)Brent Turner/BLT Productions; **60** (bkgd)CORBIS/PictureQuest, (t)CORBIS, (bl)Martin Miller, (bc)Jeff Gnass, (br)Doug Sokell/Tom Stack & Assoc.; **61** Russ Clark; **62** USGS/HVO; **63** (t)Breck P. Kent/Earth Scenes, (b)Doug Martin; **64** (Basalt)Mark Steinmetz, (Andesite)Doug Martin, (Pumice)Tim Courlas, (Granite) Doug Martin, (others)Breck P. Kent/Earth Scenes; **66** (l)Breck P. Kent/Earth Scenes, (r)Doug Martin/Photo Researchers; **67** (l r)Breck P. Kent/Earth Scenes, (c)Courtesy Kent Ratajeski & Dr. Allen Glazner, University of NC, (inset)Alfred Pasieka/Photo Researchers; **69** (inset)Aaron Haupt, (r)Robert Estall/CORBIS; **70** Paul Rocheleau/Index Stock; **71** (l)Timothy Fuller, (r)Steve McCutcheon/Visuals Unlimited; **73** (l)Icon Images, (cl)Doug Martin, (cr)Andrew Martinez/Photo Researchers, (r)John R. Foster/Photo Researchers; **74** (l)Breck P. Kent/Earth Scenes, (r)Aaron Haupt; **75** Georg Gerster/Photo Researchers, (inset)Icon Images; **77** Beth Davidow/Visuals Unlimited; **78** (l)Icon Images, (r)Breck P. Kent/Earth Scenes; **79** (l)Jack Sekowski, (r)Tim Courlas; **80** (bkgd)Y. Kawasaki/Photonica, (inset)Matt Turner/Liaison Agency; **82** Breck P. Kent/Earth Scenes; **83** Jeremy Woodhouse/DRK Photo; **86** (inset)Stephen Dalton/Animals Animals; **86–87** A.T. Willett/Image Bank/Getty Images; **88–89** S.P. Gillette/CORBIS; **90** NASA; **91** (l)Frank Rossotto/The Stock Market/CORBIS, (r)Larry Lee/CORBIS; **94** Laurence Fordyce/CORBIS; **96** Doug Martin; **97** NASA/GSFC; **98** Michael Newman/PhotoEdit, Inc.; **103** (t)Dan Guravich/Photo Researchers, (b)Bill Brooks/Masterfile; **105** (cw from top)Gene Moore/PhotoTake NYC/PictureQuest, Phil Schermeister/CORBIS, Joel W. Rogers, Kevin Schafer/CORBIS; **106** Bill Brooks/Masterfile; **108 109** David Young-Wolff/PhotoEdit, Inc.; **110** Bob Rowan/CORBIS; **116–117** Reuters NewMedia, Inc./CORBIS; **117** KS Studios; **118** Kevin Horgan/Stone/Getty Images; **119** Fabio Colombini/Earth Scenes; **123** (t)Charles O'Rear/CORBIS, (b)Joyce Photographics/Photo Researchers; **124** (l)Roy Morsch/The Stock Market/CORBIS, (r)Mark McDermott; **125** (l)Mark E.

Gibson/Visuals Unlimited, (r)EPI Nancy Adams/Tom Stack & Assoc.; **127** Van Bucher/Science Source/Photo Researchers; **129** Jeffrey Howe/Visuals Unlimited; **130** Roy Johnson/Tom Stack & Assoc.; **131** (l)Warren Faidley/Weatherstock, (r)Robert Hynes; **132** NASA/Science Photo Library/Photo Researchers; **133** Fritz Pölking/Peter Arnold, Inc; **140** (bkgd)Erik Rank/Photonica; courtesy Weather Modification Inc., **141** (l)George D. Lepp/Photo Researchers, (r)Janet Foster/Masterfile; **142** Ruth Dixon; **146–147** Andrew Wenzel/Masterfile; **151** (l)William Leonard/DRK Photo, (r)Bob Rowan, Progressive Image/CORBIS; **152** John Shaw/Tom Stack & Assoc.; **153** (tl)David Hosking/CORBIS, (tr)Yva Momatiuk & John Eastcott/Photo Researchers, (b)Michael Melford/The Image Bank/Getty Images; **154** (t)S.R. Maglione/Photo Researchers, (c)Jack Grove/Tom Stack & Assoc., (b)Fritz Pölking/Visuals Unlimited; **155** Zig Leszczynski/Animals Animals; **157** (l)Jonathan Head/AP/Wide World Photos, (r)Jim Corwin/Index Stock; **159** (t)A. Ramey/PhotoEdit, Inc., (b)Peter Beck/Pictor; **160** Galen Rowell/Mountain Light; **164** John Bolzan; **165** Chip & Jill Isenhart/Tom Stack & Assoc.; **167** Matt Meadows, Doug Martin; **170** Alberto Garcia/Saba; **171** Steve Kaufman/DRK Photo; **176–177** NASA; **178** NASA/JPL; **179** Jerry Schad/Photo Researchers; **184 187** Lick Observatory; **189** CORBIS; **191** (tr)NASA, (bl)Stephen Frisch/Stock Boston/PictureQuest; **192** NASA; **193** Timothy Fuller; **195** NASA/JPL/Northwestern University; **196** (tl)NASA/JPL, (br)Dr. Timothy Parker/JPL; **198** (l)CORBIS, (r)Erich Karkoschka, University of AZ, and NASA; **199** NASA/JPL; **200** (t)Dr. R. Albrecht, ESA/ESO Space Telescope European Coordinating Facility/NASA (b)Frank Zullo/Photo Researchers **202** NASA/JPL **204** (t)Kauko Helavuo/The Image Bank/Getty Images, (b)Charles & Josette Lenars/CORBIS; **205 208** Lick Observatory; **209** (l)CORBIS, (c)NASA/Photo Researchers, (r)NASA/Science Source/Photo Researchers; **210** Microworks/PhotoTake, NYC; **210–211** (bkgd)Doug Wilson/CORBIS; **212–213** Peter Lane Taylor/Visuals Unlimited; **213** Joanne Huemoeller/Animals Animals; **214** (l)Michael Abbey/Science Source/Photo Researchers, (r)Michael Delannoy/Visuals Unlimited; **215** (tcr)A. Glauberman/Photo Researchers, (tr)Mark Burnett, (bl bcl br)Runk/Schoenberger from Grant Heilman, (others)Dwight Kuhn; **216** (t)Bill Beaty/Animals Animals, (b)Tom & Therisa Stack/Tom Stack & Assoc.; **217** Aaron Haupt; **218** David A. Northcott/CORBIS; **219** Doug Perrine/Innerspace Visions; **220** Alvin E. Staffan; **222** (t)Kathy Talaro/Visuals Unlimited, (cl)courtesy Nikon Instruments Inc., (c)Michael Gabridge/Visuals Unlimited, (cr)Mike Abbey/Visuals Unlimited, (bl br)David M. Phillips/Visuals Unlimited; **222–223** (bkgd) David M. Phillips/Visuals Unlimited; **223** (tl)James W. Evarts, (tr)Bob Krist/CORBIS, (cl)courtesy Olympus Corporation, (cr)Mike Abbey/Visuals Unlimited, (bl)Karl Aufderheide/Visuals Unlimited, (br)Lawrence Migdale/Stock Boston/PictureQuest; **224** Don Fawcett/Photo Researchers; **225** (t)M. Schliwa/Visuals Unlimited, (b)Photo Researchers; **226** (l)George B. Chapman/Visuals Unlimited, (r)P. Motta & T. Naguro/Science Photo Library/Photo Researchers; **228** (t)Don Fawcett/Photo Researchers, (b)Biophoto Associates/Photo Researchers; **233** Dr. J.F.J.M. van der Heuvel; **234** Bruce Ayres/Getty Images; **236 237** Matt Meadows; **238** (t)Quest/Science Photo Library/Photo Researchers, (b)courtesy California Univ.; **240** NIBSC/Science Photo Library/Photo Researchers; **242** Michael Fogden/Earth Scenes; **243** Donald Specker/Animals Animals; **244–245** Jane Grushow/Grant Heilman Photography; **247** Bob Daemmrich; **249** (t)Runk/Schoenberger from Grant Heilman, (b)Klaus Guldbrandsen/Science Photo Library/Photo Researchers; **254** (l)John Fowler, (r)Richard Hamilton Smith/CORBIS; **255** KS Studios; **256** Aaron Haupt; **257** Visuals Unlimited; **258** Biophoto Associates/Science Source/Photo Researchers; **260** Matt Meadows; **262** Craig Lovell/CORBIS; **263** John Fowler; **264** David M. Phillips/Visuals Unlimited; **265** (l)Grant Heilman Photography, (r)Bios (Klein/Hubert)/Peter Arnold; **266** (t)Runk/Schoenberger from Grant Heilman, (b)Matt Meadows; **267** Matt Meadows; **268** Lappa/Marquart; **269** CNRI/Science Photo Library/Photo Researchers; **270** Biophoto Associates/Science Source/Photo Researchers; **274–275** Zig Leszcynski/Animals Animals; **276** (l)Dave B. Fleetham/Tom Stack & Assoc., (r)Cabisco/Visuals Unlimited; **278** Cabisco/Visuals Unlimited; **279** (tl)Michael Abbey/Visuals Unlimited, (others)John D. Cunningham/Visuals Unlimited; **280** (l)Matt Meadows, (r)Nigel Cattlin/Photo Researchers; **281** (l)Barry L. Runk from Grant Heilman, (r)Runk/Schoenberger from Grant Heilman; **282** (l)Walker England/Photo Researchers, (r)Tom Stack & Assoc.; **283** Runk/Schoenberger from Grant Heilman; **284** Dr. Dennis Kunkel/PhotoTake NYC; **285** (tl)Gerald & Buff Corsi/Visuals Unlimited, (bl)Susan McCartney/Photo Researchers, (r)Fred Bruenner/Peter Arnold, Inc.; **287** (l)John D. Cunningham/Visuals Unlimited, (c)Jen & Des Bartlett/Bruce Coleman, Inc., (r)Breck P. Kent; **288** (tl)Artville, (tr)Tim Fehr, (c)Bob Daemmrich/Stock Boston/PictureQuest, (bl)Troy Mary Parlee/Index Stock/PictureQuest, (br)Jeffery Myers/Southern Stock/PictureQuest; **294** Stewart Cohen/Stone/Getty Images; **296** (t)Tom McHugh/Photo Researchers, (b)file photo; **297** Monica Dalmasso/Stone/Getty Images; **298** (t)Philip Lee Harvey/Stone, (b)Lester V. Bergman/CORBIS; **300** Walker England/Photo Researchers; **302** Barry L. Runk from Grant Heilman; **303** Cabisco/Visuals Unlimited; **304–305** Ron Chapple/Getty Images; **305** Geoff Butler; **306** Stewart Cohen/Stone/Getty Images; **309** (bkgd)Jane Grushow from Grant Heilman, Special Collections, National Agriculture Library; **310** Barry L. Runk From Grant Heilman; **312** Richard Hutchings/Photo Researchers; **314** (l)Robert Maier/Animals Animals, (r)Gemma Giannini from Grant Heilman; **315** Raymond Gehman/CORBIS; **316** Dan McCoy from Rainbow; **317** (l)Phil Roach/Ipol, Inc., (r)CNRI/Science Photo Library/Photo Researchers; **318** Gopal Murti/PhotoTake, NYC; **319** Tim Davis/Photo Researchers; **320** (t)Renee Stockdale/Animals Animals, (b)Alan & Sandy Carey/Photo Researchers; **323** Tom Meyers/Photo Researchers; **324** (t)Runk/Schoenberger from Grant Heilman, (b)Mark Burnett; **325** Richard Hutchings; **326** KS Studios; **331** CNRI/Science Photo Library/Photo Researchers; **332–333** B.G. Thomson/Photo Researchers; **335** Barbera Cushman/DRK Photo; **336** (l c)Tui De Roy/Minden Pictures, (r)Tim Davis/Photo Researchers; **338** (l)Gregory G. Dimijian, M.D./Photo Researchers, (r)Patti Murray/Animals Animals; **339** (l)Darek Karp/Animals Animals, (r)Vonorla Photography; **340** (l)Joe McDonald/Animals Animals, (c)Tom McHugh/Photo Researchers, (r)Tim Davis/Photo Researchers; **341** James Richardson/Visuals Unlimited; **342** Frans Lanting/Minden Pictures; **343** (l)Dominique Braud/Earth Scenes, (c)Carr Clifton/Minden Pictures, (r)John Cancalosi/DRK Photo; **344** (cw from top)Ken Lucas/Visuals Unlimited, John Cancalosi/DRK Photo, Larry Ulrich/DRK Photo, John Cancalosi/Peter Arnold, Inc.,

Sinclair Stammers/Science Photo Library/Photo Researchers; **345** John Kieffer/Peter Arnold, Inc.; **348** (l)Kees Van Den Berg/Photo Researchers, (r)Doug Martin; **349** Peter Veit/DRK Photo; **350** (l)Mark E. Gibson, (r)Gerard Lacz/Animals Animals; **351** (l)Michael Dick/Animals Animals, (r)Carolyn A. McKeone/Photo Researchers; **352** (l)John Reader/Science Photo Library/Photo Researchers, (r)Archivo Iconografico, S.A./CORBIS; **353** Francois Ducasse/Rapho/Photo Researchers; **354** (b)Kenneth W. Fink/Photo Researchers; **354–355** Aaron Haupt; **356** (t)Oliver Meckes/E.O.S/ Gelderblom/Photo Researchers, (b)Lara Jo Regan/Saba; **357** (l)Koster-Survival/Animals Animals, (r)Stephen J. Keasemann/DRK Photo; **359** Tom Tietz/Stone/Getty Images; **360** Gregory G. Dimijian, M.D./Photo Researchers; **361** Patti Murray/Animals Animals; **362–363** Birgid Allig/Stone/Getty Images; **363** (inset)Don Mason/The Stock Market/CORBIS; **364–365** Julian Calder/CORBIS; **367** National Cancer Institute/Science Photo Library/Photo Researchers; **370** Meckes/Ottawa/Photo Researchers; **371** Aaron Haupt; **374** Matt Meadows, (inset)StudiOhio; **375** Aaron Haupt; **377** Runk/Schoenberger from Grant Heilman; **379** CC Studio/Science Photo Library/Photo Researchers; **382** Holt Studios International (Nigel Cattlin)/Photo Researchers; **383** (t b)Jack Bostrack/Visuals Unlimited, (t b)Visuals Unlimited, (cl)Cytographics Inc./Visuals Unlimited, (cr)Cabisco/Visuals Unlimited; **385** Oliver Meckes/E.O.S/ Gelderblom/Photo Researchers; **386** Andrew Syred/Science Photo Library/Photo Researchers; **390** Matt Meadows/Peter Arnold, Inc.; **391** Matt Meadows; **392** (bkgd)Science Photo Library/Photo Researchers, no credit; **393** (tl)Manfred Kage/Peter Arnold, Inc., (tr)K.G. Murti/Visuals Unlimited, (bl)Don W. Fawcett/Visuals Unlimited, Aaron Haupt; **398–399** Chris Trotman/NewSport/Corbis; **401** Geoff Butler; **405 406** KS Studios; **407** Visuals Unlimited; **409 410** KS Studios; **412** Dominic Oldershaw; **413** Bob Daemmrich; **416** Richard T. Nowitz; **418** Renee Lynn/Photo Researchers; **423** Richard Hutchings/Photo Researchers; **424** KS Studios; **425** Matt Meadows; **427** (t)Eising/Stock Food, (b)Goldwater/ Network/Saba Press Photos, (l)Ed Beck/The Stock Market/ CORBIS, (r)Tom & DeeAnn McCarthy/The Stock Market/ CORBIS; **432–433** (bkgd)Michael Pasdzior/The Image Bank/Getty Images; **433** (inset)Matt Meadows; **435** (tl)Clyde H. Smith/Peter Arnold, Inc., (tcl)Erik Sampers/Photo Researchers, (tr)Michael A. Keller/The Stock Market/ CORBIS, (tcr)Dean Conger/CORBIS, (bl)Peter Turnley/ CORBIS, (bcl)Joe McDonald/Visuals Unlimited, (bcr)Art Stein/Photo Researchers, (br)Ed Bock/The Stock Market/ CORBIS; **437** Jim Grace/Photo Researchers; **438** Mark Burnett; **439** Aaron Haupt; **440** (l)Breck P. Kent, (c)Runk/ Schoenberger from Grant Heilman, (r)PhotoTake NYC/ Carolina Biological Supply Company; **441** (b)M. McCarron, (t)C Squared Studios/PhotoDisc; **447** Geoff Butler; **448** Photo Researchers; **449 452** KS Studios; **457** Michael Newman/PhotoEdit, Inc.; **458** (t)Jeff Greenberg/PhotoEdit, Inc., (b)Amanita Pictures; **459** Amanita Pictures; **460** Sara Davis/The Herald-Sun; **465** Eamonn McNulty/Science Photo Library/Photo Researchers; **466–467** Lawerence Manning/ CORBIS; **467** John Evans; **468** David Young-Wolff/PhotoEdit, Inc.; **477** Ariel Skelley/The Stock Market/CORBIS; **479** David M. Phillips/Photo Researchers; **480** (l)Tim Davis/Photo Researchers, (r)Chris Sorensen/The Stock Market/CORBIS; **481** Science Pictures Ltd/Science Photo Library/Photo Researchers; **482** Petit Format/Nestle/Science Source/Photo Researchers; **484** (l)Jeffery W. Myers/Stock Boston, (r)Ruth Dixon; **485** (tl b)Mark Burnett, (tr)Aaron Haupt; **486** KS Studios; **487** (l)NASA/Roger Ressmeyer/CORBIS, (r)AFP/ CORBIS; **488** (t)Chris Carroll/CORBIS, (b)Richard Hutchings; **489** Matt Meadows; **490** (l)Ron Kimball Photography, (r)SuperStock, (b)Martin B. Withers/Frank Lane Picture Agency/CORBIS; **491** (l)Bob Daemmrich, (r)Maria Taglienti/The Image Bank/Getty Images; **496–497** Richard Hamilton Smith/CORBIS; **497** (l)Courtesy Broan-NuTone, (r)CORBIS; **498–499** Peter Adams/Getty Images; **500** Tom Stack & Assoc.; **501** Laat-Siluur; **502** (t)Kim Taylor/Bruce Coleman, Inc., (b)William E. Ferguson; **503** (tl)Amanita Pictures, (tr)Ken Eward/Photo Researchers, (bl)Photo Researchers, (br)Amanita Pictures; **504** (cw from top)Dan McCoy from Rainbow, Philip Dowell/DK Images, Kevin & Betty Collins/Visuals Unlimited, David Sieren/ Visuals Unlimited, Steve Callaham/Visuals Unlimited, Gerald & Buff Corsi/Visuals Unlimited, Mack Henley/Visuals Unlimited, Edward S. Ross, Douglas Peebles/CORBIS, Gerald & Buff Corsi/Visuals Unlimited, Martha McBride/Unicorn Stock Photos; **505** (t)Gail Jankus/Photo Researchers, (b)Michael P. Fogden/Bruce Coleman, Inc.; **506** (l)Larry West/Bruce Coleman, Inc., (c)Scott Camazine/Photo Researchers, (r)Kathy Merrifield/Photo Researchers; **507** Michael P. Gadomski/Photo Researchers; **509** (t)Farrell Grehan/Photo Researchers, (bl)Steve Solum/Bruce Coleman, Inc., (bc)R. Van Nostrand/Photo Researchers, (br)Inga Spence/Visuals Unlimited; **510** (t)Joy Spurr/Bruce Coleman, Inc., (b)W.H. Black/Bruce Coleman, Inc.; **511** Farrell Grehan/Photo Researchers; **512** Amanita Pictures; **513** (l)Nigel Cattlin/Photo Researchers, Inc, (c)Doug Sokel/Tom Stack & Assoc., (r)Charles D. Winters/Photo Researchers; **514** Bill Beatty/Visuals Unlimited; **516** (l)Doug Sokell/Tom Stack & Assoc., (tc)Robert C. Hermes/Photo Researchers, (r)Bill Beatty/Visuals Unlimited, (bc)David M. Schleser/Photo Researchers; **517** (cw from top)E. Valentin/ Photo Researchers, Dia Lein/Photo Researchers, Eva Wallander, Tom Stack & Assoc., Joy Spurr/Photo Researchers; **519** (l)Dwight Kuhn, (c)Joy Spurr/Bruce Coleman, Inc., (r)John D. Cunningham/Visuals Unlimited; **520** (l)J. Lotter/Tom Stack & Assoc., (r)J.C. Carton/Bruce Coleman, Inc.; **522** (t)Inga Spence/Visuals Unlimited, (b)David Sieren/Visuals Unlimited; **523** Jim Steinberg/Photo Researchers; **524** (t)Michael Rose/Frank Lane Picture Agency/CORBIS, (b)Dr. Jeremy Burgess/Science Photo Library/Photo Researchers; **526** Stephen P. Parker/Photo Researchers; **530–531** Clem Haagner/A.B.P.L./Photo Researchers; **532** WM. J. Jahoda/Photo Researchers; **533** (t)Stuart Westmorland/Photo Researchers, (c)Michael P. Gadomski/Earth Scenes, (b)George Bernard/Earth Scenes; **534** Francis Lepine/Earth Scenes; **536** (t)Roland Seitre-Bios/ Peter Arnold, Inc., (bl)Robert C. Gildart/Peter Arnold, Inc., (br)Carr Clifton/Minden Pictures; **538** Bob Daemmrich; **540** Dan Suzio/Photo Researchers; **541** (t)Tim Davis/Photo Researchers, (b)Arthur Gloor/Animals Animals; **542** Gilbert Grant/Photo Researchers; **543** John Gerlach/Animals Animals; **544** Michael P. Gadomski/Photo Researchers; **545** (bkgd)Michael Boys/CORBIS, (t)Joe McDonald/ CORBIS, (c)David A. Northcott/CORBIS, (bl)Michael Boys/CORBIS, (bc)Dennis Johnson/Papilio/CORBIS, (br)Kevin Jackson/Animals Animals; **547** (t)Ray Richardson/ Animals Animals, (tc)Suzanne L. Collins/Photo Researchers, (b)Zig Leszczynski/Earth Scenes, (bc)William E. Grenfell Jr./Visuals Unlimited; **550** (t)Geoff Butler, (b)KS Studios; **551** Matt Meadows; **552** Allen Russell/Index Stock;

Credits

553 (l)Richard Reid/Earth Scenes, (r)Helga Lade/Peter Arnold, Inc.; **554** Helga Lade/Peter Arnold, Inc.; **556** (l)George Bernard/Earth Scenes, (r)Tim Davis/Photo Researchers; **557** Arthur Gloor/Animals Animals; **558–559** Grant Heilman Photography; **560** (l)Keith Lanpher/Liaison Agency/Getty Images, (r)Richard Thatcher/David R. Frazier Photolibrary; **561** (t)Solar Cookers International, (bl)Brian F. Peterson/The Stock Market/CORBIS, (br)Ron Kimball Photography; **562** Larry Mayer/Liaison Agency/Getty Images; **565** (tr)Torleif Svenson/The Stock Market/CORBIS, (bl)Rob Williamson; (br)Les Gibbon/Cordaiy Photo Library Ltd./CORBIS; **566** Sean Justice; **567** (t)Lowell Georgia/Science Source/Photo Researchers, (cl)NASA, (c)CORBIS, (cr)Sean Sprague/Impact Visuals/PictureQuest, (bl)Lee Foster/Bruce Coleman, Inc., (br)Robert Perron; **568** Philippe Renault/Liaison Agency/Getty Images; **569** (l)NYC Parks Photo Archive/Fundamental Photographs, (r)Kristen Brochmann/Fundamental Photographs; **573** (l)Jeremy Walker/Science Photo Library/Photo Researchers, (c)John Colwell from Grant Heilman, (r)Telegraph Colour Library/FPG/Getty Images; **574** Wilford Haven/Liaison Agency/Getty Images; **575** (tl)Larry Mayer/Liaison Agency/Getty Images, (tr)ChromoSohm/The Stock Market/CORBIS, (cr)David R. Frazier Photolibrary, (br)Inga Spence/Visuals Unlimited; **576** (r)Andrew Holbrooke/The Stock Market/CORBIS, (Paint Cans)Amanita Pictures, (Turpantine, Paint thinner, epoxy)Icon Images, (Batteries) Aaron Haupt; **578** Paul A. Souders/CORBIS; **579** Icon Images; **581** Larry Lefever from Grant Heilman; **582** (t)Howard Buffett from Grant Heilman, (b)Solar Cookers International; **583** John D. Cunningham/Visuals Unlimited; **584** Frank Cezus/FPG/Getty Images; **586** Robert Cameron/Stone/Getty Images; **587** (l)Steve McCutcheon/Visuals Unlimited, (r)James N. Westwater; **588** David R. Frazier Photolibrary; **590** (inset)CORBIS/PictureQuest; **590–591** (bkgd)no credit, Stephen Frisch/Stock Boston/PictureQuest; **592–593** James L. Amos/CORBIS; **594** Fred Habegger from Grant Heilman; **595** (tr)David Nunuk/Science Photo Library/Photo Researchers, (cr)Mark Burnett, (bl)David Schultz/Stone/Getty Images, (bc)SuperStock, (br)Kent Knudson/PhotoDisc; **596** KS Studios; **597** Gary Retherford/Photo Researchers; **598** (l)Peter Steiner/The Stock Market/CORBIS, (c)Tom & DeeAnn McCarthy/The Stock Market/CORBIS, (r)SuperStock; **599** Timothy Fuller; **600** (l)Gay Bumgarner/Stone/Getty Images, (r)A. Goldsmith/The Stock Market/CORBIS; **601** (t)Matt Meadows, (others) Richard Megna/Fundamental Photographs; **602** (tl)Ed Pritchard/Stone/Getty Images, (cl bl)Kip Peticolas/Fundamental Photographs, (tr br)Richard Megna/Fundamental Photographs; **603** Rich Iwasaki/Stone/Getty Images; **604** (t)Matt Meadows, (bl br)Runk/Schoenberger from Grant Heilman, (bc)Layne Kennedy/CORBIS; **605** (t)Amanita Pictures, (b)Richard Megna/Fundamental Photographs; **606** Anthony Cooper/Ecoscene/CORBIS; **607** (tl)Russell Illig/PhotoDisc, (tcl)John D. Cunningham/Visuals Unlimited, (tcr)Coco McCoy/Rainbow/PictureQuest, (bl)Bonnie Kamin/PhotoEdit, Inc., (tr br)SuperStock; **608** (t)Grantpix/Photo Researchers, (c)Mark Sherman/Photo Network/PictureQuest, (bl)Sculpture by Maya Lin, courtesy Wexner Center for the Arts, Ohio State Univ., photo by Darnell Lautt, (br)Rainbow/PictureQuest; **609** Mark Burnett; **610 611** Matt Meadows; **612** (l)Susan Kinast/Foodpix/Getty Images, (r)Michael Newman/PhotoEdit, Inc.; **613** (l)C. Squared Studios/PhotoDisc, (r)Kip Peticolas/Fundamental Photographs; **616** Elaine Shay; **617** (l)Kurt Scholz/SuperStock, (r)CORBIS; **618–619** Joseph Sohm/ChromoSohm, Inc./CORBIS; **621** (l)Stephen W. Frisch/Stock Boston, (r)Doug Martin; **622** (t)HIRB/Index Stock, (b)Doug Martin; **623** Richard Hamilton/CORBIS; **624** John Evans; **625** (l)SuperStock, (r)Annie Griffiths/CORBIS; **628** John Evans; **630** Richard Nowitz/Phototake/PictureQuest; **632** Aaron Haupt; **633** KS Studios/Mullenix; **635** John Evans; **636** (l)Joe Sohm, Chromosohm/Stock Connection/PictureQuest, (c)Andrew Popper/Phototake/PictureQuest, (r)A. Wolf/Explorer, Photo Researchers; **637** John Evans; **638** (tl tr)Elaine Shay, (tcl)Brent Turner/BLT Productions, (tcr)Matt Meadows, (bl bcl)CORBIS, (bcr)Icon Images, (br)StudiOhio; **642 643** KS Studios; **644** CORBIS; **646** Royalty-Free/CORBIS; **649** Stephen W. Frisch/Stock Boston; **650–651** Roger Ressmeyer/CORBIS; **652** Layne Kennedy/CORBIS; **653** (t)Telegraph Colour Library/FPG/Getty Images, (b)Paul Silverman/Fundamental Photographs; **654** Bill Aron/PhotoEdit, Inc.; **655** (l)John Serrao/Photo Researchers, (r)H. Richard Johnston; **656** Tom Tracy/Photo Network/PictureQuest; **657** Annie Griffiths Belt/CORBIS; **658** Amanita Pictures; **659** (t)David Weintraub/Stock Boston, (b)James L. Amos/Peter Arnold, Inc.; **660** Dave King/DK Images; **661** Joseph Sohm/ChromoSohm, Inc./CORBIS; **662** Michael Dalton/Fundamental Photographs; **663** Swarthout & Associates/The Stock Market/CORBIS; **664** Tony Freeman/PhotoEdit, Inc.; **666** David Young-Wolff/PhotoEdit, Inc.; **667** (t)Joshua Ets-Hokin/PhotoDisc, (b)Richard Hutchings; **668** Robbie Jack/CORBIS; **670** A. Ramey/Stock Boston; **671** Mark Burnett; **672** (t)Tony Freeman/PhotoEdit, Inc., (b)Stephen Simpson/FPG/Getty Images; **674** (t)Lester Lefkowitz/The Stock Market/CORBIS, (b)Bob Daemmrich; **675** Bob Daemmrich; **676** Daniel Belknap; **677** (l)Andrew Ward/Life File/PhotoDisc, (r)NASA/TRACE; **679** Mark Burnett; **681** Joshua Ets-Hokin/PhotoDisc; **682–683** John H. Clark/CORBIS; **683** Amanita Pictures; **684** Layne Kennedy/CORBIS; **685** Dominic Oldershaw; **686** Nick Koudis/PhotoDisc; **687** Roger Ressmeyer/CORBIS; **689** (l)Tony Duffy/Allsport/Getty Images, (c)Bruce Berg/Visuals Unlimited, (r)Duomo/CORBIS; **690** Michel Hans/Allsport/Getty Images; **691** Bob Daemmrich; **692** John Evans; **693** Dennis Degnan/CORBIS; **694** David Young-Wolff/PhotoEdit, Inc.; **696** Richard Megna/Fundamental Photographs; **697** Michael Newman/PhotoEdit, Inc.; **698** (t c)David Young-Wolff/PhotoEdit, Inc., (b)Doug Martin; **699** (bkgd)David Madison/Bruce Coleman, Inc., (tl)Richard Cummins/CORBIS, (tr)Dave King/DK Images, (b)Richard T. Nowitz; **700** Peter Fownes/Stock South/PictureQuest; **701** Aaron Haupt; **702** Doug Martin; **703** Spencer Grant/PhotoEdit, Inc.; **704** Eurelios/PhotoTake NYC; **705** Scott Cunningham; **706** IT Stock International/Index Stock/PictureQuest; **707** Mark Burnett; **708** (t)Lester Lefkowitz/The Stock Market/CORBIS, (b)Firefly Productions/The Stock Market/CORBIS; **709** (l)Chuck Savage/The Stock Market/CORBIS, (c)FPG/Getty Images, (r)Doug Martin; **713** (l)Bruce Berg/Visuals Unlimited, (r)Myrleen Ferguson Cate/Photo Edit, Inc.; **714–715** Chris Knapton/Science Photo Library/Photo Researchers; **715** Matt Meadows; **716** (l c)file photo, (r)Mark Burnett; **717** (t b)Bob Daemmrich, (c)Al Tielemans/Duomo; **718** KS Studios; **719** (l r)Bob Daemmrich, (b)Andrew McClenaghan/Science Photo Library/Photo Researchers; **720** Mark Burnett/Photo Researchers; **721** Lori Adamski Peek/Stone/Getty Images; **722** Richard Hutchings; **723** Ron Kimball/Ron Kimball

Photography; **724** (t)Judy Lutz, (b)Lennart Nilsson; **726 728** KS Studios; **734** (t)Dr. Jeremy Burgess/Science Photo Library/Photo Researchers, (b)John Keating/Photo Researchers; **735** Geothermal Education Office; **736** Carsand-Mosher; **737** Billy Hustace/Stone/Getty Images; **738** SuperStock; **739** Roger Ressmeyer/CORBIS; **740** (tl)Reuters NewMedia, Inc./CORBIS, (tr)PhotoDisc, (br)Dominic Oldershaw; **741** (l)Lowell Georgia/CORBIS, (r)Mark Richards/PhotoEdit, Inc.; **746** PhotoDisc; **748** Tom Pantages; **752** Michell D. Bridwell/PhotoEdit, Inc.; **753** (t)Mark Burnett, (b)Dominic Oldershaw; **754** StudiOhio; **755** Timothy Fuller; **756** Aaron Haupt; **758** KS Studios; **759 760** Matt Meadows; **761** (t)Doug Martin, (b)Matt Meadows; **762** (t)Mary Lou Uttermohlen, (b)Doug Martin; **463** Kevin Horgan/Stone/Getty Images; **764** Aaron Haupt; **765** Sinclair Stammers/Science Photo Library/Photo Researchers; **766** KS Studios; **767** David S. Addison/Visuals Unlimited; **768** Donald Specker/Animals Animals; **770** (t)John Evans, (b)Mark Burnett; **771** John Evans; **772** Amanita Pictures; **773** Bob Daemmrich; **775** Davis Barber/PhotoEdit, Inc.; **795** Matt Meadows; **796** (l)Dr. Richard Kessel, (c)NIBSC/Science Photo Library/Photo Researchers, (r)David John/Visuals Unlimited; **797** (t)Runk/Schoenberger from Grant Heilman, (bl)Andrew Syred/Science Photo Library/Photo Researchers, (br)Rich Brommer; **798** (tr)G.R. Roberts, (l)Ralph Reinhold/Earth Scenes, (br)Scott Johnson/Animals Animals; **799** Martin Harvey/DRK Photo.

PERIODIC TABLE OF THE ELEMENTS

Columns of elements are called groups. Elements in the same group have similar chemical properties.

Element — Hydrogen
Atomic number — 1
Symbol — H
Atomic mass — 1.008
State of matter

Gas | Liquid | Solid | Synthetic

The first three symbols tell you the state of matter of the element at room temperature. The fourth symbol identifies elements that are not present in significant amounts on Earth. Useful amounts are made synthetically.

Period	1	2	3	4	5	6	7	8	9
1	Hydrogen 1 H 1.008 (Gas)								
2	Lithium 3 Li 6.941 (Solid)	Beryllium 4 Be 9.012 (Solid)							
3	Sodium 11 Na 22.990 (Solid)	Magnesium 12 Mg 24.305 (Solid)							
4	Potassium 19 K 39.098 (Solid)	Calcium 20 Ca 40.078 (Solid)	Scandium 21 Sc 44.956 (Solid)	Titanium 22 Ti 47.867 (Solid)	Vanadium 23 V 50.942 (Solid)	Chromium 24 Cr 51.996 (Solid)	Manganese 25 Mn 54.938 (Solid)	Iron 26 Fe 55.845 (Solid)	Cobalt 27 Co 58.933 (Solid)
5	Rubidium 37 Rb 85.468 (Solid)	Strontium 38 Sr 87.62 (Solid)	Yttrium 39 Y 88.906 (Solid)	Zirconium 40 Zr 91.224 (Solid)	Niobium 41 Nb 92.906 (Solid)	Molybdenum 42 Mo 95.94 (Solid)	Technetium 43 Tc (98) (Synthetic)	Ruthenium 44 Ru 101.07 (Solid)	Rhodium 45 Rh 102.906 (Solid)
6	Cesium 55 Cs 132.905 (Solid)	Barium 56 Ba 137.327 (Solid)	Lanthanum 57 La 138.906 (Solid)	Hafnium 72 Hf 178.49 (Solid)	Tantalum 73 Ta 180.948 (Solid)	Tungsten 74 W 183.84 (Solid)	Rhenium 75 Re 186.207 (Solid)	Osmium 76 Os 190.23 (Solid)	Iridium 77 Ir 192.217 (Solid)
7	Francium 87 Fr (223) (Solid)	Radium 88 Ra (226) (Solid)	Actinium 89 Ac (227) (Solid)	Rutherfordium 104 Rf (261) (Synthetic)	Dubnium 105 Db (262) (Synthetic)	Seaborgium 106 Sg (266) (Synthetic)	Bohrium 107 Bh (264) (Synthetic)	Hassium 108 Hs (277) (Synthetic)	Meitnerium 109 Mt (268) (Synthetic)

The number in parentheses is the mass number of the longest-lived isotope for that element.

Rows of elements are called periods. Atomic number increases across a period.

Series					
Lanthanide series	Cerium 58 Ce 140.116 (Solid)	Praseodymium 59 Pr 140.908 (Solid)	Neodymium 60 Nd 144.24 (Solid)	Promethium 61 Pm (145) (Synthetic)	Samarium 62 Sm 150.36 (Solid)
Actinide series	Thorium 90 Th 232.038 (Solid)	Protactinium 91 Pa 231.036 (Solid)	Uranium 92 U 238.029 (Solid)	Neptunium 93 Np (237) (Synthetic)	Plutonium 94 Pu (244) (Synthetic)

The arrow shows where these elements would fit into the periodic table. They are moved to the bottom of the table to save space.